Biology: Exploring Life

SECOND EDITION

Biology: Exploring Life

Gil Brum
California State Polytechnic University, Pomona

Larry McKane
California State Polytechnic University, Pomona

Gerry Karp
Formerly of the University of Florida, Gainesville

JOHN WILEY & SONS, INC.

New York / Chichester / Brisbane / Toronto / Singapore

iv

Acquisitions Editor *Sally Cheney*
Developmental Editor *Rachel Nelson*
Marketing Manager *Catherine Faduska*
Associate Marketing Manager *Deb Benson*
Senior Production Editor *Katharine Rubin*
Text Designer/"Steps" Illustration Art Direction *Karin Gerdes Kincheloe*
Manufacturing Manager *Andrea Price*
Photo Department Director *Stella Kupferberg*
Senior Illustration Coordinator *Edward Starr*
"Steps to Discovery" Art Illustrator *Carlyn Iverson*
Cover Design *Meryl Levavi*
Text Illustrations *Network Graphics/Blaize Zito Associates, Inc.*
Photo Editor/Cover Photography Direction *Charles Hamilton*
Photo Researchers *Hilary Newman, Pat Cadley, Lana Berkovitz*
Cover Photo *James H. Carmichael, Jr./The Image Bank*

This book was set in New Caledonia by Progressive Typographers and printed and bound by Von Hoffmann Press.
The cover was printed by Lehigh. Color separations by Progressive Typographers and Color Associates, Inc.

Recognizing the importance of preserving what has been written, it is a policy of
John Wiley & Sons, Inc. to have books of enduring value published in the
United States printed on acid-free paper, and we exert our best efforts to that end.

The paper in this book was manufactured by a mill whose forest management
programs include sustained yield harvesting of its timberlands. Sustained yield
harvesting principles ensure that the number of trees cut each year does not
exceed the amount of new growth.

Library of Congress Cataloging in Publication Data:
Brum, Gilbert D.
    Biology : exploring life/Gil Brum, Larry McKane, Gerry Karp.—2nd ed.
        p.   cm.
    Includes bibliographical references and index.
    ISBN 0-471-54408-6 (cloth)
    1. Biology.   I. McKane, Larry.   II. Karp, Gerald.   III. Title.
QH308.2.B78   1993
574—dc20   93-23383
                                                       CIP

Unit I ISBN 0-471-01827-9 (pbk)
Unit II ISBN 0-471-01831-7 (pbk)
Unit III ISBN 0-471-01830-9 (pbk)
Unit IV ISBN 0-471-01829-5 (pbk)
Unit V ISBN 0-471-01828-7 (pbk)
Unit VI ISBN 0-471-01832-5 (pbk)

Printed in the United States of America

10   9   8   7   6   5   4   3

For the Student, we hope this book helps you discover the thrill of exploring life and helps you recognize the important role biology plays in your everyday life.

To Margaret, Jan, and Patsy, who kept loving us even when we were at our most unlovable.

To our children, Jennifer, Julia, Christopher, and Matthew, whose fascination with exploring life inspires us all. And especially to Jenny—we all wish you were here to share the excitement of this special time in life.

Preface to the Instructor

Biology: Exploring Life, Second Edition is devoted to the process of investigation and discovery. The challenge and thrill of understanding how nature works ignites biologists' quests for knowledge and instills a desire to share their insights and discoveries. The satisfactions of knowing that the principles of nature can be understood and sharing this knowledge are why we teach. These are also the reasons why we created this book.

Capturing and holding student interest challenges even the best of teachers. To help meet this challenge, we have endeavored to create a book that makes biology relevant and appealing, that reveals biology as a dynamic process of exploration and discovery, and that emphasizes the widening influence of biologists in shaping and protecting our world and in helping secure our futures. We direct the reader's attention toward principles and concepts to dispel the misconception of many undergraduates that biology is nothing more than a very long list of facts and jargon. Facts and principles form the core of the course, but we have attempted to show the *significance* of each fact and principle and to reveal the important role biology plays in modern society.

From our own experiences in the introductory biology classroom, we have discovered that

- emphasizing principles, applications, and scientific exploration invigorates the teaching and learning process of biology and helps students make the significant connections needed for full understanding and appreciation of the importance of biology; and
- students learn more if a book is devoted to telling the story of biology rather than a recitation of facts and details.

Guided by these insights, we have tried to create a process-oriented book that still retains the facts, structures, and terminology needed for a fundamental understanding of biology. With these goals in mind, we have interwoven into the text

1. an emphasis on the ways that science works,
2. the underlying adventure of exploration,
3. five fundamental biological themes, and
4. balanced attention to the human perspective.

This book should challenge your students to think critically, to formulate their own hypotheses as possible explanations to unanswered questions, and to apply the approaches learned in the study of biology to understanding (and perhaps helping to solve) the serious problems that affect every person, indeed every organism, on this planet.

THE DEVELOPMENT STORY

The second edition of *Biology: Exploring Life* builds effectively on the strengths of the First Edition by Gil Brum and Larry McKane. For this edition, we added a third author, Gerry Karp, a cell and molecular biologist. Our complementary areas of expertise (genetics, zoology, botany, ecology, microbiology, and cell and molecular biology) as well as awards for teaching and writing have helped us form a balanced team. Together, we exhaustively revised and refined each chapter until all three of us, each with our different likes and dislikes, sincerely believed in the result. What evolved from this process was a satisfying synergism and a close friendship.

THE APPROACH

The elements of this new approach are described in the upcoming section "To the Student: A User's Guide." These pedagogical features are embedded in a book that is written in an informal, accessible style that invites the reader to explore the process of biology. In addition, we have tried to keep the narrative focused on *processes,* rather than on static facts, while creating an underlying foundation that helps students make the connections needed to tie together the information into a greater understanding than that which comes from memorizing facts alone. One way to help students make these connections is to relate the fundamentals of biology to humans, revealing the human perspective in each biological principle, from biochemicals to ecosystems. With each such insight, students take a substantial step toward becoming the informed citizens that make up responsible voting public.

We hope that, through this textbook, we can become partners with the instructor and the student. The biology teacher's greatest asset is the basic desire of students to understand themselves and the world around them. Unfortunately, many students have grown detached from this natural curiosity. Our overriding objective in creating this book was to arouse the students' fascination with exploring life, building knowledge and insight that will enable them to make real-life judgments as modern biology takes on greater significance in everyday life.

THE ART PROGRAM

The diligence and refinement that went into creating the text of *Biology: Exploring Life,* Second Edition characterizes the art program as well. Each photo was picked specifically for its relevance to the topic at hand and for its aesthetic and instructive value in illustrating the narrative concepts. The illustrations were carefully crafted under the guidance of the authors for accuracy and utility as well as aesthetics. The value of illustrations cannot be overlooked in a discipline as filled with images and processes as biology. Through the use of cell icons, labeled illustrations of pathways and processes, and detailed legends, the student is taken through the world of biology, from its microscopic chemical components to the macroscopic organisms and the environments that they inhabit.

SUPPLEMENTARY MATERIALS

In our continuing effort to meet all of your individual needs, Wiley is pleased to offer the various topics covered in this text in customized paperback "splits." For more details, please contact your local Wiley sales representative. We have also developed an integrated supplements package that helps the instructor bring the study of biology to life in the classroom and that will maximize the students' use and understanding of the text.

The *Instructor's Manual,* developed by Michael Leboffe and Gary Wisehart of San Diego City College, contains lecture outlines, transparency references, suggested lecture activities, sample concept maps, section concept map masters (to be used as overhead transparencies), and answers to study guide questions.

Gary Wisehart and Mark Mandell developed the test bank, which consists of four types of questions: fill-in questions, matching questions, multiple-choice questions, and critical thinking questions. A computerized test bank is also available.

A comprehensive visual ancillary package includes four-color transparencies (200 figures from the text), *Process of Science* transparency overlays that break down various biological processes into progressive steps, a video library consisting of tapes from Coronet MTI, and the *Bio Sci* videodisk series from Videodiscovery, covering topics in biochemistry, botany, vertebrate biology, reproduction, ecology, animal behavior, and genetics. Suggestions for integrating the videodisk material in your classroom discussions are available in the instructor's manual.

A comprehensive study guide and lab manual are also available and are described in more detail in the User's Guide section of the preface.

Acknowledgments

*I*t was a delight to work with so many creative individuals whose inspiration, artistry, and vital steam guided this complex project to completion. We wish we were able to acknowledge each of them here, for not only did they meet nearly impossible deadlines, but each willingly poured their heart and soul into this text. The book you now hold in your hands is in large part a tribute to their talent and dedication.

There is one individual whose unique talent, quick intellect, charm, and knowledge not only helped to make this book a reality, but who herself made an enormous contribution to the content and pedagogical strength of this book. We are proud to call Sally Cheney, our biology editor, a colleague. Her powerful belief in this textbook's new approaches to teaching biology helped instill enthusiasm and confidence in everyone who worked on it. Indeed, Sally is truly a force of positive change in college textbook publishing—she has an uncommon ability to think both like a biologist and an editor; she knows what biologists want and need in their classes and is dedicated to delivering it; she recognizes that the future of biology education is more than just publishing another look-alike text; and she is knowledgeable and persuasive enough to convince publishers to stick their necks out a little further for the good of educational advancement. Without Sally, this text would have fallen short of our goal. With Sally, it became even more than we envisioned.

Another individual also helped make this a truly special book, as well as made the many long hours of work so delightful. Stella Kupferberg, we treasure your friendship, applaud your exceptional talent, and salute your high standards. Stella also provided us with two other important assets, Charles Hamilton and Hilary Newman. Stella and Charles tirelessly applied their skill, and artistry to get us images of incomparable effectiveness and beauty, and Hilary's diligent handling helped to insure there were no oversights.

Our thanks to Rachel Nelson for her meticulous editing, for maintaining consistency between sometimes dissimilar writing styles of three authors, and for keeping track of an incalculable number of publishing and biological details; to Katharine Rubin for expertly and gently guiding this project through the myriad levels of production, and for putting up with three such demanding authors; to Karin Kincheloe for a stunningly beautiful design; to Ishaya Monokoff and Ed Starr for orchestrating a brilliant art program; to Network Graphics, especially John Smith and John Hargraves, who executed our illustrations with beauty and style without diluting their conceptual strength or pedagogy, and to Carlyn Iverson, whose artistic talent helped us visually distill our "Steps to Discovery" episodes into images that bring the process of science to life.

We would also like to thank Cathy Faduska and Alida Setford, their creative flair helped us to tell the story behind this book, as well as helped us convey what we tried to accomplish. And to Herb Brown, thank you for your initial confidence and continued support. A very special thank you to Deb Benson, our marketing manager. What a joy to work with you, Deb, your energy, enthusiasm, confidence, and pleasant personality bolstered even our spirits.

We wish to acknowledge Diana Lipscomb of George Washington University for her invaluable contributions to the evolution chapters, Judy Goodenough of the University of Massachusetts, Amherst, for contributing an outstanding chapter on Animal Behavior, and Dorothy Rosenthal for contributing the end–of–chapter "Critical Thinking Questions."

To the reviewers and instructors who used the First Edition, your insightful feedback helped us forge the foundation for this new edition. To the reviewers, and workshop and conference participants for the Second Edition, thank you for your careful guidance and for caring so much about your students.

Dennis Anderson, *Oklahoma City Community College*
Sarah Barlow, *Middle Tennessee State University*
Robert Beckman, *North Carolina State University*
Timothy Bell, *Chicago State University*
David F. Blaydes, *West Virginia University*
Richard Bliss, *Yuba College*
Richard Boohar, *University of Nebraska, Lincoln*
Clyde Bottrell, *Tarrant County Junior College*
J. D. Brammer, *North Dakota State University*
Peggy Branstrator, *Indiana University, East*
Allyn Bregman, *SUNY, New Paltz*
Daniel Brooks, *University of Toronto*

Gary Brusca, *Humboldt State University*
Jack Bruk, *California State University, Fullerton*
Marvin Cantor, *California State University, Northridge*
Richard Cheney, *Christopher Newport College*
Larry Cohen, *California State University, San Marcos*
David Cotter, *Georgia College*
Robert Creek, *Eastern Kentucky University*
Ken Curry, *University of Southern Mississippi*
Judy Davis, *Eastern Michigan University*
Loren Denny, *southwest Missouri State University*
Captain Donald Diesel, *U. S. Air Force Academy*
Tom Dickinson, *University College of the Cariboo*

Mike Donovan, *Southern Utah State College*
Robert Ebert, *Palomar College*
Thomas Emmel, *University of Florida*
Joseph Faryniarz, *Mattatuck Community College*
Alan Feduccia, *University of North Carolina, Chapel Hill*
Eugene Ferri, *Bucks County Community College*
Victor Fet, *Loyola University, New Orleans*
David Fox, *Loyola University, New Orleans*
Mary Forrest, *Okanagan University College*
Michael Gains, *University of Kansas*
S. K. Gangwere, *Wayne State University*
Dennis George, *Johnson County Community College*
Bill Glider, *University of Nebraska*
Paul Goldstein, *University of North Carolina, Charlotte*
Judy Goodenough, *University of Massachusetts, Amherst*
Nels Granholm, *South Dakota State University*
Nathaniel Grant, *Southern Carolina State College*
Mel Green, *University of California, San Diego*
Dana Griffin, *Florida State University*
Barbara L. Haas, *Loyola University of Chicago*
Richard Haas, *California State University, Fresno*
Fredrick Hagerman, *Ohio State University*
Tom Haresign, *Long Island University, Southampton*
W. R. Hawkins, *Mt. San Antonio College*
Vernon Hendricks, *Brevard Community College*
Paul Hertz, *Barnard College*
Howard Hetzle, *Illinois State University*
Ronald K. Hodgson, *Central Michigan University*
W. G. Hopkins, *University of Western Ontario*
Thomas Hutto, *West Virginia State College*
Duane Jeffrey, *Brigham Young University*
John Jenkins, *Swarthmore College*
Claudia Jones, *University of Pittsburgh*
R. David Jones, *Adelphi University*
J. Michael Jones, *Culver Stockton College*
Gene Kalland, *California State University, Dominiquez Hills*
Arnold Karpoff, *University of Louisville*
Judith Kelly, *Henry Ford Community College*
Richard Kelly, *SUNY, Albany*
Richard Kelly, *University of Western Florida*
Dale Kennedy, *Kansas State University*
Miriam Kittrell, *Kingsborough Community College*
John Kmeltz, *Kean College New Jersey*
Robert Krasner, *Providence College*
Susan Landesman, *Evergreen State College*
Anton Lawson, *Arizona State University*
Lawrence Levine, *Wayne State University*
Jerri Lindsey, *Tarrant County Junior College*
Diana Lipscomb, *George Washington University*
James Luken, *Northern Kentucky University*

Ted Maguder, *University of Hartford*
Jon Maki, *Eastern Kentucky University*
Charles Mallery, *University of Miami*
William McEowen, *Mesa Community College*
Roger Milkman, *University of Iowa*
Helen Miller, *Oklahoma State University*
Elizabeth Moore, *Glassboro State College*
Janice Moore, *Colorado State University*
Eston Morrison, *Tarleton State University*
John Mutchmor, *Iowa State University*
Jane Noble-Harvey, *University of Delaware*
Douglas W. Ogle, *Virginia Highlands Community College*
Joel Ostroff, *Brevard Community College*
James Lewis Payne, *Virginia Commonwealth University*
Gary Peterson, *South Dakota State University*
MaryAnn Phillippi, *Southern Illinois University, Carbondale*
R. Douglas Powers, *Boston College*
Robert Raikow, *University of Pittsburgh*
Charles Ralph, *Colorado State University*
Aryan Roest, *California State Polytechnic Univ., San Luis Obispo*
Robert Romans, *Bowling Green State University*
Raymond Rose, *Beaver College*
Richard G. Rose, *West Valley College*
Donald G. Ruch, *Transylvania University*
A. G. Scarbrough, *Towson State University*
Gail Schiffer, *Kennesaw State University*
John Schmidt, *Ohio State University*
John R. Schrock, *Emporia State University*
Marilyn Shopper, *Johnson County Community College*
John Smarrelli, *Loyola University of Chicago*
Deborah Smith, *Meredith College*
Guy Steucek, *Millersville University*
Ralph Sulerud, *Augsburg College*
Tom Terry, *University of Connecticut*
James Thorp, *Cornell University*
W. M. Thwaites, *San Diego State University*
Michael Torelli, *University of California, Davis*
Michael Treshow, *University of Utah*
Terry Trobec, *Oakton Community College*
Len Troncale, *California State Polytechnic University, Pomona*
Richard Van Norman, *University of Utah*
David Vanicek, *California State University, Sacramento*
Terry F. Werner, *Harris-Stowe State College*
David Whitenberg, *Southwest Texas State University*
P. Kelly Williams, *University of Dayton*
Robert Winget, *Brigham Young University*
Steven Wolf, *University of Missouri, Kansas City*
Harry Womack, *Salisbury State University*
William Yurkiewicz, *Millersville University*

Gil Brum
Larry McKane
Gerry Karp

Brief Table of Contents

Contents

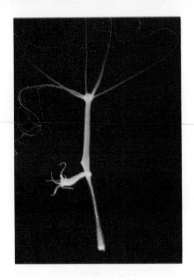

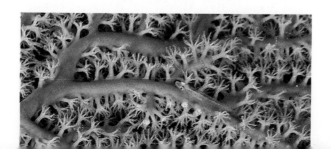

To The Student:
A User's Guide

*B*iology is a journey of exploration and discovery, of struggle and breakthrough. It is enlivened by the thrill of understanding not only what living things do but also how they work. We have tried to create such an experience for you.

Excellence in writing, visual images, and broad biological coverage form the core of a modern biology textbook. But as important as these three factors are in making difficult concepts and facts clear and meaningful, none of them reveals the excitement of biology—the adventure that un-earths what we know about life. To help relate the true nature of this adventure, we have developed several distinctive features for this book, features that strengthen its biological core, that will engage and hold your attention, that reveal the human side of biology, that enable every reader to understand how science works, that stimulate critical thinking, and that will create the informed citizenship we all hope will make a positive difference in the future of our planet.

Steps to Discovery

The process of science enriches all parts of this book. We believe that students, like biologists, themselves, are intrigued by scientific puzzles. Every chapter is introduced by a "Steps to Discovery" narrative, the story of an investigation that led to a scientific breakthrough in an area of biology which relates to that chapter's topic. The "Steps to Discovery" narratives portray biologists as they really are: human beings, with motivations, misfortunes, and mishaps, much like everyone experiences. We hope these narratives help you better appreciate biological investigation, realizing that it is understandable and within your grasp.

Throughout the narrative of these pieces, the writing is enlivened with scientific work that has provided knowledge and understanding of life. This approach is meant not just to pay tribute to scientific giants and Nobel prize winners, but once again to help you realize that science does not grow by itself. Facts do not magically materialize. They are the products of rational ideas, insight, determination, and, sometimes, a little luck. Each of the "Steps to Discovery" narratives includes a painting that is meant primarily as an aesthetic accompaniment to the adventure described in the essay and to help you form a mental picture of the subject.

STEPS TO DISCOVERY
A Factor Promoting the Growth of Nerves

Rita Levi-Montalcini received her medical degree from the University of Turin in Italy in 1936, the same year that Benito Mussolini began his anti-Semitic campaign. By 1939, as a Jew, Levi-Montalcini had been barred from carrying out research and practicing medicine, yet she continued to do both secretly. As a student, Levi-Montalcini had been fascinated with the structure and function of the nervous system. Unable to return to the university, she set up a simple laboratory in her small bedroom in her family's home. As World War II raged throughout Europe, and the Allies systematically bombed Italy, Levi-Montalcini studied chick embryos in her bedroom, discovering new information about the growth of nerve cells from the spinal cord into the nearby limbs. In her autobiography *In Praise of Imperfection*, she writes: "Every time the alarm sounded, I would carry down to the precarious safety of the cellars the Zeiss binocular microscope and my most precious silver stained embryonic sections." In September 1943, German troops arrived in Turin to support the Italian Fascists. Levi-Montalcini and her family fled southward to Florence, where they remained in hiding for the remainder of the w[...]

After the war ended, Levi-Montalcini continued research at the University of Turin. In 1946, she acce[...] an invitation from Viktor Hamburger, a leading expe[...] the development of the chick nervous system, to co[...] Washington University in St. Louis to work with him f[...] semester; she remained at Washington University [...] years.

A chick embryo and one of its nerve cells helped scientists discover nerve growth factor (NGF).

One of Levi-Montalcini's first projects was the reexamination of a previous experiment of Elmer Bueker, a former student of Hamburger's. Bueker had removed a limb from a chick embryo, replaced it with a fragment of a mouse connective tissue tumor, and found that nerve fibers grew into this mass of implanted tumor cells. When Levi-Montalcini repeated the experiment she made an unexpected discovery: One part of the nervous system of these experimental chick embryos—the sympathetic nervous system—had grown five to six times larger than had its counterpart in a normal chick embryo. (The sympathetic nervous system helps control the activity of internal organs, such as the heart and digestive tract.) Close examination revealed that the small piece of tumor tissue that had been grafted onto the embryo had caused sympathetic nerve fibers to grow "wildly" into all of the chick's internal organs, even causing some of the blood vessels to become obstructed by the invasive fibers. Levi-Montalcini hypothesized that the tumor was releasing some soluble substance that induced the remarkable growth of this part of the nervous system. Her hypothesis was soon confirmed by further experiments. She called the active substance **nerve growth factor (NGF)**.

The next step was to determine the chemical nature of NGF, a task that was more readily performed by growing the tumor cells in a culture dish rather than an embryo. But Hamburger's laboratory at Washington University did not have the facilities for such work. To continue the project, Levi-Montalcini boarded a plane, with a pair of tumor-bearing mice in the pocket of her overcoat, and flew to Brazil, where she had a friend who operated a tissue culture laboratory. When she placed sympathetic nervous tissue in the proximity of the tumor cells in a culture dish, the nervous tissue sprouted a halo of nerve fibers that grew toward the tumor cells. When the tissue was cultured in the absence of NGF, no such growth occurred.

For the next 2 years, Levi-Montalcini's lab was devoted to characterizing the substance in the tumor cells that possessed the ability to cause nerve outgrowth. The work was carried out primarily by a young biochemist, Stanley Cohen, who had joined the lab. One of the favored approaches to studying the nature of a biological molecule is to determine its sensitivity to enzymes. In order to determine if nerve growth factor was a protein or a nucleic acid, Cohen treated the active material with a small amount of snake venom, which contains a highly active enzyme that degrades nucleic acid. It was then that chance stepped in.

Cohen expected that treatment with the veno[...] ther destroy the activity of the tumor cell fracti[...] was a nucleic acid) or leave it unaffected (if [...] protein). To Cohen's surprise, treatment w[...] *increased* the nerve-growth promoting activity o[...] rial. In fact, treatment of sympathetic nerve tissu[...] venom alone (in the absence of the tumor extract[...] the growth of a halo of nerve fibers! Cohen soon d[...] why: The snake venom possessed the same nerv[...] factor as did the tumor cells, but at much higher co[...] tion. Cohen soon demonstrated that NGF was a p[...]

Levi-Montalcini and Cohen reasoned that sin[...] venom was derived from a *modified* salivary glan[...] other salivary glands might prove to be even better [...] of the protein. This hypothesis proved to be correct[...] Levi-Montalcini and Cohen tested the salivary gland[...] male mice, they discovered the richest source of NG[...] source 10,000 times more active than the tumor ce[...] ten times more active than snake venom.

A crucial question remained: Did NGF play a r[...] the normal development of the embryo, or was its abi[...] stimulate nerve growth just an accidental property o[...] molecule? To answer this question, Levi-Montalcin[...] Cohen injected embryos with an antibody against N[...] which they hoped would inactivate NGF molecules w[...] ever they were present in the embryonic tissues. The [...] bryos developed normally, with one major exception: T[...] virtually lacked a sympathetic nervous system. The [...] searchers concluded that NGF must be important dur[...] normal development of the nervous system; otherwise, [...] activation of NGF could not have had such a drama[...] effect.

By the early 1970s, the amino acid sequence of NG[...] had been determined, and the protein is now being synthe[...] sized by recombinant DNA technology. During the pas[...] decade, Fred Gage, of the University of California, ha[...] found that NGF is able to revitalize aged or damaged nerve[...] cells in rats. Based on these studies, NGF is currently being[...] tested as a possible treatment of Alzheimer's disease. For[...] their pioneering work, Rita Levi-Montalcini and Stanley[...] Cohen shared the 1987 Nobel Prize in Physiology and [...] Medicine.

Many students are overwhelmed by the diversity of living organisms and the multitude of seemingly unrelated facts that they are forced to learn in an introductory biology course. Most aspects of biology, however, can be thought of as examples of a small number of recurrent themes. Using the thematic approach, the details and principles of biology can be assembled into a body of knowledge that makes sense, and is not just a collection of disconnected facts. Facts become ideas, and details become parts of concepts as you make connections between seemingly unrelated areas of biology, forging a deeper understanding.

All areas of biology are bound together by evolution, the central theme in the study of life. Every organism is the product of evolution, which has generated the diversity of biological features that distinguish organisms from one another and the similarities that all organisms share. From this basic evolutionary theme emerge several other themes that recur throughout the book:

Relationship between Form and Function
Biological Order, Regulation, and Homeostatis
Acquiring and Using Energy
Unity Within Diversity
Evolution and Adaptation

We have highlighted the prevalent recurrence of each theme throughout the text with an icon, shown above. The icons can be used to activate higher thought processes by inviting you to explore how the fact or concept being discussed fits the indicated theme.

FORM AND FUNCTION
The length and shape of this tiger lily and that of the hummingbird's beak match, enabling the hummingbird to gather nutritious nectar from the flower base, while the flower deposits or collects pollen for reproduction.

ACQUIRING AND USING ENERGY
All organisms acquire and use energy whether by directly trapping the energy in sunlight or by harvesting energy stored in the bodies of plants or animals.

ORDER, REGULATION AND HOMEOSTASIS
Whether active or dormant, enveloped by snow or blistering heat, all organisms must maintain order and regulate internal conditions to remain alive.

UNITY WITHIN DIVERSITY
Despite the remarkable variation in different kinds of organisms on earth, all organisms are composed of one or more cells.

EVOLUTION AND ADAPTATION
Evolution has produced the astounding variety of life on earth. Like the lizard hidden on the bark of this tree, each kind of organism possesses adaptation that enable it to survive and reproduce in its particular habitat.

Reexamining the Themes

*E*ach chapter concludes with a "Reexamining the Themes" section, which revisits the themes and how they emerge within the context of the chapter's concepts and principles. This section will help you realize that the same themes are evident at all levels of biological organization, whether you are studying the molecular and cellular aspects of biology or the global characteristics of biology.

When two organisms have the same protein, the difference in amino acid sequence of that protein can be correlated with the evolutionary relatedness of the organisms. The amino acid sequence of hemoglobin, for example, is much more similar between humans and monkeys—organisms that are closely related—than between humans and turtles, who are only distantly related. In fact, the evolutionary tree that emerges when comparing the structure of specific proteins from various animals very closely matches that previously constructed from fossil evidence.

The fact that the amino acid sequences of proteins change as organisms diverge from one another reflects an underlying change in their genetic information. Even though a DNA molecule from a mushroom, a redwood tree, and a cow may appear superficially identical, the sequences of nucleotides that make up the various DNA molecules are very different. These differences reflect evolutionary changes resulting from natural selection (Chapter 34).

Virtually all differences among living organisms can be traced to evolutionary changes in the structure of their various macromolecules, originating from changes in the nucleotide sequences of their DNA. (See CTQ #7.)

REEXAMINING THE THEMES

Relationship between Form and Function

The structure of a macromolecule correlates with a particular function. The unbranched, extended nature of the cellulose molecule endows it with resistance to pulling forces, an important property of plant cell walls. The hydrophobic character of lipids underlies many of their biological roles, explaining, for example, how waxes are able to provide plants with a waterproof covering. Protein function is correlated with protein shape. Just as a key is shaped to open a specific lock, a protein is shaped for a particular molecular interaction. For example, the shape of each polypeptide chain of hemoglobin enables a molecule of oxygen to fit perfectly into its binding site. A single alteration in the amino acid sequence of a hemoglobin chain can drastically reduce the molecule's oxygen-carrying capacity.

Biological Order, Regulation, and Homeostasis

Both blood sugar levels and body weight in humans are controlled by complex homeostatic mechanisms. The level of glucose in your blood is regulated by factors acting on the liver, which stimulate either glycogen breakdown (which increases blood sugar) or glycogen formation (which decreases blood sugar). Your body weight is, at least partly, determined by factors emanating from fat cells which either increase metabolic rate (which tends to decrease body weight) or slow down metabolic rate (which tends to increase body weight).

Acquiring and Utilizing Energy

The chemical energy that fuels biological activities is stored primarily in two types of macromolecules: polysaccharides and fats. Polysaccharides, including starch in plants and glycogen in animals, function primarily in the short-term storage of chemical energy. These polysaccharides can be rapidly broken down to sugars, such as glucose, which are readily metabolized to release energy. Gram-for-gram, fats contain even more energy than polysaccharides and function primarily as a long-term storage of chemical energy.

Unity within Diversity

All organisms, from bacteria to humans, are composed of the same four families of macromolecules, illustrating the unity of life—even at the biochemical level. The precise nature of these macromolecules and the ways they are organized into higher structures differ from organism to organism, thereby building diversity. Plants, for example, polymerize glucose into starch and cellulose, while animals polymerize glucose into glycogen. Similarly, many proteins (such as hemoglobin) are present in a variety of organisms, but the precise amino acid sequence of the protein varies from one species to the next.

Evolution and Adaptation

Evolution becomes very apparent at the molecular level when we compare the structure of macromolecules among diverse organisms. Analysis of the amino acid sequences of proteins and the nucleotide sequences of nucleic acids reveals a gradual change over time in the structure of macromolecules. Organisms that are closely related have proteins and nucleic acids whose sequences are more similar than are those of distantly related organisms. To a large degree, the differences observed among diverse organisms derives from the evolutionary differences in nucleic acid and protein sequences.

The segregation of alleles and their independent assortment during meiosis increase genotype diversity by promoting new combinations of genes. But the shuffling of existing genes alone does not explain the presence of such a vast diversity of life. If all organisms descended from a common ancestor, with its relatively small complement of genes, where did all the genes present in today's millions of species come from? The answer is mutation.

Most mutant alleles are detrimental; that is, they are more likely to disrupt a well-ordered, smoothly functioning organism than to increase the organism's fitness. For example, a mutation might change a gene so that it produces an inactive enzyme needed for a critical life function. Occasionally, however, one of these stable genetic changes creates an advantageous characteristic that [increases the] fitness of the offspring. In this way, muta[tions provide] raw material for evolution and the diversi[ty of life on] earth.

One of the requirements for genes is stabi[lity . . . they must] remain basically the same from generation to [generation, or] the fitness of organisms would rapidly dete[riorate. At the] same time, there must be some capacity [for muta]tion. Alterations in genes do occur, albeit rar[ely. These] changes (mutations) represent the raw ma[terial for evolu]tion. (See CTQ #7.)

REEXAMINING THE THEMES

Biological Order, Regulation, and Homeostasis

Mendel discovered that the transmission of genetic factors followed a predictable pattern, indicating that the processes responsible for the formation of gametes, including the segregation of alleles, must occur in a highly ordered manner. This orderly pattern can be traced to the process of meiosis and the precision with which homologous chromosomes are separated during the first meiotic division. Mendel's discovery of independent assortment can also be connected with the first meiotic division, when each pair of homologous chromosomes becomes aligned at the metaphase plate in a manner that is independent of other pairs of homologues.

Unity within Diversity

All eukaryotic, sexually reproducing organisms follow the same "rules" for transmitting inherited traits. Although Mendel chose to work with peas, he could have come to the same conclusions had he studied fruit flies or mice or had he scrutinized a family's medical records on the transmission of certain genetic diseases, such as cystic fibrosis. [While] the mechanism by which genes are transmitted is [similar,] the genes themselves are highly diverse from one [species] to the next. It is this genetic difference among sp[ecies that] forms the very basis of biological diversity

Evolution and Adaptation

Mendel's findings provided a critical link in ou[r knowl]edge of the mechanism of evolution. A key tene[t of the] theory of evolution is that favorable genetic variat[ions in]crease the likelihood that an individual will survive [to re]productive age and that its offspring will exhibit these [same] favorable characteristics. Mendel's demonstration [that] units of inheritance pass from parents to offspring w[ithout] being blended revealed the means by which advanta[geous] traits could be preserved in a species over many ge[nera]tions. The subsequent discovery of genetic change by [mu]tation revealed how new genes appeared in a popula[tion,] thus providing the raw material for evolution.

SYNOPSIS

Gregor Mendel discovered the pattern by which inherited traits are transmitted from parents to offspring. Mendel discovered that inherited traits were controlled by pairs of factors (genes). The two factors for a given trait in an individual could be identical (homozygous) or different (heterozygous). In heterozygotes, one of the [two] gene variants (alleles) may be dominant over the other [, the] recessive allele. Because of dominance, the appearance [(phenotype) of the heterozygote (genotype of *Aa*) is identi]cal to that of the homozygote with two dominant alleles

Students will naturally find many ways in which the material presented in any biology course relates to them. But it is not always obvious how you can use biological information for better living or how it might influence your life. Your ability to see yourself in the course boosts interest and heightens the usefulness of the information. This translates into greater retention and understanding.

To accomplish this desirable outcome, the entire book has been constructed with you—the student—in mind. Perhaps the most notable feature of this approach is a series of boxed essays called "The Human Perspective" that directly reveals the human relevance of the biological topic being discussed at that point in the text. You will soon realize that human life, including your own, is an integral part of biology.

PART 2 / *Chemical and Cellular Foundations of Life*

◁ THE HUMAN PERSPECTIVE ▷
Obesity and the Hungry Fat Cell

FIGURE 1
Actor Robert DeNiro in (*left*) a scene from the movie *Raging Bull* and (*right*) a recent photograph.

It has become increasingly clear in recent years that people who are exceedingly overweight—that is, obese—are at increased risk of serious health problems, including heart disease and cancer. By most definitions, a person is obese if he or she is about 20 percent above "normal" or desirable body weight. Approximately 35 percent of adults in the United States are considered obese by this definition, twice as many as at the turn of the century. Among young adults, high blood pressure is five times more prevalent and diabetes three times more prevalent in a group of obese people than in a group of people who are at normal weight. Given these statistics, together with the social stigma facing the obese, there would seem to be strong motivation for maintaining a "normal" body weight. Why, then, are so many of us so overweight? And, why is it so hard to lose unwanted pounds and yet so easy to gain them back? The answers go beyond our fondness for high-calorie foods.

Excess body fat is stored in fat cells (*adipocytes*) located largely beneath the skin. These cells can change their volume more than a hundredfold, depending on the amount of fat they contain. As a person gains body fat, his or her fat cells become larger and larger, accounting for the bulging, sagging body shape. If the person becomes sufficiently overweight, and their fat cells approach their maximum fat-carrying capacity, chemical messages are sent through the blood, causing formation of new fat cells that are "hungry" to begin accumulating their own fat. Once a fat cell is formed, it may expand or contract in volume, but it appears to remain in the body for the rest of the person's life.

Although the subject remains controversial, current research findings suggest that body weight is subject to physiologic regulation in humans. Apparently, each person has a particular weight that his or her body's regulatory machinery acts to maintain. This particular value—whether 40 kilograms (80 pounds) or 200 kilograms (400 pounds)—is referred to as the person's **set-point.**

People maintain their body weight at a relatively constant value by balancing energy intake (in the form of food calories) with energy expenditure (in the form of calories burned by metabolic activities or excreted). Obese individuals are thought to have a higher set-point than do persons of normal weight. In many cases, the set-point value appears to have a strong genetic component. For instance, studies reveal there is no correlation between the body mass of adoptees and their adoptive parents, but there is a clear relationship between adoptees and their biological parents, with whom they have not lived.

The existence of a body-weight set-point is most evident when the body weight of a person is "forced" to deviate from the regulated value. Individuals of normal body weight who are fed large amounts of high-calorie foods under experimental conditions tend to have increasing amounts of weight. If these people cease their energy-rich diets, however, they return quite rapidly to their previous levels, at which point further weight loss stops. This is illustrated by actor Robert DeNiro, who reportedly gained about 50 pounds for the filming of the movie "Raging Bull" (Figure 1), and then lost the weight prior to his next acting role. Conversely, a person who is put on a strict, low-calorie diet will begin to lose weight. The drop in body weight soon triggers a decrease in the person's resting metabolic rate; that is, the amount of calories burned when the person is not engaged in physical activity. The drop in metabolic rate is the body's compensatory measure for the decreased food intake. In other words, it is the body's attempt to halt further weight loss. This effect is particularly pronounced among obese people who diet and lose large amounts of weight: Their pulse rate and blood pressure drop markedly, their fat cells shrink to "ghosts" of their former selves, and they tend to be continually hungry. If these obese individuals go back to eating a *normal* diet, they tend to regain the lost weight rapidly. The drive of these formerly obese persons to increase their food intake is probably a response to chemical signals emanating from the fat cells as they shrink below their previous size.

---

630 • PART 5 / *Form and Function of Animal Life*

◁ THE HUMAN PERSPECTIVE ▷
Dying for a Cigarette?

e average, smoking cigarettes will cut ximately 6 to 8 years off your life, han 5 minutes for every cigarette l Cigarette smoking is the greatest preventable death in the United according to a 1991 report by the for Disease Control (CDC), 0,000 Americans die each year ng-related causes. Smoking ac- 87 percent of all lung-cancer smokers are more susceptible he esophagus, larynx, mouth, bladder than are nonsmok- reased incidence of lung among smokers compared to hown in Figure 1a, and the l by quitting is shown in ffects of smoking on lung Figure 2. Atherosclero- and peptic ulcers also greater frequency than rs. For example, long- 5 times more likely to erial disease than are ysema (a condition ction of lung tissue, culty in breathing) mmation of the air- prevalent among

ger other people. sponsible for the nocent bystand- re the same air passive (invol- own; second- seriously ill rs have dou- ry infections osed to to- ng married us; 20 per- long non- ributable to inhaling other people's tobacco smoke. Another "innocent bystander" is a fetus developing in the uterus of a woman who smokes. Smoking increases the incidence of miscarriage and stillbirth and decreases the birthweight of the infant. Once born, these babies suffer twice as many respiratory infections as do babies of nonsmoking mothers.

Why is smoking so bad for your health? The smoke emitted from a burning cigarette contains more than 2,000 identifiable substances, many of which are either irritants or carcinogens. These compounds include carbon monoxide, sulfur dioxide, formaldehyde, nitrosamines, toluene, ammonia, and radioactive isotopes. Autopsies of respiratory tissues from smokers and from nonsmokers who have lived for long periods with smokers) show widespread cellular changes, including the presence of precancerous cells (cells that may become malignant, given time) and a marked reduction in the number of cilia that play a vital role in the removal of bacteria and debris from the airways.

Of all the compounds found in tobacco (including smokeless varieties), the most important is nicotine, not because it is carcinogenic, but because it is so addictive. Nicotine is addictive because it acts like a neurotransmitter by binding to certain acetylcholine receptors (page 477), stimulating postsynaptic neurons. The physiological effects of this stimulation include the release of epinephrine, an increase in blood sugar, an elevated heart rate, and the constriction of blood vessels, causing elevated blood pressure. A smoker's nervous system becomes "accustomed" to the presence of nicotine and decreases the output of the natural neurotransmitter. As a result, when a person tries to stop smoking, the sudden absence of nicotine, together with the decreased level of the natural transmitter, decreases stimulation of postsynaptic neurons, which creates a craving for a cigarette—a "nicotine fit." Ex-smokers may be so conditioned to the act of smoking that the craving for cigarettes can continue long after the physiological addiction disappears.

(a) Lung cancer deaths vs. Age — Smokers, Nonsmokers

(b) Elevation of lung cancer deaths above that for nonsmokers vs. Years after smoking stopped

*T*he "Biolines" are boxed essays that highlight fascinating facts, applications, and real-life lessons, enlivening the mainstream of biological information. Many are remarkable stories that reveal nature to be as surprising and interesting as any novelist could imagine.

◁ B I O L I N E ▷
DNA Fingerprints and Criminal Law

On February 5, 1987, a woman and her 2-year-old daughter were found stabbed to death in their apartment in the New York City borough of the Bronx. Following a tip, the police questioned a resident of a neighboring building. A small bloodstain was found on the suspect's watch, which was sent to a laboratory for DNA fingerprint analysis. The DNA from the white blood cells in the stain was amplified using the PCR technique and was digested with a restriction enzyme. The restriction fragments were then separated by electrophoresis, and a pattern of labeled fragments was identified with a radioactive probe. The banding pattern produced by the DNA from the suspect's watch was found to be a perfect match to the pattern produced by DNA taken from one of the victims. The results were provided to the opposing attorneys, and a pretrial hearing was called in 1989 to discuss the validity of the DNA evidence.

During the hearing, a number of expert witnesses for the prosecution explained the basis of the DNA analysis. According to these experts, no two individuals, with the exception of identical twins, have the same nucleotide sequence in their DNA. Moreover, differences in DNA sequence can be detected by comparing the lengths of the fragments produced by restriction-enzyme digestion of different DNA samples. The patterns produce a "DNA fingerprint" (Figure 1) that is as unique to an individual as is a set of conventional fingerprints lifted from a glass. In 1989, DNA fingerprints had already been used in more than 200 criminal cases in the United States and had been hailed as the most important development in forensic science (the application of medical facts

FIGURE 1
Alec Jeffreys of the University of Leicester, England, examining a DNA fingerprint. Jeffreys was primarily responsible for developing the DNA fingerprint technique and was the scientist who confirmed the death of Josef Mengele.

to legal problems) in decades. The widespread use of DNA fingerprinting evidence in court had been based on its general acceptability in the scientific community. According to a report from the company performing the DNA analysis, the likelihood that the same banding patterns could be obtained by chance from two *different* individuals in the community was only one in 100 million.

What made this case (known as the Castro case, after the defendant) memorable and distinct from its predecessors was that the defense also called on expert witnesses to scrutinize the data and to present

their opinions. While these experts confirmed the capability of DNA fingerprinting to identify an individual out of a large population, they found serious technical flaws in the analysis of the DNA samples used by the prosecution. In an unprecedented occurrence, the experts who earlier testified *for the prosecution* agreed that the DNA analysis in this case was unreliable and should not be used as evidence! The problem was not with the technique itself but in the way it had been carried out in this particular case. Consequently, the judge threw out the evidence.

In the wake of the Castro case, the use of DNA fingerprinting to decide guilt or innocence has been seriously questioned. Several panels and agencies are working to formulate guidelines for the licensing of forensic DNA laboratories and the certification of their employees. In 1992, a panel of the National Academy of Sciences released a report endorsing the general reliability of the technique but called for the institution of strict standards *to be set by scientists.*

Meanwhile, another issue regarding DNA fingerprinting has been raised and is hotly debated. Two geneticists, Richard Lewontin of Harvard University and Daniel Hartl of Washington University, coauthored a paper published in December 1991, suggesting that scientists do not have enough data on genetic variation within different racial or ethnic groups to calculate the odds that two individuals—a suspect and a perpetrator of the crime—are one and the same on the basis of an identical DNA fingerprint. The matter remains an issue of great concern in both the scientific and legal communities and has yet to be resolved.

◁ B I O L I N E ▷
The Fish That Changes Sex

In vertebrates, gender is generally a biologically inflexible commitment: An individual develops into either a male or a female as dictated by the sex chromosomes acquired from one's parents. Yet, even among vertebrates, there are organisms that can reverse their sexual commitment. The Australian cleaner fish (Figure 1), a small animal that sets up "cleaning stations" to which larger fishes come for parasite removal, can change its gender in response to environmental demands. Most male cleaner fish travel alone rather than with a school. Except for a single male, schools of cleaner fish are comprised entirely of females. Although it might seem logical to conclude that maleness engenders solo travel, it is actually the other way around: Being alone fosters maleness. A cleaner fish that develops away from a school *becomes* a male, whereas the same fish developing in a school would have become a female.

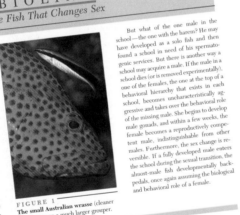

FIGURE 1
The small Australian wrasse (cleaner fish) is seen on a much larger grouper.

But what of the one male in the school—the one with the harem? He may have developed as a solo fish and then found a school in need of his spermatogenic services. But there is another way a school may acquire a male. If the male in a school dies (or is removed experimentally), one of the females, the one at the top of each behavioral hierarchy that exists in each school, becomes uncharacteristically aggressive and takes over the behavioral role of the missing male. She begins to develop male gonads, and within a few weeks, the female becomes a reproductively competent male, indistinguishable from other males. Furthermore, the sex change is reversible. If a fully developed male enters the school during the sexual transition, the almost-male fish developmentally backpedals, once again assuming the biological and behavioral role of a female.

⚠ Not all organisms follow the mammalian pattern of sex determination. In some animals, most notably birds, the opposite pattern is found: The female's cells have an X and a Y chromosome, while the male's cells have two Xs. An exception to this rule of a strict relation between sex and chromosomes is discussed in the Bioline: The Fish That Changes Sex. Although some plants possess sex chromosomes and gender distinctions between individuals, most have only autosomes; consequently, each individual produces both male and female parts.

SEX LINKAGE

For fruit flies and humans alike, there are hundreds of genes on the X chromosome that have no counterpart on the smaller Y chromosome. Most of these genes have nothing to do with determining gender, but their effect on phenotype usually *depends on* gender. For example, in females, a recessive allele on one X chromosome will be masked (and not expressed) if a dominant counterpart resides on the other X chromosome. In males, it only takes one recessive allele on the single X chromosome to determine the individual's phenotype since there is no corresponding allele on the Y chromosome. Inherited characteristics determined by genes that reside on the X chromosome are called **X-linked characteristics.**

So far, some 200 human X-linked characteristics have been described, many of which produce disorders that are found almost exclusively in men. These include a type of heart-valve defect (*mitral stenosis*), a particular form of mental retardation, several optical and hearing impairments, muscular dystrophy, and red-green colorblindness (Figure 13-8).

One X-linked recessive disorder has altered the course of history. The disease is **hemophilia,** or "bleeder's disease," a genetic disorder characterized by the inability to produce a clotting factor needed to halt blood flow quickly following an injury. Nearly all hemophiliacs are males. Although females can inherit two recessive alleles for hemophilia, this occurrence is extremely rare. In general, women who have acquired the rare defective allele are heterozygous **carriers** for the disease. The phenotype of a carrier

Bioethics Essays

*S*everal ethical issues are discussed in the Bioethics essays which add provocative pauses throughout the text. Biological Science does not operate in a vacuum but has profound consequences on the general community. Because biologists study life, the science is peppered with ethical considerations. The moral issues discussed in these essays are neither simple nor easy to resolve, and we do not claim to have any certain answers. Our goal is to encourage you to consider the bioethical issues that you will face now and in the future.

◁ B I O E T H I C S ▷
Blurring the Line between Life and Death
By ARTHUR CAPLAN
Division of the Center for Biomedical Ethics at the University of Minnesota

Theresa Ann Campo Pearson didn't have a very long life. When she died in 1992, she was only 10 days old. Despite her short life, she became the center of a very strange, sad, and wrenching ethical controversy. Theresa died because her brain had failed to form. She had anencephaly, a condition in which only the brainstem, located at the top of the spinal cord, is present. Her parents wanted to donate Theresa's organs; the courts said no. Some people found it strange that Theresa's parents, Laura Campo and Justin Pearson, did not get their way. Why not allow donation, when every day in North America a baby dies because there is no heart, lung, or liver available for transplantation?

Anencephaly is best described as completely "unabling," not disabling. Children born with anencephaly cannot think, feel, sense, or be aware of the world. Many are stillborn; the majority of the rest die within days of birth. A mere handful live for a few weeks. Theresa's parents

knew all this. But rather than abort the pregnancy, they chose to have their baby. In fact, the baby was born by Caesarean section, at least partly in the hope that it would be born alive, thereby making organ donation possible. When Theresa died at Broward General Medical Center in Fort Lauderdale, Florida, however, no organs were taken. Two Florida courts ruled that the baby could not be used as a source of organs unless she was brain-dead, and Theresa Ann Campo was never pronounced brain-dead.

Brain death refers to a situation in which the brain has irreversibly lost all function and activity. Babies born with anencephaly have some brain function in their brainstem so, while they cannot think or feel, they are alive. According to Florida law—and the law in more than 40 other states—only those individuals declared brain-dead can donate organs. The courts of Florida had no other option but to deny the request for organ donation.

One obvious solution is to change the law so that states could decide that organs can be removed upon parental consent from either those who are born brain-dead or from babies who are born with anencephaly. Another solution is to rewrite the definition of death to say that death occurs either when the brain has totally ceased to function or if a baby is born anencephalic. Do you feel that either of these changes should be made? Some may argue that medicine will fudge the line between life and death in order to get organs for transplant. Do you agree with this concern? How do you think redefining death will affect a person's decision to check off the donation box on the back of a driver's license? Do you think people may worry that if they are known to be potential donors they won't be aggressively treated at the hospital? In your opinion, would changing the definition of death to include anencephaly be beneficial or deleterious?

Like the brain, the spinal cord is composed of white matter (myelinated axons) and gray matter (dendrites and cell bodies). However, the arrangement of these types of matter is reversed in the spinal cord, compared to their arrangement in the brain: The spinal cord's white matter surrounds the gray matter (Figure 23-16).

The human central nervous system is the most complex and highly evolved assembly of matter. Among its functions are the processing of sensory information collected from both the external and internal environment; the regulation of internal physiological activities; the coordination of complex motor activities; and the endowment of such intangible "mental" qualities as emotions, creativity, language, and the ability to think, learn, and remember. (See CTQ #6.)

ARCHITECTURE OF THE PERIPHERAL NERVOUS SYSTEM

The peripheral nervous system provides the neurological bridge between the central nervous system and the various parts of the body. The peripheral nervous system is made up of paired nerves that extend into the periphery from the CNS at various levels along the body. Each nerve is composed of a large bundle of myelinated axons surrounded by a connective tissue sheath. Twelve pairs of **cranial nerves** emerge from the central stalk of the human brain, and 31 pairs of **spinal nerves** extend from the spinal cord out between the vertebrae of humans (Figure 23-16). For the most part, the cranial nerves *innervate* (supply nerves to) tissues and organs of the head and neck, whereas the spinal nerves innervate the chest, abdomen, and limbs.

Additional Pedagogical Features

We have worked to assure that each chapter in this book is an effective teaching and learning instrument. In addition to the pedagogical features discussed above, we have included some additional tried-and-proven-effective tools.

KEY POINTS

Key points follow each major section and offer a condensation of the relevant facts and details as well as the concepts discussed. You can use these key points to reaffirm your understanding of the previous reading or to alert you to misunderstood material before moving on to the next topic. Each key point is tied to a Critical Thinking Question found at the end of the chapter; together, they encourage you to analyze the information, taking it beyond mere memorization.

Plant Tissues and Organs / CHAPTER 18 • 361

▶ Many plants replenish old and dying cells with vigorous new cells. But since each plant cell has a surrounding cell wall (Chapter 7) old plant cells do not just wither and disappear when they die. Instead, dead plant cells leave cellular "skeletons" where they once lived. As a result, the longer a plant lives, the more complex its anatomy becomes. **Annuals** are plants that live for 1 year or less, such as corn and marigolds. Because they live for such a brief period, these plants do not completely replace old cells. As a result, annuals are anatomically less complex than are **biennials**—plants that live for 2 years—and **perennials**—herbs, shrubs, and trees that live longer than 2 years. Biennials (carrots, Queen Anne's lace) and perennials (rosebushes, apple trees) are able to live longer than annuals because they produce new cells to replace those that cease functioning or die, providing a continual supply of young, vigorous cells.

⬛ In this chapter, we will focus on the body construction of flowering plants, the most familiar, most evolutionarily advanced, and structurally complex of any group in the plant kingdom. All flowering plants are **vascular plants;** that is, they contain specialized cells that circulate water, minerals, and food (organic molecules) throughout the plant. Botanists divide flowering plants into two main groups: **dicotyledons,** or dicots (*di* = two, *cotyledon* = embryonic seed leaf), and **monocotyledons,** or monocots (*mono* = one). Table 18-1 illustrates the many differences that distinguish dicots from monocots and will be used as a reference throughout the chapter.

SHOOTS AND ROOTS

The flowering plant body is a study in contradictions. A typical plant grows through the soil and the air simultaneously, two very different habitats with very different conditions. As a result, the two main parts of the plant differ dramatically in form (anatomy) and function (physiology): The underground **root system** anchors the plant in the soil and absorbs water and nutrients, while the aerial **shoot system** absorbs sunlight and gathers carbon dioxide for photosynthesis (Figure 18-2). The shoot system also produces stems, leaves, flowers, and fruits. Interconnected vascular tissues transport materials between the aerial shoot system and the underground root system. These connections allow water and minerals absorbed by the root to be conducted to shoot tissues, and for food produced by the shoot to be transported to root tissues. We will discuss the various components of these two systems in more detail later in the chapter.

Over 90 percent of all plant species are flowering plants. Flowering plants are the most recently evolved plant group, having undergone rapid evolution during the past 1 million to 2 million years as environmental conditions on land became more variable. (See CTQ #2.)

TABLE 18-1

Dicot and Monocot Comparison	DICOT	MONOCOT
Embryo	2 Cotyledons	1 Cotyledon
Flowers	Parts in 4s or 5s	Parts in 3s
Leaves	Net veined	Veins parallel
Leaf anatomy	Two types of photosynthetic cells	One type of photosynthetic
Roots	One main root (tap root system)	Many main roots (fibrous root system)
Stem anatomy	Vascular bundles in rings	Scattered vascular bundles
Root anatomy	Xylem in center	Pith in center
Secondary growth	Yes	No

---

288 • PART 3 / *The Genetic Basis of Life*

the corresponding polypeptide. The cumulative effect of gradual changes in polypeptides over evolutionary time has been the generation of life's diversity.

Evolution and Adaptation

▶ Evolutionary change from generation to generation depends on genetic variability. Much of this variability arises from reshuffling maternal and paternal genes during meiosis, but somewhere along the way *new* genetic information must be introduced into the population. [...] netic information arises from mutations in existing [...] Some of these mutations arise during replication [...] occur as the result of unrepaired damage as the DN[...] "sitting" in a cell. Mutations that occur in an indi[...] germ cells can be considered the raw material on[...] natural selection operates; whereas harmful mutatio[...] duce offspring with a reduced fitness, beneficial mu[...] produce offspring with an increased fitness.

SYNOPSIS

Experiments in the 1940s and 1950s established conclusively that DNA is the genetic material. These experiments included the demonstration that DNA was capable of transforming bacteria from one genetic strain to another; that bacteriophages injected their DNA into a host cell during infection; and that the injected DNA was transmitted to the bacteriophage progeny.

DNA is a double helix. DNA is a helical molecule consisting of two chains of nucleotides running in opposite directions, with their backbones on the outside, and the nitrogenous bases facing inward like rungs on a ladder. Adenine-containing nucleotides on one strand always pair with thymine-containing nucleotides on the other strand, likewise for guanine- and cytosine-containing nucleotides. As a result, the two strands of a DNA molecule are complementary to one another. Genetic information is encoded in the specific linear sequence of nucleotides that make up the strands.

DNA replication is semiconservative. During replication, the double helix separates, and each strand serves as a template for the formation of a new, complementary strand. Nucleotide assembly is carried out by the enzyme DNA polymerase, which moves along the two strands in opposite directions. As a result, one of the strands is synthesized continuously, while the other is synthesized in segments that are covalently joined. Accuracy is maintained by a proofreading mechanism present within the polymerase.

Information flows in a cell from DNA to RNA to protein. Each gene consists of a linear sequence of nucleotides that determines the linear sequence of amino acids in a polypeptide. This is accomplished in two [...] steps: transcription and translation.

During transcription, the information spelled ou[...] the gene's nucleotide sequence is encoded in a m[...] cule of messenger RNA (mRNA). The mRNA cont[...] a series of codons. Each codon consists of three nucleoti[...] Of the 64 possible codons, 61 specify an amino acid, and [...] other 3 stop the process of protein synthesis.

During translation, the sequence of codons in [...] mRNA is used as the basis for the assembly of a cha[...] of specific amino acids. Translating mRNA mes[...] occurs on ribosomes and requires tRNAs, which se[...] decoders. Each tRNA is folded into a cloverleaf stru[...] with an anticodon at one end—which binds to a comp[...] mentary codon in the mRNA—and a specific amino acid [...] the other end—which becomes incorporated into th[...] growing polypeptide chain. Amino acids are added to t[...] appropriate tRNAs by a set of enzymes. The sequenti[...] interaction of charged tRNAs with the mRNA results in th[...] assembly of a chain of amino acids in the precise orde[...] dictated by the DNA.

Mutation is a change in the genetic message. Gen[...] mutations may occur as a single nucleotide substitution[...] which leads to the insertion of an amino acid different from [...] that originally encoded. In contrast, the addition of one or [...] two nucleotides throws off the reading frame of the ribo[...] some as it moves along the mRNA, leading to the incorpo[...] ration of incorrect amino acids "downstream" from the [...] point of mutation. Exposure to mutagens increases the rate [...] of mutation.

SYNOPSIS

The synopsis section offers a convenient summary of the chapter material in a readable narrative form. The material is summarized in concise paragraphs that detail the main points of the material, offering a useful review tool to help reinforce recall and understanding of the chapter's information.

Additional Pedagogical Features

REVIEW QUESTIONS

Along with the synopsis, the Review Questions provide a convenient study tool for testing your knowledge of the facts and processes presented in the chapter.

STIMULATING CRITICAL THINKING

Each chapter contains as part of its end material a diverse mix of Critical Thinking Questions. These questions ask you to apply your knowledge and understanding of the facts and concepts to hypothetical situations in order to solve problems, form hypotheses, and hammer out alternative points of view. Such exercises provide you with more effective thinking skills for competing and living in today's complex world.

---

224 • PART 2 / Chemical and Cellular Foundations of Life

Key Terms

zygote (p. 214)
meiosis (p. 214)
life cycle (p. 214)
germ cell (p. 214)
somatic cell (p. 214)
meiosis I (p. 216)

reduction division (p. 216)
synapsis (p. 216)
tetrad (p. 216)
crossing over (p. 216)
genetic recombination (p. 216)
synaptonemal complex (p. 218)

maternal chromosome (p. 219)
paternal chromosome (p. 219)
independent assortment (p. 219)
meiosis II (p. 219)

Review Questions

1. Match the activity with the phase of meiosis in which it occurs.

   a. synapsis
   b. crossing over
   c. kinetochores split
   d. independent assortment
   e. homologous chromosomes separate
   f. cytokinesis

   1. prophase I
   2. metaphase I
   3. anaphase I
   4. telophase I
   5. prophase II
   6. anaphase II
   7. telophase II

2. How do crossing over and independent assortment increase the genetic variability of a species?

3. Why is meiosis I (and not meiosis II) referred to as the reduction division?

4. Suppose that one human sperm contains x amount of DNA. How much DNA would a cell just entering meiosis contain? A cell entering meiosis II? A cell just completing meiosis II? Which of these three cells would have a haploid number of chromosomes? A diploid number of chromosomes?

Critical Thinking Questions

1. Why are disorders, such as Down syndrome, that arise from abnormal chromosome numbers, characterized by a number of seemingly unrelated abnormalities?

2. A gardener's favorite plant had white flowers and long seed pods. To add some variety to her garden, she transplants some plants of the same type, but with pink flowers and short seed pods from her neighbor's garden. To her surprise, in a few generations, she grows plants with white flowers and short seed pods and plants with pink flowers and long seed pods, as well as the original combinations. What are two ways in which these new combinations could have arisen?

3. Set up the meiosis template in the diagram below on a large sheet of paper. Then use pieces of colored yarn or pipe cleaners to simulate chromosomes and make a model of the phases of meiosis. (See template on opposite page)

4. Would you expect two genes on the same chromosome, such as yellow flowers and short stems, always to be exchanged during crossing over? How might they remain together in spite of crossing over?

5. Suppose paternal chromosomes always lined up on the same side of the metaphase plate of cells in meiosis I. How would this affect genetic variability of offspring? Would they all be identical? Why or why not?

Additional Readings

Chandley, A. C. 1988. Meiosis in man. *Trends in Gen.* 4:79–83. (Intermediate)

Hsu, T. C. 1979. *Human and mammalian cytogenetics.* New York: Springer-Verlag. (Intermediate)

John, B. 1990. *Meiosis.* New York: Cambridge University Press. (Advanced)

Moens, P. B. 1987. *Meiosis.* Orlando: Academic. (Advanced)

Patterson, D. 1987. The causes of Down syndrome. *Sci. Amer.* Feb:52–60. (Intermediate-Advanced)

White, M. J. D. 1973. *The chromosomes.* Halsted. (Advanced)

ADDITIONAL READINGS

Supplementary readings relevant to the Chapter's topics are provided at the end of every chapter. These readings are ranked by level of difficulty (introductory, intermediate, or advanced) so that you can tailor your supplemental readings to your level of interest and experience.

The appendices of this edition include "Careers in Biology," a frequently overlooked aspect of our discipline. Although many of you may be taking biology as a requirement for another major (or may have yet to declare a major), some of you are already biology majors and may become interested enough to investigate the career opportunities in life sciences. This appendix helps students discover how an interest in biology can grow into a livelihood. It also helps the instructor advise students who are considering biology as a life endeavor.

APPENDIX

D

Careers in Biology

Although many of you are enrolled in biology as a requirement for another major, some of you will become interested enough to investigate the career opportunities in life sciences. This interest in biology can grow into a satisfying livelihood. Here are some facts to consider:

- Biology is a field that offers a very wide range of possible science careers

- Biology offers high job security since many aspects of it deal with the most vital human needs: health and food

- Each year in the United States, nearly 40,000 people obtain bachelor's degrees in biology. But the number of newly created and vacated positions for biologists is increasing at a rate that exceeds the number of new graduates. Many of these jobs will be in the newer areas of biotechnology and bioservices.

Biologists not only enjoy job satisfaction, their work often changes the future for the better. Careers in medical biology help combat diseases and promote health. Biologists have been instrumental in preserving the earth's life-supporting capacity. Biotechnologists are engineering organisms that promise dramatic breakthroughs in medicine, food production, pest management, and environmental protection. Even the economic vitality of modern society will be increasingly linked to biology.

Biology also combines well with other fields of expertise. There is an increasing demand for people with backgrounds or majors in biology complexed with such areas as business, art, law, or engineering. Such a distinct blend of expertise gives a person a special advantage.

The average starting salary for all biologists with a Bachelor's degree is $22,000. A recent survey of California State University graduates in biology revealed that most were earning salaries between $20,000 and $50,000. But as important as salary is, most biologists stress job satisfaction, job security, work with sophisticated tools and scientific equipment, travel opportunities (either to the field or to scientific conferences), and opportunities to be creative in their job as the reasons they are happy in their career.

Here is a list of just a few of the careers for people with degrees in biology. For more resources, such as lists of current openings, career guides, and job banks, write to Biology Career Information, John Wiley and Sons, 605 Third Avenue, New York, NY 10158.

A SAMPLER OF JOBS THAT GRADUATES HAVE SECURED IN THE FIELD OF BIOLOGY*

Agricultural Biologist	Bioanalytical Chemist	Brain Function	Environmental Center
Agricultural Economist	Biochemical/Endocrine	Researcher	Director
Agricultural Extension	Toxicologist	Cancer Biologist	Environmental Engineer
Officer	Biochemical Engineer	Cardiovascular Biologist	Environmental Geographer
Agronomist	Pharmacology Distributor	Cardiovascular/Computer	Environmental Law Specialist
Amino-acid Analyst	Pharmacology Technician	Specialist	Farmer
Analytical Biochemist	Biochemist	Chemical Ecologist	Fetal Physiologist
Anatomist	Biogeochemist	Chromatographer	Flavorist
Animal Behavior	Biogeographer	Clinical Pharmacologist	Food Processing Technologist
Specialist	Biological Engineer	Coagulation Biochemist	Food Production Manager
Anticancer Drug Research	Biologist	Cognitive Neuroscientist	Food Quality Control
Technician	Biomedical	Computer Scientist	Inspector
Antiviral Therapist	Communication Biologist	Dental Assistant	Flower Grower
Arid Soils Technician	Biometerologist	Ecological Biochemist	Forest Ecologist
Audio-neurobiologist	Biophysicist	Electrophysiology/	Forest Economist
Author, Magazines & Books	Biotechnologist	Cardiovascular Technician	Forest Engineer
Behavioral Biologist	Blood Analyst	Energy Regulation Officer	Forest Geneticist
Bioanalyst	Botanist	Environmental Biochemist	Forest Manager

Study Guide

Written by Gary Wisehart and Michael Leboffe of San Diego City College, the *Study Guide* has been designed with innovative pedagogical features to maximize your understanding and retention of the facts and concepts presented in the text. Each chapter in the *Study Guide* contains the following elements.

Concepts Maps

In Chapter 1 of the *Study Guide*, the beginning of a concept map stating the five themes is introduced. In each subsequent chapter, the concept map is expanded to incorporate topics covered in each chapter as well as the interconnections between chapters and the five themes. "Connector" phrases are used to link the concepts and themes, and the text icons representing the themes are incorporated into the concept maps.

Go Figure!

In each chapter, questions are posed regarding the figures in the text. Students can explore their understanding of the figures and are asked to think critically about the figures based on their understanding of the surrounding text and their own experiences.

Self-Tests

Each chapter includes a set of matching and multiple-choice questions. Answers to the Study Guide questions are provided.

Concept Map Construction

The student is asked to create concept maps for a group of terms, using appropriate connector phrases and adding terms as necessary.

Biology: Exploring Life, Second Edition is supplemented by a comprehensive *Laboratory Manual* containing approximately 60 lab exercises chosen by the text authors from the National Association of Biology Teachers. These labs have been thoroughly class-tested and have been assembled from various scientific publications. They include such topics as

- Chaparral and Fire Ecology: Role of Fire in Seed Germination (*The American Biology Teacher*)

- A Model for Teaching Mitosis and Meiosis (*American Biology Teacher*)

- Laboratory Study of Climbing Behavior in the Salt Marsh Snail (*Oceanography for Landlocked Classrooms*)

- Down and Dirty DNA Extraction (*A Sourcebook of Biotechnology Activities*)

- Bioethics: The Ice-Minus Case (*A Sourcebook of Biotechnology Activities*)

- Using Dandelion Flower Stalks for Gravitropic Studies (*The American Biology Teacher*)

- pH and Rate of Enzymatic Reactions (*The American Biology Teacher*)

LABORATORY MANUAL

SECOND EDITION

Biology
Exploring Life

Biology:
Exploring Life

Biology: The Study of Life

To understand life you must explore the obvious and the subtle, as well as all levels in between. This translucent jellyfish represents the organismal level of biological organization. But to understand how a jellyfish or any other organism survives, grows, and reproduces, biologists must study all levels of organization, including the organs, tissues, cells, and even molecules that make up an organism, as well as the population, community and ecosystem in which the organism lives.

PART
· 1 ·

CHAPTER

◄ **1** ►

Biology: Exploring Life

**STEPS
TO
DISCOVERY**
Exploring Life—the First Step

DISTINGUISHING THE LIVING FROM THE INANIMATE

Organisms are Highly Complex and Organized

Organisms Are Composed Of Cells

Organisms Acquire And Utilize Energy

Organisms Produce Offspring Similar To Themselves

Organisms Are Built According To Genetic Instructions

Organisms Grow In Size And Change by Development

Organisms Respond To Stimuli

Organisms Carry Out A Variety Of Chemical Reactions

Organisms Maintain A Relatively Constant Internal Environment

LEVELS OF BIOLOGICAL ORGANIZATION

WHAT'S IN A NAME

Assigning Scientific Names

Classification Of Organisms

UNDERLYING THEMES OF BIOLOGY

The Relationship Between Form and Function

Biological Order, Regulation, and Homeostasis

Acquiring and Using Energy

Unity within Diversity

Evolution and Adaptation

BIOLOGY AND MODERN ETHICS

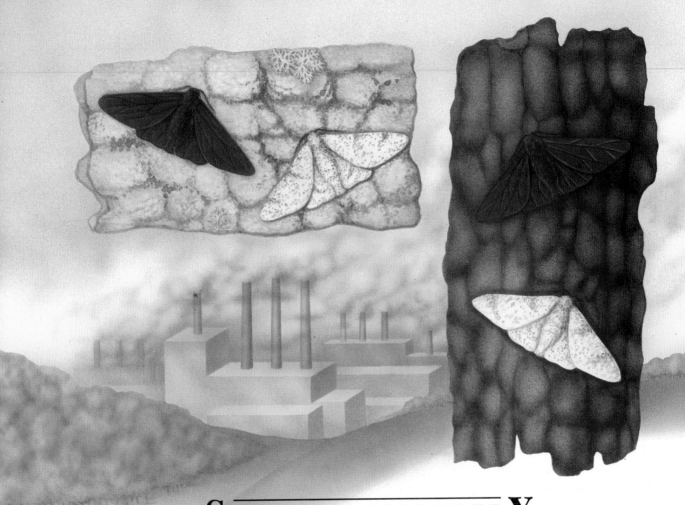

*B*iology is not magic. It does create some very impressive illusions, however, such as its amazing "disappearing acts." A leaf-shaped butterfly lands on an oakbranch and "disappears," becoming indistinguishable from the real leaves of the tree. A succulent plant growing close to the ground looks more like a rock than a living organism. Unseen by hungry animals, these organisms escape being eaten by hiding in plain sight. In England before the 1800s, where most trees were covered with light-colored lichens, the common white peppered moth was very adept at such disappearing acts. All it had to do was land on one of these mottled white tree surfaces, and the moth became virtually invisible. But then, disaster struck for these light-colored moths. The industrial revolution gained its full stride, and the white peppered moth performed a different kind of disappearing act, one that almost lasted forever.

Blackened by industrial smoke, the bright landing places of the white peppered moths changed to dark, sooty surfaces. The white peppered moths were now easily spotted by birds as they "hid" on their former sanctuaries. In these new conditions, they no longer had a competitive edge, and their numbers plunged toward the vanishing point. Yet, as the lighter-colored moths were eliminated, something unusual began to happen. Rarely seen before the industrial revolution, dark-colored peppered moths began to grow in number. Soon the population of peppered moths

Before the Industrial Revolution, dark pepper moths were eaten more frequently by birds than white moths as they rested on

was back to its former prevalence, but the new moths were dark—perfectly camouflaged on the newly blackened trees of industrialized England. Somehow the peppered moth species had "switched colors," and the species continued to survive.

To understand how the moths changed in response to their environment, we first must discuss what they *did not* do. These moths were not like chameleons: An individual moth could no more change its color to match its background than you can. It was the *species,* not the individual moths, that changed color so that, by the mid-1950s, virtually all peppered moths in industrialized Britain were dark. If they could not recognize their dark background and change colors, where did the black moths come from? The answer is *genetic variability.*

Most of the traits among the individuals in any species are similar, but there are also many genetic differences. People, for example, have two eyes, a nose, fingernails, and hundreds of other features that illustrate our similarities. Yet, we all look different from one another—even from our own parents—because of genetic variability, that is, differences in the genes possessed by different organisms. Genes are coded bits of information in cells that determine an organism's traits. Copies of these genes are passed on from parents to their offspring, who thereby inherit traits characteristic of their parents. Occasionally a spontaneous change, a *mutation,* will occur in a gene, which causes the offspring to inherit a new trait. In a population of billions of light-colored moths, for example, a few dark offspring will inevitably be produced by mutation of a pigment gene. Before the industrial revolution, however, few of these dark varieties were ever found because they were easily seen against the light surfaces and snatched up by birds. But around the mid-1800s, environmental conditions changed. The light moths had become the easy targets, and the dark moths became the "invisible" and predominant variety. Had there not been genetic variation, there would have been no dark peppered moths, and the species would now be extinct in these industrialized areas.

There is another surprising chapter in this story. If you go to industrialized England today, you will once again see the white peppered moth. Modern pollution controls have cleaned up the air, and the surfaces of the trees are once again brightly colored. The lighter moths once again have the competitive edge, and the dark moth is rarely found. The process of change continues in the peppered moth, and in all other types of organisms, according to the dictates of environmental conditions.

The case of the peppered moth illustrates how a species can change over time. The process that has changed the population of peppered moths is the same one responsible for generating the millions of species produced by three and a half billion years of biological evolution. Thus, the study of the peppered moth provides a vivid portrayal of how changes in the environment can change the genetic composition of a species. The mechanism by which new species evolve is discussed later in this chapter, and some of the overwhelming evidence for its occurrence is described in Chapter 34.

Our modern knowledge of evolution helps us understand life. It enables us to answer such childlike questions as "Why do we have houseflies?" as well as more global questions about the possibility of human extinction. Evolution enriches the study of life by making rational sense of it. It helps us understand where organisms and their properties come from and why they exist. As you progress through this book, you will find that the theme of evolution illuminates all areas of biology. It will help you piece together this information into a satisfying body of understanding that will enable you to make better sense of the world and what happens among living things.

the light bark of trees. As tree trunks were darkened by industrial smoke, the situation reversed.

You became a biologist long before you opened this book. It happens to everyone sometime during childhood, a time that marks an awakening of the fascination we have for all organisms. Children explore ant colonies and inspect sprouting seeds. They watch hypnotically as a spider spins its web. For some, the fascination grows and becomes more focused. Casual curiosity gives way to the irresistible lure of exploration, and an interesting pastime ripens into an exciting search for knowledge and discovery. These are biology's "lifers"—scientists who are engaged in a life-long adventure, a quest to make rational sense of living phenomena.

Biology is the study of life. It is a multidimensional, dynamic, creative activity that replaces mystery with understanding. No list of terms or facts can produce understanding, any more than a pile of unassembled gears and springs can explain how a clock works. Individual biological facts reveal little about how life works; they are mere threads that must be woven together by concepts or principles to arrive at an understanding of living phenomena. Biologists are detectives who use bits of information as clues to solve the complex mysteries of life. Their work also yields practical bonuses, such as controlling diseases, increasing crop yields, and proposing measures to preserve our environment.

Biology is an intellectual adventure that will take you from the landscape of molecules and cells to the global community of organisms. In distilling this information for use as a general introduction, we have struggled to retain the challenge and excitement of biology. We hope the discoveries you make in this pursuit will rekindle your fascination with biological phenomena and reaffirm your connection with all living things.

▼ ▼ ▼

DISTINGUISHING THE LIVING FROM THE INANIMATE

Most of us have little trouble identifying something as being alive or inanimate. The diverse organisms shown in Figure 1-1 are easily distinguished from nonliving environmental components. But try to define what distinguishes the two, and you may find yourself at a loss. You'll discover that many of the properties that flag an object as living cannot be found in all organisms or may be exhibited by some nonliving things. For example, you may have selected the ability to move as a basic "characteristic of life." But is a redwood tree

an inanimate object simply because you cannot observe any movement? Or, conversely, is a river alive because its movement is evident? Because of this difficulty, rather than trying to define life, biologists describe it—usually as a list of properties that characterize all living things:

- *Organization* Organisms maintain a high degree of complexity and order.
- *Cells* Organisms are composed of one or more cells.
- *Energy* Organisms acquire and use energy.
- *Reproduction* Organisms produce offspring similar to themselves.

FIGURE 1-1

Defining life. As simple as it is to identify each object in the photograph as either alive or inanimate, *justifying* your choices can be much more difficult. Most characteristics popularly associated with living things (growth, movement, reproduction, consumption of food) may also be properties of nonliving entities. Clouds and mineral crystals grow; rivers, air, and clouds move; fire consumes "food" (fuel) as it "grows" and "reproduces" giving rise to "progeny" fires. Yet none of these is alive. Defining "life" requires a combination of many properties, not one of which can, by itself, be considered *the* criterion of life.

- *Heredity* Organisms contain a genetic blueprint that dictates their characteristics.
- *Growth and development* Organisms grow in size and change in appearance and abilities.
- *Responsiveness* Organisms respond to changes in their environment.
- *Metabolism* Organisms carry out a variety of controlled chemical reactions.
- *Homeostasis* Organisms maintain a relatively constant internal environment, despite fluctuations in their external environment.

As you examine these properties in more detail in the next few pages, bear in mind that they not only "define" life, they are also inseparably linked to its success—its nonstop presence on this planet for about 3.5 billion years. Mountains crumble, continents collide, climates change, yet life on earth persists in spite of the changes (Figure 1-2).

ORGANISMS ARE HIGHLY COMPLEX AND ORGANIZED

Complexity is a measure of the number of parts that make up an object and the precision by which the parts are *organized*. An automobile is more complex than a bicycle and less complex than a space shuttle. These differences in complexity reflect the relative capabilities of these objects. Living organisms (Figure 1-3*a*) are vastly more complex than any space ship and are capable of a vastly greater variety of activities.

ORGANISMS ARE COMPOSED OF CELLS

As we will see in Chapter 5, cells are the functional units of life—all living organisms are composed of cells. For some **unicellular** organisms—those that consist of one cell—the cell *is* the organism. Most organisms, however, are **multicellular** (Figure 1-3*b*), consisting of hundreds to trillions of cells, depending on the organism's size and complexity.

(a)

(b)

FIGURE 1-2

The enduring nature of life. *(a)* This moss-covered skull testifies that life goes on in spite of the death of an individual. *(b)* This bristlecone pine has endured cold, lack of soil, and slashing winds in this same spot for thousands of years.

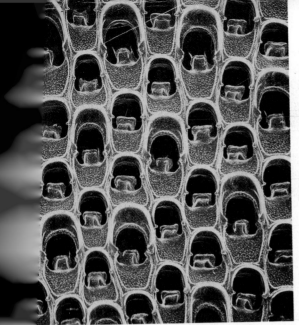

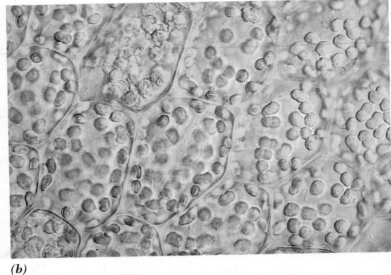

(b)

e)

(f)

FIGURE 1-3

A number of life's properties illustrated. *(a) Organization.* The skeletal remains of these marine
bryozoans reveal a high degree of order and organization. *(b) Cells.* A multicellular organism is a
cooperative constellation of cells, each of which can perform all the activities associated with life.
(c) Acquisition of energy. This spider is about to harvest energy from a fly. *(d) Reproduction.* The
resemblance within this family of swallow-tailed bee eaters is the dual product of reproduction and
heredity. *(e) Heredity.* The construction of a spiderweb is a task that requires no learning—each
spider constructs the web on the very first attempt. Like the spider's color, shape, and other physical
characteristics, web-spinning behavior is a genetically programmed trait inherited from the spider's

(c)

(d)

(g)

(h)

(i)

parents. *(f) Growth and development.* All organisms develop from a single cell. This week-old salmon embryo is acquiring the form of an adult fish but still obtains all its nutrients from the yolk sac protruding from its midsection. *(g) Responsiveness.* The final form of these sculptured tree roots reflects their ability to respond to the presence of a giant boulder. *(h) Metabolism.* The glow produced by this marine animal (ctenophore) is due to the metabolic release of energy from its food molecules, one of the thousands of chemical activities that occur in an organism. *(i) Homeostasis.* Even when exposed to subzero air temperatures, thermal homeostasis is maintained in these snow monkeys, whose body temperatures remain steady in spite of the cold.

ORGANISMS ACQUIRE AND UTILIZE ENERGY

Manufacturing and maintaining complexity requires the constant input of *energy.* Your car, for example, required energy in the manufacture and assembly of its parts; it requires additional energy to keep it in working order and to fuel your travels. Living organisms also require an input of energy to build and maintain their tissues and fuel their activities. Some organisms, such as plants, acquire the necessary energy by trapping sunlight and converting it to a form they can use. Other organisms acquire energy by feeding on the tissues of other life forms. For example, the spider in Figure 1-3*c* is about to harvest the energy stored in the body of a fly. The original source of energy for both of these animals was sunlight absorbed by plants.

ORGANISMS PRODUCE OFFSPRING SIMILAR TO THEMSELVES

New life is generated by the process of *reproduction.* Reproduction not only produces more organisms, it produces organisms that are very similar to their parent(s). Humans always produce humans, and octopuses always produce octopuses. Similar individuals are produced generation after generation, and thus the characteristics and functions of a particular kind of organism persist long after the parents have died (Figure 1-3*d*).

ORGANISMS ARE BUILT ACCORDING TO GENETIC INSTRUCTIONS

Offspring resemble their parents because they *inherit* a set of *genetic instructions* as part of the process of reproduction. The genetic instructions that we inherited from our mother and father consist of a vast collection of individual *genes* that determine the shapes of our faces, the intricate interconnections among our billions of nerve cells, and our propensity to form social units. The extent to which the genetic blueprint dictates biological activity can be appreciated by watching a spider as it plays out the intricate, genetically programmed behavior responsible for constructing a web (Figure 1-3*e*).

ORGANISMS GROW IN SIZE AND CHANGE BY DEVELOPMENT

Organisms cannot survive and perpetuate their kind without growing at some time during their life. *Growth,* an increase in size, is usually accompanied by *development,* a change in an organism's form and function. Although some simpler organisms, notably bacteria, show few developmental changes as they increase in size, virtually all other organisms change their form and capabilities dramatically as they grow (Figure 1-3*f*). An acorn develops into an oak tree; a caterpillar changes into a butterfly; a fertilized human egg develops into a person capable of contemplating his or her own nature and origin.

ORGANISMS RESPOND TO STIMULI

The ability to respond to *stimuli,* changes within an organism or in its external environment, is literally a matter of life or death. Responses may help an organism escape predators, capture prey, optimize its exposure to sunlight, move away from detrimental environmental conditions, move toward a source of water or other resources, locate mates, change growth patterns according to season, and perform many other activities necessary for survival. Some responses must be very rapid to succeed. The angler fish, for instance, responds to the presence of a smaller fish so rapidly that its strike cannot be seen; the prey seemingly disappears. Plants generally respond to stimuli more slowly than animals, but their response is just as crucial to their survival (Figure 1-3*g*).

ORGANISMS CARRY OUT A VARIETY OF CHEMICAL REACTIONS

Metabolism is the sum total of all the chemical reactions occurring within an organism. Even the simplest bacterial cell is capable of hundreds of different chemical transformations, none of which occurs to any significant degree in the inanimate world. Virtually all chemical changes in organisms require **enzymes**—molecules that increase the rate at which a chemical reaction occurs. Some enzymes participate in the breakdown of food molecules; others mediate the controlled release of usable energy; still others contribute to the assembly of substances required to build more tissue (Figure 1-3*h*).

ORGANISMS MAINTAIN A RELATIVELY CONSTANT INTERNAL ENVIRONMENT

An organism can remain alive only as long as the properties of its internal environment remain within a certain range. If, for example, an organism's cellular fluids become too salty, too warm, or too acidic, or if they retain too high a level of toxic waste products, the cells will die, and consequently the entire organism will die. Organisms possess self-regulatory mechanisms that allow them to maintain a relatively constant internal environment despite being bombarded by changing external conditions. This characteristic of life is known as **homeostasis.** The temperature of your body, for example, is normally maintained at a steady 37°C (98.6°F), even in very cold or very hot surroundings. Homeostatic control centers in your brain detect deviations from the normal temperature and "automatically" instruct the body to shiver, which generates heat, or to sweat, which cools the body. Bacteria, plants, bread mold, animals—indeed all organisms—possess homeostatic control mechanisms (Figure 1-3*i*).

> **Life defies a simple definition. Life exists only when a particular combination of properties exists in a single entity. The same combination occurs in all living organisms. (See CTQ #1.)**

LEVELS OF BIOLOGICAL ORGANIZATION

An object, animate or nonliving, is composed of relatively simple building blocks that are assembled into increasingly complex subunits, which ultimately combine to form the final complex structure. This book, for example, consists of ink and paper, which combine to form letters printed on pages (a higher level of organization). Individual letters combine to form a more complex structure (a word), which groups with other words to form the next level of complexity, a sentence. These levels of organization increase in steps until the final book is formed. Living organisms can also be placed within such a *hierarchy of organization,* in which simpler structures combine to form the more complex structures of the next level of organization, which in turn interact to form even more complex units (Figure 1-4).

All matter is built of protons, electrons, and neutrons. These three kinds of subatomic (smaller than atoms) particles combine to form many different types of atoms. Atoms, in turn, can combine with one another in a nearly infinite variety of ways to form molecules of limitless variety. In living organisms, molecules join together to form subcellular components (including structures called *organelles*) that are assembled into *cells.* Cells occupy a very special level of organization; a cell is the simplest unit in the biological hierarchy that is, in itself, alive.

The attainment of the cellular organization reveals something profound and fascinating about hierarchies of organization: *Increasing complexity not only generates a higher order of structural organization, but also creates new properties that exceed the sum of the parts used to form the structure.* To return to our book analogy, when individual letters are combined to form a word, the structure at the next level of organization, an additional property emerges —the word's meaning. No new materials were added; the new property is strictly a function of higher organization. Words are then combined to form the next level, a sentence, from which another property emerges—a statement. The levels of organization increase until the final structure—a manuscript—is created at a level complex enough for a story to be created.

Life itself is such an emergent property, a phenomenon that first appears at the level of the cell and that exceeds the sum of the parts. If not organized to form a cell, the components of a cell are incapable of generating or sustaining life. Subcellular organelles and cytoplasm removed from one another simply deteriorate—they cannot maintain their ordered state, reproduce, nor respond to stimuli. They are not alive. Yet together, as an intact cell, all the properties of life emerge.

As highly ordered as organisms are, they too are subunits of even more complex levels of organization. Individuals of the same species inhabiting the same area constitute a **population.** Different populations in a particular area interact with each other to form **communities,** which form part of a particular **ecosystem.** Ecosystems consist of the community in an area plus the nonliving environment, for example, the water, rocks, and mud, together with the bacteria, plants, and animals at the bottom of a lake. All the world's ecosystems combine to form the **biosphere.**

This hierarchy of life provides one of the foundations on which this book is organized. We begin at the atomic and molecular levels, work our way through a variety of subjects at the cellular level, turn to a discussion of organs and organ systems, then move on to an appreciation of the diverse forms of organisms on earth, culminating in a look at the ways organisms interact with each other and their environment.

> **From atoms to the biosphere, life can be investigated at different levels of complexity and organization. With each increasing step in complexity, new properties emerge that could not exist at lower states of organization. It is these emergent properties that epitomize the significance of the hierarchy of biological organization. Life itself is such a property. (See CTQ #2.)**

WHAT'S IN A NAME

The sheer complexity of organisms seems so bewildering that some people believe life to be unfathomable by the rational mind. Biologists constantly chip away at this notion by exploring and discovering real (and fully understandable) explanations for phenomena. Our ability to solve these natural mysteries would be severely handicapped, however, without a system that creates meaningful order out of the overwhelming variety of organisms to be studied. This orderly system enables us to distinguish clearly between different types of organisms, and to classify and assign names to the almost 2 million known types of living organisms, as well as their extinct ancestors. This formal system of naming, cataloguing, and describing organisms is the science of **taxonomy.**

Taxonomy perpetuates order by examining each type of organism and describing its properties, defining the set of criteria that distinguishes that particular type of organism

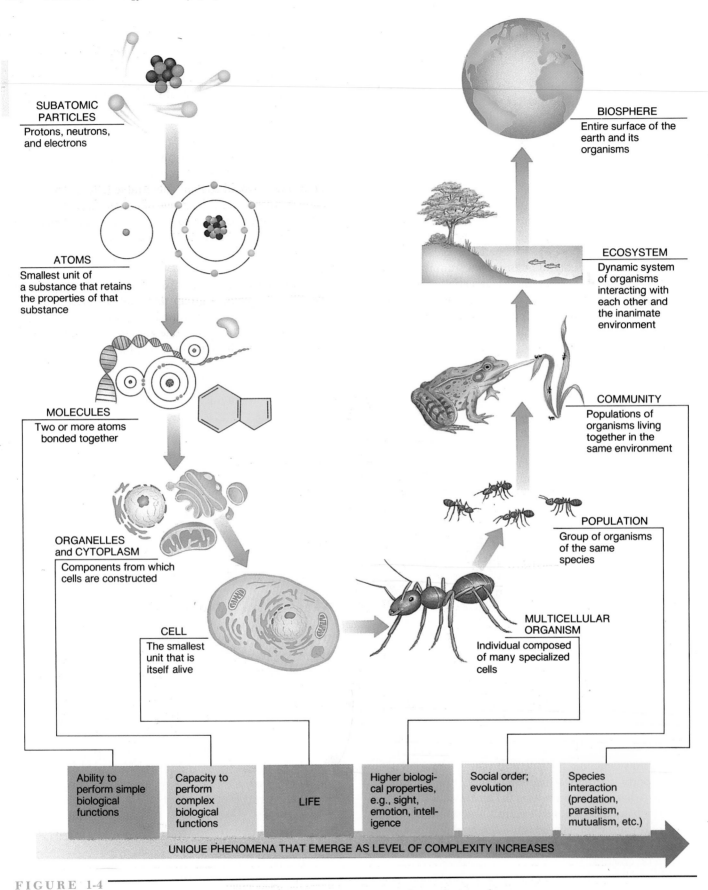

SUBATOMIC PARTICLES
Protons, neutrons, and electrons

ATOMS
Smallest unit of a substance that retains the properties of that substance

MOLECULES
Two or more atoms bonded together

ORGANELLES and CYTOPLASM
Components from which cells are constructed

CELL
The smallest unit that is itself alive

MULTICELLULAR ORGANISM
Individual composed of many specialized cells

POPULATION
Group of organisms of the same species

COMMUNITY
Populations of organisms living together in the same environment

ECOSYSTEM
Dynamic system of organisms interacting with each other and the inanimate environment

BIOSPHERE
Entire surface of the earth and its organisms

Ability to perform simple biological functions

Capacity to perform complex biological functions

LIFE

Higher biological properties, e.g., sight, emotion, intelligence

Social order; evolution

Species interaction (predation, parasitism, mutualism, etc.)

UNIQUE PHENOMENA THAT EMERGE AS LEVEL OF COMPLEXITY INCREASES

FIGURE 1-4

Levels of biological organization. Each single-step jump increases structural complexity and may also generate a unique property distinct from the structure.

from all other types. (Taxonomists have described more than 1.8 million distinct species, which represents only a fraction of the estimated total of 5 million to 50 million species alive on earth today.) From this body of descriptive information, taxonomists create the two orderly systems that are so necessary to biology:

- A *system of nomenclature* that assigns a specific name to each type of organism, a label that doesn't vary from biologist to biologist.

- A *system of classification* that groups organisms into categories according to similarities in major properties. This formal system has a dual purpose. First, it provides a standard set of criteria that can be used to ascertain the identity of an organism from its observable properties. Second, the classification scheme reveals the degree of kinship, and thus ancestral relationships, among different groups of organisms (the importance of this will become more apparent when we discuss evolution later in this chapter).

ASSIGNING SCIENTIFIC NAMES

All species are assigned a name consisting of two latinized words according to our **binomial system of nomenclature** (binomial loosely means "two names"). Latinizing these names may strike you as overly formal and difficult, but it would be infinitely more confusing if each scientific name reflected the native language of the biologist who provided the label. Some scientific names would be in English, others would be written with Chinese characters, and a few others would contain only Arabic letters. The use of Latin standardizes the language of nomenclature so that all species receive names expressed in the same language, using the same alphabet. *Homo sapiens,* for example, is recognized worldwide as the scientific name for the human species.

The first word in an organism's pair of names always identifies its *genus,* a group that contains closely related species. The second word, called the *specific epithet,* singles out from within that genus one kind of organism, the species. No two species have the same binomial name.

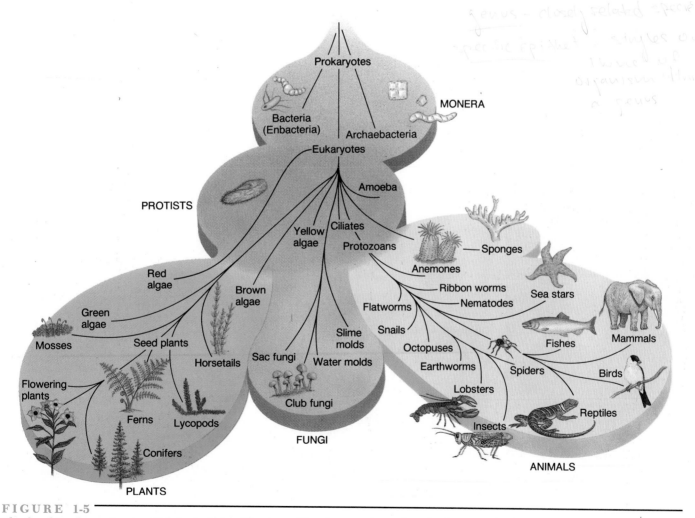

FIGURE 1-5

The five-kingdom scheme of biological classification. This diagram shows the relationship among the five kingdoms and the relative numbers of species that each contained. The animal kingdom, with its 1.3 million species, is the largest kingdom; in other words, it contains the greatest diversity.

CLASSIFICATION OF ORGANISMS

Taxonomists classify organisms by sorting them into groups according to traits that reveal *phylogenetic* relationships. This means that the organisms in the same group have a more similar ancestral history and are therefore more closely related than are organisms in different groups. ("Phylogenetic" refers to *evolutionary* relationships, revealing how recently two types of organisms shared a common ancestor.)

Phylogenetic relationships are found in all taxonomic categories. The lowest level of category is the species. Each of the millions of species contain individual organisms that are most similar to one another. The next level consists of larger groups (each one called a **genus**) containing one or more closely related species. The third level is created by grouping similar genera (plural for genus) into *families*. For example, all cats—from house cats to tigers—form a single family. The sequence continues in this manner, each level

(a)   *(b)*   *(c)*

(d)   *(e)*

FIGURE 1-6

Representative organisms from each kingdom. *(a)* Rod-shaped bacteria—Monera kingdom; *(b)* *Paramecium,* a unicellular animal-like organism—Protista (protist) kingdom; *(c)* poisonous *Amanita* mushroom—Fungi (fungus) kingdom; *(d)* Pink moccasin flower—Plantae (plant) kingdom; *(e)* a pair of young caribou—Animalia (animal) kingdom.

composed of groups formed by clustering the categories of the previous level. Related families form an *order;* related orders form a *class;* related classes form a *division* (or *phylum,* in animals); and related divisions (or phyla) form the highest level of biological classification, a *kingdom.*[1]

The currently accepted scheme of classification maintains that all organisms on earth belong to one of five kingdoms (Figures 1-5 and 1-6). The simplest organisms, bacteria and their relatives, belong to the Monera kingdom; the most complex belong to the Animal kingdom (the characteristics of each kingdom and the evolutionary relationships within each are discussed in Chapters 35 through 38).

The genus and specific epithet to which a particular type of organism belongs provide the binomial name that identifies that species. Organisms are grouped into a classification scheme that provides order and consistency to the way we view the realm of life, allowing us to communicate about biology, and giving us a better understanding of how life evolved to its current state. (See CTQ #4.)

UNDERLYING THEMES OF BIOLOGY

As you progress through your introductory biology course, you will learn about a diverse array of subjects, ranging from the atomic and molecular basis of life to the behavior of plants and animals. Yet despite such an immense spectrum, a number of general themes pervade all areas of the biological sciences. These common themes form a foundation for which much of the information gathered by biologists can be understood. As you progress from one major topic to the next in this book, you will notice that a portion of each discussion focuses on these common threads (these passages are accompanied by an icon that alerts you to the theme being emphasized). At the close of each chapter, you will have the opportunity to review how some of the material presented in the chapter exemplifies one or more of these basic themes. These themes (Figure 1-7) are

- the relationship between form and function;
- biological order, regulation, and homeostasis;
- unity within diversity;
- acquiring and using energy;
- evolution and adaptation.

THE RELATIONSHIP BETWEEN FORM AND FUNCTION

It is evident that the tools we produce in our factories serve useful functions because of their structure. A wrench is obviously better suited for loosening a nut than is a screwdriver. Organisms also manufacture "tools" and "machinery" whose functions are closely correlated with their structure. Human hands, for example, are well suited for grasping; the pointed canine teeth of a wolf are suitable for seizing prey and tearing the flesh from bone; and the canopy of leaves that forms at the top of a tree perfectly matches the plant's need to maximize its exposure to sunlight, its sole source of energy.

The relationship between form and function is evident at all levels of biological organization. Once you have read about the internal structure of the kidney in Chapter 28, for example, you will see how its form—the spatial arrangement of microscopic tubules and blood vessels—makes it possible for the organ to remove waste products, excess salts, and water from blood and to concentrate them in a liquid that is easily discharged from the body. At the molecular level, you will see how the shape of a protein (its form) allows it to carry out a particular function, whether to promote a chemical reaction, support part of a cell, or hunt down and destroy a particular virus.

BIOLOGICAL ORDER, REGULATION, AND HOMEOSTASIS

The enormous complexity needed by an organism to sustain its own life goes hand in hand with a high degree of *order,* a term that suggests precision in the arrangement of components and the management of processes. There is nothing special, for example, about the atoms that make up your body. It is the unique *arrangement* of these atoms that makes you different from every other organism on the planet. If the atoms in your body were arranged differently, they might construct a tree or perhaps a porpoise.

Maintaining biological order requires *regulation.* Without such regulation, maintaining homeostasis would be impossible. Even eating a bag of salty potato chips could be fatal—the salt concentration in your body could increase to intolerable levels, and your cells would die. Fortunately, your kidney's regulatory mechanisms remove excess salt from the bloodstream and discard it in urine.

Virtually every biological activity—from the expression of a cell's genes, to the blood pressure in the major arteries, to sexual behavior—is subject to some form of regulation. Humans, as well as most other familiar animals, have two organ systems devoted primarily to the regulation of bodily activities (these are the endocrine and nervous systems). Without biological regulation we would not be able to breathe, focus our eyes, or maintain an internal environment conducive to life. Instead, there would be chaos and disorder rather than the ordered complexity needed for life. If even a single biochemical reaction in a cell were suddenly to escape the cell's regulatory mechanisms, the product of that reaction could accumulate to higher and higher concentrations until it jeopardized the entire biochemical operations of that cell, and possibly the entire organism.

---

[1] A newly proposed taxonomic group, called a *domain,* contains kingdoms that possess similar cell types. Three domains have taxonomically solid foundations but, at the time of this printing, have not yet been officially adopted.

FORM AND FUNCTION
The length and shape of this tiger lily and that of the hummingbird's beak match, enabling the hummingbird to gather nutritious nectar from the flower base, while the flower deposits or collects pollen for reproduction.

ORDER, REGULATION AND HOMEOSTASIS
Whether active or dormant, enveloped by snow or blistering heat, all organisms must maintain order and regulate internal conditions to remain alive.

UNITY WITHIN DIVERSITY
Despite the remarkable variation in different kinds of organisms on earth, all organisms are composed of one or more cells.

ACQUIRING AND USING ENERGY
All organisms acquire and use energy whether by directly trapping the energy in sunlight or by harvesting energy stored in the bodies of plants or animals.

EVOLUTION AND ADAPTATION
Evolution has produced the astounding variety of life on earth. Like the lizard hidden on the bark of this tree, each kind of organism possesses adaptations that enable it to survive and reproduce in its particular habitat.

ACQUIRING AND USING ENERGY

Energy obtained from the environment provides the fuel needed to run the processes that make life possible. Regardless of whether an organism harvests its energy from the sun or obtains it by consuming nutrients from another organism, to run out of energy is to run out of life. Energy is needed for the biochemical reactions occurring inside cells, reactions that construct the organism and repair or replace damaged components. Energy helps maintain our constant body temperatures, whether we are walking in the desert or swimming in frigid water. Energy limitations determine the number of eggs that can be laid by an individual frog. The dynamics of energy acquisition also explain why there are more than ten times as many rabbits in the world as there are predators that feed on them. These are just a few of the countless roles of energy that make it an inescapable theme in our explorations of life.

UNITY WITHIN DIVERSITY

We share this planet with millions of other species. With such a bewildering variety of organisms, one might identify "diversity" as the central theme in biology. Despite this diversity, however, there is a remarkable amount of "sameness" among all organisms, even among such radically different organisms as humans, spiders, ferns, and bacteria.

The unity of life is particularly apparent when we examine the fundamental levels of biological organization. For example, a person and a mushroom seem to share few similarities at the level of the whole organism, but both are quite similar on the cellular and molecular levels. Both are composed of cells. Both are built from the same classes of biological molecules, as are all other organisms (Chapter 4). Both use enzymes to promote the chemical reactions of their metabolism. The universal use of enzymes (Chapter 6) to promote chemical reactions is another clear example of life's basic unity. Many of these chemical reactions are identical in all organisms.

Humans and mushrooms, like all other organisms, use the same molecule (DNA) to store genetic information. In both organisms, these instructions are contained within discrete units of genetic information, the genes. Furthermore, all organisms use the same genetic "language" to encode these instructions. The universal nature of the genetic code is a particularly striking example of biological unity, a testimony to the common ancestry of all species.

EVOLUTION AND ADAPTATION

The last of the five themes—evolution and adaptation—constitutes the most important unifying principle in biology. It explains why the biosphere is populated by millions of species rather than just one type of organism. At the same time, this principle accounts for the unmistakable unity that exists among even the most diverse organisms. The principle of **evolution** states simply that species change over

time. As a result of evolutionary changes, new species of organisms emerge while older forms of life may disappear. In other words, life evolves.

Evolution explains how all forms of life that have ever existed are members of one extended, genetically related "family" of organisms. You and that mushroom are similar in so many fundamental ways because you shared the same ancestor far back in time, and both of you retain many of the same molecular, genetic, and cellular endowments provided by that ancestor. All of the diverse species on earth are descendants of primitive cells that were the earliest life forms. It is this common ancestry that accounts for the unity of life.

Biological success for a particular type of organism depends on how well suited the organism is to its environment. An organism can survive only in environments that supply all its essential needs, and then only if the organism possesses the "equipment" needed for acquiring those environmental resources. Even then, it can succeed only if it can survive the adverse conditions to which it is exposed. In other words, an organism must be *adapted* to its particular set of environmental conditions (Figure 1-8). **Adaptations** are traits that improve the suitability of an organism to its environment.

Adaptations can be identified as part of every organism you see. The body shape and coloration of some insects (such as the peppered moths discussed earlier) better enable them to avoid detection by birds or other predators. The trunk of an elephant enables the animal to reach a supply of edible leaves on high branches that are unavailable to most of the competition (other leaf-eating, ground-based animals). The thick fur coat of an arctic fox provides the insulation needed to allow the animal to sleep on a bed of ice without expending metabolic energy keeping itself warm.

Pick out a trait that is part of your own body. The trait you have chosen is probably some type of *anatomic* (structural) adaptation, such as a grasping hand, an eyelid that protects the surface of the eye, or a foot that can balance the weight of the entire body. A few of you may have selected a property that reveals a more subtle *physiologic* trait (one concerning biological processes and functions), such as eyesight or ability to taste. Some of you may have even identified a *behavioral* adaptation, such as automatic breathing or blinking. In fact, many structural or physiologic adaptations of organisms would be useless without an accompanying behavioral adaptation. Your respiratory structures, for example, would be of little value if you didn't inherit breathing behavior as well. Spiders not only possess silk-producing glands, for example, but also inherit the "automatic" (instinctive) behavioral adaptations that direct them to weave a characteristic web and instruct them in its use for capturing prey.

Every topic we discuss in this book has an evolutionary component. As we describe certain anatomic, physiologic, or behavioral characteristics of various organisms, keep in

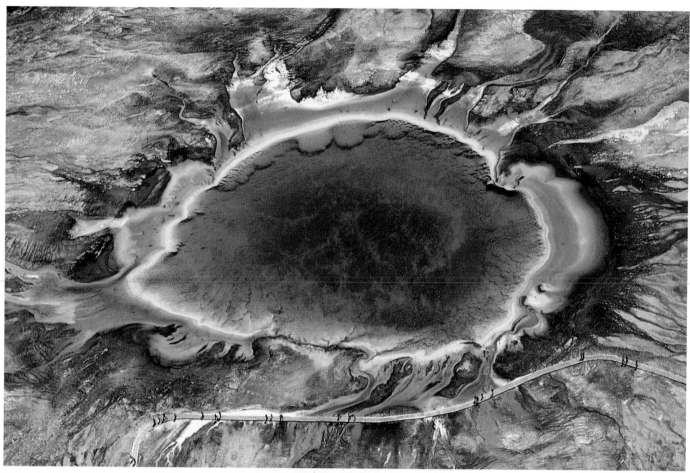

FIGURE 1-8

Adaptations to temperature differences are revealed by the concentric bands of color in this aerial photograph of a hot sulfur pool at Yellowstone National Park. The pond, which is hottest in the center and coolest at its rim, is colored by enormous populations of microorganisms concentrated in each ring-shaped thermal zone. Each distinctly pigmented type of microorganism is specifically adapted to (and limited to) one narrow temperature range.

mind how these features better suit the organism to its environment, increasing the chance for survival of both the individual and the species. We will see how some of these adaptations have become modified over evolutionary time as environmental conditions have changed and new species have evolved. For example, the tiny bones in your middle ear, which transmit sound through a recess in your skull, evolved from bones that were originally parts of the jaws of ancestral fishes that lived long before the first four-legged animal ever set forth on dry land.

Evolution is the cornerstone of our understanding of biological phenomena. The events that led to the formulation of an explanation of the mechanism by which evolution occurs constitute one of the most important stories in the natural sciences.

A Voyage That Altered Our Concept of the Origin of Species

What would you do if someone offered you a free trip around the world and all the person asked in return is that you pursue your favorite hobby during the trip? Such an opportunity presented itself to a 22-year-old university graduate named Charles Darwin when one of his professors recommended that Darwin be appointed "naturalist" for a scientific expedition around the world.

It was 1831, and Charles Darwin (Figure 1-9*a*) had just graduated from the University of Cambridge with a rather undistinguished academic record. As a young boy, he had been an avid insect and plant collector, an interest that became even more focused during his college years. Darwin's zeal is illustrated by a passage he later wrote: "One

(a)

FIGURE 1-9

Charles Darwin and the voyage of the *HMS Beagle*.
(a) Portrait of the young Charles Darwin.   **(b)** Map of the route taken by the *Beagle* during its approximately 5-year trip around the world.

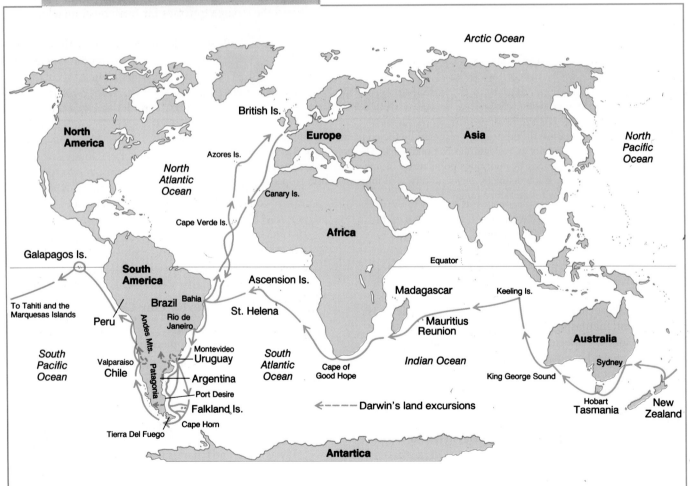

(b)

day, on tearing off some old bark, I saw two rare beetles and seized one in each hand; then I saw a third and new kind, which I could not bear to lose, so that I popped the one which I held in my right hand into my mouth. Alas, it ejected some intensely acrid fluid, which burnt my tongue so that I was forced to spit the beetle out, which was lost, as well as the third one."

Even as a college graduate, Darwin was still unsure of his lifelong profession. He had abandoned the idea of following in the footsteps of his physician father and had decided to become a clergyman in the Church of England when this "once in a lifetime" offer was proposed to him. An enthusiastic Darwin accepted the commission on the *HMS Beagle,* a small (25-meter, or 75-foot-long) three-masted surveying ship that was to map the coastlines and harbors of South America (Figure 1-9*b*). As the ship's naturalist, it was Darwin's job to collect and organize thousands of specimens from around the world.

Darwin left England with the firm belief that all of the world's plants and animals had been created directly by the hand of God. As he traveled and explored new lands, Darwin began to question this view of a static world whose life and landscape had been fixed since the time of creation. The questions began even before the *Beagle* had reached its first port of call. During the first few days, as Darwin lay in his hammock suffering from seasickness, he read the newly published *Principles of Geology,* a textbook written by Charles Lyell. Lyell's book meticulously presented evidence that the earth was much older than previously believed and that it had gradually changed over long periods by such natural forces as mountain building, erosion, volcanoes, flooding, and earthquakes.

Within the first month of the voyage, Darwin himself was able to observe evidence in support of Lyell's contentions. As the ship sailed into the harbor at the Cape Verde Islands, Darwin saw a cliff rising about 15 meters above the sea. He discovered that the cliff was composed of limestone that contained embedded seashells of the type seen along the shore. It was evident that the sea bed had been lifted upward as part of the cliffs. Later in the voyage, Darwin found marine deposits located high in the Andes Mountains, hundreds of miles from the closest shore. He saw firsthand evidence of the ability of natural forces to change the landscape after experiencing a severe earthquake off the South American coast. Darwin found that the land at one site had risen nearly a meter as a result of the quake. At a nearby site, the captain of the ship had discovered rotting mussels clinging to the rocks 3 meters above the high-tide level.

Evidence for a gradual change in the organisms living in the area was less dramatic than were the geologic observations, but they were just as convincing. A few miles inland from the east coast of Argentina, Darwin found a fossil bed containing a giant sloth, a hippopotamus-like animal, and a giant armadillo. It was clear to Darwin that these were the remnants of extinct creatures, yet they were clearly related in form to those living today. One of these creatures, Darwin wrote, was "perhaps one of the strangest animals ever discovered." It was the size of an elephant with teeth that suggested it fed by gnawing, much like modern-day rats.

Darwin saw the clear anatomic relationships between these extinct animals and those living today and wondered how it was that these ancient animals had disappeared. The captain of the *Beagle* suggested that they had been left off Noah's ark during the great flood and had been drowned as the waters covered the land. Darwin, however, considered another explanation. Perhaps they had been driven to extinction by animals that had invaded the South American continent from the north following the formation of a land bridge connecting the North and South American land masses. His proposal proved to be remarkably farsighted.

Darwin's most important observations were made as he explored the Galapagos Islands, a small archipelago (group of islands) located approximately 900 kilometers off the coast of Ecuador and named after the giant tortoises that are found only on these islands. The volcanic islands had been discovered several hundred years earlier and were a site where ships replenished their supply of fresh water and captured the strange tortoises for fresh meat on a long sea journey.

Having visited other volcanic islands along the journey, Darwin noted that the plants and animals of such habitats tended to be similar to those living on the nearby mainland. Islands off the coast of Africa, for example, contained animals similar to those found on the African mainland, whereas the Galapagos contained animals similar to those inhabiting the South American mainland. This was particularly curious for the Galapagos, since the climate and geography of the islands were much different from those of the mainland. Darwin wondered why animals of similar appearance would be living in such different habitats.

One of the most important scientific observations in history concerned a group of dull-looking birds—the now-famous Darwin's finches. Darwin observed and collected 13 species of finches that were found only on the Galapagos Islands (and one on nearby Cocos Island). Although these birds were similar to one another in overall body form, they differed in their "lifestyle" and the shape of their beaks (Figure 1-10). Some of the species lived on the ground, others in the trees. Some had strong, thick beaks adapted for crushing seeds, while others had beaks that were especially suited for feeding on flowers or insects. Among the insect eaters was the so-called woodpecker finch, which, unlike a true woodpecker that catches insects with its long tongue, digs insects out of the tree bark using a cactus spine held in its beak. Darwin wondered why animals with such different feeding habits looked so much alike.

The observations made on the Galapagos Islands were an important component in Darwin's view that species were not created in their final, unchangeable form but instead had evolved from other species. Darwin eventually concluded that all of the island finches were descendants of one

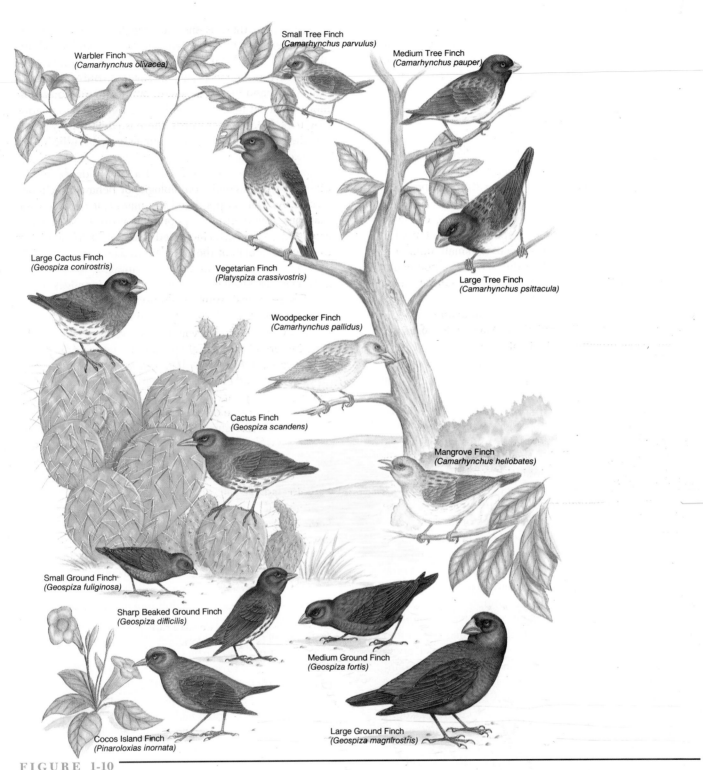

FIGURE 1-10

Darwin's finches. Darwin concluded that all 14 species of finches inhabiting the Galapagos Islands (and nearby Cocos Island) had evolved from a single ancestral finch that came to the islands from the mainland. The finches differ primarily in beak size and shape. These distinctions adapt them to different habitats and food sources. They include seed-eating ground finches; nectar-feeding finches; insect-eating finches; and one remarkable tool-using tree finch—the woodpecker finch—that bores into wood with its beak and then uses a cactus spine or twig to excavate its prey.

of the mainland species of ground finch that had drifted several hundred kilometers to one of the islands. Unlike the mainland, the islands were essentially devoid of competitors, so the immigrants were able to establish a thriving population. But how were individuals of one species able to give rise to the various species now present on the islands? In other words, how does evolution work?

Darwin's Theory of Evolution by Natural Selection

Soon after Darwin returned to England he happened to read an essay that had been written about 40 years earlier by the Reverend Thomas Malthus. Malthus pointed out that a single pair of humans have the reproductive potential to produce billions of people in a relatively small number of generations. It was evident that the size of the population was held in check as a consequence of mortality brought about by famine, war, and disease.

Darwin realized that this same potential for overpopulation was present in all populations. Any species of animal —from slow-breeding elephants to minute insects —could, given an absence of mortality, generate enough individuals in a relatively short period of time to cover the earth many times over. Yet, the numbers of individuals of most species remain relatively constant from one year to the next. Darwin concluded that there must be a "struggle for existence" among animals such that only a small percentage of those conceived actually live to maturity. Darwin was not suggesting that there was a physical struggle between individuals but rather a competition between individuals of a community and a struggle to survive the potentially adverse factors in their environment, such as lack of food or water, predation, parasitism, disease, cold, heat, flooding, and salinity.

Assuming there is such a struggle, what determines which members of the population survive and which are eliminated? If all the individuals were identical, then all would have the same chance of survival. But Darwin was aware that members of a plant or animal population are no more identical than are members of a human population (Figure 1-11). In other words, there is **variability** within a population. Variation among animals may be reflected in the color of the coat, body size, an ability to withstand high temperature, the choreography of a mating dance, or any other type of anatomic, physiologic, or behavioral characteristic. As Darwin pondered the matter, it became clear that some members of the population possessed characteristics that gave them an increased chance of survival relative to other members of the population. The survivors might have a more efficient style of gathering food, or a particularly high resistance to a common parasite, or a little faster gait. These animals would be better adapted to their environment and thus more likely to survive.

It was evident to Darwin that survival itself was not the most important consideration; rather, surviving *to reproduce successfully* was the critical factor. Those organisms that are best suited to survive tend to produce a greater number of offspring. According to Darwin, the environment "selects" those organisms that will survive and reproduce, while it eliminates those organisms that do not. Darwin called this process **natural selection.** Viewed in this way, natural selection is equivalent to *differential reproduction,* the production of more offspring by individuals that are better adapted to their environment.

Although the biologists of Darwin's time were totally ignorant of the mechanism of heredity, they knew from observation that offspring tend to inherit the characteristics

FIGURE 1-11

Genetic variation in the human population.

of their parents. Consequently, the offspring of survivors will tend to have traits that make them suited to survive and reproduce in the same environment. As a result of differential reproduction, after several generations a population will tend to collect genetic traits that make its members better adapted to its environment. Individuals with traits less suited would be less likely to successfully compete and would therefore leave fewer (if any) offspring. This change in genetic composition of a population from generation to generation is the very essence of the evolutionary process. Giraffes, for example, have long necks because individuals that happened to have longer necks were more successful in obtaining food than were their short-necked competitors. Longer-necked members of the population were more likely to survive and have offspring which, like their parents, would have longer necks.

Natural selection produces organisms that are adapted to their environment. Environments do not remain constant over long periods of time, however; climates change, fires destroy vegetation, new parasites or predators may appear, food supplies change, and so forth. Consequently, an animal that is successful within a particular habitat at one time might be poorly adapted at some other time. Darwin realized that the natural variation that exists within a population provides the basis for evolutionary change. Consider, for example, a hypothetical population of fleas that fed on the blood of the bison that roamed the American plains during the early nineteenth century. As long as the bison were plentiful, natural selection would favor those individuals who were most efficient at finding and piercing the skin of these large animals as they grazed on the prairie grasses. As the numbers of bison sharply diminished, those fleas that could locate and survive on the blood of horses or dogs might be favored by natural selection, and the population would gradually shift toward this new variety of insect. Given sufficient time, the gradual accumulation by differential reproduction of small genetic differences among individual fleas could radically transform a species' characteristics. If the populations of a species become isolated from one another so they cannot interbreed, then a strong likelihood exists that the populations will evolve into separate species. After several billion years, this divergence of species has produced the diversity of life that we see today, including the human species.

A Surprise in the Mail

Darwin returned from his voyage around the world in 1836. He spent the next few years examining the specimens he had collected, reading the writings of others, and preparing manuscripts for the journal of his voyage, including a treatise on barnacles and a book on coral atolls (coral islands surrounding a lagoon); he also pondered the implications of his thoughts on evolution. Although Darwin discussed his ideas about evolution and natural selection with his friends in the scientific community, it wasn't until 1842 that he set

them down on paper in a brief essay that was not sent out for publication. A few years later he wrote a longer essay and, at the urging of his friends, began preparing a manuscript for a book on the subject.

In June 1858, more than 20 years after the voyage of the *Beagle* and after 11 chapters of his book had been written, Darwin received a letter in the mail from Alfred Russel Wallace, a fellow naturalist who had been working in Malaysia. Contained in this letter—and in the short manuscript that soon followed—was the outline of a theory of evolution by natural selection that was virtually identical to that being formulated by Darwin. Ironically, Wallace was writing to ask Darwin if he would forward his manuscript to Charles Lyell for presentation to the Linnaean Society. Darwin was shocked by the communication and consulted with Lyell and another colleague, Joseph Hooker, as to what course of action he should take. Lyell and Hooker decided that they would present the work of both men jointly at a meeting of the Society, which they did in 1858. Rather than becoming competitors, or even enemies, Darwin and Wallace became lifelong friends. Darwin rapidly finished the remainder of the manuscript, and *The Origin of the Species* was published on November 24, 1859. The response was thunderous. All 1,250 copies of the book were sold on the first day it was available. Discussions, protests, and personal attacks followed—continuing to this very day.

A Summary of Darwin's Theory of Evolution by Natural Selection

The concept that organisms are related because of common descent had been discussed by a number of philosophers and naturalists before the nineteenth century. Charles Darwin, however, was the first to provide a feasible mechanism—natural selection—to explain how biological evolution might have occurred. Darwin arrived at this mechanism by the following logic:

1. All species have the reproductive potential to overpopulate the earth, yet populations of organisms remain relatively stable over time.

2. Consequently, a large percentage of the members of a population must die at an early age.

3. There is variation among the members of a population such that different individuals have different traits.

4. It follows that the members of a population whose traits make them better adapted to their environment at a particular time will be more likely to survive to reproductive age and to produce more offspring than will members that are less well adapted. This is the basis of natural selection.

5. Since offspring inherit the traits of their parents, those that have successful parents tend to acquire successful traits. Consequently, a population will change (evolve) over time, its members acquiring new traits that make them better adapted to a changing environment.

6. Given sufficient time, this process of evolution by natural selection can account for the formation of new species and thus for the diversity of life on earth, both past and present.

All organisms are the products of evolution, the central theme underlying biology. Evolution has generated patterns that are so recurrent among organisms, they join the list of life's fundamental themes. These themes help tie together the facts of biology to provide a deeper understanding of life. (See CTQ #5.)

BIOLOGY AND MODERN ETHICS

Our modern understanding of biological concepts and the application of these concepts is already changing the planet and all its inhabitants, and the influence continues to grow dramatically as we approach the next century. Although these changes hold tremendous promise and hope for a healthier environment for all organisms, new biological technologies come with their share of ethical controversy. Some applications seem clearly unethical, such as using technology to develop biological weapons (strains of viruses or bacteria that can incapacitate or kill people). Yet many of these issues are not so easily classified as either ethical or unacceptable. Consider the following questions:

• Should scientists be allowed to patent life?

• Do the benefits of genetic engineering outweigh the potential risks that this new technology poses?

• Should we use our biological know-how to make abortions safer and more accessible, for example, by using a "morning-after" pill that prevents pregnancy after an egg is fertilized?

• Do economic freedoms and needs justify the destruction of habitats and subsequent extinction of species?

• Although treating disease by replacing faulty genes with good ones is not so controversial, is it ethical to use the same technology to introduce genes into healthy people, genes that enhance a particular desirable characteristic, such as intelligence or athletic prowess?

• Who should decide whether or not organs can be removed from brain-dead patients so that they can be donated to save others?

• Should people with AIDS be allowed to work in healthcare professions?

These questions represent just the tip of the bioethical iceberg. An informed public is the best protection against clearly unethical applications or abuses of biological knowledge. Information and open discussion is also essential before the more debatable issues can be resolved. Regardless of how they are resolved, these bioethical decisions will permanently influence the world and all its inhabitants. To introduce you to some of these ethical dilemmas that we must all confront, we have included a series of essays that appear in several of this book's chapters. The ethical pendulums in these essays may leave you feeling indecisive, wanting some definitive piece of information that will tip the scales one way or the other. Alternatively, you may find yourself in impassioned discussions with classmates and family. As long as you can support your feelings with facts, even if those feelings are ambiguous, then you are participating in shaping the future.

REEXAMINING THE THEMES

The Relationship between Form and Function

Every component of an organism has a form that suits its function. Each enzyme has a shape that allows it to combine with a particular chemical and change it to another chemical. Similarly, an elephant's trunk suits its task of stripping leaves from tall trees and reaching water far below the elephant's head. It is much easier to understand biological concepts if the relationship between a structure and its function is firmly established.

Biological Order, Regulation, and Homeostasis

Maintaining the high degree of organization required for life depends on hundreds of homeostatic control mechanisms. Virtually every metabolic reaction in an organism needs to be regulated to prevent excesses or deficiencies of its activity. The saltiness of the internal environment must be stable or the organisms will die. All organisms have mechanisms for detecting imbalances and correcting them to maintain a stable internal environment conducive to life.

Acquiring and Using Energy

All organisms require a constant supply of energy to maintain biological order, to grow, to repair damage, to move substances within the organism, and to run other life-sustaining activities. Plants and similar organisms acquire their energy directly from the sun. Other species harvest energy from the tissues of organisms on which they feed. Regardless of the energy-acquiring strategy, the original source of energy for virtually all organisms is the sun.

Unity within Diversity

The diversity of life on earth is extraordinary; the biosphere supports at least 1.8 million species and perhaps 50 million species. Yet all organisms have fundamental similarities, such as all being composed of one or more cells, all inherit a set of genetic instructions inscribed in the same genetic "language," all possess homeostatic control mechanisms, and all use enzymes to direct metabolic activity, to name just a few. Although universals emerge at every level in the hierarchy of life, they are particularly abundant at the fundamental levels.

Evolution and Adaptation

All organisms and their traits are the products of evolution. Organisms that are well adapted to their environment survive and reproduce, while poorly adapted individuals leave far fewer offspring. As conditions change, however, different individuals are reproductively favored, and species with different characteristics emerge. The "invisible" peppered moth, the long-necked giraffe, the intelligent human, indeed all species evolved in this way.

SYNOPSIS

Organisms possess a combination of properties that is not found anywhere in the inanimate world. Organisms are highly organized. They are composed of one or more cells; they acquire and use energy; they produce offspring similar to themselves; they contain a genetic blueprint that dictates their characteristics; they grow in size and change in appearance and abilities; they respond to changes in their environment; they carry out a variety of controlled chemical reactions; and they maintain a relatively constant internal environment.

Organisms can be described at various levels of organization. At the lowest, simplest levels, life consists of subatomic particles organized into specific atoms, and atoms organized into specific molecules. At the other end of the spectrum, communities made up of various species are organized into ecosystems which, in turn, make up the biosphere. With each step-increase in complexity, new properties emerge. Life, for example, emerges at the cellular level.

While exploring life, a number of themes can be identified which reoccur at various levels of organization. These themes form a foundation for understanding some of the recurring patterns of life. Five such themes are identified throughout the text. (1) The parts of an organism, whether a molecule or an organ, have a structure that is specifically suited for their function. (2) Living objects, from cells to multicellular organisms, possess regulatory mechanisms that ensure the maintenance of a relatively constant, ordered internal state. (3) Organisms require energy to carry out the activities that make life possible. (4) Despite the enormous diversity of organisms on earth, all organisms share many common features that reveal their descent from a common ancestor. (5) Species change over time, and new species are generated. This process of biological evolution produces organisms that are well suited (adapted) to the temporary conditions of their environment.

Evolution explains both the unity and diversity of life. Darwin arrived at his theory of evolution by natural selection by considering a series of observations and conclusions about individual organisms and populations (summarized in the last paragraph of the chapter).

Key Terms

biology (p. 6)
unicellular (p. 7)
multicellular (p. 7)
metabolism (p. 10)
enzyme (p. 10)
homeostasis (p. 10)

population (p. 11)
community (p. 11)
ecosystem (p. 11)
biosphere (p. 11)
taxonomy (p. 11)
binomial system of nomenclature (p. 13)

genus (p. 14)
evolution (p. 18)
adaptation (p. 18)
variability (p. 23)
natural selection (p. 23)

Review Questions

1. Describe the various properties that characterize life, and identify some inanimate objects that exhibit one or more of these properties. Can you identify any inanimate objects that possess *all* of these properties combined?

2. Rearrange the following list from the simpler to the more complex: ecosystems, subatomic particles, cells, organs, the biosphere, populations, molecules.

3. How did Lyell's book on geology and Malthus's book on populations influence Darwin's thoughts on evolution?

4. Describe Darwin's explanation for the presence of similar-looking finches with different types of beaks and feeding habits on different Galapagos islands.

Critical Thinking Questions

1. The Nobel Prize winner Albert Szent-Gyorgyi once wrote, " . . . there is only one life and one living matter, however different its structures, colorful its functions, and varied its appearance. We are all but recent leaves on the same old tree of life and even if this life has adapted itself to new functions and conditions, it uses the same old basic principles over and over again." Discuss the meaning of this statement in light of the ideas presented in the first section of this chapter.

2. During the early history of biology, a debate ensued between the "mechanists," who believed that life would eventually be understood entirely in terms of its physical and chemical properties, and the "vitalists," who believed that life is characterized by a special property not explainable solely in terms of its physical and chemical makeup. Is the emergence of new properties at higher levels of complexity consistent with a mechanistic view, a vitalistic view, or neither? Explain.

3. Create an analogy that reveals how novel properties emerge as levels of order increase. Begin by drawing an organizational hierarchy for one of the following: composing a song, building a house, writing a novel. Identify the intangible "extra" properties that emerge as the level of complexity increases. Why would the structures be of little value without these intangible "extras"?

4. Create a classification scheme for vehicles. Begin by observing and identifying as many "species" as you can. Group similar species into "genera," and so on, until you reach the top of the taxonomic hierarchy. Ask a friend to use your scheme to identify his/her vehicle. Do all vehicles fall clearly into one group or another? (Compare this to the classification of living organisms.)

5. Identify a characteristic in yourself that represents each of the five themes identified in this book as being fundamental to exploring life.

6. Considering the process of natural selection, how does greater variation within a species strengthen the chance that the species will avoid extinction following the appearance of some new deadly environmental condition (such as the appearance of a new virus or a marked drop in temperature)?

7. The necks of giraffes were used as an example of adaptation. Explain how the combination of genetic variation, competition for food, and natural selection accounts for the development of this adaptation, and why there are no shortnecked giraffes.

Additional Readings

Asimov, I. 1980. *A short history of biology.* New York: American Museum of Science. (Introductory)

Barlow, C. 1991. *From Gaia to selfish genes: Selected writings in the life sciences.* Cambridge: MIT Press. (Advanced)

Bronowski, J. 1973. *The ascent of man.* Boston: Little, Brown. (Intermediate)

Darwin, C. 1859. *The origin of species by means of natural selection.* (recent edition) New York: Random House. (Intermediate)

Desmond, A., and J. Moore. 1992. *Darwin.* New York: Warner. (Introductory)

Gould, S. J. 1987. Darwinism defined: the difference between fact and theory. *Discover* Jan: 64–70. (Intermediate)

Mayr, E. 1991. *One long argument: Charles Darwin and the genesis of modern evolutionary thought.* Cambridge: Harvard. (Advanced)

Moorehead, A. 1969. *Darwin and the Beagle.* New York: Harper & Row. (Introductory)

Perutz, M. F. 1989. *Is science necessary: Essays on science and scientists.* New York: E. P. Dutton. (Intermediate)

The Process of Science

STEPS
TO
DISCOVERY
What is Science?

THE SCIENTIFIC PROCESS

Conducting an Experiment to Test a Hypothesis

Interpreting Experimental Results

Formulating New Hypotheses

Additional Tests, Conflicting Results, and Final Confirmation

From Hypothesis to Theory

APPLICATIONS OF THE SCIENTIFIC PROCESS

Accidents and Scientific Discovery

Testing Additives and Drugs on Animals and Humans

CAVEATS REGARDING "THE" SCIENTIFIC METHOD

BIOETHICS

Science, Truth, and Certainty:
Is a Theory "Just A Theory?"

Y ou have probably heard the time-honored saying that if you get wet on a cold day, before long you will be suffering from a cold. For many of us, our experience seems to verify this statement: We get caught in a cold rain or step in a puddle, and within a couple of days we have a cold. If this has happened to you, can you conclude that getting chilled causes colds? In other words, if two events occur close together, is that proof that the first event *caused* the second?

The answer is no. In fact, there is no relationship between getting cold and catching a cold. The reason these two events often seem to occur together is because both tend to happen with the onset of wintry weather. Actually, colds are more common in the winter because people spend more time in buildings when the weather is cold or rainy, and being indoors increases close exposure to other people, many of whom have colds and are shedding the virus. In addition, cold wintry air dries the mucus layer that lines the nasal passages, exposing the underlying cells to viral attack. Getting chilled does *not* make you more vulnerable to the common cold; however, that is just a popular myth.

Let's examine a few other time-honored beliefs plus some recent newsworthy claims:

• Shaving makes hair grow back thicker.

• Eating bean sprouts and similar "health foods" promotes good health and increases longevity. Casual observation often reveals that sprout eaters feel healthier and experience fewer health problems than do people who eat no sprouts or "health foods."

• Smoking does not cause diseases that cut short the life span. Defenders of tobacco companies claim that the long lives of some smokers proves that smoking does not cause diseases that shorten life. George Burns, for example, has

Scientists work both in the laboratory and in the field to unravel the secrets of biological activities that accompany seasonal

lived into his nineties, in spite of his many years of cigar smoking.

Each of these examples, plus countless other conclusions that find their way into our lives, suffers a serious shortcoming in reasoning: They are all products of *anecdotal evidence*—information that is based on personal experiences and testimonials. Such stories lead many people to jump to conclusions that a cause-and-effect relationship exists between events that actually may be related simply by coincidence or clustered by some underlying and less obvious set of events. Let's reexamine each of the above conclusions and see how anecdotes often lead to questionable conclusions and misinformation.

• When people shave an area of their body, the hair often grows back thicker, and the more they shave an area, the thicker the hair gets. But shaving usually begins with the onset of puberty, and the effects of puberty increases body hair growth over a period of years, so hair gradually grows in thicker, even if shaving is avoided. It is puberty, not shaving, that increases hair growth.

• People who regularly eat bean or alfalfa sprouts are generally health-conscious individuals who also tend to eat little red meat and to exercise frequently. Any or all of these factors may be responsible for the alleged better health of sprout eaters; the sprouts themselves are not responsible. In fact, Bruce Ames, a University of California researcher who invented a valuable test for detecting cancer-causing activity in various substances, has found that sprouts, mushrooms, and so-called health foods are actually loaded with chemicals that could cause cancer.

• People who believe that the continued survival of a 90-year-old smoker proves that tobacco smoke doesn't shorten life are more likely to have a false sense of security about smoking. Scientific studies (as opposed to anecdotal evidence) have proven that smoking is the leading cause of preventable death in the United States and other industrialized countries. We can explain George Burns's longevity in the same way we explain how some people survive falls out of airborne planes without parachutes: They are very lucky. We certainly would not conclude that falling out of airplanes without a parachute is harmless. Again, anecdotal evidence fails to reveal the critical information; that is, how long George Burns would live had he *not* smoked cigars?

Anecdotal evidence is the heart of misinformation. It often leads to absurdly irrational beliefs, such as black cats causing bad luck or ostriches sticking their heads in the ground to escape danger (such behavior would quickly lead to a severe ostrich shortage). Anecdotal evidence has also misled people in their attempts to understand themselves and the nature of life. One particularly prevalent anecdote asserts that the fact that we don't understand all aspects of life proves that life is a mystery beyond human understanding.

Scientists, however, believe that all phenomena in the universe have rational, verifiable explanations. We will never know everything there is to know about life and the universe, but we continue to expand our understanding by making careful observations, asking questions, and seeking answers. Yet, none of these activities alone has provided us with our current understanding of how organisms work. Observations alone, like anecdotal evidence, provide descriptive information, but they rarely reveal the mechanisms responsible for biological activities or the ways in which the intricate processes of biology affect one another. For example, the changing of seasons from warm, summer-like conditions to cold, wintry days and nights is accompanied by a profound change in the biological activity of countless organisms: Tree leaves turn brightly colored and fall; many animals hibernate, while others migrate to a warmer climate; still other animals develop a thick winter coat of fur. But is it the onset of cold weather that triggers these biological changes or the shortening of the days as winter approaches? In other words, what is the *cause* of this observed effect? Furthermore, *how* does that factor bring about the observed change?

The answers to such questions are not obtained by simple observation. Rather, they require the combined input of several different methods for acquiring and examining information, assembled into what amounts to a scientific approach. Using scientific methods provides us with a way of not only understanding how life and the universe are put together and how they interact but of verifying the accuracy of the information we discover and the explanations we propose. While this approach can help scientists evaluate and explain nature, it can also be used to solve problems in our personal lives and help us understand for ourselves.

In this chapter, we provide a number of examples that illustrate how the results of one study become the starting point of subsequent studies so that scientific knowledge builds from one confirmed explanation to another and from one generation of scientists to the next. This pattern of sequential scientific growth is revisited in the "Steps to Discovery" feature that launches each chapter. You will also become acquainted with the self-correcting nature of science, a built-in self-check that assures the eventual detection and elimination of incorrect or biased results.

The job of understanding how living things work will never be completed. Scientific findings always suggest new questions and predictions, so the path of discovery and exploration is endless. That is part of the excitement of biology—scientists will never work themselves out of a job. There will always be questions to answer.

changes, such as bird migration, leaf color changes, leaf fall, and butterfly development.

"*T*he skies were overcast, and the city was expecting heavy rain and frogs." This might have been a typical forecast in 1668 since at that time it was commonly believed that frogs developed from falling drops of rain. Although it was evident that mammals, such as pigs and dogs, were born, and that birds were hatched from eggs, the origin of smaller creatures, such as frogs and flies, was quite obscure. After all, no one had yet seen the development of an organism from a microscopic egg. The notion that most living organisms arose directly from inanimate materials was known as **spontaneous generation** and was popular among scientists and nonscientists alike for hundreds of years. It seemed particularly evident that flies arose directly from decayed meat, since everyone had observed rotting meat covered with maggots—the larval stage of flies.

One of the first skeptics of spontaneous generation was Francesco Redi, an Italian physician and naturalist. Redi was familiar with the recent accounts of the famous English physician William Harvey, who, during a dissection of one of the King's deer, had found a tiny fetus that appeared much like a miniature adult deer. Harvey had concluded that animals develop from seeds or eggs that were too small to be seen. Redi had a natural curiosity about life and was willing to question the validity of spontaneous generation.

▼ ▼ ▼

THE SCIENTIFIC APPROACH

By recognizing the possibility that spontaneous generation might be an erroneous concept, Redi had taken the first step on the path to scientific discovery. This is the way scientists typically begin an investigation; they learn about the information currently available on a subject, and they make observations that are relevant to the matter at hand. Redi had observed that maggots tended to "appear" in places where adult flies could also be found. Based both on Harvey's findings and his own observations, Redi proposed an alternative explanation for the origin of maggots: "The flesh of dead animals cannot engender worms unless the eggs of the living be deposited therein." A tentative proposal of this type is called a **hypothesis.** Two hallmarks of a good hypothesis are that

- it is consistent with observations collected up to that point; and
- its validity can be tested by experimentation (or further observation).

Redi came up with an experimental plan by which he could test his hypothesis that maggots arose only from eggs deposited by flies and not from meat (Figure 2-1).

CONDUCTING AN EXPERIMENT TO TEST A HYPOTHESIS

In his experiment, Redi put a dead snake, some fish, some eels, and a slice of veal separately into four large, wide-mouthed vessels and sealed the openings with wax. He then placed the same materials into another set of four vessels which he left open to the air.

RECORDING RESULTS OF THE EXPERIMENT

Within a matter of days, Redi observed that the decaying meat within each of the open vessels was teeming with maggots, and flies were observed coming and going at will. In contrast, the meat in the sealed vessels was in the same state of decay, but no maggots were evident.

INTERPRETING EXPERIMENTAL RESULTS

Redi concluded that the closed vessels failed to produce maggots because flies were unable to reach the meat. But Redi recognized a serious flaw in this interpretation: What if maggots failed to develop in the sealed vessels because of a lack of fresh air rather than the absence of flies? It is essential in any scientific experiment to be able to determine *cause and effect*. In Redi's initial experiment, the two sets of vessels differed by more than one **variable**—a condition that is subject to change. In this case, the two variables were the presence or absence of flies and the presence or absence of air. To determine which of these two variables was the *cause* of the results, Redi performed another experiment. He once again set up two sets of vessels containing meat, but this time both vessels were left open to the air; one set was left totally uncovered, while the other set was covered with a fine layer of gauze whose holes were too small to allow flies to pass through.

In the design of a proper experiment it is essential to allow only one condition to vary—this is the one variable whose role the investigator is attempting to evaluate (in this case, access by flies). Otherwise, it is impossible to determine which variable is causing the result. In modern scientific terminology, the gauze-covered set of vessels is called

Putrefying meat with maggots	Putrefying meat no maggots

(a)   *(b)*

FIGURE 2-1

Redi tests the validity of spontaneous generation. *(a)* In his initial experiment, Redi placed meats in two groups of vessels. The vessels in one group were left open to the air, while the vessels in the other group were sealed with wax. Maggots appeared only in the vessels exposed to air. *(b)* To eliminate the possibility that maggots failed to appear in the sealed vessels because of the absence of fresh air, Redi conducted a second series of experiments. Again he prepared two groups of vessels containing meat, but instead of sealing one group of vessels, he covered the openings with a layer of fine gauze through which flies could not pass. Once again, maggots appeared in the vessels in which flies could reach the decaying meat but not in the meat isolated from flies. (Conclusion: Flies do not spontaneously generate; it takes flies to make flies.)

the **control group** because the variable being tested—access by flies—is absent. The uncovered vessels constitute what is called the **experimental group** because they are exposed to the variable. If Redi's conclusions about his first experiment were correct, maggots would not develop in the gauze-covered vessels, even though the meat had been fully exposed to air, because that one crucial variable—access by flies—was being restricted. After a few days, Redi indeed observed that the meat in the uncovered vessel produced maggots, while the meat in the gauze-covered vessels remained free of these larvae. The results of the experiment supported his hypothesis that maggots appeared only from the deposited eggs of flies.

FORMULATING NEW HYPOTHESES

Scientists typically go beyond the specific results obtained from an experiment or a series of observations and use the data to explain a more general phenomenon. In other words, they formulate a more comprehensive hypothesis, which can be subjected to additional testing. Based on his experiments with flies, for example, Redi extended his observations, hypothesizing that *all living beings* "come from seeds of the plants or animals themselves."

Hypotheses form the foundation on which science grows, the fuel that drives new investigations. For Redi's expanded hypothesis to be correct, it would have to apply to all types of organisms, not just flies living in Florence. If someone were to demonstrate even one exception—one clear case of spontaneous generation—then Redi's hypothesis would have to be rejected or significantly modified.

ADDITIONAL TESTS, CONFLICTING RESULTS, AND FINAL CONFIRMATION

Another important characteristic of scientific findings is their repeatability. When papers are published, scientists present their methods so that other investigators will know exactly how the procedures were carried out. Redi had convinced the world that macroscopic organisms, such as frogs and flies, arose from the eggs of parents. The discovery of microorganisms about 300 years ago, however, rekindled the notion of spontaneous generation. These simpler microscopic life forms were believed to arise spontaneously from the remains of dead organisms or from other nonliving materials, substances in pond water, rain puddles, or a bowl of chicken soup. Did Redi's hypothesis hold true for microorganisms as well?

During the mid-1700s, a pair of similar experiments were performed that yielded totally opposite conclusions. In England, John Needham prepared a beef broth for growing bacteria. He briefly boiled the broth to kill any microorganisms, poured it into a test tube, and sealed the tube with a cork. Within a few days the tube was swarming with bacteria. He concluded that the bacteria must have formed from the remains of heat-killed microorganisms and proclaimed the theory of spontaneous generation to be proven. Lazzaro Spallanzani, an Italian naturalist, heard of Needham's work and performed a similar experiment, but he found no evidence of bacterial growth.

Needham's experimental results failed to pass the test of repeatability. Spallanzani believed that Needham had not been careful enough to kill all the microorganisms or to avoid contamination of his glassware. (Needham was also biased toward the concept of spontaneous generation, whereas Spallanzani was not.) Spallanzani overcame the shortcomings of Needham's experiment by boiling the broth for longer periods of time, then immediately sealing the flask. Spallanzani's critics argued that as a result of the extensive boiling procedure, he had destroyed a "vital principle" in the broth that was needed to support the spontaneous generation of organisms. More importantly, Spallanzani sealed the boiled flask so that air could not enter. Unlike Redi, he devised no way to eliminate this extra variable.

In 1860, Louis Pasteur devised an experiment in which he sterilized a preparation of nutrient broth by boiling, yet he still allowed it to remain in contact with fresh air. The

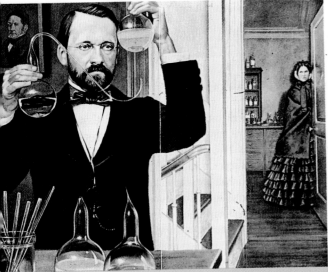

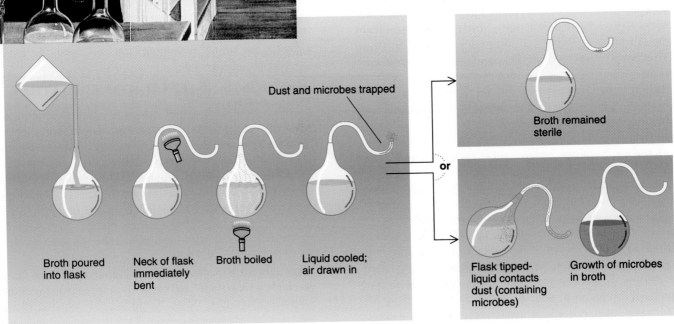

FIGURE 2-2

Pasteur's experiment disproves spontaneous generation. Pasteur poured nutrient broth into the flask, boiled it extensively, and allowed it to cool. The flask remained open to the outside air but remained free of microorganisms. After many days, Pasteur tilted the flask so that the sterilized broth would run into the neck and contact airborne particles that entered the nutrient medium. The flask soon became densely populated with microorganisms.

Dust and microbes trapped

Broth poured into flask

Neck of flask immediately bent

Broth boiled

Liquid cooled; air drawn in

or

Broth remained sterile

Flask tipped- liquid contacts dust (containing microbes)

Growth of microbes in broth

experiment was similar in principle to that performed by Redi nearly 200 years earlier. Pasteur added broth to a flask then melted the glass neck of the flask and shaped it into a long **S**-shape (Figure 2-2). The contents of the flask and the glass neck were sterilized by heat, then cooled and allowed to remain open to the outside environment. Even though fresh air could pass through the neck into the flask, all particles of matter suspended in the air settled in the "trap" of the **S**-curved tube, so no airborne organisms could reach the broth. Under these conditions, bacteria failed to grow in the flask. The absence of bacterial growth was not due to the destruction of some "vital principle" by heat, or by the absence of air; rather, the lack of growth was due to the absence of contamination by airborne bacteria. This fact was clearly established by Pasteur when he carried out the same experiment but, after several days of no growth, tilted the flask so that some of the sterilized broth would run into the **S**-shaped neck and contact the bacteria that had collected in the trap. Within 18 hours, the broth was swarming with bacteria. Pasteur had shown that living organisms do not spontaneously appear; it takes an organism to make an organism.

FROM HYPOTHESIS TO SUPPORTED THEORY

When a hypothesis has been repeatedly verified by observation and experimentation and combines with other confirmed hypotheses to help explain an important aspect of a field of investigation, the collection of related hypotheses are considered a supported **theory.** Most scientists define the term "theory" much differently from the general public, which often speaks of theories as "speculations" or "guesses." Yet, even among scientists, there is debate over the use of the term "theories." For example, should the spontaneous generation notion be called a theory? Many scientists say yes—it is a disproven theory. In this book, we will distinguish between supported theories and disproven or unsubstantiated theories. Because supported theories are backed by repeated tests, they are held with great confidence by the scientific community.

It is important to understand scientists' attitudes toward theories. It is said that a scientific theory can never be proved, only disproved. For theories to continue being accepted as correct, they must continue to be consistent with the results of new experiments or related phenomena. For instance, one of biology's most important supported theories is the theory of evolution—the theory that species change over time. The theory of evolution is no less reliable than is the atomic "theory" or the "germ theory of infectious disease." As with atoms, which cannot be directly observed, we cannot directly observe the production of new species by evolution. Biologists have gathered a tremendous amount of evidence in support of biological evolution, however, and not one major piece of evidence has ever been obtained that suggests it has not occurred. This is a very important point.

Conversely, the so-called theory of creationism proposes no testable hypotheses and thus cannot be proven right or wrong. Proponents simply accept it on faith.

The theory of evolution however, suggests many predictions that can be tested. In the words of novelist and biologist H. G. Wells, "Every mammal...is held to be descended from a reptilian ancestor. Suppose in the early Coal Measures [a period hundreds of millions of years ago], before ever a reptile existed, we found the skull of a horse or a lion. Then the whole vision of Evolution would vanish. A single human tooth...in a coal seam would demolish the entire fabric of modern biology. But never do we find any such anachronisms. The order of descent is always observed."

The scientific approach yields explanations and understanding. It is a process that combines questioning, careful observation, rigorous experimentation, and repeated verification to improve our knowledge of ourselves and the universe. (See CTQ #2.)

APPLICATIONS OF THE SCIENTIFIC PROCESS

There is another side of scientific research: It often leads to practical applications, some of which have improved our lives. Pasteur's work, which demonstrated that relatively simple measures of sterilization could prevent bacterial growth, had almost immediate practical ramifications. The British surgeon Sir Joseph Lister grasped the importance of Pasteur's work and instituted the revolutionary procedure of cleansing surgeons' hands and instruments prior to their performing an operation. The procedures produced a striking drop in the incidence of fatal postsurgical infections. Midwives and others involved in delivering babies adopted similar procedures, and the entire practice of medicine took a giant stride toward becoming safer and more effective.

ACCIDENTS AND SCIENTIFIC DISCOVERY

Most great scientific achievements are the result of creativity, diligence, and many long hours of dedicated research. However, many revolutionary discoveries were the products of unplanned events in which the observer was skilled enough to recognize the importance of the accidental occurrence. One such "accident" occurred in 1928 to Sir Alexander Fleming (Figure 2-3).

Discovering Penicillin

Fleming was a Scotsman who had studied medicine in London and was influenced by one of his professors to enter the field of bacteriology. When World War I broke out, Fleming enlisted in the army and was assigned to a bacteriologic

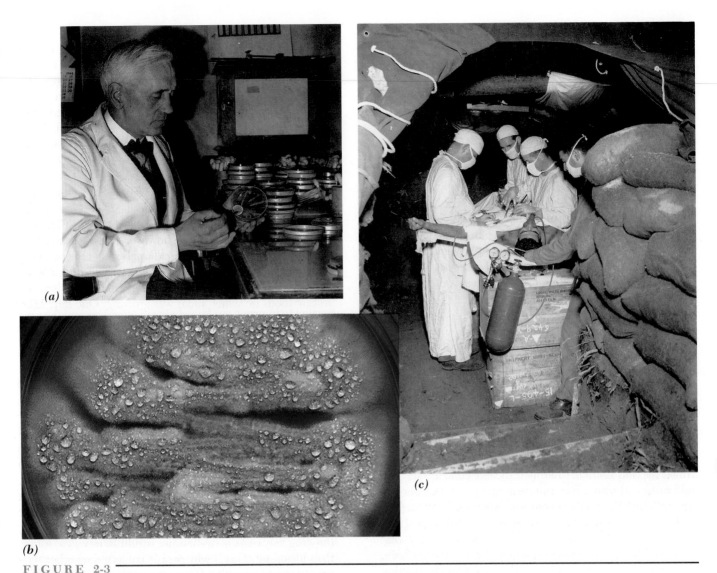

FIGURE 2-3

The history of penicillin. *(a)* Sir Alexander Fleming working in his lab. *(b)* Colony of *Penicillium* growing on the surface of culture media. The amber droplets collecting on the mold's surface contain the antibiotic penicillin. *(c)* Penicillin became available for fighting infection during World War II. Countless lives were saved by its use. The drug was so scarce, however, that it had to be "recycled." Penicillin was extracted from urine collected from patients receiving treatment and then reused.

research lab. During the war, Fleming became aware of how woefully inadequate antiseptics (such as hydrogen peroxide and carbolic acid) were in preventing wounds from becoming infected. After watching untold numbers of patients die from infected wounds, Fleming set out on a lifelong search for an agent that would effectively kill microbes while, at the same time, remain relatively nontoxic to human patients. As of 1928, only one such substance had been discovered—an arsenic-containing compound that was effective against certain bacteria, particularly the one that causes syphilis. The substance had only limited value, however.

In 1928, Fleming was working with a new antiseptic, mercuric chloride, which was much too toxic to be taken internally. Fleming's small lab in a London hospital was always cluttered with petri dishes from previous experiments. One day, while talking to a colleague, Fleming picked up one of these bacteria-inoculated dishes and noted that it was contaminated with a mold—not an uncommon occurrence in a bacteriology lab. While many researchers might have simply discarded the dish, Fleming alertly observed that in this particular case the mold was surrounded by a ring in which there were no bacterial colonies. Fleming immediately removed a bit of the mold with a scalpel and transferred it to a fresh tube of culture medium. He wanted to be sure he had a sample of this mysterious mold that could destroy bacterial cells growing in its vicinity.

Fleming cultured samples of the mold, which he identified as a member of the genus *Penicillium,* and collected samples of the medium in which the mold had been growing. These samples of mold-juice, as they were called, were found to be highly effective in killing bacteria. More importantly, the mold-juice could be injected into mice or rabbits without any evidence of toxicity. In the words of A. Maurois, Fleming's biographer, "Fleming had for a long time been hunting for a substance which should be able to kill the pathogenic microbes without damage to the patient's cell. Pure chance deposited this substance on his bench. But, had he not been waiting for 15 years, he would not have recognized the unknown visitor for what it was." Fleming then turned all of his attention away from studying bacteria toward working with mold-juice.

Determining the Value of Penicillin

Fleming enlisted a number of colleagues to purify the active substance, which he named penicillin. But penicillin resisted all early attempts to purify it. Meanwhile, Fleming began the first tests to see how effectively the mold-juice could suppress human infections. Although penicillin showed some promise when applied to a skin infection, the substance was not available in sufficient concentration to prove its real value. For the next few years, neither the scientific nor the medical community showed any interest in Fleming's substance. Remarkably, it wasn't until 1939, about the time England declared war with Germany, that Howard Florey, an Australian pathologist, and Ernst Chain, a Jewish chemist who had escaped Nazi Germany, began a concerted effort to purify penicillin at Oxford University.

As often happens in scientific research, Florey and Chain were aided by a new technique that allowed them to perform experiments that could not have been carried out earlier. The technique they used was called *lyophilization,* the technical name for "freeze-drying," the same process used to prepare instant coffee. Freeze-drying greatly facilitated the purification of unstable compounds, since cold conditions generally retard chemical deterioration. Using this technique, the mold-juice was frozen solid, and the water evaporated from the solid state under a vacuum, leaving the penicillin in a powdered state.

After tests on mice infected with various bacteria proved to be successful, the first test on a human was conducted in England in 1941. As was not uncommon at the time, the patient was dying from "blood poisoning," which had been caused by bacteria initially infecting a small scratch on his face. After several injections of purified penicillin, the patient showed dramatic improvement. Unfortunately, the supply of purified penicillin was quickly exhausted; the infection regained its ferocity and quickly claimed the patient's life.

From the Laboratory to the Factory

It was evident to the Oxford group that England did not have the facilities at the start of the war to produce sufficient quantities of this important new drug. Florey traveled to the United States carrying with him samples of the mold in hopes that he could persuade a U.S. pharmaceutical company to begin mass-producing penicillin. He handed over the samples, together with his data for purification, without any attempt to patent the process or to ensure financial gain for himself or his colleagues.

Within a matter of months, the U.S. company had greatly improved its ability to produce and harvest penicillin so that reasonable yields were being obtained. Fleming's mold was not very productive, however, and the search for a more generous species of *Penicillium* began. One company employee, nicknamed Moldy Mary, was responsible for daily trips to the markets of Peoria, Illinois, to look for rotting foods on which mold was growing. One day, Mary brought back a moldy cantaloupe that turned out to be growing a particularly productive species of *Penicillium.* Together with the new strain and improved culture conditions, the company was soon producing enough penicillin to save the lives of thousands of wounded soldiers of the Allied armies. The drug has been saving lives ever since.

In 1945, Fleming, Florey, and Chain were awarded the Nobel Prize for discovering and purifying the first **antibiotic** (a microbe-killing substance produced by a fungus or a bacterium). Subsequent research revealed that penicillin kills bacteria by interfering with the formation of their surrounding cell walls. Since human cells have no such walls, penicillin is not dangerous to the human body. Some people are allergic to penicillin, however. For these people, the drug may be fatal (sometimes within 10 minutes) rather than lifesaving. This is why people are asked whether they are allergic to penicillin before the antibiotic is prescribed.

TESTING ADDITIVES AND DRUGS ON ANIMALS AND HUMANS

Penicillin became commercially available at a time when there were very few rules and regulations governing the use of new drugs and products. That climate has changed in recent years, and most substances are extensively screened for their toxicity and *carcinogenic* (cancer-causing) properties. To conduct controlled experiments that measure these risks, biologists divide a *test population* of organisms into an experimental group and a control group. For example, consider tests to determine whether the food additive red dye #2 poses a cancer risk (Figure 2-4). A test population of rats was divided into an experimental group, which received food containing red dye #2, and a control group, which received exactly the same food but without red dye #2. All other variables (such as light, temperature, quantity of food, and amount of water) were identical for the two groups. A number of such studies demonstrated that red dye #2 did

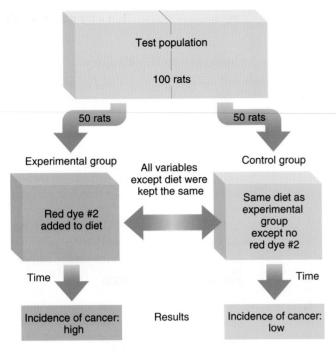

FIGURE 2-4

Does red dye #2 increase the incidence of cancer? Around the turn of the century, it was not unusual to test the health effects of food preservatives (and other chemicals) on human subjects, even children. But today not only is the use of "human guinea pigs" considered inhumane, it is also impractical since it can take 20 to 40 years for the dangerous effect of a chemical to become apparent. Because they are anatomically and physiologically similar to humans, shorter-lived rodents have now replaced human subjects in most scientific tests. Using such guidelines, scientists studied whether red dye #2 increases cancer in laboratory rats. All factors were identical in both the test group and the control group except for the addition of red dye #2 to the test group's food. Rats ingesting the dye developed an abnormally high level of cancer. Because there is a strong correlation between some causes of cancers in humans and in rats, the FDA banned red dye #2 as a food additive.

(a)   *(b)*

FIGURE 2-5

Modern biologists exploring life. Although biologists rarely become politicians or economic leaders, they nonetheless help shape the future of the world. Some scientists actively seek solutions to specific problems, such as this biologist investigating the virus that causes AIDS *(a)*. Others seek answers simply for the sake of knowledge and increased understanding—these marine biologists, for example, studying the ecology of the ocean floor *(b)*. Discoveries by such "basic researchers," however, often have hundreds of practical applications. As you will see throughout this book, scientific findings continue to dramatically influence our lives, from providing products of huge economic value to informing us how we must alter our global lifestyles in order to preserve the planet's life supporting capacity.

indeed increase the incidence of cancer in rats. Because rats and humans are biologically similar, the Food and Drug Administration (FDA) banned red dye #2 as a food additive.

Controlled experiments on people often include procedures to reduce the chance that personal bias will influence the results. Despite attempts to include as many controls as possible, research with humans is invariably open to debate. A well-publicized series of studies concerning the possible preventive and curative powers of vitamin C illustrates some of the complications involved in using human subjects. During the 1970s, one of the great chemists of the century, Linus Pauling, began proclaiming the benefits of vitamin C, first as a preventive of the common cold and later as a treatment for cancer. Pauling based his anticancer claims on work carried out in Scotland, in which 100 people with advanced colorectal cancers received large daily doses of vitamin C, while a larger control group made up of people of similar sex, age, and state of disease received the same treatment without vitamin C. The results of these studies suggested that vitamin C might prolong the life of patients with advanced cancer and might even cause some cancers to regress entirely. The pronouncements concerning the Scottish work were greeted with great controversy, pitting Pauling on one side against the disbelieving medical establishment on the other. Because of all the publicity, a group of researchers led by Charles Moertel at the Mayo Clinic in Minnesota decided to pursue a similar study. In contrast to the work in Scotland, the Mayo researchers carried out their study using a *double-blind* procedure. In a simple "blind" test, one group of patients receives a pill containing the drug to be tested, while a control group receives a *placebo,* a similar-looking pill that, unknown to them, contains no drug. When the researchers themselves are unaware of which subject is receiving the actual drug versus the placebo, the test is called a *double-blind test.* Double-blind tests eliminate the bias that may influence the results when researchers (or the subjects) strongly expect or hope for a particular result.

Moertel and his colleagues found no evidence that vitamin C prolonged the life of their patients with advanced cancer. But Pauling argued that the lack of a curative effect of the vitamin in the Mayo study was due to the patients' prior treatment by chemotherapy, which had damaged their immune systems and prevented them from responding positively to the vitamin C supplements. In contrast, the patients in the Scottish study had never received prior chemotherapy.

The public interest in megadoses of vitamin C continued, and a few years later Moertel and the Mayo group announced the results of another double-blind study, this time involving advanced colorectal cancer patients who had received no prior chemotherapy. (The clinicians felt that since this type of cancer did not respond well to chemotherapy anyway, it would not be unethical to withhold chemotherapy from the patients in this study.) Once again, the Mayo group found no benefit to patients receiving large

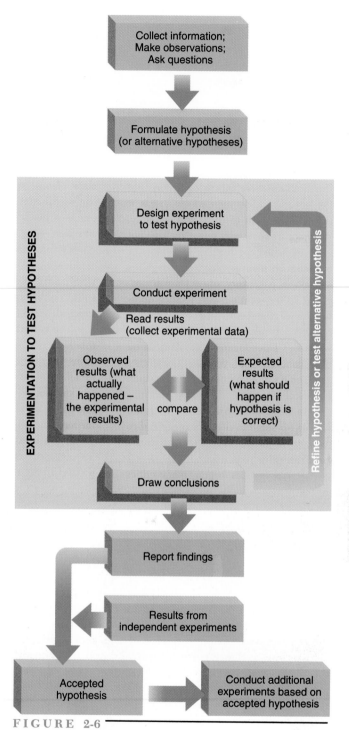

FIGURE 2-6

The scientific method. Many scientific investigations start with an observation of a phenomenon, a discussion with colleagues, or ideas triggered by listening to the presentation of a scientific paper or by reading a research article in a scientific journal. Such activities stimulate the scientist to formulate one or more hypotheses to explain a particular set of observations. These hypotheses lead to a plan for gathering experimental data that either supports a hypothesis or causes it to be rejected. Failure of one hypothesis often leads to another hypothesis, which is then tested through additional experimentation.

Science, Truth, and Certainty: Is a Theory "Just A Theory?"

By ANN S. CAUSEY
Prescott College

In recent times, science has come under fire for its apparent failure to provide what the general public increasingly insists it provide: the path to truth and certainty. We want to know why scientists have not yet established the truth of the theory of evolution by natural selection, though it's been over a century since Darwin proposed it. We want answers to such questions as was the record heat of the summer of 1988 a result of human caused global warming? Does cigarette smoking cause lung cancer? And, why can't scientists agree on a simple yes or no answer to such empirical questions?

In the absence of definite answers lies license for some to distort science and to manipulate public perception of it. Since scientists cannot state with certainty that global warming has begun, many critics have dismissed warnings of impending climate change as "just theory." Tobacco companies continue to rely on the lack of causal certainty in studies linking smoking and cancer, as they defend themselves against product-related lawsuits. And, creationists insist that until and unless Darwinian evolution is proved beyond any and all doubt, creationism should be taught alongside evolution in public schools.

The important question is why scientists have not been able to put such criticisms of scientific theory to rest once and for all. Why can't science provide the certain knowledge—the truth—that the general public seeks? The answer lies partly in the nature of science itself and partly in the problematic nature of the concept of truth. To construct the answer, let's examine these issues in more detail.

One of the primary tools of the scientific method is *inductive reasoning;* that is, reasoning from particular observations to general conclusions. The nature of induction, however, actually limits the extent of justification that any scientific "belief" can earn. The problem is that the conclusions we get through inductive reasoning are always less certain than are the premises from which they are drawn. For instance, from observations of the color of thousands of different crows, we may induce (hypothesize) that all crows are black. Of course, we can determine the color of any particular crow with certainty; yet we can never observe all crows that have ever existed or might in the future exist. Thus, our conclusion (all crows are black) based on our many observations, while highly probable, can never be absolutely certain. Of course,

---

doses of vitamin C. And, once again, Pauling responded with stinging criticism of the study. He argued that unlike the Scottish study, where patients had been kept on vitamin C for as long as they lived (as long as 12 years), the group at the Mayo Clinic had been treated with the vitamin for less than 3 months. The controversy rages on today and will continue until a definitive study on human subjects clearly proves or disproves the value of vitamin C as a therapeutic agent against cancer. Thousands of such questions, both practical and theoretical, propel scientific inquiry as modern biologists continue to investigate the world of life (Figure 2-5).

Science does not proceed along predetermined paths, and the end product cannot always be predicted. The results of scientific research often have practical applications that were never conceived of by the original researchers. (See CTQ #4.)

CAVEATS REGARDING "THE" SCIENTIFIC METHOD

The vitamin C research shows quite clearly that the results of scientific studies are often open to differences of interpretation. Readers should always be somewhat skeptical, particularly when the results are being reported second-hand in a newspaper or popular magazine.

It should also be borne in mind that the scientific method, as summarized in Figure 2-6, is only a general description of one way scientists approach their studies. Even though it is systematic, the scientific approach also allows for creativity and flexibility; there is no absolute path for discovering the answer to a question. Some scientific studies might start with an observation, others might start with a question based on previous work. Sometimes a scientific study might begin with a hypothesis to be tested, while other times an accidental discovery may trigger a finding that leads to a hypothesis. As you will see throughout

the more black crows we find, the more probable or confirmed our hypothesis is. However, no amount of observation will allow us to say with absolute certainty that *all* crows are black. And, just one observation of a non-black crow is sufficient to invalidate our hypothesis completely. Such is the nature of science: Even the best and most widely accepted hypotheses and theories generated by inductive reasoning have a built-in element of uncertainty.

Not only can we never claim absolute justification by way of certainty for any theory, we cannot even agree on what *constitutes* truth in a belief or theory. Philosophers split into three main camps on this issue. One camp, proponents of the coherence theory of truth, claims that truth is a property of a related group of consistent (or coherent) statements. Thus, the truth of any one belief depends on how well it coheres, or is consistent, with the other members of a larger generally accepted belief system. Mathematics is a good example of the coherence theory of truth in operation; from a few basic statements, an entire system of mathematical "truth" is constructed. Similarly, in science, theories

gain respectability when they are coherent with a larger body of accepted judgements.

Another camp contends that truth consists of correspondence between a belief and a fact or actual state of affairs. Thus, a belief is true if it corresponds to reality, reality consisting of some realm of facts existing objectively and independent of us. This is undoubtedly the most popular theory of truth, and one that is heavily relied on in the sciences. Most scientists do, in practice, assume that a good theory explains how things really, objectively are in nature.

Finally, the third camp insists that what makes a theory true is neither its coherence with other accepted theories nor its correspondence with some alleged objective reality, but rather its usefulness. Pragmatism, as it is called, finds true theories simply by finding those that work best in predicting, explaining, and facilitating further study. When a theory stops working, it is no longer true. We thus make our own truths as we develop and use theories. Truth, for pragmatists, is dynamic and changing; most scientists agree that these traits are essential to good theories as well.

Thus, it is true that a theory is never certain, and even truth is subject to individual interpretation. While we can never remove all elements of uncertainty from scientific theories, that uncertainty is insignificant in the face of the overwhelming amount of evidence a theory must accumulate in its favor in order to win the confidence of the scientific community. Furthermore, although scientists do not have a special claim to absolute truth, they do incorporate the important elements of coherence, correspondence, and pragmatic truth in their work. A good theory will likely be "true" in all three senses.

Thus, repeatedly verified scientific theories, such as the link between smoking and cancer, the mechanism of evolution by natural selection, and the effects of pollution on global climate, warrant a very high degree of confidence in their accuracy. Lack of scientific certainty does not justify a lack of public confidence in science; the uncertainty is part and parcel of the methodology of science itself and is essential if theories are to remain dynamic and subject to revision as new evidence accumulates.

this book, each scientific "fact" has a unique and often fascinating story of exploration and discovery.

Regardless of which path is taken, scientific knowledge grows over time, with new discoveries blossoming from the

seeds planted by previous researchers. Isaac Newton summarized this dynamic when he wrote, "If I have seen farther than other men, it is by standing on the shoulders of giants."

SYNOPSIS

Scientists believe that the world is understandable and that rational explanations exist for all phenomena. The basic objective of science is to understand life and the universe in a way that can be validated.

Scientific investigations commonly follow an investigative strategy we call the scientific method. The process includes making careful observations, learning what is known about the subject, and asking questions; formulating

hypotheses that might explain a phenomenon or answer a question; designing and conducting experiments to test the validity of hypotheses; analyzing results; drawing conclusions as to whether to accept, reject, or modify the hypotheses being tested.

When a hypothesis has been repeatedly reinforced by experimentation, it may become a supported theory, a term that indicates its widespread acceptance. Unlike a single hypothesis, a theory combines many ideas into a unifying principle. For a theory to continue to be accepted, it must remain compatible with new evidence.

Experiments must be performed with proper controls so that the effect of a single experimental variable can be determined. Controlled experiments help distinguish cause-and-effect relationships from coincidence and re-

duce the influence of human bias. The use of a control allows the investigator to determine the effect of one factor at a time.

Scientific research often leads to practical applications. This was illustrated in this chapter by the discovery and production of penicillin.

Key Terms

spontaneous generation (p. 32)
hypothesis (p. 32)
variable (p. 32)

control group (p. 33)
experimental group (p. 33)

theory (p. 36)
antibiotic (p. 37)

Review Questions

1. Why is it important to have a control group when carrying out experiments? Why was Redi's initial experiment using sealed vessels an inappropriate control?

2. How did Pasteur's use of long s-shaped-necked flasks serve as an adequate control in testing the existence of spontaneous generation of microorganisms?

3. How is it that penicillin can be used safely in humans while preventing the growth of bacteria, which are also living organisms?

4. What is meant by the term "double-blind" experiment, and why are such experiments used in science?

Critical Thinking Questions

1. Classify the following statements as products of simple observation, speculation, or the scientific approach. How might each statement that falls into the first two categories be scientifically tested?
   a. The sun rises in the east.
   b. Since the sun always rises in the east, it will rise in the east tomorrow.
   c. I took vitamins every day this year and did not catch a cold. Therefore, vitamins prevent colds.
   d. People with protein deficiencies are more vulnerable to infectious diseases.

2. Select an article from a tabloid that makes some claims of a "scientific" nature. Evaluate how scientific the claims are, using the following questions as a guide: (1) Are the claims based on observations? (2) Have the observations been made by one person or many? Are they repeatable? (3) Is the person(s) a reliable observer? (4) Are the claims testable? (5) If tests have been performed, did they involve controls?

3. What might be a practical application of each of the following scientific findings?
   a. The body responds to the presence of foreign materials by launching an immune response that attacks that foreign invader (but none other).
   b. Every living cell in a plant has the genetic capacity to become any other type of cell in that plant. In fact, one actively growing cell taken from a mature plant can grow into an exact duplicate of the plant from which it was taken.

4. Among the many discoveries of the twentieth century that have changed our lives, seven are related to biology. Select one of those listed below and research the basic scientific advances that made the discovery possible. What other practical applications have resulted from the basic scientific advance(s)? Which, if any, of the practical applications were predicted? (1) discovery of blood types, (2) development of hybrid corn, (3) use of DDT, (4) synthesis of sex hormones, (5) use of chlor-

promazine and lithium for treating mental illness, (6) discovery of structure of DNA, (7) discovery of early prehuman fossils.

5. Design a *controlled* experiment to test the hypothesis that high doses of vitamin C reduce the incidence of colds. Can you use a test population of human subjects? What are the normal limitations, and how will they affect the results?

6. How would you apply the scientific method to test the accuracy of the following statements?
    a. An ostrich sticks its head in the ground to avoid impending danger.
    b. Disasters occur in groups of three.
    c. Large doses of vitamin E improve one's energy.
    d. Walking under a ladder will bring bad luck.
    e. Camels carry water in their humps.

Additional Readings

Asimov, I. 1984. *Asimov's new guide to science.* New York: Basic Books. (Introductory)

Gibbs, A., and A. E. Lawson. 1992. The nature of scientific thinking as reflected by the work of biologists and by biology textbooks. *Am. Biol. Teacher* 54 (3):137–152. (Intermediate)

Goldberg, A. M., and J. M. Frazier. 1989. Alternatives to animals in toxicity testing. *Sci. Amer.* Aug: 24–30. (Intermediate)

Kohn, A. 1984. *Fortune or failure: missed opportunities and chance discoveries in science.* London: Blackwell. (Introductory)

MacFarlane, G. 1984. *Alexander Fleming. The man and the myth.* London: Hogarth Press. (Introductory)

Maurois, A. 1959. *The life of Sir Alexander Fleming.* London: Jonathan Cape. (Introductory)

Moberg, C. L. 1991. Penicillin's forgotten man: Norman Heatley. *Science* 253:734–735. (Intermediate)

Richards, E. 1991. *Vitamin C and cancer: Medicine or politics.* New York: St. Martin's. (Introductory)

Serafini, A. 1989. *Linus Pauling: A man and his science.* New York: Paragon House. (Introductory)

Zuckerman, H., J. R. Cole, and J. T. Bruer, eds. 1991. *The outer cycle: Women in the scientific community.* New York: Norton. (Introductory)

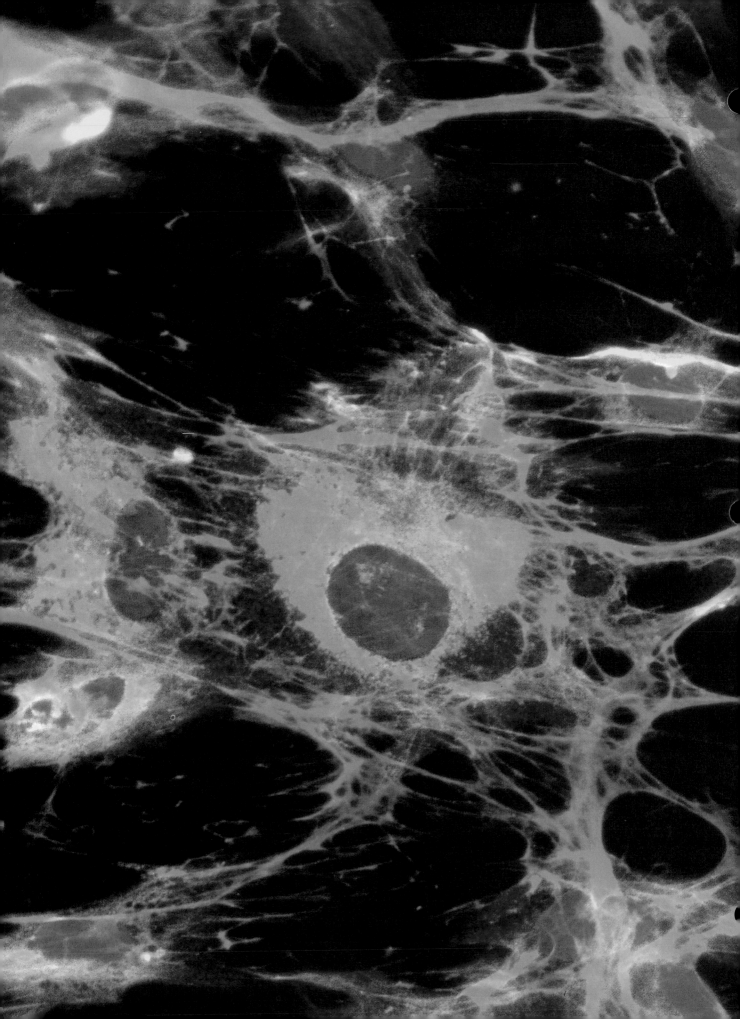

Chemical and Cellular Foundations of Life

The multitude of functions necessary for life require complexity, organization, and specialization—all of which are embodied in cells. Part of the cellular organization required for life is revealed in this photograph of a human skin cell. The green threads are microtubules, which help give the cell its shape and organization and participate in cell division. The blue threads are the cell's DNA. The remaining threads are proteins on the cell surface that enable it to interact with its environment.

PART
· 2 ·

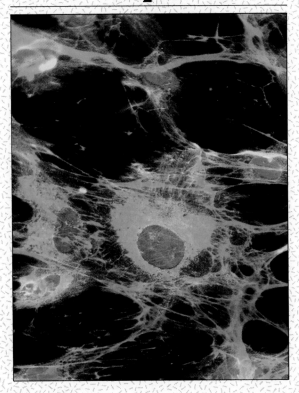

CHAPTER 3

The Atomic Basis of Life

STEPS TO DISCOVERY
The Atom Reveals Some of Its Secrets

THE NATURE OF MATTER

THE STRUCTURE OF THE ATOM

How Atoms Differ from One Another

TYPES OF CHEMICAL BONDS

Covalent Bonds

Noncovalent Bonds

THE LIFE-SUPPORTING PROPERTIES OF WATER

The Effect of Water On Nonpolar Molecules

Water as a Solvent

Capillary Action and Surface Tension of Water

The Thermal Properties of Water

ACIDS, BASES, AND BUFFERS

The Nature and Importance of pH

BIOLINE
Putting Radioisotopes to Work

THE HUMAN PERSPECTIVE
Aging and Free Radicals

STEPS TO DISCOVERY

The Atom Reveals Some of Its Secrets

*I*n the early 1890s, James Clerk Maxwell, one of the most prominent physicists of the time, authoritatively proclaimed that all of the basic principles of physics had been discovered; all that was left was to fill in the holes. Within a few years, however, the entire field of physics would undergo a revolution that altered our entire concept of the structure and properties of matter. This revolution began in 1896 as the French scientist Henri Becquerel was studying *phosphorescence*—the property of certain materials, including uranium salts, to glow in the dark after having been exposed to sunlight. A year earlier, the discovery of X-rays had been reported, and Becquerel had hypothesized that phosphorescent materials might emit X-rays after their ex-

posure to light. If this were true, he should have been able to detect the X-rays by their ability to pass through an opaque shield and to expose a photographic plate on the other side of the barrier.

On a clear day in Paris, Becquerel tried to do just that. He placed crystals of uranium salts on top of a photographic plate that had been wrapped with black paper to prevent its exposure to light. He put the package out in the sun for a few hours, then developed the plate in the darkroom, revealing a clear smudge where the "rays" from the uranium salts had penetrated the black paper and exposed the photographic film. Becquerel thought he had confirmed his hypothesis, but he wanted to repeat the experiment and

As discovered by the Curies, radioactivity results from particles hurled from the nucleus of a disintegrating atom.

prepared a similar package to be sure. The weather changed, however, and the skies became heavily overcast. Becquerel filed away the covered plate and uranium salts in a drawer, waiting for the sun to reappear. After several days of inclement weather, he impatiently developed the plate. To his astonishment, he found a very bright image of the uranium salts revealed on the film. Becquerel realized the importance of this accidental discovery. He concluded that since the "rays" emitted from the uranium salts were not a result of energy absorbed from the sun after all, they must be a spontaneous emission of the uranium itself. Becquerel had discovered a new property of matter.

Becquerel's report caught the interest of a young French physicist Pierre Curie, who had recently married Marie Sklodowska, a Polish immigrant studying physics at the Sorbonne in Paris. Marie Curie decided to conduct her doctoral thesis on the nature of Becquerel's "rays." She soon discovered that uranium was not the only source of these "rays," which she named *radioactivity*. In fact, the crude uranium ore she was working with also contained several other substances that were actually much more radioactive than uranium. She named one of these newly discovered substances *polonium*, after her native homeland. The other substance, which she called *radium*, after the Latin word meaning "ray," proved to be over a million times more radioactive than uranium. Because of the scarcity of radium in the uranium ore, it took Marie and Pierre several years, working in a small, poorly equipped shed, to extract 400 milligrams of pure radium from more than a ton of ore.

The Curies discovered both the destructive and curative properties of radium. Pierre placed a bandage containing a few crystals of radium on his arm for a few hours; the skin soon reddened, and an open sore developed, which took months to heal. This observation provided the first evidence that rays emitted from radium atoms were capable of killing living cells, and led to the use of radium to treat skin cancers. Radium crystals were placed in proximity to the cancer, causing the death of the cancerous cells and their replacement by normal tissue. Radioactive elements are still part of the arsenal used in the treatment of cancer.

The radioactive materials were affecting the health of the Curies, however, who took virtually no precautions. For example, spills of extremely radioactive materials were simply mopped up by hand with rags, which were thrown in the garbage. Consequently, Marie's hands became hardened and numb from working with the materials. For their work on discovering radioactivity, the Curies and Becquerel shared the 1903 Nobel Prize in physics, and Marie Curie won a second Nobel prize in chemistry 8 years later for her discovery of radium and polonium.

The work of the Curies also proved to be the first step in demonstrating that atoms could have an unstable structure and were not the indivisible, indestructible, minuscule spheres they were proclaimed to be. This was soon confirmed by the New Zealander Ernest Rutherford, who discovered that the rays emitted by a sample of the Curies' radium included at least two types of particles, one relatively large and positively charged, the other very small, carrying a negative charge and having very different properties. Rutherford showed that these particles were hurled out as a radioactive radium atom disintegrated with explosive force. With each explosion, the loss of particles transformed one type of atom into a different type. A radioactive uranium atom was transformed into a radioactive radium atom, which lost more particles (and discharged more energy) to become a radioactive atom of radon. The sequential deterioration continued until what had started out as an atom of uranuim was eventually transformed into an atom of nonradioactive (and thus stable) lead. In a sense, radioactivity was a fulfillment of the alchemist's dream of converting one element into another. Whereas the alchemists had tried to convert lead into gold, Rutherford had discovered that nature works in the opposite direction by spontaneously producing lead from elements vastly more valuable than gold. For his work on atomic structure and radioactive decay, Rutherford received the Nobel Prize prize in physics in 1908.

THE NATURE OF MATTER

*T*o explore life adequately, one must venture into an invisible realm—the realm of chemicals and chemical reactions, where minuscule particles combine with one another to forge everything in the universe, including that which is alive. Chemistry provides an explanation for biological properties at their most fundamental and essential level, that which ultimately accounts for what we are and what we do as living entities. Probing the chemical basis of life is a first step in understanding biology. It reveals how different particles can be arranged and rearranged to form an unlimited variety of organisms.

▼ ▼ ▼

When we look around at objects in our environment we find a seemingly endless diversity of materials. All matter, however, is composed of a limited number of basic substances, or **elements.** The diverse nature of materials stems from the variety of ways these elements can combine with one another to form more complex substances with new and different properties. The first reasonably accurate list of 28 elements was compiled in the 1780s by Antoine-Laurent Lavoisier. (His list might have grown even longer had he not been beheaded in the French revolution of 1794.) Today, scientists have identified 109 elements; 92 of these occur in nature, the remaining 17 have been synthesized by scientists. Four of these elements (carbon, hydrogen, oxygen, and nitrogen) make up over 90 percent of your body's weight.

FIGURE 3-1
Different but similar. Each of these strikingly different organisms is composed almost entirely of the same group of chemical elements.

Elements are designated by either one or two letters derived from their English or Latin names. These include carbon (symbolized by the capital letter C), hydrogen (H), oxygen (O), nitrogen (N), phosphorus (P), sulfur (S), sodium (Na, from the Latin, *natrium*), potassium (K, from the Latin, *kalium*), chlorine (Cl), magnesium (Mg), iron (Fe, from the Latin *ferrum*), and calcium (Ca). The same 12 elements (Table 3-1) mentioned above make up over 99 percent of the living matter of all the diverse organisms on earth (Figure 3-1).

There are only about 100 basic substances (elements) that can be combined in various ways to make up everything in the universe. (See CTQ #2.)

THE STRUCTURE OF THE ATOM

In 1810, John Dalton, an English schoolteacher, formulated the atomic theory of the structure of **matter,** matter being any material substance that occupies space and has weight. Dalton conceived of matter as being composed of tiny solid spheres, or **atoms** (Figure 3-2). Atoms are the smallest units of matter unique to a particular element. A century later, Ernest Rutherford began to explore the nature of the atom by allowing radioactive particles to be fired at a thin piece of gold foil. To his surprise, virtually all of the particles (over 99.9 percent) sailed right through the foil. Rutherford hypothesized that if atoms were solid like billiard balls, as previously thought, virtually all of the radioactive particles should have been deflected. He concluded that atoms consisted largely of empty space; virtually all of the weight of an atom was compacted into an extremely small volume, which he called the **nucleus.**

Based on other observations, Rutherford concluded that the compact nucleus of an atom contains one or more positively charged *subatomic* particles he called **protons.** The atomic nucleus therefore carries a positive charge—one unit of charge for each proton. Unknown to Rutherford, the nucleus of an atom also contains electrically neutral (uncharged) subatomic particles called **neutrons,** which were not discovered for another 21 years. Protons and neutrons have approximately the same mass, about $0.00000000000000000000000166$ (1.6×10^{-24}) grams. The third type of subatomic particle, called an **electron,** is about 1/1,800 the mass of a proton and is negatively charged. In Rutherford's hypothesis, the electrons were distributed in the empty space that surrounded each nucleus. He concluded that an atom is held together as a unit by the electrical attraction exerted between the oppositely charged protons and electrons.

TABLE 3-1

MOST COMMON ELEMENTS IN THE HUMAN BODY

	Percentage by weight	Atomic Number	Atomic Mass[a]
Oxygen (O)	65	8	16
Carbon (C)	18	6	12
Hydrogen (H)	10	1	1
Nitrogen (N)	3	7	14
Calcium (Ca)	2	20	40
Phosphorus (P)	1.1	15	31
Potassium (K)	0.35	19	39
Sulfur (S)	0.25	16	32
Sodium (Na)	0.15	11	23
Chlorine (Cl)	0.15	17	35
Magnesium (Mg)	0.05	12	24
Iron (Fe)	0.004	26	56

[a] Atomic number and atomic mass are discussed in the following section of the text.

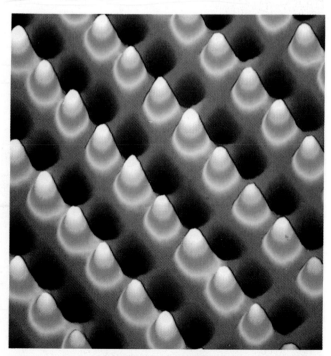

FIGURE 3-2

Photograph of rows of individual atoms of the element antimony as seen through a *scanning tunneling electron microscope.*

HOW ATOMS DIFFER FROM ONE ANOTHER

All atoms consist of the same three types of subatomic particles. An atom of one element is distinguished from those of all other elements by its **atomic number,** that is, the number of protons in its nucleus. For example, the atomic number of hydrogen is 1: All hydrogen atoms have a single proton and, conversely, all atoms with one proton are hydrogen atoms. All carbon atoms have six protons, all oxygen atoms have eight protons, and so forth (see Figure 3-4).

Atoms of the Same Element with Different-Sized Nuclei

The combined numbers of neutrons and protons in the nucleus make up the weight of an atom, or its **atomic mass.** While the number of protons is identical among all atoms of a given element, the number of neutrons can vary. Atoms having the same number of protons—the same atomic number—but different numbers of neutrons—different atomic mass—are said to be **isotopes.** Hydrogen, for example, exists as three different isotopes; all hydrogen atoms contain one proton but may have zero, one, or two neutrons. The atomic mass of these isotopes is denoted as ^1H, ^2H, and ^3H, respectively. In nature, a sample of an element will contain a mixture of various isotopes. Despite the fact that only one in every 6,000 hydrogen atoms normally contains a neutron, relatively pure *deuterium* (a hydrogen atom containing one neutron (^2H)) can be prepared by industrial procedures for use in forming "heavy water"—so called because the extra neutron makes the water more dense than the "ordinary" variety. Heavy water is incapable of supporting life. For example, seeds watered solely with heavy water will not sprout.

Hydrogen atoms with two neutrons (termed *tritium*) are **radioactive;** that is, they are physically unstable and tend to disintegrate spontaneously. Radioactive atoms such as tritium play an indispensable role in biological and medical research (see Bioline: Putting Radioisotopes to Work).

The Arrangement of Electrons in an Atom

While Ernest Rutherford was concentrating on the nucleus, a colleague of his, Niels Bohr, was studying the structural arrangement of an atom's electrons. In 1913, Bohr proposed that electrons whirl around the nucleus in orbits, not unlike the way the planets of our solar system move around the sun (Figure 3-3). Bohr proposed that each orbit—or **shell**—is situated at a specific distance from the nucleus and has a maximum number of electrons it can hold. When a shell becomes filled, any additional electrons must go into the next shell farther away from the nucleus. The innermost shell can contain only two electrons. The next two shells can contain eight each. Thus, the six electrons of the carbon atom (Figure 3-4) are distributed in two shells in a 2:4 arrangement. That is, the innermost shell holds a pair of electrons, and the second shell holds the other four electrons. Sulfur's 16 electrons form a 2:8:6 arrangement; this formation differs from that of chlorine (17 electrons) by having one fewer electron in its third shell. The electronic distribution of a number of common atoms is illustrated in

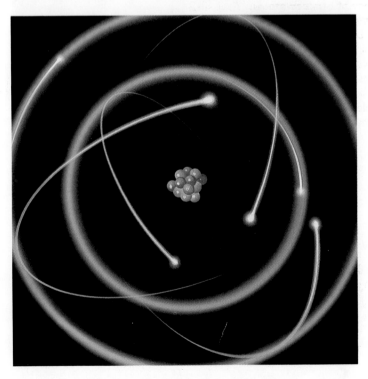

FIGURE 3-3

Simplified version of a typical atom. The nucleus of this carbon atom contains six protons and six neutrons. Six electrons orbit the nucleus in two shells. Although useful in understanding atomic structure and the chemical reactions of an atom, this simplified model (called a *Bohr atom* after the Danish physicist who proposed it) fails to show the enormous relative distance between nucleus and electrons and inaccurately represents the shape of the shells. According to modern physics, the shells consist of orbitals of spherical or dumbbell shape, in which an electron has a high probability of being located at any given instant. These orbitals are more like "clouds" encompassing the area occupied by the electrons traveling around the nucleus at about the speed of light.

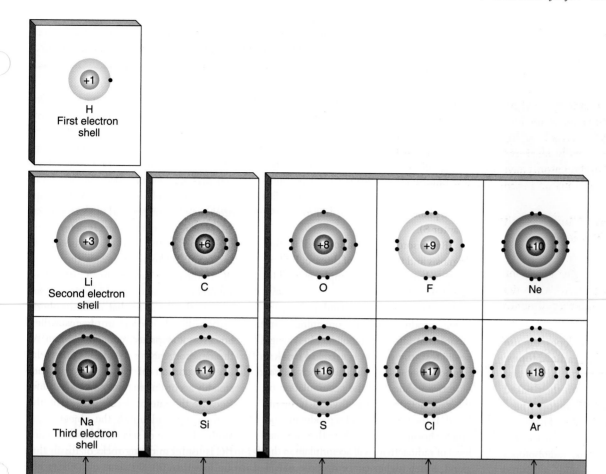

ELECTRONS NEEDED FOR ATOMS IN EACH COLUMN TO ACHIEVE STABILITY

| − 1 | + 4 | + 2 | + 1 | 0 (Inert elements) |

FIGURE 3-4

A representation of the arrangement of electrons in a number of common atoms. Each shell has a limited number of electrons it can hold. Electrons in each shell are grouped in pairs to illustrate that each orbital of a shell can hold two electrons. The number of outer-shell electrons is a primary determinant of the properties of elements. Atoms with similar number of outer-shell electrons have similar properties. Lithium (Li) and sodium (Na), for example, have one outer-shell electron, and both are highly reactive metals. Carbon (C) and silicon (Si) atoms can each bond with four different atoms. Because of its size, however, a carbon atom can bond to other carbon atoms, forming long-chained organic molecules, whereas silicon is unable to form comparable molecules. Neon (Ne), and argon (Ar) have filled outer shells, making these atoms highly nonreactive; they are referred to as inert gases.

Figure 3-4. As we will see shortly, the structure of these atoms determines their chemical reactivities.

By the 1920s, it was apparent that Bohr's model of the atom was overly simplistic and that electrons do not follow fixed circular orbits as depicted in Figure 3-3. Rather, they sweep around the nucleus in loosely defined "clouds," called **orbitals.** Orbitals are roughly defined by their boundaries, which may have a spherical or dumbbell shape.

Each orbital is capable of containing a maximum of two electrons. The innermost shell contains a single orbital, the second shell contains four orbitals (thus eight electrons), the third shell also contains four orbitals (thus eight electrons); and so forth.

Electrons contain energy; specifically, those of inner shells contain less energy than do those of outer shells. Electrons do not always stay in a particular shell but can

◁ THE HUMAN PERSPECTIVE ▷
Aging and Free Radicals

Why do humans have a maximum lifespan of approximately 100 years while our close relatives, the chimpanzees, live only about half this length of time? Many biologists believe that aging results from the gradual accumulation of damage to our body tissues, which disrupts the order that is required to maintain life. The most destructive damage probably occurs to DNA. Alterations in DNA lead to the production of faulty genetic messages that promote gradual cellular deterioration. How does cellular damage occur, and why should it occur more rapidly in a chimpanzee than a human?

Atoms are stabilized when their shells are filled with electrons. Recall that electron shells consist of orbitals, each of which can hold a maximum of two electrons. Atoms or molecules that have orbitals containing a single unpaired electron are called **free radicals.** Free radicals are often formed when a covalent bond is broken in a way that each portion keeps one half of the shared electron pair. Because free radicals are so reactive, they are ex-

tremely destructive to biological tissues. For example, water—our life-sustaining medium—can be converted into free radicals when exposed to radiation from the sun

$$H_2O \rightarrow HO \cdot + H \cdot$$

("·" indicates a free radical)

Even more common is the superoxide radical ($O_2^{\cdot-}$) which is formed when a molecule of oxygen picks up an extra electron. Superoxide radicals are inevitably formed from some of the oxygen molecules we breathe, and they may be a major contributor to the aging process.

⟳ It has been hypothesized that, compared to animals with shorter lifespans, animals with longer lifespans, such as humans, have more effective mechanisms for protecting themselves against free radicals and for repairing cellular damage that has already occurred. Several kinds of substances can provide protection from free radicals. Cells contain enzymes that can convert free radicals into noninjurious mo-

lecules. Substances called *antioxidants* can also chemically alter free radicals; these include vitamin E, C, and beta-carotene (the orange pigment in carrots and other vegetables, and the parent compound for vitamin A). It has been suggested that dietary supplements containing antioxidants may help in the fight against free-radical-induced damage. In fact, some nutritionists argue that the drop in the number of cases of stomach cancer in the United States over the past few decades is due to increased consumption of antioxidants, such BHA and BHT, which are used as preservatives in many foods. In contrast, certain dietary components, such as polyunsaturated fats (page 75), might be harmful because they may be converted to free radicals as they are metabolized. This could be the basis for a possible link between polyunsaturated fats and an increase in certain types of cancer. Free radicals have also been linked to an increased risk of cardiovascular disease because they promote the buildup of plaques within arteries.

filled the *Hindenburg* reacted with oxygen gas (O_2) present in the atmosphere, forming a cloud of H_2O and releasing enough energy to destroy the aircraft in a blazing inferno.

Oxygen can satisfy its two-electron deficit by bonding with an identical atom, forming an oxygen molecule (O_2). The two atoms of molecular oxygen are bound together by a *double bond,* in which two pairs of electrons are shared. Some elements can even form *triple bonds* composed of three pairs of electrons, as illustrated by molecular nitrogen (N_2). Quadruple bonds do not exist.

Carbon atoms—with four electrons in their outer shell—can form four single covalent bonds. Carbon, for example, can share its electrons with four atoms of hydrogen to form methane (CH_4), a gas produced by a variety of swamp-dwelling microbes as well as by the bacteria inhabiting your large intestine. Methane is also the major component of the natural gas supplied by your gas company. The

pairs of electrons that form the single bonds in a methane molecule (Figure 3-6) are shared rather equally between the carbon and hydrogen atoms. This is because neither the hydrogen nor the carbon nucleus is strongly electron attracting. In some molecules, however, the nucleus of one atom is more electron attracting than is its partner and exerts a greater pull on the shared electrons. Among the atoms most commonly present in biological molecules, nitrogen, oxygen, phosphorus, and sulfur are strongly electron attracting.

Polar versus Nonpolar Molecules

Let's examine a molecule of water. Water's single oxygen atom attracts electrons much more forcefully than do either of its hydrogen atoms. As a result, the electrons tend to be more closely associated with the oxygen atom, which becomes somewhat negatively charged. The two hydrogen atoms, which are electron deficient, become somewhat positively charged.

Molecules that contain an unequal charge distribution, such as water, are referred to as **polar molecules;** they possess distinct positive and negative regions, or *poles*. In contrast, **nonpolar molecules** lack regions of electric charge. Nonpolar molecules of biological importance consist almost exclusively of carbon and hydrogen atoms and include methane, fats, and waxes.

   The difference in structure between polar and nonpolar molecules is a key determinant of the properties of biologically important molecules. As we will see shortly, for example, the polarity within the water molecule is the reason water is able to dissolve so many different substances, a property that is essential for life.

NONCOVALENT BONDS

While covalent bonds are responsible for holding the atoms of a molecule together, noncovalent bonds occur *between* molecules (Figure 3-7) or between different parts of a large biological molecule. Noncovalent bonds are not dependent on shared electrons but rather on attractive forces between positive and negative charges. Noncovalent bonds play a key role in maintaining the intricate three-dimensional shape of large molecules, such as proteins and deoxyribonucleic acid (DNA), the genetic material found in organisms. Noncovalent bonds are also crucial for holding molecules in larger complexes together (Figure 3-7). Noncovalent bonds include *ionic bonds* and *hydrogen bonds*.

Ionic Bonds: Attractions between Charged Atoms

Some atoms are so electron-attracting that they can capture electrons from other atoms. For example, when the elements sodium (a silver-colored metal) and chlorine (a toxic gas) are mixed, the single electron in the outer shell of each sodium atom migrates to an electron-attracting chlorine atom. As a result, these two materials are transformed into sodium chloride—neither a gas nor a metal, but ordinary table salt.

$$2Na^{\cdot} + :\!\overset{\cdot\cdot}{\underset{\cdot\cdot}{Cl}}\!:\!\overset{\cdot\cdot}{\underset{\cdot\cdot}{Cl}}\!: \rightarrow 2Na:\!\overset{\cdot\cdot}{\underset{\cdot\cdot}{Cl}}\!: \rightarrow 2Na^+ + 2:\!\overset{\cdot\cdot}{\underset{\cdot\cdot}{Cl}}\!:^-$$

| Sodium metal | Chlorine gas | Transient bond while electron transfer occurs | Sodium ions | Chloride ions |

Since the chloride atom has an extra electron (relative to the number of protons in its nucleus), it has a single negative charge (Cl⁻). The sodium atom, which has lost an electron (leaving it with an extra proton relative to the number of electrons) has a single positive charge (Na⁺). Such electri-

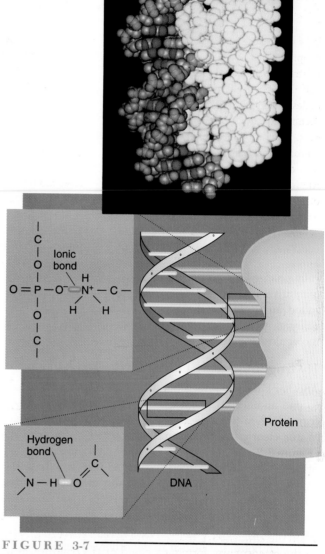

FIGURE 3-7

Noncovalent bonds play an important role in holding two or more molecules together into a complex. In the computer-simulated model depicted here, a molecule of protein (yellow atoms) is bound to a molecule of DNA by noncovalent ionic bonds. The ionic bond forms between a positively charged nitrogen atom and a negatively charged oxygen atom. The DNA molecule itself consists of two separate strands held together by noncovalent hydrogen bonds. Although a single noncovalent bond is relatively weak and easily broken, large numbers of these bonds between two molecules, as between two strands of DNA, make the overall complex quite stable.

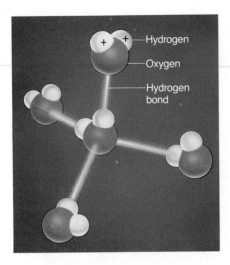

FIGURE 3-8

Hydrogen bond formation between adjacent water molecules.   Positively charged hydrogens of one polar water molecule align next to negative portions of another water molecule. These weak bonds are really attractions between oppositely charged portions of adjacent molecules. In each of the hydrogen bonds depicted, a hydrogen atom is being shared between two oxygen atoms. The length and strength of a bond are related properties. The hydrogen bond between an H and O atom is longer than the covalent bond between an H and O atom because it is a weaker bond.

cally charged atoms are called **ions.** Oppositely charged ions attract one another, forming a noncovalent linkage called an **ionic bond.** Ionic bonds often form between charged groups that are part of large, complex biological molecules (Figure 3-7).

Hydrogen Bonds: Sharing Hydrogen Atoms

Another kind of chemical bond holds polar molecules close to one another. This linkage—the **hydrogen bond**—is formed when two molecules share an atom of hydrogen. Water clearly illustrates how this attractive force works (Figure 3-8). Adjacent water molecules are drawn together at their oppositely charged regions so that, in effect, the oxygen atoms of two neighboring water molecules share a common hydrogen atom. The sharing is not equal, however, for the covalent bond that links the hydrogen to one of the oxygens is about 20 times stronger than is the noncovalent hydrogen bond that links it to the other oxygen.

Hydrogen bonding is a common characteristic of polar molecules. Hydrogen bonds help to stabilize giant protein molecules, and they also hold the two strands of a DNA molecule together (Figure 3-7). Even though individual hydrogen bonds are weak, when they are present in large numbers, as in a DNA molecule, their forces become additive and, taken as a whole, they provide a structure with considerable stability.

Atoms interact to form complexes that are held together by chemical bonds. These interactions occur because the interacting atoms achieve a greater degree of stability than they would possess independently. This increased stability is achieved when electrons are either shared between atoms or transferred completely from one atom to another. (See CTQ #4.)

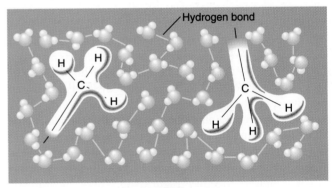

Hydrophobic interactions

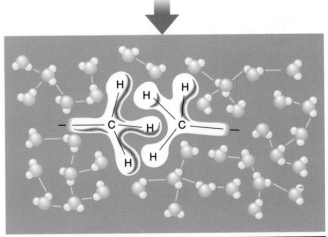

FIGURE 3-9

In a hydrophobic interaction, the nonpolar (hydrophobic) molecules are pushed together as a result of the bonds formed between the polar water molecules.

THE LIFE-SUPPORTING PROPERTIES OF WATER

Water is abundant on earth; we drink it, swim in it, and spray it on our lawns. We tend to take water for granted, even though it constitutes about 70 percent of our body weight. Life on earth is totally dependent on water, and water is probably essential to the existence of life anywhere in the universe. The life-supporting properties of water can be traced to the structure of the H_2O molecule, providing an example of the relationship between structure and function at the atomic level. We have seen how the polarity of the O—H bonds in a water molecule produces positively and negatively charged regions which, in turn, allow water molecules to form hydrogen bonds. The life-supporting properties of water stem largely from its extensive capacity to form hydrogen bonds.

THE EFFECT OF WATER ON NONPOLAR MOLECULES

If you use oil-and-vinegar salad dressings you know that you have to shake and pour the dressing rapidly or you'll end up with a lot more oil on your salad than vinegar. The reason that the oil separates from diluted vinegar, or that fat droplets form on the top of a bowl of chicken soup, has to do with the interactions between the molecules. In a bowl of chicken soup, molecules that have a polar structure, such as the amino acids derived from the meat's protein, can form hydrogen bonds with water molecules. Because of this property, polar molecules are said to be **hydrophilic,** or "water-loving" (*hydro* = water, *philic* = loving). Nonpolar molecules, such as the lipids (Chapter 4) in salad oil or chicken fat, are essentially insoluble in water because they lack charged regions that would be attracted to the poles of water molecules. Consequently, when nonpolar compounds are mixed with water, the mutual attraction between the water molecules forces the nonpolar, **hydrophobic,** or "water-fearing" (*phobia* = fearing) substances into aggregates, such as fat droplets, which minimizes their exposure to their polar surroundings (Figure 3-9). **Hydrophobic interactions** of this type play a key role in holding cell membranes together (see Figure 4-10).

WATER AS A SOLVENT

One of water's most important life-sustaining properties is its effectiveness as a solvent. A **solvent** is a substance in which another material—the **solute**—dissolves by dispersing as individual molecules or ions. The resulting product is a **solution.** Most biologically important substances, particularly polar molecules (such as sugars and amino acids) and ions, such as table salt, are highly soluble in water (Figure 3-10)—more so, in fact, than in any other solvent.

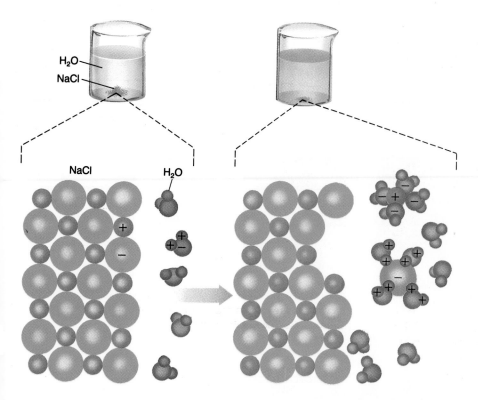

FIGURE 3-10

The solvent properties of water. Ions in the salt (NaCl) crystal attract the part of the water molecule that bears the opposite charge. The strength of this attraction exceeds the ions' attraction to their oppositely charged neighbors in the crystal, and water wins the molecular "tug of war." The ions go into solution, and the salt crystal (the solute) dissolves in the water (the solvent).

Water is such an efficient solvent that by the time a raindrop completes its fall, it has become a solution containing a number of dissolved gases, mainly nitrogen, oxygen, and carbon dioxide. The molecular oxygen (O_2) that dissolves in lakes, streams, and oceans is sufficient to supply huge communities of fish and other oxygen-dependent underwater dwellers with oxygen (Figure 3-11).

CAPILLARY ACTION AND SURFACE TENSION OF WATER

Water molecules will adhere to any hydrophilic substance. Glass, for example, has a hydrophilic surface that attracts water molecules with enough force to overcome the pull of gravity. You can see the results of this attraction if you dip a narrow glass tube into water. The water seems to "crawl" up the tube, pulled along by its attraction to the hydrophilic surface. As these outer water molecules rise, they pull along adjacent molecules to which they are hydrogen-bonded, so that the water in the center of the tube also rises. This movement of water is called **capillary action.** Capillary action plays an important role in the movement of water through the soil.

Just as water molecules adhere to hydrophilic surfaces, they adhere to one another via hydrogen bonds, a property called *cohesion.* When present at a surface, this cohesion between water molecules creates a film that resists being separated. This property of water, called **surface tension,** allows a water strider seemingly to defy gravity and to walk across the surface of a pond (Figure 3-12). Surface tension can also create problems. Our lungs, for example, consist of microscopic chambers covered by a thin film of water. As we inhale, the surface tension of this film tends to prevent the expansion of these chambers. To overcome this effect, our lungs produce a substance called *surfactant,* which lowers the surface tension of the film, aiding the breathing process.

THE THERMAL PROPERTIES OF WATER

We all know it takes a long time when we're waiting for a pot of water to boil. The reason it takes so long is that water is a very good heat absorber. Heat, or thermal energy, is measured in calories. One **calorie** is defined as the amount of heat needed to raise the temperature of 1 gram of water by 1 degree Celsius. It takes only about one-half of a calorie to raise the temperature of ethyl alcohol by a degree, and even less for most other substances. This energy-absorbing property of water is also a result of its extensive hydrogen bonding. Much of the energy that is absorbed by water as it is heated is utilized in breaking hydrogen bonds and therefore does not contribute to molecular motion (which we measure as temperature). Because of this property, organisms living in large bodies of water are protected from rapid, potentially lethal changes in body temperature despite sudden changes in the temperature of the air.

Water is also an efficient cooling agent. In order to evaporate, molecules of water must absorb a large amount of thermal energy, cooling the liquid that is left behind. This property is exploited by your body when you sweat. As the sweat evaporates, it removes thermal energy from the body surface, causing a drop in body temperature. Sweating is especially effective in dry environments that encourage rapid evaporation. In humid conditions, sweat accumulates rather than evaporating, which accounts for the common complaint, "It's not the heat, it's the humidity."

Ice Formation

As the temperature of water drops, the movement of the molecules slows down, and the molecules crowd closer to one another. In other words, as water gets colder, it becomes denser. As the temperature drops below 4°C (38°F), however, a most remarkable phenomenon occurs: The movement toward increased density reverses itself, and the

FIGURE 3-11
External gills atop this striking nudibranch (sea slug) extract enough dissolved oxygen from water to allow the animal to live permanently in the ocean without having to surface for air.

FIGURE 3-12

Surface tension of water provides enough support to allow a water strider to walk on the liquid's surface as though it were a flexible membrane.

water *expands* in volume. Expansion continues as water is cooled from 4°C to 0°C, at which point it freezes to form ice. The lowered density of ice (when compared to liquid water) explains why ice floats and why a blanket of ice covers the surface of frozen lakes and oceans. This frozen crust provides insulation against additional heat loss, allowing the water beneath the ice to stay warm enough to remain a liquid. This phenomenon plays a key role in maintaining aquatic life forms through cold winter months, preventing them from freezing to death.

Life depends on water. Because of its structure, a water molecule has the ability to form several hydrogen bonds. This enables water to force hydrophobic molecules together, to dissolve many different molecules, and to absorb large amounts of heat—three properties on which life depends. (See CTQ #5.)

ACIDS, BASES, AND BUFFERS

Protons are not only found within atomic nuclei, they are also released into the medium whenever a hydrogen atom loses an electron. Let's consider acetic acid—the major ingredient of vinegar—which can undergo the following reaction:

$$\overset{H}{\underset{H}{\overset{\cdot\cdot}{\underset{\cdot\cdot}{H:C:C}}}} \overset{\overset{\cdot\cdot}{O:}}{\underset{\overset{\cdot\cdot}{O:}}{}} \rightarrow \overset{H}{\underset{H}{\overset{\cdot\cdot}{\underset{\cdot\cdot}{H:C:C}}}} \overset{\overset{\cdot\cdot}{O:}}{\underset{\overset{\cdot\cdot}{O:^-}}{}} + \quad H^+$$

Acetic acid   Acetate   Proton
     ion (hydrogen ion)

In this reaction, the hydrogen atom of the acetic-acid molecule has transferred its electron to the adjacent oxygen atom (giving the oxygen a negative charge), while the bare proton has dissociated into the medium as a hydrogen ion (H^+). Any molecule capable of releasing a hydrogen ion is defined as an **acid.** The proton released by the acetic acid molecule in the previous reaction does not remain in the free state; instead it combines with another molecule. Possible reactions involving a proton include:

- Combination with a water molecule to form a hydronium ion (H_3O^+).

$$H^+ + H_2O \rightarrow H_3O^+$$

- Combination with a hydroxyl ion (OH^-) to form a molecule of water.

$$H^+ + OH^- \rightarrow H_2O$$

- Combination with an amino group ($—NH_2$) in a protein to form a charged amine.

$$H^+ + —NH_2 \rightarrow —NH_3^+$$

Any molecule that is capable of accepting a hydrogen ion is defined as a **base.** In the first reaction indicated above, for example, water is acting as a base by accepting a proton. Water is also capable of acting as an acid by donating a proton in the following reaction

$$H_2O \rightarrow H^+ + OH^-$$

Thus, some compounds, including water, can act as either an acid or a base, depending on conditions.

If we compare various acids, we will find that they vary considerably in how readily they give up their protons. The more easily the proton is lost, the stronger the acid. For example, hydrogen chloride, the acid produced in your stomach, is a very strong acid. Most acids produced in organisms, such as acetic acid or citric acid (the acid found in lemons and oranges), are relatively weak.

THE NATURE AND IMPORTANCE OF pH

The acidity of a solution is measured by the concentration of hydrogen ions and is expressed in terms of **pH** (Figure 3-13). The pH scale ranges from 0 to 14. Pure water has a pH of 7.0, which is the pH value of a *neutral* solution. If an acid is added to pure water, the hydrogen ion concentration increases, causing the solution to become acidic, which is measured as a *lower* pH. Conversely, if a base, such as sodium hydroxide, is added to pure water, the hydrogen ion concentration decreases, causing the solution to become *basic* (alkaline), which is measured as a *higher* pH.

↻ Most biological processes are acutely sensitive to pH, as evidenced by the devastating effects that acid rain is having on the forests in many parts of the world. The reason is that changes in hydrogen ion concentration affect the ionic state of biological molecules. For example, as the

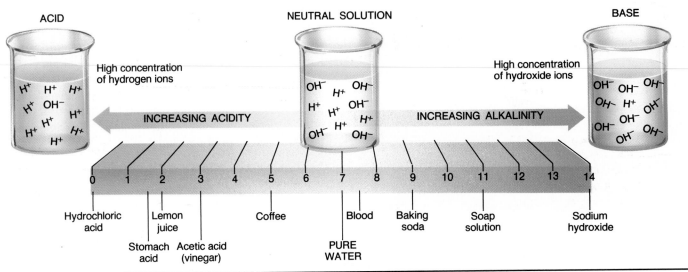

FIGURE 3-13

The pH scale. The acidity or alkalinity of a solution is indicated by a value between 0 and 14. The number is a measure of the concentration of hydrogen ions in a solution; the lower the number, the higher the H^+ concentration. The concentration of hydrogen ions in solution is always inversely related to the concentration of hydroxyl ions; as one goes up, the other goes down. The concentrations of these two ions are equal at one point on the scale, namely pH 7, which is termed a *neutral* solution. Acidic solutions have pH values below 7, alkaline (basic) solutions are greater than 7. Each unit on the scale represents a tenfold change in the hydrogen ion concentration. For example, lemon juice (pH = 2) is ten times more acidic than is vinegar (pH = 3).

hydrogen ion concentration increases, more of the —NH_2 groups in proteins become —NH_3^+, which can disrupt the shape and activity of the entire protein. Even slight changes in pH can impede the chemical reactions on which life depends. For instance, if not corrected, a drop in the pH of the blood can lead to coma and death. Organisms protect themselves from pH fluctuations with **buffers**—chemicals that couple with free hydrogen and hydroxide ions, thereby resisting changes in pH. The blood, for example, contains a certain concentration of bicarbonate ions. If the hydrogen ion concentration should suddenly rise (as occurs during exercise), the bicarbonate ions combine with the excess protons, removing them from solution.

$$HCO_3^- \ + \ H^+ \ \rightarrow H_2CO_3$$

Bicarbonate ion   Hydrogen ion   Carbonic acid

Without buffers to protect our internal fluids from becoming too acidic or alkaline, we would not survive.

The structure and function of many biologically important molecules is influenced by the concentration of hydrogen ions in the surrounding solution, which determines the solution's acidity. An excess or deficiency of hydrogen ions impairs life processes. (See CTQ #6.)

REEXAMINING THE THEMES

Relationship between Form and Function

The chemical properties of atoms are determined by their structure, most importantly, by the number of electrons present in their outermost shell. The electronic structure of an atom determines whether or not it will bond to

other atoms to form molecules, how many bonds it can form with other atoms, and whether it will gain or lose electrons to form an ion. These chemical properties of atoms form the basis for understanding the behavior of the molecules that make life possible. The life-sustaining properties of water, for example, can be traced to the molecule's tendency to form hydrogen bonds, which is the result of an unequal sharing of electrons in the covalent bonds that hold the hydrogen and oxygen atoms together.

Acquiring and Using Energy

Electrons play a central role in the capture and transfer of energy which is essential for life. For example, energy present in sunlight is captured by electrons during photosynthesis when they are boosted from an inner, lower energy shell to an outer, higher energy shell. Solar energy captured by this mechanism is converted to the energy that fuels the activities of virtually every organism on earth.

Unity within Diversity

Remarkably, all of the diverse materials that exist in the universe are composed of the same three types of subatomic particles—protons, neutrons, and electrons. This enormous diversity results from the variety of ways in which these three subatomic particles are combined to form different kinds of atoms, which in turn combine to form different kinds of molecules. Although there are only three kinds of subatomic particles and about 100 kinds of atoms (elements), there is a virtually infinite diversity of molecules.

SYNOPSIS

Organisms are composed of a small number of elements whose atoms are built from protons, neutrons, and electrons. An atom contains a positively charged central nucleus of protons and neutrons encircled by negatively charged electrons. Distinct forms of the same element that differ from each other only in their number of neutrons are called *isotopes*. Electrons are ordered into electron shells, each of which can hold a limited number of electrons. The number of protons in an atom equals the number of electrons. If an atom gains or loses an electron, it becomes a charged ion.

Covalent bonds hold atoms together to form molecules, whereas noncovalent bonds hold molecules together in larger complexes. Covalent bonds are stable partnerships formed when atoms share their outer-shell electrons, each participant gaining a filled shell. If electrons in a bond are shared unequally by the component atoms, the molecule has a polar character.

Noncovalent bonds are formed by attractions between positive and negative regions of nearby molecules. Noncovalent bonds include ionic bonds (formed between oppositely charged ionic groups) and hydrogen bonds.

Water has unique properties on which life depends. The covalent bonds that make up a water molecule are highly polarized. As a result, water is an excellent solvent, capable of forming hydrogen bonds with virtually all polar molecules. The hydrogen bonding of water molecules to one another endows water with important thermal properties. For example, water absorbs very large amounts of heat per degree of temperature rise, the evaporation of water requires very large amounts of heat, and water becomes less dense as it approaches the freezing point.

When dissolved in water, acids increase the relative proportion of hydrogen ions (which lowers pH), and bases decrease the relative proportion of hydrogen ions (which elevates pH). Neutral solutions have a pH of 7.0, the same as pure water. Some substances resist changes in pH; these buffers help protect cytoplasm and tissue fluids from pH fluctuations.

Key Terms

element (p. 50)
matter (p. 51)
atom (p. 51)
nucleus (p. 51)
proton (p. 51)
neutron (p. 51)
electron (p. 51)
atomic number (p. 52)
atomic mass (p. 52)
isotope (p. 52)

radioactive (p. 52)
shell (p. 52)
orbital (p. 53)
radioisotope (p. 54)
half-life (p. 54)
chemical bond (p. 55)
molecule (p. 55)
compound (p. 55)
covalent bond (p. 55)

free radical (p. 56)
polar molecule (p. 57)
nonpolar molecule (p. 57)
ion (p. 58)
ionic bond (p. 58)
hydrogen bond (p. 58)
hydrophilic (p. 59)
hydrophobic (p. 59)
hydrophobic interactions (p. 59)

solvent (p. 59)
solution (p. 59)
capillary action (p. 60)
surface tension (p. 60)
calorie (p. 60)
acid (p. 61)
base (p. 61)
pH (p. 61)
buffer (p. 62)

Review Questions

1. Explain what is *wrong* (if anything) with the description of each of the following key terms:

   element all atoms with the same combined number of protons and neutrons

   radioisotope the most stable form of an element

   shell always filled to its electron capacity

   ionic bond chemical partnership in which atoms share electrons

   covalent bonds may be single, double, triple, or quadruple

   hydrophobic compounds polar molecules that form hydrogen bonds with each other and dissolve in water

   pH measurement of acidity or alkalinity; the neutral pH is 0.0

2. There are three beakers on the table. One holds pure water, one holds water to which table salt (NaCl) has been added, and one holds water to which salad oil has been added. In which beaker(s) would you expect hydrogen bonds? Ionic bonds? Hydrophobic interactions?

3. Why do polar molecules, such as table sugar, dissolve so readily in water? Why does sweating help cool your body? Why does ice float on water?

4. If you were to add hydrochloric acid to water, what effect would this have on the hydrogen ion (or H_3O^+) concentration? On the pH? On the ionic charge of any proteins in solution?

Critical Thinking Questions

1. Geiger counters are instruments that detect the presence of radioactive materials. What property of radioactive elements do you suppose a Geiger counter utilizes?

2. Many different substances can be formed by combinations of atoms of the 100 naturally occurring elements. How many different types of molecules could be formed that contain only three atoms if one type of atom can be used more than once? Would all of these possibilities occur in nature? Explain why or why not.

3. Oxygen atoms have eight protons in their nucleus. How many electrons do they have? How many orbitals are in the inner electron shell? How many electrons are in the outer shell? How many more electrons can the outer shell hold before it is filled? Do all oxygen atoms have the same number of neutrons? How many neutrons does a radioactive ^{18}O atom have?

4. What types of bonds (covalent single, covalent double, ionic, or hydrogen) would you expect to find in each of the following: KCl, CH_2O, NH_4OH?

5. In 1913, the biochemist Lawrence Henderson stated that "water is the one fit substance" to support life. What evidence can you present to support Henderson's statement?

6. An enzyme in saliva, which digests starch to sugar, functions at a neutral pH. What would you expect to happen to the function of this enzyme when it enters the stomach, which has a pH of around 2.0? Why?

7. Imagine that you went outside on a hot, dry day and dipped one hand into a bowl of ethyl alcohol and the other hand into a bowl of water and then removed both hands into the air. Which hand would lose the most thermal energy? Why?

Additional Readings

Gardner, R. 1982. *Water: The life sustaining resource.* Julian Messner. New York: (Intermediate)

Pflaum, R. 1989. *Grand obsession: Madame Curie and her world.* New York: Doubleday. (Introductory)

Pais, A. 1991. *Niels Bohr's times in physics, philosophy, and polity.* Oxford: Clarendon. (Introductory)

Snyder, C. 1992. *The extraordinary chemistry of ordinary things.* New York: Wiley. (Introductory)

Yablonsky, H. A. 1975. *Chemistry.* New York: Crowell. (Introductory)

CHAPTER

4

Biochemicals: The Molecules of Life

STEPS TO DISCOVERY
Determining the Structure of Proteins

THE IMPORTANCE OF CARBON IN BIOLOGICAL MOLECULES

GIANT MOLECULES BUILT FROM SMALLER SUBUNITS

FOUR BIOCHEMICAL FAMILIES

Carbohydrates

Lipids

Proteins

Nucleic Acids (DNA and RNA)

MACROMOLECULAR EVOLUTION AND ADAPTATION

THE HUMAN PERSPECTIVE
Obesity and the Hungry Fat Cell

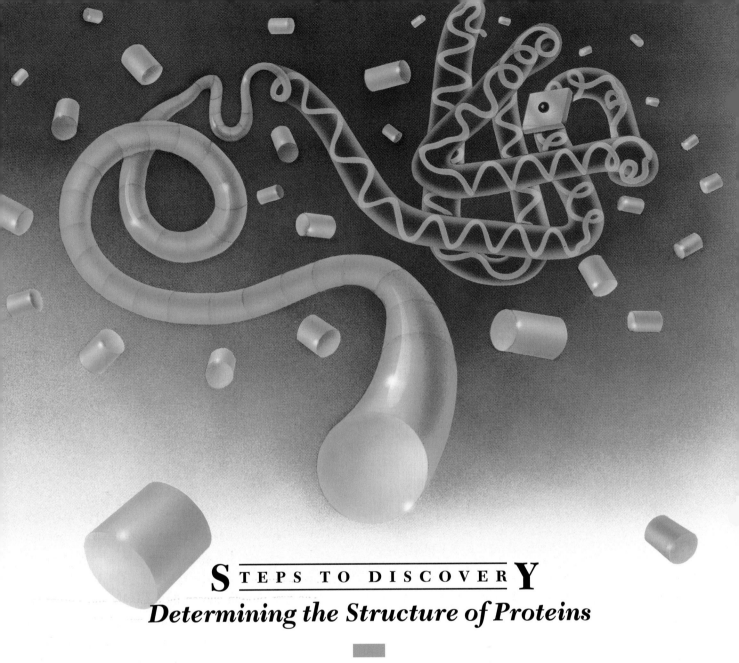

Determining the Structure of Proteins

*O*f the various types of molecules that make up the fabric of living organisms, proteins have the most complex structure and carry out the most demanding functions. At the end of World War II, in 1945, we still knew very little about the structure of these key molecules; 15 short years later, the scientific world had an accurate, detailed picture of the way these molecular giants were constructed. During this remarkable decade and a half, a handful of investigators, using very different experimental techniques, not only changed our view of proteins but dramatically improved our understanding of life at the molecular level.

Scientists in the mid-1940s knew that proteins were made up of small molecules, called *amino acids,* linked together to form chains called *polypeptides.* What they didn't know was whether or not the sequence of amino acids in a polypeptide chain was of critical importance, and if so, *how* was it important? Frederick Sanger, a young protein chemist at Cambridge University in England, began a study to determine the order of the amino acids that make up the protein insulin. Insulin was a natural choice. Not only was it one of the few proteins available commercially (it was being extracted from the pancreas of domestic pigs and used in the treatment of diabetes), it was also a much easier protein to sequence since it contained fewer amino acids than did most proteins (Sanger eventually found it was composed of only 51 amino acids).

A polypeptide, which consists of a linear chain of specific amino acids, folds to form a protein that has a complex, but precise, shape.

When Sanger began his studies, determining the sequence of amino acids linked together in a chain required a tedious, chemical process. Most importantly, only short chains of about five amino acids could be sequenced at a time. Consider an analogy in which a chain of letters is linked together forming a sequence such as

TCEJBUSETIROVAFYMSYGOLOIB.

How would you determine this sequence if you were able to identify pieces containing only five letters in a row? If you were able to cut the chain of letters at random positions, forming overlapping fragments (such as tir*ov*, *ov*afy), you should be able to figure out the entire sequence.

TCE**JB**
JBUSE
SETIR
TIROV
OVAFY
_____
TCEJBUSETIROVAFY ← the deduced sequence
(up to this point)

This is precisely the approach taken by Sanger. He chemically cut each purified polypeptide into random fragments, then *sequenced* the fragments and pieced together the linear order of amino acids in the entire chain. From start to finish, the work took Sanger and his colleagues nearly 10 years.

Sanger's sequences showed that the insulin molecule had a precise order of amino acids. Examination of other proteins showed them to have their own stable amino acid sequences, each of which differed from that of insulin. There were no predetermined rules of chemistry that dictated the sequence of amino acids in polypeptides; each sequence was unique and could only be revealed experimentally. Sanger had provided the foundation for the future understanding of protein function based on the chemical properties of their amino acids.

While Sanger was working on amino acid sequences in England, Linus Pauling of the California Institute of Technology was working on the three-dimensional organization of the amino acids in a polypeptide. Pauling utilized molecular models to determine how the amino acids could stably fit together, much like assembling pieces of a jigsaw puzzle. One day in 1948, while he was visting in England, Pauling was suffering from a bad cold and decided to stay in his room. Using a pencil, paper, and straightedge, he drew the atoms and chemical bonds of a polypeptide chain. As he was folding the paper in various positions, he discovered that when the polypeptide chain was arranged in a structure resembling a "spiral staircase," the hydrogen bonds from the amino acids of the chain fit perfectly into place, providing the chain with maximum stability. Pauling called the spiral structure an *alpha-helix* (see Figure 4-15*b*).

Meanwhile, John Kendrew and Max Perutz of Cambridge University were attempting to determine the shape of an entire protein molecule. They relied on a technique called X-ray crystallography, in which crystals of a purified substance are bombarded with X-rays. The X-rays are deflected by the molecules in the crystal and strike a photographic plate, producing a pattern of dots which, when interpreted mathematically, reveals the structure of the molecule that produced the pattern. Pauling had used X-ray crystallography in his early analysis of amino acids, but this was the first time the technique was applied to a giant, complex protein.

The first protein chosen by Kendrew and Perutz for study by X-ray crystallography was *myoglobin,* a protein that stores oxygen in muscle tissue. The picture revealed a compact molecule whose polypeptide chain was folded back on itself in a complex, irregular arrangement. It was discovered that not only do proteins have a unique amino acid sequence (as Sanger had demonstrated in the case of insulin), they also have a unique three-dimensional shape that fits their specific function—in myoglobin's case, oxygen binding. In addition, eight segments of the polypeptide chain were composed of Pauling's alpha-helixes, which together accounted for approximately 70 percent of the amino acids in the protein (see Figure 4-15*c*). Pauling was delighted. He was also delighted to receive the Nobel Prize for Chemistry in 1954. Four years later, Sanger received a Nobel Prize of his own, followed by Perutz and Kendrew in 1962.

Each cell of a living organism is a miniature chemical plant that constantly churns out an array of molecules much more varied and vastly more complex than any chemical plant built by human hands. These molecules range in size from a few atoms to gigantic structures that, if extended, would stretch the length of your finger. Among these molecules are the most complex compounds on earth, perhaps in the universe. All of these biological molecules, from the smallest to the very largest, share a common characteristic—a chemical "skeleton" composed of carbon.

▼ ▼ ▼

THE IMPORTANCE OF CARBON IN BIOLOGICAL MOLECULES

At one time, scientists referred to molecules that could be manufactured only by living organisms as *organic chemicals,* in contrast to *inorganic chemicals,* which are found in the inanimate world. Chemists eventually discovered that all of these so-called organic compounds contained the element carbon and that many of them could be synthesized in the laboratory (Figure 4-1). These developments led to the redefining of **organic** molecules as simply "carbon containing." Those organic molecules that are produced by living cells are called **biochemicals** to distinguish them from the vast array of molecules now synthesized by organic chemists.

Carbon's chemical properties make it the ideal central element on which life is based. Unlike any other type of atom, carbons have the ability to bond with one another to form long chains. These chains of carbon atoms provide the chemical backbone of biological molecules, a backbone that may be linear, cyclic, or branched:

$$C—C—C—C—C—C$$
Linear

Cyclic

Branched

Recall from the previous chapter that a carbon atom has four unpaired electrons in its outermost shell and thus is capable of forming four single covalent bonds. Consequently, even after forming bonds with other carbons to form a chain, each carbon in the skeletons depicted above is still capable of bonding to other atoms. Consider a molecule of glucose

FIGURE 4-1
The sails of these boats are made of a synthetic, carbon-containing polymer. Thousands of different organic molecules not present in living organisms are now being synthesized by chemists.

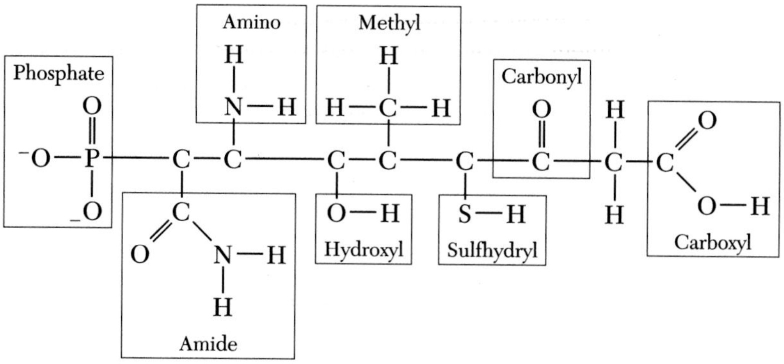

FIGURE 4-2

A gallery of functional groups attached to the carbon skeleton of a hypothetical molecule. Functional groups give organic molecules much of their chemical properties and determine the types of reactions in which the molecule can participate.

In addition to being bonded to carbon and hydrogen atoms, the carbons of a glucose molecule are also bonded to an atom that is part of a functional group. **A functional group** is a particular grouping of atoms which gives an organic molecule its chemical properties and reactivity. Some of the most common functional groups are illustrated in the hypothetical molecule depicted in Figure 4-2. Most functional groups are charged or highly polar, making organic molecules much more soluble in aqueous solutions and more chemically reactive than are those composed of only carbon and hydrogen. Functional groups account for the great diversity of organic molecules. This is illustrated by comparing similar molecules with different functional groups. Ethane (CH_3CH_3), for example, is a toxic, flammable gas. Substitute a hydroxyl group (—OH) for one of the hydrogen atoms, and the molecule (CH_3CH_2OH) becomes the intoxicating liquid, ethyl alcohol. Substitute a carboxyl group (—COOH), and the molecule (CH_3COOH) becomes acetic acid, better known as vinegar. Substitute a sulfhydryl group (—SH) and you have formed a strong, foul-smelling agent, ethyl mercaptan (CH_3CH_2SH), used by biochemists in studying enzyme reactions.

Of the 92 naturally occurring elements, only carbon has the size and bonding properties suited for forming the structural skeleton of biological molecules. The chemical properties of biological molecules are largely determined by the particular groups of atoms (functional groups) bound to the carbon skeleton. (See CTQ #2.)

GIANT MOLECULES BUILT FROM SMALLER SUBUNITS

The chemicals that form the structure and perform the functions of life are highly organized molecules. Compared to small molecules like those just mentioned, many of these organic compounds are enormous, containing hundreds to millions of carbon atoms. These enormous chemicals are called **macromolecules** (*macro* = large), and they possess special properties absent in their smaller relatives. Because of their size and the intricate shapes they can assume, these molecular giants can perform complex tasks with great precision and efficiency. Without them, complex biological activities would not be possible.

Macromolecules are constructed by assembling together small molecular subunits (Figure 4-3a), often in a linear process that resembles coupling railroad cars onto a train. This coupling process is referred to as **condensation.** Each subunit is called a **monomer** (*mono* = one; *mer* = part), and the macromolecule is referred to as a **polymer** (*poly* = many). The reverse process, in which the polymer is disassembled into its individual monomers, is called **hydrolysis** (*hydro* = water, *lysis* = split) because the bond that joins two monomers in a chain is split by the insertion of a water molecule between the two units (Figure 4-3b).

Cells produce huge macromolecules constructed from large numbers of individual subunits. The size and shape of these molecular giants contributes to their ability to perform specific tasks with rigorous precision. (See CTQ #3.)

FOUR BIOCHEMICAL FAMILIES

Macromolecules fall into four fundamental families of organic compounds: *carbohydrates, lipids, proteins,* and *nucleic acids.* The basic structure and function of each family of macromolecule is very similar in all organisms, from bacteria to humans, providing another example of the unity of life. It is not until you look very closely at the specific sequences of monomers that make up these various macromolecules that the diversity among organisms becomes apparent. The basic functions of the various macromolecules are summarized in Table 4-1.

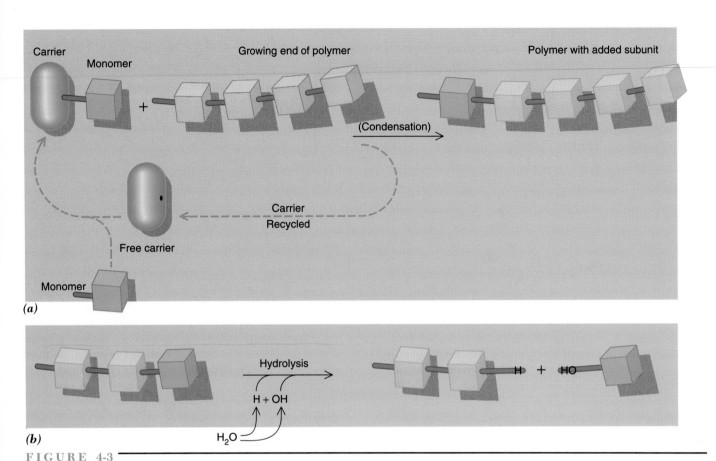

FIGURE 4-3

Monomers and polymers. *(a)* Biological macromolecules consist of monomers (subunits) linked together by covalent bonds. The assembly of macromolecules does not occur simply as a result of reactions between free monomers. Rather, each monomer is first activated by attachment to a "carrier" molecule that subsequently transfers the monomer to the end of the growing macromolecule. *(b)* Disassembly of a macromolecule occurs by hydrolysis of the bonds that join the monomers together. Hydrolysis is the splitting of a bond by water.

CARBOHYDRATES

Carbohydrates are a group of substances which includes *simple* sugars and all larger molecules constructed of sugar subunits. Carbohydrates function primarily as storehouses of chemical energy and as durable building materials for biological construction. As illustrated in Figure 4-4, glucose and fructose are six-carbon sugars, while ribose is a five-carbon sugar. Glucose and fructose are the major ingredients found in honey, which has been used as a sweetener for thousands of years, long before the cultivation of sugar cane. Of the two sugars, fructose tastes several times sweeter than glucose. You might have noticed that some beverages and other foods contain "high-fructose corn syrup." This may suggest that corn is a rich source of fructose, but this is not the case. In actuality, corn contains glucose, which is commercially converted to fructose by an industrial process that utilizes a common enzyme.

Individual sugars, called *monosaccharides*, can be covalently linked together to form larger molecules. A molecule composed of only two sugar units is a *disaccharide* (*di* = two, *saccharide* = sugar). Lactose, a disaccharide found in milk, provides a nursing baby with most of its energy. Lactose is composed of the sugars glucose and galactose. The best known (and sweetest) disaccharide is sucrose, common table sugar, which is composed of glucose and fructose. Sucrose is a major component of the sap of plants, and it carries energy from one part of the plant to another.

Polysaccharides: Macromolecular Carbohydrates

By the middle of the nineteenth century, it was known that the blood of people suffering from diabetes had a sweet taste due to an elevated level of glucose. Claude Bernard,

TABLE 4-1

MAJOR MACROMOLECULES FOUND IN LIVING SYSTEMS

Macromolecule	Constituents	Some Major Functions	Examples
Polysaccharide (large carbohydrates)	Sugars	Energy storage: physical structure	Starch; glycogen; cellulose
Lipid			
Triglycerides	Fatty acids and glycerol	Energy storage; thermal insulation; shock absorption	Fat; oil
Phospholipids	Fatty acids, glycerol, phosphate, and an R group[a]	Foundation for membranes	Plasma membrane
Waxes	Fatty acids and long-chain alcohols	Waterproofing; protection against infection	Cutin; suberin; ear wax; beeswax
Nucleic acid	Ribonucleotides; deoxyribonucleotides	Inheritance; ultimate director of metabolism	DNA; RNA
Protein	Amino acids	Catalysts for metabolic reactions; hormones; oxygen transport; physical structure	Hormones (oxytocin, insulin); hemoglobin; keratin; collagen; and a class of proteins called enzymes

[a] R group = a variable portion of a molecule.

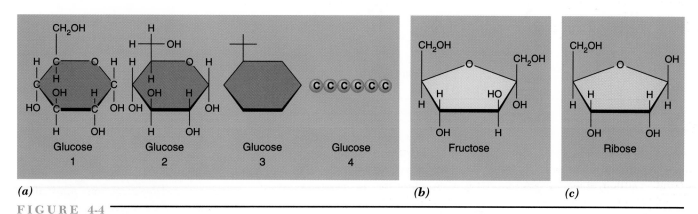

FIGURE 4-4

The structure of simple sugars. The *simple sugars* have the general formula $(CH_2O)n$, where the value of n typically ranges from three to seven. **(a)** Four structural conventions for representing glucose ($n = 6$). You might note that the carbon backbone of glucose was depicted as a straight chain on page 68, but is shown as a ring here. Sugars with five or more carbon atoms undergo a type of "self-reaction" where one end of the sugar molecule bonds with the other end, forming a closed, or ring-containing, molecule. In (1), all atoms of the molecule are shown; in (2), the carbons are omitted since their position is understood; in (3), only the skeleton is indicated, with no atomic detail shown; and (4) depicts the convention used throughout this textbook. Each ball represents a carbon. **(b)** Fructose, like glucose, is a six-carbon sugar, but it self-reacts to form a different type of ring. Note that even though glucose and fructose have the same formula ($C_6H_{12}O_6$), they have a different arrangement of atoms; they are said to be *isomers*. **(c)** Ribose is a five-carbon sugar ($n = 5$). Ribose is a major component of nucleotides and nucleic acids, which will be discussed later in the chapter.

one of the great physiologists of the period, was looking for the cause of diabetes by investigating the source of blood sugar. It was assumed at the time that any sugar present in a human or animal had to have been previously consumed in the diet. Working with dogs, Bernard found that even if the animals were placed on a diet totally lacking carbohydrates, their blood still contained a normal amount of glucose. Clearly, glucose could be formed in the body from other types of compounds.

⟳ Upon further investigation, Bernard found that glucose enters the blood from the liver. He also found that liver tissue contains an insoluble polymer of glucose he named *glycogen*. Bernard concluded that various food materials (such as proteins) were carried to the liver where they were chemically converted to glucose and stored as glycogen. Then, as the body needed sugar for fuel, the glycogen in the liver was transformed to glucose, which was released into the bloodstream to satisfy glucose-depleted tissues. This is an example of homeostasis, a concept that was first developed by Bernard and which holds that organisms maintain fairly constant internal conditions (see Chapter 22). In Bernard's hypothesis, the balance between glycogen formation

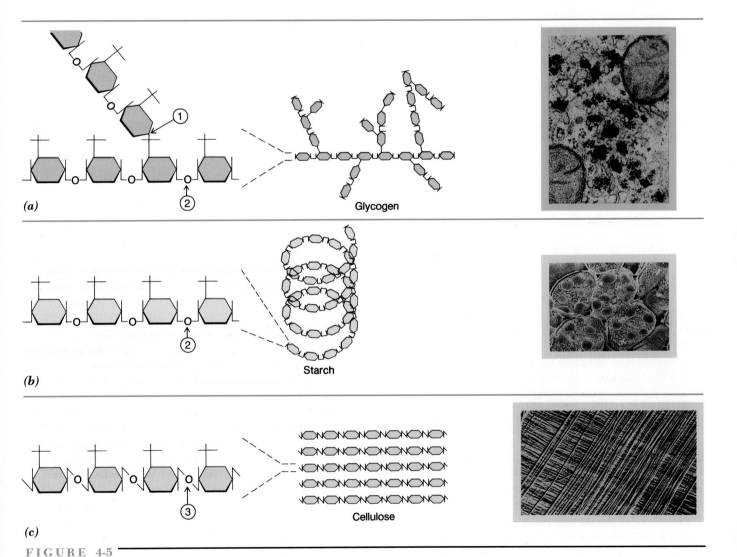

FIGURE 4-5

Three polysaccharides, identical sugar monomers, dramatically different properties. Glycogen *(a)*, starch *(b)*, and cellulose *(c)*, are each composed entirely of glucose subunits, yet their chemical and physical properties (and thus their functions) are very different due to the distinct ways that the monomers are linked together (three different types of linkages are indicated by the circled numbers). Glycogen molecules are the most highly branched; starch molecules assume a helical (spiral) arrangement; and cellulose molecules are unbranched and highly extended and are bundled together to form very tough fibers suited for their structural role. Electron micrographs show glycogen granules in a liver cell, starch grains (amyloplasts) in a plant seed, and cellulose fibers in a plant cell wall.

FIGURE 4-6

Strenuous exercise requires energy which is stored as the polysaccharide glycogen.

and glycogen breakdown in the liver was the prime determinant in maintaining the relatively constant (homeostatic) level of glucose in the blood.

Polysaccharides As Energy Stores: Glycogen and Starch Bernard's analysis proved to be correct. The molecule he named *glycogen* is a type of **polysaccharide**—a long chain of sugar units joined together as a polymer. Subsequent studies revealed that many different animals bank their surplus chemical energy in glycogen, a highly branched polysaccharide composed entirely of glucose monomers (Figure 4-5*a*). In humans, glycogen is stored and used as fuel in a wide variety of tissues, including muscles, but only the liver serves as a glucose supplier for the rest of the body. Muscles typically contain enough glycogen to fuel about 30 minutes of strenuous activity (Figure 4-6).

Another polymer of glucose is *starch*, the polysaccharide most commonly used by plants for energy storage. Potatoes and cereals, for example, consist primarily of starch. Like glycogen, starch is a polymer composed entirely of glucose monomers, but it is much less branched (Figure 4-5*b*). Even though animals don't produce starch, they can readily digest it by hydrolysis. In fact, starch is the primary source of energy in the human diet in most parts of the world.

Structural Polysaccharides: Cellulose and Chitin Two of the most important structural polysaccharides are *cellulose* and *chitin*. Cellulose is the earth's most abundant polysaccharide (Figure 4-7), forming the tough fibers present in wood and plant cell walls (Chapter 5). Cotton textiles owe their durability to the long, unbranched cellu-

lose molecules (Figure 4-5*c*), which are ideally constructed to resist pulling forces. Animals lack an enzyme capable of digesting cellulose, so this vast reserve remains unavailable to them as a direct source of energy. Ironically, cellulose is composed entirely of glucose, the same monomer found in starch, one of the most easily digested polymers. The two polysaccharides differ in the way the glucose monomers are linked together, however (Figure 4-5*b,c*). Because of this minor structural difference, the bonds of starch are readily hydrolyzed by an enzyme in our digestive tracts, while the cellulose molecules pass through intact, providing fiber that aids in the formation and elimination of feces.

Unlike animals, a variety of microorganisms possess the necessary enzyme for digesting cellulose. If not for these microscopic cellulose decomposers, the world would be permanently littered with dead bodies of plants. A number of animals are able to take advantage of the cellulose in their diet by harboring cellulose-digesting microorganisms within their digestive tract. Cows, for example, rely on bacteria and protozoa living within a special digestive chamber (the rumen) to digest the grass and hay that make up much of their diet. Even termites depend on cellulose-digesting microbes living in their stomachs to digest their meal of microscopic wood particles.

Another important structural polysaccharide—chitin—is a polymer of a nitrogen-containing sugar and is a major component of the outer covering (*exoskeleton*) of spiders, crustaceans, and insects (Figure 4-8). Chitin gives the exoskeleton a tough, resilient quality, not unlike that of certain plastics. Insects and crustaceans owe much of their biologic success to this highly adaptive polysaccharide covering.

A showcase of structural polysaccharides. The organisms in this picture reveal a few of the natural roles of polysaccharides. The fibrous polysaccharide that supports and shapes the pitcher plant is cellulose. The internal organs and skin of these frogs are secured in place by structural polysaccharides that strengthen connective tissue. Although not shown, insects that fall prey to the carnivorous pitcher plant possess external skeletons consisting of the rigid polysaccharide chitin.

F I G U R E 4-8

The body of this grasshopper is covered with a glistening outer skeleton consisting largely of the polysaccharide chitin.

LIPIDS

Lipids are a diverse group of organic molecules whose only common property is their inability to dissolve in water—a property that explains many of their varied biological functions. The hydrophobic character of lipids is revealed by their molecular structure: The carbon atoms of lipids are bonded as much to other carbon atoms as to hydrogen atoms. Lipids include fats, oils, phospholipids, steroids, and waxes.

Fats

☀ **Fats** consist of three fatty acids coupled to a single molecule of glycerol. As illustrated in Figure 4-9, **fatty acids** are long, water-insoluble chains composed primarily of —CH_2— units. Each fatty acid in a fat molecule is linked to one of the three carbons of the glycerol "backbone." Fats are very rich in chemical energy; a gram of fat contains over twice the energy content of a gram of carbohydrate (for reasons discussed in Chapter 6). Unlike carbohydrate, which functions primarily as a short-term, rapidly available energy source, fat reserves are utilized to store energy on a long-term basis. It is estimated that an average person contains about 0.5 kilograms of glycogen. During the course of a strenuous day's activity, a person can virtually deplete his or her body's entire store of glycogen. In contrast, the average person contains approximately 16 kilograms of fat and, as we all know, it can take a very long time to deplete our store of this material. The subject of fat production is discussed in The Human Perspective: Obesity and the Hungry Fat Cell.

Fats occur in either a solid or a liquid state (liquid fats are termed *oils*). In general, the greater the number of *unsaturated* (double) bonds that exist between the carbons of the fatty acid chains, the less well these long chains can be packed together. This lowers the temperature at which the lipid melts. The profusion of double bonds in vegetable oils accounts for their liquid state—both in the plant cell and on the grocery shelf—and for their label, "polyunsaturated" (many double bonds) fats. In contrast, almost all the linkages between the carbons of animal fats are single bonds; that is, they are *saturated* with hydrogens, causing the material to remain a solid at room temperature. Polyunsaturated fats in the human diet may be less likely to promote cardiovascular disease than saturated animal fats, such as those found in lard and butter, but the matter remains the subject of research.

Phospholipids

ᔕ Most **phospholipids** are similar in structure to fats except that the glycerol is attached to only two fatty acid chains instead of three. The third glycerol carbon is joined to a phosphate group, which, in turn, is linked to one of a variety of small polar groups (Figure 4-10). The end of the phospholipid containing the phosphate and polar group is soluble in water, while the opposite end containing the fatty acid tails is hydrophobic and "shuns" the aqueous medium. This bipolar nature of phospholipids allows these molecules

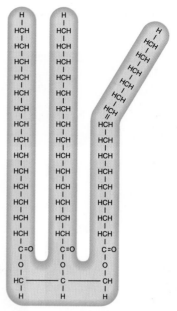

FIGURE 4-9

A fat molecule consists of three long-chain fatty acids joined to a glycerol (lower portion of the fat molecule).

to form bimolecular sheets (*bilayers*) and plays a key role in their function as a major component of cell membranes (see Chapter 5).

Steroids

Steroids are molecules that are built around a characteristic four-ringed skeleton (Figure 4-11). One of the most common steroids is *cholesterol*, a component of animal cell membranes and a precursor for the synthesis of a number of *hormones*—chemical messengers sent from one part of the body to other parts, orchestrating many of the body's processes. For example, sexual maturation is coordinated by steroid hormones: *testosterone* in males and *estrogen* in females. Cholesterol is absent from plant cells, which is why vegetable oils are considered cholesterol-free. This designation is particularly important for people with elevated levels of blood cholesterol because they are at a substantially higher risk of developing *atherosclerosis*, a dangerous "hardening of the arteries," which is a leading cause of heart disease. The presence of high cholesterol levels in the blood promotes the formation of cholesterol-containing buildup (*plaques*) on the inner lining of the arteries. Over periods of many years, these plaques can narrow the diameter of the arteries, greatly reducing blood flow and triggering the formation of blood clots, a major cause of heart attacks (see The Human Perspective, Chapter 7).

Waxes Similar in structure to fats, **waxes** contain many more fatty acids linked to a longer chain backbone. Waxes provide a waterproof covering over the leaves and stems of many plants, preventing the organism from losing precious water. Waxes also serve in construction of the honeycombs of a beehive, as a protective material in the human ear canal, and as a waterproof component spread over the feathers of birds.

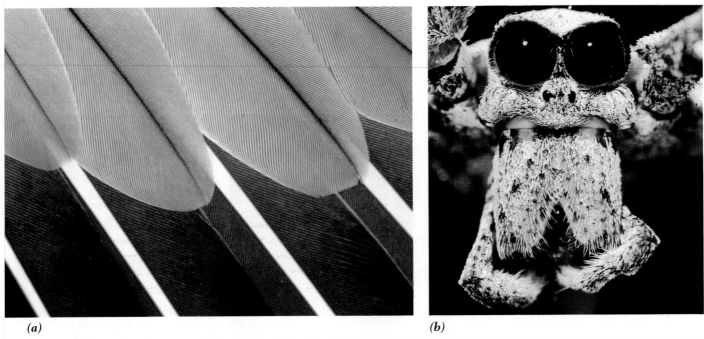

(a) *(b)*

FIGURE 4-12

This protein gallery shows two of the thousands of biological structures composed predominantly of protein. These include: *(a)* the fabric of feathers used for thermal insulation, flight, and sex recognition among birds and *(b)* the lenses of eyes, as in this net-casting spider.

FIGURE 4-13

Proteins are assembled from amino acids that are joined by peptide bonds. Each peptide bond forms by the linkage of an amino group from one amino acid and a carboxyl group from the neighboring amino acid. A string of amino acids joined by peptide bonds is called a polypeptide chain. The formation of a polypeptide is one of the most complex molecular processes in biology and involves the participation of many different components.

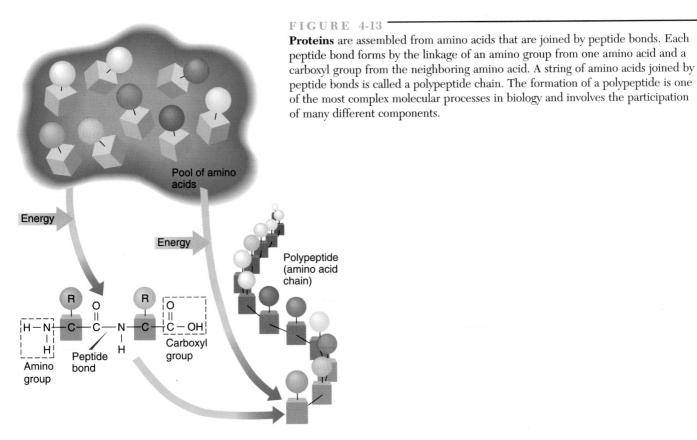

Proteins

If you were to drive off all the water from an organism's body, more than half the dried remains would consist of protein. It is estimated that the typical mammalian cell has at least 10,000 different proteins. **Proteins** are the macromolecules of the cell that "make things happen." Among their many functions, proteins determine much of what gets into and out of each cell, they regulate the expression of genes, and they form the machinery for biological movements, ranging from muscle contraction to the swimming of a sperm up the female reproductive tract. Protein constitutes the major structural component of hair, feathers, fingernails, skin, ligaments, tendons, and thousands of other biological structures (Figure 4-12). Another essential group of proteins is *enzymes,* the mediators of metabolism. Thousands of different enzymes collaborate to direct the development and maintenance of every organism.

The Building Blocks of Proteins

As we described in the beginning of the chapter, proteins are polymers made of amino acid monomers. Each protein has a unique sequence of amino acids which gives that molecule its unique properties. As a group, proteins have so many different functions because different proteins have strikingly different amino acid sequences. Many of the capabilities of a protein can be understood by examining the properties of its constituent amino acids. The same 20 amino acids are found in virtually all proteins, whether in a virus or a human. There are two aspects of amino acid structure to consider: that which is common to all of them, and that which is unique to each.

All free (unlinked) amino acids have a carboxyl group and an amino group (Figure 4-13), separated by a single carbon atom; these are the common parts of the amino acids. The remainder of each amino acid—the **R group** (Figure 4-13)—is variable among the 20 building blocks.

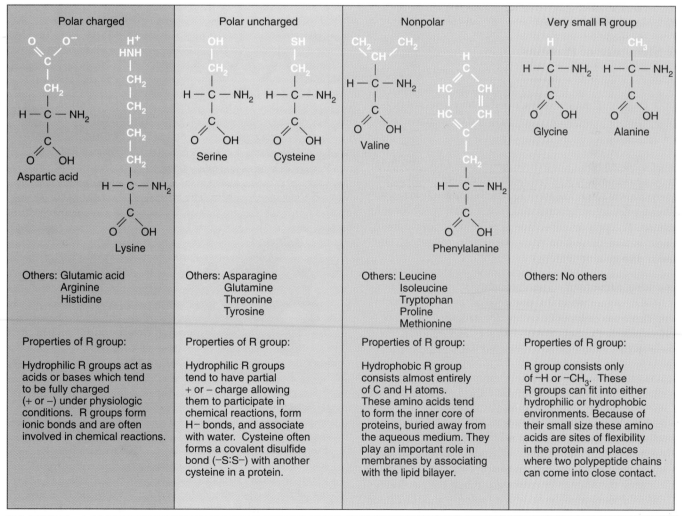

FIGURE 4-14

The structure and properties of amino acids.

Amino acids are conveniently classified on the basis of the polar (hydrophilic) versus nonpolar (hydrophobic) nature (page 59) of their R groups. They fall roughly into four categories: polar charged, polar uncharged, nonpolar, and those with a very small R group (Figure 4-14). The nonpolar amino acids tend to be clustered in the center of a protein, away from the surrounding water, giving that region an almost oil-like character. If all the amino acids are considered together, there is a large variety of reactions in which they can participate and a great many types of bonds they can form. The activities of particular amino acids, situated at particular sites within a protein, account for the function of that protein.

Proteins are synthesized by joining each amino acid to two other amino acids, forming a long, continuous, unbranched polymer called a **polypeptide** chain. The amino acids are joined by **peptide bonds**—linkages formed by joining the carboxyl group of one amino acid to the amino group of its neighbor, with the elimination of a molecule of water. The backbone of a polypeptide chain (illustrated by the gray cubes of Figure 4-13) is composed of the common parts of the string of amino acids, while the R groups (illustrated by the colored spheres of Figure 4-13) project out as "side groups."

The Structure of Proteins

Proteins are large molecules; most proteins contain at least 100 amino acids and may contain up to as many as 20,000. Protein structure is described at four levels of organization: primary, secondary, tertiary, and quaternary (Figure 4-15).

Primary Structure The *primary structure* of a protein is the specific linear sequence of amino acids that make up its polypeptide chain(s) (Figure 4-15). Different proteins have different primary structures and, consequently, different functions. The primary structure of proteins is determined by information passed on from parents to offspring, encoded in the genetic material.

Secondary Structure The secondary structure defines the spatial organization of portions of the polypeptide chain. Three major types of secondary structure are recognized: alpha helix, beta-pleated sheet, and random coil (Figure 4-15); each can be correlated with protein function.

1. In an *alpha (α) helix*, the backbone of the polypeptide assumes the form of a spiral held together by hydrogen bonds. *Alpha-keratin*, the major protein of wool, consists largely of alpha-helix, which gives the material extensibility. When a fiber of wool is stretched, the hydrogen bonds are broken, disrupting the helix and allowing the fiber to be extended. When the tension is relieved, the hydrogen bonds can reform, and the fiber snaps back to its original length.

2. In a *beta (β)-pleated sheet,* two or more sections of the polypeptide chain lie side by side, forming an accor-

dionlike sheet that is held together by hydrogen bonds. *Fibroin*, the major protein of silk, is composed largely of pleated sheets stacked on top of one another. Unlike the fibers of wool, those of silk cannot be stretched because the polypeptide chains are already extended.

3. Any portion of a polypeptide chain not organized into a helix or a sheet is said to be in a *random coil*. For example, the sites in the myoglobin (Figure 4-15)

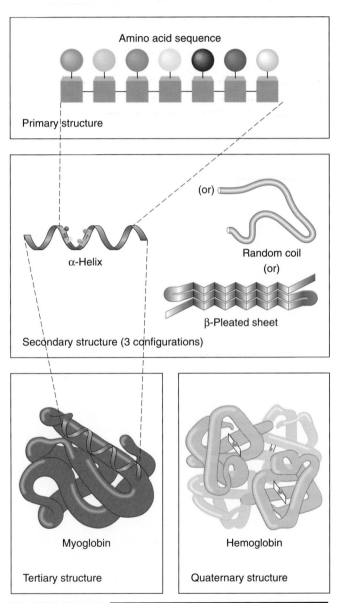

FIGURE 4-15

Four levels of protein structure. Primary structure describes the specific sequence of amino acids of a polypeptide chain. Secondary structure describes the conformation of a portion of the polypeptide chain. Tertiary structure describes the manner in which an entire polypeptide chain is folded. Proteins consisting of more than one polypeptide chain have quaternary structure, which describes the way the chains are arranged within the protein.

where the polypeptide makes a sharp turn are regions of random coil. The random coils tend to be the most flexible portions of a polypeptide and often represent sites of greatest activity.

Tertiary Structure The *tertiary structure* describes the shape of an entire polypeptide chain. Like myoglobin, the polypeptide chains of most proteins are folded and twisted to form a globular-shaped molecule with a complex internal organization (Figure 4-15c). Each protein has a precise shape that enables it to carry out a precise function. The polypeptide chain is held in this complex shape largely by noncovalent bonds (ionic bonds, hydrogen bonds) and hydrophobic interactions. Proteins are not rigidly fixed structures but flexible molecules capable of relatively large-scale, internal movements, called **conformational changes.** These changes in shape contribute to their various activities. The movement of your body, for example, results from the additive effect of millions of conformational changes taking place within the proteins of your muscles.

Quaternary Structure Many proteins are composed of more than one polypeptide chain. The spatial arrangement of the combined chains describes these proteins' *quaternary structure.* Hemoglobin, for example, is an iron-containing blood protein consisting of four polypeptide chains (Figure 4-15d), each of which can bind and transport a single molecule of oxygen.

Determination of Protein Shape

How does the complex, folded form of a protein take shape within a cell? The first important insights into this question were obtained in the late 1950s by Christian Anfinsen and his co-workers at the National Institutes of Health. Anfinsen and his colleagues were working on an enzyme called *ribonuclease.* In these experiments (Figure 4-16), purified preparations of ribonuclease were treated with agents that caused the protein molecules to unfold, losing their secondary and tertiary structure. This type of disruption is known as **denaturation.** It was found that once the denaturing agents were removed from the preparation, the disorganized ribonuclease molecules *spontaneously* reformed themselves into active enzyme molecules. The results of these experiments indicated that the three-dimensional shape of a protein may "spring" directly from the protein's primary structure. In other words, once a given linear sequence of amino acids is strung together, the protein generally spontaneously folds into the proper shape.

The importance of amino acid sequence in determining the form (and function) of a protein is dramatically illustrated by the consequences of changing just one amino acid in hemoglobin. Substituting a hydrophobic, uncharged amino acid (valine) for a hydrophilic, negatively charged amino acid (glutamic acid) distorts the protein and seriously impairs its ability to carry oxygen. Under certain conditions, the distorted hemoglobin elongates red blood cells into a sickle shape (Figure 4-17) that clogs vessels. This potentially fatal disease is called *sickle cell anemia,* and it is a chronic inherited anemia that occurs primarily in people of African descent.

NUCLEIC ACIDS (DNA AND RNA)

One of the properties distinguishing the living from the inanimate is the ability of the living to reproduce offspring that have characteristics similar or identical to those of their parent(s). The instructions required to "build" a new indi-

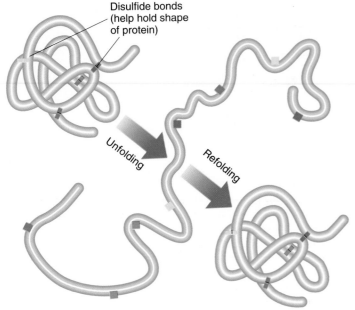

Disulfide bonds (help hold shape of protein)

Unfolding

Refolding

FIGURE 4-16

Self-folding. When the polypeptide chain of a ribonuclease molecule is experimentally unfolded, it is capable of spontaneously reforming its natural three-dimensional shape when the agents that caused denaturation are removed. This experiment provided the first evidence that it is primarily the sequence of amino acids in a polypeptide that dictates the way the polypeptide becomes folded to form the active molecule.

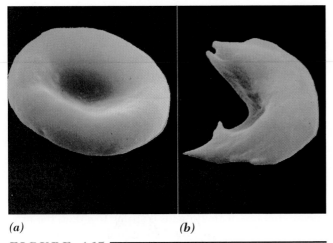

(a)   *(b)*

FIGURE 4-17
The effects of a single amino acid change in a polypeptide chain of hemoglobin. *(a)* A normal human red blood cell. *(b)* A red blood cell from a person with sickle cell anemia. The cell becomes sickle shaped when the level of oxygen in the blood drops. These abnormally shaped cells can clog small blood vessels, causing pain and life-threatening crises.

vidual are transmitted to the offspring encoded in the polymer **deoxyribonucleic acid (DNA)**—the material of which genes are made. DNA is primarily a storehouse of genetic information (Figure 4-18). The hereditary messages stored in the DNA blueprint govern cellular activities through the formation of molecules of **ribonucleic acid (RNA)**. Both DNA and RNA are **nucleic acids**—macromolecules constructed as a long chain (strand) of **nucleotide** monomers. For the time being, we will describe the basic structure of nucleic acids, using RNA as the representative molecule. We will describe the more complex three-dimensional structure of DNA, along with the story of its discovery, in Chapter 14.

Each nucleotide in a strand of RNA consists of three parts: (1) a five-carbon sugar called *ribose*, (2) a *phosphate* group, and (3) a *nitrogenous (nitrogen-containing)* base (Figure 4-19a). While the sugar and phosphate groups are identical in all nucleotides, there are four different types of nitrogenous bases in RNA: uracil, cytosine, guanine, and adenine. Nucleotides are covalently joined to one another through sugar–phosphate linkages forming a single linear strand of RNA (Figure 4-19b).

◯ Nucleic acid structure and function illustrate the importance of biological order at the molecular level. The

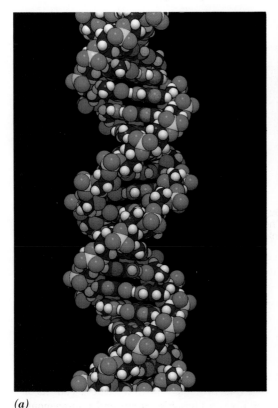

(a)   *(b)*

FIGURE 4-18
DNA: The material of the genes. *(a)* A structural model of a DNA molecule. *(b)* Even complex behaviors, such as the defense of territory and mating rituals of these Japanese cranes, are encoded within the DNA that makes up the animals' genes.

information in a nucleic acid is encoded in the ordered linear sequence of its nucleotides. There is a flow of information in a cell from DNA to RNA to protein. The sequence of nucleotides in the DNA of the genes determines the sequence of nucleotides in a molecule of RNA, which in turn determines the sequence of amino acids in a protein. Consequently, a change in the sequence of nucleotides in the DNA can lead to a change in the sequence of amino acids in the corresponding protein. This is the basis for the single change in amino acid sequence in the sickle cell version of hemoglobin.

Nucleotides are not only important as monomers in nucleic acids, they have important functions in their own right. Most of the energy being put to use at this very instant within your body is derived from the nucleotide **adenosine triphosphate (ATP).** The structure of ATP and its key role in cellular metabolism is discussed in Chapter 6.

Just as different organisms are adapted to different habitats, and different organs in your body are adapted for different physiologic activities, different types of macromolecules are adapted to perform different chemical functions. Carbohydrates are sugar-containing molecules that provide readily available energy and durable building materials. Lipids are a varied group of energy-rich, water-insoluble molecules; they function in long-term energy storage and as insoluble walls within cellular membranes. Proteins derive their properties from a precise sequence of their amino acid subunits; they carry out most of the diverse, precision-requiring biological activities. Nucleic acids are composed of strands of nucleotide subunits whose precise sequence serves as a storehouse of genetic information. (See CTQ #5.)

MACROMOLECULAR EVOLUTION AND ADAPTATION

III▶ Recall from Chapter 1 that adaptations are traits that improve the suitability of an organism to its environment. Macromolecules are biochemical adaptations that are subject to natural selection and evolutionary change in the same way as are other types of characteristics. Some macromolecules evolve very slowly. Cellulose, for example, is a structural component of cell walls in virtually all plants from the simplest algae to the most complex trees. Similarly, glycogen is almost universally present as the storage form of glucose in animals. In contrast, proteins show marked changes from organism to organism, both in form and function; most of the proteins in your body are not found in simpler animals, such as sponges and worms. For example, while hemoglobin is the oxygen-carrying protein of all vertebrates (animals with backbones), from fish to mammals, many other animals utilize a totally different protein for this same function. Octopus blood, for example, contains a bright blue, copper-containing protein called *hemocyanin* which is totally unrelated to hemoglobin.

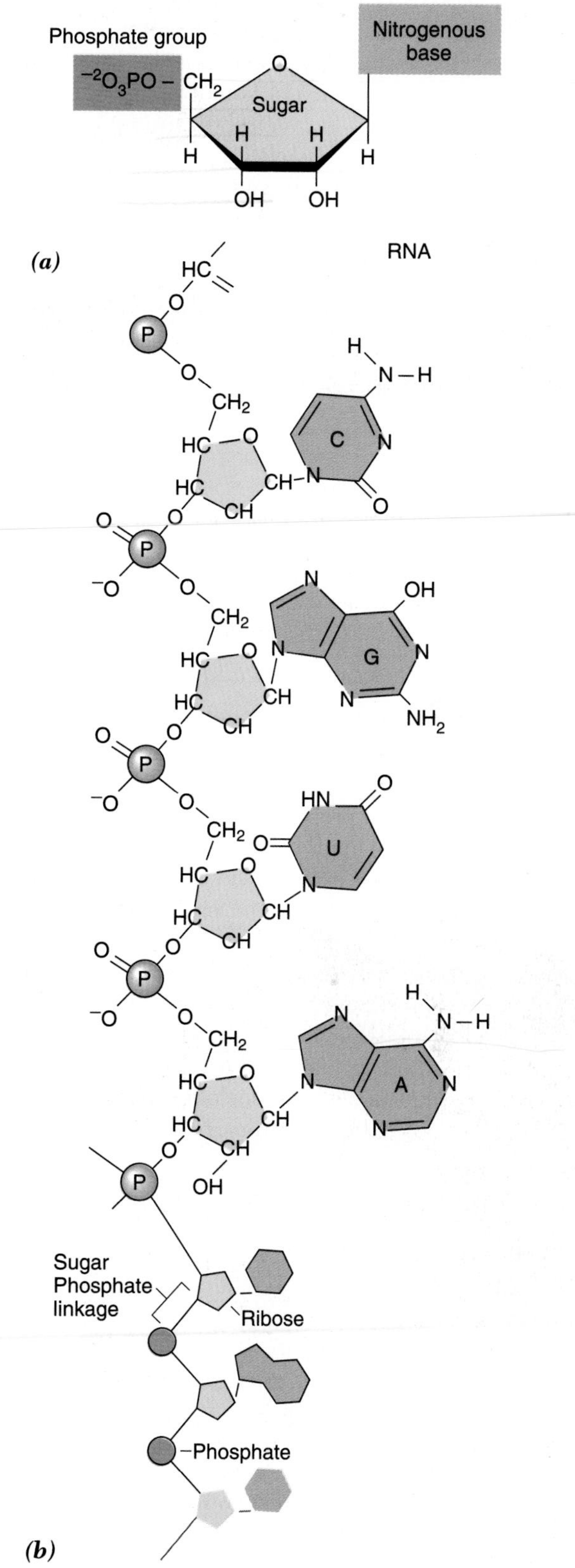

FIGURE 4-19

Nucleotides and nucleic acids. (*a*) A nucleotide consists of a sugar (ribose in RNA), a phosphate group, and a nitrogenous base (either uracil, cytosine, guanine, or adenine in RNA). (*b*) Nucleotides are joined to one another by sugar–phosphate linkages to form long strands of nucleic acids.

When two organisms have the same protein, the difference in amino acid sequence of that protein can be correlated with the evolutionary relatedness of the organisms. The amino acid sequence of hemoglobin, for example, is much more similar between humans and monkeys—organisms that are closely related—than between humans and turtles, who are only distantly related. In fact, the evolutionary tree that emerges when comparing the structure of specific proteins from various animals very closely matches that previously constructed from fossil evidence.

The fact that the amino acid sequences of proteins change as organisms diverge from one another reflects an underlying change in their genetic information. Even though a DNA molecule from a mushroom, a redwood tree, and a cow may appear superficially identical, the sequences of nucleotides that make up the various DNA molecules are very different. These differences reflect evolutionary changes resulting from natural selection (Chapter 34).

Virtually all differences among living organisms can be traced to evolutionary changes in the structure of their various macromolecules, originating from changes in the nucleotide sequences of their DNA. (See CTQ #7.)

REEXAMINING THE THEMES

Relationship between Form and Function

The structure of a macromolecule correlates with a particular function. The unbranched, extended nature of the cellulose molecule endows it with resistance to pulling forces, an important property of plant cell walls. The hydrophobic character of lipids underlies many of their biological roles, explaining, for example, how waxes are able to provide plants with a waterproof covering. Protein function is correlated with protein shape. Just as a key is shaped to open a specific lock, a protein is shaped for a particular molecular interaction. For example, the shape of each polypeptide chain of hemoglobin enables a molecule of oxygen to fit perfectly into its binding site. A single alteration in the amino acid sequence of a hemoglobin chain can drastically reduce the molecule's oxygen-carrying capacity.

Biological Order, Regulation, and Homeostasis

Both blood sugar levels and body weight in humans are controlled by complex homeostatic mechanisms. The level of glucose in your blood is regulated by factors acting on the liver, which stimulate either glycogen breakdown (which increases blood sugar) or glycogen formation (which decreases blood sugar). Your body weight is, at least partly, determined by factors emanating from fat cells which either increase metabolic rate (which tends to decrease body weight) or slow down metabolic rate (which tends to increase body weight).

Acquiring and Utilizing Energy

The chemical energy that fuels biological activities is stored primarily in two types of macromolecules: polysaccharides and fats. Polysaccharides, including starch in plants and glycogen in animals, function primarily in the short-term storage of chemical energy. These polysaccharides can be rapidly broken down to sugars, such as glucose, which are readily metabolized to release energy. Gram-for-gram, fats contain even more energy than polysaccharides and function primarily as a long-term storage of chemical energy.

Unity within Diversity

All organisms, from bacteria to humans, are composed of the same four families of macromolecules, illustrating the unity of life—even at the biochemical level. The precise nature of these macromolecules and the ways they are organized into higher structures differ from organism to organism, thereby building diversity. Plants, for example, polymerize glucose into starch and cellulose, while animals polymerize glucose into glycogen. Similarly, many proteins (such as hemoglobin) are present in a variety of organisms, but the precise amino acid sequence of the protein varies from one species to the next.

Evolution and Adaptation

Evolution becomes very apparent at the molecular level when we compare the structure of macromolecules among diverse organisms. Analysis of the amino acid sequences of proteins and the nucleotide sequences of nucleic acids reveals a gradual change over time in the structure of macromolecules. Organisms that are closely related have proteins and nucleic acids whose sequences are more similar than are those of distantly related organisms. To a large degree, the differences observed among diverse organisms derives from the evolutionary differences in their nucleic acid and protein sequences.

SYNOPSIS

Carbon is the central element in all organic compounds. Carbon's versatile properties include its ability to form four covalent bonds; to link up with other carbons to form straight, branching, or ring-shaped chains; to attach itself to various functional chemical groups; and to form single, double, or triple bonds. Its versatility makes the extraordinarily complex molecules on which life depends possible.

Some biochemicals are small molecules; others are polymers (macromolecules) made up of smaller building blocks. Macromolecules constitute the fabric of an organism and conduct its needed functions. Most macromolecules are polymers constructed by linking together the same class of subunits—monomers—into long chains. Macromolecules are disassembled by hydrolysis.

There are four families of biochemicals. Carbohydrates include simple sugars and larger molecules (polysaccharides) that are formed by linking together many sugars. Some polysaccharides store energy (e.g., glycogen in animals and starch in plants), others are structural polysaccharides (e.g., cellulose and chitin). Glycogen, starch, and cellulose are all polymers of glucose but, because of differ-ences in the way the sugars are linked, these polysaccharides have different properties. Lipids are a diverse array of hydrophobic molecules, including fats, which are composed of fatty acids and glycerol; phospholipids, which contain a phosphate group; steroids, which are constructed of four rings linked together; and waxes, which contain a number of fatty acids. Fats are primarily energy stores, phospholipids are major components of membranes, steroids are found in animal membranes and hormones, and waxes form waterproof coverings. Proteins consist of chains of 20 or so different amino acids, which differ by the structure of their R group. The properties of the R groups of particular amino acids give the protein its particular activity. The sequence of amino acids in a protein determines the shape of the protein, which in turn determines the protein's biological role. Some proteins have a structural role; others, such as hormones, antibodies, hemoglobin, and enzymes, have functional roles. Nucleic acids are informational molecules that consist of strands of nucleotide monomers. Both types of nucleic acids, DNA and RNA, consist of four different types of nucleotides. The precise sequence of nucleotides in a chain determines the information content of the molecule. Changes in the sequence of nucleotides in DNA, which occur during evolution, lead to alterations in the sequence of amino acids in the corresponding protein.

Key Terms

organic (p. 68)
biochemical (p. 68)
functional group (p. 69)
macromolecule (p. 69)
condensation (p. 69)
monomer (p. 69)
polymer (p. 69)
hydrolysis (p. 69)
carbohydrate (p. 70)
polysaccharide (p. 73)

lipid (p. 75)
fat (p. 75)
fatty acid (p. 75)
phospholipid (p. 75)
steroid (p. 75)
wax (p. 75)
set-point (p. 76)
protein (p. 79)
R group (p. 79)

polypeptide chain (p.80)
peptide bond (p. 80)
conformational change (p. 81)
denaturation (p. 81)
deoxyribonucleic acid (DNA) (p. 82)
ribonucleic acid (RNA) (p. 82)
nucleic acid (p. 82)
nucleotide (p. 82)
adenosine triphosphate (ATP) (p. 83)

Review Questions

1. Why are macromolecules described as polymers?

2. Why is a potato more fattening than an equivalent weight of spinach, when both are composed largely of carbohydrates?

3. Discuss how carbon's ability to form four chemical bonds is critical to life.

4. A popular myth states that camels store water in their humps. In actuality, a camel's hump functions in storing fat. From what you know about hydrolysis, would utilizing this stored fat produce more water or use it up?

5. Complete the following table:

Compound	Class of Compound	Monomer(s)	Function
———	disaccharide	———	———
ribonuclease	———	———	———
———	fat	———	———
DNA	———	———	———

Critical Thinking Questions

1. Frederick Sanger worked out the amino acid sequence of beef insulin. How would you predict this sequence would compare to the insulin found in buffaloes? To that found in fish? Why?

2. Silicon is similar to carbon in that it has four electrons in its outer shell; however, it is less suitable for forming biological molecules than is carbon. Research other properties of silicon and discuss why this is so.

3. Discuss the ways in which the formation of macromolecules from smaller molecular subunits is analogous to the manufacture of automobiles on an assembly line. In each case, what are the advantages of constructing large, complex units from smaller subunits?

4. Knowing that diabetes is characterized by a high level of glucose in the blood and that diabetics are treated by injections of insulin, what effect would you expect insulin to have on glycogen levels in the liver?

5. Using their chemical composition, structure, and functions, develop a classification system for the following "species" of organic molecules: monosaccharides, di-saccharides, polysaccharides, fats, phospholipids, waxes, steroids, DNA, RNA, and proteins. For example, you might begin by placing the nitrogen-containing molecules (DNA, RNA, proteins) in one group, and those without nitrogen (all of the others) in a second group. You would then subdivide each group until you had classified each of the molecules.

6. Termites harbor colonies of microorganisms in their digestive tracts. Why do you suppose this is adaptive? What would happen if you were to treat termites with drugs that killed all of these microorganisms?

7. Explain why macromolecules, but not small organic molecules, can be used to trace evolutionary changes in organisms.

8. Bacteria are known to change the kinds of fatty acids they produce as the temperature in which they are living changes. What types of changes in the fatty acids would you expect as the temperature drops, and why would this be adaptive?

Additional Readings

Burrows, G. D. , et al., eds. 1988. *Handbook of eating disorders.* New York: Elsevier. (Intermediate)

Field, H. L., ed. 1987. *Eating disorders throughout the life span.* Chicago: Greenwood. (Intermediate)

Fruton, J.S. 1972. *Molecules and life.* New York: Wiley. (Advanced)

Schulz, G. E., and R. H. Schirmer. 1990. *Principles of protein structure.* New York: Springer-Verlag. (Advanced)

Judson, H.F. 1980. *The eighth day of creation.* New York: Simon & Schuster. (Intermediate)

Lehninger, A. L. 1982. *Principles of biochemistry.* New York: Worth. (Intermediate)

Richards, F. M. 1991. The protein folding problem. *Sci. Am.* Jan: 54–63. (Intermediate)

Stryer, L. 1988. *Biochemistry.* 3d ed., New York: Freeman. (Advanced)

Voet, D., and J. G. Voet. 1990. *Biochemistry.* New York: Wiley. (Advanced)

Scientific American October 1985. "The Molecules of Life." (Special Edition) (Intermediate)

CHAPTER

5

Cell Structure and Function

STEPS TO DISCOVERY
The Nature of the Plasma Membrane

DISCOVERING THE CELL

BASIC CHARACTERISTICS OF CELLS

TWO FUNDAMENTALLY DIFFERENT TYPES OF CELLS

THE PLASMA MEMBRANE: SPECIALIZED FOR INTERACTION WITH THE ENVIRONMENT

THE EUKARYOTIC CELL: ORGANELLE STRUCTURE AND FUNCTION

The Nucleus: Genetic Control Center of the Cell

Membranous Organelles of the Cytoplasm: A Dynamic Interacting Network

Mitochondria and Chloroplasts: Acquiring and Using Energy in Cells

The Cytoskeleton: Providing Support and Motility

JUST OUTSIDE THE CELL

Cell Walls

Intercellular Junctions

EVOLUTION OF EUKARYOTIC CELLS: GRADUAL CHANGE OR AN EVOLUTIONARY LEAP?

THE HUMAN PERSPECTIVE
Lysosome-Related Diseases

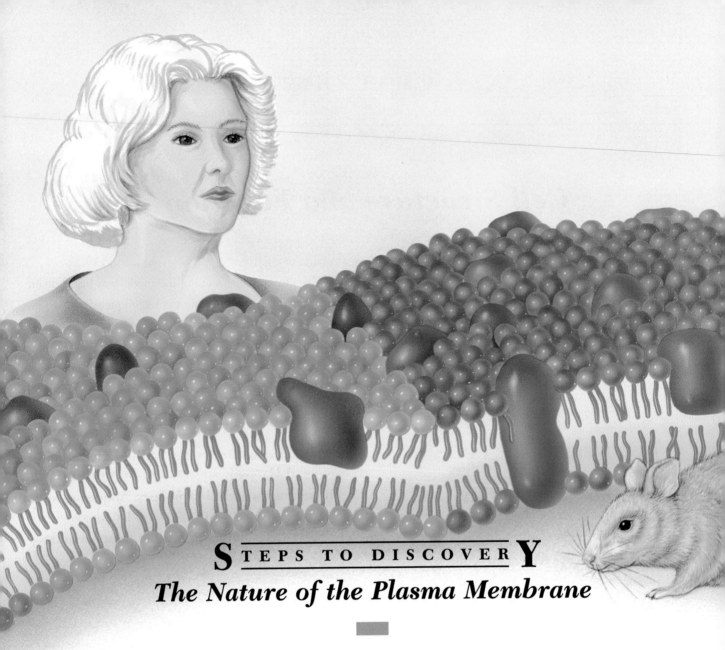

The Nature of the Plasma Membrane

*O*bserving the parts of a cell through a light microscope is like looking for cars and people through the window of an airplane at 35,000 feet. In both situations, the objects are simply too small to be seen. Before the advent of the electron microscope, information on cell structure depended largely on indirect techniques and subsequent interpretations. This was particularly true of the plasma membrane —the delicate structure at the outer edge of all cells; it is so thin that its presence is not revealed by the light microscope.

In the 1890s, a German physiologist, Ernst Overton, thought he could obtain information about the structure of a cell's outer boundary layer by analyzing the types of substances that passed from the outside environment through

the "invisible" barrier into the cell. Overton placed living plant cells into solutions containing various types of solutes. He found that the more nonpolar the solute, the better it was able to penetrate the cell boundary and enter the cell. Overton concluded that the dissolving power of a cell's outer layer matched that of a fatty oil. He hypothesized that the cell possessed a "lipid-containing membrane" that separated its living contents from the outer world.

In 1925, two Dutch scientists, E. Gorter and F. Grendel, designed an experiment to answer two questions: (1) Does the plasma membrane contain lipid? (2) If so, how much? Gorter and Grendel extracted the lipid present in human red blood cells and concluded that the amount of lipid in each cell was just enough to form a layer two mole-

By fusing the membranes of a mouse and human cell, scientists discovered the fluid state of the lipid bilayer which allows

cules thick—a **lipid *bi*layer.** This conclusion assumed that all of the lipid of the cell was part of the plasma membrane. This is not an unreasonable assumption to make when it comes to human red blood cells since they are essentially a "bag of dissolved hemoglobin" with virtually no internal structure, not even a nucleus.

The lipid bilayer was soon shown to be composed of phospholipids (page 75). Phospholipids have a "split personality"—a hydrophilic end (the phosphate and polar R group) that can form bonds with water, and a hydrophobic end (the two nonpolar fatty acid tails) that shuns the watery medium. As a result of their structure, phospholipids become aligned into a bilayered sheet, with the hydrophilic ends of the lipid molecules facing out toward the water, and the fatty acid tails facing inward toward each other and away from the water (see Figure 4-10). Gorter and Grendel concluded that cells were surrounded by a lipid bilayer that acts as a barrier to protect the cell's internal contents. A plasma membrane is more than just lipid, however; it also contains protein. In 1935, James Danielli of Princeton University and Hugh Davson of University College in London proposed a model for membrane structure that became the focal point of experimentation for 30 years. In the Davson–Danielli model, as it became known, the protein was present as a layer of globular molecules on both sides of the lipid bilayer, not unlike two slices of bread surrounding a double layer of cheese in a sandwich.

Nearly 40 years later, in 1972, S. Jonathan Singer and Garth Nicolson of the University of California proposed a new model of plasma membrane structure, which they named the *fluid-mosaic model* (see Figure 5-4). According to this model, the phospholipid bilayer of a membrane exists in a liquid (*fluid*) phase, having a viscosity similar to that of light machine oil. In the fluid-mosaic model, proteins are embedded in the membrane, penetrating into or passing completely through the lipid bilayer. What led Singer and Nicolson to view the membrane so differently from the earlier models? The answer relates back to the late 1960s, when a new procedure, termed *freeze-fracturing,* was developed. In this procedure, tissue that had been frozen solid was struck with a knife blade and fractured into two pieces, and the exposed surfaces then examined with an electron microscope. Many of the cracks produced by the knife edge ran directly through the interior of plasma membranes. Examination of the interior of these membranes revealed proteins that were deeply embedded in the lipid bilayer or even extended completely through the membrane (see Figure 5-4). The scattered protein particles seen in the interior of the plasma membrane gave rise to the "mosaic" component of the fluid-mosaic model.

The "fluid" component of the model was based on an earlier experiment employing the technique of *cell fusion.* In these experiments, cells taken from humans and mice were fused to form single cells containing both human and mouse cell nuclei surrounded by a common cytoplasm and a continuous human-mouse plasma membrane (depicted in the illustration that accompanies this essay). Locations of human and mouse membrane proteins were mapped at various times after fusion. At the instant of fusion, the two types of proteins were located in their respective separate portions of the membrane. Within 40 minutes, however, the membrane proteins from the two cells were completely intermixed. This experiment was the first to suggest that membrane proteins were not necessarily fixed in place but were capable of diffusing laterally within the membrane. For this to happen, the lipid bilayer of the membrane must be in a fluid state. Since its initial proposal, the fluid-mosaic model has been confirmed for virtually all cell membranes, regardless of their location in the cell or the type of organism in which they are found, once again emphasizing the unity of life.

The Singer–Nicolson model of membrane structure has ramifications far beyond the structure of the plasma membrane. Just as the discoveries of atomic structure led to an understanding of the way atoms interact (Chapter 3), the Singer–Nicolson model led to an appreciation of the dynamic quality of membranes, which, as we will see, participate in nearly all cellular activities.

proteins to move from one membrane to another.

*I*n these times of burgeoning computer technology, we have grown accustomed to hearing how smaller and smaller microchips are able to store and process greater and greater amounts of information. A 75-volume encyclopedia, for instance, can now be stored on a single chip and "read" by a computer in about 1 second. As remarkable as this may be, there is another microscopic package that can accomplish this feat and even more—the cell. In addition to storing an enormous amount of information—from the contours of a thumb print to the rate of hair growth—the cell is the center of life itself and is responsible for screening, sheltering, organizing, and coordinating a multitude of life-sustaining chemical reactions.

▼ ▼ ▼

DISCOVERING THE CELL

The invention of the microscope around 1600 sparked one of the greatest revolutions in the history of science.[1] By the mid-1600s, a handful of pioneering scientists had uncovered a new world that would have never been revealed had they relied on observations available to the naked eye. The discovery of cells is generally credited to Robert Hooke, an English microscopist who, at age 27, was awarded the position of curator of the Royal Society, England's foremost scientific academy. Three years later, in 1665, Hooke published *Micrographia,* a book in which he described his microscopic observations on such far-ranging subjects as fleas, feathers, and cork. Hooke wondered why stoppers made of cork were so well suited to holding air in a bottle. While viewing thin slices of cork that he cut with his pen knife, Hooke saw rows of tiny compartments resembling a honeycomb (Figure 5-1). He called these compartments "cells" because they reminded him of the cells that housed monks living in a monastery.

Meanwhile, Anton van Leeuwenhoek, a Dutchman who earned a living selling clothes and buttons, was spending his spare time grinding lenses and constructing *simple* (single-lensed) microscopes of remarkable quality. For 50 years, Leeuwenhoek sent letters to the Royal Society of London describing his microscopic observations—along with a rambling discourse on his daily habits and the state of his health. Leeuwenhoek was the first to examine a drop of pond water and, to his amazement, to observe the teeming microscopic "beasties" that darted back and forth before his

FIGURE 5-1

The discovery of the cell. Microscope used by Robert Hooke, with lamp and condensor for illumination of the object. (Inset) Hooke's drawing of a thin slice of cork, showing the honeycomb-like network of "cells."

eyes. He was also the first to describe various forms of bacteria which he obtained from water in which pepper had been soaked and from scrapings of his own teeth. His initial letters to the Royal Society describing this unseen world were met with such skepticism that the Society dispatched its curator, Robert Hooke, to confirm the observations. Hooke did just that, and Leeuwenhoek, an untrained amateur scientist, was soon a worldwide celebrity, receiving visits in Holland from Peter the Great of Russia and the Queen of England.

As microscopes gradually improved, biologists began to see a consistent pattern: Cells were present in all types of plants and animals. In 1838, Matthias Schleiden, a German lawyer-turned-botanist, proposed that all plants were constructed of cells. The following year, Theodor Schwann, a German zoologist, extended this generalization to include animals. Summarizing the new observations gained from studying the microscopic structure of life, Schwann proposed the first two tenets of the **cell theory.**

---

[1] The properties and uses of microscopes are discussed in Appendix B at the back of the book.

1. *All organisms are composed of one or more cells.*
2. *The cell is the basic organizational unit of life.*

   More than 10 years later, after studying cell reproduction, Rudolf Virchow, a German physician, proposed the third tenet of the cell theory

3. *All cells arise from preexisting cells.*

The cell theory is one of the greatest unifying concepts in biology; no matter how diverse organisms appear, all are made up of one or more cells. The cell is the most fundamental structure to harbor and sustain life as well as to give rise to new life. The evolution of the first primitive cells was one of the most crucial—and poorly understood—steps in the entire course of biological evolution on this planet. For all of these reasons, understanding the cell is one of the cornerstones of biology.

Technological invention often triggers new scientific advancements. The invention of the light microscope empowered biologists with the ability to see a multitude of previously unseen life forms and enabled them to investigate the fundamental unit of life—the cell. (See CTQ #2.)

example, have been kept in culture (growing in laboratory dishes) for decades. These cultured cells have all the necessary regulatory mechanisms to maintain a homeostatic condition. They take up nutrients, digest them, and excrete waste products; they take up oxygen and release carbon dioxide; they maintain a particular water and salt content; they are capable of growth, reproduction, and movement; they respond to external stimulation; they expend energy to carry out their activities; they inherit a genetic program from their parents and pass it on to their offspring; and, finally, they die. These are the characteristics of life, and they are all exhibited by individual cells. Cells come in a variety of shapes and sizes, ranging from enormously extended nerve cells that connect your spinal cord to your big toe (while very long, these cells have a very small diameter), to minute bacteria so small that more than a thousand would be needed to fill the dot in the letter *i*.

Cells are the fundamental units of life, exhibiting all of the basic properties of life. (See CTQ #3.)

BASIC CHARACTERISTICS OF CELLS

Cells can be removed from a plant or animal and kept alive and healthy outside the organism. Certain human cells, for

TWO FUNDAMENTALLY DIFFERENT CLASSES OF CELLS

All cells are either prokaryotic or eukaryotic (Figure 5-2), distinguishable by the types of internal structures, or **organelles,** they contain. The existence of two distinct classes of cells, without any known intermediates, repre-

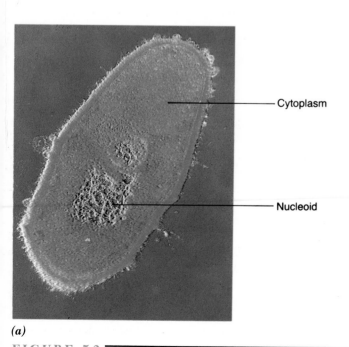

(a)

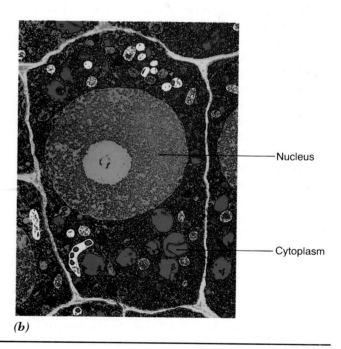

(b)

FIGURE 5-2

A comparison of prokaryotic *(a)* and eukaryotic *(b)* cells, the two fundamental cell types. *(a)* A bacterial cell showing the cytoplasm devoid of visible organelles and the DNA housed within an ill-defined nucleoid. *(b)* A plant root tip cell showing the varied cytoplasmic organelles and the membrane-bound nucleus, that houses the cell's DNA.

sents one of the most fundamental evolutionary discontinuities that exists in the biological world. Both types of cells—prokaryotic and eukaryotic—are bounded by a similar plasma membrane, and both may be surrounded by a nonliving cell wall. In other respects, the two classes of cells are very different. **Prokaryotic cells** (*pro* = before, *karyotic* = nucleus) are the simpler type; they contain much less genetic information and house it in a *nucleoid*, a rather poorly demarcated region of the cell that lacks a membrane to separate it from the surrounding **cytoplasm.** Prokaryotic cells are essentially devoid of organelles (other than ribosomes) and are capable of much less complex activities than are eukaryotic cells. Prokaryotic cells are found in prokaryotic organisms, all of which are bacteria (members of the kingdom Monera; see Chapter 1). Prokaryotic cells are thought to represent the vestiges of an early stage in the evolution of life which was present before the more complex eukaryotes appeared.

Eukaryotic cells (*eu* = true) are larger and much more structurally and functionally complex than are prokaryotic cells. As the name suggests, the genetic information is housed in a true **nucleus**—one surrounded by a complex membranous envelope. The cytoplasm outside the nucleus is filled with a variety of membranous and nonmembranous organelles that are specialized for various activities. Eukaryotic cells make up eukaryotic organisms, including fungi, protists, plants, and animals.

Nearly all cells—prokaryotic and eukaryotic—are microscopic; that is, they are too small to be seen without a microscope. Prokaryotic cells seldom reach diameters greater than just a few micrometers (μm), while most eukaryotic cells range in diameter from only 10 μm to 100 μm (Figure 5-3). Each micrometer is one-millionth of a meter. In general, any organism larger than 100 μm in diameter is composed of more than one cell. There are a number of reasons for the small size of cells. The larger a cell's volume,

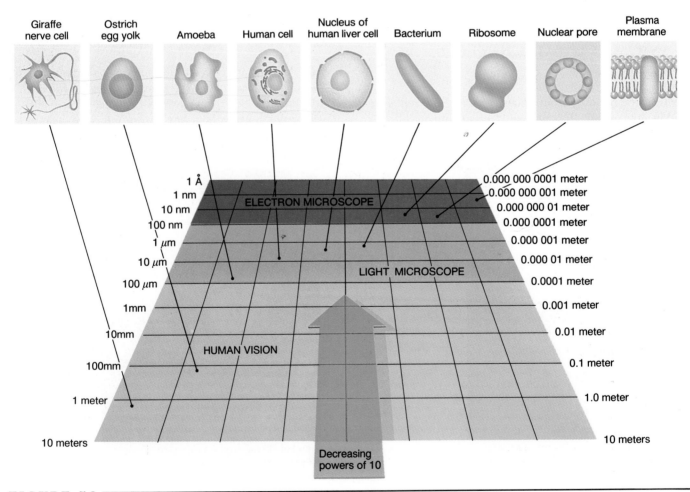

FIGURE 5-3

Relative sizes of cells and cell components. Each unit of measurement is one-tenth as large as the preceding unit. While the huge ostrich egg is technically a single cell, the living portion is present only as a microscopically thin disk located on one edge of the large inert yolk mass.

the more difficult it becomes for that cell to receive the oxygen and nutrients necessary to sustain its metabolic needs. This problem arises because larger cells have *relatively* less plasma membrane surface across which nutrients and oxygen can enter the cell. This problem of surface and volume is dealt with in detail in Chapter 22. Another limitation on cell size stems from the fact that a single nucleus can only support the activities of a limited volume of active cytoplasm. Cells that grow to large sizes, such as those of many protists (see Figure 1-6*b*), typically have more than one nucleus.

Two distinct classes of cells have evolved: a structurally simpler prokaryotic type characteristic of all bacteria, and a more complex eukaryotic type that constitutes all other living organisms. The simpler plan limits function, restricting life to the unicellular level. (See CTQ #4.)

THE PLASMA MEMBRANE: SPECIALIZED FOR INTERACTION WITH THE ENVIRONMENT

All cells are surrounded by an exceedingly thin **plasma membrane** of remarkably similar construction in all organisms—from bacteria to mammals. Among its numerous functions, the plasma membrane (1) forms a protective outer barrier for the living cell and (2) helps maintain the internal environment of the cell by regulating the exchange of substances between the cell and the outside world (Chapter 7). As we discussed earlier, the plasma membrane consists of a fluid, phospholipid bilayer that is pierced by proteins (Figure 5-4). In addition, the plasma membranes of animal cells contain cholesterol, a steroid that generally increases the fluid nature of the lipid bilayer.

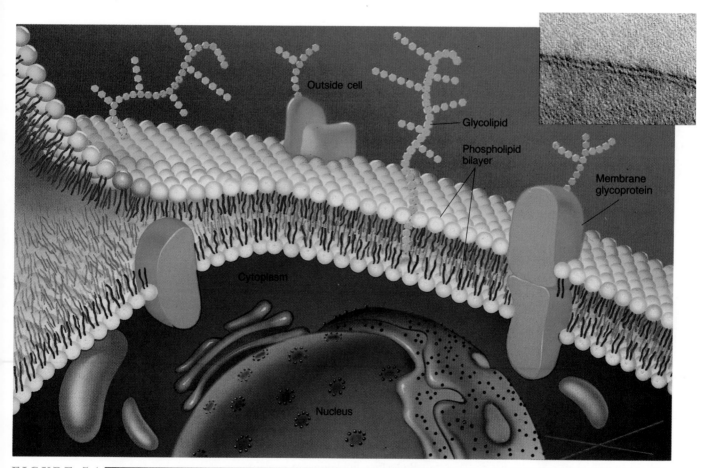

FIGURE 5-4

The structure of the plasma membrane. All cells have a plasma membrane of similar structure. Proteins penetrate into or through the lipid bilayer, with their hydrophobic surfaces forced into proximity with the fatty acid tails of the phospholipids. The hydrophilic portions of the membrane proteins project out into the external environment and inward into the cytoplasm. Chains of sugars are covalently bonded to the outer surface of most, if not all, of the proteins, and to a small percentage of the phospholipids. Membranes appear in the electron micrograph (inset) as three-layered structures having two dense outer layers and a less dense middle layer.

Membrane proteins provide a good example of the relationship between amino acid structure and protein function. The outer surface of membrane proteins consists predominantly of nonpolar amino acids (see Figure 4-14), which form hydrophobic interactions (page 59) with the nonpolar tails of the membrane's phospholipids (Figure 5-4). This facilitates the insertion of these proteins into the hydrophobic interior of the membrane, much like pegs can be inserted into the holes of a pegboard. Those portions of the membrane proteins that protrude beyond the lipid bilayer generally contain attached carbohydrates, making them *glycoproteins*. The carbohydrate groups are present as short, branched hydrophilic chains (Figure 5-4). In contrast to the carbohydrates discussed in Chapter 4 (glycogen, starch, etc.), which are polymers of a single sugar, the carbohydrates of the membrane are composed of a number of different sugars. The sequences of sugars and the branching patterns of the carbohydrate chains vary from one glycoprotein to another. Because of their variable structure, the carbohydrate chains of membrane glycoproteins can engage in *specific* interactions with other molecules. The basis of a person's blood type (A, B, AB, or O), for example, is determined by the particular sugars in the carbohydrate chains present on the glycoproteins at the surface of the person's blood cells.

The plasma membrane is able to recognize and interact with certain substances in the cell's environment. These specific interactions are mediated by those parts of the glycoproteins that protrude from the outer surface of the membrane. Membrane proteins having this role are termed **receptors.** In eukaryotes, each type of specialized cell has its own set of receptors, which allows the cell to respond to particular ions, hormones, antibodies, and other circulating molecules.

All cells are bounded by a plasma membrane which separates the living and nonliving worlds. In this capacity, the plasma membrane regulates the exchange of materials between the cell and its environment. (See CTQ #5.)

TABLE 5-1
THE PRIMARY FUNCTIONS OF CELL COMPONENTS OF EUKARYOTIC CELLS

Component	Primary Functions
Plasma membrane	Boundary of cell, exchange with environment
Nucleus	Storage of hereditary information, control of cell activities
Nuclear envelope	Exchange between nucleus and cytoplasm
Nucleolus	Ribosome synthesis
Nuclear matrix	Support, synthetic machinery
Endoplasmic reticulum (ER)	Synthesis of protein, steroids, lipids; Storage of Ca^{2+}; detoxification
Ribosomes	Sites of protein synthesis
Mitochondria	Chemical energy conversions for cell metabolism
Chloroplasts (plant cells only)	Conversion of light energy into chemical energy, storage of food and pigments
Golgi complex	Synthesis, packaging, and distribution of materials
Lysosomes	Digestion, waste removal, discharge
Central vacuoles (plant cells only)	Storage, excretion
Microfilaments, microtubules, intermediate filaments	Cellular structure, movement of internal cell parts, cell movement
Cilia and flagella	Locomotion, production of currents
Vesicles	Storage and transport of materials
Cell wall (plant cells only)	Protection, fluid pressure, support
Intercellular junctions	Cell-to-cell adhesion, occlusion, communication

THE EUKARYOTIC CELL: ORGANELLE STRUCTURE AND FUNCTION

The unity and diversity of life is revealed at the cellular level, just as it was at the atomic and biochemical levels (Chapters 3 and 4). You have several hundred different types of cells in your body, each recognizably different from the others. Yet virtually all of these different cells are composed of the same types of organelles. The primary functions of all the components found in and around eukaryotic cells are summarized in Table 5-1. Figure 5-5 shows a "generalized" plant and animal cell, each containing a combination of organelles typically found in eukaryotic cells. Keep in mind that there is really no such thing as a "generalized" or "typical" cell. The appearance and distribution of cellular organelles vary greatly from a nerve cell, to a bone cell, to a gland cell. The analogy might be made to a variety of orchestral pieces: All are composed of the same notes, but varying arrangement gives each its unique character and beauty.

Organelles are specialized compartments inside the cell, in which specific activities take place without interference from other events. Viewed in this way, organelles maintain order within a cell, preventing biochemical chaos. A eukaryotic cell could no more operate without its organelles than a restaurant could operate without a kitchen, dining room, restrooms, and garbage bin. The dimensions of some cellular organelles are shown in Figure 5-3.

THE NUCLEUS: GENETIC CONTROL CENTER OF THE CELL

Typically, the most prominent structure in a eukaryotic cell is the nucleus (Figure 5-6). The nucleus is the residence of DNA, the cell's genetic material. Genetic instructions leave the nucleus via a form of RNA and enter the cytoplasm, where they direct the synthesis of specific proteins. Because of its role as genetic headquarters, the nucleus is often thought of as the "control center" of the cell. The nucleus contains chromosomes, nucleoli, and the nucleoplasm, and is surrounded by the nuclear envelope.

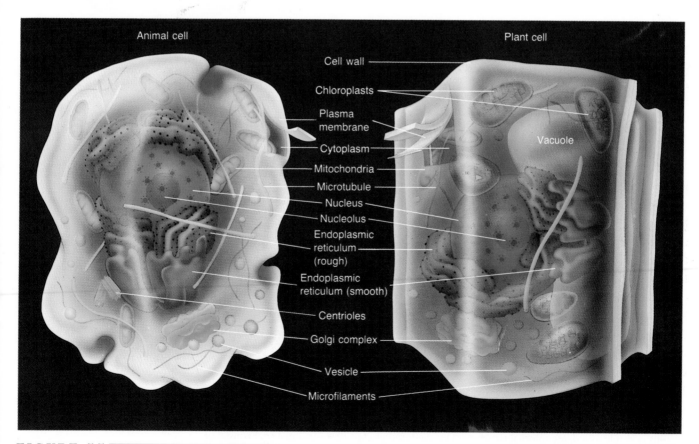

FIGURE 5-5

Generalized structure of eukaryotic cells. Both plant and animal cells have a plasma membrane, nucleus, and cytoplasmic organelles. They differ in some types of organelles and by the presence of cell walls just outside of plant cells, which animal cells do not have.

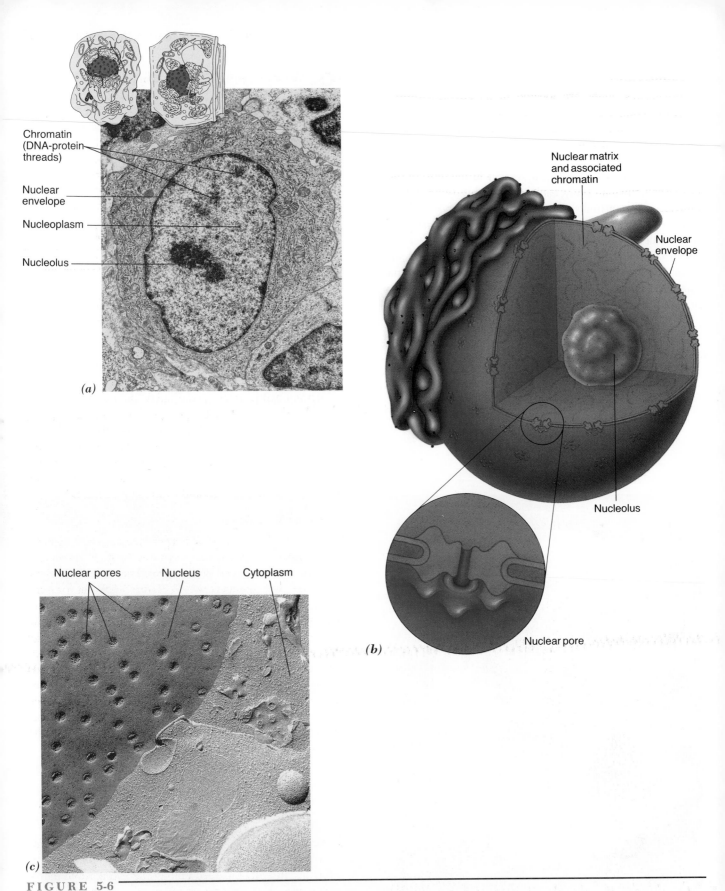

Chromatin
(DNA-protein
threads)

Nuclear
envelope

Nucleoplasm

Nucleolus

(a)

Nuclear matrix
and associated
chromatin

Nuclear
envelope

Nucleolus

Nuclear pore

(b)

Nuclear pores   Nucleus   Cytoplasm

(c)

FIGURE 5-6

The nucleus. *(a)* An electron micrograph of a section through the nucleus of a eukaryotic cell. The nucleolus is visible as a dense internal structure. The scattered clumps of stained material consist of DNA-protein fibers (chromatin) that make up the chromosomes. *(b)* Cut-away diagram showing the nuclear envelope that bounds the nuclear compartment and the nucleolus. *(c)* Using the technique of freeze-fracture (discussed in the chapter opener), the interior of cell membranes can be examined. In this freeze-fractured preparation, the pores in the nuclear envelope are readily visible.

Chromosomes

An organism's genes are located in its **chromosomes,** structures composed of DNA molecules and bound proteins. The number of chromosomes in the nucleus varies according to species; for example, a parasitic roundworm (*Ascaris*) has only four chromosomes, whereas the common sunflower has 34, and humans have 46. During most of the life of a cell, the material that makes up the chromosomes is unraveled to form highly elongated threads, called **chromatin.** It is only when the cell gets ready to divide that the threads become packaged into microscopically visible chromosomes (Figure 5-7).

Nucleoli

The nucleus often contains one or more dense-looking structures called **nucleoli** (Figure 5-6*a*). If you look closely at an electron micrograph of a single nucleolus, you will see tiny granules. These granules are the precursors of **ribosomes,** particles consisting of RNA and protein that serve as sites of protein synthesis in the cytoplasm. The two subunits that make up each ribosome are formed in the nucleoli and then squeezed through the nuclear pores into the cytoplasm, where they combine to form functioning ribosomes.

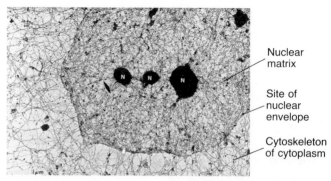

FIGURE 5-8

The nuclear matrix. The interconnecting fibers of the nuclear matrix help maintain the shape of the nucleus and provide sites for the attachment of enzymes involved in nuclear activities. Several nucleoli (N) are also shown.

Nucleoplasm

The **nucleoplasm** is the semifluid substance in the nucleus that contains proteins, granules, and an interconnecting network of fibers, termed the **nuclear matrix** (Figure 5-8). In addition to maintaining the shape of the nucleus, the nuclear matrix is thought to organize the contents of the nucleus and to provide attachment sites for enzymes involved in the duplication and expression of the DNA.

Nuclear Envelope

The **nuclear envelope** is a complex structure containing a double membrane that separates the nuclear contents from the rest of the cell (Figure 5-6). The nuclear envelope is studded with complex pores (Figure 5-6*c*) that form the passageway for materials entering or exiting the nucleus. For instance, proteins needed for making new DNA are synthesized in the cytoplasm and then transported through the nuclear pores to DNA assembly sites in the nucleus. Conversely, the genetic messages needed for synthesizing those proteins are formed in the nucleus and then transported through nuclear pores to the cytoplasm. Thus, the nuclear envelope is an important gateway for regulating traffic and maintaining essential differences between the two major regions of the cell.

MEMBRANOUS ORGANELLES OF THE CYTOPLASM: A DYNAMIC INTERACTING NETWORK

The cytoplasm of most eukaryotic cells is filled with membranous structures that extend to every nook and cranny of the cell's interior. Included among the cytoplasmic membranes are small spherical containers—or **vesicles**—of varying diameter, long interconnected channels, and flattened membranous sacs. These membranes illustrate the unity and diversity of cellular structure. While the membranes of the cytoplasm have the same basic structure as the plasma membrane, the particular proteins embedded in the

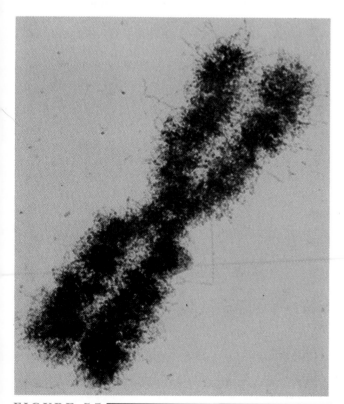

FIGURE 5-7

A chromosome. During cell division in a eukaryotic cell, each chromosome becomes greatly compacted to form the type of structure seen in this electron micrograph. The chromatin fibers that make up the chromosome are visible at its edges.

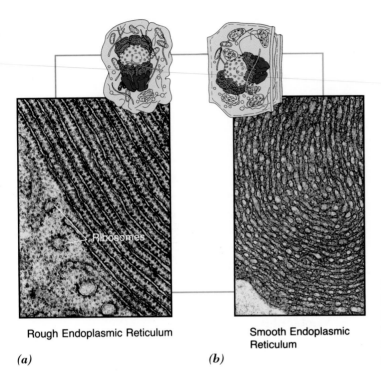

Rough Endoplasmic Reticulum

(a)

Smooth Endoplasmic Reticulum

(b)

FIGURE 5-9

Endoplasmic reticulum. *(a)* Portion of a pancreatic cell showing the rough endoplasmic reticulum (RER) where digestive enzymes are assembled. *(b)* Portion of a cell from the testis showing the smooth endoplasmic reticulum (SER) where steroid hormones are synthesized.

lipid bilayer vary from one part of the cell to another. These membrane proteins give each organelle many of its specialized functions.

Although the membranes of the cytoplasm may appear disconnected in an electron micrograph, in reality they form a highly interdependent membranous network composed of the endoplasmic reticulum, the Golgi complex, lysosomes, and vesicles.

Endoplasmic Reticulum

Throughout the cytoplasm is an elaborate system of folded, stacked, and tubular membranes known as the **endoplasmic reticulum,** or simply the **ER.** The ER occurs in two different forms: rough ER and smooth ER. **Rough ER (RER)** appears bumpy (rough) in electron micrographs because of its many attached ribosomes (Figure 5-9*a*). ER lacking ribosomes is called **smooth ER (SER)** (Figure 5-9*b*). SER is generally more tubular, whereas RER is usually composed of flattened sacs of membranes.

The membranes of the ER divide the cytoplasm into two compartments: one within the confines of the ER membranes, and one outside the ER membranes. As a result, materials confined within the ER space are segregated from the remainder of the cytoplasm. Moreover, the continuous membranes of the endoplasmic reticulum create a system of passageways that allows materials to be channeled to different locations within the cell.

Functions of the RER. The ribosomes located on the outer surface of the RER membranes are sites where pro-

teins are assembled, one amino acid at a time. The newly synthesized protein passes through the ER membrane and is then segregated within the ER compartment. These proteins have specific destinations: Some are exported out of the cell, some end up as membrane proteins, and some become part of other cytoplasmic organelles. Interest is currently focused on how these proteins are sorted and targeted to their specific destinations. The RER is most highly developed in cells that export *(secrete)* large quantities of protein, including the cells of the pancreas (which produce digestive enzymes) and the salivary glands (which produce salivary proteins). The RER is also the primary site of the synthesis of membrane phospholipids.

Functions of the SER. Depending on the particular type of cell, the SER may participate in various functions. The SER may be responsible for the synthesis of steroids; sex hormones of the gonad, for example, are SER products. The SER may also destroy toxic materials in the liver. For example, ingestion of a barbiturate, such as phenobarbitol, is followed by a rapid increase in the SER of a person's liver cells, leading to the destruction of these potentially toxic drug molecules. The SER also regulates calcium ion concentration. In many cells, the controlled release of calcium ions from the SER acts to trigger physiologic responses, including muscle contraction (Chapter 26).

Golgi Complex

In 1898, an Italian biologist, Camillo Golgi, was working with a new type of metallic stain when he discovered a dark

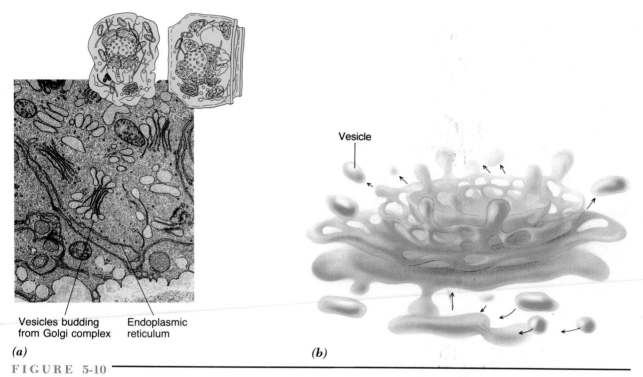

Vesicle

Vesicles budding from Golgi complex

Endoplasmic reticulum

(a)

(b)

FIGURE 5-10

The Golgi complex. *(a)* Electron micrograph of the Golgi complex of a plant cell. Vesicles are seen budding from the lateral edges of the Golgi sacs. *(b)* Diagrammatic view of a Golgi complex. As with most organelles, the number of Golgi complexes varies from cell to cell. For example, dividing plant cells contain many Golgi complexes clustered near the area where a new cell wall is being manufactured. A single cell of the human pancreas, churning out enzymes needed to digest three meals a day, may contain thousands of Golgi complexes.

yellow network near the nuclei of stained nerve cells. This network was later named the **Golgi complex** and helped earn its discoverer the 1906 Nobel Prize. The Golgi complex remained a center of controversy between those who believed that the structure existed in living cells and those who believed it was an *artifact*—an artificial structure formed during preparation for microscopy. The electron microscope finally verified the existence of the Golgi complex as a true organelle—one composed of flattened, membranous sacs and associated vesicles (Figure 5-10). The sacs are arranged in an orderly pile, resembling a stack of "hollow" pancakes. The Golgi complex is a way station in the movement of materials from the ER through the cell (Figure 5-11). The Golgi membranes themselves are thought to form by the fusion of small vesicles that arrive from the RER. The contents of these vesicles, including the proteins destined for export, are chemically modified in the Golgi complex. For example, the insulin that controls the level of sugar in your blood is originally synthesized in the RER as a much larger protein than that ultimately secreted; the larger, precursor protein is cut down to length in the Golgi complex. Similarly, the sugars that determine your blood type are added as the glycoproteins pass through the Golgi complex. The Golgi complex is also the site of

synthesis of many of the complex polysaccharides that make up the plant cell wall, which we will discuss later in the chapter.

Materials processed in the Golgi complex are packaged into vesicles that bud from the lateral edges of the Golgi sacs (Figure 5-10 and 5-11). Some vesicles remain in the cell, storing important chemicals until they are needed; others move to the cell surface and discharge their contents to the outside by a process called **exocytosis.** Exocytosis occurs when the membrane of a vesicle fuses at some point with the overlying plasma membrane, thus opening the vesicle and allowing its contents to be released into the medium.

The Secretory Pathway. We have seen in this discussion that the Golgi complex forms a crucial link in a *secretory pathway* that carries materials from their site of synthesis in the ER to the cell's exterior (Figure 5-11). In addition to carrying enclosed materials, the secretory pathway provides a mechanism for building plasma membrane since the membrane of a cytoplasmic vesicle becomes *incorporated* into the plasma membrane during exocytosis (Figure 5-11). This is, in fact, how the plasma membrane is formed—from vesicles whose membranes were originally produced in the ER.

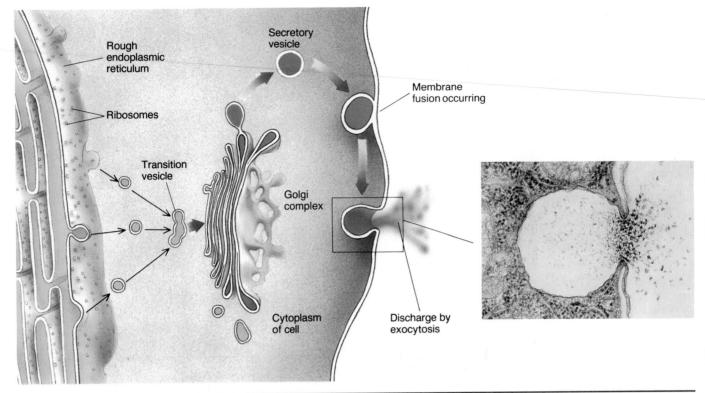

FIGURE 5-11

The secretory pathway. Proteins destined for secretion (such as the digestive enzymes of the pancreas or the proteins of salivary mucus) are synthesized on ribosomes situated on the outer surface of the ER membranes. The proteins are then passed into the ER compartment, packaged in a transition vesicle, and transferred to the Golgi complex, where the proteins are concentrated and often modified before being sent on their way in cytoplasmic vesicles. The vesicles may be stored in the cytoplasm before fusing with the plasma membrane (exocytosis), allowing the vesicle contents to be discharged into the external medium (shown in inset).

The studies that identified the various steps along the secretory pathway were carried out primarily in the 1960s in the laboratory of George Palade of Yale University, one of the corecipients of the 1974 Nobel Prize in Medicine. Palade and his colleagues were working with pancreatic cells that manufacture and export large quantities of digestive enzymes destined for service in the small intestine. They prepared thin slices of pancreatic tissue, placed the slices into a medium containing radioactive amino acids, and then waited for varying amounts of time before the cells were killed and examined. Palade and his co-workers found that the pancreatic cells took up the labeled amino acids, which were rapidly incorporated by the ribosomes of the RER into newly synthesized digestive enzymes. As a result, radioactivity rapidly appeared within the RER (Figure 5-12). If they waited for about 20 minutes before killing the cells, however, much of the radioactivity had moved away from the RER and into the Golgi complex. After an additional 20 minutes or so, radioactivity appeared in material discharged from the cell. Based on these experiments, the researchers concluded that secretory proteins were synthesized in the RER and passed through the Golgi complex on their way out of the cell.

Lysosomes

Some of the vesicles that bud from the Golgi complex remain in the cytoplasm as **lysosomes**—storage vesicles that contain dozens of dangerously powerful hydrolytic enzymes capable of digesting virtually every type of macromolecule in the cell. The membrane surrounding the lysosomes keeps these lethal enzymes safely sequestered (see The Human Perspective: Lysosome-Related Diseases).

Why do cells house such potentially injurious materials? Lysosomes play a key role in maintaining constancy and order within the cell; they function as digestion chambers for destroying dangerous bacteria and foreign debris, as well as sites for disposing of impaired or worn-out organelles. In some organisms, including protists and sponges, lysosomes digest trapped food particles. In a few cases, an organism's own cells are deliberately digested by

◁ THE HUMAN PERSPECTIVE ▷
Lysosome-Related Diseases

Almost from the time of the discovery of lysosomes it was proposed that their malfunction might be a major cause of various diseases. For example, a miners' disease known as *silicosis* results from the uptake of silica fibers by wandering scavenger cells in the lungs. The fibers become enclosed within lysosomes but cannot be digested; instead, they cause the lysosomal membrane to leak, spilling the contents of digestive enzymes into the cell and damaging the tissue of the lungs. A similar result occurs when asbestos fibers are taken up by scavenging cells, resulting in the disease *asbestosis*. Both conditions can be debilitating and may even be fatal. Certain types of inflammatory diseases, such as rheumatoid arthritis, are believed to result, in part, from the release of lysosomal enzymes from the cell into the extracellular space, causing damage to materials in the joints.

One of the functions of anti-inflammatory drugs, such as steroids, is to stabilize the membrane of lysosomes, thus preventing their rupture.

In 1992, several papers were published that implicated lysosomes in the development of Alzheimer's disease. This type of senility is characterized by the formation of plaques, located outside of nerve cells, which contain a protein called *beta-amyloid,* a small fragment of a larger protein. Researchers have now found that the conversion of the larger protein to the small, apparently harmful, fragment occurs in the lysosomes of the nerve cells. This finding suggests that Alzheimer's disease may occur as the result of some defect in the lysosomal processing of the large precursor protein. While this matter is still highly speculative, it opens up the possibility that treatment of the disease might be achieved by inhibiting the activity of certain lysosomal enzymes.

Just as there are diseases resulting from excessive lysosomal activity, there are also serious consequences associated with a lack of lysosomal enzymes. These rare genetic disorders are called *lysosomal storage diseases* (of which about 30 are known) and are characterized by the buildup within the tissues of the particular macromolecule that the missing enzyme would normally hydrolyze. Tay-Sachs disease, the most familiar lysosomal storage disease, is characterized by severe mental retardation and death by about age 5. Damage to the brain results from an accumulation of certain plasma membrane lipids in the child's nerve cells. The lethal nature of lysosomal storage diseases illustrates the importance of these hydrolytic enzymes in cell metabolism.

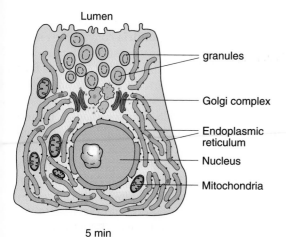
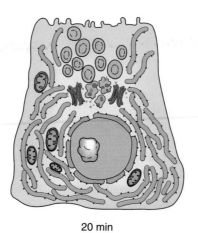
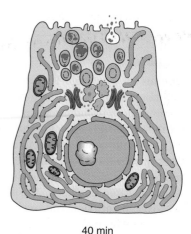

Lumen

granules

Golgi complex

Endoplasmic reticulum

Nucleus

Mitochondria

5 min

20 min

40 min

FIGURE 5-12

Experimental demonstration of the secretory pathway. When a secretory cell is given radioactively labeled amino acids, the amino acids are immediately incorporated into proteins in the rough ER. After about 20 minutes, the radioactively labeled proteins have moved onto the Golgi complex, and within about 40 minutes the label is seen discharged into the external medium. Radioactivity is indicated by the red spots.

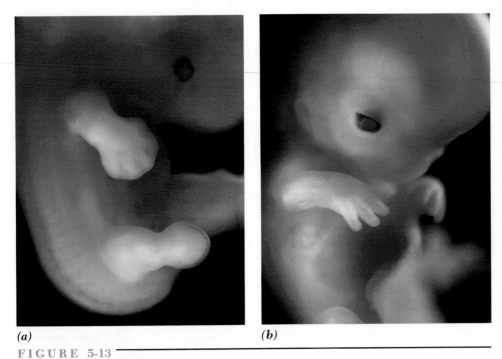

(a)   *(b)*

FIGURE 5-13

Carving human fingers. *(a)* At an early embryonic stage, the human hand is paddle-shaped. *(b)* The fingers are "carved" out of the paddle by the death of the intervening cells. Cell death results from the release of lysosomal enzymes.

lysosomal enzymes. Your hand, for example, began as a paddle-shaped structure. When you were an early embryo, your fingers were carved out of the "paddle" due to the death of the intervening cells by digestive enzymes released from lysosomes (Figure 5-13). Lysosomes also play a key role in digesting "unwanted" tissues in a tadpole as it undergoes metamorphosis into a frog, or in a caterpillar as it transforms into a butterfly.

Vacuoles

Vacuoles are essentially large vesicles; that is, large, membrane-bound, fluid-containing sacs. Vacuoles are particularly prominent in mature plant cells, where they may occupy more than 90 percent of the cell's volume (see Figure 5-5). The plant vacuole is surrounded by a single membrane that governs which materials are exchanged between the cytoplasm and the fluid inside the vacuole. In addition to containing water, the fluid in the plant vacuole may contain gases (oxygen, nitrogen, and/or carbon dioxide), acids, salts, sugars, crystals, pigments that account for some of the colors of flowers and leaves, and even toxic wastes. Unlike animals, who have elaborate excretory systems for expelling toxic wastes, plants apparently "excrete" such substances into their own vacuoles, safely partitioning the toxins from the plant's cytoplasm. Plant vacuoles also maintain high internal water pressure, which aids in the support of the plant (discussed in Chapter 7).

MITOCHONDRIA AND CHLOROPLASTS: ACQUIRING AND USING ENERGY IN THE CELL

Mitochondria can be likened to miniature "power plants" located within the cytoplasm of eukaryotic cells. Whereas power plants convert the energy stored in energy-rich fuels (such as oil and coal) into electricity—a form usable by the consumer, **mitochondria** convert the energy stored in energy-rich macromolecules (such as fats and polysaccharides) into adenosine triphosphate (ATP)—a form usable by the cell in running virtually all of its immediate activities (Chapter 6).

Mitochondria are typically sausage-shaped (Figure 5-14*a*), although their contours change as they flow through the cytoplasm. Regardless of shape, each mitochondrion is constructed of two membranes—an *outer membrane* surrounding an elaborately folded *inner membrane* (Figure 5-14*b,c*). Just as the plasma membrane is specialized for interacting with external substances, and the RER membrane is specialized for synthesizing and packaging proteins, the inner mitochondrial membrane is specialized for energy transfer. The inner mitochondrial membane illustrates another important feature of membranes: their ability to organize molecules into a stable, ordered array. The inner mitochondrial membrane contains dozens of different components arranged in the precise spatial order required for the formation of a cell's ATP (Chapter 9). The

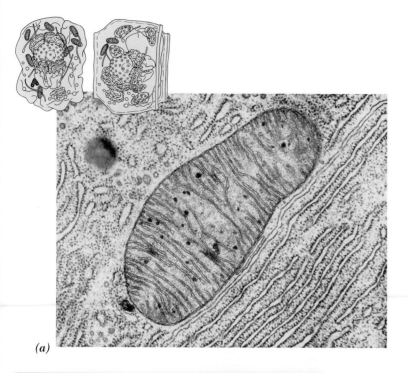

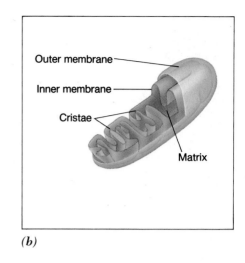

(b)

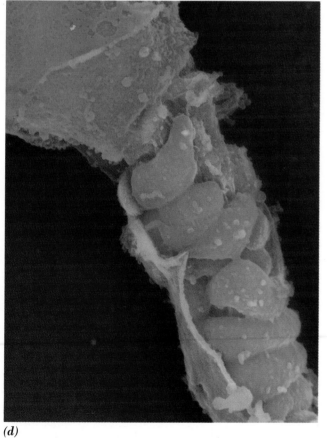

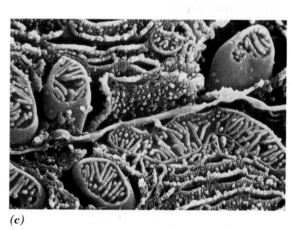

(c)

(a)

(d)

FIGURE 5-14
Mitochondria. Electron micrographs **(a,c,d)** and drawing **(b)** of mitochondria showing the outer and inner membranes. The labyrinth of convolutions created by the inner membrane forms the cristae, which project into an internal, semifluid compartment, the matrix. The matrix also contains DNA and ribosomes. The electron micrograph in part **(d)** shows the way in which mitochondria (orange structures) are packaged in the middle portion of a mammalian sperm.

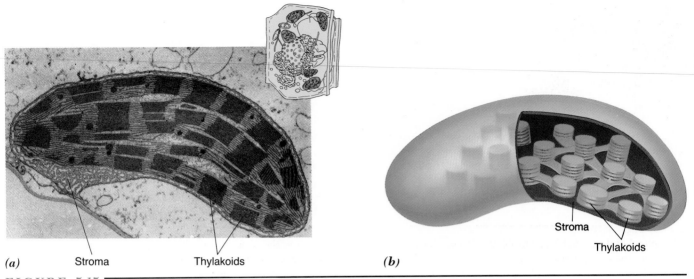

(a) Stroma   Thylakoids   *(b)*   Stroma   Thylakoids

FIGURE 5-15

Chloroplasts. *(a)* An electron micrograph, and *(b)* a three-dimensional diagram of a chloroplast. A chloroplast contains an outer and inner membrane as well as stacks of membranous thylakoids that are contained in the central stroma. The stroma also contains DNA and ribosomes.

labyrinth of convolutions created by the inner membrane forms the **cristae,** which project into the **matrix,** an internal, semifluid compartment (Figure 5-14*b,c*)

As with other organelles, the number and location of mitochondria in a cell depend on the cell's activities. Mitochondria are particularly numerous in muscle cells, which reflects the high energy requirements of muscle contraction. Often, mitochondria associate with fat-containing oil droplets, from which they derive the raw materials for energy production. A particularly striking arrangement of mitochondria is seen in many sperm cells, where the mitochondria are distributed in a spiral in the middle of the cell (Figure 5-14*d*). The movements of the sperm are powered by the ATP produced in these mitochondria. Since the mitochondria in a human sperm remain outside the fertilized egg, all the mitochondria of the offspring are derived from the mother. This is not an inconsequential matter since mitochondria contain a small amount of genetic material, which is thus inherited maternally.

Chloroplasts (Figure 5-15) are found only in plants and certain photosynthetic protists and are sites of photosynthesis, a complex process during which light energy is captured and used to construct complex biochemicals (Chapter 8). Virtually all life on earth depends on the energy captured by photosynthesis. Each chloroplast is bounded by a double-membrane envelope and contains an elaborate internal system of flattened membranous discs called **thylakoids,** in which light-capturing pigments, such as chlorophyll, are contained. The stacks of thylakoids are located in the central **stroma.** It is the green-colored chlorophyll of the chloroplasts that give plants their color.

Both mitochondria and chloroplasts contain small circular DNA molecules and ribosomes, which function as sites for the synthesis of certain proteins of these organelles. The presence of this system for storing and translating genetic information has sparked an interesting hypothesis concerning the origin of these organelles, which we will discuss later in the chapter.

THE CYTOSKELETON: PROVIDING SUPPORT AND MOTILITY

The human skeleton consists of hardened parts of the body which support the soft tissues and play a key role in mediating body movements. The cell also has a "skeletal system" —a **cytoskeleton** (*cyto* = cell), with some analogous functions. The cytoskeleton serves as an internal framework that supports the shape of the cell, organizes its contents, and provides the machinery for moving cells and their organelles. The cytoskeleton includes three components—microtubules, microfilaments, and intermediate filaments—which are typically interconnected to form an elaborate interactive network of fibers (Figures 5-8 and 5-16*a*). The elements of the cytoskeleton are a dynamic group of structures capable of rapid and dramatic reorganization. Unfortunately, this property cannot be appreciated by viewing images in a photograph which are fixed in time.

Microtubules are hollow, cylindrical structures whose wall is composed of subunits made of the protein *tubulin* (Figure 5-16*b*). **Microfilaments** (Figure 5-16*c*) are solid, thinner and more flexible than are microtubules and are polymers of the contractile protein *actin*. **Intermediate**

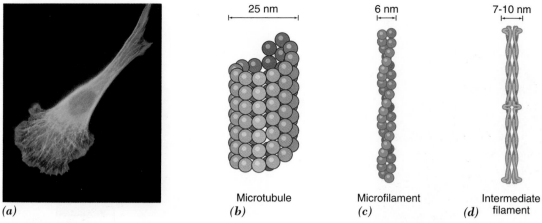

25 nm

6 nm

7-10 nm

Microtubule
(b)

Microfilament
(c)

Intermediate
filament
(d)

(a)

FIGURE 5-16

The cytoskeleton. *(a)* This fibroblast cell was caught in the act of moving over the surface of a culture dish. The cell has been stained to reveal the distribution of microfilaments and microtubules. The rounded edge of the cell is leading the way; the clusters of microfilaments at the leading surface are sites where the forces required for movement are generated. *(b-d)* Microtubules, microfilaments, and intermediate filaments are composed of different types of protein subunits, that become arranged in characteristic patterns.

filaments (Figure 5-16*d*) are tough, ropelike fibers composed of a variety of proteins. One type of intermediate filament, which is present only in nerve tissue, accumulates in deteriorating regions of the brain of Alzheimer's disease patients.

The structures that make up the cytoskeleton function in two interrelated activities:

1. as a scaffold, providing structural support, maintaining the shape of the cell, and organizing the internal contents of the cytoplasm;

2. as part of the machinery required for the movement of materials within the cell or for the movement of the cell itself.

An example of the first activity, structural support, is seen in the long "tentacles" that project from the "body" of certain unicellular organisms (Figure 5-17*a*). These fragile processes are supported in their extended position by an internal skeleton of microtubules (Figure 5-17*b*). We will now take a closer look at the role of cytoskeletal elements in cellular movements.

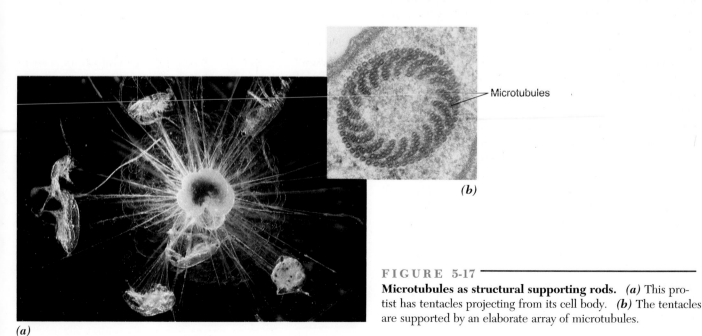

Microtubules

(b)

FIGURE 5-17

Microtubules as structural supporting rods. *(a)* This protist has tentacles projecting from its cell body. *(b)* The tentacles are supported by an elaborate array of microtubules.

(a)

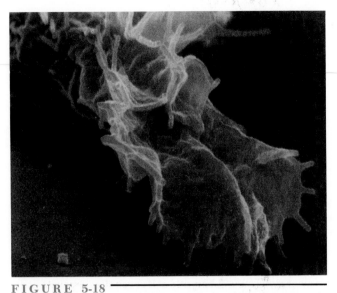

FIGURE 5-18

The leading edge of a moving cell may contain ruffles, which are formed by the contractile activity of microfilaments located just beneath the plasma membrane.

The Movement of Cells from One Place to Another

At one time, when you were an early embryo, cells that went on to form the pigmentation in the skin of your arm and the nerve cells in your brain were very close neighbors. These two types of cells end up in very different locations in the adult because pigment cells migrate away from the developing brain, across the entire width of the embryo. Even as you read this book, cells are leaving your bloodstream by squeezing through the walls of blood capillaries and making their way through the surrounding tissue. We know relatively little about how cells are able to carry out these remarkable movements. This is particularly unfortunate because the movement of single cancer cells away from the site of a tumor is one of the primary causes of the spread of malignant diseases.

One way to learn about the function of a cell structure is to destroy it selectively and study the consequences. If migrating cells are treated with drugs that cause microfilaments to fall apart into their subunits, cell locomotion halts. Studies of this type have drawn attention to microfilaments as important structures in cell locomotion. The microfilaments that drive cell movement tend to be concentrated in a thin layer just beneath the plasma membrane. In fact, the plasma membrane at the leading edge of a moving cell undergoes a dramatic ruffling activity (Figure 5-18) due to the contractile activity of the underlying microfilaments.

Cilia and Flagella. Microscopic organisms typically dart through their aqueous environment, powered by tiny locomotory organelles called cilia and flagella, which are really two versions of the same structure. **Cilia** are short, hairlike organelles that project from the surface of many small eu-

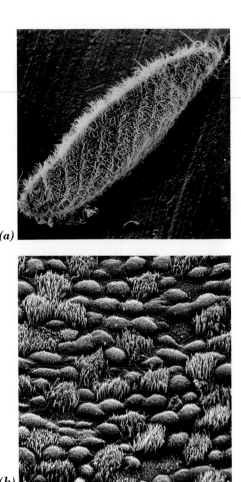

(a)

(b)

FIGURE 5-19

Cilia. *(a)* The numerous cilia that cover the external surface of this protist, *Paramecium*, propel the cell through the water while creating currents that channel food into its "mouth." *(b)* The coordinated beating of the cilia lining the mammalian trachea moves mucus and its trapped dust particles out of the respiratory tract into the throat.

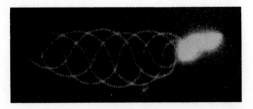

FIGURE 5-20

The undulating beat of the flagellum of a sea urchin sperm. Although this cell has only a single flagellum, the multiple-exposure photograph shows it in several different positions.

karyotic organisms; they act like oars to propel the organism through the water (Figure 5-19*a*). In larger organisms, such as mammals, cilia often line the surfaces of various tracts (Figure 5-19*b*), where they help propel materials through the channel. Generally, cilia occur in large numbers and are densely packed.

Flagella are longer than cilia and are present in fewer numbers. Flagella beat with an undulating motion that pushes or pulls the cell through the medium. In addition to powering microscopic organisms, including flagellated protozoa and algae, flagella provide the propulsive force for the movement of sperm in most animals (Figure 5-20).

▐▶ Cilia and flagella are powered by microtubules. The microtubules are arranged in a specific pattern called a *9 + 2 array* — nine pairs of microtubules encircling two central microtubules (Figure 5-21). This same 9 + 2 array is found in the cilia and flagella of virtually every eukaryotic organism, from fungi to humans, another of many reminders of the unity of life. We must presume that the 9 + 2 array is critical to ciliary and flagellar function, otherwise it would surely have undergone some evolutionary modification over the 2 billion years that eukaryotic cells have existed on earth.

Cilia and flagella move in a whiplike motion that requires that these fibers bend along their length. Bending results from the sliding of microtubules past one another (Figure 5-21). This sliding is actively accomplished by the movement of small arms that connect each pair of microtubules with its neighbor. The importance of the movable arms became apparent when it was discovered that people suffering from certain types of sterility produced sperm devoid of these arms. Further examination indicated that these people also suffered from respiratory ailments as a result of the absence of arms on their respiratory tract cilia.

Eukaryotic cells contain a variety of organelles that are specialized for particular activities. Many of these organelles are bound by cellular membranes; other organelles are nonmembranous. Eukaryotic cell organelles separate complex activities that could interfere, compete, or somehow disrupt the orderly sequence of reactions essential to each biological task. Compartmentalized organelles also enable cells to run many complex processes simultaneously. (See CTQ #6.)

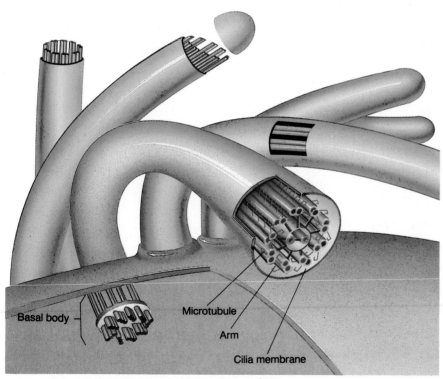
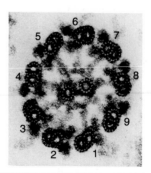

FIGURE 5-21

Cilia structure and movement. The 9 + 2 array of microtubules is shown in the cutaway sketch and accompanying electron micrograph. The bending of a cilium results from the sliding of one pair of microtubules over its neighboring pair. Those microtubules at one edge of the cilium (such as numbers 5 and 6 in the inset) move outward, while microtubules at the opposite edge (such as numbers 9 and 1), move inward. Sliding is powered by the movement of small arms that connect adjacent microtubules. A cilium moves when the arms projecting from one pair of microtubules "walks" along the neighboring pair of microtubules. The base of each cilium (or flagellum) contains a specialized structure called a *basal body*.

JUST OUTSIDE THE CELL

Even though the living portion of a cell ends at the plasma membrane, most cells produce extracellular materials which, as the name suggests, remain just outside the cell. Most animal cells secrete proteins and polysaccharides that form a thin **extracellular matrix** around their outer surface, which mediates interactions between different cells, and between cells and their nonliving environment. A few cell types, such as those found in bone and cartilage, are surrounded by an extensive extracellular matrix that gives these tissues their hardness and durability. Collagen—the most abundant protein in the human body—is a major constituent of the extracellular matrix.

CELL WALLS

In the beginning of this chapter we described how Robert Hooke was the first person to have observed a cell. In actual fact, the compartments he described were the empty cell walls of dead cork tissue which had originally been pro-duced by the living cells they once surrounded. Cell walls, of one type or another, are present in bacteria, fungi, and plants but are absent in animals. These walls form a rigid outer casing that provides support, slows dehydration, and prevents a cell from bursting when internal pressure builds due to an influx of water. We will focus our discussion on the cell walls of plant cells (Figure 5-22).

You might not think that plant cell walls and reinforced concrete have much in common, but they are both built of materials having a similar basic function. Concrete is an amorphous (nonstructured) material designed to resist crushing (compression forces). The steel rods used to reinforce the concrete are designed to help the concrete resist being pulled apart (tension forces). Plant cell walls have a similar structure. The steel rods of reinforced concrete are analogous to the cellulose molecules of a plant cell, which become bundled together to form thicker cables, called *microfibrils* (Figure 5-22). Microfibrils enable the cell wall to resist forces that might pull it apart. The microfibrils are embedded in an amorphous polysaccharide matrix that is formed in the Golgi complex and secreted outside the cell.

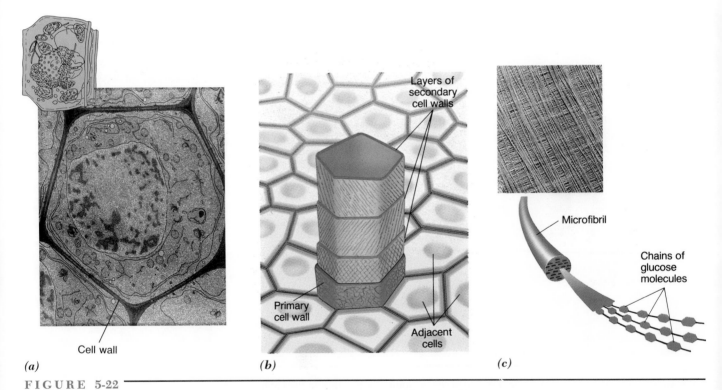

(a)   *(b)*   *(c)*

FIGURE 5-22

Plant cell walls. *(a)* Electron micrograph of a plant cell surrounded by its cell wall. *(b)* A diagrammatic plant cell wall, telescoped to reveal the primary cell wall, layers of the secondary cell wall, and the arrangement of cellulose microfibrils. *(c)* A surface view of the layers of parallel microfibrils in a secondary cell wall. Each microfibril is a complex of cellulose chains, held together by hydrogen bonds between adjacent glucose molecules to form flat, sheetlike strips. Cellulose molecules are assembled from glucose monomers by proteins embedded in the plasma membrane. Plant cell walls provide support and resist external forces.

The matrix of the cell wall holds the fibrils together; at the same time, it resists forces that might crush the enclosed cell. The various polysaccharides of the wall form a porous matrix that allows passage of dissolved gases and nutrients into the cell.

When a "young" plant cell is formed at the end of cell division it becomes surrounded by a thin (0.1 μm) **primary cell wall** composed largely of cellulose. The primary cell wall provides a firm, protective casing for the fragile internal cell, but it is not absolutely confining. As the young cell grows, the pressure exerted against the surrounding wall increases, causing the microfibrils to slide over one another, allowing the cell to increase in volume. When cell growth slows or ceases, the cell wall may become modified and thickened by the addition of other materials. This strengthened **secondary cell wall** greatly increases the cell's resiliency. Plant cells with secondary cell walls are found clustered in stems, as well as in the stalks that support leaves, flowers, and fruits. Secondary walls are often impregnated

with a hardening substance called *lignin*. Lignified secondary wall layers are responsible for the strength of wood and natural fibers, providing the means by which trees can grow hundreds of meters taller than the tallest animal.

INTERCELLULAR JUNCTIONS

Animal cells often interact with one another, forming **intercellular junctions** that can be seen in the electron microscope. The function of each type of junction is correlated with its structure (Figure 5-23).

* ***Anchoring junctions*** *or adhering junctions contain adhesive materials between the plasma membranes* The cells of tissues subjected to pulling or stretching forces, such as the uterine cervix or the skin, are typically "welded" together by anchoring junctions. A rare and potentially fatal disease called *pemphigus* results from deterioration of the junctions that anchor the

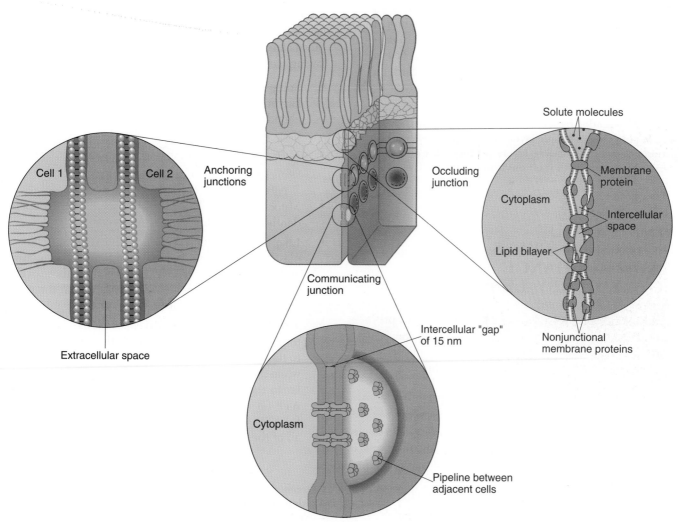

FIGURE 5-23

Intercellular junctions. This epithelial cell contains anchoring, occluding, and communicating junctions.

cells of the skin. The disease is characterized by severe blistering due to fluid leaking into the loosened skin tissue.

- ***Occluding junctions*** *are sites where adjacent plasma membranes make direct contact*   Occluding junctions prevent materials from moving between the cells. Junctions of this type are found between the cells lining the capillaries in the brain, for example. Because of these junctions, many drugs are unable to pass from the bloodstream into the tissues of the brain. Pharmaceutical companies are working on drugs that are better able to penetrate this so-called blood–brain barrier.

- ***Communicating junctions*** *contain pipelines between adjoining plasma membranes*   Communicating junctions join cells in such a way that materials are able to move directly through a channel from the cytoplasm of one cell into the cytoplasm of the adjoining cell. Communicating junctions often function to coordinate activities occurring in a sheet or mass of cells, as evidenced by the muscle tissue of your heart. Each heartbeat is initiated by direct electrical stimulation of only a small part of the heart (known as the *pacemaker*). The muscular wall of the heart contracts as a result of an ionic current that spreads rapidly through the muscle cells via communicating junctions.

Cells are surrounded by secreted materials that form structures ranging from a simple layer to a complex cell wall or junction. These surrounding layers protect cells and may govern interactions with their environment. (See CTQ #8.)

EVOLUTION OF EUKARYOTIC CELLS: GRADUAL CHANGE OR AN EVOLUTIONARY LEAP?

■■▶ Eukaryotic cells evolved about 2 billion years ago, presumably from simpler, prokaryotic ancestors. With specialized cytoplasmic organelles and a nucleus that sequestered genetic material, eukaryotic cells were more highly organized than were prokaryotes and were capable of conducting more complex functions. One of the questions that has intrigued biologists for decades is how the transition from the prokaryotic to eukaryotic state took place. Since there are no organisms living today that are intermediate in complexity between prokaryotes and eukaryotes, the question has been the subject of considerable speculation. Two modern hypotheses—the membrane invagination hypothesis and the endosymbiosis hypothesis—offer possible explanations.

The **membrane invagination hypothesis** (Figure 5-24*a*) proposes that the plasma membrane of the ancestral prokaryotic cell gradually folded in on itself, forming pockets, which pinched off from the surface membrane to form organelles in which particular enzymes or cellular materials could be concentrated. Over long periods of time, these simple organelles became highly specialized, forming the complex organelles found in modern eukaryotic cells.

It is easy to see how membrane invagination could have formed the endoplasmic reticulum, Golgi apparatus, and even the nucleus. However, the evolution of complex DNA-containing organelles, such as mitochondria and chloroplasts, is usually explained by another theory, the **endosymbiosis hypothesis** (Figure 5-24*b*), which proposes that the mitochondria and chloroplasts of eukaryotic cells were originally derived from small prokaryotic cells that took up residence *inside* a larger cell. (The name of this hypothesis uses the word *symbiosis*, a term that refers to a close association between different kinds of organisms.) For example, chloroplasts would have derived from a small photosynthetic bacterium, and mitochondria from a nonphotosynthetic bacterium that had evolved the machinery to use oxygen in the formation of ATP (Chapter 9). Convincing evidence in behalf of the endosymbiosis hypothesis has been assembled by Lynn Margulis of the University of Massachusetts. Among the evidence she cites are the following facts:

- Some bacteria form symbiotic partnerships with eukaryotic cells today. These bacteria resemble mitochondria in structure and carry out the same steps in oxygen utilization as do mitochondria.

- Mitochondria and chloroplasts contain their own nucleic acids, including a circular strand of DNA which resembles the chromosomes of prokaryotic cells.

- Mitochondria and chloroplasts can divide independently of the cell in which they reside.

- Like an independent cell, mitochondria and chloroplasts contain their own ribosomes whose structure and drug sensitivity more closely resemble those of the ribosomes of prokaryotic cells than those in the cytoplasm of eukaryotoic cells.

- The thylakoid membranes of chloroplasts contain photosynthetic pigments and proteins that are very similar to those in the membranes of photosynthetic bacteria, while the inner mitochondrial membrane is very similar to the plasma membrane of most bacteria where ATP formation takes place. The enzymes responsible for ATP synthesis, for example, are virtually identical in all of these membranes.

Evidence indicates that prokaryotic cells evolved before —and later gave rise to—eukaryotic cells. How this transition took place is currently the focus of vigorous scientific investigation and debate. (See CTQ #9.)

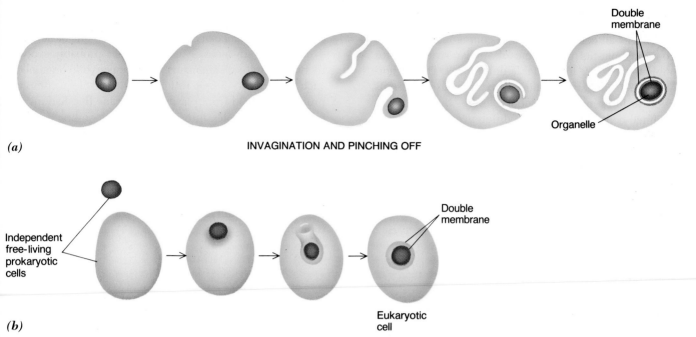

(a)

INVAGINATION AND PINCHING OFF

Double membrane

Organelle

Independent free-living prokaryotic cells

Double membrane

(b)

Eukaryotic cell

FIGURE 5-24

Proposals for the origin of eukaryotic cells. *(a)* Invagination hypothesis. *(b)* Endosymbiosis hypothesis.

R E E X A M I N I N G   T H E   T H E M E S

The Relationship between Form and Function

The structure of every cell component is well-suited to its function. For example, a plant cell wall provides protection and support for the fragile cell it surrounds. The cellulose microfibrils within the wall provide resistance to pulling (tension) forces, while the amorphous matrix in which the microfibrils are embedded provides resistance to crushing (compression) force. This type of composite structure is analogous to that of reinforced concrete; the steel cables resist tension, while the concrete resists compression. Reinforced concrete and cell walls allow office buildings and trees to tower over the horizon.

Biologic Order, Regulation, and Homeosatasis

Like an entire organism, a cell remains alive only as long as the composition of its internal medium is maintained within a tolerable range. Each of a cell's organelles plays some role in maintaining this internal stability. For example, the plasma membrane prevents a cell from losing essential ingredients; the nuclear envelope helps maintain essential differences between the composition of the nucleus and cytoplasm; mitochondria provide the chemical energy necessary to fuel the cell's activities and to prevent its downhill slide to greater disorder; lysosomes digest various materials, including invading bacteria whose presence would threaten the cell; plant cell walls help maintain a cell's volume; and a plant cell's central vacuole serves as a storage site that removes toxic substances from the cell itself.

Acquiring and Using Energy

Two cellular organelles—the mitochondrion and the chloroplast—are the principal players in life's energy balance. Chloroplasts trap light energy from the sun and convert it to chemical energy stored in carbohydrate. Mitochondria convert the energy stored in fats and polysaccharides into ATP, which is used to run cellular activities. In both organelles, the inner membranes hold the components necessary for energy transfer.

Unity within Diversity

⚠ Whereas all cells share certain common features—a surrounding plasma membrane of similar structure, for example—there are also notable differences. The dichotomy between prokaryotic and eukaryotic cells represents diversity at life's most fundamental level—the cell. Plant and animal cells also have basic differences; chloroplasts, cell walls, and large central vacuoles are all present in plants, for example, but absent in animals. Even within an individual plant or animal, cellular diversity is evident from the various cell types specialized for particular functions. Even though these various cells contain the same kinds of organelles, their form and function can be very different.

Evolution and Adaptation

▮▶ The appearance of eukaryotic cells was one of the great evolutionary advances. Eukaryotic cells probably evolved by two separate mechanisms. Some eukaryotic organelles, such as the endoplasmic reticulum and the Golgi complex, probably evolved from plasma membrane that had invaginated into the cell's interior, whereas membrane-bound mitochodria and chloroplasts were probably derived from prokaryotic organisms that took up residence inside a larger host cell. Eukaryotic organelles remain relatively similar in structure over long periods of evolution. For example, the 9 + 2 array of mirotubules in cilia and flagella is seen across the entire spectrum of eukaryotic organisms, from protists to mammals.

SYNOPSIS

The cell theory has three parts. 1) All organisms are composed of one or more cells; 2) the cell is the basic organizational unit of life; and 3) all cells arise from preexisting cells.

Cells are either prokaryotic or eukaryotic. Prokaryotic cells are found only among bacteria (kingdom Monera); all other organisms are composed of eukaryotic cells. Both types of cells are surrounded by a fluid-mosaic plasma membrane consisting of a lipid bilayer and embedded proteins. Unlike prokaryotic cells, eukaryotic cells contain a true membrane-bound nucleus and a variety of distinct cytoplasmic organelles.

Eukaryotic cells contain similar organelles, but their number, form, and distribution vary from one cell type to another. The plasma membrane serves as a barrier between the cell and the outside world and regulates exchanges between the two. The plasma membrane contains receptors that recognize and interact with specific substances in the external medium. The nucleus is bound by a complex nuclear envelope and houses the chromosomes, which contain highly elongated DNA molecules and bound protein. Genetic messages move into the cytoplasm through pores in the nuclear envelope. The cytoplasm contains an interconnected, interfunctional network of membranous organelles. Membrane proteins, secretory proteins, and the proteins of some organelles are synthesized on ribosomes bound to the outer surface of the ER membranes, passed through the ER membrane to the internal space, and then sent off in vesicles to the Golgi complex. Once in the Golgi complex, materials are often modified

and then packaged for transport either to specific vesicles, such as lysosomes, or for discharge outside the cell during exocytosis. Lysosomes contain a variety of hydrolytic enzymes that are utilized in digesting macromolecules and aging cytoplasmic organelles. Plant cells often contain large vacuoles that store various substances. Mitochondria are specialized for the transfer of energy from stored macromolecular fuels to ATP for immediate use in the cell. Chloroplasts, which are not found in animal cells, are organelles in which light energy is captured and used to manufacture biochemicals during photosynthesis. The cytoskeleton of microtubules, microfilaments, and intermediate filaments maintains the shape of a cell, organizes its internal components, and provides the machinery necessary for cell movement. Some eukaryotic cells have cilia or flagella that are employed in the movement of single cells as well as the movement of materials along the surface of various tracts.

Cells secrete materials that form various types of extracellular coverings. The most complex extracellular layers are the walls that surround plant cells. These walls contain bundles of cellulose fibers embedded in an amorphous polysaccharide matrix. Animal cells may form specialized intercellular junctions that function in cell–cell attachment, occlusion of the space between cells, and intercellular communication.

The evolution of eukaryotic cells from prokaryotic ancestors is explained by two hypotheses. Many of the cytoplasmic organelles of a eukaryotic cell may derive from invaginated membrane, while mitochondria and chloroplasts are probably derived from prokaryotic symbionts.

Key Terms

lipid bilayer (p. 89)
cell theory (p. 90)
organelle (p. 91)
prokaryotic cell (p. 92)
eukaryotic cell (p. 92)
nucleus (p. 92)
cytoplasm (p. 92)
plasma membrane (p. 93)
receptor (p. 94)
chromosome (p. 97)
chromatin (p. 97)
nuclear envelope (p. 97)
nucleoli (p. 97)
ribosome (p. 97)
nucleoplasm (p. 97)
nuclear matrix (p. 97)

vesicle (p. 97)
endoplasmic reticulum (ER) (p. 98)
rough ER (RER) (p. 98)
smooth ER (SER) (p. 98)
Golgi complex (p. 99)
exocytosis (p. 99)
lysosome (p. 100)
vacuole (p. 102)
mitochondria (p. 102)
cristae (p. 104)
matrix (p. 104)
chloroplast (p. 104)
thylakoid (p. 104)
stroma (p. 104)
cytoskeleton (p. 104)

microtubule (p. 104)
microfilament (p. 104)
intermediate filament (p. 104)
cilia (p. 106)
flagella (p. 107)
extracellular matrix (p. 108)
primary cell wall (p. 109)
secondary cell wall (p. 109)
intercellular junction (p. 109)
anchoring junction (p. 109)
occluding junction (p. 110)
communicating junction (p. 110)
membrane invagination hypothesis (p. 110)
endosymbiosis hypothesis (p. 110)

Review Questions

1. List the similarities and differences between prokaryotic and eukaryotic cells.

2. Prepare a chart comparing the similarities and differences between plant and animal cells.

3. Why do the surfaces of membrane proteins contain both hydrophobic and hydrophilic regions?

4. Which cytoplasmic organelles synthesize steroid hormones? Which synthesize proteins to be secreted? Which store calcium? Which store hydrolytic enzymes? Which synthesize plasma membrane proteins? Which store toxic waste products in plant cells? Which synthesize polysaccharides for plant cell walls?

5. Try to convince a friend that the mitochondria in their cells are derived from bacteria.

Critical Thinking Questions

1. Gorter and Grendel used human red blood cells for their determination of a cell's lipid content. Would their results have differed if they had used a salivary gland cell instead? In what way? What do you think they would have been able to conclude about plasma membrane structure?

2. In 1674, Anton van Leeuwenhoek used a single glass lens (a "simple" microscope) to look at a drop of lake water. He discovered a new world, one of tiny organisms he called animalcules ("little animals"). Imagine you are van Leeuwenhoek. Write a letter to the Royal Society of London in which you to try to convey your feelings about your momentous discovery.

3. When Leeuwenhoek looked at what he saw through his microscopes, he concluded that they were living organisms. Most of what he saw we now know are single-celled organisms. List the evidence Leeuwenhoek could have used in drawing the conclusion that the things he saw were indeed alive.

4. Given that both single-celled prokaryotes and single-celled eukaryotes carry out all of the life functions, why do you think biologists describe the former as "simple" and the latter as "complex?" Compare your answer to that of a professional biologist.

5. What changes would occur in the interaction of cells with external substances if enzymes are used to remove: (1) the carbohydrate chains that project from the outer surface of the plasma membrane; (2) the polar amino acids that project from the outer surface of the plasma membrane; (3) the phospholipids that make up much of the plasma membrane?

6. The individual cells found in living organisms exhibit the characteristics that distinguish living from nonliving things (see Chapter 1). Complete the chart below by listing the appropriate cellular components (organelles) in the right-hand column. You may use a component more than once.

Characteristic	Cellular Components
maintaining order	
acquiring and using energy	
reproduction/ heredity	
growth/ development	
responsiveness to environment	
metabolism	
homeostasis	

7. Suppose you wanted to test the hypothesis that cellulose fibers were synthesized in the Golgi complex. (They are actually synthesized at the plasma membrane.) How might you go about testing this hypothesis?

8. How do the materials secreted by cells in multicellular organisms contribute to the life functions of homeostasis and responsiveness for those organisms?

9. Fossil evidence indicates that prokaryotic cells evolved before eukaryotic cells. If you did not know this, what evidence from the structure of these two main types of cells would lead you to the same conclusion?

Additional Readings

Alberts, B., et al. 1989. *Molecular biology of the cell*, 2d ed. New York: Garland. (Intermediate)

Becker, W. and D. W. Deamer. 1991 *The world of the cell*, 2d ed. San Francisco: Benjamin-Cummings. (Intermediate)

Darnell, J., H. Lodish, and D. Baltimore. 1990. *Molecular cell biology*, 2d ed. New York: Freeman. (Advanced)

Karp, G. 1984. *Cell biology*. New York: McGraw-Hill. (Intermediate)

Kleinsmith, L. J., and V. M. Kish. 1988. *Principles of cell biology*. New York: Harper & Row. (Intermediate)

Loewy, A. G., et al. 1991. *Cell structure and function*, 3d ed. Philadelphia: Saunders. (Advanced)

Marx, J. 1992. Boring in on β-Amyloid's Role in Alzheimer's. *Science* 255:688–689. (Intermediate).

Prescott, D. M. 1988. *Cells*. Boston: Jones and Bartlett. (Intermediate)

Sharon, N. and Lis, H. 1993. Carbohydrates in cell recognition. *Sci. Amer.* Jan:82–89. (Intermediate)

Sheeler, P. and D. Bianchi. 1987. *Cell and molecular biology*. New York: Wiley. (Intermediate)

Wolfe, S. L. 1993. *Molecular and cell biology*. Belmont, Ca: Wadsworth. (Advanced)

CHAPTER

◄ 6 ►

Energy, Enzymes, and Metabolic Pathways

**STEPS
TO
DISCOVERY**
The Chemical Nature of Enzymes

ACQUIRING AND USING ENERGY

The Laws of Thermodynamics

Energy is Transferred During
Chemical Reactions

ENZYMES

Why Enzymes are Such Effective Catalysts

The Effect of Temperature and pH on
Enzyme Activity

METABOLIC PATHWAYS

Oxidation and Reduction: A Matter of Electrons

ATP: The Energy Currency of Life

Directing Metabolic Traffic

**MACROMOLECULAR EVOLUTION
AND ADAPTATION OF ENZYMES**

BIOLINE

Saving Lives by Inhibiting Enzymes

THE HUMAN PERSPECTIVE

The Manipulation of Human Enzymes

Hexokinase

Alcohol
dehydrogenase

Yea

The Chemical Nature
of Enzymes

*I*n 1779, the French Academy of Science offered 1 kilogram (2.2 pounds) of gold to anyone who could unravel the mystery by which sugars present in grape juice are converted to alcohol during the formation of wine. The prize was never collected.

A hundred years later, two conflicting explanations for the nature of alcohol formation (a process known as *fermentation*) prevailed. On one side of the argument were the biologists, led by Louis Pasteur, a biologist working on behalf of the French wine industry. Pasteur correctly believed that the conversion of sugars to alcohol was a process

carried out by living yeast cells. In fact, he applied this scientific fact to the faltering French wine industry, thereby restoring it to its previous station as the best winemaker in the world, simply by manipulating the organisms used for fermentation. In adopting this position, however, Pasteur hypothesized that fermentation required a "vital force" that could only be supplied by an intact, highly organized, living organism. Pasteur rejected the notion that life processes, such as fermentation, could be reduced to simple chemical reactions.

On the other side of the argument were the organic

Yeast cells provide the enzymes that ferment the sugar in grapes to form the alcohol found in wine.

chemists of the period, including the aggressive debater Justus von Liebig, who ridiculed the suggestion that yeast cells were responsible for fermentation. Liebig and his fellow chemists proposed that fermentation was simply an organic reaction that occurred on its own, no different from those reactions they had been studying in the test tube. The battle lines were drawn, biologists versus chemists.

Then came Eduard Buchner. Buchner was both a chemist—teaching chemistry at a German university—and a biologist—working in his brother's bacteriology laboratory. In 1897, 2 years after Pasteur's death, Buchner was preparing "yeast-juice,"—an extract made by grinding yeast cells with sand grains and then filtering the mixture through filter paper. Buchner was planning to use this "yeast-juice" in a series of medical studies, but he wanted to preserve it for later work. After trying to preserve the extract with antiseptics and failing, Buchner attempted to protect his cell juice from spoilage by adding sugar, the same procedure used to preserve jams and jellies. Instead of preserving the solution, the sugar produced gas, which bubbled continuously for days. Buchner's scientific curiosity was aroused by this unexpected occurrence, and he kept the solution instead of discarding it. Analysis revealed that fermentation was occurring, producing alcohol and bubbles of carbon dioxide. All this was taking place without a single living yeast cell in the "soup," a severe blow to Pasteur's hypothesis that fermentation required living organisms.

Buchner had accidentally discovered that chemical agents (later identified as *enzymes*) promote biological reactions, and can do so *outside* of the cells in which they were originally produced. Investigators could now purify enzymes and study their activities without interference from other cellular ingredients. The door to modern biochemistry had swung wide open, raising the question: What was the nature of these remarkable molecular mediators?

One prominent scientist to turn his attention to the study of enzymes was the German chemist Richard Willstater, who had earlier won a Nobel Prize for his work on the structure of chlorophyll, the central plant pigment in photosynthesis. Willstater set out to purify plant enzymes but found that his most active preparations contained so little material they could not be chemically identified. Today, we understand that enzymes are so efficient that they are active at extremely low concentrations—too low to be characterized using 1920s' methods. To Willstater, it was bewildering. Because he was unable to detect any protein in the purified enzyme preparations, Willstater erroneously concluded that enzymes could not be made of protein.

Then in 1926, James Sumner, an American biochemist, prepared the first crystals of an enzyme (urease), which he had purified from the seeds of a tropical plant; he determined that the crystals were indeed composed of protein. He also demonstrated that protein-denaturing treatments destroyed the preparation's enzymatic activity and concluded that the enzyme was a protein. Sumner's finding was not greeted with much acclaim. Willstater argued that the crystals may have been partly composed of protein but that the enzyme was only a minor contaminant of the preparation and it was not being detected in the chemical analysis. A few years later, however, another American, John Northrop, crystallized a number of different enzymes (including pepsin, a digestive enzyme of the stomach) and showed conclusively that all were made of protein. As a result of their work, Buchner won the Nobel Prize in chemistry in 1907, while Sumner and Northrop shared the same prize in 1946.

*E*very organism is a precisely coordinated collection of chemicals, each of which is created through biochemical reactions. Every detail of your body, from the hair on your head down to the nails on your toes, is the product of such reactions. Your eyes, for example, owe their color to pigments manufactured by metabolic activity. The chemical changes that construct these pigments don't happen automatically or haphazardly; they require the tight supervision provided by enzymes, which may be thought of as "metabolic traffic directors." Enzymes act on prepigment chemicals, changing them step by step until they become the molecules that ultimately produce eye color. Most of your inherited characteristics developed because your cells produced enzymes that encouraged the biochemical construction of each particular trait. In this sense, an organism is the product of its unique combination of enzymes (Figure 6-1). Most series of reactions—whether they lead to the production of pigment or of some other material—also require a source of chemical energy. The topic of energy is a good starting point in a discussion of metabolism.

▼ ▼ ▼

ACQUIRING AND USING ENERGY

A living cell bustles with activity. Materials constantly enter and exit the cell; genetic instructions flow from the nucleus to the cytoplasm, where proteins and other substances are synthesized or degraded. To maintain such a high level of activity a cell needs energy. For most of us, the word "energy" conjures up a rather vague concept of being energetic, such as playing baseball, feeling enthusiastic, or making an effort. For scientists, **energy** has a more precise definition: It is the capacity to do work; that is, the capacity to change or move something. There are several different forms of energy. Chemical energy can change the structure of molecules, mechanical energy can move objects, light energy can boost electrons to an outer shell, thermal energy (heat) can increase the motion of molecules, and electrical energy can move electrically charged particles.

Energy exists in two general states: potential and kinetic. **Potential energy** is stored energy; it has the potential to perform work if allowed. **Kinetic energy,** which literally means "energy in motion," is energy expended in the process of performing work. Each form of energy (chemical, mechanical, light, thermal, and electrical) can exist in either state. The glycogen stored in your muscle and liver cells, for example, contains abundant potential energy. When the glycogen molecules are broken down and metab-

(a)

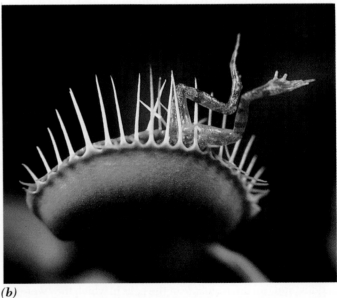

(b)

FIGURE 6-1

A sampler of enzyme performance. The properties and activities of every organism depend on enzymes. *(a)* The unusual color of this blue frog is due to a pigment manufactured by enzymes, a set different from that of his green counterparts. *(b)* Digestive enzymes secreted by the Venus-flytrap will disassemble the body of this imprisoned frog. Other enzymes will reassemble the molecules released from the frog's flesh into plant tissue.

FIGURE 6-2

Biological transfers of energy. *(a)* Energy is captured for the biosphere by photosynthesis. Plant leaves transform sunlight into chemical energy stored in the plant's tissues primarily in the form of cellulose and starch. In this illustration, the potential energy stored in the cellulose (structure shown) that makes up a match becomes kinetic energy when the cellulose molecules are "burned." Animals that feed on plants, such as a hare *(b)*, utilize the chemical energy in starch to fuel their diverse energy-requiring activities. Plant-eating organisms (herbivores), such as the hare, serve as energy sources for other animals, this lynx for example *(c)*. In each case, however, most of the potential energy of the food matter escapes back into the environment as unusable energy.

olized, the energy appears as kinetic energy and is used by the body to contract muscles, to move fluid through various tracts and vessels, to produce body heat, and myriad other activities.

THE LAWS OF THERMODYNAMICS

Much of what we know about energy is summarized in two basic laws of nature—the laws of thermodynamics:

- **The first law of thermodynamics** is the *law of energy conservation*. Simply put, it states that energy can

neither be created nor destroyed. Energy can be converted from one form to another, however. For example, electrical energy is converted to mechanical energy when we plug in a clock or turn on the blender; chemical energy is converted to heat and light energy when fuel is burned in an oil heater. Living organisms are also capable of energy conversion. Green plants perform *photosynthesis*, a process by which solar energy is converted into chemical energy that is stored in organic chemicals, such as starch or cellulose (Figure 6-2a). The next time you eat a piece of fruit or meat, remember that the chemical energy you are receiving was originally derived as light energy from the sun.

- **The second law of thermodynamics** expresses the concept that events in the universe proceed in a predictable "downhill" direction, from a higher energy level to a lower energy level. Rocks fall off cliffs to the ground, but they never spontaneously lift themselves back to the higher energy level at the top of the cliff. The second law of thermodynamics carries with it the notion that nature is "wasteful." Any time energy is exchanged, some *usable* energy is inevitably lost. In other words, the energy *available to perform additional work* decreases. For example, when a hare browses on the leaves of a tree (Figure 6-2*b*), or a lynx preys on the hare (Figure 6-2*c*), most of the chemical energy in the food is inevitably lost to the animal having the meal. What happens to this additional energy? A significant portion is "wasted" by increasing the random movements of atoms and molecules. This so-called unusable energy has a special name—entropy.

The Concept of Entropy

Entropy can be measured in terms of an increased state of randomness or disorder in the universe, which is often a result of the release of heat. Entropy can be illustrated by innumerable familiar activities. Suppose you were to walk into a library and search for books by three of your favorite authors. You find the books, take them to a desk, browse through them, and leave them on a table. You have just increased the entropy of the universe. Prior to your visit, the books were in an ordered location; there is only one proper place for each book in the library. In contrast, after you left the library, the books were left on a table that was selected essentially at random; you could just as well have left them in any one of a hundred or more places. As a result of your activities, the library has become more disordered; that is, its entropy has increased. Another example of entropy is shown in Figure 6-3.

⟳ One of the themes we continue to stress throughout this book is the complexity and order of living organisms. If every event increases the disorder in the universe, how is it possible for organisms to maintain their state of order and, even more perplexing, how did complex organisms ever evolve in the first place? The fact that entropy in the universe always increases does not mean that *every part* of the universe has to become more and more disordered. On the contrary, each organism represents a temporary departure from the relentless march toward disorganization.

FIGURE 6-3

A victim of entropy. A house left on its own for a number of years will spontaneously fall into a state of disorder, increasing the entropy of the universe. All organized structures require a constant input of energy to maintain their complexity and order; otherwise, they slowly deteriorate. The same is true for organisms; without the input of energy, they die and decompose.

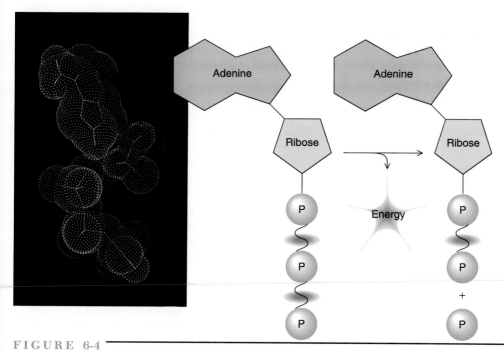

FIGURE 6-4

ATP. ATP is a nucleotide consisting of a sugar (ribose), a nitrogenous base (adenine), and three phosphate groups joined to each other. When hydrolyzed, the molecule is split into two products: ADP and inorganic phosphate (P_i). The squiggly lines connecting the second and third phosphate groups represent the bonds that can be broken when ATP is hydrolyzed. Space-filling model of ATP generated by a computer. Each ball represents an atom (white = hydrogen, green = carbon, blue = nitrogen, yellow = phosphorus).

Maintaining a state of low entropy requires the input of energy, which, in keeping with the first law of thermodynamics, lowers the energy content of the remainder of the universe. Consider just one molecule of DNA located in one cell in your liver. That cell has dozens of different enzymes whose sole job it is to patrol the DNA, looking for damage and repairing it. Without this expenditure of energy, the ordered arrangement of nucleotides in the cells of your body would literally disappear overnight. Energy is expended to maintain order at every level of biological organization—from molecules to ecosystems.

The evolution of complex organisms from simpler forms has also been driven by the input of energy. Energy is required to build and maintain the organisms on which natural selection operates. As organisms became larger and more complex, they required more energy to develop, survive, and reproduce; this energy is ultimately derived from the sun.

ENERGY IS TRANSFERRED DURING CHEMICAL REACTIONS

◮ It is easy to understand how a moving sledgehammer releases mechanical energy as it slams through a cement sidewalk or how a pot of water absorbs thermal energy as it becomes hotter and hotter. Understanding how energy is released or absorbed during the transformation of one type of chemical compound into another is a bit more complex. Let us begin with one of the most important compounds found in all living organisms, the key compound of energy metabolism, *adenosine triphosphate* or *ATP* (Figure 6-4). ATP is a nucleotide (Chapter 4); it is made up of a sugar (ribose), a nitrogenous base (adenine), and three phosphate groups linked to one another. ATP is the molecule in which all cells temporarily store the chemical energy used to run virtually all of a cell's activities. These small, usable amounts of energy are stored in a form that is instantly accessible as the energy is needed. For this reason, ATP is often described as the "energy currency" of the cell. As we will see, organisms spend their ATP on various energy-requiring tasks, such as the assembly of macromolecules, the contraction of muscles, and the buildup of ions on opposite sides of the plasma membrane.

The amount of ATP present in a cell at a given moment is surprisingly small. The average bacterial cell, for example, has fewer than 5 million ATP molecules—only enough to sustain the cell's activities for a second or two. Similarly, the human body has only enough ATP "on hand" to last about 20 seconds. Consequently, ATP supplies must be continually replenished from the energy stored in its large organic molecules, particularly the polysaccharides and fats, which maintain the energy reserves of the cell. In keeping with the comparison of ATP to currency, a cell's energy-rich macromolecules can be likened to money banked in a savings account, while its ATP supply can be thought of as money in the cell's "pocket," available to be "spent" on its needs at that very instant.

ATP is broken down in a cell by the following hydrolytic reaction

$$ATP + H_2O \rightarrow ADP + P_i\text{(inorganic phosphate)}$$
$$+ \text{energy}$$

It is a highly *favored* reaction, that is, a reaction that will proceed spontaneously toward the formation of products, in this case ADP and P_i. The reason this reaction is so highly favored is that ATP has considerably more energy than do ADP and P_i; therefore, when ATP is hydrolyzed the contents of the test tube attain a lower energy level. A similar principle underlies the use of common explosives, such as nitroglycerin $[C_3H_5(NO_3)_3]$ and trinitrotoluene (TNT) $[C_7H_5(NO_2)_3]$. The explosive reaction of these high-energy molecules is driven by the production of one of the most stable, lowest energy molecules in the universe — N_2 gas — the most common component of the atmosphere.

Highly favored chemical reactions, such as ATP hydrolysis or the explosion of TNT, are referred to as **exergonic** because they occur spontaneously with the release of energy. Whereas the energy released by a TNT explosion is simply released into the environment, the energy released by ATP hydrolysis is utilized by the cell to accomplish useful work. This will be illustrated in the following discussion.

Unlike ATP hydrolysis, many chemical reactions that occur in a cell are not favored at all; they are described as thermodynamically *unfavorable.* An example of a thermodynamically unfavorable reaction is the synthesis of the amino acid glutamine

$$\text{glutamic acid} + \text{ammonia} \rightarrow \text{glutamine}$$

Such reactions, which lead to the formation of products with more energy than the original reactants, are referred to as **endergonic.** The most important endergonic reaction in biology is the splitting of water (H_2O) into its component elements, hydrogen and oxygen; this is the fundamental reaction that occurs during photosynthesis. In contrast to exergonic reactions, endergonic reactions, such as glutamine formation or the splitting of water, do not occur spontaneously but require an input of external energy. As we will see later in the chapter, this is a crucial factor of metabolism. First, let us examine how endergonic reactions can be made to occur.

The Use of Energy-Releasing Reactions to Drive Energy-Requiring Reactions

If the conversion of glutamic acid into glutamine cannot occur spontaneously, how does a cell produce this essential

amino acid? The answer was revealed in 1947, when it was shown that the formation of glutamine is *coupled* to the hydrolysis of ATP. **Coupling** is accomplished by utilizing two sequential reactions, both of which are favored.

First reaction:
  glutamic acid + ATP → glutamyl phosphate + ADP
Second reaction:
  glutamyl phosphate + ammonia → glutamine + P_i

Net reaction:
  glutamic acid + **ATP** + ammonia → glutamine
          + **ADP** + **P_i**

The situation is analogous to the scene depicted in Figure 6-5, where a mallet driven down onto a platform (an energy-releasing reaction) is used to lift another object (an energy-requiring reaction). ATP hydrolysis is used to drive

FIGURE 6-5

A mechanical analogy for the use of ATP in driving endergonic reactions. The energy released as the mallet strikes the base is used to drive the uphill reaction as the weight is moved upward against the force of gravity to ring the bell.

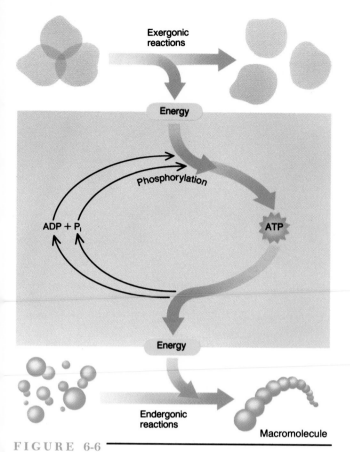

FIGURE 6-6

Producing and "spending" the cell's energy currency.
Endergonic and exergonic reactions are coupled by ATP, the
common denominator in energy production and utilization. Exergonic reactions generate ATP, whereas endergonic reactions,
such as macromolecular assembly, utilize ATP.

virtually every endergonic process within the cell, including
chemical reactions such as the formation of glutamine, the
assembly of macromolecules (Figure 6-6), the concentration of ions across a membrane (Chapter 7), and the movement of filaments in a muscle cell (Chapter 26). In most
coupled reactions, the phosphate group is transferred from
ATP to an acceptor—such as glutamic acid, a sugar, or
often a protein—and is subsequently removed in a second
step. The molecules that make all of this possible are
enzymes.

**Every time energy is exchanged, as occurs during chemical
reactions, a portion of that energy is "lost" for further use.
As a result, events in the universe tend to proceed toward a
state of lower energy. Life is a highly organized state. Maintaining such a high degree of organization demands a constant input of energy. (See CTQ #2.)**

ENZYMES

Even though a biochemical reaction is thermodynamically
favored, it usually takes place at extremely slow rates.
Herein lies the value of enzymes. Enzymes are biological
catalysts—substances that greatly increase the rate of
particular chemical reactions. An enzyme may cause a reaction to proceed billions of times faster than it would
otherwise occur. Consider, for example, what would happen if you were to add a teaspoon of glucose to a test tube,
seal the tube, and sterilize the solution. The glucose would
remain essentially intact for years. If you were to add bacteria to the solution, however, the sugar molecules would be
taken into the cells and degraded in seconds by the enzymes
of the bacteria.

Because of their powerful catalytic activity, enzymes
are effective in small numbers. Even a tiny bacterial cell has
space for hundreds of different enzymes, all catalyzing different reactions. Just as importantly, enzymes help maintain biological order by converting reactants into specific
products needed by the cell. In contrast, when these same
reactions are attempted in the laboratory by organic chemists without the benefit of enzymes, many different byproducts are usually produced. In a cell, formation of unwanted
byproducts would disrupt the progress of other reactions
and kill the cell. The importance of enzymes can be appreciated when one considers that such devastating diseases as
phenylketonuria (PKU) and Tay-Sachs, which lead to severe mental retardation and infant death, result from the
body's deficiency of a single enzyme.

WHY ENZYMES ARE SUCH EFFECTIVE
CATALYSTS

We have seen how certain reactions, such as glucose breakdown or ATP hydrolysis, are thermodynamically favored
because they lead to the formation of lower-energy products. But why don't these reactions occur at significant rates
in the absence of enzymes? The reason is that the reactant
molecules—the starting substances—must contain a certain amount of energy to become transformed into product
molecules—the new substances. This required energy is
called **activation energy.** In Figure 6-7, the activation
energy is represented by the height of the barrier. Enzymes
act by lowering a reaction's activation energy—the amount
of energy needed by molecules to undergo a reaction. We
will return to the matter of activation energy below.

The question still remains: Why are enzymes such effective catalysts? How is it that an enzyme can cause a
reaction to occur ten times a second when that same reaction might occur only once every hundred years in the
enzyme's absence? A simple analogy may clarify the answer.

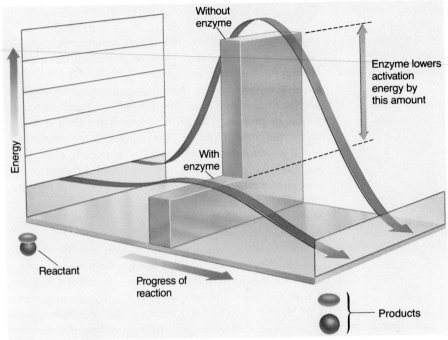

FIGURE 6-7

Activation energy and enzymes. Activation energy is the energy required to initiate a chemical reaction—to overcome the reactants' tendency to resist change. This energy barrier must be hurdled before a chemical reaction can occur. Enzymes reduce activation energy, so a reaction proceeds at a much faster rate than would occur in their absence.

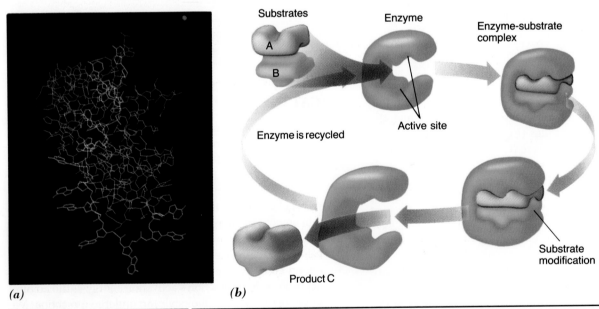

(a)   *(b)*

FIGURE 6-8

An enzyme in action. *(a)* The closeness of the enzyme–substrate fit is realistically revealed by this computer-generated version showing the RNA substrate (in green) being attacked and disassembled by the enzyme ribonuclease (in purple). *(b)* Substrates A and B are enzymatically altered to form product C. The reaction occurs following the binding of the substrates within the enzyme's active site. Enzymes are recycled—following completion of a reaction and the dispersal of the product, the enzyme binds a new pair of substrates and catalyzes another reaction.

Suppose you were to place a handful of nuts and bolts into a bag and shake the bag for 15 minutes. It is very unlikely that any of the bolts would have a nut firmly attached to its end when you stopped shaking the bag. In contrast, if you were to pick up a bolt in one hand and a nut in the other, you could guide the bolt into the nut very rapidly. In this analogy, enzymes are the guiding hands, and reactants are the nuts and bolts. Enzymes bind reactants—termed **substrates** when an enzyme is involved—increasing the likelihood that they will be converted to products.

The area on the enzyme that binds the substrate(s) is called the **active site,** and it is often situated within a groove or cleft of the enzyme. The active site and the substrate(s) have complementary shape enabling them to bind together with a high degree of precision, like pieces of a jigsaw puzzle (Figure 6-8*a*). In addition to binding the substrate(s), the active site contains a particular array of amino acids whose presence lowers the activation energy the substrates need to undergo a reaction. There are a number of ways this can happen (Figure 6-9). These include:

- *Substrate orientation*   Like the connection between nuts and bolts, many reactions require that two different substrate molecules collide with each other in just the right way. Enzymes hold their substrates in a particular orientation that forces them together in the proper relationship.

- *Physical stress*   Once a substrate molecule is in the "grip" of an enzyme, certain bonds within the substrate molecule may be placed under physical stress, increasing the likelihood that the bond will rupture.

- *Changes in substrate reactivity*   Refer back to Figure 4-14 and notice the variety in the structures of amino acid R groups. When these R groups on the enzyme's amino acids come into close proximity with a part of the substrate, they can change the charge of the substrate, alter the distribution of electrons within the substrate's bonds, tie up surrounding water molecules, and cause other changes that increase the reactivity of the substrate.

The interaction between enzyme and substrate is not a rigid one, as would occur, for example, between a lock and key. Rather, enzymes are built with an internal flexibility that allows them to undergo **conformational changes,** that is, changes in shape. As a result of conformational changes in the enzyme, the fit between an enzyme and substrate often improves after the substrate initially binds to the active site. In addition, conformational changes invariably occur within an enzyme while a substrate is being chemically transformed into a product. Once the reaction has taken place, the product(s) leaves the enzyme, which is then ready to bind a new set of substrates. This cycle of substrate binding, chemical modification, and product release is illustrated in Figure 6-8*b*.

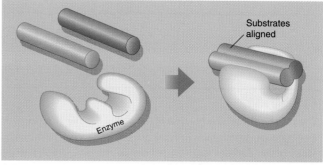

(a)

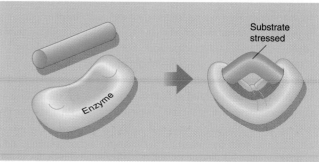

(b)

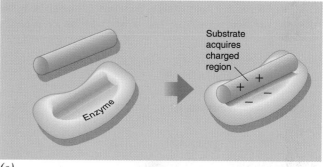

(c)

FIGURE 6-9

Three processes by which enzymes modify substrates: *(a)* maintaining precise substrate orientation, *(b)* physical stress, and *(c)* change in substrate reactivity.

The complementarity in the shapes of the active site of the enzyme and substrate accounts for the high degree of *specificity* of enzyme activity. Specificity is a term used to describe the high degree of selectivity and precision with which biological molecules interact; it is a key ingredient in maintaining biological order. Regarding enzyme activity, for example, only the proper substrate fits into the active site of an enzyme. As a result, the course of an enzymatic reaction is not affected by the hundreds of other types of molecules present in the cell at the time. However, the activity of an enzyme can be affected by adding a substance whose structure is very similar to that of a substrate (see Bioline: Saving Lives by Inhibiting Enzymes).

◁ B I O L I N E ▷
Saving Lives by Inhibiting Enzymes

You may be alive today because of the ability of chemicals to inhibit enzyme activity. Many of the diseases that ravaged our ancestors are treatable today with chemicals that selectively block the action of enzymes present in disease-causing bacteria but not in people. Such inhibitors may be safely introduced into an infected person's body, to inhibit bacterial metabolism without impeding the person's metabolism.

Many of these inhibitors function because their structure is similar enough to the enzyme's normal substrate that they can compete with the substrate in binding to the enzyme's active site. Unlike the substrate, however, the inhibitor cannot be converted to products and, while in the active site, prevents normal substrate bind-

ing (Figure 1a). In this state, the enzyme ceases to function.

Sulfa drugs—compounds that have saved countless human lives—provide an example of this type of **competitive inhibitor.** These agents are given to people to fight bacteria that cause diseases ranging from urinary bladder infections to pneumonia. The drugs block the ability of bacteria to transform a particular chemical in their cells—*para-aminobenzoic acid* (PABA)—to the essential coenzyme *folic acid.* The structural similarity between sulfa drugs and PABA (Figure 1b) creates metabolic confusion in the bacteria. Although many critical reactions in the human body also depend on folic acid, we cannot naturally manufacture this coen-

zyme. Instead, all of our folic acid is obtained as a vitamin in our diet. For this reason, we are not affected by the sulfa drugs in the same way as the bacteria, which must make their own folic acid. As a result, we survive the treatment unimpaired, but the bacteria do not.

Penicillin is another drug that owes its lifesaving antibacterial activity to its ability to compete with the substrate of a bacterial enzyme that directs formation of the cell wall of the bacterium. Penicillin is even more effective than the sulfa drugs: Once penicillin enters the active site of the enzyme instead of the normal substrate, it forms a covalent bond with the enzyme, permanently inactivating it. Without their cell walls, bacteria responsible for such diseases as pneumonia and syphilis are readily destroyed.

Enzymes may also be inactivated by heavy metals, such as lead, silver, mercury, and arsenic. In contrast to competitive inhibitors, these poisonous substances bear no resemblance to the enzyme's substrates but bind nonspecifically to many proteins. Some of these inhibitors inactivate enzymes by altering their shape; others bind in place of required cofactors. Since these metals act as inhibitors of essential human enzymes, as well as those of bacterial cells, they are not particularly useful for treating human diseases. Before the widespread use of penicillin during the 1940s, however, arsenic-containing solutions were commonly employed as "home remedies" and may have been responsible for the gradual poisoning of large numbers of people who used these drugs. The bacteria responsible for syphilis are particularly sensitive to an arsenic compound developed by the microbiologist Paul Ehrlich in 1910. The drug—which was the first compound discovered that would kill bacteria without also killing the human taking the drug—was the 606th arsenic compound that Ehrlich had methodically synthesized in the laboratory and tested.

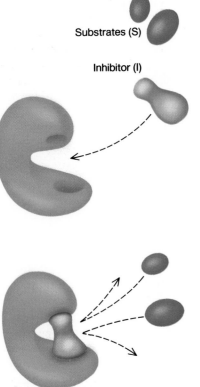

Substrates (S)

Inhibitor (I)

Enzyme-inhibitor complex
(active site blocked)

(a)

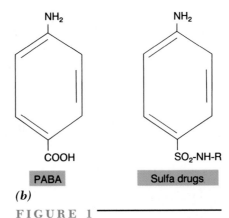

NH₂

COOH

PABA

NH₂

SO₂-NH-R

Sulfa drugs

(b)

FIGURE 1

Competitive inhibition. *(a)* Competitive inhibition of enzyme activity is due to the structural similarity betwen the inhibitor and the substrate(s). The enzyme's active site recognizes both the inhibitor (I) and the substrate (S), but the inhibitor cannot be converted to products. *(b)* Due to its similar structure, the sulfa drug acts as a competitive inhibitor of PABA, the enzyme's normal substrate.

Cofactors: Nonprotein Helpers

Although enzymes are proteins, many enzymes utilize nonprotein helpers, called **cofactors,** which are required for the enzyme to carry out its function. Depending on the enzyme, the cofactor may be an organic molecule, termed a **coenzyme,** or a metal atom. Cofactors typically carry out chemical activities for which amino acids are not well suited. Thus, they add to the repertoire of chemical reactions that enzymes can catalyze. The importance of coenzymes is illustrated by your daily need for vitamins. For the most part, these nutritional supplements act as essential coenzymes that you cannot synthesize for yourself.

THE EFFECT OF TEMPERATURE AND pH ON ENZYME ACTIVITY

Increases in temperature boost enzyme activity for the same reason that heat accelerates chemical reactions—they increase molecular motion. As a result, more reactant molecules possess the activation energy needed to be converted to products. Above a critical temperature, however, the rate of an enzymatic reaction rapidly decreases because enzymes suffer heat damage. Excessive heat disrupts hydrogen bonding and other forces that stabilize the shape of a protein. Enzymes change shape, lose solubility, and coagulate (thicken), as illustrated by the protein in a cooked egg (Figure 6-10). Proper cooking destroys *(denatures)* the proteins of many disease-causing microbes residing in food. This is one of the reasons why it is safer to eat cooked meats and eggs than raw ones. Pasteurization (controlled heating) of milk and other dairy foods has dramatically reduced the spread of serious milk-borne diseases, such as tuberculosis. High temperatures can also destroy human enzymes, however, which is one reason why high, prolonged fevers can be life-threatening and should be taken seriously.

Enzyme activity is also altered by changes in pH. A rise or fall in H^+ concentration can change the charge of many of the amino acids, thus affecting the structure of the active site. An enzyme that functions optimally at neutral pH (7.0) will usually be inactivated when its environment becomes either too basic or too acidic. Not all enzymes have their optimum near pH 7, however. Enzymes that act in an acidic environment, such as within a lysosome or the stomach of a mammal, have a structure conducive to functioning at very low pH.

Now that we have described the structure and properties of enzymes, we can examine how these remarkable biological catalysts are functionally linked into pathways to form molecules needed in the construction of cellular materials.

Life demands precisely orchestrated chemical activity. Cells govern metabolic activities by producing specific enzymes that vastly increase the rate of specific chemical reactions. Reactions promoted by enzymes generate only the desired products, which prevents metabolic chaos. (See CTQ #3.)

METABOLIC PATHWAYS

The formation of complex biochemicals within a cell occurs in stepwise fashion, similar to the way an automobile or appliance is built on an assembly line. Each biochemical is either assembled or dismantled by an orderly sequence of chemical reactions which comprise a **metabolic pathway.** Each pathway is catalyzed by a series of specific enzymes in which the product(s) of one reaction become the substrate(s) of the following reaction. The substances formed along the pathway (**b**, **c**, and **d** below) are called **metabolic intermediates.**

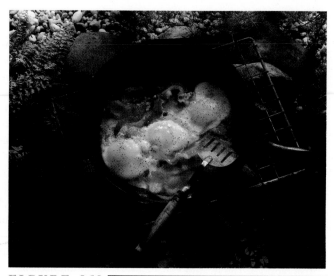

FIGURE 6-10

Denaturation. The protein in the bacon and eggs has been denatured by heat.

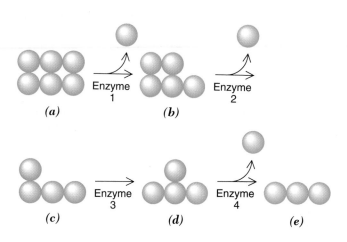

Many of the same metabolic pathways are present in virtually all modern organisms—from bacteria to mammals to trees—indicating that they arose at an early stage in biological evolution and have been retained for billions of years.

Metabolic pathways fall into two categories, depending on whether the products are more or less complex than the original substrates. **Catabolic pathways** (illustrated by the diagram above) degrade complex compounds into simpler molecules, releasing building materials and chemical energy. Therefore, catabolic pathways are exergonic (energy releasing). In contrast, **anabolic pathways** lead to the biosynthesis of complex molecules from simpler components. Since molecular construction requires energy, anabolic reactions are said to be endergonic (energy absorbing). But there is more to the differences between anabolism and catabolism than just molecular complexity and energy transfers. A key feature of both types of pathways is the transfer of electrons. In order to understand these electron transactions, we must first consider the processes of oxidation and reduction.

OXIDATION AND REDUCTION: A MATTER OF ELECTRONS

Oxidation and reduction describe the relationship between atoms and electrons. For example, we saw in Chapter 4 how the electrons of a covalent bond may be shared unequally. When a carbon atom is covalently bonded to a hydrogen atom, the carbon atom has the strongest pull on the shared electrons. Because of its greater "possession" of electrons, carbon atoms bonded to hydrogen atoms are said to be in a *reduced* state. In contrast, if a carbon atom is bonded to a more electron-attracting atom, such as an oxygen or a nitrogen atom, the electrons are pulled *away* from the carbon atom; the carbon is said to be in an *oxidized* state. But the state of **reduction** or **oxidation** of a carbon atom is not an all-or-nothing condition. Since carbon has four outer-shell electrons it can share with other atoms, it can exist in various *oxidation states*. This is illustrated by a series of one-carbon molecules (Figure 6-11), ranging from the fully reduced methane (CH_4) to the fully oxidized carbon dioxide (CO_2). The oxidation state of an organic molecule is a measure of its energy content. The compounds that we use as chemical fuels to run our furnaces and automobiles are highly reduced organic molecules, such as natural gas (CH_4) and petroleum derivatives. Energy is released when these molecules are burned in the presence of oxygen, converting the carbons to more oxidized forms, primarily carbon dioxide gas. The energy-rich fuel molecules in a living organism are its carbohydrates and fats. Carbohydrates are rich in chemical energy because they contain strings of (H—C—OH) units. Fats contain even greater energy per unit weight because they contain strings of more reduced (H—C—H) units. Oxidation of these cellular fuels provides organisms with their usable energy.

Transferring Electrons during Chemical Reactions

We used the terms reduced and oxidized in the previous sections to describe the electronic state of a carbon atom. These terms are also used to describe a *change* in that state.

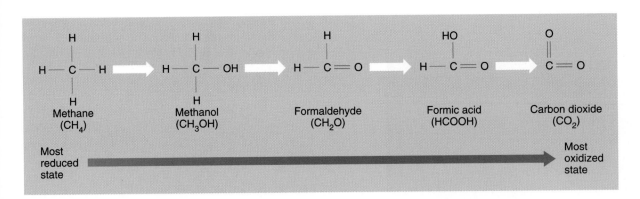

— Covalent bond in which carbon atom has greater share of electron pair
— Covalent bond in which oxygen atom has greater share of electron pair

FIGURE 6-11

The oxidation state of a carbon atom. The oxidation state of a given carbon atom depends on the other atoms to which it is bonded. Each carbon atom can form a maximum of four bonds with other atoms. This series of simple, one-carbon molecules illustrates the various oxidation states in which the carbon atom can exist. In its most reduced state, the carbon is bonded to four hydrogens (forming methane); in its most oxidized state it is bonded to two oxygens (forming carbon dioxide). The relative oxidation state of an organic molecule can be roughly determined by counting the number of hydrogen versus oxygen (or nitrogen) atoms per carbon atom.

A molecule is said to be *oxidized* when it loses one or more electrons, whereas it is said to be *reduced* when it gains one or more electrons. Electrons lost during oxidation are not simply released free into the medium, however; instead, they are immediately captured by another molecule, which is therefore reduced in the process. Consequently, whenever one substance is oxidized, another must be reduced.

We can illustrate these features by examining one of the key reactions of the anabolic pathway of photosynthesis. Don't be concerned with the chemical structures of the molecules in these reactions. Your attention is best directed to the red carbons and the transfer of the electrons indicated by the circles.

In this reaction, a pair of electrons (together with a proton) are transferred from NADPH (nicotinamide adenine dinucleotide phosphate) to the substrate DPG. NADPH is an important molecule in many anabolic reactions that occur in all types of organisms, from bacteria to humans. NADPH functions primarily as an electron donor, raising the energy level and state of reduction of the substrate. A cell's reserve of NADPH is referred to as its **reducing power,** which is one measure of a cell's energy supply. As a result of the transference of electrons in the above reaction, the NADPH becomes oxidized to $NADP^+$, while the DPG becomes reduced to PGAL. This reduction is evidenced by the red carbon; before the reaction, the carbon was bonded to an oxygen, and after the reaction, it is bonded to a hydrogen. In this reaction, energy has been transferred in the form of electrons from the coenzyme NADPH to the substrate, leaving the substrate with a higher energy content.

DIRECTING METABOLIC TRAFFIC

Just as organisms must adapt to changes in external conditions, cells must respond to changes in internal conditions. One of the mechanisms by which cells maintain an ordered internal environment is by directing materials into the metabolic pathways that best satisfy the cell's needs at that particular moment. Glucose, for example, can be directed along several different metabolic routes. It can be dismantled into carbon dioxide and water to generate energy; it can be linked to other glucose molecules to form a polysaccharide; it can be partially disassembled into fragments used to build lipids; or it can be modified to form amino acids or nucleotides. Even the simplest cell can accurately "evaluate" these needs and direct the fate of glucose (and the thousands of other molecules for which multiple options exist). How can cells "decide" which of these options best satisfies its needs? The answer is they can't, but there is no need for them to because a cell has built-in mechanisms that regulate enzyme activities according to the cell's needs.

Changing an enzyme's activity is usually accomplished by modifying the enzyme in a way that alters the shape of its active site. Two of the most common mechanisms for doing so are covalent modification and feedback inhibition.

Altering Enzyme Activity by Covalent Modification

During the mid 1950s, Edmond Fischer and Edwin Krebs of the University of Washington were studying *phosphorylase*, an enzyme found in muscle cells, which disassembled glycogen into its glucose subunits. The enzyme could exist in either an inactive or an active form. Fischer and Krebs prepared a crude extract of muscle cells and found that inactive enzyme molecules in the extract could be converted to active ones simply by adding ATP to the test tube. Further analysis revealed a second enzyme in the extract— a "converting enzyme," as they called it—that transferred a phosphate group from ATP to one of the 841 amino acids that make up the glycogen-disassembling enzyme. The presence of the phosphate group altered the shape of the active site of the enzyme molecule and increased its catalytic activity. Subsequent research has shown that covalent modification of enzymes, as illustrated by the addition of phosphates, is a general mechanism for activating (or inactivating) enzymes. Enzymes that transfer phosphate groups to other proteins are called **protein kinases,** and are involved in regulating such diverse activities as hormone action, cell division, and gene expression. The discovery of protein kinases earned Krebs and Fischer the Nobel Prize in 1992.

Altering Enzyme Activity by Feedback Inhibition

Feedback inhibition is a mechanism whereby a key enzyme in a metabolic pathway is temporarily inactivated when the concentration of the end product of that pathway —an amino acid, for example—becomes elevated. This is illustrated by the simple pathway depicted in Figure 6-12 in which two substrates A and B are converted to the end product E. As the concentration of the product E—an amino acid, for example—rises, it binds to the enzyme BC, causing a conformational change in the enzyme that decreases the enzyme's activity.

Feedback inhibition is analogous to a refrigerator or air conditioner whose function is to generate cold air rather than a chemical product. When the environment becomes too cold, the cold air acts on the thermostat to switch off production of additional cold air. In the case of feedback inhibition, the end product E does not bind to the enzyme's active site but to a separate "feedback site" on the large enzyme molecule. Like the sequential collapse of a row of dominoes, the binding of the end product to the feedback site sends a "ripple" through the protein, producing a de-

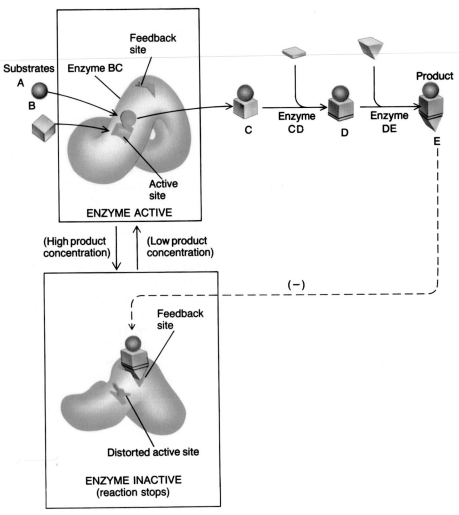

FIGURE 6-12

Metabolic control by feedback inhibition. When concentrations of the product E are low, the first enzyme (BC) is active, and the pathway proceeds to completion. As the end product (E) accumulates, it binds to the enzyme's feedback (or *allosteric*) site, changing the shape of the enzyme's active site, decreasing further production of the end product (E).

fined change in the shape of the active site, thereby decreasing the enzyme's activity. Feedback inhibition prevents an organism from wasting resources by continuing to produce compounds that are not needed at the time and may even become toxic if allowed to accumulate to high levels.

Control of Enzyme Synthesis

Although covalent modification and feedback inhibition of enzyme activity may provide immediate inactivation of a metabolic pathway, these activities still allow energy to be wasted on synthesizing unneeded enzymes. Organisms possess additional control mechanisms that prevent such metabolic extravagance by inhibiting the continued production of unneeded enzymes. For example, the enzymes

for digesting the sugar lactose are not manufactured when there is no lactose to digest. The mechanism by which cells avoid squandering their resources, making unneeded enzymes, is described in Chapter 15.

The molecules of biological importance are usually quite different from one another structurally and cannot be converted from one to another in a single step. Rather, such conversions require a series of chemical reactions, each catalyzed by a different enzyme, forming a metabolic pathway. Certain key reactions in these pathways include the transfer of electrons from one substrate to another, a process that changes the energy level of the substrates and products. (See CTQ #4.)

◁ THE HUMAN PERSPECTIVE ▷
The Manipulation of Human Enzymes

Recent advances in recombinant DNA technology (Chapter 16) have allowed investigators to isolate an individual gene from human chromosomes, to alter its information content, and to synthesize the modified protein with its altered amino acid sequence. One protein that has been the subject of such manipulation is tissue plasminogen activator (TPA), a key enzyme in the complex process by which blood clots are naturally dissolved in the human body. Human TPA is one of the first proteins to be produced commercially by the biotechnology industry, and its synthesis is finding widespread application in hospital emergency rooms around the world.

Most heart attacks result from blood clots that become lodged in the coronary arteries—the vessels that carry fresh oxygenated blood to the heart. TPA is a "clot-buster," an enzyme that, if injected very soon after a heart attack, dissolves the blood clot that is blocking the flow of blood to the heart tissues. The enzyme presently being marketed is identical to that produced naturally within the human body. There are certain problems with the natural form of the enzyme, however. For example, the blood contains a TPA inhibitor that binds to an inhibitor site on the enzyme, decreasing activity at the active site. In response to this problem, a synthetic version of the enzyme has been produced which lacks a number of amino acids in the inhibitor-binding site, thus preventing the

body's natural TPA inhibitor from binding to the synthetic enzyme.

A much more drastically modified version of TPA has been produced by combining the catalytic portion of the TPA molecule with a totally different protein (an antibody molecule). The antibody portion of this "hybrid enzyme" acts to target the molecule directly to the blood clot with which it binds very tightly. This brings the TPA portion of the hybrid enzyme into very close proximity with its intended substrate. In the laboratory, this hybrid enzyme is a very effective clot dissolver, but whether or not such modified enzymes are safe for human use remains to be determined.

MACROMOLECULAR EVOLUTION AND ADAPTATION OF ENZYMES

▐▶ Enzymes subject to feedback inhibition probably began as simple catalytic proteins without a separate feedback site. Over time, natural selection favored the survival of organisms whose enzymes were inhibited by metabolic products because these organisms made more efficient use of available nutrients than did their counterparts lacking such regulatory mechanisms. This, in turn, led to the appearance of enzymes with sites to which metabolic end products could bind.

"Fingerprints" left by evolution can also be revealed by comparing the same enzyme in similar organisms that live under very different environmental conditions. Consider two species of bacteria, one living in hot springs in Yellowstone Park at temperatures above 90°C, and another living in a nearby pond fed by cool spring water. The enzymes present in these two types of bacteria have their optimal activity at drastically different temperatures. For example, the purified, DNA-synthesizing enzyme from the bacteria that lives in the hot springs is active in the test tube at 90°C, while the same type of enzyme from the other bacteria is totally inactive at this temperature. Differences in primary, secondary, and tertiary structure provide added stability to the enzyme having the higher temperature optimum. Biotechnologists have taken advantage of the temperature stability of the DNA-synthesizing enzyme from hot springs bacteria to develop a revolutionary new technique that allows them to produce large amounts of DNA from a miniscule amount of starting material (Chapter 16). Another example of the scientific manipulation of enzymes is discussed in the accompanying Human Perspective box.

Enzymes are adaptations, just as are fingers in humans and feathers in birds. Just as natural selection leads to the survival of species with more efficient anatomic traits, it also leads to survival of species with more efficient enzymes. (See CTQ #6.)

REEXAMINING THE THEMES

Relationship between Form and Function

The ability of an enzyme to catalyze a specific reaction derives from its structure, most notably from the position of particular amino acids within the active site. The active site of an enzyme has a precise molecular shape that selectively binds to a specific substrate and converts the molecule to a product. If the shape of the active site should be altered, even slightly (as occurs, for example, when the feedback site of an enzyme is occupied by an end product), the activity of the enzyme may be greatly decreased.

Biological Order, Regulation, and Homeostasis

Just as organisms must maintain a stable, ordered internal environment to remain alive, so too must cells. Cells accomplish this, in part, by regulating the activity of enzymes that govern critical steps in metabolic pathways. Covalent modification and feedback inhibition provide two of the best examples of biological regulation at the molecular level. During covalent modification, the activity of an enzyme is changed by the attachment of a small chemical group, such as a phosphate, by another enzyme. Feedback inhibition operates when a particular end product of a metabolic pathway reaches an elevated concentration, which promotes its binding to a key enzyme in the pathway, preventing further production of the product.

Acquiring and Using Energy

Energy is the central topic of this chapter. Here, we have defined energy, described how it is converted from one form to another, discussed how some of it inevitably raises the level of disorder in the universe, and seen how the direction taken by chemical reactions can be explained by a shift from a higher to a lower energy state. Most of the body's energy is stored in energy-rich macromolecules. Energy released by the disassembly of these macromolecules is stored temporarily in ATP, whose subsequent hydrolysis drives reactions that would otherwise be unfavored.

Unity within Diversity

No matter how different their form and function, all cells use ATP as energy currency. The same enzymes and metabolic pathways use ATP to drive endergonic reactions in organisms as diverse as bacteria and humans, illustrating that not only are the types of biochemicals remarkably similar among diverse organisms (Chapter 4), so too are the enzymes and the metabolic pathways that form these biochemicals.

Evolution and Adaptation

Enzymes are adaptations and, therefore, are the result of natural selection. For example, organisms that live at markedly different temperatures have enzymes that are active at temperatures that correspond to that of their normal environment. The DNA-synthesizing enzyme from bacteria living in certain hot springs, for example, is active at temperatures above 90°C, while the DNA-synthesizing enzyme from most bacteria is totally inactive at such high temperatures.

SYNOPSIS

Energy is the capacity to do work. Energy can occur in various forms, including chemical, mechanical, light, electrical, and thermal, which are interconvertible. Whenever an exchange of energy occurs, the total amount of energy in the universe remains constant, but there is a loss of usable energy, that is, energy available to do additional work. This loss of usable energy, called entropy, results from an increase in the randomness and disorder of the universe. Living organisms are systems of low entropy, maintained by the constant input of external energy—energy ultimately derived from the sun. All spontaneous (favorable) reactions proceed to a lower state of energy. The hydrolysis of ATP is

an example of a favorable reaction. Many unfavorable reactions that would normally fail to occur in a cell are driven by coupling them to ATP hydrolysis.

Enzymes are proteins that vastly accelerate the rate of specific chemical reactions by binding to the reactant(s) and increasing the likelihood that they will be converted to products. Enzymes act by lowering the activation energy—the energy required by reactants to undergo a reaction. To do this, reactants may be held in the proper orientation, subjected to physical stress, and made more reactive by interacting with amino acids in the protein. The specificity and catalytic activity of enzymes are due to the complementary structure of the active site that binds the substrate(s), enabling the enzyme to exert its influence. Many enzymes also contain nonprotein "helpers," which may be organic coenzymes or metals. Enzyme structure—and thus function—is affected by temperature, pH, and the presence of inhibitors.

Enzymatic reactions are organized into metabolic pathways in which the product(s) of one reaction becomes the substrate(s) of a subsequent reaction. Catabolic pathways lead to less complex products, with the accompanying release of energy. In contrast, anabolic pathways build more complex products at the expense of cellular energy. Many reactions include the transfer of electrons from one molecule to another. When a molecule receives one or more electrons it becomes reduced; when a molecule gives up one or more electrons it becomes oxidized. Electrons carry with them chemical energy; thus, the more reduced an organic molecule, the higher its energy content. The activity of many key enzymes is under cellular control and can be altered by interaction with certain metabolic end products or by covalent modification, such as the addition of a phosphate group.

Key Terms

energy (p. 118)
potential energy (p. 118)
kinetic energy (p. 118)
first law of thermodynamics (p. 119)
second law of thermodynamics (p 120)
entropy (p. 120)
exergonic (p. 122)
endergonic (p. 122)
coupling (p. 122)

catalyst (p. 123)
activation energy (p. 123)
substrate (p. 125)
active site (p. 125)
conformational change (p. 125)
competitive inhibitor (p. 126)
cofactor (p. 127)
coenzyme (p. 127)
metabolic pathway (p. 127)

metabolic intermediate (p. 127)
catabolic pathway (p. 128)
anabolic pathway (p. 128)
reduction (p. 128)
oxidation (p. 128)
reducing power (p. 129)
protein kinase (p. 129)
feedback inhibition (p. 129)

Review Questions

1. Contrast potential and kinetic energy, using both a biological and a nonbiological example.

2. Describe how enzymes decrease activation energies; how they are able to bind only specific substrates; why they may need cofactors; and why their activity is sensitive to temperature and pH.

3. What is the relationship between enzyme activity and each of the following:

   a. *increasing temperature (two answers)*
   b. *cold*
   c. *the presence of a structural analog of the substrate (an analog is a molecule with a similar shape)*
   d. *lead poisoning*
   e. *extreme acidity*

4. Why does NADPH provide a cell with reducing power?

Critical Thinking Questions

1. Suppose that Eduard Buchner had added sugar to his yeast-juice preparation and failed to observe the occurrence of fermentation. Would this have proven that fermentation could only occur in a living yeast cell. Why or why not?

2. Using the energy principles discussed in the beginning of this chapter, explain why it is a more efficient use of agricultural land for people to eat grains rather than meat from grain-fed cattle.

3. Lead was formerly used in the manufacture of house paints. Young children who live in homes where lead paint still exists unintentionally ingest flakes of it, which may result in serious and permanent brain damage. Using your knowledge of enzymes, explain how even minute quantities of lead can have such disastrous consequences.

4. Study the diagram (*right*) and identify which compounds (A–K) would *not* be produced under each of the following conditions. Explain why.

   (1) Compound E binds to and inhibits enzyme 2 when E is present in high concentrations.
   (2) A compound is added that competes with compound F for enzyme 7 but cannot be converted to compound G.
   (3) Enzyme 5 is inactivated by a change in pH.
   (4) Mercury is added and contacts enzymes 1–10.

5. How is it that feedback inhibition can block the formation of a particular product, such as an amino acid, even though it binds to only one enzyme in the entire pathway?

6. The enzyme lactase digests lactose, a disaccharide found in milk, to glucose and galactose, two monosaccharides. When lactase is not present, lactose is not digested and can cause intestinal disturbances in humans. Lactase is normally present in young children and in adults who continue to eat a diet that includes milk products, but it is not present in adults who do not consume milk. What do you think was the adaptive value of this enzyme for primitive humans?

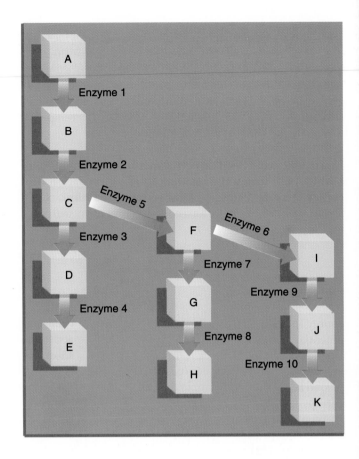

Additional Readings

Cutnel, J. D., and K. W. Johnson, 1992. *Physics.* New York: Wiley. (Introductory)

Kornberg, A. 1989. Never a dull enzyme. *Ann. Revs. Biochem.* 58:1–30. (Intermediate)

Kornberg, A. 1991. *For the love of enzymes: The odyssey of a biochemist.* Cambridge: Harvard. (Intermediate)

Laszlo, P. 1986. *Molecular correlates of biological concepts: A history of biochemistry.* Comprehensive Biochemistry,

Vol. 34A. New York: Elsevier. (Advanced)

Lehninger, A. L. 1982. *Principles of biochemistry.* New York: Worth. (Intermediate)

Stryer, L. 1988. *Biochemistry.* 3d ed., New York: Freeman. (Advanced)

Voet, D., and J. G. Voet. 1990. *Biochemistry.* New York: Wiley. (Advanced)

Movement of Materials Across Membranes

STEPS
TO
DISCOVERY
Getting Large Molecules Across Membrane Barriers

MEMBRANE PERMEABILITY

DIFFUSION: DEPENDING ON THE RANDOM MOVEMENT OF MOLECULES

Conditions For Diffusion Across Membranes
Osmosis
Facilitated Diffusion

ACTIVE TRANSPORT

Generating Ionic Gradients And Storing Energy

ENDOCYTOSIS

Phagocytosis
Pinocytosis

THE HUMAN PERSPECTIVE

LDL Cholesterol, Endocytosis, and Heart Disease

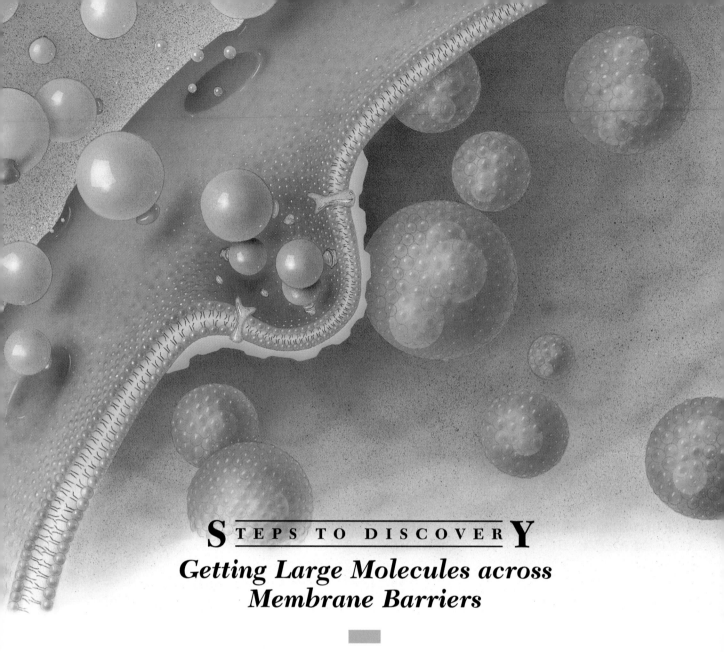

Getting Large Molecules across Membrane Barriers

*F*or many years, a question that puzzled cell biologists was how large molecules were able to get across an intact plasma membrane. Your blood, for example, delivers a variety of large molecules to all your cells. These molecules pass from the bloodstream across the plasma membrane, and into the cytoplasm of the cells in which they are used. It was not until the 1970s that this process came to be understood at the molecular level, primarily as a result of the work of Michael Brown and Joseph Goldstein at the University of Texas.

Brown and Goldstein stumbled into the area of mem-

brane transport as a result of their interest in a genetic disease called *familial hypercholesterolemia* in which people suffer from extremely high levels of blood cholesterol. Cholesterol is a hydrophobic molecule. Because it is virtually insoluble in water, cholesterol is transported through the circulation in special packages called low-density lipoproteins (LDLs). LDLs are spheres that contain about 1,500 cholesterol molecules packaged in a sac of phospholipids and protein that keeps the enclosed cholesterol suspended in the blood. Cells take up these large LDLs and

Cholesterol-containing particles enter a cell after binding to receptors situated in coated pits on the cell surface. The particles

liberate the cholesterol molecules, which—depending on the type of cell—may be used for various functions, including assembling cell membranes, forming steroid hormones, or producing bile in the liver.

Brown and Goldstein began their studies on cholesterol by examining cells growing in cultures, thus avoiding the complications of working with whole animals. The investigators used cells called *fibroblasts*, which remove cholesterol-containing LDLs from the culture medium in which they are growing. Brown and Goldstein incubated the fibroblasts with a preparation of LDL that contained radioactive atoms (the radioisotope acted as a marker to reveal the location of the labeled LDLs). They found that the radioactively labeled LDLs bound tightly to the surface of the fibroblast, suggesting that the surface of these cells contained *receptor* molecules with binding sites that specifically fit the LDLs. Within 10 minutes of attachment, virtually all of the bound LDL had entered the cell and was being converted to cholesterol for use in cellular metabolism. The LDL receptors then returned to the plasma membrane where they could bind additional LDLs.

These early findings raised an important question: How do these giant LDL particles get into the cell? To answer this question, Brown and Goldstein, together with Richard Anderson, prepared LDLs that were chemically bonded to iron particles (which are easily seen in the electron microscope). When they examined the cells under the microscope, they were surprised to find that the iron-containing LDL particles were not spread uniformly over the fibroblast surface, but were concentrated into special sites. These sites are called *coated pits* because the membrane is indented to form a pit that is coated on its inner surface by a

layer of bristly looking protein. This finding suggested that the coated pits were sites where LDL receptors became concentrated. The investigators envisioned the LDL receptors as proteins that spanned the membrane. As these proteins moved around within the membrane, they would inevitably bump into a coated pit. When this happened, the receptors would bind to the bristly coating and remain in the pit. Viewed in this way, the coated pits were acting as "molecular traps" that ensnared the LDL receptors, concentrating them within small regions of the membrane. When they waited longer periods of time before killing and examining the cells, the scientists found that the coated pits eventually bulged inward, forming pouchlike structures, which then pinched off from the remainder of the membrane, forming bristly coated cytoplasmic vesicles (see Figure 7-11). The bristly protein and the LDL receptors then returned to the membrane, while the LDL was broken down by lysosomal enzymes, releasing free cholesterol molecules, which passed into the cytoplasm for use by the cell.

Within a few years, a dozen or so different receptors were discovered which acted in a similar manner to the LDL receptor, suggesting that this was a general mechanism for the specific ingestion of a variety of large, blood-borne molecules or particles. For their work on the role of the LDL receptor, as well as their important discoveries concerning cholesterol metabolism, Brown and Goldstein were awarded the 1985 Nobel Prize for Medicine.

are enclosed in vesicles pinched off from the plasma membrane.

As we have seen in previous chapters, a living cell is a bastion of complexity and order—a microscopic package that contains thousands of different molecules whose concentrations are maintained by carefully controlled reactions. A cell's internal order is protected by the plasma membrane—a structure so thin it would take approximately 40,000 of them stacked one on top of the other to equal the thickness of one page of this book. The plasma membrane keeps essential substances inside the cell and regulates the passage of materials between a cell and its environment. The basis for these functions becomes apparent when we consider the plasma membrane's structure.

▼ ▼ ▼

MEMBRANE PERMEABILITY

We described in Chapter 5 how the plasma membrane is a mosaic of glycoproteins embedded in a fluid, bilayered sheet of phospholipids (Figure 7-1; see also Figure 5-4). The two major components of the plasma membrane—its proteins and lipid bilayer—affect whether substances enter or leave a cell. On the one hand, the lipid bilayer acts primarily as a barrier, blocking the passage of most molecules based primarily on their insolubility in lipid. On the other hand, membrane proteins provide an alternate pathway across the membrane for a substance that would be unable to penetrate a hydrophobic lipid sheet. Unlike the lipid bilayer, which allows any lipid-soluble material to pass, membrane proteins are able to *control* the entrance and exit of "selected" substances. Just as enzymes bind specific substrates and catalyze their reaction, these membrane proteins bind specific solutes and promote their movement across the membrane.

Together, the phospholipid bilayer and embedded proteins make the plasma membrane **selectively permeable;** that is, it allows certain substances to cross, while restricting the passage of others. Selective permeability enables a cell to import and accumulate essential molecules to concentrations high enough for normal metabolism. It also enables the cell to export wastes and other substances that might interfere with metabolism. The appearance of a lipid-containing membrane that allowed a primitive "cell" to retain its nutrients and other essential materials must have been one of the most critical steps in the early evolution of life on earth.

If we are to understand the selective permeability of the plasma membrane, we need to consider the means by which individual molecules can traverse the membrane. There are basically two means for such movement: passively, by diffusion, or actively, by a transport process requiring an energy boost from the cell.

Normal cell function requires a dynamic barrier between the cell and its surroundings in order to prevent dilution of cell contents, to safeguard contents against external conditions, and to maintain a concentration of essential molecules and ions inside the cell different from that in the surrounding fluid. The plasma membrane accomplishes these functions by selectively controlling the materials entering and exiting the cell. (See CTQ #2.)

DIFFUSION: DEPENDING ON THE RANDOM MOVEMENT OF MOLECULES

Diffusion is the passive movement of molecules in a gas or liquid from a region where they are present at higher concentration to a region where they are present at lower concentration; in other words, *down a concentration gradient.* Diffusion is readily demonstrated by dropping colored dye into a glass of water. The dye molecules are localized at first in one region, which appears dark in color. After time, however, the color spreads until the dye molecules are distributed rather evenly throughout the glass. Let's examine why this happens.

All molecules exist in a state of continuous, random (that is, unpredictable) movement. If molecular move-

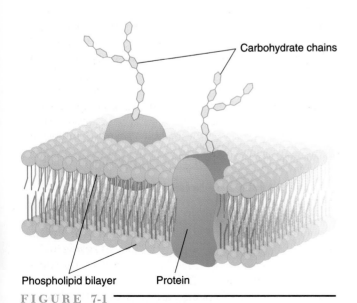

Carbohydrate chains

Phospholipid bilayer   Protein

FIGURE 7-1

The structure of the plasma membrane.

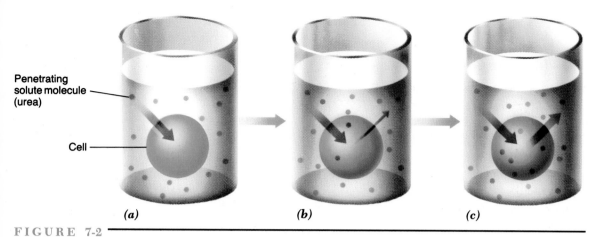

FIGURE 7-2

Diffusion of a solute through the plasma membrane. In this example, a cell is dropped into a solution containing a solute (urea) capable of penetrating the membrane. Initially, the concentration of solute is much greater outside the cell than inside, resulting in the rapid movement of solute into the cell *(a)*. After the solute enters the cell, some of the molecules will diffuse back across the membrane to the outside, but the *net* movement (as indicated by the larger arrow) will still be into the cell *(b)*. Eventually, the concentration on the two sides of the membrane will be equal, and the movement of solute will occur at the same rate in both directions *(c)*.

ments are random, why is there a predictable diffusion of molecules from a region of high concentration to one of low concentration? The answer can best be visualized using an analogy. Consider what would happen if two people were to go into a room with a line painted down the center, and each person were to stand at equal distances from the line on opposite sides. Then, one person were to drop ten bouncing rubber balls on his side of the line and the other were to drop only a single ball on her side. Even though the movement of each ball occurred at random, the chances would be very high that, after the balls had come to rest, there would be more than one ball on the second person's side and fewer than ten balls on the first person's side. In other words, a *net* movement of balls from the side of initial high concentration to the side of initial low concentration is likely to occur. Thus, even though the movement of an individual ball (or molecule) is not predictable, the collective action of a large number of them is quite predictable. In the case of molecules (Figure 7-2), which, unlike rubber balls, never come to rest, the ultimate outcome will be a state of equilibrium in which the concentration of molecules will be equal on both sides of the membrane. When this occurs, molecules will be crossing back and forth across the membrane at equal rates; there will no longer be any *net* movement of molecules. This process is known as *simple diffusion*. Diffusion plays a key role in the movement of many substances within the body. When you breathe, for example, oxygen moves by diffusion from the lungs into the bloodstream. Then, when the oxygen-rich blood moves through the tis-

sues, oxygen again moves by diffusion out of the blood into the cells where it is utilized.

CONDITIONS FOR DIFFUSION ACROSS MEMBRANES

Two qualifications must be met before a substance can diffuse across a plasma membrane. First, the substance must be present at a higher concentration on one side of the membrane than on the other, and second, the membrane must be permeable to (able to be penetrated by) the substance. The *rate* of diffusion of a given substance across the plasma membrane—that is, the number of molecules that will diffuse through a given section of membrane per unit time—depends on a number of conditions. In general, the lipid bilayer is much more permeable to nonpolar molecules and to small molecules than to ions, polar molecules, or large molecules. Some cells are quite permeable to specific ions (Na^+, K^+, or Ca^{2+}) because they contain special channels that allow the ions to pass unimpeded across the lipid bilayer. These ion channels consist of clusters of membrane proteins that form a "doughnut-shaped" complex that spans the membrane (Figure 7-3). The ions pass through the center of the "doughnut." Most ion channels are equipped with *gates* that may be either open or closed. The opening and closing of these gates play a key role in the movement of an impulse along the membrane of a nerve cell (discussed in Chapter 23). Two types of diffusion of particular importance in biology are osmosis and facilitated diffusion.

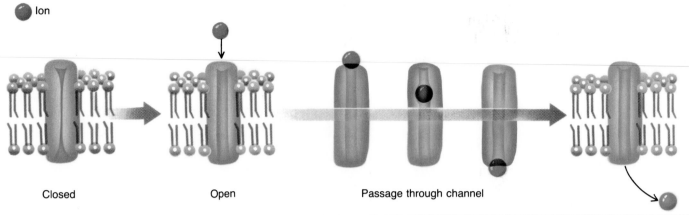

Ion

Closed — Open — Passage through channel

FIGURE 7-3
Ion channels consist of pores situated within the center of membrane proteins or within clusters of such proteins. Most ion channels possess a "gate" that can exist in either an open or a closed conformation, depending on conditions in the cell.

OSMOSIS

Even though water is highly polar, it diffuses readily across the lipid bilayer of the plasma membrane because it is a very small molecule and is able to squeeze between neighboring phospholipids. The diffusion of water through a membrane has a special name—**osmosis.** In understanding osmosis, let us consider what might happen to a sea urchin egg under different environmental conditions. Sea urchin eggs (a favorite of both embryologists and gourmet diners) are spawned into sea water, where they are fertilized and develop into microscopic swimming larvae. The concentration of dissolved substances (solutes) in a sea urchin egg is virtually identical, or **isotonic,** to that of the surrounding sea water. In such isotonic (*iso* = equal) solutions, the movement of water into and out of a cell is equal; therefore, the egg neither gains nor loses water by osmosis (Figure 7-4*a*).

Now let us consider what might happen if the same egg should be caught in a shallow tidepool on a hot day at a very low tide. As the water in the tidepool evaporates, its solute concentration rises and, conversely, its water concentration decreases. As a consequence of the difference in water concentration between the egg and its surroundings, water will move out of the cell by osmosis, causing the cell to shrink. In other words, cells shrink when surrounded by **hypertonic** solutions—solutions with a higher solute concentration (*hyper* = more) than that found inside the cell (Figure 7-4*b*).

Finally, let us consider what might happen if the sea urchin egg drifts into an *estuary,* a site along the coastline where a freshwater river empties into the sea. Under these conditions, the egg would find itself in a medium of diluted sea water, which would have a lower solute concentration

—and, thus, a higher water concentration—than that of the egg. In this case, the external medium is said to be **hypotonic** to that of the egg (*hypo* = less), and water will move by osmosis into the egg, causing the egg to swell (Figure 7-4*c*).

To summarize, water always moves across a membrane by osmosis from the hypotonic compartment (one with a lower solute concentration) to the hypertonic compartment (one with a higher solute concentration). It makes no difference if the types of solute molecules on the two sides of the membrane are completely different—the only important factor is the total solute concentration on each side.

▐▌▶ Osmosis is a fact of life with which cells—and entire organisms—have to live. In some circumstances, osmosis is a challenge that natural selection has been able to overcome. For example, as organisms moved from the sea to inhabit fresh waters, they generally moved from an environment that was isotonic to their bodily fluids into one that was very hypotonic. These organisms evolved adaptations that allowed them to expel the excess water that flooded into their bodies (Chapter 28). In other circumstances, osmosis has facilitated certain bodily functions. Your digestive tract, for example, produces several liters of fluid, which is reabsorbed osmotically by the cells that line your intestine. If this reabsorption process did not occur, as happens in cases of extreme diarrhea, you would face the prospect of rapid dehydration.

Plants utilize osmosis in other ways. The movement of water is affected not only by solute concentration, but also by pressure. For example, when a plant cell is placed in a hypotonic solution, water diffuses into the cell by osmosis. Because the cell has a rigid wall that resists expanding, however, the plant cell will not swell to the point of burst-

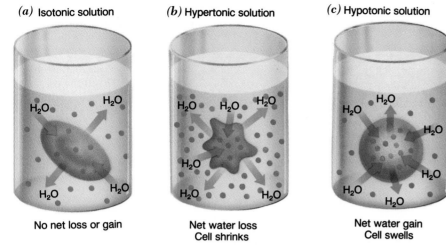

(a) Isotonic solution

H_2O H_2O
H_2O H_2O

No net loss or gain

(b) Hypertonic solution

H_2O H_2O H_2O
H_2O
H_2O H_2O

Net water loss
Cell shrinks

(c) Hypotonic solution

H_2O
H_2O
H_2O
H_2O H_2O
H_2O

Net water gain
Cell swells

FIGURE 7-4

The effects of isotonic, hypertonic, and hypotonic solutions on osmosis. *(a)* A cell placed in an isotonic solution (one containing the same solute concentration as the cell) neither swells nor shrinks because it gains and loses equal amounts of water. *(b)* A cell in a hypertonic solution (one containing a higher solute concentration than the cell) soon shrinks because of a net loss of water by osmosis. *(c)* A cell placed in a hypotonic solution (one having a lower solute concentration than the cell) swells because of a net gain of water by osmosis.

ing, as would an animal cell. Instead, the internal water pressure builds as water continues to diffuse into the plant cell. Eventually, the water pressure builds high enough inside the cell to stop the influx of more water by osmosis, even though the solute concentration inside the cell remains higher than the surrounding solution. Plant cells utilize the water pressure that develops due to osmosis to maintain their rigidity. This internal pressure, called **turgor pressure,** provides support for nonwoody plants or for the nonwoody parts of plants, such as the leaves. When plant cells lose water, as occurs when they are placed in a hypertonic solution, the cell shrinks away from its cell wall, a process termed **plasmolysis** (Figure 7-5). The loss of water due to osmosis (or evaporation in a dry, terrestrial habitat) causes plants to lose their support and wilt.

FACILITATED DIFFUSION

As we mentioned earlier, the diffusion of a substance across a membrane always occurs from a region of higher concentration to a region of lower concentration, but the penetrating molecules do not always move through the membrane unaided. Many plasma membranes contain **carrier proteins** that bind specific substances and facilitate their diffusion across the membrane. This process, termed **facilitated diffusion** (Figure 7-6), might be likened to a ferry

moving people across a river. Like the lipid bilayer, the river is a barrier to traffic, while the ferry provides a specialized avenue across the barrier. The ferry can move its cargo in both directions but, as long as there are more people waiting on the north side than on the south side of the river, there will be a net flow of traffic from north to south. Similarly, in facilitated diffusion, as long as there are more molecules of a substance on the outside of the cell, there will be a net flow of those molecules into the cytoplasm. Facilitated diffusion is particularly common in the inward movement of sugars and amino acids, substances that would be unable to penetrate the hydrophobic lipid bilayer directly. Just as enzymes are specific for particular substrates, carrier proteins are specific for particular solutes; a protein that facilitates diffusion of glucose, for example, will not carry amino acids.

If a specific solute is present at higher concentration on the outside of the plasma membrane than in the cytoplasm, and the membrane is permeable to that particular solute, then there will be a net movement of the solute into the cell without the need for the cell to expend energy. If the membrane is *not* permeable to the solute, then water will move in the direction of higher solute concentration. Such osmotic movements of water are critical to life. (See CTQ #3.)

142 • P A R T 2 / *Chemical and Cellular Foundations of Life*

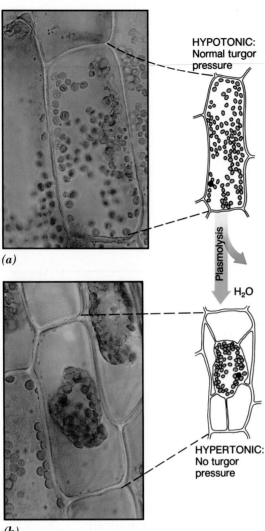

HYPOTONIC:
Normal turgor
pressure

(a)

Plasmolysis

H_2O

HYPERTONIC:
No turgor
pressure

(b)

FIGURE 7-5

The effects of osmosis in plants. Aquatic plants living in fresh water are surrounded by a hypotonic environment. Water therefore tends to flow into the cells, creating high water (*turgor*) pressure *(a)*. If the plant is placed in a hypertonic solution, such as seawater, the cell loses water, and the plasma membrane pulls away from the cell wall, a process known as *plasmolysis* *(b)*. If the cells of land plants evaporate large amounts of water to dry air, water pressure plunges, and the plant wilts.

ACTIVE TRANSPORT

A frog sitting in a shallow freshwater pond has internal fluids with a solute concentration 40 to 50 times greater than its aqueous environment, which is not much different from pure water. How is this frog able to maintain a salt concentration that is so much higher than its surrounding medium? It can do so because its skin contains proteins capable of "pumping" salts from the environment into the body *against* a very large concentration gradient (Figure 7-7). Movement of substances against a concentration gradient—that is, from a region of lower concentration to one of higher concentration—is called **active transport.** This process requires the input of considerable chemical energy, which is either directly or indirectly supplied by the hydrolysis of ATP.

GENERATING IONIC GRADIENTS AND STORING ENERGY

One of the most widely occurring active transport proteins is the *sodium–potassium pump,* a protein found in all types of eukaryotes. This protein transports sodium ions out of cells and potassium ions into cells, both against steep concentration gradients (Figure 7-8). Human nerve cells, for example, have internal concentrations of potassium that are 30 times greater than that of the extracellular fluid, due to the pumping activities of this protein. The sodium–potassium pump is more than just a transport protein, it is also an enzyme—an *ATPase.* As an ATPase, the protein hydrolyzes ATP, utilizing the energy to move ions against their concentration gradients. During this reaction, the phosphate group released by ATP hydrolysis becomes transiently linked to the transport protein, changing the shape

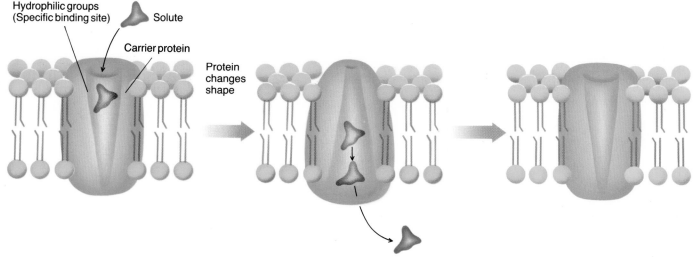

Hydrophilic groups
(Specific binding site)

Solute

Carrier protein

Protein
changes
shape

FIGURE 7-6

Facilitated diffusion. Facilitated diffusion is responsible for the uptake of many small hydrophilic molecules, such as sugars and amino acids. Facilitated diffusion is initiated when a solute molecule binds to a carrier protein at the outer surface of the membrane. The binding of the solute is thought to trigger a conformational change within the carrier protein, exposing the solute molecule to the inner surface of the membrane. The solute molecule is then free to diffuse into the cytoplasm down its concentration gradient. Once the solute diffuses into the cytoplasm, the carrier protein snaps back to its original conformation, ready to bind another solute molecule. Like simple (unaided) diffusion, facilitated diffusion can only occur in a direction from higher concentration to lower concentration and does not require the input of energy.

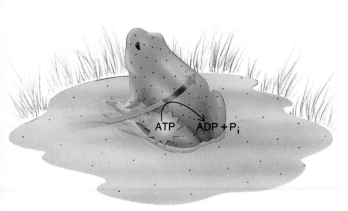

ATP   ADP + P$_i$

FIGURE 7-7

An example of active transport. Animals that live in freshwater habitats tend to lose salts from their body to their environment by diffusion. Some of this salt is regained in their diet, but most freshwater animals, including this frog, possess mechanisms to take back salts from their environment against a concentration gradient. This process requires the input of energy and is called *active transport*. Active transport proteins in this frog are located in the plasma membranes of its skin.

of the protein and forcing the ions through the membrane (Figure 7-8). With the exception of maintaining an elevated temperature, a resting person utilizes more ATP in pumping ions than in any other activity.

Recall from Chapter 6 that energy can be stored in several forms. ATP serves as a storage of chemical energy, some of which is expended in forming ionic gradients, such as those of sodium and potassium. Ionic gradients are themselves a form of stored (potential) energy since, given the opportunity, these ions will spontaneously diffuse back across the membrane, thereby eliminating the concentration difference. Like the flow of water over a dam, the directed movement of ions back across the membrane is a form of kinetic energy that can be used to accomplish work. For example, when you eat a food that is rich in starch, such as a potato, the glucose molecules derived from the starch are transported from the intestine into the bloodstream — across the intervening intestinal cell — *against a concentration gradient* (Figure 7-9). In this case, glucose molecules are transported by a membrane protein that also carries sodium ions. The "downhill" movement of sodium ions into the cell provides the energy used to *cotransport* the "uphill" movement of glucose molecules. The movement of nerve impulses back and forth between your brain and the other

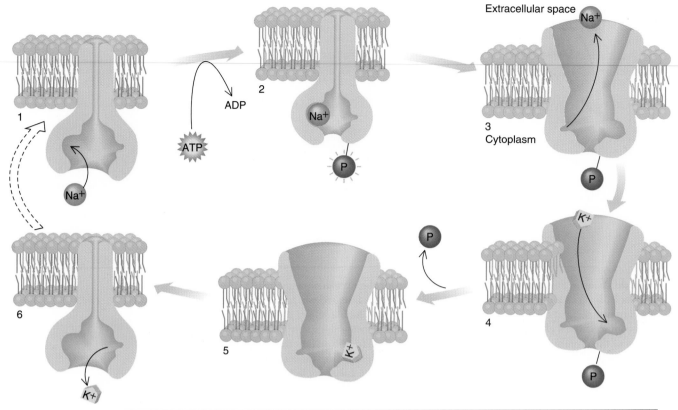

FIGURE 7-8

Schematic concept of the mechanism of the sodium–potassium pump. Sodium ions (1) bind to the protein on the inside of the membrane; ATP is hydrolyzed and the phosphate produced is linked to the protein (2), causing a change in the shape (conformation) of the protein (3), allowing sodium ions to be expelled to the external space. Potassium ions then bind to the protein (4), followed by the removal of the phosphate group (5) which causes the protein to "snap" back to its original conformation, moving the potassium ions to the inside of the cell (6). Unlike facilitated diffusion, the changes in the shape of the protein are driven by energy from ATP hydrolysis, which allows the transport system to move these ions against their concentration gradient.

parts of your body also depends on the movement of sodium ions across a membrane (Chapter 23). As we will see in the following chapters, the formation of ATP in mitochondria and chloroplasts also results from energy stored in ionic gradients.

The plasma membrane utilizes energy to transport solutes into regions of higher concentration, a process vital to the survival and normal functioning of a cell. This ability enables cells to set up concentration gradients necessary for generating the energy to drive many energy-requiring activities. (See CTQ #5.)

ENDOCYTOSIS

Both diffusion and active transport can move smaller-sized solutes directly through the plasma membrane, but what of the uptake of materials that are too large to penetrate a membrane, regardless of its nature? We saw in the last

chapter how materials stored in cytoplasmic vesicles can be released from cells by exocytosis (page 99). As described in the chapter opener, cells also carry out a reverse process, called **endocytosis,** in which large molecular weight materials are enclosed within invaginations of the plasma membrane, which subsequently pinch off to form cytoplasmic vesicles. The ability of the plasma membrane to perform this remarkable activity depends on the dynamic, fluid nature of the membrane. There are two forms of endocytosis:

1. **phagocytosis** (*phago* = eating), in which a cell ingests large particles, such as bacteria or pieces of debris, and

2. **pinocytosis** (*pino* = drinking), in which a cell ingests liquid and/or dissolved solutes and small suspended particles.

PHAGOCYTOSIS

An example of phagocytosis is pictured in Figure 7-10*a*. In Figure 7-10*b*, a unicellular amoeba engulfs its microscopic

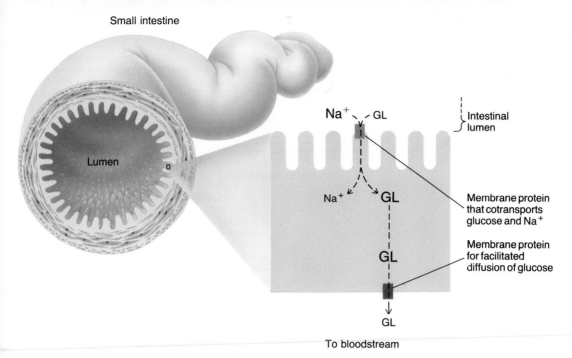

Small intestine

Lumen

Na⁺ — GL

Intestinal lumen

Na⁺   GL

Membrane protein that cotransports glucose and Na⁺

GL

Membrane protein for facilitated diffusion of glucose

GL

To bloodstream

FIGURE 7-9

An example of the use of energy stored in an ionic gradient. The sodium–potassium pump establishes a steep sodium gradient by pumping sodium ions out of the cell. This gradient represents a storehouse of potential energy that can be put to use in various ways. In this case, the gradient is employed to move glucose molecules into an intestinal cell against a concentration gradient. The glucose molecules exit the opposite end of the cell by facilitated diffusion. The relative size of the "Gl" and "Na⁺" letters indicates the directions of the concentration gradients.

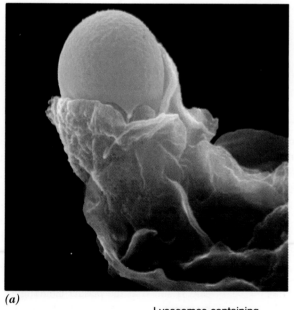

(a)

FIGURE 7-10

Cell uptake by phagocytosis. *(a)* Cell ingesting a synthetic particle. *(b)* A food particle becomes enclosed within an invagination of the plasma membrane, which subsequently pinches off to form a large membranous vesicle (called a *food vacuole*). Food matter is digested by enzymes discharged from lysosomes that fuse with the vesicle membrane. The digested food is then absorbed into the amoeba's cytoplasm.

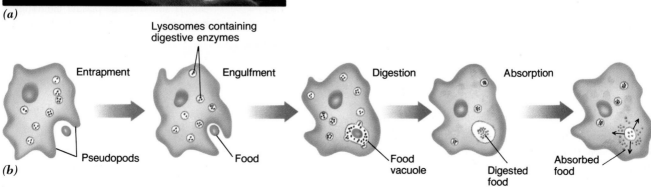

Lysosomes containing digestive enzymes

Entrapment   Engulfment   Digestion   Absorption

Pseudopods   Food   Food vacuole   Digested food   Absorbed food

(b)

◁ THE HUMAN PERSPECTIVE ▷
LDL Cholesterol, Endocytosis, and Heart Disease

We saw in the beginning of the chapter how cholesterol is transported in protein-containing particles called LDLs, which are taken into cells by receptor-mediated endocytosis. This is a critical process since excess LDLs—those that are not ingested by cells and thus remain in the bloodstream—can build up as plaques on the walls of blood vessels and eventually cause *atherosclerosis* (narrowing of the arteries) or heart attacks (Figure 1).

In its severe form, the rare genetic disease *familial hypercholesterolemia* strikes about 1 in every million individuals. Children with this disease have six to ten times the normal level of blood cholesterol and often suffer heart attacks before age 15. Goldstein and Brown, the scientists depicted in the chapter opener, discovered the reason for this inherited condition: These individuals lack LDL receptors; consequently, their cells are not able to remove cholesterol from the blood. This genetic condition has provided the strongest evidence that elevated blood cholesterol levels *alone* are sufficient to predispose a person to cardiovascular disease.

It is evident why these genetically deficient individuals would suffer from atherosclerosis and heart attacks, but why do some people who possess *normal* LDL receptors develop these same diseases, albeit at an older age? Evidence strongly suggests that the root of the problem still lies in elevated blood cholesterol levels which, in many people can be traced to diet (along with genetic factors, smoking, and a lack of exercise).

⟳ Cholesterol is required by all animal cells as a component of their cell membranes. When a cell's cholesterol needs are

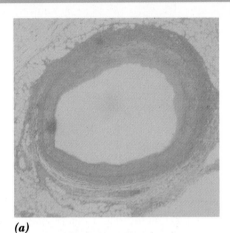

(a)

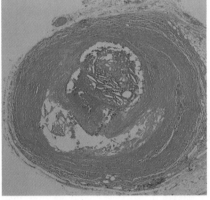

(b)

FIGURE 1 ────

Comparison of a normal coronary artery *(a)* and an artery whose channel is almost completely closed as the result of atherosclorosis *(b)*.

met, the production of LDL receptors is shut down, and the receptors present are gradually removed from the cell's surface. Evidence suggests that people with a diet high in cholesterol (or saturated fats, which lead to the formation of cholesterol) quickly provide their cells with their needed allotment of cholesterol. As a result, the cells stop producing LDL receptors and stop removing cholesterol from the blood, leading to a rise in blood cholesterol levels and an increased risk of atherosclerosis.

These findings have produced an experimental strategy for lowering blood cholesterol in individuals with particularly high levels. The key to the strategy is to inhibit the synthesis of cholesterol by the cells of the body, particularly those of the liver. This can be accomplished by the prescription drug mevinolin (trade name *Lovastatin*), a compound derived from a fungus that inhibits a key enzyme in cholesterol synthesis. As the cells produce less cholesterol, they must take up more cholesterol from the blood, which is accomplished by an increased production of LDL receptors. The more LDL receptors, the more cholesterol that is taken up, and the less cholesterol that remains in the blood to become deposited on the walls of arteries. Recent evidence suggests mevinolin may actually reverse arterial disease.

The breakthroughs achieved by Goldstein and Brown have paved the way to understanding—and hopefully treating—other receptor-related diseases, including diabetes. Don't be surprised if you soon hear that the medical treatment for a disease is aimed at the plasma membrane's receptors.

meal, a bacterium, gradually enclosing the food particle in a large vesicle. Once the bacterium is enclosed, the vesicle fuses with a lysosome, allowing the hydrolytic enzymes of the lysosome to digest the bacterium. After the macromolecules have been hydrolyzed, the products (sugars, amino acids, etc.) of the digested bacterium are transported though the vesicle membrane and utilized by the amoeba. A similar mechanism is employed by your white blood cells when defending your body against invading microorganisms; the cells engulf the microorganisms in a vesicle and destroy them.

PINOCYTOSIS

Most cells are not capable of phagocytosis, but virtually all eukaryotic cells are capable of pinocytosis. This form of endocytosis is often initiated by the interaction of dissolved or suspended molecules with specific receptors on the plasma membrane; it is known as **receptor-mediated endocytosis,** which we introduced at the beginning of the chapter. Included within this group are receptors for hormones, enzymes, and blood materials, such as LDL (see The Human Perspective: LDL Cholesterol, Endocytosis,

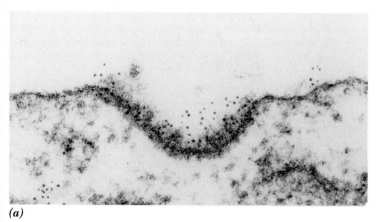

(a)

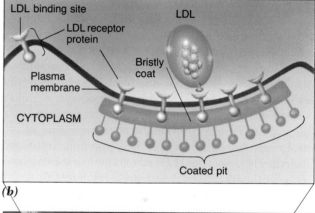

(b)

FIGURE 7-11 ━━━━━━━━

Binding and uptake of LDL. *(a)* LDL particles (indicated by black dots on photo, left) bind to receptors situated in coated pits in the plasma membrane. *(b)* The receptors are held in the coated pit by their attachment to a bristly protein on the inner surface of the membrane. Following binding to the LDL receptor, the LDL particles are taken into the cell as part of a bristly-coated vesicle. *(c)* Events that occur following LDL uptake. LDL receptors are returned to the plasma membrane while cholesterol molecules are released from the LDL by lysosomal enzymes.

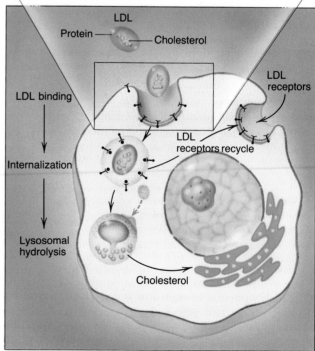

(c)

and Heart Disease). Certain viruses and bacteria take advantage of this cellular activity by binding to a surface receptor and gaining entry to the cell by receptor-mediated endocytosis (Figure 7-11). The cytoplasmic vesicle formed during endocytosis normally fuses with a lysosome whose hydrolytic enzymes degrade the vesicle's contents. The digested products of the vesicle then move across the surrounding membrane into the fluid of the cytoplasm.

Cells must sometimes take in molecules or particles that are too large to pass through the membrane by diffusion or active transport. The fluid nature of the plasma membrane enables cells to surround large molecules or particles completely and to fuse with itself to engulf the enclosed substances. (See CTQ #6.)

REEXAMINING THE THEMES

Relationship between Form and Function

The structure of the plasma membrane—consisting of a fluid lipid bilayer and a mosaic of embedded proteins—accounts for many of the cell's life-sustaining functions. For example, the lipid bilayer acts as a barrier, preventing essential molecules from leaking out of the cell. The fluidity of the lipid bilayer gives the membrane the necessary flexibility it needs to facilitate such dynamic processes as endocytosis, which requires the membrane to invaginate and pinch off vesicles. Proteins provide the membrane with selectivity. Protein receptors bind specific extracellular substances, including hormones and LDL cholesterol, or act as passageways for the transmembrane movement of specific ions or dissolved solutes.

Biological Order, Regulation, and Homeostasis

The concentration of many substances within a cell is maintained by transport activities of the plasma membrane. Sodium ions, for example, are maintained at a low concentration inside a cell by the expulsion of these ions by membrane transport. Cholesterol levels, both inside the cell and in the blood, are also determined by the activity of membrane proteins. When cholesterol levels in a cell are low, the cell produces additional LDL receptors, which then import more cholesterol from the blood. When cellular cholesterol levels are high, LDL receptors are removed from the plasma membrane, causing less cholesterol to be taken up.

Acquiring and Using Energy

One of the essential uses of energy in living organisms is the active transport of materials across cell membranes. Active transport of sodium and potassium ions is accomplished by conformational changes brought about by the transfer of a phosphate group from ATP to one of the amino acids in the transport protein. This is an example of one of the common mechanisms by which the chemical energy in ATP is used to power an otherwise energetically unfavorable reaction. Although energy is used to establish these ionic gradients, the energy is not "wasted" since the gradient itself is a form of potential energy. The energy stored in the gradient, for example, is used to move other substances across a membrane against their concentration gradient. This is illustrated by the movement of sugar molecules in your intestine.

SYNOPSIS

The cell's plasma membrane is a dynamic boundary that protects the internal environment of the cell while directing the exchange of materials between the cell and the extracellular medium. The barrier properties of the membrane stem largely from the lipid bilayer, which is impermeable, except to very small polar molecules (such as water and carbon dioxide) or to lipid-soluble substances, such as steroids. Proteins promote the passage of selected molecules. Together, the lipids and proteins make the membrane selectively permeable.

Diffusion is the passive movement of a substance from a region of higher concentration to a region of lower concentration. Diffusion depends solely on the random movements of molecules and does not require the input of energy. The rate of diffusion depends on the polarity and size of the molecule and the magnitude of the concentration gradient. The diffusion of water across a membrane, called osmosis, is of great importance in the lives of diverse organisms. During osmosis, water moves from a compartment of lower solute concentration to one of higher solute concentration. Some membranes have special channels that allow ions to diffuse across the membrane without having to traverse the lipid bilayer directly. Membranes may also possess carrier proteins that bind to specific solutes and facilitate the diffusion process.

Active transport is a process in which substances are moved against a concentration gradient by mem- **brane proteins at the expense of cellular energy.** The establishment of gradients by active transport provides a storage of energy the cell can use to drive various energy-requiring activities.

Endocytosis is the uptake of materials in the extracellular environment by invaginations of the plasma membrane. The material becomes enclosed within a cytoplasmic vesicle, which generally fuses with a lysosome, allowing digestive enzymes to disassemble the material. Endocytosis is categorized as either phagocytosis—the uptake of relatively large particles, such as bacteria or debris—or pinocytosis—the uptake of smaller materials or fluid. Pinocytosis is often initiated by the interaction of material (such as LDL) with specific membrane receptors.

Key Terms

selectively permeable (p. 138)
diffusion (p. 138)
osmosis (p. 140)
isotonic (p. 140)
hypertonic (p. 140)

hypotonic (p. 140)
turgor pressure (p. 141)
plasmolysis (p. 141)
carrier protein (p. 141)
facilitated diffusion (p. 141)

active transport (p. 142)
endocytosis (p. 144)
phagocytosis (p. 144)
pinocytosis (p. 144)
receptor-mediated endocytosis (p. 147)

Review Questions

1. Hexane (C_6H_6), a commercial organic solvent, and glucose ($C_6H_{12}O_6$) contain six carbon atoms each. Which one would move more rapidly across the membrane by simple diffusion? Why?

2. Why does water stop moving into a plant cell even if the cell remains hypertonic to its environment?

3. In comparing simple diffusion, facilitated diffusion, and active transport, which of these processes is being described by the following phrases:

   a. penetrating molecules move in only one direction across the membrane;
   b. requires the use of ATP hydrolysis;
   c. rate of penetration is related to lipid solubility;
   d. involves a membrane protein.

4. Sea urchins are isotonic to the surrounding sea water, while most fishes are extremely hypotonic. What type of osmotic challenge, if any, would each of these animals face?

Critical Thinking Questions

1. If you were studying viruses and discovered one whose protein coat resembled the protein found in an LDL particle, what conclusion might you draw about the way the virus entered the cell?

2. If you were designing instruments to look for signs of life on Mars, would you include a search for a selectively permeable membrane, such as that found in living cells

on Earth? Adopt a position on this question and defend it.

3. How does osmosis explain each of the following?
    (1) a red blood cell swells and bursts when placed in pure water
    (2) a red blood cell shrinks when placed in a concentrated salt solution

4. Amino acids are found to enter most cells rapidly, even if the cells are treated with drugs that block ATP formation. Can you conclude that amino acids are capable of rapid diffusion across the lipid bilayer? Why or why not?

5. For each case given below, indicate whether it illustrates simple diffusion, facilitated diffusion, or active transport.
    (1) Some marine algae contain much higher concentrations of iodine than does the surrounding sea water.
    (2) A single-celled freshwater organism uses a special organelle to pump excess water out of its cytoplasm.
    (3) A single-celled freshwater organism is hypertonic to the surrounding water, and water moves into the organism.
    (4) Urea, a small molecule that is relatively soluble in lipids is produced inside of cells. It moves into the surrounding medium, which contains little or no urea.

6. Use the terms below to prepare a concept map of the various means by which materials are transported across the cell membrane. [To make a concept map, the concepts (terms listed) are arranged graphically to represent relationships among the concepts. Concepts are enclosed by boxes, circles, or ovals and are connected by lines indicative of relationships. Connecting words are written on the lines.]

    passive transport; active transport; diffusion; osmosis; facilitated diffusion; endocytosis; phagocytosis; pinocytosis.

7. Recall from Chapter 6 that events in the universe tend to proceed spontaneously in the direction that increases entropy (disorder) in the universe. Do you think that diffusion is an example of this principle? Why or why not?

Additional Readings

Alberts, B., et al. 1989. *Molecular biology of the cell.* New York: Garland. (Intermediate)

Brown, M. S., and J. L. Goldstein. 1984. How LDL receptors influence cholesterol and atherosclerosis. *Scientific American* Nov: 58–66. (Intermediate)

Darnell, J., H. Lodish, and D. Baltimore. 1990. *Molecular cell biology*, 2d ed. New York: Freeman. (Advanced)

Lienhard, G. E., et al. 1992. How cells absorb glucose. *Scientific American* Jan: 86–91. (Intermediate)

Myant, N. B. 1990. *Cholesterol metabolism, LDL and the LDL receptor.* New York: Academic Press. (Advanced)

Stein, W. D. 1990. *Channels, carriers, and pumps: An introduction to membrane transport.* San Diego: Academic Press. (Advanced)

CHAPTER

◄ 8 ►

Processing Energy: Photosynthesis and Chemosynthesis

STEPS TO DISCOVERY
Turning Inorganic Molecules Into Complex Sugars

Turning Inorganic Molecules into Complex Sugars

*I*n the late 1930s, Martin Kamen, having recently received his doctorate in nuclear physics, arrived at the University of California at Berkeley, the site where the original cyclotron had recently been constructed. A cyclotron is an instrument that accelerates subatomic particles to high speeds and directs them at the nuclei of atoms. The subsequent crash changes the target atom's structure. Kamen's goal was to use the cyclotron to produce new isotopes, such as ^{11}C, a radioactive isotope of carbon. Soon after his arrival, Kamen met Sam Ruben, a chemist at Berkeley, who pointed out the possibilities for using a radioactive carbon atom for working out the steps of metabolic pathways. One of the least understood, and most important, metabolic pathways was that

of photosynthesis. At the time, scientists knew that the process of photosynthesis converted carbon dioxide and water to carbohydrates, such as glucose, but they knew very little about the steps by which such a remarkable conversion took place. How are these inorganic molecules converted to the "food" substances on which virtually all life on earth depends? Kamen and Ruben set out to provide the answer.

The scientists planned to grow either barley plants or cultures of photosynthetic algae in the presence of radioactive carbon dioxide ($^{11}CO_2$). After a short period of time, they would kill the cells with acidified, boiling water, which would extract the small polar organic molecules that form the intermediate compounds in the photosynthetic path-

Water from the soil and carbon dioxide from the air combine during photosynthesis to produce carbohydrates and oxygen,

way. By determining the chemical identity of the radioactive compounds, Kamen and Ruben felt they could assemble a diagram of the metabolic pathway. The approach was sound, but the technical problems were overwhelming.

The first problem was presented by the isotope itself. ^{11}C has a half-life of only 21 minutes, meaning that within 21 minutes of the onset of the experiment, only half the starting amount was left; the other half had already disintegrated. Before the experiment was an hour old, only 15 percent of the original radioactivity remained. Because of the isotope's short life span, the entire procedure—from producing the isotope at the cyclotron, to carrying out the experiment, to measuring the radiation with a counter—had to be performed impossibly fast. Kamen and Ruben found themselves running across campus in the middle of the night. The second problem was the lack of a reliable method for separating radioactive organic molecules from one another. After 3 years of research, no substantial progress on the project had been made.

Then, in 1940, Kamen and Ruben discovered a longer-lived isotope of carbon. They found that bombarding nitrogen atoms with neutrons produced a new isotope of carbon—^{14}C—which has a half-life of 5,700 years. Unfortunately, just as they were gearing up for an onslaught on photosynthesis using the new isotope, World War II broke out. Kamen was assigned to work on war-related isotope research; Ruben was put to work on the development of chemical weapons and was killed in a laboratory accident.

By the end of the war, the second problem that had plagued Kamen and Ruben—how to separate organic molecules from one another—was solved with a surprisingly simple technique. Archer Martin and Richard Synge, a pair of British biochemists, discovered that they could separate closely related organic compounds with a technique they called *paper chromatography*. In this procedure, a sample containing a mixture of compounds is applied to a piece of filter paper, and one end of the paper is submerged in a solvent. The solvent then migrates through the paper in much the same way a paper towel "sucks up" water. As the solvent moves through the applied sample, it dissolves the organic molecules and carries them along at different speeds—the smaller, more soluble compounds moving fastest. At the end of the "run," all the compounds lie in different locations on the paper strip. The isolated spots can then be cut out of the strip and dissolved in a solvent to obtain purified preparations of each compound in a sample.

Following the war, another group of scientists at Berkeley picked up where Ruben and Kamen had been forced to stop. Headed by the biochemist Melvin Calvin, this team used the approach of Ruben and Kamen. Armed with ^{14}C and paper chromatography, however, their work was much more fruitful. Calvin and his colleagues, James Bassham and Andrew Benson, grew algae in the presence of $^{14}CO_2$, killed the cells after a time, and extracted the labeled compounds with boiling alcohol. They reasoned that a very brief exposure to the radioactive carbon dioxide would allow radioactive carbon to become incorporated into the initial compounds formed during photosynthesis. Exposing the cells to ^{14}C a little longer allowed the next compounds in the pathway to become radioactive, and so on, until all the intermediates in the photosynthetic pathway were labeled. Using this approach, the researchers attempted to trace the route by which carbon travels from carbon dioxide to carbohydrate.

In their initial studies, they found that even short exposures of about 5 seconds to labeled carbon dioxide produced a number of radioactive compounds (see Figure 8-10). When they shortened the exposure to 2 seconds, one radioactive spot appeared, which corresponded to the first compound in the pathway formed after carbon dioxide was "fixed" to another molecule within the cells. The exposures were lengthened until the team had identified all the compounds in what Calvin concluded was a *circular* pathway—the route from carbon dioxide to carbohydrates. For his work in unraveling the pathway of the photosynthetic conversion of carbon dioxide to carbohydrate, Calvin received the Nobel Prize in Chemistry in 1961.

which is released into the air.

AUTOTROPHS AND HETEROTROPHS: PRODUCERS AND CONSUMERS

*L*ife is powered by the sun. The energy that gives a cheetah its incredible speed or a bird its power to fly does not originate within the planet, but travels 93 million miles from the sun. The energy of sunlight, however, becomes energy for life only after it has been transformed into chemical energy by the process of **photosynthesis.** The cheetah, for example, gets its food by eating an antelope, which in turn ate grasses and other plants that had photosynthesized. Virtually all organisms depend on photosynthesis for supplies of energy-rich food. For this reason, photosynthesis might be considered the most important set of chemical reactions in biology.

▼ ▼ ▼

Life on earth can be divided into two groups: autotrophs and heterotrophs. **Autotrophs** (*auto* = self, *troph* = feeding) are organisms that use carbon dioxide as their sole source of carbon. They convert simple inorganic molecules, such as carbon dioxide and water, into all of the complex, energy-rich organic materials needed to sustain their life. This feat requires the input of external energy derived from a nonliving source. Most autotrophs harness the energy from the sun; these are called *photosynthetic autotrophs*. A number of autotrophic bacteria, however, have evolved an alternate mechanism for capturing external energy. They acquire the energy from inorganic substances through *chemosynthesis*, a process discussed in more detail at the end of the chapter.

Although all organisms depend on the metabolic activity of autotrophs, only plants, some protists, and some bac-

(a)

FIGURE 8-1

Life is centered around photosynthesis. *(a)* Plants supply energy for virtually all organisms by transforming light energy into chemical energy. Every year, plants convert about 600 billion tons of carbon dioxide into sugars, a weight equivalent to about 1,000 times the combined weight of all people living on earth today. *(b)* Tea fields growing near Mt. Fuji, Japan. Photosynthetic plants are essential to human existence as well as to the existence of virtually all other animals. Although about 3,000 species of plants have been cultivated throughout human history, today only 15 plant species supply most of the world's human population with food.

(b)

teria are categorized as autotrophs (Figure 8-1). The remainder of organisms—more than 80 percent of living species—are unable to carry out either photosynthesis or chemosynthesis. These organisms are called **heterotrophs** (*hetero* = other). Since heterotrophs cannot manufacture complex organic molecules from inorganic precursors, they must consume preformed organic materials initially produced by the autotrophs. Thus, autotrophs not only provide themselves with food, but they provide food to a vast array of heterotrophs as well—you included. It is estimated that each year plants convert about 600 billion tons of carbon dioxide into organic nutrients and release about 400 billion tons of oxygen into the environment. Much of this activity is accomplished by single-celled algae living in the thin upper layer of the world's sensitive—and increasingly more polluted—oceans.

Organisms depend on their environment to provide them with the raw materials and energy necessary for life. Autotrophs derive their energy and materials from inorganic sources and provide the resources on which heterotrophs depend. This process is not always a one-way street, however, because heterotrophs help provide the inorganic resources (carbon dioxide) needed by the autotrophs. (See CTQ #2.)

AN OVERVIEW OF PHOTOSYNTHESIS

In the early 1600s, it was generally accepted that plants, which obviously did not prey on other organisms, absorbed all of their food and nutrients from the soil alone. This was the common belief until Jan Baptista van Helmont, a Belgian physician, conducted a simple experiment that disproved this hypothesis of "soil-eating" plants. After planting a 5-pound willow tree in 200 pounds of dry soil, van Helmont added only rainwater to the soil for a period of 5 years. When he reweighed the soil and the willow tree, the tree had gained 164 pounds, yet the soil had lost only 2 ounces—not 164 pounds, as would be expected of a soil-eating plant. Van Helmont concluded that the willow's weight gain must have come from the added water, a conclusion we now know is only partly correct.

Willows, like most other plants, gather the carbon needed to build their tissues from the air, not the soil. Through photosynthesis, the carbon in atmospheric carbon dioxide is used to form the backbone of new organic compounds that are fashioned into the cellular material required for growth. The soil provides only a few minerals needed for growth, which accounts for the 2-ounce weight loss measured by van Helmont.

Photosynthesis in eukaryotes takes place inside chloroplasts (page 104), organelles whose structure is ideally suited to their function. Chloroplasts contain an elaborate array of membranes in which clusters of light-absorbing pigments are embedded. These pigments capture solar energy, which is used to transform carbon dioxide into energy-rich sugars. These sugars not only provide a store of chemical energy but serve as the raw materials for producing thousands of other molecules needed by the plant. Oxygen, which is vital to energy utilization by both plants and animals, is released as a byproduct of photosynthesis.

Expressing the overall process in a chemical equation makes photosynthesis appear rather simple:

$$CO_2 + H_2O \xrightarrow[\text{chlorophyll}]{\text{light energy}} (CH_2O) + O_2$$
$$\underset{\text{carbohydrate}}{\text{Unit of}}$$

You might conclude from this equation that the energy in light is used to split carbon dioxide, releasing molecular oxygen (O_2) and transferring the carbon atom to a molecule of water to form a unit of carbohydrate (CH_2O). In fact, this was the prevailing line of thought as late as 1941, the year Ruben and Kamen reported on an experiment using a specially labeled isotope of oxygen (^{18}O) as a replacement for the common isotope (^{16}O). In this experiment, one group of plants was given labeled $C^{18}O_2$ and "regular" water, while the other group was given "regular" carbon dioxide and labeled $H_2{}^{18}O$. The investigators asked a simple question: Which of these two groups of plants released labeled $^{18}O_2$ gas? The results showed that those plants given the labeled water produced labeled oxygen, while the plants given the labeled carbon dioxide produced "regular" oxygen. Con-

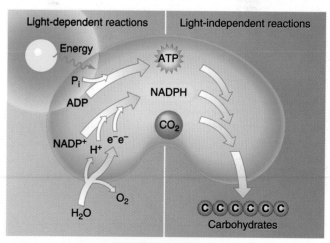

FIGURE 8-2 ————

Photosynthesis: A two-stage process. Photosynthesis is divided into light-dependent reactions and light-independent (dark) reactions. During the light-dependent reactions, the energy of sunlight provides the power to generate ATP and NADPH from ADP, P_i, $NADP^+$, and water. Oxygen gas, derived from water, is given off as a byproduct. In the light-independent reactions, ATP and NADPH from the light-dependent reactions provide the energy and electrons to convert low-energy carbon dioxide molecules into energy-rich carbohydrates.

trary to popular belief, it wasn't carbon dioxide that was being split into its two atomic components, but water.

Botanists—biologists who specialize in the study of plant life—group the reactions of photosynthesis into two general stages: light-dependent reactions and light-independent reactions (Figure 8-2). During the **light-dependent reactions,** energy from sunlight is absorbed and converted to chemical energy, which is stored in two energy-rich molecules: ATP and NADPH. This process involves the splitting of water. During the second stage of photosynthesis, the **light-independent reactions** (or "dark reactions"), light is not required; carbohydrates are synthesized from carbon dioxide using the energy stored in the ATPs and NADPHs formed in the light-dependent reactions.

Life depends on the nonliving environment. Photosynthesis brings together light energy and simple inorganic molecules (H₂O and CO₂) to produce living material by a complex series of light-dependent and light-independent reactions. This new living material is the source of nutrients that support virtually every organism on earth. (See CTQ #3.)

THE LIGHT-DEPENDENT REACTIONS: CONVERTING LIGHT ENERGY INTO CHEMICAL ENERGY

Chloroplasts are large organelles, by organelle standards. Their role in photosynthesis was revealed in 1881 in an ingenious demonstration by German biologist Theodor Engelmann. Engelmann showed that when the cells of an aquatic plant were illuminated, actively moving bacteria would collect outside the cell near the site of the large ribbonlike chloroplast. The bacteria were utilizing the minute quantity of oxygen released by photosynthesis in the chloroplast to stimulate their energy metabolism.

Recall from Chapter 5 that chloroplasts contain stacks of membrane discs, or thylakoids, surrounded by a semifluid stroma. The light-dependent reactions take place within the thylakoid membranes. They begin when the energy in sunlight is captured by light-absorbing pigments.

PHOTOSYNTHETIC PIGMENTS: CAPTURING LIGHT ENERGY

☀ Before we can understand how photosynthetic pigments capture light energy, we need to consider a few basic properties of light. Light energy travels in packets called **photons.** The energy content of a photon depends on the wavelength of the light: the shorter the wavelength, the higher the energy content. Since a photon cannot be divided into smaller packets of energy, a molecule that absorbs light must absorb an entire photon. Consequently, each pigment molecule absorbs only those wavelengths whose energy matches that needed to boost an electron to a higher orbital (page 54). As we will see below, the fate of these "photoexcited" electrons is a key aspect of photosynthesis.

Plants are able to absorb a wide range of wavelengths because they contain a variety of pigment molecules that have different structures and absorption properties. As a group, these pigments absorb light energy with wavelengths between 400 nanometers (or 400 billionths of a meter), violet light, and 700 nanometers, red light, which repre-

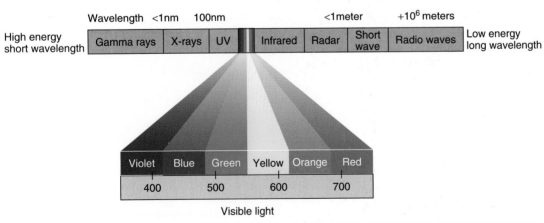

FIGURE 8-3

Solar energy for life. The wavelengths of radiation reaching the earth range from short-wavelength (high-energy) gamma rays, to extremely long-wavelength (low-energy) radio waves. Our eyes can detect only a small segment of the radiation spectrum, the portion we call *visible light* (from 380 to 750 nanometers). Photosynthetic pigments are also sensitive to visible light. They selectively absorb wavelengths between 400 and 700 nanometers.

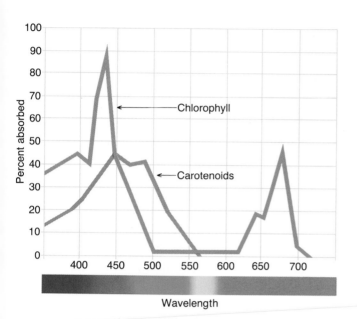

FIGURE 8-4

Absorption spectrum of photosynthetic pigments. Most absorption of light energy by photosynthetic pigments takes place at the peaks of each graph; the valleys indicate wavelengths that are reflected by the pigments. All plants use chlorophylls and carotenoids as photosynthetic pigments. Chlorophylls absorb primarily violet-blue and orange-red wavelengths, while carotenoids absorb wavelengths only in the violet-blue range of the spectrum. Differences in absorption from one chlorophyll molecule to the next depend largely on differences in the proteins to which the pigments are bound.

sents only a small fraction of the electromagnetic spectrum that reaches the earth (Figure 8-3).

There are two major groups of light-capturing pigments in plant chloroplasts: chlorophylls and carotenoids. **Chlorophylls** are complex, magnesium-containing compounds that absorb light of red and blue wavelengths (Figure 8-4). Plants are typically green because chlorophyll, which is the predominant plant pigment, does not absorb light of green wavelengths. Instead, this light is reflected to our eyes, which is why we see the green color. **Carotenoids** absorb light of blue and green wavelengths (Figure 8-4). Yellow, orange, and red wavelengths are reflected by carotenoids, producing the characteristic orange and red colors of carrots, oranges, tomatoes, and the leaves of some plants during the fall (Figure 8-5). As the leaves stop producing chlorophyll in the cooler weather, the gold and red carotenoids become more visible, which accounts for the brightly colored landscapes of autumn.

FIGURE 8-5

Fall colors. As chlorophyll is broken down in autumn, the underlying red, orange, and yellow carotenoids in the leaf chloroplasts are exposed.

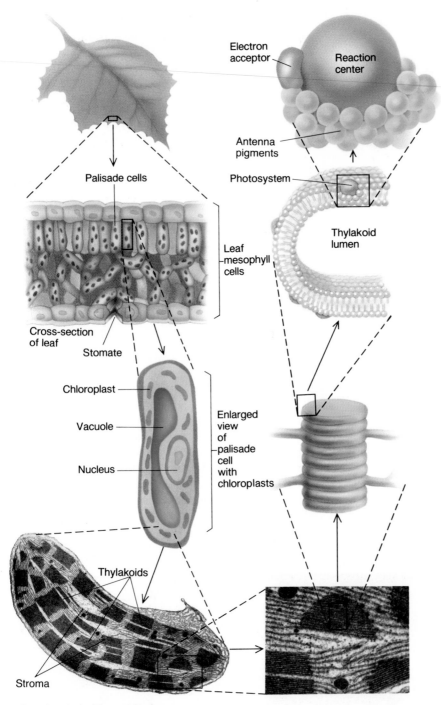

Electron acceptor

Reaction center

Antenna pigments

Photosystem

Palisade cells

Thylakoid lumen

Leaf mesophyll cells

Cross-section of leaf

Stomate

Chloroplast

Vacuole

Nucleus

Enlarged view of palisade cell with chloroplasts

Thylakoids

Stroma

* Greatly enlarged for emphasis

FIGURE 8-6

Organization for photosynthesis is evident at many different levels. Photosynthesis takes place in the chloroplasts of plant cells. A leaf cell may contain as many as 60 chloroplasts. The thylakoid system of the chloroplast is the site of the light-dependent reactions, whereas the stroma is the site of the light-independent reactions. Photosynthetic pigments are precisely arranged in the thylakoid membranes to form light-harvesting photosystems. Within a photosystem, light is absorbed by antenna pigments, and the energy is transferred to the reaction-center pigment, boosting an electron to a higher energy level from which it is passed to an electron-acceptor molecule. This starts a chain of chemical reactions that leads to the formation of ATP and NADPH and, ultimately, energy-rich carbohydrates.

ORGANIZATION OF PHOTOSYNTHETIC PIGMENTS

Capturing light energy and converting it into usable chemical energy requires an ordered biological structure, which can be seen at all levels of organization from the arrangement of individual pigment molecules to the entire leaf (Figure 8-6). Photosynthetic pigments are embedded in the thylakoid membranes of chloroplasts in clusters called **photosystems.** A single chloroplast contains several thousand photosystems. Each photosystem has an ordered structure in which a special chlorophyll molecule, called the **reaction center,** is surrounded by 250 to 350 **antenna pigments** including both chlorophyll and carotenoid molecules. Antenna pigments are so named because they gather light energy of wavelengths varying between 400 and 700 nanometers, and then channel this energy to the reaction center, much as an antenna receives and relays radio and television waves. As a result, the reaction center always ends up with the absorbed energy.

The membranes of the chloroplast contain two types of photosystems, **Photosystem I** and **Photosystem II**—distinguishable by their reaction-center pigments (Figure 8-7). The reaction center of Photosystem I is referred to as **P700**—"P" standing for "pigment," and "700" standing for the wavelength of light that this particular chlorophyll molecule absorbs most strongly. The reaction center of Photosystem II is referred to as **P680,** for comparable reasons.

When sunlight strikes the thylakoid membrane, the energy is absorbed simultaneously by the antenna pigments of both Photosystems I and II and passed to the reaction centers of both photosystems. Electrons of both reaction-center pigments are boosted to an outer orbital, and each photoexcited electron is then passed to an "open" **electron acceptor;** that is, a molecule that will receive the electron. The transfer of electrons out of the photosystems leaves the two reaction-center pigments missing an electron and, thus, are positively charged. After losing their electrons, the reaction centers of Photosystem I and II can be denoted as

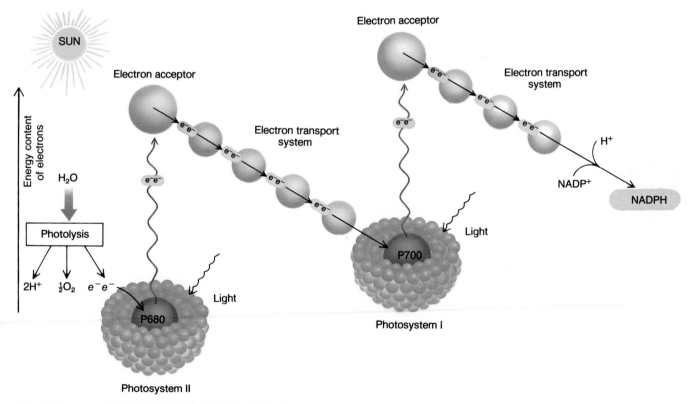

FIGURE 8-7

Electron flow from water to NADP⁺. Photoexcited electrons from the P700 and P680 reaction centers are passed to electron acceptors, which in turn transfer electrons to an electron transport system. The electron vacancy created by the absorption of a photon by P680 is filled with an electron released during photolysis. The electron vacancy in $P700^+$ is filled with electrons that originated in P680. The photoexcited electrons from P700 pass down a second electron transport system until they react with $NADP^+$ and H^+ to form NADPH. The overall consequence of the process is to remove low-energy electrons from water, boost them to a higher energy level, and transfer the energized electrons to $NADP^+$, providing the cell with reducing power.

P700$^+$ and P680$^+$, respectively. Let's look at the subsequent events occurring at each photosystem separately.

OBTAINING ELECTRONS BY SPLITTING WATER

The positively charged P680$^+$ has a very strong attraction for electrons—strong enough to pull tightly held (low-energy) electrons from water (Figure 8-7), splitting the molecule. The magnitude of this statement becomes evident when you consider that the splitting of water in a laboratory requires the use of a strong electric current or temperatures approaching 2,000°C. Yet, a plant cell can accomplish this feat on a snowy mountainside using the small amount of energy of visible light.

The splitting of water is termed **photolysis** (*photo* = light, *lysis* = splitting) and can be expressed as

$$\text{energy} + H_2O \rightarrow \tfrac{1}{2}O_2 + 2e^- + 2H^+$$

This reaction generates three different products: (1) oxygen produced by photolysis is released into the atmosphere; (2) electrons that travel to the P680$^+$ reaction center of

Photosystem II, returning the system to the uncharged (P680) state; and (3) protons that are released into the thylakoid lumen (see Figure 8-8). We will return to the fate of these protons shortly. But first, let's consider the fate of the electrons that had been transferred earlier from P680, producing the electron-deficient P680$^+$ reaction center.

TRANSPORTING ELECTRONS AND FORMING NADPH

Each electron lost by P680 following absorption of a photon is passed through the "hands" of a number of electron carriers that together form an **electron transport system** (Figure 8-8). The components of the electron transport system are precisely arranged within the thylakoid membrane to facilitate the passage of electrons through the system. Electron transport systems provide one of the best examples of the role of membranes in organizing components (proteins, lipids, pigments, etc.) in spatially precise arrays. As a result of its position in the membrane, each carrier in the system is able to receive electrons from the previous member and to transfer them to the next member,

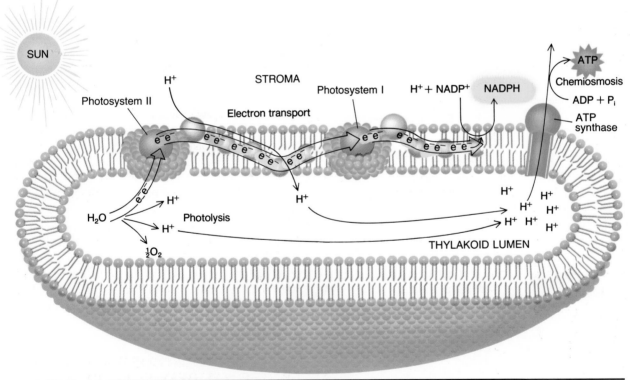

FIGURE 8-8

Photosynthesis and the thylakoid membrane. The carriers that transport electrons during the light-dependent reactions of photosynthesis are embedded within the thylakoid membrane. The transport of energized electrons "downhill" along an electron transport system and the splitting of water during photolysis both lead to the concentration of protons inside the thylakoid lumen. This produces a concentration gradient of protons across the thylakoid membrane with sufficient energy content to drive the phosphorylation of ADP to ATP as protons diffuse across the membrane through special channels in the ATP synthase. This process is called *noncyclic photophosphorylation*. Measurements indicate that approximately four ATPs are formed per molecule of oxygen released.

releasing energy in the process. This process of shuttling electrons along an electron transport chain is analogous to the passage of buckets from one hand to another along a bucket brigade.

At the end of the electron transport chain lies $P700^+$, the electron-deficient reaction center of Photosystem I, which accepts the electron from the last carrier in the chain. When $P700^+$ accepts an electron, it returns to the uncharged state (P700), at which time it is ready to absorb another photon. But what happens to the electrons that had been previously removed from P700 to create the electron-deficient $P700^+$ of Photosystem I? These electrons are passed through another set of electron carriers and on to $NADP^+$ to form NADPH (Figure 8-7, 8-8).

We have now followed the process in which electrons flow from water to Photosystem II to Photosystem I to $NADP^+$. The NADPH formed in this process provides the cell with reducing power, which is a form of stored energy (page 129). This process is called **noncyclic photophosphorylation** because electrons move in one direction, from water to $NADP^+$. But the flow of electrons and the storage of energy in NADPH is only half the story. The term "photophosphorylation" describes the fact that light energy is also used to drive a reaction in which ATP is formed, the subject of the next section.

MAKING ATP

☀ Recall from Chapter 6 that spontaneous reactions proceed to a state of lower energy, releasing energy as they occur. Electron transport is a spontaneous process in which electrons drop to lower and lower energy levels, releasing energy that is used to manufacture ATP (discussed at greater length in Chapter 9). For decades, a question of intense interest was: How is the "downhill" process of electron transport coupled to the "uphill" process of ATP formation? In 1961, the British physiologist Peter Mitchell formulated what seemed like a radical hypothesis to explain this phenomenon. In essence, Mitchell proposed that the energy released as the electrons fell to lower and lower energy levels during electron transport produced a *proton gradient* across the thylakoid membrane. A proton gradient is one in which hydrogen ions (H^+) are present at higher concentration on one side of the membrane than on the other side. Recall from Chapter 7 that an ionic gradient, whether Na^+ or H^+, represents a store of potential energy. In chloroplasts, the H^+ gradient is used to drive the energy-requiring (endergonic) reaction in which ADP reacts with P_i to form ATP. Mitchell called his mechanism for ATP formation **chemiosmosis.** Initially, Mitchell's hypothesis met with skepticism because it lacked direct evidence. However, experimental support for the proposal accumulated over the next decade to an overwhelming degree, culminating in Mitchell's receiving the Nobel Prize in 1978.

There are two matters to consider in this description of ATP formation: (1) How is a proton gradient formed across

the thylakoid membranes of chloroplasts? and (2) How does this gradient lead to the formation of ATP?

Storing Energy in a Proton Gradient

We partially answered the question of how a proton gradient is established earlier in the chapter. Recall that the splitting of water (photolysis) releases protons into the thylakoid lumen (Figure 8-8). Since the thylakoid membrane is highly impermeable to protons, the protons remain at a higher concentration within the thylakoid lumen than in the stroma, thus forming a proton gradient across the membrane.

Protons are also released into the thylakoid lumen as a result of electron transport. If you look closely at the electron transport chain between Photosystem II and Photosystem I in Figure 8-8, you will notice that protons move into the thylakoid membrane at its outer surface and are discharged into the thylakoid lumen at the inner surface of the membrane. This movement of protons from the stroma to the thylakoid lumen accompanies electron transport, contributing to the proton gradient across the thylakoid membrane.

Using the Stored Energy in the Formation of ATP

The second question remains: How does the energy stored in a proton gradient lead to the formation of ATP? ATP synthesis requires a large enzyme complex called **ATP synthase,** whose shape resembles a ball sitting on a stalk (Figure 8-8). Each part of this complex has a role to play. The stalked portion of the enzyme crosses through the thylakoid membrane and contains an internal channel or "proton pore" that allows the protons massed in the lumen to flow through the thylakoid membrane toward the region of lower concentration in the stroma. The flow of protons through these channels is analogous to water flowing through a bathtub drain: Water trapped in the tub escapes through the drain, flowing to a lower, more stable energy level. The active site of the enzyme—where ADP and P_i come together to form ATP—is located in the ball-shaped portion of the complex at the end of the proton channel. The "downhill" movement of protons through the channel drives the "uphill" synthesis of ATP at the enzyme's active site. The mechanism by which this happens remains one of the central questions in cell biology.

CYCLIC PHOTOPHOSPHORYLATION: AN ALTERNATE PATHWAY FOR ENERGY CONVERSION

In addition to noncyclic photophosphorylation, plants have a second mechanism to form ATP by chemiosmosis. This mechanism is called **cyclic photophosphorylation** (Figure 8-9). Only Photosystem I participates in cyclic photophosphorylation, a process that involves neither photolysis nor the formation of NADPH.

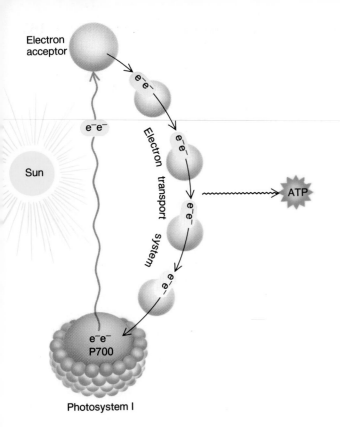

Electron acceptor

Sun

Electron transport system

ATP

e⁻ e⁻
P700

Photosystem I

FIGURE 8-9

Electron flow during cyclic photophosphorylation. During cyclic photophosphorylation, light energy is absorbed by Photosystem I. Photoexcited electrons from the P700 reaction center are transferred to an electron acceptor and passed down an electron transport system which returns the electron to the P700⁺ pigment. During this cyclic flow of electrons, protons are shuttled across the membrane, generating a proton gradient used to form ATP. The electron carriers that participate in cyclic photophosphorylation include some of the same molecules used in noncyclic electron flow. Cyclic photophosphorylation does not involve Photosystem II and does not generate NADPH.

During cyclic photophosphorylation, light energy is absorbed by antenna pigments and funneled into the P700 reaction center. Electrons from P700 are passed to an electron acceptor and then relayed through a number of electron carriers, which return the electrons back to the P700⁺ reaction center. As the electrons move along this circular electron transport pathway, a pair of protons are picked up on the outer surface of the thylakoid membrane and deposited into the thylakoid lumen, forming the proton gradient necessary to generate ATP by chemiosmosis. The differences between cyclic and noncyclic photophosphorylation are summarized in Table 8-1.

Cyclic photophosporylation is likely the means by which the earth's first autotrophic bacteria converted light energy into chemical energy. Even today, the only means of making ATP for some prokaryotic bacteria is through cyclic photophosphorylation. However, the failure of the cyclic pathway to generate NADPH limits its ability to provide

TABLE 8-1

SUMMARY OF CYCLIC AND NONCYCLIC PHOTOPHOSPHORYLATION

Cyclic	Noncyclic
1. Light is absorbed by pigments in Photosystem I.	1. Light is absorbed by pigments in Photosystems I and II.
2. Energized electrons from P700 pass to an electron acceptor.	2. Energized electrons from P680 of Photosystem II and energized electrons from P700 of Photosystem I pass to separate electron acceptors.
3. Energized electrons are relayed down an electron transport system and returned to the original P700⁺ reaction center.	3. Splitting of water fills the electron vacancy in P680⁺, releasing protons and oxygen.
4. ATPs are generated a proton gradient as a result of the cyclic flow of electrons.	4. P680's electron acceptor passes energized electrons to an electron transport system that relays them to the P700⁺ reaction center of Photosystem I.
	5. The electron acceptor for P700 passes energized electrons to another electron transport chain that shuttles them to NADP⁺, forming NADPH.
	6. ATPs are generated by a proton gradient as a result of a noncyclic flow of electrons and the release of protons as water is split.

chemical energy for larger autotrophic organisms. This was probably the major selective pressure that led to the evolution of the noncyclic pathway.

When light is absorbed by chlorophyll molecules, water molecules are split, and electrons are boosted to a higher energy state. These high-energy electrons are used to form NADPH and ATP, which provide the energy required to convert carbon dioxide to carbohydrate. The capture of light energy and its conversion to chemical energy require the precise arrangement of chlorophyll and other components within the thylakoid membrane. (See CTQ #4.)

THE LIGHT-INDEPENDENT REACTIONS

Since both ATP and NADPH are forms of chemical energy, why do photosynthesizers go on to manufacture another form of chemical energy—energy-rich sugars—during the synthesis reactions? The answer is that neither ATP nor NADPH can be stored or moved from one part of a plant to another, but sugars can. Transport and storage of energy-rich compounds are critical to the survival of a multicellular plant. For example, glucose produced in leaves is converted to the disaccharide sucrose and moved to growing regions of roots and stems where photosynthesis does not take place. Sucrose is an ideal transport molecule since it is highly stable and can be rapidly disassembled and used to construct other organic compounds. Thus, the function of the light-independent reactions is to use the readily available energy of ATP and the energy-rich electrons of NADPH to manufacture glucose. Carbon dioxide is gathered from air to supply carbon for glucose construction.

To date, three variations of light-independent reactions have been identified in plants—C_3, C_4, and *CAM synthesis*. Table 8-2 lists common plants that utilize each of these variations. All three modes of carbohydrate synthesis share the same pathway for producing glucose; they differ in the way carbon dioxide is first combined with a carbon-acceptor molecule. As we will see, in certain environments, this seemingly small difference has given some plants an adaptive edge over others.

C_3 PLANTS

C_3 synthesis—and indeed all three types of synthetic pathways—begins with **carbon dioxide fixation,** the combining of carbon dioxide with a carbon-acceptor molecule. When he began his studies on the incorporation of $^{14}CO_2$ by photosynthetic algae, one of Melvin Calvin's first objectives was to determine the nature of the carbon dioxide acceptor—a goal that proved to be rather elusive.

Calvin and his colleagues noticed that when the cells were exposed to $^{14}CO_2$ for very short periods of time—5 seconds or less—one spot appeared that contained most of the radioactivity (Figure 8-10). This spot contained the three-carbon compound, phosphoglyceric acid (PGA). Calvin concluded that the one-carbon compound carbon dioxide reacted with a two-carbon acceptor to form the three-

TABLE 8-2
PLANTS WITH C_3, C_4, AND CAM SYNTHESIS REACTIONS

C_3	C_4	CAM
Legumes (bean, pea)[a]	Sugar cane[a]	Pineapple[a]
Wheat[a]	Corn[a]	Cactus
Oats[a]	Sorghum[a]	Jade plant
Barley[a]	Crabgrass	Lillies
Rice[a]	Russian thistle (tumbleweed)	Agave (tequila)[a]
Bluegrass	Bermuda grass	Some orchids

[a] Agricultural crops.

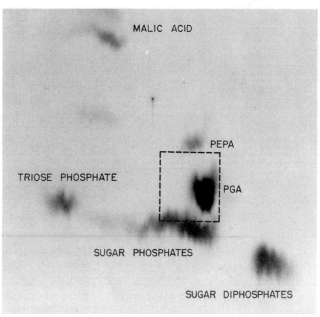

FIGURE 8-10

Unraveling the metabolic pathway for photosynthesis. A photograph of a chromatogram obtained after incubating algal cells with $^{14}CO_2$ for 5 seconds. Each of the spots corresponds to a different organic molecule containing one or more radioactive carbon atoms derived from the labeled carbon dioxide. During short incubations, the spot (boxed) corresponding to the compound phosphoglyceric acid (PGA) was most heavily labeled. PGA is the first stable compound into which carbon dioxide becomes incorporated.

carbon PGA, hence the designation **C_3 pathway,** also known as the *Calvin–Benson cycle.* A search for the mysterious two-carbon compound began. Further studies revealed, however, that carbon dioxide is not fixed to a two-carbon molecule but to a five-carbon molecule—**ribulose biphosphate (RuBP)**—by the enzyme **RuBP carboxylase,** nicknamed "Rubisco" (Figure 8-11). However, the six-carbon molecule formed by this union is so unstable that it "immediately" splits into two molecules of PGA, thus accounting for the early PGA spot on the chromatogram.

PGA is not a high-energy molecule. In fact, two molecules of PGA contain less energy than does a single molecule of RuBP, the actual carbon dioxide acceptor. This explains why carbon dioxide fixation is spontaneous and does not require energy from the light-dependent reactions. However, producing high-energy molecules from PGA requires the energy from ATP and electrons from NADPH formed by the light-dependent reactions. These compounds are used to reduce each PGA to a more energy-rich molecule, phosphoglyceraldehyde (PGAL), in a two-

step reaction, the first step using ATP, and the second using NADPH (Figure 8-11). (The second of these reactions—DPG to PGAL—was used as an example of an oxidation–reduction reaction on page 129.) Two PGAL molecules can ultimately give rise to one glucose by the C_3 pathway depicted in Figure 8-11. The vast majority of all plants use the C_3 pathway exclusively for constructing glucose, making it the predominant synthetic pathway among the earth's photosynthesizers.

C_4 PLANTS: ADAPTATIONS TO A HOT, DRY ENVIRONMENT

An alternate means by which plants can fix carbon dioxide was uncovered by Hugo Kortschak in the early 1960s. While working with sugar cane in Hawaii, Kortschak was surprised to find that when these plants were provided ^{14}C-labeled carbon dioxide, radioactivity first appeared in organic compounds containing four carbons rather than three, hence the name **C_4 synthesis.** It soon became apparent that these

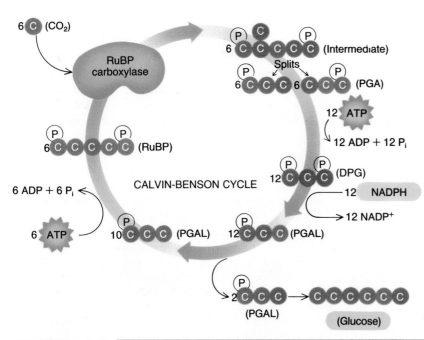

FIGURE 8-11

The Calvin-Benson Cycle (C_3 pathway). During the formation of glucose from carbon dioxide in C_3 plants, six carbon dioxide molecules are combined with six RuBPs to form six unstable, six-carbon carbohydrates. Each carbohydrate immediately splits into two, producing a total of 12 molecules of PGA. The PGAs are converted to DPGs by the transfer of phosphate groups from ATP. This is one of the places where the ATPs produced by the light-dependent reactions are utilized. NADPHs formed by the light-dependent reactions are used in the following step in which the 12 DPGs are reduced to 12 PGALs by the transfer of electrons. Two of the PGALs are "drained" away to form a molecule of glucose, whereas the remaining 10 PGALs (representing a total of 30 carbon atoms) recombine to regenerate six molecules of RuBP. The regeneration of six RuBPs is the same number of acceptor molecules with which we began.

◁ THE HUMAN PERSPECTIVE ▷
Producing Crop Plants Better Suited to Environmental Conditions

Biologists are currently investigating whether some plants are naturally able to shift among C_3, C_4, and/or CAM pathways, depending on conditions, and whether it is possible to genetically engineer "flexible" plants that can shift between two, or possibly all three, of these synthetic pathways. Shifting between different pathways could give plants a selective advantage over those with only one pathway. For example, a "flexible" plant could be in the CAM pathway when conditions are very hot and dry

and then shift into C_3 or C_4 following irrigation or a rainfall, enabling it to grow many times faster than it could if it remained in CAM.

So far, scientists have not found a plant capable of switching among all three synthetic pathways. But a few species have been described as natural C_3–C_4 intermediates, and preliminary results indicate that some intermediates show higher C_4 photosynthetic rates. Other investigations reveal natural C_3–CAM intermediates.

As for genetically engineered "flexible" plants, scientists have engineered C_3–C_4 hybrid plants. Unfortunately, none of the hybrids have shown increased rates of photosynthesis over the C_3 level as of yet. But investigations will continue, for if biologists can engineer biochemically flexible crop plants, much of the world's hunger problems could be solved— assuming there is no corresponding increase in human population.

plants utilize a different mechanism for carbon dioxide fixation, giving them an adaptive advantage over C_3 plants in hot, dry habitats. At high temperatures (45°C, 103°F or higher), plants that utilize C_4 synthesis produce as much as six times more carbohydrate than do C_3 plants. C_4 plants also survive better than do C_3 plants in environments in which water is limited and soil nutrients are scarce. This is

why crabgrass, a C_4 plant, tends to take over a lawn, displacing the domestic C_3 grasses originally planted.

The adaptive advantage of C_4 plants can be traced to the way they fix carbon dioxide which, in turn, is related to the structure of their leaf (Figure 8-12). RuBP carboxylase, the enzyme that fixes carbon dioxide in C_3 plants, requires relatively high concentrations of carbon dioxide before it

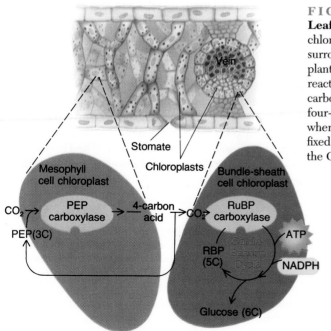

FIGURE 8-12

Leaf anatomy and C_4 synthesis. Plants that undergo C_4 synthesis have chloroplasts in their *mesophyll* cells and in large *bundle-sheath* cells surrounding the leaf veins. This unique chloroplast distribution enables C_4 plants to photosynthesize up to six times faster than can C_3 plants. The C_4 reactions take place in different leaf cells. Carbon dioxide fixation by PEP carboxylase takes place in the chloroplasts of mesophyll cells. The resulting four-carbon products are moved to the chloroplasts in bundle-sheath cells, where the carbon dioxide is released. The released carbon dioxide is then fixed to RuBP by RuBP carboxylase and carbohydrates are synthesized by the C_3 pathway.

can bind this substrate. This is important since carbon dioxide constitutes only about 0.03 percent of the gas content of the atmosphere. C_4 plants have a different carbon dioxide-fixing enzyme—**PEP carboxylase**—which is able to bind carbon dioxide at very low concentrations. Consequently, carbon dioxide levels can be very low in a leaf of a C_4 plant, even much lower than normal air concentrations, and carbon dioxide will still be linked to its acceptor.

How does this information fit with our earlier statement that all plants use the same cyclic pathway to produce glucose? The answer is explained by a C_4 plant's structure and function. In C_4 plants, PEP carboxylase fixes carbon dioxide in the leaf's *mesophyll cells* (Figure 8-12). The four-carbon products are then shipped to *bundle-sheath cells* lying deeper in the leaf. Once in the chloroplast of the bundle-sheath cells, the carbon dioxide is split from the four-carbon carrier, generating high levels of dissolved carbon dioxide—up to 100 times that in the mesophyll cell. This elevated carbon dioxide level is now sufficient to drive fixation by the less efficient RuBP carboxylase in the bundle sheath, and glucose is produced through the same reactions that operate in C_3 plants (Figure 8-11). It appears, therefore, that the C_4 pathway evolved as a mechanism to "get around" the problems resulting from the inability of RuBP carboxylase to bind carbon dioxide at low concentrations.

In hot, dry environments where water loss is critical, C_4 plants possess a major survival advantage. Carbon dioxide enters the leaves of all plants through small pores, called **stomates.** When a stomate opens for carbon dioxide uptake, however, water vapor escapes through the same pore. Since C_4 plants are able to fix carbon dioxide even at extremely low carbon dioxide concentrations, stomates can be kept very narrow or closed altogether, reducing water loss, without affecting their rate of carbon dioxide fixation. In contrast, when C_3 plants save water in hot, dry environments by closing their stomates, they drastically reduce their rate of carbon dioxide fixation and subsequent growth.

Another survival advantage of C_4 plants stems from their resistance to **photorespiration**—a phenomenon whereby high levels of oxygen actually interfere with photosynthesis by binding to RuBP carboxylase and inhibiting its carbon dioxide-fixing activity. Photorespiration can cut glucose production in C_3 plants by as much as 50 percent during peak sunlight hours. C_4 plants are less susceptible to photorespiration because they are able to photosynthesize under lower carbon dioxide and oxygen levels.

So why haven't vigorously growing C_4 plants taken over all the earth's habitats? When C_4 plants are introduced by people into some environments, the consequences to the native flora have indeed been swift. But C_4 plants, such as sugar cane and kudzu, do better than C_3s only where it is warm or dry (Figure 8-13). In cold environments, C_3s do better than C_4s because they are less sensitive to low temperature.

CAM PLANTS: FURTHER ADAPTATIONS TO A HOT, DRY ENVIRONMENT

Less than 5 percent of plants have another biochemical adaptation that allows them to survive in very hot, dry desert habitats. These plants, called **CAM plants,** utilize the same enzyme for carbon dioxide fixation as do C_4 plants, but conduct the light-dependent reactions and CO_2 fixation at different times rather than in different cells. (CAM is an acronym for *crassulacean acid metabolism,* named for plants of the family Crassulaceae in which it was first discovered.) While C_3 and C_4 plants open their stomates and fix carbon dioxide during the daytime, CAM plants, such as cactus and pineapple, open their stomates at night, when the rate of water vapor loss is greatly reduced. Throughout the night, more and more carbon dioxide is fixed and the products stored in the cell's vacuole. At sunrise, stomates close (conserving water), and sunlight powers ATP and NADPH production via the light-dependent reactions. At the same time, the products of carbon dioxide fixation are gradually removed from the vacuole and converted to carbohydrates. Attempts to manipulate C_3, C_4, and CAM plants in the laboratory are discussed in "The Human Perspective: Producing Crop Plants Better Suited to Environmental Conditions."

FIGURE 8-13

Kudzu, a C_4 plant, is a familiar sight growing along the highways in the southeastern United States. Introduced from Asia, kudzu is able to overgrow many indigenous species.

New life is built mostly from air and water. The original source of carbon to build new organic molecules needed for growth and reproduction is carbon dioxide gas obtained from the nonliving environment. Carbon becomes incorporated into organic molecules during the light-independent reactions of photosynthesis, forming the life-sustaining bridge across which carbon passes from the realm of the nonliving into that of the living. (See CTQ #6.)

◁ B I O L I N E ▷
Living on the Fringe of the Biosphere

In 1977, Robert Ballard and fellow scientists of the Woods Hole Oceanographic Institute guided a small research submarine through the pitch-black depths of the Pacific Ocean near the Galapagos Islands. Ballard's group was looking for cracks in the ocean floor that had appeared in photographs taken by cameras suspended at the end of long cables. There, at a depth of more than 2,500 meters (8,000 feet), the researchers discovered a flourishing community of diverse organisms (Figure 1*a*), including sea anemones, clams, mussels, blind crabs, and giant-sized red tube worms that (Figure 1*b*) had no mouths and were many times larger than their closest known relatives. All of these strange marine creatures were crowded near "chimneys" that spewed scalding, black water. The scientists crowded into the small research vessel were amazed at having discovered one of the most extraordinary and productive ecosystems on earth—lying at the very fringe of the biosphere. Animals living on the deep ocean bottom generally must depend on a trickle of organic material that filters through the water from the surface. How is it possible, then, that such an abundant fauna could live in such a hostile environment?

The cracks in the ocean floor are termed **hydrothermal vents.** These are sites where frigid sea water seeps into red-hot fissures and becomes heated to more than 350°C (662°F). The scalding water

(a)   *(b)*

FIGURE 1

mixes with hydrogen sulfide released from the earth's core, producing a habitat capable of supporting the growth of dense clouds of chemosynthetic bacteria that depend on hydrogen sulfide as an energy source. These bacteria form the diet of barnacles, clams, mussels, and small worms, which filter the microscopic organisms from the sea water. These filter feeders, in turn, are eaten by larger crabs and fishes. A number of animals, including the giant red tube worm, house the chemosynthetic bacteria *inside their tissues*, thereby receiving a direct supply of food. Remarkably, the organisms of the hydrothermal vents form a self-sufficient community independent of life elsewhere on earth.

CHEMOSYNTHESIS: AN ALTERNATE FORM OF AUTOTROPHY

Although the predominant means of generating energy-rich organic molecules is through photosynthesis, a number of bacteria depend on chemosynthesis for a supply of chemical energy. In photosynthesis, electrons and protons come from the splitting of water, while in **chemosynthesis,** they are stripped from reduced inorganic substances, such as ammonia (NH_3) and hydrogen sulfide (H_2S). The chemical equation for chemosynthesis using hydrogen sulfide is:

$$CO_2 + 2\ H_2S \rightarrow (CH_2O) + H_2O + 2\ S$$

The removal of electrons (oxidation) provides the energy needed for the formation of NADPH and ATP. Until recently, chemosynthetic bacteria did not receive much attention from scientists, but this changed in 1977 when a whole new type of community was discovered that depended entirely on chemosynthesis of microscopic prokaryotes. This community is discussed in the Bioline entitled "Living on the Fringe of the Biosphere."

Light is not the only source of energy for autotrophs. The energy contained in some inorganic molecules is harvested by a few types of autotrophs. In some habitats devoid of light, these autotrophs supply the energy that supports an entire community of organisms. (See CTQ #8.)

REEXAMINING THE THEMES

Relationship between Form and Function

Differences in the growth of C_3 and C_4 plants in different environmental conditions are based on differences in the structure of the leaves and the carbon dioxide-fixing enzymes of these plants. C_4 plants are able to flourish in warm, dry habitats partly because their leaves separate parts of photosynthesis into two specialized types of cells, each with a different carbon dioxide-fixing enzyme. PEP carboxylase, whose active site is able to bind carbon dioxide at low levels, is present in the outer mesophyll cells, while RuBP carboxylase, a less efficient enzyme but an important component of the pathway leading to carbohydrate synthesis, is present in bundle-sheath cells lying deeper in the leaf. C_3 plants do not use PEP carboxylase, nor do they separate reactions in different cells.

Biological Order, Regulation, and Homeostasis

Chloroplast thylakoids probably contain the most highly ordered components of any cellular membrane. Within these thin membrane sheets is an array of proteins and pigments capable of harvesting light energy and converting it to chemical energy of ATP and NADPH. Each of the components required for the light-dependent reactions —from the elements of the two photosystems to those of the electron transport chains—must be held in a precise orientation within the thylakoid membrane; otherwise, electrons could not pass along a defined sequence of carriers, and protons could not be transported across specific sites of the thylakoid membrane.

Acquiring and Using Energy

Photosynthesis (and chemosynthesis) are the only means by which the energy used and lost by living organisms is replaced, allowing life to continue. The energy present in sunlight drives endergonic reactions in the chloroplast, which lead to the synthesis of ATP and NADPH. The energy stored in these compounds is subsequently used in the synthesis of carbohydrate and other organic materials.

Unity within Diversity

Organisms can be divided into two groups based on their ability or inability to manufacture their own organic materials from "scratch," that is, from inorganic precursors such as carbon dioxide and water. Organisms with this ability are autotrophs—life's "producers"—while organisms lacking this ability are heterotrophs—life's "consumers." Organisms as different as the chemosynthetic bacteria living in the cracks in the ocean floor, the flagellated protists that form a scum over ponds during the summer, and the trees in the forest are all autotrophs. Conversely, organisms as different as disease-causing bacteria, mushrooms, baker's yeast, and mammals are all heterotrophs.

Evolution and Adaptation

Plants living in hot, dry environments are frequently threatened with severe water loss. Saving water requires that the plants close or reduce the size of their stomates during the day, which reduces the carbon dioxide supply to the chloroplasts. C_4 and CAM plants have evolved adaptations that get around these problems. Both groups possess a carbon dioxide-fixing enzyme that can operate at low carbon dioxide levels. The C_4 plant separates fixation of atmospheric carbon dioxide and carbohydrate synthesis into different cells of the leaf, whereas CAM plants carry out these two processes at different times of the day.

SYNOPSIS

Organisms are categorized as autotrophs if they manufacture organic nutrients from carbon dioxide and other inorganic precursors; they are categorized as heterotrophs if they depend on organic materials synthesized by others. Most autotrophs are photosynthesizers, converting light energy into chemical energy, which is then used to manufacture carbohydrates from carbon dioxide and water.

During the light-dependent reactions, photons of light are absorbed by pigments embedded in the thylakoid membranes. The pigments are associated into clusters, called photosystems, each of which includes several hundred antenna pigments that absorb varying wavelengths of light and a single reaction-center pigment that ultimately receives the energy. There are two types of photosystems, I and II, which operate in conjunction with each other. When energy is absorbed by the reaction-center pigments, electrons are boosted to an outer orbital from which they are transferred to an electron acceptor, leaving the reaction-center pigments positively charged. The reaction center of Photosystem II ($P680^+$) is sufficiently electron attractive to pull electrons from water, splitting the water molecules into protons, electrons, and oxygen atoms. The protons pass into the inner space of the thylakoid; the oxygen atoms form molecular oxygen, which is released as a byproduct; and the electrons pass to $P680^+$, filling the electron vacancy. The electrons previously lost by P680 pass along a chain of electron carriers to the electron-deficient reaction center ($P700^+$) of Photosystem I, filling its electron vacancy. The electrons previously lost by P700 pass along a chain of electron carriers to $NADP^+$, forming NADPH. During the process of electron transport, protons are moved across the thylakoid membrane into the lumen.

The protons that accumulate in the thylakoid lumen as the result of the splitting of water and electron transport produce a proton gradient—a form of stored energy that is used to drive the reaction in which ADP is converted to ATP. ATP formation occurs as a result of the movement of protons back across the thylakoid membrane through a channel in an ATP-synthesizing complex. This entire process in which electrons pass from water to NADPH and ATP is formed by a proton gradient is called noncyclic photophosphorylation.

During the light-independent reactions, the chemical energy stored in NADPH and ATP is used in the synthesis of carbohydrates from carbon dioxide. There are three variations in these synthetic pathways. The majority of plants utilize C_3 synthesis in which carbon dioxide is fixed by RuBP carboxylase to a five-carbon compound, RuBP, forming an unstable six-carbon compound, which immediately splits into two molecules of PGA. NADPH and ATP are used to convert PGA molecules to glucose, regenerating RuBP to participate in another round of the cycle. In C_4 and CAM plants, carbon dioxide is initially fixed by a different enzyme (PEP carboxylase) to a three-carbon compound. PEP carboxylase has the ability to bind carbon dioxide at much lower carbon dioxide concentrations than can RuBP carboxylase. This allows C_4 and CAM plants to flourish under hot, dry conditions where they can keep their stomates closed, which greatly lowers the carbon dioxide concentration but prevents water loss.

Chemosynthesis practiced by certain bacteria utilizes energy obtained by oxidizing inorganic substrates, such as ammonia and hydrogen sulfide to form NADPH and ATP needed in the synthesis of carbohydrates.

Key Terms

photosynthesis (p. 154)
autotroph (p. 154)
heterotroph (p. 155)
light-dependent reaction (p. 156)
light-independent reaction (p. 156)
photon (p. 156)
chlorophyll (p. 157)
carotenoid (p. 157)
photosystem (p. 159)
reaction center (p. 159)
antenna pigment (p. 159)

Photosystem I (p. 159)
Photosystem II (p. 159)
P700 (p. 159)
P680 (p. 159)
electron acceptor (p. 159)
photolysis (p. 160)
electron transport system (p. 160)
noncyclic photophosphorylation (p. 161)
chemiosmosis (p. 161)
ATP synthase (p. 161)
cyclic photophosphorylation (p. 161)

carbon dioxide fixation (p. 163)
C_3 pathway (p. 164)
ribulose biphosphate (RuBP) (p. 164)
RuBP carboxylase (p. 164)
C_4 synthesis (p. 164)
PEP carboxylase (p. 166)
stomate (p. 166)
photorespiration (p. 166)
CAM plant (p. 166)
chemosynthesis (p. 167)
hydrothermal vent (p. 167)

Review Questions

1. Rearrange the order of the following terms to match the correct sequence of reactions during photosynthesis:

glucose formation	PGA production
splitting of water	chemiosmosis
carbon dioxide fixation	absorption of light
noncyclic flow	PGAL production

2. Contrast the functions of the light-dependent and light-independent reactions. Why do the light-independent reactions depend on the light-dependent reactions having already occurred?

3. How does light absorption by the Photosystem II reaction center contribute to the splitting of water? How does the splitting of water contribute to the formation of ATP?

4. Contrast the basic roles of Photosystems I and II in generating NADPH.

5. Describe the adaptations in C_4 and CAM plants which allow them to flourish in hot, dry environments.

Critical Thinking Questions

1. If Melvin Calvin had been working with a C_4 plant, what type of compound would he have found most heavily labeled after administering $C^{18}O_2$ for 2 seconds? What if he had been working with a CAM plant?

2. Considering the second law of thermodynamics (Chapter 6), which states that usable energy becomes lost whenever energy is exchanged, would you expect there to be a greater number of heterotrophic or autotrophic individuals in the world? How does your conclusion fit with the statement that approximately 80 percent of all species are heterotrophs?

3. Technological advances and scientific discoveries often go hand in hand. List the technological advances that made it possible for scientists to unravel the mysteries of photosynthesis, and explain how each advance contributed to our knowledge of photosynthesis.

4. Suppose you discovered a mutant plant having thylakoids lacking Photosystem II. How would this affect carbohydrate production? Why? Do you suppose this plant would be able to produce any ATP from converted light energy?

5. Organization is one of the characteristics of all living things. Explain how organization is reflected in the structure of plant cells and the biochemical processes that constitute photosynthesis.

6. Construct a diagram that illustrates the flow of carbon from carbon dioxide in the atmosphere to heterotrophic organisms and back to the atmosphere.

7. If you were trying to engineer C_3 plants genetically to make them more efficient and able to survive better in hot, dry climates, on which enzyme would you focus? What changes would you want to develop in this enzyme?

8. Suppose somebody were to tell you that all life is dependent on energy from the sun and that the "sudden death" of the sun would quickly extinguish all life forms. Would you agree with their conclusion? Why or why not?

Additional Readings

Alberts, B., et al. 1989. *Molecular biology of the cell.* New York: Garland. (Intermediate)

Cone, J. 1991. *Fire Under the Sea: The Discovery of the Most Extraordinary Environment on Earth—Volcanic Hot Springs on the Ocean Floor.* New York: Morrow. (Intermediate)

Govindjee, and W. J. Coleman. 1990. How cells make oxygen. *Sci. Amer.* Feb: 50–58. (Intermediate)

Gregory, R. P. 1989. *Photosynthesis.* London: Chapman and Hall. (Intermediate)

Kaharl, V. A. 1991. *Water baby: The story of Alvin.* Oxford. (Alvin is the name of the submersible vessel that discovered life in the hydrothermal vents) (Introductory)

Kamen, M. D. 1986. A cupful of luck, a pinch of sagacity. *Ann Rev. Biochem.* 55:1–34. (Introductory)

Prebble, J. N. 1981. *Mitochondria, chloroplasts, and bacterial membranes.* New York: Longman. (Advanced)

Stryer, L. 1988. *Biochemistry.* 3d ed., New York: Freeman. (Advanced)

Youvan, D. C., and B. L. Marrs. 1987. Molecular mechanisms of photosynthesis. *Sci. Amer.* June: 42–50. (Intermediate)

Processing Energy: Fermentation and Respiration

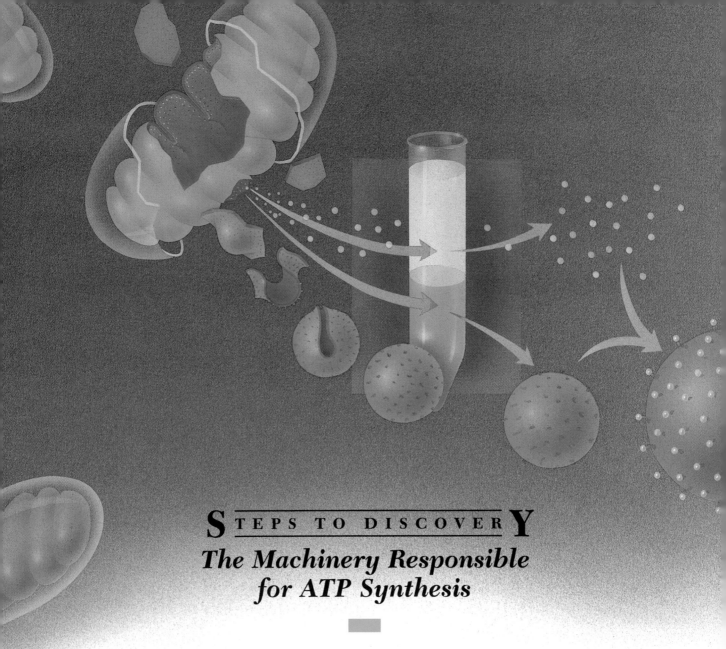

With the improvement of light microscopes in the latter part of the nineteenth century, biologists observed thread-like particles within the cytoplasm of various types of cells. They named the organelle "mitochondrion," from the Greek meaning "thread granule." Despite their prevalence in cells, the function of mitochondria remained a mystery until the 1940s, when scientists perfected techniques to remove and isolate various organelles from cells. The isolation of mitochondria depended on the development of a new instrument—the "ultracentrifuge." This device could spin tubes in a circular path at speeds high enough to gen-

erate centrifugal forces (forces that pull an object outward when it is rotating around a center) over 100,000 times the force of gravity. Cells were first broken open to release their contents, which were placed in a tube and spun at various speeds. Larger organelles, such as cell nuclei, would settle at the bottom of the tube at slower speeds. The supernatant (the liquid remaining in the upper portion of the tube) containing smaller particles was then removed and recentrifuged *at higher speeds*, causing the mitochondria to settle at the bottom. The supernatant was then discarded, and the purified mitochondria were resuspended.

When mitochondria are broken, the inner membranes form vesicles (red spheres) and particles (yellow dots).

In 1948, using these techniques, Eugene Kennedy and Albert Lehninger, then at the University of Chicago, purified mitochondria and demonstrated that these organelles were the cells' "chemical power plants." These purified mitochondria were able to oxidize organic compounds (such as fatty acids) and to use the released energy to make ATP.

The mitochondrion is a complex organelle composed of many parts. Which part was responsible for synthesizing ATP? Efraim Racker and his colleagues in New York attempted to answer this difficult question. According to Racker, "We had no new ideas, so we did what I call 'instrumental research': You have no ideas, use a new instrument. At about that time, Dr. Nossal in Australia had designed a mechanical shaker for breaking up yeast cells with glass beads. So we used this machine to break up mitochondria."

When Racker's team subjected the suspension of broken mitochondria to high-speed centrifugation, they divided the material into two portions: one portion consisted of membrane vesicles that went to the bottom of the centrifuge tube, the other consisted of material that remained in the liquid supernatant. The isolated vesicles at the bottom could oxidize various organic substrates, but they were unable to make ATP. When mixed with the liquid supernatant, however, these same vesicles acquired full ATP-generating capacity. It seemed that the supernatant contained some factor needed by the vesicles for ATP production. Racker called this factor F_1.

Meanwhile, in Boston, Humberto Fernandez-Moran of Massachusetts General Hospital was examining mitochondrial membranes using a new electron-microscopic technique called negative staining, whereby objects are brightly outlined against a dark background. This makes very small objects consisting of macromolecules much easier to see. In this case, the technique revealed never-before-seen rows of particles protruding from the inner mitochondrial membrane. Each particle (see Figure 9-9) was attached to the membrane by a thin stalk.

Racker then used the electron microscope to examine the supernatant formed from disrupted mitochondria to see if his F_1 factors could be visualized. Racker observed particles in the supernatant that appeared identical to the stalked particles Fernandez-Moran had discovered attached to the mitochondrion's inner membrane. Furthermore, when Racker added the F_1 particles to the membrane vesicles formed from disrupted mitochondria, the particles became attached to the membranes by "little stalks." Racker concluded that the F_1 particles released during mitochondrial disruption were the same ones as those seen on the inner wall of mitochondria. After a series of biochemical studies, Racker discovered that these particles were in fact the sites in mitochondria where ATP was synthesized. He called the entire complex (sphere, stalk, and membrane-base piece) *ATP synthase.*

Similar types of stalked particles have been found attached to the thylakoid membranes of chloroplasts and the plasma membrane of bacteria (which have no mitochondria). In all of these organisms, the stalked particles are the sites where energy from substrate oxidation drives the synthesis of ATP. Once again, the unity of life reveals the common evolutionary origins of all organisms. Bacteria, plants, and animals are all descended from the same ancestor, one from which they inherited their energy-generating mechanisms.

When mixed, particles reattach to the vesicles to form complexes capable of ATP formation.

*I*t may have been the most catastrophic single occurrence in the history of life on earth. It was presumably responsible for the deaths of billions of individual organisms and spelled extinction for countless species. Creatures that had prevailed for millions of years disappeared forever, banished by one of the most universally poisonous substances the world had ever known. This unwelcome toxic intruder is still around today; in fact, it saturates today's biosphere. We are referring to molecular oxygen (O_2).

Molecular oxygen gradually appeared on earth beginning about 3 billion years ago. It was produced by a new type of microorganism—an evolutionary innovator that changed the face of the planet even more dramatically than would the eventual reign of human beings billions of years later. These devastating microbes were *cyanobacteria* (formerly known as "blue-green algae"), the prokaryotes that "invented" oxygen-evolving photosynthesis.

Up to that time, none of the earth's organisms required molecular oxygen, and the gas was not present in the atmosphere. The world was populated by **anaerobes** —organisms that utilized *oxygen-free (anaerobic)* metabolic processes to disassemble nutrients and obtain energy. This was the ancient order on the early oxygen-devoid earth. Following the rise of the cyanobacteria, however, the oceans and the atmosphere were infused with so much molecular oxygen that the old order was overthrown, forever altering the history of life on earth. The success of the cyanobacteria's energy-acquiring strategy catapulted these microbes to the top of the evolutionary hierarchy, where they remained for more than 2 billion years.

When it first appeared, molecular oxygen fatally oxidized the cytoplasm of all but a few types of microorganisms. Eventually, however, species evolved that not only withstood the poisonous effects of molecular oxygen but actually became dependent on it, using oxygen to increase their metabolic efficiency enormously. Organisms dependent on oxygen are called **aerobes.** These evolutionary innovators—the first users of oxygen—were the pioneers of an evolutionary line that ultimately produced all oxygen-using organisms, including humans.

Living with oxygen is still a dangerous business, however. As we saw on page 56, molecular oxygen can pick up an extra electron, becoming transformed into the highly destructive superoxide radical. The survival of aerobes depends on protective enzymes that disarm these free radicals by converting them to water the instant they are formed. Without this protection you inherited from early aerobic prokaryotes, the oxygen in your next breath would bring certain death instead of sustaining your life.

Perhaps after reading this chapter, you will have a new appreciation of breathing (Figure 9-1). Breathing draws in life-supporting oxygen and expels the waste product, carbon dioxide. Molecular oxygen is needed for you to release the energy stored in the chemical fuel you consume as food. Without oxygen, the metabolic furnace that powers your life would be extinguished.

FERMENTATION AND AEROBIC RESPIRATION: A PREVIEW OF THE STRATEGIES

⬢ One of the foundations of biology is the evolutionary unity of organisms. Even such distantly related organisms as a bacterium, a garden weed, and a human being share many common biological characteristics. We now come to another of these biological "universals." Recall from previous chapters that reduced organic molecules represent a storehouse of chemical energy. All organisms harvest chemical energy by oxidizing these organic molecules, generating

FIGURE 9-1

Oxygen—a precious commodity. Without a supply of oxygen from the tank, this scuba diver would only be able to remain submerged for a minute or two.

raw materials, and releasing energy that is trapped momentarily in the form of ATP. Moreover, they accomplish this shuffling of chemical energy from one compound to another by virtually the same set of metabolic reactions that arose at a very early stage of evolution. The starting point for most of these metabolic reactions is glucose.

Glucose is an energy-rich molecule, but it cannot directly power biological activities. In other words, a glucose molecule in a cell is like a person carrying around a $500 bill and trying to use it to purchase a few gallons of gasoline: Both the cell and the gasoline purchaser require smaller denominations. Therefore, cells convert energy-rich molecules, such as glucose, to other compounds containing more usable quantities of energy, most often ATP.

☀ The *complete* oxidation of a gram of glucose releases 3,811 calories. Not all organisms can take full advantage of glucose's energy content, however. The extent to which glucose is dismantled to release its energy is the basis for distinguishing between two fundamental processes: fermentation and aerobic respiration. As we will see shortly, **fermentation** is an incomplete breakdown of glucose, which occurs in the absence of oxygen and extracts only a small portion of the sugar's energy content. In contrast, during **aerobic respiration,** each glucose molecule is *completely* disassembled, step by step, with pairs of high-energy electrons being stripped from the substrate and transferred to molecular oxygen.[1] Electrons removed from the substrate are passed through a series of membrane-embedded electron carriers that comprise an electron transport system similar to that utilized during photosynthesis, which we discussed on page 160. The energy released during electron transport is used in the formation of ATP.

Regardless of the strategy (fermentation or aerobic respiration), tapping the energy stored in glucose begins with **glycolysis** (*glyco* = sugar, *lysis* = split), a universal pathway in which glucose is split into two fragments, both of which are converted to pyruvic acid, which contains a storehouse of readily available energy. We will now take a closer look at the process of glycolysis.

If all the energy in an energy-rich molecule were released in one step, most of the energy would be lost as heat and could not be used to power biochemical reactions. A multistep, energy-harvesting pathway enables cells to release the energy gradually to form molecules capable of delivering smaller, usable amounts of energy, most often as ATP. (see CTQ #2.)

---

[1] Cellular respiration should be distinguished from the term "respiration" as it applies to inhaling and exhaling. Nearly all organisms, even those that have never taken a breath, rely on aerobic respiration as their primary energy-harvesting strategy. These organisms include animals, plants, protists, and bacteria—every organism on earth, with the exception of a few prokaryotes and animals (such as intestinal parasites) that live in conditions where oxygen is not available. Clearly, "respiration" implies more than just breathing.

GLYCOLYSIS

In 1905, two British chemists, Arthur Harden and W. J. Young, were studying glucose breakdown by yeast cells, a process that generates bubbles of carbon dioxide. Harden and Young noted that the carbon dioxide bubbling eventually slowed and stopped, even though there was plenty of glucose left to metabolize. Apparently, some component of the broth was being exhausted. After experimenting with a number of substances, the chemists found that adding inorganic phosphates started the reaction going again. They concluded that the reaction was exhausting phosphate, the first clue that phosphate played a role in metabolic pathways. It would be several decades before biochemists would understand that the inorganic phosphate was being used to form ATP, which was subsequently used in glucose disassembly. This is illustrated by the first reaction of glycolysis.

THE REACTIONS OF GLYCOLYSIS

Glycolysis begins with the linkage of two phosphate groups to glucose at the expense of two molecules of ATP (Figure 9-2). This process of phosphate addition, called *phosphorylation,* "activates" the sugar—in this case, making it reactive enough to be split into two fragments, each containing three carbons and a phosphate. (Keep in mind that, because of the splitting of glucose, two molecules of each reaction product are formed from a single molecule of glucose. This is indicated in Figure 9-2.) In this case, the loss of a pair of ATPs can be considered the cost of getting into the glucose-oxidation business.

The next two reactions are particularly important because they have the potential to generate ATP. The first of these reactions is the conversion of phosphoglyceraldehyde (PGAL) to 1,3-diphosphoglycerate (DPG). You needn't be concerned with the chemical structures of the molecules in these reactions; your attention is best directed to the red carbons and the transfer of the electrons, indicated by the circle.

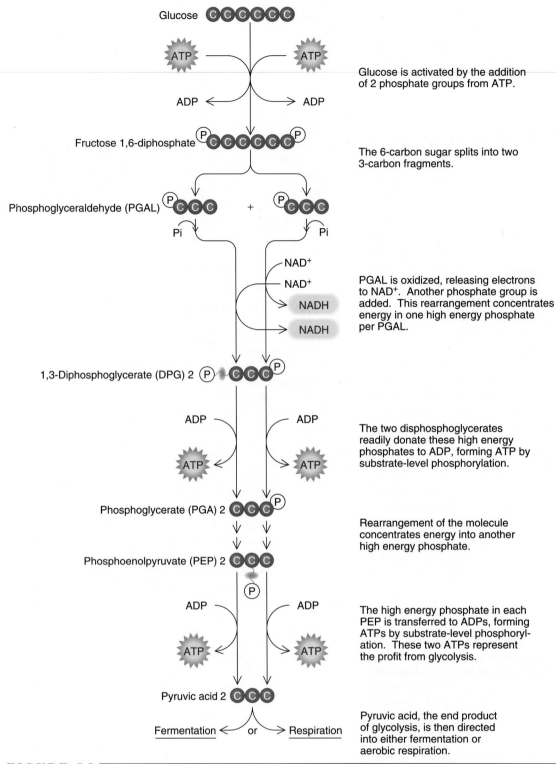

Glucose is activated by the addition of 2 phosphate groups from ATP.

The 6-carbon sugar splits into two 3-carbon fragments.

PGAL is oxidized, releasing electrons to NAD⁺. Another phosphate group is added. This rearrangement concentrates energy in one high energy phosphate per PGAL.

The two disphosphoglycerates readily donate these high energy phosphates to ADP, forming ATP by substrate-level phosphorylation.

Rearrangement of the molecule concentrates energy into another high energy phosphate.

The high energy phosphate in each PEP is transferred to ADPs, forming ATPs by substrate-level phosphorylation. These two ATPs represent the profit from glycolysis.

Pyruvic acid, the end product of glycolysis, is then directed into either fermentation or aerobic respiration.

FIGURE 9-2

A condensed version of glycolysis. Although some of the ten reactions have been omitted for clarity, the overall activities of the pathway are evident—the oxidation of glucose to two molecules of pyruvic acid, with the release of two NADHs and a profit of two ATPs. The overall reaction can be written: Glucose + 2 ATPs + 2 NAD⁺ → 2 pyruvic acid + 4 ATPs + 2 NADH. Note: each ball represents a carbon atom.

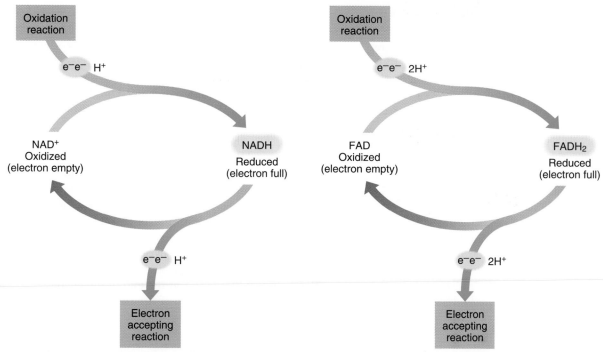

FIGURE 9-3

Electron carriers in action. A number of enzymes oxidize substrates by removing a pair of electrons, which is transferred to NAD$^+$, forming NADH. The electrons are only stored temporarily in NADH. As we will see, during fermentation, the electrons stored in NADH are transferred to another substrate, thereby regenerating NAD$^+$. During aerobic respiration, the electrons are passed along an electron transport system and used to generate ATP. (NADH may also be converted to NADPH, whose electrons are used in anabolic pathways rather than to form ATP.) As we will see later in the chapter, FAD (flavin adenine dinucleotide) is another coenzyme that accepts electrons from substrates and passes the electrons along an electron transport system, thereby generating ATP.

NAD$^+$ *(nicotinamide adenine dinucleotide)* is a coenzyme for a number of enzymes that oxidize substrates. This key compound of metabolism is derived from the vitamin niacin, a substance readily obtained from various types of meats and leafy vegetables. In the reaction shown above, the enzyme removes "high-energy" electrons (and a proton) from the PGAL molecule and transfers them to NAD$^+$, forming NADH (Figure 9-3), which is now free to leave the enzyme. In respiring organisms, these high-energy electrons stored in NADH will be "cashed in" for ATP at a later stage in the process.

In the next reaction of glycolysis, a phosphate group is transferred from DPG to ADP to form a molecule of ATP. Once again, you needn't be concerned with the chemical structures shown in this reaction, for it is the transfer of the phosphate group (indicated by the circle) that is important here:

This "direct" route of ATP formation is referred to as **substrate-level phosphorylation,** because a phosphate group is transferred directly from a substrate molecule to ADP *and does not require an electron transport chain.* A second substrate-phosphorylation step occurs later in glycolysis during the formation of pyruvic acid (Figure 9-2).

We can now look back over glycolysis and total up the energy profits. The two substrate-level phosphorylation steps produce four ATPs for each glucose entering the pathway. Because two ATPs must be spent just to get glycolysis "rolling," however, the *net* ATP yield is only two. In addition to a net gain of two ATPs, glycolysis also generates two pairs of energized electrons (carried in two NADHs).

The molecular remains of glycolysis are two molecules of pyruvic acid, each containing three carbons. This is the point where anaerobic (oxygen-independent) fermentation and aerobic (oxygen-dependent) respiration diverge. They differ in the way they solve a common problem: how to regenerate the NAD$^+$ molecules that were converted to NADH during the oxidation of PGAL. In all organisms, the reserves of available NAD$^+$ are replenished by transferring electrons from NADH to another molecule, regenerating NAD$^+$. If NADH donates its electrons directly to an organic substrate, which is simply excreted as a waste product, the process is called *fermentation.* If, instead, NADH passes its electrons to an electron transport chain in which oxygen is the final electron acceptor, the process is called *aerobic respiration.* If the regeneration of NAD$^+$ could not occur, the cell would rapidly run out of its small supply of this coenzyme, and neither glycolysis nor the subsequent energy-generating process could proceed.

Glucose contains a large amount of chemical energy. Before its energy can be extracted, however, glucose molecules must first be activated by phosphorylation. Activation requires adding energy. *Glycolysis* accomplishes this activation, and extracts a small amount of usable energy in the process. (See CTQ #3.)

FERMENTATION

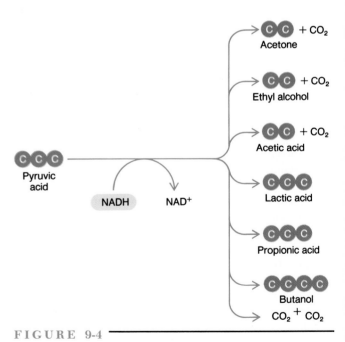 During fermentation, the electrons from NADH are transferred either to pyruvic acid, the end product of glycolysis, or to a compound formed from pyruvic acid. The end products of fermentation (Figure 9-4) vary according to the organism but always lead to the regeneration of NAD$^+$. In the most familiar type of fermentation—**alcoholic fermentation**—yeast cells convert pyruvic acid to ethyl alcohol, the alcohol consumed in alcoholic beverages. Alcoholic beverages differ primarily in the raw materials used in the fermentation process. Wines are produced by fermenting fruits, particularly grapes, whereas beers are produced by fermenting malted cereals, such as barley. Both wine and beer production are ancient arts; a recipe for making beer, for example, was found inscribed on stone tablets from Mesopotamia that dated back 9,000 years. By itself, fermentation is only able to produce a liquid containing about

15 percent alcohol. Those alcoholic products containing higher quantities of alcohol, such as whiskey or brandy, are produced by *distillation,* whereby the alcohol is evaporated from the fermentation liquid and then recondensed at higher concentration. The role of microorganisms in producing alcohol and other commercial products of fermentation is discussed in the Bioline: The Fruits of Fermentation.

In 1907, the British biochemist Frederick Hopkins demonstrated that when isolated frog muscles were caused to contract in an anaerobic environment, the muscles produced large amounts of lactic acid. Lactic acid was being produced by fermentation of pyruvic acid. **Lactic acid fermentation** also occurs in human muscle cells when they undergo such strenuous activity that they cannot obtain enough oxygen to oxidize pyruvic acid fully (see "The Human Perspective: The Role of Anaerobic and Aerobic Metabolism in Exercise," at the end of the chapter).

To reiterate, all types of fermentation are anaerobic processes. During the early stages of life on earth, when oxygen had not yet appeared, glycolysis and fermentation were probably the primary metabolic pathways by which energy was extracted from sugars by primitive prokaryotic cells. Today, many organisms still live in environments lacking oxygen, as in deep soil or inside the body of another animal, and glycolysis and fermentation remain the only

FIGURE 9-4

Some metabolic products of fermentation. Pyruvic acid may be converted to several end products, depending on the organism and the presence or absence of oxygen. In all cases, the final compound contains electrons donated by NADH. This recycles the coenzyme NAD$^+$, allowing it to continue its essential role as electron acceptor in glycolysis.

◁ B I O L I N E ▷
The Fruits of Fermentation

FIGURE 1 ———

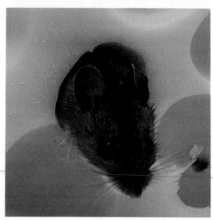

FIGURE 2 ———

Biological events that take place in the absence of oxygen have been creating fortunes and livelihoods for people since the dawn of recorded history. The metabolic byproducts of fermentation, for example, make our diets more interesting, protect us from diseases, retard food spoilage, and provide us with alcoholic beverages (Figure 1). Alcoholic drinks are the products of converting pyruvic acid to ethyl alcohol by brewer's yeast, *Saccharomyces cerevisiae*. Alcoholic fermentation also releases carbon dioxide, which, if not allowed to escape, becomes trapped in the liquid,

producing the natural carbonation of beer and champagne. The same process, alcoholic fermentation by yeast cells, elevates dough during the leavening of bread and pastries. In fact, until the mid-1800s, commercial bakeries used yeast left over from the brewing of beer to leaven their pastries. The dried baker's yeast familiar to modern cooks contains living cells that quickly convert the sugar in dough to ethyl alcohol and carbon dioxide. The carbon dioxide forms the expanding bubbles that cause the dough to rise; the dough literally inflates with the gas. The alcohol is driven off by baking, which explains why you can eat bread without fear (or hope) of inebriation.

Bacteria generate a wide variety of valuable fermentation products (see Figure 9-4), of which lactic acid is perhaps the most common. This slightly sour acid imparts flavor to yogurt, rye bread, and some cheeses (Figure 2). It also lowers the pH of food below a level tolerable by many spoilage microbes. Dairy products that contain lactic acid are therefore more resistant to spoilage than is the milk from which they were made. Swiss cheese is produced by other types of bacteria that convert pyruvic acid to propionic acid and carbon dioxide. The acid imparts a nutty flavor to the cheese, whereas the gas creates the large bubbles that form its characteristic holes. Vinegar is a product of acetic-acid-

producing bacteria that grow in apple cider or grape mash. The tangy taste of pickles, sauerkraut, and olives reflects the presence of lactic and acetic acids, microbial byproducts that provide a flavorful twist to these foods. The process of making soy sauce requires an 8- to 12-month fermentation period before the mash of soy beans acquires its characteristic flavor.

Acetone and other industrially produced solvents are also byproducts of fermentation, as is isopropyl (rubbing) alcohol. Fermentation may be an inefficient means for an organism to harvest biological energy, but it has proven a very efficient source of valuable products.

methods of obtaining energy.[2] The total ATP yield of such organisms is a mere two molecules per glucose oxidation. Most of the potential energy of glucose is left in the waste products that are simply discarded, such as ethyl alcohol or

lactic acid. In fact, more than 90 percent of glucose's chemical energy is untapped by glycolysis and fermentation. (The flammability of ethyl alcohol, which can be used as a fuel in automobiles, testifies to the high energy content of this fermentation product.)

---

[2] A few bacteria in these oxygen-devoid environments use anaerobic respiration rather than fermentation. Anaerobic respiration resembles a less efficient form of aerobic respiration that uses a terminal electron acceptor such as nitrate or sulfate rather than oxygen.

In oxygen-poor environments, cells utilize a less efficient metabolic pathway that allows them to continue to harvest a limited amount of energy. (See CTQ #4.)

◁ THE HUMAN PERSPECTIVE ▷
The Role of Anaerobic and Aerobic Metabolism in Exercise

Most of you have tried lifting a barbell or doing "push-ups." You may have noticed that the more times you repeat the exercise, the more difficult it becomes until you are no longer able to perform the activity. The failure of your muscles to continue to work can be explained by oxidative metabolism. Skeletal muscles (the muscles that move the bones of the skeleton) consist of at least two types of fibers (insets): "fast-twitch" fibers—which can contract very rapidly—and "slow-twitch" fibers—which contract more slowly. Fast-twitch fibers are nearly devoid of mitochondria, which indicates that these cells are unable to produce much ATP by aerobic respiration. In contrast, slow-twitch fibers contain large numbers of mitochondria—sites of aerobic ATP production. These two types of skeletal muscle fibers are suited for different types of activities. Lifting weights or doing push-ups depends primarily on fast-twitch fibers, which are able to generate more force than are their slow-twitch counterparts. Fast-twitch fibers produce nearly all of their ATP anaerobically as a result of glycolysis. The problem with producing ATP by glycolysis is the rapid use of the fiber's available glucose (stored in the form of glycogen) and the production of an undesirable end product, lactic acid. Let's consider this latter aspect further.

FIGURE 1

Skeletal muscles (*insets*) contain a mix of fast-twitch fibers (darkly stained) and slow-twitch fibers (lightly stained). The muscles of world-class weightlifters (*top photo*) typically have a higher-than-average percentage of anaerobic, fast-twitch fibers, while world-class marathon runners (*bottom photo*) tend to have a higher-than-average percentage of aerobic, slow-twitch fibers.

Recall that the continuation of glycolysis requires the regeneration of NAD⁺, which occurs by fermentation. Muscle cells regenerate NAD^+ by reducing pyruvate—the end product of glycolysis—to lactic acid. Most of the lactic acid diffuses out of the active muscle cells into the blood, where it is carried to the liver and converted back to glucose. Glucose produced in the liver is released into the blood, where it can be returned to the active muscles to continue to fuel the high levels of glycolysis. Not all of the lactic acid is transported to the liver, however; some of it remains behind in the muscles. The buildup of lactic acid and the associated drop in pH within the muscle tissue may produce the pain and cramps that accompany vigorous exercise and, together with the depletion of glycogen stores, accounts for the sensation of muscle fatigue.

If, instead of trying to use your muscles to lift weights or do push-ups, you were to engage in an "aerobic" exercise, such as jumping jacks or fast walking, you would be able to continue to perform the activity for much longer periods of time without feeling muscle pain or fatigue. Aerobic exercises, as their name implies, are exercises designed to allow your muscles to continue to perform aerobically; that is, to continue to produce the necessary ATP by electron transport. Aerobic exercises depend largely on the contraction of the slow-twitch fibers of your skeletal muscles; these muscles are able to generate less force but can continue to function for long periods of time, due to the continuing aerobic production of ATP without the corresponding buildup of lactic acid. Aerobic exercise is initially fueled by the glucose molecules stored as glycogen in the muscles themselves, but after a few minutes, the muscles depend on glucose and fatty acids released into the blood by the liver. The longer the period of exercise, the greater the dependency on fatty acids. After 20 minutes of vigorous aerobic exercise, it is estimated that about 50 percent of the calories being consumed by the muscles are derived from fat. Aerobic exercise, such as jogging, fast walking, swimming, or bicycling, is one of the best

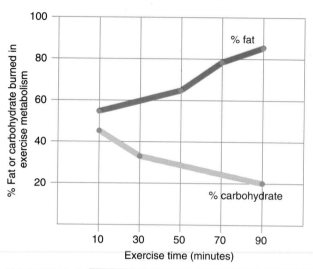

FIGURE 2

ways of reducing the body's fat content (Figure 2).

The ratio of fast-twitch to slow-twitch fibers varies from one particular muscle to another. For example, postural muscles in the back that are needed to allow a person to stand consist of a higher proportion of slow-twitch fibers than do arm muscles used to throw or lift an object. The ratio of fast-twitch to slow-twitch fibers in a particular muscle is genetically determined—a factor that may play a role in allowing a particular individual to excel in certain types of physical activities. For example, among world-class athletes, those who excel in activities that require short bursts of exertion, such as weight lifters (top photo) or sprinters, tend to have a higher proportion of fast-twitch fibers in their muscles than do long-distance runners (bottom photo), who excel in events that require endurance (Table 9-1).

TABLE 9-1

TYPICAL MUSCLE FIBER COMPOSITION IN ELITE ATHLETES REPRESENTING DIFFERENT SPORTS AND IN NONATHLETES

Sport	*Slow-Twitch Fibers* (%)	*Fast-Twitch Fibers* (%)
Distance running	60–90	10–40
Track sprinters	25–45	55–75
Shot-putters	25–40	60–75
Nonathletes	47–53	47–53

From Scott K. Powers and Edward T. Howley, *Exercise Physiology: Theory and Application to Fitness and Performance.* Copyright © 1990 Wm. C. Brown Communications, Inc., Dubuque, Iowa. All Rights Reserved. Reprinted by permission.

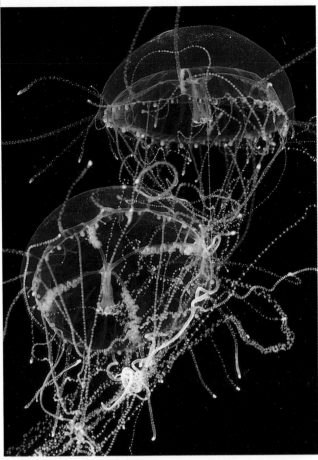

FIGURE 9-11
A luminescent jellyfish utilizes chemical energy stored in NADH to emit light.

some of the energy present in the original glucose molecule inevitably is lost, the formation of 36 to 38 ATPs represents an energy-capture efficiency of about 40 percent (the actual percentage varying with the conditions in the cell at the time). Compare this to a typical automobile engine that runs at less than 25 percent efficiency.

A metabolic ledger (Figure 9-10) pinpoints the source of each ATP. A balanced equation for the complete oxidation of glucose in a eukaryotic cell, for example, would be written

$$C_6H_{12}O_6 + 6\ O_2 + 36\ ADP + 36\ P_i \rightarrow 6\ CO_2 + 6\ H_2O + 36\ ATP$$

This equation represents the *potential* number of ATPs each molecule of glucose provides an organism. Electrons (and energy) are also needed for other purposes, however, such as biosynthesis of new molecules, movement, making sounds, or, in a more unusual example, light generation

(Figure 9-11). Although these activities reduce the number of ATPs produced, their energy is not wasted because it is used to accomplish biological work.

The laws of thermodynamics apply to the living world. The chemical energy originally stored in glucose is converted by cells to the energy stored in proton gradients and then to the energy stored in ATP. Not all of the original available energy is used to form ATP, however; some of the energy is lost in the process. (See CTQ #7.)

COUPLING GLUCOSE OXIDATION TO OTHER PATHWAYS

Thus far, we have limited our discussion of energy to that which is obtained by glucose oxidation. But you have probably never sat down to a meal of pure glucose. What about the hamburger and french fries that may be fueling your biological processes right now? How is that energy utilized?

The respiratory pathway we have described in this chapter (glycolysis, the Krebs cycle, and electron transport) not only extracts energy from glucose but is also the central pathway utilized for the breakdown of a diverse variety of materials (Figure 9-12). For example, the protein in your hamburger is converted to amino acids in your intestine and absorbed into the bloodstream. These amino acids are carried to the liver, where they are converted to molecules that are part of the pathways described in this chapter. The three-carbon amino acid alanine, for example, can be converted to pyruvic acid simply by removing its amino (NH_2) group. The pyruvic acid can then be fed into the Krebs cycle for complete oxidation. Similarly, the long fatty acid chains that make up fat molecules in your french fries are broken down into two-carbon units that enter the Krebs cycle as acetyl CoA molecules.

A pyruvic acid or acetyl CoA molecule is treated exactly the same by an enzyme, regardless of whether it came from glucose degradation or from some other carbohydrate, fat, or protein. Thus, any biochemical that can be degraded or converted to an intermediate of these central respiratory pathways can be completely oxidized for its energy.

We have just seen how the breakdown products of all types of materials are fed into glycolysis and the Krebs cycle. This principle also works in reverse. That is, virtually any compound that an organism must synthesize can be manufactured by diverting metabolic intermediates of glycolysis or the Krebs cycle into the appropriate biosynthetic pathway. For example, much of the surplus acetyl CoA produced in your body when glucose is abundant will be diverted to fat synthesis for energy storage.

The energy used to form more complex materials, such as proteins, fats, and nucleic acids, is derived largely from the ATP generated by electron transport. Many of these

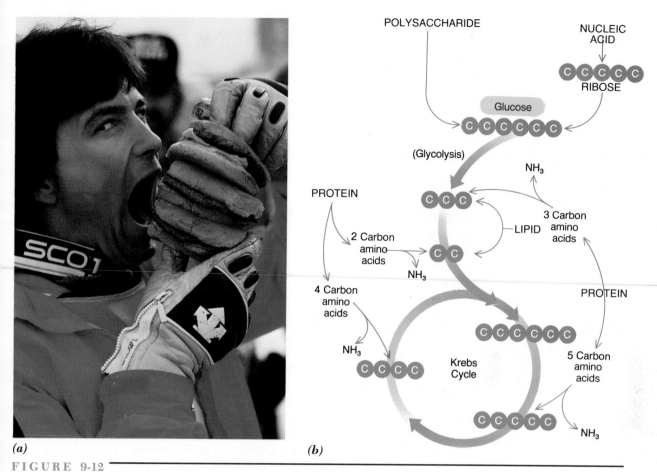

(a)   *(b)*

FIGURE 9-12

Some common denominators of metabolism. *(a)* Most meals contain a variety of macromolecules. The meat and cheese in this hamburger are rich in protein and fat, while the bun is rich in polysaccharide. *(b)* Each of the catabolic pathways by which proteins, lipids, carbohydrates, and nucleic acids are broken down produce metabolic intermediates that are channeled into glycolysis or the Krebs cycle, where they are completely oxidized. Alternatively, the intermediates of glycolysis and the Krebs cycle provide raw materials that are diverted into anabolic pathways, leading to the synthesis of macromolecules.

materials, particularly fats and other lipids, are more reduced than the intermediates from which they are built. Reduction of these intermediates is accomplished by electrons donated by NADPH. A cell's reservoir of NADPH represents its *reducing power* (page 129). Whereas NADPH is formed directly during photosynthesis (see Figure 8-7), humans and other nonphotosynthetic organisms must build a pool of reducing power by transferring electrons from NADH to NADP$^+$:

$$NADH + NADP^+ \rightarrow NAD^+ + NADPH$$

We can see, therefore, that not all of the NADH generated by glucose oxidation is used to make ATP. When energy is abundant, the production of NADPH is favored, providing an ample supply of electrons needed for biosynthesis of new macromolecules, such as proteins and lipids, which are essential for growth. When energy resources are scarce, however, most of the high-energy electrons of NADH are "cashed in" for ATP, whereas NADP$^+$ gets just

enough electrons for the minimal biosynthesis needed to maintain the status quo. As a result of this regulation, an organism integrates its metabolic processes into a "traffic pattern" that can be directed according to the demands of the cell.

Now that we have described how nutrients are dismantled and their energy content extracted, we can understand how this energy can be used to fuel human exercise. This is the subject of The Human Perspective.

Metabolic pathways, like roads, intersect one another. Glycolysis and the Krebs cycle are central pathways in all types of organisms. A diverse array of biochemicals are converted to compounds that can be fed into these pathways where they are dismantled and their energy content put to use. Conversely, glycolysis and the Krebs cycle provide the cell with raw materials for the construction of other types of molecules. (See CTQ #8.)

REEXAMINING THE THEMES

Relationship between Form and Function

The structure of a mitochondrion is correlated with its role in energy metabolism. The inner mitochondrial membrane contains a highly ordered array of enzymes and electron carriers that must be kept in exact position for the removal of electrons from coenzymes and for their passage from carrier to carrier and, ultimately, to molecular oxygen. The impermeability of the inner mitochondrial membrane maintains the proton gradient. In contrast, each ATP synthase contains an internal channel that allows protons to move through the membrane and drive the reaction by which ATP is synthesized. The central compartment of the mitochondrion contains the soluble enzymes that catalyze the reactions of the Krebs cycle. The electrons removed during these reactions are then passed to the carriers in the surrounding inner mitochondrial membrane.

Biological Order, Regulation, and Homeostasis

The needs of a cell change from moment to moment. For example, the fat cells of a person ingesting a large number of calories are likely to be engaged in the synthesis of fat molecules, whereas these same cells may be engaged in the breakdown of fat during times when calorie intake is low. At certain times, the pathways of glycolysis and the Krebs cycle may run primarily in a direction that releases energy, while at other times they may be running primarily in a direction that produces intermediates for the construction of macromolecules. The rates of activity of the various metabolic pathways in a cell are closely regulated so that they meet the needs of the cell at the time.

Acquiring and Using Energy

Glycolysis, the Krebs cycle, and the electron transport system are the three primary pathways by which cells extract energy from organic molecules. Energy extraction is accomplished primarily through the removal of energized electrons. Lipids, proteins, carbohydrates, and nucleic acids are broken down either into glucose or into metabolic intermediates that can be fed into these oxidative pathways. As the molecules are oxidized, the energy is conserved as usable packets in the form of ATP and/or NADPH. Cells that rely on fermentation are able to extract only a small percentage of the energy contained in their organic substrates. In contrast, aerobic respiration is much more "profitable" because high-energy electrons are allowed to drop to a much lower energy state by combining with molecular oxygen. Because no energy transfer can be 100 percent efficient, some useful energy is inevitably lost in the process.

Unity within Diversity

Remarkably, the chemical reactions and metabolic pathways described in this chapter are found in virtually every living cell, from the simplest bacterium to the most complex mammal. These pathways constitute a cell's most basic metabolic activity, providing usable intermediates, chemical energy (ATP), and reducing power (NADPH).

Evolution and Adaptation

The metabolic pathways described in this chapter did not appear "overnight"; they undoubtedly evolved step by step. Glycolysis, which is anaerobic, was probably one of the first metabolic pathways to appear, providing small amounts of ATP for use by primitive prokaryotic cells. Fermentation must have been an early accompaniment to glycolysis because it allowed the regeneration of NAD^+ needed for glycolysis to continue. Then, the cyanobacteria appeared and permeated the atmosphere with molecular oxygen. At first, cells must have evolved safeguards against this toxic substance, but they eventually acquired enzymes that could take advantage of oxygen's presence. Organisms that could utilize oxygen would have been able to obtain much greater quantities of energy from their nutrients. These aerobes would have been rapidly selected over their anaerobic competitors, forcing the anaerobes into habitats lacking oxygen, such as deep in the soil, where they can still be found today.

SYNOPSIS

During the early stages of life on earth, the atmosphere was devoid of oxygen and the earth was populated with anaerobic organisms. With the evolution of the cyanobacteria, the earth's atmosphere and bodies of water became infused with oxygen. This paved the way for a new breed of organisms that not only withstood the toxic effects of the substance but also possessed metabolic pathways that took advantage of the ability of oxygen to attract electrons and, in the process, to extract energy from organic substrates.

The first stage in glucose disassembly is glycolysis. Located in the fluid phase of the cytoplasm, glycolysis converts glucose to two molecules of pyruvic acid (a three-carbon molecule), four ATPs (by substrate level phosphorylation), and two NADHs (which can be "cashed in" for up to six additional ATPs). On the negative side of the ledger, glycolysis costs the cell two ATPs to initiate the pathway by activating the glucose molecule before it is split into two fragments.

Under anaerobic conditions, cells carry out fermentation as a means to regenerate NAD$^+$ from the NADH formed during glycolysis. At the end of fermentation, more than 90 percent of glucose's chemical energy remains in discarded end products, such as ethyl alcohol or lactic acid.

Aerobic respiration continues the disassembly of glucose in the presence of oxygen. The two pyruvic acid molecules generated from each glucose by glycolysis are completely oxidized to six carbon dioxide molecules. Located in the matrix of the mitochondria in eukaryotes (or the soluble cytoplasm of prokaryotes), the Krebs cycle generates reduced coenzymes for biosynthesis and energy production. It yields two ATPs by substrate-level phosphoryl-ation, eight NADHs, and two FADH$_2$s, per glucose oxidized, or a potential yield of 30 ATPs. The high-energy electrons stored temporarily in NADH and FADH$_2$ are used to generate ATP (three ATPs per NADH, and two per FADH$_2$) by passage down an electron transport system embedded in the inner mitochondrial membrane. The electrons are passed to carriers having a greater affinity for electrons until eventually they are passed to molecular oxygen, leading to the formation of water. As the electrons pass down the electron transport system, the energy released is stored in the form of a proton gradient across the ion-impermeable inner mitochondrial membrane. As protons move down their concentration gradient through the channel in the ATP synthase and into the active site located in the spherical (F_1) portion of the complex, they somehow provide the energy necessary to drive the phosphorylation of ADP. NADH is also used in the formation of NADPH, providing the cell with reducing power.

Glycolysis, the Krebs cycle, and electron transport are central metabolic pathways that provide virtually all aerobic cells with energy and raw materials. A diverse array of materials, ranging from proteins to lipids, are broken down into compounds that can be fed into glycolysis or the Krebs cycle and further metabolized. Conversely, intermediates from these central pathways can be diverted into the formation of various types of biochemicals, depending on the needs of the cell at the time.

Key Terms

anaerobe (p. 174)
aerobe (p. 174)
fermentation (p. 175)
aerobic respiration (p. 175)

glycolysis (p. 175)
substrate-level phosphorylation
  (p. 177)
alcoholic fermentation (p. 178)

lactic acid fermentation (p. 178)
Krebs cycle (p. 180)
cytochrome oxidase (p. 184)

Review Questions

1. Under what conditions do muscle cells form lactic acid and why is this adaptive?

2. What is the role of NAD$^+$ and FAD in the Krebs cycle reactions?

3. What is the role of the proton gradient and why would its function be disrupted by dinitrophenol?

4. Rank the following substances from the least electron attracting to the most electron attracting: NAD$^+$, glucose, molecular oxygen, cytochrome oxidase.

5. Rank the following compounds in terms of energy content: pyruvic acid, glucose, carbon dioxide, lactic acid, PGAL.

Critical Thinking Questions

1. Why do you suppose the isolated vesicles in the supernatant of disrupted mitochondria in Racker's experiments were able to oxidize glucose?

2. Burning organic molecules (i.e. fuels) in an internal combustion engine is only about 25 percent efficient (about 25 percent of the potential chemical energy in the molecules is transformed into usable energy). In contrast, the release of energy from glucose molecules by cells in the process of respiration is about 40 percent efficient. What accounts for the greater efficiency of respiration, compared to combustion?

3. If you were to place some sugar in a small metal dish and expose it to the flame of a Bunsen burner, it would break down into carbon dioxide and water without being activated. Why is activation not necessary when sugar is burned, and why is it necessary when sugar is broken down during respiration?

4. When glucose is broken down during anaerobic respiration, only about 5 percent of the potential energy of glucose is transferred to ATP. Where is the remainder of the energy? (Remember: The first law of thermodynamics states that the total energy present when a process begins must equal the total energy present at the end of the process.)

5. Why would an organism with respiratory equipment ever resort to fermentation? Why can't you—as a human being—switch to fermentation to sustain yourself indefinitely in similar circumstances?

6. Compare fermentation and aerobic respiration by completing the table below:

	Fermentation	Aerobic Respiration
initial compound		
final products		
part of cell where it occurs		
net ATP molecules produced		
involves glycolysis (y/n)		
involves Krebs cycle (y/n)		
involves electron transport chain?		

7. Consistent with the laws of thermodynamics, not all of the energy stored in glucose is converted to ATP; some of the energy is "lost" in the process. Where does this "lost" energy go? Does it have any benefits, or is it entirely wasted?

8. How do you personally benefit from the fact that your metabolic pathways are interconnected? How does this fact benefit people who have had to subsist on starvation diets for long periods of time?

Additional Readings

Bursztyn, P. G. 1990. *Physiology for sportspeople: a serious user's guide to the body.* New York: Manchester University Press. (Intermediate)

Darnell, J., H. Lodish, and D. Baltimore. 1990. *Molecular cell biology.* 2d ed. New York: Freeman. (Advanced)

Holmes, F. L. 1991. *Hans Krebs: The formation of a scientific life.* New York: Oxford. (Introductory)

Kalckar, H. 1991. 50 years of biological research—from oxidative phosphorylation to energy requiring transport. *Ann. Rev. Biochem.* 60:1–37. (Intermediate)

Krebs, H. 1982. *Reminiscences and reflections.* Oxford. (Introductory)

Powers, S. K., and E. T. Howley. 1990. *Exercise physiology.* W. C. Brown. (Intermediate-Advanced)

Racker, E. 1976. *A new look at mechanisms in bioenergetics.* New York: Academic. (Intermediate)

Stryer, L. 1988. *Biochemistry.* 3d ed. New York: Freeman. (Advanced)

Cell Division: Mitosis

STEPS
TO
DISCOVERY
Controlling Cell Division

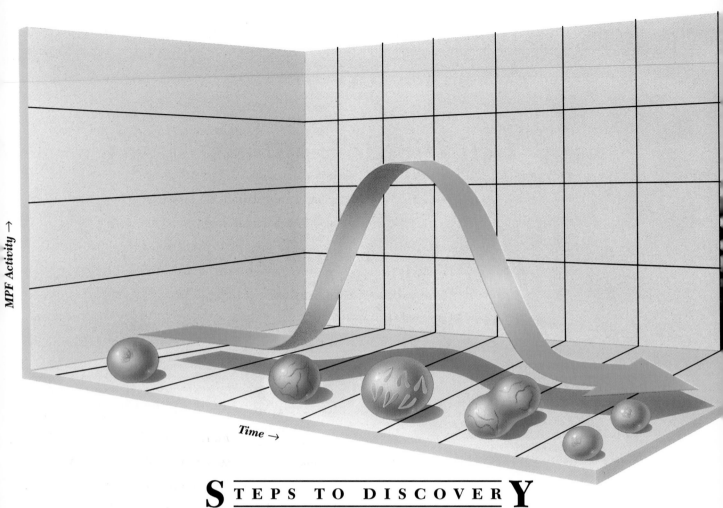

S TEPS TO DISCOVER Y
Controlling Cell Division

Cells reproduce—that is, they generate more of themselves—by dividing in two. Many of the events that occur during cell division have been known for decades, but the mechanism that triggers cell division has been shrouded in mystery until very recently. Research into the factors that control whether a cell is "resting" or dividing has enormous practical implications, particularly in combating cancer, a disease that results from a breakdown in a cell's ability to regulate its own division. Loss of this regulatory ability produces cells that divide out of control and form a continuously growing malignant tumor.

In 1970, Potu Rao and Robert Johnson of the Univer-

sity of Colorado devised an ingenious way of examining whether or not cells contain factors that trigger cell division. The scientists wanted to know what would happen if a cell that was just about to divide were to fuse with a cell that had divided recently, and would not divide again for many hours. To find out, they fused two cells to form a single "giant" cell that contained the nuclei and cytoplasm of both original cells. Consider the possible consequences of fusing these two types of cells: Two nuclei, one about to divide and the other totally unprepared to divide, are brought together in cytoplasm made up of a mixture of the two fused cells. Does the cytoplasm contain regulatory factors that affect

Cell division is triggered as MPF activity rises to its peak level.

the nuclei? If so, does the cytoplasm donated by the nondividing cell contain factors that can *block* the dividing nucleus from continuing its activities? Or, does the cytoplasm from the dividing cell contain factors that *stimulate* division activities in the nucleus derived from the nondividing cell?

Rao and Johnson found evidence for the latter alternative. As we will discuss later in the chapter, before a cell normally divides, its chromosomes become transformed from an "invisible" state to a condensed (packaged) state, in which they are readily visible under the microscope. In Rao and Johnson's experiment in which a dividing and a nondividing cell were fused, the chromosomes donated by the nondividing cell underwent condensation—a process that would not normally have occurred in that nucleus for many hours. The scientists concluded that the cytoplasm of the dividing cell contained one or more regulatory factors that triggered the nondividing nucleus to behave as if it were getting ready to divide.

Over the next few years, evidence accumulated that showed that the triggering agent for this behavior is a protein named *mitosis promoting factor* (MPF). This same protein was found in a wide variety of eukaryotic organisms, from yeast to humans. MPF purified from one species could be injected into the cells of a distant species, causing the injected cell to divide. Because of its widespread occurrence, MPF is thought to have appeared very early in the evolution of eukaryotic organisms and was retained as a key cellular component as more and more complex eukaryotes evolved.

If MPF triggers cell division, it seemed likely that it would be found in very small concentrations in nondividing cells, then jump to high levels in cells just before cell division, then return again to very low levels just after cell division is completed. The first evidence for such an oscillation in the level of MPF activity came in 1978 from a study conducted in the laboratory of Dennis Smith at Purdue University. Smith and his colleagues had been studying the first few cell divisions in the life of a frog; that is, the first divisions after the egg had been fertilized. These investigators found that as cells got ready to divide, the level of MPF activity rose dramatically; it then fell to undetectable levels after division, only to rise once again as the next ensuing division approached. These results indicated that MPF was acting like an "alarm clock"—whenever its activity rose to a high enough level it would set off the events that led to cell division. The question was: How was MPF able to do this?

Much of the progress toward answering that question has come from recent studies by James Maller and his colleagues at the University of Colorado. One of the events that accompanies a cell's preparation for division is the addition of large numbers of phosphate groups to a variety of the cell's proteins, some of which are found in the chromosomes. MPF is a protein kinase—an enzyme that adds phosphates to proteins. Injecting MPF into a cell induces an immediate increase in protein phosphorylation.

Perhaps the most widely accepted hypothesis based on these and other observations proposes that nondividing cells contain the proteins that initiate cell division, but they are present in an inactive state. Before they can participate in cell division, they must first be activated by the addition of a phosphate by MPF. Thus, MPF provides the major activating stimulus by phosphorylating key proteins needed for cell division.

Y ou were once just a fertilized egg—the product of the union of gametes: a sperm from your father and an egg from your mother. From this inauspicious beginning, you have grown into an organism consisting of trillions of cells. How did this remarkable transformation take place?

Recall the third tenet of the cell theory: New cells originate only from other living cells. The process by which this occurs is called **cell division.** For a multicellular organism like yourself, countless divisions of the fertilized egg result in an organism of astonishing cellular complexity and organization. Furthermore, cell division continues throughout life. Biologists estimate that more than 25 million cells are undergoing division each second of your life. This enormous output of cells is needed to replace those cells that have aged or died. Old, worn blood cells, for example, are removed and replaced by newcomers at the rate of about 100 million per minute. Not surprisingly, then, anything that blocks cell division, such as exposure to heavy doses of radiation, can have tragic effects. For example, many people who valiantly worked to seal the damaged nuclear reactor at Chernobyl, in the former Soviet Union, died because their bodies were unable to produce new, healthy blood cells (Figure 10-1). Ironically, the same type of destructive radiation is also used as a treatment for cancer because it selectively destroys rapidly dividing cells, such as those of a tumor.

Each dividing cell is called a **mother cell,** and its descendants are appropriately called **daughter cells.** There is a reason for using these "familial" terms. The mother cell transmits copies of its hereditary information (DNA) to its daughter cells, which represent the next cell generation. In turn, the daughter cells can become mother cells to their own daughter cells, passing along the same genes they inherited from their mother to yet another cellular generation. For this reason, cell division is often referred to as *cellular reproduction.*

Cell division is more than just a means of reproducing more cells; it is the basis for reproducing more *organisms.* Cell division, therefore, forms the link between a parent and its offspring; between living species and their extinct ancestors; and between humans and the earliest, most primitive cellular organisms.

▼   ▼   ▼

TYPES OF CELL DIVISION

Despite the great diversity of living organisms and the types of cells they contain, there are only three basic types of cell division, which are distinguished primarily by the way the genetic material is partitioned between the daughter cells. Prokaryotic cells partition their genetic material by a simple process involving membrane growth, whereas eu-

FIGURE 10-1

Testing for radiation in a village near the Chernobyl nuclear power plant in the former Soviet Union.

karyotic cells employ a complex process of nuclear division —either *mitosis* or *meiosis*. In all three of these processes, the cell first prepares for division by duplicating (*replicating*) its genetic material, a process discussed in detail in Chapter 14. The mother cell later splits itself in such a way that each daughter cell receives at least one complete set of hereditary instructions, as well as a portion of the cytoplasm.

CELL DIVISION IN PROKARYOTES

You will recall from earlier chapters that prokaryotes lack a nuclear membrane (page 92). The DNA of prokaryotic cells is attached directly to the cell's plasma membrane, providing a mechanism by which the replicated copies can be precisely distributed to the daughter cells (Figure 10-2). At the beginning of the division process, the replicated DNA molecules are attached to slightly different points on the plasma membrane. A new section of plasma membrane

grows between these attachment points, separating the DNA molecules. A partitioning plasma membrane and cell wall then develop between the DNA molecules to form two daughter cells. This process is called **prokaryotic fission.**

CELL DIVISION IN EUKARYOTES

The division of a eukaryotic cell into daughter cells occurs in two stages: First, its nuclear contents are divided by either *mitosis* or *meiosis*. This is followed by a second step, **cytokinesis,** in which the cell is actually split into two. Cell division in eukaryotes is much more complex than is prokaryotic fission for a number of reasons:

1. Eukaryotic cells are much larger than are prokaryotic cells and contain a diverse array of organelles, including a membrane-bound nucleus. The cytoplasmic organelles must be sorted out more or less equally between the daughter cells, and a new nuclear envelope must be reassembled for each daughter cell nucleus.

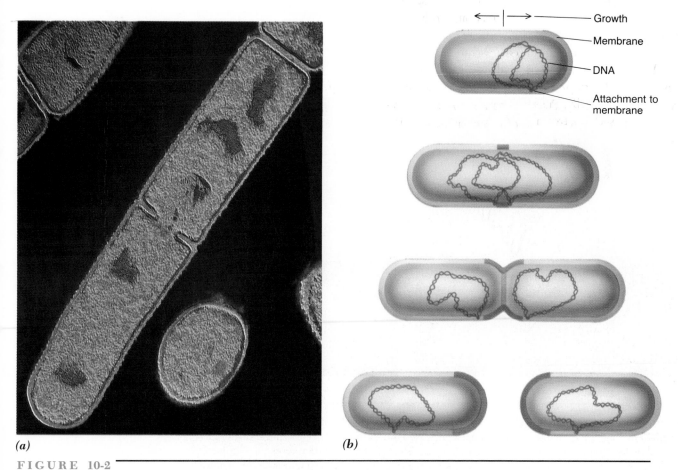

(a)   *(b)*

F I G U R E 10-2

Cell division in prokaryotes. *(a)* Photograph of a bacterial cell in the final stages of cell division. The DNA has already been separated, and the crosswall that will separate the cells is partially manufactured. *(b)* Stages in the process of prokaryotic fission. The DNA is attached directly to the cell's plasma membrane. Separation of the duplicated DNA molecules is accomplished by growth of the plasma membrane between the points of attachment.

2. Eukaryotic cells house much greater amounts of DNA than do prokaryotic cells. The quantity of DNA in a human cell, for example, is roughly 700 times that found in the most complex prokaryote.

3. Eukaryotic cells contain anywhere from 2 to more than 1,000 chromosomes, each of which consists of a molecule of DNA associated with a complex variety of proteins. In contrast, prokaryotic cells contain a single "chromosome," consisting essentially of a "naked" DNA molecule; that is, one devoid of permanently associated proteins.

As we will see below, eukaryotic cells are able to deal with such large amounts of chromosomal material by organizing it into specialized "packages" that can be distributed to their daughter cells. During mitosis or meiosis, each "packaged" chromosome appears as two thick "rods," called **chromatids,** which are connected to one another in an indented region called the **centromere** (inset, Figure 10-3). The two "sister" chromatids that make up each chromosome are genetic duplicates of each other—exact copies that were constructed at an earlier stage, when the DNA was duplicated. Each chromosome of a dividing cell has a

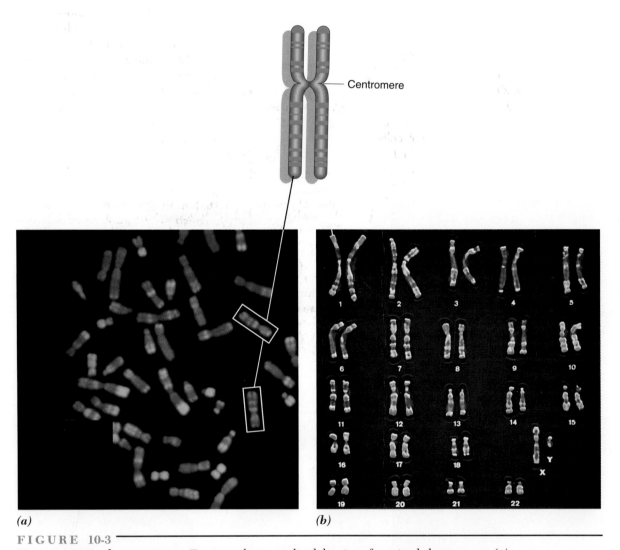

Centromere

(a) *(b)*

FIGURE 10-3

Human mitotic chromosomes. Top inset shows a stylized drawing of a stained chromosome. *(a)* Photograph of a cluster of mitotic chromosomes that spilled out of the nucleus of a single dividing human cell. This diploid set of 46 chromosomes has been stained with a dye that gives the chromosomes a banded pattern (see top inset). The two chromatids that make up each chromosome can be distinguished from each other. As discussed in the text, diploid cells contain pairs of homologous ("lookalike") chromosomes. One pair of homologous chromosomes is indicated by the boxes. *(b)* The chromosomes of a human male. In this figure, called a *karyotype,* homologous chromosomes are paired and arranged according to number (size). If the chromosome preparation had been made from the cells of a female, two X chromosomes would be seen, instead of an X and Y.

characteristic shape and size and can be identified from other chromosomes. Let's look more closely at the chromosomes of a dividing cell.

Chromosome Number: Haploid and Diploid

In eukaryotic cells, chromosomes occur in pairs (Figure 10-3a,b); each chromosome has a partner that is virtually identical in appearance. The two similar-shaped chromosomes are called **homologues** (*homo* = same, *log* = writing) because each one has the same sequence of genes along its length as the other. (You will recall from Chapter 1 that genes are the blueprints that determine our heritable traits, such as eye color and hair texture.) Homologous chromosomes can be recognized by their identical size, shape, and appearance (Figure 10-3b). Even though homologues have the same genes—in the sense that they determine the same trait—they may have different *versions* of those genes. For example, consider a gene for height in plants. One homologue may code for a tall plant, whereas the other may code for a short one. Both chromosomes have a gene for height at the same location, but each may produce a different version of the trait.

Most eukaryotic cells carry two complete sets of homologues—a **2N** number of chromosomes; such cells are said to be **diploid** (*di* = two, *ploid* = multiple). In diploid cells, one set of homologous chromosomes was originally contributed by each parent during sexual reproduction. Every cell in your brain, for example, has 46 chromosomes—a set of 23 from your father and a homologous set of 23 from your mother. Put differently, the cells in your body contain 23 pairs of homologous chromosomes. Each species has a characteristic diploid number, ranging from 2 in the horse roundworm and *Penicillium* fungus to a whopping 1,262 in Adder's tongue fern. Our closest relative, the chimpanzee, has 48 chromosomes, most of which are indistinguishable in appearance from our own (Chapter 13).

In contrast, **haploid** cells contain only one set of homologues—a **1N** number of chromosomes. The haploid number for humans is 23. Except for haploid sperm or eggs produced by your testes or ovaries, the cells in your body are diploid.

Mitosis: The Production of Two Identical Nuclei

The name "mitosis" comes from the Greek word *mitos,* meaning "thread." The name was first used in the 1870s to describe the threadlike chromosomes that appeared to "dance around" the cell just before it divided in two. **Mitosis** is a process of nuclear division in which duplicated chromosomes are separated from one another, producing two nuclei, each with one copy of each chromosome. Mitosis is usually accompanied by cytokinesis, resulting in *two* daughter cells with genetic potential identical to each other and to the mother cell from which they arose (Figure 10-4, left column). Mitosis, therefore, maintains the chromosome number and generates new cells for the growth, maintenance, and repair of an organism. Mitosis can take place in either diploid or haploid cells, the latter occurring in plants and a few animals (including male bees known as drones).

In some plants and animals, mitotic cell divisions are also a means of producing an entirely new organism through *asexual reproduction*—reproduction that does not involve the union of male and female gametes but involves the growth of offspring from the cells of a single parent (discussed further in Chapter 31). The offspring produced by asexual reproduction have exactly the same genes as does their parent.

Meiosis: The Production of Four Haploid Nuclei

Unlike mitosis, **meiosis** occurs only in diploid cells and produces *four* daughter nuclei, each containing a haploid (1N) number of chromosomes. A mother cell undergoing meiosis is able to produce four daughters by duplicating its own chromosomes and then proceeding through two consecutive nuclear divisions (Figure 10-4, right column). The nuclear divisions are usually accompanied by a corresponding division of the cytoplasm, resulting in four haploid cells. Haploid (1N) cells produced by meiosis form **gametes**—reproductive cells that fuse during fertilization, forming a diploid (2N) cell with a genetic potential different from that of either parent. The events that take place during meiosis, and their importance in sexual reproduction, are discussed in the next chapter.

For life to be perpetuated, organisms must reproduce. Since all organisms are composed of one or more cells, cell division is the foundation of reproduction. Prokaryotic cells have a simpler structure and divide by a simpler mechanism than do eukaryotic cells. In eukaryotic cells, there are two types of cell division, depending on whether the nucleus divides by mitosis or meiosis. Mitosis produces nuclei of equivalent genetic constitution to that of the original parent cell, whereas meiosis leads to the formation of nuclei with half the number of chromosomes of the original parental cell. (See CTQ #2.)

THE CELL CYCLE

The life of a cell begins with its formation from a mother cell by division and ends when the cell divides to form daughter cells of its own or when it dies. The stages through which a cell passes from one cell division to the next constitute the **cell cycle** (Figure 10-5).

Based on cellular activities visible with a light microscope, biologists divide the cell cycle into two phases: the "M phase" (*M* = mitotic) and the "interphase." The **M phase** includes (1) the process of mitosis, whereby chromosome separation occurs (which will be discussed in more detail later in the chapter); and (2) **cytokinesis,** whereby the entire cell is physically divided into two daughters. The M phase occupies only a small portion of the cell cycle since the separation of the chromosomes usually takes only 30 minutes to an hour, whereas the life of a cell may extend

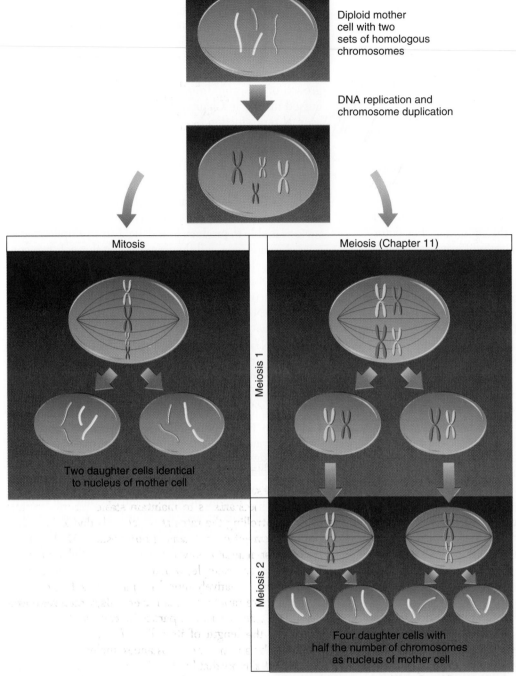

Diploid mother cell with two sets of homologous chromosomes

DNA replication and chromosome duplication

Mitosis

Meiosis (Chapter 11)

Meiosis 1

Two daughter cells identical to nucleus of mother cell

Meiosis 2

Four daughter cells with half the number of chromosomes as nucleus of mother cell

FIGURE 10-4

Schematic comparison of mitosis versus meiosis. DNA replication and the duplication of the chromosomes occur similarly prior to both mitosis and meiosis. When a cell enters either mitosis or meiosis, the chromosomes become condensed and visible in the light microscope as rodlike structures, each consisting of two chromatids. Mitosis produces two daughter nuclei that have exactly the same number of chromosomes as does the nucleus of the mother cell. In meiosis, however, the mother cell divides *twice*, producing four daughter nuclei, each with *half* the number of chromosomes of the nucleus of the mother cell. (Meiosis is discussed in detail in Chapter 11.)

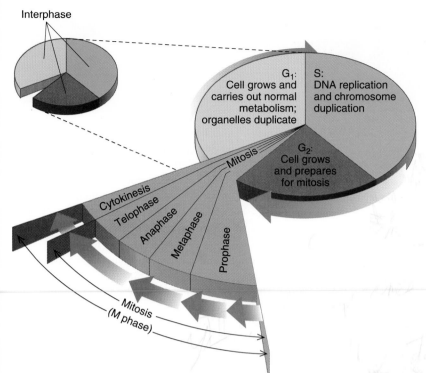

Interphase

The eukaryotic cell cycle. This diagram of the cell cycle indicates the stages through which a cell passes from one division to the next. The cell cycle is divided into two stages: M phase and interphase. M phase includes the successive stages of mitosis and cytokinesis. Interphase is divided into G_1, S, and G_2 phases—S phase being equivalent to the time of DNA synthesis. G_1 typically lasts 6 to 10 hours; S lasts 3 to 5 hours; and M less than 1 hour. G_1 is highly variable, lasting from a few minutes to months.

over many years. **Interphase** takes up the remainder of the cycle. During interphase, the cell grows in volume, and normal metabolic functions, such as glucose oxidation or the production of proteins for export, are carried out.

Although M phase is the period when the contents of a cell are actually divided, numerous preparations for an upcoming mitosis occur during interphase, including the important events involved in the replication of its DNA. The interphase period of the cell cycle is generally divided into three parts—G_1, S, and G_2 (Figure 10-5)—based on the timing of DNA replication. G_1 (Gap$_1$) is the period following mitosis and preceding the onset of DNA synthesis. S (Synthesis) is the period during which all the DNA in the chromosomes is replicated (Chapter 14). G_2 (Gap$_2$) is the period following DNA synthesis and preceding the subsequent mitosis. During G_2, the cell makes its final preparations for the upcoming division. Of the three phases, G_1 is the most variable. For example, in adult humans, the rapidly dividing cells that line the intestines remain in the G_1 phase for only about 2 hours, whereas it may take months for the very slowly dividing cells that make up the liver to move from mitosis through G_1 to S phase. Some types of cells, most notably nerve and muscle cells, do not divide under any circumstances. These cells can be considered to be "permanently" arrested in G_1 and, thus, will never enter S phase.

REGULATING THE CELL CYCLE

We described in Chapter 1 how organisms utilize homeostatic mechanisms to maintain stable internal conditions. Controlling the rates at which cells divide is an important element in maintaining homeostasis. Each type of tissue must maintain its own characteristic cellular makeup. Blood cells, for example, which are needed in great numbers and have relatively short life spans, must be produced at much more rapid rates than liver cells, which have long lives. The rate at which a particular cell divides is determined by the length of its cell cycle, which, in turn, is regulated by a number of substances, including hormones and growth factors that bind to the outer surface of the cell and either trigger or block cell division, as well as substances produced within the cell itself, such as MPF. For various reasons, these controls sometimes break down, causing some cells to divide independently. This malfunction can lead to one of the most dreaded types of diseases—cancer (see The Human Perspective: Cancer: The Cell Cycle Out of Control).

Maintaining homeostasis also requires that cells be able to change their rate of division depending on present conditions. For example, under ordinary conditions, a skin-producing cell takes about 20 hours to progress through interphase prior to producing two new skin cells by mitosis.

◁ THE HUMAN PERSPECTIVE ▷
Cancer: The Cell Cycle Out of Control

CANCER! The name alone strikes fear in all of us—and for good reason. Cancer is a leading cause of death in the Western world, second only to heart disease. Humans are not the only kind of organism that develops cancer. Humans, frogs, chickens, mice, and even plants—indeed all multicellular organisms—are candidates for cancer. And, contrary to what many believe, cancer is not a new disease. Cancerous lesions have been found in Egyptian mummies and even in dinosaur bones.

✪ Cancer is a disease that results from uncontrolled cell divisions. Normal cells may divide very rapidly, as occurs, for example, among liver cells following the removal of a portion of the liver. These normal cells are closely regulated, however, and they stop dividing when the liver has been returned to its normal size. In contrast, cancer cells continue to grow and divide indefinitely. For some reason, they no longer respond to the normal metabolic checks and balances that would otherwise limit and coordinate their growth with other cells. To fuel this unbridled growth, cancer cells out-compete surrounding cells for energy and nutrients. Since cancerous cells invade and spread to other tissues, many organs of the body can become adversely affected.

Why would a perfectly normal cell begin dividing wildly? We don't yet know the answer to this question, but it is becoming increasingly apparent that the genes involved with the cell cycle play an important role. Normal cells become transformed into cancer cells when something happens to certain genes, converting them to **oncogenes,** which causes them either to change their level of activity or to produce proteins with altered amino acid sequences (that is, mutant proteins). This is the basis of the action of *carcinogens* (cancer-causing agents), such as cigarette smoke, ultraviolet radiation, X-rays, and more than 1,000 known chemicals, including numerous pesticides, household products, and food additives. Carcinogens act by causing alterations in the DNA.

The study of oncogenes has led to a better understanding of the genes that control normal cellular growth and division. The first oncogene discovered was subsequently shown to code for a protein kinase—an enzyme that phosphorylates other proteins (page 129). (For their discovery of this oncogene in the mid 1970s, J. Michael Bishop and Harold Varmus of the University of California were awarded the 1989 Nobel Prize for Medicine.) There are now several dozen known oncogenes, and the list continues to grow. Included on the list are genes that code for (1) growth factors—that is, substances that bind to

cell receptors and stimulate the cell to divide; (2) receptors for growth factors—that is, the cell-surface protein that binds the growth factor and mediates its response; (3) regulatory proteins that bind to genes involved in cell growth; and (4) a number of protein kinases. These same genes and proteins are found not only in humans, or even mammals, but in virtually all eukaryotes. Since all of these proteins are normally involved in cell growth and division, it is easy to understand how changes in these proteins could make cells less responsive to the body's growth-control mechanisms.

Our understanding of the factors that control the cell cycle has received a boost in recent years with the purification of mitosis promoting factor, MPF—the subject of the chapter-opening essay. MPF is a protein kinase, as are the proteins encoded by some oncogenes. In fact, some of the proteins phosphorylated by MPF may be the same proteins that, when mutated, transform a normal cell into a cancerous one. A great deal of research is currently focused on how the activity of a single protein, MPF, can trigger such a complex process as mitosis. When we understand how mitosis and the cell cycle are regulated, we will be closer to winning one of the greatest medical battles ever waged.

If you cut yourself, however, the interphase period is reduced, allowing rapid cell replacement for healing.

Most cells play two critical roles: They conduct the various cellular functions that are characteristic of their cell type, and they divide. In general, a cell spends much more time conducting metabolic functions than it spends in cell division. However, cell division demands that virtually all cell components and activities be devoted to the process. As a result, cell division stands out as a distinct stage in the life of a cell. (See CTQ #3.)

THE PHASES OF MITOSIS

Even though interphase occupies nearly all of the cell cycle, it is the "dance of the chromosomes" during mitosis that has captured the attention of biologists for over a century. Mitosis is a continuous process. For the sake of discussion, however, we will divide mitosis into four sequential phases: prophase, metaphase, anaphase, and telophase (Figure 10-6). Each phase is defined by the behavior of the chromosomes.

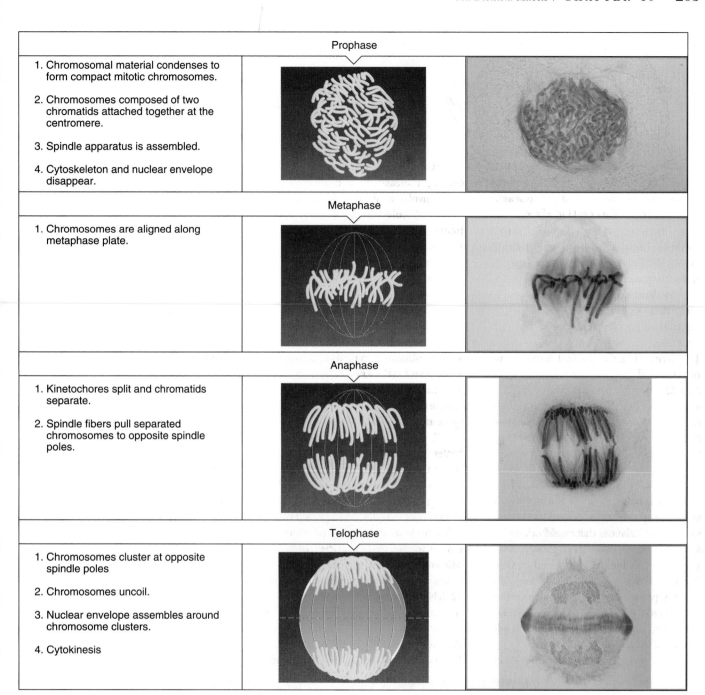

Prophase

1. Chromosomal material condenses to form compact mitotic chromosomes.

2. Chromosomes composed of two chromatids attached together at the centromere.

3. Spindle apparatus is assembled.

4. Cytoskeleton and nuclear envelope disappear.

Metaphase

1. Chromosomes are aligned along metaphase plate.

Anaphase

1. Kinetochores split and chromatids separate.

2. Spindle fibers pull separated chromosomes to opposite spindle poles.

Telophase

1. Chromosomes cluster at opposite spindle poles

2. Chromosomes uncoil.

3. Nuclear envelope assembles around chromosome clusters.

4. Cytokinesis

F I G U R E 10-6
The phases of mitosis.

PROPHASE: PREPARING FOR CHROMOSOME SEPARATION

The first phase of mitosis, **prophase** (*pro* = before), is the longest phase and includes a number of complex activities, including chromosome condensation and formation of the spindle apparatus (Figure 10-6).

Chromosome Condensation

Each chromosome contains a single molecule of DNA which, together with its associated protein, is spread throughout the nucleus of an interphase cell. In this extended state, DNA is well suited for interaction with various enzymes and regulatory proteins required for gene expression and replication. As cell division approaches, however, these remarkably long DNA-protein fibers undergo a coiling process, in which they are packaged into compact **mitotic chromosomes** (Figures 10-3, 10-6, 10-7)— structures that are ideally suited for the upcoming separation process.

The importance of chromosome coiling becomes clear when you consider that the DNA of a single human chromosome is 30,000 times longer in the extended state than when it is coiled. Imagine the molecular chaos that would reign inside a human cell during the separation of the duplicates of 46 *uncoiled* chromosomes. Inevitable tangling would pull the uncoiled DNA to pieces, forming a jumble of DNA fragments that would be unevenly sorted into the daughter cells. The result would be daughter cells with an incomplete set of genetic blueprints. No species could survive disorder of this magnitude.

As the condensation of the chromatin nears completion, it becomes evident that each mitotic chromosome consists of two identical, attached chromatids (Figure 10-7). These sister chromatids are visible evidence of chromosome duplication, a process that occurred earlier during interphase.

Formation of the Spindle Apparatus

In eukaryotes, the precise separation of duplicated mitotic chromosomes into two daughter cells requires the activity of a cellular "machine" called the **spindle apparatus,** which has no counterpart in prokaryotic cells. The spindle apparatus is constructed primarily of cytoskeletal materials. Recall from Chapter 5 that the cytoskeleton has the dual function of providing support and moving particles within a cell. The spindle apparatus consists of bundles of microtubules, called **spindle fibers,** which are organized to form both a supportive scaffolding and a chromosome-pulling machine (Figure 10-8). Remarkably, the microtubules that make up the spindle apparatus are formed from the same subunits that a few minutes earlier might have been part of a different microtubule having an entirely different function. This is analogous to demolishing a brick building and then using the same bricks to construct a new office building—all within a matter of minutes. The rapid assembly of the spindle apparatus during mitosis, and its even more rapid disassembly following mitosis, illustrates the dynamic nature of cytoskeletal elements.

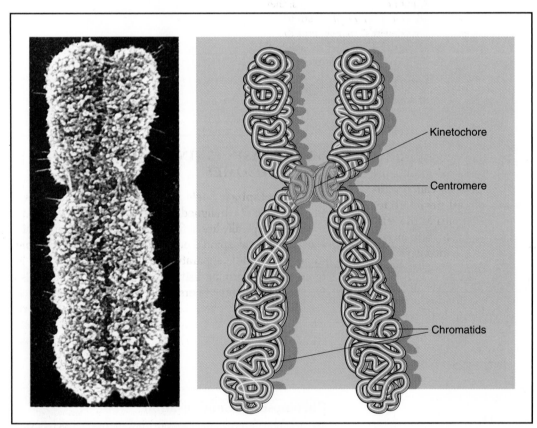

Kinetochore

Centromere

Chromatids

FIGURE 10-7

The structure of a mitotic chromosome.   During mitosis, the DNA-protein fibers coil, as is indicated in both the drawing and the electron micrograph. Each chromatid remains distinct but is joined to the other in the indented region of the chromosome (the *centromere*). When the centromere is examined under the electron microscope, it is seen to contain a dense, protein-containing structure called the *kinetochore*.

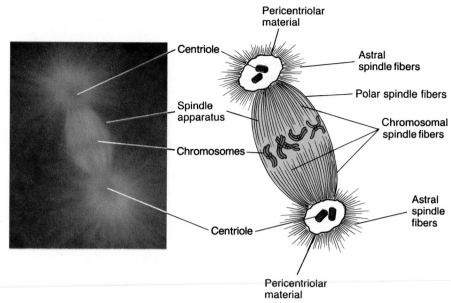

Pericentriolar material

Centriole

Astral spindle fibers

Polar spindle fibers

Spindle apparatus

Chromosomal spindle fibers

Chromosomes

Centriole

Astral spindle fibers

Pericentriolar material

Spindle apparatus of an animal cell in metaphase of mitosis. Each spindle pole contains a pair of centrioles surrounded by a shapeless pericentriolar material from which the microtubules which make up the spindle fibers radiate. Spindle fibers can be categorized as astral fibers, chromosomal fibers, or polar fibers, each having its own distinct functions. *Astral spindle fibers* radiate like a starburst around the centrioles; they probably help position the spindle apparatus in the cell. *Polar spindle fibers* stretch completely across the cell from pole to pole, forming "guidewires" that support the spindle apparatus during its operation. Shorter *chromosomal spindle fibers* attach to the kinetochores of each mitotic chromosome. When the kinetochores split at the start of anaphase, the sister chromatids are pulled to opposite poles by the action of the chromosomal spindle fibers.

Spindle fibers radiate out from the two poles (ends) of the spindle apparatus. In animal cells, the poles contain a pair of peculiar, pinwheel-shaped structures called **centrioles,** surrounded by dense, unstructured **pericentriolar material (PCM)** (see Figure 10-8). Ironically, the relatively conspicuous centrioles seem to have little or no function in the process of mitosis, while the inconspicuous PCM appears to play a key role in governing the assembly of the microtubules of the spindle apparatus. This hypothesis is supported by studies on dividing plant cells that possess a mitotic apparatus that lacks centrioles entirely but contain material resembling the PCM found in animals. You might also note that the growing ends of all of the various microtubules depicted in Figure 10-8 are embedded in the PCM.

Assembly of the spindle apparatus takes place outside the nucleus at the same time the chromosomes are becoming compacted inside the nucleus. Toward the end of prophase, the nuclear envelope surrounding the compacted chromosomes becomes fragmented and disappears from view. At about the same time, the nucleolus disappears, and the chromosomes become attached to the ends of certain spindle fibers. By the end of prophase, the chromosomes have moved to the center of the cell.

METAPHASE: LINING UP THE CHROMOSOMES

During **metaphase** (*meta* = middle), the chromosomes of the dividing cell are aligned in a plane (called the *metaphase plate*) that typically lies at the cell's "equator"; that is, midway between the spindle poles (Figure 10-6). The chromosomes and spindle apparatus of the metaphase cell are interconnected, forming a structure depicted in Figure 10-8. One group of spindle fibers extends to a specialized attachment site on each chromosome, called the **kinetochore,** which is located in the narrowed centromere region of the chromosome (Figure 10-7).[1] During metaphase, the kine-

---

[1] When duplicated chromosomes were first observed in the light microscope in the late 1800s, the two chromatids were seen to be attached at a point where the chromosome was pinched inward, forming a "waist." This indented region was called the *centromere*. Later, when chromosomes were observed in the electron microscope, the point where the chromatids were attached to one another revealed a specialized structure called the *kinetochore*. These two terms are often confused; the kinetochore is a physical structure embedded within the body of the chromosome, whereas the centromere is that region of the chromosome marked by the indentation. The kinetochores are located within the centromere.

tochores of all the chromosomes are aligned within the metaphase plate, and the "arms" of the chromatids extend out in various directions (Figure 10-6). The alignment of kinetochores and their attachment to the spindle fibers ensure precise separation of sister chromatids during the next stage, anaphase.

ANAPHASE: SEPARATING THE CHROMATIDS

Anaphase (*ana* = backward) begins with the sudden, synchronous separation of the kinetochores present on sister chromatids. Separated chromatids (which, after separation, are called chromosomes) now move away from each other, toward opposite spindle poles (Figure 10-6). Since each daughter cell receives one copy of every chromosome, the two daughters are genetically identical to each other as well as to the mother cell from which they arose.

TELOPHASE: PRODUCING TWO DAUGHTER NUCLEI

Telophase (*telo* = end) begins when the chromosomes reach their respective spindle poles (Figure 10-6). The events of telophase are virtually the reverse of those that occur during prophase: The chromosomes uncoil, the spindle apparatus is dismantled, nucleoli reappear, and nuclear envelopes form around each chromosome cluster. Telophase ends when cytokinesis is completed and two separate daughter cells are produced.

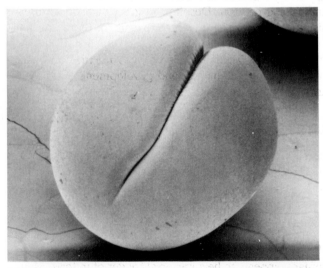

FIGURE 10-9

One cell splits into two. During cytokinesis, a frog egg divides into two cells as microfilaments contract and draw in the plasma membrane. Eventually, the microfilaments pull the edges of the plasma membrane together, pinching the cell's cytoplasm in two.

Mitosis is an intricate process that follows similar steps among diverse eukaryotic cells. These steps include packaging the chromosomal material to facilitate precise chromosome separation; assembling the machinery necessary for chromosome separation; and moving the duplicated chromosomes to opposite ends of the cell. (See CTQ #4.)

CYTOKINESIS: DIVIDING THE CELL'S CYTOPLASM AND ORGANELLES

Although mitosis is very similar in plant and animal cells, the way in which the cytoplasm is divided by cytokinesis is very different. If you watch a sea urchin or frog egg undergoing its first cell division, for example, you will see the first hint of cytokinesis as an indentation of the cell surface during late anaphase. Depending on the type of cell, the indentation may occur at one end (as in the frog egg depicted in Figure 10-9) or at points completely encircling the cell. As time passes, the indentation deepens and becomes a *cleavage furrow* that moves through the cytoplasm, pinching the cell in two. The plane of the furrow always lies in the same plane previously occupied by the metaphase chromosomes, thereby ensuring that the separated chromosomes will be partitioned in different cells.

Cytokinesis in animal cells is the result of contractions of a ring of microfilaments (page 104), which assembles just beneath the plasma membrane. As the ring contracts, it pulls the overlying membrane inward, constricting the cell, much like a purse string closes the diameter of a purse's opening. The ring of microfilaments disassembles after constriction is completed, and the subunits (composed of the protein actin) are put to use in the formation of microfilaments of the cytoskeleton.

Formation of a contractile ring would not be possible in plant cells because the rigid plant cell walls cannot be pulled inward. In a plant cell, cytokinesis begins with the formation of a **cell plate** between the daughter nuclei (Figure 10-10). The cell plate is formed by secretion vesicles containing polysaccharides produced by the nearby Golgi complex (see Chapter 5). As these vesicles accumulate in the center of the dividing cell, their membranes fuse together. The released polysaccharide forms the primary cell wall between the new cells, and the membranes of the fusing vesicles become incorporated into the plasma membranes of the adjoining daughter cells.

The exact separation of duplicated chromosomes during mitosis is complicated because it demands great precision. Dividing the cell's cytoplasm and organelles into daughter cells requires less precision, however, because equal amounts of these components are not necessary. How a dividing cell partitions its cytoplasm and organelles depends on whether or not the cell is surrounded by a cell wall. (See CTQ #5.)

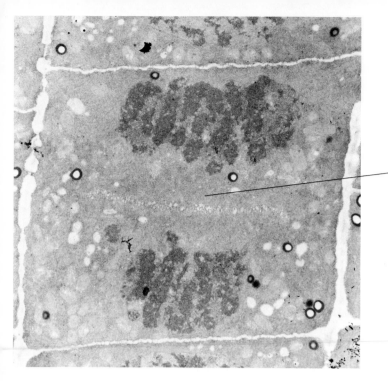

Formation of a
cell plate

FIGURE 10-10

Cytokinesis in an onion root cell. Secretion vesicles
from nearby Golgi complexes become aligned midway between newly formed daughter nuclei. The membrane of the
secretion vesicles will fuse to form the plasma membranes
of the two daughter cells, and the contents of the vesicles
will provide the material that forms the cell plate that will
separate these cells. Additional materials for constructing a
cell wall for each daughter cell will be delivered later by
other secretion vesicles.

REEXAMINING THE THEMES

Relationship between Form and Function

The dramatic differences in the structure of the chromosomes and the cytoskeleton between interphase and M
phase of the cell cycle reflect their different functions in
these two stages. During interphase, the chromosomes are
in an extended state, which allows them to send messages
that direct cellular activities. During mitosis, these same
chromosomes are compacted into a state where they could
not possibly function as cell directors but can readily be
moved around and separated into daughter cells. Similarly,
the cytoskeleton of the interphase cell maintains cell shape
and movement of cytoplasmic organelles and vesicles. During M phase, cytoskeletal elements are disassembled, and
the same subunits are reused to construct a spindle apparatus, required for chromosome movement, and a contractile ring, required for cytokinesis in dividing animal cells.

Biological Order, Regulation, and Homeostasis

The cellular makeup of the complex tissues of multicellular plants and animals is maintained by strict controls
over the rates of cell division. Many substances are involved
in regulating cell cycles, including hormones and growth
factors that bind to the outer surface of the cell, as well as
substances (such as MPF) produced within the cell itself.
Some substances stimulate cell division, while others inhibit
the process. Mutations in any one of a number of different
genes can abolish growth control and lead to the formation
of cancer.

Unity within Diversity

Every organism consists of one or more cells capable of
undergoing division. Despite the great diversity of organisms and the types of cells they contain, there are only
three basic types of cell division: prokaryotic fission, mitosis,
and meiosis. These three processes provide the underlying
basis for the reproduction and development of all organisms.

Evolution and Adaptation

The existence of similar processes occurring by similar
mechanisms and utilizing similar molecules provides some
of the strongest evidence for the evolutionary relationship
of all organisms. Mitosis, for example, follows a similar sequence of steps among diverse eukaryotes, utilizing similar
types of cellular organelles and regulatory mechanisms.
MPF, which is present in organisms as diverse as yeast and
humans, appears to be a universal trigger of mitosis. Many
of the oncogenes—genes thought to control cell growth
and division—are also found in diverse organisms. It appears that the mechanisms that regulate cell division in
eukaryotes appeared at an early stage of evolution and have
been retained ever since.

SYNOPSIS

Cells arise from other living cells by cell division. Cell division enables organisms to grow (by increasing the number of cells), to reproduce, and to repair or replace damaged or worn tissues.

There are three basic types of cell division: prokaryotic fission, mitosis, and meiosis. In prokaryotic cells, duplicate copies of the DNA attach to different sites on the plasma membrane; they are subsequently separated from each other by growth of the membrane and formation of a partitioning membrane and cell wall.

Eukaryotic cells divide their nuclear contents by the complex processes of either mitosis or meiosis. Mitosis produces genetically identical cells as part of the process of growth, repair, and asexual reproduction. Mitosis occurs in both diploid and haploid cells. Meiosis produces haploid daughter cells containing one-half the number of chromosomes as the diploid mother cell. The daughter cells from meiosis eventually develop into sex cells (gametes) for sexual reproduction.

The cell cycle is the sequence of stages through which a cell progresses from one division to the next. Most of the cell cycle is spent in interphase, a period divided into G_1, S, and G_2 stages, with S representing the period of DNA replication. The remainder of the cell cycle (M phase) includes mitosis and cytokinesis.

Mitosis is divided into several stages, based on the activities relating to the preparation and separation of the chromosomes. During prophase, the chromosomes undergo a coiling process in which the DNA-protein fibers become highly compacted and the cell assembles a mitotic spindle apparatus consisting of microtubules. During metaphase, the chromosomes are aligned at the center of the cell, attached to spindle fibers. Each chromosome consists of identical chromatids attached at their joined kinetochores. During anaphase, the chromatids split apart and move to opposite poles as separate chromosomes. During telophase, the cell returns to the interphase state.

Cytokinesis in animal cells occurs by the contraction of a ring of microfilaments, which splits the mother cell in two. In plant cells, the mother cell is divided by formation of an intervening cell plate resulting from the fusion of Golgi-derived vesicles.

Key Terms

cell division (p. 196)
mother cell (p. 196)
daughter cell (p. 196)
prokaryotic fission (p. 197)
cytokinesis (p. 197)
chromatid (p. 198)
centromere (p. 198)
homologue (p. 199)
diploid (2N) (p. 199)
haploid (1N) (p. 199)

mitosis (p. 199)
meiosis (p. 199)
gamete (p. 199)
cell cycle (p. 199)
M phase (p. 199)
interphase (p. 201)
oncogene (p. 202)
mitosis promoting factor (MPF) (p. 202)
prophase (p. 203)
mitotic chromosome (p. 204)

spindle apparatus (p. 204)
spindle fiber (p. 204)
centriole (p. 205)
pericentriolar material (PCM) (p. 205)
metaphase (p. 205)
kinetochore (p. 205)
anaphase (p. 206)
telophase (p. 206)
cell plate (p. 206)

Review Questions

1. Complete the figure by filling in the terms in their correct locations.

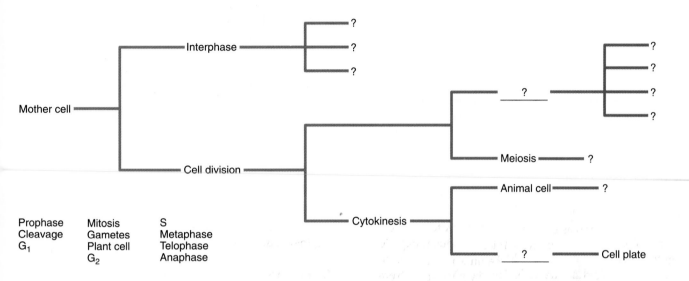

Prophase   Mitosis   S
Cleavage   Gametes   Metaphase
G_1 Plant cell Telophase
           G_2 Anaphase

2. How does the process of chromosome coiling facilitate cell division? During what stage does it occur? During what stage does DNA replication occur?

3. Using pieces of colored yarn or crayons, set up the mitosis template (below) on a large sheet of paper and then physically run through the phases of mitosis.

4. How would you describe the genetic relatedness of sister chromatids? Of two chromosomes that were split apart during anaphase and moved to different daughter cells? Of the two members of a pair of homologous chromosomes?

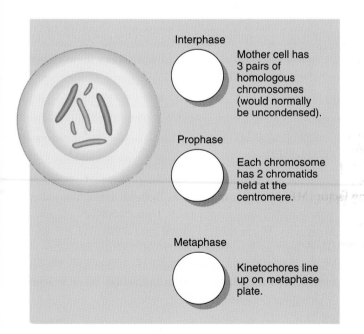

Interphase

Mother cell has 3 pairs of homologous chromosomes (would normally be uncondensed).

Prophase

Each chromosome has 2 chromatids held at the centromere.

Metaphase

Kinetochores line up on metaphase plate.

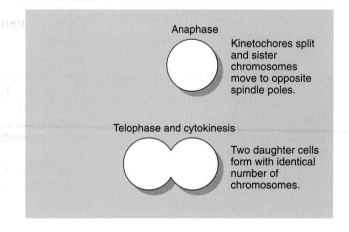

Anaphase

Kinetochores split and sister chromosomes move to opposite spindle poles.

Telophase and cytokinesis

Two daughter cells form with identical number of chromosomes.

Critical Thinking Questions

1. What do you expect the result would be if a zygote contained mutant MPF genes that produced a deficient MPF protein? Would you expect a liver cell or skin cell that had mutant MPF genes to be susceptible to developing into a cancer? Why or why not?

2. Complete the table below comparing mitosis and meiosis.

	Mitosis	Meioisis
chromosome number of mother cells		
number of cell divisions		
number of daughter cells produced		
chromosome number of daughter cells		
genetic composition of daughter cells (compared to mother cells)		

3. The graph opposite shows the results of measuring the DNA content of a suspension of cells by labeling the DNA with a fluorescent dye and then measuring the fluorescence of each cell as it passes through a sensitive detector. Identify the stage (or stages) of the cell cycle indicated by the letters A, B, and C.

4. Taxol is a drug used in the treatment of certain cancers. The drug acts by stabilizing microtubules, that is, pre-

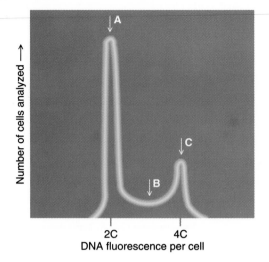

venting their disassembly. What effect would you expect taxol to have on cell division? Why? Why would taxol be useful in cancer treatment?

5. Explain how the manner in which cytokinesis occurs in animal cells and plant cells demonstrates the principle that structure and function are complementary.

6. The length of the cell cycle varies greatly from one cell type to another and for the same cell under different conditions. What do such differences tell you about the life span of the cell? In which parts of the human body would you expect to find cells with the short life cycles?

7. Cell division is usually blocked by the addition of cytochalasin and other drugs that cause the disassembly of actin filaments. Why would you expect these drugs to have this effect?

Additional Readings

Brooks, R. et al., eds. 1989. The cell cycle. *J. Cell Sci. Suppl.* 12. (Advanced)

Glover, D. M. et al., 1993. The centrosome. *Sci. Amer.* June:62–68. (Intermediate)

Hyams, J. S., and B. R. Brinkley, eds. 1989. *Mitosis: Molecules and mechanisms.* New York: Academic. (Advanced)

Marx, J. 1989. The cell cycle coming under control. *Science* 245:252–255. (Intermediate)

Marx, J. 1991. The cell cycle: Spinning farther ahead. *Science* 252:1490–1492. (Intermediate)

McIntosh, J. R., and K. L. McDonald. 1989. The mitotic spindle. *Sci. Amer.* Oct:48–56. (Intermediate)

Murray, A. W., and M. W. Kirschner. 1991. What controls the cell cycle. *Sci. Amer.* Mar:56–63. (Intermediate)

Pardee, A. B., et al. 1989. Frontiers in biology: The cell cycle. *Science* 246:603–640 (a collection of six research review papers). (Advanced)

CHAPTER
‹ 11 ›

Cell Division: Meiosis

**STEPS
TO
DISCOVERY**
Counting Human Chromosomes

THE HUMAN PERSPECTIVE

Dangers That Lurk in Meiosis

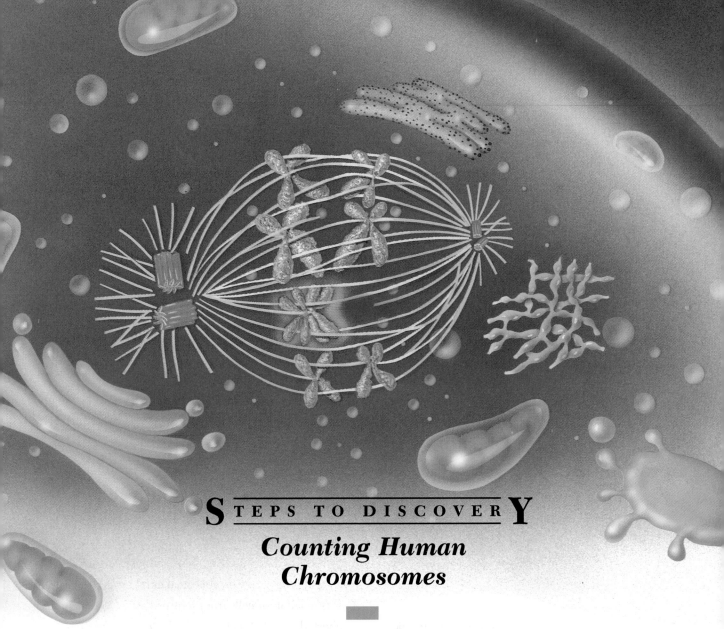

. . . *"Thus man has 48 chromosomes."* This quote was taken from a biology book published in 1952. You might recall from the last chapter, however, that the diploid number of chromosomes in humans was stated to be *46*, not 48. But before you condemn the authors of this book for their inaccuracy, consider that, up until 1955, biologists accepted the "fact" that human cells contained 48 chromosomes.

Why did it take so long to discover the right number? Prior to the 1950s, chromosome numbers were determined primarily by examining sections of tissue in which cells were occasionally "caught" in the process of cell division. Trying to count several dozen chromosomes crammed together within a microscopic nucleus is like trying to count the number of rubber bands present in a cellophane package without opening the package. It is actually quite remarkable that biologists got as close as 48! Before 1922, when Theophilus Painter, a geneticist at the University of Texas, arrived at this number, previous guesses had ranged from 8 to more than 50; 48 chromosomes then became the accepted value for more than 30 years until a remarkably simple discovery changed everything.

In 1951, times were tough in academia, and many new Ph.D.'s couldn't find teaching positions. Having just completed his doctorate on insect chromosomes at the University of Texas, T. C. Hsu fell into this distressing category. Reluctantly, Hsu accepted a postdoctoral research position

When a pair of homologous chromosomes fail to separate during meiosis, the resulting daughter cells will have an abnormal

in Galveston working on mammalian chromosomes, which were notoriously difficult to study. After several months of frustration, he was examining a new batch of cells with a microscope when, in Hsu's words "I could not believe my eyes when I saw some beautifully scattered chromosomes in these cells. I did not tell anyone, took a walk around the building, went to the coffee shop, and then returned to the lab. The beautiful chromosomes were still there. I knew they were real."

He tried to repeat the work, but his preparations regained their "normal miserable appearance." For many months after that, every attempt to discover what he had done "wrong" to get that beautiful preparation failed. Finally, Hsu tried pretreating the cells with a hypotonic (more dilute) salt solution. The cells expanded like balloons and exploded onto the slide, releasing the chromosomes and spreading them out so each was separated from its neighbors (review Figure 10-3). The earlier preparation of cells must have been accidentally washed with a dilute saline solution to which someone had failed to add enough salt. Since then, Hsu's hypotonic technique for treating cells has become a standard part of preparing chromosomes for microscopic examination. Ironically, Hsu did not use his technique to reexamine the question of the human diploid number; Painter had been one of his mentors at the University of Texas, and Hsu accepted the 48-chromosome count.

Within 3 years, however, Albert Levan and Jo Hin Tijo, working in the United States and Sweden, used the new hypotonic pretreatment technique on human cells treated with a drug called colchicine. Colchicine disassembles the mitotic spindle apparatus so that cells "freeze" in their metaphase configuration, providing many more mitotic cells to observe. Levan and Tijo carefully counted the chromosomes of these mitosis-arrested cells and found only 46. They repeated their observations and cautiously concluded that, at least in lung cells, the human diploid number was 46. The number was soon confirmed by other investigators on other human cell types and has been the accepted value ever since.

Meanwhile, Jerome Lejeune, a French clinician, had been studying children with Down syndrome for many years. Such children (discussed in the Human Perspective box) are characterized by a short stocky stature, distinctive folds of the eyelids (which gave rise to the earlier name "mongolism"), and mental retardation. After hearing a lecture by Jo Hin Tijo, who described how chromosomes could be counted using newly developed techniques, Lejeune decided to examine the chromosomes from a few of his Down syndrome patients. These children showed such a wide range of abnormalities that some alteration in the chromosomes was very likely responsible. Lejeune had never been involved in this type of research, however, and, in fact, did not even possess a microscope. He finally located one that had been discarded by the bacteriology lab. The microscope had been used so much that it would not remain in focus. In order to use it, Lejeune inserted a piece of tinfoil from a candy wrapper to hold the focusing knob in place. In 1959, Lejeune and two colleagues who had helped in cell preparation published a two-page paper indicating that the cells of nine different patients with Down syndrome all had 47 chromosomes, rather than 46.

Lejeune's paper opened the door to a new field of medical genetics. It was soon followed by a number of other reports in which patients with various types of disorders were shown to have an abnormal number of chromosomes. Examination of cells from fetuses that had spontaneously aborted revealed that many had three or more full sets of chromosomes in their cells. Thus, an abnormal fetal chromosome number was discovered to be a common cause of miscarriage. These insights into the effects of chromosome abnormalities revealed how important it was that the cells of developing embryos contain precisely the correct number of chromosomes. It soon became obvious that the formation of gametes with an abnormal number of chromosomes could be traced to a defect occurring during meiosis—the subject of this chapter.

number of chromosomes.

*E*ach organism is a living showcase, possessing thousands of inherited features. When you glance in a mirror, you can see dozens of inherited characteristics—the length and thickness of your eyelashes, the width of your smile, the depth of a dimple. The genetic blueprint containing the plans for your construction is contained in only 46 chromosomes, which are carried in virtually every cell of your body. Within these 46 chromosomes, there are actually two similar sets of genetic instructions: One haploid set of 23 chromosomes is derived from your father's sperm, and the other haploid set of 23 chromosomes originated from your mother's egg. At conception, the two sets of homologous chromosomes unite at fertilization to form a diploid cell, the **zygote.** The zygote and all of its progeny cells have dutifully duplicated those 46 chromosomes each time they divided, eventually growing into a remarkable cell collection—you.

But your parents are also collections of cells containing 46 chromosomes. How did they manufacture gametes (sex cells—sperm or eggs) with only 23 chromosomes? They could not have done so by mitosis, since mitosis produces cells with exactly the same number of chromosomes as the mother cell (see Figure 10-4). Rather, they were able to do so through **meiosis,** a type of nuclear division that divides the number of chromosomes in half, thereby forming haploid gametes for sexual reproduction.

▼ ▼ ▼

MEIOSIS AND SEXUAL REPRODUCTION: AN OVERVIEW

◢ Reproduction may be the most important characteristic that distinguishes organisms from nonliving objects. The sequence of events that occur in the life of an organism from the time it is formed by reproduction until it produces offspring itself makes up the **life cycle** of that organism. The life cycles of nearly all eukaryotic organisms include sexual reproduction, a process that is just as important to the survival of trees and worms as it is to humans. Meiosis is a critical step in all forms of sexual reproduction because it generates haploid nuclei in cells that eventually form gametes. Gametes are produced differently in animals and plants, although meiosis is a key step in their production in both groups (Figure 11-1). In animals, meiosis produces haploid daughter cells that are directly transformed into sperm or eggs. In plants, meiosis produces haploid daughter cells called *spores,* which then divide by *mitosis* to grow into a multicellular haploid plant that produces the

gametes—either male pollen cells or female eggs. (Sexual reproduction in plants and animals is discussed further in Chapters 20 and 31, respectively.) Haploid gametes, in turn, come together at fertilization to generate a diploid zygote (a fertilized egg). Thus, meiosis and fertilization are critical events in the life cycle of sexual organisms, since both processes change the number of chromosome sets.

Sexually reproducing organisms are made up of two types of cells: germ cells and somatic cells. **Germ cells** are those cells that will undergo meiosis to form the gametes used in sexual reproduction. (In this case, the word "germ" does not refer to a disease-causing organism.) Germ cells are found in the *gonads* of animals—the female ovaries and the male testes. In most plants, germ cells are located in the female ovaries and the male anthers. All remaining cells that make up the body of an animal or plant—that is, all cells except the germ cells—are called **somatic cells.** Somatic cells divide only by mitosis, never by meiosis.

THE IMPORTANCE OF MEIOSIS

In 1883, Edouard van Beneden, a Belgian biologist, noted that the body cells of the roundworm *Ascaris* contained four large chromosomes, but the male and female gametes had but two chromosomes apiece. In 1887, August Weismann, a German biologist, proposed the existence of some type of "reduction division" that occurred in both the ovaries and the testes, allowing the chromosome number to be divided in half during the process of gamete formation. This reduction division was soon shown to constitute the first part of meiosis.

Consider for a moment what would happen if meiosis did not occur and the gametes contained the same number of chromosomes as did the somatic cells. Every time two gametes came together at fertilization, the chromosome number of the offspring would be twice that of their parents. For humans, the first generation of offspring would have 92 chromosomes (46 from the sperm plus 46 from the egg); the second generation would have 184; the third, 368; and so forth. Cells would soon become overloaded with surplus chromosomes and die after just a few generations. Thus, for organisms that reproduce sexually, the formation of reproductive cells containing only *one* set of chromosomes (rather than two sets, as in somatic cells) is crucial for ensuring a stable chromosome number for a species from generation to generation. The role of meiosis in maintaining a constant number of chromosomes in the cells of parents and offspring is another example of a mechanism that maintains biological order.

▮▶ The importance of meiosis goes beyond maintaining the chromosome number, however. As we will see later in the chapter, meiosis also builds *genetic variability* in a species. That is, meiosis increases the variety in characteristics among individuals that make up a species' population. A glance at a crowd of people instantly reveals the enormous variability that can exist within the human species. Different individuals may be better suited for different habitats.

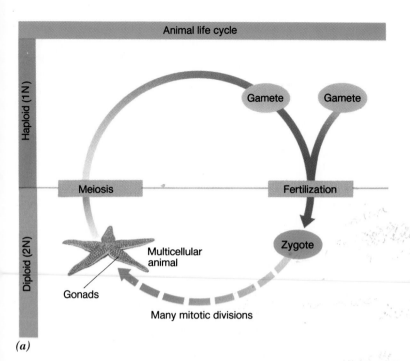

(a)

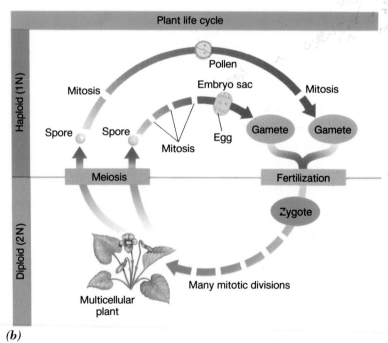

(b)

For example, shorter members of a population might be best suited for living in a forest whose trees have very low branches, while darker-skinned individuals may be better suited for living in bright, sunny climates, and so forth. A greater range of characteristics in a population improves the chances that some individuals will survive environmental changes, thereby increasing the likelihood that the species will be perpetuated.

Meiosis is not only important for the perpetuation of a species; it is also necessary for the formation of new species. Together with the appearance of new genetic characteristics by mutation, meiosis is one of the underlying mecha-

nisms that ensure genetic change in a population over time (discussed further in Chapters 33 and 34). If meiosis had never appeared, life on earth may well have been limited to bacteria and simple, unicellular eukaryotes.

Genetic variability increases during meiosis because chromosomes and genes are reshuffled to form new combinations before they are distributed to haploid daughter cells. Consequently, each gamete has a unique genetic composition. When gametes combine during fertilization, the organism produced is different from all others of the species. This explains why you *resemble* members of your family, yet you do not look *exactly* like any of them. The one exception to genetic variability among humans is identical twins; since identical twins develop from the same zygote, they have the same genes.

Life on earth has persisted over billions of years because species have the capacity to change. Change is facilitated by sexual reproduction, which requires meiosis as a step in the path toward formation of gametes. Meiosis provides genetic diversity that facilitates biological change while maintaining chromosomal numbers from one generation to the next. (See CTQ #2.)

THE STAGES OF MEIOSIS

There are many similarities between meiosis and mitosis (Chapter 10). Even the names of the stages are the same (prophase, metaphase, anaphase, and telophase). There are also some very important differences, however. This is not unexpected, considering that mitosis maintains chromosome number, whereas meiosis divides the number in half.

Like mitosis, meiosis is a continuous process—one that progresses through similar steps in all eukaryotic organisms. Biologists divide meiosis into stages based primarily on differences in chromosomal activity. Like mitosis, meiosis is preceded by an interphase period in which the chromosomes are duplicated. Unlike mitosis, in which the chromosomes are divided between two daughter nuclei, meiosis includes *two* sequential divisions—*meiosis I* and *meiosis II*—so that the chromosomes are distributed among four nuclei, rather than two. The entire sequence of meiotic stages is shown in Figure 11-2.

MEIOSIS I

Meiosis I is called the **reduction division** of meiosis because the nuclei of the two daughter cells contain half the number of chromosomes as does the nucleus of the mother cell. This reduction in the number of chromosomes is achieved by separating the members of each pair of homologous chromosomes into different nuclei. As a result, the daughter cells from meiosis I are haploid; that is, they contain only one complete set of chromosomes instead of two sets as in the diploid mother cell. In order to ensure that each of the daughter cells has only one member of each pair of homologues, an elaborate process of chromosome pairing occurs that has no counterpart in mitosis. During the period in which the homologous chromosomes are paired, some very important chromosome choreography takes place that increases genetic variability.

Prophase I

Unlike the prophase of mitosis, which usually lasts only a few minutes, the prophase I stage of meiosis can take hours, days, and sometimes even years. In humans, for example, all of the potential eggs that can ever be formed during the life of a woman have already entered prophase I by the time of her birth. Most of these germ cells will remain "stuck" in this stage of meiosis for decades, which is not without possible consequences.

Prophase I begins when the diffuse threads of the interphase chromosomes begin coiling. When the chromosomes have reached a stage of partial compaction, the members of each homologous pair recognize each other and become aligned together along their entire length. This process of chromosomal alignment is called **synapsis.** Since each chromosome is made up of two chromatids, a synapsed pair of homologous chromosomes is called a **tetrad,** which is a unit of four chromatids (Figure 11-3).

Early observers of meiosis noticed that the homologous chromosomes of the tetrads actually become wrapped around each other while they are synapsed (Figure 11-3). In 1909, F. A. Janssens proposed that this interaction might result in the breakage and exchange of pieces between different chromatids of the tetrad. Janssens' hypothesis proved to be basically correct; he had predicted the process now called **crossing over.**

Crossing Over and Genetic Recombination: An Increase in Genetic Variability Recall from page 199 that homologous chromosomes have identical sequences of genes along their length, although the corresponding genes on the two chromosomes may code for different characteristics of a particular trait. Crossing over between homologous chromatids can produce gene combinations different from those originally found in either chromatid. Let's see how crossing over can produce new gene combinations in homologous chromosomes with genes for plant height at one location and genes for flower color at another. These chromosomes are depicted in Figure 11-4.

In this example, one chromosome (on top) directs the development of a tall individual with yellow flowers, whereas its homologue produces a short plant with white flowers. Crossing over can *reshuffle* these genes (Figure 11-4)—a process known as **genetic recombination.** In our example of plant chromosomes, crossing over and recombination team the gene for a tall plant with the gene for white flowers on one chromatid and the gene for a short plant with the gene for yellow flowers on the homologous chromatid.

Genetic recombination greatly increases the genetic variability of the offspring by producing gametes with novel combinations of traits. Notice that the recombined genes are found on only two of the chromatids; the original gene combinations remain on the two chromatids that did not participate in crossing over. Genetic recombination during meiosis is one of the biological processes responsible for generating the diversity of life on earth.

The Molecular Basis of Genetic Recombination Crossing over rearranges chromosomal material, producing chromatids that contain sections of both paternal and maternal origin (Figure 11-4). Diagrams and photographs of

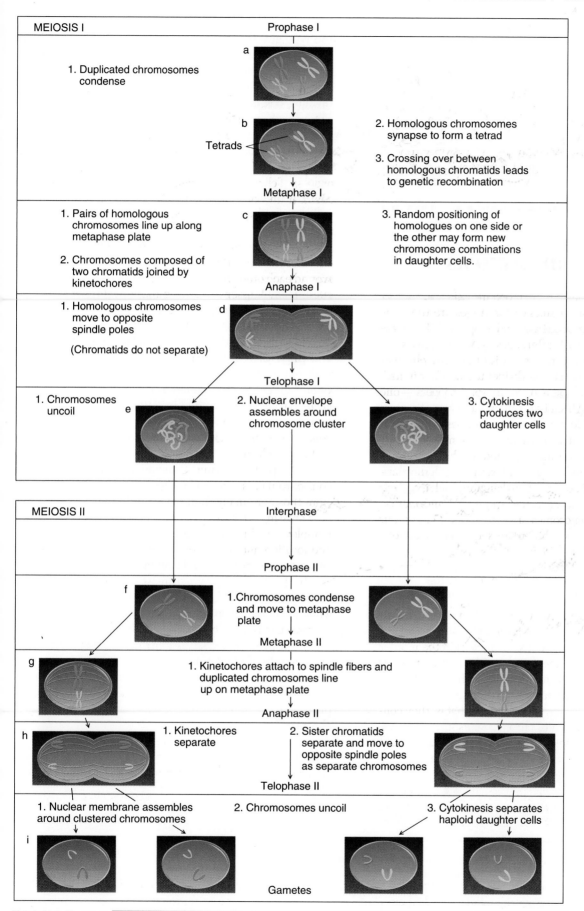

FIGURE 11-2
The stages of meiosis.

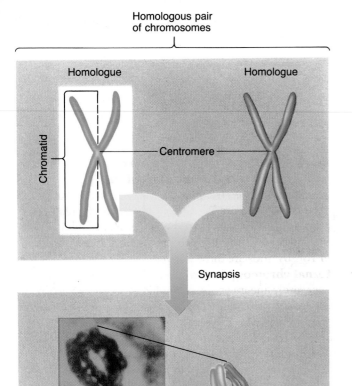

Homologous pair
of chromosomes

Homologue   Homologue

Chromatid

Centromere

Synapsis

Chiasmata

Tetrad

FIGURE 11-3

Homologous chromosomes forming a tetrad after synapsis. The points where the two homologous chromosomes are in contact are called *chiasmata*. These are sites where crossing over is believed to have occurred at an earlier stage.

this type give the erroneous impression that large sections of two chromatids simply break off and are physically exchanged with each other. Consider for a moment what this would entail. Because it is compacted, each chromatid of a dividing cell is so thick that it must contain thousands of DNA fibers lying side by side, as depicted in Figure 10-7. Are all of these fibers broken simultaneously? What about the genes contained in these DNA fibers? Recall what happens when just one alteration occurs in a gene that codes for a globin protein—the result can be sickle cell anemia (page 81). The breakage of an entire compacted chromatid would split *hundreds* of genes. Could all of these genes be reunited precisely with their missing segment when the two chromatids fused during crossing over? The answer is undoubtedly no.

⬿ Let us go back to an earlier stage of prophase I and reexamine the process of synapsis more closely. As the homologous chromosomes come together during synapsis, they can be seen in the electron microscope to be physically attached to a ladderlike structure called the **synaptonemal complex.** Each lateral element of the "ladder" is associated with one of the chromosomes of the tetrad, so that the homologues are not actually pressed against each other. Instead, they are separated by a space of about 100 nanometers (Figure 11-5a). The space in the middle of the ladder is not empty but contains loops of DNA extending from each of the chromosomes (Figure 11-5b). This suggests that the synaptonemal complex acts as a physical scaffold to which the chromosomes adhere, and that genetic recombination results from exchanges between the DNA loops extending into the center of the ladderlike structure. Viewed in this way, crossing over does not require the breakage and reunion of thick, compact meiotic chromatids but of individual DNA molecules.

Metaphase I and Anaphase I

Near the end of prophase I, the nucleolus and nuclear envelope become dispersed; spindle fibers become con-

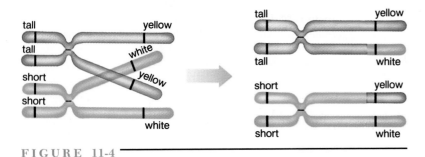

tall   yellow
tall
                              white
short   yellow
short
                              white

tall   yellow

tall   white

short   yellow

short   white

FIGURE 11-4

Genetic recombination. Meiotic chromosomes showing the two duplicates (sister chromatids) joined to one another. The sites (*loci*) on the chromosomes for the genes that govern two plant traits—height and flower color—are indicated. After crossing over, each chromosome contains one chromatid with the original combination of genetic characteristics and one chromatid with a mixture of the maternal and paternal characteristics.

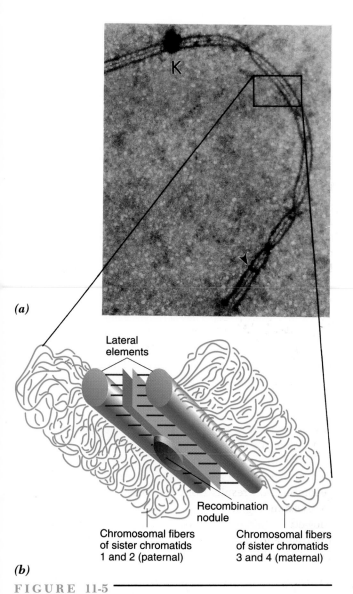

(a)

(b)

Lateral elements

Recombination nodule

Chromosomal fibers of sister chromatids 1 and 2 (paternal)

Chromosomal fibers of sister chromatids 3 and 4 (maternal)

FIGURE 11-5

Crossing over and the synaptonemal complex. *(a)* During prophase I of meiosis, homologous chromosomes become paired with one another and may appear as parallel fibers joined by the kinetochore (K). *(b)* Closer examination shows that the homologous chromosomes are tightly associated with a ladderlike structure, the synaptonemal complex. The dense granules (*recombination nodules*) seen between the "rungs of the ladder" are thought to contain the enzymatic machinery required for genetic recombination.

nected to the kinetochores of the chromosomes; and the tetrads are moved to the metaphase plate (page 205). In metaphase I, the tetrads become aligned in such a way that the two homologous chromosomes of each tetrad are situated on opposite sides of the plate (Figure 11-2*c*). Anaphase I begins when the spindle fibers attached to each chromosome shorten, pulling homologous chromosomes of each tetrad to opposite spindle poles (Figure 11-2*d*). Unlike in anaphase of mitosis, the kinetochores do not split during

anaphase I of meiosis. Each homologous chromosome is therefore still made up of two chromatids joined by their fused kinetochores. As a result of anaphase I, each daughter cell now has half the number of chromosome sets of its mother cell.

Independent Assortment of Homologues: An Additional Increase in Genetic Variability When homologous chromosomes become aligned at the metaphase plate, each homologue can line up on one side of the plate or the other, depending entirely on chance. Hence, all the chromosomes originally derived from one parent (**maternal chromosomes** are those derived from the mother and **paternal chromosomes** are those derived from the father) very rarely line up on the same side of the metaphase plate. Since each pair of homologous chromosomes lines up independently of other pairs, the chromosomes of each parent are sorted independently during meiosis I (Figure 11-6). This **independent assortment** of homologues reshuffles homologous chromosomes to form new chromosome combinations in the daughter cells.

Let's look at the numbers. If the diploid number of chromosomes of a species happened to be four (two pairs of homologous chromosomes), four (2^2) different types of daughter cells could be produced (Figure 11-6*a*). If the diploid number is six, there can be eight (2^3) different daughter cells (Figure 11-6*b*). In each case, there are 2^n number of genetically different daughter cells, where *n* is the haploid number of chromosomes. In a human, there are more than 8 million (2^{23}) possible chromosome combinations, with 23 chromosomes from your father and 23 chromosomes from your mother. Thus, excluding crossing over, the possibility that two parents would produce two identical offspring in two different pregnancies would be about one in 70 million ($2^{23} \times 2^{23}$). Add to this the gene exchanges that occur during crossing over, and it is easy to see how meiosis vastly increases the genetic variability of a species. This is the evolutionary advantage of sexual reproduction.

Telophase I and Cytokinesis

Telophase I of meiosis (Figure 11-2*e*) produces less dramatic changes than does telophase of mitosis. Although in many cases the chromosomes undergo some uncoiling, they do not reach the extremely extended state of the interphase nucleus. Similarly, the nuclear envelope may or may not reform during telophase I. Telophase I is followed by cytokinesis, which splits the cell into two daughter cells. Because DNA does not replicate during the interphase between cytokinesis and prophase II, the period separating meiosis I and II (called *interkinesis*) is usually very brief.

MEIOSIS II

Since sister chromatids become separated into different nuclei during **meiosis II,** the events in meiosis II are similar to those that occur during mitosis. During prophase II, the chromosomes shorten and thicken as they journey toward

Distribution of two pairs of homologous chromosomes by independent assortment during meiosis

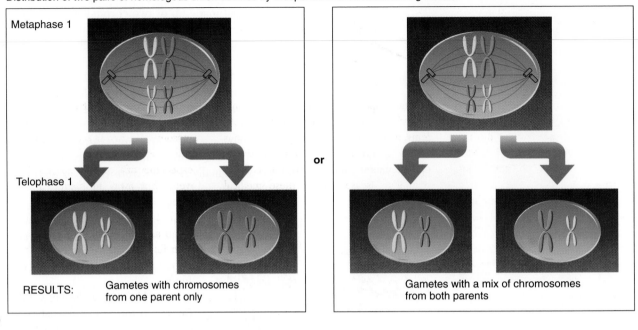

(a)

Distribution of three pairs of homologous chromosomes by independent assortment

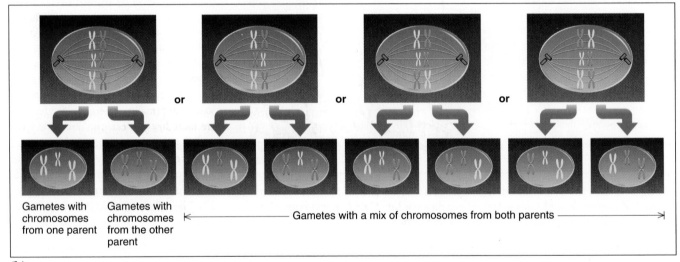

(b)

FIGURE 11-6

Independent assortment of homologous chromosomes during metaphase I of meiosis.
Paired homologous chromosomes line up randomly along the metaphase plate. Because each pair of homologous chromosomes lines up independently of other pairs, new chromosome combinations are often produced in the daughter nuclei, leading to increased genetic variability in a species. (To follow independent assortment, the homologues originally derived from one parent are colored blue, and those from the other parent are orange.) **(a)** In a cell with two pairs of homologous chromosomes, there are four possible combinations of chromosomes and, thus, four genetically different daughter nuclei that can form. Either both chromosomes from the same parent go into the same daughter cell, or the chromosomes are mixed, and each daughter cell receives one chromosome from each parent. **(b)** In a cell with three pairs of chromosomes, eight (2^3) genetically different daughter nuclei can form, not including the added genetic variability introduced by crossing over.

◁ THE HUMAN PERSPECTIVE ▷
Dangers That Lurk in Meiosis

On occasion, a mistake occurs during meiosis. Homologous chromosomes may fail to separate from each other during meiosis I, or sister chromatids may fail to come apart during meiosis II. When either of these situations occurs, gametes are formed that contain an abnormal number of chromosomes—either an extra chromosome or a missing chromosome. What happens if one of these gametes happens to fuse with a normal gamete and forms a zygote with an abnormal number of chromosomes? In most cases, the zygote develops into an abnormal embryo that dies in the womb, causing a miscarriage. In a few cases, however, the zygote develops into an infant whose cells have an abnormal chromosome number. These infants usually develop characteristic deformities.

The most common congenital disorder resulting from a meiotic mistake is Down syndrome (Figure 1), a condition in which individuals have three copies of chromosome number 21 in their cells, rather than two. Chromosome 21 is the smallest chromosome (see Figure 10-3*b*), containing an estimated 1,500 genes. Consequently, you might think that having a bit of extra genetic information would not be harmful. Yet, as we mentioned earlier in the chapter, Down syndrome is characterized by serious problems. These include mental impairment, alteration in some body features, frequent circulatory problems, increased susceptibility to infectious diseases, a greatly increased risk of developing leukemia, and the early onset of Alzheimer's disease. All of these medical problems are thought to result from the abnormal activity of genes located on chromosome 21. Seventy years ago, children with Down syndrome had an average life expectancy of about 10 years; most of them were shut away in mental institutions, where they withered and died. Today, most of these children are being raised at home and are encouraged to develop to their full potential; many attend "regular" schools and grow up to become working adults.

The likelihood of having a child with Down syndrome rises dramatically with the age of the mother, from 0.05 percent for mothers 19 years of age to nearly 2 percent for women over the age of 45. There are several possible explanations for this age-related increase. Some geneticists believe that the increased risk results primarily from aging of the germ cell as it remains for longer periods in the ovary. Another possibility is that younger females are more likely to abort a fetus bearing a chromosomal abnormality, whereas older females are more likely to carry an abnormal fetus to full term. Regardless of the reasons, older pregnant women are encouraged to undergo tests that examine the fetal cells for chromosomal abnormalities (Chapter 17). The effects of other chromosome abnormalities are described in Chapter 13.

FIGURE 1

Chris Burke, star of the television program *Life Goes On*, **was born with** Down syndrome.

---

the cell's metaphase plate (Figure 11-2*f*). In metaphase II, the chromosomes line up on the metaphase plate, and the kinetochores attach to the spindle fibers (Figure 11-2*g*). Anaphase II begins when the kinetochores split and the newly independent chromosomes are pulled toward opposite spindle poles (Figure 11-2*h*). During telophase II, a nuclear envelope is assembled around each of the four clusters of chromosomes, generating four nuclei. The two cells then divide by cytokinesis to produce a total of four haploid daughter cells (Figure 11-2*i*). Each daughter cell contains one-half of the chromosomes of the original diploid cell (although each chromosome may be heavily altered due to crossing over) and, therefore, has *one* complete set of hereditary instructions. The separation of chromosomes during meiosis must occur with great precision; the consequences of a failure in the process are considered in The Human Perspective: Dangers That Lurk in Meiosis.

Like mitosis, meiosis requires precision in order to separate duplicated chromosomes exactly. Because meiosis produces daughter cells that contain half the number of chromosomes as does the dividing cell, two sequences of cell division are required. (See CTQ #3.)

MITOSIS VERSUS MEIOSIS REVISITED

The major differences in the processes of mitosis and meiosis were summarized in Figure 10-4. In short, mitosis produces two daughter nuclei that are genetically identical to each other and to the parent cell, whereas meiosis produces four daughter nuclei that are genetically different from one another, each containing half the number of chromosomes of the parent cell. We have also discussed the different roles of mitosis and meiosis in the life cycles of plants and animals. Mitosis generates the nuclei of the body's cells, while meiosis generates the nuclei that ultimately become incorporated into the gametes.

Now let's go beyond these general comparisons and see how the differences between mitosis and meiosis can make a difference in the life of a specific organism. Suppose you were to visit a local pond, remove an eyedropper full of pond water, and put it in a vial. When you get back to your biology lab, you find you have captured a single ciliated protist that you identify as a member of the genus *Paramecium*. You place the contents of the vial into a container of purified water and add some algae or yeast to provide the organism with food. Over the next few weeks, your "domesticated" ciliate seems to thrive, dividing repeatedly and producing large numbers of genetically identical offspring by asexual reproduction (mitosis and cytokinesis). Periodically, you remove a certain volume of your culture and transfer it to fresh surroundings with fresh algae or yeast so that the culture remains unpolluted.

You might think that you could keep a culture of *Paramecium* alive indefinitely this way, but soon the organisms would begin to divide more slowly and to appear abnormal under the microscope; eventually, they would die off. This was the finding of Tracy Sonneborn of Indiana University, who studied the longevity of cultures of ciliates that reproduced solely by mitosis. A similar phenomenon has been observed when *cells* of humans and various animals are grown in culture. The cells divide mitotically for about 50 generations and then gradually lose their ability to divide, becoming unhealthy-looking, and eventually dying.

In contrast, meiosis seems to have the power to rejuvenate cells. While human cells growing in culture cannot undergo meiosis, the ciliates swimming in a pond can. Suppose that, instead of beginning with a single individual ciliate, you were able to capture a few dozen of these protists.

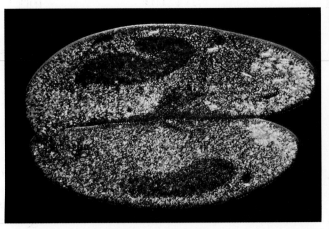

FIGURE 11-7

These protists are engaged in conjugation. A haploid nucleus from each individual passes into the other cell and fuses with the haploid nucleus of the recipient cell, forming a diploid nucleus with a new genetic constitution.

Although ciliated protists do not produce gametes that fuse to form a zygote, they do carry on a type of sexual reproduction called *conjugation* (Figure 11-7). During conjugation, two individuals of different strains of a species come together; the nucleus of each individual undergoes meiosis; and one haploid nucleus from each member of the conjugated pair travels across a bridge into the other member cell. There, it fuses with a haploid nucleus in the recipient, forming a diploid nucleus. The two ciliates then separate from each other, and each swims away.

Although this type of sexual reproduction does not produce additional offspring, it does generate individuals with a new genetic composition. Some ciliates that have undergone conjugation are found to die rather rapidly; presumably, these individuals have become genetically unfit for their environment. Others, however, appear to do much better than the original stocks, suggesting that these individuals possess a genetic constitution that makes them better adapted to their surroundings. This example illustrates the advantage of meiosis. It generates genetically diverse individuals required for the perpetuation of a species.

Because of its capacity to generate offspring with a unique genetic composition, meiosis is important for the survival of a species. (See CTQ #5.)

REEXAMINING THE THEMES

Relationship between Form and Function

Genetic recombination requires that the DNA molecules of different chromatids be broken and precisely reunited. This process is facilitated by the synaptonemal complex, a scaffoldlike structure that allows the DNA molecules of homologous chromosomes to be brought into close enough proximity for crossing over to occur. At the same time, the synaptonemal complex prevents the chromatids from becoming intertwined, which would result in genetic chaos.

Biological Order, Regulation, and Homeostasis

Maintaining biological order requires that the cells of a species maintain a constant number of chromosomes. The importance of chromosome number is dramatically illustrated by Down syndrome, a human disorder resulting from the presence of just one extra chromosome. Meiosis is *the* process by which the number of chromosomes in sexually reproducing organisms is cut in half, ensuring that the number of chromosomes will remain constant from one generation to the next.

Unity within Diversity

Sexual reproduction occurs throughout the four eukaryotic kingdoms, and, in every case, the formation of haploid gametes requires meiosis. In the more than 1 million different kinds of sexually reproducing organisms that exist, the process of meiosis is remarkably similar. At the same time, meiosis is largely responsible for generating the genetic variability that, through natural selection, has led to the remarkable diversity of life on earth.

Evolution and Adaptation

There could be no sexual reproduction without meiosis. If reproduction occurred solely by mitosis, offspring would be virtually identical to their parents. Biological evolution by natural selection depends on the genetic variability produced by independent assortment and crossing over. It is not unreasonable to conclude that, without meiosis, life on this planet may well have been limited to bacteria and a smattering of primitive, single-celled protists.

SYNOPSIS

Unlike mitosis, which maintains chromosome number, meiosis divides the number of chromosomes in half by separating the homologues of a diploid mother cell into four haploid daughter nuclei. The haploid products ultimately become incorporated into gametes for sexual reproduction. The fusion of two haploid gametes restores the diploid number of chromosomes in the zygote, thus maintaining a stable number of chromosomes from generation to generation.

Meiosis includes two sequential nuclear divisions —meiosis I and meiosis II—without an intervening period of replication. The first division is called the reduction division because it separates homologous chromosomes into two different daughter nuclei. The second division separates attached chromatids, producing a total of four haploid nuclei.

Genetic variability in the daughter cells (when compared to the original mother cell) is ensured by two events: independent assortment and genetic recombination. Genetic recombination results from crossing over, which occurs during prophase I of the first meiotic division. In crossing over, homologous portions of maternal and paternal chromatids are exchanged with each other, thus producing chromosomes that contain both maternal and paternal regions. Since both homologues in a pair are separated (assorted) independently of each other during anaphase I, daughter cells receive mixtures of both maternal and paternal chromosomes.

Genetic recombination and independent assortment during meiosis increase the genetic variability of a species. Genetic variability enhances the chances that some members of the species will survive inevitable environmental changes.

Key Terms

zygote (p. 214)
meiosis (p. 214)
life cycle (p. 214)
germ cell (p. 214)
somatic cell (p. 214)
meiosis I (p. 216)

reduction division (p. 216)
synapsis (p. 216)
tetrad (p. 216)
crossing over (p. 216)
genetic recombination (p. 216)
synaptonemal complex (p. 218)

maternal chromosome (p. 219)
paternal chromosome (p. 219)
independent assortment (p. 219)
meiosis II (p. 219)

Review Questions

1. Match the activity with the phase of meiosis in which it occurs.

   a. synapsis
   b. crossing over
   c. kinetochores split
   d. independent assortment
   e. homologous chromosomes separate
   f. cytokinesis

   1. prophase I
   2. metaphase I
   3. anaphase I
   4. telophase I
   5. prophase II
   6. anaphase II
   7. telophase II

2. How do crossing over and independent assortment increase the genetic variability of a species?

3. Why is meiosis I (and not meiosis II) referred to as the reduction division?

4. Suppose that one human sperm contains x amount of DNA. How much DNA would a cell just entering meiosis contain? A cell entering meiosis II? A cell just completing meiosis II? Which of these three cells would have a haploid number of chromosomes? A diploid number of chromosomes?

Critical Thinking Questions

1. Why are disorders, such as Down syndrome, that arise from abnormal chromosome numbers, characterized by a number of seemingly unrelated abnormalities?

2. A gardener's favorite plant had white flowers and long seed pods. To add some variety to her garden, she transplants some plants of the same type, but with pink flowers and short seed pods from her neighbor's garden. To her surprise, in a few generations, she grows plants with white flowers and short seed pods and plants with pink flowers and long seed pods, as well as the original combinations. What are two ways in which these new combinations could have arisen?

3. Set up the meiosis template in the diagram below on a large sheet of paper. Then use pieces of colored yarn or pipe cleaners to simulate chromosomes and make a model of the phases of meiosis. (*See template on opposite page*)

4. Would you expect two genes on the same chromosome, such as yellow flowers and short stems, always to be exchanged during crossing over? How might they remain together *in spite of* crossing over?

5. Suppose paternal chromosomes always lined up on the same side of the metaphase plate of cells in meiosis I. How would this affect genetic variability of offspring? Would they all be identical? Why or why not?

Additional Readings

Chandley, A. C. 1988. Meiosis in man. *Trends in Gen.* 4:79–83. (Intermediate)

Hsu, T. C. 1979. *Human and mammalian cytogenetics.* New York: Springer-Verlag. (Intermediate)

John, B. 1990. *Meiosis.* New York: Cambridge University Press. (Advanced)

Moens, P. B. 1987. *Meiosis.* Orlando: Academic. (Advanced)

Patterson, D. 1987. The causes of Down syndrome. *Sci. Amer.* Feb:52–60. (Intermediate-Advanced)

White, M. J. D. 1973. *The chromosomes.* Halsted. (Advanced)

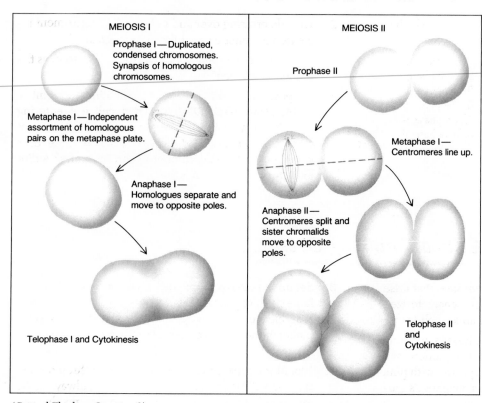

MEIOSIS I

Prophase I—Duplicated, condensed chromosomes. Synapsis of homologous chromosomes.

Metaphase I—Independent assortment of homologous pairs on the metaphase plate.

Anaphase I— Homologues separate and move to opposite poles.

Telophase I and Cytokinesis

MEIOSIS II

Prophase II

Metaphase I— Centromeres line up.

Anaphase II— Centromeres split and sister chromalids move to opposite poles.

Telophase II and Cytokinesis

(Critical Thinking Question 3)

genotypes and phenotypes (and their expected ratios) when there is an equal chance of acquiring either of the two alleles. A Punnett square of each of Mendel's monohybrid crosses predicts a 3 : 1 ratio of dominant to recessive phenotypes, just as Mendel had observed (Figure 12-5).

The ratios of genotypes and phenotypes predicted by Punnett squares are a matter of *probability;* they are validated by experiments only if the number of offspring is large enough to eliminate chance variations. The same is true for any random combination of events. For example, if you flipped a coin three times, and it happened to land on "heads" all three times, the results could be attributed to chance variation. In other words, three flips are not enough to conclude that the probability of heads is 100 percent. But if you flipped the coin 100 times, the likelihood of its landing on heads every time would be very, very small. Instead, the frequency of obtaining heads would likely be close to 50 percent. In genetics, the larger the number of offspring examined, the closer the results should approach the predicted ratio.

Punnett squares are convenient for determining the genotypes of the offspring when only one or two traits are involved, but they soon become unwieldy when genetic crosses involve many traits. Although Punnett squares are valuable because they offer a visual portrait of gamete combinations, it is easier to reach the same predictions by simple multiplication. Keep in mind that *the probability of two independent events occurring together is the product of the probability of each occurring alone.* In the case of garden peas, the chance of a gamete acquiring the recessive allele for wrinkled seeds from a heterozygous parent is one in two, or 1/2. In a cross between heterozygotes the probability of a zygote receiving both alleles for wrinkled seeds is one in four, or 1/4, since each gamete has a probability of 1/2 for having this allele (1/2 × 1/2 = 1/4). Thus, on the average, one of every four offspring of two heterozygous parents will be homozygous recessive (wrinkled seeds); the other three will show the dominant phenotype (round seeds). This brings us back to the 3 : 1 ratio Mendel obtained from the thousands of genetic crosses he performed.

Hereditary Patterns in Humans

Since the principles of dominance and segregation of alleles apply to all sexually reproducing diploid organisms, not just garden peas, Mendel's findings can help you explain hereditary patterns in humans. A dimple in the chin (Figure 12-6*a*), a "widow's peak" hairline (Figure 12-6*b*), achondroplastic dwarfness, webbed fingers, and the absence of fingerprints are all examples of traits determined primarily by dominant alleles (though other factors may also play a role). A person with just *one* of these dominant alleles will develop the corresponding characteristic, regardless of the second allele. A child of parents both of whom are heterozygous for one of these traits has a 3 in 4 probability (75 percent chance) of exhibiting the same dominant phenotype as that found in the parents. This does not mean that individuals with a dimple in their chin are three times more numerous than persons without them, however. In fact, for

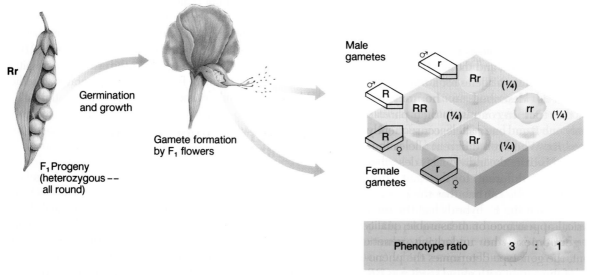

FIGURE 12-5

The Punnett square method of determining probability ratios. Following self-fertilization of a heterozygous pea plant, each F$_1$ gamete will contain either an *R* (round) or *r* (wrinkled) allele. In the Punnett square representation, all possible male gametes are listed along the top, and all female gametes along the side. The possible combinations of alleles following fertilization are shown in the boxes, each box representing the genotype created by the union of these two alleles. The genotype ratio is 1 : 2 : 1 *(RR : Rr : rr),* meaning that there are twice as many *Rr* boxes as either *RR* or *rr* boxes. The 3 : 1 phenotype ratio explains why recessive (wrinkled) characteristics reappeared in one-fourth of Mendel's F$_2$ progeny after disappearing from the F$_1$ generation.

(a)   *(b)*

FIGURE 12-6

Dominant alleles. The dimple in Cary Grant's chin *(a)* and Marilyn Monroe's widow's peak hairline *(b)* result from dominant alleles.

some traits, the recessive phenotype is far more common than is the dominant phenotype. After all, most of us *do* have fingerprints, even though it is a recessive condition. In this case, the dominant allele is so rare that nearly everyone is homozygous recessive, and therefore has a complete set of fingerprints.

The Test Cross

As we saw earlier, outward appearance of a dominant phenotype cannot reveal whether an individual is homozygous or heterozygous for that trait. A pea plant with purple flowers, for example, may be homozygous or heterozygous for the trait of flower color. Mendel concluded that if his concept of dominant and recessive factors were correct, then it should be possible to determine genotype by crossing an individual in question with a mate that shows the recessive trait. This strategy is known as a **test cross,** and it reveals the presence of a "hidden" recessive allele if the individual is heterozygous for that trait, as explained in Figure 12-7.

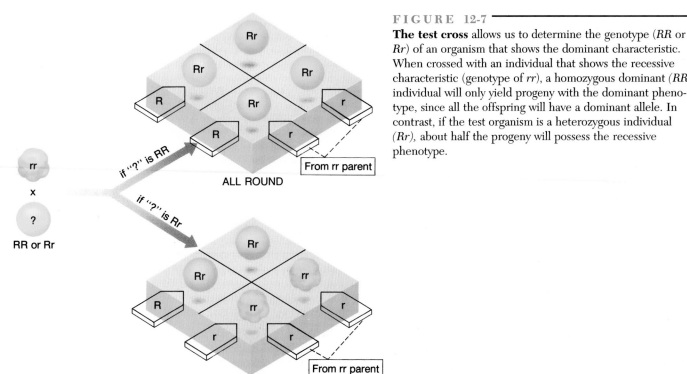

FIGURE 12-7

The test cross allows us to determine the genotype (*RR* or *Rr*) of an organism that shows the dominant characteristic. When crossed with an individual that shows the recessive characteristic (genotype of *rr*), a homozygous dominant (*RR*) individual will only yield progeny with the dominant phenotype, since all the offspring will have a dominant allele. In contrast, if the test organism is a heterozygous individual (*Rr*), about half the progeny will possess the recessive phenotype.

CROSSING PLANTS THAT DIFFER BY MORE THAN ONE TRAIT: MENDEL'S LAW OF INDEPENDENT ASSORTMENT

Once Mendel understood how a single trait was inherited, he was ready to examine the transmission of two traits simultaneously. To do this, he performed a series of **dihybrid crosses**—crosses between plants that differed from one another in two distinct traits, such as seed shape and seed color. Mendel crossed true-breeding plants having round, yellow seeds (genotype of *RRYY*) with plants having wrinkled, green seeds (*rryy*). As expected, the seeds of the F_1 offspring showed only the dominant characteristics (Figure 12-8); that is, seeds of a round, yellow phenotype (*RrYy* genotype). When he allowed these F_1 hybrids to self-fertilize, however, four different types of seeds were found among the next generation. Most of the seeds (approxi-

mately 9 out of 16) were round and yellow, while approximately 1 out of 16 were wrinkled and green. In addition, two types of seeds had a combination of traits not seen in either parent—round, green seeds and wrinkled, yellow seeds (both represented by approximately 3/16 of the population). Thus, the ratio of these four phenotypes was approximately 9:3:3:1.

In order to account for these results, Mendel made a final conclusion: The segregation of the pair of alleles for one trait had no effect on the segregation of the pair of alleles for another trait; that is, alleles for different traits segregate *independently* of one another. Just because an F_1 individual inherited one *R* and one *Y* allele from one gamete did not mean that those two alleles must remain together when that F_1 plant formed its own gametes. Rather, either of the alleles for seed shape could team up with either of the alleles for seed color (Figure 12-8). This would produce

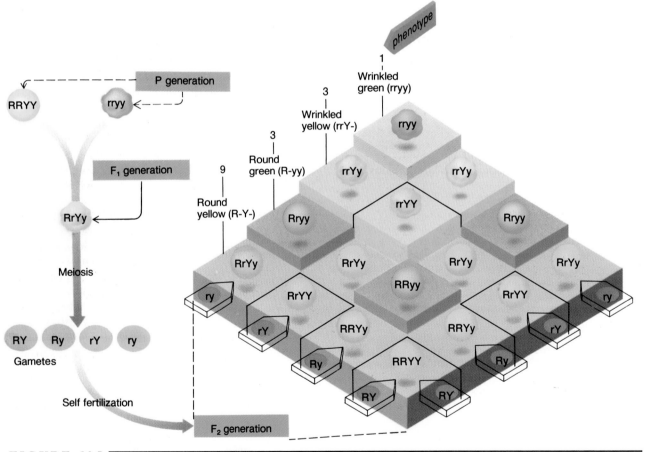

FIGURE 12-8

Mendel's dihybrid crosses. When plants that breed true for round, yellow (*RRYY*) seeds are crossed with those producing wrinkled, green (*rryy*) seeds, the F_1 progeny are heterozygous (*RrYy*) for both seed shape and color. Independent assortment allows either *R* or *r* to combine with either *Y* or *y*, generating gametes with four different possible combinations of alleles (*RY, Ry, rY, ry*). Self-fertilization of an F_1 plant produces an F_2 generation with the nine genotypes and four phenotypes shown in the Punnett square. When Mendel performed this experiment with hundreds of peas, he obtained ratios very close to the 9:3:3:1 phenotype pattern predicted here. Mendel also determined the genotypes of the F_2 progeny by allowing them to self-fertilize and then examining their offspring. The F_2 generation consisted of all nine genotypes in a ratio that approximated the predicted 1:2:1:2:4:2:1:2:1.

gametes with four possible combinations (*RY, Ry, rY, ry*), each in equal proportion. This conclusion is referred to as Mendel's **law of independent assortment** and, as we saw in the previous chapter, derives directly from the fact that homologous chromosomes become aligned independently in the metaphase plate during meiosis I.

Because the two traits are independent of each other, the probability of acquiring any phenotypic combination of both can be calculated by multiplying the probability of inheriting each phenotype alone:

Probability of Acquiring Each Characteristic

Round seeds = 3/4   Yellow seeds = 3/4
Wrinkled seeds = 1/4   Green seeds = 1/4

Probability of Combination of Characteristics

Round, yellow seeds = 3/4 × 3/4 = 9/16
Round, green seeds = 3/4 × 1/4 = 3/16
Wrinkled, yellow seeds = 1/4 × 3/4 = 3/16
Wrinkled, green seeds = 1/4 × 1/4 = 1/16

Through the use of carefully designed experiments, meticulous gathering of data, and insightful interpretation, Mendel discovered the "rules" by which genetic characteristics are transmitted. He correctly formulated why the patterns of transmission were observed and how they could be applied to predict the genetic outcome of particular matings. (See CTQ #3.)

MENDEL AMENDED

Even though Mendel's "laws" provided the foundation on which modern genetics was built, several exceptions to Mendel's principles were soon discovered as the science of genetics matured. For example, not all traits have two alternative forms of appearance or display the simple dominant–recessive relationship exhibited by the seven traits examined by Mendel. Furthermore, just as Mendelian patterns of inheritance are found in all types of sexually reproducing eukaryotic organisms, the same types of non-Mendelian patterns also appear in diverse organisms. The reason for the nearly universal application of these principles is simple: Regardless of the organism, genetic information is encoded in the same types of material and is expressed by the same types of molecular processes. This will become more apparent in the following section.

INCOMPLETE DOMINANCE AND CODOMINANCE

At first glance, some traits seem to follow the notion of "blending" rather than Mendelian inheritance. For example, when red-flowered snapdragons are crossed with white-flowered snapdragons, all of the progeny are pink, as though the traits had blended. But the next generation of plants clearly demonstrates that blending has *not* occurred.

Self-pollination of pink snapdragons produces an interesting mix of offspring—50 percent pink, 25 percent red, and 25 percent white. If blending had occurred, red and white varieties could never have been recovered from pink parents. How does this happen?

The answer is evident if we first introduce the concept of "gene products." Genes are not abstract factors that pass from parents to offspring; rather, they are portions of chromosomes that dictate the formation of specific proteins. This relationship between genes and proteins was first glimpsed in 1908 by a British physician, Archibald Garrod, who found that individuals suffering from the genetic disease *alcaptonuria* were missing a key enzyme essential for metabolizing certain amino acids. The defect, which does not result in serious health problems, is detected because the urine of the affected individual turns black upon exposure to air. This curious symptom results from the oxidation of a compound that is absent in the urine of normal individuals. In one of biology's most farsighted proposals, Garrod suggested that genes carried information directing the manufacture of specific enzymes.

⬥ Although Garrod's proposal proved to be true, it wasn't until biologists turned to less conspicuous organisms—including fruit flies and bacteria—that the universality of Garrod's conclusions came to be accepted. We can illustrate Garrod's principle using one of the traits studied by Mendel. The gene product for the purple flower allele in pea plants is an enzyme that catalyzes the formation of a purple pigment. The allele for white flowers is a variant of this gene which directs the formation of an inactive enzyme. This explains why purple is dominant over white in garden pea flowers: As long as there is one purple flower allele in the cells, enough pigment will be produced to turn the flower purple. The white color appears only in the homozygous recessive configuration because of the flower's inability to produce the purple pigment. Many genes control their traits in such a fashion; the recessive allele is simply unable to direct formation of an active gene product.

The case of snapdragons is a bit different, however. The red allele directs the formation of an active enzyme that produces red pigment, while the white allele cannot generate an active enzyme. Therefore, when two copies of the red allele are present, enough enzyme is manufactured to produce a true red phenotype. When only one red allele is present, however, only enough enzyme is manufactured to produce a paler (pink) color. This phenomenon is known as **incomplete** (or **partial**) **dominance.** Another example of this phenomenon is found in people who are heterozygous for the gene that causes familial hypercholesterolemia (page 146). Heterozygotes for this gene have a markedly elevated blood-cholesterol level, but one not nearly as high as that of homozygotes.

Heterozygous individuals are also distinguishable in cases of **codominance,** but, in these cases, the phenotype is not an intermediate (diluted) form. Rather, codominance means that both alleles are expressed simultaneously and are unmodified in the heterozygous individual. A classic example of codominance is seen in the ABO blood group antigens—proteins that are found on the surface of red blood cells and are used to determine human blood type (page 94). The codominant alleles for A and B, written as I^A and I^B, direct the blood cell to manufacture the A antigen and the B antigen, respectively, giving the individual type AB blood. How, then, do people acquire type O blood? The answer lies in multiple alleles.

MULTIPLE ALLELES

The human ABO blood group system can be used to illustrate another departure from classical Mendelian genetics —the existence of more than two possible alleles for a single trait. Although a single diploid individual can possess only two alleles, the existence of additional alleles increases the number of possible genotypes in the population. There are three different alleles in the ABO system, for example: I^A, I^B, and I^O. The I^O allele is a variant that, like the allele for white flowers in pea plants, directs the formation of a defective protein. A homozygous person with two I^O alleles would lack both A and B antigens. These individuals are blood type O. An allele such as I^O is recessive to both of the alternative alleles, which is sometimes why it is written as "*i*." Therefore, an $I^A I^O$ heterozygous pair would produce the same phenotype as would the homozygous $I^A I^A$—that is, a person with type A blood. (These atypical genetic symbols were coined many years ago by a pioneering Austrian physician—Karl Landsteiner—and remain in use today.)

COMBINED EFFECTS OF GENES AT DIFFERENT LOCI

Not only did Mendel's seven pairs of alleles segregate independently, they also *functioned* independently of one another. In many cases, however, the expression of a pair of alleles is *influenced* by the genotype at other *loci* (plural of

locus), or sites, on the chromosomes. This relationship is illustrated by human pigmentation. Human beings have separate genes for hair color, eye color, and skin color. Let's consider a person who has a genotype that dictates brown hair, brown eyes, and a relatively dark complexion. All of these phenotypes require the production of a dark pigment (known as *melanin*) which, in turn, requires the activity of a certain enzyme encoded by a gene at the A locus. If a person carries two alleles *(aa)* that produce an *inactive* enzyme, their hair, eyes, and skin will be devoid of color, regardless of the genotypes at the hair, eye, and skin color loci. Individuals who acquire this pair of recessive alleles *(aa)* are called *albinos.* Albinism (Figure 12-9) is an example of **epistasis,** a process whereby a pair of recessive alleles at one locus masks the effect of genes at other loci.

PLEIOTROPY

The product of one gene can have far-reaching secondary effects on many diverse characteristics, a condition called **pleiotropy.** The inherited disease *cystic fibrosis* is one example of this phenomenon. The diverse (and serious) symptoms of this disease result from the production of an abnormal membrane protein that causes thickened mucus, which impairs the functioning of the digestive, sweat, and respiratory mucous glands. This, in turn, produces widespread disorder within the body. For example, the digestive enzymes are poorly secreted, causing improper digestion and absorption of food; the sweat contains abnormally high levels of salt; and the mucus that lines the airways is too thick to be propelled out of the lower respiratory tract by its ciliated lining. This viscous mucus, laden with trapped microbes, settles into the lungs. As a result, pneumonia and other lung infections are the most frequent causes of death in those afflicted with cystic fibrosis (Chapter 17).

POLYGENIC INHERITANCE

If human height were determined by simple dominance at a single locus, people would come in two sizes—tall and short (dwarf)—much like the pea plants studied by Mendel. Height in peas is described as **discontinuous** because the phenotypes fall into two distinct categories. But many traits, such as human height and skin coloration, show **continuous variation** within the population. Such traits are determined by a number of different genes present at different loci, rather than by a single gene. This phenomenon is known as **polygenic inheritance** (Figure 12-10). The overall expression of the polygenic trait of skin color represents the sum of the influences of many contributing genes. Two people who differ in just one pair of alleles show only slight color differences, for example, as compared to two people who differ in several of the polygenic determinants.

Many of the traits currently being studied by geneticists are polygenic. These include animal and plant traits selected by breeders for improving livestock and crops, as

FIGURE 12-9

Albinism occurs throughout the vertebrates, including these crocodilians.

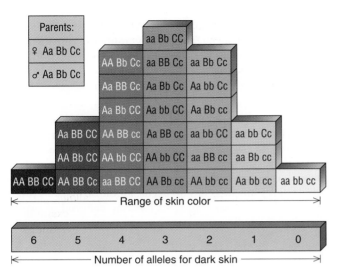

FIGURE 12-10

Polygenic inheritance and skin color. People don't come in two contrasting colors: dark and light. Most of us lie somewhere between these extremes. There are, at least, three genes—*A*, *B*, *C*—located at three different loci, that determine the amount of pigment in human skin. At each locus, there are two possible alleles of the gene, one for maximum pigment (the capital letter), and one for no pigment (the lowercase letter). This figure shows the 27 genotypes and 7 phenotypes that could be produced by two people who are heterozygous for all three genes (*AaBbCc*). The 27 genotypes are indicated within each of the boxes; the 7 different phenotypes are indicated by the 7 columns indicating different shades of darkness. The darkness of skin is determined by the total number of alleles for dark pigment (capital letters) in the genotype. The darkest phenotype would have all 6 genes represented by the dominant allele, while the lightest would have all six genes represented by the recessive allele.

well as such elusive human characteristics as susceptibility to cardiovascular disease and athletic prowess. Polygenic inheritance poses particular challenges to geneticists because the outcome of an experimental cross is the sum of many genotypic variables.

ENVIRONMENTAL INFLUENCES

Nearly all phenotypes are subject to modification by the environment. A genetically tall person, for example, will not achieve his or her maximum height potential if deprived of adequate nutrition. Furthermore, two organisms with the same genotype may have different phenotypes in different environments. For example, a light-skinned person living in a sunny, warm climate is likely to be considerably darker skinned than would a genetically similar individual living in a cold, overcast climate.

Environmental alteration of phenotype can affect single-gene traits as well as polygenic traits. An excellent example is the two-color pattern of the Siamese cat (Figure 12-11). This type of cat has only one gene for hair color, and the genotype ($c^s c^s$) at this locus is the same in all of the cat's cells, including those that produce light fur and those that make dark fur. The *siamese* allele (c^s) directs the synthesis of a temperature-sensitive variant of an enzyme that manufactures the dark pigment. This enzyme is operative only on the cooler areas of the body (feet, snout, tip of the tail, and ears). The rest of the cat's body is warm enough to inactivate the enzyme, causing the fur to remain light.

Each of the traits studied by Mendel was determined by two alternate alleles, one of which was dominant over the other. Following the rediscovery of Mendel's work, geneticists learned that not all traits followed such simple patterns of inheritance. The basis of these non-Mendelian patterns can be understood when interpreted in the light of more recent research on the nature of gene products. (See CTQ #4.)

MUTATION: A CHANGE IN THE GENETIC STATUS QUO

▐▶ It is fitting to end this chapter with the subject of mutation because, in a sense, we have been dealing with it all along. **Mutations** are rare but permanent changes in a gene that may alter a gene's product. Thus far, we have been describing the inheritance of alleles—alternate forms of a gene. Alleles arise by mutations in an existing gene. The original allele is termed the *wild type*, to distinguish it from the newer *mutant* form. If there were no mutations, there would be no new alleles, *and there would be no biological evolution*. This may seem overly dramatic, so let us explore it a bit further.

FIGURE 12-11

A Siamese cat provides a visual portrayal of the influence of environment on phenotype.

The segregation of alleles and their independent assortment during meiosis increase genotype diversity by promoting new combinations of genes. But the shuffling of existing genes alone does not explain the presence of such a vast diversity of life. If all organisms descended from a common ancestor, with its relatively small complement of genes, where did all the genes present in today's millions of species come from? The answer is mutation.

Most mutant alleles are detrimental; that is, they are more likely to disrupt a well-ordered, smoothly functioning organism than to increase the organism's fitness. For example, a mutation might change a gene so that it produces an inactive enzyme needed for a critical life function. Occasionally, however, one of these stable genetic changes creates an advantageous characteristic that increases the fitness of the offspring. In this way, mutation provides the raw material for evolution and the diversification of life on earth.

One of the requirements for genes is stability; genes must remain basically the same from generation to generation or the fitness of organisms would rapidly deteriorate. At the same time, there must be some capacity for genes to change; otherwise, there would be no potential for evolution. Alterations in genes do occur, albeit rarely, and these changes (mutations) represent the raw material of evolution. (See CTQ #7.)

REEXAMINING THE THEMES

Biological Order, Regulation, and Homeostasis

Mendel discovered that the transmission of genetic factors followed a predictable pattern, indicating that the processes responsible for the formation of gametes, including the segregation of alleles, must occur in a highly ordered manner. This orderly pattern can be traced to the process of meiosis and the precision with which homologous chromosomes are separated during the first meiotic division. Mendel's discovery of independent assortment can also be connected with the first meiotic division, when each pair of homologous chromosomes becomes aligned at the metaphase plate in a manner that is independent of other pairs of homologues.

Unity within Diversity

All eukaryotic, sexually reproducing organisms follow the same "rules" for transmitting inherited traits. Although Mendel chose to work with peas, he could have come to the same conclusions had he studied fruit flies or mice or had he scrutinized a family's medical records on the transmission of certain genetic diseases, such as cystic fibrosis. Although the mechanism by which genes are transmitted is universal, the genes themselves are highly diverse from one organism to the next. It is this genetic difference among species that forms the very basis of biological diversity

Evolution and Adaptation

Mendel's findings provided a critical link in our knowledge of the mechanism of evolution. A key tenet in the theory of evolution is that favorable genetic variations increase the likelihood that an individual will survive to reproductive age and that its offspring will exhibit these same favorable characteristics. Mendel's demonstration that units of inheritance pass from parents to offspring without being blended revealed the means by which advantageous traits could be preserved in a species over many generations. The subsequent discovery of genetic change by mutation revealed how new genes appeared in a population, thus providing the raw material for evolution.

SYNOPSIS

Gregor Mendel discovered the pattern by which inherited traits are transmitted from parents to offspring. Mendel discovered that inherited traits were controlled by pairs of factors (genes). The two factors for a given trait in an individual could be identical (homozygous) or different (heterozygous). In heterozygotes, one of the gene variants (alleles) may be dominant over the other, recessive allele. Because of dominance, the appearance (phenotype) of the heterozygote (genotype of *Aa*) is identical to that of the homozygote with two dominant alleles

(*AA*). An individual must possess two recessive alleles (*aa*) to exhibit the recessive phenotype. Although a particular recessive characteristic may disappear in heterozygotes for a generation or more, it can eventually reappear because the allele for that characteristic remains intact as it is passed from parents to offspring. Monohybrid crosses between heterozygous parents produce three times as many offspring with the dominant characteristic as with the recessive characteristic.

Mendel's law of segregation explained the results of monohybrid crosses. Paired alleles segregate from each other during formation of the gametes. Each gamete receives only one of the two alleles, and a gamete has an equal chance of receiving either allele.

Mendel's law of independent assortment explained the results of dihybrid crosses. Dihybrid crosses between heterozygotes produce plants with four phenotypes, two of which were present among the parental plants. The ratios of the four phenotypes are approximately 9:3:3:1, which is exactly what would be expected if the alleles of the two genes are combining in a random manner, as predicted by simple probability equations (or by the Punnett square method). These findings led Mendel to conclude that the segregation of the pair of alleles for one trait had no effect on the segregation of the pair of alleles for another trait.

Not all inherited traits are transmitted according to Mendelian patterns. (1) Some alleles show incomplete dominance, whereby the pair of alleles "dilute" each other's effect in heterozygous individuals. (2) Both codominant alleles are fully expressed in heterozygous individuals. (3) Multiple (more than two) alleles for a particular trait can combine to form more than three genotypes and two phenotypes. (4) Alleles present at one locus can effect the expression of genes at a different locus. (5) A single gene may be pleiotropic, meaning that it can have many effects. (6) One trait may be due to multiple genes at several loci; this is known as polygenic inheritance. (7) Tracking the fate of alleles from generation to generation can be complicated by the effect of environment on phenotype. (8) New alleles can emerge by mutations—changes in the genetic message. Such changes have produced a rich diversity of living organisms.

Key Terms

heredity (p. 232)
monohybrid cross (p. 233)
P generation (p. 233)
F_1 generation (p. 233)
dominant (p. 234)
recessive (p. 234)
F_2 generation (p. 234)
allele (p. 234)
homozygous (p. 235)

heterozygous (p. 235)
phenotype (p. 235)
genotype (p. 235)
law of segregation (p. 235)
Punnet square (p. 235)
test cross (p. 237)
dihybrid cross (p. 238)
law of independent assortment (p. 239)

incomplete (partial) dominance (p. 239)
codominance (p. 239)
epistasis (p. 240)
pleiotropy (p. 240)
discontinuous variation (p. 240)
continuous variation (p. 240)
polygenic inheritance (p. 240)
mutation (p. 241)

Review Questions

1. Why can't the law of independent assortment be discussed without an initial understanding of the law of segregation?

2. How do Mendel's findings explain how a recessive genetic disorder, such as sickle cell anemia, may skip over one, or even two, generations?

3. How does each of the following increase genetic diversity: (a) meiosis, (b) fertilization, (c) mutation?

4. If a person has very light hair and skin but dark eyes, what might you conclude about the genotype of his or her melanism locus (*A*)?

Critical Thinking Questions

1. Do any of the non-Mendelian patterns of inheritance discussed in the chapter provide a useful hypothesis for understanding the fact that one identical twin may not display the same magnitude of disease as his or her twin? Explain your answer.

2. Why was it necessary for Mendel to begin his work by self-fertilizing his original stocks of pea plants? How would Mendel's conclusions have differed had he found that one of his parental plants produced offspring with two different characteristics? What might you conclude about the genotype of this particular plant? How might this genotype have arisen?

3. In peas, yellow seed color is dominant over green. The chart below shows the results of crossing pea plants of known phenotypes. Using the letters Y for yellow and y for green, identify the genotypes of each parent plant.

	Offspring	
Parents	*Yellow Seeds*	*Green Seeds*
yellow × green	104	100
yellow × yellow	156	54
green × green	0	112
yellow × green	92	0
yellow × yellow	89	0

4. Do the non-Mendelian patterns of inheritance, such as incomplete and codominance, multiple alleles, pleiotropy, and polygeny negate Mendel's Laws of Inheritance? How did Mendel's laws contribute to biologists' understanding of non-Mendelian patterns? Which of the non-Mendelian patterns would have been most helpful to Mendel and others had Mendel found it in his pea plants?

5. Design a mating experiment that would demonstrate that incomplete dominance is not evidence of genetic blending. Explain how your proposed experiment proves that genes remain intact, even in organisms with the intermediate phenotype.

6. Suppose Mendel had made a trihybrid cross between a pea plant that was tall with purple flowers and green seeds (genotype *TTPpyy*) with a pea plant that was tall with white flowers and green seeds (genotype *Ttppyy*). What fraction of the progeny would have the genotype *TT*? What fraction would have white flowers? What fraction would have the phenotype of dwarf with purple flowers and green seeds? What would be the ratio of genotypes if the parental plant with purple flowers was test-crossed? (Try to answer these questions without using a Punnett square.)

7. In his theory of evolution by natural selection, Darwin assumed that individuals in a population vary, but he could not explain the origin of those variations because the principles of genetics had not yet been discovered. If Darwin were to appear to you today, how would you explain the origin of variations and the relationship of mutations to evolution?

Additional Readings

Brennan, J. R. 1985. *Patterns of human heredity: An introduction to human genetics.* Englewood Cliffs, NJ: Prentice-Hall. (Intermediate)

Carlson, E. A. 1973. *The gene: A critical history,* 2d ed. Philadelphia: Saunders. (Intermediate-Advanced)

Gardner, E. J., M. J. Simmons, and D. P. Snustad. 1991. *Principles of genetics,* 8th ed. New York: Wiley. (Intermediate-Advanced).

Mendel, G. 1866. *Experiments in plant hybridization* (English translation). Cambridge: Harvard University Press (1951). (Intermediate)

Suzuki, D. T., et al., 1989. *Introduction to genetic analysis,* 4th ed. New York: Freeman. (Intermediate-Advanced)

Wender, P. H., and D. F. Klein. 1981. *Mind, mood, and medicine.* New York: Farrar, Straus & Giroux. (Introductory)

Young, P. 1988. *Schizophrenia.* New York: Chelsea House. (Introductory)

CHAPTER
◄ 13 ►

Genes and Chromosomes

**STEPS
TO
DISCOVERY**
The Relationship Between Genes and Chromosomes

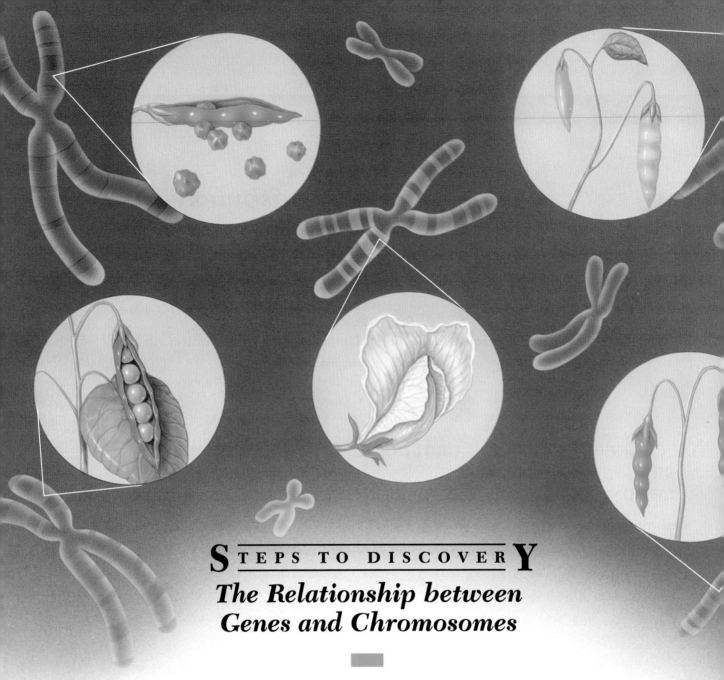

*B*y the 1880s, a number of prominent European biologists were closely following the activities of cells and using rapidly improving light microscopes to observe newly discernable cell structures. None of these scientists was aware of Mendel's work, which had demonstrated the particulate nature of hereditary factors. Yet, they realized that whatever it was that governed inherited characteristics would have to be passed from cell to cell and from generation to generation. This, by itself, was a crucial realization; all the genetic information needed to build and maintain a human

being, for example, had to fit within the boundaries of a single cell.

Observations of dividing cells revealed that the cytoplasmic contents were merely distributed among the two daughter cells, and often not very evenly. In contrast, the cell's nuclear material was always evenly divided. The contents of the nucleus become organized into thick, darkly stained strands, inspiring biologists in 1888 to give them the name "chromosomes," meaning "colored bodies."

During cell division, each chromosome appeared out

The genes for five of the traits studied by Mendel—seed shape, seed color, flower color, pod color, and pod shape—are located

of the formless nuclear mass, split down the middle into two separate entities (each of which passed into separate cells), and then disappeared again. Careful observations of the sizes and shapes of the mitotic chromosomes suggested that a chromosome that disappeared after one division seemed to be the same structure that reappeared before the next. Even though the chromosome could not be seen in the nucleus of the nondividing cell, it was presumed to be present in an "invisible" state. Just as cells arise only from preexisting cells, it appeared that chromosomes arose only from preexisting chromosomes.

Attention then turned to the gametes and the material present in these cells which might be responsible for endowing the offspring with its genetic inheritance. Even though the contribution from the male was just a tiny cell (a sperm), it was as genetically important as the much larger egg, contributed by the female. What was it that these two very different cells had in common? The most apparent feature was the nucleus and its chromosomes.

During the 1880s, two German biologists, Theodor Boveri and August Weismann, turned their attention to the chromosomes. Boveri discovered that sea urchin eggs possessing abnormal numbers of chromosomes (a result of his experimental treatments), developed into abnormal embryos. Boveri concluded that the orderly process of normal development is "dependent upon a particular combination of chromosomes; and this can only mean that the individual chromosomes must possess different qualities." Weismann focused on the germ cells and concluded that the number of chromosomes in these cells must be divided precisely in half prior to the formation of gametes. If not, the chromosome number would double every time an egg fused with a sperm at fertilization. Each generation would have twice the number of chromosomes as the previous generation, which obviously did not occur.

The discovery and confirmation of Mendel's work in 1900 had an important influence on the work of the cell biologists of the time. Whatever their physical nature, the carriers of the hereditary units would have to behave in a manner consistent with Mendelian principles. In 1903, Walter Sutton, a graduate student at Columbia University, published a paper that pointed directly to the chromosomes as the physical carriers of Mendel's genetic factors. Sutton followed the chromosomes during the formation of sperm cells in the grasshopper. He found that for each chromosome in one of the diploid body cells of a grasshopper, there was another chromosome in the cell that matched it perfectly in size and shape. In other words, the nucleus of diploid cells contained homologous pairs of chromosomes, a feature that correlated perfectly with the pairs of inheritable factors uncovered by Mendel. When he examined cells at the beginning of meiosis, Sutton found that the members of each pair of homologous chromosomes were joined together until they separated at the first meiotic division. This was the physical basis for Mendel's proposal that genes exist in pairs that segregate from each other upon formation of the gametes. Sutton concluded that genes are situated on chromosomes.

The reduction division observed by Sutton explained several of Mendel's other findings: Gametes could contain only one version (allele) of each gene; an equal number of gametes with each allele would be formed; and two gametes that united at fertilization would produce an individual with two alleles for each trait. But many questions remained unanswered. For instance, how were the genes organized within the chromosomes, and could the location of specific genes be determined?

at specific sites on particular chromosomes.

*T*hey have been called biological computers and living strings of pearls. Such comparisons diminish the importance of their two-pronged biological role, however, which is absolutely essential to life. In the first of these roles, they retain all of the information for the genetic traits of a species that are passed from generation to generation; in the second, they control the molecular activity in cells so that biological order prevails. They reside deep inside a cell's interior, dictating which inheritable features to manufacture and, in so doing, instructing an organism to develop into itself. They are the same cellular structures that the nineteenth-century European microscopists first saw "dancing" around within the nucleus just before a cell divided. They are the bearers of the genes. They are the chromosomes.

▼ ▼ ▼

THE CONCEPT OF LINKAGE GROUPS

Although Mendel provided convincing evidence that inherited traits were governed by discrete factors, his studies were totally unconcerned with the physical nature of these factors or their location within the organism. Mendel was able to carry out his entire research project without ever observing anything under a microscope. But genes are physical structures residing within the chromosomes of living cells. Mendel had shown that each trait was governed by two copies of a gene. These copies were located at the same locus on each homologous chromosome (Figure 13-1). If the individual were heterozygous for a particular trait, the homologous chromosomes would contain different alleles at the corresponding sites. If homozygous, the homologous chromosomes would contain identical alleles.

As clearly as Sutton saw the relationship between chromosomal behavior and Mendel's results on pea plants, he saw one glaring problem. Mendel had examined the inheritance of seven traits and found that the manner in which each of these characteristics was inherited was independent of the inheritance of the others. This formed the basis of his law of independent assortment. But if genes are packaged together on chromosomes, like pearls on a string, then each package of genes should be passed from each parent to its offspring, just as intact chromosomes are passed. In other words, genes on the same chromosome should act as if they are *linked* to one another; that is, they should be part of the same **linkage group.**

Let's reconsider the $9:3:3:1$ ratio from the F_1 dihybrid cross ($RrYy \times RrYy$) discussed on page 238. Suppose the genes for seed color and seed shape lay near each other on the same chromosome; the two dominant alleles should stay together during gamete formation, as should the two recessive alleles. This should reduce the number of possible gametes from four (RY, Ry, rY, and ry) to two (RY and ry). Such crosses should yield progeny showing two possible phenotypes instead of four.

How is it that all of Mendel's seven traits demonstrated independent assortment? Were they all on different linkage groups; that is, on different chromosomes? As it turned out, the garden pea has seven different pairs of homologous chromosomes, and each of the traits on which Mendel re-

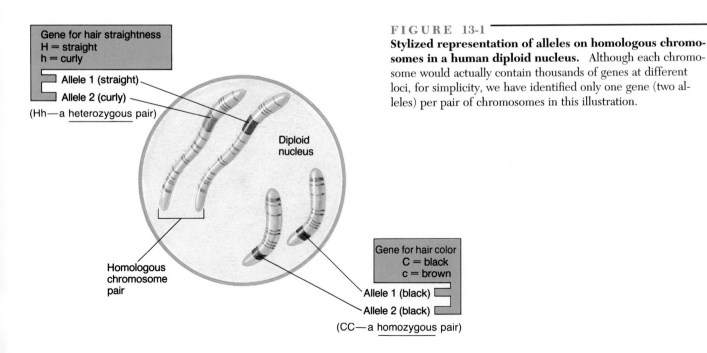

FIGURE 13-1

Stylized representation of alleles on homologous chromosomes in a human diploid nucleus. Although each chromosome would actually contain thousands of genes at different loci, for simplicity, we have identified only one gene (two alleles) per pair of chromosomes in this illustration.

Gene for hair straightness
H = straight
h = curly

Allele 1 (straight)
Allele 2 (curly)
(Hh—a heterozygous pair)

Diploid nucleus

Homologous chromosome pair

Gene for hair color
C = black
c = brown

Allele 1 (black)
Allele 2 (black)
(CC—a homozygous pair)

ported either occurred on a different chromosome or was so far apart on the same chromosome that it acted independently. Whether Mendel owed his findings to good fortune or simply to a lack of interest in any traits that did not fit his predictions is unclear.

Sutton's prophecy of linkage groups was soon transformed from speculation to fact. Within a couple of years, two traits (flower color and pollen shape) in sweet peas were shown to be linked. Other evidence would soon follow.

Because genes are physically connected to one another, like pearls on a strand, the various alleles on a single chromosome tend to remain linked together as they pass from cell to cell and from parents to offspring. (See CTQ #1.)

LESSONS FROM FRUIT FLIES

Genetic investigations on domestic plants or animals can be enormously time-consuming. Experimental matings yield results only when offspring appear, often a year or more after the parents mate. Analysis of subsequent generations requires additional years of study. Geneticists needed an organism that was easy to maintain, produced multiple generations every year, and was governed by the same "rules" of genetic inheritance exhibited by more complex organisms. They found just such an organism swarming around rotting fruit. Fruit flies (of the species *Drosophila melanogaster*) could easily be raised by the thousands in small

laboratory bottles, completing their entire reproductive cycle in 10 days. In 1909, the fruit fly seemed like the perfect organism to Thomas Hunt Morgan of Columbia University, as he began the research that initiated the "golden age of genetics," a period which has yet to end.

GENETIC MARKERS ALONG THE CHROMOSOME

There was one major disadvantage in beginning work with this insect: There was only one type of fly available, the *wild type* (Figure 13-2a). Inheritance in fruit flies, pea plants, or any other organism can be charted only if there is detectable evidence of genetic differences. Whereas Mendel had simply purchased his different genetic strains of peas from seed merchants, Morgan had to generate his own strains. Morgan expected that variants, or *mutants*, from the wild type might appear if he bred sufficient numbers of flies. Several months and thousands of flies later, he found his first mutant, a fruit fly with white eyes instead of the normal red ones.

By 1915, Morgan and his students had found 85 different mutants. As expected, some of these mutant alleles assorted independently; others did not (such as the linked characteristics illustrated on page 250). Taken together, the mutant alleles belonged to four different linkage groups, one of which contained very few genes (only two had been discovered by 1915). This discovery correlated perfectly with the finding of four different chromosomes in the cells of *Drosophila*, one of which was very small (Figure 13-2). There was little remaining doubt that genes resided on chromosomes.

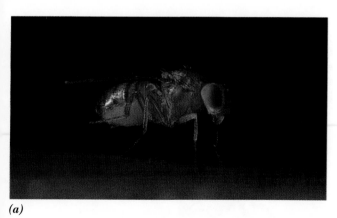

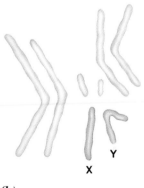

(a)   *(b)*

FIGURE 13-2

***Drosophila melanogaster*, the geneticist's ally.** *(a)* These petite fruit flies reproduce rapidly in bottles, dining on yeast that grows on a medium placed in the bottle. Each female produces hundreds of offspring, each of which can reproduce before they are 2 weeks old. Results from 1 year of *Drosophila* experimentation would require 30 years to obtain using an organism that reproduced only once a year. Note the red eyes and the other normal (wild-type) characteristics. *(b)* Fruit flies have four pairs of chromosomes, one of which is very small. The two dissimilar homologues are the sex chromosomes that determine gender.

CROSSING OVER: GENE EXCHANGE

Morgan and his students soon discovered that alleles in a linkage group occasionally segregated independently, as though on different chromosomes, ending up in separate gametes. The genes for body color and wing length, for example, reside on the same chromosome (Figure 13-3). Therefore, crossing a heterozygous gray-bodied, long-winged fly (*BbWw*) with a black-bodied, short-winged mate (*bbww*) should yield only two phenotypes, the same as the parents (Figure 13-4). Yet, contrary to these predictions, in about one-sixth of the progeny the two alleles in a pair appeared to have traded places with each other on homologous chromosomes. These offspring showed neither of the predicted phenotypes. Instead, they inherited a combination that would be possible only if the alleles were no longer connected, for example, a gray-bodied and short-winged fly (*Bbww*). In 1911, Morgan offered an explanation for this case of incomplete linkage—**crossing over.** Crossing over is the exchange of genetic segments between homologous chromosomes during meiosis and is described in Figure 13-5.

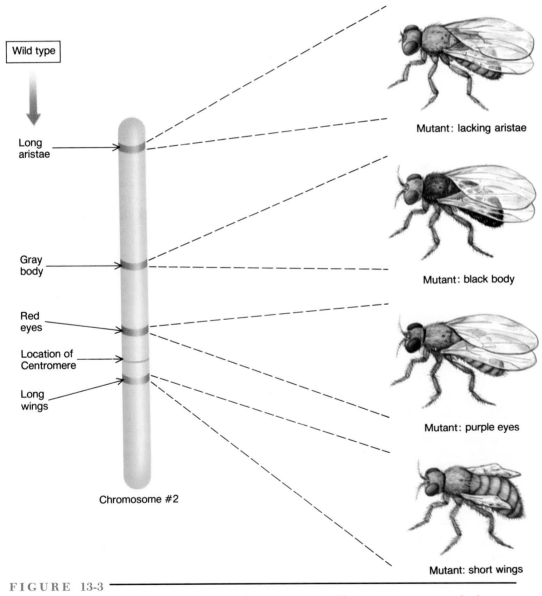

FIGURE 13-3

The mutation that produced each mutant phenotype pictured here occurs at a particular location on one of the chromosomes. Each wild-type characteristic is printed next to its location on the chromosome; the corresponding mutant characteristic is illustrated on the right.

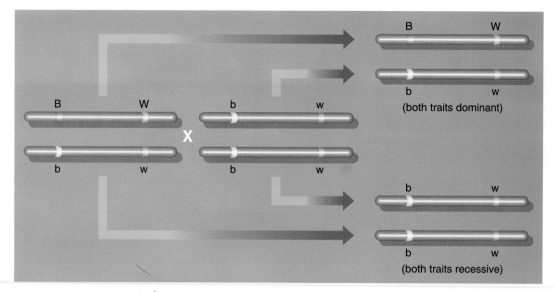

BbWw x bbww ➡ BbWw or bbww (*no offspring with combination of recessive and dominant traits*)

FIGURE 13-4

The pattern of inheritance if linkage were 100 percent complete. If two alleles on the same chromosome always remained together, as predicted by Mendel's law of independent assortment, then all of the offspring in the case depicted here would resemble one or the other parent. They would be either gray-bodied and long-winged (*BbWw*) *or* black-bodied and short-winged (*bbww*).

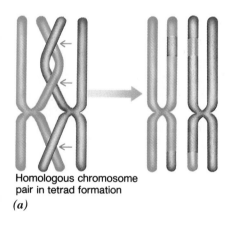

Homologous chromosome pair in tetrad formation

(a)

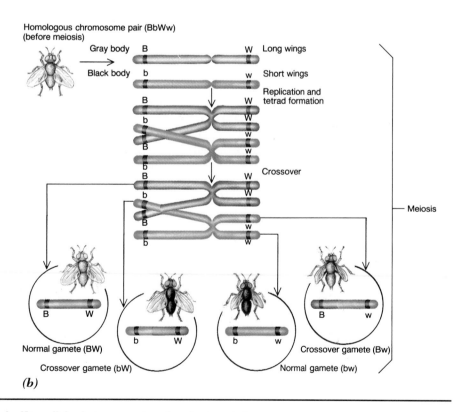

(b)

FIGURE 13-5

Crossing over provides the mechanism for reshuffling alleles between maternal and paternal chromosomes. *(a)* Tetrad formation during synapsis, showing three possible crossover intersections (*chiasmata,* indicated by red arrows). *(b)* Simplified representation of a single crossover in a *Drosophila* heterozygote (*BbWw*) at chromosome number 2, and the resulting gametes. If one of the crossover gametes participates in fertilization, the offspring will have a combination of alleles that was not present in either parent.

MAPPING THE CHROMOSOME

Studie on fruit flies also revealed that specific genes are arranged along a particular chromosome in a predictable linear sequence, much like towns along a highway. All individuals of a given species have chromosomes with the same sequence of genes. Since scientists cannot "drive" along a chromosome, recording the locations of genes and the distances that separate them, they had to discover an indirect way of determining gene location. Crossing over supplied this indirect approach.

We noted earlier that genes on a chromosome are like pearls along a strand. Consider what would happen if you were to pick up a linear strand of pearls, close your eyes, and randomly cut the strand once with a scissors. It is more likely that two pearls located close to each other will remain *together* on one of the cut pieces, than would two pearls located much farther apart. The same principle applies to crossing over, which involves *random* physical breakage of the chromosome. The more room between two sites on the chromosome for breakage to occur, the more likely a break will occur between those two sites. Since alleles of two genes at opposite ends of a chromosome have more room between them than do alleles of two genes that lie close to each other, the former are more likely to be separated during crossing over than are the latter. In other words, the crossover frequency between two genes reflects their relative distance from each other—the greater the distance, the greater the frequency. This information is used to construct genetic maps that show relative distances between genes on a chromosome. In *Drosophila,* for example, alleles of the linked genes for body color and wing length (Figure 13-3) become observably detached from each other (Figure 13-5) in about 180 of every 1,000 offspring (180/1,000 = 0.18 crossover frequency). To geneticists, 1 percent crossing is equated to one *map unit.* The alleles w and b are thus 18 map units apart (180/1,000 × 100 = 18% = 18 map units).

Mapping not only reveals relative distances between different genes but also delineates the sequence of those genes along a chromosome. We have already indicated that the alleles for body color (b) and wing length (w) show a crossover frequency of 18 percent. Let's consider a third "genetic marker"—pr, the allele for purple eyes—that shares the same chromosome as b and w (Figure 13-3). Mating experiments show that the crossover frequency between w and pr is 12 percent and the crossover frequency between b and pr is 6 percent. Only one sequence of these three genes along the chromosome is consistent with these data. That sequence is shown in the following genetic map:

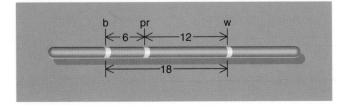

Using this technique, hundreds of gene loci have been mapped on this one fruit fly chromosome alone. The precise positioning of specific genes along specific chromosomes of a species represents another example of the high degree of order that exists in biological systems. Consider what would happen if the linear arrangement of genes differed from one individual to the next. As you know by now, diploid organisms contain homologous chromosomes derived from two different parents. If these homologous chromosomes had a different arrangement of genes, then crossing over between homologs would create chromosomes that were lacking certain genes. Offspring inheriting these chromosomes would be genetically deficient.

Crossover frequencies have been used to map genes on a diverse variety of organisms, from yeast to mice. As we will see in Chapter 17, genes on human chromosomes have also been extensively mapped—using very different techniques. Each month or so, researchers announce the discovery of the chromosomal location of a gene responsible for a different inherited disease, such as cystic fibrosis or muscular dystrophy. With each announcement comes the hope that the underlying cause of the disease will be determined and a new, more effective treatment forthcoming.

THE USE OF MUTAGENIC AGENTS

During the early period of genetics, the search for mutants was a slow and tedious procedure that depended on the *spontaneous* appearance of altered genes. In 1927, Herman Muller, a former undergraduate in Morgan's laboratory, discovered that flies subjected to sublethal doses of X-rays exhibited 100 times the spontaneous mutation frequency as did their counterparts who were not exposed to X-rays. This finding had several important consequences. On the practical side, the use of *mutagenic* (mutation-causing) agents, such as X-rays, provided a great increase in the number of mutants available for research using a variety of laboratory organisms, ranging from bacteria and yeast to mice. This discovery also pointed out the hazard of the increasing use of radiation in the industrial and medical fields, since most mutations are harmful. The work earned Muller the Nobel Prize in 1949.

GIANT CHROMOSOMES

In 1933, Theophilus Painter of the University of Texas rediscovered a phenomenon that had been observed many years earlier: giant chromosomes in certain insect cells (Figure 13-6). For example, cells from the larval salivary gland of *Drosophila* contain chromosomes that are about 100 times thicker than are the chromosomes found in most other *Drosophila* cells. During larval development, these salivary gland cells cease dividing, yet the cells keep growing in size. DNA replication continues, providing the additional genetic material needed to support the high level of secretory activity of these enormous cells. The duplicated DNA strands remain attached to each other in perfect side-by-

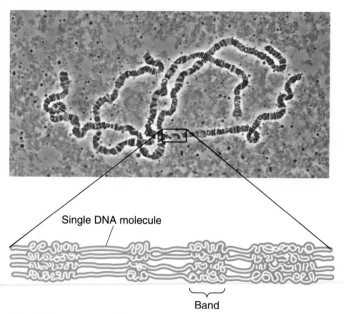

FIGURE 13-6

Giant chromosomes of the fruit fly—genes on display.
These giant polytene chromosomes from the salivary gland of
larval fruit flies show thousands of distinct, darkly staining
bands. Many of the bands have been identified as the loci of
particular genes. The inset shows how polytene chromosomes
consist of a number of individual DNA strands. The bands on
the chromosomes correspond to those sites where the DNA is
more tightly compacted.

side alignment (see inset of Figure 13-6), producing giant
chromosomes with as many as 1,054 more copies of the
DNA than are found in normal chromosomes. These unusual *polytene chromosomes,* as they are called, are rich in
visual detail, revealing about 6,000 dark bands when stained
and examined microscopically.

Painter realized that these giant chromosomes might
provide the opportunity to obtain a visual portrait of the
chromosomes of a species that had been mapped by breeding experiments over the past 20 years. Could a correlation
be made between the genetic maps obtained by crossing-
over analyses and the chromosomal banding patterns?
Painter and others found that it could. Many of the bands
correlated with specific genes, providing visual confirmation of the entire mapping procedure.

**Over the past 90 years, one organism, the fruit fly, has re-
vealed more about the chromosomal organization of genes
and their role in directing biological activities than has any
other organism. The primary value of fruit flies to biologists
stems from the great variety of mutants that have been isolated. Each new genetic mutant tells us something about the
activity of a normal gene product whose existence would
otherwise have gone undetected. (See CTQ #3.)**

SEX AND INHERITANCE

Genetically, humans are a lot like flies and, therefore,
most of the genetic breakthroughs revealed by *Drosophila*
apply to humans as well as to most other organisms. We
even resemble fruit flies in the fundamentals of sex; an
animal's gender is determined by specific chromosomes.

GENDER DETERMINATION

The chromosomes of males and females of the same animal
species typically look identical, with the exception of one
chromosomal pair. This pair is called the **sex chromosomes** to distinguish it from the other pairs of
chromosomes—the **autosomes**—which are identical in
both genders. In both fruit flies and mammals, the female's
cells harbor a pair of identical sex chromosomes, called **X
chromosomes,** whereas the male's cells possess one X
chromosome and a smaller **Y chromosome** (illustrated for
Drosophila in Figure 13-2 and in humans in Figure 10-3).

During gamete formation in humans and other mammals, the X and Y chromosomes synapse and then separate
at the first meiotic division. Accordingly, half the sperm
produced by males will contain the Y chromosome, and the
other half will possess the X chromosome. Since all gametes
produced by an XX female will contain an X chromosome,
the sex of each offspring depends on whether the egg is
fertilized by an X-bearing sperm (forming an XX *female*
offspring) or a Y-bearing sperm (forming an XY *male* offspring). Gender in humans is thus determined by the male
(Figure 13-7). King Henry VIII of England, who disposed
of many wives because they failed to bear him a son, should
have pointed his lethal finger squarely at himself. It was *his*
royal gametes that determined the gender of his progeny!

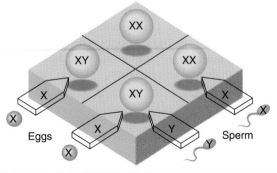

FIGURE 13-7

Sex determination in *Drosophila*. As in humans, in *Drosophila,* males have an X and a Y chromosome, while females
have two X chromosomes. Males produce two types of gametes
in equal numbers, while females produce only one. As depicted
in this Punnett square, the sex of each offspring is determined
by the chromosome composition of the fertilizing sperm.

◁ B I O L I N E ▷
The Fish That Changes Sex

In vertebrates, gender is generally a biologically inflexible commitment: An individual develops into either a male or a female as dictated by the sex chromosomes acquired from one's parents. Yet, even among vertebrates, there are organisms that can reverse their sexual commitment. The Australian cleaner fish (Figure 1), a small animal that sets up "cleaning stations" to which larger fishes come for parasite removal, can change its gender in response to environmental demands. Most male cleaner fish travel alone rather than with a school. Except for a single male, schools of cleaner fish are comprised entirely of females. Although it might seem logical to conclude that maleness engenders solo travel, it is actually the other way around: Being alone fosters maleness. A cleaner fish that develops away from a school *becomes* a male, whereas the same fish developing in a school would have become a female.

FIGURE 1

The small Australian wrasse (cleaner fish) is seen on a much larger grouper.

But what of the one male in the school—the one with the harem? He may have developed as a solo fish and then found a school in need of his spermatogenic services. But there is another way a school may acquire a male. If the male in a school dies (or is removed experimentally), one of the females, the one at the top of a behavioral hierarchy that exists in each school, becomes uncharacteristically aggressive and takes over the behavioral role of the missing male. She begins to develop male gonads, and within a few weeks, the female becomes a reproductively competent male, indistinguishable from other males. Furthermore, the sex change is reversible. If a fully developed male enters the school during the sexual transition, the almost-male fish developmentally backpedals, once again assuming the biological and behavioral role of a female.

Not all organisms follow the mammalian pattern of sex determination. In some animals, most notably birds, the opposite pattern is found: The female's cells have an X and a Y chromosome, while the male's cells have two Xs. An exception to this rule of a strict relation between sex and chromosomes is discussed in the Bioline: The Fish That Changes Sex. Although some plants possess sex chromosomes and gender distinctions between individuals, most have only autosomes; consequently, each individual produces both male and female parts.

SEX LINKAGE

For fruit flies and humans alike, there are hundreds of genes on the X chromosome that have no counterpart on the smaller Y chromosome. Most of these genes have nothing to do with determining gender, but their effect on phenotype usually *depends on* gender. For example, in females, a recessive allele on one X chromosome will be masked (and not expressed) if a dominant counterpart resides on the other X chromosome. In males, it only takes one recessive allele on the single X chromosome to determine the individual's phenotype since there is no corresponding allele on the Y chromosome. Inherited characteristics determined by genes that reside on the X chromosome are called **X-linked characteristics.**

So far, some 200 human X-linked characteristics have been described, many of which produce disorders that are found almost exclusively in men. These include a type of heart-valve defect (*mitral stenosis*), a particular form of mental retardation, several optical and hearing impairments, muscular dystrophy, and red-green colorblindness (Figure 13-8).

One X-linked recessive disorder has altered the course of history. The disease is **hemophilia,** or "bleeder's disease," a genetic disorder characterized by the inability to produce a clotting factor needed to halt blood flow quickly following an injury. Nearly all hemophiliacs are males. Although females can inherit two recessive alleles for hemophilia, this occurrence is extremely rare. In general, women who have acquired the rare defective allele are heterozygous **carriers** for the disease. The phenotype of a carrier

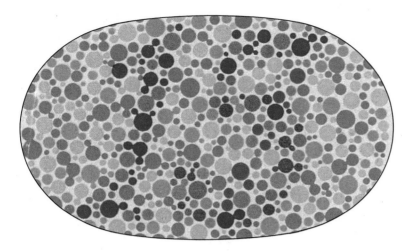

FIGURE 13-8

Distinctively male. There are hundreds of X-linked recessive characteristics that appear in males but rarely in females, who would have to have the recessive allele on both X chromosomes. Red-green colorblindness is such a trait. Males with this defect in vision cannot see the number "15" in the above pattern.

discloses no suggestion of the presence of the recessive allele because the normal allele on the other X chromosome directs formation of enough blood-clotting factor to assure a normal phenotype. Since the Y chromosome has no allele for producing the clotting factor, a boy who inherits the defective allele from his heterozygous mother will develop hemophilia.

The mutant hemophilia gene in the royal family of England can be traced to Queen Victoria, who transmitted the recessive allele to four of her children—a son (who eventually died of hemophilia) and three daughters (who were heterozygous carriers). The disease plagued several royal families when male descendants married Victoria's daughters; their descendants were, in turn, afflicted, in some cases killing the heir to the throne. The son of Nicholas, the last Russian czar, was a hemophiliac, having acquired the disease from his mother, Alexandra, a granddaughter of Queen Victoria (Figure 13-9). Desperate to save their son, Nicholas and Alexandra put their faith in the infamous monk Rasputin, who convinced them he was able to cure the young Alexis. It has been argued that Rasputin's influence on the monarchy contributed to the overthrow of the czar during the Russian Revolution. In case you are wondering about members of the current royal family of England, they are descended from Edward VII, one of Victoria's sons who did *not* inherit her "tainted" X chromosome.

The path of inheritance of specific traits is best revealed through a diagram called a **pedigree,** such as that depicted in Figure 13-10. Pedigrees reveal that X-linked recessive characteristics tend to skip a generation. When a male with an X-linked recessive characteristic has children, none of them shows evidence of the trait (except for the unlikely circumstance where his mate also carries the defective allele) because all male progeny receive a normal allele from the mother. All daughters will be carriers. If a carrier and a normal mate conceive offspring, the chance that a male will inherit the sex-linked characteristic is 50

percent, versus 0 percent for female offspring; half of the daughters will be carriers.

Although nearly all known sex-linked disorders are associated with a gene located on the X chromosome, a small number of genes are carried on the Y chromosome and determine **Y-linked characteristics.** In mammals, male reproductive anatomy is Y-linked, the Y chromosome carrying genetic information that instructs the embryonic gonads to develop into testes. The testes then produce the hormone testosterone which, in turn, directs the embryo to develop other male reproductive structures. In the absence of a Y chromosome, a female will develop (Chapter 17).

FIGURE 13-9

The last Russian royal family. Alexis, great-grandson of Queen Victoria, is seen in front of his mother in this family photo taken in 1914.

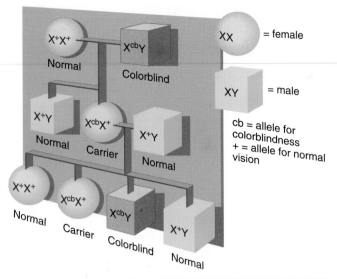

FIGURE 13-10

The "generation gap." Pedigrees show the tendency of X-linked traits—in this case, colorblindness—to skip a generation, producing a phenotypic gap: a generation of normal-visioned men and normal-visioned, carrier women, but no individuals showing the abnormal characteristic. Hemophilia and other X-linked traits follow this basic inheritance.

BALANCING GENE DOSAGE

The amount of gene product manufactured by a cell is influenced by the "gene dosage"—the number of copies of the allele that directs its synthesis. Increasing gene dosage usually intensifies the expression of the trait. It would seem, then, that a mammalian female who is homozygous for a trait on the X chromosome would express that trait with twice the intensity as would a male who has only one copy of the gene. Yet, this does not happen; most X-linked traits are expressed more or less equally in both sexes. This puzzling observation was explained by Mary Lyon, a British geneticist, who suggested that one of the X chromosomes in each cell of females was *inactivated* during embryonic development. The inactivated chromosome forms a dense mass in the interphase nuclei of female cells (Figure 13-11*a*), which is absent in males. Thus, like the cells in males, each cell in a female has only one *functional* copy of each X-linked gene.

This phenomenon may appear to contradict what you have just learned about sex-linked traits. If the cells of females have only one active X chromosome, why don't females show X-linked diseases such as hemophilia? They would if the same X chromosome were inactivated in all of the female's cells. But this isn't the case. Inactivation occurs randomly in the cells of early embryos. In about half the cells, one X remains active, while in the other cells, the homologous X prevails, creating two genetically distinct types of cells scattered randomly throughout the embryo.

(a)

(b)

FIGURE 13-11

Chromosome in retirement. (*a*) The inactivated X chromosome in the nucleus of a woman's cells appears as a darkly staining structure, named a *Barr body* after its discoverer. (*b*) Random inactivation of either X chromosome in different cells during early embryonic development creates a mosaic of tissue patches that descend from each cell after inactivation. These patches are visually evident in calico cats, which are heterozygous for fur color—the allele for black coat color residing on one X chromosome, and that for yellow on the other X. This explains why male calico cats are virtually non-existent since all cells in the male will have the same color allele.

The millions of cells that descend from each of these early embryonic cells form a patch of tissue that may be genetically and phenotypically different from adjacent tissue patches in which the other X chromosome directs the action. As a result, every female is a *mosaic* of genetically different tissue patches, some of which may have an active dominant allele, and others that may have an active recessive allele for a particular gene. X-chromosome inactivation appears to occur in all mammals and is reflected in the patchwork coloration of the fur of some mammals, including the calico cat (Figure 13-11*b*). Pigmentation genes in humans are not located on the X chromosome, hence the absence of "calico women." Mosaicism due to X inactivation can be demonstrated in women, however. For example, if a narrow beam of red or green light is shone into the eyes of a woman who is a heterozygous carrier for red-green colorblindness, patches of retinal cells with defective color vision can be found interspersed among patches with normal vision.

Sex and chromosomes are closely related. Most animal species have distinct male and female individuals whose gender and sexual characteristics are determined by the chromosomes they inherit. Many nonsexual genetic traits are also carried on the sex chromosomes, which produces striking differences between males and females in the frequency of many genetic characteristics. (See CTQ #5.)

ABERRANT CHROMOSOMES

In addition to mutations that alter the information content of a single gene, chromosomes may be subjected to more extensive alterations that typically occur during cell division. Pieces of a chromosome may be lost or exchanged between nonhomologous chromosomes, or extra segments may appear (Table 13-1). Since these **chromosomal**

TABLE 13-1
MODIFICATIONS IN CHROMOSOMAL STRUCTURE

Type of Alteration	Example of How Change May Occur	Some Possible Effects ● Favorable ○ Harmful
Deletion		Rarely favorable; perhaps elimination of detrimental genes
		Loss of critical genes is lethal; disrupts chromosome separation during meiosis
Duplication		Provides raw material for evolution of new proteins as part of a family of related proteins
		May interfere with chromosome separation; may disrupt gene function if duplication occurs within a gene
Inversion		Increases genetic diversity by changing gene positions
		Reduced fertility; loss of control of gene expression
Translocation		Enormous genetic changes may generate rapid evolutionary advances
		May activate genes that cause cancer; reduce fertility; may result in gain or loss of part or whole chromosome

aberrations follow chromosomal breakage, their incidence is increased by exposure to agents that damage DNA, such as a viral infection, X-rays, or exposure to chemicals that can break the DNA backbone.

The consequence of a chromosomal aberration depends on the circumstances involved. If the aberration occurs during a mitotic cell division, the effects on the organism are generally minimal since only a few cells of the body will usually be affected. On rare occasions, however, a daughter cell that inherits the aberration may be transformed into a malignant cell, which can grow into a cancerous tumor.

▐▶ Many chromosome aberrations occur during meiosis, particularly as a result of abnormal crossing over, and can be transmitted to the next generation. When an aberrant chromosome is inherited through a gamete, all cells of the offspring will have the aberrant chromosome, and the individual generally does not survive embryonic development. However, like single-gene mutations, chromosomal aberrations may occasionally confer advantageous traits. These structural modifications in chromosomes produce large-scale genetic changes that may propel evolution forward in giant steps. There are several types of chromosomal aberrations (Table 13-1), including:

- *Deletions.* A **deletion** occurs when a portion of a chromosome is lost. Forfeiting a portion of a chromosome usually results in a loss of critical genes, producing severe consequences. The first correlation between a human disorder and a chromosomal deletion was made in 1963 by Jerome Lejeune, the same French geneticist who had earlier discovered the chromosomal basis of Down syndrome (Chapter 11). Lejeune discovered that a baby born with a variety of facial malformations and abnormal development of the lar-

ynx (voice box) was missing a portion of chromosome number 5. A defect in the larynx caused the infant's cry to resemble the sound of a suffering cat. Consequently, the scientists named the disorder *cri du chat syndrome,* meaning cry-of-the-cat syndrome.

- *Duplications.* A **duplication** occurs when a portion of a chromosome is repeated. Duplications have played a very important role in biological evolution. Many proteins are present as families consisting of a variety of closely related molecules. Consider the adult hemoglobin molecule illustrated in Figure 4-15*d*, which consists of two alpha-globin chains and two beta-globin chains. Examination of the amino acid sequence of alpha- and beta-globin chains reveals marked similarities, indicating that they both evolved from a single globin polypeptide present in an ancient ancestor. The human globin family contains several other members that form the hemoglobin molecules of embryos and infants. The amino acid sequences of these globins are also closely related to each other and to the alpha and beta forms.

The first step in the evolution of a protein family is thought to be the duplication of the gene. Over time, the two copies of the gene evolve in different directions, generating polypeptides with different, but related, amino acid sequences. In some instances, the polypeptides encoded by a given family of genes have evolved divergent functions, even though their amino acid sequences still show their ancestral relationship. Growth hormone and prolactin (a hormone that stimulates milk production), for example, are pituitary hormones that evoke completely different responses in different target cells, yet their amino acid sequences indicate that they have evolved from a common ancestral gene.

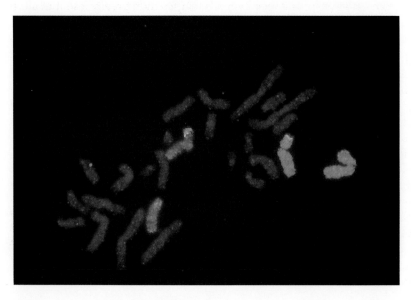

FIGURE 13-12

Translocation. Using special fluorescent stains, investigators can demonstrate when a piece of one chromosome breaks away and becomes attached to another chromosome. In this case, chromosome number 12 (green) and chromosome number 7 (red) have exchanged pieces.

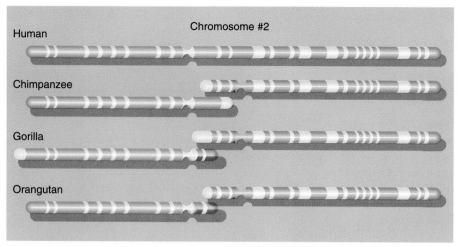

FIGURE 13-13

Translocation and evolution. Translocation may have played an important role in the evolution of humans from their apelike ancestors. If the only two ape chromosomes that have no counterpart in humans are "fused," they match human chromosome number 2, band for band.

Inversions. Sometimes a chromosome is broken in two places, and the segment between the breaks becomes resealed into the chromosome in reverse order. This aberration is called an **inversion** and can interfere with the way that neighboring genes are expressed. It is possible, for example, for an inversion to disrupt the orderly expression of an oncogene (page 202), leading to the formation of a malignant tumor.

Translocations. When all or a piece of one chromosome becomes attached to a nonhomologous chromosome, the aberration is called a **translocation** (Figure 13-12). Certain translocations increase the incidence of cancer in humans. The best-studied example is the *Philadelphia chromosome,* which is found in individuals with certain forms of leukemia, a malignancy of the white blood cells. The Philadelphia chromosome, which is named for the city in which it was discovered in 1960, is a shortened version of human chromosome number 22. For years, it was believed that the missing segment represented a simple deletion, but with improved techniques for observing chromosomes, the missing genetic piece was found translocated to another chromosome (number 9). This change in the positions of genes causes the cell to become malignant. Chromosome number 9 contains a gene that codes for a protein kinase (page 129) which plays a role in cell proliferation. As a result of translocation, one small end of this protein is replaced by about 600 extra amino acids encoded by a gene carried on the translocated piece of chromosome number 22. This new, greatly enlarged protein apparently retains the catalytic activity of the original version but is no longer subject to the cell's normal regulatory mechanisms. In other words, it cannot be turned off. As a result, the cell is no longer subject to normal regulatory mechanisms that prevent one type of cell from overgrowing the body and it continues to proliferate. The result is a cancerous transformation. Several other types of cancer are also associated with chromosomal aberrations (see The Human Perspective: Chromosome Aberrations and Cancer).

Translocations also play an important role in evolution, generating large-scale changes that may be pivotal in the branching of separate evolutionary lines from a common ancestor. Such a genetic incident probably happened during our own recent evolutionary history. A comparison of the 23 pairs of chromosomes in human cells to the 24 pairs of chromosomes in the cells of chimpanzees, gorillas, and orangutans reveals a striking similarity (Figure 13-13). But how did humans "lose" the twenty-fourth pair of chromosomes? A close examination of human chromosome number 2 reveals that we didn't really lose it at all. If the two ape chromosomes that have no counterpart in humans were to fuse together to form a single chromosome, they would form a perfect match, band for band, with human chromosome number 2 (Figure 13-13). At some point during the evolution of humans, an entire chromosome was translocated to another, creating a single fused chromosome, reducing the haploid number from 24 to 23.

Chromosomes are the repository of an organism's genetic endowment, and different parts of a chromosome contains different information. The maintenance of a well-ordered, smoothly functioning organism requires virtually all of its inherited information. It is not surprising, therefore, that alterations in a cell or organism's chromosomal composition almost invariably lead to a disruption of normal activities. (See CTQ #6.)

◁ THE HUMAN PERSPECTIVE ▷
Chromosome Aberrations and Cancer

Malignant cells can often be identified by the abnormal structure of their chromosomes. In some cases, as with the Philadelphia chromosome, alterations in chromosome structure are thought to play a role in the original transformation of a normal cell to a cancerous state. Careful analysis of the banding patterns of chromosomes (see Figure 10-3) from human malignant cells indicates that the cancer-causing aberrations are not random but tend to occur in approximately 100 specific bands. It is presumed that these bands contain genes that play a special role in the formation of malignant tumors.

We have already discussed one group of cancer-causing genes, called *oncogenes.* When *activated* ("turned on"), oncogenes cause cells to proliferate out of control, leading to tumors (Chapter 10). More recently, we have discovered a different group of genes, called **tumor-suppressor genes** (or anti-oncogenes), whose normal role is to *block* the formation of cancerous cells. Both copies of a tumor-suppressor gene in a cell must be knocked out in order to eliminate their protective function, allowing the transformation of the cell to the malignant state.

Several examples of this phenomenon have been documented. For instance, a rare childhood cancer of the eye, called *retinoblastoma,* occurs in a small number of families. Examination of normal cells from children suffering from retinoblastoma indicated that one of the thirteenth pair of homologous chromosomes was missing a small piece from the interior portion of the chromosome. The missing piece normally contains the *retinoblastoma gene* which, when altered or absent, predisposes the child to the eye cancer. The cancer itself develops only in those individuals where a retinal cell accidently loses the second copy of the gene on the homologous chromosome, leaving the individual with a cell completely lacking the tumor-suppressive capacity. Recent evidence suggests that the retinoblastoma gene directs formation of a protein that inhibits cell division; the protein loses this inhibitory function when it is phosphorylated. Tumor-suppressor genes have been implicated in a variety of common malignancies and are of widespread importance.

The discovery of tumor-suppressor genes has important clinical implications. Rather than using toxic chemotherapeutic agents to kill malignant cells, it might be possible to treat cells with the protein encoded by the tumor-suppressor gene, a protein that might stop tumor growth, as it normally does in the body. Identification of the products and functions of tumor-suppressor genes is one of the foremost goals of cancer biologists.

POLYPLOIDY

We saw in Chapter 11 how the presence of an extra chromosome due to a failure of chromosome separation during meiosis can cause abnormal development, leading to Down syndrome. On occasion, an entire set of chromosomes fails to separate during meiosis, resulting in a zygote with an increased number of chromosomal sets—a condition called **polyploidy.** In most higher animal species, polyploidy is lethal. In humans, for example, a high percentage of miscarriages are due to a polyploid fetus. Plants are much more tolerant of polyploidy. In fact, half of all known plant species are polyploid or contain some polyploid tissue. It is important that polyploid plants have an even number of chromosomal sets (4N, 6N, etc.) so that every chromosome has a chance to pair with a homologue during meiosis. As a result, gametes receive equal shares of chromosomes, and the plant retains fertility. Plants that have odd-numbered sets (such as triploid—3N—plants) have meiotic problems and tend to be sterile. Such species may still survive by asexual reproduction, as discussed below.

▶ Polyploidy in plants may confer advantages over their diploid counterparts, especially when the extra set of chromosomes is the result of hybridization of two closely related plant species. The polyploid hybrid is often hardier than either of the diploid species since it combines the adaptations of both species (for example, disease resistance of one and drought resistance of the other). In addition to enhanced hardiness, polyploid plants have several other adaptive advantages over their diploid relatives, such as increased yields and production of fruit with special qualities. Many of our crop plants, including wheat, strawberries, and potatoes, are polyploid. The development of new polyploid strains of wheat, rice, and corn has helped raise the worldwide level of food production. Some seedless fruits, including bananas, are produced by triploid plants. Although these plants are usually sterile—hence, their inability to make seeds—they can be propagated vegetatively by cuttings or grafts. Polyploids of pears, apples, and grapes produce giant fruit that show commercial promise.

> **Some organisms may possess more than a diploid number (two sets) of chromosomes. In some cases, these extra sets disrupt the normal chromosomal balance in a cell and prove disastrous. In many plants, however, extra sets of chromosomes provide an opportunity for enhanced genetic variation and genetic vigor. (See CTQ #8.)**

REEXAMINING THE THEMES

Relationship between Form and Function

Chromosomes are the physical carriers of the genes. The structure of chromosomes correlates with their genetic function and helps explain the basis of the principles of inheritance discovered by Mendel and others. Chromosomes provide a discrete location for each gene and a mechanism for transmitting copies of genes to offspring. The existence of chromosomes as pairs of homologs explains why each body cell has two copies of each allele. The separation of homologous chromosomes during meiosis accounts for the presence of only one of a pair of alleles in a gamete.

Biological Order, Regulation, and Homeostasis

The chromosomes of a higher plant or animal may contain tens of thousands of genes, yet each particular gene is predictably located at a particular site (locus) on that chromosome in virtually every member of that species. This ordered arrangement allows the positions of genes to be mapped by crossing over and other techniques. If different individuals of a species had chromosomes with different arrangements of genes, then crossing over between homologous chromosomes would result in the loss of genes in the gametes and death of the ensuing offspring. Similarly, the loss of parts of a chromosome, or rearrangements within or between chromosomes, disrupts the orderly expression of genes and can lead to biological disorders, including abnormal development and cancer.

Unity within Diversity

The chromosomes of all eukaryotic organisms are constructed in a similar manner and consist of linear arrays of genes comprising linkage groups. Crossing over occurs during meiosis in all sexually reproducing diploid organisms; therefore, crossover frequencies can be used to map the relative distances between genes in a wide variety of organisms. While chromosomes have the same internal construction in all eukaryotes, their number, size, and shape vary greatly among diverse organisms. Fruit flies, for example, have a haploid number of 4 chromosomes, whereas humans have 23. Both fruit flies and humans have one pair of sex chromosomes, and in both organisms, X-linked recessive characteristics are passed from mothers to male offspring.

Evolution and Adaptation

The duplication of genes within a chromosome is one of the most important cellular processes underlying the process of evolution. Once a gene is duplicated, the two copies of the gene tend to evolve in different directions, often generating distinct proteins with related amino acid sequences. In some cases, the two proteins retain similar functions, as occurs among the various polypeptides that can form a hemoglobin molecule. In other cases, the two proteins evolve different functions, as occurs with the pituitary hormones prolactin and growth hormone.

Because evolution depends on changes in genes, and genes are carried on chromosomes, changes in chromosomes can provide a direct visual display of the evolutionary process. Although mammalian chromosomes don't exist in a giant polytenic form, they can be stained to reveal banding patterns that can be used to identify chromosomes and to provide an indication of genetic relatedness, as between apes and humans, whose chromosomes are nearly identical.

SYNOPSIS

Genes reside on chromosomes in fixed linear orders. The physically connected set of alleles on the same chromosome constitutes a linkage group—a group of alleles that do not assort independently.

Linked genes don't always stay linked. Crossing over during meiosis shuffles alleles between homologous chromosomes, creating new combinations that are not found in the parents. Crossing over increases genetic diversity beyond that resulting from independent assortment.

The positions of genes on a chromosome can be mapped by analysis of crossing over data. The farther apart two genes are from each other on a chromosome, the more likely they will be separated by crossing over between the two. Genetic maps of chromosomal loci can be constructed by determining the frequency with which alleles become "unlinked" during meiosis.

Gender is determined by sex chromosomes in most animals. In humans (and fruit flies), females possess X chromosomes (XX), whereas males possess an X and a Y chromosome (XY). X-linked recessive characteristics, such as hemophilia and muscular dystrophy, are normally expressed when males inherit a recessive allele on the X chromosome from their mother. In female mammals, one of the X chromosomes is inactivated in each cell. Inactivation of one of the X chromosomes occurs randomly during embryonic development so the cells of adult females are a mosaic of X-linked alleles.

Chromosomes are subject to alterations that change their genetic content. Chromosomal aberrations include the loss of a segment (a deletion), the acquisition of extra genes (a duplication), a reversal of the order of genes in part of a chromosome (an inversion), and the transfer of a chromosome (or a part of a chromosome) to a nonhomologous chromosome (a translocation). Each of these aberrations can have serious effects on the cell and even the whole organism.

Key Terms

linkage group (p. 248)
crossing over (p. 250)
sex chromosome (p. 253)
autosome (p. 253)
X chromosome (p. 253)
Y chromosome (p. 253)

X-linked characteristic (p. 254)
carrier (p. 254)
pedigree (p. 255)
Y-linked characteristic (p. 255)
chromosomal aberration (p. 258)
deletion (p. 258)

duplication (p. 258)
inversion (p. 259)
translocation (p. 259)
tumor-suppressor gene (p. 260)
polyploidy (p. 260)

Review Questions

1. The traits of one well-studied organism fall into seven distinct linkage groups. How many chromosomes are in the *somatic* cells of this organism?

2. Draw the processes of synapsis, crossing over, and segregation, and indicate the severity of the consequences if different individuals of a species were to have a different order of genes on their chromosomes.

3. Why do X-linked traits tend to skip a generation? Under what circumstances would this *not* happen? (For instance, how could a colorblind man have a child that is also colorblind?)

4. What is the likelihood that Nicholas would acquire hemophilia, knowing that his grandmother, Queen Victoria, was a carrier of the mutant allele?

5. If you were offered a calico kitten, why would you not have to ask whether it is a male or female?

Critical Thinking Questions

1. Sutton was able to provide visual evidence for Mendel's law of segregation. Why would it have been impossible for him visually to confirm or refute Mendel's law of independent assortment?

2. Draw a simple map showing the gene order and relative distances among the genes X, Y, and Z, using the data below:

Crossover Frequency	Between These Genes
36%	X–Z
10%	Y–Z
26%	X–Y

3. Make a list of reasons why the fruit fly has been an ideal organism to use in genetic studies.

4. Alleles on opposite ends of a chromosome are so likely to be separated by crossing over between them that they segregate independently. How would we ever know that these two genes belong to the same linkage group? (Hint: If A is linked to B, and B is linked to C, then A is linked to C.)

5. Below is a map of the human X-chromosome, showing the location of known genes. Why are the traits associated with these genes said to be "sex-linked" even though most of them have nothing to do with determining the sex of individuals? Why do such traits often skip a generation? What circumstances must occur for the traits not to skip a generation?

6. Genes are often compared to beads on a string; each bead is independent of the other beads. How do results of studies on translocations of chromosomes demonstrate that genes are not independent of other genes?

7. Why do you suppose the Philadelphia chromosome causes only one type of cancer—leukemia—rather than other types as well, such as colon cancer or skin cancer?

8. In a diagram, show how a plant with a 3N chromosome number could arise? 4N number? Why are polyploids more likely to survive mitosis than meiosis? Which would you expect to have more chance of reproducing, the 3N or the 4N organism? Why? Can you think of at least one reason why polyploids are scarcer among animals than among plants?

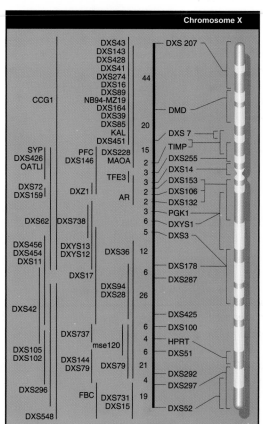

Source: *National Institutes of Health Human Genome; New York Times,* Tuesday, October 6, 1992.

Additional Readings

Edey, M. A., and D. C. Johanson. 1989. *Blueprints: Solving the mystery of evolution.* Boston: Little Brown (Introductory)

Marx, J. 1991. How the retinoblastoma gene may inhibit cell growth. *Science* 252:1492. (Intermediate)

Moore, J. A. 1972. *Heredity and development.* 2d ed. New York: Oxford. (Intermediate)

Weinberg, R. A. 1988. Finding the anti-oncogene. *Sci. Amer.* Sept:44–51. (Intermediate)

Wills, C. 1989. *The wisdom of the genes.* New York: Basic Books. (Intermediate)

CHAPTER
‹ 14 ›

The Molecular Basis of Genetics

STEPS TO DISCOVERY
The Chemical Nature of the Gene

THE HUMAN PERSPECTIVE
The Dark Side of the Sun

S TEPS TO DISCOVERY

The Chemical Nature of the Gene

*B*y the end of the 1930s, biologists were well versed in the alignment of genes along chromosomes and the ways these genes were reshuffled during meiosis and transmitted from one generation to the next. By 1940, the major question confronting geneticists was very simple: What is the identity of the genetic material?

Ironically, the genetic substance had been hiding in "plain sight" for over 70 years. DNA was discovered in 1869 by the Swiss physician Friedrich Miescher, who first purified DNA (which he called "nuclein") from pus cells obtained from discarded surgical bandages and later from salmon sperm. He chose these unlikely cells because they had large nuclei surrounded by very little cytoplasm.

In 1923, the German biochemist Robert Feulgen developed a procedure that specifically stained DNA in tissues. Observing stained cells under the microscope, Feul-

gen demonstrated the presence of DNA in mitotic chromosomes. Since protein had also been shown to be present in chromosomes, a new question emerged: Was it the DNA or the protein that carried the hereditary instructions? The answer seemed obvious. During the first half of the century, the complexity, versatility, and importance of proteins had been determined. In contrast, DNA was thought to be composed of a monotonous repeat of only four nucleotide subunits (page 270). A molecule of such simple construction could hardly be considered a candidate for the master molecule that directs life itself!

A few years later, Fred Griffith, a meticulous, soft-spoken bacteriologist who worked at the British Ministry of Health, was conducting research on techniques that would distinguish different strains of pneumococcus, the bacterium that causes pneumonia. Griffith was studying two ge-

A DNA timeline ties the findings of studies on pea plants, fruit flies, and bacteria that led to the discovery of DNA as the

netically stable strains of pneumococcus. One strain produced capsules that surrounded the cell, enabling the bacteria to evade the defense system of both a mouse and a human, thereby causing pneumonia. The other strain produced no capsule but was identical to its counterpart in every other way. These nonencapsulated bacteria were quickly destroyed by the host's defenses before they could cause disease. Griffith grew a batch of the encapsulated strain and heated the cells until they died. As expected, injections of these heat-killed bacteria into mice were harmless, as were injections of the live, nonencapsulated strain.

In 1928, Griffith injected both these preparations into the same mouse; remarkably, the mouse contracted pneumonia and died. Furthermore, the dead mouse contained live, encapsulated bacteria, even though none had been injected into it. Since there was no possibility that the heat-killed bacteria had been brought back to life, Griffith concluded that the presence of the dead encapsulated cells had *transformed* the nonencapsulated cells into an encapsulated strain. The transformed bacteria and their progeny continued to produce capsules; thus, the change was permanent and *inheritable*. In 1928, however, the scientific community had yet to recognize that bacteria possessed genes, and the genetic significance of Griffith's results went unappreciated.

Meanwhile, a physician named Oswald Avery decided to give up his clinical practice and shift his attentions to bacteriologic research. In 1928, Avery was 51 years old and an expert on the immunological properties of pneumococcal capsules. At first, Avery was skeptical of Griffith's results and had a young scientist in his laboratory attempt to confirm them. Not only did Avery's colleague confirm the findings, but he discovered that the transformation of nonencapsulated cells to the encapsulated state did not require the use of a host animal (such as a mouse). Rather, transformation could be accomplished in a culture dish simply by adding a soluble extract of the encapsulated cells to the medium in which the nonencapsulated cells were growing. For the next decade, Avery was preoccupied in purifying the substance responsible for transformation and determining its chemical nature.

Extensive chemical analysis by Avery and his colleagues Colin MacLeod and Maclyn McCarty eventually identified DNA as the transforming substance. In addition, an enzyme (*DNase*) that selectively destroys DNA was the only enzyme the scientists found that was capable of abolishing the substance's transforming activity; protein-degrading enzymes had no effect. In 1944, Avery's group published a paper on their findings in the *Journal of Experimental Medicine*. The paper was written with scrupulous caution and made no dramatic statements that genes were made of DNA rather than protein. Avery was determined not to repeat the embarassing mistake of Richard Willstater (Chapter 6), who had claimed that enzymes

could not be made of protein simply because he could not detect proteins in his most active preparations.

The paper drew remarkably little attention. Maclyn McCarty, one of the three coauthors, describes an incident in 1949 when he was asked to speak at Johns Hopkins University along with Leslie Gay, who had been testing the effects of the new drug Dramamine for the treatment of seasickness. The large hall was packed with people and "after a short period of questions and discussion following [Gay's] paper, the president of the Society got up to introduce me as the second speaker. Very little that he said could be heard because of the noise created by people streaming out of the hall. When the exodus was complete, after I had given the first few minutes of my talk, I counted approximately 35 hardy souls who remained in the audience because they wanted to hear about pneumococcal transformation or because they felt they had to remain out of courtesy." But Avery's awareness of the potential of his discovery was revealed in a letter he wrote in 1943 to his brother Roy, also a bacteriologist:

> If we are right, & of course that's not yet proven, then it means that nucleic acids are not merely structurally important but functionally active substances in determining the biochemical activities and specific characteristics of cells — & that by means of a known chemical substance it is possible to induce predictable and hereditary changes in cells. This is something that has long been the dream of geneticists. . . . Sounds like a virus — may be a gene. But with mechanisms I am not now concerned—one step at a time. . . . Of course the problem bristles with implications. . . . It touches genetics, enzyme chemistry, cell metabolism & carbohydrate synthesis—etc. But today it takes a lot of well documented evidence to convince anyone that the sodium salt of deoxyribose nucleic acid, protein free, could possibly be endowed with such biologically active & specific properties & that evidence we are now trying to get. It's lots of fun to blow bubbles,—but it's wiser to prick them yourself before someone else tries to.

Why were Avery's findings so broadly overlooked? And what finally convinced the world that DNA was the genetic material?

genetic material.

*I*f ever there was a fleeting period of discovery that changed the face of biology, it was the period between the early 1940s and the mid-1960s. Virtually all that you will read about in this chapter was discovered in that scant period of about 25 years — a period of biological revolution. Before this revolution, we knew that genes were carried on chromosomes, and we knew the rules by which these genes were transmitted from generation to generation. But we knew little else. By the time the dust had settled, we understood how it all worked.

The revolution began with Avery's discovery of DNA as the substance responsible for transforming one type of pneumococcus into another. Many articles and passages in books have dealt with the reasons why Avery's findings were not met with greater acclaim. Part of the reason may be due to the subdued manner in which the paper was written and the fact that Avery was a bacteriologist, not a geneticist. Some biologists were persuaded that Avery's preparation was contaminated with miniscule amounts of protein and that the contaminant, not the DNA, was the active transforming agent. Others questioned whether studies on bacteria had any relevance to the field of genetics, and they viewed the phenomenon of transformation as a bacterial peculiarity. In other words, Avery's findings were ahead of their time.

△ During the years following the publication of Avery's paper, the climate in genetics changed in a very important way. The existence of the bacterial chromosome was recognized, and a number of prominent geneticists turned their attention to these prokaryotes. These scientists believed that knowledge gained from the study of the simplest cellular organisms would shed light on the mechanisms that operate in the most complex plants and animals. In addition to being much simpler than fruit flies, pea plants, or mice, bacteria are haploid, so dominant traits cannot hide the presence of recessive alleles. Bacteria also have the advantage of being unicellular and capable of rapidly increasing in number; when grown under rich nutrient conditions, a culture of bacteria can double every 20 minutes. Millions of genetically identical cells can be grown in small containers in a matter of hours.

With bacteria as their research tool, geneticists began to hunt for clues at the most basic level of life — the molecular level. As the quest for answers drew scientists into the molecular domain, a new field of science was created: **molecular biology.** By 1950, the emerging field of molecular biology still had a central question to answer: What was the chemical nature of the gene?

▼ ▼ ▼

CONFIRMATION OF DNA AS THE GENETIC MATERIAL

Seven years after the publication of Avery's paper on bacterial transformation, Alfred Hershey and Martha Chase of the Cold Spring Harbor Laboratories in New York turned their attention to an even simpler system — **bacteriophage** — viruses that infect bacterial cells. By 1950, researchers recognized that every virus had a genetic program. The genetic material was injected into the host cell, where it directed the formation of new virus particles inside the infected cell. Within a matter of minutes, the infected cell broke open, releasing new bacteriophage particles, which infected neighboring host cells.

It was clear that the genetic material directing the formation of viral progeny had to be either DNA or protein because these were the only two molecules the virus contained. Hershey and Chase reasoned that the virus's genetic material must possess two properties: First, if the material were to direct the development of new bacteriophage during infection, it must pass into the infected cell. Second, it must be passed on to the next generation of bacteriophage. The molecular biologists prepared two batches of bacteriophage to use for infection. One batch contained radioactively labeled DNA (^{32}P-DNA); the other batch contained radioactively labeled protein (^{35}S-protein). Since DNA lacks sulfur (S) atoms, and protein usually lacks phosphorus (P) atoms, these two radioisotopes provided specific labels for the two types of macromolecules.

The course of infection of a bacterial cell by a bacteriophage containing labeled DNA and protein is illustrated in Figure 14-1. Of the two labeled molecules, only the ^{32}P-DNA entered the infected cell and was passed on to the next generation of viruses. By 1952, the scientific community finally accepted DNA as the genetic material responsible for storing an individual's hereditary message.

The first important step in the newly emerging field of molecular biology was to convince the scientific community that DNA was the genetic material. This was ultimately accomplished using bacteriophage, the simplest system known to contain a genetic program. (See CTQ #2.)

THE STRUCTURE OF DNA

Before Hershey and Chase published their findings, DNA was far from a preoccupation of biologists. In fact, only one group of biologists — located at King's College in England — was working full-time in 1951 trying to determine the structure of DNA. One member of this group, Rosalind Franklin, was busy firing X-rays through DNA crystals, trying to learn how the atoms of the DNA molecule were arranged. This is the same technique (X-ray crystallogra-

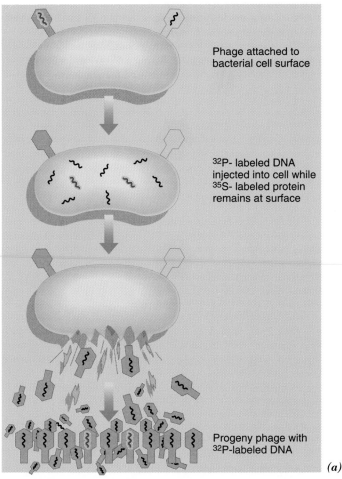

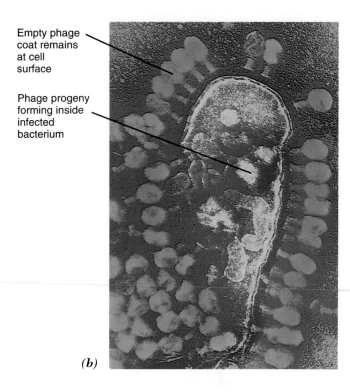

Empty phage coat remains at cell surface

Phage progeny forming inside infected bacterium

(b)

Phage attached to bacterial cell surface

^{32}P- labeled DNA injected into cell while ^{35}S- labeled protein remains at surface

Progeny phage with ^{32}P-labeled DNA

(a)

FIGURE 14-1

The Hershey-Chase experiment. *(a)* Bacterial cells were infected with bacteriophage that contained radioactively labeled protein (blue) or radioactively labeled DNA (red). Hershey and Chase found that none of the labeled protein entered the infected cells (it remained outside the bacteria in the viral coats). In contrast, labeled DNA entered the cells and was recovered in the viral progeny. *(b)* Micrograph showing a bacterial cell infected with bacteriophage. New viruses are being formed within the cell, and empty viral coats are seen attached to the outer cell surface.

phy) that played such an important role in determining the shape of proteins (page 67). Although Franklin didn't interpret her pictures correctly, a couple of other scientists eventually did.

By 1953, the stage was set for the central event in the DNA story, a crowning achievement by two young scientists at Cambridge University in England whose names were about to become household words—James Watson and Francis Crick. Revealed in the seemingly shapeless smudges of Franklin's X-ray photographs was the very essence of the DNA molecule—a helix. You will recall from Chapter 4 that DNA consists of nucleotides, each of which contains a five-carbon sugar called **deoxyribose,** a phosphate group, and a nitrogenous (nitrogen-containing) base (Figure 14-2). All nucleotides in DNA contain the same sugar and phosphate, but they differ in their nitrogenous base. Two of the bases, adenine (A) and guanine (G), are

purines, while the other two bases, cytosine (C) and thymine (T), are *pyrimidines* (Figure 14-3).[1] Using molecular-shaped cutouts of these nucleotides, Watson and Crick constructed a helical model of the DNA molecule which, 40 years later, remains the centerpiece of molecular genetics.

Nucleotides are monomers that become linked together to form a long strand of nucleic acid polymer. Each nucleotide has a polarized structure (Figure 14-2). One edge, where the phosphate is located, is called the 5′ end (pronounced 5-prime end), while the other edge is called the 3′ end. We saw in Chapter 4 how a strand of nucleic acid is formed by linkages between the sugar of one nucleotide

---

[1] It might help to remember the acronym "PurAG," signifying that A and G are purines. Purines have a shorter name than pyrimidines, but a more complex molecular structure (two rings, rather than one).

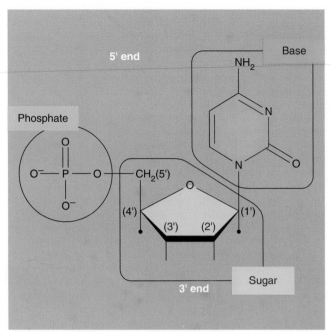

FIGURE 14-2
The structure of a nucleotide. Nucleic acids (DNA and RNA) are composed of repeating nucleotide units. Each nucleotide in DNA consists of the sugar deoxyribose, a nitrogenous base, and a phosphate group. Each nucleotide is polarized, having a 5′ end and a 3′ end. These numbers are based on the system used for numbering the carbons of the sugar.

FIGURE 14-3
Nitrogenous bases in DNA. There are four different nucleotides in DNA, depending on the nitrogenous base linked to the sugar. Adenine and guanine are purines; thymine and cytosine are pyrimidines.

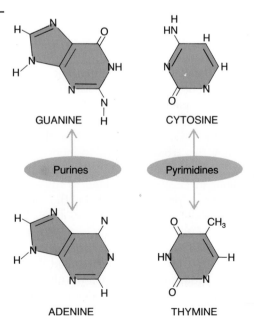

and the phosphate group of the adjoining nucleotide. Since each of the stacked nucleotides in a strand faces the same direction, the entire strand has a direction. In this regard, a DNA strand is like a line of people waiting to get into a theatre: One end—identifiable by a face—is the front of the line, while the other end—identifiable by the back of a head—is the rear of the line. For a strand of nucleic acid, one end is the 3′ end, the other is the 5′ end. These distinctions are useful in understanding the mechanisms for encoding and utilizing genetic information.

THE WATSON-CRICK PROPOSAL

The Watson-Crick model of DNA structure first proposed in 1953 is shown in Figure 14-4. A few of its main elements are listed below.

- A DNA molecule is composed of two chains of nucleotides that coil around each other to form a double helix.

- The two chains of a helix run in opposite directions, like two lines of people standing side by side, but facing opposite directions.

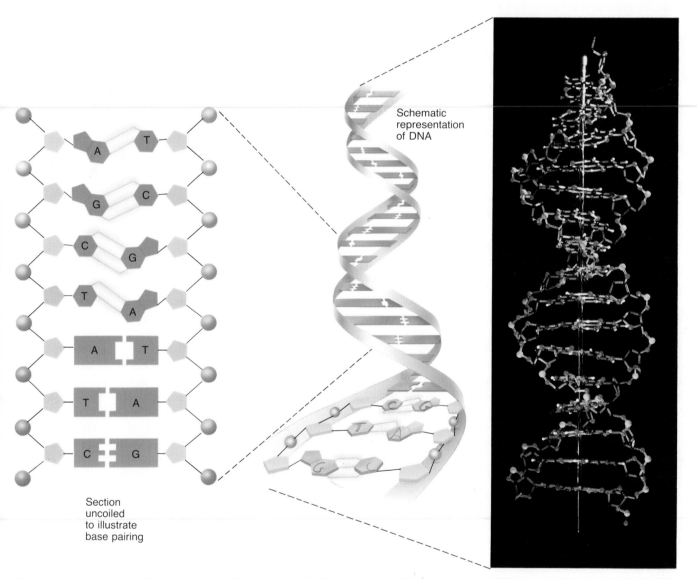

Schematic representation of DNA

Section uncoiled to illustrate base pairing

FIGURE 14-4

The double helix. Base pairing creates a double-stranded molecule that twists into a helix, resembling a spiral staircase. The nitrogenous bases of the nucleotides interact with one another by forming hydrogen bonds. The nucleotide sequence in the two strands is complementary. Guanine (G) binds to cytosine (C) by three hydrogen bonds, whereas adenine (A) binds to thymine (T) by two hydrogen bonds. Consequently, if you were given the sequence in one strand, you could predict the nucleotide order in the other strand. The three base pairs at the lower end of the left side of the figure depict the way nucleotides are illustrated in many of the following figures. The model on the right shows the spatial relationship between the two strands of the DNA molecule.

- The backbone (—sugar—phophate—sugar—phosphate—) of each chain is located on the outside of the molecule, whereas the bases project toward the center, giving the molecule the appearance of a spiral staircase.

- The two chains are held together by hydrogen bonds that form between the nitrogenous bases. Since individual hydrogen bonds are quite weak, the two strands of the double helix can separate during certain essential biological activities.

- An adenine on one chain always pairs with a thymine on the other chain, while a guanine on one chain always pairs with a cytosine on the other chain. Watson and Crick came to this conclusion about **base pairing** based on the way the molecular cutouts of the nucleotides fit with one another and on earlier findings in the late 1940s by Erwin Chargaff of Columbia University. Chargaff had studied the composition of the four different nucleotides in DNA molecules isolated from a variety of organisms. He found that in any given sample of DNA, the percentage of nucleotides that contained adenine always equaled the percentage that contained thymine (A = T). Similarly, the percentage of nucleotides containing guanine always equaled the percentage containing cytosine (G = C). Watson and Crick were the first to grasp the significance of Chargaff's "rules" and, in so doing, laid the cornerstone for their model of DNA structure.

If someone were to tell you the sequence of nucleotides along one strand of a DNA molecule, you could immediately tell him or her the sequence of bases in the other strand. Whenever the person mentioned a G, you would know that a C was situated across the way in the other strand. This relationship is known as **complementarity.** G is said to be complementary to C; one entire strand in a DNA molecule is complementary to the other stand. As we will see later on, the concept of complementarity is of overriding importance in nearly all the activities in which nucleic acids are involved.

Form and function go hand in hand. Once DNA was discovered to be a helical molecule consisting of two complementary strands of nucleotides held together by hydrogen bonds, biologists were finally poised to understand how such a molecule could participate in all of the activities required of genetic material. (See CTQ #3.)

DNA: LIFE'S MOLECULAR SUPERVISOR

From the time biologists first considered DNA as the genetic material, DNA was expected to fulfill three primary functions:

1. **Storage of genetic information.** DNA is the molecular "blueprint," a stored record of precise instructions that determine all the inheritable characteristics an organism can exhibit.

2. **Self-duplication and inheritance.** Since DNA contains the entire blueprint for an organism, it has to contain the information for its own *replication* (duplication). DNA replication provides the means by which genetic instructions can be transmitted from one cell to its daughter cells or from one organism to its offspring.

3. **Expression of the genetic message.** For many years, scientists had suspected that individual genes carried the information for specific proteins. Some mechanism had to exist by which the information stored in a gene was actually put to use to direct the synthesis of a specific polypeptide.

The elucidation of the structure of DNA by Watson and Crick provided an instantaneous stimulus for determining how such a structure facilitated these three essential functions. We will consider the structure of DNA as it relates in turn to each of them.

FUNCTION 1: STORAGE OF GENETIC INFORMATION

The term "information" is difficult to define and is subject to many different interpretations. In biology, information usually refers to a kind of "instruction manual" situated within the DNA that makes up the genes. *The information content of the DNA is encoded in the precise order (linear sequence) of its nucleotides.* The nucleotides of a strand of nucleic acid are like chemical "letters" of an alphabet; they encode genetic information, just as the sequence of printed letters on this page encode the factual information you are now learning. Unlike the 26 letters in the English language, however, the genetic "alphabet" consists of only four letters (G, C, A, and T, corresponding to the four types of nucleotides in DNA). But four "letters" are all that are needed to "write" an unlimited variety of genetic messages. For example, a portion of the double helix only ten nucleotides long can be arranged in more than a million different sequences. Imagine the number of possible sequences you could form with human DNA, which contains about 3 billion base pairs!

With the publication of the Watson-Crick model, the gene was no longer seen as just a vague portion of a chromosome; it had become a specific stretch of nucleotides within a highly elongated DNA molecule. Organisms are highly three-dimensional—from their entire bodies, to the smallest subcellular organelles, to the proteins that bestow them with life. Yet, all of this three-dimensional structure is encoded in a one-dimensional array of nucleotide building blocks. The linear sequence of nucleotides dictates a linear sequence of amino acids in a polypeptide chain, which folds into a complex three-dimensional protein. Virtually all else in biology follows from this organization.

THE ORGANIZATION OF DNA IN CHROMOSOMES IN PROKARYOTES AND EUKARYOTES

The bacterial chromosome consists of a single, circular molecule of DNA, typically about 1 mm long, packaged into a cell that is only about 1 μm in length. One millimeter of DNA is sufficient to encode approximately 3,000 genes, of which about 1,000 have been localized in the chromosome of the common intestinal bacterium *Escherichia coli* (*E. coli*). Although the packaging of the bacterial chromosome into such a small cell is a remarkable feat (as evidenced by the photograph in Figure 14-5), it pales by comparison to that in the packaging of a eukaryotic cell. In every nucleated cell in your body, about 2 meters (6 feet) of DNA are packed into the nucleus, a sphere vastly smaller than the dot on the letter *i*. Moreover, a cell's DNA is not simply stuffed into the nucleus but is present in an orderly arrangement that allows the molecule to direct protein synthesis, to duplicate without tangling, and to coil periodically into compacted chromosomes in preparation for cell division.

The differences in structure between eukaryotic and prokaryotic chromosomes are summarized in Figure 14-5. Eukaryotic chromosomes contain a rich supply of proteins, including a group of small basic proteins called **histones.** Histones are described as basic because they contain large amounts of the basic amino acids arginine and lysine (see Figure 4-14). The positive charges on the R groups of these amino acids form ionic bonds with the negative charges of

DNA's phosphate groups, which facilitates the packaging process. Two loops of DNA are always wrapped around a central cluster of eight histone molecules, forming a unit called a **nucleosome** (Figure 14-6). Not only does the DNA wrap around the histone "spools," it strings the nucleosomes together like beads on a necklace (Figure 14-6).

The relationship between histones and DNA is highly ordered and constant throughout all eukaryotic kingdoms. It appears that the nucleosome has worked well as a DNA-organizing unit for more than a billion years. Once again, the unity of life supplies evidence for the evolutionary kinship of all organisms as descendants of the same early ancestor.

Winding the DNA around nucleosomes shortens its length by approximately one-sixth. The nucleosomal filament is shortened even further by its coiling into thicker fibers, analogous to a coiled telephone cord, which, in turn, are bent into "looped domains" (Figure 14-6). Taken together, these structural properties of the chromosome allow the chromosomal material to fit into the tiny nucleus of a nondividing cell. The preparation of duplicated chromosomes for separation during mitosis requires an additional series of compaction steps, as illustrated in Figure 14-6. The highly coiled DNA of the interphase chromosome condenses around a "scaffold" (Figure 14-7) composed of structural proteins to form a mitotic chromosome. This final packaging step allows a dividing cell to parcel out identical allotments of the genetic material.

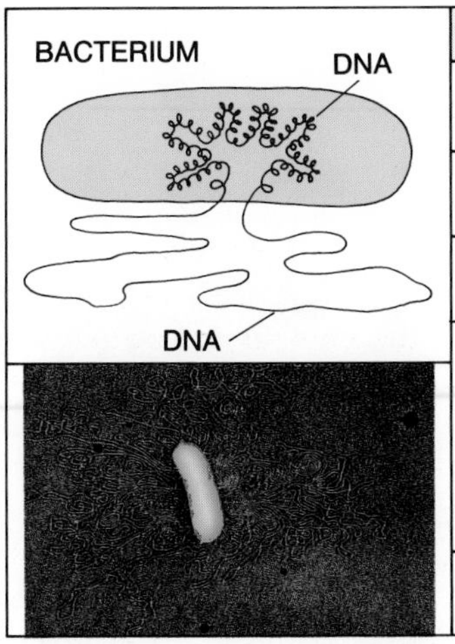

BACTERIUM	CHARACTERISTIC	PROKARYOTIC	EUKARYOTIC
	Configuration of DNA	Circular	Linear (open-ended)
	Length (average)	1000 μm	1.8 meters (in humans)
	Number of chromosomes per cell	1	At least 2, up to 1,262
	Associated proteins	Enzymes and regulatory proteins transiently associated	Histones "permanently" associated. Also transiently associated enzymes and regulatory proteins
	DNA housed in nucleus	No	Yes

FIGURE 14-5

The prokaryotic chromosome is a model of both complexity and simplicity. It is a single circular DNA molecule having a diameter about 1,000 times that of the cell in which it is tightly packed. The adjacent table compares the chromosomes of prokaryotic and eukaryotic cells. Whereas the DNA of prokaryotes is only transiently associated with various proteins, the DNA of eukaryotic cells occurs in permanent association with structural proteins called *histones* (see Figure 14-6). The photo is of *Escherichia coli* broken open and spilling out its chromosome.

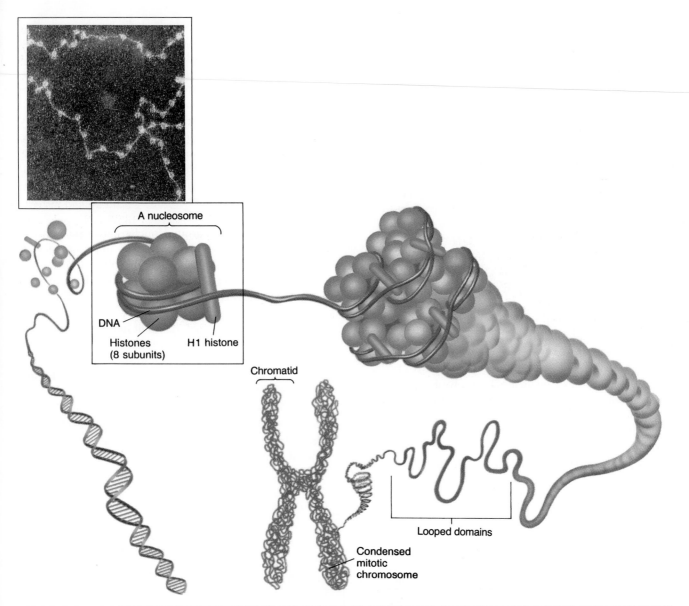

FIGURE 14-6

Packaging DNA into nucleosomes. Current model depicting packaging of DNA into condensed chromatin. The nucleosome (boxed area) is the fundamental packing unit. Each nucleosome consists of eight histone molecules encircled by approximately two turns of DNA. A different type of histone (called H1) locks the nucleosome complex together so that the DNA cannot unwind from the histone core. The nucleosomes are then coiled into thicker fibers that bend into "looped domains." These thickened strands can coil even further, forming the arms of a condensed mitotic chromosome. Nucleosomes are visible in this electron micrograph of uncondensed chromatin. The nucleosomes are the "beads" on the DNA "string."

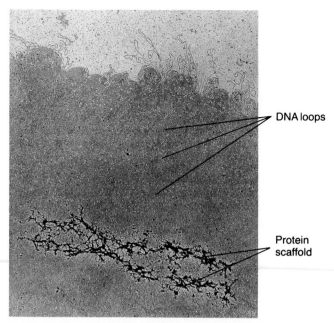

each "daughter" molecule. As a result of replication, two new DNA molecules are formed, each containing precisely the same genetic message as that stored in the original molecule. This is the mechanism by which genetic instructions can be passed on from generation to generation.

The Mechanism of Replication

As in the case of other metabolic processes, much of the pioneering work on replication was carried out on bacterial cells. The first important steps in unraveling the mysteries of replication were the purification and analysis of the enzyme **DNA polymerase** by Arthur Kornberg and his colleagues at Washington University in 1957. DNA polymerase is the enzyme that moves along each template strand of the open helix, reading the nucleotide in the template and

FIGURE 14-7

The protein scaffold of a mitotic chromosome. When mitotic chromosomes are treated so as to remove the histones, the freed DNA is seen to form giant loops that are attached at their base to the protein scaffold (which can be seen to retain the shape of the mitotic chromosome).

FUNCTION 2: PASSAGE OF GENETIC INFORMATION TO DESCENDANTS THROUGH DNA REPLICATION

The ability to reproduce is one of the most fundamental properties of all living systems. This process of duplication can be observed at several levels: Organisms duplicate by asexual or sexual reproduction; cells duplicate by cellular division; and the genetic material duplicates by **replication.**

The initial publication of Watson and Crick's proposal on the structure of DNA was accompanied by a proposed explanation of how such a molecule might replicate. The scientists suggested that during replication the hydrogen bonds holding the two strands of the DNA helix were sequentially broken, causing the gradual separation of the strands, much like the separation of two halves of a zipper. Each of the separated strands, with its exposed nitrogenous bases, would then serve as a **template** (a mold or physical pattern), directing the order in which complementary nucleotides become assembled to form the complementary strand. When complete, the process would have generated two identical molecules of double-stranded DNA, each containing one strand from the original DNA molecule and one newly synthesized strand (Figure 14-8). This form of DNA synthesis is called **semiconservative replication** because half the original DNA molecule is conserved in

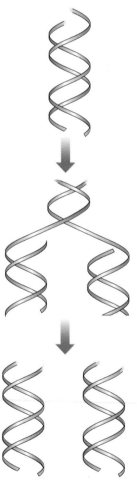

FIGURE 14-8

Semiconservative DNA replication. During replication, the double helix unwinds, and each of the strands serves as a template for the assembly of a new complementary strand. Following replication, each new DNA molecule consists of one strand from the original duplex and one newly constructed strand. This arrangement, which was first predicted by Watson and Crick in their first publication of the structure of DNA, is called *semiconservative replication.*

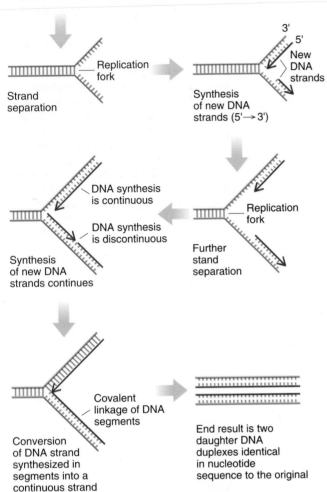

3' end
5' end

Original DNA duplex

Replication fork

Strand separation

Synthesis of new DNA strands (5'→3')

3'
5'
New DNA strands

DNA synthesis is continuous

DNA synthesis is discontinuous

Synthesis of new DNA strands continues

Replication fork

Further stand separation

Covalent linkage of DNA segments

Conversion of DNA strand synthesized in segments into a continuous strand

End result is two daughter DNA duplexes identical in nucleotide sequence to the original

FIGURE 14-9

The mechanism of replication. DNA consists of two strands that run in opposite directions. One of the strands runs in the 3' to 5' direction, and the other runs in the 5' to 3' direction. DNA polymerase molecules are only capable of moving along a template in one direction, toward the 5' end of the template. Consequently, polymerase molecules (and their associated proteins) move in opposite directions along the two strands. As a result, the two newly assembled strands also grow in opposite directions, one growing toward the replication fork, and the other growing away from it. One strand is assembled in continuous fashion, the other in segments that must be joined together by an enzyme.

covalently joining the complementary nucleotide onto the end of the new strand, which thereby grows in length.

It was initially thought that DNA polymerase molecules would move along both template strands toward the *replication fork,* the site where the helix was unzipping. If this were the case, both of the new strands would simply grow continuously in length by addition of nucleotides at the end near the fork. It was soon shown, however, that polymerase molecules move in opposite directions along the two template strands—toward the replication fork on one side, and away from the fork on the other side (Figure 14-9). As a result, the two new strands must be assembled in different ways. The new strand that grows toward the replication fork is constructed by continuous assembly, while the new strand that grows away from the replication fork is assembled by discontinuous assembly—in pieces that are subsequently linked together.

The process of replication is actually much more complex than that shown in Figure 14-9, and a number of proteins besides DNA polymerase are involved. For example, proteins are needed to unwind the helix, to keep the strands separated, and to join segments of DNA together into a continuous strand. Many of these proteins are clustered together to form a giant enzyme complex that moves along the template strand, much as a locomotive moves along a railroad track.

In bacteria, replication begins at one site, the **origin,** and progresses outward in both directions, as indicated in Figure 14-10*a*. At 37°C, the replication machinery of *E. coli* moves along the DNA at an astounding rate of 850 nucleotides per second, reading the template and incorporating a complementary nucleotide at each step along the way. In contrast, the chromosomes of eukaryotic cells contain much more DNA than do those of bacteria, so it would take days to replicate the DNA of a large chromosome if there were only one origin of replication. Instead, replication of the DNA of a plant or animal chromosome begins at many sites simultaneously, proceeding outward from each site in both directions (Figure 14-10*b*).

The Accuracy of Replication

The maintenance of biological order and stability from one generation to the next requires that the process of replication occur with a minimal number of mistakes. Imagine that you are a DNA polymerase molecule moving along a template, selecting complementary nucleotides out of a bag simply by their shape. How many mistakes do you think you would make? What do you think the consequences of picking a mismatched base pair would be? Remarkably, the DNA polymerase of bacteria makes a mistake only about once in every billion nucleotides it incorporates, which is less than once every 100 cycles of replication. One of the reasons for this extraordinary accuracy is that DNA polymerase is one of a handful of proteins that is actually two enzymes in one; it has one active site for polymerization, and another active site for "proofreading." If the first active site happens to incorporate a noncomplementary

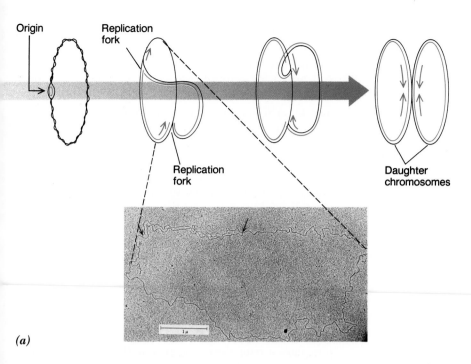

(a)

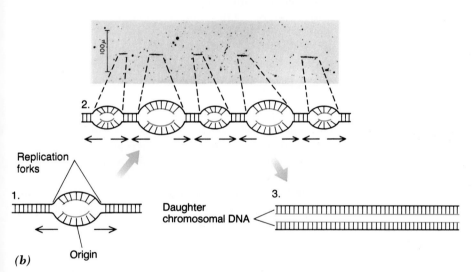

(b)

FIGURE 14-10
Comparison of bacterial and eukaryotic chromosomal replication. In both types of cells, replication occurs at "forks" that travel away from each other in opposite directions. Replication is therefore bidirectional. *(a)* The circular chromosome of bacteria has only two replication forks. Chromosomal duplication is completed when the forks meet each other halfway around the circle. The chromosome in the photo has completed about one-sixth of the process. *(b)* In eukaryotes, replication begins at many points along the chromosome, each site forming two replication forks that travel away from each other until they meet another fork. Five distinct replication sites are apparent in the photo of a single DNA molecule from a mammalian cell. The dark lines are areas with newly replicated DNA.

monomer, the mistake is immediately recognized by the second active site, which removes the incorrect nucleotide. If the mistake happens to slip by the "proofreading" mechanism, it may result in a permanent change, or genetic mutation, in the information content of the DNA. The consequences of a mutation are discussed later in the chapter.

FUNCTION 3: GENETIC EXPRESSION FROM DNA TO RNA TO PROTEIN

An organism is the manifestation of its particular constellation of proteins, each of which is made up of one or more polypeptide chains. A single gene encodes the information for a single polypeptide chain. But just how does the linear order of nucleotides in the DNA lead to the assembly of a linear order of amino acids in a polypeptide? During the late 1950s and early 1960s, it was discovered that the relationship between DNA and protein involves an intermediate—RNA. In other words, genetic information in a cell flows from DNA to RNA to protein.[2] This flow of encoded information is depicted in Figure 14-11. The cell first *transcribes* (copies) a gene's encoded instructions into a molecule of RNA, which is then sent to the "construction site" (ribosomes). These instructions direct the activities at the site, telling the "workers" (various proteins and RNA molecules) which polypeptide to build. The workers must be able to *translate* the instructions in the RNA into the exact gene product ordered for construction.

---

[2] This flow is reversed under certain circumstances—from RNA to DNA — as occurs in a cell infected by HIV, the virus that causes AIDS (Chapter 30).

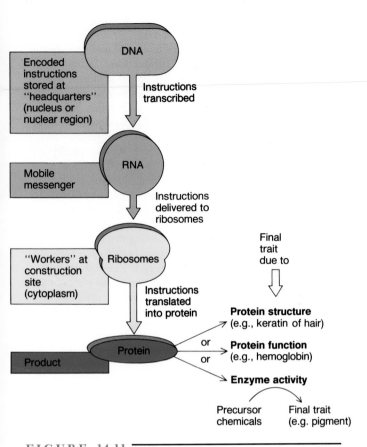

FIGURE 14-11

A summary of the flow of genetic information from DNA to RNA to protein.

Two questions of particular importance will occupy most of the remainder of this chapter. First, how is the genetic information stored in DNA transcribed into RNA? Second, how is the information contained in the RNA translated into the corresponding polypeptide? In order to understand the answers to these questions, we must reconsider the structure of RNA (originally described on page 82). RNA differs structurally from DNA in three ways.

- The nucleotides in RNA contain *ribose,* a sugar that has one more oxygen atom than does DNA's sugar (*deoxyribose*). This is an important difference because it allows enzymes to distinguish between the two types of nucleotides.

- As in DNA, RNA has four distinct nitrogenous bases, but one of them — *uracil* — is unique to RNA. Uracil replaces thymine as the base that is complementary to adenine.

- RNA is typically a single-stranded molecule, so its bases are generally exposed and available for interaction with other molecules.

Transcribing the Message

The information stored in DNA is carried to the ribosomes by an RNA molecule aptly called **messenger RNA (mRNA).** RNA molecules are assembled by **transcription,** a process that, in some ways, resembles the synthesis of a new DNA strand during replication. During transcription (Figure 14-12), the double helix temporarily separates, and a complementary strand of mRNA assembles along one of the single DNA strands that acts as a template. **RNA polymerase,** the enzyme that directs the process, distinguishes between deoxyribose- and ribose-containing nucleotides, polymerizing only the latter type into the growing chain. The enzyme also distinguishes between the two strands of the DNA molecule, selecting the **sense strand;** that is, the strand with the appropriate sequence for protein synthesis.

In this way, genetic information stored in a gene is transcribed into an mRNA molecule that contains the DNA's message by virtue of its complementary sequence of nucleotides. This mobile "messenger" leaves the DNA template carrying the information to the ribosomes, where it directs the synthesis of a specific polypeptide. Messenger RNA is only one of three major types of RNA synthesized by cells. As we will see shortly, the other two types — transfer RNA and ribosomal RNA — are also involved in protein synthesis, but their function is very different from mRNA.

The Genetic Code

In a sense, the genetic basis of life is a matter of "reading, copying, and following instructions" on a molecular scale. Just as a reader of a sentence deciphers an encoded message by translating a linear string of letters into a meaningful thought, cells use a language — a *genetic code* — which they translate into genetic characteristics. As you translate the line of characters in this sentence, you do so by recognizing groups of letters (words) that have specific meanings. In constructing proteins, the linear array of nucleotides in mRNA is also translated in groups, called **codons,** which could be considered molecular "words." Each codon is three nucleotides long and specifies the insertion of one — and only one — of the 20 different amino acids at a specific point in the polypeptide being built. With a four-letter genetic alphabet (A, G, C, and U), 64 triplet combinations (4 × 4 × 4) are possible; 64 combinations are more than enough to assign at least one unique codon to each of the 20 different amino acids.

By 1961, the general properties of the genetic code were known, but one great task remained: breaking the code itself. What is the "language" that cells use to instruct ribosomes to insert the correct amino acids in the proper sequence in a growing polypeptide chain? At the time, most experts believed it would take 5 to 10 years to decipher the entire code. But the codebreakers received a terrific boost from the development of techniques used to synthesize "artificial mRNAs" having known nucleotide sequences. The first of these artificial mRNAs, which was synthesized and tested by Marshal Nirenberg and Heinrich Matthaei of

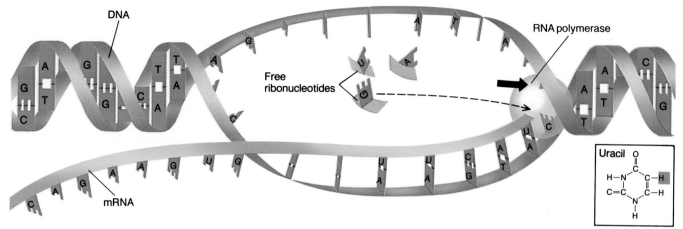

FIGURE 14-12

Transcription: dispatching the molecular messenger. Only one strand of DNA (called the "sense strand") encodes information in the message sent to the ribosomes. During transcription, the RNA polymerase binds to the DNA, the double helix is temporarily separated, the RNA polymerase recognizes the sense strand as the proper template, and the enzyme then assembles a complementary strand of RNA which grows toward its 3′ end. (The uracil in RNA—see inset—that replaces thymine in DNA differs by the single chemical group shaded for emphasis.)

the National Institutes of Health, contained nucleotides with only one type of nitrogenous base—uracil. When these synthetic RNA strands (called "poly-U") were added to a test tube containing a bacterial extract with all 20 amino acids plus the materials necessary for protein synthesis (ribosomes and various soluble factors), the system followed the artificial messenger's instructions and manufactured a polypeptide. This polypeptide turned out to be polyphenylalanine, a polymer of the amino acid phenylalanine. Nirenberg had thus shown that the codon UUU specifies "phenylalanine."

Over the next 4 years, synthetic mRNAs were used to test all 64 possible codons to determine which amino acids would be incorporated. The result was the universal decoder chart, or "genetic code," shown in Figure 14-13. (Instructions for reading the chart are provided in the accompanying figure legend.) The chart is "universal" because regardless of the type of cell—whether a bacterium, yeast, mushroom, redwood tree, or human—the same codons specify the same amino acids.[3] The codon CAC, for example, will always specify the insertion of histidine at the corresponding point in the polypeptide being assembled. The universality of the genetic code is powerful evidence that all organisms on earth have evolved from a common

ancestor at some early point in the history of life on this planet.

Translating the Message

Let us return to a natural mRNA molecule containing a spectrum of different codons. The presence of a particular codon triplet in mRNA orders the insertion of the corresponding amino acid in the growing polypeptide chain. The next codon in the mRNA would specify the insertion of the next amino acid in the polypeptide being synthesized. In this way, the entire message is read—codon by codon—until the polypeptide chain is completely assembled according to the instructions originally encoded in the DNA.

The mRNA-directed assembly of a polypeptide is called **translation** and is the most complex synthetic activity occurring in a cell. Translation is the process of protein synthesis; it requires mRNA, amino acids, numerous enzymes, ribosomes, and chemical energy in the form of ATP (and GTP). It also requires another type of RNA that decodes mRNA's encoded message (written in codons) and translates it into the language of proteins (amino acids). This "molecular decoder" is called **transfer RNA (tRNA).**

Transfer RNAs are small molecules (as RNAs generally go), and each one is able to fold in a characteristic manner (Figure 14-13, right panel, Figure 14-14). The function of tRNAs is closely correlated with their three-dimensional shape. Once it has folded, one end of each tRNA molecule contains a unique sequence of three nucleotides which can form base pairs with one of the mRNA codons. The nucleotide triplet of the tRNA that recognizes and attaches to an mRNA codon is called the **anticodon.** On the side opposite

---

[3] Minor variations have been discovered in a few microorganisms. In *Paramecium,* for example, UAG = glutamine rather than "stop." Mitochondria, which have their own protein-synthesizing machinery, also have some minor differences in codon recognition. It is believed that these alterations evolved from the standard genetic code.

◁ THE HUMAN PERSPECTIVE ▷
The Dark Side of the Sun

We owe our lives to the rays of the sun. The energy in these rays are captured by photosynthetic protists and plants and used to manufacture complex organic molecules upon which all heterotrophic organisms—you included—depend. The sun is also important in less tangible ways: We thrive on warm, sunny days; we spend our vacations on sunlit beaches; and we become depressed when the weather turns dark and "gloomy." But the sun also emits a constant stream of ultraviolet rays that ages and mutates the cells of our skin. Ultraviolet radiation damages DNA by causing adjacent thymine bases to become covalently bonded to one another, forming a "dimer."

**UV damaged DNA
(contains T-T dimer)**

— C-C-T-A-T-T-A-G-C-A —
⋯⋯ ⋯⋯⋯⋯ ⋯⋯ ⋯
— G-G-A-T-A-A-T-C-G-T —

To repair the damage, the region containing the dimer must be cut out of the damaged strand, and the original nucleotides replaced.

⟳ The hazardous effects of the sun are most dramatically illustrated by the rare recessive genetic disorder, *xeroderma pigmentosum (XP)*. Victims of XP possess a deficient repair system that is unable to remove segments of DNA damaged by ultraviolet light. Because of their genes, children with this disease are forced to live in a continually dark environment. Even brief exposure to the ultraviolet rays in sunlight increases the danger of severe skin damage and promotes the formation of skin tumors and other fatal cancers. Sleeping during the day behind blackened windows, playing indoors under conditions of low interior light, going outside only at night when there is no possibility of exposure to ultraviolet light—these children can die from

enjoying what most of us take for granted, a day in the sun.

Before you conclude that because you don't have xeroderma pigmentosum you don't have anything to worry about from exposure to the sun, consider the following statistics. Over 600,000 people develop one of three forms of skin cancer every year in the United States; most of these cases are attributed to overexposure to the sun's ultraviolet rays. Fortunately, the two most common forms of skin cancer—basal cell carcinoma and squamous cell carcinoma—rarely spread to other parts of the body and can usually be excised in a doctor's office. Both of these types of cancer arise from cells that form the bulk of the epidermis, the outer layer of the skin (Chapter 26).

Malignant melanoma, the third type of skin cancer, is a potential killer. Unlike the other types of skin cancer, melanomas

(a)

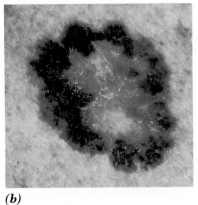
(b)

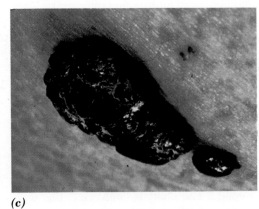
(c)

FIGURE 1 ▬
Stages in the growth of a melanoma. These malignant skin lesions are often characterized by rapid growth, an irregular boundary, variegation in color, and a tendency to become crusty and bleed.

the anticodon on the tRNA is the site that carries a specific amino acid. Amino acids are covalently linked to tRNAs by a special set of enzymes (referred to as "charging enzymes" in Figure 14-18).

Once the amino acid is linked to the tRNA, base pairing between the tRNA's anticodon and the mRNA's codon brings the corresponding amino acid into position for in-

corporation into the polypeptide being synthesized. For example, the codon UCU in mRNA binds only with the anticodon AGA of the tRNA carrying the amino acid serine. Therefore, UCU in mRNA instructs the cell to insert serine at that point in the newly forming polypeptide. In this way, tRNA bridges the language gap between the nucleotide codon and the amino acid.

develop from pigment cells in the skin. They may arise in an existing, nonmalignant mole, or they may appear where no previous mole is present. Every year, nearly 30,000 Americans are diagnosed with melanoma, and the number is climbing at an alarming rate due to the increasing amount of time people have been spending in the sun over the past few decades. Many scientists predict that the rate of melanoma will climb even more rapidly in the future if the UV-absorbing ozone layer of the atmosphere continues to deteriorate.

If a melanoma is detected and removed at an early stage, when it is small and has not yet penetrated into the deeper layers of the skin, the prognosis is very good. Unlike most tumors, melanomas appear on the surface of the skin where they can be observed, so there is no reason for such cancers to go undetected. Everyone should know the warning signs of a melanoma, which are depicted in Figure 1.

Light-skinned individuals are particularly susceptible to developing melanoma, as are those with close relatives who have had the disease; those who received a severe, blistering sunburn as a child; or those who live in sunnier regions. For example, the highest incidence of melanoma is found in Australia, which has a sunny, tropical climate and is populated largely by light-skinned individuals of northern-European descent. Similarly, in the United States, the incidence of melanoma in Arizona is over twice that found in Michigan. The best way to avoid developing melanoma is to avoid sunbathing (Figure 2) and to wear sunblock creams when you plan to spend time in the sun. Most important: Avoid serious sunburns and be sure your children do the same.

FIGURE 2

Four codons warrant special mention. Three (UGA, UAG, and UAA) have no corresponding amino acid (Figure 14-13). These triplets, called **stop codons** (or *nonsense codons*), spell "stop!" when mRNA is being translated into protein. They are used to terminate synthesis at the completion of a polypeptide. Another codon, AUG (specifying the amino acid methionine), is the "start" codon. AUG always appears at the beginning of the coding portion of an mRNA and initiates the assembly of amino acids into the polypeptide.

Translation occurs at ribosomes—complex particles composed of several dozen different proteins and a number of different RNA molecules. The RNAs of the ribosome, called **ribosomal RNAs (rRNAs),** constitute the third major type of RNA made by cells. Whereas mRNAs carry encoded information, and tRNAs serve to decode this information, rRNAs are primarily structural molecules functioning as a scaffold to which the various proteins of the ribosome can attach themselves. The proteins that make up the ribosome have varied functions: Some play a structural role in holding the particle together; others bind to the mRNA or tRNAs; still others act as enzymes involved directly in protein synthesis.

The ribosome is a nonspecific component of the translation machinery—sort of a "workbench"—in that any ribosome can serve as a translation site for any mRNA. This is why bacterial cells can be used as "pharmaceutical factories" to churn out proteins encoded by human mRNAs (Chapter 16). A functioning ribosome consists of a large and a small subunit (Figure 14-15). The ribosome assembles from its subunits at the start of polypeptide synthesis and then disassembles when synthesis has been completed.

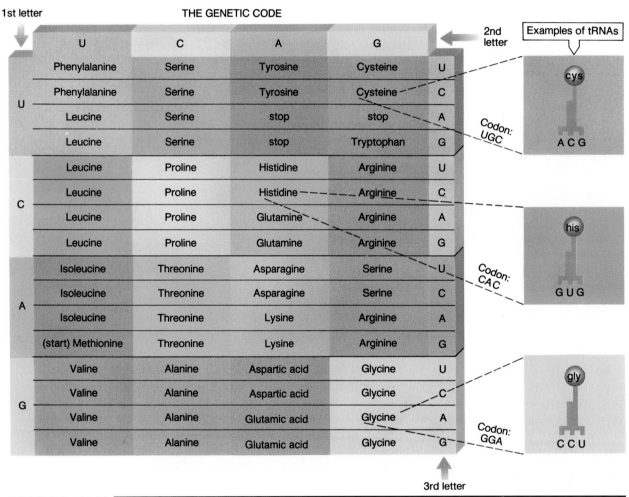

FIGURE 14-13

The genetic code. The genetic code is a universal biological language. The correlation between codon and amino acid indicated in this decoder chart is the same in virtually all organisms. To use the chart to translate the codon UGC, for example, find the first letter (U) in the indicated row on the left. Follow that row to the right until you reach the second letter (G) indicated at the top, then find the amino acid that matches the third letter (C) in the row on the right. UGC specifies the insertion of cysteine. Each amino acid (except two) has several codons that order its insertion. These are genetic "synonyms," backup systems that reduce the danger of lethal mutations disrupting the cell. A change in a single nucleotide in the codon's third position, for example, would change which amino acid would be incorporated into the polypeptide, unless the new codon were synonymous with the original (which happens frequently). In that case, the same amino acid would be incorporated, and the mutation would have no phenotypic effect. As discussed in the following section, decoding in the cell is carried out by tRNAs, a few of which are illustrated in the right side of the figure.

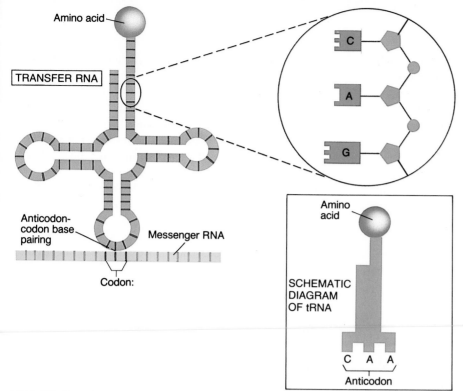

FIGURE 14-14 ━━━━━━━━

Molecular decoder. Transfer RNA molecules translate one "language" (a three-nucleotide codon "word") into another (a specific amino acid) by using anticodons. Each anticodon in a tRNA recognizes a single codon in mRNA. Similarly, each tRNA carries a specific amino acid. In this way, a particular tRNA associates the correct amino acid with its corresponding codon in mRNA. The "cloverleaf" structure illustrates how the anticodon base pairs with its complementary codon. The attachment of a specific amino acid to its corresponding tRNA is catalyzed by a set of amino acid "charging" enzymes.

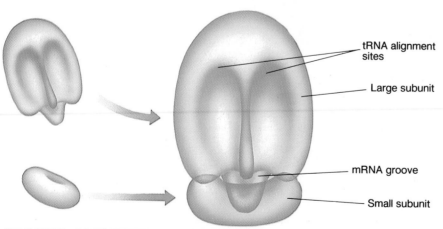

FIGURE 14-15 ━━━━━━━━

Assembly of a functional ribosome. Two subunits, one large and one small, fit together to form a groove for mRNA and create sites for accepting two tRNAs at a time. A component of the large subunit catalyzes the formation of a peptide bond that connects the two aligned amino acids. Thus, ribosomes are more than mere "workbenches" on which proteins are synthesized, they are more like "workers" that help assemble the protein.

An assembled ribosome has a groove through which the mRNA molecule can travel. The ribosome also has sites that position two tRNA molecules so that their amino acids end up adjacent to each other on the large subunit. The large subunit contains a factor that covalently links the adjacent amino acids. As the ribosome moves along the mRNA strand, amino acids are incorporated into the growing polypeptide chain in the order specified by the mRNA. The incorporation of each amino acid is accompanied by the hydrolysis of GTP (a high-energy compound similar to ATP), which provides the energy driving the assembly process. A simplified version of the events of translation is depicted in Figure 14-16. (Many of the enzymes and accessory molecules have been omitted for clarity.) The three stages of translation are initiation, chain elongation, and chain termination.

Step 1: Initiation   The process of translation begins when the small subunit of a ribosome binds to the mRNA near its 5′ end. This binding always occurs at the initiation codon, AUG. The binding of the small subunit is followed by the attachment of the first tRNA, the *initiator tRNA*, whose anticodon is UAC. The initiator tRNA always brings a methionine as the first amino acid of the assembling polypeptide. The large subunit soon joins the complex, and the assembly of the polypeptide begins. The binding of the ribosome to the AUG codon fixes the "reading frame," assuring that translation begins with the correct nucleotide.

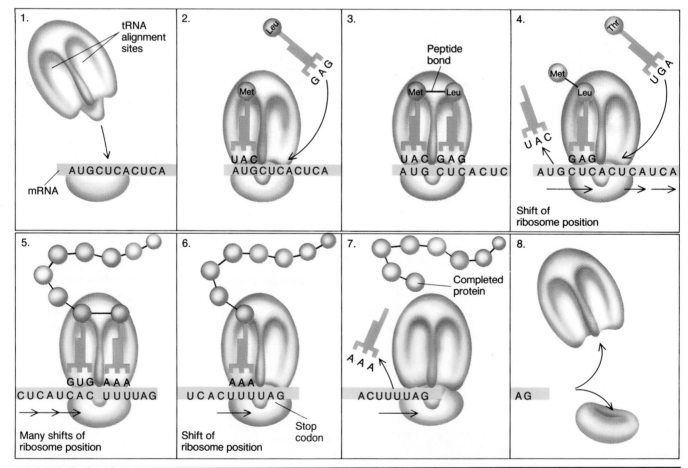

FIGURE 14-16

General steps in translation.   Each of the three major steps in translation—initiation, elongation, and termination—is discussed in the text. (The mRNA in the figure is only a short portion of the entire molecule, which, for most proteins, would exceed 600 nucleotides in length.) The methionine inserted by the initiator tRNA as the first amino acid is usually clipped from the polypeptide by an enzyme.

If the message were initiated one or two nucleotides over, all the remaining triplets would be incorrectly read, and the wrong amino acids would be inserted, producing a totally useless polypeptide. The brackets below show the incorrect codons that would be produced if initiation began two nucleotides over from the proper site.

correct ⟶   met   leu   his   pro

mRNA ⟶ — AUGCUGCAUCCA —

incorrect ⟶   ala   ala   ser

Steps 2–5: Chain Elongation

With the initiator tRNA in place, a second free site remains on the ribosome, where another tRNA can align with the next codon in the mRNA (which is CUC, in this illustration). CUC pairs with the anticodon GAG of the tRNA carrying the amino acid leucine. The two amino acids become aligned next to each other, and the amino acid attached to the first tRNA is enzymatically transferred to the second amino acid, forming a covalent (peptide) bond. The first tRNA, which has lost its amino acid cargo, then departs, and the ribosome moves down the mRNA by three nucleotides, bringing the third codon into position. A tRNA molecule with a complementary anticodon binds to the third codon, orienting its amino acid next to the previous one, and a peptide bond now forms between them. The growing polypeptide chain is now three amino acids long. The second tRNA is then released, and the ribosome moves down the mRNA, bringing the fourth codon into position. This process continues, adding amino acids in the proper sequence until the entire polypeptide chain is synthesized according to the original genetic instructions.

Steps 6–8: Chain Termination

The completion of a polypeptide chain is signaled by the presence of a stop codon in the mRNA strand. Since these triplets do not specify amino acids, their presence produces a region on the mRNA to which no tRNA can bind. Therefore, the amino acid inserted just before the stop codon becomes the terminal member of the chain.

Protein synthesis occurs very rapidly, and its efficiency is increased by simultaneous translation of each mRNA by numerous ribosomes. As soon as one ribosome has translated the first few codons, the mRNA initiation site is again available for another ribosome to become attached. As a result, ribosomes are often found in chains held together by an mRNA strand. This complex is called a **polysome** (Figure 14-17). Since amino acids are being incorporated at each ribosome along the polysome, a single mRNA may be generating dozens of identical polypeptide chains simultaneously.

The ordered flow of genetic information from its stored form in DNA to its transcription into RNA and its expression as a specific protein is summarized in Figure 14-18.

Within 20 years after Avery's first indication of the importance of DNA, scientists had worked out the mechanism by which genetic information is stored, replicated, and used to run cellular activities—which is nothing less than the molecular basis of life itself. In so doing, these scientists have delivered a deeply satisfying message: The universe inside every living organism is within our cognitive reach. Life is explainable in physical and chemical terms, without the need to evoke mystical forces. (See CTQ #5.)

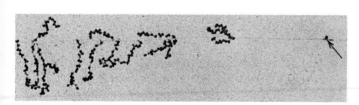

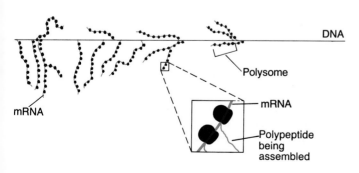

FIGURE 14-17

Gene expression—caught in the act. In bacteria, ribosomes attach to an mRNA molecule as it is being synthesized and begin translation (as shown in this photograph). In both prokaryotic and eukaryotic cells, a single mRNA is translated by a number of ribosomes that follow each other along the mRNA. The complex of ribosomes held together by an mRNA molecule is called a polysome. Colored lines emerging from the ribosomes depicted in the box indicate that each ribosome is synthesizing a polypeptide chain.

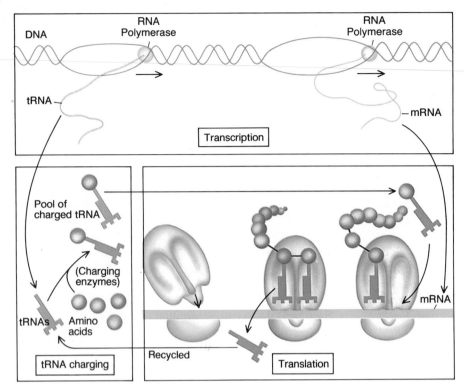

FIGURE 14-18

Overview of protein synthesis. All the processes depicted here occur in a single cell. The steps are compartmentalized for clarity. The genetic information in DNA ultimately dictates the amino acid sequence in protein, explaining why cells are biologically "obedient" to their genes. A charged tRNA is a tRNA with an attached amino acid, and a charging enzyme is an enzyme that attaches a specific amino acid to the appropriate tRNA (one with the appropriate anticodon).

THE MOLECULAR BASIS OF GENE MUTATIONS

If genetic information is stored in the linear sequence of nucleotides along the strands of a DNA molecule, it follows that changes in that sequence will alter the DNA's information content. This is the basis of a **gene mutation.** As we saw in Chapte 4, sickle cell anemia is a genetic disease that results from the substitution of one amino acid (a valine) for another (a glutamic acid) in the oxygen-carrying protein hemoglobin. Looking back at Figure 14-13, notice the various codons for glutamic acid and valine. All codons for both amino acids begin with G, but the second nucleotide for the glutamic acid codons is A, while that for valine is U. Herein lies the basis for the gene mutation that causes sickle cell anemia. (Remember, codons represent letters in the mRNA.) A change from CTC to CAC in the DNA produces a change from GAG to GUG in the mRNA codon, which, in turn, produces a change from a glutamic acid to a valine in the corresponding polypeptide.

The substitution of one base for another in the DNA is called a **point mutation** because it only affects a single "point" in the gene. This type of mutation can occur during replication, when an incorrect nucleotide is incorporated into the growing DNA chain, or it may occur as the result of damage to the DNA of a nonreplicating cell. The likelihood of mutation is greatly increased if the DNA happens to come into contact with a **mutagen**—a chemical or physical agent that induces genetic changes.

A more serious mutation occurs to a gene when one or two nucleotides are added or deleted because such an alteration throws off the reading frame of the rest of the mRNA molecule, changing all the codons "downstream" from the point of change. As a result, the message is read in the wrong sequence, generating a useless protein that contains a string of incorrect amino acids beyond the shift point. This type of *frameshift mutation,* as it is called, is responsible for certain cases of human *thalessemia,* a type of anemia (a deficiency of red blood cells) resulting from the production of abnormal hemoglobin molecules.

Any agent capable of causing gene mutations is also a potential carcinogen (cancer-causing agent) since the alteration of certain types of genes (including oncogenes and tumor-suppressor genes) can lead to the transformation of a normal cell into a malignant cell. Because of this relationship between mutation and cancer, large numbers of substances can be screened for their carcinogenicity by determining whether or not they are capable of causing mutations. To carry out this test—called the *Ames test,* after Bruce Ames of the University of California, who developed the procedure—a culture of bacteria is simply exposed to the chemical in question, and the number of detectable mutant cells produced is counted. This number is then compared to the number of mutants that appear in a control culture treated the same way but without exposure to the chemical being tested. An increased mutation rate in the chemically treated bacteria flags the substance as a potential human carcinogen. This procedure has detected cancer-causing potential in many substances to which people are frequently exposed, including materials in hair dyes, cured meats, artificial food colors, and cigarette smoke. Ames has become a controversial figure in the current debate over the danger of carcinogenic environmental pollutants by determining that many of the "natural" foods we eat, such as sprouts and mushrooms, contain rather high levels of compounds capable of causing mutations.

Chemicals are not the only agents capable of causing mutations: DNA is also very susceptible to damage by radiation. In fact, one of the most common mutagenic agents is ultraviolet radiation—the subject of The Human Perspective: The Dark Side of the Sun.

Mutagens do not only arise in the external environment; they are also produced in large numbers within the body as the result of normal metabolic reactions. To appreciate the magnitude of the mutation "problem," consider the estimate that, on the average, several thousand bases are lost from the DNA of *every human cell every day!* How is the DNA of a cell able to maintain a nucleotide sequence necessary for life while absorbing such molecular punishment? The answer is that cells contain a diverse array of DNA **repair enzymes** that patrol the DNA, searching for alterations and distortions they can recognize and repair. In this regard, a cell might be likened to a car driving around at all times with a team of mechanics inside who monitor and repair problems as they develop, while the car carries out all of its normal activities. DNA repair systems provide another example of the mechanisms that maintain the order and complexity of the living state.

▶▶ While genetic mutations are generally thought to be harmful, resulting in decreased fitness, some mutations will inevitably be beneficial, providing an individual with an increased chance to survive and produce offspring. As we discussed in Chapter 12, mutations are the raw material of evolution since they introduce new genetic information into a population. Without these spontaneous changes in the DNA, the level of genetic variability would be greatly limited, and evolution, as we know it, could not have occurred.

Life on earth is exposed to a variety of destructive agents that can act on DNA and alter its information content. If organisms lacked mechanisms to repair genetic alterations, life could not continue. Yet, if all of these alterations were repaired, there would be no mutation and, thus, no variation or biological evolution. (See CTQ #6.)

REEXAMINING THE THEMES

Relationship between Form and Function

Although relatively simple, the structure of DNA explains a great deal of biological function. Most importantly, it reveals how information—genetic instructions—can be stored in the linear sequence of nucleotides. Elucidation of the structure of DNA immediately suggested a likely mechanism of replication; the two strands would separate, and each would act as a template for the assembly of a complementary strand. The *translation* of the information in DNA into a polypeptide chain is more complex. The assembly of a polypeptide requires the intervention of a tRNA decoder —a molecule whose structure allows it to translate nucleotide languages into amino acid languages. Each tRNA carries an anticodon at one end, which interacts with an mRNA codon, and an amino acid at the other end, which is incorporated into the assembling polypeptide.

Biological Order, Regulation, and Homeostasis

Biological information is encoded in highly ordered sequences of nucleotides. The deletion of a single nucleotide in a DNA molecule can lead to the formation of a messenger RNA molecule that will direct the assembly of a meaningless string of incorrect amino acids. The ordered sequence of nucleotides in the DNA is maintained by a highly accurate mechanism of replication, which includes a molecular proofreading system, and a battery of repair enzymes that patrol the genetic material, discovering and repairing nucleotide damage. Maintaining the stability of genetic information can be thought of as a type of molecular homeostasis.

Unity within Diversity

Nowhere is the unity and shared ancestry of life better revealed than in the genetic code. Pick any three nucleotides to make up a codon, and they spell the same amino acid (or stop message) in a bacteriophage, a human, or a lilac bush. Introduce human mRNAs for insulin into a bacterial cell, and that bacterium begins producing human insulin indistinguishable from that made by your own pancreas. The diversity of life is also derived from the genetic code. Over periods of time, mutations cause changes in the sequence of nucleotides in a gene, thereby changing the codons of the gene and the amino acids incorporated into the corresponding polypeptide. The cumulative effect of gradual changes in polypeptides over evolutionary time has been the generation of life's diversity.

Evolution and Adaptation

▮▶ Evolutionary change from generation to generation depends on genetic variability. Much of this variability arises from reshuffling maternal and paternal genes during meiosis, but somewhere along the way *new* genetic infor-

mation must be introduced into the population. New genetic information arises from mutations in existing genes. Some of these mutations arise during replication; others occur as the result of unrepaired damage as the DNA is just "sitting" in a cell. Mutations that occur in an individual's germ cells can be considered the raw material on which natural selection operates; whereas harmful mutations produce offspring with a reduced fitness, beneficial mutations produce offspring with an increased fitness.

S Y N O P S I S

Experiments in the 1940s and 1950s established conclusively that DNA is the genetic material. These experiments included the demonstration that DNA was capable of transforming bacteria from one genetic strain to another; that bacteriophages injected their DNA into a host cell during infection; and that the injected DNA was transmitted to the bacteriophage progeny.

DNA is a double helix. DNA is a helical molecule consisting of two chains of nucleotides running in opposite directions, with their backbones on the outside, and the nitrogenous bases facing inward like rungs on a ladder. Adenine-containing nucleotides on one strand always pair with thymine-containing nucleotides on the other strand, likewise for guanine- and cytosine-containing nucleotides. As a result, the two strands of a DNA molecule are complementary to one another. Genetic information is encoded in the specific linear sequence of nucleotides that make up the strands.

DNA replication is semiconservative. During replication, the double helix separates, and each strand serves as a template for the formation of a new, complementary strand. Nucleotide assembly is carried out by the enzyme DNA polymerase, which moves along the two strands in opposite directions. As a result, one of the strands is synthesized continuously, while the other is synthesized in segments that are covalently joined. Accuracy is maintained by a proofreading mechanism present within the polymerase.

Information flows in a cell from DNA to RNA to protein. Each gene consists of a linear sequence of nucleotides that determines the linear sequence of amino

acids in a polypeptide. This is accomplished in two major steps: transcription and translation.

During transcription, the information spelled out by the gene's nucleotide sequence is encoded in a molecule of messenger RNA (mRNA). The mRNA contains a series of codons. Each codon consists of three nucleotides. Of the 64 possible codons, 61 specify an amino acid, and the other 3 stop the process of protein synthesis.

During translation, the sequence of codons in the mRNA is used as the basis for the assembly of a chain of specific amino acids. Translating mRNA messages occurs on ribosomes and requires tRNAs, which serve as decoders. Each tRNA is folded into a cloverleaf structure with an anticodon at one end—which binds to a complementary codon in the mRNA—and a specific amino acid at the other end—which becomes incorporated into the growing polypeptide chain. Amino acids are added to their appropriate tRNAs by a set of enzymes. The sequential interaction of charged tRNAs with the mRNA results in the assembly of a chain of amino acids in the precise order dictated by the DNA.

Mutation is a change in the genetic message. Gene mutations may occur as a single nucleotide substitution, which leads to the insertion of an amino acid different from that originally encoded. In contrast, the addition of one or two nucleotides throws off the reading frame of the ribosome as it moves along the mRNA, leading to the incorporation of incorrect amino acids "downstream" from the point of mutation. Exposure to mutagens increases the rate of mutation.

Key Terms

molecular biology (p. 268)
bacteriophage (p. 268)
deoxyribose (p. 269)
base pairing (p. 272)
complementarity (p. 272)
histone (p. 273)
nucleosome (p. 273)
replication (p. 275)
template (p. 275)

semiconservative replication (p. 275)
DNA polymerase (p. 275)
origin (p. 276)
messenger RNA (mRNA) (p. 278)
transcription (p. 278)
RNA polymerase (p. 278)
sense strand (p. 278)
codon (p. 278)
translation (p. 279)

transfer RNA (tRNA) (p. 279)
anticodon (p. 279)
stop codon (p. 281)
ribosomal RNA (rRNA) (p. 281)
polysome (p. 285)
gene mutation (p. 286)
point mutation (p. 286)
mutagen (p. 286)
repair enzyme (p. 287)

Review Questions

1. How is complementary base pairing important to inheritance and to the expression of genetic messages?

2. Describe at least two similarities and two differences between the processes of DNA replication and transcription.

3. If a portion of DNA had the nucleotide sequence AGCAGGCAGC, would you be able to predict the next nucleotide in the chain? Why or why not?

4. Why are tRNA molecules, rather than mRNA molecules, considered the decoders?

5. Considering the importance of maintaining the correct reading frame during translation, what would be the effect of (1) a single nucleotide deletion from the DNA, or (2) the loss of three successive nucleotides?

Critical Thinking Questions

1. The use of DNase was an important part of Avery's studies, pointing to DNA as the transforming principle. DNase is purified from the pancreas, an organ that produces a variety of digestive enzymes. With this in mind, can you think of any other possible explanations for Avery's results? Is there any control that Avery might have run to eliminate this possibility?

2. If you were teaching biology to college students in 1942, would you tell them that the genetic material was protein or nucleic acid? What evidence would you have used to support your position? If you were still teaching in 1952, what would you tell them, and what evidence would you draw on?

3. How do you think the discovery of the molecular structure and function of DNA affects the argument between vitalists and mechanists? (See Critical Thinking Question, #2, Chapter 1.)

4. Suppose that protein, not DNA, were the genetic material. How would the results of the Hershey-Chase experiment have been different?

5. Using the genetic code chart in Figure 14-13, construct the polypeptide chain coded for in a strand of DNA with the following sequence of bases: TACGGATCGCCTACG. (Remember to include the mRNA sequence.)

6. If mutations play an important role in evolution, why are many scientists concerned about the mutagenic effects of x-rays, radiation from nuclear power plants, chemicals that cause mutations, etc.?

7. What would be the effect on the offspring if a DNA polymerase were absolutely foolproof in its proofreading activity? What would be the long-term effect on biological evolution?

8. Suppose the first synthetic polynucleotide that Nirenberg synthesized had been poly A. What type of polypeptide would have been assembled in his experiment? What if he had been able to synthesize the polynucleotide AUAUAUAU . . . by polymerizing the dinucleotide AU? What polypeptide would that polynucleotide encode?

Additional Readings

Alberts, B., et al. 1989. *Molecular Biology of the Cell.* 2d ed. New York: Garland. (Intermediate-Advanced)

Ames, B. N., et al. 1987. Ranking possible carcinogenic hazards. *Science* 236:271–280. (Intermediate-Advanced)

Dubos, R. J. 1976. *The professor, the institute, and DNA: Oswald T. Avery.* New York: Rockefeller University Press. (Intermediate)

Jaroff, L. 1990. Special report on skin cancer. *Time,* July 23: 68–70. (Introductory)

Jaroff, L. 1993. Happy birthday, double helix. *Time,* March 15:56–59. (Introductory)

Judson, H. F. 1979. *The eighth day of creation.* New York: Simon & Schuster. (Intermediate)

McCarty, M. 1985. *The transforming principle: Discovering that genes are made of DNA.* New York: Norton. (Intermediate)

Prescott, D. 1988. *Cells.* Boston: Jones and Bartlett. (Intermediate)

Radman, M., and R. Wagner, 1988. The high fidelity of DNA duplication. *Sci. Amer.* Feb:40–47. (Intermediate)

Watson, J. D. 1969. *The double helix.* New York: Dutton. (Introductory)

Watson, J. D., and F. H. C. Crick, 1953. Molecular structure of nucleic acids. A structure of deoxyribose nucleic acid. *Nature* 171:737–738. (The original paper describing the structure of DNA.)

Orchestrating Gene Expression

STEPS
TO
DISCOVERY
Jumping Genes: Leaping into the Spotlight

◁ B I O L I N E ▷
RNA as an Evolutionary Relic

◼▶ Which came first, the protein or the nucleic acid? Evolutionary biologists have been arguing this point for decades. The dilemma arose from the seemingly non-overlapping functions of these two types of macromolecules. Nucleic acids store information, whereas proteins catalyze reactions. With Cech's discovery of ribozymes, it became apparent that one type of molecule—RNA—could do both.

These findings have led to speculation that neither DNA nor protein existed during an early stage in the evolution of life. Instead, RNA molecules performed double duty; they served as genetic material, and they catalyzed necessary enzymatic reactions. Only at a later stage in evolution were these activities "turned over" to DNA and protein, leaving RNA to function primarily as a "go-between" in the flow of genetic information.

How feasible is this proposition? So far, only a few chemical reactions have been found to be catalyzed by RNA *in the cell*. Many biologists were shocked when it was demonstrated in 1992 that the "enzyme" that catalyzes the reaction in which amino acids are polymerized into polypeptides was actually an RNA molecule present within the large subunit of the ribosome (p. 282). It had long been assumed that this reaction was catalyzed by one of the ribosome's proteins. Several research groups are now exploring the catalytic *potential* of RNA. These investigators are modifying RNAs in various ways and searching for previously unknown catalytic properties. It has recently been shown that some of these modified RNAs can carry out RNA replication—the duplication of other RNA molecules. Since self-duplication is a fundamental property of life, this finding provides support for the existence of an ancient "RNA world."

In addition to fortifying our understanding of evolution, the practical considerations of enzymatically active RNAs are particularly exciting. Competing patent applications have been filed in both the United States and Australia that capitalize on the ability of ribozymes to recognize specific sequences in other RNA molecules and to cut the RNA strand. These applicants hope to be able to synthesize ribozymes that can recognize and destroy certain mRNAs, such as those produced by the AIDS virus, while leaving the rest of the cell's RNA unharmed. The first ribozyme to be subjected to clinical trials will probably be aimed at the herpes virus responsible for human skin lesions.

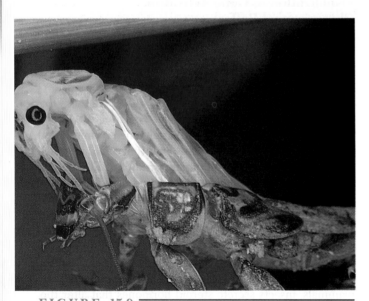

FIGURE 15-9
The metamorphosis of this stone fly was induced by the steroid hormone ecdysone.

products of the newly transcribed genes. These new proteins would normally participate in those activities that carry the insect beyond the larval stage.

REGULATING GENE EXPRESSION AT THE RNA PROCESSING LEVEL

The presence of introns within the boundaries of genes introduces serious obstacles in the path of gene expression. Consider what would happen if these intervening sequences were present in an mRNA molecule being translated. We saw in the last chapter that a ribosome moves along an mRNA from the initiation codon to the stop codon, producing a continuous polypeptide chain. The presence of intervening sequences in mRNA would cause the message to be translated into a polypeptide with disruptive stretches of amino acids within its midst. By 1977, when introns were discovered, a number of eukaryotic mRNAs had already been isolated and sequenced, and it was evident that they contained an unbroken "coding sequence." It seemed obvious, therefore, that even though a *gene* may contain noncoding introns, the mRNA formed from that gene does not. Subsequent investigations revealed that the entire split gene (including introns and exons) is transcribed into a giant RNA molecule, called a *primary transcript*, which is subsequently *processed* into the mature mRNA.

RNA processing removes those segments of the primary transcript that correspond to the introns (Figure 15-10). But investigations of the RNA processing reactions revealed further unexpected revelations.

In 1982, Thomas Cech of the University of Colorado was studying the processing of RNA in a ciliated protozoan. Cech discovered that RNA molecules possessed the ability to catalyze their own processing. Self-processing is not a simple reaction: The RNA has to cut itself into pieces; those pieces that are not needed must be eliminated; and the remaining pieces must be joined together to form the final mRNA product (Figure 15-10). At first, Cech was skeptical about his results, but he soon confirmed his findings. The concept that RNA molecules, as well as proteins, could possess enzyme activity emerged. The question was whether this was some peculiar property of a few RNAs in a strange protozoan or a universal process.

Further studies have revealed that RNA plays a key role in RNA processing in all eukaryotes, but not in the manner first discovered by Cech. Apparently, as a gene is being transcribed, the RNA transcript becomes associated with particles that contain both RNA and protein. These particles move along the primary transcript, cutting the RNA chain, removing the intron sequences, and joining *(splicing)* the exon sequences together (Figure 15-10). As anticipated by the work on the protozoan, the RNAs in the particles carry out the cutting and splicing reactions; the proteins play only a supportive role. Because of their enzymelike properties, these catalytic RNAs are referred to as **ribozymes**.

For his unexpected discovery (as well as further investigations), Cech was awarded the 1989 Nobel Prize in Chemistry. The discovery that RNA molecules can have catalytic activities has also led to a revision in our thinking about the origin of life (see Bioline: RNA as an Evolutionary Relic). The fact that proteins are not life's only catalysts illustrates once again that science is not a fixed, unchangeable body of knowledge. Rather, through the scientific method, our concepts are constantly revised as new information becomes available.

Alternative Processing

Cells may exert control over gene expression by processing RNA transcripts differently. For example, in one type of cell, a primary transcript may be processed into a cytoplasmic mRNA molecule, whereas in another type of cell, the transcript is simply degraded in the nucleus without ever producing a translatable mRNA. Cells can also process the same transcript in different ways; consequently, the same gene can have different forms of expression. An example of this *alternative processing* is provided in Figure 15-11.

REGULATING GENE EXPRESSION AT THE TRANSLATIONAL LEVEL

Just because a mature messenger RNA is able to escape from the nucleus and enter the cytoplasm does not guaran-

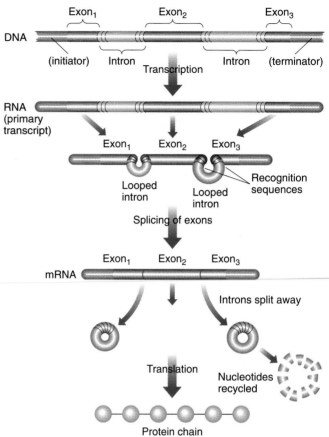

FIGURE 15-10

Editing a gene's message. As an RNA polymerase moves along a split gene, it generates a primary transcript whose length is equivalent to that of the gene itself. The final mRNA product is produced by RNA processing, whereby the sections corresponding to the introns are removed from the RNA transcript, and the remaining sections corresponding to the exons are spliced (joined) together. RNA processing must occur with absolute precision so that the nucleotides that code for necessary amino acids are not accidentally removed from the exons. Specific nucleotide sequences identify the beginning and the end of each intron to be removed. These signals pinpoint the location where the RNA is to be cut. As a result, the mRNA formed by the splicing process contains a continuous sequence of codons which specifies the sequence of amino acids for a polypeptide chain.

tee that it will be translated immediately. The cell regulates gene expression at the translational level as well. In some situations, mRNAs are temporarily "masked" by proteins, so they cannot be translated. Consequently, huge amounts of mRNA for certain proteins can accumulate in dormant cells, such as an unfertilized egg awaiting activation by a sperm. Once the dormant cell is activated, the mRNAs are "unmasked" by removal of the blocking proteins, and synthesis of the corresponding polypeptides takes place.

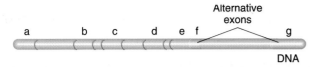

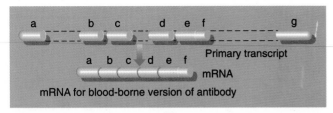

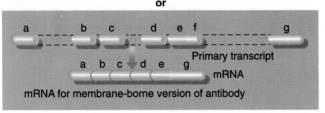

FIGURE 15-11

Orchestrating gene expression by alternative processing.
A single region of the DNA can be processed in more than one way so as to generate different mRNA molecules. The case depicted here concerns antibodies: disease-fighting proteins that occur (1) as blood-borne proteins and (2) as part of the plasma membrane of certain white blood cells. These two forms of an antibody—the blood-borne and the membrane-borne forms—have different amino acids at one end of the polypeptide. The two forms of the protein are translated on different mRNAs that are derived from the same primary transcript as a result of alternate processing. The two different mRNAs have different terminal sections (as indicated by the differently colored portions of the message).

Cells also regulate gene expression by controlling the life span of mRNAs. For example, protecting specific mRNAs from enzymatic degradation increases the number of times the mRNA can be translated into a polypeptide. Prolactin—the hormone that triggers milk production in mammary glands—operates by this mechanism. The cells of the mammary glands of a nursing mother must produce tremendous quantities of milk proteins in a sustained manner over a period of months or even years. It would be wasteful for these cells continually to synthesize and destroy the mRNAs that carry the message for milk proteins. Prolactin stabilizes these mRNAs, thus augmenting their translation. Withdrawal of prolactin decreases the stability of these mRNAs, which quickly deteriorate, causing milk protein production to stop almost immediately.

Understanding the mechanisms of gene control in eukaryotes is a primary goal of biological investigators. Not only does an understanding of gene control contribute to our comprehension of life on one of its most fundamental levels, but it also helps us learn how and why these controls malfunction in ways that lead to human disease.

Before a polypeptide can be synthesized in a eukaryotic cell a gene must be transcribed; the RNA transcript must be cut and spliced; and the mRNA product must enter the cytoplasm and become associated with a ribosome to be translated. Each of these activities is complex and serves as a step where gene expression can be regulated. (See CTQ #5.)

REEXAMINING THE THEMES

The Relationship between Form and Function

Regulation of gene expression at the transcriptional level requires gene regulatory proteins (such as the lac repressor or the testosterone receptor), whose shape can precisely recognize one specific sequence of nucleotides in the DNA among the millions of possible sequences present. While different regulatory proteins are shaped to recognize different sequences, many of them have "fingers" that fit into the grooves of the DNA at the site of recognition.

Biological Order, Regulation, and Homeostasis

Cells contain much more genetic information than they can possibly use at any one time. Bacterial cells typically express only those genes that code for proteins that are useful under existing environmental conditions. Eukaryotic cells typically express only those genes that are appropriate for that particular cell type. A host of different regulatory mechanisms intervene at various points along the path leading from gene to polypeptide, ensuring that only the appropriate proteins are manufactured.

Acquiring and Using Energy

In addition to preventing the breakdown of biological order, gene regulation in prokaryotic cells helps avoid needless waste of energy (and chemical resources) by preventing the synthesis of enzymes and metabolic products that are not useful at that time. For example, it would be a waste of energy and resources for a bacterial cell to produce enzymes that metabolize lactose when there is no lactose in the medium.

Unity within Diversity

Gene regulatory proteins that bind to specific DNA sequences regulate gene expression at the transcriptional level in all cells. The binding of these proteins either activates or represses the transcription of a nearby gene. Prokaryotic and eukaryotic cells have fundamentally different mechanisms by which these regulatory proteins act, however. Prokaryotes possess operons in which a sequence of adjacent genes is controlled by a single operator site. Large, multicellular eukaryotes may contain hundreds of different cell types, each with its own unique set of gene regulatory proteins.

Evolution and Adaptation

▐▶ The discovery that genes contain introns (noncoding, intervening sequences) has led to an important hypothesis concerning the mechanism by which new genes can arise during evolution. According to this hypothesis, each exon represents a distinct ancestral polypeptide. As the exons moved around from one place to another in the chromosomes, they came together in new combinations. Some of these combinations led to the formation of new genes whose "combined" polypeptides had new or improved functions, which gave the organism a selective advantage.

SYNOPSIS

All cells regulate which part of their genetic endowment is expressed.   Every cell has far more genes than it uses at any particular time. Cells possess mechanisms that determine which genes are turned on and which are turned off at a given time. Consider, for example, a bacterial cell living in a medium whose composition suddenly changes, or a eukaryotic cell transforming from an undifferentiated stage into a liver cell, or a cell in your reproductive tract being exposed to higher levels of a hormone. In each of these cases, the cell responds by activating new genes and repressing the expression of others.

In prokaryotes, the fundamental regulatory mechanism is the operon.   Operons are clusters of structural genes (and their control elements O, P, and R) that typically code for different enzymes in the same metabolic pathway. Since all of the structural genes are transcribed into a single mRNA, their expression can be regulated in a coordinated manner. The level of gene expression is controlled by a key metabolic compound, such as the inducer lactose, which attaches to a protein repressor and changes its shape. This event alters the ability of the repressor to bind to the operator site on the DNA, thereby allowing transcription.

Regulation of gene expression in eukaryotes is more complex than in prokaryotes.   Eukaryotic cells have much more DNA than do prokaryotes, and much of this DNA is never expressed. Most of the nonexpressed DNA lies between genes, but some of it (in the form of introns) lies within the coding regions of the genes. Introns require their removal from the RNA transcript during processing, but they are thought to have increased the rate of evolution by facilitating the formation of new genes by exon shuffling.

Eukaryotic gene expression is regulated primarily at three levels.   (1) Gene regulation at the transcriptional level determines whether or not a gene will be transcribed and, if so, how often. Gene regulatory proteins, such as the testosterone-receptor complex, are capable of binding to specific DNA sites and controlling the rate of transcription of nearby genes. (2) Genes are transcribed into large primary transcripts that must be cut and spliced to form the mature mRNA. Control of these activities constitutes regulation at the processing level. (3) Control of gene expression at the translational level centers on whether or not an mRNA is translated and how long the mRNA will survive.

Key Terms

gene regulatory protein (p. 295)
clone (p. 296)
totipotent (p. 297)
induce (p. 298)
repress (p. 299)
operon (p. 299)
structural gene (p. 299)

promoter (p. 299)
operator (p. 299)
repressor (p. 299)
regulatory gene (p. 299)
cell differentiation (p. 302)
intron (p. 303)
exon (p. 303)

transcriptional-level control (p. 304)
processing-level control (p. 304)
translational-level control (p. 304)
testosterone receptor (p. 304)
chromosomal puff (p. 305)
RNA processing (p. 307)
ribozyme (p. 307)

Review Questions

1. Describe the cascade of events responsible for the sudden changes in gene expression in a bacterial cell following the addition of the milk sugar lactose.

2. Which of the following statements are true? (1) The regulatory gene produces the promoter. (2) The inducer binds to the operator site. (3) The repressor binds to DNA in the presence of lactose. (4) The structural genes are transcribed sequentially into one long mRNA.

3. Cite an example of an RNA molecule forming hydrogen bonds between complementary nucleotides in the same molecule.

4. Describe the different levels at which gene expression is regulated to allow a beta-globin gene with the following structure to direct the formation of a protein that accounts for over 95 percent of the protein of the cell.

exon #1/intron/exon #2/intron/exon #3

5. How are the functions of a bacterial lac repressor and a human testosterone receptor similar? How are they different?

6. What is the primary difference between a primary transcript and the mRNA to which it gives rise?

Critical Thinking Questions

1. Suppose you were studying tobacco plants and found a gene that coded for a protein very similar to that of a bacterial cell that infects these plants. Can you suggest two alternate mechanisms by which these proteins could be so similar between these widely divergent organisms?

2. What are the adaptive advantages to organisms of selective gene expression that can change over time?

3. What is the advantage of clustering structural genes so that all the enzymes for a metabolic pathway are regulated together rather than independently?

4. In eukaryotes, only about 1 percent of the DNA ever codes for mRNA that is subsequently translated into proteins. In prokaryotes, the figure is above 90 percent. Explain how gene structure and gene expression in eukaryotes account for this difference. What are the advantages of the eukaryotic system? What are the disadvantages?

5. Hormones dramatically alter gene expression, as exemplified by the distinctions between men and women, which are fundamentally the result of sex hormones selecting different genes for activation. Create a model by which one hormone might temporarily activate a gene and another hormone might turn off that gene later in development. Your model can use any of the mechanisms discussed in this chapter.

Additional Readings

Beardsley, T. 1991. Smart genes. *Sci. Amer.* Aug:86–95. (Intermediate-Advanced)

Gibbons, A. 1991. Molecular scissors: RNA enzymes go commercial. *Science* 251: 521. (Intermediate)

Hoffman, M. 1991. Brave new (RNA) world. *Science* 254: 379. (Intermediate)

Keller, E. F. 1983. *A feeling for the organism: The life and work of Barbara McClintock.* New York: W. H. Freeman. (Introductory-Intermediate)

McKnight, S. L. 1991. Molecular zippers in gene regulation. *Sci. Amer.* April:54–64. (Intermediate)

Ptashne, M. 1989. How gene activators work. *Sci. Amer.* Jan:40–47. (Intermediate)

Rhodes, D. and Klug, A. 1993. Zinc fingers. *Sci. Amer.* Feb:56–65. (Intermediate)

Singer, M. and P. Berg. 1991. Genes and genomes. University Science Books. (Advanced)

CHAPTER
◄ 16 ►

DNA Technology: Developments and Applications

**STEPS
TO
DISCOVERY**
DNA Technology and Turtle Migration

GENETIC ENGINEERING

Genetically Engineered Cells and Human Proteins

Genetically Engineered Cells and Industrial Products

Recombinant Organisms at Work in the Field

Genetic Engineering of Domestic Plants

Genetic Engineering of Laboratory and Domestic Animals

Controversy over the Perceived Dangers of Genetic Engineering

DNA TECHNOLOGY I: THE FORMATION AND USE OF RECOMBINANT DNA MOLECULES

Tools for Assembling Recombinant DNA Molecules

Amplification of Recombinant DNAs by DNA Cloning

Expression of a Eukaryotic Gene in a Host Cell

DNA TECHNOLOGY II: TECHNIQUES THAT DO NOT REQUIRE RECOMBINANT DNA MOLECULES

The Separation of DNA Fragments by Length

Enzymatic Amplification of DNA

USE OF DNA TECHNOLOGY IN DETERMINING EVOLUTIONARY RELATIONSHIPS

BIOLINE

DNA Fingerprints and Criminal Law

THE HUMAN PERSPECTIVE

Animals That Develop Human Diseases

BIOETHICS

Patenting a Genetic Sequence

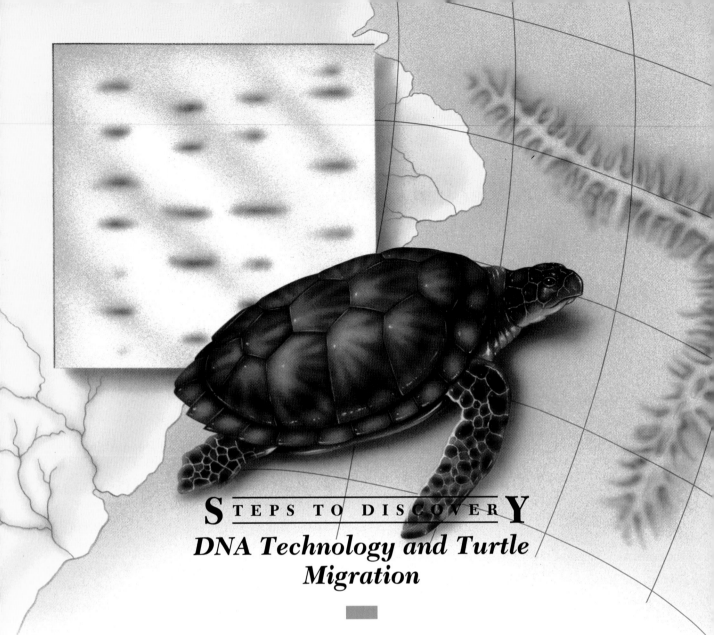

S TEPS TO DISCOVERY
DNA Technology and Turtle Migration

A well-studied population of green turtles lives most of the year off the coast of Brazil. Every winter, the adults from this population begin a month-long journey that takes them from Brazil to Ascension Island, a remote site in the mid-Atlantic Ocean. There, the turtles mate; the females lay their eggs on the beaches; and the animals return to their feeding grounds 2,000 kilometers (1,240 miles) away. Why don't these turtles simply lay their eggs in Brazil? Why do they make such a long, and seemingly unnecessary, journey?

About 20 years ago, Archie Carr, of the University of Florida, offered an explanation. An authority on sea turtles, Carr based his hypothesis on the theory of continental drift, which states that the earth's continents were not always

located in the same places as they are today. Rather, at one time—more than 100 million years ago—all of the continents were situated in relatively close proximity (see Chapter 35).

Carr further based his hypothesis on *natal homing*—the phenomenon whereby a green turtle always returns to the beach where it was hatched. According to Carr, a green turtle travels from Brazil to Ascension Island to breed because it was hatched on this island, as were its ancestors, and their ancestors, and so on, going back millions of years. In the distant past, however, the trip from the turtle's feeding grounds to its nesting grounds was very short because the continents were close together. As the continents gradually moved apart, natural selection favored animals that could

By using electrophoresis to compare the DNA patterns from different populations of sea turtles, biologists determined

make longer and longer journeys back to the place where they were hatched. What had once been a short commute eventually became a long sea odyssey.

A few years after Carr proposed his hypothesis, he was challenged by the evolutionary biologist Stephen Jay Gould. Gould argued that it was too much to expect that over a period of 100 million years conditions would remain suitable, year after year, for turtles to return to nesting grounds on Ascension Island. All it would take to break the link would be a series of "bad" years when conditions prevented a generation of turtles from making their long journey back home. If they could not reach Ascension Island, the turtles would have to start colonies at other sites, or perish as a population. Gould argued that the colonization of the Ascension Islands' breeding grounds was a relatively recent event.

You may be wondering what any of this has to do with DNA technology—the subject of this chapter. In fact, the techniques used to create and study DNA molecules can be used to probe other biological questions that are far removed from the subject of molecular biology, such as the mystery of turtle migration. As you will see in this chapter, the nucleotide sequence of a DNA molecule can be readily determined. This technique of DNA sequencing allows investigators to compare the genetic relatedness among various populations of animals on the basis of similarities in their nucleotide sequences, which brings us back to the green turtles.

Ascension Island is not the only site where green turtles go to breed; other populations nest on the beaches of Florida and Venezuela. If Carr's hypothesis is correct, and animals have returned to the same hatching site for hundreds of thousands of generations, then populations with different nesting grounds should have remained reproductively isolated from each other over a very long period of time. Reproductive isolation, in turn, leads to genetic divergence that should be reflected in differences in the nucleotide sequences in the DNA of members of different populations. How similar are the DNA sequences from animals that nest in these different parts of the Atlantic Ocean? A former graduate student of Carr's named Anne Meylan,

working with Brian Bowen and John Avise, two evolutionary biologists from the University of Georgia, recently investigated this question. The team indeed found distinct differences in the DNA from one population of turtles to another, indicating that they had been reproductively isolated from one another. But the differences were not nearly as great as one would expect if the isolation had continued for millions of years. Rather, the populations are estimated to have remained distinct for a few tens of thousands of years, at most.

The results suggest that the various turtle populations investigated are all descended from a population of ancestral turtles that once laid their eggs on a particular beach in the Atlantic Ocean. Over time, a number of descendants of these ancestral turtles failed to make it back to the beach where they were hatched. Instead, they made a "wrong turn" and ended up at a new nesting site, which is probably how the Ascension Island breeding ground became established. Regardless of how it came into being, the connection between Brazil and Ascension Island appears to be of relatively recent origin, which still leaves us wondering why these turtles travel so far to carry out their breeding activities. We can see from this example how biologists are able to use the recent advances in DNA technology as a tool for determining the evolutionary relationships among organisms—whether among populations of the same species or among different species that are only distantly related.

the approximate time the populations had separated.

Watchers of Wall Street were stunned. The company's stock began to sell the moment the opening bell sounded. Regarded by many as a novelty stock, it was the first company of its kind to be listed on the stock exchange, and its only products were unmarketable medicines years away from federal approval. Yet the stock caused a frenzy of buying and selling. By the end of its first day in 1976, it was worth $80 million!

The company was Genentech, the world's first stockholder-backed *biotechnology* organization, a company devoted to applying some of the recent lessons of molecular biology to manufacturing products of value in medicine, agriculture, and industry. Genentech was created as an offshoot of the scientific breakthroughs of a group of geneticists who have created new ways of putting organisms to work for people and the environment. In doing so, these geneticists have captured the attention of industrial leaders, members of the news media, and even representatives of the legal community. An important step in this "biotechnology revolution" came in 1980, when the U.S. Supreme Court ruled that "life" could be patented; that is, a company (or an individual scientist) can literally own the rights to the new life forms it "invents" in its laboratories. This issue is discussed further in Bioethics: Patenting a Genetic Sequence.

▼ ▼ ▼

GENETIC ENGINEERING

For thousands of years, people have been selectively breeding organisms to increase their value. They do so by mating individuals with desirable characteristics with each other to produce improved progeny. Such artificial breeding programs may even cross species lines, as illustrated by the emergence of a mule from the mating of a female horse and a male donkey. Although sterile, mules possess a useful combination of the genetic characteristics of both their parental species. Modern *genetic engineering* goes one step further, however; an organism's genotype can be modified according to human design by introducing genes that have never been present in the chromosomes of that particular species. This feat is accomplished using **recombinant DNA** molecules—molecules that contain DNA sequences derived from different biological sources that have been joined together in the laboratory. The techniques by which specific pieces of DNA are isolated, recombined, and intro-duced into host cells are discussed in the second section of this chapter. Let us first describe some of the medical, agricultural, and industrial applications to which this technology is being put to use.

GENETICALLY ENGINEERED CELLS AND HUMAN PROTEINS

The earliest successes in genetic engineering were achieved by the creation of strains of bacteria that would act as microscopic "factories," churning out molecular products specified by a newly acquired gene. Today, genetically engineered cells are used to manufacture biological products ranging from drain cleaners to medicines. One biotechnologist has referred to these cells as "the largest nonunion workforce in the world."

Insulin was the first usable human protein to be produced in bacterial cells. The gene was first synthesized chemically, nucleotide by nucleotide, in 1978 and was subsequently introduced into bacterial cells, which produced the human insulin molecule and released it into the culture medium. Even though it is produced in bacterial "factories," the product of the human gene is identical to human insulin and is used to treat tens of thousands of diabetics, who suffer from a deficiency in insulin.

Prior to genetic engineering, human proteins with medical uses were either unavailable or were harvested in very low quantities from whole organs, blood, or "nonengineered" cells grown in the laboratory. The protein being sought was usually present at very low levels amidst a sea of thousands of different, unwanted proteins. Separating the desired protein from the unwanted proteins was very expensive, and adequate purity was impossible to achieve. Just consider the number of hemophiliacs who received the AIDS virus from contaminated blood products and you can see the danger inherent in treating patients with these substances.

Human proteins are now being produced safely and efficiently through the use of genetically engineered cells. Dozens of proteins are awaiting approval by regulatory agencies, while others have become available in doctors' offices and hospitals throughout the world. For example, the protein tissue plasminogen activator (TPA) is used widely in emergency rooms to dissolve blood clots that block coronary arteries, causing heart attacks. Patients treated quickly with TPA are much less likely to suffer permanent damage to heart tissue than are untreated patients. Similarly, hemophiliacs can now receive pure blood-clotting factors without the risk of infection from blood-borne viruses, and children who are unusually short for their age can receive recombinant human growth hormone to help them grow to a normal height. Other products currently produced by recombinant DNA technology are listed in Table 16-1. Virtually any protein manufactured by any organism is a candidate for production by genetically engineered microbes or cultured cells.

◁ BIOETHICS ▷
Patenting a Genetic Sequence
By ARTHUR CAPLAN
Division of the Center for Biomedical Ethics at the University of Minnesota

Scientists at the National Institutes of Health (NIH) in Washington, D.C., have been very busy recently. They have been rummaging through cells obtained from a collection of human brains. By extracting the DNA from these cells, they can locate the genes that permit a stew of chemicals to become a bit of brain. The rapid progress being made in mapping these genes has set some other brain cells working— those located in the heads of the lawyers, venture capitalists, Wall Street analysts, corporate executives, and government officials who are wondering who will own these gene maps.

In 1991, the NIH scientists planned to publish some of their initial discoveries in the prestigious journal *Science.* Just before they did, the attorneys responsible for patents at the NIH got wind of the scientists' publication plans. The lawyers realized that if the newly discovered genetic

sequences appeared in print without a patent having been filed, then anyone could use them without having to pay fees or royalties. Consequently, the NIH attorneys postponed publication until they could file patent applications on the genetic sequences.

Many members of the scientific community were—and remain—outraged at the decision to seek patents on the information stored in brain cell genes. They feel that information about genetic language ought to be freely available to anyone who wishes to use it. Others feel that, while morally high-minded, it is doubtful whether this view can prevail. The free exchange of information may prove to be no competition when the stakes are in the tens of billions of dollars. Whoever controls knowledge about the genetic makeup of humans, plants, and animals will have a huge edge in getting products to the mar-

ket. Many people feel that the lawyers at the NIH know this and that is why they are seeking patents even though their scientists have got only a partial list of the genetic sequences in brain cells. Furthermore, without practical plans for applying this new knowledge, the patent may not be granted so soon. Demonstrating utility is one of the key requirements for getting a patent.

The patenting of genetic information seems inevitable, however. If that is the case, then even though the NIH is probably a bit premature, it might be onto something important. Do you feel it would be best if no proprietary ownership were granted over the genetic code? Assuming the huge sums of money to be made from licensing the genetic code are cycled back to the NIH to fund future biomedical research, how do you feel about the leading U.S. scientific institution owning the code?

TABLE 16-1
SOME PRODUCTS OF RECOMBINANT DNA TECHNOLOGY

Product	Activity
Interferons	Fight viral infection; boost the immune system; possibly effective against melanoma (a form of skin cancer) and some forms of leukemia; may help relieve rheumatoid arthritis.
Interleukin 2	Activates the immune system and may help in treating immune-system disorders. Although it produces serious side effects, the drug is proving valuable in treating kidney cancer.
Tumor necrosis factor (TNF)	Attacks and kills cancer cells. Presently being used in the first experimental attempt to treat human cancer by introducing cells carrying foreign genes (page 648).
Erythropoietin	Stimulates red blood cell production; may be used to combat anemia.
Beta-endorphin	The body's "natural morphine"; used to treat pain.
Metabolic enzymes	Perform a multitude of services, from catalyzing chemical reactions in the pharmaceutical industry to replacing defective human enzymes.
Vaccines (e.g., hepatitis B)	Stimulate the body's immunity to protect against disease-causing viruses and bacteria.

GENETICALLY ENGINEERED CELLS AND INDUSTRIAL PRODUCTS

Fuel shortages may be relieved by genetically engineered microbes that inexpensively store solar energy in the chemical bonds of combustible organic compounds, providing a virtually inexhaustible fuel source. Genetic engineers are attempting to improve on the efficiency of natural systems that already accomplish this energy conversion (Figure 16-1*a*). In addition, genetic engineers have created "oil-producing" microorganisms that generate highly combustible organic compounds that are equivalent to petroleum in their potential energy. Another supplier of energy is cellulose—the earth's most abundant renewable resource. Each year, well over 3.5 billion tons of paper, cardboard, and other cellulose-based products are manufactured in the United States (15 tons for every person in the country). Nearly 900 million tons of this material ends up as refuse. Microbes can convert this huge resource to one of two valuable fuels: ethanol (the combustible alcohol found in "gasohol") or methane (natural gas, Figure 16-1*b*).

(a)

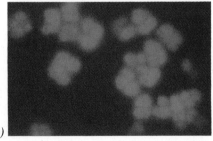

(b)

FIGURE 16-1

Two natural fuel producers. *(a)* Generally considered a nuisance that chokes waterways, this aquatic fern is being used to generate combustible fuel, methane. *(b)* Bacteria like these green fluorescent stained cells are especially prolific methane producers. Most biological fuel generators depend either on these kinds of cells or on different microbes that produce flammable ethanol.

RECOMBINANT ORGANISMS AT WORK IN THE FIELD

Some genetically engineered organisms must first make their way out of the lab and into the world before they can perform the task for which they were designed. One such organism has been constructed to reduce food loss when crops are subjected to freezing temperatures. Certain bacteria of the genus *Pseudomonas* that normally reside on plants produce a protein that acts as "seed crystals" around which water freezes; these bacteria are referred to as "ice-plus" cells (Figure 16-2). Freezing water expands and forms sharp crystals that crush cells and slash membranes. This type of cellular damage nourishes the bacteria residing on the plants, while possibly destroying the crop. The bacterial molecule that induces such water solidification is called the *ice-nucleating protein*.

Through recombinant DNA technology, a strain of bacteria has been developed that lacks the gene for producing the ice-nucleating protein. Called the "ice-minus" bacterium, this strain of *Pseudomonas* fails to induce ice formation, even when temperatures drop below $-4°C$ (25°F). When sprayed on plants, the engineered ice-minus bacteria compete with normal ice-forming bacteria, displacing them and reducing crop loss. In 1987, after a lengthy court battle over the risks involved in releasing genetically engineered organisms in the environment, ice-minus bacteria (under the trade name Frostban) were sprayed on a plot of strawberry plants in California, becoming the first genetically engineered microorganism to be legally released into the environment.

Genetically engineered microbes can also increase crop yields in other ways. We add fertilizer to plants to provide them with nitrogen-containing compounds for use in manufacturing protein. Some crops, such as soybeans and peas, are able to manufacture their own nitrogen-containing compounds from the nitrogen gas (N_2) in the air—a process known as *nitrogen fixation*. It is not the plant cells themselves that are able to "fix" N_2 but bacteria of the genus *Rhizobium* that live in nodules on the plant's roots. Plants that harbor these bacteria are called *legumes*. Genetically engineered strains of *Rhizobium* have increased soybean yields by 50 percent. Similarly, genetic engineers are trying to develop a strain of *Azotobacter*, a free-living, nitrogen-fixing bacterium, which would grow on the roots of nonleguminous plants, such as corn, freeing these plants from their need for applied nitrogenous fertilizer.

Genetic engineers are also releasing their microbial work force against pollution and toxic wastes, which pose a serious threat to the biosphere. Many of today's toxic pollutants are new synthetic compounds. Organisms capable of decomposing these chemicals have not yet had a chance to evolve. Consequently, the compounds accumulate to dangerous levels in the environment. Genetic engineers are trying to accelerate evolution in the laboratory by developing bacteria that can quickly degrade several types of poi-

FIGURE 16-2

The ice crystals seen on these strawberry leaves form around a bacterial *ice-nucleating protein.* This frost damage could have been prevented if the plants had been sprayed with the genetically engineered "ice minus bacterium."

sonous compounds; an oil spill, for example, would be one big meal for these bacteria (Figure 16-3).

GENETIC ENGINEERING OF DOMESTIC PLANTS

Biotechnologists are also attempting to modify the genetics of multicellular organisms. For example, genetic engineering can be used to improve plants by producing crop species that are more resistant to drought, disease, poor soil conditions, and chemical pesticides and herbicides. It has been known for many years that plants inoculated with a mild form of a virus may become resistant to more devastating strains. While the mechanism of plant "immunity" is unclear, the phenomenon has led to successful attempts by biotechnologists to induce viral resistance in certain crop plants. For example, Roger Beachy and his colleagues at Washington University introduced a gene into tobacco plants that codes for the outer coat protein of tobacco mosaic virus (TMV). Once incorporated into the chromosomes of the plants, the tobacco plants exhibited a marked resistance to TMV infection. The same technique has been used to protect tomato plants and potato plants from similar viruses.

▐▶ Genetic engineers are also developing plants that produce insect-killing toxins—"one bite and the pest dies." The best-studied protein toxins are produced by the bacterium *Bacillus thuringiensis* (BT). Different strains of the bacterium produce different varieties of the toxin, which are effective against different insect pests, including the tobacco hornworm and the gypsy moth. In 1986, a plant that had been genetically engineered with a BT toxin gene suc-

cessfully passed a field test, growing undamaged by pests that ravaged the "unprotected" plants close by. More recently, however, agricultural researchers are finding that insect pests are becoming resistant to the toxin produced by the genetically engineered plants, just as they would have become resistant if the substance had simply been sprayed over the field in the form of a pesticide. Researchers are hoping to stay one step ahead of the insects' evolutionary adaptations by engineering variations in the toxin's structure so that the insects do not have an opportunity to become immune.

FIGURE 16-3

Oil-eating bacteria. Workers are spraying a fertilizer solution on this oil-contaminated shore to stimulate the activity of oil-eating bacteria. This procedure greatly accelerates the rate of oil removal.

In another effort, genetic engineers have recently developed a new strategy to improve the quality of tomato plants. "Mushy" tomatoes result from the presence of an enzyme (polygalacturonase) that breaks down the plants' cell walls. In order to keep the tomatoes firm, farmers pick the fruits before they are ripe, redden them artificially, and then transport them to market. A tastier product could be marketed if the tomatoes were allowed to ripen on the vine and then kept from becoming mushy during transport. This is now possible through genetic engineering. Recall from Chapter 14 that only one strand of the DNA (the *sense strand*) is transcribed by an RNA polymerase; this strand contains the information for the encoded polypeptide. Consider what might happen if the other strand (the *antisense strand*) were also transcribed. The cell would contain RNA molecules with complementary sequences that have the potential to bind to one another, forming a double-stranded RNA molecule that could not be translated into the encoded protein. Recombinant DNA technologists

have been able to create genes in which the "wrong" strand is transcribed, including the gene that codes for the enzyme that causes mushy tomatoes. When this "reversed" gene is introduced into the cells of a tomato plant, the antisense RNA from the foreign gene interferes with translation of the sense RNA from the original gene, and production of the troublesome enzyme is reduced by 99 percent. The result is firm, "mush-resistant" tomatoes.

Taken together, these various approaches to improving the quality of crop plants provide an unparalled opportunity to increase the quantity of food that can be produced on a given amount of land. However, we should not lose sight of the fact that technological solutions often create unforseen problems that require new technological solutions. In the 1950s and 1960s, for example, plant geneticists created new strains of certain plants, particularly rice and wheat, which grew much more rapidly than did the original strains and led to much greater crop yields in many underdeveloped countries. Some of these countries became grain exporters,

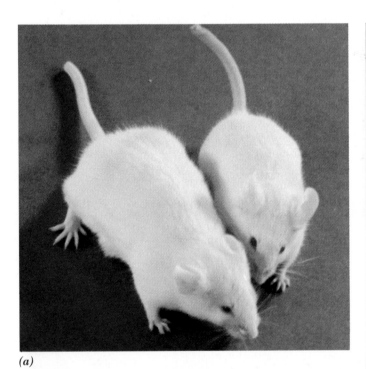

(a)

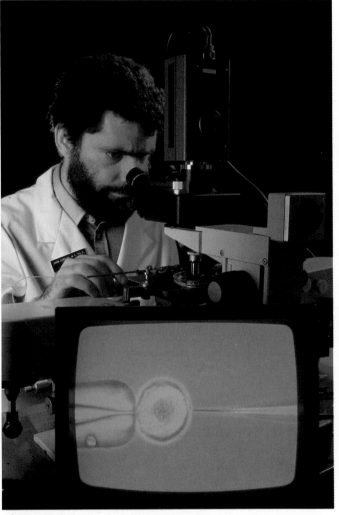

(b)

FIGURE 16-4

Transgenic mice. *(a)* A pair of littermates at 10 weeks of age. The larger mouse carries copies of rat growth hormone genes in all of its cells, which has caused this transgenic mouse to be much larger than its littermate, who lacks the rat gene. *(b)* The procedure by which a fertilized egg is injected with foreign DNA containing a gene that will become integrated into the egg's chromosomes. The video screen reveals the image seen by the researcher injecting the DNA. The egg is seen near the middle of the screen; the glass tube on the left holds the egg in place by suction, while the glass micropipette on the right contains the DNA to be injected.

rather than grain importers. This "Green Revolution," as it has been called, was not without problems. For example, the new strains relied heavily on the use of artificial fertilizers, which may be difficult to obtain and can contribute to serious pollution problems. Furthermore, increases in human population growth have virtually obliterated the gains of the Green Revolution, illustrating that no technological advance in agriculture will be successful until worldwide population control has been achieved.

GENETIC ENGINEERING OF LABORATORY AND DOMESTIC ANIMALS

In 1982, a litter of mice was born that included individuals unlike any mouse that had ever been born before; the chromosomes of these mice contained rat genes! Within a few weeks, those mice containing the rat genes were huge, compared to their littermates (Figure 16-4a). The increased size was not due to the fact that rats are larger than mice but to the fact that the foreign genes that had become integrated into the mouse chromosomes coded for the protein growth hormone (GH). When these rat GH genes were activated, the mice produced excess quantities of growth hormone, which led to their increased size and weight.

How did this rat gene get into the chromosomes of unborn mice? In order to accomplish this feat, Ralph Brinster of the University of Pennsylvania and Richard Palmiter of the University of Washington isolated fertilized mouse eggs from the reproductive tracts of female mice. They injected the nucleus of each egg with about 600 copies of the rat GH gene (Figure 16-4b). Some of these injected DNA fragments became integrated into the chromosomes of the injected eggs. These eggs were then implanted in the reproductive tracts of "surrogate mothers," where they developed normally to term. Because the foreign DNA was integrated into the chromosomes, it was passed on during mitosis to every cell of the fetus. When this first litter of genetically engineered mice reached maturity, the GH genes passed into the gametes and were transmitted to the next generation of offspring, which again exhibited exceptional growth. Animals that possess genes of a different species, such as these mice, are called **transgenic animals.**

These early experiments with transgenic mice have laid the groundwork for the development of genetically engineered livestock carrying genes that improve their food value. For example, pigs born with foreign growth hormone genes incorporated into their chromosomes (Figure 16-5) grow much leaner than do control animals lacking the genes. The meat of the former animals is leaner because the excess growth hormone stimulates the conversion of nutrients into protein rather than fat. Transgenic animals are also playing an important role in clinical laboratories (see The Human Perspective: Animals That Develop Human Diseases).

FIGURE 16-5

Transgenic "pork." A pair of young pigs who have developed from eggs that had been injected with foreign genes. Transgenic pigs containing foreign growth hormone genes produce leaner meat.

◁ THE HUMAN PERSPECTIVE ▷
Animals That Develop Human Diseases

(a)

(b)

FIGURE 1

(a) This woman suffers from Waardenburg's syndrome, a rare genetic disorder characterized by deafness, mismatched, widely spaced eyes, and a white forelock of hair. *(b)* The Splotch mouse shown on the left is an animal model of Waardenburg's syndrome. This mouse has a white forelock and tiny eyes compared with a normal mouse (right).

Animal models are laboratory animals that are susceptible to a particular human disease. Animal models provide one of the most important tools in evaluating the effectiveness of new drugs and therapies against human disorders. For example, a certain strain of rabbit has a defective gene for the LDL receptor (page 146) and, consequently, has very high blood-cholesterol levels. Another example is illustrated in Figure 1. Animal models are an important tool in the development of drugs for treating the corresponding condition in humans. But where do we turn when there are no animals with disorders similar to ours? Recent advances in molecular biology have opened the door to the possibility of "creating" animal models.

Sickle cell anemia is a disease from which no animal other than humans suffers—at least this was the case before 1990. In order to produce an animal model for sickle cell anemia, transgenic mice that carried the sickle cell version of the human globin gene were developed. Two mice were born that produced relatively high levels of the sickle cell type of hemoglobin in their red blood cells, and some of these cells exhibited the characteristic sickled shape. For unknown reasons, however, the mice showed no evidence of either anemia or of an elevated rate of red blood cell destruction, both characteristics of the human disease. These mice probably failed to produce enough of the human protein to trigger the disease's symptoms. Whatever the reason, the procedure is being refined, and animal models for the disease will likely be developed in the near future. Similar work is under way for other human genetic diseases, and it is hoped that a new era may be dawning in which experimental drugs can be more rapidly and safely tested on genetically engineered animal models.

Transgenic animals (and transgenic plants) have also been tested as "living factories" for the production of human proteins. As a result of pioneering work at the University of Edinburgh in Scotland, transgenic sheep have been developed which carry human genes for either Factor IX (a blood-clotting factor absent in some hemophiliacs) or alpha-antitrypsin (a protein used in the treatment of emphysema, a lung disease in which patients cannot inhale sufficient air). In both cases, the human gene is joined to a sheep gene that codes for a milk protein. DNA containing the fused genes is then injected into fertilized sheep eggs, which develop into sheep whose milk contains sufficient quantities of the human protein to be purified for clinical use. The next time you see farm animals grazing on the countryside, you may be looking at a living pharmaceutical plant!

CONTROVERSY OVER THE PERCEIVED DANGERS OF GENETIC ENGINEERING

In spite of its successes, releasing genetically engineered organisms into the environment has elicited considerable opposition from groups concerned that such practices might upset the balance of ecosystems or may create other unforseen catastrophies. For example, the release of ice-minus bacteria elicited a great deal of opposition. Some local farmers feared an accidental contamination of their fields with these engineered bacteria. Most of the opposition arose from a public fear of genetic engineering in general, a technology many suspected would lead to a world full of harmful genetic "creations." Even though the ice-minus bacterium showed no evidence of being harmful, injunctions, court battles, and even vandalism plagued the project. Finally, the project directors decided to isolate a naturally occurring ice-minus bacterium to use in place of the genetically engineered strain.

Use and release of genetically engineered organisms is subject to careful scrutiny by regulatory agencies and by the scientific community itself. Even though thousands of experiments have been carried out without any apparent danger, concern over this new technology is still prevalent among the general public. Perhaps the answer to realizing the benefits of biotechnology without promoting fear in the general community is to keep the general population better informed and to create well-publicized regulatory agencies to monitor the development and use of genetically engineered organisms. Regulatory controls should ensure public safety while also guarding against an atmosphere that retards progress in biotechnology.

In addition to working with plants and animals, genetic engineers are studying ways to cure inherited human diseases by "gene therapy"—implanting an effective gene in place of a defective or missing one. We will return to this provocative new development in medicine in Chapter 17.

In the meantime, let's turn to some of the specific techniques that have made these practical applications of genetic engineering possible.

Scientists can now isolate specific genes, which can be incorporated into the chromosomes of host cells to create new, genetically engineered life forms. Using this technology, cells can be converted into living pharmaceutical plants; crop plants and domestic animals can be modified to provide additional food for a hungry human population; and it is hoped that one day, individuals suffering from debilitating genetic diseases may be cured. (See CTQ #2.)

DNA TECHNOLOGY I: THE FORMATION AND USE OF RECOMBINANT DNA MOLECULES

All of the biotechnological successes described thus far depended on the introduction of "new" genes into a recipient "host" cell. This feat is not as simple as it may sound and actually required years of research before it was successfully accomplished. First, a recombinant DNA molecule must be formed by inserting, or **splicing,** the gene in question into a larger molecule of DNA. This DNA then acts as a transport vehicle, or *vector,* carrying the desired gene into the host cell. Common vectors include viruses and bacterial plasmids. Viruses carry the foreign gene into a host cell during an infection; the viral DNA, together with the foreign gene, then becomes integrated into the host cell's chromosome. **Plasmids** (Figure 16-6) are small, circular DNA molecules found in bacteria, which are separate from the main bacterial chromosome. Plasmids often contain genes that make bacterial cells resistant to antibiotics such as ampicillin and tetracycline, which are commonly used to treat bacterial infections.

TOOLS FOR ASSEMBLING RECOMBINANT DNA MOLECULES

One of the first steps in constructing most recombinant DNA molecules is to fragment large DNA molecules into smaller pieces. This is accomplished by **restriction enzymes,** special DNA-cutting enzymes that are found in bacteria. Unlike eukaryotic DNA-digesting enzymes (such as the DNase produced by your pancreas), restriction enzymes are *sequence specific,* meaning that they recognize specific DNA sequences four to six nucleotides long and make their incision within that sequence (Figure 16-7). Since particular nucleotide sequences of this length occur quite frequently simply by chance, they appear in all types of DNA—viral, bacterial, plant, and animal—and are thus susceptible to fragmentation by these enzymes. Bacteria

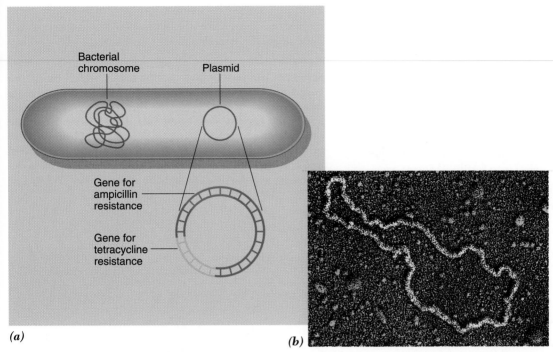

(a)

(b)

FIGURE 16-6 ————————————————————————————

A bacterial plasmid.   In addition to its chromosome, many bacteria contain small circular strands of DNA, called plasmids. The plasmid shown in part **(a)** contains genes that make the bacterial cells resistant to two antibiotics, tetracycline and ampicillin (a relative of penicillin). Consequently, a bacterium with this plasmid can live in the presence of these antibiotics. This particular plasmid is commonly used in recombinant DNA experiments. The photograph **(b)** is an electron micrograph of a bacterial plasmid.

FIGURE 16-7 ————————

Restriction enzymes are the molecular biologist's DNA-cutting scissors. In the case depicted here, a restriction enzyme called *Eco* R1 cuts both strands of a DNA double helix at a particular DNA sequence (in this example, after every "TTAA" sequence). This restriction enzyme makes staggered cuts that generate DNA fragments possessing "sticky ends."

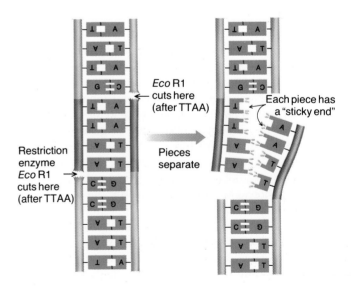

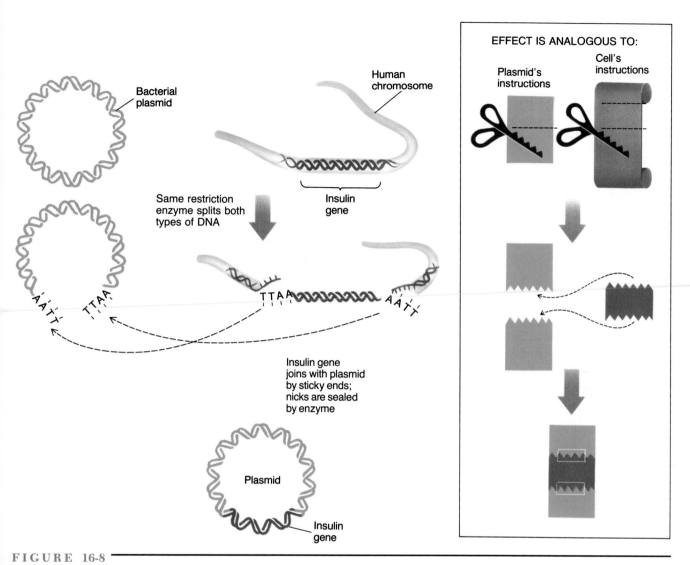

EFFECT IS ANALOGOUS TO:

Plasmid's instructions

Cell's instructions

FIGURE 16-8

Formation of recombinant DNA molecules. In this example, both plasmid DNA and human DNA containing the insulin gene are treated with the same restriction enzyme so that the DNA from the two sources will have complementary "sticky ends." As a result, the two DNA molecules become joined to each other and are then covalently sealed by DNA ligase, forming a recombinant DNA molecule.

use their restriction enzymes to destroy ("restrict") the DNA of invading viruses, while genetic engineers have used the enzymes as molecular "scissors" to "cut and paste" DNA molecules obtained from different sources.

As illustrated in Figure 16-7, many restriction enzymes generate a "staggered cut" in the DNA because the two strands of the double helix are nicked at different sites along their length. Staggered cuts leave short, single-stranded tails that act as "sticky ends" that can bind with a complementary single-stranded tail on another DNA molecule to restore a double-stranded structure.

Since a particular restriction enzyme always cuts a DNA molecule at the same site in a sequence, it creates the same sticky ends regardless of the organism that donated the DNA. For example, the sticky ends of a genetic segment

cut from a human chromosome readily adhere to the complementary single-stranded tails of a bacterial plasmid that has been cut with the same enzyme (Figure 16-8). When these two DNA preparations are mixed together, the complementary sticky ends join the isolated human genetic segment to the plasmid DNA, much as an extra paragraph may be taped into a sheet of instructions. As long as both papers are cut with scissors that make the same pattern, the cuts fit, and the pieces can be joined together. The "tape" that permanently bonds the joined DNA fragments together is **DNA ligase,** an enzyme that seals the gaps in the DNA backbone by forming a covalent bond (page 55). The result is a recombinant DNA molecule—in this case, a bacterial plasmid containing a gene that codes for a human protein. The first recombinant DNA molecule—a frog gene joined to a bacterial plasmid—was formed in this manner in 1973 by Stanley Cohen of Stanford University and Herbert Boyer of the University of California, marking the birth of modern genetic engineering.

AMPLIFICATION OF RECOMBINANT DNA BY DNA CLONING

Before the recombinant DNA molecule formed by the procedure discussed above can be put to much use, it is necessary to generate a relatively large quantity of the gene(s) needed. This DNA amplification process can be accomplished by a procedure called **DNA cloning** (Figure 16-9). The first step in this procedure is to get the recombinant DNA molecule into a host bacterial cell. This is surprisingly easy since bacteria readily take up DNA from their medium.

Once inside a bacterium, the plasmid DNA molecule replicates in each cell before the cell divides in two; therefore the number of recombinant DNA molecules increases, as does the number of bacterial cells. A geneticist can begin with a single recombinant plasmid inside a single bacterial cell and, after a period of time, millions of copies of the plasmid will be formed. Once the desired amount of DNA replication has been reached, the amplified recombinant DNA can be purified (Figure 16-10) and used in other procedures.

EXPRESSION OF A EUKARYOTIC GENE IN A HOST CELL

⚠ In addition to serving as cellular "copying machines" for generating large quantities of recombinant DNA, host bacterial cells can transcribe and translate a human gene residing within a bacterial plasmid, producing large amounts of high-quality human protein. This process requires that the DNA encoding a human protein be placed in the proper position next to a bacterial promoter (otherwise, the bacterial RNA polymerase cannot attach to the DNA and transcribe it). The human DNA used in these processes must also be free of introns (page 303) since bacterial cells do not have the machinery required to remove the intervening sequences from the transcribed message.

Biotechnologists are not restricted to using bacterial cells as such living pharmaceutical "factories." Similar techniques have been developed in which mammalian cells serve as hosts for human genes. These cells generally prove to be better suited than do bacteria for manufacturing human gene products and, like bacteria, can be grown to high density in large vats, where they secrete human proteins (such as those listed in Table 16-1) into the medium.

Work involving recombinant DNA technology has once again revealed the basic unity of life. Enzymes isolated from bacterial cells are used to prepare fragments of bacterial, viral, or eukaryotic DNA and to join them into a single,

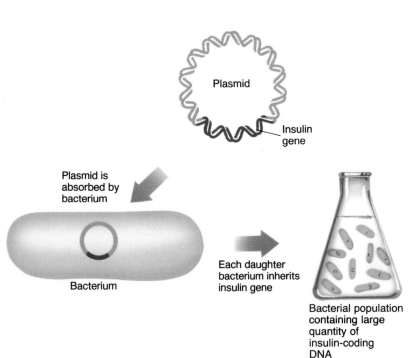

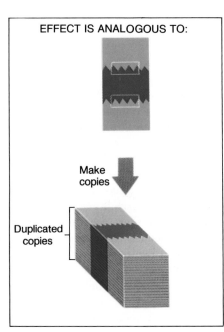

EFFECT IS ANALOGOUS TO:

Make copies

Duplicated copies

Plasmid

Insulin gene

Plasmid is absorbed by bacterium

Bacterium

Each daughter bacterium inherits insulin gene

Bacterial population containing large quantity of insulin-coding DNA

FIGURE 16-9

DNA cloning produces large quantities of a particular recombinant DNA molecule inside bacterial cells. The recombinant molecules are taken up by the bacterial cells and become amplified as the bacterial cells proliferate.

FIGURE 16-10

Purifying DNA. This centrifugation tube can be seen to contain two separate bands. One band represents plasmid DNA carrying a foreign DNA segment that has been cloned within the bacteria, while the other band contains chromosomal DNA from these same bacteria. The two types of DNA are separated during centrifugation. The researcher is attempting to remove the DNA from the tube with a needle and syringe. The DNA in the tube is made visible by using a fluorescent dye.

continuous DNA molecule. These recombinant DNA molecules can be introduced into either a bacterial or a eukaryotic cell, where they direct the formation of the same gene product as that which is in the cells of the species from which they were taken.

Recombinant DNA molecules form the cornerstone of the recent advances in genetic engineering. Formation and use of recombinant DNA have required the development of a new arsenal of molecular techniques in which specific DNA segments can be isolated, amplified, and inserted into suitable vectors. (See CTQ #3.)

DNA TECHNOLOGY II: TECHNIQUES THAT DO NOT REQUIRE RECOMBINANT DNA MOLECULES

The examples of biotechnology described in the beginning of this chapter require the formation of a recombinant DNA molecule so that a foreign gene can be expressed in a host cell. Several other techniques that are also of great importance to the "biotechnology revolution" do not require the use of recombinant DNA. These techniques are also important for understanding human genetics, which will be discussed in the next chapter.

THE SEPARATION OF DNA FRAGMENTS BY LENGTH

As you will recall, the DNA molecules present in eukaryotic chromosomes are much larger than DNA molecules that make up the circular chromosome of a bacterial cell. When molecular biologists treat a sample of eukaryotic DNA with a restriction enzyme, thousands of different DNA pieces of varying lengths are generated. These pieces are called **restriction fragments.** It is important for most purposes to separate restriction fragments so that individual fragments can be identified. For example, suppose you have just treated a sample of human DNA with a restriction enzyme and you want to isolate any restriction fragments that contain part of the insulin gene. The easiest way to separate a large variety of restriction fragments is by a technique called **gel electrophoresis** (Figure 16-11), whereby DNA fragments move through a porous gel in response to an electric current. While the negative charge of the DNA fragment causes the DNA fragment to move, it is the length of the fragment that determines the actual speed of its movement. The smaller the DNA fragment, the faster it can "work its way" through the pores in the gel. Consequently, smaller fragments move farther down the gel than do larger fragments.

After electrophoresis, it is possible to determine where in the gel a particular fragment, such as one containing the insulin gene, is located. One DNA fragment is distinguished from others by virtue of differences in its nucleotide sequence. The search for specific DNA fragments in a giant "haystack" of different DNAs requires a **DNA probe** —a single-stranded DNA possessing a sequence complementary to the fragments being sought. The probe is like a guided missile, capable of searching out and binding to a particular target. In this case, the target in the gel is a single-stranded fragment of the insulin gene. Since the DNA probe contains radioactive atoms, an investigator can find the insulin gene by locating the radioactive emissions given off by the bound probe. The radioactive band(s) in the gel can be detected after electrophoresis by pressing the gel against a piece of X-ray film.

Determining DNA Nucleotide Sequences

Reported in 1965, the first nucleic acid to be sequenced was a strand of transfer RNA (tRNA) that was 77 nucleotides long and obtained from bacteria. It had taken approximately 7 years of painstaking work to sequence this nucleic acid, earning Robert Holley of Cornell University a Nobel Prize. Today, a fragment of DNA several times this length can be sequenced within an hour using a totally automated apparatus. Attempts to determine the complete sequence of two multicellular eukaryotes—the fruit fly *Drosophila* and a nematode worm—are now under way, and plans to launch a full-scale offensive to determine the entire nucleotide sequence of human DNA has been the subject of great plans and debate for several years. This undertaking is discussed in the Bioline box in Chapter 17.

ENZYMATIC AMPLIFICATION OF DNA

Within the past few years, a new technique has become available for amplifying a specific DNA fragment (such as the insulin gene), which does not require the use of a bacte-

rial cell. As we noted on page 131, the DNA-synthesizing enzyme (DNA polymerase) from hot springs bacteria is stable at temperatures of 90°C. Biotechnologists have taken advantage of this enzyme to develop a technique called **polymerase chain reaction (PCR),** whereby the temperature is raised far above that which would destroy the activity of other DNA polymerases. Using PCR, a single region of DNA, present in vanishingly small amounts, can be amplified cheaply, rapidly, and extensively.

To carry out PCR, a sample of DNA is mixed with the temperature-resistant DNA polymerase, along with short, synthetic DNA fragments (called *primers*) that are complementary to DNA sequences at either end of the region of the DNA to be amplified. The mixture is then heated above

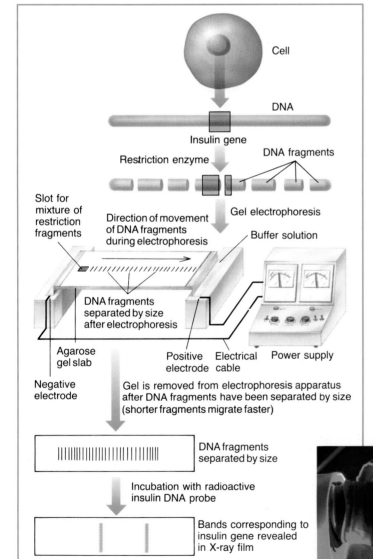

(a)

(b)

FIGURE 16-11

Separation of DNA restriction fragments by gel electrophoresis. *(a)* In the experiment outlined here, the researcher is attempting to determine the size of the restriction fragment(s) containing the insulin gene. To carry out gel electrophoresis, a thin slab of a gel (made of the polysaccharide agarose, obtained from seaweed) is prepared, and the sample containing the mixture of DNA fragments is placed in a small slot, or "well," located at one end of the gel. The slab (and the DNA sample) is then subjected to an electric current. Since DNA molecules have many negatively charged phosphate groups, they move through the porous gel toward the oppositely charged positive pole. But the gel itself acts like a sieve so that smaller fragments are able to make their way through the molecular obstacle course more easily than larger fragments. Consequently, the smaller the fragment, the faster it moves. After a period of time, the fragments become separated according to their length. In order to locate the DNA sequences that make up the insulin gene, the separated fragments are incubated in a liquid containing a radioactive insulin DNA probe that binds to the complementary DNA fragment(s). The radioactive band(s) can be located by laying the DNA-containing sheet onto X-ray film in a lightproof container. The radioactive particles expose the X-ray film, revealing the location of two distinct fragments that make up the insulin gene. *(b)* In *(a)*, specific bands of DNA were located following electrophoresis using a radioactive probe. In *(b)*, the DNA in the gel is revealed following electrophoresis by using a fluorescent dye that stains all of the DNA fragments.

75°C, which is hot enough to cause the DNA molecules in the sample to separate into their two component strands (Figure 16-12). The mixture is then cooled to allow the primers to bind to the two ends of the single-stranded target DNA. The polymerase then binds to the primers and selectively copies the intervening region, forming a new complementary strand. The temperature is raised once again, causing the newly formed and the original DNA strands to separate from each other. The sample is then cooled to allow the synthetic primers in the mixture to bind to the ends of the target DNA, which is now present at twice the original amount. This cycle is repeated over and over again, each time doubling the amount of the specific region of DNA that is flanked by the bound primers. Billions of copies of this one specific region can be generated in just a few hours.

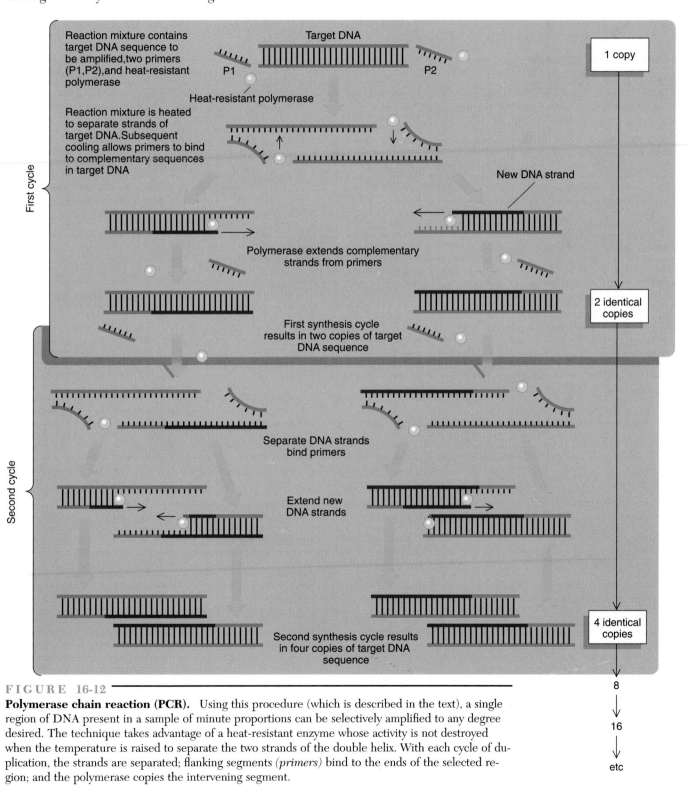

FIGURE 16-12
Polymerase chain reaction (PCR). Using this procedure (which is described in the text), a single region of DNA present in a sample of minute proportions can be selectively amplified to any degree desired. The technique takes advantage of a heat-resistant enzyme whose activity is not destroyed when the temperature is raised to separate the two strands of the double helix. With each cycle of duplication, the strands are separated; flanking segments *(primers)* bind to the ends of the selected region; and the polymerase copies the intervening segment.

◁ B I O L I N E ▷
DNA Fingerprints and Criminal Law

On February 5, 1987, a woman and her 2-year-old daughter were found stabbed to death in their apartment in the New York City borough of the Bronx. Following a tip, the police questioned a resident of a neighboring building. A small bloodstain was found on the suspect's watch, which was sent to a laboratory for DNA fingerprint analysis. The DNA from the white blood cells in the stain was amplified using the PCR technique and was digested with a restriction enzyme. The restriction fragments were then separated by electrophoresis, and a pattern of labeled fragments was identified with a radioactive probe. The banding pattern produced by the DNA from the suspect's watch was found to be a perfect match to the pattern produced by DNA taken from one of the victims. The results were provided to the opposing attorneys, and a pretrial hearing was called in 1989 to discuss the validity of the DNA evidence.

During the hearing, a number of expert witnesses for the prosecution explained the basis of the DNA analysis. According to these experts, no two individuals, with the exception of identical twins, have the same nucleotide sequence in their DNA. Moreover, differences in DNA sequence can be detected by comparing the lengths of the fragments produced by restriction-enzyme digestion of different DNA samples. The patterns produce a "DNA fingerprint" (Figure 1) that is as unique to an individual as is a set of conventional fingerprints lifted from a glass. In fact, DNA fingerprints had already been used in more than 200 criminal cases in the United States and had been hailed as the most important development in forensic medicine (the application of medical facts

FIGURE 1

Alec Jeffreys of the University of Leicester, England, examining a DNA fingerprint. Jeffreys was primarily responsible for developing the DNA fingerprint technique and was the scientist who confirmed the death of Josef Mengele.

to legal problems) in decades. The widespread use of DNA fingerprinting evidence in court had been based on its general acceptability in the scientific community. According to a report from the company performing the DNA analysis, the likelihood that the same banding patterns could be obtained by chance from two *different* individuals in the community was only one in 100 million.

What made this case (known as the Castro case, after the defendant) memorable and distinct from its predecessors was that the defense also called on expert witnesses to scrutinize the data and to present

their opinions. While these experts confirmed the capability of DNA fingerprinting to identify an individual out of a huge population, they found serious technical flaws in the analysis of the DNA samples used by the prosecution. In an unprecedented occurrence, the experts who had earlier testified *for the prosecution* agreed that the DNA analysis in this case was unreliable and should not be used as evidence! The problem was not with the technique itself but in the way it had been carried out in this particular case. Consequently, the judge threw out the evidence.

In the wake of the Castro case, the use of DNA fingerprinting to decide guilt or innocence has been seriously questioned. Several panels and agencies are working to formulate guidelines for the licensing of forensic DNA laboratories and the certification of their employees. In 1992, a panel of the National Academy of Sciences released a report endorsing the general reliability of the technique but called for the institution of strict standards *to be set by scientists.*

Meanwhile, another issue regarding DNA fingerprinting has been raised and hotly debated. Two geneticists, Richard Lewontin of Harvard University and Daniel Hartl of Washington University, coauthored a paper published in December 1991, suggesting that scientists do not have enough data on genetic variation within different racial or ethnic groups to calculate the odds that two individuals—a suspect and a perpetrator of the crime—are one and the same on the basis of an identical DNA fingerprint. The matter remains an issue of great concern in both the scientific and legal communities and has yet to be resolved.

In addition to its use in the amplification of specific DNA fragments, PCR is able to generate large amounts of DNA from minute starting samples. For example, PCR has been used in criminal investigations to generate quantities of DNA from a spot of dried blood left on a crime suspect's clothing or even from the DNA present in part of a single hair follicle left at the scene of a crime (see Bioline: DNA Fingerprinting and Criminal Law). In 1992, PCR was used to amplify DNA extracted from the exhumed bones of a man who had drowned several years earlier while swimming off the coast of Brazil. Some people claimed that the man was Dr. Josef Mengele, the infamous Nazi doctor who had carried out sadistic experiments on inmates at the Auschwitz concentration camp. Many skeptics doubted the identification, however. To confirm the man's identity, DNA from the bones was isolated, amplified by PCR, and compared to a DNA sample donated by Mengele's son, using DNA fingerprinting. The results were conclusive: The drowning victim was indeed Josef Mengele, closing the chapter on one of the most intensive international manhunts in history.

DNA fragments generated by restriction enzymes have applications beyond genetic engineering. Separation and analysis of DNA restriction fragments form the basis for DNA fingerprinting, constructing detailed genetic maps, and determining evolutionary similarity among species. (See CTQ #4.)

USE OF DNA TECHNOLOGY IN DETERMINING EVOLUTIONARY RELATIONSHIPS

▮▶ In the beginning of this chapter, we saw how DNA sequencing techniques were used to shed light on an eco-logical question concerning turtle migration. The technique has also been used by evolutionary biologists. Mutations lead to changes in the nucleotide sequence of DNA—changes that serve as the raw material for biological evolution. Since mutations in DNA occur continuously over time, the longer the amount of time that passes since two organisms diverged from a common ancestor, the greater the differences in their DNA sequences. The analogy might be made to two people who happen to meet on a bike trip through Europe and then set out on their bikes independently to see different parts of the continent. The longer the time that passes since the departure from their point of common origin, the greater the distance the cyclists are likely to be separated. Similarly, the more time that passes after two organisms have diverged from a common ancestor, the greater the differences in their DNA sequences.

An evolutionary question of interest to humans is: Who is our closest living relative? According to sequencing studies of a large region of DNA near the beta-globin gene (see Figure 15-6) carried out by Michael Miyamoto and his colleagues at Wayne State University, the answer is chimpanzees. While this answer may not be surprising, Miyamoto's group made another, unexpected finding. It has generally been assumed that chimpanzees and gorillas are more closely related to each other than either is to the human species. However, Miyamoto found that there were fewer differences in the DNA nucleotide sequences between humans and chimpanzees (a 1.6 percent nucleotide sequence difference) than between chimpanzees and gorillas (a 2.1 percent difference). According to these data, chimpanzees are more closely related to humans than they are to gorillas.

Until recently, biologists had to rely solely on phenotypic differences among organisms to assess evolutionary relatedness. With advances in DNA technology, evolutionists can now focus directly on differences in genotypes, the source of biological variation. (See CTQ #5.)

REEXAMINING THE THEMES

Unity within Diversity

⬙ The entire technology used in the formation and use of recombinant DNAs is based on the universality of the genetic code. DNAs from such diverse sources as humans, viruses, and bacteria can be spliced together to form a single, continuous DNA molecule. This recombinant DNA can then be introduced into a host cell (usually either bacterial or mammalian). Once in the host cell, the human DNA will direct the formation of a human protein that is indistinguishable from the same protein produced by a human cell.

Evolution and Adaptation

▮▶ Techniques are now available to alter the genetic composition of any organism, creating, in effect, a new type of life form. The advent of this new age of biotechnology reveals the power of science to alter genetic inheritance and even the course of evolution itself. At the same time, techniques described in this chapter are being used to study the natural course of evolution. Comparisons of DNA sequences between diverse organisms provide a measure of the evolutionary relatedness of the organisms. This experimental approach rests on the assumption that organisms that have diverged more recently from a common ancestor will have more similar DNA sequences than will organisms that diverged from one another at a more distant point in the past.

SYNOPSIS

Genetic engineering and the "biotechnology revolution" has depended on the formation of recombinant DNA. Recombinant DNA molecules contain portions derived from different sources and joined together in the laboratory. Recombinant DNAs have been used to manufacture human proteins with clinical applications, such as insulin, growth hormone, and plasminogen activator. In such cases, a gene encoding the desired product is spliced into a bacterial plasmid, and the recombinant molecule is introduced into a bacterial or a mammalian host cell, which transcribes and translates the foreign gene, releasing the gene product into the medium.

Genetic engineers have developed organisms that carry recombinant DNA. The recombinant DNA encodes a product that alters the organism's characteristics. For example, genetic engineers have developed plants that contain a bacterial gene that codes for an insect-killing protein, microbes that can produce enzymes to digest oil spills, crops and livestock with improved nutrient value, and animals that are capable of manufacturing human proteins of clinical value.

The biotechnology revolution has been built on the development of new techniques in DNA technology. The isolation of particular genes of interest in medicine and agriculture is achieved by using restriction enzymes— bacterial DNA-digesting enzymes that recognize a particular stretch of nucleotides. Once the DNA is enzymatically fragmented, and the gene of interest is isolated, the gene is spliced into a suitable vector, such as a bacterial plasmid, and the recombinant molecule is amplified either by DNA cloning or by PCR. When amplified by DNA cloning, the plasmid is introduced into bacterial cells, where it is replicated as the bacterial cells proliferate. When amplified by PCR, a particular gene is experimentally and selectively targeted for replication by a heat-resistant DNA polymerase.

DNA fragments can be separated from one another, and their nucleotide sequences determined. Created by treatment of a DNA sample with a restriction enzyme, DNA fragments are separated by gel electrophoresis, whereby the DNAs move through a sievelike gel in response to an electric current. The pattern of fragments generated by electrophoresis forms the basis for comparing DNA from different individuals of a species. For example, distinguishing the pattern of restriction fragments from different humans forms the basis for the forensic technique of DNA fingerprinting.

Comparing nucleotide sequences of DNAs obtained from individuals of different species provides a measure of the evolutionary relatedness of the species. Based on this technique, chimpanzees appear more closely related to humans than to gorillas.

Key Terms

recombinant DNA (p. 314)
transgenic animals (p. 319)
splicing (p. 321)
plasmid (p. 321)

restriction enzyme (p. 321)
DNA ligase (p. 323)
DNA cloning (p. 324)
restriction fragment (p. 325)

gel electrophoresis (p. 325)
DNA probe (p. 325)
polymerase chain reaction (PCR) (p. 326)

Review Questions

1. Distinguish between the following: bacterial plasmid and chromosome; conventionally produced insulin and insulin produced by genetically engineered *E. coli;* restriction enzyme and human pancreatic DNase; transgenic animal and a "normal" animal; a sense strand and an antisense strand of DNA; amplification by cloning and amplification by PCR.

2. Why is the production of "sticky ends" important in recombinant DNA technology?

3. Describe three ways crop plants might be improved by genetic engineering.

4. How are investigators able to use PCR to amplify one particular gene, while all the other parts of the DNA remain at their original amount?

5. Why do larger fragments of DNA migrate more slowly than do smaller fragments during gel electrophoresis?

Critical Thinking Questions

1. Would you expect that the ability of a sea turtle to return to the beach where it was hatched is somehow built into the animal's genes? That is, is natal homing an instinctive behavior? How might you test whether or not such behavior is learned from other turtles? Can you think of any clues a turtle could use to navigate across the ocean to a particular island where it was hatched?

2. Given the many benefits of genetic engineering illustrated in this chapter, why are some people opposed to it? List at least four arguments against genetic engineering. For each, develop a counterargument. After considering these pros and cons, what is your position on genetic engineering, and why?

3. Compare the mechanism for bacterial transformation discovered by Frederick Griffith (page 266) and the production of human insulin inside an "engineered" bacterial cell. What are the similarities and differences?

4. Until recently, bacteria that cause tuberculosis (TB) were susceptible to a variety of antibiotics. Now, some of these bacteria have developed resistance to many drugs. Most resistant strains of bacteria have genes that code for enzymes that destroy antibiotics, but the TB-resistant bacteria are the result of a deletion mutation. Can you suggest a mechanism by which a deletion could lead to drug resistance?

5. The graph below shows the percentage of nucleotide substitutions between the DNA and three proteins (fibrin, hemoglobin, and insulin) of the cow, sheep, and pig. Which pair of animals is most closely related?

Which of the four types of molecules shows the greatest rate of change? the least? What can you conclude from these data about the rate of change of different genes?

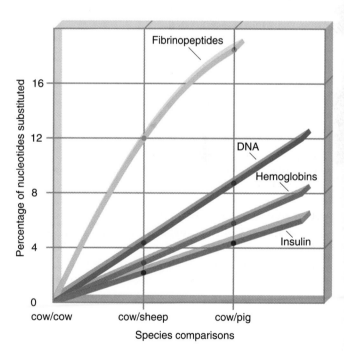

6. Do you harbor any fears about recombinant DNA technology? If so, what steps would you want taken to allay those fears?

Additional Readings

Barinaga, M. 1992. Knockout mice offer first animal model for CF. *Science* 257:1046–1047. (Intermediate)

Barton, J. H. 1991. Patenting life. *Sci. Amer.* May:84–91. (Intermediate)

Gibbons, A. 1992. Moths take the field against biopesticide. *Science* 254:646. (Introductory)

Kessler, D.A., et al. 1992. The safety of foods developed by biotechnology. *Science* 256:1747–1832. (Intermediate)

Kolberg, R. 1992. Animal models point the way to clinical trials. *Science* 256:772–773. (Intermediate)

Lewin, R. 1989. New look at turtle migration mystery. *Science* 243:1009. (Intermediate)

Lewontin, R. C., and D. L. Hartl. 1991. Population genetics in forensic DNA typing. *Science* 254:1745–1750. (Advanced)

McElfresh, K.C., et. al. 1993. DNA-based identity testing in forensic science. *Bioscience* March:149–157. (Intermediate)

Moffat, A. S. 1991. Making sense of antisense. *Science* 253:511. (Intermediate)

Moffat, A. S. 1992. Plant biotechnology explored in Indianapolis. *Science* 254:25. (Intermediate)

Moffat, A. S. 1992. Transgenic animals may be down on the pharm. *Science* 254:35–36. (Intermediate)

Neufeld, P. J., and N. Coleman. 1990. When science takes the witness stand. *Sci. Amer.* May:46–53. (Intermediate)

Roberts, L. 1991. Fight erupts over DNA fingerprinting. *Science* 254:1721–1723. (Intermediate)

Suzuki, D., and P. Knudson. 1990. *Genethics: The clash between the new genetics and human values.* Cambridge, MA: Harvard Univ. Press. (Intermediate)

Human Genetics: Past, Present, and Future

STEPS TO DISCOVERY

Developing a Treatment for an Inherited Disorder

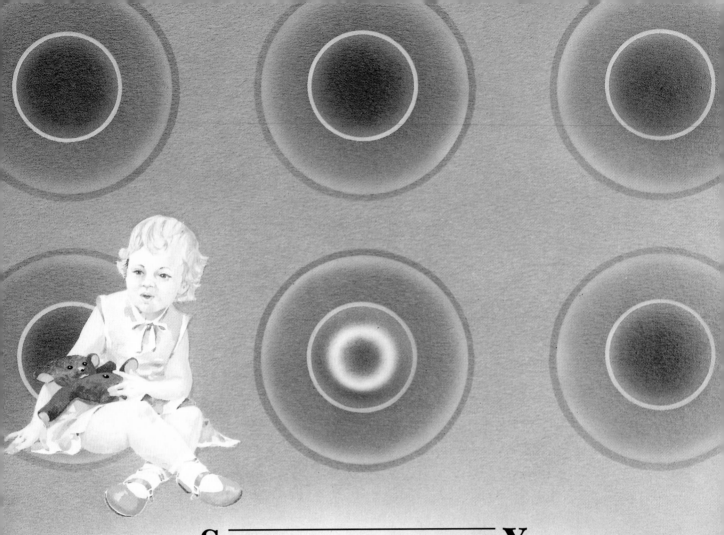

S TEPS TO DISCOVER Y

Developing a Treatment for an Inherited Disorder

*I*n 1934, Asbjorn Folling, a Norwegian physician, reported on a study of two mentally retarded infants. The infants' mother had complained that her children emitted a continual musty odor. Chemical tests of the infants' diapers revealed the presence of high levels of phenyl ketones, toxic compounds formed from the metabolic breakdown of an essential amino acid, phenylalanine. Because of the presence of these compounds, the childrens' disease was called *phenylketonuria (PKU)*. Subsequent research indicated that the condition was caused by the inherited deficiency of phenylalanine hydroxylase, an enzyme that normally converts phenylalanine to another amino acid, tyrosine. As a result of this deficiency, the blood of these infants contained high levels of phenylalanine, which caused mental retardation.

In 1953, working at a British children's hospital, Evelyn Hickmans and co-workers Horst Bickel and John Gerrard, hypothesized that if the crippling effects of PKU were due to the presence of high levels of phenylalanine in the blood, then it might be possible to treat PKU children by restrict-

Phenylketonuria is detected in newborn infants when phenylalanine levels are high enough to promote the growth of

ing their dietary intake of phenylalanine. The researchers tested their hypothesis on a 2-year-old PKU victim who was "unable to stand, walk, or talk . . . and spent her time groaning, crying, and banging her head." The child was put on a diet containing only enough phenylalanine to support the synthesis of vital proteins. Over the next few months, the little girl improved dramatically; she learned to stand and climb on chairs, and she stopped crying and banging her head. Other studies soon followed, confirming the benefits of a low-phenylalanine diet for infants born with PKU. For treatment to be effective, however, early diagnosis is essential so that the infant can be placed on the diet before permanent damage to the nervous system occurs.

In 1961, Robert Guthrie of the University of Buffalo published a one-page "letter" outlining a procedure by which newborn infants could easily be screened for PKU. Guthrie's procedure took advantage of the fact that the blood of newborn infants with PKU contains about 30 times the level of phenylalanine as does that of normal infants. Using the Guthrie test, a drop of blood from a newborn infant is dried on a small piece of filter paper, which is then added to a well in a culture dish containing a low concentration of bacteria that require phenylalanine to grow. When the filter paper contains blood from an infant with PKU, the high phenylalanine content promotes the growth of the bacteria, producing a visible "halo" around the well. In contrast, blood from a normal infant does not promote the growth of the bacteria, so no halo will be observed.

Since the development of the Guthrie test, the vast majority of infants born in the United States and other western countries are automatically screened for PKU. Those infants who test positively for the inherited condition (about 1 in 18,000 newborns) can be placed on the prescribed diet. Once these children reach a certain age, their nervous systems are no longer susceptible to damage by high levels of phenylalanine in their blood, and they can begin eating a normal diet.

Although early diagnosis and treatment had virtually eliminated this rare form of mental retardation, PKU has returned to the spotlight in recent years. Successful screening and dietary treatment for PKU have allowed children with the disorder to develop into normal adults, who are having children of their own. Even though a high level of phenylalanine in the blood has little effect on the mother, it can produce terrible damage to the developing fetus. Damage to the fetus is best prevented by the mother's return to her strict, low-phenylalanine diet *before* she becomes pregnant to ensure that the developing baby will have a safe environment throughout the entire gestation period. This potentially serious problem illustrates the importance of genetic counseling for people at risk of having children with genetic diseases.

Dietary treatment of PKU prevents development of the effects of the disease without altering the root of the genetic problem itself. It can be said that a diet low in phenylalanine changes a person's phenotype without altering his or her genotype. In this chapter, we will examine new approaches in the treatment of genetic disorders where the genotype, rather than the phenotype, is the target for modification.

bacteria around blood samples, producing a halo.

*H*uman genetics is stocked with tales of tragedy and triumph. One strange tragedy comes in the form of a bizarre disease called Lesch-Nyhan syndrome. Victims of this syndrome, who are almost all boys, begin to mutilate themselves during their second year of life by biting off their lips and fingertips. Although the behavior was once attributed to lunacy, or "demonic possession," scientific examination of pedigrees of victims and controlled analysis revealed that the underlying defect was not a dysfunctional family life or some childhood trauma but something completely biological—a deficiency in a single enzyme, one needed for normal purine metabolism. Without this enzyme (hypoxanthine-guanine phosphoribosyltransferase, or HGPRT) children exhibit a buildup of uric acid in their blood and urine. In fact, the urine of these children often contained "orange sand," the precipitated uric acid crystals. The disorder was inherited from the children's mothers, carriers who showed no signs of the disorder.

Sadly, the triumph in this tragedy has yet to be realized, although through genetic counseling potential parents can discover their probability of having a baby with the syndrome. If a woman is pregnant, some of the methods described in this chapter can determine whether the incubating fetus is a victim of Lesch-Nyhan syndrome. Someday we may score a real triumph over this genetic disease by developing the ability to replace the faulty gene with a fully functional copy. Such *gene therapy* is the frontier that holds the most promise for genetic researchers, for physicians, and most importantly, for people who must cope with the consequences of acquiring one of the many diseases caused by defective human genes. Altogether, over 3,500 distinct genetic disorders have been described, 15 of which are described in Table 17-1.

▼ ▼ ▼

THE CONSEQUENCES OF AN ABNORMAL NUMBER OF CHROMOSOMES

We saw in Chapter 11 how homologous chromosomes separate during meiosis I and sister chromatids separate during meiosis II. On a rare occasion, homologous chromosomes fail to come apart during meiosis I or sister chromatids fail to detach during meiosis II. If either of these types of meiotic **nondisjunction** occurs, gametes with either an extra chromosome or a missing chromosome are formed (Figure 17-1). Normal development depends on the proper number of chromosomes. If one of these abnormal gametes

fuses with a normal gamete, the resulting zygote usually develops into an abnormal embryo that dies in the womb and is miscarried. In a few cases, however, the zygote develops into a viable offspring, whose cells have an abnormal chromosome number. The most common disorder resulting from meiotic nondisjunction is Down syndrome, which, as discussed in Chapter 11, occurs in individuals having three homologues of chromosome number 21 in their cells, rather than two, which is why the disease is also called *trisomy 21*.

AN ABNORMAL NUMBER OF SEX CHROMOSOMES

Occasionally, a baby is born with an abnormal number of sex chromosomes due to meiotic nondisjunction. A zygote with three X chromosomes (XXX) develops into a relatively "normal" female, although she will likely have subaverage intelligence and experience menstrual irregularities. A zy-

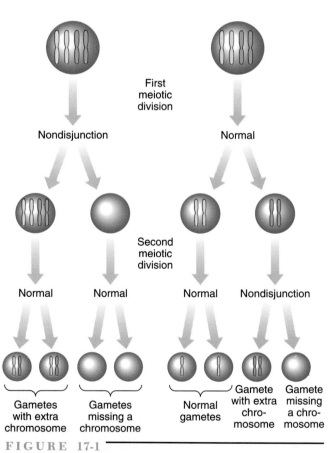

FIGURE 17-1

Meiotic nondisjunction occurs when chromosomes fail to separate from each other during meiosis. If the failure to separate occurs during the first meiotic division, all of the gametes will have an abnormal number of chromosomes. If nondisjunction occurs during the second meiotic division, only two of the four gametes will be affected.

gote with only one X chromosome and no second sex chromosome (X0) develops into a female with *Turner syndrome*, in which genital development is arrested in the juvenile state; the ovaries fail to develop; and body structure is slightly abnormal.

Males may also develop with abnormal numbers of sex chromosomes. Since a Y chromosome is male determining, persons with at least one Y chromosome develop as males. A male with an extra X chromosome (XXY) will develop *Klinefelter syndrome*, which is characterized by mental retardation, underdevelopment of genitalia, and the presence of feminine physical characteristics (such as breast enlargement). Alternatively, a zygote with an extra Y (forming an XYY male) develops into a man who appears normal in every way except that he will probably be taller than average. Considerable controversy has developed surrounding claims that XYY males tend to exhibit more aggressive, antisocial, and criminal behavior than do XY males, but this correlation has never been proven.

Human cells typically possess two copies of each gene, one on each homologue, except for those genes on the sex chromosomes for which a cell typically has only one functional copy. The well-being of many human cells depends on this "gene dosage"; when it is altered by the absence of a chromosome or the presence of an extra chromosome, the normal balance within the cells is disturbed. (See CTQ #2.)

TABLE 17-1
GENETIC DISORDERS

Genetic Disorder	Cause	Nature of Illness	Incidence	Inheritance
Down syndrome	Extra chromosome number 21	Mental retardation, body alterations	1 in 800	Sporadic
Klinefelter syndrome	Male with extra X chromosome	Abnormal sexual differentiation	1 in 2,000	Sporadic
Cystic fibrosis	Abnormal chloride transport	Complications of thickened mucus	1 in 2,500 Caucasions	Autosomal recessive
Huntington's disease	Unknown	Progressive neurological degeneration	1 in 2,500	Autosomal dominant
Duchenne muscular dystrophy	Deficient muscle protein dystrophin	Progressive muscle degeneration	1 in 7,000 males	X-linked
Sickle-cell anemia	Abnormal beta-globin	Weakness, pain, impaired circulation	1 in 625; mostly African descent	Autosomal recessive
Hemophilia	Deficiency in one of a number of clotting factors	Uncontrolled bleeding	1 in 10,000 males	X-linked
Phenylketonuria	Deficiency in enzyme phenylalanine hydroxylase	Mental retardation	1 in 18,000	Autosomal recessive
Tay-Sachs syndrome	Deficiency in enzyme acetylhexosaminidase	Deposition of fatty materials in brain, infant death	1 in 3,000 Ashkenazic Jews	Autosomal recessive
Lesch-Nyhan syndrome	Deficiency in enzyme HGPRT	Mental retardation, self-mutilation	1 in 100,000 males	X-linked
Galactosemia	Deficiency in enzyme galactose transferase	Mental retardation, digestive problems	1 in 60,000	Autosomal recessive
Xeroderma pigmentosum	Deficiency in a UV repair enzyme	Sensitivity to sunlight, cancer	1 in 250,000	Autosomal recessive
Severe combined immunodeficiency (SCID)	Deficiency in enzyme adenosine deaminase or others	Absence of immune defenses	Extremely rare	Autosomal recessive
Ehlers-Danlos syndrome	Deficiency in collagen-hydroxylating enzyme	Abnormal connective tissues, joint problems	1 in 100,000	Autosomal recessive
Familial hypercholesterolemia	Deficiency in LDL receptors	Cardiovascular disease	1 in 100,000	Autosomal recessive

DISORDERS THAT RESULT FROM A DEFECT IN A SINGLE GENE

Every gene encodes a product, either a polypeptide or an RNA (such as a transfer RNA), each of which has some function. Some gene products play a relatively minor role in the life of an organism; others are critical. When genes become altered so that they produce inactive products, the consequences are often serious. One of the first steps involved in studying the properties of a defective gene is determining the chromosome on which the gene is located and the relative position of the gene within that chromosome. This process of gene mapping has undergone a revolution in recent years, made possible by the DNA technology discussed in Chapter 16.

NEW TECHNIQUES FOR MAPPING DEFECTIVE GENES

Geneticists map genes along a chromosome by determining the frequency of crossing over (page 252); the closer two genes reside on a chromosome, the less likely their alleles will become separated from one another during crossing over. Using this mapping procedure, the different forms (alleles) of each gene act as **genetic markers,** serving as signposts that help reveal the relative locations of other genetic sites along the chromosome (see Bioline: Mapping the Human Genome). Genes that are known to exist in several allelic forms, such as those that determine blood type, are called **polymorphic genes,** and they make the best markers.

In 1978, a group of geneticists gathered to describe their research on the inheritance of diseases within large human families. The scientists were disturbed that there were so few known polymorphic genes that could be used as genetic markers to locate disease-causing genes within human chromosomes. During a presentation, two members of the audience—David Botstein of MIT and Ronald Davis of Stanford University—began talking together and announced that they knew a way to provide genetic markers located throughout the chromosomes. The markers Botstein and Davis had in mind were not genes but DNA fragments.

⬡ As you will recall, treatment of DNA with a restriction enzyme produces a collection of different-sized fragments (page 325). The restriction fragments generated from the DNA *of a given individual* form a precisely defined set of fragments. When restriction fragments from different people (even siblings) are compared, the patterns of fragments are similar, but not identical, due to differences in nucleotide sequences from one person to the next. Let's look more closely at these genetic differences among humans.

Recent analysis of human DNA molecules has revealed the presence of short, repeating sequences that lie adjacent to each other in clusters within the vast stretches between genes. One such cluster of these repeating sequences is outlined by the colored boxes in Figure 17-2*a*. In this illustration, the two DNA molecules represent comparable portions of two homologous chromosomes from a man with a wife and two daughters (Figure 17-2*b*). Note the different number of the repeating sequences (indicated by the boxes) on the two DNA molecules. Differences of this type are very common and, in fact, account for much of the genetic variation that occurs in the human population. While these differences play no apparent role in biological function, they have proven invaluable as markers for locating human genes.

Differences in the number of repeating sequences in a cluster produce restriction fragments of different length after cleavage by a restriction enzyme. In the example illustrated in Figure 17-2*a*, corresponding segments of DNA located on two homologous chromosomes are cleaved into different-sized fragments—1,400 versus 2,400 base pairs in length—because of the differences in the number of times a small sequence happens to be repeated. Such differences are called **restriction fragment length polymorphisms (RFLPs),** or simply "riflips." RFLPs produce distinctive differences in banding patterns following electrophoresis (Figure 17-2*c*–*d*) which can be used to identify individuals or to track the inheritance of particular genes.

The first scientist to search for RFLPs was Raymond White, then a recent arrival at the University of Massachusetts and a colleague of Botstein's. By 1979, White and co-worker Arlene Wyman had identified the first RFLP (Figure 17-2*d*). White and Wyman examined the DNA from a large Mormon family chosen because the genealogical relationships among the members of the family were well established. When DNA from 56 different people in the study group was subjected to digestion by the same restriction enzyme, a labeled probe identified fragments of 30 different sizes among the population. It was like finding 30 different alleles for a gene among 56 related individuals. In order to illustrate how this type of variability can be put to use in gene mapping, we will describe the use of RFLPs in locating the gene responsible for cystic fibrosis.

Locating the Gene Responsible for Cystic Fibrosis

Cystic fibrosis is the most common debilitating inherited disease among Caucasians (about 1 per 2,500 newborns), but it is almost nonexistent among members of other races. Among a variety of symptoms, victims of cystic fibrosis (CF) produce a thickened, sticky mucus that is very hard to propel out of the airways leading from the lungs. As a result, these individuals typically suffer from chronic lung diseases, including potentially fatal infections. Children with cystic fibrosis once faced near-certain death before the age of 5, but recent therapies to help clear congested airways (Figure 17-3) and antibiotics to fight infections have allowed these patients to live into their adult years.

In the summer of 1989, a press conference was held to announce the isolation of the gene responsible for cystic

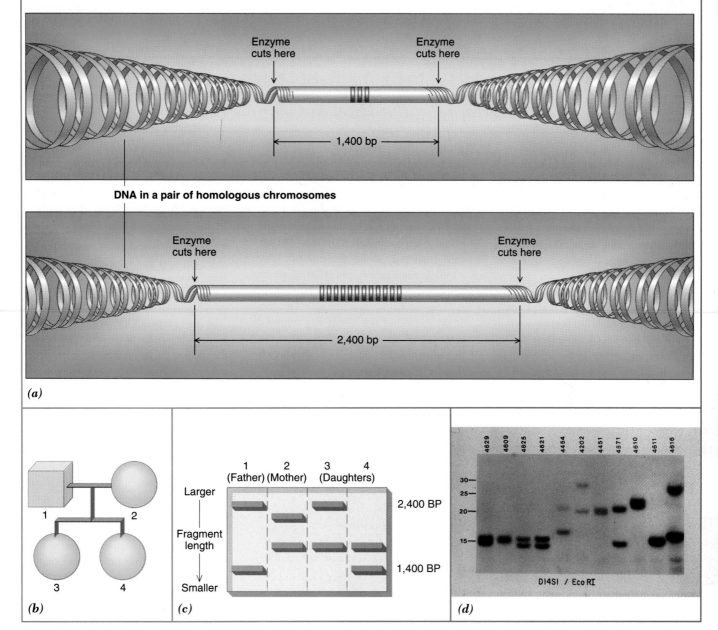

Enzyme
cuts here

Enzyme
cuts here

1,400 bp

DNA in a pair of homologous chromosomes

Enzyme
cuts here

Enzyme
cuts here

2,400 bp

(a)

(b)

(c)

1
(Father)
2
(Mother)
3   4
(Daughters)

Larger

Fragment
length

Smaller

2,400 BP

1,400 BP

(d)

D14S1 / Eco RI

FIGURE 17-2

Restriction Fragment Length Polymorphisms (RFLPs). *(a)* Homologous segments of DNA
from a pair of homologous chromosomes from one individual. The arrows represent sites in the DNA
that are attacked by a particular restriction enzyme. The colored boxes denote short, repeating sequences in the DNA. In this example, the DNA on one chromosome has only three repeats of this sequence, whereas the DNA on the homologous chromosome has 12. When these two DNA molecules
are treated with the restriction enzyme, one DNA molecule yields a 2,400 base pair fragment, while
the other DNA molecule yields a fragment of only 1,400 base pairs. This is an example of a restriction
fragment length polymorphism (RFLP). *(b)* A short pedigree of four family members: a father,
mother, and their two daughters. The father's DNA is shown in part *(a)*. *(c)* The pattern of DNA
fragments from the four family members. Each lane (corresponding to a different person) has two bands
—one from each homologous chromosome. The two bands in lane 1 identify the DNA fragments of
1,400 and 2,400 base pairs generated by enzyme treatment of the father's DNA from part *(a)*. The
DNA fragments of the mother are different from those of the father, reflecting differences in the nucleotide sequence of her DNA. Each of the daughters has inherited one chromosome from her
mother and one from her father, generating the patterns in lanes 3 and 4. *(d)* The first demonstration
by Wyman and White that RFLPs could be used to detect genetic differences in a human population.
The number at the top of each lane stands for a particular individual. Note the marked differences
among these banding patterns. Those lanes that have only one dense band are from individuals whose
two homologous fragments either (1) are similar enough in size so that the bands merge into a single
thicker band or (2) happen to be homozygous for this particular RFLP, so that both homologues produce the same-sized fragments.

◁ B I O L I N E ▷
Mapping the Human Genome

In 1986, a group of scientists from around the world gathered at the Cold Spring Harbor Laboratories in Long Island, New York, to discuss the possibility that all of the genetic information stored in human chromosomes might be decoded as the result of a giant, international scientific collaboration. Within a few years, governmental agencies in the United States, Europe, and Japan had decided to proceed with the effort, which has become known as HGP, the Human Genome Project. The term "genome" refers to the information stored in all the DNA of a single set of chromosomes. The human genome contains approximately 3 billion base pairs. If each base pair in the DNA were equivalent to a single letter on this page, the information contained in the human genome would produce a book approximately 1 million pages long!

The goal for the Human Genome Project is to sequence the DNA of all 24 human chromosomes (22 autosomes, plus the X and Y sex chromosomes) by the year 2005. Centers committed to achieving this goal have been established throughout the United States and other countries. During the first few years of the project, researchers will prepare highly detailed genetic maps of each human chromosome, with markers located every 100,000 or so bases. They will then isolate overlapping fragments of DNA that can be arranged in an order that spans the entire length of each chromosome. By the time these efforts are completed, DNA sequencing techniques should be substantially improved so that the isolated DNA fragments can eventually be sequenced in less time for less money than is currently possible. Advances in biocomputer technology will

also be required to enable researchers to sift through the massive amounts of data in order to come up with meaningful conclusions about the functions of gene products.

The overall cost for the project is estimated at approximately $3 billion—$1 for each base pair to be determined. Why spend so much money and effort simply to determine a list of four nucleotide letters running a million pages in length? The goal of the project is not simply to compile a sequence of letters but to learn what the sequence of letters "stands for." It is hoped that the Human Genome Project will

- Reveal the amino acid sequence of each polypeptide encoded in our genes. Using reverse genetics (page 342), biologists can then learn about the probable function of these polypeptides. This information could shed light on virtually every aspect of human biology, from the biochemical basis of memory to the gene products responsible for a cell's loss of growth control and consequent malignancy.

- reveal detailed information about the sequences involved in controlling gene expression.

- identify those genes that cause or contribute to human diseases— including complex diseases such as cancer, hypertension, Alzheimer's, and mental illness—and determine the changes in sequence that trigger the disease.

- reveal insights into evolution at the molecular level by comparing nucleotide and amino acid sequences in humans to those of more distant relatives. Studies of this type will also reveal which gene products are unique to humans, helping to explain how we are biologically different from—and similar to—other organisms.

- allow investigators to receive samples of DNA containing any particular gene, thereby providing an unlimited opportunity for continuing research in human molecular biology.

- provide biotechnologists with the opportunity to manufacture previously unknown human proteins that have potential medical value.

The Human Genome Project is not without its critics. Some scientists believe the project is too costly, at least at our current level of DNA sequencing expertise, and it will siphon money from other research endeavours. In addition, most of the work will be tedious and noncreative —not the kind of project attractive to most scientific minds. Sydney Brenner, one of England's pioneering molecular biologists, facetiously suggested that the work would be suitable punishment for convicts of a penal colony. Other critics argue that knowledge of human DNA sequences opens the door to further invasion of our privacy. It is feasible, for example, that insurers or employers could demand information about the genetic makeup of potential employees or insurees to determine their predisposition toward particular illnesses. Still others fear that knowledge about the nucleotide sequences responsible for certain genetic characteristics could lead unscrupulous governments to attempt to alter the human genome in desirable directions. Since it is inevitable that the human genome will be sequenced, it is up to all of us to remain vigilant and ensure that this genetic information is used for the betterment of the human species and not misused by those who feel they can profit from it.

FIGURE 17-3

Coping with cystic fibrosis requires dislodging as much of the thickened mucus from the airways as possible. The standard therapy for the disease is to pound the person's back. In 1992, a new experimental treatment was introduced in which the patient inhales a mist containing the DNA-digesting enzyme, DNase, which is manufactured using recombinant DNA technology. The enzyme degrades the DNA that contributes to the viscosity of the mucus. The DNA present in the mucus is derived from disintegrating inflammatory cells that move into the respiratory tract.

fibrosis. For the first time, victims of cystic fibrosis saw a glimmer of light at the end of a long, dark tunnel; there was hope that a cure for their crippling disease might be developed. To understand how the search for RFLPs was a critical factor in the isolation of the CF gene, consider the following analogy: Imagine that you live in a neighborhood where the outsides of the houses are similar and there are no addresses to help you locate your house. Each block has a single commercial building at one corner, each housing a different business. One block has a laundry, another a grocery store, and so forth. You are fortunate to have a small coffeehouse on your corner, where you can have a cup of coffee and read your biology text. Every day when you come home, you drive along the main street until you see the coffeehouse then turn right and proceed to the eighth house, which is where you live.

The similar-looking houses are analogous to the genes that lie along the chromosome. Just as the houses are virtually identical from one street to the next, human genes, for most purposes, are identical from one person to the next. In contrast, the commercial properties, which are analogous to

RFLPs, are highly polymorphic—they vary from block to block. Once you have located a commercial property that is a given distance from the particular house you live in, you can always find your house (or CF gene) using the commercial property (or linked RFLP) as a guidepost. In other words, RFLPs serve as genetic markers for locating genes whose positions would otherwise be undetectable.

The goal of the researchers searching for the CF gene was to find a RFLP that was very close to the CF gene itself. How does a gene mapper know when he or she has identified such a RFLP? Remember that two genes located very close to each other are very unlikely to become separated from each other during crossing over. Consequently, if a particular RFLP and the CF gene are very close to each other, then all of the individuals in a family that are actually stricken with the disease will probably have the same version of the RFLP marker (Figure 17-4). Other families with the disease will have different versions of the RFLP. If a researcher finds a RFLP that is invariably present in those members of a family with the disease, then it follows that this RFLP is close to the gene causing the disease. Once a RFLP that resided within a million or so nucleotides of the CF gene had been identified, investigators used other techniques to move from the RFLP into the unknown, adjoining regions of the DNA until they arrived at the CF gene itself.

▶ Once the gene responsible for cystic fibrosis had been located and its nucleotide sequence determined, samples of DNA from a variety of afflicted individuals, their parents, and other members of the population were analyzed. It turned out that 70 percent of the alleles responsible for cystic fibrosis in the United States contained the same genetic alteration—they were all missing three base pairs of DNA that coded for a phenylalanine at the 508th position within the polypeptide chain encoded by the CF gene. Individuals with alleles containing this small deletion were generally of Northern European descent, and it is likely that all of these individuals are descendants of a single Northern European who lived many generations ago.

The isolation of a disease-causing gene is a major step in understanding the underlying basis of the disease and may provide a foundation for new and innovative treatments. This is illustrated by some of the research described below which followed the announcement that the CF gene had been isolated.

Reverse Genetics: Working Backward from Genotype to Phenotype

Mendel began his studies with pea plants distinguishable by height and seed color; Morgan began his studies with fruit flies having different-colored eyes; geneticists working on sickle cell anemia learned decades ago that this disease was characterized by abnormal hemoglobin. In each of these cases, the investigator began with a relatively clear phenotypic difference. But what is the cause of traits such as cystic fibrosis, which affects many different organs in various ways? More specifically, what is the product of the CF gene, whose alteration has such devastating effects?

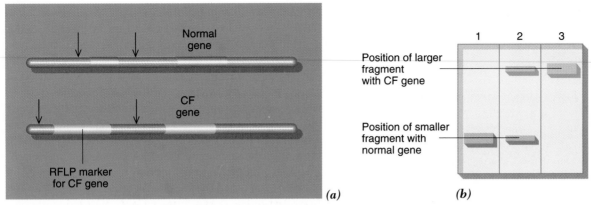

FIGURE 17-4

Linkage of the cystic fibrosis gene with a genetic marker. (*a*) The DNA of a heterozygous carrier for CF will have one normal allele and one mutant allele. The two alleles, located on homologous chromosomes, will inevitably be linked to different-sized restriction fragments. In this hypothetical example, the CF gene is closely linked to the green-colored RFLP marker. In this case, any member of the family who carries the CF allele will also have the longer restriction fragment. (*b*) Suppose two heterozygotes with the same homologues shown in (*a*) were to marry and have children. The children would be expected to have three different restriction fragment patterns. Lane 1 shows the DNA from a child who is homozygous normal and thus has two copies of the smaller fragment; lane 2 is from a heterozygous child who has one long and one short fragment and would be a carrier like the parents; and lane 3 is from a child who suffers from the disease and would have two copies of the longer fragment. Once a nearby marker is found, such as the green-colored RFLP shown in (*a*), molecular biologists can work their way from the marker to the gene being sought.

➡ Molecular biologists are now able to carry out a form of "reverse genetics." First, they isolate the gene and determine its nucleotide sequence. Then, they deduce the amino acid sequence of the protein using the universal genetic code and compare it to the amino acid sequence of other proteins whose function is known. Proteins of related (*homologous*) amino acid sequence have evolved from a common ancestral protein and often have a similar function. The amino acid sequence deduced for the CF gene suggested that the gene product was a membrane protein involved in the movement of chloride ions across the plasma membrane. Recall from Chapter 7 that water flows passively from regions of lower to regions of higher salt concentration. The blockage of chloride movement in people with CF is thought to have a secondary effect on the movement of water, causing a decreased water content—and thus an increased viscosity—in bodily secretions. The presence of DNA also contributes to the high viscosity of the mucus found in CF patients (see Figure 17-3).

CYSTIC FIBROSIS: A LESSON IN EVOLUTION

Why should one ethnic group have such a high incidence of a disease that is rare or nonexistent in other groups? One answer is simply by chance; the alleles for cystic fibrosis may simply have increased in frequency along with the rapid expansion of the Caucasian population over the past hundred years or so. It is also possible that some selective pressure has existed to maintain the CF alleles at a high frequency in the Caucasian population. Clearly, there is no selective advantage for homozygotes with cystic fibrosis, but there may be some advantage for the vastly larger number of heterozygous CF carriers. One recent hypothesis suggests that CF heterozygotes may be more resistant to cholera (a disease that causes the loss of body water and death by dehydration) than are homozygotes that lack a CF allele. Recall that the CF gene product is a chloride-transport protein that affects the movement of water in the body. A person having one CF allele would be expected to have a reduced capacity for chloride transport, which might result in a reduced loss of body water during a cholera infection.

The fact that a large percentage of the CF alleles in the American population are probably derived from a single Northern European ancestor illustrates how alleles can spread from one population to another as individuals move from place to place. As we will see in Chapter 33, evolution is characterized by a change in the percentages of alleles of a species. The movement of individuals from one population to another is one way to initiate such change.

Recent advances in DNA technology have led to techniques designed to locate specific disease-causing alleles on particular chromosomes, to determine the functions of the products encoded by these genes, and to offer hope that a cure for these devastating genetic disorders may be on the horizon. (See CTQ #3.)

PATTERNS OF TRANSMISSION OF GENETIC DISORDERS

The transmission of genetic characteristics in humans follows the same basic "Mendelian rules" described for pea plants and fruit flies. Disorders that result from mutations in specific genes are usually divided into three categories—autosomal recessive, autosomal dominant, and X-linked—depending on whether or not one defective allele is sufficient to cause the disease and on the type of chromosome on which the gene appears.

DISORDERS THAT RESULT FROM AUTOSOMAL RECESSIVE ALLELES

Recall that autosomes include all chromosomes except the sex chromosomes (X and Y). Since each cell has two copies of each autosome, a recessive mutation on one autosome will be masked by the presence of a normal allele on the homologous autosome. These heterozygotes are referred to as *carriers* because they harbor the mutant gene without being adversely affected, in most cases. You are probably a carrier for as many as five to ten lethal recessive alleles. Fortunately, recessive alleles only produce a disease phe-notype when they are present on both homologues; that is, in the homozygous recessive condition (Figure 17-5). Consequently, these disorders are usually rare. If two heterozygotes (carriers) have children, however, on the average, one in four of their offspring will suffer from the abnormal phenotype. Many homozygous recessive conditions lead to the death of the fetus and subsequent miscarriage, leaving the parents unaware that they are carriers of the defective alleles. Other homozygous recessive disorders, such as cystic fibrosis, phenylketonuria, and sickle cell anemia, lead to the birth of an infant with serious medical problems.

DISORDERS THAT RESULT FROM AUTOSOMAL DOMINANT ALLELES

Some genetic diseases result from *dominant* mutations (Figure 17-6) in which the defective gene codes for a product that directly causes the diseased condition. In other words, rather than the normal gene overriding the defective one, as in recessive conditions, the tables are turned, and the normal gene remains a "helpless bystander." Disease-causing dominant mutations can reside on either an autosome or a sex chromosome. Huntington's disease is an example of a condition that results from such an *autosomal dominant mutation*.

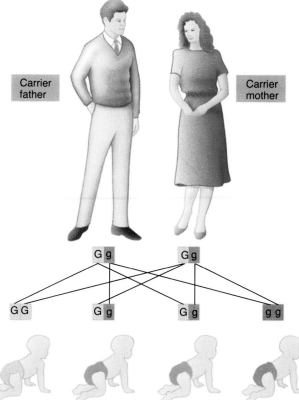

Normal   Carrier   Carrier   Affected

FIGURE 17-5

Inheritance of an autosomal recessive disorder. For these traits (including sickle cell anemia, phenylketonuria, and cystic fibrosis), both parents are carriers of the defective allele (indicated as g), but they do not show its ill effects due to the presence of the dominant normal allele (indicated as G). Each offspring has a one in four chance (25 percent) of inheriting two copies of the defective gene, resulting in the disease. Each offspring has a two in four chance of being a carrier and a one in four chance of inheriting two normal alleles.

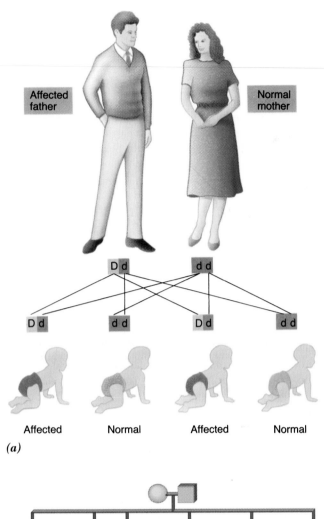

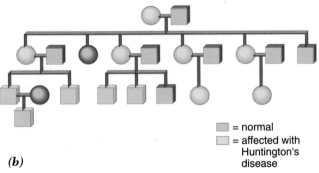

= normal

= affected with
Huntington's
disease

(b)

FIGURE 17-6

Inheritance of an autosomal dominant disorder. *(a)* For these traits, only one parent carries the defective allele (indicated as D), and that person exhibits the disease. On average, half the offspring will inherit the defective allele and develop the disorder. *(b)* Pedigree of a family with Huntington's disease.

Huntington's Disease

In 1872, Dr. George Huntington published his only scientific paper. It was titled "On Chorea," and it described a strange neurological disease that had plagued a large family living in New York. Victims of the disease exhibited no symptoms until they were 35 to 50 years old, at which time they began to show signs of unusual involuntary movements, loss of memory, depression, and irrational behavior. The disease progressed steadily and unerringly toward dementia, loss of motor control, and finally death. The underlying biochemical defect causing the neurological deterioration was then, and remains, a mystery. Huntington noted, however, that if one parent had the disease, "one or more offspring almost invariably suffer from it, if they live to adult age. But if the children go through life without it, the thread is broken and the grandchildren and great-grandchildren are free from the disease." Although unaware of it, Huntington was describing a trait that was inherited as an autosomal dominant mutation. Unlike autosomal recessive diseases, such as cystic fibrosis and PKU, only one parent has to have the defective allele for a child to inherit an autosomal dominant disease.

Huntington's disease received media attention when legendary songwriter Woody Guthrie died from the disease in 1967 (Figure 17-7a). Until very recently, children of Huntington's victims (such as Guthrie's son, Arlo) found themselves in a horrible position. Knowing that they faced a 50 percent risk of developing the disease, they had to wait until they were well along in their adult life before learning their destiny and possibly that of their children. Every time they forgot something or tripped over their own feet, they were reminded of their genetic inheritance and wondered whether it was the first sign of the onset of the disease.

Today, another name is associated with Huntington's disease—Columbia University's Nancy Wexler, psychologist and head of the Hereditary Diseases Foundation (Figure 17-7b). Nancy Wexler, herself at risk for Huntington's disease, has been the driving force behind an intensive research program aimed at isolating the gene responsible for the disease and determining the underlying malfunction. Wexler has managed to bring together researchers from several laboratories in an unprecedented collaborative hunt for the gene responsible for the disease.

The search for the Huntington's gene began in earnest in 1982, using RFLP technology. Armed with probes for 13 RFLPs and DNA samples from a small Iowa family with a history of Huntington's disease, James Gusella of Harvard University began his study of linkage analysis. The chance that any of the 13 RFLPs would be closely linked to the Huntington's gene was remote; Gusella's primary goal in this initial round of experiments was simply to work out the techniques needed for the analysis. Instead, Gusella hit the jackpot. The third RFLP he tested was not only on the same chromosome as the Huntington's gene (chromosome number 4) but was within about 5 million base pairs of the gene (the entire chromosome has about 200 million base pairs of

(a) *(b)*

FIGURE 17-7

(a) Woody Guthrie and *(b)* Nancy Wexler. Dr. Wexler is shown in front of a wall containing a pedigree of the inhabitants in the community on the shore of Lake Maracaibo, Venezuela.

DNA). For his good fortune, Gusella earned the nickname "Lucky Jim."

▶▶ Soon after his early finding, Gusella switched to the analysis of DNA samples that had been collected by Nancy Wexler during a series of trips to a number of remote fishing villages dotting the lagoons of Lake Maracaibo in Venezuela (Figure 17-7b). The residents of Lake Maracaibo were plagued with a strange malady they called *El Mal* ("the bad"), their term for Huntington's disease. According to their story, El Mal could be traced to a European sailor who had visited the area nearly 200 years earlier. By the 1980s, the Huntington's allele had spread rapidly through the local population; over 150 people were afflicted with the disease, and 1,100 more were at 50 percent risk. The Venezuelan community provided an ideal human pedigree for researchers to study, and the search began for a RFLP marker that was very close to the defective gene.

For ten years, over a dozen labs followed one lead after another, each leading to a dead end. Finally, in March, 1993, the gene was isolated and, appropriately, it was James Gusella's lab that found it. Sequencing studies revealed the basis for the defect. The Huntington's gene includes a region containing a repeating sequence of the trinucleotide, CAG, which specifies the amino acid glutamine. Whereas the normal gene contains about 10 to 20 copies of the trinucleotide, the altered gene that causes Huntington's disease contains over 40 copies of the triplet. The effect of these extra repeats is under intense investigation.

DISORDERS THAT RESULT FROM X-LINKED ALLELES

Some genetic disorders for which the defective allele is recessive are expressed in men even when only one copy of the gene is inherited. As we discussed on page 254, these conditions are due to X-linked alleles—faulty genes that lie on the X chromosome. X-linked disorders include hemo-philia, red–green colorblindness, congenital night blindness, ichthyosis (skin hardens to scalelike consistency), some forms of anemia (low hemoglobin content in the blood), and muscular dystrophy.

Muscular Dystrophy

Muscular dystrophy is an X-linked gene that strikes about 1 in every 4,000 males and is characterized by progressive muscle deterioration. In its severest form, known as *Duchenne muscular dystrophy*, the patient is confined to a wheelchair by age 12 and often dies of cardiac failure by age 20.

The gene responsible for muscular dystrophy was discovered piece by piece during the late 1980s. Unlike the other genes that had been sought, the gene causing muscular dystrophy is unusually large—spanning approximately 2 million base pairs of DNA, or nearly 0.1 percent of a person's total supply of genetic information! Most of this DNA consists of senseless introns, but the encoded protein, *dystrophin*, is also exceptionally large. Studies indicate that dystrophin contributes to the structure of the muscle cell's surface; in its absence, the membrane weakens and gradually deteriorates, accounting for the progression from seemingly normal muscles in an infant to the progressive loss of muscular tissue over the years. Curiously, a mouse strain that fails to produce dystrophin has been isolated, yet these animals do not exhibit the same type of muscle-wasting disease seen in humans.

The discovery of the dystrophin gene as the cause of muscular dystrophy has suggested a possible cure for the disease. What if the muscles of these sick children could somehow receive a supply of dystrophin? Could the deterioration be prevented or even reversed? In 1991, Peter Law, formerly of the University of Tennessee, conducted experiments with this goal in mind. In his study, 32 young boys with muscular dystrophy were injected with healthy, undifferentiated human muscle cells to determine whether these cells would participate in the formation of dystrophin-containing muscle tissue. The cells were injected into several muscles of the leg. Initial results, which have been met with controversy, indicate that the affected muscles were strengthened, but whether or not such a procedure can serve as an effective treatment remains uncertain.

COMPLEX DISEASES

Before you conclude that studies on the genetics of human disease only apply to those persons born with rare congenital disorders, consider the following: Many common diseases, including cancer, atherosclerosis, diabetes, Alzheimer's, manic depression, and even alcoholism, have a strong genetic component. For example, we have discussed in earlier chapters how mutations in oncogenes (page 202) or tumor suppressor genes (page 260) can lead to the development of cancer. As a result, those individuals born with mutations in certain genes are predisposed to develop particular types of cancer. This is the reason why women whose mothers have had breast cancer are at much higher risk of developing the disease than are members of the general population. Similarly, a man whose father develops prostate cancer is at increased risk of doing the same. It is becoming increasingly evident that the more we learn about human genetics, the better we will be able to understand the complex diseases that threaten our health.

Most genetic diseases are the result of single-gene defects and follow simple Mendelian patterns of inheritance, allowing risk assessments to be calculated based on simple probability laws. The genetic basis of more common diseases, including cancer, is much more complex, making risk assessments more difficult. (See CTQ #5.)

SCREENING HUMANS FOR GENETIC DEFECTS

The discovery of RFLPs (and their use in locating defective genes) has opened new vistas in the field of medical diagnostics. Couples can be tested so that they no longer have to wonder if they might be carriers for cystic fibrosis or sickle cell anemia or if their unborn fetus will be afflicted with the disease. A person whose mother or father is dying of Huntington's disease can now learn with 95 percent certainty whether or not he or she will be stricken with the disease. It may soon be possible to determine whether an infant has a predisposition for developing a particular form of cancer, heart disease, or diabetes. Let's examine how geneticists screen DNA in order to make these types of determinations.

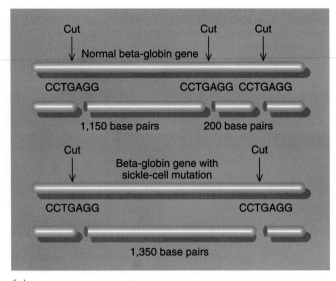

(a)

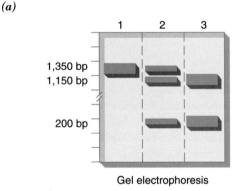

Gel electrophoresis

(b)

FIGURE 17-8

Screening for the sickle cell gene. *(a)* Restriction enzyme cuts the normal beta-globin gene at the enzyme's recognition sites (arrows), creating two fragments. The mutation responsible for sickle cell anemia changes the nucleotide sequence of the DNA in a way that eliminates a recognition site for the restriction enzyme. The sickle cell allele is therefore cut into only one fragment by this enzyme. This difference in restriction fragment length is readily revealed through electrophoresis, forming a basis for screening the population for the mutant allele. Part *(b)* shows the pattern produced by DNA from a person with sickle cell anemia (lane 1), a heterozygous carrier of the disease (lane 2), and a person who is homozygous normal (lane 3).

- *Sickle cell anemia*   Sickle cell anemia is an autosomal recessive disease that primarily afflicts individuals of central African ancestry. Sickle cell anemia is caused by a specific alteration in the beta-globin gene, which happens to lie directly within one of the sites on the DNA normally recognized by a restriction enzyme. DNA containing the normal globin allele is cut at this site, whereas DNA containing the sickle cell allele is not. As a result, RFLP analysis can determine if a person lacks the defective gene entirely; is a heterozygous carrier of the disease; or has two copies of the defective gene and, thus, the disease (Figure 17-8).

- *Cystic fibrosis*   Approximately 1 in 25 Caucasian Americans are carriers of the cystic fibrosis gene; thus, approximately 1 in 625 (25 × 25) Caucasian couples are at risk of having a child with the disease. Earlier, we noted that in 70 percent of the cases of CF in the United States, the alleles responsible are missing three particular nucleotides. Consequently, any screening test based on this single type of gene alteration would miss 30 percent of the carriers and more than half the couples at risk. Fortunately, screening tests have been developed that identify several other frequently occur-

ring types of gene alterations, so we can now screen for—and successfully detect—at least 90 percent of all abnormal CF genes.

- *Huntington's disease*   Even though the gene for Huntington's disease has just been isolated, a person at risk for the disease can usually determine whether or not he or she carries the defective allele. The current test for Huntington's disease depends on obtaining DNA from a close relative who has the disease. Analysis of this DNA can specifically identify the RFLP that is closely linked to the defective gene in the particular family being studied. If a person at risk in the family carries this same RFLP in his or her DNA, then the probability is very high that he or she has the same allele for Huntington's disease and will develop the disease.

SCREENING FETAL CELLS FOR GENETIC DISORDERS

Two commonly performed procedures allow physicians to obtain a sample of cells of an unborn fetus. These cells can then be screened for genetic deficiencies.

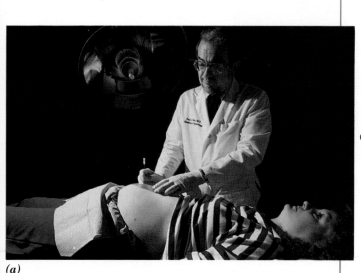

(a)

FIGURE 17-9 ⸺

Sampling cells from the fetus. *(a)* Amniocentesis. *(b)* By employing the techniques of amniocentesis, or chorionic villus sampling, geneticists can remove a sample of fetal cells to determine whether the child will be born with a detectable chromosome abnormality (which can be determined by microscopic examination of the chromosomes of a fetal cell) or a defect in a gene (which can be determined by the ability of the fetal cells to grow in various media or by RFLP analysis).

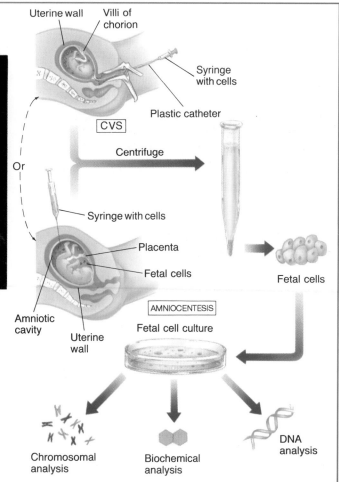

(b)

◁ THE HUMAN PERSPECTIVE ▷
Correcting Genetic Disorders by Gene Therapy

During the 1970s, a young boy named David captured the attention of the American public as "the boy in the plastic bubble" (Figure 1). The "bubble" was a sterile, enclosed environment in which David lived nearly his entire life. David required this extraordinary level of protection because he was born with a very rare inherited disease called *severe combined immunodeficiency disease* (SCID), which left him virtually lacking an immune system—the system that protects us from invading pathogens. The bubble protected David from viruses or bacteria that might infect and kill him, but it also kept him from any direct physical contact with the outside world, including his parents. In approximately 25 percent of cases, SCID results from the hereditary absence of a single enzyme, adenosine deaminase (ADA).

Gene therapy is the prospect by which a patient is cured by altering his or her genotype. The goal is to provide the patient with a normal gene, capable of pro-

FIGURE 1

- **Amniocentesis** is a procedure in which a hypodermic needle is inserted through the body wall of the pregnant woman into the fluid-filled amniotic sac that surrounds and cushions the growing fetus, and a sample of fluid is withdrawn (Figure 17-9*a*). Fetal cells that have been shed into the amniotic fluid are isolated and cultured in the laboratory for 1 or 2 weeks and are then subjected to genetic analysis. Amniocentesis is not usually performed before the fifteenth week of pregnancy, which leaves a relatively small amount of time in which the mother can choose to have a safe abortion should the tests reveal serious genetic problems in the fetus.

- **Chorionic villus sampling (CVS)** is a newer procedure, whereby a small sample of tissue is removed from the developing placenta. The fetal cells contained in this tissue sample can then be subjected to genetic analysis. Since CVS can be performed as early as the ninth week of pregnancy, it provides the mother with much more time to decide whether or not to terminate the pregnancy.

Both types of prenatal screening procedures carry some risk (0.5 to 1 percent) to the life of the fetus. Newer, safer techniques are being developed that will enable physicians to obtain fetal cells from the blood of the mother.

ducing the missing gene product. SCID is an excellent candidate for the development of such therapy for a number of reasons. First, there is no cure for the disease, which inevitably proves fatal at an early age. Second, SCID results from the absence of a single gene product (ADA). The ADA gene has been isolated and cloned and thus is available for treatment. Finally, the cells that normally express the ADA gene are white blood cells that can easily be removed from a patient, genetically modified, and then reintroduced into the patient by transfusion.

In 1990, a 4-year-old girl suffering from SCID became the first person authorized by the National Institutes of Health and the Food and Drug Administration (FDA) to receive gene therapy. In September 1990, the girl received a transfusion of her own white blood cells which had been genetically modified to carry normal copies of the ADA gene. It was hoped that the modified white blood cells would provide the girl with the necessary armaments to ward off future infections. Since white blood cells have a limited lifetime, the procedure must be repeated periodically to maintain the patient's immune capacity. At the time of this writing, the child is attending school and doing well; her immune system is apparently functioning normally.

Intensive effort is currently focused on isolating the *stem cells* that give rise to both red and white blood cells. If these stem cells can be isolated, and their genotype modified, patients with genetic blood cell diseases, such as SCID, will need only a single treatment since the genetically engineered stem cells will continue to provide healthy blood cells throughout the person's lifetime.

Cystic fibrosis is also a good candidate for gene therapy because the gene has been isolated and because the worst symptoms of the disease are caused by cells that line the airways and, therefore, are accessible to substances that can be delivered by inhalation of an aerosol (as in Figure 17-3). It is hoped that a similar aerosol approach can be used to deliver a normal CF gene directly to the defective cells of the respiratory tract of CF patients. Scientists are optimistic, based on the results of recent studies from Ronald Crystal's laboratory at the National Institutes of Health, in which the normal CF gene, attached to the DNA of a cold virus, was delivered to the lung cells of rats. The researchers found that the added CF gene was integrated into the chromosomes of the rat cells, and the human protein was produced for at least 6 weeks. Tests with humans will probably begin once the safety of the delivery procedure can be confirmed.

Another approach is being considered as a therapy for diseases that result when an abnormal product is carried in the blood. Hemophilia, for example, results from a deficiency of a particular clotting factor in the blood. The clotting factor is normally produced by the liver, but the source of the protein is not important to its function; its presence in the blood alone is required. Cells from the deep layer of the skin have been isolated and genetically modified so that they carry extra genes for this clotting factor. These cells might be reintroduced into the skin of a hemophiliac, where they will secrete the clotting factor into the blood, thus compensating for the abnormal clotting factor. Trial studies for this procedure are currently being carried out on dogs.

It is important to note that all of the procedures discussed above, as well as all of those being contemplated, involve the modification of *somatic cells*—those cells of the body that are not on the path to gamete formation. Modification of somatic cells will affect only the person being treated, and the modified chromosomes cannot be passed on to future generations. This would not be the case if the germ cells of the gonads were modified since these cells are part of the line that produces gametes. Thus far, the consensus is that no studies involving the modification of human germ line cells will be performed. Such studies would present risks for the genetic constitution of future generations and raise serious ethical questions about scientists tampering with human evolution.

Even though fetal cells are present in very small amounts in the mother's blood, techniques are available to separate them from the mother's cells and to amplify the fetal DNA by PCR (page 326). It will probably be several years before this procedure will be available, however.

The ability to obtain fetal cells allows medical geneticists to make several types of determinations (Figure 17-9b):

1. *Detection of chromosomal abnormalities* Amniocentesis and CVS are most commonly performed on older pregnant women who, because of their age, have an increased likelihood of giving birth to a child with Down syndrome. The determination is made by counting the number of chromosomes in the fetal cells in search of an extra chromosome number 21, which is readily detected by karyotype analysis (see Figure 10-3). The presence of some chromosomal aberrations can also be detected by analysis of fetal cells.

2. *Detection of metabolic deficiencies* Tests for over 200 genetic disorders that affect metabolic processes can also be carried out on cultured fetal cells. If fetal cells are missing a particular enzyme—such as HGPRT, whose deficiency causes Lesch-Nyhan syndrome—the condition may be revealed by the cells' inability to grow in certain culture media.

3. *Detection of defective alleles*   With the development of RFLP analysis, more and more tests will become available for screening the DNA of fetal cells for genetic disorders that are not readily detected by standard microscopic or biochemical tests, such as cystic fibrosis and Huntington's disease.

ETHICAL CONSIDERATIONS

One of the byproducts of the search for the genes responsible for human diseases has been the development of techniques for diagnosing genetic disorders. These techniques have been hailed as modern medical "miracles." Consider, for example, how many people carrying a pair of PKU alleles have been saved from severe mental retardation. At the same time, serious unforseen ethical and psychological considerations have emerged from our recent breakthroughs. Consider the following situations.

- You have a 50 percent chance of developing Huntington's disease. Would you want to know as a young adult that your body will eventually (and inevitably) deteriorate and you will become mentally deranged? Or would you rather live with the uncertainty, holding onto the hope that you will not contract the disease? If you choose the latter route, would you have children, knowing that you might be passing on the defective gene? In fact, the test for Huntington's disease has been available for a number of years, and genetic counselors have been surprised at the relatively small number of people who have come forward to learn of their genetic fate.

- You find out that both you and the person whom you plan to marry carry the gene for cystic fibrosis. You were planning to get married and have a large family. Do you break off the relationship? Do you marry and have children, accepting the possibility that one out of four will have a disease that will be emotionally draining and expensive to treat and, in all likelihood, end in a premature death? If your wife becomes pregnant, do you submit to fetal testing and abort a fetus that carries a pair of defective alleles? Knowing that you face the possibility of having dependents with costly medical problems, should you be forced to reveal this information on applications for health or life insurance? Do insurance companies have the right to be informed of a person's genetic status?

- You learn that you are a heterozygous carrier of the sickle cell gene, which means that a portion of your hemoglobin molecules are abnormal. You have always wanted to be a pilot, and your doctor tells you this will not affect your ability to fulfill that goal, but you find there are barriers against sickle cell carriers in the airline industry because of the belief that these individuals will be adversely affected if a drop in cabin air pressure should occur. Do you inform the airline of your genetic status before accepting a job?

It is evident that DNA-based diagnostic tests have created new questions for which answers are not readily available. The greatest problems stem from the fact that these tests predict the presence or likelihood of diseases for which there are no cures. Given this circumstance, some argue it would be better not to employ such tests at all. Since current tests screen for relatively rare conditions, they have not had a major impact on most of our lives. But as we develop diagnostic tests to detect an individual's predisposition to develop far more common complex diseases, such as cancer, Alzheimer's, and heart disease, the problems will be compounded, and the entire matter of genetic screening is likely to come under careful scrutiny.

Just as a typographical error in a sentence can be detected by a proofreader, a change in the nucleotide sequence of a gene can be detected by screening procedures that use recently developed techniques in molecular biology. While these techniques offer great promise in identifying genetic abnormalities at an early age, they raise serious ethical considerations. (See CTQ #7.)

REEXAMINING THE THEMES

Biological Order, Regulation, and Homeostasis

Most of the disorders discussed in this chapter result from a defect in a single gene, producing a severe breakdown of the body's normal functions. The fact that a deficiency in one gene product can have such devastating consequences illustrates an organism's need for a high degree of order. A deficiency in one process often leads to secondary failures, triggering a chain reaction that produces the visible consequences, such as life-threatening infections, muscle deterioration, digestive failure, or neurological degeneration.

Unity within Diversity

None of the diseases described in this chapter has any known counterpart in other mammals. This fact may be explained by our ignorance of animal afflictions, but it is more likely attributed to differences that exist even among closely related organisms. For example, a defect in the dystrophin gene has also been discovered in mice. But even though these mutant mice fail to produce dystrophin, the animals don't exhibit the deteriorative muscle wasting so characteristic of humans with muscular dystrophy.

Unity and diversity are also revealed by examination of nucleotide sequences in different people. While most sections of the genome are very similar from one person to the next, reflecting the genetic homogeneity within a species, large variations exist in the number of times certain sequences are repeated. These variations allow researchers to map genes by RFLP analysis and to develop various DNA screening procedures.

Evolution and Adaptation

Differences between species that arise during evolution result from underlying changes in the species' genetic composition. Such changes occur as new alleles arise by mutation and then spread throughout a species' population. In this chapter, we have seen two examples of how the apparent immigration of a single individual has promoted the spread of a particular allele into a new population—the cystic fibrosis allele in the United States, and the Huntington's allele in Venezuela. Since both alleles are deleterious, one would expect the selective disadvantage of possessing these alleles to restrict their spread. However, the spread of deleterious alleles has probably occurred more readily in recent years in human populations than in those of other organisms because of the explosive increase in the population in the past few hundred years and the medical advances in the treatment of inherited disorders. In the case of cystic fibrosis, it has been suggested that heterozygous carriers of the CF allele have a selective advantage over others in that they are more resistant to the effects of cholera.

SYNOPSIS

Human life is very sensitive to the proper number of chromosomes. Gametes containing an abnormal chromosome number are produced following nondisjunction during meiosis. In humans, zygotes having an abnormal chromosome number usually die in the womb or grow into individuals who are born with characteristic abnormalities, such as Down syndrome.

Recent advances in gene mapping have allowed geneticists to locate the chromosomal position of genes, even when the gene product has not been characterized. Gene mapping takes advantage of differences in DNA nucleotide sequences from one person to another—differences that result in variations (polymorphisms) in the length of restriction fragments (RFLPs) produced when human DNA is treated with restriction enzymes. The location of the gene in question is discovered by identifying a specific RFLP that is always present in the same members of a family that have a particular allele of the gene. Since the particular RFLP and the particular allele occur together in the relatives, the two must be closely linked.

Once a gene has been isolated, the amino acid sequence of the polypeptide can be deduced and the function of the polypeptide usually determined by comparing its amino acid sequence to that of proteins of known function. This is the technique used to determine that the CF gene codes for a chloride channel, providing insight into the molecular basis of the disease.

Genetic disorders are transmitted from generation to generation in several well-defined patterns. Disorders resulting from autosomal recessive alleles, such as cystic fibrosis and sickle cell anemia, occur only when both members of a pair of homologous autosomes carry the recessive alleles. In a heterozygous carrier, the normal allele produces a normal gene product that compensates for the defective product produced by the recessive allele. Disorders that result from autosomal dominant alleles, such as Huntington's disease, occur when only one member of a pair of homologous autosomes carries the defective allele. The abnormal gene product encoded by the allele causes the disorder, despite the presence of the normal product from the other homologue. Disorders that result from X-linked alleles, such as hemophilia and muscular dystrophy, occur almost exclusively in males. The allele producing the defective gene product is located on the X chromosome, and there is no compensating locus for the allele on the Y.

Recent advances in molecular genetics have initiated changes in the potential treatment of genetic-based disorders. The use of restriction enzymes and RFLP mapping has led to the development of techniques to screen fetuses and adults for the presence of certain genetic disorders and the *potential* to screen individuals for a genetic predisposition to developing cancer, heart disease, or mental illnesses. Gene therapy has moved from the drawing boards to clinical trials, as researchers are attempting to change the genotypes of somatic cells of individuals with certain genetic disorders, including SCID and cystic fibro-sis. In addition, a 15-year project has begun to sequence the entire human genome. Information gained from this endeavor is expected to increase vastly our understanding of the genes that control all of our functions, to help in the treatment of genetic-based diseases, to provide an opportunity to manufacture many new protein products, and to help us better understand the course of human evolution.

Key Terms

nondisjunction (p. 336)
genetic marker (p. 338)
polymorphic gene (p. 338)

restriction fragment length
   polymorphism (RFLP) (p. 338)
amniocentesis (p. 348)

chorionic villus sampling (CVS) (p. 348)
gene therapy (p. 348)

Review Questions

1. Draw pedigrees that show the patterns by which autosomal recessive, autosomal dominant, and X-linked disorders are inherited.

2. Why does a person with an XXY chromosome abnormality develop into a male rather than a female? Which sex would you predict a person with XXXXY would exhibit?

3. Why do RFLPs make such useful genetic markers?

4. Give two different molecular bases for the occurrence of RFLPs. (Hint: check Figures 17-2a and 17-8a.)

5. Why do researchers believe that the majority of people carrying a CF allele have one particular Northern European ancestor? Why is it so difficult to identify all CF carriers by genetic screening?

6. Compare the extent of our knowledge about the molecular and physiological basis of PKU, CF, Huntington's disease, and muscular dystrophy.

Critical Thinking Questions

1. Diet soft drinks and other foods containing the artificial sweetener aspartame (trademark Nutrasweet) carry a warning for people with PKU. Aspartame is a dipeptide. Why do you suppose this warning might be necessary?

2. If the chromosome carrying a gene for Lesch–Nyhan syndrome were involved in nondisjunction, how might this affect the occurrence of the condition in female offspring? Explain your answer using what you have learned about gene dosage.

3. If you were a member of Congress and were asked to vote to appropriate money for the Human Genome Project, would you vote for it or against it? Why?

4. Suppose you have just isolated the gene responsible for causing Huntington's disease. What steps might you take to determine why the gene causes such devastating neurological damage?

5. How would you explain to someone why it may be easier to find a cure for muscular dystrophy than for atherosclerosis?

6. In this chapter, we noted that a person having a parent with Huntington's disease can learn with about 95 percent certainty whether he or she will develop the condition. Why do you suppose this person cannot learn this with 100 percent certainty?

7. In the 1960s, genetic screening for sickle cell anemia, followed by genetic counseling, was tried in a small town in Greece in which the disease was prevalent. The program failed because people were unwilling to inform potential marriage partners if they were carriers, in case this would reduce their chances of marrying. When a similar program was proposed for screening African Americans for the disease, it failed because of charges that the program was racially motivated. What implications do the ethical and sociological questions raised by these cases have for the future of genetic diagnosis and screening?

Additional Readings

Baskin, Y. 1984. *The gene doctors.* New York: Morrow. (Introductory)

Bishop, J. E., and M. Waldholz. 1990. *Genome.* New York: Simon & Schuster. (Introductory)

Collins, F. S. 1992. Cystic fibrosis and therapeutic implications. *Science* 256:774-779. (Intermediate)

Culliton, B. J. 1991. Gene therapy on the move. *Nature* 354:429. (Intermediate)

Davies, K. 1991. Mapping the way forward. *Nature* 353:798–799. (Advanced)

Levitan, M. 1988. *Textbook of human genetics.* New York: Oxford University Press. (Advanced)

Maddox, J. 1991. The case for the human genome. *Nature* 352:11–14. (Intermediate)

Marx, J. 1992. Gene therapy for CF advances. *Science* 255:289. (Intermediate)

Nichols, E. K. 1988. *Human gene therapy.* Cambridge, MA: Harvard University Press. (Introductory)

Roberts, L. 1992. The Huntington's gene quest goes on. *Nature* 258:740-741. (Intermediate)

Thompson, L. 1992. Stem-cell gene therapy moves toward approval. *Science* 255:1072. (Intermediate)

Thompson, L. 1993. The first kids with new genes. *Time* June 7:50–53. (Introductory)

Verma, I. M. 1990. Gene therapy. *Sci. Amer.* Nov:68-84. (Intermediate)

Vogel, F., and A. G. Motulsky. 1987. *Human genetics.* New York: Springer-Verlag. (Intermediate)

Watson, J. D. 1990. The human genome project: past, present, and future. *Science* 248:44–49. (Intermediate)

Wingerson, L. 1990. *Mapping our genes.* New York: Dutton. (Introductory)

White, R., and J.-M. Lalouel. 1988. Chromosome mapping with DNA markers. *Sci. Amer.* 258:40–48. (Intermediate)

Form and Function of Plant Life

One of the most
strikingly beautiful and complex
plant structures is the flower. Flowers
capture the attention of passing animals
(mostly insects). As animals gather
nectar, they transfer sperm-containing
pollen from flower to flower for the plant's
sexual reproduction.

CHAPTER

◄ 18 ►

Plant Tissues and Organs

**STEPS
TO
DISCOVERY**
Fruit of the Vine and the French Economy

BIOLINE

Nature's Oldest Living Organisms

THE HUMAN PERSPECTIVE

Agriculture, Genetic Engineering, and Plant Fracture Properties

◁ B I O L I N E ▷
Nature's Oldest Living Organisms

FIGURE 1

FIGURE 2

Middle Ages to experience the glory of the Renaissance; you will survive tragic world wars, hear the first report of people walking on the moon, and experience the beginning of the age of computers and biotechnology.

The scenario is intriguing but impossible. Humans cannot live that long; vital cells simply wear out and are not replaced. But perennial plants are continually able to replace worn-out cells with new ones, maintaining a constant supply of young, efficient cells. Unlike humans, perennials have the potential to live forever.

For many years, the bristlecone pine (*Pinus longaeva*) has been the reigning "oldest living organism," with a documented age of 4,900 years (see Figure 1). In fact, a bristlecone pine actually lived through all of the milestones in the development of human civilization discussed above. Strangely, these ancient plants do not live in a rich environmental setting. Instead, they grow on the wind-beaten, desertlike slopes of the White Mountains in eastern California.

The reign of the bristlecone is currently being threatened by an unimposing, common neighbor, the creosote bush (*Larrea divaricata*) (Figure 2). Some creosote bushes have been growing identical clones of themselves for an estimated 11,000 years; the bristlecones are middle

aged in comparison. The clones form a ring of new individuals that surround the parent plant, each clone having developed asexually from a branch that contacted the soil surface. A number of botanists argue that the creosote's clone production is analogous to the way the bristlecone produces rings of new secondary tissues year after year. Either way—producing rings of new secondary tissues or rings of new individuals—the bristlecone pine and the creosote bush perpetuate cells over thousands of years, cells that originated from a single individual plant. If more botanists begin to agree that perpetuation by cloning is the same as perpetuation by secondary growth, the common creosote bush will likely assume the throne as the world's "oldest living organism."

Imagine living for 5,000 years. Imagine having been born around 3000 B.C., at the beginning of human civilization. You are just starting to practice agriculture, build cities, develop art, and scratch out the first forms of writing. During your 5,000-year lifetime, you will live through the grandeur of the Roman Empire and see it fall into ruin; you will survive the constraints of the

---

all plant organs; and the ground tissue system provides support and storage continuity.

THE DERMAL TISSUE SYSTEM

The dermal tissue system (*derma* = skin) has two critical, yet opposing, functions: The epidermis must form a barrier to protect the plant's delicate internal tissues, yet it must allow for exchanges of essential materials between the plant and its surrounding environment. In herbaceous plants,

such as peas, the dermal tissue system is a single sheet of cells; it is the only thing that stands between the fragile interior cells and the often hostile external environment. For protection, the outer cell walls of the shoot epidermis are covered by a **cuticle,** a waxy layer that retards water loss and helps prevent dehydration. The cuticle is such an effective water barrier that it has been used to protect some of our valued possessions. For example, the cuticle of the Brazilian wax palm is harvested to make carnuba wax for safeguarding furniture and cars (Figure 18-7).

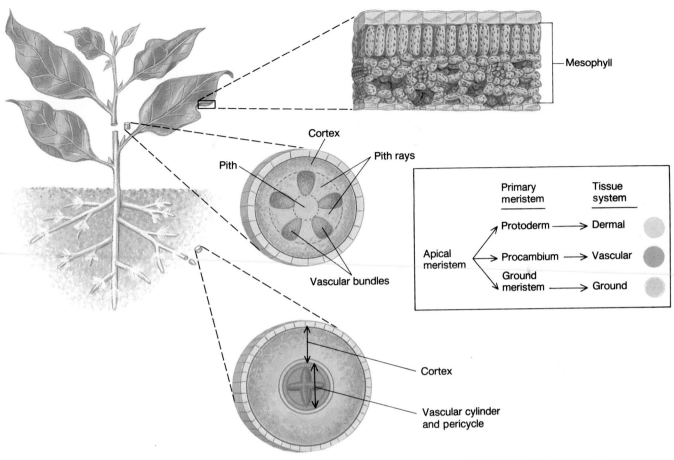

— Mesophyll

Cortex

Pith

Pith rays

Vascular bundles

Primary meristem		Tissue system
Apical meristem	Protoderm →	Dermal
	Procambium →	Vascular
	Ground meristem →	Ground

Cortex

Vascular cylinder and pericycle

FIGURE 18-6

Plant tissue systems. Three tissue systems run continuously throughout the body of a plant: the dermal tissue system, the vascular tissue system, and the ground tissue system. The ground tissue system found in the stem consists of the cortex (the area between the epidermis and vascular tissues), pith (the area from inside the vascular tissues to the stem or root center), and pith rays between the vascular tissues; the mesophyll found in leaves; and the cortex found in roots.

FIGURE 18-7

The cuticle that protects Brazilian wax palm leaves also protects automobile finishes from damage.

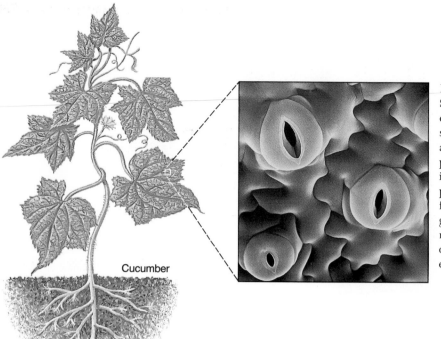

Cucumber

FIGURE 18-8

Stomatal pores between adjacent guard cells enable gases to diffuse into and out of air spaces between internal cells. Three stomates are shown in the photograph. Each stomatal pore is flanked by two guard cells. Carbon dioxide mainly diffuses inward through the stomatal pore for photosynthesis, while water vapor diffuses out. As internal turgor pressure increases, guard cells bend and enlarge the size of the stomatal pore; decreasing turgor pressure narrows or closes the stomatal pore, blocking further gas exchange.

The epidermis of the shoot is riddled with microscopic pores called **stomates** (Figure 18-8), which allow the exchange of gases between the plant and the outside environment. The stomate is flanked by two specialized **guard cells** which regulate the rate of gas diffusion—carbon dioxide for photosynthesis, oxygen for aerobic respiration, and water vapor for cooling. By regulating water vapor, guard cells not only moderate temperature, they also help prevent dehydration in plants that grow in hot, dry environments.

FIGURE 18-9

Creating the effect. A sculptured epidermis (inset) gives marigold petals their velvety texture. Bright pigments stored in the vacuoles of these epidermal cells give the petals their brilliant orange and yellow color.

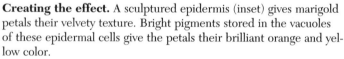

(a)

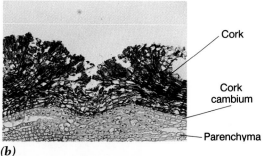

Cork

Cork cambium

Parenchyma

(b)

FIGURE 18-10
The periderm not only protects internal tissues during secondary growth but allows for gas exchange through lenticels. *(a)* Birch periderm with thousands of elongate lenticels on the surface. *(b)* Section through a single lenticel.

In general, the cells in the dermal tissue system are referred to as *epidermal cells.* Some epidermal cells, such as those found in begonias, become meristems, asexually producing buds that grow into complete new plants. In the petals of many colorful flowers, the vacuoles of epidermal cells are engorged with water-soluble pigments that help attract animals to the flower for sexual reproduction (see Figure 18-9 and Chapter 20).

In woody plants, the **periderm** takes over the protective and regulating functions of the epidermis when the epidermis has become damaged as the stem and root increase in width. The periderm is commonly referred to as "bark," but it is actually only the outer part of bark. **Bark** includes all those plant tissues outside the wood (secondary xylem) and is made up of many distinct tissues, including the vascular cambium, phloem, cells in the cortex, cork cambium, and cork. The periderm is composed only of cork cells, the cork cambium, and a layer of parenchyma cells immediately inside the cork cambium (Figure 18-10).

Like sclerenchyma, cork cells continue to function even when the cell is dead. Each cell is composed of layers of secondary walls that are impregnated with a waterproof wax called suberin. These waxy cell layers effectively seal off internal tissues, protecting them against excess water-vapor loss, disease, extreme weather, and foraging insects. To allow oxygen from the atmosphere to reach living cells beneath the periderm, random eruptions in the periderm, called **lenticels,** form air channels that allow the transfer of carbon dioxide and oxygen (Figure 18-10).

THE VASCULAR TISSUE SYSTEM

The vascular tissue system forms an internal circulatory network—the "veins and arteries" of a vascular plant. This tissue system contains two types of complex tissues: **xylem** and **phloem.** Water and dissolved minerals are carried primarily through the xylem; food (carbohydrates produced during photosynthesis) and other organic chemicals syn-

Pits

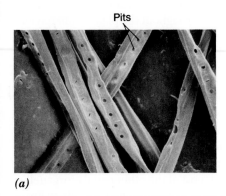

(a)

Perforation

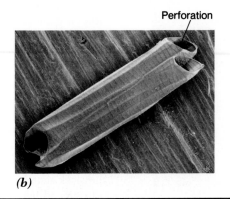

(b)

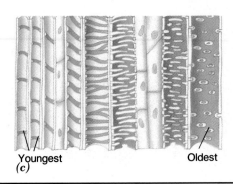

Youngest   Oldest
(c)

FIGURE 18-11
Conducting cells in the xylem transport water and dissolved minerals. They include *(a)* long, thin tracheids that are aligned precisely so that the openings (pits) of one match up with those of other tracheids, and *(b)* vessel members that have pits and perforated ends. Some vessel members *(c)* have characteristic cell wall patterns, the result of being stretched during stem and root growth.

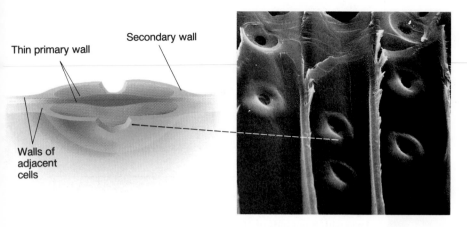

Thin primary wall

Secondary wall

Walls of adjacent cells

thesized by the plant (such as hormones, amino acids, and proteins) move through the phloem.

Xylem: Water and Mineral Transport and Support

The main conducting cells of the xylem are **tracheids** and **vessel members** (Figure 18-11), both of which are dead cells; water and minerals flow through their hollow interiors. The walls of each cell are riddled with pits—small holes in the secondary wall that are adjacent to thin depressions in the primary wall (Figure 18-12). Xylem cells line up so that the pits of one cell match up exactly with those of a neighbor, forming a pit pair. Water flows from one cell to the next through these pit pairs. Opposite ends of vessel members also have a **perforation plate**—an area where

all or large portions of the end wall are absent (Figure 18-11). Vessel members stack one on top of another to form open **vessels** through which water readily travels (Figure 18-13).

Secondary xylem is called **wood.** New secondary xylem cells are produced by the vascular cambium. Toward the end of one growth period (usually at the end of summer), the inside of xylem tracheids and vessel members gets progressively smaller. When water becomes plentiful again (usually not until the following spring), new tracheids and vessel members with large diameters are formed. The abrupt transition from xylem cells with narrow diameters to those with large diameters forms a distinct line, the border of a **growth ring** (Figure 18-14). Because one growth ring is usually produced each year, the number of growth rings is often used to estimate the age of a tree. Nevertheless, many botanists have abandoned using the term "annual ring" because plants sometimes have more than one growth spurt and, therefore, form more than one growth ring in a single year.

Phloem: Food Transport

Phloem runs parallel with the xylem, transporting food and other organic chemicals throughout the plant. Unlike xylem tracheids and vessel members, however, phloem conducting cells function as living cells. In flowering plants, phloem conducting cells are called **sieve-tube members** because the clusters of pores in their cell walls resemble a strainer or sieve (Figure 18-15). These porous areas are located at opposite ends of the sieve-tube member and are called **sieve plates.** A row of sieve-tube members, stacked sieve plate to sieve plate, forms a sieve tube.

Although sieve-tube members are *technically* alive, they lack a nucleus and most of the other cellular organelles of a typical living cell. As these conducting cells mature, their organelles degenerate, leaving only a veneer of cytoplasm against the inside walls and a delicate mesh of protein that extends from sieve-tube member to sieve-tube member. To remain alive and active, sieve-tube members are continually nourished by an adjacent parenchyma cell

Pits

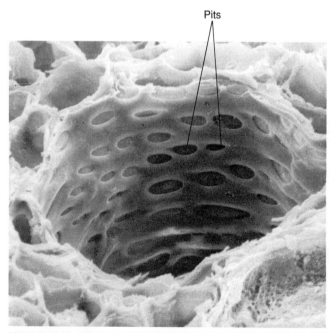

FIGURE 18-13
A vessel pipeline is formed by stacks of vessel members. Pits allow lateral transport.

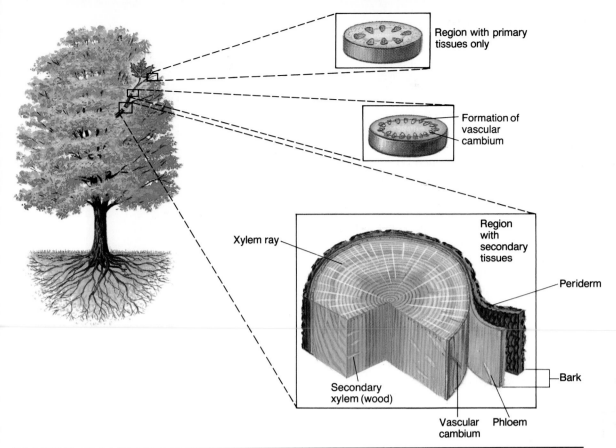

Region with primary tissues only

Formation of vascular cambium

Region with secondary tissues

Xylem ray

Periderm

Secondary xylem (wood)

Vascular cambium

Phloem

Bark

FIGURE 18-14

Sections of a stem and trunk of a tree. Although primary growth continues from apical meristems at the tips of stems, most of the tissue in a tree is secondary xylem, or wood. The vascular cambium divides to produce secondary xylem toward the inside and secondary phloem toward the outside.

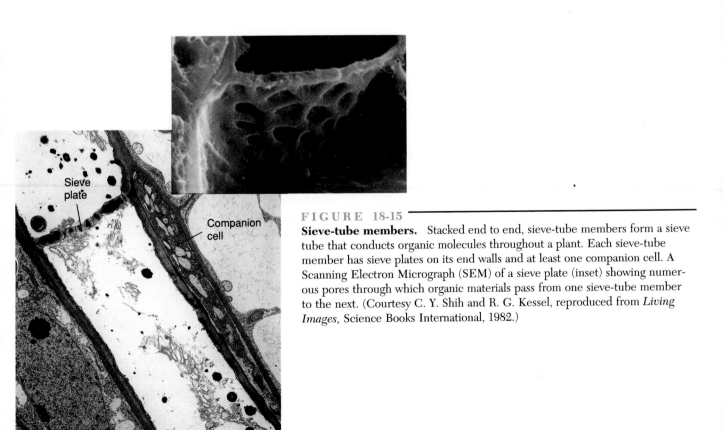

Sieve plate

Companion cell

FIGURE 18-15

Sieve-tube members. Stacked end to end, sieve-tube members form a sieve tube that conducts organic molecules throughout a plant. Each sieve-tube member has sieve plates on its end walls and at least one companion cell. A Scanning Electron Micrograph (SEM) of a sieve plate (inset) showing numerous pores through which organic materials pass from one sieve-tube member to the next. (Courtesy C. Y. Shih and R. G. Kessel, reproduced from *Living Images*, Science Books International, 1982.)

called a **companion cell.** Sieve-tube members and companion cells are interdependent; if one dies, its partner dies soon after.

THE GROUND TISSUE SYSTEM

With protection and exchange functions performed by the dermal tissue system, and long-distance transport accomplished by the vascular tissue system, all remaining plant functions are carried out in the ground tissue system, which is composed of a mixture of cells with different functions (see Figure 18-6). The ground tissue system is made up of cells in the **cortex** (the region between the epidermis and the vascular tissues), the **pith** (the area from inside the vascular tissues to the center of the stem or root), and **pith rays** (the areas between bundles of vascular tissues). The photosynthetic cells between the upper and lower epidermis of leaves are also part of the ground tissue system.

☀ The ground tissues are the principal sites of photosynthesis and food and water storage in a plant. Ground tissues also contain collenchyma and sclerenchyma, which provide support for shoot structures.

Plant cells and tissues are organized into three tissue systems, each of which is continuous throughout the entire plant body. Each tissue system has the same unique set of functions in all plant organs. (See CTQ #4.)

PLANT ORGANS

All the plant cells, tissues, and tissue systems we have discussed thus far are organized into four plant organs: the *stem, root, leaf,* and *flower.* By necessity, photosynthesis is restricted to aerial organs that are exposed to light, whereas water and minerals are absorbed by root tissues that grow in the water and mineral reservoirs found in soil.

THE STEM: GROWTH, SUPPORT, AND CONDUCTION

The **stem** is an exquisite example of nature's bioengineering. Not only do stems physically support all of a plant's leaves, flowers, fruits, and even its other stems, they produce the structures they support. The stems of herbaceous plants consist of only primary stem tissues, whereas the stems of woody plants consist of both primary and secondary stem tissues.

Primary Stem Tissues

At the tip of every stem is a dome-shaped apical meristem (Figure 18-16). The end of each stem is divided into three regions:

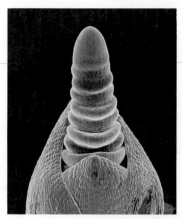

FIGURE 18-16

Apical meristem. A scanning electron micrograph of a stem tip shows the dome-shaped apical meristem and upwardly arched juvenile leaves. Above each young leaf is a developing axillary bud.

- The **meristematic region** contains cells that divide for primary growth. Young leaves and juvenile axillary buds (a bud that is directly above each leaf on a stem) begin developing in this region.
- The **region of elongation** lies just below the meristematic region. This is where cells enlarge and lengthen as they differentiate.
- The **region of maturation,** which lies immediately below the region of elongation, is where nearly all cells have completed enlarging and differentiating.

Monocot and dicot flowering plants differ in the way their stems' vascular tissues are arranged (refer back to Table 18-1). Dicot vascular bundles are organized in distinct rings (Figure 18-17), whereas the more numerous monocot vascular bundles are scattered throughout the ground tissues. Monocots and dicots differ in another important way. Perennial and biennial dicots form cambia for secondary growth, whereas monocots do not.

Secondary Stem Tissues

The tallest organisms on earth are those plants that experience secondary growth. In crowded habitats, taller plants have access to greater amounts of sunlight for photosynthesis than do shorter, shaded plants. A tall plant is able to grow so large because its diameter (and thus the plant's supportive strength) increases as the plant ages. Thus, as apical meristems produce cells that elongate stems, making the plant taller, the vascular and cork cambia increase the stem's support and transport capacities by adding new cells that increase the diameter of the stem. All new secondary vascular cells match up with those in the primary xylem and phloem to maintain a continuous connection of vascular tissues throughout even the tallest plant.

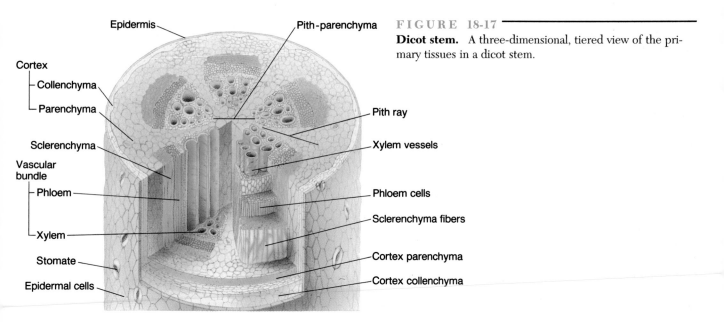

FIGURE 18-17

Dicot stem. A three-dimensional, tiered view of the primary tissues in a dicot stem.

The vascular cambium produces two arrangements of conducting cells: One allows substances to move up and down the stem, while the other conducts materials laterally through secondary xylem and phloem rays (Figure 18-14). In plants, the term "ray" is used to describe any structure that runs parallel to a radius line from the stem center to the epidermis.

THE ROOT: GROWTH, ABSORPTION, AND CONDUCTION

Much of a plant lies hidden beneath the ground. The extensive nature of a plant's root system is illustrated by a single, tiny rye plant. A rye plant only 25 centimeters tall (10 inches) has more than 14 million roots, which, if placed end to end, would stretch over 609 kilometers (380 miles). Knowing this, you might assume that a massive plant like a giant sequoia would have even more roots than would the rye; surprisingly, this is not the case. The sequoia has fewer roots than does the rye plant, but the entire root system of a giant sequoia is much larger than is that of the rye. This unexpected disparity exists because the rye and giant sequoia have different types of root systems. The giant sequoia has a **tap root system** (Table 18-1), which is similar to a carrot, with its one main root and many smaller branches, called **lateral roots** (Figure 18-18a). Like the giant sequoia, most dicots have a tap root system. In contrast, the rye is a monocot and has a **fibrous root system,** which is composed of many main roots—more than 140 for rye (Figure 18-18b). In both tap and fibrous root systems, main roots produce lateral roots; lateral roots form more lateral roots; and so on, until their origin can't be traced any further. This repeated branching of roots generates many

root tips—locations of **root hairs,** extended surface cells through which water and minerals are absorbed (Figure 18-19).

Sometimes roots arise in unexpected places—on stems or even leaves. Roots that develop directly from shoot tissues form an **adventitious root system.** For example, corn plants develop prop roots from the base of their stems (Figure 18-18c), which support the shoot when it is burdened with heavy clusters of fruits (what we call a "corn cob").

Primary Root Tissues

Unlike stem apical meristems, root apical meristems are not located at the very tip of each root (Figure 18-20). Instead, the tip of a root has a **root cap**—a protective cellular "helmet" that surrounds delicate meristematic cells and shields them from abrasion as the root grows through the soil. The root's apical meristem continuously regenerates its root cap as cells are worn away or pierced by sharp soil particles. Root cap cells also help the root penetrate the soil by manufacturing and secreting a gelatinlike substance that helps lubricate the root. This secretion also creates a favorable habitat for certain soil bacteria that supply the plant with essential elements, especially nitrogen.

In addition to forming its own protective cap, the root apical meristem produces all the cells that will differentiate into the root's primary tissue systems. As in stems, roots have a *meristematic region,* a *region of elongation,* and a *region of maturation* (Figure 18-20). Root hairs form in the region of maturation and are the principle site of water and mineral absorption.

Roots contain two important tissues that are not found in stems—an endodermis and a pericycle—both of which

Tap root system   Fibrous root system   Adventitious root system

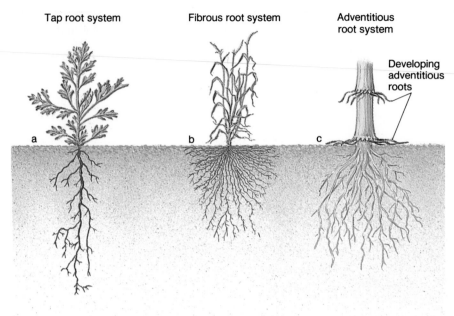

Developing adventitious roots

FIGURE 18-18

Three types of root systems in plants.
(a) A tap root system consists of one main root with many branching lateral roots.
(b) A fibrous root system has more than one main root, each with many branching lateral roots. Fibrous root systems are generally shallow and spread out from the stem base. *(c)* In an adventitious root system, roots arise from organs other than the root itself, mainly from stems or leaves. Adventitious roots develop when plant cuttings are made.

are cylinders of cells that encircle the vascular tissues (Figure 18-20 and 18-21). The **endodermis** is the innermost layer of the cortex. Early in its development, each endodermal cell is encircled by a band of waxy suberin, forming the **Casparian strip.** Water-repelling Casparian strips of adjacent endodermal cells are aligned, creating a "gasket" that seals the space between adjacent cells so that water and minerals can only pass through living endodermal cells. By changing their solute concentration, endodermal cells control the uptake of water and minerals and prevent water from leaking out of the root and back into the sometimes dry or salty soil.

FIGURE 18-19

Root hairs give this radish seedling its fuzzy appearance. The enlargement (inset) shows the root cells beginning to expand to form root hairs for absorbing water and minerals from the soil.

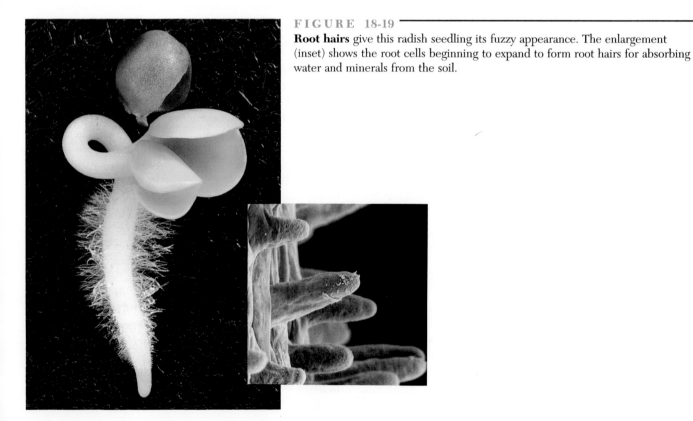

◁ THE HUMAN PERSPECTIVE ▷
Agriculture, Genetic Engineering, and Plant Fracture Properties

Attempts to breed or genetically engineer plants with larger fruits, stems, or leaves for improved productivity could fail if the fracture properties of plant structures is not considered. As a plant's structures increase in size, they are more likely to fracture. In order to produce large fruits, stems, or leaves that will not fracture, they would have to be so "tough" (filled with supportive cells and tissues) that they may not be palatable.

Some botanists study the strength of plant structures as they are subjected to varying stresses in an attempt to reach some general conclusions about the dura-

bility of plant parts; such conclusions may have agricultural and horticultural applications. For example, the shoot system of a plant must sometimes withstand slashing winds and pelting rains. The way a plant resists fracturing under such conditions is the result of its mechanical design, from the cellulose in its cell walls, to specialized support and strengthing cells, to complexes of support tissues. Under normal conditions, the parallel arrangement of sclerenchyma fibers in a leaf petiole (such as a celery stalk) resists damage. If winds are strong enough to inflict damage, however, the cellular structure and arrange-

ment control the fracture in a way that reduces injury, an advantage to an agricultural field of celery on a windy day.

Fracture studies point out a basic principle: The smaller the plant or plant part, the greater its inherent toughness and structural integrity. This principle can be of great importance as biologists try to engineer crops that can survive in regions that are exposed to severe weather, greatly expanding the world's agricultural area. Increasing the food supply is the first step in solving world hunger.

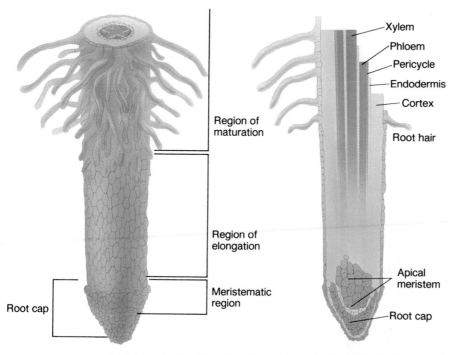

FIGURE 18-20

The three regions of the root apex. The meristematic region of active cell division. The region of elongation, where cells enlarge in length as well as width and begin differentiating. The region of maturation, where most cells, including root hairs, have completed differentiation. Water and minerals are absorbed only in the region of maturation.

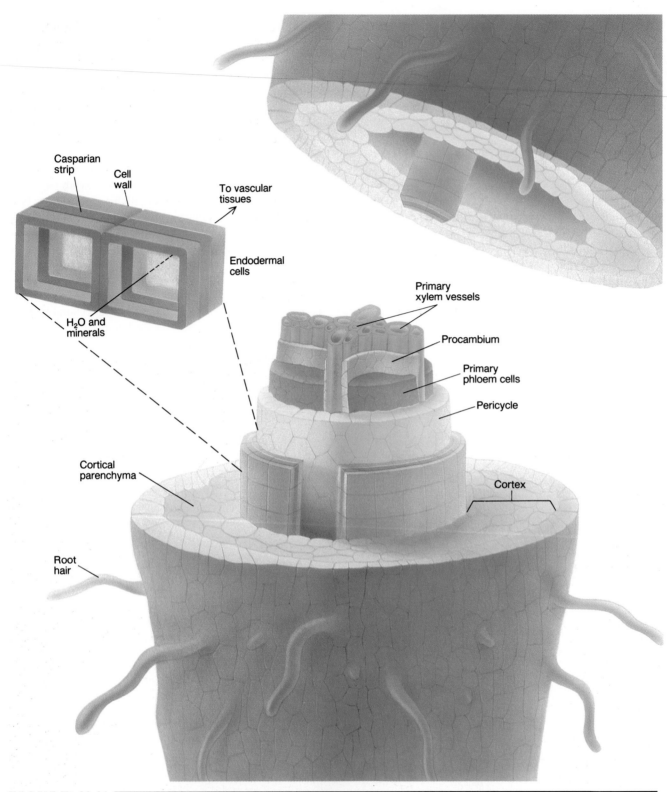

FIGURE 18-21

A three-dimensional, tiered view of the primary tissues found in a dicot root. The innermost layer of cortex cells is the endodermis. A Casparian strip encircles each endodermal cell, enabling endodermal cells to control water movement. The pericycle is a cylinder of cells that lies immediately inside the endodermis. The pericycle produces lateral roots or forms part of the vascular cambium during secondary growth.

Like the endodermis, the **pericycle** is one cell layer thick. The pericycle lies immediately inside the endodermis (Figure 18-21) and divides to form lateral roots. In a perennial or a biennial plant, the pericycle also contributes to the formation of a vascular cambium in the root for secondary growth.

Secondary Root Tissues

If there is secondary growth in the stem, there will also be secondary growth in the root. As in the stem, secondary tissues arise from the vascular and cork cambia. Except for the absence of a pith in the root, secondary growth in the root and stem appear very similar.

THE LEAF: THE PLANT'S PRIMARY PHOTOSYNTHETIC ORGAN

Leaves are the plant's "solar-collectors," "energy generators," and "energy transmission lines," all rolled into one. Leaves capture solar energy, house chloroplasts for converting the energy in sunlight into chemical energy during photosynthesis, and transport materials through an extensive network of interconnecting vascular tissues.

The leaves of flowering plants typically are made up of a flattened **blade** that collects sunlight for photosynthesis, and a **petiole**—a stalk that connects the blade to the stem. The part of a plant stem where one or more leaves are attached is called the **node;** regions between nodes are **internodes.** Directly above the spot where a leaf joins the stem is an **axillary bud**—an underdeveloped cluster of cells that will either grow into a new stem with additional leaves and more axillary buds or will develop into flowers.

Leaves are classified as one of two types, depending on whether or not the blade is divided (Figure 18-22). A **simple leaf** has a single, undivided blade, although, in some cases the blade is deeply indented (lobed) almost to the central, main vein. In contrast, the blade of a **compound leaf** is divided into a number of clearly separated *leaflets.*

(a)   *(b)*

(c)   *(d)*

FIGURE 18-22

Simple versus compound leaves. *(a)* Apple *(Malus)* and *(b)* sycamore *(Platanus)* trees both have simple leaves because each blade is undivided, even though the leaves of sycamore are clearly lobed. *(c)* Roses *(Rosa)* and *(d)* poison ivy *(Toxicodendron)* have compound leaves with leaflets.

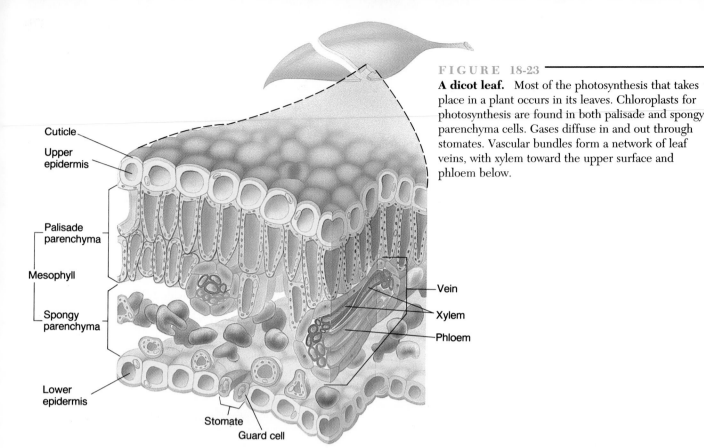

FIGURE 18-23

Cuticle

Upper epidermis

Palisade parenchyma

Mesophyll

Spongy parenchyma

Lower epidermis

Stomate

Guard cell

Vein

Xylem

Phloem

A dicot leaf. Most of the photosynthesis that takes place in a plant occurs in its leaves. Chloroplasts for photosynthesis are found in both palisade and spongy parenchyma cells. Gases diffuse in and out through stomates. Vascular bundles form a network of leaf veins, with xylem toward the upper surface and phloem below.

Leaves have an upper and lower epidermis that sandwich the leaf's **mesophyll**—the layers of cells found between the two epidermises (Figure 18-23). In dicot leaves, the mesophyll contain two types of cells: column-shaped **palisade parenchyma** toward the top, and loosely packed **spongy parenchyma** toward the bottom. Air spaces between these mesophyll cells form corridors for diffusion of carbon dioxide for photosynthesis.

The extensive vascular leaf tissues form veins (Figure 18-24), which create a characteristic **venation** pattern for

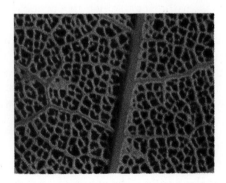

FIGURE 18-24

An extensive network of vascular tissue is revealed when all tissues except the xylem are cleared from a leaf.

different plants. Monocots have leaves with larger veins that run parallel to each other, whereas dicots have highly branched venation (Table 18-1).

THE FLOWER: THE SITE OF SEXUAL REPRODUCTION

The balance between form and function is most elegantly displayed in flowers, which are the site of sexual reproduction. Each part of a flower participates in the process of reproduction by either producing gametes or by influencing the transfer of gametes so that fertilization can take place. Following fertilization, all or part of the flower develops (ripens) into a fruit that contains seeds with newly formed plant embryos. The fruit protects these embryos and helps disperse them to new habitats. Because of their complexity and central role in reproduction, we devote an entire chapter (Chapter 21) to the structure and functions of flowers and fruits.

Dividing essential functions among different organs greatly improves the effectiveness of each. Leaf construction facilitates photosynthesis, flower color and form improves sexual reproduction, stems produce new growth and support aerial structures, and roots absorb water and minerals. (See CTQ #6.)

REEXAMINING THE THEMES

Relationship between Form and Function

⟲ The relationship between form and function is found not only at the more familiar organ level but at the cellular level as well. Specialized cells that help support aerial stems, leaves, flowers, and fruits all have reinforced cell walls. The shape and characteristics of guard cells enable these cells to change shape, thereby regulating the flow of carbon dioxide and water vapor. The hollow interiors of xylem conducting cells form after the cell dies, allowing water and minerals to flow from one cell to the next. The various shapes of thick-walled sclereids enable these cells to carry out specific functions.

Acquiring and Using Energy

☀ The basic plant structure is a consequence of the way plants acquire and use energy. The relatively few structures needed to gather light, water, and carbon dioxide for photosynthesis include leaves, which provide an enlarged surface area for absorbing radiant energy from the sun; stems, which contain vascular tissues that connect with vascular tissues in the root to supply necessary water and minerals and to distribute the products of photosynthesis; stomate pores, which allow gaseous carbon dioxide and oxygen to diffuse to photosynthetic cells; and cells with chloroplasts, the cellular site of photosynthesis.

Evolution and Adaptation

▐▶ Flowering plants are the most evolutionarily advanced plant group. Judgments about whether organisms are "primitive" or "advanced" are based on many criteria, including, but not limited to, fossil record data, similarities and differences in structures, and the means by which efficient structures carry out specific functions. Flowering plants have a number of structures that are more efficient at completing their tasks than are those found in other plant groups. Examples include the flower, for sexual reproduction; fruits, for protecting and disseminating offspring; sieve-tube members, for translocating organic molecules; and vessel members, for transporting water and dissolved minerals.

SYNOPSIS

Flowering plants are the most evolutionarily advanced and structurally complex group in the plant kingdom. Whether they are dicots or monocots, flowering plants form a branched root system that absorbs and conducts water and dissolved minerals, and helps support the aerial shoot system. The plant's shoot produces stems with leaves and axillary buds that develop into more stems or into flowers for sexual reproduction. Leaves are adapted for photosynthesis.

Plants can be classified by their growth pattern. Short-lived, annual plants experience only primary growth, whereas longer-lived biennials and perennials experience both primary and secondary growth simultaneously. Primary growth from meristems increases the length of stems and roots, whereas secondary growth from cambia increases stem and root width as older or dead cells are replaced with new, active cells.

Long-lived perennials replace old or damaged cells with new cells from meristems and cambia. As a result, many plants live for hundreds, or even thousands, of years. Growth in width from cambia produces wood (secondary xylem) for increased support and water and mineral transport, as well as an outer protective periderm.

All flowering plants contain specialized vascular tissues. The xylem conducts water and minerals, and the phloem transports food and other organic molecules.

Plant tissues form groups of cells that perform the same function. Meristems (apical and cambia) divide by mitosis to produce complex vascular tissues (xylem and phloem) and simple tissues (parenchyma, collenchyma, sclerenchyma). The cells that make up a plant's tissues vary in shape, size, and cell wall characteristics, enabling the tissues to carry out specific functions.

Some plant cells—mainly xylem and sclerenchyma—function as dead cells. The unique characteristics of the surrounding cell wall enables some plant cells to continue functioning even after the cell dies.

Parenchyma cells are the most prevalent type of cell in many plants. Versatile parenchyma cells carry out many functions, including photosynthesis, storage, and the ability to change into any other type of plant cell.

The organs of a flowering plant are interconnected by three tissue systems that extend throughout the plant's body. The outer epidermis forms the protective dermal tissue system. The internal network of conducting cells forms the vascular tissue system of xylem and phloem, which is surrounded by cells that make up the ground tissue system.

Each of the four plant organs—root, stem, leaf, flower—is adapted for a different function. Roots absorb water and minerals, store food reserves, and help support the shoot system. Stems produce leaves and axillary buds, which develop into new stems with more leaves and more axillary buds or into flowers for sexual reproduction. Most photosynthesis takes place in the plant's leaves and stems.

Key Terms

Review Questions

1. Name an important adaptation that all shoot dermal tissue system cells have in common but is entirely absent in the root's epidermis. Why is this adaptation restricted to the shoot? Why is it so critical to the survival of a land plant?

2. Plants contain cells (such as xylem vessels and sclerenchyma fibers) that continue to function even after the living part of the cell is dead. List the types of functions these dead cells perform. What characteristics do dead, functioning cells have in common that enable them to continue to function even after the cell dies?

3. How are the terms "periderm" and "bark" related? Why do most botanists avoid using the term "bark"?

4. Why are parenchyma cells sometimes called "the most important type of cell in a plant"?

5. Using the following list of terms, identify with an "S" those that are found only in the shoot system, with an "R" those that are found only in the root system, and with a "B" those found in both the shoot and root systems:

apical meristem	collenchyma	sclerenchyma
parenchyma	node	vessel member
bundle sheath	Casparian strip	axillary bud
companion cell	lenticel	vascular cambium

6. Why do biennials and perennials experience both primary and secondary growth simultaneously, while annual plants only experience primary growth?

7. List five differences and five similarities between the tissues in the stem and the tissues in a root.

8. What is the function of the Casparian strip on endodermal cells? Why don't stem cells have such a strip?

9. How are flowers and stems similar? From an evolutionary point of view, what is the significance of the similarity?

10. List the structural adaptations land plants have evolved to help reduce the amount of valuable water they lose to the surrounding air.

Critical Thinking Questions

1. Imagine that you have been asked to help identify the source of destruction of the grape vines in the French vineyards. When you started working on the project, it was not known that the damage was being caused by an insect. How would you go about determining the source of destruction? How would you differentiate damage caused by insects (or some other animal) from that caused by other environmental factors, such as shortage of rainfall, nutrient deficiency, lack of pollinators, and so on? Choose five environmental factors that could account for the devastation. For each, design a test to verify whether or not that factor was responsible for the damage. In nature, often more than one factor may cause damage. How would you test for a combination of factors?

2. The longer a group of organisms exists, the greater the time for new species to evolve from that group. It seems logical to assume, then, that older groups of organisms would have the greatest species diversity. But the most *recently* evolved plant group (the flowering plants) contains the greatest number of species—over 90 percent of all plant species. How do you explain this apparent contradiction? List some factors (both environmental and biological) that may have triggered such rapid evolution in flowering plants.

3. Businesses use organizational charts to explain the working relationships among personnel. Prepare an organizational chart for a flowering plant, showing how it is organized to carry out the life functions.

4. Using the characteristics described for the various types of cells found in flowering plants, construct a dichotomous key that could be used to identify cells microscopically.

5. Sieve-tube members in the phloem and vessel members in the xylem are both transport cells. What features do these cells have in common that aid in transport, and how are they different? Why wouldn't all transport cells have the same structure?

6. Suppose you could design a flowering plant. What modifications would you make to the roots, stems, leaves, or flowers to create a plant that would be adapted to each of the following habitats: a desert, a rocky intertidal zone, the surface of a freshwater pond?

7. On a nature walk, you find a herbaceous plant that has leaves with parallel veins and flowers with 12 petals. There are no fruits or seeds, so it is impossible to determine whether the embryo has one or two cotyledons. Why would you suspect that the plant is a monocot? What other characteristics would you look for to verify your hypothesis?

8. Plants are annuals, biennials, or perennials. What is the adaptive value for each of these life span strategies? Compare a desert, seashore, and mountain environ-ment. Would one strategy have increased survivability and reproduction over the others in each of these different environments?

Additional Readings

Esau, K. 1977. *Anatomy of seed plants.* New York: Wiley. (Advanced)

Evans, M., R. Moore, and K. Hasenstein. 1986. How roots respond to gravity. *Sci. Amer.* Dec:112–119. (Introductory)

Hohn, R. 1980. *Curiosities of the plant kingdom.* New York: Universe Book. (Introductory)

Raven, P., R. Evert, and S. Eichhorn. 1992. *Biology of plants.* New York: Worth. (Intermediate)

Simpson, B. and M. Conner-Ogorzaly. 1986. *Economic botany: Plants in our world.* New York: McGraw-Hill.

CHAPTER
◄ 19 ►

The Living Plant:
Circulation and Transport

STEPS
TO
DISCOVERY
Exploring the Plant's Circulatory System

XYLEM: WATER AND MINERAL TRANSPORT

Absorption of Water and Minerals

Transpiration

PHLOEM: FOOD TRANSPORT

Transport of Organic Substances

Translocation Through the Phloem

Pressure-Flow Mechanism For Phloem Transport

BIOLINE
Mycorrhizae and Our Fragile Deserts

THE HUMAN PERSPECTIVE
Leaf Nodules and World Hunger: An Unforeseen Connection

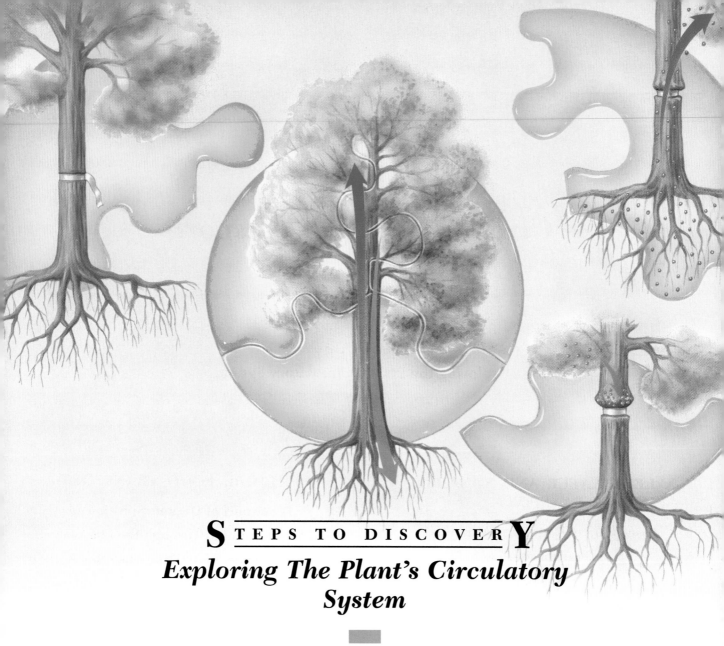

*F*or 3 centuries, biologists have known that blood surges through the human body through vessels. In 1661, Marcello Malpighi, a professor of medicine at the University of Bologna, described the network of the smallest of these vessels, the capillaries. But Malpighi's interest extended beyond the circulatory system of humans; he was curious about how materials moved through the bodies of *all* organisms, both animals and plants.

Malpighi suspected that, like humans, larger plants must have some type of a circulatory system for carrying materials from one part to another; otherwise, how could materials move across long distances, such as between deep roots and the leaves on the tips of branches? Unlike animal research, little microscopic work had been done on plant tissues, so very little was known about how materials could be transported through a plant. Malpighi decided to test whether the bark of a tree contained the plant's circulatory system. He removed a complete strip of bark from around a tree's trunk, a technique called *girdling*. After a period of time, he observed that the bark *above* the girdle swelled; eventually, the tree died. Because the bark swelled only above the girdle, Malpighi concluded that girdling had

By stripping off a ring of bark, scientists solved the puzzle of how materials flow through plant tissues—essential nutrients

blocked the downward movement of materials. The blockage caused these materials to accumulate above the girdle, which then caused the bark to swell; if the materials had been moving upward, the bark *below* the girdle would have swelled. The fact that the tree eventually died led Malpighi to conclude that the materials that move downward through the bark are essential to the plant's life. But just what were these materials?

In 1928, two British plant physiologists, T. Mason and E. Maskell, repeated Malpighi's girdling experiment. After girdling the tree's trunk, the researchers measured the rate of water-vapor loss from the tree's leaves. Mason and Maskell knew from earlier studies that most of the water that moves through a plant is lost by evaporation from the leaves. By measuring water-vapor loss from the leaves of a girdled plant, the researchers could determine whether the *main* pathway of water movement was through the bark or through tissues interior to the bark which were not destroyed by girdling. As did Malpighi, Mason and Maskell observed that the bark swelled only above the girdle. But their study determined that girdling did not change the rate of water-vapor loss from the plant's leaves. Based on these results, Mason and Maskell concluded that the bulk of water must move up through tissues interior to the bark. To explain the swelling above the girdle, the researchers concluded that other materials move *down* through the bark itself. By the 1920s, botanists had named the distinctive tissue within the bark the *phloem* and the tissue interior to the bark the *xylem*. Since most of the water that moves through a plant is released as water vapor from leaves, Mason and Maskel concluded that water circulates mostly upward through the xylem.

During the 1940s, a new technique was developed that enabled researchers to label organic compounds with radioactive carbon and, consequently, to trace the movement of these molecules through an organism. To trace the movement of sugars produced during photosynthesis, for example, plants were exposed to radioactive carbon dioxide, which was absorbed and incorporated into photosynthesized sugars. Scientists then traced the movement of the labeled sugars and discovered that they were concentrated completely in the phloem tissues (in the bark) rather than in the xylem. The predominant direction of movement was downward through the phloem from the leaves—the centers of photosynthesis—to all remaining plant tissues. Later studies using labeled compounds also revealed that virtually all organic molecules produced by a plant circulate through the phloem.

Nearly 300 years since Malpighi's experiments, scientists had finally proven that plants—like most animals—have a circulatory system. But in contrast to animals, water and minerals are transported in plants through tissues that are separate from those through which organic molecules are transported.

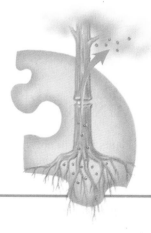

flow down through the bark, while water streams up through internal tissues.

Sometimes the English language gets in the way of our understanding biology. For example, on the one hand, being "animated" means being energetic and lively, like an animal. On the other hand, "vegetating" means being sluggish, dull, and passive —in other words, plantlike. But as we will discover in this and subsequent chapters, plants are not inactive at all. Hidden behind their immobility and stillness is not only the biochemical bustle of myriad metabolic reactions but also a number of surprisingly "animated" behaviors:

- *Competition* Some plants, such as mints and manzanitas, engage in battle with others, competing for space and environmental resources. And, like some animals, many of these plants use biochemical warfare tactics to defeat their rivals.

- *Movement* Although their responses are often slower than are those of most animals, some plants, like the Venus flytrap, move fast enough to catch living, motile prey.

- *Body temperature regulation* Like warm-blooded animals, some plants, such as a snow plant, store the heat produced from their own metabolism, melting their snowy habitats so that water is easily available.

- *Periods of dormancy* Like hibernating animals, many plants (sycamores, walnuts, and aspens, for example) become dormant during cold or dry periods; these plants monitor environmental conditions and resume growth and reproduction when favorable conditions return.

- *Circulation* Like many animals, most plants contain a network of cells that forms a "circulatory system" that distributes materials throughout the plant body. Unlike animals, however, plants have no muscles or heart to pump materials through their circulatory system. Instead, plants rely on the physical properties of water and concentration gradients to transport needed materials, the subject of this chapter.

▼ ▼ ▼

XYLEM: WATER AND MINERAL TRANSPORT

Plants take in and transport far greater amounts of water than do most animals. A single sunflower plant, for example, will take in more than 17 times as much water as you will in 1 day. Plants demand such large amounts of water because many lose up to 98 percent of the water they absorb to the

air as water vapor. Water vapor is lost when plants expose moist leaf cells to the air to take up gaseous carbon dioxide for photosynthesis. When water is exposed to air, it evaporates; the drier the air, the faster the rate of evaporation.

The loss of water vapor from plant surfaces is called **transpiration.** Most water vapor is transpired from leaf surfaces through open stomates; the cuticle prevents evaporation from the surfaces of epidermal cells (page 366). Unless the plant's roots absorb enough water to replace the amount lost during transpiration, the plant will wilt and eventually die from dehydration. Recall from Chapter 8 that some plants have evolved biochemical and anatomic adaptations to hot, dry environments (C_4 and CAM synthesis) which help them reduce water-vapor loss. Transpiration will be discussed in more detail later in the chapter.

ABSORPTION OF WATER AND MINERALS

In the previous chapter, we learned that plants absorb water and minerals through root hairs found near root tips. Since water and minerals are usually widely dispersed in the soil, plants often develop extensive root systems with many root tips to mine the soil for these scant resources. In addition, more than 90 percent of vascular plants form root "partnerships" with soil fungi. In these fungus-root associations, called **mycorrhizae** (*myco* = fungus, *rhiza* = root), fungal filaments often form a tangled sheath around the root, with some filaments penetrating into the root's cortex cells (Figure 19-1a).

Mycorrhizal associations increase a plant's ability to extract water and minerals from soil because fungal filaments extend further into the surrounding soil than does the plant's root system. Some mycorrhizal alliances are so critical that young seedlings that fail to develop mycorrhizae perish from dehydration or nutrient deficiency (see Bioline: Mycorrhizae and Our Fragile Deserts). Furthermore, experiments have shown that plants for whom mycorrhizae are not a matter of life and death (citrus, some pines, and ornamental plants, such as eucalyptus and sugar maple) grow more efficiently with mycorrhizae than without them. Mycorrhizal partnerships are mutually beneficial: The plant gains greater supplies of water and minerals than it could absorb on its own, and the heterotrophic fungus absorbs food stored in the plant's root. Some studies reveal that plants with mycorrhizae may also be less susceptible to certain diseases caused by soil-borne pathogens. Furthermore, there is evidence that mycorrhizae enable some plants to grow better in soils that could not otherwise support favorable growth.

Mycorrhizal hyphae can extend up to 8 meters (25 feet) out from the root. Sometimes the same fungus penetrates the roots of different plants, forming linkages through which materials are exchanged, even among unrelated plants. In addition to water and minerals, carbohydrates and hormones can be exchanged in this manner. Some field biologists are currently investigating the extent of mycor-

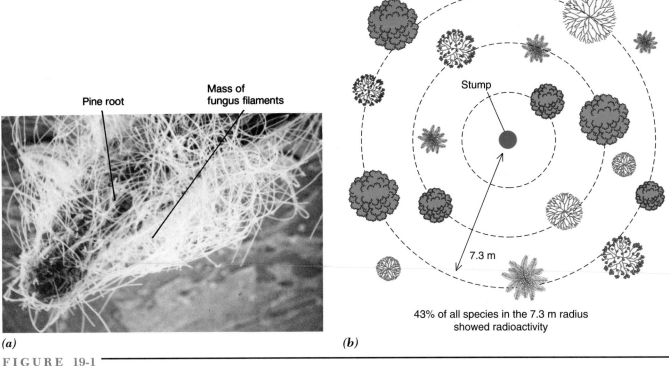

Pine root

Mass of fungus filaments

Stump

7.3 m

43% of all species in the 7.3 m radius showed radioactivity

(a)   *(b)*

FIGURE 19-1

Mycorrhizae: An intimate association with mutual benefits. The filaments of a soil fungus join with the roots of plants to form mycorrhizae. The fungus absorbs some of the food (carbohydrates, amino acids) stored in the root, and the plant receives water and minerals absorbed by widespread fungal filaments. *(a)* In most mycorrhizae, a mass of fungal filaments forms a tangled sheath around the root tip, with some of the filaments penetrating root cells. *(b)* Scientists discovered that the roots of many plants in a North Carolina forest were interlinked by mycorrhizal filaments that extended from plant to plant. In this study, researchers added radioactive calcium and phosphorus to a freshly cut stump of a red maple tree. The radioactive materials quickly spread through 43 percent of the species within a radius of 7.3 meters, indicating that, because of mycorrhizae linkages, the roots of these forest plants form a single, widespread functional unit for absorbing water and minerals over very broad areas.

rhizal interconnections in ecosystems. For example, F. Woods and K. Brock added radioactive calcium and phosphorus to a freshly cut maple tree stump; within 8 days, 19 other types of trees and shrubs within a radius of 7.3 meters contained the labeled elements (Figure 19-1*b*). Woods and Brock suggested that because mycorrhizae interconnect root systems, the root mass of a forest may act as a single functional unit. This possibility is considered further in Chapter 41.

Avenues of Water Uptake

Water enters a plant's roots by osmosis, whereas dissolved minerals are actively transported into root cells through the cell membrane. Once inside the root, water travels through the root cortex in one of two ways: along cell walls or from cell to cell via connecting passages called **plasmodesmata** (Figure 19-2). Once water reaches the endodermis, however, aligned, waxy Casparian strips prevent further movement of water between or along cell walls. Water then

moves through the endodermal cells as minerals are actively transported into interior cells; that is, water flows along an osmotic gradient. To help visualize this phenomenon, consider what happens when you place a limp celery stalk in a bowl of water to make it crisp. Water from the bowl flows into the celery stalk along an osmotic gradient; that is, the concentration of solutes in the stalk is greater than that in the bowl so water flows into the stalk. Water continues to follow the inbound path of actively transported minerals into a conducting cell in the xylem. Once inside the xylem, water and minerals become part of a stream that conducts them through the plant.

Mineral Absorption and Requirements for Normal Growth

In the nineteenth century, two independent researchers, J. Sachs and W. Knop, grew plants by immersing their roots in a solution of inorganic salts rather than soil (a technique now referred to as **hydroponics**). The studies proved for

◁ BIOLINE ▷
Mycorrhizae and Our Fragile Deserts

Hot, dry, and desolate. That is the image most of us have of the desert. In fact, *Webster's New World Dictionary* defines a desert as "arid, barren land . . . incapable of supporting any considerable population without an artificial water supply." Although this description may aptly characterize some areas of the Sahara Desert in Africa, it certainly does not describe the deserts of North America. In the Colorado Desert of southeastern California, for example, biologists have identified more than 1,200 plant species, 200 species of vertebrate animals, and numerous insects and invertebrates.

Some people believe that because desert plants and animals can withstand unusually harsh conditions, they are "tough" enough to withstand virtually any kind of abuse, from hordes of off-road vehicles and motorcycles to the many disturbances incurred from the construction of roads and power-transmission lines. Unfortunately, this perception of "toughness" is a false one. In reality, the desert is easily damaged and is very slow to recover from even small disturbances; 1,000-year-old Indian trails and campsites are still clearly visible, proving that virtually no natural recovery has taken place after thousands of years.

The California deserts are a valuable recreational, industrial, and biological re-

source. More than 100 of the plant and animal species found in these deserts are listed as rare, endangered, or threatened with extinction. With continued fast-paced development and heavy abuse of the deserts, these species cannot be saved without care, habitat protection, and efforts to rehabilitate damaged habitats.

In 1970, a local electrical company built a power-transmission line in a portion of the Mojave Desert in California, cutting through one of the few remaining populations of the Mojave ground squirrel, an endangered species. The wide power line road and path of disturbance destroyed the surrounding plants, preventing the ground squirrels on one side of the line from crossing over and breeding with squirrels on the other side. (Ground squirrels scamper from plant to plant not only to find food but also to hide from predators. When a ground squirrel reaches an area without plants, it turns around.) This segregation quickly lowered the squirrels' reproduction rate, causing the number of Mojave ground squirrels to plummet. In an effort to reverse this trend, field biologists tried to reseed the disturbed area with desert plants a number of times, but all efforts failed. Even when seeds and seedlings of the desert plants were repeatedly planted and irrigated, none survived more than a few weeks. Investigators eventually discov-

ered that the damaged soils lacked the fungus that naturally forms mycorrhizae with desert plant roots. Without mycorrhizae, the young plants were incapable of forming a root system large enough to gather sufficient water supplies, and they quickly died of dehydration. When the researchers added mycorrhizal fungi to the disturbed soils, however, many of the young plants lived. Unfortunately, the costs of injecting several square miles of disturbed soil with mycorrhizal fungi is so astronomical that the remaining population of Mojave ground squirrels will simply have to survive on its own.

A number of important lessons were learned from the Mojave ground squirrel studies. The electrical company vowed not to scrape the plants and desert soil to one side when constructing a power line, removing critical habitats and essential microorganisms (like the mycorrhizal fungus) in the soil. The company also pledged to adopt new construction practices that cause fewer environmental disturbances. And, perhaps most importantly, the power company agreed to be much more careful in the future in planning new power line corridors or construction sites in order to avoid tampering with the habitats of any endangered species.

---

the first time that plants could receive all their needs from sunlight and inorganic elements. Since that time, researchers have used hydroponics to grow plants in solutions in which only one element is missing at a time. From these studies, botanists have identified 16 **essential nutrients** —elements that are absolutely necessary for plants to complete their life cycle and for normal growth and development to occur (Table 19-1). Nine of these essential nutrients are needed in very large amounts; as such, they are called **macronutrients.** The seven remaining **micronu-**

trients are needed only in small amounts. Among their many roles in plant metabolism, essential nutrients are used to construct organic compounds, to activate enzymes, and to contribute to turgor pressure in plant cells. A plant deficient in an essential nutrient displays characteristic symptoms, some of which are described in Table 19-1.

Minerals are actively transported into roots, even when mineral concentration is lower in the root than it is in the soil or surrounding solution. Once inside the root, dissolved minerals are actively transported through cortical paren-

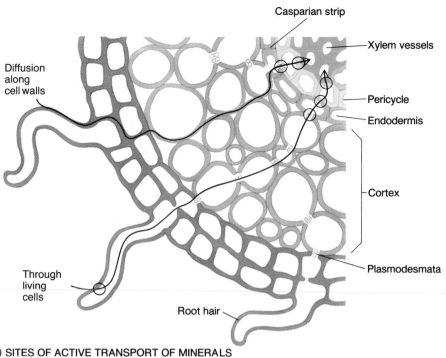

Casparian strip

Xylem vessels

Pericycle

Endodermis

Cortex

Plasmodesmata

Diffusion along cell walls

Through living cells

Root hair

○ SITES OF ACTIVE TRANSPORT OF MINERALS

FIGURE 19-2

Roadmap to the xylem. Water enters roots by osmosis, whereas most dissolved mineral ions are actively transported into root cells. Water moves either through living cortex cells or along cortex cell walls but is prevented from moving between endodermal cells by adjoining, waxy Casparian strips. Minerals are actively transported through the endodermis, the pericycle, the procambium, and into xylem conducting cells, creating an osmotic gradient. Water flows along this gradient into the xylem.

chyma cells, the endodermis, the pericycle, and, finally, into the xylem (Figure 19-2). Because of this one-way active transport path, minerals become concentrated in the xylem.

As minerals travel through the xylem, some diffuse out into surrounding root and stem tissues; the rest continue to flow within the xylem stream to the leaves. Minerals not used in the leaf are discharged into the descending phloem stream. Some minerals travel back to the root, where they are either used by root cells or reenter the ascending xylem stream.

▐▶ Some plants, particularly legumes (peas, beans, clover, alfalfa) form **root nodules**—swellings that house nitrogen-fixing (capable of converting nitrogen gas to ammonia or nitrate) *Rhizobium* bacteria (Figure 19-3). As in the case of mycorrhizal associations, both organisms benefit from this association: The plant receives a supply of nitrogen (an

FIGURE 19-3

Root nodules. Legumes, such as this snow pea *(a)* commonly form root nodules, *(b)*. The lumplike root nodules house nitrogen-fixing bacteria. The bacteria benefit from this alliance by harvesting some of the plant's carbohydrate stores, and the plant benefits by receiving nitrogen from the bacteria in a form it can use.

(a)

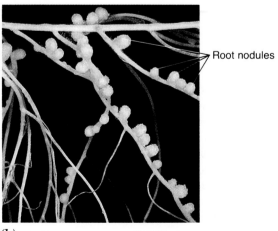

Root nodules

(b)

TABLE 19-1

PRIMARY USES OF THE 16 ESSENTIAL PLANT NUTRIENTS

Nutrient	Function	Percentage Dry Weight[a]	Deficiency Symptoms
Macronutrient			
Carbon (C)	Major component of organic molecules	45.0	Severely impaired growth
Oxygen (O)	Major component of organic molecules	45.0	Severely impaired growth
Hydrogen (H)	Major component of organic molecules	6.0	Severely impaired growth
Nitrogen (N)	Component of amino acids, nucleic acids, chlorophyll, and coenzymes	1.0–4.0	Yellowing of older leaves; plant spindly
Potassium (K)	Component of proteins and enzymes; helps regulate opening and closing of stomata	1.0	Mottled, yellow older leaves with die back at tips and margins
Calcium (Ca)	Component of middle lamella; enzyme cofactor; influences cell permeability	0.5	Youngest leaves die back at tip and margins; deformed leaves
Phosphorus (P)	Component of ATP, ADP, proteins, nucleic acids, and phospholipids	0.2	Youngest leaves marked with purple
Magnesium (Mg)	Component of chlorophyll; enzyme activator; amino acid and vitamin formation	0.2	Yellowing between leaf veins of older leaves; leaves may turn orange
Sulfur (S)	Component of proteins and coenzyme A	0.1	Yellowing of young leaves; veins bright red; stunted
Micronutrient			
Iron (Fe)	Catalyst for chlorophyll synthesis; component of cytochromes	0.01	Yellowing between veins of youngest leaves
Chlorine (Cl)	Osmosis and photosynthesis	0.01	Wilting of leaf tips; yellow leaves with bronze color
Copper (Cu)	Enzyme activator and component; chlorophyll synthesis	0.0006	Dark green leaves with black spots at tips and margins
Manganese (Mn)	Enzyme activator; component of chlorophyll	0.005	Black spots on veins in youngest leaves
Zinc (Zn)	Enzyme activator; influences synthesis of hormone, chloroplasts, and starch	0.002	Reduced internodes; leaf margins puckered
Molybdenum (Mo)	Essential for nitrogen fixation	0.0001	Yellowing between veins of older leaves with black spots
Boron (B)	Important to flowering, fruiting, cell division, water relations, hormone movement, and nitrogen metabolism	0.002	Black spots on young leaves and buds

[a] Percentage Dry Weight: Proportion of weight after all water has been removed.
Note: To help remember the macronutrients, use the phrase "C. HOPK(*i*)NS Ca*r is an* MG";
to help remember the micronutrients, use the phrase "A Fe*stive* MoB *comes in* (= CuMnZn) Cl*apping*"

essential macronutrient), and the bacteria receive food, water, and living space. Mutually beneficial interactions like mycorrhizae and root nodules impart a survival advantage to both participants; as a result, such relationships are a common outcome of evolution even between very different kinds of organisms (this principle of coevolution is discussed in Chapter 34).

Nitrogen can mean the difference between success or disaster of agricultural crops. Nitrogen is a key element in many vital organic compounds, including nucleic acids and proteins; of the nine macronutrients, only carbon, oxygen,

and hydrogen are needed in greater quantities than is nitrogen (Table 19-1). Plants absorb nitrogen only in the form of nitrate (NO_3) or ammonia (NH_3) but not as nitrogen gas (N_2), the form of nitrogen found in our atmosphere.

⟳ As we mentioned in the previous paragraph, nitrogen gas is converted to ammonia or nitrate during **nitrogen fixation**, 90 percent of which is accomplished by nitrogen-fixing bacteria (the other 10 percent is performed by lightning). Nitrogen-fixing bacteria are found in the soil or in root nodules, where the plant receives a direct supply of fixed nitrogen. Root-nodule bacteria fix more nitrogen than the plant can actually use. Excess ammonia is then released into the soil, a natural means of enriching croplands. In fact, many farmers have discovered that they can improve productivity by rotating crops every other year—one year they plant clover or alfalfa, which produces root nodules that enrich the soil with nitrogen; the next year they plant corn, a plant that does not form root nodules. The corn grows vigorously in the nitrogen-rich soil. If corn were planted year after year, it would quickly deplete nitrogen supplies, and productivity would plummet.

Root Pressure and Guttation

As root cells pump minerals into the xylem by active transport, the concentration of dissolved particles increases, compared to the concentration in surrounding cells. This increase causes water to be drawn into the xylem by osmosis. The continuous influx of water can produce a positive **root pressure** that pushes water and dissolved minerals up a xylem column. Under certain conditions, root pressure builds high enough to force water and minerals completely out the tips of leaves, a process called **guttation** (Figure 19-4). Guttation only happens at night when there is no transpiration; as soon as transpiration begins in the morning, root pressure plunges and guttation stops. Guttation also occurs only in small plants growing in soils saturated with water. A drop of water on the tip of a leaf of a small plant in the early morning may be the result of either guttation or dew formation. A dew drop forms when water from the air condenses on a cool leaf tip, whereas a guttation drop forms as internal water is forced out of a plant.

Researchers have demonstrated that water cannot be pushed up a column to a height greater than 10 meters (about 30 feet, the height of a medium-size tree). Thus, root pressure is inadequate for moving water and minerals up plants that are taller than 10 meters and does not explain how water and minerals move through a plant during the daytime.

TRANSPIRATION

Why is it that when you put a cut flower in a vase of water, sometimes the flower still wilts? Doesn't the water in the vase simply replace the water being lost from leaves, stems, and petals through transpiration? It depends. In order for the water in the vase to replace transpired water, there must be a continuous connection between the water molecules in the xylem and the water in the vase; that is, water molecules in the xylem must be *cohesively bonded* to water molecules in the vase. If the flower was cut off the plant on a warm afternoon when transpiration was high, or if you waited some time before putting the flower in a vase, cohesive bonding between the xylem water and the water in the vase has been prevented.

If you could have observed the xylem under a microscope as you cut the flower from the plant, you would have seen that the tension of the water in the xylem channel from high transpiration caused the water to "snap up" the minute

the stem was cut, similar to the way a stretched rubber band retracts in two directions if it is cut. The snapped water column creates an air pocket in the xylem channel. (Air pockets will also form in the xylem if you wait a while before putting the flower in a vase because the flower continues to lose water through transpiration.) When the flower is put in the vase, the air pocket separates the xylem water from the water in the vase, preventing cohesive bonding between them. The flower continues to lose water through transpiration. When it eventually uses up all the water in the xylem, the flower wilts and dies, even though there is plenty of water in the vase. The solution, as many florists will tell you, is to recut the stem of the flower under water before putting the flower in a vase. This removes the section of the stem with the air pocket and allows cohesive bonds to form between liquid water molecules in the xylem and the vase. In this case, as the flower loses water vapor, transpired water is replaced with the water from the xylem, which is replaced with water from the vase.

Transpiration Pull Mechanism for Transport

Plant physiologists have now shown that the primary mechanism for water and mineral translocation is **transpiration**

FIGURE 19-4

Forcing the issue. Glistening droplets of water and dissolved minerals are forced out of leaf tips by positive root pressure during guttation.

pull—a process triggered by water-vapor loss. Transpiration pull results from a chain of events that starts when leaves begin absorbing solar radiation in the morning. Sunlight heats the plant's leaves, causing more water to evaporate from moist cell walls. The evaporated water is immediately replaced with water from inside the cell, which is replaced with water from neighboring cells deeper in the leaf, which, in turn, is replaced with water in the xylem (Figure 19-5).

Recall from Chapter 4 that liquid water molecules are unique in their ability to cling to one another by strong cohesive bonds. In the xylem, liquid water molecules are cohesively bonded to one another, forming an unbroken water chain that extends all the way from the leaf vein, through the stem, and down into the root. Because of these strong bonds, as water molecules exit the xylem, they pull adjacent water molecules up, which, in turn, pull more

adjacent water molecules up, and so on, down the xylem column to the roots below. In other words, cohesive bonds pull water up the xylem. This process is transpiration pull.

The constant pull between cohesively bonded water molecules creates a tension in the xylem that is transmitted down the entire water column. As transpiration increases during the day, the tension in the xylem builds, as more and more water molecules exit the xylem to replace the transpired water. But why don't the cohesive bonds simply snap from the tension, pulling liquid water molecules apart and breaking the continuous water column? The reason has to do with the strength of the cohesive bond *and* the narrow width of xylem tracheids and vessels (page 370). In narrow xylem cells, many water molecules are both cohesively bonded to other water molecules *and* adhesively bonded to xylem cell walls. This combination of cohesive and adhesive bonds enables a column of water to withstand tension ten times greater than that needed to pull water to the top of the tallest tree.

Peak periods of transpiration can build such tremendous tension in the xylem that the diameter of a large tree trunk may actually shrink! Because some water molecules in the xylem adhere to xylem cell walls as well as cohere to other water molecules, high tension pulls tracheid and vessel walls inward, enough to reduce the diameter of a tree measureably as the day progresses.

Regulating Transpiration Rates

The difference in concentration of water vapor between the air spaces *inside* a leaf and that found in the air *outside* a leaf determines the rate of transpiration; that is, the speed at which water vapor will be lost. Simply put, the greater the difference in concentration, the faster the rate of transpiration; the smaller the difference, the slower the rate (Figure 19-6). Since the air inside a leaf remains saturated with water vapor, there is almost always a difference between the air inside and the air outside a leaf. The only time there would not be a difference is when it is raining, foggy, or very humid outside.

Under normal daytime conditions, a leaf could transpire its entire water content in about 1 hour. Unless transpired water is replaced just as quickly, the leaf (and plant) will wilt and die. Thus, for plants growing in dry environments, the ability to control transpiration is a matter of life and death.

Because shoot surfaces are covered with a waxy cuticle, less than 10 percent of transpired water is lost directly from the surface of epidermal cells, unless something damages the cuticle layer. One reason biologists are so concerned about acid rain is that it dissolves the cuticle, destroying one of the plant's natural adaptations for controlling water-vapor loss. Acid rain also kills soil bacteria (which supply plants with needed nutrients) and increases the rate of leaching of nutrients from the soil—two more reasons why it is so important to control acid rain.

Most water-vapor loss occurs through open stomates—the pores formed between adjacent guard cells. You will

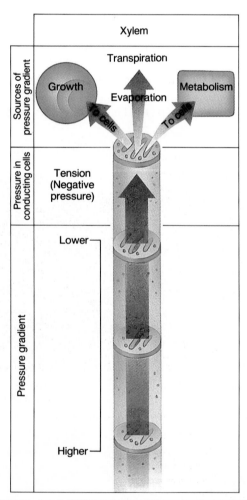

FIGURE 19-5

Xylem transport. Tension (negative pressure) builds in the xylem as water evaporates during transpiration or is used by cells for metabolism and growth. As water exits the xylem, cohesively bonded water molecules are pulled up.

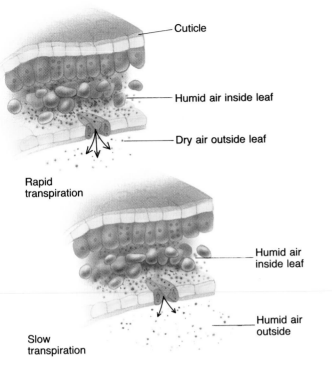

FIGURE 19-6

The rate of transpiration. The greater the difference in concentration of water-vapor molecules between the air spaces inside a leaf and that of the surrounding outside air, the greater the rate of transpiration. Water vapor is always high inside the leaf (unless it is wilted); changes in the humidity of the surrounding air affects the rate of water-vapor loss.

recall from Chapter 18 that stomates open to allow carbon dioxide to diffuse into the leaf for photosynthesis; when stomates open, water vapor usually flows out. Guard cells are able to change the size of a stomatal pore, thereby changing the rate of transpiration. The ability of guard cells to narrow, or even close, a stomatal pore, can save a plant from wilting or dying from dehydration. Plants have evolved other adaptations for regulating transpiration rate, three of which are illustrated in Figure 19-7.

Water and dissolved minerals are transported through the xylem. Since plants often transpire large quantities of water vapor from leaf surfaces, water and minerals flow up through the xylem, from the roots to the leaves. (See CTQ #2.)

PHLOEM: FOOD TRANSPORT

Substances are continually moving through the vascular tissues of a plant. The rate at which food moves through the phloem can reach speeds up to 100 cm/hour, or roughly 0.0013 miles/hour—about the speed of a tired snail. This is about one-fifth the rate at which water and minerals move through the xylem. Although phloem transport is slower than xylem transport, it is still much faster than cytoplasmic

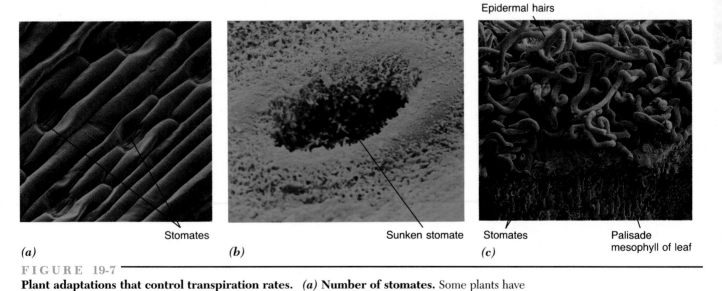

(a) Stomates   *(b)* Sunken stomate   Stomates   *(c)*   Epidermal hairs   Palisade mesophyll of leaf

FIGURE 19-7

Plant adaptations that control transpiration rates. *(a)* **Number of stomates.** Some plants have many thousands of stomates per each square centimeter of leaf surface. Plants adapted to growing in moist habitats have stomates on both the top and bottom surfaces; plants adapted to arid areas have fewer stomates, which are restricted to the cooler, lower surfaces where the vapor gradient between the inside and outside of the leaf is less. *(b)* **Position of stomates.** Recessed stomates help some plants adapted to arid habitats conserve water. This placement doesn't affect carbon dioxide diffusion but retards water-vapor loss by forming pockets of high humidity. Sunken stomates also protect the plant from the drying effects of wind. *(c)* **Epidermal hairs.** Dense epidermal hairs not only reflect light, reducing the leaf's heat load, but they also slow down transpiration by shielding stomate openings from drying winds.

streaming or diffusion of substances from cell to cell. Such rates would be too slow to distribute adequate supplies of nutrients throughout a plant.

TRANSPORT OF ORGANIC SUBSTANCES

It wasn't until 40 years ago that investigators actually identified which substances were being transported through the phloem. Before then, no matter how carefully researchers tried to extract the sap from phloem, the sieve tubes collapsed and plugged up, preventing researchers from getting a pure sample of phloem sap. Oddly enough, a common plant pest called *aphids* provided scientists with the "tool" they needed to collect pure phloem sap. Not only did the aphids enable botanists to verify what they had suspected all along—that food and other organic substances are transported through the phloem—but the pest also allowed researchers to measure how fast these substances move through phloem tubes.

➔ Aphids are insects that gather on the growing stem tips and leaves of many plants in early spring. Their hollow, needlelike mouthparts (stylets) enable aphids to bore directly into the phloem and feed on the plant's juices. The moment the aphid's stylet penetrates a sieve tube, its body balloons out, as sap gushes from the phloem transport cells; sometimes the pressure in the sieve tube is so high that it forces sap completely through the insect's digestive tract and out its anus, forming small droplets of honeydew (Fig-

ure 19-8*a*). But researchers did not want to analyze the contents of honeydew drops because the drops would be "contaminated" with materials from the aphids' digestive system. To collect "uncontaminated" phloem sap, investigators simply allowed aphids to tap into the phloem, then quickly anesthetized the insects and cut off their bodies, leaving the stylet still tapped into the phloem. The pure sap that exuded from the remaining stylets (Figure 19-8*b*) contained only materials flowing through the phloem—10 to 25 percent sugar (mainly sucrose), small amounts of amino acids, other nitrogen-containing substances, and plant hormones—all molecules synthesized by the plant. A few dissolved minerals were found in the sap as well. After nearly 350 years, the puzzle of what materials flow through the phloem was finally solved.

TRANSLOCATION THROUGH THE PHLOEM

Phloem sap always flow from *sources* to *sinks* (Figure 19-9*a*). Sources are those areas of a plant where sugars and other organic substances are synthesized (such as a photosynthesizing leaf cell) or stored (such as in parenchyma tissues). Sinks are plant tissues, such as actively growing meristems, developing fruits and seeds, or storage tissues, that use organic molecules. (Notice that storage tissues can be both sources and sinks. They are sources when they export their stored molecules, and they are sinks when they import molecules to build reserves.)

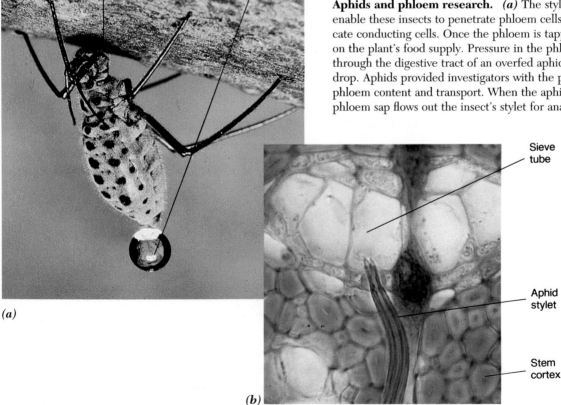

Stylet   Honeydew drop

(a)

Sieve tube

Aphid stylet

Stem cortex

(b)

FIGURE 19-8

Aphids and phloem research. *(a)* The stylet mouthparts of aphids enable these insects to penetrate phloem cells without damaging delicate conducting cells. Once the phloem is tapped *(b)*, the aphid feeds on the plant's food supply. Pressure in the phloem forces phloem sap through the digestive tract of an overfed aphid, forming a honeydew drop. Aphids provided investigators with the perfect tool for studying phloem content and transport. When the aphid's body is removed, phloem sap flows out the insect's stylet for analysis.

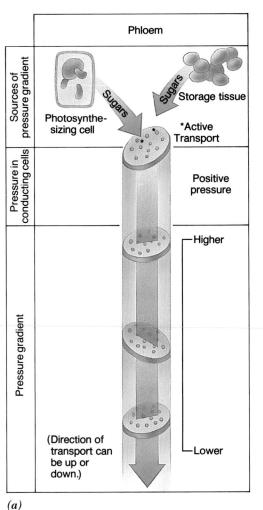

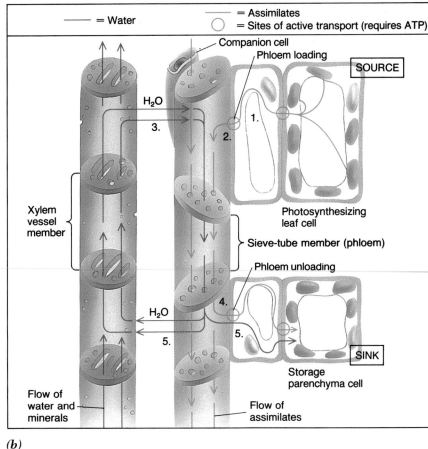

(a)   *(b)*

FIGURE 19-9

Phloem transport by pressure flow. *(a)* Sugars and organic molecules are actively loaded into sieve elements, creating an osmotic gradient that draws water into phloem transport cells, thereby increasing internal pressure at the point of active loading. At the other end of a phloem tube, pressure falls as molecules are actively unloaded and water flows out of the phloem. This pressure gradient pushes materials through the phloem from sources to sinks. *(b)* Phloem sap flows from sources to sinks. (1) A photosynthesizing leaf cell is a source because it produces sugars. (2) Sugars are actively transported into a sieve tube during phloem loading. (3) As sugar concentration increases in the sieve tube, water is drawn in from nearby xylem vessels. This influx of water raises the pressure in the sieve tube at the source. (4) At sinks (in this example, a storage parenchyma cell is a sink because it stores sugars), sugars are actively unloaded, lowering sugar concentration in the sieve tube and (5) causing water to exit the phloem tube and be recycled back to the xylem or drawn into surrounding cells. The pressure differences between sieve-tube members at sources (high pressure) and sinks (low pressure) initiate the flow of an assimilate stream.

Because phloem is so delicate and therefore difficult for researchers to work with, there is still some debate over exactly how materials are translocated through the phloem. Biologists have proposed and tested many hypotheses. So far, the greatest evidence supports the "pressure-flow hypothesis" of transport, which we discuss below. But the quest for understanding organic-molecule transport has not been abandoned. In fact, some studies suggest there may be more than one mechanism involved in phloem transport, illustrating the ongoing nature of scientific investigations.

PRESSURE-FLOW MECHANISM FOR PHLOEM TRANSPORT

To initiate **pressure flow,** higher pressures are created at sources when organic molecules are actively transported into sieve-tube members; this is called **phloem loading** (Figure 19-9*b*). Phloem loading increases the concentration of solutes inside the sieve tube and creates an osmotic gradient that causes water to flow in from nearby xylem cells. This process builds water pressure in the sieve tube at the source. At the sink end of the sieve tube, molecules are actively transported out of the phloem. This **phloem un-**

◁ THE HUMAN PERSPECTIVE ▷
Leaf Nodules and World Hunger: An Unforeseen Connection

Roots are not the only plant organ that form associations with microorganisms. Several hundred species of tropical plants develop bacteria leaf nodules, including an important crop plant, yams (*Dioscorea*). Because leaf nodules occur on less than 0.07 percent of plants, they have been mostly overlooked by investigators. In fact, it still isn't clear which bacteria are involved or whether these bacteria play a beneficial role for the plant. The work that has been done has made one thing certain: Leaf nodule bacteria, unlike root nodule bacteria, are not nitrogen-fixers. If leaf nodules are so rare and have no clear benefits, why study them at all?

The bacteria in leaf nodules are unique in that they are passed from one plant generation to the next. As the ovule develops into a seed, bacteria from the parent plant are transported to the embryo. When the embryo germinates, the transported bacteria immediately infect the leaves of the growing seedling. The ability of these plants to coexist permanently with resident leaf bacteria and to transfer bacteria from generation to generation may hold a key to solving some critical human problems. For example, yams are cultivated in many parts of the world for their edible tubers (fleshy roots). If leaf nodule bacteria on yams were genetically engineered to fix nitrogen or to deliver systemic pesticides, they could be transferred from one year's crop to the next, increasing yam production while reducing the need for fertilizers or pesticides, neither of which are available in many regions that depend on yams as a food staple. Of course, this cannot happen until more basic research is done on plant leaf nodules.

Sometimes studying rare, seemingly unimportant problems may sound like a useless and wasteful pursuit. We often hear criticisms of government-supported research projects that, on the surface, seem like a waste of our tax dollars. But studying rare phenomena may also provide new scientific breakthroughs that can solve important human problems, such as world hunger, in the case of leaf nodule bacteria. Shutting the door on conducting research also shuts the door on reaching new understanding of the world around us.

loading reduces solute concentration in the phloem, causing water to flow out of the phloem and into the sink or back to the xylem. As a result, water pressure drops at a sink. Since all sources and sinks are connected by columns of sieve-tube members, higher water pressure at sources pushes the phloem sap to lower water pressures at sinks (Figure 19-9*b*). In other words, pressure differences cause organic molecules to flow from sources to sinks. The ATP that is needed for phloem loading and unloading is not produced by the sieve-tube members because, as we discussed in Chapter 18, sieve-tube members do not contain any mitochondria. The ATP needed for loading and unloading is provided by adjacent companion cells—cells that also supply all the other metabolic needs of the sieve-tube member. Only by working together do sieve-tube members and companion cells create conditions necessary to power assimilate flow.

Organic substances that are synthesized by a plant move through the phloem. Since plants usually synthesize more sugar through photosynthesis than they do any other organic molecule, and since photosynthesis takes place predominantly in leaves, sugars and other organic substances generally flow downward through the phloem, from leaves to stems and roots. (See CTQ #5.)

REEXAMINING THE THEMES

Relationship between Form and Function

The structure of vascular plant tissues enables them to translocate materials from one region of the plant to another. Water and minerals flow through hollow, lifeless xylem cells. Their thick walls are able to withstand the high tensions that build during transpiration, the driving force behind the movement of water and minerals up to the top of the tallest trees. In the phloem, both sieve-tube members and companion cells create the pressure necessary to cause assimilates to flow from a source to a sink. Because of their interdependence, sieve-tube members and companion cells must join together. Finally, the juxtaposition of xylem and phloem in the same vascular bundle allows water to circulate back and forth between xylem and phloem; water is the basic medium for both mineral and assimilate flow.

Unity within Diversity

⚠ All multicellular organisms that grow larger than a few layers of cells face a similar problem: how to distribute essential materials throughout all of the cells in the body. Although higher plants and animals may be very different, they have all evolved very similar adaptations for solving the problem of moving materials from one region to another.

Both have specialized vascular tissues that form a network of tubes through the body for distributing materials, and both rely on water as the primary transport medium.

Evolution and Adaptation

▶ Water supply is probably the most important environmental factor that affects the growth, distribution, and health of any organism. Plants grow in many kinds of habitats, from those that are soaking wet all of the time (a tropical rainforest), to those that receive intense radiation and scant amounts of rainfall (a desert). Plants in dry, hot environments must control water-vapor losses; otherwise, the plant can easily lose more water than is available from its environment. Guard-cell behavior, stomate number and placement, the presence of epidermal hairs, reflective surfaces, and photosynthetic pathways are all adaptations plants have evolved for regulating water-vapor loss.

SYNOPSIS

A vascular plant's circulatory system includes two sets of conduits. Water and dissolved minerals are conducted through the xylem; organic substances synthesized by the plant are translocated through the phloem.

Water and dissolved minerals are absorbed through root hairs or mycorrhizae fungal filaments. Interconnecting Casparian strips on endodermal cells in the root direct water through the endodermis to xylem vascular tissues and prevent water from leaking out of roots as the soil dries.

Most of the water that is absorbed evaporates from leaf cell surfaces and exits through stomates during transpiration. As water vapor is lost during transpiration, water from the xylem replaces the lost water vapor. This builds a strong tension in the xylem that *pulls* water and minerals through the columns of hollow conducting cells.

The faster the rate of transpiration, the stronger the tension, and the more accelerated the flow of the xylem stream.

Products of biochemical reactions (sugars from photosynthesis, hormones, amino acids, and so on) are transported through living sieve tubes of the phloem. Organic substances are actively loaded into the phloem at sources and are unloaded at sinks, where they are used or stored. Loading at sources causes water from nearby xylem to flow into the sieve tube, generating pressure that pushes the phloem sap to sinks by pressure flow.

The roots of many plants establish mutually beneficial partnerships with soil microorganisms. Many plants form mycorrhizae associations with soil fungi, greatly improving their ability to absorb water and minerals. Some plants form lump-shaped root nodules housing nitrogen-fixing bacteria that directly supply roots with usable nitrogen.

Key Terms

transpiration (p. 386)
mycorrhizae (p. 386)
plasmodesmata (p. 387)
hydroponics (p. 387)
essential nutrient (p. 388)

macronutrient (p. 388)
micronutrient (p. 388)
root nodule (p. 389)
nitrogen fixation (p. 390)
root pressure (p. 391)

guttation (p. 391)
transpiration pull (p. 391)
pressure flow (p. 395)
phloem loading (p. 395)
phloem unloading (p. 395)

Review Questions

1. Using an "X" for xylem and a "P" for phloem, indicate where each of the following is found:

    transpiration pull   micronutrients
    calcium   sugars
    pressure flow   organic molecules
    sieve tubes   vessel members

2. Where appropriate, put an "R" for root nodule or an "M" for mycorrhize.

    _____ a. root and fungus partnership
    _____ b. supplies nitrogen to plants
    _____ c. root and bacteria partnership
    _____ d. creates large water-absorption surface area
    _____ e. is mutually beneficial

3. How can you determine whether a drop of water on the tip of a leaf is dew (condensation from the air) or water that has been forced out by guttation?

4. List the essential nutrients that are:
    a. components of plant organic molecules
    b. needed for chlorophyll formation
    c. used to construct plant cell walls

5. Why are roots considered one of the main "sinks" in an actively growing plant? List three sources and three sinks in a typical plant.

6. What happens to the water in sap that flows down a stem and into the root, the site of water and mineral absorption?

7. Why is flow rate of water and minerals through the xylem several times faster than is the flow rate of phloem sap?

Critical Thinking Questions

1. Why does a girdled tree eventually die? Why doesn't the tree die immediately?

2. (*a*). The table below shows the number of stomates for several species of land plants. What is the adaptive value of the differences between the numbers of stomates on the upper and lower leaf surfaces?

Number of Stomates per Square NM

Plant Species	Upper Surface	Lower Surface
sunflower	175	325
cottonwood	89	132
olive	0	625
birch	0	237

(*b*). In some plants, stomates are found only in the bottom depressions of the epidermis. What advantage would this have?

3. Most plants will die if the water content of their cells drops below 60 percent. Discuss some of the possible causes of death from dehydration.

4. Transpiration rate is affected by many environmental conditions, including temperature, wind, light, relative humidity, and water availability in the soil. For the following environments, rank each of these factors, from most important to least important, in *increasing* the rate of transpiration: tropical rainforest, desert, and arctic tundra (refer to Chapter 40 for descriptions). Justify your rankings.

5. Girdling showed that the area just above the removed bark would accumulate sugars and swell up and that the loss of water from leaves above the girdling was not affected. Explain how these observations led scientists to a correct description of the functions of the xylem and phloem.

Additional Readings

Allen, Michael F. 1991. *The ecology of mycorrhizae.* New York: Cambridge University Press. (Intermediate)

Brum, G., R. Boyd, and S. Carter. 1983. Recovery rates and rehabilitation of powerline corridors. In *Environmental effects of off-road vehicles,* R. Webb and H. Wilshire (eds.). New York: Springer-Verlag. (Intermediate)

Cronshaw, J. 1981. Phloem structure and function. *Ann. Rev. Plant Physiol.* 32:465–484. (Intermediate)

Raven, P., R. Evert, and S. Eichhorn. 1992. *Biology of plants.* New York: Worth. (Introductory)

Taiz, L., and E. Zeiger. 1991. *Plant physiology.* Reading, MA: Benjamin/Cummings. (Advanced)

CHAPTER
◄ 20 ►

Sexual Reproduction of Flowering Plants

STEPS TO DISCOVERY
What Triggers Flowering?

FLOWER STRUCTURE AND POLLINATION

Flower Diversity

FORMATION OF GAMETES

Production of Female Gametes

Production of Male Gametes

FERTILIZATION AND DEVELOPMENT

Embryo Development

Seed Development

Fruit Development

FRUIT AND SEED DISPERSAL

GERMINATION AND SEEDLING DEVELOPMENT

Seed Viability and Seed Dormancy

Seedling Development

Seedling Survival and Growth to Maturity

ASEXUAL REPRODUCTION

AGRICULTURAL APPLICATIONS

BIOLINE
The Odd Couples: Bizarre Flowers and Their Pollinators

THE HUMAN PERSPECTIVE
The Fruits of Civilization

During springtime, plants in the Mojave Desert of California begin to flourish following the cold winter season and brief, warm spring rains. After a period of rapid vegetative growth, during which plants produce new stems, leaves, and roots, the plants promptly shift to reproductive growth, triggering the development of a profusion of desert flowers. What activates this shift from vegetative to reproductive growth? Oddly enough, this dramatic transition is not set in motion by plant stems, where the flowers are produced. Instead, researchers have discovered that the internal trigger for flowering originates elsewhere in the plant.

By the 1930s, botanists had clearly demonstrated that, in many plants, flowering is regulated by seasonal changes in the photoperiod—the relative length of day to night during a 24-hour period. For example, plants such as spinach are called *long-day plants* because they flower only after being exposed to the lengthening days of spring. Conversely, other plants, like chrysanthemums, bloom only after exposure to the progressively shorter days of autumn; thus, they are called *short-day plants*. (*Day-neutral plants* are not affected by photoperiod.) Although it was easy for botanists to document that changes in photoperiod induced flowering, it was not as easy to reveal how plants were able to detect photoperiod changes or to determine which part of the plant actually detected the change.

Part of the puzzle was solved in 1938 through the work

Increasing daylength (the yellow bar in the boxes) triggers the leaves of desert creosote bush and sand verbena to synthesize a

of two botanists, Karl Hamner and James Bonner. The scientists reasoned that because leaves are well adapted to absorbing light energy for photosynthesis, perhaps they are also adapted to monitoring photoperiod. Hamner and Bonner chose a short-day plant, the cocklebur (*Xanthium*), to test their hypothesis because this plant blooms within only 2 weeks following exposure to daylight. They exposed only a single leaf of the plant to the correct photoperiod, while the rest of the plant's stems and leaves remained exposed to longer periods of light. By the end of the experiment, the entire plant had bloomed, even though only one leaf had been exposed to the proper day length. Hamner and Bonner concluded that the flowering "signal" had to have originated in the plant's leaves and that some impulse or transmissible substance had acted on the plant's stems to promote a shift from vegetative bud development to flower bud development. But whether the trigger was an impulse or a substance could not be determined from these studies.

Part of the answer on the nature of the trigger came from research done in 1958 by Jan A. Zeevaart in the Netherlands. Zeevaart removed the leaves of a short-day plant, exposed the detached leaves to the correct photoperiod, and then grafted a single leaf onto short-day plants that had been maintained on long days. A control group of short-day plants received no grafts and were maintained on long days. None of the plants in the control group bloomed, whereas all the plants that had received leaf grafts produced flowers. Like Hamner and Bonner, Zeevaart concluded that the flowering signal originates in the leaf and that the signal must be a chemical that is transmitted from the leaf to the stem, where flowers develop.

In 1968, R. W. King, L. T. Evans, and I. F. Wardlaw, three Australian plant physiologists, devised an experiment to establish the speed at which the flowering signal travels from the leaves to the stem. The scientists exposed the leaves of a series of plants to the correct photoperiod and then removed all of the leaves from the plants, each of which had retained the exposed leaves for differing lengths of time. Plants that had retained their leaves for only a very brief time following exposure to the correct photoperiod failed to flower, indicating that a minimum length of time was needed for the trigger to travel to the stem. But how much time? Since the experiment was designed so that there was a succession of plants, each of which had retained leaves for increasing periods of time, the investigators were able to determine the shortest time necessary for the trigger to travel to the stem by identifying the plant that had retained leaves for the shortest period yet still bloomed.

The results of this study revealed that the flowering signal traveled at a rate equal to that reported by other researchers for materials traveling through the plant's phloem tissues, an average of 100 centimeters/hour. King, Evans, and Wardlaw deduced that the flowering signal was an organic chemical that travels through the phloem from the leaf to the stem (recall from Chapter 19 that all organic molecules are transported through the phloem). When this chemical reaches the stem, it triggers axillary buds to shift from vegetative growth to flower development. The exact nature of the chemical is still being investigated.

chemical that travels to stems where it induces flowering in the Spring.

*I*t could be a scene from a horror movie. A victim is lured to a pool by a powerful force and then knocked into the water. Although the victim frantically attempts to climb out, the sides of the pool are so high and slippery it is impossible for him to get a firm hold. Eventually, he discovers a narrow passageway in the back of the pool, squeezes through it, and escapes. But the ordeal is far from over.

As the victim emerges, he discovers that two bulging sacs have been attached to his back. As he tries to flee the scene, the victim is lured to yet another pool, unable to resist the same powerful force that enticed him to the first one. The events that follow are the same: The victim is dunked into the pool; he makes several unsuccessful attempts to climb out the sides; and he eventually escapes through a narrow exit. This time, however, the two sacs are gone.

This is not a scene from a movie but a real-life drama between a male euglossine bee (the "victim") and a bucket orchid flower, *Coryanthus leucocorys* (the "captor"). During its ordeal, the bee inadvertently transfers the orchid's male gametes from one flower to another. Like other flowering plants, *Coryanthes* houses male gametes in **pollen grains**—the contents of the bulging sacs attached to the bee's back. The bucket orchid is a marvelously complicated flower that baits and then captures male orchid bees by dunking them into a pool of water (Figure 20-1). When the bee escapes through a narrow channel in the back of the flower, bags of pollen grains are attached to or removed from the insect's back, transferring gametes between flowers.

Male bees are attracted to the bucket orchid not only because of its alluring fragrance but also because the flower produces a waxy substance the male bee uses to make its own sexy "perfume," which he uses to attract female euglossine bees. Thus, the sexual reproduction of both the orchid and the bee are inseparably intertwined; each organism depends on the other to form a critical link in a chain that leads to the sexual reproduction of its species.

▼ ▼ ▼

FLOWER STRUCTURE AND POLLINATION

The flower is the center of sexual activity for all flowering plants. Flowers come in an array of sizes, shapes, colors, color patterns, and arrangements, each representing an adaptation that promotes pollen transfer for sexual reproduction. The transfer of pollen grains between flowers is called **pollination.** Many plants, like *Coryanthes*, rely on insects for pollination; others are adapted to utilizing wind, water,

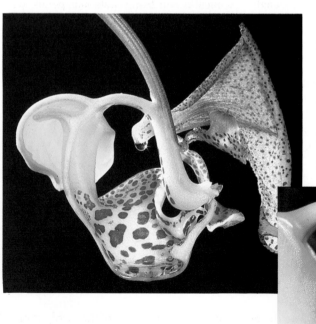

FIGURE 20-1

Glamorous and alluring, the bucket orchid (*Coryanthes leucocorys*) has two tiny spouts that alternately drip water into a floral cup, forming a small pool. The flower's fetching and disorienting aroma lures male euglossine bees to the spouts. Eventually, a drop of water knocks the befuddled bee into the pool. The steep, slippery sides of the flower prevent the bee from escaping, except through a narrow groove in the back of the flower. As the bee departs, the flower either deposits or retrieves bags of pollen grains from the bee's back. By capturing and then loading and unloading sacs of pollen, the plant succeeds in transferring sperm from flower to flower for sexual reproduction.

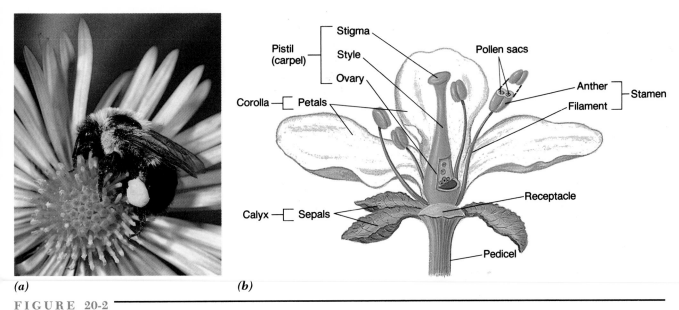

FIGURE 20-2

Structures of a typical flower. *(a)* Flowers produce the plant's gametes for sexual reproduction and attract pollinators that transport male gametes from flower to flower. *(b)* The anthers of the stamen form pollen, which eventually produce the sperm gametes. The pistil is the female part of a flower. Within the pistil, the stigma collects pollen; the style provides a channel that directs the growth of a pollen tube to the egg (which is produced in the ovary) and, following fertilization, the ovary develops into a fruit that protects, nourishes, and helps disperse developing embryos contained in its seeds.

or other types of animals, such as birds and even a few mammals. Most animal-pollinated flowers produce food (usually nectar and pollen) that attracts the animal to the flower. However, a few plants employ some rather unusual tactics to accomplish pollination (see Bioline: The Odd Couples).

To understand how flowers accomplish sexual reproduction, we need to investigate the structure of the flower itself. A flower, such as the one illustrated in Figure 20-2, is a cluster of highly modified leaves attached to a shortened stem, called the **pedicel.** The flower parts emerge from a **receptacle**—the widened end of the pedicel which forms the flower base. A typical flower is made up of four groups of modified leaves—*sepals, petals, stamens,* and a *pistil*—each functioning in some way to achieve sexual reproduction.

Generally, the **sepals** surround and protect the flower bud as it develops. The collection of individual sepals make up the plant's calyx. To discourage hungry animals, the sepals of some flowers contain a distasteful latex, a milky white fluid that quickly deters animals from biting into the nutrient-rich flower bud. Bright or patterned **petals** distinguish the flower from its surroundings, catching the attention of pollen-transferring animals. The petals collectively make up the flower's corolla. Although some petal colors and markings are invisible to humans, they are readily spotted by the animals they attract (Figure 20-3). Petals are not

the only structures that attract animals; sometimes, the sepals, clumps of flowers, and even brightly colored leaves attract animals for pollination (Figure 20-4).

Although it is easy to see that sepals and petals are really modified leaves (they are leaf-shaped and have prominent veins), the stamens and pistil of the flower do not appear leaflike at all, but they too are modified leaves. These structures are the product of extensive evolutionary modifications of spore-producing leaves in the ancestors of flowering plants—modifications that enabled these leaves to manufacture and house the plant's gametes. The **stamens** are the flower's male reproductive structures; they produce pollen grains, which contain sperm. Most stamens are made up of a slender stalk—the **filament**—with a swollen end, called the **anther.** Pollen grains are produced inside the anther. When pollen matures, the anthers split open, releasing the pollen.

The **pistil** consists of one or more **carpels**—modified leaves that produce egg gametes. Each pistil has three parts:

• the **stigma,** or sticky area to which pollen adheres;

• the **style,** or tube that connects the stigma to the ovary; and

• the **ovary,** or enlarged bottom part of the pistil, where eggs are produced and embryos and seeds develop.

In plants adapted to utilizing wind for pollination, the stigma is often enlarged and feathery, creating a large sur-

Many plants enlist an army of "middlemen" of various sizes, shapes, and behavior for pollination. Usually, the plant entices pollinators by providing "rewards," such as nectar, pollen, oils, or plant tissue. But some plants seem to resort to evil and manipulative means to reach this end, as in the following cases:

CASE 1: THE IMPOSTOR

The dull-red, wrinkled flowers of the South African milkweed, *Stapelia gigan-*

tea, look and smell like putrefying flesh, deceiving female blowflies into laying her eggs on them. In doing so, the duped female blowfly carries pollen from flower to flower. But the floral hoax has grave consequences for the blowfly. All of the fly's maggots perish from lack of proper food because blowfly maggots can live only on animal flesh, not flower tissue—no matter how foul it is.

CASE 2: OPEN HOUSE

The elongated petal of the Dutchman's pipe (*Aristolochia clematitis*) is covered with millions of tiny grains of wax. Gaining a footing on such a surface is like trying to walk down a mountain covered with greased ball bearings. Even with hooks and suckers, the gnats and flies attracted to the flower cannot gain a foothold on this slick surface. As soon as an insect lands, it slips down the flower tube, past a forest of downward-pointing hairs, and finally

crashes into a cluster of stigmas (part of the female reproductive structure) at the bottom of the chamber. "Visitors" are prevented from leaving this hairy forest for at least 2 or 3 days. During that time, the stamens (the male reproductive structures) mature and shower the insects with pollen. Then the flower tilts over, and the retaining hairs shrivel. The insects walk out, usually only to be quickly lured into captivity by yet another Dutchman's pipe flower, completing the pollination cycle.

(a)

(b)

ports pollen among them. The flower is just as successful in arousing the amorous attention of the male wasp as is the genuine female of his own species!

CASE 4: R.I.P.—MURDER AMONG THE LILIES

The South African waterlily attracts hundreds of eager hoverflies, bees, and beetles to its flowers, using an abundance of nutritious pollen as bait. The insects soon learn that waterlilies are a rich source of food. Lily flowers first go through a female stage when no pollen is offered and then go into the male stage. But by the time pollen is offered, the mortal deed has already been done, for hidden below the bountiful stamens float the remains of previous visitors.

The deceased visitors happened upon the lily flower on the first day the flower opened, a day in which the flower was in its female stage. In their feeding frenzy, the insects carrying pollen from other lily flowers in their male stage failed to notice there was no pollen in this flower. The immature stamens form a ring of slick columns surrounding a watery pool. Once the in-

sects land and start crawling over the stamens in search of pollen, it is too late. Unable to get a grip, the insects helplessly slide down the slippery stamens into the pool and drown. At night, the flower closes, and the pollen collected on the insects from visits to other lily flowers washes off the corpses and settles to the bottom of the pool, where the stigma is located. Pollination is completed. By morning, the murky burial pool is hidden by a blanket of mature stamens, the flower now luring insects with a bounty of pollen. In its male stage, the lily flower is safe, but insects cannot tell the difference between flowers that offer pollen and those that offer certain death, a fact on which the lily has staked its sexual reproduction.

CASE 3: THE CASE OF MISTAKEN IDENTITY

The flowers of the orchid *Ophrys speculum* take advantage of the powerful sex drive and energetic nature of male wasps. To a male wasp, the orchid looks and smells like a female of his own species. The illusion is perfect; as the male wasp "copulates" with a number of orchids, it trans-

---

◀ **FIGURE 20-3**

A bee's-eye view. Humans and bees see different wavelengths of light, as these two photos of a marsh marigold (*Caltha palustris*) illustrate. **(a)** To humans, this marsh marigold is almost uniformly yellow. **(b)** To bees, the petals have distinct lines that converge in the dark center where the nectar (and stamens and pistils) are found. These nectar guides absorb ultraviolet (UV) wavelengths, while the remaining parts reflect UV wavelengths (humans do not see UV wavelengths, but bees do).

face that gleans pollen from the air (Figure 20-5*a*). In flowers that utilize animals for pollination, the stigma is shaped in such a way that it contacts the pollen-laden areas of the animal's body. For example, the stigma of the monkey flower depicted in Figure 20-5*b* remains wide open until a bee deposits pollen on it. The stigma then closes, reducing inbreeding by preventing pollen produced from the same flower from being deposited on the stigma surface. In most cases, pollen must physically and chemically fit the stigma before a pollen tube will grow down the style and into the ovary to deliver sperm to the egg. This "recognition" of pollen helps ensure that only sperm from members of the same species will fertilize the egg.

☀ Many flowers have **nectaries**—glands that secrete sugary nectar that attracts animals for pollination. Some animals collect both nectar and protein-rich pollen. As an animal laps up nectar and collects pollen, some of the pollen will adhere to its body. As the animal moves from flower to

(a)

(b)

(c)

FIGURE 20-4

Flower power. In most cases, only the petals of a flower are brightly colored to help attract animal pollinators, but some plant species do not produce flowers with petals or have very inconspicuous flowers. In these cases, other structures have become modified to attract pollinators. *(a)* Although they look like petals, flannel bushes *(Fremontedendron)* have brightly colored sepals that attract animals. *(b)* An African daisy is not a single flower but a miniature bouquet of tiny, densely packed flowers that forms a *composite*. Each of the outer *ray flowers* has long petals that attract animals. The petals of the inner, inconspicuous *disk flowers* are barely visible. *(c)* In poinsettias, it isn't the flower that attracts animal pollinators but a cluster of colored leaves located just below the inconspicuous flowers.

(a)

Stigma

(b)

FIGURE 20-5

Improving the odds. *(a)* Wind-pollinated plants, like this grass, produce tremendous quantities of pollen and form large feathery stigmas. Together, these characteristics improve the chances of pollination. *(b)* This two-lipped monkey flower stigma *(Mimulus)* "licks" pollen off the heads of entering bees and then quickly closes to prevent pollen from its own stamens from entering as the bee exits.

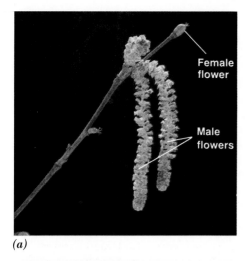

FIGURE 20-6

Separating the sexes. *(a)* Monoecious plants, such as this oak, have both male and female flowers (the male flowers are in a pendulous, or suspended, group, and the female flower is at the tip of the stem). Dioecious plants, such as meadow rue *(Thalictrum)*, produce male flowers on one plant *(b)* and female flowers on another *(c)*. Separate-sexed flowers reduce the chances of fertilization between sperm and eggs from the same individual, thereby increasing genetic variability.

(a)

(b)

(c)

flower, pollen is carried from the animal's body to the flower's stigma, completing pollination and setting the stage for fertilization.

FLOWER DIVERSITY

A flower that contains all four groups of modified leaves—sepals, petals, stamens, and pistil—is a **complete flower,** as opposed to an **incomplete flower,** which lacks one or more of these four basic parts. Roses and daisies are examples of complete flowers, whereas dogwood and palms produce incomplete flowers. Some plants produce flowers with only stamens (male flowers) or only pistils (female flowers). A flower that lacks either stamens or pistils, or both, is called an **imperfect flower,** as opposed to a **perfect flower,** which has both stamens and pistils. If separate male and female flowers are produced on one plant, the plant is said to be **monoecious** (*mono* = one, *ecious* = house) (Figure 20-6*a*), whereas **dioecious** (*di* = two) plants produce male and female flowers on separate individuals (Figure 20-6*b*). All monecious and dioecious plants produce imperfect flowers. The tassels of male flowers and the separate cobs of female flowers make corn a monoecious plant, whereas eggplant and other squashes are dioecious plants because

each individual produces either male or female flowers. Producing different-sexed flowers on different plants guarantees that eggs cannot be fertilized by sperm from the same individual, thus promoting genetic variability.

> **A flower is a multipurpose structure for sexual reproduction. The flower produces sexual gametes; facilitates pollination; and serves as the site of fertilization and embryo, seed, and fruit development. (See CTQ # 1.)**

FORMATION OF GAMETES

Gamete formation always begins with meiosis (Chapter 11). In plants, special diploid mother cells inside the anthers and ovaries of flowers divide by meiosis to form haploid *spores*. Mother cells in the anthers divide to form **microspores,** whereas mother cells in the ovary produce **megaspores.** (The prefixes "micro" and "mega" do not necessarily refer to the size of the spores but to the size of the gametes produced; the male sperm is smaller than the female egg.)

Before gametes are formed, both microspores and megaspores divide by mitosis to form a haploid, multicellular gametophyte, which produces the gametes for sexual reproduction. Megaspores develop into **embryo sacs,** the female gametophytes; microspores develop into pollen grains, the male gametophytes.

PRODUCTION OF FEMALE GAMETES

Female gametophytes form within **ovules**—round masses of cells that are attached by a thin stalk to the inner surfaces of the flower's ovary. The outer cells of the ovule form one or two protective layers (the integuments) that leave a small opening, the *micropyle,* through which the pollen tube grows, to deliver sperm.

As illustrated in Figure 20-7, the production of the egg gamete begins when a diploid mother cell inside the ovule undergoes meiosis. Three of the four resulting megaspores degenerate. The nucleus of the surviving megaspore divides three times by mitosis to produce the female gametophyte, which contains eight haploid nuclei. (The first mitotic division produces two nuclei; the second, four; and the third, eight.) The eight nuclei become partitioned into seven cells, one of which contains two nuclei; only one of the cells—that nearest the micropyle—becomes the egg. The cell that contains two nuclei is called an **endosperm mother cell,** which later proliferates into the **endosperm,** an important tissue that nourishes the embryo and the growing seedling.

PRODUCTION OF MALE GAMETES

As embryo sacs develop in the ovary, pollen grains form in the anthers of stamens (Figure 20-8). Mother cells in the anther divide by meiosis to produce four microspores, each of which divides once by mitosis, forming a two-celled pollen grain. One of the cells—the *generative cell*—is enclosed within the other—the *tube cell*; the generative cell will divide by mitosis to produce two sperm. Pollen grains are surrounded by a tough wall that protects the male gametophyte while being transported to a stigma. The design of the outer walls of some pollen grains helps them adhere to animals (Figure 20-9*a*) or remain airborne during pollination (Figure 20-9*b*).

> The production of plant gametes begins with meiosis and the formation of four haploid spores. Each spore divides by mitosis to form a many-celled structure that produces either an egg or sperm gamete. (See CTQ #2.)

FERTILIZATION AND DEVELOPMENT

Even after pollination is completed, the plant's male and female gametes remain physically separated. The pollen grain lies on top of the stigma, while the egg is buried deep within the ovary. But when pollen lands on a stigma, the

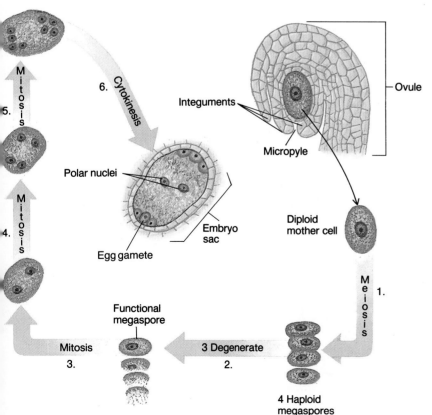

FIGURE 20-7

Formation of the female gametophyte is a six-step process that takes place inside an ovule. (1) A diploid mother cell divides by meiosis to produce four haploid megaspores. (2) Three megaspores degenerate, leaving a single functional megaspore. (3–5) The nucleus of the surviving megaspore divides by mitosis three times to form an eight-nucleate gametophyte. (6) Cytokinesis (page 206) partitions the nuclei into seven cells, producing the mature embryo sac with a single egg gamete.

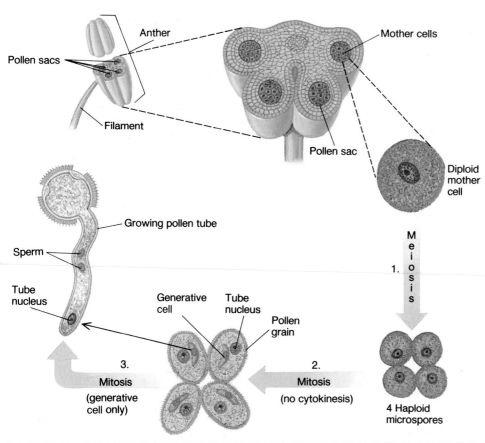

FIGURE 20-8

Formation of the male gametophyte is a three-step process that takes place inside the anthers of stamens. (1) A mother cell divides by meiosis to produce four haploid microspores. (2) Each microspore divides by mitosis to a pollen grain containing a tube cell and a generative cell. (3) When deposited on a stigma, the tube cell forms a pollen tube, while the generative cell divides to form two sperm.

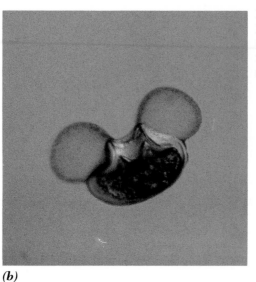

FIGURE 20-9

Pollen form and function. (*a*) Sculpturing of cotton pollen helps them adhere to insects. (*b*) Bladderlike wings of pine pollen (*Pinus nigra*) help keep them airborne.

(*a*)

(*b*)

tube cell of the microgametophyte forms a pollen tube that grows down through the stigma, style, and ovary tissues (Figure 20-10), eventually worming its way through the micropyle of an ovule. Two sperm are discharged; one fuses with the egg to produce a diploid zygote, while the other fuses with the diploid endosperm mother cell to form a triploid (3N) *primary endosperm cell*. Because both sperm from a pollen grain fuse with cells in the embryo sac, this is sometimes called "double fertilization," although technically there is only one fertilization—the fusion of egg and sperm to form the zygote.

Fertilization sparks dramatic transformations in the flower, including the following:

- the zygote develops into the embryo;
- the primary endosperm cell forms nourishing endosperm tissue;
- the ovule is transformed into a seed; and
- the ovary develops into a fruit.

Together, the endosperm, seed, and fruit nourish and protect the embryo as it develops (Figure 20-11) and during its journey to a new habitat.

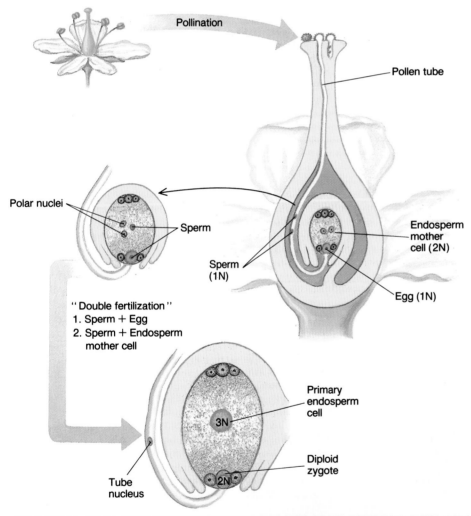

FIGURE 20-10

Pollination and fertilization. The beginnings of a new generation. Pollen is transferred from anthers to a stigma during pollination. The pollen grain grows a pollen tube down the style, into the ovary, and through the micropyle of an ovule. The generative cell divides to produce two sperm; one fertilizes the egg to form a zygote, the other fuses with the two nuclei of the endosperm mother cell to form a triploid primary endosperm cell.

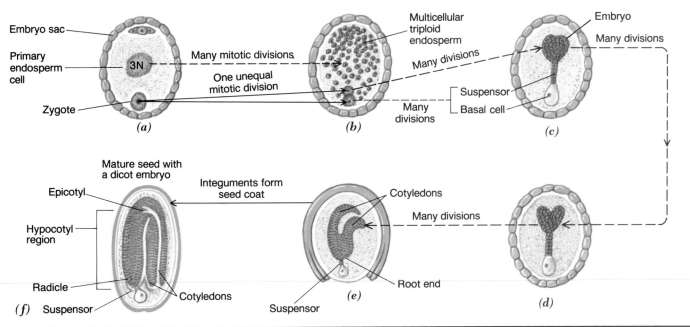

FIGURE 20-11

Development of a dicot embryo, *Capsella*. *(a)* A zygote and primary endosperm cell form following fertilization. *(b)* The triploid (3N) primary endosperm cell divides many times by mitosis to produce nutritive endosperm. The first mitotic division of the zygote produces different-sized daughter cells. *(c)* The smaller cell divides many times to produce the plant embryo. Divisions of the larger, basal cell produce the suspensor. *(d–f)* Continued mitotic divisions leads to the formation of a mature dicot embryo.

EMBRYO DEVELOPMENT

A newly formed zygote divides by mitosis to produce two different-sized daughter cells (Figure 20-11b). The smaller cell eventually grows into the plant embryo, while the larger, basal cell produces the *suspensor*—a column of cells that conducts nutrients to the growing embryo (Figure 20-11c).

The plant embryo is a relatively simple structure, consisting of only three parts: one or two cotyledons, an epi-

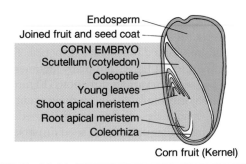

FIGURE 20-12

Corn (*Zea mays*) fruit (kernel) with mature embryo. A corn kernel is a combination of fruit and seed. The ovary of the pistil forms an outer "fruit + seed" coat. The single cotyledon digests and absorbs endosperm stored outside the embryo.

cotyl, and a hypocotyl. The **cotyledons,** or seed leaves, are the first structures to begin photosynthesis when the seedling emerges from the soil. In most dicots, the endosperm (food) is located within the cotyledons and is part of the embryo; in most monocots, the endosperm surrounds the embryo and is part of the seed. As their names indicate, dicot embryos have two cotyledons, whereas monocot embryos have only one (Figure 20-12). The **epicotyl** is the portion of the embryo above the cotyledons (*epi* = above, *cotyl* = cotyledon); it develops into the shoot system of a plant. The **hypocotyl** is the portion of the embryo below the cotyledons (*hypo* = below). At the tip of the hypocotyl is the **radicle,** which eventually develops into most or all of the plant's root system.

SEED DEVELOPMENT

Following fertilization, the ovule develops into a seed, which contains the embryo, endosperm, and a seed coat (Figure 20-13). The seed coat forms from the outer cell layers of the ovule. The integuments become tough and hard, protecting the embryo from abrasion or other dangers as it is dispersed to new habitats. Some seed coats have projections that help absorb water, attach the seed to the fur of an animal for dispersal, or protect the embryo from hungry animals.

Bean fruit with seeds

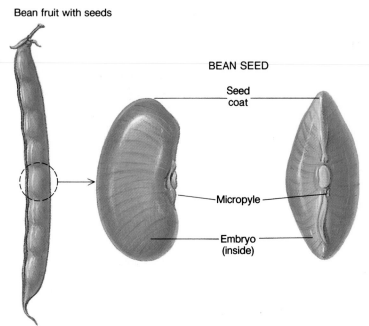

BEAN SEED

Seed coat

Micropyle

Embryo (inside)

FIGURE 20-13
Seed. The seed is a life-sustaining capsule that houses, nourishes, and protects the embryo after separation from its parent and during dispersal.

FRUIT DEVELOPMENT

With sexual reproduction completed, those parts of the flower that were involved in pollination (sepals, petals, stamens, stigma, and style) wither. As embryos and seeds develop, the ovary of the pistil matures into the *fruit*. Many animal species, including humans, rely on plant fruits for nutrition. In fact, fruits have had a profound effect on human life, perhaps even sparking the development of human civilization (see The Human Perspective: The Fruits of Civilization).

Whether a fruit is green or brightly colored, juicy or dry, soft or hard, all (or part) of it formed from a ripened ovary. Some so-called vegetables, such as cucumbers, tomatoes, squash, string beans, and grains, technically are fruits because they develop from an ovary. There are three categories of fruits (Figure 20-14)—simple, aggregate, and multiple—depending on whether a fruit develops from the ovary of one pistil of a single flower **(simple fruits),** from many pistils in a single flower **(aggregate fruits),** or from the pistils of a number of flowers **(multiple fruits).**

Flower structure and function change following pollination and fertilization. The parts of the plant involved in pollination (petals, sepals, nectaries, and stamens) wither as the pistil undergoes a dramatic transformation during embryo, seed, and fruit development. (See CTQ #3.)

FRUIT AND SEED DISPERSAL

▶▶ In just one flowering season, a single plant can produce hundreds, thousands, and even millions of seeds, each containing an embryo capable of growing into a new plant. If a parent plant simply dropped all of its seeds onto the ground directly beneath its stems and branches, the resulting seedlings would compete with each other and with the parent plant for limited space, light, water, and nutrients as they grew. Such intense competition would lead to widespread malnourishment and massive numbers of deaths. *Seed dispersal* reduces competition by separating seeds. Dispersal also scatters seeds to new areas where conditions may be more favorable for growth and development.

With such clear advantages, it is not surprising that natural selection has resulted in the proliferation of seeds and fruits that are adapted to enhance dispersal. The fruits of wild cucumber, witch hazel, and many legumes literally explode, flinging seeds in various directions (Figure 20-15). However, the majority of plants rely on some agent—wind, water currents, or animals—to disperse their seeds (Figure 20-16). Plants that utilize animals for seed dispersal recruit a broad range of species, from worms to bears. In plants that develop juicy fruits, the seeds are often consumed along with the fruit but pass undamaged through the animal's digestive system and are scattered along with the animal's feces. For example, the seeds of the giant Saguaro cactus are dispersed by birds that eat the fruit and seeds simulta-

Simple fruits

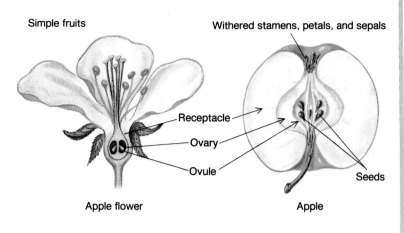

Receptacle

Withered stamens, petals, and sepals

Ovary

Ovule

Seeds

Apple flower

Apple

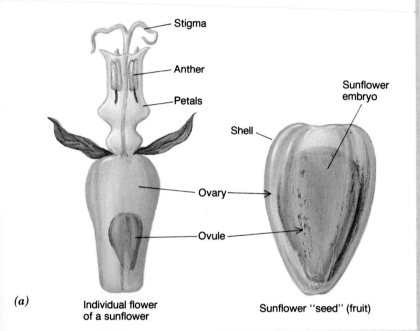

Stigma

Anther

Petals

Shell

Sunflower embryo

Ovary

Ovule

Individual flower of a sunflower

Sunflower "seed" (fruit)

(a)

Aggregate fruit

Separate carpels (pistils) form separate dry fruits

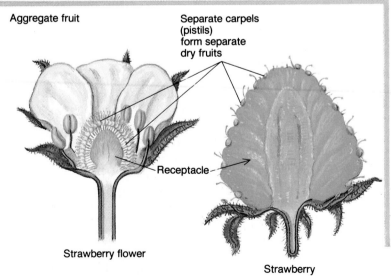

Receptacle

Strawberry flower

Strawberry

(b)

Multiple fruits

Single fruits developed from different flowers

Pineapple

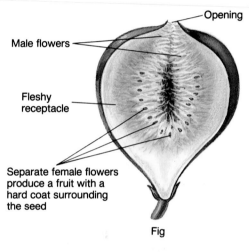

Opening

Male flowers

Fleshy receptacle

Separate female flowers produce a fruit with a hard coat surrounding the seed

Fig

(c)

FIGURE 20-14

Development of three main fruit types. *(a)* Whether fleshy (apple) or dry (sunflower), simple fruits develop from one pistil. *(b)* Aggregate fruits, such as strawberries, are formed from separate pistils in the same flower. *(c)* Multiple fruits develop as the ovaries of several, often densely packed flowers mature together. Multiple fruits form a single fruit (such as a pineapple) or develop from many flowers that are physically joined by other flower tissues. A fig, for example, is an enlarged receptacle filled with many tiny simple fruits, each of which developed from a separate flower.

◁ THE HUMAN PERSPECTIVE ▷
The Fruits of Civilization

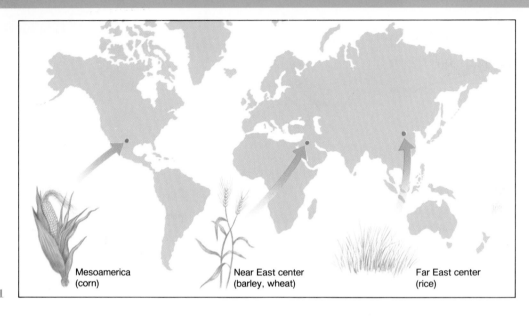

Mesoamerica (corn)　　Near East center (barley, wheat)　　Far East center (rice)

FIGURE 1

As humans, our existence depends on fruits and seeds, but our dependency goes beyond the obvious apple, orange, or peanut butter sandwich. Flour for making bread is ground grain, the fruit of grasses; we feed domestic livestock fruits and grains for meat production. In fact, most of our food comes directly or indirectly from fruits and seeds.

The same was true for our early ancestors. Early humans traveled in nomadic tribes, forced to follow seasonal changes in plant growth in order to harvest enough food to feed their members. The perpetual search for food occupied most of the time and energy of every member of the tribe. There was no time for contemplation, no time or energy to develop art or science or even to begin the first scratchings of a written language. So how did these intellectual and civilized pursuits begin? Human civilization may owe its existence to one of the simplest kinds of fruits—grains.

Historians have identified three geographic regions where human civilization flourished and then radiated out to other areas of the world (Figure 1). Each of these centers of civilization had at least one naturally growing grain that likely provided the nutritional foundation for the development of civilization. The Near East Center had wheat and barley, the Far East Center had rice, and the Middle America Center had corn. Because of the abundance of these grasses and their high nutritional content, a few members of each tribe could gather, tend, and cultivate enough grain to feed the entire group. In addition, grass grains could be stored for long periods without spoiling, eliminating the need to travel from place to place to find a fresh supply of food. Some anthropologists speculate that as more and more members of the tribe were freed from the oppression of an endless, exhausting search for food, they had time to contemplate their surroundings and to develop the innovations that allowed civilization to progress. The development of plant and animal husbandry (production and care of domestic plants and animals) practices freed even more people to specialize in other activities. Rice terraces, such as these in the Phillipines (Figure 2) greatly expand food production in mountainous regions. Eventually, language, arts, and sciences began to flourish, leading to the social and technological advances we enjoy today.

FIGURE 2

FIGURE 20-15

Exploding fruits disperse their own seeds. This common vetch (*Vicia sativa*) produces a power-packed legume that flings seeds 3 meters (10 feet) from the plant.

(a)   *(b)*

FIGURE 20-16

Wind dispersal. Airborne seeds and fruits often have wings or plumes that help them remain aloft, enabling them to travel further. *(a)* Plumes of milkweed, *(b)* wings of swamp maple.

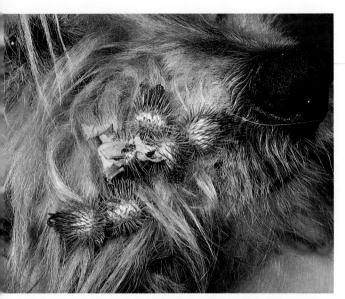

FIGURE 20-17
Hitchhiking fruits, such as the burrs of common Burdock (*Arctium minus*) have hooks, spines, or barbs to "bum a ride" to new habitats by attaching to the skin or fur of animals.

neously. Saguaro seeds pass undamaged through the bird's digestive system as it flies about and are released along with the bird's guano (excrement). In some cases, rodents gather and stockpile nuts and seeds in hiding places; the uneaten seeds grow into new plants.

Seeds can be dispersed over very long distances— across oceans, to other continents or to distant islands— carried on a bird's feathers or in mud adhering to the feet of migrating birds. (Some of the original plants that grew on the Hawaiian Islands were dispersed there in this way.) Barbs, hooks, spines, or sticky surfaces that temporarily fasten fruits or seeds to an animal's fur aid dispersal of some plants to new locations (Figure 20-17).

> **Fruits and seeds protect the plant embryo and aid in the dispersal of offspring to new habitats as well as reduce competition among offspring or competition between offspring and the parent plant. (See CTQ #4.)**

GERMINATION AND SEEDLING DEVELOPMENT

When environmental conditions are favorable (proper temperature, sufficient water, and adequate oxygen, for example), a seed germinates and a new generation of plants begin to grow and develop. During **germination,** the radicle of the hypocotyl is usually the first structure to thrust through the seed coat. The radicle then forms a root that anchors the new plant and begins absorbing water and minerals from the soil. As the root system develops, the epicotyl begins growing into the shoot system; photosynthesis begins when the shoot system emerges from the soil surface.

SEED VIABILITY AND SEED DORMANCY

The embryos of different plant species remain viable (capable of germinating) for different periods of time. The *seed viability* of some species, such as willows, is only a few days, whereas the seeds of other plants may remain viable for months or even years. For example, oriental-lotus seeds that were recovered from ancient tombs germinated after 3,000 years!

A seed may fail to germinate for one of two reasons: (1) the embryo is dead (that is, the seed is not viable), or (2) the embryo is "resting," a condition known as **seed dormancy.** Seeds may remain dormant if there is a lack of adequate water; the embryo is underdeveloped; hormones are blocking germination; the seed requires a cold period; or the radicle is unable to break through a tough seed coat.

SEEDLING DEVELOPMENT

A young plant's (seedling's) battle for survival begins instantly. Producing a root system to obtain water is more important to survival than is breaking through the surface for photosynthesis. Nutrients stored in the seed can usually last for several days after germination, but without an immediate supply of moisture the seedling will quickly die of dehydration. Most plants establish mycorrhizae associations with soil fungi during the initial growth of the root system (page 386), improving their ability to absorb water and minerals.

Seedling development is slightly different for monocots than for dicots. To illustrate these differences, we discuss a representative example of each: a dicot bean seedling (*Phaseolus*), and a monocot corn seedling (*Zea*).

Bean Seedling Development

During bean germination, the radicle emerges and grows downward, producing many lateral roots from a single main root (Figure 20-18). Unequal cell growth causes the bean hypocotyl to bend, forming a hook that nudges its way up through the soil, carving a channel so that the delicate apical meristem, cotyledons, and young leaves can be drawn along without injury. Once the hypocotyl breaks the soil surface, it straightens, exposing the cotyledons and epicotyl to sunlight.

Up to this point, the seedling has depended on the nutrients stored in the endosperm of its cotyledons. Once the cotyledons and young leaves are exposed to light, however, the plant manufactures chlorophyll and other pigments, and photosynthesis begins. As food reserves are used up, the cotyledons wither and fall.

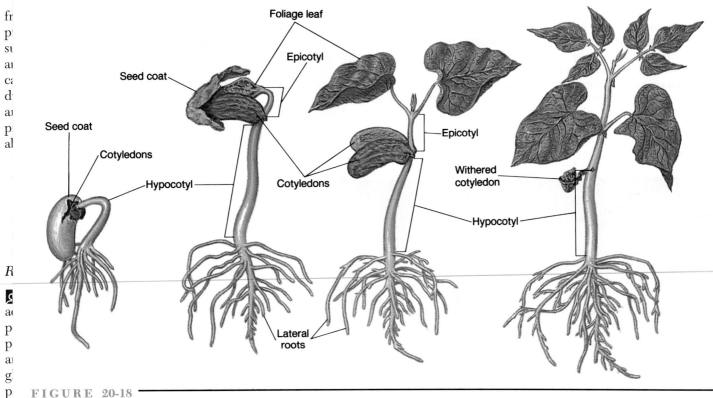

FIGURE 20-18

Seed germination and seedling development of a dicot, bean (*Phaseolus vulgaris*).

Corn Seedling Development

Food reserves usually aren't stored in the single cotyledon of a monocot embryo. Instead, the cotyledon absorbs food from surrounding endosperm tissues, which fuels germination and initial seedling growth.

When a corn kernel germinates, the root tip breaks through the surrounding tissues, as the epicotyl forces its way through a protective sheath (the *coleoptile*) and then up through the soil (Figure 20-19). As in dicots, once young monocot leaves are exposed to sunlight, photosynthesis begins, and the young plant begins its life of autotrophic independence.

SEEDLING SURVIVAL AND GROWTH TO MATURITY

A seedling's life does not depend solely on its ability to absorb water and minerals or to photosynthesize enough food to support its own growth and development. Often, the seedling must also be able to outcompete other plants for nutrients and light as well as avoid the many hazards of life, including continuous onslaughts of plant-eating animals, attacks by hundreds of disease-causing microbes, fluctuations in temperature, slashing winds, and a host of other life-threatening environmental conditions. The few seed-

lings that survive establish a new generation of plants. The ability of a plant to survive depends on a number of growth and development strategies and behaviors that are regulated by plant hormones, the subject of the next chapter.

A new plant begins its independent life after it germinates; it develops a root system for absorbing water and minerals and a shoot system for photosynthesis. (See CTQ #5.)

ASEXUAL REPRODUCTION

Flowering plants reproduce new individuals both sexually and asexually (without the fusion of gametes and resulting embryo, seed, and fruit development). Asexual, or *vegetative reproduction,* results from mitosis and consequently produces offspring that are always genetically identical to the parent plant.

Some plants whose stems grow over the soil surface, such as strawberries and spider plants, form asexual offspring from horizontal stems—*runners* (or *stolons*). If a plant's stems grow at or below the soil surface (as in the case of crab grass and bamboo), asexual offspring are formed

◁ THE HUMAN PERSPECTIVE ▷
Saving Tropical Rain Forests

FIGURE 1

Satellite photographs reveal a perpetual cloud of smoke covering the forests of Brazil—a cloud the size of the state of Texas. Beneath the smoke, one of the earth's most critical life-sustaining habitats is being consumed. Every minute, 100 acres of tropical rain forest disappear, burned or hacked away by people seeking new land for cultivation, by subsistence farmers whose lives depend on the crops grown on small plots of land (FIGURE 1). But the livelihoods of these farmers are doomed to a few years of marginal success; they then must clear more forest for their crops.

Nowhere is life more abundant or diverse than in the tropical rain forests, where more than half the world's species of plants and animals reside. But the abundance of life in the natural rain forest does not translate into lush crop growth once the forest has been destroyed. In fact, cleared rain forest land is paradoxically unproductive. Because the forest abounds with so much life, litter does not accumulate. Dead organisms are quickly consumed by other organisms. Consequently, little or no humus collects in the soil to condition it, and precious little nitrogen or phosphorous seeps into the ground to fertilize it. In other words, the topsoil is thin and poor. With the rain forest cleared away, the area can no longer support a rich abundance of life, and what few nutrients are present in the soil rapidly erode away with each rainfall. Within a few years of clearing, the land is abandoned by its cultivators, who then burn off more rain forest in their need to grow food for survival (FIGURE 2).

Deforestation in the tropics has rapidly accelerated since World War II. New

highways penetrate into the deep forests, providing easy access by timber harvesters, land developers, cattle ranchers, cash-crop planters, and oil and mineral exploration teams. Each year, 27 million acres of tropical forest disappear, along with the habitats that support the most diverse range of species on earth. One of the most distressing aspects of tropical rain forest destruction is the extinction of thousands of species. Although extinction is part of the natural history of life, deforestation has accelerated the rate by 10,000 times, compared to the extinction rate that existed before humans. Extinction claims the lives of three to four species *every day!*

The tropics provide humans with most of our edible plant species (a typical breakfast of bananas, orange juice, corn-flakes and sugar, coffee or hot chocolate, and hashed-brown potatoes consists entirely of tropical plants). More than 40 percent of our medicines, from pain-relieving aspirin to treatments for malaria, have come from tropical plants. For example, the rain forest's rosy periwinkle provides a life-saving drug used for treating otherwise fatal cases of lymphocytic leukemia. Cures for cancer and AIDS may be hiding in plants on the brink of extinction, plants that could disappear before they are ever discovered.

Destruction of the tropical rain forests also alters the earth's climate. Ecologists refer to tropical rain forests as "rain machines" because they recycle water back into the atmosphere (roots draw up the ground water, which evaporates from leaves, returning about 50 percent of rainwater to the atmosphere). Deforested areas receive much lower amounts of rainfall; nearly 100 percent of the water runs off into streams, most of it flowing all the way to the the ocean. Some lush tropical areas have become desertlike following deforestation.

Deforestation also plays a very important role in accelerating global warming. The forests are critical to the balance of carbon dioxide on earth. The plants of the forest remove carbon dioxide from the atmosphere for photosynthesis. In burning the rain forests, we are not only destroying our major carbon-dioxide "sink," we are also turning the trees into carbon dioxide (about 1 billion tons of carbon dioxide are ejected into the atmosphere by deforestation). Because carbon dioxide in the atmosphere traps heat, this ecological "combination punch" will likely hasten global warming which, in turn, will trigger unexpected and unwelcome environmental changes.

Already half the rain forests have disappeared; at the current rate of destruction, virtually all the earth's rain forests may be eliminated (along with more than half the earth's species) within the next 50 years. What will the world be like then?

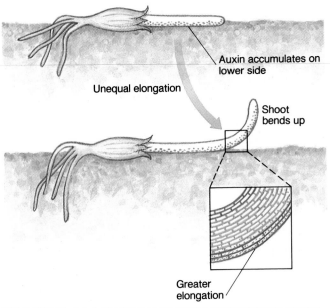

Auxin accumulates on
lower side

Unequal elongation

Shoot
bends up

Greater
elongation

FIGURE 21-2

Which way is up? When a seedling is placed on its side, auxin accumulates on the lower side of the shoot, increasing cell elongation there and causing the shoot to grow up.

- Induce cells to differentiate into secondary xylem, thereby increasing water and mineral transport and adding reinforcement for greater support.
- Promote the development of flowers and fruits.

The first (and most common) auxin to be isolated was **indoleacetic acid (IAA)**. As you will recall from the chapter opening, researchers knew long ago that the center of auxin production in grasses was in the tips of their coleoptiles. It took the collection of more than 10,000 oat coleoptile tips to obtain a single microgram of the hormone, confirming the fact that plant hormones are indeed active in *extremely* small amounts.

In addition to their growth-promoting functions, auxins sometimes inhibit development. For example, auxins inhibit the development of axillary buds, producing a familiar growth pattern, known as **apical dominance** (Figure 21-4). As horticulturists and gardeners will tell you, pinching off the tips of a plant's main stems stimulates the production of more leafy branches. Removing the tips of stems also removes the apical meristem, a center of IAA production. Normally, IAA moves down the stem and inhibits axillary buds from growing into new leafy stems. By removing the center of IAA production, you allow axillary buds to begin to develop into new stems with more leaves and more axillary buds.

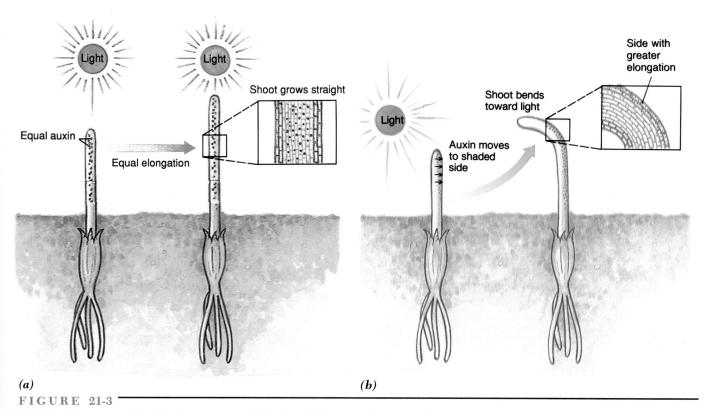

(a)   *(b)*

FIGURE 21-3

Light and stem growth. *(a)* Shoots grow straight when light comes from above because auxin is evenly distributed, resulting in equal cell elongation on all sides. *(b)* Shoots grow toward light because auxin is transported to the shaded side, where it increases cell elongation.

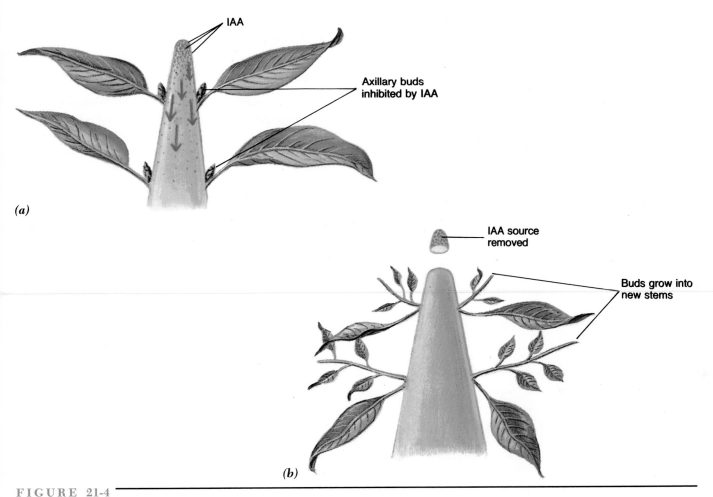

FIGURE 21-4

Apical dominance. *(a)* Produced by apical meristems, IAA moves down the stem and inhibits axillary bud growth. *(b)* When the shoot tip is clipped, the source of IAA is removed, and axillary buds grow into new stems.

Over the past 40 years, hundreds of synthetic auxins have been manufactured in chemical laboratories. Synthetic auxins (in fact, synthetic versions of all plant hormones) are called *growth regulators*. The active ingredient in Agent Orange, the defoliant (herbicide) that was used extensively in Vietnam, is a potent growth regulator that disrupts normal plant growth. Agent Orange was intended to defoliate densely populated jungles that provided enemy cover. Unfortunately, the herbicide also contains the contaminant dioxin, which is now known to be one of the most toxic substances for humans and has been linked to leukemia, miscarriages, birth defects, and liver and lung disorders.

Experiments confirm that other plant hormones play a role in many of the effects produced by auxins. As more research is completed, it becomes increasingly clear that hormonal control of plant growth and development is not simply the result of a single hormone acting alone but of a combination of hormones that triggers specific responses.

GIBBERELLINS

At the same time that researchers were attempting to isolate IAA during the 1920s, several Japanese scientists were investigating the cause of a fungal disease found in rice. Rice infected with a fungus (*Gibberella fujikuroi*) grew taller than did noninfected plants. As a result, the plants' stems weakened, and the plants collapsed. In 1926, E. Kurosawa purified the chemical in the fungus which stimulated cell elongation and division in infected rice; he named the substance **gibberellin** after the genus name of the fungus. During the 1940s and 1950s, intensive studies in Japan, the United States, and England revealed a number of growth-promoting chemicals, all of which had similar molecular structures to gibberellin. Today, more than 70 compounds are classified as gibberellins.

When gibberellins are applied to dwarf plants, normal stem growth is restored (Figure 21-5), suggesting that dwarf plants lack the gene needed to manufacture their own sup-

FIGURE 21-5

Gibberellin-enhanced growth. Applications of gibberellins to a plant cause skyrocketing vertical growth due to rapid enlargement of internode cells.

ply of gibberellins. Applications of gibberellins stimulate both cell division and cell elongation. But normal stem growth is not regulated by gibberellins alone; auxins and cytokinins are also involved.

▓▶ In addition to affecting stem growth, gibberellins are critical to embryo and seedling development. In all monocot seeds (corn, orchids, and grasses), gibberellins stimulate mobilization of food reserves stored in the endosperm. Without such mobilization, stored energy and nutrients would not be available to the embryo; without energy and nutrients, the embryo cannot grow, the seed cannot germinate, and a seedling cannot grow into a new plant. Gibberellins also inhibit seed formation, stimulate pollen-tube

growth, increase fruit size, initiate flowering, and end periods of dormancy in seeds and axillary buds.

CYTOKININS

In the 1950s and 1960s, researchers discovered that rapid growth could be stimulated in plant embryos and plant tissue cultures by adding coconut milk and yeast extract to the culture growth solution. Later studies isolated the active ingredients, which were named **cytokinins** because they stimulated rapid growth by promoting cell division (cytokinesis). Injuries to plants (such as a cut or a torn branch) are quickly healed as a result of the production of cytokinins.

In addition to repairing damaged tissues, cytokinins also play a role in regulating normal stem and root growth. Investigators have found that applying cytokinins stimulates cell division in stems and inhibits cell division in roots. These opposite effects on cell division in stems and roots is apparently related to the presence of auxins and gibberellins; but just how varying levels of these hormones control promotion or inhibition of cell division is still not known.

Cytokinins also delay **senescence**—the aging and eventual death of an organism, organ, or tissue. In leaves, cytokinins slow senescence by delaying the breakdown of chlorophyll. In flowers and fruits, cytokinins delay senescence by increasing nutrient transport.

ETHYLENE GAS

Some enterprising entrepreneurs have made a great deal of money from an unusual plant hormone, **ethylene gas**. The entrepreneurs' advertisement claims that when green fruits are placed in a "specially designed bowl," they will ripen two or three times faster than normal. This claim is true, but it would also be true for any closed container, including an ordinary paper bag. A covered bowl or closed bag traps ethylene gas that is released from ripening fruit. Ethylene gas hastens fruit ripening, and as fruits ripen, they produce more and more ethylene gas. Enclosing fruits sets up a positive feedback system (ethylene breeds more ethylene), causing fruits to ripen faster and faster. Production of ethylene gas by overripe fruit explains why "one rotten apple can spoil the whole barrel."

The discovery that ethylene gas promotes fruit ripening has dramatically lowered the cost of shipping fruits to markets. Tomatoes, bananas, apples, oranges, and many other fruits are usually picked when they are still green (reducing the chance of spoilage), transported in ventilated crates (to prevent ethylene gas buildup), and then gassed with synthetic ethylene at distribution centers to promote last-minute ripening.

In addition to its role in fruit ripening, ethylene gas stimulates stems to grow thicker in response to physical stress (Figure 21-6). Trees that have been staked while growing (and thus remain protected from the bending

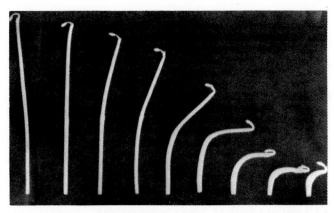

FIGURE 21-6

Ethylene gas and the emergence of seedlings. Stresses from pushing through the soil stimulate ethylene gas production in young seedlings. The bigger the obstacle, the greater the stress, and the more ethylene gas produced. This photo shows that increasing level of ethylene (from left to right) causes the stems of pea seedlings to thicken and curve. Thickening and curving of stems aid a seedling as it forces its way around obstacles and through the soil in its struggle to emerge.

stresses caused by winds) usually have weaker trunks and stems than do plants allowed to bend in the wind. The production of ethylene gas in response to physical stresses would be favored by natural selection since it increases survivorship under natural conditions, which, in turn, increases the plant's chances of living long enough to reproduce.

High concentrations of ethylene gas cause abnormal stem bending and leaf curling, discolor flowers, and inhibit root growth. Ethylene gas interacts with other plant hormones to control leaf fall (ethylene gas + abscisic acid), to stimulate flower production (ethylene gas + auxins), and to control the ratio of male to female flowers on some monoecious plants (ethylene gas + gibberellins).

ABSCISIC ACID

Not all plant hormones promote growth. **Abscisic acid (ABA)** almost always *inhibits* growth by either reducing the rate of cell division and elongation or halting the processes altogether. Abscisic acid was given its name because early investigators believed that ABA was involved in **abscission** —the process that causes flowers, fruits, and leaves to fall. While originally thought to be associated with both dormancy and abscission, more recently, abscisic acid has been shown to have little or no role in either process.

◑ ABA-suspended growth can lead to plant dormancy, which can save a plant's life during periods of severe environmental conditions. Repressed growth better enables dormant plants to withstand the stresses of dry, hot periods

or periods of freezing temperatures. During fall and winter, abscisic acid levels remain high in the axillary buds of deciduous trees—plants that drop their leaves during unfavorable growth periods. High abscisic acid levels counteract growth-promoting auxins and gibberellins, preventing cell division and elongation at a time when delicate, growing tissues would normally be exposed to lethal temperatures. ABA concentration drops at the onset of favorable growing conditions, releasing buds from their dormancy and causing plants to erupt with new growth (Figure 21-7). Alternating periods of dormancy and active growth which are keyed to environmental changes are examples of order and regulation in plants.

Under drought conditions, the concentration of ABA in guard cells may increase up to 40 times. Increased ABA concentration causes stomates to close, thereby reducing water loss and averting wilting and dehydration. Because ABA is active during periods of stress (triggering stomate closure during periods of water stress or promoting plant dormancy during periods of temperature stress), it is sometimes referred to as "the stress hormone."

OTHER PLANT HORMONES

Experimental evidence suggests there may be more than just five groups of plant hormones. For example, some studies suggest the presence of a flower-stimulating hor-

FIGURE 21-7

Deciduous plants burst out with new growth after a period of dormancy, when conditions become favorable for growth.

mone. In one experiment, two plants were grafted together; both were kept completely in the dark, except for the tip of one leaf on one plant. When this leaf tip was exposed to the critical photoperiod, both plants bloomed, suggesting that a hormone may have been involved. The hormone was transported to stems via the phloem, where it triggered flower-bud development. Since both experimental plants produced flowers, the hormone must have traveled to the stems of the grafted plant as well, inducing flower development. This elusive flowering chemical has yet to be isolated, although it already has been named **florigen.**

In addition to the possibility of a flowering hormone, some investigations suggest the existence of a root-growth hormone. Like the flowering hormone, the root-forming hormone has not yet been isolated.

Recently, scientists have discovered that some plants produce chemicals that act like the growth hormones of insects. These plant chemicals mimic the insect's hormones so closely that they disrupt the insect's normal growth and development, preventing sexual reproduction of the insect.

Plant growth and development is regulated by five groups of hormones—auxins, gibberellins, cytokinins, ethylene gas, and abscisic acid—although more hormones may be discovered in the future. Each hormone affects the plant differently; however, many growth and development responses result from the combined affects of two or more hormones. (See CTQ #3.)

TIMING GROWTH AND DEVELOPMENT

⟳ In many parts of the world, spring triggers a lavish surge of plant growth and development. Seeds are jolted out of dormancy; ghostly deciduous trees erupt with new leaves; and branches become crowded with masses of blossoms. But not all plants bloom or germinate in early spring. Flowering or germination is delayed in some plants until late spring, summer, or even fall. Clearly, plants are able to synchronize growth and development with the changing seasons; some, with great precision. In fact, Indian tribes in Arizona based their annual calender on the punctual flowering and fruiting of the saguaro cactus; a new year was marked by the appearance of the first saguaro fruits.

In addition to seasonal cycles, plants also have daily cycles. For example, the flowers of some plants open only at certain times of the day—morning glory (*Ipomoea purpurea*) at 4:00 A.M., fireweed (*Epilobium angustifolium*) at 6:00 P.M., nightshades (*Datura*) at 7:00 P.M., and the queen of the night (*Selenicereus grandiflora*) at 9:00 P.M. In the mid 1700s, Carolus Linnaeus, the Swedish scientist who introduced the binomial system for naming organisms (see page 13) established a "flower clock" in the Botanical Gardens of Uppsala, using plants that bloomed throughout the day and night to indicate the time of day.

Just how do plants monitor the time of day or the season of the year? Plants are able to detect and respond to seasonal environmental changes by monitoring changes in temperature and **photoperiod**—the relative length of daylight to dark in a 24-hour period. Plants use an internal *biological clock* to measure the time of day. Together, temperature, photoperiod, and an internal biological clock regulate cell metabolism and hormone production so that plant growth and development remain synchronized with environmental changes, enabling plants to live in some of the earth's harshest environments.

TEMPERATURE AND GROWTH

Plants grow in virtually all parts of the world, including Death Valley, California, where the world's highest air temperature was recorded at 57.8°C (136°F), and Siberia, where the world's lowest recorded temperature was −87°C (−126°F).

Each plant species has a particular range of temperatures it can tolerate, including an optimum temperature at which it grows best, and maximum and minimum temperatures it can tolerate for only limited periods time. Thus, where a plant grows is partly determined by the prevailing temperatures and the duration of temperature extremes. For example, saguaro cactus seedlings cannot survive more than 19 hours of freezing temperatures, so the saguaro stops growing at the Haulapi Mountains of Arizona. Any further north, freezing temperatures persist for longer than 19 hours, and the saguaro could not survive.

Because plant responses are sensitive to temperature, wide temperature fluctuations or prolonged exposure to extreme temperature can threaten a plant's life. Because a dormant plant's metabolism is significantly reduced, however, the plant is able to withstand greater temperature extremes for longer periods. On the days that the highest and lowest air temperatures were recorded in Death Valley and Siberia, most of the plants were dormant. Had they been actively growing, vulnerable meristems and primary tissues would have been killed, jeopardizing the life of the entire plant.

▐▶ Dormancy requires a number of preparations, including building up stores of food, water, and other nutrients and growing protective structures (such as axillary bud scales) to shield delicate tissues. Plants must be able to detect the approach of unfavorable growing conditions to prepare for dormancy as well as to detect the end of unfavorable growing conditions so that they can resume growth as favorable conditions return. Because temperature is closely tied to other weather factors (such as relative humidity, rainfall, and snowfall), gradual temperature changes act as a predictor of upcoming weather. By responding to temperature trends, plants synchronize growth and development to upcoming weather.

Temperature pattern plays another role in synchronizing plant activity. Some plants must be exposed to a certain period of cold before they will flower or before their seeds will germinate. The promotion of flowering by cold is called *vernalization*. Many stone fruit trees—plants that produce fruit that have pits (peaches, apricots, plums)—and apple trees fail to flower (and, therefore, fail to produce fruit) unless they are exposed to cold temperatures for precise numbers of hours (ranging from 250 to more than 1,200 hours, depending on the species and the latitude and altitude of the environment). Similarly, many seeds germinate only after they are exposed to near-freezing temperatures for several weeks. Plant sensitivity to cold is an adaptation favored by natural selection. A requisite, minimum cold period is an indicator that bleak winter conditions have passed and is not just a brief cold spell. If a plant were to resume growth following *any* length of cold, it might do so soon after at the onset of winter, freezing all new tissues and threatening the plant's life.

LIGHT AND GROWTH

In addition to responding to temperature cues, many plants react to variations in photoperiod. In the Northern Hemisphere, the period of daylight gradually lengthens from December until late June and then gradually shortens from late June until early December. Photoperiods are a very reliable indicator. On any given day of the year, the length of daylight and darkness is nearly constant from one year to the next in a particular region. This constancy explains the precise blooming of the saguaro cactus each year in the southwest deserts of North America.

The ability of organisms to respond to day length is known as **photoperiodism**. Photoperiodism enables plants to induce flowering at times when pollinators are active; to initiate dormancy as conditions gradually become less favorable; to correlate seed germination with periods favorable for seedling growth; and to control stem and root development of an emerging seedling. But how do plants measure photoperiod?

Phytochrome

In the 1950s, a group of plant scientists at the U.S. Department of Agriculture in Beltsville, Maryland, conducted a series of studies that exposed seeds and other plant parts to specific wavelengths of light and measured the plant responses. The Beltsville group observed that red light (660 to 680 nanometers) induced certain responses, while far-red light (700 to 740 nanometers) inhibited or nullified responses (Figure 21-8). These observations, together with results from studies on the affects of length of darkness on

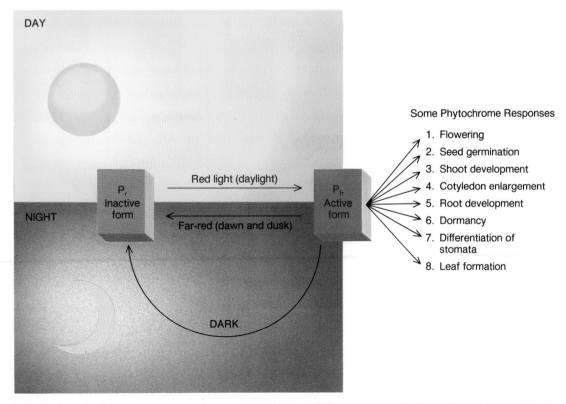

FIGURE 21-8

Phytochrome and plant development. P_{fr}, the active form of phytochrome, is formed in daylight when P_r absorbs red light and converts to P_{fr}. P_{fr} reverts back to inactive P_r at dawn, dusk, and in the dark and is gradually destroyed by cell enzymes. The length of the dark period largely determines how much active P_{fr} remains to stimulate or inhibit a number of plant growth responses.

plant responses, led the Beltsville researchers to propose that plants contained a blue-green pigment that exists in two, interconvertible forms. They named this alternating pigment **phytochrome** (*phyto* = plant, *chrome* = color).

After many studies, scientists verified that phytochrome contributes to the plant's ability to monitor photoperiod. One form of phytochrome absorbs red light at 660-nanometer wavelengths and is called P_r (phytochrome red). The other form of phytochrome absorbs far-red light at 730 nanometers and is called P_{fr} (phytochrome far-red). P_{fr} is the active form of phytochrome; it promotes or inhibits photoperiodic responses.

Phytochrome flips back and forth between its two forms as it absorbs light: When P_r absorbs red light, it immediately converts to the P_{fr} form; when P_{fr} absorbs far-red light, it immediately converts to P_r. P_{fr} also gradually converts into P_r in the dark (refer to Figure 21-9). These back-and-forth conversions between the two forms of phy-

tochrome enables the plant to measure the length of daylight and darkness during a 24-hour period.

At sunrise, red light predominates, triggering the conversion of phytochrome to the P_{fr} form. Thus, the first appearance of P_{fr} marks the beginning of a new day. Since red light predominates throughout the day, P_{fr} accumulates. At sunset, far-red light predominates, so phytochrome is converted to the P_r form; the appearance of P_r denotes the end of daylight. In other words, P_{fr} accumulates during daylight, and P_r accumulates in the dark. The proportion of P_{fr} to P_r is the means by which plants chemically measure the length of daylight and darkness. But, as researchers have discovered during their experiments, even a brief flash of light during the dark hours converts P_r back to P_{fr}, thereby changing the P_{fr} to P_r ratio, and, ultimately, the photoperiodic response.

The means by which phytochrome stimulates or blocks photoperiodic responses are still being investigated. Because phytochrome is not transported from cell to cell it is

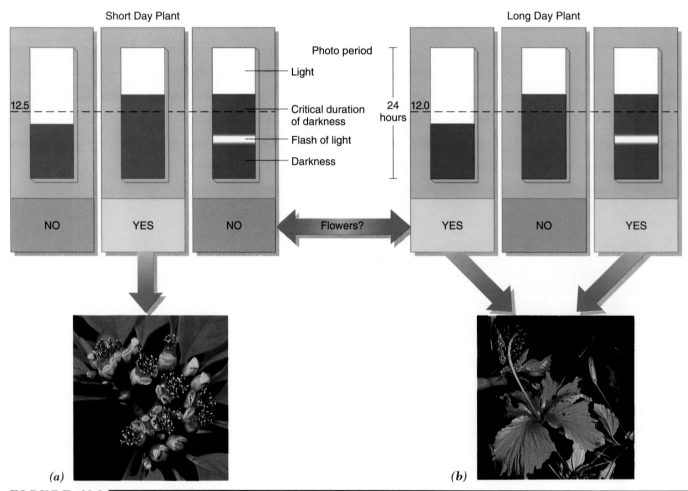

FIGURE 21-9

Photoperiod and flowering. Short-day and long-day plants flower in response to changes in photoperiod. *(a)* The poinsettia is a short-day plant that flowers when the day length becomes *less* than 12.5 hours of daylight (critical day length). *(b)* The hibiscus is a long-day plant that flowers when the day length becomes *greater* than 12 hours (critical day length).

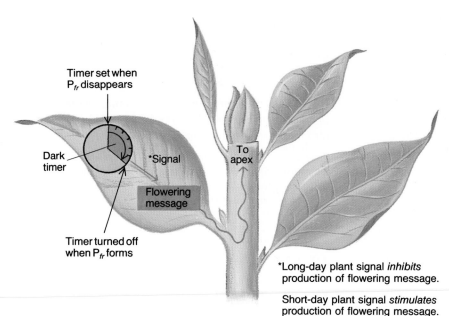

FIGURE 21-10
Regulating flowering in long-day and short-day plants. Leaf cells have a "dark timer" that measures the length of darkness. When the night reaches a critical length, the signal either inhibits (in long-day plants) or stimulates (in short-day plants) the production of one or more hormones that convey a flowering message to the stem, promoting flower bud development.

not a hormone, although evidence suggests that phytochrome probably controls the activity of plant hormones. For example, studies suggest that phytochrome is embedded in cell membranes and controls the passage of hormones into the cell, thereby altering cell activity.

Photoperiod and Flowering

■▶ It would be advantageous for plants to time flowering so that the process coincides with pollinator activity as well as with favorable growing conditions that last long enough for flowers, fruits, and seeds to mature before severe weather returns. Natural selection favors those plants that have adaptations that enable them to monitor photoperiod so that reproduction is optimized.

Plants are classified into three groups, depending on how changes in day length affect their flowering. **Short-day plants**, such as poinsettias, rice, morning glory, chrysanthemum, and ragweed, flower when the length of daylight grows shorter than some critical photoperiod, usually in late summer or fall (Figure 21-9*a*). **Long-day plants** flower when the length of daylight becomes longer than some critical photoperiod, usually in the spring and early summer (Figure 21-9*b*). Larkspurs, spinach, wheat, lettuce, mustard, and petunias are examples of long-day plants. **Day-neutral plants** flower independently of day length. Most day-neutral plants flower only when they are mature enough to do so, although other factors, such as temperature and water supply, may also affect flowering. Day-neutral plants include carnations, roses, snapdragons, sunflowers, and, as most gardeners know, many common weeds. Many agriculturally important plants, such as corn, beans, and tomatoes, are also day-neutral plants.

It is easy to assume mistakenly that short-day plants will always bloom when days are short, and long-day plants will only bloom when days are long. In fact, early researchers

developed this terminology by noting that some plants required shorter days to bloom, while others required either longer days or no specific day length. But further experimentation revealed that the difference between short- and long-day plants depended on whether the daylength was shorter or longer than some *critical photoperiod* (Figure 21-10). In long-day plants, the critical photoperiod represents a *minimum* day length; in short-day plants the critical photoperiod represents a *maximum* day length. For example, hibiscus (*Hibiscus syriacus*) and rice (*Oryza sativa*) are plants with an identical critical day length of 12 hours. However, the hibiscus is a long-day plant because it must be exposed to a day length *longer* than 12 hours in order to flower (12 hours is the *minimum* day length), whereas rice is a short-day plant because it must be exposed to a day length *shorter* than 12 hours in order to flower (12 hours is the *maximum* day length).

⟳ As you might expect, phytochrome plays a role in inducing flowering in both short-day and long-day plants. But after a great deal of research, investigators discovered there was more to photoperiod and flowering than just phytochrome conversions. Experiments suggest that the formation and disappearance of P_r and P_{fr} apparently sets a natural timing system within plant cells. But instead of measuring the length of daylight, as the terms *long-day* and *short-day* suggested, the timer actually measured the length of darkness.

The cell's "dark timer" works like a stopwatch. The timer starts running when P_{fr} begins to disappear at sunset and stops when P_{fr} begins forming at sunrise. If the dark timer has run for the critical length of time, then a message (perhaps an unknown hormone) is sent from the leaves to the shoot apex, stimulating axillary buds to develop into flowers (Figure 21-10). Long-day and short-day plants differ in whether the timer triggers (short-day plants) or inhibits (long-day plants) the formation of this flowering message.

Phytochrome and Shoot Development

The moment a seedling nudges its way through the soil surface, light begins affecting the way the seedling grows and develops (for example, auxins cause shoots to bend). As a shoot emerges, daylight induces P_{fr} conversion. The appearance of P_{fr} stimulates hormonal changes that cause the shoot to straighten; stimulate leaf enlargement, and begin chlorophyll synthesis for photosynthesis—normal shoot growth and development for a seedling emerging from the soil.

But even before a seedling nudges through the soil surface, the *absence* of light has already had an impact on shoot development. In buried seeds, phytochrome remains in the P_r form. The active P_{fr} form that triggers normal growth regulators is lacking, so the shoot elongates rapidly; leaves remain small and underdeveloped; the shoot hook (if present) remains bent; and the plant remains colorless, devoid of chlorophyll. These features are associated with a condition called **etiolation**. Plants kept in the dark, such as the grass under a board, will also become etiolated because of an absence of P_{fr}.

Shade affects stem growth similarly. A plant growing in the shade receives mostly far-red light, red light having been absorbed by the leaves of surrounding taller plants. Far-red light triggers the conversion of phytochrome to P_r, reducing the amount of P_{fr} present. A lower P_{fr} to P_r ratio boosts stem elongation, increasing a plant's chances of growing out from under the shade.

Phytochrome and Seed Germination

Some seeds do not germinate in the dark, which means that they will not germinate under the ground either. When these seeds are exposed to light, however, they germinate. This is why you must continually pull weeds from your garden: The seeds of many weedy plants germinate only after exposure to light, which is just what they get when you turn the soil. Experiments demonstrate that even a short flash of red light induces light-requiring seeds to germinate. This is a clue that phytochrome is involved. The red light shifts P_r to P_{fr}, removing the germination block.

BIOLOGICAL CLOCKS

Plant growth and development are controlled not only by changes in temperature and light but also by internal biological clocks. Recall that the "dark timer" we mentioned earlier measures the length of night, launching flowering in some plants. Other plant activities are also controlled by biological clocks. For example, the leaves of the silk tree (*Albizzia*), as well as those of other legumes, are oriented horizontally during daylight but fall into a more compact vertical "sleeping" position at night. Even when kept in continuous light and at constant temperature, the leaves fall into the "sleeping" position at exactly the same time each day. This rhythm is said to be **endogenous** (controlled internally), rather than **exogenous** (controlled by outside environmental changes). The sleep movements of leaves follow a circadian (*circa* = about, *diem* = day) cycle, recurring at 24-hour intervals. Other endogenous circadian plant activities include flower opening and closing and nectar and flower odor production.

By monitoring changes in temperature and light, plants adjust their growth and development to coincide with favorable environmental periods. (See CTQ #4.)

PLANT TROPISMS

Several environmental factors change the rate of production or distribution of growth-regulating hormones, thereby changing the direction of plant growth. Such responses are called **tropisms**. Tropisms are stimulated by light, gravity, contact with objects, chemicals, temperature gradients, wounding, and the presence of water. The three most common tropisms are phototropism, gravitropism, and thigmotropism.

PHOTOTROPISM

A familiar tropism occurs in many houseplants. If left in one spot, houseplants gradually turn their leaf surfaces toward a light source, often a nearby window. Since light is the environmental stimulus, the response is called a **phototropism** (*photo* = light). Shoot bending, which we discussed in the chapter opener, is another example of phototropism. Experiments reveal that a light-absorbing, yellow pigment—a flavoprotein—promotes the movement of auxins away from the light source, increasing cell elongation in the cells on the shaded side (Figure 21-1 and 21-4).

GRAVITROPISM

When the pull of gravity causes changes in a plant's growth, the response is called **gravitropism**. Gravity indicates which way is down and which way is up, helping to direct growth of roots and shoots in the right direction. Shoot growth away from the gravitational force is called **negative gravitropism**, whereas root growth toward the gravitational force is called **positive gravitropism**.

Researchers believe that plants detect the direction of gravity by small particles inside their cells which are pulled to the bottom of the cell by gravity. The change in position of these particles apparently redirects the transport of growth hormones, generating the gravitropic response. In one experiment, a growing shoot was placed on its side. Within 15 minutes, the lower side of the shoot contained twice as much auxin than did the upper side. This unequal distribution caused the stem to bend upward (see Figure 21-3). Once upright, the particles settled into their original position inside the cell; auxin concentration became equally distributed on all sides; and the shoot grows straight up.

↻ The root cap also plays an important role in controlling the direction and rate of root growth. If the root cap is experimentally removed, the root grows faster than normal but not downward. This suggests that the root cap produces a growth inhibitor that slows the production and elongation of certain root-tip cells so that growth is always downward. When a root is placed horizontally, the inhibitor accumulates in the cells on the lower side, reducing their growth relative to cells on the top side and producing downward bending (Figure 21-11). Unlike stems, root bending is probably not controlled by auxin; some experimental evidence suggests that abscisic acid may in fact be the growth inhibitor in roots.

THIGMOTROPISM

⟳ A change in plant growth that is stimulated by contact with another object is called a **thigmotropism**. For example, many climbing plants have stems and branches that are too weak to support their own weight so they rely on other objects to keep them upright. Using tendrils, some plants either wrap around a supporting structure (Figure 21-12*a*) or produce pads at the tips of branches that fasten the plant to a wall, fence post, or some other solid object for support (Figure 21-12*b*).

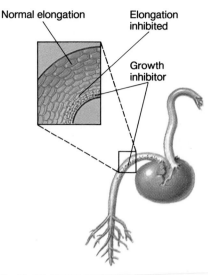

FIGURE 21-11

Which way is down? Gravity causes a growth-inhibiting hormone to accumulate on the lower side of a root, slowing cell elongation relative to cells on the upper side. This differential elongation causes the root to turn downward.

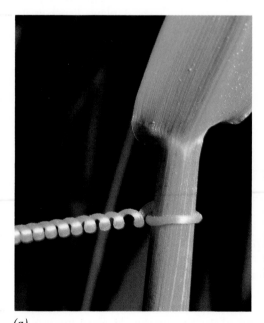

(a)

(b)

FIGURE 21-12

Some climbing vines have very weak stems and rely on a sturdier plant or some other solid object for support. The tendril of a pea plant grows around a corn stem for support *(a)*, while expanded pads of a Virginia creeper fasten its growing stems to a wall *(b)*.

Considered together—the diversity of plant tropisms, the mechanisms by which plants monitor and respond to changes in environmental conditions, and the production of hormones that coordinate growth and development—all help dispel a common misconception that plants are a static, comparatively unresponsive life form. The diversity of plant growth and development adaptations, the complexity by which plants are regulated, and the precision by which plants operate, reveal just the opposite: Plants are dynamic organisms; they continually respond and adjust to their surroundings in ways that promote survival and reproduction.

Tropisms are the result of changes in the size or shape of many cells or of changes in internal water pressure within a few strategically placed cells. Changes in a plant's internal water pressure bring about rapid responses, whereas long-term adjustments require permanent modifications in cell structure. (See CTQ #6.)

REEXAMINING THE THEMES

Relationship between Form and Function

Hormones control and coordinate plant growth and development. The production of hormones enables plants to adjust their form so that resulting structures will be better able to carry out their functions. Shoots growing upward, roots growing downward, stems and leaves growing toward light, flowering at certain times of the year, seedling stems increasing thickness and bending as they push through the soil, and tendrils and pad production to support stems are all examples of the modification of plant form as a result of hormone production to increase function.

Biological Order, Regulation, and Homeostasis

The regulation of growth and development in plants follows a hierarchy of controls, the combination of which enables plants to germinate, grow, develop, and produce flowers and fruits at the most optimum times of the year.

Genes are controlled by hormones. In turn, the production and distribution of hormones are affected by the environment.

Evolution and Adaptation

Plant seeds can find their way to any number of habitats, each with slightly different environmental circumstances. For example, a seed may be dispersed to the side of a mountain or on flat ground. To survive, a developing seedling would have to grow differently in each of these habitats. On a mountainside, the shoot cannot grow perpendicular to the surface, as it would on flat ground, and the root could not grow directly down. A plant disperses its embryos when they are in a very underdeveloped form, which enables them to adjust to particular environmental circumstances.

SYNOPSIS

Plant growth and development are the result of three interacting levels of control. The intracellular level contains fixed genetic controls, which are regulated by intercellular hormonal controls, which, in turn, are affected by various aspects of the environment. Environmental factors greatly influence plant growth and development, helping adjust a plant's form to its immediate surroundings.

Effects of plant hormones vary with hormone concentration, target tissue, time of year, developmental stage of the plant, and type of plant. Most growth and development responses are the result of combinations of plant hormones.

In general, four of the five groups of plant hormones (auxins, gibberellins, cytokinins, and ethylene gas) promote growth, whereas the fifth (abscisic acid) inhibits growth. Recent research suggests that there may be other plant hormones besides these five.

Many plants synchronize dormancy and periods of active growth with changing seasons by responding to variations in temperature and day length and by using an internal clock. Dormancy is often critical for a plant to survive periods of unfavorable growing conditions.

Interconvertible forms of light-absorbing phytochrome enable plants to monitor changes in photoperiod. Monitoring photoperiod helps plants synchronize flowering, seed germination, and renewed vegetative growth to favorable periods of environmental conditions.

Key Terms

auxin (p. 427)
indoleacetic acid (IAA) (p. 430)
apical dominance (p. 430)
gibberellin (p. 431)
cytokinin (p. 432)
senescence (p. 432)
ethylene gas (p. 432)
abscisic acid (ABA) (p. 433)
abscission (p. 433)

florigen (p. 434)
photoperiod (p. 434)
photoperiodism (p. 435)
phytochrome (p. 436)
short-day plant (p. 437)
long-day plant (p. 437)
day-neutral plant (p. 437)
etiolation (p. 438)
endogenous (p. 438)

exogenous (p. 438)
tropism (p. 438)
phototropism (p. 438)
gravitropism (p. 438)
negative gravitropism (p. 438)
positive gravitropism (p. 438)
thigmotropism (p. 439)

Review Questions

1. List five reasons why it has been so difficult for plant researchers to work with and determine the functions of plant hormones.

2. How do the characteristics of etiolated seedlings improve their chances of emerging from the soil?

3. Match the hormone with its function (multiple matches are possible).

   _____ auxins   a. fruit ripening
   _____ cytokinins   b. apical dominance
   _____ gibberellins   c. senescence
   _____ ethylene gas   d. stimulate stem elongation
   _____ abscisic acid   e. seed dormancy

4. Growth can be the result of cell division or cell enlargement (mainly elongation). In plants, most growth results from cell elongation; a mature plant cell may be ten times larger than a cell produced by the apical meristem. Since all plant cells have a surrounding cell wall, what must happen in order for plant cells to be able to elongate? What force is involved?

5. Consider a young vine struggling to grow in a dense tropical rain forest. Which tropisms are likely be involved if the vine is to grow successfully and develop to maturity?

6. Are plant tropisms irreversible?

7. List three hormones that *promote* growth. What is different about the way the hormones increase growth?

8. Phytochrome is a plant pigment, yet it differs from those involved in photosynthesis, such as chlorophyll. How is phytochrome similar to photosynthetic pigments, and how is it different?

9. If you remove the apex of a stem, will growth in length of that stem stop altogether? Explain.

10. Explain how photoperiodism works. What is the adaptive advantage to photoperiodism for plants?

Critical Thinking Questions

1. Explaining the results of their experiments on light and stem bending, the Darwins reasoned that a growth signal was produced in the coleoptile tip. According to the Darwins, the signal traveled down the shoot and triggered the cells below to enlarge differentially which, in turn, caused the young shoot to bend toward light. Describe two ways in which cells could "enlarge differentially" and cause a stem to bend toward light. How can you determine which way is correct?

2. One molecule of a plant hormone can cause 150,000 glucose molecules to be converted to cellulose. What does this tell you about how plant hormones work? How are they similar to enzymes? Different from enzymes?

3. A botanist gave wheat seedlings a dose of X-rays sufficient to prevent cell division but which left the seedlings able to function normally otherwise. He gave half of the irradiated seedlings gibberellin, but not the other half. The length of the seedlings is plotted in the graph below. Which of the following conclusions is supported by the experiment, and why? (1) gibberellins stimulate cell elongation only; (2) gibberellins stimulate cell division only; (3) gibberellins stimulate both cell elongation and cell division.

4. The seeds of some plants do not germinate in the spring, even though temperatures are generally warm and water is plentiful—two conditions that are favorable for growth. Suggest what might be preventing these seeds from germinating. Would there be any advantage to delaying germination past spring?

5. Discuss some of the ways you can use your new knowledge of plant growth and development to improve the growth of plants in your garden.

6. The discovery of auxin began with the study of tropisms by Darwin and others. Knowledge of auxins has helped scientists understand tropisms. Summarize the significant events in this history by means of a diagram that illustrates the interaction between these two lines of research.

7. Colonizing space presents many difficult challenges, not the least of which is how plants can be grown for food and for providing a quality environment for the human inhabitants. Discuss some of the problems you would anticipate in trying to grow plants in a space station. For each problem, try to suggest a solution.

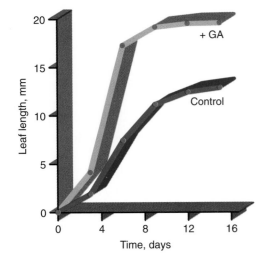

Additional Readings

Darwin, C. 1881. *The power of movement in plants.* New York: Appleton-Century-Crofts. (Intermediate)

Davies, P. 1987. *Plant hormones and their role in plant growth and development.* Dordrecht, Netherlands: Martinus Nijhoff. (Advanced)

Gregory, P. J., J. V. Lake, and D. A. Rose (eds.). 1987. *Root development and function.* Society for Experimental Biology. Press Syndicate of the Univ. of Cambridge, Cambridge, England. (Advanced)

Halstad, T., and D. Dilcher. 1987. Plants in space. *Ann. Rev. Plant Physiol.* 38: 317–345. (Intermediate)

Taiz, L., and E. Zeiger. 1991. *Plant physiology.* Reading, MA: Benjamin Cummings. (Advanced)

Torrey, J. 1985. The development of plant biotechnology. *Amer. Sci.* 73: 354–363. (Intermediate)

Van Overbeek, J. 1968. The control of plant growths. *Sci. Amer.* July: 75–81. (Introductory)

Zimmermann, M. H., and C. L. Brown. 1975. Trees: *Structure and Function.* Springer-Verlag, New York. (Intermediate)

Salisbury, F. B., and C. W. Ross. 1991. *Plant Physiology,* 4th ed. Wadsworth Publishing Co., Belmont, Ca. (Advanced)

Steeues, T. A., and I. M. Sussex. 1989. *Patterns in plant development.* 2nd ed. Prentice Hall, Inc. New Jersey. (Intermediate)

Form and Function of Animal Life

Of any group
of organisms, animals display the
most complex array of form and function.
This praying mantis has specialized
appendages for flight, walking, capturing
and devouring victims, and mate-seeking
and reproduction, as well as dozens of
other complex activities.

PART
· 5 ·

An Introduction to Animal Form and Function

STEPS
TO
DISCOVERY
The Concept of Homeostasis

*I*n 1843, at the age of 30, Claude Bernard moved from a small French town, where he had been a pharmacist and an aspiring playwright, to Paris, where he planned to pursue his literary career. Bernard's plans changed, however, when he enrolled in medical school, and he went on to become one of the leading physiologists of the nineteenth century. During his prestigious career, Bernard studied body temperature, stomach juices, the role of nerves in controlling the diameter of blood vessels, and the function of the liver, but he focused particular attention on the fluids that bathe the body's internal tissues. Bernard carefully monitored the temperature, acidity, and salt and sugar concentrations within the blood under various conditions and found that the body's fluids resisted change, providing cells with a stable, ordered environment. In his own translated words: "It is the fixity of the internal environment which is the condition of free and independent life. . . . All the vital mechanisms, however varied they may be, have only one object, that of preserving constant the conditions of life in the internal environment."

Even in extreme environments, humans maintain a constant body temperature. The regulatory center is the hypothalamus

Bernard's work provided the foundation for a diverse array of studies of the mechanisms by which the body maintains its internal constancy. For example, in the 1880s, Charles Richet in France and Isaac Ott in the United States independently determined that one part of the brain, the hypothalamus, played a key role in maintaining body temperature. If this part of the brain of laboratory animals was damaged, these animals lost their ability to hold their body temperature constant in the face of increasing or decreasing environmental temperatures. In 1912, Henry Barbour of Yale University developed a technique to selectively warm or cool this part of the brain. Barbour accomplished this feat by implanting fine silver tubes into the hypothalamus, and then circulating water of particular temperatures through the tubes. Cooling the hypothalamus below normal body temperature caused the animals to increase heat production, just as if they had been outside on a cold winter night. Warming the hypothalamus evoked the opposite response; the animals reacted by losing heat, just as if they had been exposed to a hot desert sun. The results of these studies suggested that one of the functions of the hypothalamus was to determine if the body's temperature was rising or falling and then to elicit an appropriate response that would return the temperature to its normal value.

One of the most influential physiologists of the twentieth century was Walter Cannon of Harvard University. Working in a field hospital in France during the fierce trench warfare of the last few months of World War I, Cannon was struck by the body's ability to withstand the terrible trauma resulting from severe wounds and to restore the orderly conditions necessary for survival. After the war, Cannon returned to Harvard and turned his attention to studying the mechanisms by which the body "fights" to maintain the internal environment. In 1926, he coined the term "homeostasis" (*homeo* = sameness, *stasis* = standing still; balance). Cannon did not mean to suggest that the body was stagnant or incapable of change but, rather, that it had the capacity to respond in dynamic ways to situations that threatened to disrupt its internal stability.

Cannon summarized his views on homeostasis in a book entitled *The Wisdom of the Body*. In his book, Cannon noted that humans could be exposed *for short periods* of time to dry heat above 115°C (239°F), without raising their body temperature, or to high altitudes with greatly reduced oxygen, such as when flying in an airplane or climbing a mountain, without showing serious effects of oxygen deprivation. He also noted that the body could resist disturbances that arise from within. For example, a person running for 20 minutes produces enough heat to "cause some of the albuminous substances of the body to become stiff, like a hard-boiled egg" and enough lactic acid (the acid of sour milk) to "neutralize all the alkali contained in the blood." Yet, the body becomes neither overheated nor poisoned.

The concept of homeostasis identified by Bernard and Cannon pervades virtually every aspect of the physiologic sciences. As we consider the function of major organs, such as the lungs, heart, kidney, and liver, we will see how each of these organs helps maintain a stable, ordered internal environment that promotes our health and well-being. We will also examine some of the dangers that lurk should these homeostatic controls fail.

of the brain.

Go without water on a hot summer day and you build up a thirst that has to be quenched. Spend a cold winter night in a bed with inadequate covers and you sleep huddled into a ball, trying to keep warm. Stay up all night studying for an exam and you find yourself so tired you can barely stay awake. All of these common events in our lives illustrate cases where our bodies are "telling" us that we need to take some action to maintain the stability of our internal environment. Most homeostatic activities take place without the need for these types of behavioral responses. For example, without conscious effort, our bodies "automatically" prevent themselves from becoming too salty or dilute, too hot or too cold; our bodies prevent waste products from accumulating to dangerous levels and stop oxygen or ATP concentrations from dropping below required amounts. All organisms have homeostatic mechanisms that enable them to respond to changes that threaten to upset their internal environment. In animals, these responses are carried out by teams of cells, tissues, organs, and organ systems.

▼ ▼ ▼

HOMEOSTATIC MECHANISMS

As you are reading this chapter or listening to your professor's lecture in class, you are probably in an environment where the temperature is actively maintained at a relatively constant value. Maintaining a constant room temperature requires more than just a heater or an air conditioner. The system requires

1. a *sensor* that continually monitors the temperature in the room, sending information about the air temperature to
2. a *control unit, or thermostat,* which is set at a particular temperature. If the information from the sensor indicates that the temperature in the room is either cooler or warmer than that for which the thermostat is set, a signal is transmitted to
3. an *effector*—the furnace or air conditioner that generates hot or cold air. When the air in the room is restored to the temperature set by the thermostat, the effector is turned off, until the temperature again deviates from the thermostat's set-point.

The operation of this temperature-control system is an example of a **negative-feedback mechanism.** The term "feedback" indicates that a change in a property is "fed back" to a control unit, inducing a subsequent change. Feedback is "negative" when a change in the property, such as air temperature, *reverses* further changes. For example, a rise in room temperature triggers a response to lower the temperature, while a drop in room temperature triggers the opposite response.

⟳ Maintaining a constant *body* temperature requires an analogous set of biological components (Figure 22-1) and provides one of the best examples of how homeostatic mechanisms maintain the constancy of the internal environment. Your body remains at a temperature of 37°C (98.6°F) as a result of

1. *sensors* (called *temperature receptors*) located both at the body surface and deep within the body's interior; sensors monitor the temperature of both the external and internal environments. Information from the temperature receptors is passed to
2. a *control unit,* or *integrator,* (known as the body's "thermostat") located within the hypothalamus of the brain, which is normally "set" at 37°C. When the thermostat receives information that the body temperature is *deviating* from 37°C, signals are sent to the body's
3. *effectors,* which consist of widely scattered muscles and glands. If, for example, the temperature receptors detect a rise in body temperature, the brain initiates a response that causes skin glands to release sweat, cooling the body back down to 37°C (Chapter 28).

Negative-feedback mechanisms operate by reversing change. In contrast, **positive-feedback mechanisms** operate when a change in the body feeds back to *increase* the magnitude of the change. Blood clotting is an example of positive feedback. Your blood contains proteins that have the potential to form a blood clot that can plug a wound in a blood vessel which can be a life-saving response. The clotting of the blood at the site of an injured vessel begins on a very small scale. First, a small amount of clotted protein is deposited. As the clot forms, it induces a response that triggers the deposition of additional clotted protein, which, in turn, triggers even more deposition. As a result of this positive-feedback mechanism, what began as a small response is rapidly amplified so that vessel damage can be controlled quickly, halting any further loss of blood and maintaining internal stability.

We have described several examples of homeostasis and the role of negative- and positive-feedback mechanisms, but we have yet to discuss the parts of the body responsible for these physiologic activities. Look back for a moment at Figure 1-4 and the discussion of the hierarchy of biological organization. We saw in earlier chapters how atoms, molecules, and organelles are organized into increasingly complex biological structures, including cells, which are the fundamental units of life. Let us now continue farther up the ladder of biological organization and see how the cells of an animal carry out their activities in concert

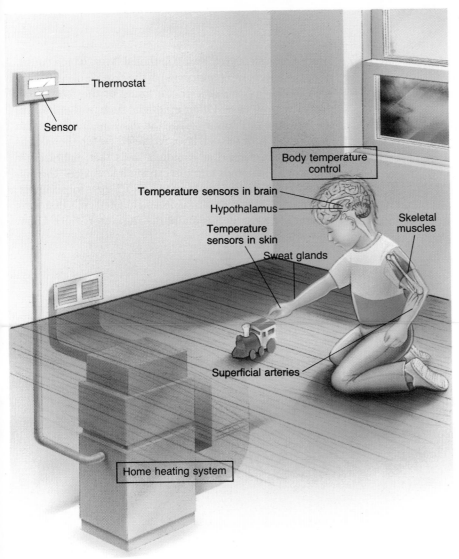

Thermostat

Sensor

Body temperature
control

Temperature sensors in brain

Hypothalamus

Temperature
sensors in skin

Sweat glands

Skeletal
muscles

Superficial arteries

Home heating system

FIGURE 22-1

Controlling temperature levels by negative feedback. The temperature within the room and the body temperature of the child are maintained at a relatively constant value by an analogous set of elements that include temperature sensors, thermostats, and effectors. In the regulation of body temperature, the temperature sensors are located in the skin and brain; the thermostat is in the hypothalamus; and the effectors are the sweat glands, which produce sweat, and the skeletal muscles, which cause shivering.

with one another to meet the many needs of the entire animal.

Maintaining a stable internal environment requires a diverse array of sensors, integrators, and response mechanisms that detect and reverse any tendency toward physiologic disorder. (See CTQ #2.)

UNITY AND DIVERSITY AMONG ANIMAL TISSUES

The human body contains several hundred recognizably different types of cells, each of which works with other types

of cells to accomplish a common function. The simplest of these teams of cells is a **tissue**—an organized group of cells with a similar structure and a common function. Tissues are the "fabrics" from which complex animals (and plants) are constructed. The use of the word "fabric" to describe tissues is not just a recent invention of biology writers; the term is derived from *tissu*, an Old English word meaning a "fine cloth." Like intricately woven fabric, tissues are the products of finely organized biological designs and not mere aggregates of cells randomly packed together in a unit. These living fabrics form the foundation of the multicellular organism.

Despite the tremendous diversity that exists in the types of cells present within a single animal (and even greater diversity found among different animals), there is a striking overall unity of function. All of the diverse cells found among animals can be classified as part of one of four fundamental types of tissues of which all organs and organ systems are composed. Each of your organs, for example, is composed entirely of epithelial, connective, muscle, and/or nervous tissue.

EPITHELIAL TISSUE

Epithelial tissue (Figure 22-2) is present as an **epithelium**—a sheet of tightly adhering cells that lines the spaces of the body, such as the outer edge of an organ or the inner lining of a blood vessel or duct. The surface of the entire body is covered by an epithelial layer, which constitutes the outer layer (*epidermis*) of the skin. Some epithelia, such as that of the skin or the lining of the mouth, are primarily protective. Other epithelia, such as the lining of the intestine or lungs, regulate the movement of materials from one side of the cell layer to the other. Epithelial cells are often specialized as glandular cells that manufacture and *secrete* materials into the extracellular space. The epithelial lining of the trachea (Figure 22-2*b*), for example, contains mucous-secreting cells and ciliated cells that work together to move debris out of the respiratory tract. Other

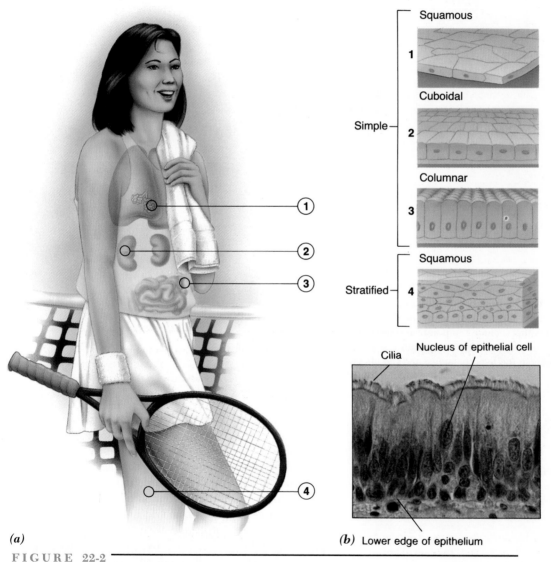

Squamous

Cuboidal

Columnar

Simple

Squamous

Stratified

Cilia

Nucleus of epithelial cell

(a)

(b) Lower edge of epithelium

FIGURE 22-2

Epithelial tissue. *(a)* Epithelial tissues are categorized as either *simple* or *stratified* depending on whether they consist of one or more than one layer of cells. Epithelia are also categorized as either squamous (flattened), cuboidal, or columnar depending on the shapes of the cells. Lung tissue is an example of a simple squamous epithelium; kidney tubules contain a simple cuboidal epithelium; the inner lining of the small intestine is a simple columnar epithelium; and the outer part of the skin is a stratified epithelium. *(b)* A stained section showing the ciliated epithelum that lines the inner surface of the trachea (windpipe). Even though the nuclei are found at different levels, this epithelium is only one cell layer thick; it is called a pseudostratified epithelium.

(a)

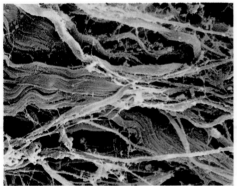

(b)

FIGURE 22-3

Connective tissue. *(a)* Connective tissues are quite diverse in structure and function and include bone, cartilage, blood, and tendons. All of these tissues are categorized by extensive extracellular materials. *(b)* Bundles of extracellular collagen fibers give connective tissues their strength.

secretory epithelia release hormones, oils, sweat, milk, and digestive enzymes. The structure and functions of epithelia will be discussed in greater detail in later chapters on the skin, respiratory tract, digestive tract, and kidney.

CONNECTIVE TISSUE

The body is held together by **connective tissues** (Figure 22-3), which consist of a loosely organized array of cells

surrounded by a nonliving extracellular matrix. The properties of connective tissue are due largely to the extracellular matrix, which contains proteins and polysaccharides secreted by the "entrapped" cells. For example, collagen, the most common component of the extracellular matrix (and the most abundant protein in the human body) is an inelastic molecule that provides resistance to pulling forces. Fibers of collagen (Figure 22-3*b*) provide tendons and ligaments with the strength to connect muscles to bones and bones to one another. Looser arrangements of collagen protect delicate structures, such as the eyeball, much like packing material prevents breakage of fragile glassware. Connective tissues may also contain elastic fibers that provide some tissues, such as skin and vocal cords, with the capacity to stretch and recoil. Other examples of connective tissues include skeletal elements composed of cartilage and bone; sheets (*mesenteries*) that support the visceral (internal) organs; the transparent outer layer (*cornea*) of the eyeball; fat (*adipose*) deposits; and the deeper layer (*dermis*) of the skin. Blood is also a type of connective tissue; blood cells are surrounded by an extracellular matrix with a fluid consistency. The structure and functions of connective tissues will be discussed in greater detail in the chapters on the skin, skeleton and circulatory system.

MUSCLE TISSUE

Muscle tissue (Figure 22-4) consists of muscle cells — cells that contain elongated protein filaments that slide over one another, causing the muscle cells to shorten (contract). When muscle tissue contracts, it generates the forces needed for motion — either to move the animal (or its parts) or to propel substances within the body. Muscle tissue is present in large masses capable of moving the largest bones of the body and is the major element of the heart. Muscle tissue can also be found scattered more subtly throughout the body's internal organs. Among other consequences, when such muscle tissue contracts, blood vessels constrict; food moves through the digestive tract; and urine is emptied from the bladder. The various types of muscle tissue will be described in Chapter 26.

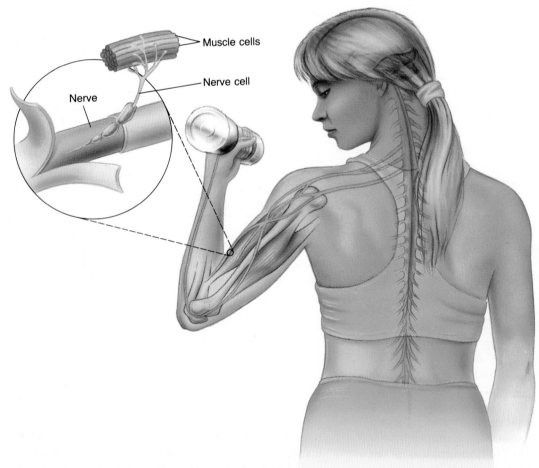

FIGURE 22-4

Muscle and nerve tissues. The structure of one type of muscle tissue is exemplified by a large skeletal muscle of the arm. Other types of muscle tissue are found in the heart and scattered less noticeably throughout the internal organs of the body. Nerve tissue is concentrated within the brain and spinal cord, but also extends from these central neural organs into most of the distant regions of the body.

NERVE TISSUE

Nerve tissue (Figure 22-4) consists of highly elongated nerve cells—cells that are specialized for long-distance communication within an animal. Communication is accomplished when a signal is sent along the membrane of a nerve cell to its terminal end and is relayed to another cell—a nerve cell, muscle cell, or gland cell—thereby evoking a response. Nerve tissue is responsible for coordinating such diverse bodily activities as breathing, sweating, sexual responses, and defecation, as well as providing some animals with memory, emotions, and consciousness. The structure and function of nervous tissues will be discussed in Chapter 23.

Physiologic activities are accomplished by groups of similar cells working together as tissues. Four broad types of tissues are distinguished, but numerous subtypes can be identified. (See CTQ #3.)

UNITY AND DIVERSITY AMONG ORGAN SYSTEMS

Animals as diverse as sponges, whales, earthworms, and anteaters have common needs: All animals must obtain oxygen, digest food, eliminate wastes, protect themselves from microbes and predators, and so forth. Each of these basic needs are met by **organs**—structures that are composed of different tissues working together to perform a particular function. A major function of the kidney, for example, is the removal of metabolic waste products from the blood. The elimination of waste products from the body requires additional organs, including the bladder and a series of tubular pathways (the ureter and urethra) that lead to the body surface. All of the various organs that work together to perform a common task, such as the elimination of waste products, make up an **organ system.**

⚠ Common needs often lead to similar, but independent, evolutionary solutions. Even though an earthworm and a whale are so distantly related that their most recent common ancestor *may* have been a single-celled protist, these two animals possess organ systems that have similar overall functions. In the following chapters, we will examine each type of organ system found in animals, focusing on human systems. Despite similarities in overall function, such as uptake of oxygen or elimination of waste products, the organ systems of different animals may be constructed very differently.

▐▶ Keep in mind that each animal's environment is a critical factor in determining the types of homeostatic mechanisms that will be adaptive for that animal. For example, sea animals must cope with very salty surroundings and, consequently, must possess mechanisms to prevent the loss of water by osmosis. In contrast, animals inhabiting a fresh-water stream live in a highly dilute environment that has very little available salt. As a result, freshwater animals must possess adaptations to eliminate water that floods their body by osmosis. In addition, freshwater animals must cope with marked fluctuations in water temperature, oxygen availability, and the risk of the evaporation of their home. Terrestrial (land) animals face even harsher conditions. They must cope with rapid changes in climate, a shortage of external water, a propensity to lose water to the surrounding air, and the lack of a medium to help support their body weight. Each of these different environments presents different physiologic challenges and selects for different types of homeostatic responses.

TYPES OF ORGAN SYSTEMS

Figure 22-5 provides an introduction to the types of organ systems we will encounter in the following chapters. We will summarize them briefly here.

- *Digestive systems* break down (*digest*) the macromolecules present in food. The breakdown products are absorbed across the lining of the digestive tract, providing the body with small-molecular-weight nutrients that supply energy and building materials.

- *Excretory/osmoregulatory (urinary) systems* accomplish two interrelated functions: They rid the body of waste products, the unusable byproducts of metabolism, and they maintain balanced internal concentrations of salt and water. Waste products include potentially toxic nitrogen-containing compounds, most notably ammonia, which must be either eliminated directly or modified to a less toxic compound, such as urea, and then eliminated. Similarly, many biological processes, ranging from enzyme activity to nerve-impulse conduction, are very sensitive to the concentration of dissolved salts, which must be strictly regulated.

- *Respiratory systems* absorb the oxygen animals need to oxidize the organic molecules that fuel their biological activities. The respiratory system may also expel carbon dioxide, a waste product of metabolism. Animals that live in water and absorb the oxygen dissolved in their aquatic medium have different types of respiratory organs than do animals that breathe air.

- *Circulatory systems* transport materials inside an animal's body. Food absorbed by digestive systems and oxygen absorbed by respiratory systems must be distributed to those locations in the body where they are needed; wastes generated by cell metabolism must be carried to a site of elimination. Most animals, particularly those of larger size, distribute materials via a system of branched vessels to all parts of the body.

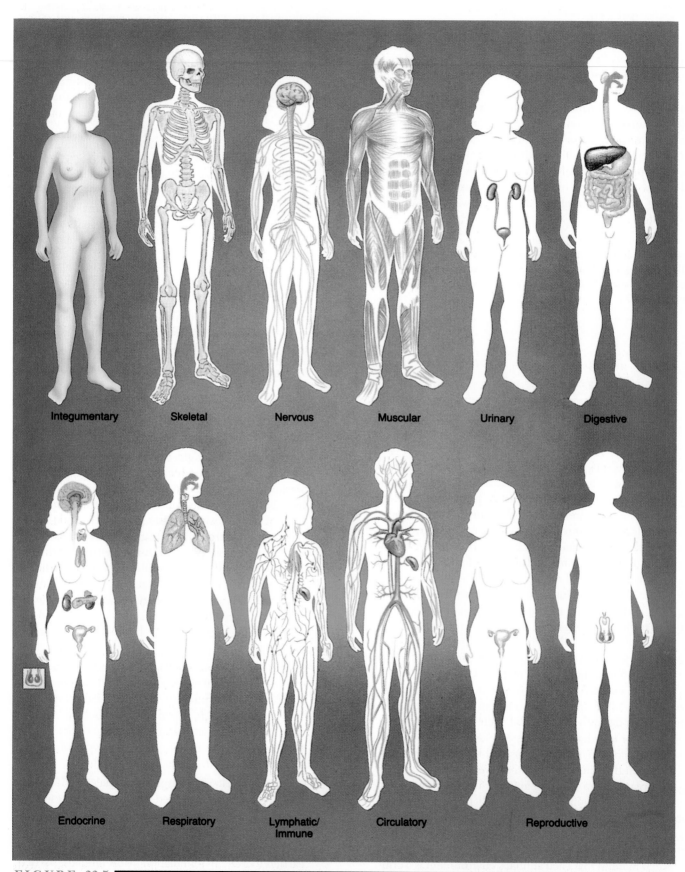

Integumentary   Skeletal   Nervous   Muscular   Urinary   Digestive

Endocrine   Respiratory   Lymphatic/ Immune   Circulatory   Reproductive

FIGURE 22-5

The types of organs systems found in the human body.

- *Immune systems* protect animals from foreign substances and invading microorganisms. The body's immunological defenses consist of various cells that recognize and ingest foreign materials as well as soluble blood proteins, called *antibodies,* which specifically bind to foreign materials and inactivate them.

- *Integumentary systems* cover the surfaces of animals and provide a barrier between the animal's body and the external environment. These systems protect the animal from dehydration, provide physical support, prevent invasion by foreign microorganisms, and help regulate internal temperature.

- *Skeletal and muscular systems* work together to provide both support and movement. Most animals possess rigid skeletal structures that provide support and maintain body shape. In addition, skeletal structures are often involved in *movement* (a shift in position of a part of the body) and *locomotion* (a shift of the entire animal from one place to another). Movement is often accomplished as rigid skeletal elements are pulled in one direction or another by the contraction of attached muscles. Locomotor structures, such as wings, legs, or flagella, typically project from the body as *appendages* (Figure 22-6) that act upon the external environment.

- *Reproductive systems* produce gametes (sperm or eggs) for fertilization and the subsequent development of new individuals of a species. All organisms have a limited life span. Individual bacteria live for minutes, a Galapagos tortoise for hundreds of years, and a giant redwood tree for thousands of years. Yet, these *species* have survived for millions of years—the products of continual reproduction. Reproduction also supplies the means by which variability among individuals in an animal population is generated and is the basis for biological evolution.

The Nervous And Endocrine Systems: Regulating Bodily functions

The multitude of physiologic processes that proceed simultaneously within an animal must be continually monitored and regulated. If, for example, oxygen levels in the body's tissues drop, mechanisms are triggered that increase the uptake of oxygen from the environment. Or, if an animal is confronted by a potential predator, a protective response is triggered that increases the animal's chance for survival. Two types of systems—the nervous and endocrine systems —regulate and coordinate bodily functions.

- *Nervous systems* are networks of nerve cells *(neurons)* that receive information concerning changes in the external and internal environment, integrate the information, and send out directives to the body's muscles and glands to respond in an appropriate manner. Information moves along the pathways of the nervous system in the form of impulses traveling along the plasma membranes of the nerve cells. These impulses provide a mechanism by which all parts of the body can rapidly communicate with one another. Information enters the nervous system via a series of *sensory* structures that detect changes in the external and internal environment.

- *Endocrine systems* also regulate and coordinate many of an animal's internal activities. Endocrine systems consist of a disconnected network of glands that release chemical messengers (hormones) into the blood. Hormones circulate through the bloodstream and ultimately interact with their particular target cells, triggering specific responses. While responses triggered by the nervous system tend to occur rapidly, those triggered by the endocrine system occur more slowly and often include metabolic changes. Endocrine responses include changes in the level of glucose in the blood, metamorphosis of a caterpillar into a butterfly, and sexual maturation (Figure 22-7).

THE COST OF RUNNING THE BODY'S BUSINESS

The operation of each physiologic system requires the expenditure of energy. An animal takes in chemical energy that is stored in food. Energy-containing macromolecules enter the digestive system, where they are disassembled; the energy-rich products are then absorbed into the body, where they are made available to all of the body's cells. The biggest consumers of energy are the muscular and nervous systems. Even when you are at rest, muscular contraction is at work, pumping blood through your vessels, food through your digestive tract, and air into your lungs. Maintaining an elevated body temperature can be the most costly of all activities, consuming approximately 90 percent of the energy you expend when you are at rest. If your physical activity should increase, muscle activity takes a much larger share of your energy supply.

As a group, animals have a common set of physiologic needs, which are met by organ systems that fall into distinct categories, according to function. The structure of the organs, their mechanism of action, and their organization into systems is highly varied among diverse animals. (See CTQ #4.)

THE EVOLUTION OF ORGAN SYSTEMS

Much of what we know about the evolutionary relationship among animals is derived from studies of fossil remains left behind by individuals living millions of years ago. Fossils almost invariably consist of the hardened skeletal parts of

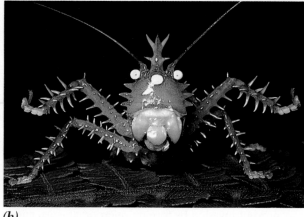

(a)   (b)

FIGURE 22-6

A gallery of animal appendages. *(a)* A giant Pacific octopus with a shark caught in the suction grip of its tentacles. *(b)* An inhabitant of the South American rain forest, this katydid exhibits a number of appendages including legs, antenna, and even its mouthparts which are derived from modified embryonic appendages. *(c)* The wings of this African cape gannet are appendages used for flight. *(d)* The appendages of a chimpanzee are similar in structure and function to our own arms and legs.

ancient animals; they reveal very little direct information about any of the other systems introduced in this chapter. Consequently, our knowledge about the evolution of most organ systems is based largely on comparative studies of living animals.

■▶ This approach always raises the question of origin. Are the similarities in an organ between two distantly related living animals due to the fact that the organ evolved from a similar organ present in a common ancestor (in which case, the organs are said to be *homologous*) or from independent courses of evolution? Two organs with similar functions often resemble each other, even though they are not de-

rived from a common ancestral organ, because they have evolved in response to similar types of selective pressures. In such cases, the organs are said to be *analogous*. For example, the excretory organs of both a human and a lobster consist of microscopic tubules. While the tubules may look superficially similar in both organisms, they have evolved independently to meet a similar physiologic need. This phenomenon is an example of "convergent evolution" (Chapter 33).

When biologists compare the organ systems of invertebrates—animals lacking backbones—they often find a progression in anatomic complexity from simpler

(a)   (b)

FIGURE 22-7

The result of hormones. This emperor fish is transformed from a juvenile *(a)* to a sexually mature adult *(b)* as the result of hormones secreted by the fish's gonads.

(c) *(d)*

animals, such as sea anemones and flatworms, to more complex animals, such as beetles and octopuses. For example, the nervous system of a sea anemone consists of a diffuse net of nerve cells, whereas that of an octopus includes a large, complex brain that coordinates the animal's intricate behavior.

A comparison of most organ systems in vertebrates—animals that possess backbones—provides some of the clearest insight into how natural selection can lead to the modification of structures to meet different physiologic challenges. In some cases, a particular structure has undergone a transformation from one form and function to another. For example, the bones situated just behind your

eardrum are derived from bones that formed part of the jaws of your vertebrate ancestors. The evolutionary movement of vertebrates from the water onto the land required certain changes in the way sound vibrations were transmitted from the environment to the sensory receptors in the ear. Fortuitously, one of the bones used to support the jaws of fishes was no longer needed as a jaw brace in the early amphibians. Instead, this bone became "pressed into service" as an ear bone (Figure 22-8a). The other two bones in the middle ear of mammals are derived from bones that were previously part of the jaws of our amphibian ancestors (Figure 22-8b). We can see from this description that, during the course of vertebrate evolution, these bones under-

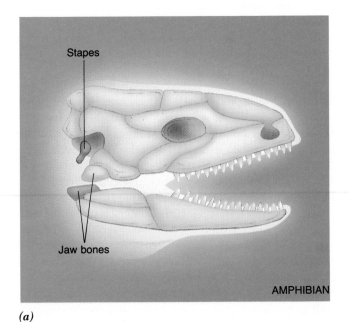

(a)

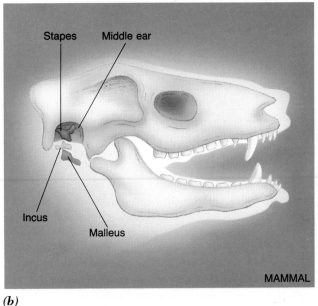

(b)

FIGURE 22-8

The evolution of the bones of the ear. *(a)* The middle ear of amphibians contains a single bone, the stapes, which can be traced to a bone in the skull of ancestral fishes. *(b)* Mammals have three bones in the middle ear—the incus, malleus, and stapes (often called the anvil, hammer, and stirrup, respectively). The incus and malleus have evolved from bones that are present in the upper and lower jaws of our amphibian ancestors.

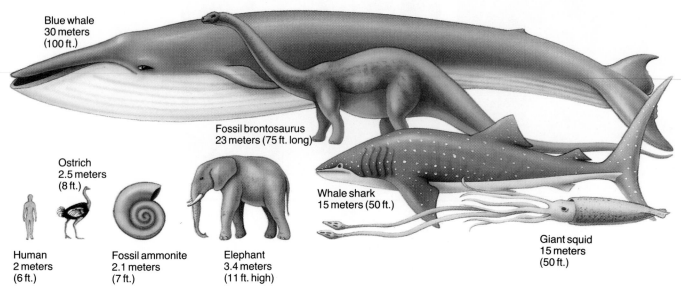

FIGURE 22-9

Animals are found in many different sizes.

went a dramatic change in position, shape, and function. What were once food-gathering structures evolved into transmitters of sound waves.

The characteristics of each organ system within an animal can be understood by considering the animal's ancestry and the environmental challenges with which the system must cope. (See CTQ #6.)

BODY SIZE, SURFACE AREA, AND VOLUME

Multicellular animals span a range of size more than five orders of magnitude (100,000 times), from microscopic, aquatic rotifers (100 micrometers long) to the blue whale (30 meters long). As biologists measured physiologic activities in animals of widely different dimensions (Figure 22-9), it became apparent that the levels of these activities did not increase in a proportional manner. For example, when an elephant and a mouse are compared *per gram of body weight*, the elephant utilizes less oxygen, eats less food, and has a greater skeletal mass than does the mouse. These lack of proportionalities are known as **scaling effects.**

Many scaling effects in biology can be explained on the basis of unequal changes in surface area and volume as animals increase or decrease in size. Surface area, which might be measured in units from *square* micrometers to *square* meters, is the area that covers the outside of the body. Volumes of these same animals would then be mea-

sured in units ranging from *cubic* micrometers to *cubic* meters. (We have italicized the words "square" and "cubic" to emphasize that the surface area of an animal is a function of the square of the animal's length, width, and height, while the volume is a function of the cube of these dimensions.) As the dimensions of an object—whether it be a simple sphere or a complex animal—increase, the surface area of the object increases to less of a degree than does its volume (Figure 22-10). In other words, the larger the animal, the smaller the **surface area/volume ratio (SA/V).** This relationship has very important physiologic consequences.

An animal's surface is the boundary between itself and its external environment. All materials and energy (food, water, respiratory gases, waste products, heat) that pass between the animal and its environment must cross the body surface, which includes internal surfaces, such as the digestive tract and the lungs, as well as the external skin. The rate at which this exchange can occur is directly proportional to the surface area available to be crossed. In contrast, the mass, or weight, of an animal is proportional to its volume; the greater the volume, the more food and oxygen the animal needs, and the more waste materials it produces.

Imagine, for a moment, that an animal, let's say a frog, absorbs most of its oxygen across its body surface. As the frog grows larger, its *need* for oxygen will increase in proportion to its volume, while its *ability to absorb* oxygen will only increase in proportion to its surface area. As a result, the frog's need for oxygen will outstrip its ability to provide this vital substance. As we will see later on, larger animals have evolved special adaptations to counteract this scaling problem. For example, the respiratory organs of large ani-

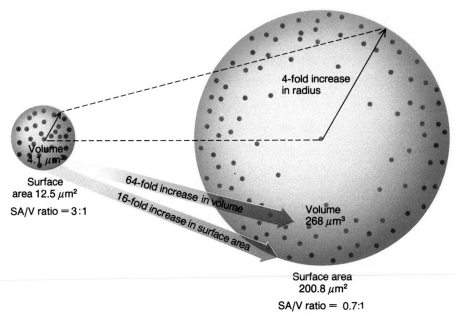

FIGURE 22-10

Relationship between surface area and volume. Quadrupling the size of a sphere increases its surface area 16 times and its volume 64 times. Suppose the spheres were single-celled organisms—floating protozoa, for example. The plasma membrane of the larger cell would have to provide nutrients and oxygen (shown as blue dots) for a mass of living protoplasm that was 64 times greater than the protoplasm of the smaller cell, but through a surface only 16 times as large. Without special mechanisms, the large cell would not be able to maintain the same concentration of nutrients and oxygen as would the smaller cell.

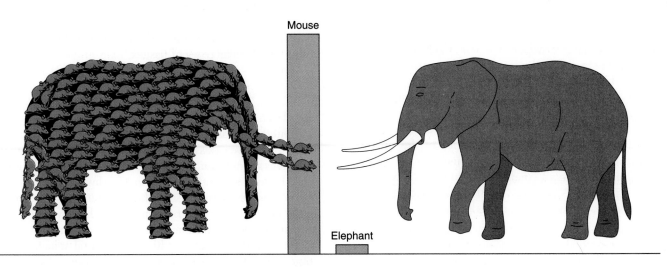

FIGURE 22-11

The mouse and the elephant. An average mouse weighs about 25 grams, compared to an average elephant that weighs nearly 4 million grams. Because of their difference in size, the elephant has a much smaller surface area/volume. This difference is reflected in the metabolic rates (volume of oxygen consumed per gram of body weight per hour) of these animals—as indicated by the height of the bars, the metabolic rate of the mouse is 24 times greater than that of the elephant.

mals contain extensive folds, or partitions, that greatly increase the area across which oxygen can be absorbed. Similarly, the lining of a large animal's digestive tract contains folds and projections that greatly increase the surface area available for absorbing nutrients. Reducing surface area may also be adaptive. For example, people tend to curl up or huddle together with friends when cold, in an attempt to decrease the surface area across which heat is lost.

But why does an elephant eat less food and utilize less oxygen *per gram of body weight* than does a mouse (Figure 22-11)? At least part of the explanation can be traced to the fact that elephants have a much smaller surface area/volume ratio than do mice. Most of the energy that a mouse or elephant obtains from its food is utilized to maintain its high body temperature. The loss of body heat, however, depends on the body's surface area; on a cool night, a small mouse will lose its body heat much more rapidly than will a large elephant. Consequently, the mouse must take in *relatively* more calories in its food and produce relatively more heat (as measured by its oxygen consumption) than does the elephant.

The level of various physiologic activities exhibited by different animals is often understood when their relative surface area/volume ratio is considered. (See CTQ #7.)

REEXAMINING THE THEMES

Relationship between Form and Function

As we will see throughout this section of the book, the structures of various tissues, organs, and organ systems are closely correlated with their functions. For example, an epithelium may separate compartments of different solute concentration because this tissue consists of a continuous sheet of tightly adhering cells. Similarly, connective tissue can function to protect and support parts of the body because it contains extracellular materials that have strength, hardness, elasticity, and/or resiliency.

Biological Order, Regulation, and Homeostasis

The basic function of most organ systems is to maintain the constancy of the internal environment. Each system has a particular role to play in this overall "drama," from maintaining a stable, elevated temperature, to keeping the salt and water level of body fluids at a relatively constant concentration, to eliminating wastes, to protecting the body from foreign invasion. Two systems—the nervous and endocrine systems—coordinate the activities of the other organ systems. The nervous and endocrine systems must collect information about conditions within the body and send out appropriate messages to the various effectors, whose responses maintain the ordered state.

Acquiring and Using Energy

All of the various organ systems require energy to fuel their activities. Virtually every activity—from contracting a muscle to thinking about the answer to a question on an exam—requires energy. Chemical energy is derived from the food we eat, which must be disassembled by our digestive system. The products of digestion are then absorbed through the lining of our digestive tract and carried by the circulatory system to the sites where the energy is needed.

Unity within Diversity

Animals exhibit remarkably diverse sizes, shapes, and internal body plans, yet they display an equally remarkable physiologic unity. Virtually all animals have the same types of organ systems, even if these systems have evolved independently. The presence of similar organ systems reflects the fact that all animals have similar needs that are basic to maintaining life. Similarly, even though animals possess a diverse array of cells, these cells are organized into a small number of similar types of tissues.

Evolution and Adaptation

Organ systems are adapted to the particular challenges presented by an animal's environment. For example, fishes living in the sea possess an osmoregulatory system that maintains water balance in the face of water loss by osmosis, while their close relatives living in a freshwater stream possess an osmoregulatory system that maintains water balance in the face of water gain by osmosis. Each body part of an animal has evolved from a structure that was present in an ancestor and was then modified by natural selection. In some cases, a structure can undergo dramatic changes in form and function, as organisms adapt to new conditions. For example, the movement of vertebrates from water to land was accompanied by a shift in certain bones from the jaws to the middle ear.

SYNOPSIS

Multicellular animals must conduct three critical functions in order to maintain a homeostatic state. The body must (1) receive information from sensors that detect changes; (2) pass the information to a control unit, typically located in the brain, which is "set" to maintain a certain value; and (3) send out signals to effectors (muscles and glands) to initiate a response that restores conditions to that which is set by the control unit. Maintaining a property at a constant level requires negative feedback. In certain cases, homeostasis requires that a change in the body is amplified, rather than reversed. This can be accomplished by positive feedback, as illustrated by the rapid formation of a blood clot, which prevents further loss of blood.

Multicellular animals have similar types of organ systems that meet common needs. These include a digestive system, which disassembles food matter and provides the body with nutrients; an excretory/osmoregulatory system, which eliminates metabolic waste products and maintains a balanced concentration of salt and water; a respiratory system, which absorbs oxygen and often expels carbon dioxide; a circulatory system, which transports materials from place to place in the body; an immune system, which protects against foreign substances and invading microorganisms; an integumentary system, which covers the animal; skeletal and muscular systems, which facilitate support, movement, and locomotion; and a reproductive system, which produces sperm and eggs.

The nervous and endocrine systems control bodily activities. These two systems coordinate the activities of the other organ systems by collecting information about conditions in both the internal and external environment and providing signals to effectors that carry out a particular response.

Although diverse in form and function, cells are organized into four basic types of tissues. Epithelial tissues consist of sheets of cells that act as linings. Their functions include protection, exchange, and secretion. Connective tissues consist of cells surrounded by a nonliving extracellular matrix. As skeletal materials, connective tissues provide support and facilitate movement; as ligaments and tendons, they connect parts of the body and resist stretching; as the cornea, they provide a transparent layer for vision; and as the blood, they distribute materials from place to place. Muscle tissue provides the force for movement of materials within the body and movement of attached skeletal elements. Nerve tissue forms a communication network that transmits information used in coordinating bodily activities.

Animals with markedly different surface area/volume ratios face different physiologic challenges. As an animal (or series of animals) increases in size, its surface area increases as a function of the square of its linear dimensions, while its volume increases as a function of the cube of those dimensions. Consequently, surface area/volume ratio decreases as the size of an animal increases. Exchange of substances between the animal and its environment are determined by the body's surface area. Larger animals require special adaptations of their respiratory and digestive tracts in order to provide sufficient surface area to facilitate the absorption of the large amounts of oxygen and nutrients needed to sustain life. Smaller, warm-blooded animals tend to gain and lose heat more rapidly than do larger ones.

Key Terms

negative-feedback mechanism (p. 450)
positive-feedback mechanism (p. 450)
tissue (p. 451)
epithelial tissue (p. 452)

epithelium (p. 452)
connective tissue (p. 453)
muscle tissue (p. 454)
nerve tissue (p. 455)

organ (p. 455)
organ system (p. 455)
scaling effect (p. 460)
surface area/volume ratio (p. 460)

Review Questions

1. Describe the basic components your body employs to maintain homeostasis. Describe these components in connection with the maintenance of a constant body temperature.

2. Compare and contrast analogous and homologous organs; negative- and positive-feedback mechanisms; tissues and organs; connective tissue and epithelial tissue; and the functions of the digestive system and the excretory system.

3. Contrast the basic mechanism of operation of the nervous and endocrine systems in regulating bodily functions.

4. Consider an animal whose shape is a perfect cube. When it hatched, the animal measured 1 centimeter along each side; when it was fully grown, it measured 10 centimeters along each side. What are the differences in surface area and volume between the animal's hatched and fully grown state? How does its surface area/volume ratio differ at these two stages of life? What physiologic changes would you expect to find as the animal grows?

Critical Thinking Questions

1. In the Steps to Discovery vignette, several activities were mentioned that have the potential to disrupt the internal stability of our bodies but are kept from doing so by homeostatic mechanisms. Can you think of any other activities in which you engage that have this potential? Explain your answer.

2. When the concentration of carbon dioxide in your blood increases, the pH decreases. This stimulates the respiratory center in your brain. You then breathe faster and deeper, expel more carbon dioxide from your lungs, and the pH increases. In this manner, your body maintains a relatively constant blood pH. Identify the following parts of this system: sensor, control unit, effectors. Is this an example of negative or positive feedback? Explain.

3. You are cutting an apple with a sharp knife and accidentally cut your finger deeply. Which of the four types of tissues did the knife pass through? Support your answer.

4. How does the organization of flowering plants compare with that of higher animals, such as the human? What is there in the life styles of flowering plants and higher animals that could account for differences? What tissues in plants are analogous to the four animal tissue types? Do flowering plants have organ systems?

5. Select any five of the organ systems surveyed in this chapter and discuss the effects on your body if one of these systems should suddenly stop operating.

6. Discuss the different types of challenges that would face an animal living on land compared to one living in the ocean and the adaptations required in the excretory/osmoregulatory, skeletal/muscular, and respiratory systems of both animals to meet those challenges.

7. According to Jonathan Swift, the author of *Gulliver's Travels*, once Gulliver landed in Lilliputia and became accepted by the tiny emperor of that land, it was deemed that he should be provided with a daily allowance of meat and drink sufficient for the support of 1,728 Lilliputians. This value was determined by mathematicians, who calculated Gulliver's volume as equal to 1,728 times their own. Was this the appropriate diet for Gulliver? Why or why not?

Additional Readings

Benison, S., A. C. Barger, and E. Wolfe. 1987. *Walter B. Cannon: The life and times of a young scientist.* Cambridge, MA: Harvard University Press. (Introductory)

The following are intermediate–advanced level histology and physiology texts that describe all of the organ systems covered in this section.

Eckert, R. 1988. *Animal physiology: Mechanisms and adaptations.* 3rd ed. New York: W. H. Freeman.

Fawcett, D. 1986. *Bloom and Fawcett: A textbook of histology,* 11th ed. Philadelphia: Saunders.

Fox, S. I. 1984. *Human physiology.* Dubuque, IA: W. C. Brown.

Guyton, A. C. 1992. *Human physiology and mechanisms of disease.* Philadelphia: Saunders.

Vander, A. J., J. H. Sherman, and D. S. Luciano. 1990. *Human physiology: The mechanisms of body function,* 5th ed. New York: McGraw-Hill.

Weiss, L., Ed. 1988. *Cell and tissue biology: A textbook of histology,* 6th ed. Baltimore: Williams & Wilkins.

CHAPTER

23

Coordinating the Organism: The Role of the Nervous System

STEPS TO DISCOVERY

A Factor Promoting the Growth of Nerves

NEURONS AND THEIR TARGETS

Neurons: Form and Function

GENERATING AND CONDUCTING NEURAL IMPULSES

The Membrane Potential of a Neuron "At Rest"

Action Potentials: Triggered by a Reduction in Membrane Potential

NEUROTRANSMISSION: JUMPING THE SYNAPTIC CLEFT

Neurotransmitters: Chemicals that Carry the Message

Synapses: Sites of Information Integration

THE NERVOUS SYSTEM

The Reflex Arc: Shortening Reaction Time

More Complex Circuits

ARCHITECTURE OF THE HUMAN CENTRAL NERVOUS SYSTEM

The Brain

The Spinal Cord

ARCHITECTURE OF THE PERIPHERAL NERVOUS SYSTEM

The Somatic and Autonomic Divisions of the Peripheral Nervous System

THE EVOLUTION OF COMPLEX NERVOUS SYSTEMS FROM SINGLE-CELLED ROOTS

BIOLINE

Deadly Meddling at the Synapse

THE HUMAN PERSPECTIVE

Alzheimer's Disease: A Status Report

BIOETHICS

Blurring the Line Between Life and Death

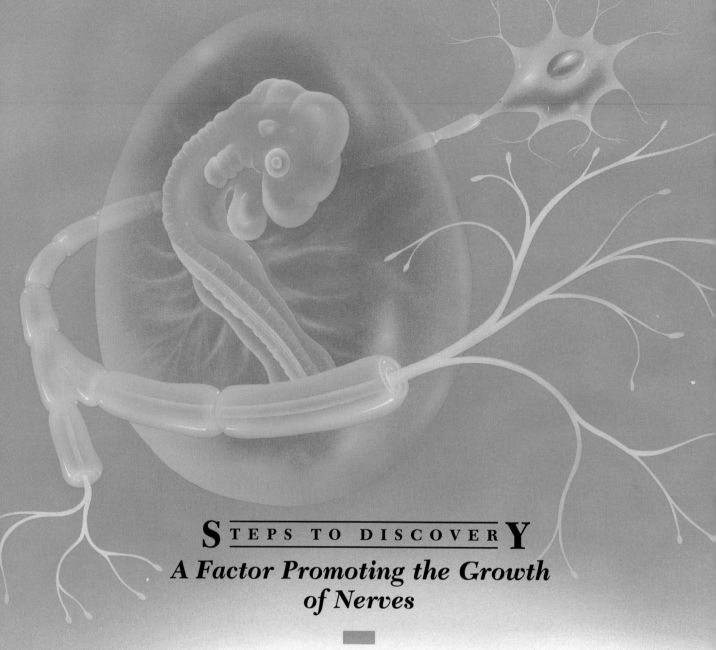

A Factor Promoting the Growth of Nerves

*R*ita Levi-Montalcini received her medical degree from the University of Turin in Italy in 1936, the same year that Benito Mussolini began his anti-Semitic campaign. By 1939, as a Jew, Levi-Montalcini had been barred from carrying out research and practicing medicine, yet she continued to do both secretly. As a student, Levi-Montalcini had been fascinated with the structure and function of the nervous system. Unable to return to the university, she set up a simple laboratory in her small bedroom in her family's home. As World War II raged throughout Europe, and the Allies systematically bombed Italy, Levi-Montalcini studied chick embryos in her bedroom, discovering new information about the growth of nerve cells from the spinal cord into the nearby limbs. In her autobiography *In Praise of*

Imperfection, she writes: "Every time the alarm sounded, I would carry down to the precarious safety of the cellars the Zeiss binocular microscope and my most precious silver-stained embryonic sections." In September 1943, German troops arrived in Turin to support the Italian Fascists. Levi-Montalcini and her family fled southward to Florence, where they remained in hiding for the remainder of the war.

After the war ended, Levi-Montalcini continued her research at the University of Turin. In 1946, she accepted an invitation from Viktor Hamburger, a leading expert on the development of the chick nervous system, to come to Washington University in St. Louis to work with him for one semester; she remained at Washington University for 30 years.

A chick embryo and one of its nerve cells helped scientists discover nerve growth factor (NGF).

One of Levi-Montalcini's first projects was the reexamination of a previous experiment of Elmer Bueker, a former student of Hamburger's. Bueker had removed a limb from a chick embryo, replaced it with a fragment of a mouse connective tissue tumor, and found that nerve fibers grew into this mass of implanted tumor cells. When Levi-Montalcini repeated the experiment she made an unexpected discovery: One part of the nervous system of these experimental chick embryos—the sympathetic nervous system—had grown five to six times larger than had its counterpart in a normal chick embryo. (The sympathetic nervous system helps control the activity of internal organs, such as the heart and digestive tract.) Close examination revealed that the small piece of tumor tissue that had been grafted onto the embryo had caused sympathetic nerve fibers to grow "wildly" into all of the chick's internal organs, even causing some of the blood vessels to become obstructed by the invasive fibers. Levi-Montalcini hypothesized that the tumor was releasing some soluble substance that induced the remarkable growth of this part of the nervous system. Her hypothesis was soon confirmed by further experiments. She called the active substance **nerve growth factor (NGF).**

The next step was to determine the chemical nature of NGF, a task that was more readily performed by growing the tumor cells in a culture dish rather than an embryo. But Hamburger's laboratory at Washington University did not have the facilities for such work. To continue the project, Levi-Montalcini boarded a plane, with a pair of tumor-bearing mice in the pocket of her overcoat, and flew to Brazil, where she had a friend who operated a tissue culture laboratory. When she placed sympathetic nervous tissue in the proximity of the tumor cells in a culture dish, the nervous tissue sprouted a halo of nerve fibers that grew toward the tumor cells. When the tissue was cultured in the absence of NGF, no such growth occurred.

For the next 2 years, Levi-Montalcini's lab was devoted to characterizing the substance in the tumor cells that possessed the ability to cause nerve outgrowth. The work was carried out primarily by a young biochemist, Stanley Cohen, who had joined the lab. One of the favored approaches to studying the nature of a biological molecule is to determine its sensitivity to enzymes. In order to determine if nerve growth factor was a protein or a nucleic acid, Cohen treated the active material with a small amount of snake venom, which contains a highly active enzyme that degrades nucleic acid. It was then that chance stepped in.

Cohen expected that treatment with the venom would either destroy the activity of the tumor cell fraction (if NGF was a nucleic acid) or leave it unaffected (if NGF was a protein). To Cohen's surprise, treatment with the venom *increased* the nerve-growth promoting activity of the material. In fact, treatment of sympathetic nerve tissue with the venom alone (in the absence of the tumor extract) induced the growth of a halo of nerve fibers! Cohen soon discovered why: The snake venom possessed the same nerve growth factor as did the tumor cells, but at much higher concentration. Cohen soon demonstrated that NGF was a protein.

Levi-Montalcini and Cohen reasoned that since snake venom was derived from a *modified* salivary gland, then other salivary glands might prove to be even better sources of the protein. This hypothesis proved to be correct. When Levi-Montalcini and Cohen tested the salivary glands from male mice, they discovered the richest source of NGF yet, a source 10,000 times more active than the tumor cells and ten times more active than snake venom.

A crucial question remained: Did NGF play a role in the normal development of the embryo, or was its ability to stimulate nerve growth just an accidental property of the molecule? To answer this question, Levi-Montalcini and Cohen injected embryos with an antibody against NGF, which they hoped would inactivate NGF molecules wherever they were present in the embryonic tissues. The embryos developed normally, with one major exception: They virtually lacked a sympathetic nervous system. The researchers concluded that NGF must be important during normal development of the nervous system; otherwise, inactivation of NGF could not have had such a dramatic effect.

By the early 1970s, the amino acid sequence of NGF had been determined, and the protein is now being synthesized by recombinant DNA technology. During the past decade, Fred Gage, of the University of California, has found that NGF is able to revitalize aged or damaged nerve cells in rats. Based on these studies, NGF is currently being tested as a possible treatment of Alzheimer's disease. For their pioneering work, Rita Levi-Montalcini and Stanley Cohen shared the 1987 Nobel Prize in Physiology and Medicine.

*I*magine that you are sitting at your desk reading a book when, out of the corner of your eye, you catch a glimpse of a dark furry-looking object resting on your right forearm. If you are like most people, within a second you will have flicked the object off your arm with the back of your left hand. You will then get out of your chair, bend over, and try to get a better look at the insect or spider you presume to have wounded. When you find the object, you identify it as a clump of thread that must have come loose from the sweater you're wearing. You smile, realize your heart is beating a little faster, sit back down, and return to your book.

The events we have just described—the glimpse of an object, the "instantaneous" determination of a threat, the quick muscular response by your hand, the curiosity as to the nature of the object, the increase in your heart rate, the humor you find in the events that have occurred, and the desire to return to your book—are all a direct result of activities taking place in your nervous system. So too is the visualization of these events in your "mind's eye."

This one brief sequence of events points out many of the functions of the nervous system. Most importantly, the nervous system communicates information from one part of the body to another and, in so doing, regulates the body's activities and maintains homeostasis. Your very survival depends on continuous neural activity. For example, without orders from the nervous system, you would be unable to contract the muscles that draw air into your lungs; you would be unable to activate your sweat glands to release fluids needed to lower your body temperature; and you would be unable to chew, swallow, or send food along your digestive tract.

The nervous system also controls your basic drives (such as hunger, thirst, and sexual desires) and emotional responses (including anger and fear). Consciousness itself is derived from neural activity. Despite the recent advances in cell and molecular biology, we are still very far from understanding the underlying neural basis of thought, learning, memory, perception, and behavior. One thing is certain: The operation of the nervous system will remain a fertile ground for investigation for many years to come.

▼   ▼   ▼

NEURONS AND THEIR TARGETS

Each nerve cell, or **neuron,** is specialized for conducting messages, in the form of moving *impulses,* from one part of the body to another. Messages can be sent along these cellular "transmission lines" at speeds of over 100 meters per second (225 miles per hour). The effect of the impulse depends on two properties: the nature of the neuron, and the type of target cell that responds to the neuron.

- *Nature of the neuron.* Some neurons, called **excitatory neurons,** stimulate their target cells into activity; others, called **inhibitory neurons,** oppose a response, encouraging target cells to remain at rest.
- *Nature of the target cell.* Only three basic types of cells can respond *directly* to an arriving impulse. They are

1. muscle cells, which respond to excitatory stimulation by contracting, thereby exerting force;
2. gland cells, which respond to excitation by secreting a substance;
3. other nerve cells, which may generate impulses of their own, thereby relaying the message to another target cell.

NEURONS: FORM AND FUNCTION

Describing the form of a "typical" nerve cell is like trying to describe a "typical" human personality. Your body contains more than 100 billion neurons, and no two are exactly alike. Nonetheless, all neurons are composed of the same basic parts (Figure 23-1), which allow them to collect, conduct, and transmit impulses.

The form of a neuron is readily correlated with its function. The nucleus of the neuron is located within an expanded region, called the **cell body,** which is the metabolic center of the cell and the site where most of its material contents are manufactured. Extending from the cell bodies of most neurons are a number of miniscule extensions, called **dendrites,** which receive *incoming* information from external sources, typically other neurons. Also emerging from the cell body is a single, more prominent extension, the **axon,** which conducts *outgoing* impulses away from the cell body and toward the target cell. Impulses are generally initiated in the region where the cell body merges into the axon. While some axons may be only a few micrometers in length, others extend for considerable distance. The neurons that carry impulses from a giraffe's spinal cord to its legs, for example, may extend 3 meters, placing them among the longest cells in the animal world.

Most axons split near their ends into smaller processes, each ending in a **synaptic knob**—a specialized site where impulses are transmitted from neuron to target cell. Some cells in your brain may end in thousands of synaptic knobs, allowing these brain cells to communicate with thousands of potential targets.

Neuroglia: The Supporting Cast

Only about 10 percent of the cells in the human nervous system are neurons; the rest are **neuroglial cells.**

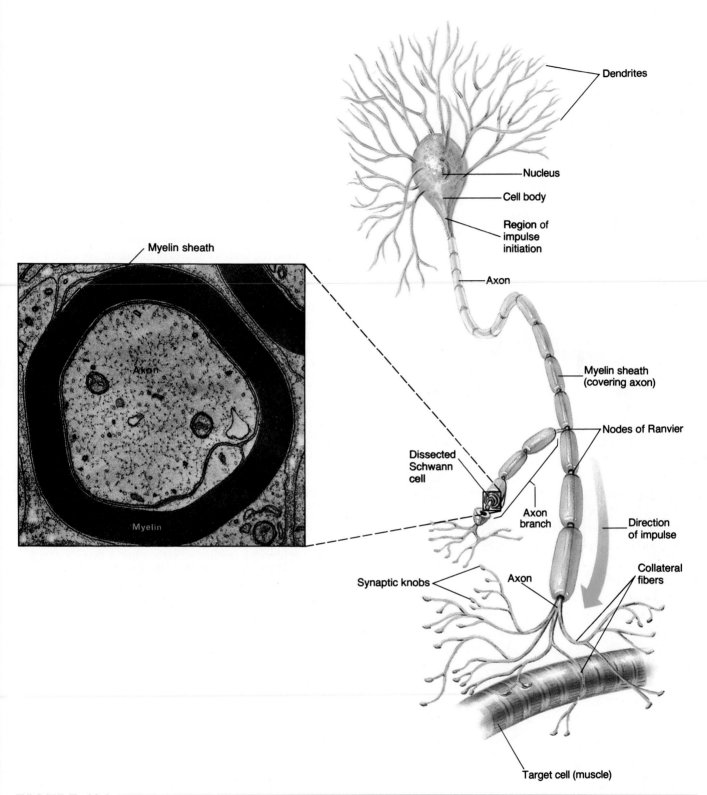

Myelin sheath

Dendrites

Nucleus

Cell body

Region of impulse initiation

Axon

Myelin sheath (covering axon)

Nodes of Ranvier

Dissected Schwann cell

Axon branch

Direction of impulse

Synaptic knobs

Axon

Collateral fibers

Target cell (muscle)

Axon

Myelin

FIGURE 23-1

Anatomy of a neuron. Information enters the nerve cell through a branching network of dendrites. The incoming signals impinge on the cell body, which contains the nucleus and the cell's synthetic machinery. The cell body merges with the elongated axon, which often branches into numerous smaller processes, each of which ends in a synaptic knob. Most vertebrate axons are wrapped in a myelin sheath composed of Schwann cells. The inset shows an electron micrograph of a cross section through an axon surrounded by a myelin sheath.

These "accessory cells" include **Schwann cells,** which consist almost wholly of plasma membrane, with very little cytoplasm (see inset, Figure 23-1). These cells wrap themselves around axons, forming a layered jacket of cell membranes, called a **myelin sheath.** Since cell membranes consist predominantly of lipids, which are poor conductors of electricity, the myelin sheath functions as living "electrical tape," insulating the axon against electrical interference from its neighbor. The insulation is not complete, however; tiny, naked gaps remain exposed between adjacent Schwann cells. These uninsulated gaps, called the **nodes of Ranvier,** help speed nerve impulses along an axon by a "leap-frog" mechanism that enables the impulses to skip from gap to gap. This phenomenon is discussed in more detail later in the chapter.

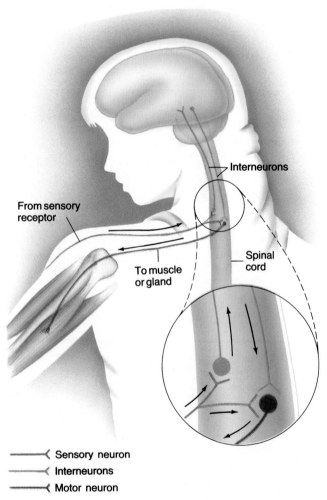

— ⊰ Sensory neuron
— ⊰ Interneurons
— ⊰ Motor neuron

FIGURE 23-2 ──────────

Three types of neurons. Sensory neurons carry impulses from the periphery (sensory receptors) to the CNS (as indicated by the spinal cord). The various pathways that run up and down the human brain and spinal cord are composed of large numbers of interneurons, which are located entirely within the brain or spinal cord. Motor neurons carry impulses from the CNS to effector cells in the periphery.

Classification of Neurons According to Function

Nerve cells can be grouped into three classes—sensory neurons, interneurons, and motor neurons—depending on the direction impulses are carried and the type of cells to which they are functionally linked (Figure 23-2). **Sensory neurons** carry information about changes in the external and internal environments *toward* the central nervous system (the brain and spinal cord). Once they enter the central nervous system (CNS), impulses from sensory neurons are transmitted to **interneurons,** which transmit impulses from one part of the CNS to another. Interneurons route incoming or outgoing impulses, integrating the millions of messages constantly racing through the CNS. Outgoing impulses are carried by **motor neurons,** which stretch from the CNS to the body's muscles or glands whose contraction or secretion may be stimulated or inhibited.

Neurons are highly specialized, cellular communication systems, equipped for receiving, conducting, and transmitting information. Neurons regulate biological function by stimulating or inhibiting actions in target cells. (See CTQ #2.)

GENERATING AND CONDUCTING NEURAL IMPULSES

Propagation of an impulse along a neuron is often compared to the conduction of a pulse of electricity along a wire. But this analogy fails to take into account basic differences: Electricity is a flow of electrons along a wire, but a nerve impulse involves neither electrons nor a flow of charged particles along an axon. Rather, the nerve impulse occurs as the result of the movement of ions *across* the plasma membrane, rather than along its length. Why do ions move across the plasma membrane? And how can ionic movement across the membrane at one point lead to an impulse that speeds along the entire length of a neuron to an awaiting target cell? The answers to these questions will be revealed as we compare a "resting" neuron to one that has been triggered to conduct an impulse.

THE MEMBRANE POTENTIAL OF A NEURON "AT REST"

The concentrations of specific ions on the two sides of the plasma membrane of a resting neuron—one that is not conducting an impulse—are very different (Figure 23-3). The concentration of potassium ions (K^+) is approximately 30 times higher inside the cell than outside, while the concentration of sodium (Na^+) and chloride (Cl^-) ions are approximately 10 to 15 times higher outside the cell than inside. (These *ionic gradients* are established by the sodium–potassium pump, discussed on page 142.) You might expect that as a result of their concentration gradients, potassium ions would diffuse out of the cell and sodium ions would diffuse inward. But the ability of ions to move across a membrane is not automatic; rather, it depends on the permeability of the cell membrane. Recall from Chapter 7 that ions move across the plasma membrane through specific channels. Nerve cell membranes have two types of channels: *leak channels,* which are always open, and *gated channels,* which can be either open or closed. In the resting state, potassium ions diffuse out of a cell through potassium leak channels. In contrast, since the nerve cell lacks sodium leak channels, the plasma membrane is virtually impermeable to sodium ions.

Potassium ions are positively charged. The movement of positive charges out of the cell leaves the inside of the membrane more negatively charged than the outside. This separation of positive and negative charge is called a **potential difference,** or **voltage.** The voltage across a cell membrane can be measured by inserting microscopic electrodes into a nerve cell. The **resting potential**—the potential difference when the cell is at rest—measures approximately − 70 millivolts. The negative value, due largely to the outward diffusion of potassium ions, indicates the negativity of the inside of the cell, relative to the outside. In the resting state, the membrane is said to be *polarized.*

ACTION POTENTIALS: TRIGGERED BY A REDUCTION IN MEMBRANE POTENTIAL

Physiologists first learned about membrane potentials in the 1930s from studies on the giant axons of the squid. These axons, which are approximately 1 millimeter in diameter, carry impulses at high speeds, enabling the squid to escape rapidly from predators. If the membrane of a resting squid axon is stimulated by poking it with a fine needle or jolting it with a very small electric current, the axon responds by opening the gates of some of its sodium channels, allowing a number of sodium ions to move into the cell. This movement of positive charges into the cell reduces the

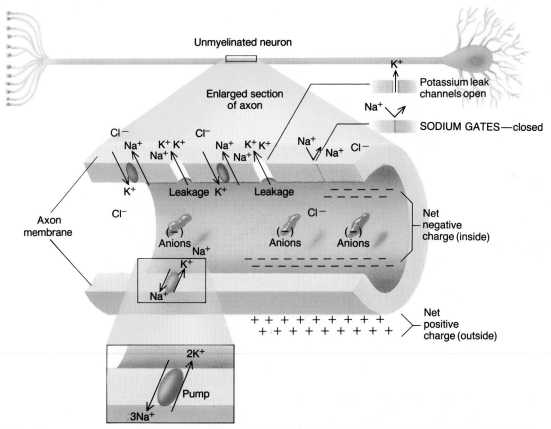

FIGURE 23-3

Formation of a membrane potential in an unmyelinated, "resting" neuron. An ATP-driven pump maintains a steep ionic gradient across the membrane by pumping sodium ions (Na⁺) out of the cell in exchange for potassium ions (K⁺), which are pumped inward (the lower square shows the details of the pump). The membrane of the resting neuron is permeable to potassium and impermeable to sodium (insets at upper right). As a result, positively charged potassium ions diffuse out of the cell through potassium leak channels, resulting in a separation of charge (a *membrane potential*), with the inside of the cell negative relative to the outside. The negativity of the inside of the cell is maintained largely by negatively charged anions, including protein molecules, bicarbonate ions, and phosphate groups.

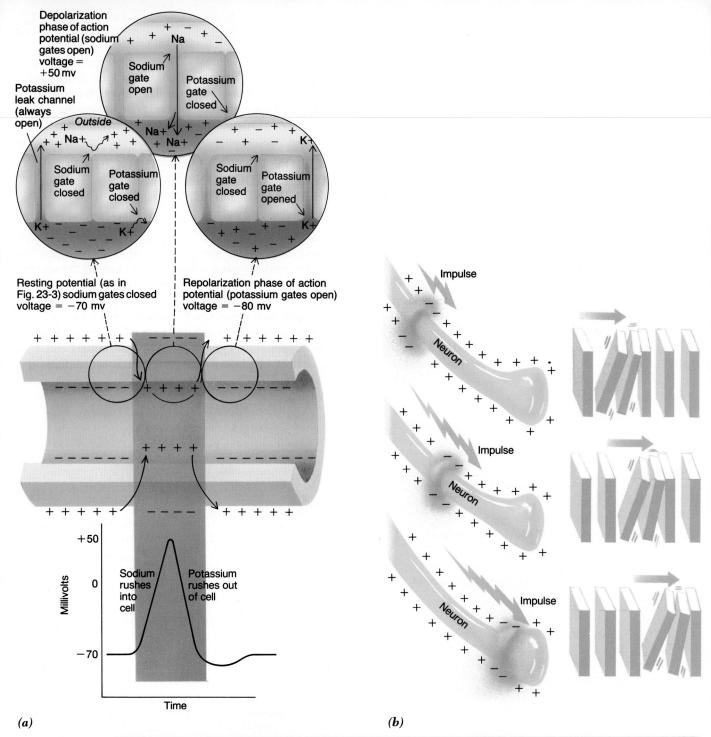

FIGURE 23-4

Formation and propagation of an action potential. *(a)* When depolarization of the membrane exceeds the threshold value, the membrane's sodium gates open and allow positively charged sodium ions to move into the cell. The influx of sodium causes a fleeting reversal in the polarity of the membrane potential, typically from the resting value of − 70 millivolts to + 50 millivolts. This charge reversal is known as an action potential. Within a brief fraction of a second, the sodium gates close and the potassium gates open, allowing potassium to diffuse across the membrane and reestablish an even more negative potential at that location (− 80 millivolts) than that of the resting potential. Almost as soon as they open, the potassium gates close, leaving the potassium leak channels as the primary path of ion movement across the membrane, and reestablishing the resting potential. *(b)* **Propagating an action potential as an impulse.** The disruption in polarity of the membrane that accompanies the action potential at one site triggers the opening of the sodium gates at an adjacent site, initiating an action potential farther along the axon. This process repeats itself from point to point, causing an impulse to race along the axon to the end of the neuron, much like falling dominoes, but with one important difference: Each "domino" along the neuron rights itself immediately after falling.

membrane voltage, making it less negative. Since the reduction in membrane voltage causes a decrease in the polarity between the two sides of the membrane it is called a *depolarization*.

If the stimulus causes the membrane to depolarize by only a few millivolts, say from -70 to -60 millivolts, the membrane rapidly returns to its resting potential as soon as the stimulus has ceased (Figure 23-3; 23-4*a* left circle). If the stimulus is great enough, however, the membrane becomes depolarized beyond a certain point, called the **threshold.** When this happens, a new series of events is launched. The change in voltage causes the gates on the sodium channels to swing open, and sodium ions flood into the cell (Figure 23-4*a*, middle circle). As a result of the inflow of sodium ions, the membrane potential briefly reverses itself (Figure 23-4*a*, lower plot), becoming a positive potential of about $+50$ millivolts. Then, as the membrane potential reaches its peak positive value, the sodium gates close again, and the gated potassium channels open (Figure 23-4*a*, right circle). As a result, potassium ions flood out of the cell, and the membrane potential swings back to a negative value (Figure 23-4*a*, lower plot). The large negative potential causes the gated potassium channels to close, leaving only the potassium leak channels open. As a result, the membrane is *repolarized* and returns to its resting state. These collective changes in membrane potential are called an **action potential;** the entire sequence occurs in a few milliseconds (thousandths of a second).

The movements of ions across the plasma membrane of nerve cells forms the basis for neural communication. Certain *local* (topically applied) anesthetics, such as Xylocaine and Novocaine, the anesthetic used primarily by dentists, act by closing the gates of ion channels in the membranes of nerve cells. As long as these ion channels remain closed, the affected neurons are unable to generate action potentials and cannot inform the brain of the painful insults being experienced by your gums or teeth. The next time you are in a dentist's chair listening to the sound of the drill, think of the millions of plugged ion channels in the sensory neurons leading from the roots of your teeth to your brain.

Propagation of Action Potentials as an Impulse

The action potential that occurs at one point along the neuron is the "spark" that creates a neural impulse. Like falling dominoes, the propagation of action potentials along the entire length of the axon is the result of a chain reaction (Figure 23-4*b*). An action potential at one site of a membrane induces a depolarization at the adjacent site of the membrane farther along the neuron, initiating an action potential at that site. This new action potential stimulates the next region of the membrane, and the chain reaction continues, producing a wave of action potentials that travels along the entire length of the excited neuron. Once a wave of action potentials is triggered, the wave passes down the entire length of the neuron without any loss of intensity, arriving at its target cell with the same strength it had at its point of origin.

Speed Is of the Essence

When you accidentally stumble, the only thing that keeps your face from hitting the floor is your ability to thrust your foot or hands forward fast enough to arrest your forward plunge—faster than gravity can plaster you against the ground. The speed of your response is even more remarkable when you realize that neural impulses have to travel to the central nervous system (notably, the spinal column) so they can be routed to the arm and leg muscles that prevent you from losing your race against gravity. Think of how often the speed of your response has saved you from personal injury—jumping out of the path of an oncoming car, maneuvering your car when you are cut off by another vehicle, withdrawing your hand from a hot object fast enough to prevent serious injury. The speed of all these actions depends on how fast a nerve impulse reaches a muscle and stimulates it to contract. The speed an impulse travels depends on (1) the diameter of the axon, and (2) whether or not the axon is jacketed in a myelin sheath.

In general, the larger a neuron's diameter, the faster the neuron conducts impulses. This is the adaptive advantage of giant axons in squids and other invertebrates; the larger diameter of these animals' axons increases the speed at which the animals can escape danger. Vertebrate evolution improved on this utility with the adaptation of the myelin sheath. Composed almost entirely of lipid-containing membranes, the myelin sheath is ideally suited to preventing the passage of ions across the plasma membrane. As a result, action potentials can only occur in the unwrapped nodes between the Schwann cells (the nodes of Ranvier). An action potential at one node is strong enough to trigger another action potential at the next node. Consequently, a nerve impulse skips along myelinated neurons from node to node, rather than taking the slower continuous route along the membrane. This "hopping" type of propagation is called **saltatory conduction,** after the Latin "saltare," meaning "to leap." Impulses are able to travel along a myelinated axon at speeds up to 120 meters per second, which is nearly 20 times faster than the speed of impulses of an unmyelinated neuron of the same diameter.

The importance of myelination is dramatically illustrated by multiple sclerosis, a disease that results from the gradual deterioration of the myelin sheath that surrounds axons in various parts of the nervous system. The disease usually begins in young adulthood; victims will experience weakness in their hands, difficulty in walking, and/or problems with their vision. The disease is characterized by progressive muscular dysfunction, often culminating in permanent paralysis.

Because of the ionic gradients generated across its plasma membrane, a resting neuron is always primed and ready. A slight depolarizing stimulus may trigger a self-propagating wave of activity that sweeps down the length of the axon, undiminished in intensity. This wave constitutes a nerve impulse—the mechanism of communication within the nervous system. (See CTQ #3.)

NEUROTRANSMISSION: JUMPING THE SYNAPTIC CLEFT

Neurons are linked with their target cells at specialized junctions called **synapses.** Careful examination of a synapse reveals that the two cells do not make direct contact but are separated from each other by a narrow gap of about 20 to 40 nanometers. This gap is called the **synaptic cleft** (Figure 23-5). Somehow the presynaptic neuron's impulse must "jump" across this cleft in order to affect the postsynaptic target cell.

The first indication that neural messages are carried across the synaptic cleft by chemicals came from an ingenious experiment conducted by Otto Loewi in 1921 (Figure 23-6). The design of the experiment (for which Loewi was awarded the Nobel Prize) came to the scientist in a dream. The heart rate of a vertebrate is regulated by the balance of input from two opposing (antagonistic) nerves, each consisting of a large number of neurons. Loewi isolated a frog's heart together with both nerves. When he stimulated the inhibitory (*vagus*) nerve, a chemical was released from the heart preparation into a salt solution, which was allowed to drain into a second, isolated heart. The rate of the second heart slowed dramatically, as though its own inhibitory nerve had been activated. The substance responsible for inhibiting the frog's heart (and the human heart) was later identified as *acetylcholine,* the first *neurotransmitter* to be discovered.

NEUROTRANSMITTERS: CHEMICALS THAT CARRY THE MESSAGE

Acetylcholine is only one of a number of different chemicals (Table 23-1) that act as **neurotransmitters**—molecules released from neurons, stimulating or inhibiting target cells. These substances, stored in numerous membrane-bound packets called **synaptic vesicles** found inside the synaptic knob, have no influence as long as they remain

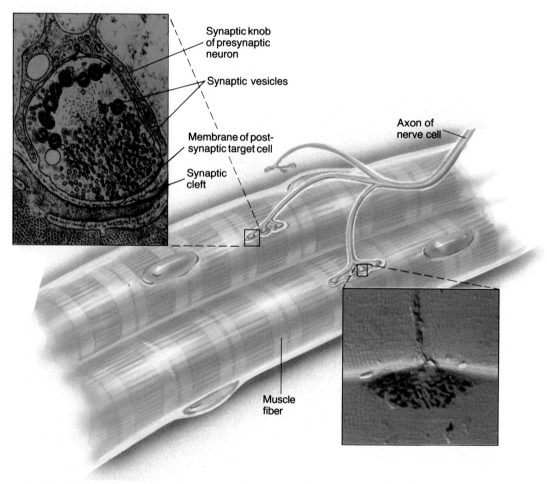

Synaptic knob
of presynaptic
neuron

Synaptic vesicles

Membrane of post-
synaptic target cell

Synaptic
cleft

Axon of
nerve cell

Muscle
fiber

FIGURE 23-5

Synaptic junction between a neuron and an effector cell. Each synaptic knob abuts the target cell membrane very closely; the synaptic vesicles within the knob are indicated in the micrograph in the upper-left box.

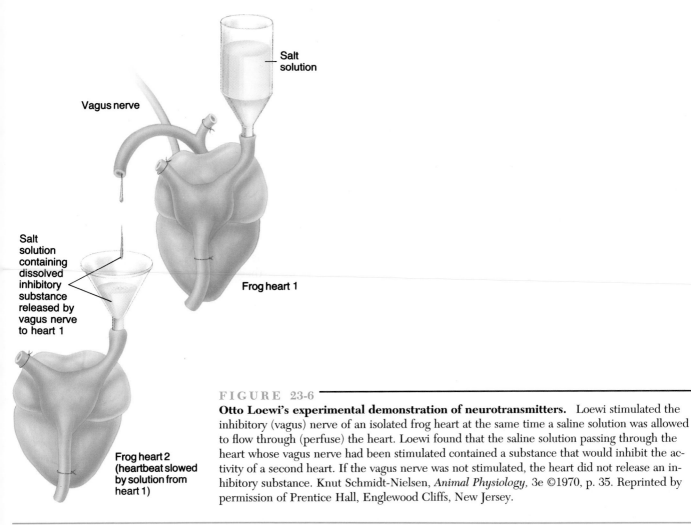

Salt
solution

Vagus nerve

Salt
solution
containing
dissolved
inhibitory
substance
released by
vagus nerve
to heart 1

Frog heart 1

Frog heart 2
(heartbeat slowed
by solution from
heart 1)

FIGURE 23-6

Otto Loewi's experimental demonstration of neurotransmitters. Loewi stimulated the inhibitory (vagus) nerve of an isolated frog heart at the same time a saline solution was allowed to flow through (perfuse) the heart. Loewi found that the saline solution passing through the heart whose vagus nerve had been stimulated contained a substance that would inhibit the activity of a second heart. If the vagus nerve was not stimulated, the heart did not release an inhibitory substance. Knut Schmidt-Nielsen, *Animal Physiology,* 3e ©1970, p. 35. Reprinted by permission of Prentice Hall, Englewood Cliffs, New Jersey.

TABLE 23-1

SOME NEUROTRANSMITTERS AND THEIR EFFECTS

Neurotransmitter	*+ or −*[a]	*Most Common Target Cells*	*Predominant Effect*
Acetylcholine	+ −	Voluntary muscles Heart muscle	Stimulates muscle contraction. Increases threshold of contraction.
Glycine	−	Motor neurons to voluntary muscles	Raises threshold of excitation, checking uncontrolled muscle contraction.
Dopamine	−	Neurons that produce acetylcholine	Prevents overactivity of neurons that activate muscles. (Deficiencies result in uncontrolled muscle contractions of Parkinson's disease.)
Norepinephrine (noradrenaline)	+	Neurons of central nervous system responsible for arousal, attention, and mood; involuntary muscles (e.g., heart); glands	Increases alertness and attention; heightens readiness for muscular activity.
GABA	−	Motor neurons to voluntary muscles	Prevents uncontrolled muscle contraction.
Serotonin	−	Neurons in the brain that maintain wakefulness	Induces sleep; may modulate mood.

[a] "+" = excitatory; "−" = inhibitory.

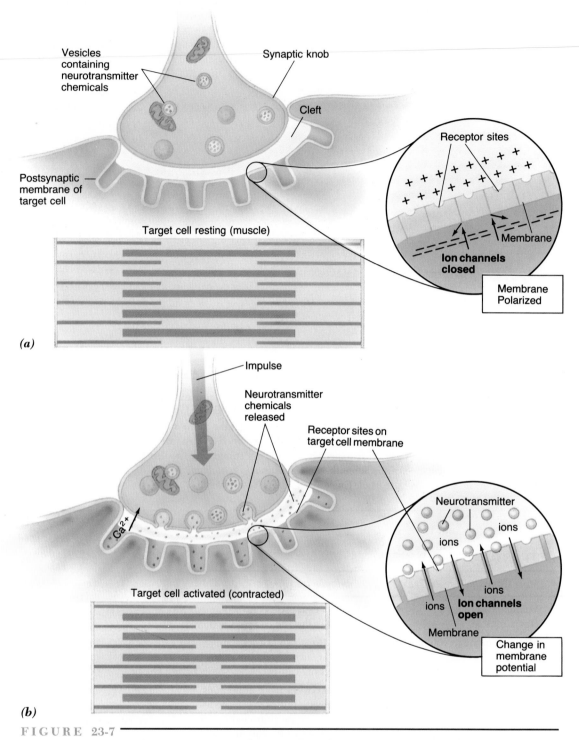

FIGURE 23-7

Synaptic transmission between a neuron and target cell (in this case a muscle fiber). *(a)* Neuron at rest. The synaptic cleft contains no neurotransmitter, the receptor sites remain empty, and the target remains unaffected by the neuron. *(b)* When the impulse reaches the tip of a neuron, calcium ions enter the synaptic knob, neurotransmitter molecules are released and bind to receptor proteins on the postsynaptic neuron, sodium ion channels swing open, and the muscle fiber contracts.

packaged in the neuron (Figure 23-7*a*). When an impulse reaches the end of a presynaptic neuron, however, it triggers an opening of calcium ion channels in that part of the membrane, allowing calcium ions to flow *into* the synaptic knobs of the cell (Figure 23-7*b*). Calcium ions are a potent inducer of exocytosis, the discharge of materials out of a cell (page 99). As calcium ions flow into the cell, the membranes of a number of synaptic vesicles fuse with the plasma membrane, and the vesicles spill their contents of neurotransmitter molecules into the synaptic cleft. The discharged neurotransmitter molecules then diffuse across the narrow gap and bind specifically to *receptor proteins* in the membrane of the postsynaptic target cell. The interaction between a neurotransmitter and its specific receptor can have one of two opposite effects depending on the target cell:

1. it can decrease the membrane potential (*depolarize* the membrane), which will *excite* the target cell, making it more likely to respond, as in Figure 23-7, or

2. it can increase the membrane potential (*hyperpolarize* the membrane) which will *inhibit* the target cell, making it less likely to respond.

If you are running to class, for example, neurons stimulate the muscles in your legs to contract by releasing acetylcholine, which *decreases* the membrane potential of the voluntary muscle cells in your leg, making it easier to excite the cells. A different neurotransmitter, *norepinephrine*, is released by nerves that stimulate heart contraction.

If the synaptic story ended here, the target cell would be unable to recover from the chemical message. Neurochemical excitation at the synapse would lock an activated target into a perpetually excited state. This is not the case, however. Target cells are prevented from maintaining a perpetual state of excitation in at least two ways: by enzymes that destroy neurotransmitter molecules almost as soon as they react with their receptors, and by enzymes that transport neurotransmitter molecules back to the neuron that orginally released them—a process called *reuptake*. Because of the destruction and/or reuptake of neurotransmitter molecules, the effect of each impulse lasts no more than a few milliseconds; order is maintained.

Neurobiologists have discovered over 30 different chemicals that act as neurotransmitters, some of which are described in Table 23-1. Most of these neurotransmitters act within the brain alone and, as we will discuss later in the chapter, some neurotransmitters, such as dopamine, can have dramatic effects on our emotional state.

Unfortunately, neurons do not always function properly. Things can go dreadfully wrong at the synapse. When the synaptic cleft is occupied by chemicals that interfere with neurotransmitters, for example, the resulting disorder can lead to paralysis or even death (see Bioline: "Deadly Meddling at the Synapse").

SYNAPSES: SITES OF INFORMATION INTEGRATION

Synaptic transmission does not operate on the basis of "one impulse in, one impulse out." In fact, a single impulse transmitted by a single neuron rarely initiates a response in a target cell because it fails to exceed the target's threshold. Activating a target cell requires a number of excitatory signals which, added together (*summated*), exceed the threshold. Summation can result from (1) simultaneous arrival of multiple stimuli from several adjacent neurons (Figure 23.8) or (2) a virtual "nonstop" barrage of impulses from just one neuron. If every neuron that spontaneously "goes off" were to generate a response from its target, the resulting chaos would disrupt the body's homeostasis. It would be similar to a car alarm's sensitivity being set so high that every little breeze would activate the siren.

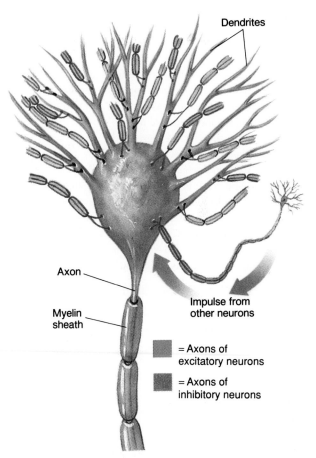

Dendrites

Axon

Myelin sheath

Impulse from other neurons

= Axons of excitatory neurons

= Axons of inhibitory neurons

FIGURE 23-8

Multiple synapses on a single neuron can run into the thousands. Whether or not the neuron is activated and relays the message depends on whether or not the combined stimulation of the excitatory and inhibitory neurons exceeds the membrane threshold of the target neuron.

◁ B I O L I N E ▷
Deadly Meddling at the Synapse

Nerve poisons! Few substances have acquired such notorious reputations as this group of chemicals, and none deserves the reputation more. Tiny quantities of these substances can sabotage the transmission of impulses across synaptic junctions, interfering with essential motor activity. A number of mechanisms lead to these neurological disasters; the following neurotoxins exemplify just a few.

CURARE

For centuries, hunters in South America have enhanced their predatory skills by dipping their weapons in extracts prepared from tropical plants. For example, a small dart from a blowgun becomes a missile of death when coated with curare, a powerful nerve poison contained in one such botanical extract. Curare blocks the acetylcholine receptor sites on the membrane of muscle cells. As a result, muscles are paralyzed in a state of relaxation and fail to contract in response to neural commands, even though the synaptic cleft is saturated with acetylcholine. Death from suffocation occurs quickly once the muscles used for breathing stop functioning. Curare is now used under clinically-controlled conditions (under the name *Tubocurarine*) as a mus-

cle relaxant to prevent muscle damage that can occur from overcontraction as the result of tetanus or during electroconvulsive shock therapy.

BOTULISM

The bacterium *Clostridium botulinum* releases one of the most potent toxins known; a few ounces is enough to kill every person on earth. A person who eats food containing this toxin (usually a result of improperly canned food) is attacked by a poison that spreads through the bloodstream from the intestine to all the neurons in the body. The toxin prevents the release of acetylcholine from motor neurons. Its effect is therefore similar to curare, only 50 times more potent. The symptoms of botulism include double vision, loss of coordination, and eventual fatal paralysis.

ORGANOPHOSPHATES AND NERVE GAS

Some neurotoxins block the enzymatic destruction of acetylcholine so that the neurotransmitter remains in the synapse, preventing the muscles from returning to their relaxed state once stimulated. Victims suffer sustained paralysis; their muscles re-

main in a state of permanent contraction —just the opposite of the cause of death by curare. Nerve gas exerts its lethal effect in this way, as do organophosphate pesticides, which can be as deadly to humans as they are to the insects for which they are intended.

TETANUS

Another type of paralysis occurs if a wound becomes infected by a common soil bacterium. This disease is known as *tetanus*. As the bacteria grow in the wound, they release a powerful neurotoxin that seeps into the bloodstream and blocks inhibitory synapses on motor neurons throughout the body. The removal of inhibitory influences on these neurons creates a hyperexcitable state; consequently, even low levels of excitatory impulses set off unchecked muscle spasms throughout the body. Although the jaw muscles are the first to be affected (which is why tetanus is commonly known as "lockjaw"), the toxin eventually strikes all of the body's voluntary muscles. Muscular spasms in the back bend the spine into an exaggerated arch, which can snap the backbone. The victim eventually loses the ability to breathe and dies of asphyxiation.

Such hypersensitization leading to accidental activation of target cells is further prevented by the presence of inhibitory neurons and their neurotransmitters, which increase the threshold; that is, they decrease the response sensitivity of the target. Most target cells are influenced by impulses from hundreds of different excitatory and inhibitory neurons (Figure 23-8). If the combined input from all the incoming synapses is enough to trigger a response, the cell performs its activities. Until then, the target cell remains at rest.

Taken as a group, synapses are more than just connecting sites between adjacent neurons; they are key determi-

nants in the orderly routing of impulses throughout the nervous system. The billions of synapses that exist in a complex mammalian nervous system act like gates stationed along the various pathways; some pieces of information are allowed to pass from one neuron to another, while other pieces are held back or rerouted in some other direction. Such synaptic integration allows us to focus on the book we are reading or the music we are listening to, while simultaneously ignoring all of the distracting background noise with which we are constantly bombarded. We will return to the importance of synapses when we discuss the basis of learning and memory later in the chapter.

Now that we have described the form and function of neurons and the way they transmit information to other cells, we will examine how nerve cells are organized into more complex neural structures.

Neurons accomplish their regulatory functions by passing information—in the form of chemical transmitter substances—to other cells, across synaptic clefts. The interaction between neurotransmitters and membrane receptors then initiates or inhibits a response in the target cell. Synapses determine the paths by which information travels through the nervous system. (See CTQ #4.)

THE NERVOUS SYSTEM

The nervous system of a vertebrate is divided into two major divisions: the central nervous system and the peripheral nervous system. The **central nervous system (CNS)** consists of two major parts: the **brain,** which is the center of neural integration, and the **spinal cord,** which contains billions of neurons that run to and from the brain and also mediates many of the body's reflex responses. All neurons, or parts of neurons, situated outside the central nervous system are part of the **peripheral nervous system,** "peri" meaning "around the edge," as in *peri*meter. The peripheral nervous system connects the various organs and tissues of the body with the brain and spinal cord. The neurons of the peripheral nervous system are grouped into **nerves**—"living cables" composed of large numbers of individual neurons bundled together in parallel alignment, together with their supporting cells (Figure 23-9). All incoming and outgoing impulses are routed through the CNS, which functions as a centralized "command and control center." The simplest example of neural centralization is the reflex arc.

THE REFLEX ARC: SHORTENING REACTION TIME

The fastest motor reactions to stimuli are reflex responses. A **reflex** is an involuntary response to a stimulus—a response that occurs "automatically" and requires no conscious deliberation or awareness of the stimulus. Reflex responses occur so rapidly because the impulse travels the shortest route possible: from the site of the stimulus, through the central nervous system, to the responding effector. The chain of neurons that mediate a reflex make up a **reflex arc.** One of the simplest reflex arcs—that which is responsible for your foot jumping forward when the doctor strikes the area below your knee with a rubber hammer—is illustrated in Figure 23-10*a*.

A reflex arc begins with a **sensory receptor**—a cell that responds to a change in its environment. In many cases, the receptor is the tip of a sensory neuron that is specialized to respond to a stimulus. For example, the *stretch receptor* responsible for the knee-jerk response is a sensory neuron whose end is wrapped around a muscle fiber embedded within a muscle of the thigh. When the tendon of the knee is tapped, the attached muscle is stretched, activating the receptor, which generates a neural impulse. Some of the terminal processes of the sensory neuron end in the spinal cord (Figure 23-10*a*) in synapses with motor neurons leading directly back to the thigh muscle. Impulses traveling back along these motor neurons cause the muscle to contract, and the leg extends forward.

The *stretch reflex* just described didn't evolve to help doctors evaluate the state of your nervous system. Rather, the reflex is an adaptive response that helps you maintain your posture and balance. The same stretch reflex that keeps you upright also works for the mountain goats pictured in Figure 23-10b.

Your day is full of adaptive reflex responses. For example, an overambitious sip of steaming coffee triggers a reflex activation of the muscles in your tongue, jaw, and mouth.

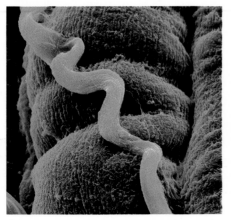

(a)

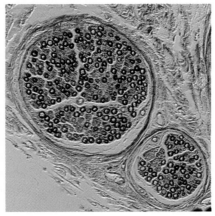

(b)

FIGURE 23-9

Nerves. *(a)* A nerve winding across the surface of muscle fibers. *(b)* In this cross section, the individual neurons in the nerve are seen bound together in a sheath of connective tissue.

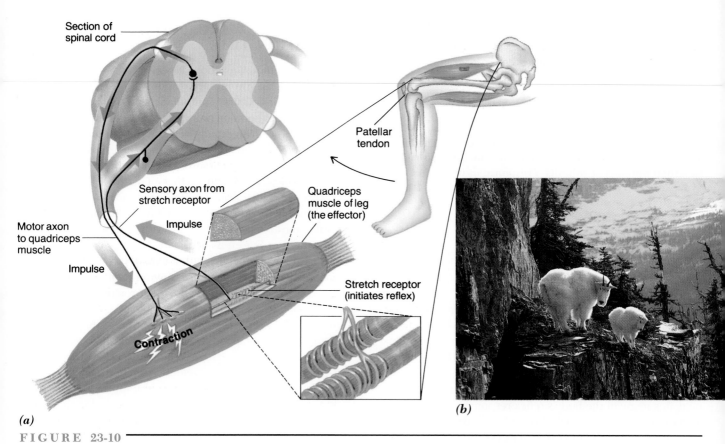

(a)

(b)

<u>FIGURE 23-10</u>

The reflex arc that mediates the knee-jerk response. *(a)* A reflex arc includes a receptor, which generates a nerve impulse after it is activated by some change in the environment; a sensory neuron, which carries the impulse to the CNS; a motor neuron, which sends a command signal back to the periphery; and an effector (either a muscle or gland), which provides the response. In the case of the knee-jerk response, the receptor is actually the end of a sensory neuron, which is wrapped around a special muscle fiber situated within the quadriceps muscle of the thigh. When the muscle fiber is stretched, it activates the sensory neuron, sending impulses to the spinal cord. In this stretch reflex, impulses are transmitted directly from the sensory neuron to the motor neuron, causing the quadriceps muscle to contract and extend the leg forward. Unlike this stretch reflex, most reflexes include one or more interneurons in the arc. *(b)* These mountain goats depend on stretch reflexes to maintain their footing on steep trails.

The tongue automatically jumps to the back of the mouth, which may open and discharge the liquid before your brain can consider the reaction of the other people at the table. Other reflexes include regulation of blood pressure, control of pupil size in response to changes in light intensity, and withdrawal of the hand when it encounters a sharp or hot object. Reflexes are involuntary responses that occur very rapidly in a *stereotyped* manner; that is, the reflex is the same every time the same simple stimulus is encountered.

MORE COMPLEX CIRCUITS

Not all of the sensory neurons from the stretch receptors in the thigh muscle terminate at a motor neuron which returns to the muscle. Many sensory neurons synapse with interneurons of the spinal cord. These interneurons carry impulses upward into the brain, where we perceive the stimulus (such as the hammer striking our knee) and send commands to our muscles. Our ability to stand upright, for example, depends on sensory input reaching the brain from many different muscles. Similarly, the motor neurons that contract the thigh muscle are not activated solely by a simple reflex arc. Rather, a single motor neuron may be covered with synaptic knobs from thousands of different interneurons. As a result, motor neurons leading to the thigh muscle can be activated by many unrelated activities, such as willingly kicking a football or automatically jumping out of the way of an oncoming truck.

Each neuron is a link in a chain, joined to other links by synapses. Neurons are organized into functional circuits that allow information to be routed in meaningful ways through the billions of cells that make up a complex nervous system. The simplest circuits consist of reflex arcs, whereby sensory information can direct motor activity without the participation of higher neural centers. (See CTQ #5.)

ARCHITECTURE OF THE HUMAN CENTRAL NERVOUS SYSTEM

The central nervous system performs the most complex neural functions. It collects information about the internal and external environment, "sorts" through the information, relays impulses of its own along different pathways to various parts of the brain and spinal cord, and then acts on the information by sending command messages to peripheral effectors. The human CNS is the most complex, highly organized structure found on earth. We will now take a closer look at the two main components of the human CNS: the brain and the spinal cord.

THE BRAIN

Although the brain constitutes only about 2.5 percent of your body weight, it consumes 25 percent of your body's oxygen supplies while generating the ATP needed to fuel its activities. If not replenished, the brain's oxygen content would be exhausted in about 10 seconds. Within only a few minutes, the damage to the brain would be irreversible.

The human brain (Figure 23-11) is a mass of nearly 1.5 kilograms (3 pounds) of gelatinlike tissue. It contains a darker outer region—**gray matter**—in which the cell bodies and dendrites of the brain's neurons reside. Gray matter is rich in neuron-to-neuron synapses, places where

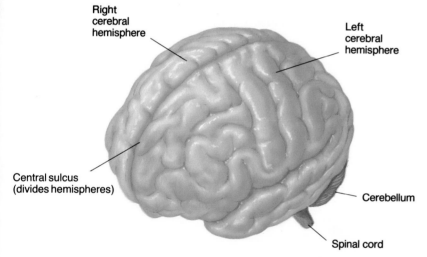

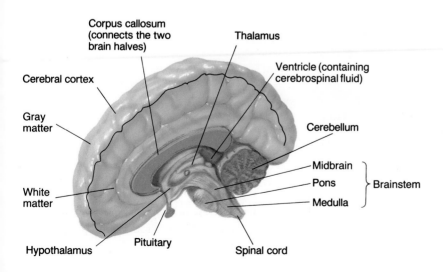

FIGURE 23-11

Two views of the brain. The intact human brain reveals mostly cerebrum, which is divided into a right and left cerebral hemisphere. Bisecting the brain reveals some of the brain's structural complexity. The medulla, pons, and midbrain constitute the brainstem, which controls visceral functions, such as breathing and cardiovascular activity. The thalamus is primarily a way station for sensory information that passes to the cerebrum, while the hypothalamus is one of the brain's homeostatic control centers. The right and left halves of the cerebral cortex are connected by a thick mass of nerve fibers that form the corpus callosum. The entire brain is covered by the meninges and bony cranium (not shown).

neural associations are made. The inner region of the brain consists largely of **white matter** composed of myelinated axons; its whitish cast is provided by the light-colored myelin sheaths, which insulate the neurons. The tissue of the brain surrounds a series of distinct but interconnected chambers, called **ventricles,** which are filled with a protein-rich liquid, the **cerebrospinal fluid,** which cushions and nourishes the brain. This fluid also surrounds the delicate brain (and spinal cord), cushioning it against injury. The brain and its surrounding fluid is encased in a complex, protective, watertight sheath, called the *meninges,* which, in turn, is enclosed within a protective, bony case, the *cranium.*

The brain can be divided in various ways, depending on different criteria. For our purposes, we will restrict the discussion to three functional groups: (1) the cerebrum, (2) the cerebellum, and (3) a series of interrelated networks that form the brainstem, limbic system, and reticular formation.

The Cerebrum

The cerebrum is the most prominent part of the human brain (Figure 23-11). Its two halves, called **cerebral hemispheres,** are generally associated with "higher" brain func-

tions, such as speech and rational thought. Actually, these functions are attributes of the **cerebral cortex,** the outer, highly wrinkled layer of the cerebrum (*cortex* = rind). Every cubic inch of this thin layer of gray matter contains 10,000 miles of interconnecting neurons. The convolutions (wrinkles) in the cerebrum increase the surface area of the cortex without enlarging the space required to house it. Convolutions in the cerebral cortex are presumed to be an evolutionary sign of higher cerebral capabilities, such as intellect; artistic and creative abilities; and a greater capacity for language, learning, and memory, compared to those of the smooth-brained, lower vertebrates, such as amphibians and reptiles.

Each cerebral hemisphere is composed of four lobes —temporal, frontal, occipital, and parietal—each of which has a unique set of functions (Figure 23-12). The two cerebral hemispheres are connected by the *corpus callosum*—a thick cable made up of hundreds of millions of neurons— which allows the left and right hemispheres to communicate with each other. As we will discuss in the Human Perspective box in Chapter 24, most of the sensory information from the right side of the body is transmitted through the corpus callosum to the left cerebral hemisphere, and vice versa.

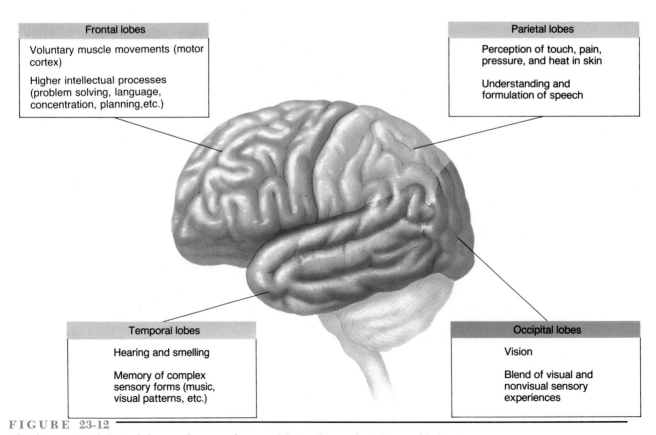

Frontal lobes

Voluntary muscle movements (motor cortex)

Higher intellectual processes (problem solving, language, concentration, planning,etc.)

Parietal lobes

Perception of touch, pain, pressure, and heat in skin

Understanding and formulation of speech

Temporal lobes

Hearing and smelling

Memory of complex sensory forms (music, visual patterns, etc.)

Occipital lobes

Vision

Blend of visual and nonvisual sensory experiences

FIGURE 23-12

The four major lobes of the cerebrum and some of their roles in physiology and behavior.

Memory Imagine, for a moment, that you are standing in front of the entrance to your former high school. You walk through the doors, through the halls, and into the room where you took one of your more memorable classes. Your former teacher is standing in the front of the class, and you can hear the sound of her voice. A minute earlier, none of this was on your mind, and now you have "directed" your brain to recall a series of specific, ordered images about which you can consciously reminisce. How is this possible? Scientists cannot yet answer this question, but they can provide some interesting insights. The types of images you have just brought to "mind" can also be evoked by electrical stimulation of various parts of the cerebral cortex. When this procedure is performed, the subject may see an image of something they thought they had long forgotten, or hear a voice, or smell an aroma. The information is "stored" in the cerebral cortex, probably in the pattern of neural circuits formed by the billions of neurons that make up this portion of the brain.

Scientists have also learned about memory by studying individuals who have suffered specific brain injuries that have affected their memory. During the 1950s, a young man (known as H. M.) came to the attention of Brenda Milner of the Montreal Neurological Institute. A portion of the man's brain had been removed during surgery in an attempt to stop severe seizures. The operation had unexpected and tragic results. H. M. was able to remember events from his childhood and adolescence, but he was unable to memorize any new information. For example, if H. M. were given a number to remember, he could do so only as long as he focused all of his attention on it. As soon as he lost his attention, he would forget the number, as well as any recollection of ever being told about it. Over and over, H. M. would greet as strangers the same doctors and researchers he worked with on a daily basis. Perhaps most tragic, H. M. was not able to grasp the reason for his problem because he could not remember what he was told.

The case of H. M. revealed for the first time the dual nature of human memory. H. M. could remember events from the past, similar to your having remembered your high-school classroom, because his *long-term memory* remained basically intact. In contrast, *short-term memory* allows us to retain a piece of information only for a matter of seconds or minutes. Short-term memory can be considered "working memory" since we use it to carry out our daily activities. Remembering a phone number from a telephone book until we get a chance to write it down is an example of short-term memory. As we go about our activities, certain pieces of information from our short-term memory become processed subconsciously in some unknown way so that they become part of our long-term memory. Short- and long-term memory apparently involve different types of neural processes and different parts of the brain.

Studies on H. M.—and others with similar conditions—have indicated that short-term memory is processed in the temporal lobes of the cerebrum and the hippocampus, a part of the limbic system discussed below. These are the same regions of the brain that degenerate in individuals with Alzheimer's disease (see The Human Perspective: Alzheimer's Disease: A Status Report). Like H. M., Alzheimer's victims often vividly remember events from the distant past but are unable to remember what has happened to them a few minutes earlier. Long-term memory appears to be more diffusely located in the cerebrum, as evidenced by the fact that stimulation to many different parts of the cerebral cortex will evoke memories of past events.

Learning is a change in behavior that results from prior experience. Learning occurs as the nervous system processes sensory stimuli and uses the information to make changes in the nervous system that lead to new types of responses. Virtually all animals are capable of modifying their behavior as the result of prior experience. Most of what we know about the cellular mechanisms responsible for learning have come from studies of simpler animals, particularly the sea slug, *Aplysia,* a marine mollusk whose nervous system consists of approximately 10,000 unusually large neurons. These studies have been carried out largely by Eric Kandel and his colleagues at Columbia University.

As a sea slug moves across the sea bottom or the surface of a rock, water passes through its siphon and over the gills. If the siphon is gently stimulated, the animal responds by retracting its gills. This *gill-withdrawal reflex* is a protective mechanism. But if the stimulation is due to wave action, or contact with a rock, there would be no reason for gill withdrawal. In fact, repeated gentle contact with the siphon soon leads to a cessation of the gill withdrawal response; the behavior has been *habituated.* Habituation is a simple form of learning which is "remembered" for several days.

The gill-withdrawal reflex in *Aplysia* is mediated by a simple reflex arc. Studies have shown that, following habituation, action potentials continue to be generated in the sensory neuron leading out of the siphon, but these impulses are not transmitted to the target motor neuron leading to the gills. This is because successive impulses in the sensory neuron trigger the release of a decreasing quantity of neurotransmitter, which, in turn, evokes a smaller stimulus to the motor neuron. You might conclude that the sensory neuron is simply running out of transmitter substance, but this is not the case. Recall that neurotransmitter release is mediated by an influx of calcium ions at the tip of the neuron. The underlying cause of habituation of the gill-withdrawal reflex is a regulated decrease in the number of calcium channels that open in response to successive action potentials reaching the tip of the sensory neuron. From this finding, we can conclude that, in *Aplysia,* the memory of recent events is stored in a short-term change in individual neurons; if one stops touching the siphon for a period of time, the sensory neuron reverts to its original state in which the maximum number of calcium channels open upon stimulation.

The studies on *Aplysia* indicate that learning is accompanied by structural modifications of synapses. Studies on

◁ THE HUMAN PERSPECTIVE ▷
Alzheimer's Disease: A Status Report

First it robs victims of their memory of recent events, followed by a loss of simple reasoning and the ability to feed or clean oneself. Finally, it takes the victim's life. Alzheimer's disease strikes about 5 percent of people age 65 or older, and possibly as many as 50 percent of those 85 or older. This insidious disease destroys nerve cells in the brain, particularly those that release acetylcholine. The hippocampus, a region of the brain important for memory, is affected most severely. The severe effects of Alzheimer's disease appear in brain scans of afflicted persons (Figure 1).

The brains of Alzheimer's patients show two microscopic abnormalities not found in normal brains. The cytoplasm of degenerating neurons contains strange tangled fibrils, and the space outside the neurons contains dense plaques. The tangled fibrils consist of disorganized cytoskeletal proteins, while the plaques consist of fragments of a membrane protein called beta-amyloid. Two questions that are the subject of intense current interest are whether or not the tangles or the plaques cause Alzheimer's disease, and why either of these abnormal formations develop.

Many neurobiologists believe that individuals who suffer from Alzheimer's disease have sustained genetic damage that has caused their nerve cells to produce abnormal gene products. Evidence for this belief comes from a rare genetic disorder in which victims develop all of the charac-

teristics of Alzheimer's at a much earlier age than the disease normally appears. This form, termed *familial Alzheimer's disease (FAD)*, is characterized by the same tangled fibrils and plaques seen in individuals with *sporadic* (noninherited) cases. More than one gene may be responsible for FAD. One of these genes was isolated in 1991 and was shown to code for the pre-

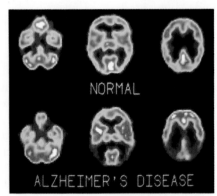

FIGURE 1

Characteristic signs of Alzheimer's disease. These PET (positron-emission tomography) scans, which depict levels of metabolic activity within living tissue, reveal marked differences between the brain of a normal person and one suffering from Alzheimer's disease.

cursor of the beta-amyloid protein found in the plaques that characterize the disease. In studies of two separate families, individuals afflicted with FAD were found to have alterations in this particular gene which changed the amino acid sequence of the protein. This finding strengthens the belief that the amyloid plaques are a cause of the disease rather than simply an effect. Researchers have now successfully introduced this mutant gene into mice in an attempt to induce an Alzheimer's-like disease in laboratory animals and to establish an animal model for possible treatments. This is particularly important since current treatments have had little success. As we discussed in the opening pages of the chapter, NGF may eventually prove a useful treatment.

Researchers are also attempting to develop diagnostic procedures that would detect Alzheimer's disease at a very early stage, even before symptoms develop. A number of biotechnology companies are working on diagnostic procedures based on substances that might appear in the blood of people with the disease. Now that a gene for FAD has been isolated, the possibility also exists for developing genetic screening tests. The same ethical question arises in this case as in the case of Huntington's disease (page 350): Who among us would want to learn that we will ultimately be stricken with a horrible, degenerative disease for which no treatment is available?

more complex animals have corroborated this view, indicating that as new tasks are learned, physical changes (such as the loss or gain of dendrites or the phosphorylation of membrane proteins) may occur at key synapses, which weaken or strengthen the connections between the neurons involved in the response. In humans, these synaptic changes occur primarily in the cerebrum.

Language Communication between members of a species occurs throughout the animal world. Fireflies communicate by flashes of light; gypsy moths communicate by the release of airborne chemicals; and birds and dolphins communicate by vocalization. Nowhere does the complexity of language approach that used by humans, however. Children learn to communicate verbally at a very early age, long before they are capable of learning complex motor activities. Studies with infants as early as a few months old have convinced most researchers that humans are born with an innate, genetically determined ability to learn languages without formal instruction. For most people, especially right-handed individuals, this ability is localized in the left frontal lobe of the cerebrum. (In 10 to 20 percent of people, the center is localized in either the right frontal lobe or in both frontal lobes.) The localization of the language center in the left frontal lobe can be dramatically revealed by anesthetizing this part of the brain, leaving the person awake and alert but unable to speak.

The importance of the left frontal lobe in language abilities has been confirmed in studies of deaf children who learn to communicate via sign language at an early age. Sign language is a form of communication which is performed by motor activities of the hands. You might expect that, like other motor activities, control over signing would emanate from the motor areas of the brain, but researchers have found that it is actually the language center that controls the motions used in sign communication.

Motor Activities Within the cerebral cortex, some regions are specialized for receiving sensory input, and other regions are specialized for initiating motor output. The locations and functions of the sensory portions of the cortex will be discussed in the next chapter, together with the sensory organs themselves. For now, we will concentrate on the **motor cortex** (Figure 23-13)—the paired strips of gray matter found in the frontal lobe of the brain, where most of the impulses that command our muscles emanate. Stimulation of various parts of the motor cortex results in contraction of specific groups of muscle fibers. Those parts of the body which perform the most intricate and delicate movements, including the fingers, lips, tongue, and vocal cords, have the greatest representation in the motor cortex (Figure 23-13). As a result, stimulation of just a few cells may produce reactions ranging from a slight movement of a finger to a major movement of the torso.

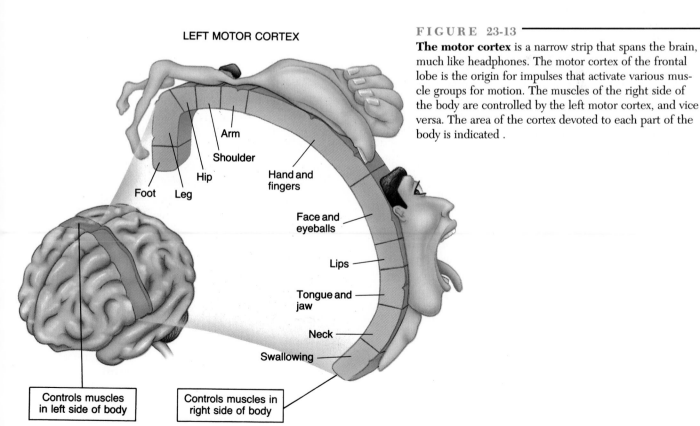

LEFT MOTOR CORTEX

Arm
Shoulder
Hip
Foot Leg
Hand and fingers
Face and eyeballs
Lips
Tongue and jaw
Neck
Swallowing

Controls muscles in left side of body

Controls muscles in right side of body

FIGURE 23-13

The motor cortex is a narrow strip that spans the brain, much like headphones. The motor cortex of the frontal lobe is the origin for impulses that activate various muscle groups for motion. The muscles of the right side of the body are controlled by the left motor cortex, and vice versa. The area of the cortex devoted to each part of the body is indicated .

The Cerebellum

The cerebellum is a bulbous structure (Figure 23-11) that receives information from receptors located in the muscles, joints, and tendons, as well as from the eyes and ears. The amount of information converging on the cerebellum is revealed by examining the huge number of synaptic contacts made by some of these cells; some individual cerebellar neurons receive input from as many as 80,000 other neurons!

Vast amounts of sensory information can be processed within a fraction of a second. Messages are then sent to the motor cortex of the cerebrum, where they are used in directing such complex motor activities as playing a musical instrument, writing, or shooting a basketball (Figure 23-14). For this reason, a person who sustains cerebellar damage will have difficulty performing smooth, coordinated movements; motor activities lose their subconscious basis, and the person may have to think about each movement that would otherwise be performed automatically.

The Brainstem, Reticular Formation, and Limbic System

The brainstem forms the central stalk of the brain (Figure 23-11) and is responsible for regulating most of the involuntary, visceral activities of the body, such as breathing and swallowing, as well as maintaining heart rate and blood pressure. Consequently, permanent damage to the brainstem usually leads to coma or death. Whereas the cerebral cortex is the most recently evolved part of the vertebrate brain, the *brainstem* is probably the oldest part; it makes up the bulk of the brain of lower vertebrates, such as fish and amphibians.

Most of the sensory information streaming into the cerebral cortex is routed through the *thalamus*—the part of the brain located just beneath the cerebrum (Figure 23-11). The thalamus also coordinates outgoing motor signals. Associated with the thalamus is the *reticular formation,* which is composed of several interconnected sites in the core of the brainstem that selectively arouse conscious activity, producing a state of wakeful alertness. The reticular formation can be activated by any number of internal factors or external stimuli—the sound of an approaching horn, a flashlight shining in your eyes, or the middle-of-the-night realization that you forgot to study for tomorrow morning's biology exam. When the reticular system fails to maintain arousal, the brain falls asleep and stays asleep until the reticular formation is activated by sensory input from, say, an alarm clock.

The reticular formation is no larger than your little finger. Nonetheless, it does much more than simply keep you awake and alert (no small task itself). The reticular formation helps you cope with the millions of impulses that assault the brain every waking second (Figure 23-15). It screens out the trivial signals, while allowing vital or unusual signals to pass through and alert the mind.

FIGURE 23-14
Michael Jordan about to score two.

Beneath the thalamus (Figure 23-11) is the **hypothalamus** (*hypo* = below), a portion of the brain that is only about the size of the tip of your thumb. Despite its small size, the hypothalamus is critical for maintaining homeostasis. The hypothalamus regulates body temperature, keeping it around 37°C (98.6°F). It also helps control blood pressure, heart rate, and the body's urges, such as hunger, thirst, and sex drive. The hypothalamus also controls the nearby *pituitary gland* (Figure 23-11), which is a key component of the endocrine system (discussed in Chapter 25).

The hypothalamus and the thalamus comprise part of an interconnected group of structures buried in the core of the brain. This complex is called the **limbic system.** As discussed above, one part of the limbic system, the hippocampus, is an important center for short-term memory; without it, you would have already forgotten how this sentence began. The limbic system is also associated with pleasure and joy, pain and fury, and an ability to balance emotions. Stimulation of different parts of the limbic system with electrodes can induce immediate anger or euphoria, sexual arousal, or deep relaxation.

The importance of the limbic system in regulating our emotions is dramatically illustrated by "Julie," an expatient of Vernon Mark, a neurosurgeon at Harvard Medical School. Since childhood, Julie had suffered from epileptic seizures, which initiated outbursts of violent behavior. In one episode, at age 18, Julie was overcome by a seizure in a movie theatre. She then went to the restroom, where she stabbed a woman who accidentally bumped against her. Mark identified the *amygdala,* a part of the limbic system, as the site in the brain that was being electrically disturbed during Julie's seizures. When this site was stimulated with an electrode, the patient became unresponsive for a few seconds, then reacted violently by smashing her fists against the wall or swinging an object she held in her hand. Mark removed the abnormal brain tissue, and Julie never again demonstrated an uncontrolled rage. The importance of the limbic system in controlling our emotions is also revealed by mood-altering drugs.

The Effect of Cocaine on the Limbic System When certain parts of the limbic system are stimulated—either artificially (by electrodes) or naturally (by incoming impulses)—a strong feeling of pleasure is elicited. This feeling results from the release of the neurotransmitter dopamine, which binds to certain neurons in the "pleasure centers" of the limbic system. Normally, the effects of dopamine are short-lived because the neurotransmitter is rapidly removed from the synaptic cleft by the quick reuptake into the neuron that released it. However, cocaine, a compound extracted from the leaves of the South American coca plant (*Erythroxylum coca*) and inhaled by most users, interferes with the reuptake of dopamine. The sustained presence of dopamine in the synaptic clefts of the limbic center produces a feeling of euphoria—a "high" that lasts for several minutes. The high is followed by a "crash," in which the person feels depressed, irritable, and anxious. Since the fastest way to relieve the unpleasant effects of the crash is to inhale more cocaine, the drug can rapidly become addictive.

In addition to causing mood elevations, cocaine also elevates a person's heart rate, blood pressure, body temperature, and blood sugar. The drug has also been known to trigger seizures and heart failure, the latter of which accounts for nearly 1,000 deaths per year in the United States alone.

Opium Derivatives and Endogenous Opiates Morphine and heroin are structurally related molecules, called *opiates*, derived from the opium poppy. Like cocaine, opiates interfere with the transmission of impulses by neurons of the limbic system. Unlike cocaine, opiates act by binding to specific receptors situated in the postsynaptic membrane of neurons found in several areas of the brain. Among their targets, opiates bind to neurons in the pain pathways of the limbic system, blocking the perception of pain.

The discovery of opiate receptors in the early 1970s raised a key question: Why should neurons of the brain possess receptors that specifically bind substances derived from an opium poppy? One likely explanation was that the brain produced one or more substances that were similar in structure to opiates and that bound to the same receptors. After an intense search carried out by a number of laboratories, two classes of peptides were discovered. These peptides, called *endorphins* (a contraction for "endogenous morphinelike substances") and *enkephalins* (from the Greek meaning "in the head"), normally bind to receptors located on neurons that are part of pain-transmitting pathways. Like morphine or heroin, endorphins and enkephalins interfere with the delivery of information through this pathway to the brain, thereby blocking the sensation of pain. Even strenuous exercise or painful workouts may lead to the release of endorphins and enkephalins, causing a feeling of euphoria. This phenomenon is commonly referred to as "runner's high." Other psychoactive drugs have also been found to act by binding to receptors for neurotransmitters in the brain. For example, nicotine binds to acetylcholine receptors; lysergic acid diethylamide (LSD) binds to serotonin receptors; and mescaline (extracted from the peyote cactus) binds to norepinephrine receptors.

FIGURE 23-15
The limbic system allows this student to focus on her book and to ignore the surrounding noise.

All of the spinal nerves contain both sensory and motor neurons, which separate from one another a short distance after leaving the spinal cord, forming two distinct roots (Figure 23-16). Sensory impulses enter the spinal cord through *dorsal roots*, while motor impulses leave the spinal cord through *ventral roots*.

The sensory neurons that form part of the cranial and spinal nerves provide the CNS with the sensory input needed for conscious awareness and the ability to respond to changes in the body and in the external environment. The motor neurons present in these nerves carry messages to the body's muscles and glands. These outbound motor neurons constitute two distinct divisions of the peripheral nervous system: the somatic and the autonomic divisions.

THE SOMATIC AND AUTONOMIC DIVISIONS OF THE PERIPHERAL NERVOUS SYSTEM

The **somatic division** of the peripheral nervous system carries messages to the skin and to those muscles that move the skeleton, such as the major muscles of the head, trunk, and limbs. These movements generally follow voluntary orders from the brain, although stereotyped reflex movements, such as the knee-jerk reflex, are also mediated by the somatic division.

↻ The **autonomic division** controls the involuntary, homeostatic activities of the body's internal organs and blood vessels. Autonomic motor neurons regulate heart rate, blood-vessel diameter, respiratory rate, glandular secretions, excretory functions, digestion, sexual responses, and so forth. Although this system's operations are largely involuntary, many autonomic functions can be influenced by conscious control. Biofeedback training, for example, teaches how to modify certain types of autonomic activity to help reduce blood pressure or to lower physiological stress, responses that are classically considered beyond voluntary control.

The Sympathetic and Parasympathetic Divisions of the Autonomic Nervous System

The autonomic division can be further divided into two parts: the **sympathetic system** and the **parasympathetic system** (Figure 23-17). Most organs of the body receive neurons from both of these systems, which evoke opposite (*antagonistic*) responses. Your heart rate, for example, is precisely regulated by the balance that is maintained between continuous stimulatory influences from sympathetic neurons and inhibitory influences from parasympathetic neurons.

A survey of the specific responses evoked by the autonomic division (Figure 23-17) reveals the different adaptive functions of the sympathetic and parasympathetic systems. When effector organs are stimulated by the sympathetic system, an adaptive response is triggered which better enables the body to cope with stressful situations, as might occur if you were suddenly confronted with a person coming toward you with a weapon. In such situations, the heart

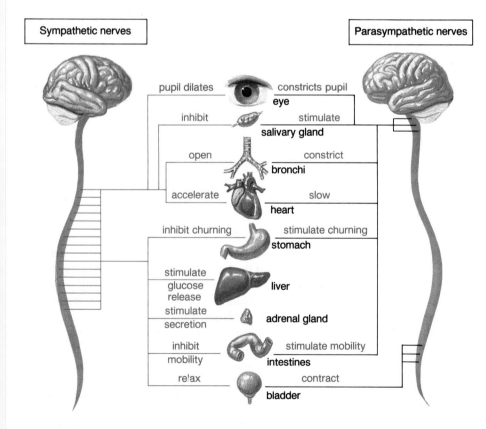

Sympathetic nerves

Parasympathetic nerves

pupil dilates — eye — constricts pupil
inhibit — salivary gland — stimulate
open — bronchi — constrict
accelerate — heart — slow
inhibit churning — stomach — stimulate churning
stimulate glucose release — liver
stimulate secretion — adrenal gland
inhibit mobility — intestines — stimulate mobility
relax — bladder — contract

FIGURE 23-17

The two divisions of the autonomic nervous system control most of the same internal organs but under opposite circumstances. The effects of the parasympathetic system predominate during normal, relaxed activity. During times of stress, the sympathetic system predominates, eliciting physiological changes that increase the level of performance to better adapt to the stressful situation. Parasympathetic neurons release acetylcholine at synapses with their target organs, while most sympathetic neurons release norepinephrine.

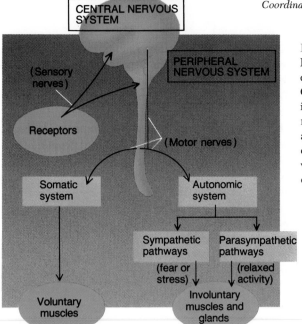

FIGURE 23-18

Interplay between the central and peripheral nervous systems. Receptors relay information about the internal and external environment to the CNS by way of sensory (or *afferent*) neurons. The CNS responds to this input by sending impulses to target cells through motor (or *efferent*) neurons. Motor nerves are part of two systems. (1) the somatic system, which acts on voluntary muscles; and (2) the autonomic system, which either excites or inhibits involuntary muscles and glands. The autonomic system is divided into the sympathetic and parasympathetic pathways, which typically elicit opposite responses in target organs.

rate increases, and blood is shunted away from the periphery toward the lungs for added oxygen uptake and toward the skeletal muscles, whose energy demands soar during periods of activity. The liver is stimulated to release additional glucose into the blood, while the activity of the digestive organs is reduced. The level of metabolism is increased, causing a rise in body temperature, which triggers the release of fluids by the sweat glands. Taken together, these changes constitute the "fight-or-flight" response, whereby the body is physiologically prepared either to confront the danger or to beat a fast retreat. Sympathetic neurons also stimulate the adrenal gland, which releases hormones that further prepare the body to meet the crisis at hand. This teamwork is one example of the interrelatedness of the nervous and endocrine systems (Chapter 25).

In contrast, the parasympathetic system handles the mundane, "housekeeping" functions of the body, such as digestion and excretion (Figure 23-17). For example, emptying the urinary bladder is a reflex that is mediated by the parasympathetic system and can be overridden by voluntary control. A person who has suffered a spinal-cord injury loses this voluntary control and thus cannot prevent voiding urine whenever the stretch receptors in the bladder wall launch the bladder-contraction reflex. The functional relationship between the central and peripheral nervous system is summarized in Figure 23-18.

The central nervous system receives information about conditions in the external environment and governs the activities of the body's organs by receiving and sending impulses over the neurons that make up the peripheral nervous system. (See CTQ #7.)

THE EVOLUTION OF COMPLEX NERVOUS SYSTEMS FROM SINGLE-CELLED ROOTS

Many people tend to think that a large and sophisticated brain is the ultimate achievement of evolution and that humans therefore reside at this "evolutionary pinnacle." Yet, the nervous systems of simpler animals (some of which lack even the earliest traces of a brain) are just as effective in enabling these organisms to survive and reproduce as is the large brain of humans. Our "advanced" brains are merely one adaptive strategy. Some animals, notably sponges, survive without a single neuron, as do all plants, fungi, protists, and monerans. That is not to say that these organisms lack neuron-like communication. Even unicellular organisms respond to the presence of external stimuli. For example, a chemical detected by a bacterial cell activates or inhibits flagella at the opposite end of the cell. In this way, the cell moves toward desirable chemicals, such as nutrients, and away from harmful substances, such as toxins. Impulses are relayed from "sensors" at one end of the cell to the flagella at the other end of the bacterial cell by a wave of depolarization similar to that responsible for neural impulse transmission. From these simple beginnings, more complicated nervous systems evolved.

Figure 23-19 illustrates some of the stages in the evolution of complex nervous systems. Even though there is remarkable diversity in the design and complexity of these various nervous systems, they are all built from the same type of nerve cells that utilize the same basic mechanism for generating, conducting, and transmitting nerve impulses. The simplest multicellular nervous systems are found in cnidarians: hydras, jellyfish, and sea anemones. These animals possess a *nerve net*—a diffuse system of individual nerve cells that gives these animals the ability to move around slowly, to seize and paralyze prey with their stinging tentacles, and to shove the incapacitated victim into the mouth (Figure 23-20*a*).

Compare the anemone's nerve net with the primitive nervous system of the flatworm in Figure 23-19. Unlike the anemone, which is a sedentary animal, the flatworm slowly creeps over a surface, its head end leading the way. This type of directed movement is correlated with the evolutionary beginnings of **cephalization**—the concentration of nerve cells near the anterior end of the animal, forming a primitive "brain." In most animals, the brain is located near the area of greatest concentration of sense organs, which is situated in the head. As evolution generated more and more complex animal forms, this tendency toward cephalization became more pronounced, and the remainder of the central nervous system became more complex as well.

The brains of flatworms and earthworms provide so little control that the worm can lose its head and still continue to live while regenerating a replacement. The brains of insects, lobsters, and spiders play a more prominent and indispensable role in integrating the responses of various parts of a complex body. The most complex brain of any invertebrate belongs to the octopus. An octopus can distinguish among relatively similar objects and can accomplish such complex tasks as removing a stopper from a jar (Figure 23-20*b*). Some parts of the octopus's brain are concerned with memory and decision-making activities. If, for example, an octopus's higher brain centers are damaged, the animal may still be able to attack a crab placed in its aquar-

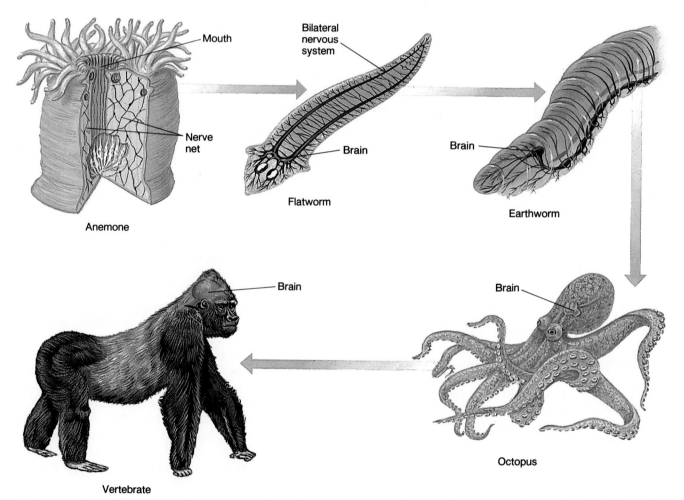

FIGURE 23-19

Evolution of the nervous system. (The shaded arrows indicate increasing neural complexity, *not* an evolutionary pathway.) The simplest nervous systems are found in cnidarians, such as the sea anemone, which possess diffuse nerve nets but lack a central system. The trend toward cephalization (the development of control centers in the head) first appears in flatworms and becomes successively more pronounced in the earthworm, octopus, and vertebrate.

(a)   *(b)*

FIGURE 23-20

Animals with nervous systems of contrasting complexity. *(a)* The anemone's predatory skills rely on a simple system of sensory neurons that launch a feeding response when they detect prey. In this instance the prey is a sea star that has made contact with the anemone's tentacles. *(b)* An octopus has a highly complex nervous system that allows the animal to figure out how to dine on a crab trapped inside a stoppered jar.

ium, but if the crab were to hide under a rock, the octopus might "forget" that food is nearby. Without its normal brain function, the octopus illustrates a case of "out of sight, out of mind."

▮▶ It is likely that sea anemones, flatworms, earthworms, and octopuses are part of a single evolutionary pathway and that the increased complexity of their nervous systems occurred by stepwise advances of the same basic plan. Even though vertebrates evolved along a different pathway, their nervous systems show unmistakable similarities to those of complex invertebrates. The nervous system of an octopus and a human, for example, both consist of a peripheral portion, where impulses initiate and terminate, and a cen-

tral portion, where impulses are received, sorted, and then relayed to the appropriate destination, illustrating once again, that similar physiological needs often lead to similar evolutionary solutions.

All organisms possess an inherent ability to respond to stimuli; this ability formed the basis for the evolution of primitive nervous systems. The evolution of more active, anatomically complex animals was accompanied by the evolution of more complex nervous systems that are capable of mediating more complex physiological and behavioral responses. (See CTQ #8.)

REEXAMINING THE THEMES

Relationship between Form and Function

▨ Both the single nerve cell and the entire nervous system are structurally adapted for their tasks of collecting and disseminating information. This can be illustrated by comparing the living components of the nervous system to that of a telephone network. A telephone system contains a

collecting device at one end (a mouthpiece); a long, thin, transmission line to carry coded information over a distance; and a speaker at the other end to deliver information that can be interpreted by the target listener. Similarly, nerve cells contain a network of dendrites that collect information from various sources, a cablelike axon that propagates the information in coded form, and synaptic knobs that transmit the information to a variety of targets. Like a telephone network, a nervous system contains a central communications center, where incoming messages from scattered sites are sorted, and outgoing messages are routed to selected target sites.

Biological Order, Regulation, and Homeostasis

The nervous system is more than just a communications network; it is the control center for regulating bodily function. The nervous system monitors the environment, integrates sensory input, and sends out command signals to initiate appropriate responses when changes in the internal or external environment are detected. If your body becomes too warm, nerve cells stimulate your sweat glands to secrete a cooling layer of water; if the concentration of oxygen in your blood becomes too low, nerve cells stimulate your breathing muscles to increase their rate of contraction; if you are threatened by a hungry predator, your nervous system will send out directives either to outsmart the beast or to beat a fast retreat.

Unity within Diversity

While the architecture and complexity of nervous systems are highly diverse, all nervous systems are built from very similar types of nerve cells that conduct and transmit impulses by the same basic mechanisms. The complexity of the human nervous system (as well as that of other "higher" animals) can be traced to the intricate circuitry of our system rather than to an increased complexity of the units that make up the system. Similarly, the nervous systems of virtually all animals, from flatworms to mammals, consist of a peripheral portion in which impulses are initiated and terminated, and a central portion in which impulses are received and sorted, and then relayed to the appropriate destinations.

Evolution and Adaptation

The basic mechanism responsible for neural activity can be found in the simplest unicellular organisms— bacteria—where communication from one part of the cell to another is accomplished by waves of depolarization that pass along the plasma membrane. With the evolution of multicellular animals, communication became more complex; specialized cells evolved that stretched considerable distances across the body. In its simplest form (as exemplified by hydras and sea anemones), the nervous system consists of a simple, diffuse net, with no evidence of centralization. With the evolution of more complex animals came a central concentration of nerve tissue, usually near the anterior end of the animal, close to the sense organs of the head.

SYNOPSIS

The nervous system communicates information from one part of the body to another and, in doing so, regulates the body's activities and maintains homeostasis. Nerve cells (or neurons) are specialized in that they have dendrites, for collecting information; a cell body that houses the nucleus and most of the cellular machinery; an axon, for conducting information in the form of impulses; and terminal synaptic knobs, for transmitting information to a target cell (a muscle cell, gland cell, or another neuron).

Neural impulses result from sequential changes in the distribution of ions across the plasma membrane. Molecular pumps in the plasma membrane of a neuron maintain steep gradients of sodium (Na^+) and potassium (K^+) across the membrane. In the resting neuron, potassium ions leak outward across the membrane creating a resting potential. Should the neuron receive a stimulus that causes the potential to drop (depolarize) past the threshold, an action potential is automatically triggered, whereby gated sodium channels open, allowing an inward movement of sodium ions and a reversal in the potential. This change triggers the closing of the sodium channels and the opening of the gated potassium channels, causing the potential to return to a negative value. When the gated potassium channels close, the membrane returns to its resting state. Once triggered, a wave of action potentials passes down the length of the membrane without losing intensity. In vertebrates, the speed of neural impulses is greatly increased by wrapping the axons in a lipid-containing myelin sheath. The sheath is punctuated along its length by nodes, and the impulse travels along the axon by skipping from node to node.

An impulse is transmitted from a neuron to its target cell by the release of a chemical-transmitter substance that diffuses across the narrow synaptic cleft between the two cells. Neurotransmitter substances are stored in vesicles within the synaptic knobs of a neuron. When an impulse arrives at the synaptic knob, calcium ions flow into the neuron, triggering the fusion of the membrane of a portion of the vesicles with the plasma membrane and releasing the neurotransmitter into the synaptic cleft. When the neurotransmitter binds to a specific receptor on the target cell membrane, it may either excite (depolarize) or inhibit (hyperpolarize) the target cell. In most cases, one excitatory transmission is not sufficient to stimulate a target cell to respond. Several excitatory impulses may become additive if they arrive at the target cell membrane close enough together in either time or space. Synapses act as gates that route information in the proper directions through the nervous system. Moreover, synapses can change over time, as occurs during learning and adapting to new situations.

The nervous system of vertebrates consists of a central nervous system (brain and spinal cord) and a peripheral nervous system (cranial and spinal nerves). Sensory neurons carry information about changes in the environment from the periphery toward the CNS, while motor neurons carry command signals to the body's muscles and glands. Interneurons convey impulses within the CNS.

The fastest motor reactions to stimuli are mediated by reflex arcs, which consist of at least one sensory and one motor neuron. This type of simple circuit provides the shortest possible route for the flow of information through the CNS. Information also passes to and from higher centers in the CNS, which mediate more complex responses.

The human brain consists of several distinct parts, including the cerebrum, the cerebellum, and the brainstem. The cerebrum is covered by the cerebral cortex, a layer of gray matter that houses the sites of higher brain functions, such as speech and rational thought. The cerebellum, in conjunction with the motor cortex of the cerebrum, controls complex motor activities. The brainstem controls involuntary visceral activities, such as breathing and heart rate. Those parts of the brain that control wakefulness, emotions, and homeostatic functions (such as temperature control) are also associated with the brainstem.

The spinal cord receives sensory information via dorsal roots of its spinal nerves and sends out motor information through the ventral roots of these nerves. Motor commands may be sent via the somatic division, which controls voluntary muscle activity, or the autonomic division, which controls involuntary activity. The autonomic division is divided into a parasympathetic branch responsible primarily for "housekeeping" activities, such as digestion and excretion, and a sympathetic branch primarily responsible for readying the body for meeting stressful situations. Many organs of the body receive stimulation from both parasympathetic and sympathetic neurons, which have opposing influences on the organ.

Key Terms

nerve growth factor (NGF) (p. 467)
neuron (p. 468)
excitatory neuron (p. 468)
inhibitory neuron (p. 468)
cell body (p. 468)
dendrite (p. 468)
axon (p. 468)
synaptic knob (p. 468)
sensory neuron (p. 470)
interneuron (p. 470)
motor neuron (p. 470)
neuroglial cell (p. 470)
Schwann cell (p. 470)
myelin sheath (p. 470)
nodes of Ranvier (p. 470)
potential difference (p. 471)
voltage (p. 471)

resting potential (p. 471)
threshold (p. 473)
action potential (p. 473)
saltatory conduction (p. 473)
synapse (p. 474)
synaptic cleft (p. 474)
neurotransmitter (p. 474)
synaptic vesicle (p. 474)
central nervous system (CNS) (p. 479)
brain (p. 479)
spinal cord (p. 479)
peripheral nervous system (p. 479)
nerve (p. 479)
reflex (p. 479)
reflex arc (p. 479)
sensory receptor (p. 479)
gray matter (p. 481)

white matter (p. 482)
ventricle (p. 482)
cerebrospinal fluid (p. 482)
cerebral hemisphere (p. 482)
cerebral cortex (p. 482)
learning (p. 483)
motor cortex (p. 485)
hypothalamus (p. 486)
limbic system (p. 486)
cranial nerve (p. 489)
spinal nerve (p. 489)
somatic division (p. 490)
autonomic division (p. 490)
sympathetic system (p. 490)
parasympathetic system (p. 490)
cephalization (p. 492)

Review Questions

1. Describe the various parts of a myelinated neuron and the role of each part in neural function.

2. Describe the effects of depolarization and hyperpolarization of a target cell on the inhibition or excitation of that cell.

3. What is the source of the energy needed to fire an impulse along a neuron? Why doesn't the strength of the impulse diminish as it travels along the neuron?

4. Describe the components required to mediate a simple reflex response, such as the knee-jerk. How can these same components become part of more complex circuits that involve conscious decisions, such as kicking a football?

5. Compare and contrast the following: somatic and autonomic divisions; sympathetic and parasympathetic systems; resting potential and action potential; the role of leak channels versus gated channels in generating an action potential; the roles of sodium and potassium gated channels; cranial and spinal nerves; myelinated and nonmyelinated axons.

Critical Thinking Questions

1. Suppose that antibodies against NGF injected into a chick embryo had no visible effect on the development of the animal's nervous system. What might you conclude about the role of NGF in development? How could you explain the results of experiments in which NGF injected into an embryo causes enlargement of the sympathetic nervous system?

2. Neurons typically have many short, highly branched dendrites and one axon, which is longer and thicker than the dendrites. How do these differences in structure reflect the differences in function between dendrites and axons?

3. Tetrodotoxin is a toxin present in the spines of puffer fish; it has the capability of blocking the function of gated sodium channels. What effect do you suppose this substance would have on the formation of nerve impulses? On the state of muscle contraction in the body? Why?

4. Transmission of nerve impulses is often referred to as an "electrochemical" phenomenon. Explain what is meant by this term.

5. Most actions controlled by simple reflex arcs involve protective responses or actions that must be completed frequently. Why is it adaptive that these responses do not depend upon the central nervous system?

6. What effects would you predict would result from severe injury to the following parts of the central nervous system: medulla, cerebellum, ventral root of a spinal nerve to the arm, reticular formation, motor cortex, motor cortex on the right side of brain, lower two-thirds of the spinal cord?

7. Explain how the existence of two autonomic systems permits fine-tuning of behavior.

8. Much of what is known about the functioning of the human nervous system has been learned by studying simpler animals, such as the snail, squid, and grasshopper. Why is it possible to learn about the human nervous system from studying these organisms that appear very different from us?

Additional Readings

Alkon, D. A. 1989. Memory storage and neural systems. *Sci. Amer.* July:42–50. (Intermediate)

Aoki, C., and P. Siekevitz. 1988. Plasticity in brain development. *Sci. Amer.* Dec:56–64. (Intermediate)

Kalil, R. E. 1989. Synapse formation in the developing brain. *Sci. Amer.* Dec:76–85. (Intermediate)

Kandel, E., and J. Schwartz. 1985. *Principles of neural science*, 2d ed. New York: Elsevier. (Intermediate–Advanced)

Levi-Montalcini, R. 1988. *In praise of imperfection.* New York: Basic Books. (Introductory)

Marx, J. 1991. Mutation identified as a possible cause of Alzheimer's disease. *Science* 251:876–877. (Intermediate)

Robertson M. 1992. Alzheimer's disease and amyloid. *Nature* 356:103. (Intermediate)

Selkoe, D. J. 1991. Amyloid protein and Alzheimer's disease. *Sci. Amer.* Nov:68–78. (Intermediate)

Tuomanen, E. 1993. Breaching the blood-brain barrier. *Sci. Amer.* Feb:80–84. (Intermediate)

Wallace, B. C. 1991. *Crack cocaine.* New York: Brunner/Mazel. (Introductory)

The brain. 1990. Time-Life Books. (Introductory)

CHAPTER

24

Sensory Perception: Gathering Information about the Environment

STEPS TO DISCOVERY

Echolocation: Seeing in the Dark

THE RESPONSE OF A SENSORY RECEPTOR TO A STIMULUS

SOMATIC SENSATION

VISION

The Lens: Focusing Images on the Retina

The Retina: A Living Projection Screen

HEARING AND BALANCE

Translating Vibrations into Sound

Translating Fluid Movements into a Sense of Balance

CHEMORECEPTION: TASTE AND SMELL

Taste

Smell

THE BRAIN: INTERPRETING IMPULSES INTO SENSATIONS

Determining the Nature and Location of the Stimulus

EVOLUTION AND ADAPTATION OF SENSE ORGANS

BIOLINE

Sensory Adaptations to an Aquatic Environment

THE HUMAN PERSPECTIVE

Two Brains in One

Echolocation:
Seeing in the Dark

*F*or hundreds of years, naturalists have wondered how bats are able to fly through caves in total darkness or through wooded fields at night, capturing fast-flying insects they can't possibly see. Near the end of the eighteenth century, Lazarro Spallanzani, an Italian biologist, caught a number of bats in a cathedral belfry, impaired their vision, and let them go. Returning to the belfry at a later time, Spallanzani found the bats back in their roost at the top of the cathedral. Not only did the bats return home, but their stomachs were full of insects, indicating that they had been able to capture food without their sense of sight. In contrast, bats whose ears had been plugged with wax were reluctant even to take flight and frequently collided into objects when they tried to fly. Hanging objects in front of the bats' mouths

also impeded the animals' flying abilities. Spallanzani was baffled: How could an animal's ears and mouth be of more importance than its eyes in guiding flight?

The question remained unanswered until 1938, at which time, Donald Griffin, an undergraduate at Harvard University, brought a cage of bats to one of his physics professors, George Pierce. Pierce had developed an apparatus for converting high-frequency sound waves into a form that could be heard by the human ear. After placing the cage of bats in front of the sound detector, Griffin and Pierce became the first people to "hear" the bursts of sounds emitted by bats. When the bats were set free in the physics laboratory, these same high-frequency sounds could be heard as a bat flew in a straight direction toward

By placing invisible barriers in the path of porpoises, biologists discovered that these animals use echolocation to navigate

the microphone. For Griffin, this was the initial step in a long, illustrious career centered around studying the role of sound in animal navigation.

The mechanism bats use to fly through a pitch-black obstacle course relies on the same principles first employed by the Allied navy to hunt for German submarines during World War II. That mechanism is sonar. The basic principle of sonar is simple: Sound waves are emitted from a transmitter; they bounce off objects in the environment; and the "echoes" are detected by an appropriate receiver. The elapsed time between the emission of the sound pulse and its return provides precise information about the location and shape of the object. Most bats employ a remarkably compact and efficient sonar system to detect insect prey and to avoid colliding with objects during flight. High-frequency sounds—those beyond the range of the human ear—are produced by the bat's larynx (voicebox), and the echoes are received by the animal's large ears. Impulses from the ears are then sent to the brain, where they are perceived as a detailed mental picture of the objects in the environment. Griffin coined the term "echolocation" to describe this phenomenon.

Echolocation by bats is an adaptation appropriate for their nocturnal (nighttime) peak of activity. Using echolocation, a bat flying at night can detect and avoid a wire as thin as 0.3 millimeters in diameter or can home in on a tiny, moving insect with unerring accuracy. These feats become even more impressive when you realize that the echo returning to the bat from such objects is about 2,000 times less intense than is the sound emitted. Moreover, a bat can pick out this faint echo in a crowd of other bats, each sending its own pulses of sound waves.

During the early 1950s, studies by Winthrop Kellogg, a marine biologist at Florida State University, and others, demonstrated that bats were not the only mammals capable of navigating by echolocation. One day, Kellogg and Robert Kohler, an electronics engineer, were out in a boat in the Gulf of Mexico when they saw a school of porpoises swimming toward them. The scientists lowered a microphone connected to a speaker and tape recorder into the water. As the porpoises swam toward them, all that could be heard above water was the sound of the animals exhaling through their blowholes. " . . . but the underwater listening gear told a very different story." According to Kellogg, the animals were emitting sequences of underwater clicks and clacks "such as might be produced by a rusty hinge if it were

around objects.

opened slowly. . . . By the time the group was about to make its final dive, the crescendo from the speaker in our boat had become a clattering din which almost drowned out the human voice."

To study the echolocating capabilities of these animals, Kellogg persuaded Florida State University to build a special "porpoise laboratory" at its nearby marine station. There, investigators discovered that porpoises possessed a sophisticated sonar system they could use to distinguish between similarly shaped objects, to avoid invisible barriers (such as sheets of glass), to swim through an elaborate obstacle course, to locate food, and to pick up objects off the bottom of their enclosure.

It may surprise you to learn that you probably have some capability to echolocate as well. Over the past hundred years or so, many reports have been written about the remarkable ability of many blind people to detect the presence of obstacles in their path without the use of a cane. The first carefully controlled experiment on this subject was conducted in 1940 by Michael Supa and Milton Kotzin, a pair of graduate students at Cornell University, one of whom was blind himself. Subjects that were either blind or sighted but blindfolded were asked to walk down a long wooden hallway. They were told that a large, fiberboard screen *might* be placed at some random site along their way and were asked to report when they detected the presence of the screen. The subjects were able to detect the screen within 1 to 5 meters of its presence. In those cases when the screen was absent, none of the subjects reported its presence. In contrast, when the subjects' ears were tightly plugged, they collided with the screen in every trial. The authors of the experiment concluded that the echo of the sound emitted from the subjects' footsteps on the floor provided information about the location of the screen. A later report documented the case of a young blind boy who was able to avoid obstacles while riding a bicycle by making clicking sounds and listening to the echoes.

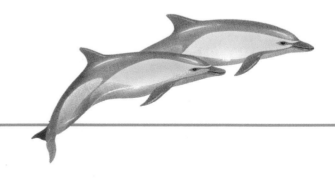

*C*onsider for a moment what it would be like if you were suddenly deprived of information from your sense organs. Our concept of the world is largely based on information gathered by our eyes, ears, nose, mouth, and the surface of our skin. Without our sense organs, we could not find food, avoid danger, or communicate with others. We would know absolutely nothing about the outside world. Our sense organs also provide us with many of the pleasures in life—the taste of food, the sounds of music, and the touch of another person. Collecting sensory data satisfies a basic human need. In fact, in experiments where people are placed in an environment that is devoid of sensory stimulation—a dark, silent, constant-temperature chamber—the subjects quickly become restless and irrational and often begin to hallucinate.

We also depend on information gathered by the sense organs in our internal environment that are required to maintain homeostasis. Without information from sensors located in muscles, joints, tendons, and internal organs, we could not walk, stand, or digest food; we would have no awareness of our body's state of well-being. Without information from sensors that detect chemicals and pressure in the walls of our arteries, we would not be able to maintain proper cardiovascular or respiratory function. In fact, very little in our body would work properly.

The properties of sense organs vary widely among different animals. Our "picture" of the outside world is shaped largely by our sense of sight, but many other animals rely on very different types of sensory organs (Figure 24-1). Consider, for example, the shark, which can find prey buried in the sand by detecting tiny electric potentials generated by the muscles of the buried animal; the bloodhound, which can follow the trail of a specific person hours after he or she has passed; or the rattlesnake, which can strike in total darkness, guided only by the heat emitted from its warm-blooded prey.

▼ ▼ ▼

THE RESPONSE OF A SENSORY RECEPTOR TO A STIMULUS

The first step in the chain of events leading to sensory awareness is the interaction between a *stimulus* and a *sensory receptor*. A stimulus is any change in the internal or external environment, such as a cold breeze or an empty stomach, and a sensory receptor is a cell that can be activated by that change. Some receptors, such as those of the eyes and ears, are specialized cells that respond to a stimulus and transmit the information to a sensory neuron. Other receptors, such as the stretch receptors of the muscles (page 479), are part of the sensory neurons themselves. Each type of stimulus activates only one type of sensory receptor (Table 24-1), which include:

- *Mechanoreceptors,* which respond to mechanical pressure and detect motion, touch, pressure, and sound.
- *Thermoreceptors,* which detect changes in temperature and react to heat and cold.
- *Chemoreceptors,* which are activated by specific chemicals, such as those that induce a particular taste or smell. Chemoreceptors also monitor concentrations of critical nutrients (glucose, amino acids) or respiratory gases (oxygen and carbon dioxide) in the blood.
- *Photoreceptors,* which respond to light.
- *Pain receptors,* which respond to excess heat, pressure, and irritating chemicals or chemicals that are released from damaged or inflamed tissue.

The structure of each sensory receptor enables the receptor to respond specifically to a particular type of stimulus, such as light of certain wavelengths or particular chemicals. The interaction between a stimulus and a receptor elicits some change in the receptor, which, in turn, evokes an alteration in the ionic permeability of the plasma membrane. This change alters membrane potential at the site of stimulation, which may trigger an action potential in a sensory neuron. For example, photoreceptors located in the retina of your eye contain membrane pigments that absorb light energy. Absorption of light energy causes a change in the molecular conformation of the pigment, leading to a change in membrane voltage. This alteration can trigger an impulse in a neuron of the optic nerve, generating electrical activity in those parts of the brain concerned with vision. The result is the perception of light.

We learn about conditions in the external environment and within our own bodies through receptors, which are activated by specific types of stimuli: light, pressure, chemicals, and so forth. The perception of a stimulus occurs in the brain. (See CTQ #2.)

SOMATIC SENSATION

Somatic sensation (*soma* = body) is a sense of the physiological state of the body. The information is gathered by somatic sensory receptors located in the skin, skeletal

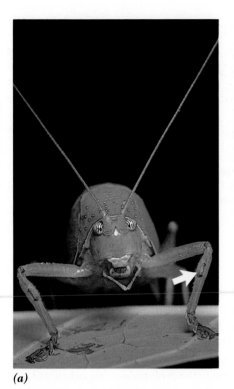 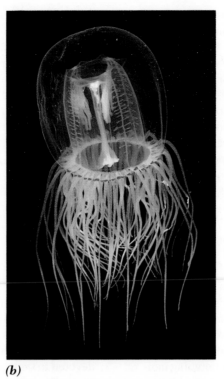 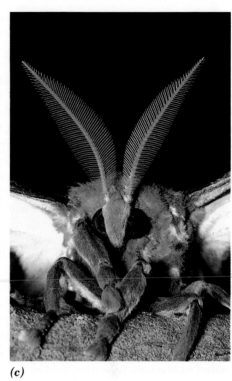

(a)   *(b)*   *(c)*

FIGURE 24-1

A gallery of invertebrate sense organs. *(a)* The hearing organ of this katydid is located on its legs. Sound produces vibrations of the exoskeleton. *(b)* Motion detectors *(statocysts)* in the rim of the bell of this jellyfish allow the animal to maintain its orientation as it swims. *(c)* Chemoreceptors in the antennae of this male atlas moth allow it to detect the smell of a female moth of the same species from a distance of 11 kilometers.

TABLE 24-1

TYPES OF SENSORY RECEPTORS

Example of Receptor	*Example of Function*
1. *Mechanoreceptors*	
Pacinian corpuscle in deep layers of skin	Sense of deep touch
Stretch receptors in skeletal muscles	Maintenance of posture
Hair cells of vestibular apparatus and cochlea of inner ear	Sense of balance and hearing
2. *Thermoreceptors*	
Ends of sensory neurons in skin	Monitor external temperature
Neurons in hypothalamus	Monitor internal temperature
3. *Chemoreceptors*	
Receptor cells in arteries	Monitor blood oxygen level
Receptor cells in taste buds	Sense of taste
Receptor cells in olfactory epithelium	Sense of smell
4. *Photoreceptors*	
Rod cells in retina	Low-light vision
Cone cells in retina	bright-light, color vision
5. *Pain receptors*	
Ends of sensory neurons	Awareness of tissue damage

muscles, tendons, and joints. Somatic sensory receptors inform the CNS about pressure, stretch, temperature, and whether or not a stimulus is intense or threatening— judgments that often lead to the perception of pain. Information from these receptors is required for bodily movements and for maintaining homeostasis.

Although the skin is not considered a specialized sensory organ like the eye or the ear, it is generously endowed with several types of sensory receptors, all of which are the tips of sensory neurons (Figure 24-2). Different nerve endings are specialized to respond to warmth, cold, painful stimuli, or pressure (touch). Some of the pressure receptors of the skin, such as the Pacinian corpuscle shown in Figure 24-2, contain a multilayered capsule that resembles an onion. The capsule insulates the sensitive nerve ending so that weak stimuli, such as pressure exerted by the clothes you are wearing, fail to trigger neural impulses.

Information gathered by somatic sensory receptors in the skin travels through the central nervous system until it reaches two narrow strips of cortex, one on each side of the brain (Figure 24-3). These strips are called the **somatic sensory cortex.** The map of the **somatic sensory cortex** has been obtained by studying individuals undergoing neurosurgery. For example, touching a person on the fingers, stomach, and toes produces electrical activity in three distinct and predictable areas of the brain's sensory cortex. Conversely, if a neurosurgeon directly stimulates a few cells in this part of the brain using an electrified probe, a person

reports feeling warmth or tingling in the corresponding part of the body. Those parts that are most sensitive to stimulation and have the greatest number of sensory receptors, such as the fingers and lips, are represented by a disproportionately large number of neurons in the somatic sensory cortex (Figure 24-3). The pathways by which tactile (touch) information is processed in the brain is discussed in The Human Perspective: Two Brains in One.

Somatic sensory receptors gather information on the state of the body, which is used primarily to direct bodily movements and to maintain homeostasis. (See CTQ #3.)

VISION

◗◗▶ Vision is a sense that is based on light rays reflected into our eyes from objects in the external environment. To most people, vision is the richest form of sensory input. The complexity of our eyes and the large amount of brain tissue devoted to sight attest to the evolutionary importance of this sense. We inherited a keen sense of sight from our primate ancestors who lived in trees and got around by jumping from branch to branch. One miscalculation could result in a long fall to the ground. In this environment, natural selection favored those animals with better eyesight.

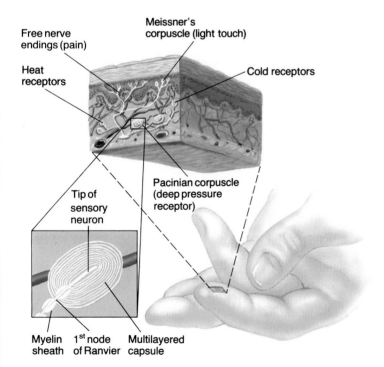

FIGURE 24-2

Human skin is the residence of a variety of somatic sensory receptors that respond to mechanical, chemical, and thermal stimuli. These receptors are particularly numerous in the skin of the fingertips, lips, and genitals, where they provide heightened sensitivity. Several mechanoreceptors, such as the Pacinian corpuscle, consist of a free sensory nerve ending surrounded by a capsule whose layers of connective tissue resemble the layers of an onion. The contents of the capsule suppress slight mechanical stimuli, preventing initiation of a neural impulse. In contrast, Meissner's corpuscles, which are located closer to the surface and lack the multilayered capsule, are activated by light touch.

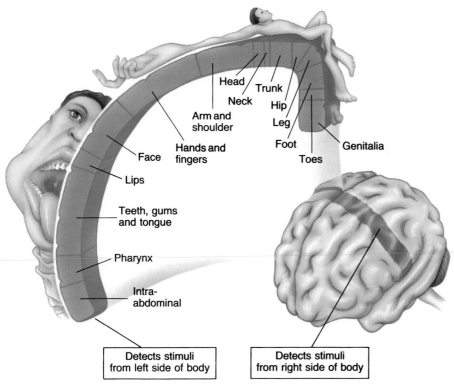

RIGHT SOMATIC SENSORY CORTEX

FIGURE 24-3

The somatic sensory cortex. This strip of gray matter is located just behind the motor cortex (see Figure 23-13) on both sides of the brain. Information from sensory receptors in the skin, muscles, tendons, and joints is transmitted to the somatic sensory cortex, where it is processed and used in directing bodily movements and maintaining homeostasis. The area of cortex devoted to each part of the body is indicated.

The main component of the human eye—the eyeball (Figure 24-4)—is roughly spherical in shape and is situated within a protective socket of the skull. The wall of the eyeball can be divided into three complex layers, each with distinct functions.

1. The outer layer consists of the sclera in the rear and the cornea in the front. The **sclera** is a tough, connective-tissue capsule that protects the eye, whereas the **cornea** is a delicate, transparent window, whose remarkable structure lets light rays pass into the eye unobstructed.

2. The middle layer consists of the choroid in the rear and the iris in the front. The **choroid** is a dark tissue that contains large numbers of blood vessels which supply nutrients and oxygen to all parts of the eye. The **iris** is a disklike, muscular structure with a circular opening, the **pupil,** through which light passes into the eyeball's interior. Pigments in the iris block the passage of stray light and give the eye its distinctive coloring.

3. The inner layer is evident only in the rear of the eye, where it forms the **retina,** a blanket of photoreceptors that provides the screen on which visual images are projected.

THE LENS: FOCUSING IMAGES ON THE RETINA

The interior of the eyeball contains the **lens** (Figure 24-4), which focuses incoming light rays onto the retina. The focusing abilities of the lens are derived from the glasslike proteins of which it is composed. Contraction of the *ciliary muscles* changes the shape of the lens so that objects at different distances from the eye can be focused on the retina. If this change in lens shape (called *accommodation*) did not occur, the image of nearby objects would be focused behind the retina and would appear blurry. As we age, changes in the structure of the lens proteins makes the lens less elastic. Consequently, our eyes become less able to focus on close objects, which explains why so many people need reading glasses after the age of about 45. In some older individuals, the lens becomes opaque, causing a *cataract*, which can be treated surgically.

THE RETINA: A LIVING PROJECTION SCREEN

The retina contains two different types of photoreceptor cells, the *rods* and *cones* (Figure 24-4, inset), names that

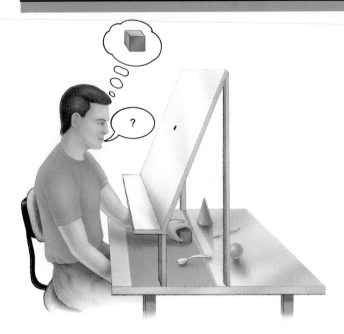

FIGURE 1

Apparatus for testing the functions of the corpus callosum. In this experiment, the subject can pick up the objects on the table with one hand or the other or can touch an object with the feet but cannot see the objects being manipulated.

As you will recall from the previous chapter, the largest part of the human brain is divided into right and left cerebral hemispheres, which are connected to one another by the corpus callosum, a "cable" approximately 200 million nerve cells thick (see Figure 23-11). During the first half of the twentieth century, scientists observed that accidental damage to the corpus callosum in people who suffered from epileptic seizures led to dramatic improvement in the person's epileptic condition. In the 1940s, William Van Wagenen, a neurosurgeon at the University of Rochester, attempted to treat a group of epileptic patients by surgically severing their corpus callosum. The patients improved and showed virtually no ill effects from the operation. In fact, neurological and psychological tests indicated no significant loss of brain function.

Roger Sperry of the California Institute of Technology had studied the neural basis of consciousness and was convinced that earlier investigators had failed to discover the behavioral effects of severing the corpus callosum because they hadn't examined the function of each hemisphere independently of the other. In the 1960s, Sperry devised a procedure for testing these split-brain patients so that the func-

tions of each half of the cerebrum could be analyzed separately. He controlled experimental conditions so that the sense organs of "split-brain" patients were stimulated on only one side of their body. For example, Sperry would ask the subjects to pick up an object with only one hand while sitting at an apparatus that prevented them from seeing the object (Figure 1). If the subjects picked up a cube with their right hand and were asked to name the object, they were able to say it was a cube. They could also write the word "cube" on a piece of paper. If they picked up the same object with the left hand, however, the subjects were unable to identify the object either verbally or in writing. This was clearly different from a "normal person" with an intact corpus callosum, who had no difficulty identifying the object or writing the object's identity, regardless of which hand held it.

Sperry had an explanation for this seemingly strange behavior in split-brain subjects: The speech and writing centers of the brain are located in the left cerebral hemisphere, which receives information from the sense organs of the right side of the body. Consequently, when the subjects picked up an object with their right hand, sensory information about the shape of the

object was sent to the left cerebral hemisphere which, because it contains the speech and writing center, allowed the subjects to describe the object. In contrast, when the same object was picked up by the left hand, the information went to the right cerebral hemisphere, which had no way of "informing" the speech center in the left hemisphere of the experience. Consequently, the subjects could not verbalize the shape, often guessing wildly at its identity. Yet, they could accurately point to a picture of a cube among a gallery of different-shaped objects, indicating an awareness of the object's identity. Remarkably, they might even draw a picture of a cube and, at the same time, verbally guess that they were holding a sphere or a rod. Such responses were accompanied by signs of frustration from the subjects, indicating that the right side of the brain recognized that this was an erroneous verbal response. Sperry concluded that, without an intact corpus callosum, each hemisphere is unaware of the sensory and mental experience of its partner. In recognition of his work on the brain, Sperry was awarded the 1981 Nobel Prize in Medicine and Physiology.

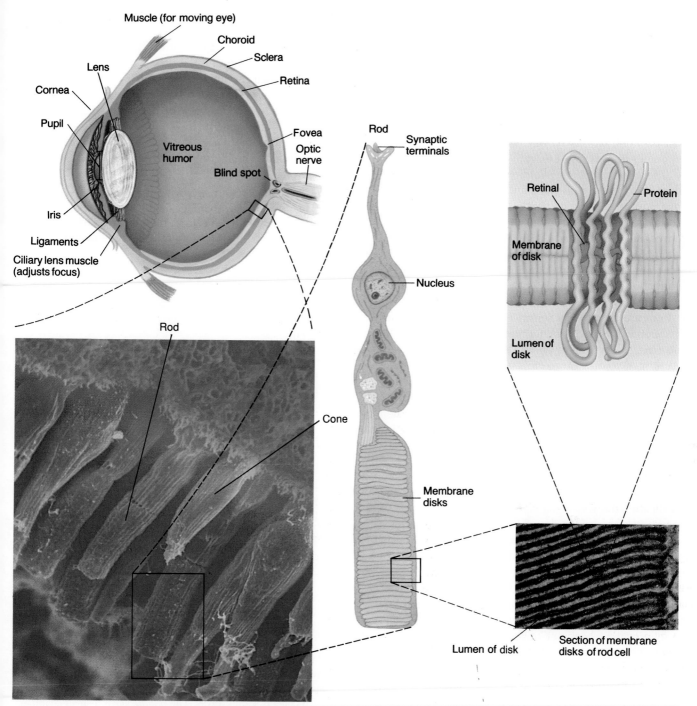

FIGURE 24-4

The human eye.   Light enters through the cornea and then travels through the pupil, the hole in the middle of the colored iris. Pupil size changes to regulate the amount of light that enters the eye. Light rays are focused on the retina by the lens, whose shape is changed by *ciliary lens muscles* acting on ligaments. The retina has two kinds of photoreceptors: rods and cones. Activation of the photoreceptors initiates impulses, which are conducted along neurons that make up the *optic nerve*. The light-absorbing pigments of the photoreceptors are embedded in membranous disks of the photoreceptor cells. The acute light sensitivity of rods allows us to see in very dim light. Cones allow us to see color. The *fovea* is the site in the retina of the greatest concentration of cones. The hole through which the optic nerve exits the inner surface of the eye lacks photoreceptors and creates a small "blind spot" in the visual field. The rear chamber of the eye is filled with a viscous fluid, the *vitreous humor.* Upper right inset courtesy of J. David Rawn, *Biochemistry,* ©1989, p. 1066. Reprinted by permission of Prentice Hall, Englewood Cliffs, New Jersey.

reflect their microscopic shape. Both types of cells contain pigment molecules that absorb incoming light. Each pigment molecule consists of a protein linked to a molecule of *retinal,* which is synthesized from vitamin A (thus making carrots, which are rich in vitamin A, good for your eyesight). Like the light-absorbing pigments of plant chloroplasts (Chapter 8), photopigments of the rods and cones are embedded in membrane disks. *Rhodopsin,* the best-studied visual pigment, worms its way back and forth across the membranes of rods (most magnified inset of Figure 24-4). The absorption of a single photon of light (the smallest unit of light energy) can cause an alteration in the shape of the retinal portion of a single pigment molecule, triggering a sequence of changes that culminate in the closure of several hundred ion channels and a change in membrane potential of about 1 millivolt. The additive changes in potential that result from the absorption of a number of photons lead to the initiation of an action potential in a sensory neuron of the *optic nerve* leading to the brain.

Rods are much more sensitive to light than are cones and function primarily under low-light conditions. The human retina contains over 100 million rods, which are concentrated toward the peripheral regions of the retina. Rods provide us with night vision, which is characterized by a lack of color and sharpness but a heightened sensitivity. The high sensitivity of the rods can be appreciated when you walk out of bright sunshine into a darkened theater. At first you are virtually blind because only your cones—which respond to bright light—are functioning; the pigment molecules of the rods have been temporarily inactivated by the bright light. As you spend more time in the dark theater, the rod pigments are reactivated, and your eyes become *dark-adapted,* enabling you to see the outlines of people in the theater. The dilation of the pupil, which allows more light to reach the retina, is another response of the dark-adapted eye. Even with these dark adaptations, humans have relatively poor night vision compared to nocturnal animals, such as cats and owls. Many nocturnal animals have a mirrorlike layer *(tapetum)* behind their retina that reflects light back out of the eye, giving the rods a "second chance" to absorb light rays. This mirrorlike layer is the reason the eyes of many animals appear to glow at night (see Figure 24-10).

Unlike rods, cones are relatively insensitive to light and are essentially useless under conditions of low light intensity. Cones are concentrated in the center of the human retina, where they provide a highly detailed image of the visual field. Unlike rods, cones also provide information on the wavelength of the light, which we perceive as color. Three different types of cones are distinguished by the structure of their pigments and the wavelengths of light they absorb. It is thought that the color we perceive in a particular part of the visual field depends on the ratio of impulses generated by the blue-, green-, and red-absorbing cones in the corresponding area of the retina. The various types of colorblindness can be explained as a loss of one of the three types of cones.

Impulses triggered by the rods and cones of the retina travel to the **visual cortex** of the occipital lobe of the brain (Figure 23-12). The visual cortex was mapped with considerable precision around the turn of the century by a Japanese physician, Tatsuji Inouye. Inouye interviewed soldiers from the Russo-Japanese War who had recovered from bullet wounds to the back of the head. He was able to correlate blind spots in the soldier's visual field with the location of brain damage determined by the entrance and exit wound of the high-velocity Russian bullets.

The eye collects light rays from the external environment and focuses them as an image on the blanket of photoreceptors that comprise the retina. Impulses from the receptors are interpreted by the brain as a detailed mental picture. (See CTQ# 4.)

HEARING AND BALANCE

The human ear contains two separate sensory systems, one that governs hearing, and the other that allows us to maintain our balance. The sensory receptors for both of these senses are **hair cells**—mechanoreceptors that lie in the interior portion of the ear (Figure 24-5a) and contain very fine "hairs." When these hairs are bent or displaced by movements or pressure, a change in the membrane potential of the hair cell is generated, which may launch an impulse along a sensory neuron carrying the information to the brain. Depending on the part of the ear in which the impulses are generated, we may perceive a particular sound or a feeling of motion.

FIGURE 24-5
The human ear. *(a)* The *outer ear* consists of the ear flap (or *pinna*) and the channel leading to the tympanic membrane, which collects sound waves. The *middle ear* is a small chamber composed of three interconnected ear bones (often called the hammer, anvil, and stirrup) which transmit and amplify sound waves. The *inner ear* contains the cochlea (and its receptors for hearing) and the vestibular apparatus (and its receptors for balance). Motion of the fluid in the cochlea is initiated when the stirrup pushes in at the *oval window.* Waves are transmitted through the cochlear fluid until they reach the *round window,* where they are dissipated. The first electron micrograph shows several rows of hair cells along a portion of the cochlea, and the second micrograph shows a single hair cell. The hair cells translate vibrations into neural impulses that travel to the hearing centers of the brain. *(b)* Pete Townshend of the "Who" has experienced severe hearing loss as a result of exposure to amplified music.

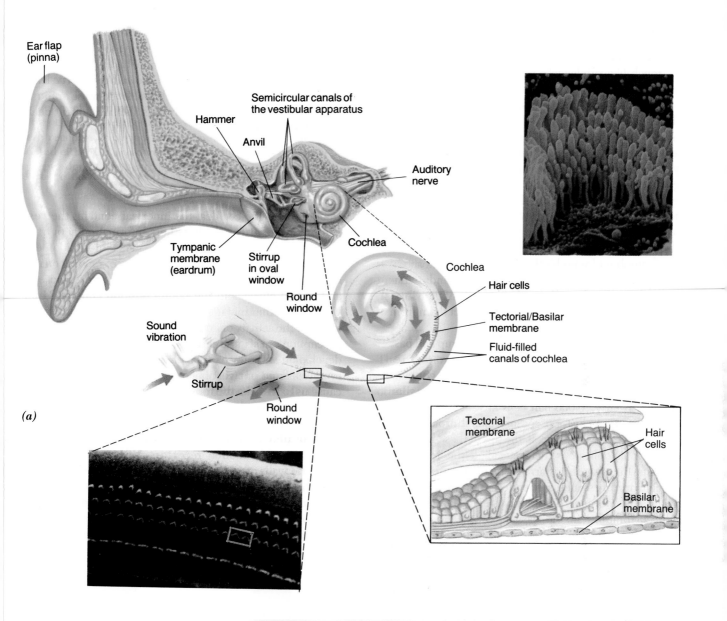

(a)

(b)

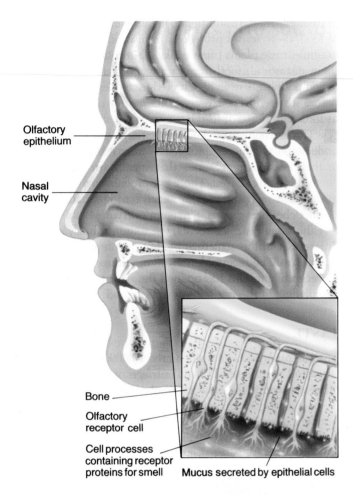

Olfactory epithelium

Nasal cavity

Bone

Olfactory receptor cell

Cell processes containing receptor proteins for smell

Mucus secreted by epithelial cells

FIGURE 24-8

Olfaction. The olfactory receptor cells are located in patches within the inner lining of the nasal cavity. The receptor proteins themselves are located on processes that extend into a layer of mucus that lines the nasal chamber. Activation of the receptor cells sends impulses along fibers of the olfactory nerve into a nearby portion of the brain, called the olfactory bulb. Information is then transferred from the olfactory bulb to the olfactory centers of the cerebral cortex, where odor is perceived.

taste bud through an open pore and interact with sensory hairs projecting from each receptor cell. There are only four *primary tastes:* salty, sweet, sour, and bitter (Figure 24-7*b*). These primary tastes become combined in various ways to give us our rich discrimination of flavors. The perception of each primary taste can be elicited by chemicals with very different structures. For example, the sweetest substance yet discovered is thaumatin, a protein isolated from the berry of an African shrub. Thaumatin is several hundred times sweeter than the dipeptide aspartame (Nutrasweet), which is much sweeter than is the disaccharide sucrose (table sugar). Even though taste and smell constitute distinct senses, the taste of our food is greatly affected by its smell. This becomes evident when you suffer from a bad cold; you lose the ability to discriminate among subtle taste differences.

SMELL

The sensory receptors for smell are located within the olfactory epithelium that lines the interior wall of the nasal cavity (Figure 24-8). Even humans, whose sense of smell is relatively poor compared to other mammals (such as mice or dogs), can discern thousands of different types of sub-

stances and can detect them at extremely low concentrations. Methyl mercaptan, for example, the substance added to natural gas (which is otherwise odorless) to give it the characteristic aroma that alerts us to gas leaks, can be detected at a concentration of one part methyl mercaptan per billion parts of air. For many substances, a single molecule impinging on an olfactory receptor cell can trigger an impulse to the CNS.

Our senses of taste and smell derive from chemoreceptors present in the mouth and the lining of the nasal cavities, respectively. Our perception of a taste or smell results from the particular combination of chemoreceptors activated. (See CTQ# 6.)

THE BRAIN: INTERPRETING IMPULSES INTO SENSATIONS

In the previous chapter, we mentioned that all impulses traveling along a neuron exhibit the same strength. Consequently, neurons cannot conduct stronger impulses in re-

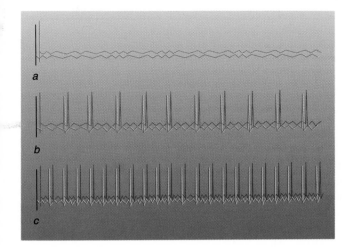

F I G U R E 24-9

The strength of a stimulus is generally encoded in the frequency of the impulses carried by the neuron. These recordings, taken from a living neuron, show a neuron at rest *(a)*, after exposure to a weak stimulus *(b)*, and after exposure to a stronger stimulus *(c)*. Each spike represents an action potential moving across a point on the neuron's membrane.

sponse to increased intensity of stimulation. Yet, we are clearly capable of detecting differences in the strength of a stimulus. If you put your fingers into a bathtub of water, for example, you will learn more than whether or not the water is hot. You are able to perceive a *range* of thermal differences, from unpleasantly cold to scalding hot.

The ability to make sensory discriminations depends on several factors. For example, a stronger stimulus, such as scalding water, activates more nerve cells than does a weaker stimulus, such as warm water. It also activates "high-threshold" neurons that would remain at rest if the stimulus were weaker. Stimulus strength may also be *encoded* in the frequency by which action potentials are launched down a particular neuron; the stronger the stimulus, the greater the frequency (Figure 24-9).

DETERMINING THE NATURE AND LOCATION OF A STIMULUS

Why does an impulse create pain when delivered from a pain receptor, whereas the exact same type of impulse creates sight when delivered by a neuron from a photoreceptor? Furthermore, how do we know where the pain is felt, even without looking at the part of the body experiencing it? Just think about the last time a doctor drew a blood sample from your thumb. Even without looking at the damage, you know what has happened and the exact spot where the needle entered. Your brain has been notified of the disturbance and translates the incoming impulses as "sharp pain in the tip of the left thumb." The neurons triggered by this stimulus carry no such message, however; all they can transmit is an action potential sweeping along the neuron. The nature and exact location of the disturbance are determined by the specific portions of the brain that are stimulated. Sensory input from the eyes, ears, nose, and somatic sensory receptors of the skin travel to separate locations within the cerebral cortex (see Figure 23-12). These are the sites in the brain where our perceptions of sight, hearing, smell, and touch are derived. Any time the cells in a particular portion of the brain are activated, the same sensation is perceived in the same location in the body; the brain perceives the sensation, not the finger. This phenomenon is readily demonstrated by electrode stimulation during open-skull surgery. If the electrode stimulates a particular site in the brain, the patient will experience a very real sensation of a sharp object piercing his or her unassaulted thumb, for example, just as though the finger were really being pierced.

The type of sensation perceived by the brain depends on three factors: (1) the specific region of the cerebral cortex that is stimulated; (2) the number of brain cells that are stimulated; and (3) the frequency with which impulses arrive along various neurons. (See CTQ# 7.)

EVOLUTION AND ADAPTATION OF SENSE ORGANS

Humans don't use their eyes to help them hear or their ears to help them see. Unlike a nervous system or a digestive system, whose parts are functionally linked, sense organs function and evolve relatively independent of one another. As a result, sense organs exhibit a much greater diversity in form and function among various animals than do most other types of physiological structures. For example, sense organs that detect light have appeared over and over again throughout the course of evolution, producing animals with a great variety of light-sensitive organs (Figure 24-10). Yet, despite the diversity among sense organs, obvious similari-

◁ B I O L I N E ▷
Sensory Adaptations to an Aquatic Environment

▶ In this chapter, we have focused on human sense organs, which are adapted for life in a terrestrial environment. A brief examination of some of the sense organs found in fishes will illustrate how aquatic environments select for sense organs having very different properties.

VISION

If you have ever tried to find an object on the bottom of a pool, you are aware that humans don't see very well underwater. Our poor underwater eyesight can be blamed in part on our eyes and in part on the limitations that water places on vision. For a vertebrate living in a terrestrial habitat, light rays are bent (*refracted*) as they pass from air across the cornea into the watery tissue of the eye; this bending of light is part of the focusing process. In contrast, light rays are not bent as they pass from the aquatic environment into the eye of a fish. Consequently, virtually all of the focusing by the fish's eye must be accomplished by the lens, which is much harder and more spherical than is the lens of a terrestrial vertebrate. Because of its hardness, the shape of a fish's lens cannot be changed by muscles; instead, light rays are focused by a back and forth motion of the lens within the eye. The reason we see so poorly underwater is that our cornea is essentially eliminated as a focusing element. Furthermore, light rays cannot travel nearly as far through water as they can through air; thus, fish cannot use their eyes as distance receptors as do terrestrial vertebrates.

HEARING

A fish's ears consist exclusively of an inner-ear compartment; there is no outer or middle ear. In most fishes, sounds are probably transmitted to the receptors in the inner ear by vibrations of the bones of the skull, which is a relatively insensitive mechanism. In fact, the structure of a fish's inner ear suggests that the ear primarily provides information on body position, not hearing. It is thought that the original function of the vertebrate ear was to maintain body equilibrium; it became a hearing organ secondarily.

CHEMORECEPTION

Due to the limitations imposed on their visual sense, most fishes rely primarily on their sense of smell. Substances dissolved in the water inform a fish of the presence of food, a potential mate, a toxic substance, or even a predator. For example, some fishes that travel in large schools emit an *alarm substance* from their skin in the presence of danger, which alerts other fish to disperse.

MECHANORECEPTION

Water is an excellent conductor of vibrations or pressure disturbances. Fishes have evolved a unique type of sensory system— the *lateral line*—to monitor such stimuli. The system consists of canals that run down the flank of the fish and across its head. The canals contain sensory cells with hairlike projections that are activated by the flow of water. The lateral line can be thought of as a system for sensing "distant touch." Using this system, fishes can detect waves hitting a distant shore or waves generated by a distant boat. Closer to home, the system can inform the fish of objects in its immediate surrounding which affect the flow of water around the animal. For example, under normal circumstances, a fish will avoid a finger placed in its midst. If the lateral line system is covered so that the fish cannot receive sensory information, the finger is ignored, and the fish can be gently pushed with it.

ELECTRORECEPTION

Certain eels, catfish, and rays are able to deliver a strong electric shock—up to 600 volts—to anyone who happens to come too near. Other fishes have much weaker electric organs that are used not as weapons but as sensory systems and a means of communication. These fish generate an electric current that flows from the tail to the head. Objects that come within the fish's electric field will interfere with the flow of electric current through the water; this change is detected by sense organs in the fish's head. Impulses from these sensory receptors provide information on the nature and location of objects in the water. These electrolocation systems are particularly important to species that live in murky waters, where vision is highly limited.

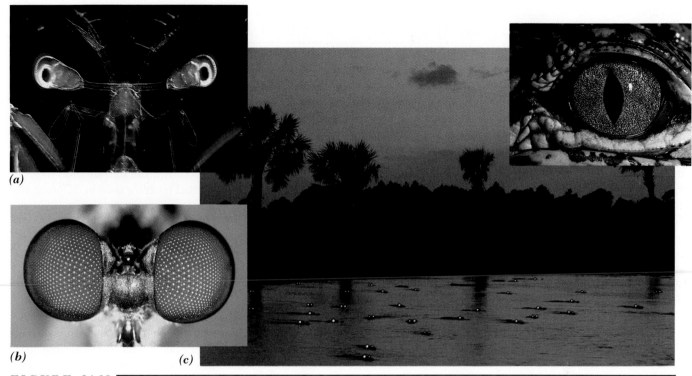

(a)

(b)   *(c)*

FIGURE 24-10

Some striking adaptations for photoreception. *(a)* Stalked eyes of a mantis shrimp. *(b)* The compound eyes of a robber fly are composed of thousands of individual lenses that give the fly an extremely broad visual field. *(c)* Only their eyes betray the silent nocturnal approach of this flotilla of alligators closing in on an observer's spotlight. A special layer of tapetum cells in the alligator's retina reflects light, creating the telltale orange-red glow. (Inset: An alligator's eye in daylight.)

ties exist among many of the components, particularly the receptor cells. For example, all photoreceptors contain pigments capable of absorbing light and opening ion channels in the membrane. On the surface, the sound-receptor organs of a grasshopper (Figure 24-1*a*) and a human could not be more different: The organ is constructed out of a portion of the exoskeleton on the grasshopper's leg, while the human organ is embedded deep within the skull. Yet, both sound detectors contain a mechanoreceptor cell that depends on vibrations of a part of the body surface. Grasshoppers sense vibrations of their thin, outer exoskeleton, while humans detect vibrations of their thin eardrum.

▮▶ Consider for a moment the different types of challenges facing animals that live in different environments. An earthworm or mole that lives underground is exposed to an entirely different set of environmental stimuli than is a fish or shrimp that lives in the sea. Similarly, an aquatic animal faces entirely different conditions than does a lizard or a bird living in a terrestrial habitat. Since sense organs monitor changes in an animal's environment, they must be able to respond to the types of stimuli to which the animal is exposed. In fact, sense organs play a key role in adapting an animal to a particular environment. For example, earthworms have chemoreceptors that measure the acidity of the soil; fish have mechanoreceptors that detect vibrations in the water; and the eyes of a lizard are adapted to focusing light rays from the air (rather than from water). The role of the environment in shaping the properties of sense organs is discussed in the accompanying Bioline: Sensory Adaptations to an Aquatic Environment.

Similarly, each animal has sense organs that are adapted to the animal's particular mode of existence in that environment. For example, an owl, which feeds in the dark, has such a keen sense of *hearing* that it can attack and kill a mouse while blindfolded. In contrast, a hawk, which feeds during the day, has such a keen sense of vision that it can spot a mouse while flying hundreds of feet in the air. Echolocation in bats is one of the best examples of sensory adaptation. Bats live in a world of darkness, roosting in pitch-black caves and picking minute insects out of the air while flying at night. Eyes are of minimal use to bats and play a minimal role in their lives. In contrast, a bat's ears are unaffected by darkness, making echolocation an ideal primary sense.

The type of sense organs an animal possesses can be correlated with the animal's habitat and mode of existence. Despite the diverse structure of sense organs, the receptor cells often share common features that may reflect common ancestry and/or common selective pressures. (See CTQ# 8.)

REEXAMINING THE THEMES

Relationship between Form and Function

Sensory organs are devices whose structure allows them to detect a particular type of stimulus. The human ear, for example, contains a large flap for collecting sound waves, an external channel for focusing sound waves on the eardrum, and an elaborately constructed mechanical apparatus for amplifying remarkably small vibrations. A closer look at the actual receptor cells within a sense organ reveals a similar relationship between structure and function: Receptors in the ear contain clusters of "hairs" that are sensitive to bending; receptors in the eye contain membrane pigments whose structure allows them to absorb particular wavelengths of light; and receptors in the nose and mouth contain proteins whose structure allows them to bind specific molecules in the air or food.

Biological Order, Regulation, and Homeostasis

Maintaining the order and constancy of the internal environment requires sense organs that monitor conditions both inside and outside an animal's body. If the environment becomes too warm, too salty, or too acidic, sensors scattered around the body inform the central nervous system of these changing conditions, providing the information necessary for the body to initiate an appropriate homeostatic response.

Unity within Diversity

Regardless of the type of animal, photoreceptors contain light-sensitive pigments; chemoreceptors contain proteins that combine with specific chemicals; mechanoreceptors contain elements that become deformed when pressure is applied; and so forth. Yet, despite this unity, the organs that house the actual receptor cells in different animals are often very different in structure. For example, the hearing organ of a grasshopper, the eye of a fly, or the olfactory organs of a moth, all have very different structures than do their counterparts in the human body.

Evolution and Adaptation

Each sense organ has evolved in response to the selective pressures at work in the particular environment in which an animal lives. For example, humans and most other mammals possess sense organs that are adapted to a terrestrial environment. Our eyes and ears are adapted to collecting sound waves from the air, which is why we have such a terrible time seeing and hearing underwater. Similarly, humans evolved as a species that was active during the day and slept at night. Natural selection favored individuals whose eyes were adapted for conditions of relatively bright light, which is why we see sharp, color images during daylight hours and only dim shadows on a dark, moonless night. In contrast, the eyes and ears of mammals that live underwater, or that have their peak activity at night, function optimally under aquatic or nocturnal conditions, respectively.

SYNOPSIS

Sense organs collect information about conditions in the external and internal environment and pass the information on to the central nervous system. Each type of sensory receptor is specialized to respond to a particular type of stimulus. The interaction between a stimulus and a receptor induces a change in the receptor, which leads to a voltage change across the receptor cell plasma membrane, which may initiate an action potential in a sensory neuron leading to the CNS.

Humans rely on several different types of sensory stimuli. Somatic sensation monitors conditions on the surface and within the body, providing information required for movement and homeostasis. Vision is a sense that is based on light rays reflected into our eyes from objects in the external environment. Light rays pass through an outer, transparent cornea and are focused by a glasslike lens onto a living "screen" that contains photoreceptors. The brain interprets impulses from these receptors as a visual image of the environment. The retina contains two types of receptor cells: rods, which help provide an image under conditions of low light levels, and cones, which provide a highly detailed, multicolored image when environmental light levels are high. The inner ear contains a pair of sense organs—the cochlea and the vestibular apparatus—which provide us with the sense of hearing and balance, respectively. Both senses derive from hair cells whose membrane voltage changes when sensory hairs projecting from the receptor cell are displaced. Displacement of hairs in the cochlea occurs when sound vibrations are transmitted from the environment into the basilar membrane of the cochlea, while displacement of hairs in the vestibular apparatus results from fluid movements that occur when the head is moved. Taste and smell are senses that depend on the stimulation of chemoreceptors located in taste buds on the tongue and in the inner lining of the nasal cavity, respectively. In both cases, the chemoreceptor cells contain membrane protein receptors that combine with specific chemicals—either as dissolved food matter (taste) or as airborne particles (smell).

The strength, location, and nature of the stimulus are interpreted by the brain. Stimulus intensity may be coded in the number of neurons stimulated and the frequency of impulses propagated. The cerebral cortex contains different regions devoted to somatic sensation, sight, hearing, and smell. The location and nature of the stimulus are determined by the particular sites within these areas that ultimately receive the impulses. In humans, the two cerebral hemispheres function relatively independently of each other and communicate across the corpus callosum.

Key Terms

mechanoreceptors (p. 500)
thermoreceptors (p. 500)
chemoreceptors (p. 500)
photoreceptors (p. 500)
pain receptors (p. 500)
somatic sensation (p. 500)
somatic sensory cortex (p. 502)
sclera (p. 502)

cornea (p. 503)
choroid (p. 503)
iris (p. 503)
pupil (p. 503)
retina (p. 503)
lens (p. 503)
hair cell (p. 506)
tympanic membrane (p. 508)

cochlea (p. 508)
basilar membrane (p. 508)
tectorial membrane (p. 508)
vestibular apparatus (p. 508)
semicircular canal (p. 508)
taste (p. 509)
smell (p. 509)
olfaction (p. 509)

Review Questions

1. Trace the events that occur in the eye and the central nervous system that allow you to see the outlines of people seated in a dark theater.

2. How do you know where a stimulus is coming from in the body? Or the strength of the stimulus? Or its nature?

3. Why wouldn't a person with a severed corpus callosum notice any loss of abilities in his or her daily activities?

4. Name the sensory structure(s) in humans that contain(s): hair cells, photoreceptors, mechanoreceptors, thermoreceptors, chemoreceptors, olfactory receptors, semicircular canals, tympanic membrane, and somatic receptors.

Critical Thinking Questions

1. What control might the graduate students working on blind or blindfolded human subjects have performed to be certain that sound reception was the basis for avoidance of the screen?

2. The Chinese describe the senses as "the gateway to the mind." Is this an accurate description? Explain your answer.

3. Diagnose the area of injury in each of the following cases: (1) a stroke patient is paralyzed on the right side and cannot speak or write; (2) following an accident, a truck driver can no longer perceive distance; (3) after surgery, a patient can pick out specific shapes but cannot name them correctly.

4. How, and why, would each of the following affect a person's vision? (1) cataracts in both eyes; (2) injury to the ciliary muscles; (3) vitamin A deficiency; (4) loss of elasticity of the lens; (5) damage to the visual cortex of the brain; (6) blindness in one eye.

5. Otosclerosis is a disease that limits the movement of the bones in the middle ear. Why would this condition affect hearing? Can you devise some surgical procedure that might correct the problem?

6. What is the justification for lumping the senses of taste and smell together as "chemoreception?"

7. How can you explain the facts that: (1) images are projected onto the retina upside-down, but we "see" objects right side up; (2) people who become able to see after being blind from birth do not immediately have a three-dimensional view of the world; (3) we can be fooled by optical illusions?

8. Which sense(s) would you expect would be highly developed in each of the following, and why?: a hawk, a monkey, an owl, a bat, a mole, a salmon.

Additional Readings

Barinaga, M. 1991. How the nose knows: Olfactory receptor cloned. *Science* 252:209–210. (Intermediate–Advanced)

Borg, E., and S. A. Counter. 1989. The middle-ear muscles. *Sci. Amer.* Aug:74–81 (Intermediate)

Freeman, W. J. 1991. The physiology of perception. *Sci. Amer.* Feb:116–125. (Intermediate)

Glickstein, M. 1988. The discovery of the visual cortex. *Sci. Amer.* Sept:118–127. (Introductory)

Griffin, D. R. 1974. *Listening in the dark.* New York: Dover. (Introductory)

Hudspeth, A. 1983. The hair cells of the inner ear. *Sci. Amer.* Jan:54–66. (Intermediate)

Kandel, E., and J. Schwartz. 1985. *Principles of neural science.* New York: Elsevier. (Intermediate–Advanced)

Kellogg, W. N. 1961. *Porpoises and sonar.* Chicago: University of Chicago Press. (Introductory)

Konishi, M. 1993. Listening with two ears. *Sci. Amer.* April:66–73. (Intermediate)

Parker, D. 1980. The vestibular apparatus. *Sci. Amer.* Nov:118–130. (Intermediate)

Shreeve, J. et al. 1993. The mystery of sense. *Discover* June:35–85. (Intermediate)

Stryer, L. 1987. The molecules of visual excitation. *Sci. Amer.* July:42–50. (Intermediate)

CHAPTER
◄ 25 ►

Coordinating the Organism: The Role of the Endocrine System

STEPS
TO
DISCOVERY
The Discovery of Insulin

BIOLINE
Chemically Enhanced Athletes

THE HUMAN PERSPECTIVE
The Mysterious Pineal Gland

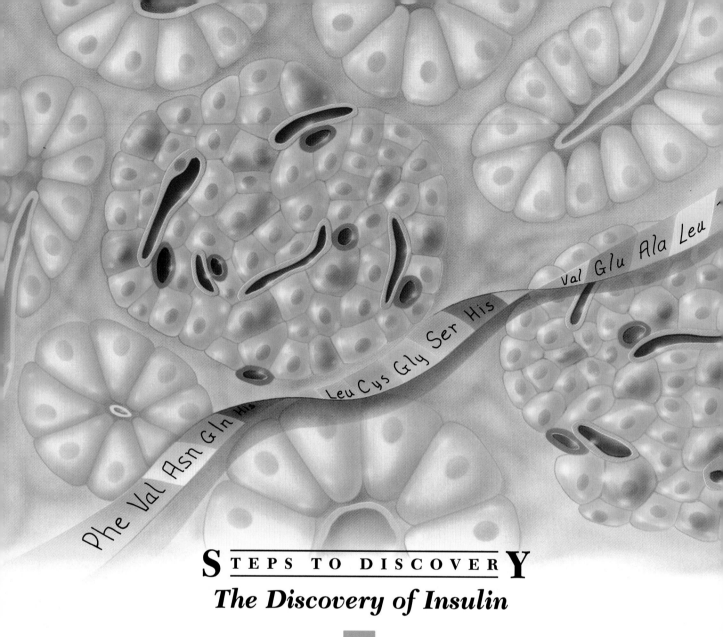

Phe Val Asn Gln His Leu Cys Gly Ser His Leu Val Glu Ala Leu

STEPS TO DISCOVERY

The Discovery of Insulin

Diabetes remains a common and potentially fatal disease, one that can destroy the kidneys, damage the heart, disrupt circulation (leading to the amputation of toes and feet), and cause blindness. Yet, diabetes is treatable; today, most diabetics lead normal lives. Until the 1920s, however, diabetes claimed its victims rapidly and horribly. Diabetics excreted enormous quantities of urine; possessed an unquenchable thirst; and generally "wasted away" due to their inability to utilize the nutrients they consumed. Today, daily injections of a simple protein, insulin, help control diabetes.

The story begins in 1869, when Paul Langerhans, a German doctoral student, discovered that the pancreas had two very different types of secretory cells. One group, the *acinar cells,* produced digestive enzymes that were shipped through ducts from the pancreas to the small intestine. The other group of cells was clustered into islands (later named

the *islets of Langerhans*), which possessed neither ducts nor any detectable avenues for the export of the substance these glands secreted; their function was unknown.

Nearly 20 years later, two German physiologists, Oscar Minkowski and Joseph von Mering, attempted to determine the functions of the pancreatic digestive enzymes by removing the pancreas from a dog. Following removal of the glands, the dog, who had previously been house-trained, began urinating frequently on the floor. Minkowski and von Mering were aware that frequent and large-volume urination was one of the key symptoms of diabetes, as was a high level of sugar in the urine. They analyzed the dog's urine and found that it had an elevated sugar content. Removal of the pancreas had made the dog diabetic. The scientists traced the induced effect to the absence of a product from the islets of Langerhans. Yet, for the next 30

A stained section of the pancreas reveals islands of endocrine cells (reddish), which include cells that produce insulin. The

years, all attempts to isolate the product of these cells proved unsuccessful. This would soon change when, in 1920, a young doctor lay in bed struggling with insomnia. Unable to sleep, he devised a plan.

The doctor was Frederick Banting, a 29-year-old Canadian surgeon. As part of his preparation for a lecture he was to deliver to a medical class at the University of Western Ontario, Banting had just read a paper describing an autopsy on a person who had died from an obstruction of the major duct leading from the pancreas. The acinar cells of the pancreas had degenerated, leaving the islets of Langerhans unharmed. As Banting tossed around in bed, the contents of the article circulated through his mind. On a pad of paper lying next to his bed he wrote: "Diabetes. Ligate [tie off] pancreatic ducts of dogs. Keep dogs alive till acini degenerate leaving islets. Try to isolate the internal secretion of these to relieve glycosuria [sugar in the urine]." Banting had concluded (as had others) that the inability to isolate the product of the islets of Langerhans was due to the destruction of the substance by the digestive enzymes of the acinar cells. Banting intended to tie off the pancreatic duct, allowing the acinar cells to degenerate. He hoped that this would allow the antidiabetes factor of the Langerhans cells to be extracted in an undigested state.

Many research projects begin with a scientist reading a paper that arouses his or her curiosity or sparks an idea for a new experimental approach to a nagging question. But Banting was not a research scientist, and he had no lab. Instead, he traveled to the nearby University of Toronto to discuss his plan with John Macleod, an authority on carbohydrate metabolism and diabetes. Banting asked to work in Macleod's laboratory, but Macleod resisted initially; he didn't think the experiment would work, nor did he have confidence in Banting's abilities. After all, even experienced researchers had failed to isolate the secretion of the islets. Banting was persistent, however, and he finally convinced Macleod to give him a bit of lab space and a few dogs to work with. Macleod also asked his physiology class for a volunteer to help out in the summer research project. A 21-year-old medical student named Charles Best came forward. Macleod soon left for an extended holiday in his homeland of Scotland, leaving Banting and Best to carry out what he thought would be a fruitless project.

Together, the two inexperienced researchers worked out a technique to ligate the duct of the pancreas; they then waited several weeks to allow the dog's gland to degenerate. Meanwhile, the researchers prepared several test subjects —dogs whose pancreas had been removed, causing them to become diabetic. Toward the end of July, Banting and Best prepared an extract from the ligated, shrivelled pancreas and injected the extract into a diabetic dog. The urine and blood of the dog showed a dramatic decrease in sugar content. After repeating the procedure several times over the next few weeks, the researchers noted a clear pattern: Injection of the pancreatic extract temporarily relieved the symptoms of diabetes.

By the time Macleod returned in September, Banting and Best were convinced that they were on the verge of a treatment for diabetes. Macleod remained skeptical, however, insisting that the researchers conduct additional experiments. Slowly, the findings were confirmed, and Macleod turned his entire lab over to the study of the pancreatic secretion. He invited James Collip, a visiting biochemist, onto the project to help develop improved procedures for extracting the antidiabetic substance, which the group had named *insulin*. By the end of 1921, the group's work had not yet been published, but the newspapers got wind of the story, and insulin became front-page news around the world.

In January 1922, a 12-year-old boy lay dying of diabetes in a Toronto hospital bed. The Toronto group extracted insulin from beef pancreas, and the boy's physician injected the insulin into the patient, producing astounding results. After receiving daily injections of the extract, the boy left the hospital and resumed a normal life. In the words of M. Bliss, Banting's biographer, "Those who watched the first starved, sometimes comatose, diabetics receive insulin and return to life saw one of the genuine miracles of modern medicine. They were present at the closest resurrection of the body that our secular society can achieve. . . ." Within the year, insulin was being prepared commercially by Eli Lilly, a U.S. pharmaceutical company, and thousands of diabetics were soon receiving the life-saving medication.

In 1923, the Nobel Prize in Medicine and Physiology was awarded to Banting and Macleod, a choice that probably caused more controversy than any other presented. Banting immediately announced that he would share his prize money with Best, whom he felt should have been the corecipient, rather than Macleod. Macleod, not to be outdone, announced that he would share his prize money with Collip. Macleod and Banting remained bitter enemies for the rest of their lives.

ribbon illustrates part of the amino acid sequence of an insulin molecule.

*T*hings could have gone terribly wrong. For days, the child received no calcium in her diet. The absence of this essential mineral in the blood could have triggered a number of life-threatening breakdowns: Neurons could have failed to release their neurotransmitters; muscles might have locked up in painful spasms; blood could have failed to clot, resulting in fatal hemorrhages. But these disasters never materialized; they were prevented by four small glands buried in the child's neck. The pea-sized glands sensed the drop in blood calcium and sounded a molecular alarm, releasing tiny amounts of a chemical—a *hormone*—that stirred the body into corrective action. Some calcium was withdrawn from the bones, the body's calcium bank, and was poured into the bloodstream. At the same time, the hormone slowed excretion of the mineral, while enhancing its absorption from the intestine. The deficit was managed without misfortune.

Yet, the child was not out of danger. There was still the risk that a backlash from these calcium-boosting measures might elevate the concentration of calcium to dangerously high levels in the blood. Fortunately, before that could happen, another gland began releasing its hormone, one that prevents calcium excesses by *reversing* the effects of the first hormone. The constant increase and decrease in the output of these two glands maintains a stable concentration of calcium in the blood, even during temporary deficiencies. This is the essence of homeostasis; without such a system, wild vacillations in calcium concentrations would cause a breakdown in biological order and quickly eradicate any chance of life.

THE ROLE OF THE ENDOCRINE SYSTEM

Like the nervous system (Chapter 23), the **endocrine system** is a communications network, one that uses chemical messengers, called hormones, rather than electrical impulses, to carry information. The hormone molecules are secreted by "ductless" glands into the tissue fluids surrounding the gland. From there, the hormones diffuse into the bloodstream and are carried to distant parts of the body, where they interact with specific *target cells*—cells that have receptors for that particular hormone. When hormone molecules bind to the receptors of a target cell, the hormone evokes a dramatic change in the target cell's activity. For example, your gonads secrete steroid hormones—testosterone, if you are a male, and estrogen, if you are a female. These hormones circulate in the blood, but only target cells, such as those that make up your reproductive tract, bind the hormone and are affected by it. If these hormone molecules should stop being produced, as occurs in women during menopause, the reproductive tract loses its ability to carrry out reproductive activities.

 Like the nervous system, the endocrine system is also a command-and-control center. Animals face a constant threat of disruptions. Fluctuations in hundreds of critical chemicals occur every minute and must be corrected quickly before they disrupt homeostasis and interfere with

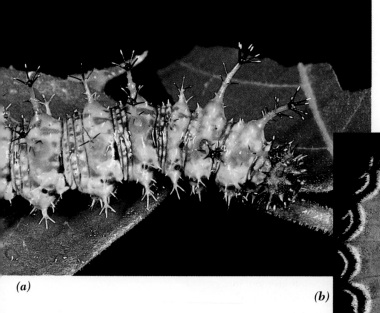

(a)

(b)

FIGURE 25-1

An effect of hormones. The larva *(a)* of this Malaysian butterfly is transformed into an adult *(b)* as the result of hormones secreted by endocrine glands within the insect's body.

vital processes. The endocrine system is responsible for detecting many such variations and releasing hormones whose actions help maintain a stable internal environment. Unlike the nervous system, which evokes rapid responses that typically require muscle contraction, the endocrine organs stimulate slower responses that require changes in metabolic activities of target cells.

Not all endocrine functions maintain homeostasis. Some hormones evoke changes that move the body in a new direction, away from a stable, homeostatic condition. For example, hormones "tell" an animal when to become reproductively mature, when to produce sperm, when to release eggs, and, in mammals, when to expel a fully developed fetus and produce milk. In some cases, hormones have the power to transform an animal into a different form that bears little resemblance to its former appearance (Figure 25-1).

The endocrine system regulates bodily functions through the action of hormones—chemical messengers that circulate in the blood and act on specific target cells, changing their metabolic activity. (See CTQ #2.)

THE NATURE OF THE ENDOCRINE SYSTEM

The vertebrate endocrine system consists of a collection of disconnected glandular tissues positioned around the body (Figure 25-2). Some of these tissues exist as physically separate glands, including the thyroid and adrenal glands. Others are components of a larger organ that has multiple functions, such as the pancreas or gonad. Regardless of their location, all endocrine cells secrete hormones that travel to their targets via the bloodstream.

THE DISCOVERY OF HORMONES

The existence of hormones was first glimpsed in 1902 by two English physiologists, W. M. Bayliss and E. H. Starling, in their study on the control of pancreatic secretions. In addition to secreting insulin and other hormones, the pancreas produces digestive juices that pass through ducts into the intestine. These juices only flow when food materials enter the intestine, which suggested to biologists at the time that nerves informed the pancreas of the arrival of food in

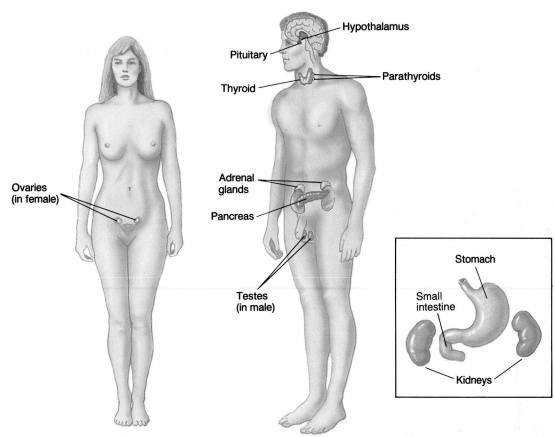

FIGURE 25-2

The major endocrine glands, plus some endocrine cells that are dispersed in nonendocrine organs (inset). Endocrine glands are defined as ductless glands; their secretions diffuse into the bloodstream, where they are carried to a distant target cell. (The endocrine activities of the digestive tract, kidneys, and gonads are discussed in Chapters 27, 28, and 31, respectively.)

the intestine. Bayliss and Starling found otherwise. In one of their numerous experiments, they joined together the bloodstreams of two dogs and observed that the pancreas of both animals would secrete their juices when only one animal was being fed. The researchers concluded that a *chemical* message was being sent from the intestine to the pancreas via the circulatory system. They called this blood-borne chemical messenger *secretin*. Bayliss and Starling further suggested that secretin was probably just one of a number of hormones that acted as chemical messengers within the body.

TYPES OF HORMONES

In 1970, fewer than 30 hormones had been identified. Today, some 200 possible hormones produced by various cells scattered throughout the body are being investigated. Some of the best-studied human hormones are listed in Table 25-1. Most of the known hormones fall into two broad categories: (1) amino acids, derivatives of amino acids, peptides, or proteins; and (2) steroids, all of which are constructed around the same type of multiringed, molecular skeleton (see Figure 4-11).

⚠ Similar types of molecules serve as hormones in most animals, although they may evoke very different types of responses. For example, estrogen causes a thickening of the wall of the human uterus, whereas ecdysone, a molecule of similar structure, causes a lobster to shed its outer skeletal covering. Not only do hormones have a similar structure in diverse animals, their mechanism of action at the cellular and molecular level is virtually identical in all animals, as we will discuss later in the chapter.

TABLE 25-1

THE MAJOR ENDOCRINE GLANDS AND THEIR HORMONES

Gland	Hormone	Regulates
Anterior pituitary	Growth hormone (GH)	Growth; metabolism
	Thyroid-stimulating hormone (TSH)	Thyroid gland secretions
	Adrenocorticotropic hormone (ACTH)	Adrenal cortex secretions
	Prolactin	Milk production
	Gonadotropic hormones: Follicle-stimulating hormone (FSH); Luteinizing hormone (LH)	Production of gametes and sex hormones by gonads
Posterior pituitary	Oxytocin	Milk secretion; uterine motility
	Antidiuretic hormone (ADH) or vasopressin	Water excretion; blood pressure
Adrenal cortex	Cortisol	Metabolism
	Aldosterone	Sodium and potassium excretion
Adrenal medulla	Epinephrine and norepinephrine	Organic metabolism; cardiovascular function; stress response
Thyroid gland	Thyroxine (T-4)	Energy metabolism; growth
	Triiodothyronine (T-3)	
	Calcitonin	Calcium in blood
Parathyroids	Parathyroid hormone (PTH)	Calcium and phosphate in blood
Ovaries	Estrogen and progesterone	Reproductive system; growth and development; female secondary sex characteristics
Testes	Testosterone	Reproductive system; growth and development; male secondary sex characteristics; sex drive
Pancreas	Insulin and glucagon	Metabolism; blood-glucose concentration
Kidneys (see Chapter 28)	Renin	Adrenal cortex; blood pressure
	Erythropoietin	Erythrocyte production
Gastrointestinal tract (see Chapter 27)	Gastrin	Secretory activity of stomach-small intestine; liver; pancreas; gall bladder
	Secretin	
	Cholecystokinin	Release of digestive enzymes
Pineal gland	Melatonin	Biological cycles; sexual maturation

"Reins and Spurs": Regulating Hormone Concentration

⟳ Hormones exert their effects in extraordinarily low concentrations, and their targets are acutely sensitive to even slight endocrine variations. In many cases, the precise concentrations of a hormone are regulated by:

- *Negative feedback of the hormone itself.* In many cases, the concentration of a hormone directly influences its own production. Some endocrine cells produce hormones only when the levels of those hormones in the blood drop below the homeostatic level (Figure 25-3). The drop in concentration of a particular hormone directly or indirectly spurs the gland to secrete more; consequently, the hormone concentration in the blood rises. Elevated concentrations have the opposite effect: They pull the reins in on hormone secretion before an excess amount can accumulate.

- *Negative feedback based on the hormone's effect.* Some glands are sensitive to the changes their hormones elicit. For example, the calcium-sensitive glands we discussed earlier in the chapter stabilize concentrations of calcium in the blood. In this form of negative feedback, the gland responds to the eventual *effect* (in this case, the concentration of blood calcium) rather than to concentrations of the hormone itself.

- *Inactivation or removal.* Hormones do not persist indefinitely. Most are removed by the kidneys and excreted; others are enzymatically inactivated or transformed into other compounds. If it is not replaced by continued secretion, the hormone slowly diminishes in concentration.

In many cases, endocrine cell activity is regulated by concentrations of blood-borne substances that stimulate or inhibit the production and secretion of hormones. (See CTQ #3.)

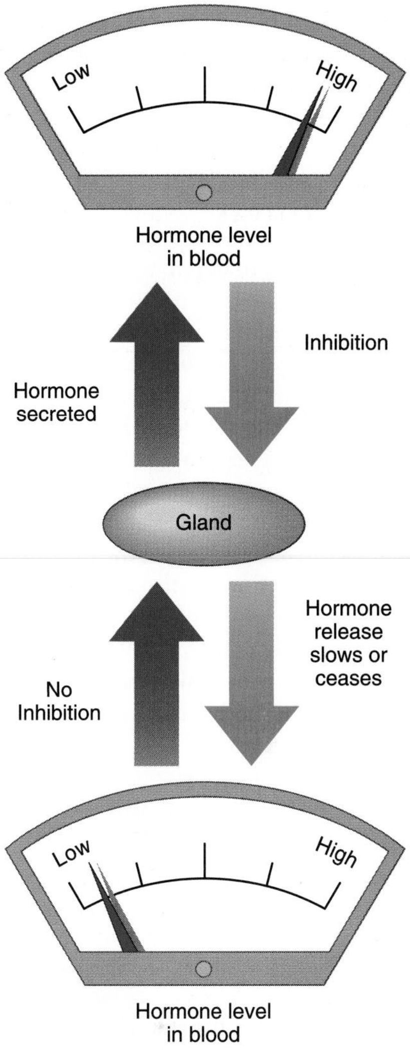

FIGURE 25-3

Negative feedback provides one mechanism for regulating hormone concentration in the blood. Higher concentrations of a hormone decrease the production and release of that hormone, reducing levels to the homeostatic value. As the concentration falls, production and release of the hormone increases, raising levels to the homeostatic value.

A CLOSER LOOK AT THE ENDOCRINE SYSTEM

Although endocrine cells have recently been found almost everywhere in the human body, a few organs stand out as major sources of hormones. In this section, we will examine each of these major endocrine glands (Table 25-1). We will begin with those parts of the endocrine system that have a close relationship with the nervous system. The nervous and endocrine systems work hand in hand as an integrated team, supervising bodily functions. In their role as regulatory systems, the nervous and endocrine systems collaborate on some activities and work independently on others.

⟳ Throughout the animal kingdom, from flatworms to mammals, biologists have discovered the existence of cells that look very much like neurons but act like endocrine cells. The form and function of these so-called **neurosecretory cells** make them ideally suited to serve as a "bridge" between the two major regulatory systems of the body. Neurosecretory cells look and behave like neurons in many ways (see Figure 25-5). Like neurons, neurosecretory cells receive information from other nerve cells and have elongated axons that conduct neural impulses. When the impulses reach the tip of the axon, they stimulate the release of stored materials from vesicles. Unlike neurons, however, the substance released from the tip of a neurosecretory cell is a hormone, which diffuses into the bloodstream, rather than a transmitter, which diffuses across a synaptic cleft. For example, oxytocin, the hormone that induces labor during childbirth, is a product of neurosecretory cells in the brain.

THE HYPOTHALMIC-PITUITARY CONTROL PATHWAY

⟳ The endocrine system has no true central coordinating system comparable to the brain of the nervous system. The endocrine system's pituitary gland has been called the "master gland," however, because it controls the activities of so many other endocrine elements. The master gland has a master of its own—the hypothalamus—which, you will recall from Chapter 23, is part of the brain (see Figure 25-2). The hypothalamus and the pituitary gland are functionally related, as illustrated in Figure 25-4. Information about the state of the body arrives at the hypothalamus from two sources: (1) blood flowing through the hypothalamus provides information on the concentration of various chemicals in the bloodstream; and (2) sensory information arriving by neurons provides information on the conditions of various parts of the body. This information is used to control pituitary secretions.

The pituitary is actually two glands in one: the **posterior pituitary** and the **anterior pituitary.** These two parts of the pituitary gland are controlled by the hypothalamus in very different ways, but both mechanisms utilize neurosecretory cells (Figure 25-5).

1. The posterior pituitary manufactures no hormones of its own but stores and secretes two hormones that are produced by neurosecretory cells that originate in the hypothalamus (Figure 25-5*a*).

2. The anterior pituitary is a true endocrine gland, manufacturing at least six separate hormones. These pituitary hormones are secreted in response to *releasing factors* that are produced by neurosecretory cells of the hypothalamus (Figure 25-5*b*). These releasing factors are produced in extremely small amounts; their purification and characterization were monumental tasks, requiring the extraction of hundreds of tons of sheep brains. The work was conducted by Roger Guillemin of Baylor University and Andrew Schally of Tulane University and earned the scientists the 1977 Nobel Prize.

   When the hypothalamus senses a need for one of the anterior pituitary hormones, an impulse passes down the axon of the neurosecretory cell, and the appropriate releasing factor is discharged into tiny blood vessels. The blood then carries the releasing factor to the nearby anterior pituitary, where it triggers the secretion of pituitary hormones. The hypothalamus also produces *inhibiting factors* that prevent excessive secretion of specific hormones by the anterior pituitary.

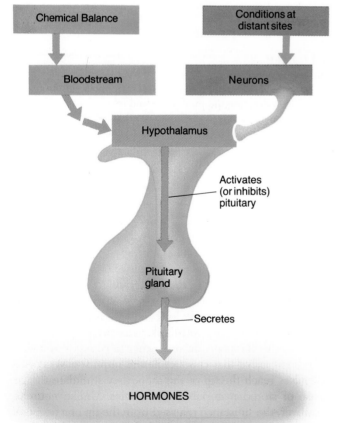

FIGURE 25-4

The major neuroendocrine connection.   Input from both the bloodstream and the nervous system influences the hypothalamus in its regulation of the secretory activity of the pituitary gland. The anatomic nature of the hypothalmic-pituitary relationship is illustrated in Figure 25-5.

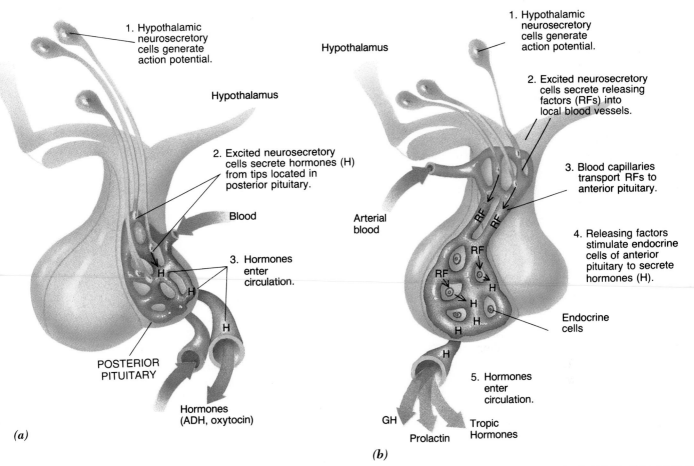

1. Hypothalamic neurosecretory cells generate action potential.

Hypothalamus

Hypothalamus

2. Excited neurosecretory cells secrete hormones (H) from tips located in posterior pituitary.

Blood

3. Hormones enter circulation.

H

H

H

POSTERIOR PITUITARY

Hormones (ADH, oxytocin)

(a)

1. Hypothalamic neurosecretory cells generate action potential.

2. Excited neurosecretory cells secrete releasing factors (RFs) into local blood vessels.

3. Blood capillaries transport RFs to anterior pituitary.

Arterial blood

RF   RF

4. Releasing factors stimulate endocrine cells of anterior pituitary to secrete hormones (H).

RF

RF

H

H   H

H   H

H

Endocrine cells

5. Hormones enter circulation.

GH

Prolactin

Tropic Hormones

(b)

FIGURE 25-5

The pituitary-hypothalmic linkage. *(a)* The posterior pituitary is a storage depot for two hormones —ADH and oxytocin—which are produced in the cell bodies of hypothalmic neurons. Hormones are enclosed within vesicles in the cell body and are transported down the length of the axon to the posterior pituitary; from there, the hormones are released into the bloodstream. Once in circulation, these hormones are carried to distant sites, where they act on target cells. *(b)* Secretion of the various hormones of the anterior pituitary is controlled by releasing and inhibiting factors, which are secreted from the hypothalamus by neurosecretory cells into local blood capillaries that carry the factors to the anterior pituitary. Hormones produced and secreted by the anterior pituitary circulate through the bloodstream and act on target cells around the body.

Hormones of the Posterior Pituitary

The hormones released by the posterior pituitary (Figure 25-5a) are both peptides. The first, **antidiuretic hormone (ADH)**, acts on the kidneys to reduce the loss of water during urine formation (Chapter 28). In females, the second hormone, **oxytocin,** triggers uterine contractions during childbirth. Oxytocin also stimulates the release of milk when the breast is stimulated by nursing.

Hormones of the Anterior Pituitary

Of the six known hormones produced by the anterior pituitary (Figure 25-6), four are **tropic hormones;** that is, they regulate the activity of other endocrine glands. The remaining two are direct-acting hormones. One of these, **prolactin,** promotes the production of milk by mammary glands in the breast. The other, **growth hormone (GH),** encourages body growth by stimulating bone elongation and accelerat-

ing fat breakdown and protein synthesis in a wide variety of target cells.

Growth hormone (also called *somatotropin*) is produced predominantly during childhood and adolescence and plays a key role in promoting normal body growth. When the anterior pituitary ceases its GH output, overall growth stops. This typically occurs sooner in girls than in boys, accounting for the shorter average height of women, compared to men. Interestingly, most growth takes place while sleeping, when bursts of GH secretions usually occur. Extremes in height (giantism or certain kinds of dwarfism) are due to overproduction or underproduction of GH during childhood (Figure 25-7). Children who fail to grow at normal rates due to a deficiency of GH secretion can now be treated with growth hormone that is produced by recombinant DNA technology (Chapter 16). Excessive secretion of GH *in an adult,* which is typically due to a pituitary tumor,

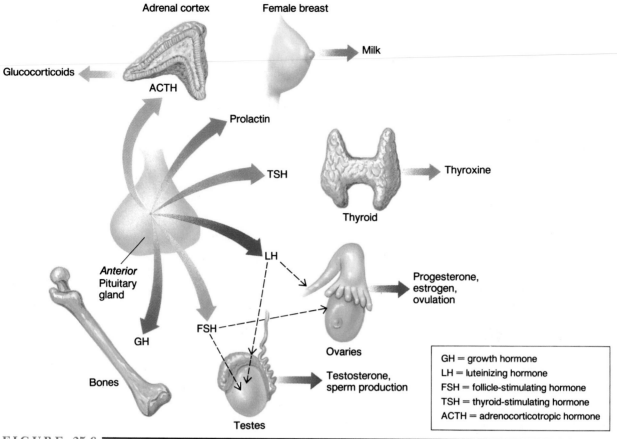

Adrenal cortex

Female breast

Glucocorticoids

Milk

ACTH

Prolactin

TSH

Thyroxine

Thyroid

Anterior
Pituitary
gland

LH

Progesterone,
estrogen,
ovulation

FSH

GH

Ovaries

Bones

Testosterone,
sperm production

| GH = growth hormone |
| LH = luteinizing hormone |
| FSH = follicle-stimulating hormone |
| TSH = thyroid-stimulating hormone |
| ACTH = adrenocorticotropic hormone |

Testes

FIGURE 25-6

Hormones of the anterior pituitary. Four of the six hormones (ACTH, TSH, LH, and FSH) are tropic hormones, which stimulate other endocrine glands to produce hormones. Of the two direct acting hormones, prolactin stimulates milk production, while GH stimulates protein synthesis and the growth of bones.

leads to a condition known as *acromegaly,* in which certain parts of the body, including the hands, feet, jaw, and nose, become enlarged (Figure 25-8). Because it also produces increased muscle mass, GH has been used by some bodybuilders and athletes (see Bioline: Chemically Enhanced Athletes).

The four tropic hormones of the anterior pituitary stimulate endocrine glands to produce other hormones. **Adrenocorticotropic hormone (ACTH)** stimulates the cortex (outer layer) of the adrenal glands to secrete its hormones (glucocorticoids and mineralocorticoids). **Thyroid-stimulating hormone (TSH)** activates secretion of thyroid hormones by the thyroid gland. The other two tropic hormones act on the gonads and are called **gonadotropins.** Both gonadotropins are identical in males and females. They are **follicle-stimulating hormone (FSH)** and **luteinizing hormone (LH).** FSH and LH promote gamete development and stimulate gonads to produce sex hor-

mones, as discussed in Chapter 31. In the following discussion, we will see how two of the tropic hormones—ACTH and TSH—influence their target glands.

THE ADRENAL GLANDS

Perched atop each kidney is an adrenal gland (*ad* = upon, *renal* = kidney). Like the pituitary gland, each adrenal gland is essentially composed of two independent glands that produce very different hormones (Figure 25-9).

The Adrenal Cortex

The gland's outer layer—the **adrenal cortex**—secretes a family of steroid hormones. These include:

- *Glucocorticoids,* which promote the conversion of amino acids to glucose and its uptake by the brain. The most important glucocorticoid is *cortisol* (also known

FIGURE 25-7
Circus side shows often included a pituitary dwarf and giant.

(a) *(b)* *(c)* *(d)*

FIGURE 25-8

Acromegaly. Photos were taken at *(a)* age 9 years, *(b)* age 16 years, *(c)* age 33 years, and *(d)* age 52 years. The effects of excess growth hormone production in the adult are seen by age 33 as an enlarged lower jaw and nose and thickened lips, features which become exaggerated with age.

◁ B I O L I N E ▷
Chemically Enhanced Athletes

In 1896, Ellery Clark leaped nearly 21 feet to win the long-jump event in the Olympic Games; in the 1936 Olympic Games, Jesse Owens won the event with a jump of more than 26 feet; today, the world record is more than 29 feet. While modern athletes are superior to any in history, their achievements have been tainted over the past decade by the discovery that too many athletes rely on drugs to maximize their physical potential. The disqualification of gold-medal winner Ben Johnson from the 100-meter dash in the 1988 Olympics, and his subsequent explusion from competition for life in 1993, exemplifies the severity of the problem. The reason for Johnson's explusion: Highly sensitive tests detected the presence of illegal drugs in the athlete's urine.

Topping the list of banned substances taken by some athletes are steroids, commonly referred to as "anabolic" steroids because they tend to increase biosynthesis (anabolism), especially protein synthesis.

The most common of these steroids is the male sex hormone testosterone and its synthetic derivatives. These drugs help build muscle, restore energy, and enhance aggressiveness. Their use has grown so common, even among high-school athletes, that suppliers of black-market drugs distribute printed advertising brochures and order forms at local gyms.

The dangers of steroid abuse are considerable. A steroid surplus in the blood, produced by taking external sources of testosterone, causes the body's own testosterone production to drop due to negative feedback and causes the gonadal cells that produce the male hormone to waste away. Other side effects include lowered sperm counts, penis atrophy, reduced resistance against infectious disease, and potentially serious heart irregularities.

When a man's body is flooded with testosterone, some of the hormone is converted to the female hormone estrogen,

often stimulating female-type breast enlargement. Women taking testosteronelike steroids experience the opposite effects: The male hormones stimulate the development of facial hair and other masculine features. Prolonged exposure to steroids suppresses the immune system in both sexes, compromising the body's ability to defend itself against many infectious diseases. Steroid supplements also increase the likelihood of cancer.

In recent years, growth hormone has become commercially available through recombinant DNA technology and is being taken by some athletes to increase muscle mass. Excess growth hormone is broken down within the body within an hour or so, so that it is virtually undetectable by testing procedures. Like steroids, excess growth hormone carries with it serious health risks, including irreversible overgrowth of certain bones (a characteristic of acromegaly, illustrated in Figure 25-8), arthritis, and enlargement of the heart.

---

as hydrocortisone), whose level is highest during periods of stress, such as that which occurs following a severe physical injury or a period of emotional trauma. Secretion of cortisol helps maintain normal blood glucose levels and fuels brain activity. Cortisol secretion is stimulated by ACTH released from the pituitary. At very high concentrations, cortisol suppresses the body's normal response to injury, including inflammation. Consequently, cortisol and related synthetic compounds are highly effective in treating persistent inflammation, such as that caused by arthritis or bursitis, and severe allergies or asthma. Glucocorticoids may have serious side effects, however, including suppression of the body's ability to fight infections and inducement of hypertension and ulcers, and, thus, must be used with great caution. Extended periods of stress may take their toll on the body as a result of the increased secretion of cortisol.

- *Mineralocorticoids*, which regulate the concentrations of sodium and potassium in the blood. The most important mineralocorticoid is *aldosterone*, which stimulates the kidney to reabsorb sodium into the blood, maintaining homeostatic levels of this important ion. If the level of sodium in the blood rises, it reduces aldosterone secretion, increasing excretion of the ion in the urine.

◑ The adrenal cortex is essential for survival. Without aldosterone, for example, the body quickly loses sodium ions needed for neuron activity, muscular contraction, and blood-pressure stability. Without glucocorticoids, glucose concentrations in the blood plummet, crippling cellular energy metabolism. A person whose adrenal cortex fails to produce sufficient levels of these hormones suffers from *Addison's disease,* which is characterized by extreme weakness, weight loss, and impaired heart and kidney function and is treated by administering synthetic adrenal cortical steroids. John F. Kennedy suffered from Addison's disease prior to the campaign for the presidency in 1960; diagonosis and treatment of the condition transformed a sick, weak-looking candidate into a healthy, vigorous campaigner.

The Adrenal Medulla

The "fight-or-flight" response discussed on page 491 includes activation of the **adrenal medulla,** the core of the adrenal gland (Figure 25-9). When stimulated by sympathetic nerve fibers, the adrenal medulla secretes two hormones: **epinephrine** (also called adrenalin) and **norepinephrine** (or noradrenalin). These hormones jolt the body into readiness to escape or confront an emergency. When confronted with an emergency, a person's metabolic rate increases and heart rate accelerates; additional glucose and oxygen are shunted to voluntary muscles, increasing muscular performance; blood vessels to the skin constrict, helping reduce the loss of blood through injured tissues; and red blood cells pour into the bloodstream from their reserves in the spleen, enhancing oxygen transport and replacing blood cells that are lost during bleeding. Epinephrine and norepinephrine also inhibit activities that are of little help in an emergency, such as contraction of muscles in the digestive tract.

THE THYROID GLAND

In the early twentieth century, many people found themselves afflicted with enormous swellings in the front of their necks. This condition, known as *simple goiter,* is the result of an enlargement of the **thyroid gland,** a butterfly-shaped gland that lies just in front of the windpipe (trachea) in humans (Figure 25-10). The swelling is often accompanied by lethargy, hair loss, slowed heart beat, and mental sluggishness. Goiter initially occurred frequently in geographic regions where natural sources of dietary iodine were scarce. The link between iodine deficiencies and the disease was not established until 1916, at which time it was shown that adding iodine to the diet, such as in iodized table salt, prevented the formation of goiters.

Why should iodine be related to thyroid function? The reason becomes evident when we examine the structure of two hormones produced by the thyroid gland. These hormones—**thyroxin** and **triiodothyronine**—are amino acids that contain iodine. Together, they are called *thyroid hormone.* When dietary iodine is inadequate, the thyroid cannot produce enough thyroid hormone. Since thyroid gland activity is regulated by negative feedback through the pituitary hormone TSH, the low concentration of thyroid hormone in the blood cannot feed back to inhibit TSH

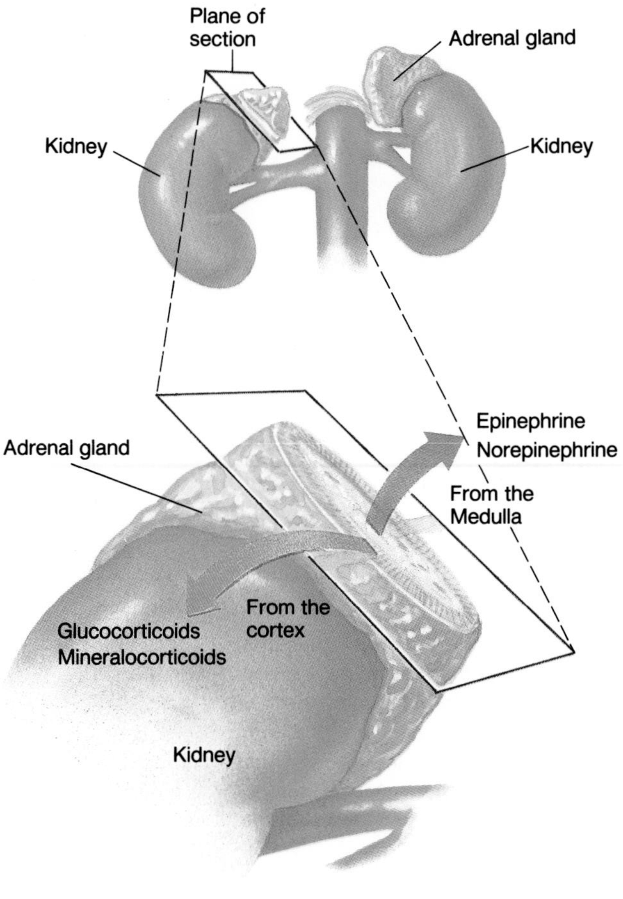

FIGURE 25-9

The adrenal glands are located in the fatty tissue adjacent to the anterior ends of the kidneys. The adrenal cortex produces an array of steroid hormones, including glucocorticoids and mineralocorticoids, and a small amount of sex hormones, while the adrenal medulla produces two related amino acid derivatives: epinephrine and norepinephrine.

The thyroid and parathyroid glands produce hormones that regulate metabolic rate and blood calcium levels respectively.

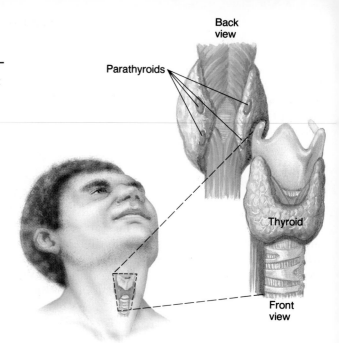

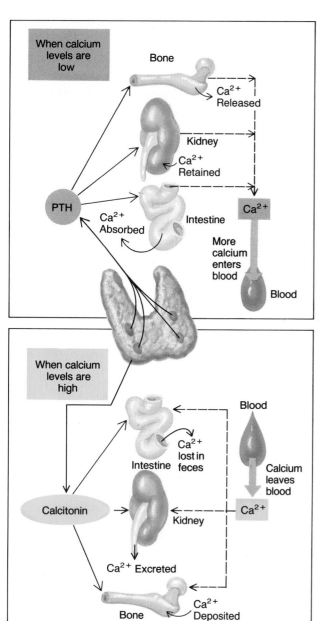

F I G U R E 25-11

Balancing calcium. A low calcium (Ca^{2+}) concentration in the blood stimulates PTH secretion by the parathyroid glands (shown in green) and inhibits calcitonin secretion by the thyroid gland (in red). PTH leads to the demineralization of bone; the retention of calcium by the kidney; and the absorption of calcium by the intestine. All of these responses trigger an increase in blood calcium levels. A high concentration of calcium has the opposite effects, decreasing calcium levels in the blood.

production. As a result, the high levels of TSH cause the thyroid to enlarge into a large lump, or goiter, in its perpetual "attempt" to produce enough thyroid hormone.

☀ Normally, thyroid hormone enhances oxidation in the mitochondria of various target cells, which increases energy availability and metabolic rate. Consequently, a person with a lowered output of thyroid hormone—whether due to a decreased availability of iodine or to a sluggish thyroid gland—experiences fatigue and low energy levels. Such thyroid hormone deficits are called *hypothyroidism.* Extreme hypothyroidism during childhood produces *cretinism,* a type of dwarfism that is characterized by severe mental deficiency. Cretinism can be prevented by early treatment with thyroid hormone supplements. The same treatment is effective in correcting the sluggish metabolic rate of adult hypothyroid victims. An excess of thyroid hormone *(hyperthyroidism)* may lead to *Grave's disease,* which is characterized by hyperactivity, weight loss, nervousness, insomnia, irritability and, in extreme cases, *exophthalmia,* in which the eyeballs bulge from the sockets.

After decades of studying thyroid function, a third thyroid hormone, **calcitonin,** was discovered in the early 1960s. Calcitonin regulates blood calcium levels in cooperation with another set of endocrine structures, the parathyroid glands.

THE PARATHYROID GLANDS

You probably think of your bones as stable, inert structures. In reality, the inorganic salts that give bones their hardness are continually being dissolved and redeposited by cells embedded in the bone tissue. This is the major reason why calcium is an essential dietary substance; it is needed to replace the calcium that is removed from bone and lost in the urine. Maintaining the calcium concentrations in the blood requires the cooperation of the thyroid gland and four tiny **parathyroid glands.** The parathyroids were discovered only at the end of the last century, when it was shown that the muscular spasms and convulsions that sometimes followed thyroid removal were due to the loss of the adjoining parathyroid glands, not the thyroid.

🔄 When calcium levels in the blood are low, the parathyroid glands secrete **parathyroid hormone (PTH),** which acts on the bones, kidneys, and intestines to restore normal calcium concentrations (Figure 25-11, *top*). Under PTH influence, calcium is withdrawn from bones, and the released mineral is absorbed into the bloodstream; kidneys retain calcium; vitamin D is activated; and calcium absorption from the intestines is enhanced. As blood calcium levels normalize, negative feedback decreases the level of PTH secretion. In contrast, if blood calcium levels should rise to abnormally high levels, the thyroid gland is stimulated to secrete calcitonin, which exerts the opposite effects of PTH (Figure 25-11, *bottom*).

THE PANCREAS

The pancreas is predominantly an *exocrine* gland; that is, a gland that secretes its products into a duct. The exocrine products of the pancreas are digestive enzymes. The pancreas also contains tiny endocrine centers, called **islets of Langerhans,** which secrete several protein hormones into the blood. One of these hormones—insulin—is secreted when the concentration of glucose in the blood begins to exceed normal levels, usually as sugar floods the bloodstream following a meal (Figure 25-12, *left*). Insulin acts on numerous organs of the body to stimulate the cellular uptake of glucose, which is necessary in initiating the utilization of the sugar. Insulin also directs the conversion of surplus glucose into glycogen for storage in the liver and muscles. This conversion prevents the loss of surplus sugar since excess sugar that remains in the blood is excreted in urine.

Insulin can do too good a job, however, and deplete the blood of glucose. When the concentration of blood sugar begins to drop below normal, the islets of Langerhans alter their secretory priorities and increase the secretion of **glucagon,** another pancreatic hormone (Figure 25-12, *right*). Glucagon promotes glycogen breakdown in cells where it is stored and elevates glucose concentration in the blood, especially during times of stress, when increased cellular and physical activity is likely to expend greater amounts of energy.

As we discussed in the chapter opener, a deficiency in insulin production can lead to *diabetes mellitus;* complications can include cardiovascular damage, kidney failure, blindness, and susceptibility to life-threatening infections. Because they are unable to take up glucose from the blood, the cells of untreated diabetics must turn to other sources of energy, such as protein and fat reserves. Consequently, some diabetics may become emaciated. Diabetics may also be very thirsty or dehydrated because increased blood glucose levels promote frequent urination.

In approximately 15 percent of the cases, diabetes appears during childhood as the result of the destruction of the insulin-producing cells of the pancreas, which apparently results from either a viral infection or an attack by the person's own antibodies (Chapter 30). In these cases, which are classified as *juvenile-onset,* or *Type 1,* diabetes, the person produces little or no insulin and must be treated by daily injection of the hormone. Advanced delivery systems that continually release metered supplies of the hormone are beginning to replace daily injection routines, thereby avoiding undesirable fluctuations in the hormone's concentration associated with periodic insulin shots.

The majority of diabetics are classified as having the less severe, *adult-onset,* or *Type 2,* form of the disease, in which insulin levels may be normal, but the target cells fail to respond to the hormone because of insulin receptor abnormalities. Even though Type 2 diabetics may not ex-

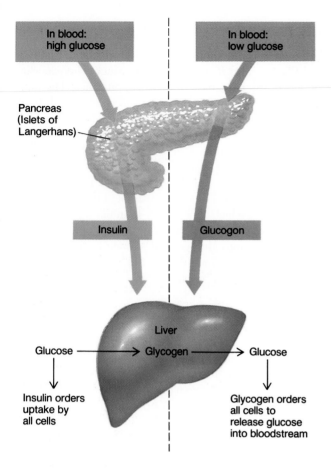

In blood: high glucose

In blood: low glucose

Pancreas (Islets of Langerhans)

Insulin

Glucogon

Liver

Glucose → Glycogen → Glucose

Insulin orders uptake by all cells

Glycogen orders all cells to release glucose into bloodstream

FIGURE 25-12

Control of blood sugar concentration by the islets of Langerhans in the pancreas. When blood glucose levels are high, insulin is secreted, promoting glucose uptake and storage by the liver and muscle cells as glycogen. When blood glucose levels are low, glucagon is secreted, promoting the breakdown of glycogen into glucose and its subsequent release into the bloodstream.

hibit extreme urinary and metabolic derangements, it is still essential that they take steps to maintain normal blood-sugar levels, or life-threatening complications may develop. Type 2 diabetes can often be controlled if the patient adheres to strict dietary recommendations.

GONADS: THE TESTES AND OVARIES

When stimulated by pituitary gonadotropins, the gonads—the male **testes** and the female **ovaries**—secrete the powerful steroid sex hormones **testosterone** and **estrogen,** respectively. These substances promote the development of the reproductive tracts and secondary sex characteristics that distinguish genders, such as deeper voices and facial hair, which is stimulated by testosterone, and breast enlargement, which is triggered by estrogen. The activities and regulation of the sex hormones are discussed in more detail in Chapter 31. The role of another endocrine structure that affects maturation of the reproductive organs is discussed in The Human Perspective: The Mysterious Pineal Gland.

PROSTAGLANDINS

Unlike the hormones described thus far, **prostaglandins** are secreted by endocrine cells scattered throughout the body; their activities seem as diverse as are their sources.

Prostaglandins are modified fatty acids and are quite different in structure from other hormones. Prostaglandin activities affect reproduction, digestion, respiration, neurological function, pain perception, inflammation, and blood clotting. Especially high concentrations of prostaglandins are found in *semen,* the fluid discharged during male orgasm. The name "prostaglandin" is derived from the presumption that these substances were produced in the prostate gland, which is situated at the base of the male urethra. The prostaglandins in semen cause uterine muscles to contract, which helps propel sperm toward the egg. In women, the contraction of uterine muscles responsible for cramping during menstruation has also been traced to prostaglandins. Other prostaglandins promote inflammation and stimulate pain receptors. The pain-relieving and anti-inflammatory effects of aspirin and ibuprofen are probably due to the ability of these drugs to inhibit prostaglandins.

Many of the body's endocrine glands are part of a hierarchical control system. The hypothalamus sits at the top of the hierarchy, collecting information from the nervous system and the bloodstream and sending out commands via neurosecretory cells to the pituitary gland. In turn the pituitary gland releases tropic hormones that control various peripheral endocrine tissues. Other endocrine glands are regulated by blood-borne substances whose concentration is directly affected by the hormone's action. (See CTQ #5.)

◁ THE HUMAN PERSPECTIVE ▷
The Mysterious Pineal Gland

Embedded deep within the brain is a tiny, pinecone-shaped organ, the **pineal gland,** whose function has been a source of speculation for over 2,000 years. As late as the seventeenth century, the philosopher Descartes designated the pineal gland as the "seat of the soul." One of the first important findings regarding the biological function of the pineal gland was made in 1898 by Otto Heubner, a German physician, who reported on a 4-year-old boy who had undergone premature puberty and then died. Autopsy results indicated that the boy had died from a tumor of the pineal gland. This finding, and similar reports from others, led Heubner to propose that the pineal gland normally produces a hormone that suppresses sexual development during childhood. If production of that chemical should cease—if the gland is destroyed by a tumor, for example—the inhibition is removed, and sexual maturation occurs prematurely.

Heubner's hormone was not isolated and identified until 1958. At that time, Aaron Lerner of Yale University purified the substance from approximately 200,000 cattle pineal glands. He named the hormone *melatonin.* Although melatonin is still thought to be a factor involved in inhibiting the onset of sexual maturation in mammals, another role of the hormone has captured the recent attention of researchers. For years, it was known that the pineal gland in amphibians and reptiles was a light-sensitive organ, whose hormonal secretions controlled the darkening of the skin. The pineal gland of many amphibians and reptiles is directly exposed to light, whereas the pineal gland of mammals is covered by a thick layer of cranial bone. Yet, a number of studies have indicated a relationship between light exposure and the activity of a mammal's pineal gland. For example, when rats are exposed to constant illumination, the level of melatonin synthesis rapidly drops as much as 80 percent, compared to animals kept in continuous darkness.

In humans, melatonin concentrations in the blood are highest at night and lowest in the day. Daily variations in the synthesis of melatonin may, in turn, regulate a variety of human daily rhythms, including sleep, motor activity, and brain waves. Studies of people who fly across several time zones indicate that it takes several days for the normal rhythm of melatonin to reestablish itself, suggesting that the familiar problem of "jet lag" may be due to the time needed to reset the pineal "clock." The pineal gland has also been implicated as a cause of seasonal depression, or seasonal affective disorder syndrome (SADS). As winter approaches, and the days become shorter, some people become tired, sad, and depressed, only to rebound when the days become longer in the spring. The onset of the "winter blues" correlates with a seasonal rise in melatonin production and, in some cases, can be treated by exposing depressed subjects to periods of bright light, which decreases the pineal's output of melatonin.

ACTION AT THE TARGET

In order for a cell to respond as a target to a particular hormone, the cell must have a protein receptor whose structure allows it to bind that hormone specifically. Depending on the particular hormone, the receptor may be located at the cell surface or within the cytoplasm. The position of the receptor within the cell is a key factor in determining the mechanism of action of the corresponding hormone.

CELL SURFACE RECEPTORS

Most of the hormones that act via cell surface receptors are water-soluble substances (amino acids, peptides, and proteins) that cannot passively diffuse through the plasma membrane and enter the cytoplasm. (The prostaglandins are an exception to this rule; despite their lipid-soluble structure, prostaglandins also act at the cell surface.) All of these hormones act without entering the target cell.

The hormone itself can be considered a "first messenger." It binds to a receptor on the outer surface of the target cell's plasma membrane (Figure 25-13), promoting a change at the membrane and the release of a **"second messenger,"** which enters the cytoplasm and actually triggers the response. The best-studied and most widespread second messenger is a small molecule called *cyclic AMP.* We will illustrate this type of mechanism using the pancreatic hormone glucagon.

When glucagon binds to a glucagon receptor located at the surface of the plasma membrane, the hormone changes the shape of the receptor, which activates an enzyme located on the inner surface of the membrane (Figure 25-

14a). This enzyme, called *adenylate cyclase,* converts ATP to cyclic AMP (cAMP), one of the most universally important molecules associated with cellular regulation. Cyclic AMP then diffuses into the cytoplasm, where it activates protein kinases—enzymes that attach phosphates to other enzymes (page 129). When glucagon binds to a liver cell, causing the synthesis of cAMP, the activated protein kinase activates an enzyme that splits glycogen into its glucose monomers. As a result, glucagon secretion increases the concentration of glucose in the blood. The involvement of two messengers—a hormone and cAMP—amplifies the original signal. The binding of one glucagon molecule at the cell surface promotes the synthesis of thousands of cAMP molecules inside the cell. As a result of this amplification, a very small concentration of hormone in the blood can produce a rapid, massive response within a target cell.

But how does such a mechanism explain the specificity of hormones? If glucagon, calcitonin, parathyroid hormone, epinephrine, the pituitary-releasing factors, and several other hormones do no more than increase the cytoplas-

mic cAMP concentration in target cells, why don't they all cause the body to change in the same way? There are two parts to the answer to this question. First, different target cells have different hormone receptors on their surface. If the only receptors on a cell happen to be epinephrine receptors, for example, only epinephrine will cause the cAMP-mediated response. Second, cAMP has different effects in different target cells, depending on which enzymes are present in that cell. For example, when liver cell enzymes are activated by cAMP, glucose is released into the bloodstream, but when the proteins found in the cells of the adrenal cortex are activated by cAMP, glucocorticoids are produced and released.

CYTOPLASMIC RECEPTORS

Steroid hormones have a very different mechanism of action (Figure 25-14b). These hydrophobic hormones diffuse through the plasma membrane, where they bind to a specific cytoplasmic receptor molecule. The receptor–hormone complex then enters the nucleus, where it becomes a gene regulatory protein that binds to a particular DNA sequence. Binding of the receptor–hormone complex activates the genes responsible for the hormone-induced changes (page 305). Again, the specificity of the receptor molecule determines which cells will respond as targets to particular hormones. Only those cells that have the appropriate hormone receptors can be triggered by a particular hormone. Thyroid hormones also act via cytoplasmic receptors, but their precise mechanism of action remains a subject of research.

A target cell responds to a particular hormone because it has a receptor that is capable of binding that hormone as well as the machinery to respond to the hormone-receptor complex. The type of response triggered by the hormone depends on the unique array of components present in the target cell. (See CTQ #6.)

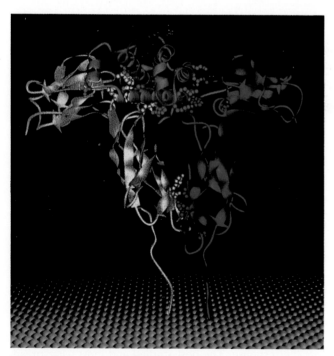

FIGURE 25-13

A hormone-receptor complex. Computer model of human growth hormone (red) bound to two receptor molecules (blue and green), which are projecting outward from the plasma membrane (the pebbled reddish surface). This binding event generates a signal that is transmitted through the membrane and activates the cell as illustrated in Figure 25-14.

EVOLUTION OF THE ENDOCRINE SYSTEM

◗◗▶ Among the simpler invertebrates, such as flatworms, roundworms, and echinoderms (sea stars and sea cucumbers), hormones are produced primarily by neurosecretory cells, suggesting that the first endocrine cells to evolve were modified neurons. It is a relatively small evolutionary step from a typical neuron secreting a chemical transmitter substance into a synaptic cleft to a neurosecretory cell secreting a hormone into the bloodstream. Yet, the evolution of the neurosecretory cell provided new opportunities to coordinate events occurring in the outside world with the animal's own internal activities. Information relayed from surface

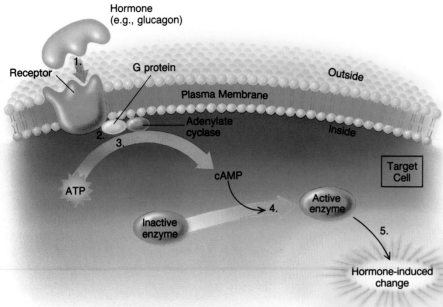

(a)

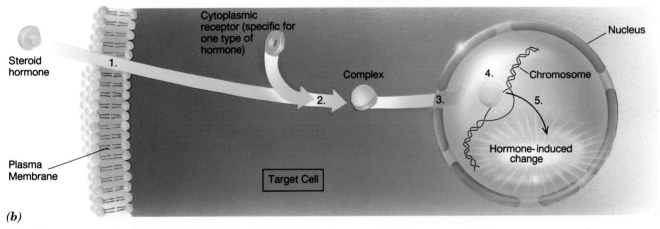

(b)

FIGURE 25-14

Mechanisms of hormone action. **(a) Second messenger model:** (1) Hormone binds to surface receptor. (2) Binding activates a G protein, which activates adenylate cyclase. (3) Activated adenylate cyclase converts ATP to cAMP (a "second messenger"). (4) cAMP activates (or inhibits) specific enzymes. (5) Activated enzymes produce specific changes in the cell. **(b) *Steroid hormones:*** (1) Hormone readily passes through the plasma membrane and (2) reacts with a protein receptor molecule. (3) The hormone-receptor complex enters the nucleus and (4) attaches to a specific site on the chromosome. (5) Attachment activates those genes responsible for the hormone-induced change, such as the genes required for construction of a hair shaft by a previously dormant hair follicle on an adolescent boy's chin.

receptors could be transmitted through the nervous system to neurosecretory cells; chemical messengers could then regulate various internal physiological activities. As invertebrates became more complex, they developed more elaborate endocrine systems that included separate endocrine glands as well as neurosecretory cells. Arthropods (such as lobsters, spiders, and insects) and mollusks (especially squids and octopuses) have complex endocrine systems analogous to those of vertebrates.

△ The endocrine glands and their hormones are quite similar among all vertebrates, from fish to mammals. Insulin and glucagon regulate blood-sugar levels in a catfish and in a cat; thyroid hormones influence the metabolic rate of dogs and dolphins; epinephrine induces a life-saving fight-or-flight response in people and pigeons. There are exceptions to this unity, however. For example, prolactin maintains the balance of salt and water concentrations in lower vertebrates but has shifted in function to the control of milk production in mammals.

The chemical structure of many different polypeptide hormones is remarkably similar among different species. For example, even though evolution has created different functions for FSH, LH, and TSH, these hormones share a similar amino acid sequence in vertebrates and are undoubtedly derived by evolution from a single protein that was present in some prevertebrate ancestor.

While the mechanisms by which hormones act have remained remarkably uniform among diverse animals throughout their evolution, the types of metabolic processes that are regulated by hormones vary greatly from one type of animal to another. (See CTQ #7.)

REEXAMINING THE THEMES

Relationship between Form and Function

Ⓝ Neurosecretory cells essentially are neurons that secrete hormones rather than neurotransmitters. These cells combine the structure and function of both neurons and endocrine cells; in doing so, they form a bridge between the two great regulatory systems of the body. As part of the nervous system, neurosecretory cells receive impulses from other neurons and conduct their own impulses down an axon. As part of the endocrine system, neurosecretory cells respond to stimuli by secreting hormones into the bloodstream.

Biological Order, Regulation, and Homeostasis

○ Along with the nervous system, the endocrine system is the primary regulatory agency within an animal's body. Whereas the nervous system is geared for rapid responses, such as those requiring muscle contraction and movement, the endocrine organs regulate slower responses that require changes in the metabolic activities of target cells. Endocrine organs regulate such diverse activities as protein and polysaccharide synthesis, body growth, reproductive maturation, blood-cell formation, and the ionic content of the blood.

Acquiring and Using Energy

☀ One of the simplest measures of energy utilization is metabolic rate: the amount of oxygen consumed per unit of body weight per minute. Metabolic rate can be directly related to energy utilization. In vertebrates, metabolic rate is affected by a number of hormones including those that affect glucose metabolism (e.g., glucagon, epinephrine, insulin, and glucocorticoids) and those that affect mitochondrial function (thyroid hormone). For example, daily doses of thyroid hormone can transform a person with a hypothyroid condition from a lethargic, sluggish individual to one with a normal metabolic rate.

Unity within Diversity

△ Even though different animals possess different hormones that evoke different responses, the basic structure and mechanism of action is very similar in all species. For example, both insects and mammals produce steroid hormones that bind to the DNA and activate nearby genes. Yet, a steroid hormone in an insect may cause metamorphosis from a pupa to an adult, while a steroid hormone in a mammal may cause the reabsorption of sodium from the blood or the development of breasts. Similarly, neurosecretory cells are found throughout the animal kingdom, but they regulate totally different types of functions in different animals.

Evolution and Adaptation

▐▶ The endocrine system provides some of the best evidence for the common ancestry of all vertebrates. Consider that a hormone, such as insulin, has virtually the same structure and function from one end of the vertebrate scale to the other. Common ancestry explains how you and a catfish have acquired so many similar hormones.

SYNOPSIS

Hormones are chemical messengers that are secreted into the bloodstream via ductless glands of the endocrine system. Hormones act on specific target cells that possess receptors for that hormone, eliciting a specific metabolic response in the target cell. Some hormones maintain homeostatic conditions; others cause irreversible bodily changes, including metamorphosis.

Hormone concentrations are often regulated by negative feedback. As the concentration of the hormone (or its effect) increases, the output of the hormone decreases. The effects of hormones are short-lived, because hormones are regularly excreted or enzymatically destroyed.

The double-lobed pituitary is the "master gland" of the endocrine system. The hormones of the posterior pituitary—antidiuretic hormone (ADH) and oxytocin—are products of neurosecretory cells which originate in the hypothalamus. These neurosecretory cells release their hormone secretions from their tips following the arrival of neural impulses. ADH regulates water reabsorption in the kidney, and oxytocin regulates uterine contractions and the release of milk from the mammary glands. The anterior lobe of the pituitary consists of endocrine cells that produce at least six hormones: four tropic hormones (ACTH, FSH, LH, FSH), prolactin, and growth hormone (GH). The secretion of each of these hormones occurs in response to factors released by neurosecretory cells from the hypothalamus. Prolactin induces milk production, and growth hormone stimulates body growth.

The adrenal cortex secretes steroid hormones that regulate sugar metabolism and stabilize sodium and potassium concentrations in the blood. **The adrenal medulla** secretes epinephrine, which boosts metabolic activity and prepares the body to cope with perceived danger, emergencies, or stressful situations.

The thyroid gland secretes thyroid hormone, which regulates the body's overall rate of metabolism, and calcitonin, which lowers the concentration of calcium in the blood. This latter effect is countered by the parathyroid hormone, which helps balance blood calcium concentrations.

The pancreas secretes insulin and glucagon. Insulin promotes glucose absorption by cells and conversion to glycogen, decreasing blood glucose. Glucagon promotes glycogen breakdown, increasing blood glucose.

Prostaglandins are produced by secretory cells that are scattered throughout the body; they regulate activities associated with reproduction, blood clotting, uterine contraction, inflammation, and pain perception.

Hormone action depends on the position of the receptor within the target cell. Most hormones bind to specific receptor molecules on the target cell's surface, causing the release of a second messenger (most often cyclic AMP) into the cytoplasm. Cyclic AMP activates a protein kinase, which, in turn, modifies the activities of particular enzymes, depending on the cell. Steroid hormones bind to cytoplasmic receptor molecules. The resulting complex attaches to specific DNA sequences in the chromosomes and turns on specific genes, whose products trigger the response.

Key Terms

endocrine system (p. 520)
neurosecretory cell (p. 523)
pituitary gland (p. 524)
posterior pituitary (p. 524)
anterior pituitary (p. 524)
antidiuretic hormone (ADH) (p. 525)
oxytocin (p. 525)
tropic hormone (p. 525)
prolactin (p. 525)
growth hormone (GH) (p. 525)
adrenocorticotropic hormone (ACTH) (p. 526)

thyroid-stimulating hormone (TSH) (p. 526)
gonadotropin (p. 526)
follicle-stimulating hormone (FSH) (p. 526)
luteinizing hormone (LH) (p. 526)
adrenal cortex (p. 526)
glucocorticoid (p. 526)
mineralocorticoid (p. 528)
adrenal medulla (p. 529)
epinephrine (p. 529)
norepinephrine (p. 529)
thyroid gland (p. 529)

thyroxin (p. 529)
triiodothyronine (p. 529)
parathyroid gland (p. 531)
parathyroid hormone (PTH) (p. 531)
islets of Langerhans (p. 531)
glucagon (p. 531)
testes (p. 532)
ovaries (p. 532)
testosterone (p. 532)
estrogen (p. 532)
prostaglandin (p. 532)
second messenger (p. 533)

Review Questions

1. Many hormones have antagonistic hormones that reverse their effects. Name an antagonist for each of the following hormones, and describe how each pair works together: parathyroid hormone, insulin, and prostaglandin (as a promoter of inflammation).

2. Why is cAMP called a "second messenger"? What is the first messenger? How is it possible that a variety of different hormones can all utilize the same second messenger and evoke markedly different responses?

3. How does the anterior pituitary differ from the posterior pituitary? (Consider the different hormones secreted by these two parts of the pituitary gland and the relationship between the secretion of these hormones and the hypothalamus.)

4. Name two organs that produce steroid hormones, and describe the effects of the hormones from these two sources.

Critical Thinking Questions

1. If Banting and Best had injected far too strong an extract from a ligated pancreas during their first trials on dogs, what effect do you think the injection of excess insulin would have had on these diabetic animals?

2. Both the nervous and endocrine systems perform the same general function—transfer of information—but by different mechanisms. Complete the chart below, comparing these two systems.

Characteristic	Nervous System	Endocrine System
type of message		
message carried by		
area affected by message		
speed of responses		
duration of responses		

3. Why is it important to regulate the concentrations of hormones precisely?

4. In one of Bayliss and Starling's early experiments on the control of pancreatic secretions, the researchers attempted to cut the nerves between the intestine and the pancreas. Why do you suppose they would have carried out this experiment? What results do you think they would have found when their test animals were fed?

5. Describe two major links between the endocrine and nervous systems.

6. Explain how the endocrine system achieves feedback, amplification, and specificity.

7. How are hormones evidence of the unity of life? What evidence is there for a common origin for nervous and endocrine systems? What reasons can you suggest to explain why the two diverged to form separate systems?

Additional Readings

Atkinson, M. A., and N. K. Maclaren. 1990. What causes diabetes. *Sci. Amer.* July:62–71. (Intermediate)

Bliss, M. 1982. *The discovery of insulin.* Chicago: Univ. of Chicago Press. (Introductory)

Guillemin, R., and R. Burges. 1972. The hormones of the hypothalamus. *Sci. Amer.* Nov:24–33. (Intermediate)

Hadley, M. 1988. *Endocrinology,* 2d ed. Englewood Cliffs, NJ: Prentice Hall. (Intermediate)

Lehrer, S. 1979. *Explorers of the body (Banting & Best).* New York: Doubleday. (Introductory)

Linder, M. E., and A. G. Gilman. 1992. G proteins. *Sci. Amer.* Dec:108–115. (Advanced)

Reiter, R. J., ed. 1984. *The pineal gland.* New York: Raven. (Advanced)

Turner, C. D., and J. T. Bagnara. 1976. *General endocrinology,* 6th ed. Philadelphia: Saunders. (Intermediate—Advanced)

Weissmann, G. 1991. Aspirin. *Sci. Amer.* Jan:84–91. (Intermediate)

Witzmann, R. 1981. *Steroids: Keys to life.* New York: Van Nostrand-Reinhold Co. (Intermediate)

CHAPTER

◄ 26 ►

Protection, Support, and Movement: The Integumentary, Skeletal, and Muscular Systems

STEPS TO DISCOVERY
Vitamin C's Role in Holding the Body Together

THE HUMAN PERSPECTIVE
Building Better Bones

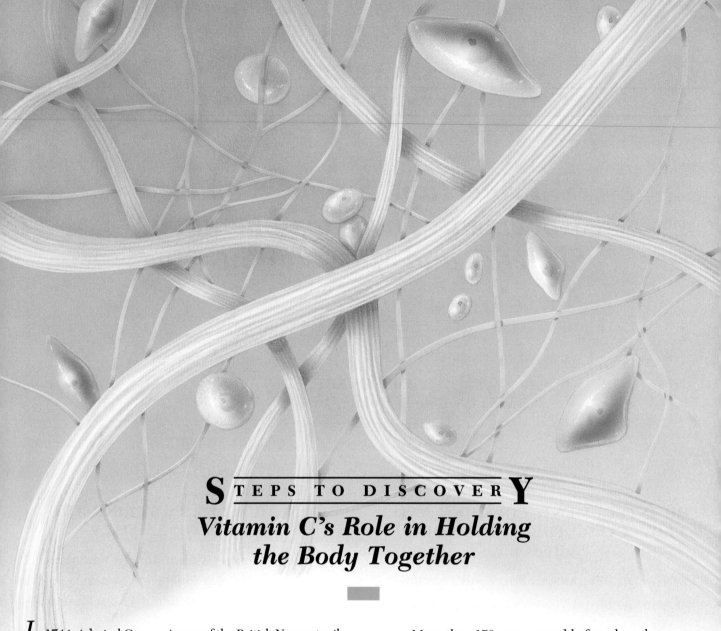

Vitamin C's Role in Holding the Body Together

In 1744, Admiral George Anson of the British Navy set sail with approximately 1,000 men aboard ship. When he returned to England less than a year later, only 144 men remained alive. More than 85 percent of the crew had died of a sailor's disease called *scurvy*. Three years later, James Lind, a surgeon in the British Navy, published a treatise on scurvy, concluding that the dreaded disease was caused by an imbalanced diet. Sailors on these expeditions subsisted almost exclusively on preserved meats and fish. Lind discovered that the addition of citrus fruits, such as lemons and limes, to the sailors' diets would totally prevent the disease (it also earned British sailors the nickname "limeys").

More than 150 years passed before the substance responsible for preventing scurvy was isolated, first from lemon juice, then from cabbage, and, finally, from the adrenal gland of laboratory animals. The structure and properties of the substance, which had been given the name *vitamin C*, were described in the late 1920s and early 1930s, primarily by the Hungarian-born biochemist Albert Szent-Gyorgyi and the British carbohydrate chemist Walter Haworth. Together, Szent-Gyorgyi and Haworth named vitamin C *ascorbic acid*, for its antiscurvy activity. They both received Nobel prizes for their work in 1937.

Although vitamin C and scurvy are obvious topics in a

Loose connective tissue containing scattered cells (fibroblasts), thinner elastic fibers, and thick collagen bundles that are held

discussion of nutrition, it is less evident how they relate to the subjects of this chapter, namely skin, bones, and muscles. All of these parts of the body contain a large amount of connective tissue. Victims of scurvy typically suffer from inflamed gums and tooth loss, poor wound-healing, brittle bones, and internal bleeding. All of these are consequences of serious defects in the formation and maintenance of connective tissues throughout the body. The properties of connective tissues are determined largely by the properties of the *extracellular matrix,* which contains polysaccharides and proteins secreted by cells into their surroundings. The strength and resiliency of connective tissues often depend on a single protein, collagen. Examination of the connective tissues from laboratory animals suffering from scurvy revealed a marked reduction in the number of collagen fibers that normally fill the spaces between the cells. It appeared that there was some relationship between ascorbic acid and the formation of collagen fibers.

Over the next few decades, a number of laboratories turned their attention to the structure and synthesis of collagen. The great strength of the collagen fibers in the extracellular matrix depends in large part on the formation of cross-linking chemical bonds that bind the collagen polypeptides together into strong fibers. The formation of these cross-links requires that some of the amino acids (specifically lysine and proline) in the collagen polypeptide chains are first modified by the addition of a hydroxyl (—OH) group. During the 1960s, scientists found evidence that ascorbic acid deficiency, which is the cause of scurvy, decreases the amount of cross-linking that occurs between collagen polypeptides, thereby weakening the entire fabric of the body's connective tissues. Subsequent research revealed that ascorbic acid is, in fact, a coenzyme that is

required by the enzymes which add hydroxyl groups to the amino acids. A person suffering from scurvy could not modify his or her amino acids to produce the cross links that strengthen collagen fibers.

Ehlers-Danlos syndrome (EDS) is another condition that has been traced to an inability to add hydroxyl groups to collagen. This inherited disorder is characterized by poor wound-healing, tissues that bruise and bleed, extremely flexible joints, and highly extensible skin (which led some EDS sufferers to work as "rubber men" in circus side shows). In 1972, Sheldon Pinnell and his co-workers at Harvard University discovered that a common form of EDS was due to a mutation in an autosomal recessive gene (page 343) that ordinarily directed fibroblasts to produce the enzyme that added a hydroxyl group to the amino acid lysine. A person who is homozygous recessive for this gene fails to form collagen molecules with normal lysine cross-links, a condition that leads to the various symptoms of EDS. Ehlers-Danlos syndrome is less severe than is scurvy, however, because other cross-links (those involving the amino acid proline) are still able to form. In scurvy victims, neither lysine nor proline crosslinks are formed, creating a much weaker collagen fiber.

together by covalent cross linkages.

Skin, bones, and muscles constitute more than 65 percent of your body mass and largely determine your physical appearance, from your body stature to your facial features. Your skin, bones, and muscles protect you from the environment, support you against the effects of gravity, and enable you to move toward food and away from danger. We will begin our discussion of these systems with a look at an animal's first line of defense: its outer body surface.

▼ ▼ ▼

THE INTEGUMENT: COVERING AND PROTECTING THE OUTER BODY SURFACE

The **integument** is the outer body covering of an animal; in vertebrates, it is the **skin.** Depending on the functions it performs, the integument may be soft, flexible, and permeable (as in an earthworm or a frog) or coarse, stiff, and impermeable (as in a lizard or a fish). Whatever its nature, the integument is strategically located at the boundary between the living animal and its environment. Consequently, the integument must act as a protective barrier; it helps shield the individual's delicate, moist, internal tissues from a changing and often harsh environment that might otherwise infect the body with bacteria, freeze the body's fluids, evaporate the body's water, or mutate the body's genes.

THE HUMAN INTEGUMENT: FORM AND FUNCTION

Your skin is a biological cooperative of the four tissue types: epithelial, connective, muscle, and nerve (Figure 26-1). Thus, skin is an organ. In fact, it is the largest organ of your body. Human skin consists of two distinct layers: the outer **epidermis** and the inner **dermis.** These layers have very different structures that reflect their different functions.

The Epidermis: An Outer, Protective Layer

The epidermis is a protective epithelium, approximately 0.1 millimeter thick. It consists of many layers of cells that are formed by mitosis in the deepest layer of the epidermis and then move toward the body surface. As they approach the surface, the cells become flattened, and their cytoplasm becomes filled with filaments of the tough, resistant protein *keratin* (page 801). By the time they reach the surface, the cells have been transformed into a dead, outer layer of keratin, making your skin airtight, watertight, and resistant to bacteria and most chemicals. Pigmentation of the skin is due to the presence of dark granules that eventually reside within the dead epidermal cells.

Fingernails and toenails are remnants of the *cuticle*— the narrow, sensitive flap of living tissue that is located at the back rim of the nail. As the cells of the outer layer of the cuticle die and form nails, they acquire a special hardness due to additional covalent linkages that form between the keratin molecules.

The Dermis: A Complex, Inner Layer

Beneath the epidermis lies the dermis (Figure 26-1), which consists of dense connective tissue and a rich supply of blood vessels, nerve fibers, and smooth muscle cells. The border between the dermis and the epidermis is characterized by hills and valleys, which increase the hand's gripping ability and form the basis of a person's fingerprints. Bundles of dermal collagen fibers give the skin its strength and cohesion as a continuous thin layer. Elastic fibers provide the skin with the elasticity that allows it to snap back when stretched. The dermal blood vessels provide nutrients and oxygen to the overlying epidermis, which lacks its own blood supply. These vessels also play a key role in maintaining the body's constant temperature by carrying warm blood to the body's surface, where heat can be lost to the environment. Blood flow into the dermis can range from a bare trickle, when heat must be conserved during cold external conditions, to 50 percent of the body's blood supply when heat loss is needed to cool the body. The pinkness of the skin in light-skinned individuals is due to blood flow through the dermis. In dark-skinned individuals, the presence of the dermal vessels is obscured by overlying epidermal pigmentation.

Hair: Protecting and Insulating the Skin

Most mammals have considerably more hair (fur) than do humans. A thick layer of hair provides an excellent cover for protecting the body against abrasion and insulating it against heat loss. Human evolution, which probably occurred in a warm, tropical environment, was accompanied by the loss of body hair, which is thought to have been a result of natural selection favoring those individuals who were better able to lose excess body heat.

Hairs consist of dead, keratinized cells similar to those of the outer layer of the skin. Each hair is formed within a living **follicle** (Figure 26-1). Attached to each follicle is a small, smooth muscle that, when contracted, causes the hair to "stand on end." In furrier mammals, such as bears, the contraction of these smooth muscles increases the insulation value of the coat. In humans, contraction of these muscles simply causes an indentation of the skin, producing "goose bumps."

The Skin's Various Secretions

Your skin contains large numbers of glands, whose secretions find their way to the surface of your body. All skin

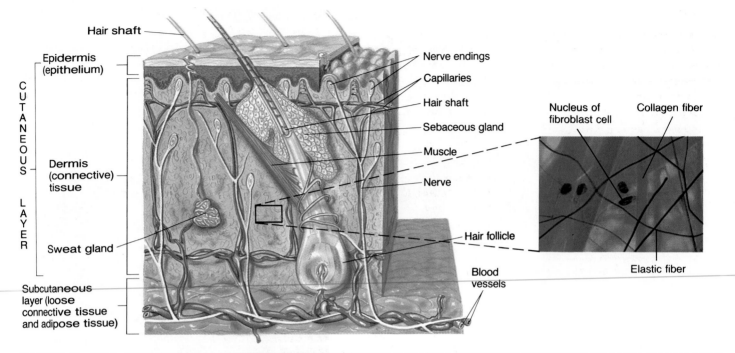

FIGURE 26-1

Human skin. The outer epidermis consists of a superficial layer of dead cells that cover the underlying living epithelial cells. Below the epidermis lies the dermis, where connective tissue predominates. Embedded in the dermis are blood vessels, muscles, nerves, and the basal portions of glands and hair follicles, which form in the epidermis and then sink into the deeper layers of the skin. Hairs are formed from epidermal cells that are generated deep within the follicle. The cells become filled with keratin; they then die and become part of the *hair shaft*. As new layers of dead, keratinized cells are added at its base, the hair is forced upward, increasing in length. Skin (the *cutaneous layer*) is firmly secured to the underlying layer (the *subcutaneous layer*) by connective tissue. Inset shows the connective tissue of the dermis contains scattered cells (fibroblasts) and extracellular collagen and elastic fibers.

glands are **exocrine glands;** that is, they release their secretory products through a duct (in contrast to the "ductless" endocrine glands discussed in Chapter 25). The glands of the skin include two broad types: sebaceous glands and sweat glands. **Sebaceous glands** produce a mixture of lipids *(sebum)* that oil the hair and skin, keeping them pliable. **Sweat glands** are distributed over most of the skin's surface, where they secrete a dilute salt solution, whose evaporation cools the body. A second type of sweat gland is restricted to just a few sites, including the anal and genital regions, nipples, and armpits. These glands appear after puberty and secrete a fluid that contains protein and other organic molecules. Although this fluid is odorless as it is initially secreted, the molecules are eventually broken down by skin bacteria into products that give the human body some of its characteristic odors.

As you can see, the skin is a dynamic, living organ. Its importance is dramatically evident when you consider that burns to as little as 20 percent of the body can be fatal if not treated rapidly. The cause of death in such burn cases is dehydration, which results from the loss of water through the damaged, previously waterproof, body cover.

EVOLUTION OF THE VERTEBRATE INTEGUMENT

All vertebrate integuments are built according to the same two-layered, epidermal–dermal plan, but the two layers can have a very different structure and function, depending on the habitat and lifestyle of the particular animal (Figure 26-2).

The earliest vertebrates were jawless, bottom-dwelling fishes that were clothed in heavy, bony armor that protected them from predators. The plates of bone that formed the fish's armor shields were located in the dermis; hence, they are called *dermal bone*. During subsequent evolution, fishes moved away from the ocean bottom, becoming more buoyant and mobile. The thick plates of dermal bone along the sides of the body were no longer adaptive and became reduced to the thin, familiar bony *scales* that are scraped away when a fish is "cleaned."

As vertebrates moved out of the water and onto the land, the integument became adapted to terrestrial habitats. The bony scales of the ancestral fishes were lost, and the dermis became a more fibrous, flexible layer. In

(a)

(b)

FIGURE 26-19

Examples of Animal locomotion. *(a)* Cuttlefish swimming, *(b)* a basilisk lizard running over the water's surface, and *(c)* lacewings flying.

(c)

EVOLUTION OF THE SKELETOMUSCULAR SYSTEM

▥▶ The skeleton within your body can be traced back through a series of fossils to the skeletons of primitive fishes that lived more than 400 million years ago. Earlier, we noted that these fishes were covered with plates of bone that formed within the dermis of the skin. While nearly all of this dermal bone has disappeared over the course of vertebrate evolution, a number of the bones in your jaw and skull

are derived from these dermal bony plates present in ancestral fish. Other parts of your skull and jaws, as well as the bones in your middle ear, are derived from bones that once supported the gills of these ancient fish (Chapter 39). In contrast, the bones in your limbs have more recent evolutionary roots; limbs first appear in the fossil record in primitive amphibians, soon after vertebrates made their way onto the land. The bones of these limbs were derived from bones that supported the fins of the fishes from which these amphibians evolved.

⚠ The skeletal system of vertebrates probably arose in primitive fishes, whereas the muscles of vertebrates reveal an even earlier origin. Actin and myosin, the major proteins of vertebrate muscle tissue, are found in virtually all eukaryotes, from yeast and fungi to trees and mammals. Muscle cells, which are specialized for contraction, can be found in the most primitive multicellular animals—the sponges—and thus must have appeared very early in the course of animal evolution.

Next time you dine on a trout and are carefully pulling the bony skeleton away from the meat, you might remember that this is the form from which the skeletal muscles in your body probably evolved—as a "wall" of muscle tissue on each side of a fish's body. As vertebrates moved out of the water, and their skeletons became reorganized by natural selection in many different ways, the solid wall of skeletal muscle became divided into discrete muscle masses capable of moving individual bones. The shape and position of the muscles of an animal are closely correlated with the peculiarities of its skeleton and the specialized movements it makes.

Muscle tissue is present throughout the animal kingdom, whereas bone is unique to vertebrates. The particular shapes and arrangements of the muscles and bones of an animal can be correlated with the types of movements the animal performs. (See CTQ #7.)

REEXAMINING THE THEMES

Relationship between Form and Function

⬚ Bones are shaped to carry out particular functions; their internal architecture reflects the specific stresses to which they are subjected. Bones are moved by skeletal muscles, whose cells contain an elaborate array of overlapping protein filaments that convert these cells into pulling machines. A key component in this machinery is myosin, a long, rod-shaped molecule with bulbous heads that project from one end. The receptor sites in the actin filaments fit perfectly with the myosin heads. The bending of the myosin heads creates the power stroke that causes the actin filaments to slide over the myosin rods, shortening the entire length of the cell.

Biological Order, Regulation, and Homeostasis

⟳ Contraction of a skeletal muscle cell is regulated by the concentration of calcium ions in the myofibrils. In the relaxed state, calcium ions are sequestered within the membranous compartments of the SR. When an impulse speeds along a motor neuron and reaches a muscle cell, the SR membranes become permeable to calcium ions; the calcium concentration around the contractile filaments rises; calcium ions bind to the thin filaments, exposing sites on the actin molecules to attachment of the myosin bulbs; and the fiber shortens. When nerve impulses stop arriving, calcium ions are once again sequestered, and contraction ceases.

Acquiring and Using Energy

☀ The large amount of energy required to fuel muscle activity derives from one simple molecular activity: bending the heads on the myosin molecules that push the actin filaments toward the center of the sarcomeres of the muscles that are contracting. ATP molecules bind to the myosin heads before they attach to the actin filaments and are subsequently hydrolyzed as the actin filaments are slid over the myosin. This mechanism of contraction is so efficient that up to 50 percent of the energy released by ATP hydrolysis is converted to mechanical energy, the remainder is released as heat.

Unity within Diversity

⚠ All types of vertebrate integuments are built of the same two layers—an epidermis and a dermis—yet the properties of the integument vary greatly from one animal to another. The bones of vertebrates have a similar molecular composition and a similar microscopic construction, yet the overall shape of each bone varies greatly. Even within a given person, individual bones range from the tiny elements of the wrist to the massive, long bones of the leg. Similarly, the muscles of vertebrates (and all animals) are built of the same types of contractile proteins and operate by a very similar molecular mechanism, yet the particular movements they promote vary greatly both within an animal and among diverse species.

Evolution and Adaptation

▐▶ A specific organ, derived from a particular part of an early embryo, can become modified dramatically over the course of evolution as it meets the needs of different types of animals living in different types of habitats. The vertebrate integument, for example, ranges from thick and scaly in most fishes, to moist and permeable in most amphibians, to dry and horny in most reptiles, to feather covered in birds and hair covered in mammals. Even among related animals, there is great variation. For example, gorillas are covered with hair, while humans are almost devoid of these epidermal derivatives.

SYNOPSIS

An animal's body surface is covered by a protective integument. In humans and other vertebrates, the skin is composed of an outer epidermis and an inner dermis. The outer, protective layer of the human epidermis consists of dead, keratinized cells that are continually sloughed and replaced. The dermis contains connective tissue, which provides support and skin cohesion, and blood vessels, which nourish the skin and play a role in heat conservation or heat loss. Sebaceous glands keep the skin pliable. Sweat glands secrete a dilute salt solution, whose evaporation cools the body. The properties of the vertebrate integument vary, depending on the type of animal and the habitat in which it lives.

Skeletal systems provide the rigidity that is required for support and movement. Some animals use only internal water pressure to support the body; others have a rigid exoskeleton that is external to the animal's living tissues. This may be a calcified shell or a hardened cuticle of protein and chitin. Still other animals have an internal endoskeleton: Sponges have needlelike spicules, sea stars and sea urchins have calcified plates, and vertebrates have bones.

The human skeleton consists of 206 bones, each with a unique shape that allows it to perform a particular function. Most bones arise in the embryo as cartilaginous structures. They are then converted to bone by the deposition of calcium phosphate to the protein–polysaccharide matrix which surrounds the living cells. Bones are typically hollow and contain both compact and spongy regions. The axial skeleton includes the skull, which protects the brain; the vertebral column, which protects the spinal cord; and the rib cage, which protects the organs of the chest cavity.

The appendicular skeleton consists of the paired pectoral and pelvic girdles, which attach the limbs to the axial skeleton, and the bones of the limbs themselves. The bones of the skeleton are connected by joints that possess varying degrees of flexibility.

Muscles are composed of specialized cells that contract (shorten) and generate a pulling force. Vertebrate muscle is divided into skeletal, smooth, and cardiac types. Skeletal muscle responds to voluntary commands and is primarily responsible for moving portions of the skeleton. Skeletal muscles are composed of large, multinucleate cells (fibers) containing myofibrils that have a markedly striated appearance. The structural unit of the myofibril is the sarcomere, which contains overlapping sets of thick (myosin-containing) and thin (actin-containing) filaments. When a muscle fiber is activated to contract, the thin filaments of the sarcomeres slide over the thicker filaments. The force required for this motion is fueled by ATP hydrolysis and is generated by the bending movements of the bulbous heads of the myosin molecules when they are attached to the adjacent, thin filaments. The attachment of the myosin heads to the actin molecules of the thin filaments is triggered by calcium ions, which are released from the sarcoplasmic reticulum of the muscle cell following the arrival of a nerve impulse. The strength by which a muscle contracts depends on the number of fibers that contract, which depends on the number of neurons that carry impulses into the muscle. Smooth muscles, which lack striations, mediate involuntary movements, such as the constriction of blood vessels and the closure of the pupil of the eye. Cardiac muscle, which is striated, makes up the wall of the heart.

Key Terms

integument (p. 542)
skin (p. 542)
epidermis (p. 542)
dermis (p. 542)
exocrine gland (p. 543)
sebaceous gland (p. 543)
sweat gland (p. 543)
hydrostatic skeleton (p. 544)
exoskeleton (p. 545)
cuticle (p. 545)
endoskeleton (p. 546)
bone (p. 546)
osteocyte (p. 546)
compact bone (p. 546)
spongy bone (p. 546)

red marrow (p. 546)
yellow marrow (p. 546)
cartilage (p. 548)
chondrocyte (p. 548)
axial skeleton (p. 548)
skull (p. 548)
vertebral column (p. 548)
rib cage (p. 548)
appendicular skeleton (p. 548)
pectoral girdle (p. 548)
scapulae (p. 548)
clavicle (p. 548)
pelvic girdle (p. 548)
pelvis (p. 548)
joint (p. 550)

ligament (p. 551)
synovial cavity (p. 551)
muscle fiber (p. 553)
myofibril (p. 553)
sarcomere (p. 553)
thin filament (p. 555)
thick filament (p. 555)
actin (p. 555)
myosin (p. 555)
transverse (T) tubule (p. 556)
sarcoplasmic reticulum (SR) (p. 557)
smooth muscle (p. 558)
cardiac muscle (p. 558)
intercalated disk (p. 558)

Review Questions

1. Compare and contrast skeletal muscle and cardiac muscle; skeletal muscle and smooth muscle; osteoclasts and osteoblasts; outer and inner layers of the human epidermis; sebaceous and sweat glands; endocrine and exocrine glands; compact and spongy bone; red and yellow marrow; hydrostatic skeleton and exoskeleton; cartilage and bone.

2. What properties do collagen and cellulose molecules have in common? How is the role of collagen in bone similar to that of cellulose in a plant cell wall?

3. Describe how the integument has changed over the period of vertebrate evolution. Contrast the appendicular skeleton of a fish with that of a human. Contrast the musculature of a fish and a human.

4. Describe the events that occur in a muscle fiber following the arrival of an excitatory nerve impulse. Carry the discussion through to the point where the fiber returns to its original, relaxed state.

Critical Thinking Questions

1. Explain, in terms of enzyme activity, why Ehlers-Danlos syndrome would produce less severe effects than would extreme ascorbic acid deficiency.

2. In addition to protecting the body against the environment, the skin has to receive information from and exchange materials with the environment. How is the skin organized and structured to perform these somewhat contradictory functions?

3. What would you expect to remain if a chicken leg bone were soaked in an acid, such as vinegar, for several days? How would it differ from an untreated bone? What if, instead of acid, you subjected the bone to very high heat in a dry oven?

4. Consider two terrestrial animals, one with an exoskeleton (such as an insect) and the other with an endoskeleton (such as a dog). As these two animals increase in size, which skeleton would you expect to represent a greater and greater percentage of overall body weight? What effect do you think this might have on the size each of these animals could attain?

5. If smooth, instead of striated, muscle were attached to the bones of the skeleton, what effects would this have on the movement of the body?

6. Imagine that your right elbow is resting on the table and your right hand is lifting a heavy weight. Considering the role of the biceps in bending your arm (see Figure 26-10), how do the relative positions of the fulcrum, load, and force differ in this case, compared to the example of standing on your tiptoes (Figure 26-18)?

7. Why are all large animals with exoskeletons aquatic? Why are there no insects as large as elephants (or even dogs)? Why are the largest animals with an exoskeleton (the whales) aquatic?

Additional Readings

Alexander, R. M. 1983. *Animal mechanics.* London: Blackwell. (Intermediate–Advanced)

Diamond, J. 1993. Building to Code. *Discover* May:92–98. (Intermediate)

Gans, C.. 1980. *Biomechanics: An approach to vertebrate biology.* Ann Arbor: Univ. of Michigan. (Intermediate–Advanced)

Hildebrand, M. 1989. Vertebrate locomotion: An introduction. *BioScience* 39:764–765. (Intermediate)

Huxley, H. 1969. The mechanism of muscular contraction. *Science* 164:1356–1366. (Intermediate–Advanced)

Langley, L. L., I. R. Telford, and J. B. Christensen. 1980. *Dynamic anatomy and physiology,* 5th ed. New York: McGraw-Hill. (Intermediate)

Levin, R. M. 1991. The prevention of osteoporosis. *Hosp. Pract.* May:77–97. (Advanced)

Luciano, D. S., A. J. Vander, and J. H. Sherman. 1978. *Human structure and function.* New York: McGraw-Hill. (Intermediate)

Weiss, L. ed. 1988. *Cell and tissue biology: A textbook of histology.* Baltimore: Williams & Wilkins. (Advanced)

Processing Food and Providing Nutrition: The Digestive System

STEPS TO DISCOVERY

The Battle against Beri-beri

Unlike so many other human diseases, beri-beri was probably not an ancient scourge, but primarily a product of the Industrial Revolution. The first clearly documented cases of this nervous disorder, which is characterized by fatigue, muscle deterioration, and possible paralysis, appeared in Asia in the nineteenth century. The disease became prevalent among prisoners and soldiers stationed in the Dutch East Indies in the 1880s. The Dutch government dispatched a team of scientists to look into the problem. Among the members of the team was a medical officer named Christiaan Eijkman.

During the 1880s, etiology (the study of disease) was dominated by the findings of Louis Pasteur and Robert Koch, who had been instrumental in proving that diseases are often caused by "germs" that grow in the body. Unfortunately, Pasteur's and Koch's contemporaries believed that this "germ theory" applied to all diseases; that is, they believed that all diseases were attributed either to bacterial infections or to the toxins produced by bacteria. Eijkman spent 4 fruitless years trying to isolate the bacterium responsible for beri-beri.

One day in 1896, a sudden development provided an unexpected breakthrough in Eijkman's research. For no apparent reason, the chickens that Eijkman was using as

Experiments on chickens revealed that rice kernels contain a vitamin required for normal metabolism. The red bar

experimental animals developed a nerve disease whose symptoms resembled those of human beri-beri. Many of the animals died, but after 4 months, the chickens that had survived the disorder had recovered completely. Upon investigating the matter, Eijkman discovered that the chickens began to recover after a new animal keeper had stopped feeding them leftovers from the military hospital, which consisted largely of polished rice—rice that had been processed by a steam mill until the outer hulls had been removed.

Eijkman used this information in a dietary experiment. He fed some of the chickens a diet of polished rice; these chickens soon developed symptoms of the disease. In contrast, control animals that were fed whole rice remained healthy. Furthermore, the afflicted group could be cured if they were fed either whole rice or polished rice to which the outer hulls had been added. Eijkman concluded that the disease was not due to a bacterial infection but to a dietary deficiency, providing the first evidence that a disease could be caused by the absence of some trace component of the diet. The idea that a disease could result from a dietary deficiency did not initially gain widespread acceptance, however. Eijkman and a colleague, Gerrit Grijns, tried to isolate the substance from the rice hulls which corrected the deficiency, but all they learned was that the factor could be extracted from the hulls in water.

In 1911, Casimir Funk, a chemist working at the Lister Institute in London, succeeded in purifying a substance from the hulls of rice. Funk believed the substance was the same ingredient that Eijkman had found to reverse the symptoms of beri-beri. The substance was an amine (one containing an amino [—NH_2] group), which led Funk to coin the word "vitamine," meaning an *amine* that was *vital* to life. Later work showed that the substance crystallized by Funk was not, in fact, the same one that was active against beri-beri. The name caught on, however, and remained the common term used to describe organic substances that are required by the body in trace amounts. Most vitamins, in fact, contain no amine groups.

Finally, in 1926, two Dutch chemists, B. C. Jansen and W. Donath, working in Eijkman's old lab in the East Indies, developed a procedure for purifying the anti-beri-beri factor from rice bran. Crystals of the substance were sent back to Holland, where Eijkman confirmed that this single chemical compound was effective against the nervous disorder exhibited in birds. However, when Jansen and Donath determined the chemical formula for the substance, they overlooked an important feature: the presence of a sulfur atom. This oversight set back the effort to determine the structure of the compound, which had been named vitamin B_1. The presence of the sulfur atom wasn't discovered until 1932; the correct structure was published in 1936 by Robert Williams, an American chemist who had been working on the problem for over 20 years. Within a year, Williams had worked out a complex procedure for synthesizing the compound, which he named thiamin. Soon, thiamin was being manufactured and became available as a widespread vitamin supplement.

The last major step in the story of vitamin B_1 was the discovery of its biological action. In 1937, two German biochemists, K. Lohmann and P. Schuster, found that thiamin was a coenzyme involved in the reaction that converts pyruvate to acetyl CoA prior to its entry into the Krebs cycle (page 181). Later investigations revealed that virtually all vitamins act in conjunction with enzymes in carrying out one or more crucial metabolic reactions. It was the *failure* to catalyze these reactions that led to the symptoms of the deficiency diseases, such as beri-beri. For his work in establishing the existence of dietary deficiency diseases, Eijkman was awarded the 1929 Nobel Prize in Medicine and Physiology.

and bent arrow pinpoint the reaction that is blocked when the substance is absent from the diet.

*F*or some animals, obtaining food is a relatively simple task. Tapeworms, for example, have no mouth or digestive tract; they simply attach to the wall of an animal's intestine and absorb digested nutrients across their outer body surface. For these animals, eating is not necessary. Unfortunately, as humans, we can't enjoy the same advantage. We can't recline in a bathtub of oatmeal for breakfast and chicken soup for lunch and simply soak up the nutrients. Any such endeavor would only end in starvation. Even if the nutrients could penetrate the epithelial layer of the skin and enter the bloodstream, most of the molecules would be too large to cross the plasma membranes and enter the cells. The molecules would only be wasted since the value of food is unleashed only inside living cells. To meet this requirement for life, most animals possess a team of specialized organs that constitute the **digestive system.**

▼ ▼ ▼

THE DIGESTIVE SYSTEM: CONVERTING FOOD INTO A FORM AVAILABLE TO THE BODY

At what point does food enter the body? Most people would answer, "as soon as you put it in your mouth," or "when you swallow it." Yet, eating does not actually introduce food into your body. When substances are in the stomach or intestines, they are still outside of you, just as your finger poking through the hole of a doughnut remains outside the pastry. The **digestive tract,** or *gut*, is actually a tubelike continuation of the animal's external surface into or completely through its body. The walls of the tract might be likened to an absorbant version of the skin, one that forms a barrier between the external environment and the internal tissues of the body. Because of this barrier property, the digestive tract can safely provide residence for a large number of bacteria that would be dangerous to the interior, living tissues of the body.

To enter the body itself, nutrients must be absorbed across the epithelium that lines the digestive tract. **Digestion** prepares food to do just that. Digestion is the process of disassembling large food particles into molecules that are small enough to be absorbed by the cells that line the digestive tract. Ultimately, these molecules enter the cytoplasm of every cell in the body, where the nutritive value of food is harvested. Eating only initiates the digestive process, launching food on a journey through tunnels and chambers, where digestion and absorption occur.

An animal's diet consists largely of macromolecules—proteins, polysaccharides, lipids, and nucleic acids—that have been synthesized to the specifications of the organism being eaten. To be of use to an animal, these macromolecules must first be disassembled (digested) and then absorbed into the animal's body, where the nutrients can be used for synthesis of new macromolecules. Digestion and absorption are functions of the digestive system.

THE HUMAN DIGESTIVE SYSTEM: A MODEL FOOD-PROCESSING PLANT

The human digestive tract is approximately 9 meters (30 feet) long. It consists of the *mouth, esophagus, stomach, small intestine, large intestine,* and *anus,* plus a variety of accessory organs (Figure 27-1). Each part of the digestive tract is structurally adapted to carry out a particular phase of the digestive process. In fact, the human digestive tract is a model food-processing plant for the stepwise disassembly and absorption of food material. During its journey through the digestive tract, ingested food matter is mixed with various fluids; churned and propelled by the musculature of the wall; broken down by enzymes that are secreted by various glands; and absorbed by cells that line the digestive channel. The indigestible residues are then eliminated from the tract through the anus. All of these complex processes are regulated by the coordinated action of both the nervous and endocrine systems (Chapters 23 and 25).

The wall of the digestive tract (Figure 27-1) is composed of several layers, including an inner glandular epithelium (*mucosa*), layers of circular and longitudinal smooth muscle, and a connective tissue sheath (*serosa*). Glandular cells in the epithelium secrete mucus, enzymes, ions, and other substances. Contraction of the muscle layers help break up congealed food matter, which is mixed with secreted fluids and moved through the tract. We will begin our journey through the digestive system as food enters the first part of the digestive tract: the mouth.

THE MOUTH AND ESOPHAGUS: ENTRY OF FOOD INTO THE DIGESTIVE TRACT

Digestion of food begins in the **mouth,** where food is cut and ground by the teeth. This action makes the food matter easier to swallow and increases its access to digestive enzymes (see Bioline: Teeth: A Matter of Life or Death). While in the mouth, the macerated food is mixed with **saliva:** the secretion that initiates chemical digestion. Saliva is produced by three pairs of **salivary glands,** whose ducts open into the mouth. (It is the salivary glands that become swollen during a mumps virus infection.) Saliva contains the

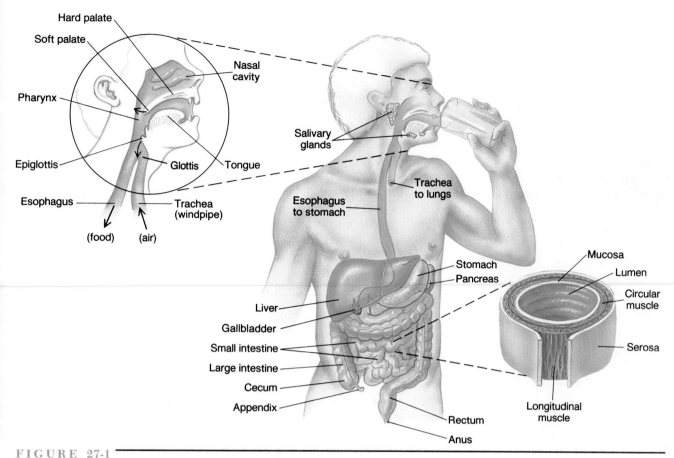

The human digestive system. In addition to the tubular digestive tract, the digestive system includes accessory organs (the pancreas, liver, and gallbladder) that aid digestion. The mucosa and musculature of the digestive tract are shown in the intestinal cross section. The inset depicts structures involved in swallowing. During swallowing, the *soft palate* elevates and closes off the nasal cavity, while
the *glottis* (the opening to the windpipe and lungs) is sealed by the *epiglottis,* leaving the esophagus
the only open passageway for the mass of chewed food (bolus).

enzyme amylase that initiates digestion of starch and helps
protect teeth from decay by breaking down and dislodging
starchy food particles that get trapped between teeth. Saliva
also contains *mucin,* the major protein of mucus, which acts
as a lubricant. This feature is best appreciated when we try
to swallow something that has not been adequately covered
by the slippery fluid. Mucin also binds the macerated food
together into a cohesive mass, called a **bolus.**

Swallowing is essentially a two-step process. The first
step is voluntary: The tongue pushes the bolus to the back
wall of the oral cavity (the *pharynx*), where it stimulates
sensory receptors that initiate the second step. This step
involves a series of reflex, muscular contractions that push
the bolus into the **esophagus,** the tubular channel that
leads to the stomach. During the swallowing process, the
openings to the respiratory and nasal passages are automatically closed to ensure that food is kept out of these nondi-
gestive pathways (see inset, Figure 27-1). This explains why

you can't breathe while swallowing. The swallowing reflexes
occasionally fail to maintain the proper sequence, and food
or liquids accidentally enter the airways. The explosive
coughing reflex that is triggered by such misdirection protects the respiratory tract, which gracefully accepts only
gases (Chapter 29).

The walls of the esophagus contain muscle layers that
contract in a rhythmic manner, sending successive waves of
contraction, or **peristalsis,** down its length (Figure 27-2).
Peristalsis constricts the channel of the esophagus, pushing
the bolus ahead of the traveling wave and through a thickened muscular band, called the **cardiac sphincter** (see
Figure 27-5), and into the stomach. The cardiac sphincter
opens automatically as the wave reaches the bottom of the
esophagus and then quickly closes so that the acidic contents of the stomach do not back up and injure the esophagus. Occasionally, this backup occurs anyway, causing
"heartburn" or "acid stomach," which is felt neither in the

◁ B I O L I N E ▷
Teeth: A Matter of Life and Death

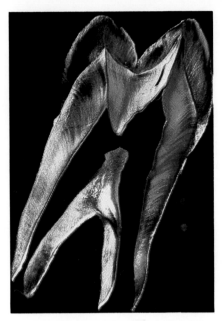

FIGURE 1

The elephant's heart was in excellent condition. So were the animal's other vital organ systems. The giant was free of infection and disease. Yet, she was dying, slip-ping toward the same end that has claimed her kind for thousands of generations. Although surrounded by food, the elephant was starving to death.

The coarse diet of plants on which an elephant dines steadily grinds away the animal's teeth. During the elephant's lifetime, its worn-out *molars* (which grind food) are periodically replaced so that eating can continue. Only six sets of teeth can be produced, however; by the age of 60, the last set has usually been ground flat. The otherwise healthy elephant can no longer chew its food and literally starves to death.

🜨 Teeth provide an excellent example of the relationship between form and function. Animals whose diets include large amounts of plant material (such as that of the elephant) require teeth that can crush the tough cell walls, releasing the intracellular nutrients. The teeth must have broad-ridged surfaces that grind together, as do our molars (Figure 1). Dogs, cats, and other carnivores that feed almost exclusively on meat have little need for teeth with grinding surfaces since there are no cell walls in their diets to crush. In these animals, grinding surfaces have been replaced by sharp, cutting edges that rip and slice the food into smaller pieces. The knifelike, cutting edges of the front teeth (*incisors*) are specialized for cutting off bites of food. Four pointed teeth, called *canines* (or "fangs"), are instrumental in securing and killing prey. The reduction in the length of canine teeth during human evolution reflects our different strategy for capturing and handling prey. Primitive humans used weapons and their hands rather than their teeth for this purpose.

🜨 Animals that swallow their food whole do not use their teeth to assist digestion. For example, snakes have backward-curving teeth that are used for capturing and holding prey. The sharp teeth of sharks provide a fascinating, if not chilling, example of teeth that are adapted strictly for tearing off chunks of meat to be swallowed without being chewed. Birds, which lack teeth altogether, have a muscular grinding organ, called the *gizzard* (Figure 27-8c). With the help of swallowed stones, the gizzard macerates food as it passes through the digestive tract.

---

heart nor in the stomach, but *near* the heart, in the esophagus.

THE STOMACH: A SITE FOR STORAGE AND EARLY DIGESTION

The **stomach** continues the mechanical breakdown of food solids and limited enzymatic digestion, which were begun in the mouth and esophagus. The average human stomach can hold and store about a liter (nearly a quart) of material. The stomach contents are churned to a pastelike consistency and are mixed with *gastric juice* (*gastro* = stomach), forming a solution called **chyme.** Gastric juice is produced by secretory cells that are located in pits in the wall of the stomach (Figure 27-3). *Hydrochloric acid* (HCl), one of the compounds that makes up gastric juice, lowers the pH of the stomach contents to around 2.0. This extremely acidic environment kills most microbes found in food, including many of those that could cause illness. Stomach acid also denatures (opens up) highly folded protein chains so that they can be more easily attacked by protein-digesting enzymes.

Although most enzymatic digestion occurs in the small intestine, protein digestion begins in the stomach with the action of the enzyme *pepsin.* To prevent the stomach from becoming its own next meal, the protein-digesting enzyme is secreted as an inactive precursor, *pepsinogen,* which is converted to an active pepsin molecule by the hydrochloric acid of the stomach. The living tissue that lines the stomach is protected from both acid and pepsin by a thick layer of alkaline mucus, which is secreted by specialized glandular cells found in the stomach wall. When protective mechanisms fail, the stomach may begin digesting portions of itself, causing painful and dangerous *peptic ulcers.* For decades, peptic ulcers have been treated by administration of antacids. Persons suffering from these lesions, however,

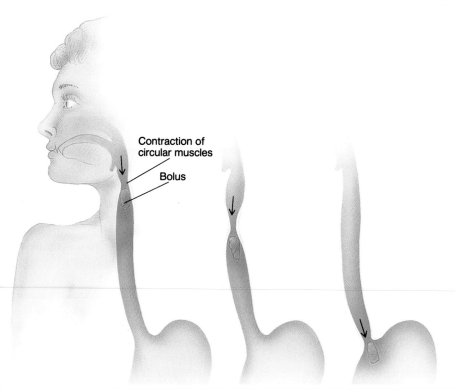

FIGURE 27-2

Movement by peristalsis. Contraction of a layer of circular muscle in the wall of the tubular esophagus sweeps down the length of the organ, pushing the bolus into the stomach. Similar types of peristaltic waves force ingested material through the entire length of the digestive tract.

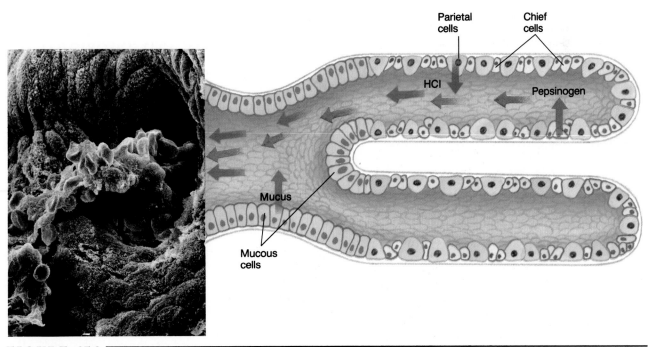

FIGURE 27-3

Gastric juice combines the secretory products from several different types of gland cells, which are located in pits (inset) that open into the stomach chamber. Within the wall of the pit are *mucous cells* that secrete mucus, *chief cells* that secrete pepsinogen (the inactive precursor of the enzyme pepsin), and *parietal cells* that secrete hydrochloric acid (HCl).

have found this therapy ineffective; while antacids help to alleviate symptoms, over 95 percent of patients experience a relapse within two years. Within the past few years, a number of studies have shown that, while stomach acidity may be a factor in the development of peptic ulcers, the primary cause is infection by the acid-resistant bacterium *Helicobacter pylori.* Ulcer patients treated with antibiotics that kill the bacteria are much less likely to suffer relapses than those treated solely with antacids. Don't be surprised if, one day, a vaccine becomes available to prevent ulcers. According to one estimate, the eradication of *H. pylori* could cut as much as 80 percent of the cost of gastrointestinal medicine.

Hunger and the Control of Gastric Secretions

Food-seeking behaviors, such as those that overcome many of us on a late-night outing, usually follow sensations of *hunger:* the body's message to your conscious mind that nutrients need to be replenished. During the hours following a meal, the concentration of glucose in the blood is high enough to stimulate neurons, called *glucoreceptors,* in the hypothalamus. Activated glucoreceptors inhibit the brain's *hunger center,* which is also located in the hypothalamus. The sensation of hunger reoccurs when glucose levels fall slightly, signaling glucoreceptors to release their inhibitory

grip on the hunger center. Yet, glucoreceptors cannot be the sole factor responsible for regulating hunger since the sensation subsides as soon as our stomachs become full, long before glucose from the meal is absorbed into the bloodstream. This rapid inhibition of hunger is probably the direct result of stomach distension, or swelling, which stimulates pressure-sensitive sensory neurons in the stomach wall.

The control of the secretion of gastric juices illustrates the complex communication that occurs during physiological processes (Figure 27-4). The first phase of gastric secretion is stimulated by nerve impulses that reach the stomach from the brain as a result of the smell, taste, or even thought of food. When food actually enters the stomach, two new types of signals are generated, which lead to a marked increase in the secretion of gastric juice. One of the signals is carried by sensory neurons from the stomach to the brainstem, which responds by sending impulses down autonomic motor fibers. This stimulates the digestive-gland cells of the stomach wall to release their products. The other signal is a chemical message that is sent by the hormone gastrin, which is released by endocrine cells located in the stomach lining. The message is carried locally in the blood vessels of the stomach wall to the stomach's glandular cells, triggering these cells to release gastric juices. By sending these nervous and endocrine signals, the stomach ensures that gastric juices will be secreted when they are necessary.

Only a few small molecules, such as aspirin and alcohol, enter the bloodstream through the stomach wall. This explains the rapid onset of their effects. Most nutrients are not absorbed until they enter the small intestine.

THE SMALL INTESTINE: A SITE FOR FINAL DIGESTION AND ABSORPTION

Peristaltic waves moving along the wall of the stomach repeatedly push small quantities of chyme through a muscular valve (the *pyloric sphincter*) and into the **small intestine** (Figure 27-5). The small intestine consists of 6 to 7 meters (about 21 feet) of highly coiled muscular tubing, about 2.5 centimeters (1 inch) in diameter. During their stay in the small intestine, macromolecular food substances are digested into small, organic molecules, such as simple sugars, amino acids, and nucleotides, all of which are absorbed into the bloodstream.

Digestion of materials within the small intestine requires the cooperative activities of several major organs and their secretions, including intestinal juice, pancreatic secretions, and emulsifying materials from the liver and gallbladder.

Intestinal Secretions

When the inflow of chyme from the stomach stretches the intestinal wall, the action triggers a neural reflex response, whereby the cells of the intestinal lining secrete *intestinal juice* and mucus. Under normal conditions, the small intestine secretes about 2 to 3 liters of intestinal juice per day.

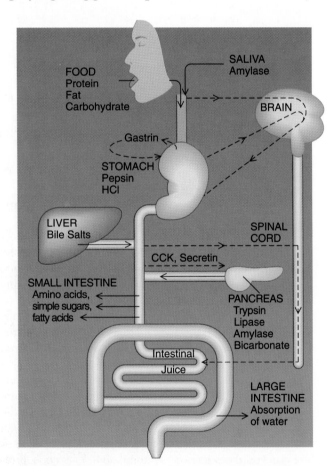

FIGURE 27-4

Control of the processes of digestion by the nervous and endocrine system.

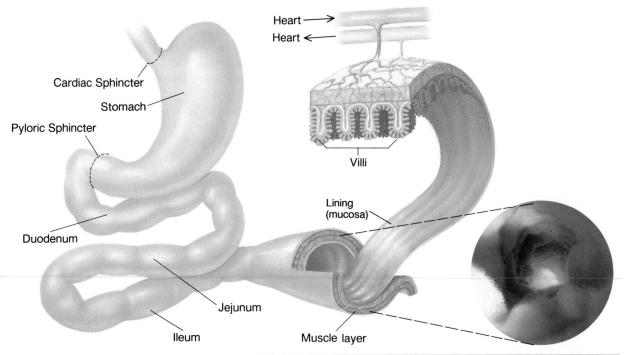

Heart →
Heart ←

Cardiac Sphincter

Stomach

Pyloric Sphincter

Villi

Duodenum

Lining (mucosa)

Jejunum

Ileum

Muscle layer

FIGURE 27-5

Anatomy of the small intestine. Approximately the first 30 centimeters of the small intestine forms a C-shaped curve, called the *duodenum*. The *jejunem* comprises the next 1.5 meters of the small intestine, and the *ileum* constitutes the remaining 2.5 meters of the tube. The surface area of the intestinal lining is increased by the folding of the intestine and by the presence of villi, which are shown in the inset photo. The average person sends about 725 kilograms (330 pounds) of solid food and 700 liters (650 quarts) of ingested liquids through this tunnel every year.

This fluid is needed to dissolve the molecules for digestion and to facilitate absorption across the intestinal epithelium. Cholera, one of the most dreaded human diseases, results from a bacterial toxin that greatly increases the fluid released by these intestinal cells. When stimulated by the cholera toxin, the intestine can pour out over 1 liter of fluid per hour, most of which is simply lost through the digestive tract as diarrhea. Unless this loss of fluid is made up by drinking or intravenous entry, patients are at high risk of dying from dehydration. The bacteria that cause cholera are usually acquired by drinking water that is contaminated by the feces of other people with the disease.

Pancreatic Secretions

As we discussed in Chapter 25, the pancreas (review Figure 27-1) is a gland that manufactures digestive enzymes and discharges them directly into the *duodenum*, the first region of the small intestine. The pancreas also releases sodium bicarbonate, an alkaline substance that helps neutralize the severe acidity of the chyme entering the small intestine from the stomach. Within the pancreatic secretion are enzymes that digest all four major types of macromolecules:

- *Trypsin, chymotrypsin,* and other such proteolytic enzymes degrade proteins to amino acids.

- *Nucleases* break down nucleic acids to nucleotides and nitrogenous bases.

- *Carbohydrases* hydrolyze complex carbohydrates to disaccharides or simple sugars.

- *Lipases* split triglycerides into fatty acids and glycerol.

The secretion of pancreatic enzymes and bicarbonate is stimulated by two hormones: *cholecystokinin (CCK)* and *secretin* (Figure 27-4). These hormones are secreted into the blood by endocrine cells in the wall of the small intestine in response to the inflow of chyme from the stomach.

The enzymes produced by the pancreas are potentially dangerous molecules since they have the ability to destroy most of the body's own materials. The pancreas is protected from some of these enzymes by two mechanisms: (1) synthesis of the enzymes in an inactive state, and (2) synthesis of the enzymes along with the inhibitors that keep them inactive until they reach the intestine. The reality of the threat is revealed by a rare condition called *pancreatitis,* in which the duct from the pancreas to the intestine becomes blocked, and the secretions accumulate to high concentrations. This accumulation can cause the pancreas to digest itself rapidly with its own secretions, a condition that can prove fatal.

Liver and Gallbladder Secretions

The common approach to removing baby oil or motor grease from your hands or animal fat from a pan is to wash your hands or the pan with soap and water. Soap contains detergent molecules that are both hydrophobic and hydrophilic—one end is soluble in fat, and the other end is soluble in water. Because of their structure, detergents can surround fat molecules and suspend them in water. As a result, the grease or fat comes off your hands or the pan and becomes *emulsified* (suspended) in the surrounding water. A similar approach is taken by your digestive system in dealing with fats in your diet. Suppose you eat a pizza topped with double cheese. In order for the fat molecules in the cheese to be efficiently hydrolyzed by the pancreatic lipases in the intestine, large globules of fat must be broken apart into much smaller clusters. This process is accomplished by **bile salts,** which are produced in the liver and stored in the **gallbladder** (see Figure 27-1), a small sac that empties its contents into the intestine through a duct. Bile salts are similar in structure to the detergents present in soap. In the presence of bile salts, fat globules are reduced to stable, microscopic droplets that can be more efficiently attacked by lipid-digesting enzymes.

In addition to bile salts, the fluid of the gallbladder also contains the breakdown products of hemoglobin molecules of aging red blood cells that are pulled from the circulation and destroyed by the liver. Diseases that affect liver function, such as hepatitis, often interfere with the normal processing and discharge of these products, which remain in the blood and cause the skin to take on a yellowish, or *jaundiced,* appearance.

Absorption of Products Across the Small-Intestinal Wall

The first step in the absorption of the small, digested food molecules is their movement from the lumen (the space inside a tube) of the small intestine into the epithelial cells that form its lining. The inner surface of the intestinal wall has a specialized structure that makes it ideal for efficient absorption. The "velvety" texture of the intestinal lining is provided by fingerlike projections, called **villi** (Figures 27-5, 27-6), which protrude from the entire surface. Villi increase the absorptive surface of the intestine in the same way that the texture of terry-cloth towels enables them to soak up much more water than can smooth cloth towels.

Each villus (singular of villi) is covered with smaller projections, called **microvilli** (inset, Figure 27-6), which further increase the surface area of the small intestine enormously. Together, intestinal villi and microvilli create an interior surface area that is 150 times greater than the surface of your entire skin. Packed inside your abdomen is an intestinal surface equivalent to the surface area of a tennis court!

Each villus is laced with a rich network of capillaries surrounding a single, centrally located lymphatic vessel known as a **lacteal.** The lacteal absorbs the products of lipid digestion, such as the fatty acids that are produced by lipase digestion of the cheese on the pizza you consumed hours earlier. From the lacteal, microscopic fat droplets are transported through a series of lymphatic vessels that eventually drain into a large vein in the neck (Figure 27-6 and Chapter 28). Most of these fat molecules are removed from the bloodstream by adipose (fat) cells scattered throughout the body, adding to the body's fat content.

What about the remainder of the ingredients in that pizza, such as the glucose in the starchy dough, and the amino acids in the protein-rich pepperoni? Most nonfatty nutrients diffuse directly into the blood capillaries of the intestinal villi, where they are carried to the liver and are removed from the bloodstream. The liver is the body's primary metabolic regulatory center, producing waste products, controlling blood glucose levels, and releasing substances into the bloodstream, as needed by the body's tissues.

Some nutrients undergo the last stages of digestion as they pass across the plasma membrane of the epithelial cells. Lactase, for example, is an enzyme in the plasma membrane which severs the disaccharide lactose (milk sugar) into two simple sugars. Many people cannot drink milk or eat dairy products without developing severe diarrhea because they lack the enzyme lactase. Consequently, milk sugar remains undigested in the large intestine, upsetting this organ's osmotic balance by creating a gradient that draws an excessive amount of water into the intestinal lumen.

THE LARGE INTESTINE: PREPARING THE RESIDUE FOR ELIMINATION

By the time digested food material has made its long journey to the end of the small intestine, virtually all of its nutrients have been removed, along with most of the water. The nutrient-depleted chyme is now propelled by peristalsis into the next part of the digestive tract, the **large intestine** (Figure 27-7).

The most important functions of the large intestine are the reabsorption of water from the digestive tract and the conversion of the remaining contents into a mass, called *feces.* Pressure-sensitive neurons detect when solids accumulate in the terminal portion of the large intestine (the **rectum**) and respond by provoking a *defecation reflex:* Impulses from the large intestine travel to the spinal cord and back to the muscles of the rectum, causing the muscles to contract and force the feces out through the anus. Because one of the two anal sphincters is under voluntary control, we can consciously delay expulsion of feces until an appropriate time.

Projecting from the large intestine is a short, blind (dead-ended) pouch, the *cecum,* from which a small blind tube, the *appendix,* extends (Figure 27-7). The cecum and appendix have no known function in humans. They are

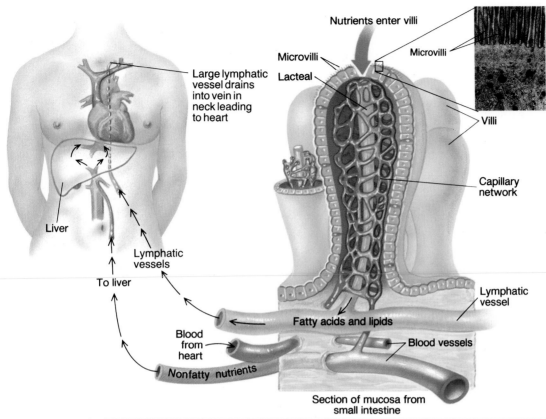

FIGURE 27-6

Structure and function of the small intestine's villi. Nutrients pass from the lumen of the small intestine into capillaries and a central lacteal located within each villus, a projection of the intestinal lining. Tiny projections, called microvilli, extend from the end of each epithelial cell in the villi, further increasing the absorbing surface. Substances absorbed from the small intestine into the bloodstream are carried to the liver, which detoxifies many harmful molecules and stores various nutrients, such as glucose. From the liver, nutrients enter the general circulation.

vestigial reminders of our evolutionary descent from mammals whose enlarged ceca (plural of cecum) aided digestion of plant material. Inflammation of the appendix leads to *appendicitis*, which, if untreated, can cause the appendix to rupture, spilling bacteria into the abdominal cavity and creating a potentially fatal infection.

The Large Intestine as a Home for Bacteria

Huge numbers of bacteria reside in the healthy human digestive tract. In fact, the number of resident microbes in the large intestine is so large that bacteria constitute almost half the dry weight of human feces. These intestinal bacteria metabolically attack organic substances in chyme and use them as nutrients, often producing unpleasant-smelling byproducts. These organisms are not freeloaders, however; they manufacture vitamin K, biotin, folic acid, and other nutrients we absorb and utilize. Although these services are

not required when the diet is well balanced, intestinal bacteria contribute enormously to our well-being by competing with potentially dangerous microbes for the body's limited space and nutrients. This becomes apparent when we destroy our normal bacterial flora by the extended use of antibiotics, which can cause digestive dysfunction (such as extended bouts of diarrhea) and yeast infections.

In providing nutrients for the body, the digestive system must grind ingested materials, disassemble macromolecules, absorb fluid and digested products, and eliminate undigestible residues. These processes occur in a stepwise manner, as food material passes through a series of chambers whose walls are specialized for secreting, churning, absorbing, and sending neural and hormonal signals to associated glands to release digestion-related substances. (See CTQ #3.)

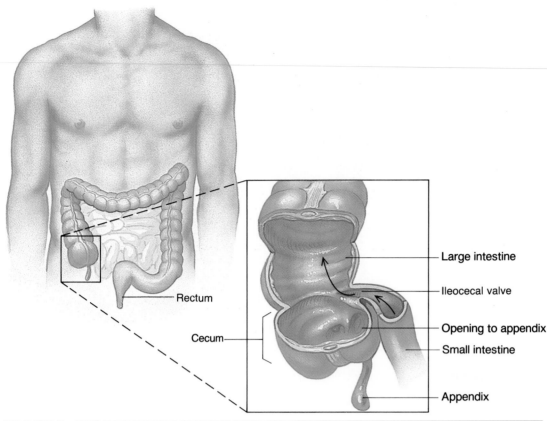

FIGURE 27-7
The large intestine receives unabsorbed wastes from the small intestine through the *ileocecal valve*.
The large intestine in humans is about 1.5 meters long; the final 18 centimeters is the *rectum*. The
large intestine, except for the rectum, is called the *colon*. The short cecum and appendix in humans is
all that remains of larger chambers that housed cellulose-digesting micro-organisms in our ancestors.

EVOLUTION OF DIGESTIVE SYSTEMS

▮▶ As with most other organ systems, the evolution of increasingly complex animals was accompanied by the appearance of more elaborate digestive systems. The most primitive form of digestion in animals is best revealed by observing the process in many of the larger protists. Figure 27-8a shows one ciliated protozoan engulfing another ciliated protozoan by phagocytosis (page 145), a process known as **intracellular digestion.** Such prey are incorporated into food vacuoles and digested by lysosomal enzymes; the molecular products are then absorbed into the cytoplasm. Many of the simpler animals, including sponges, sea anemones, and flatworms also utilize intracellular digestion. Food is taken into the body of the animal, where it is engulfed by cells and disassembled within food vacuoles.

Most animals—including arthropods, annelids, and vertebrates—utilize **extracellular digestion,** whereby food material is hydrolyzed within a digestive tract by enzymes that are secreted into the tract by various digestive glands. The small-molecular digestive products are then absorbed, and the undigested residue, such as arthropod exoskeletons or plant cell walls, passes out of the body.

Some of the simpler animals—including jellyfish, sea anemones and flatworms—have an **incomplete digestive system** (Figure 27-8b), which consists of a blind digestive chamber with only one opening to the outside environment. This opening serves two functions: It is both an entrance for food and an exit for undigested residues. In contrast, most animals require the specialization afforded by a **complete digestive system,** which includes a tube with openings at both ends: a mouth for entry and an anus for exit (Figure 27-8c). In humans, food moves sequentially through different compartments, each region performing its specialized tasks of maceration, digestion, absorption, dehydration, and elimination. Complete digestive tracts can be thought of as "disassembly lines"—assembly lines that operate in reverse.

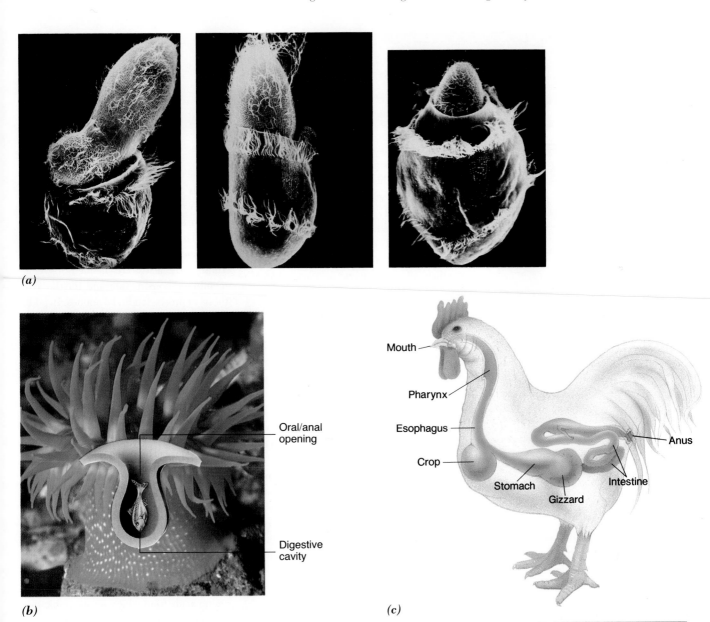

FIGURE 27-8
Digestive diversity. *(a)* Intracellular digestion. The protozoan *Didinium* engulfs a larger *Paramecium* for dinner. *(b)* Incomplete digestive tract of a sea anemone; food enters and residues exit through the same opening. *(c)* Complete digestive tract of a bird. Food is pushed to the back of the mouth (pharynx) and swallowed into the esophagus; from there, it passes into the crop, where it is stored. Food is ground, with the help of swallowed stones, in the bird's muscular gizzard.

ADAPTATIONS TO DIFFERENT DIETARY BEHAVIORS

■■▶ An animal's feeding apparatus and digestive system are adapted to the type of food the animal regularly consumes. The human digestive system is specialized for digesting a variety of foods, ranging from chunks of meat to fibrous vegetables and fruits. Many animals have more specialized diets. For example, a number of mammals, including ant-eaters and pangolins, feed exclusively on ants (Figure 27-9*a*). Each of these ant-eating species has powerful claws that can dig up anthills; elongated snouts that can extend into an ant nest; and long, sticky tongues that trap ants. Many aquatic animals, including barnacles, clams, and blue whales, feed on tiny organisms suspended in the water (Figure 27-9*b*). These *filter feeders* have some type of straining device or sticky, mucous-covered surfaces that screen or trap tiny food particles suspended in the water.

◁ THE HUMAN PERSPECTIVE ▷
Human Nutrition

FIGURE 1

Controversy continues to rage over such topics as the impact of dietary sugar, cholesterol, and saturated fats on our health. Nutritionists generally agree, however, that the healthiest diets are those that balance carbohydrates, triglyceride lipids (fats and oils), and proteins (Figure 1). Foods that provide these three groups of molecules should also contain enough energy, organic building blocks, vitamins, and minerals to satisfy the needs of the average person.

CALORIES AND ENERGY

☀ You may recall from the discussion of oxidative metabolism that subunits from all of the various macromolecules (amino acids, sugars, fatty acids, and nucleotides) are fed into the same metabolic pathways that provide ATP for the cell (page 188). Consequently, the body can fulfill its energy needs from any of these macromolecules. It is estimated that an average per-

son engaged in a relatively sedentary lifestyle requires about 2,500 calories per day to maintain his or her body at a stable level. A person who engages in frequent strenuous activity, such as a professional athlete, may need over 6,000 calories per day. Since people differ in their metabolic set-points (page 76), the number of calories that will maintain one person's weight may cause another person either to gain or lose pounds.

CARBOHYDRATES

Carbohydrates provide the most readily available form of glucose and, therefore, the most rapid, readily available form of usable energy. This is why runners, for example, eat large amounts of pasta (which is high in carbohydrates) the night before a marathon. Glucose is the "all-purpose" energy source; it is also the only one that can be used by all brain and nerve cells. When these cells are deprived of energy, the resulting temporary *hypoglycemia* (low blood sugar) may cause unclear thinking, clumsiness, depression, or muscle tremors due to diminished neurological function.

Not all carbohydrates are easily digestible. Cellulose, for example, the polysaccharide of plant cell walls, resists disassembly in the human digestive tract. Yet, cellulose (such as that found in celery stalks) promotes health by providing bulk-fiber, which assists in the formation and elimination of feces. Low-cellulose diets may cause constipation and have been linked to colon cancer. For this reason, foods that are rich in both polysaccharides and undigestible bran, such as fruits, grain products, and legumes, are recommended as part of a balanced diet. In contrast, the simple carbohydrates that are found in sugary products, such as candy and soft drinks, provide calories but lack the fiber necessary for healthy digestion.

LIPIDS

Because of their highly reduced state, fats and oils are rich sources of energy (page 128). Two fatty acids, linolenic and linoleic acid, are *essential fatty acids,* since they are needed but cannot be manufactured in the body and are thus required in the diet. These fatty acids are needed for cell-membrane construction and for the synthesis of prostaglandin hormones. In addition to being fattening, diets rich in saturated fats and cholesterol (found in butter, animal fats, and eggs) may predispose susceptible individuals to circulatory disease by increasing the likelihood of deposition of lipids in the walls of arteries. In contrast, unsaturated fats, which are found in most vegetable oils, may increase the risk of cancer because breakdown of the fatty acids in these molecules produces free radicals (page 56). Nutritionists generally agree that an ideal diet should be low in fat; that is, fat should provide less than 30 percent of food calories.

PROTEINS

Dietary protein is needed to supply the amino acids from which we assemble our enzymes, antibodies, hormones, and various other types of proteins. Protein can be obtained from virtually any food, including meat, cheese, eggs, and vegetables. We can manufacture all but eight amino acids metabolically. The absence of even one of these eight *essential amino acids* prevents the synthesis of all proteins. All of these required amino acids can be synthesized by plants and microorganisms. In fact, it is thought that during the early evolution of animal life, the enzymes needed for the synthesis of these particular amino acids were lost because these molecules were present in adequate supply within the diet. Plant proteins are *incomplete* because they provide essential amino acids but not in the proportions necessary for proper nutrition. For this reason, vegetarians are advised to combine vegetables and grains in particular ways to ensure proper amino acid intake.

The prolonged absence of any of the essential amino acids in the diet can lead to severe protein deficiency and a condition known as *kwashiorkor,* which is one of the world's most serious health problems. We are all too familiar with pictures of listless, malnourished children with swollen bellies (a result of water retention) and arms and legs composed of just skin and bones. These symptoms are a result of the world's shortage of dietary protein.

VITAMINS

An organic compound is designated a *vitamin* when it is needed in trace amounts for normal health but cannot be synthesized by the body's own metabolic machinery and thus must be obtained in the diet. As humans, we must acquire 13 vitamins from our diet, or we run the risk of suffering vitamin-deficiency disorders, some of which can be fatal. A well-balanced diet normally provides all the vitamins needed.

Vitamins are usually divided into two groups, based on their solubility properties. Vitamins in the first group are water soluble; these include eight different vitamins of the B complex and vitamin C. Vitamins in the second group are insoluble in water but soluble in oil; these include vitamins A, D, E, and K. Most of these vitamins function as coenzymes that assist essential enzymatic reactions (Table 27-1). The required daily allowance of each vitamin is relatively low because these molecules are not consumed in the reactions they assist so each molecule is used again and again. Furthermore, the small amount that is required replaces that which is normally excreted.

Given this information, is there any value to a person with a balanced diet taking vitamin (and mineral) supplements?

TABLE 27-1
DIETARY ESSENTIALS IN HUMAN NUTRITION: VITAMINS

Designation	Major Mode of Action	Major Sources[a]	Symptoms of Deficiency[b]
Retinol (A)	Part of visual pigment, maintenance of epithelial tissues	Egg yolk, butter, fish oils; conversion of carotenes[c]	Nightblindness, corneal and skin lesions, reproductive failure
Calciferol (D)	Ca and P absorption, bone and teeth formation	Fish oils, livers; irradiation of sterols[c]	Rickets, osteomalacia
Tocopherols (E)	Antioxidant	Vegetable oils, green, leafy vegetables	In animals: muscular degeneration, infertility, brain lesions, edema[d]
Vitamin K	Synthesis of blood coagulation factors	Green, leafy vegetables; bacterial synthesis	Slowed blood coagulation
Thiamine (B_1)	Energy metabolism–decarboxylation	Whole grains, organ meats	Beri-beri, polyneuritis
Riboflavin (B_2)	Hydrogen and electron transfer (FAD)	Whole grains, milk, eggs, liver	Cheilosis, glossitis, photophobia
Nicotinic acid (niacin)	Hydrogen and electron transfer (NAD, NADP)	Yeast, meat, liver[e]	Pellagra
Pyridoxine (B_6)	Amino acid metabolism	Whole grains, yeast, liver	Convulsions, hyperirritability
Pantothenic acid	Acetyl group transfer (CoA)	Widely distributed	Neuromotor and gastrointestinal disorders
Biotin	CO_2 transfer	Eggs, liver; bacterial synthesis	Seborrheic dermatitis
Folic acid	One-carbon transfer	Leafy, green vegetables, meat	Anemia
Cobalamine (B_{12})	One-carbon synthesis; molecular rearrangement	Animal products, esp. liver; bacterial synthesis	Pernicious anemia
Ascorbic acid (C)	Hydroxylations, collagen synthesis	Citrus, potatoes, peppers	Scurvy

[a] Most vitamins, especially of the B group, occur in a multitude of foodstuffs and in all body cells.
[b] A variety of symptoms occur with certain vitamin deficiencies; vitamin deficiencies are frequently of a multiple nature, and symptoms similar to those described may have their origin in conditions not related to nutrition.
[c] Certain carotenes, found in green and yellow vegetables, are precursors of vitamin A. Certain sterols, including 7-dehydrocholesterol, which is synthesized in the body, are precursors of vitamin D.
[d] No well-defined syndrome is described for humans.
[e] Niacin is one of the end products of normal tryptophan metabolism.
Source: From P. D. Sturkie, *Basic Physiology,* New York, Springer-Verlag, 1981, p. 345.

This question continues to be debated, and no sweeping, satisfactory answer is currently available. Scientific studies in human nutrition are notoriously difficult to carry out in a controlled manner (as exemplified by the debate over cancer and vitamin C, page 39), and the effects of various eating habits may not become apparent for decades. On the one hand, there is no current scientific evidence to suggest that taking *large* amounts of vitamins is helpful, and there is clear evidence that taking *large* amounts of vitamins A and D can be injurious and even fatal. (There are documented cases, for example, of hungry individuals dying from consuming a bear's liver, which contains high concentrations of fat-soluble vitamins.) On the other hand, some biochemists and nutritionists argue that the capability of vitamins C and E to destroy free radicals (page 56) makes them valuable supplements *in moderate doses.* Clearly, the subject of vitamin supplements is an area in which more controlled research studies are required.

MINERALS

A dietary supply of *minerals* (Table 27-2) is just as important for proper nutrition as is that of vitamins. Without calcium and magnesium, for example, large numbers of enzyme-mediated reactions would simply not occur. Calcium is also needed for bone growth and muscle function. Iron forms the functional core of cytochromes, which are needed by all cells for aerobic respiration, and of hemoglobin, the oxygen-transporting protein in blood. Phosphorus is needed for ATP and nucleic acid synthesis. Sodium, potassium, and chloride are required for maintaining osmotic balance and for propagating nerve impulses. A number of other minerals, including iodine, copper, zinc, cobalt, molybdenum, manganese, and chromium, are required in such tiny amounts that they are referred to as *trace elements.*

TABLE 27-2

DIETARY ESSENTIALS IN HUMAN NUTRITION: MINERALS[a]

Designation	Major Functions	Major Sources	Symptoms of Deficiency[b]
Calcium (Ca)	Bone and teeth, nervous reactions, enzyme cofactor	Dairy products, leafy, green vegetables	Calcium tetany, demineralized bones
Phosphorus (P)	Bone and teeth, intermediary metabolism	Dairy products, grains, meat	Demineralized bones
Magnesium (Mg)	Bone, nervous reactions, enzyme cofactor	Whole grains, meat, milk	Anorexia, nausea, neurological symptoms
Sodium (Na)	Maintenance of osmotic equilibrium and fluid volume	Table salt[c]	Weakness, mental apathy, muscle twitching
Potassium (K)	Cellular enzyme function	Vegetables, meats, dried fruits, nuts	Weakness, lethargy, hyporeflexia
Chlorine (Cl)	Maintenance of fluid and electrolyte balance	Table salt[c]	[d]
Iron (Fe)	Hemoglobin, myoglobin; respiratory enzymes	Meat, liver, beans, nuts, dried fruit	Anemia
Copper (Cu)	Enzyme cofactor (cytochrome-c-oxidase)[e]	Nuts, liver, kidney, dried legumes, raisins	Anemia, neutropenia, skeletal defects
Manganese (Mn)	Enzyme cofactor, bone structure, reproduction	Nuts, whole grains	[d]
Zinc (Zn)	Enzyme cofactor (carbonic anhydrase)[e]	Shellfish, meat, beans, egg yolks	Growth failure, delayed sexual maturation
Iodine (I)	Thyroid hormone synthesis	Iodized table salt, marine foods	Goiter
Molybdenum (Mo)	Enzyme cofactor (xanthine oxidase)[e]	Beef kidney, some cereals and legumes	[d]
Chromium (Cr)	Regulation of carbohydrate metabolism (glucose tolerance factor)	Limited information available	[d]

[a] A human need for the following trace elements is possible but has not been unequivocally established: selenium (Se), fluorine (F), silicon (Si), nickel (Ni), vanadium (V), and tin (Sn). The need for sulfur (S) is satisfied by ingestion of methionine and cystine, and for cobalt (Co) by vitamin B$_{12}$.
[b] Except for Ca, Fe, and I, dietary deficiency in humans is either unlikely or rare.
[c] Many processed foods contain considerable amounts of sodium chloride.
[d] No specific deficiency syndrome described in humans.
[e] Examples of activity as enzyme cofactors.
Source: From P. D. Sturkie, *Basic Physiology,* New York, Springer-Verlag, 1981, p. 344.

(a)

(b)

(c)

FIGURE 27-9

Feeding specialists. *(a)* Anteaters trap their tiny prey on the sticky surface of their tongue. *(b)* Barnacles have modified appendages that strain the surrounding water, collecting microscopic, organisms. *(c)* Some sea cucumbers—relatives of sea stars—collect microbes on their feathery tentacles, which are periodically placed in the mouth.

Some filter feeders, such as sea cucumbers, extend feathery or highly branched tentacles that harvest microscopic bits of food (Figure 27-9c). Filter feeders usually feed *continuously*, while most animals feed *discontinuously*, taking in food matter only when it is convenient and available.

Some animals paradoxically obtain nutrients from fibrous plants that they alone cannot digest. Cattle, antelopes, buffalo, giraffes, and other such *ruminant* animals possess additional stomach chambers, one of which is heavily fortified with cellulose-digesting microorganisms (Figure 27-10). In this chamber, called the *rumen*, bacteria and protozoa break down the otherwise indigestible cellulose fibers and use them for growth of more microorganisms. This growing crop of microorganisms is then digested by the animal as the contents of the rumen travel through the rest of the digestive tract. Horses and other grazing animals that lack rumens are able to utilize plant cellulose because they possess an elongated *cecum*, a sac that extends from the large intestine. The cecum serves as a fermenting vat in which microbes disassemble cellulose into products that are digestible by the animal.

Animals procure food in various ways: They strain it from their aqueous environment, extract it from mud, scrape it from rocks, swallow it whole or in large pieces, and so forth. The structure and function of an animal's digestive system are adapted to the particular feeding habits of the animal. (See CTQ #5.)

FIGURE 27-10

The cape buffalo; an example of a ruminant.

REEXAMINING THE THEMES

Relationship between Form and Function

🔄 The human digestive tract is a model food-processing plant for the stepwise disassembly of food and absorption of nutrients. Each part of the tract is ideally suited for particular activities. For example, the mouth contains a set of differentiated teeth that are specialized for cutting, tearing, and grinding food material to facilitate its digestion. The esophagus, with its thick, circular bands of muscle tissue, is specialized for pushing mouthfuls of food matter into the stomach. The pouchlike stomach is specialized as a food storage site, and its muscular walls churn the food matter into a liquified slurry. The small intestine, with its tremendous number of villi and microvilli, is ideally suited for absorbing small-molecular-weight nutrients that have been generated by the digestive enzymes that pour in from the pancreas. The water-reabsorbing properties of the large intestine (and its large population of bacteria) make this organ suited for preparing fecal residues for elimination.

Biological Order, Regulation, and Homeostasis

🔄 The passage of food through the human digestive tract is not a continuous activity; rather, it occurs periodically. The activities that take place in various parts of the digestive tract must be timed to correspond to the presence of food matter in each specific part of the tract at a particular time. Timing and coordination of digestive activities is controlled by the cooperative activities of both the nervous and endocrine systems. When food enters the mouth, stomach, and intestine, it triggers reflex responses and hormonal secretions, which initiate specific secretory and muscular re-

sponses. Food seeking behavior is initiated by sensations of hunger, which occur in response to decreased blood glucose and lack of pressure in the stomach.

Acquiring and Using Energy

☀️ The energy that is expended in fueling our activities is derived primarily from polysaccharides and lipids that are present in the food we eat. These macromolecules must first be broken down into their subunits, which are then absorbed across the wall of the digestive tract into the bloodstream. These breakdown products eventually reach all the individual cells of the body, where the energy is harvested by oxidation. While carbohydrates and lipids provide the most readily available sources of chemical energy, proteins and nucleic acids in our diets can also be broken down and oxidized for their energy content.

Evolution and Adaptation

▶ Digestive tracts show a graded increase from simple to more complex systems. Simpler, multicellular animals, such as sea anemones and flatworms, have an incomplete digestive tract. In addition, they utilize intracellular digestion, whereby microscopic food particles are phagocytized and enzymatically broken down within food vacuoles. Most animals have a complete digestive tract which is open at both ends and allows food matter to pass in a single direction along a continuous "disassembly line," where it is subjected to stepwise processing. Even though animals may have similar digestive tracts, each digestive system is specifically adapted to the particular feeding habits of the animal. For example, elephants have huge, replaceable molars for grinding tough vegetation. Birds, which lack teeth, have a large crop for storing food, and a muscular gizzard for grinding it. Mammals that graze on grass often possess a chamber, such as a rumen or cecum, that harbors cellulose-digesting microorganisms.

SYNOPSIS

An overview of digestion. Large, complex food substances are dismantled into molecules small enough to pass through plasma membranes and enter cells, where their nutritional value is unleashed. A complete digestive system, such as that of humans, consists of a continuous digestive tract and various accessory organs. Ingested material is forced through the entire tract by peristalsis—waves of contraction of the layers of muscle tissue in the wall of the tract. The muscular and secretory activities that occur at different sites along the digestive tract are coordinated by the actions of both the nervous and endocrine system.

Each part of the digestive tract is specialized for particular activities that occur in a stepwise fashion as food matter is pushed along the tract. The mouth macerates food and covers it with lubricating mucus. Swallowing combines voluntary and involuntary responses that push the food bolus into the esophagus, where it is moved by peristalsis into the stomach and churned into chyme. Glands in the stomach wall secrete a fluid that contains acid and protein-digesting enzymes. Chyme passes into the small intestine, where its acidity is neutralized by bicarbonate ions from the pancreas; its macromolecules are digested by enzymes also produced in the pancreas. Most of the fluid in which the food matter is suspended is derived from secretions of the wall of the small intestine itself. The fat that is present in the food is emulsified in the form of microscopic droplets by bile salts that are produced in the liver and stored in the gallbladder. As the various macromolecules are enzymatically digested, their component subunits are absorbed across the wall of the intestinal villi into either blood capillaries or lymphatic lacteals. Most of these nutrients are carried to the liver, where they are removed from the bloodstream. Those materials that cannot be digested and absorbed pass into the large intestine, where the remaining water is reabsorbed and the insoluble residues are compacted (together with bacteria) into feces, which are eliminated.

Most digestive systems consist of a continuous tube that is open at both ends. Because it is continuous with the external environment, the space within the digestive tract is actually a part of the outside world. Although most digestive systems have a similar overall plan, each has special adaptations that are appropriate to the feeding habits of the particular animal.

A healthy human diet must contain a variety of components. Chemical energy is most readily supplied by carbohydrates and fats, but it can also be provided by proteins and nucleic acids, which also can be degraded through a cell's oxidative pathways. Of the 20 amino acids incorporated into proteins, eight cannot be synthesized from other compounds and must be obtained from the diet. Protein deficiency is the world's major cause of malnutrition. Two fatty acids are also required as dietary elements. Vitamins are organic compounds that function primarily as coenzymes but cannot be synthesized by the body; thus, they are required in the diet. The diet must also supply a number of minerals. Some of these, such as calcium, magnesium, iron, sodium, and potassium, play major roles in the body and must be present at high levels in the diet. In contrast, the roles of trace elements, such as iodine, copper, and zinc, are more limited.

Key Terms

digestive system (p. 568)
digestive tract (p. 568)
digestion (p. 568)
mouth (p. 568)
saliva (p. 568)
salivary glands (p. 568)
bolus (p. 569)
esophagus (p. 569)

peristalsis (p. 569)
cardiac sphincter (p. 569)
stomach (p. 570)
chyme (p. 570)
small intestine (p. 572)
bile salts (p. 574)
gallbladder (p. 574)
villi (p. 574)

microvilli (p. 574)
lacteal (p. 574)
large intestine (p. 574)
rectum (p. 574)
intracellular digestion (p. 576)
extracellular digestion (p. 576)
incomplete digestive tract (p. 576)
complete digestive tract (p. 576)

Review Questions

1. Why isn't food considered to be inside the body immediately after it is swallowed? With this consideration in mind, is human digestion intracellular or extracellular? At what point do nutrients actually enter the body?

2. Trace the fate of a mouthful of food through the entire digestive tract. Describe the changes that occur in each portion of the digestive tract, and discuss the activities of saliva, gastric secretions, intestinal secretions, pancreatic secretions, bile, and intestinal bacteria.

3. How are the various activities described in your answer to the previous question regulated by the endocrine and/or nervous system?

4. Compare and contrast the sites of the preliminary digestion of starch, protein, and fats; the role of CCK and secretin; the effects of high versus low blood glucose concentrations on hunger; the role of most vitamins, compared to minerals, such as calcium and iron; the role of molars in an elephant and a dog.

Critical Thinking Questions

1. Suppose you were studying nutrition and found that laboratory rats that were fed on a diet of raw carrots remained healthy, but those that were fed on a diet consisting only of cooked carrots developed a condition that caused them to lose their hair. What conclusion might you draw about this substance? What kind of experiment would you run to confirm your conclusion? What controls would you use? If you attempted to extract the substance from carrots, how could you determine which of your extracts contained the necessary ingredient?

2. Both mechanical and chemical processes are involved in the breakdown of most foods into small molecules that can be absorbed by cells lining the small intestine. What is the role of each type of activity, and why are both necessary?

3. What effects might each of the following have on an individual: mumps infection; eating slowly; removal of a cancerous stomach; secretion of excess stomach acid; reversal of peristalsis; gallstones; removing a portion of the small intestine; appendectomy; colostomy (removal of the colon); long treatment with antibiotics?

4. Why do you think elevating your head when you sleep may help prevent heartburn in the middle of the night? How do you suppose people are able to swallow food even when they are standing on their heads? Do you think these two observations contradict each other? If so, how can you resolve the apparent contradiction?

5. How do teeth reflect the varied diet of humans? How do human teeth differ from the teeth of carnivores, such as lions? Herbivores, such as deer?

Additional Readings

Alper, J. 1993. Ulcers as an infectious disease. *Science* 260:159–160. (Intermediate)

Campbell-Platt, G. 1988. The food we eat. *New Sci.* May:1–4. (Introductory)

Davenport, H. W. 1978. *Physiology of the digestive tract,* 4th ed. Chicago: Year Book Medical Publishers. (Intermediate–Advanced)

Hamilton, E. 1982. *Nutrition: Concepts and controversies.* Menlo Park: West. (Introductory)

Makhlouf, G. M. 1990. Neural and hormonal regulation of function in the gut. *Hospital Practice* Feb:79–98. (Advanced)

Moog, F. 1981. The lining of the small intestine. *Sci. Amer.* May:154–176. (Intermediate)

Scrimshaw, N. S. 1991. Iron deficiency. *Sci. Amer.* Oct:46–53. (Intermediate)

Schutz, Y., et al. 1991. Unsolved and controversial issues in human nutrition. *Experentia* 47:166–193. (Intermediate–Advanced)

Readings from Scientific American. 1978. *Human nutrition.* New York: W. H. Freeman. (Intermediate–Advanced)

CHAPTER
◄ 28 ►

Maintaining the Constancy of the Internal Environment: The Circulatory and Excretory Systems

STEPS TO DISCOVERY
Tracing the Flow of Blood

THE HUMAN CIRCULATORY SYSTEM: FORM AND FUNCTION

Blood Vessels

The Human Heart: A Marathon Performer

Composition of the Blood

EVOLUTION OF CIRCULATORY SYSTEMS

THE LYMPHATIC SYSTEM: SUPPLEMENTING THE FUNCTIONS OF THE CIRCULATORY SYSTEM

EXCRETION AND OSMOREGULATION: REMOVING WASTES AND MAINTAINING THE COMPOSITION OF THE BODY'S FLUIDS

The Human Excretory System: Form and Function

Evolution and Adaptation of Osmoregulatory and Excretory Structures

THERMOREGULATION: MAINTAINING A CONSTANT BODY TEMPERATURE

Gaining Heat to Keep the Body Warm

Losing Heat to Keep the Body Cool

BIOLINE
The Artificial Kidney

THE HUMAN PERSPECTIVE
The Causes of High Blood Pressure

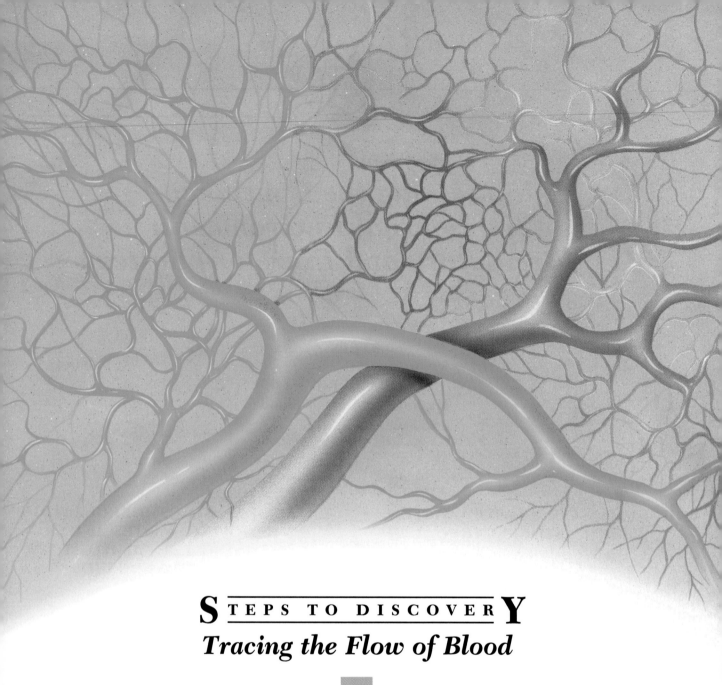

Tracing the Flow of Blood

*I*n 1628, a book entitled *An Anatomical Treatise on the Movement of the Heart and Blood in Animals* was written in Latin by an English physician and published in Germany. It was a small book, published on cheap, crumbling paper and filled with typographical errors. Yet, it has been hailed as the most important work ever published in the field of physiology. In this book, William Harvey, the son of a prosperous English merchant and a graduate of the University of Padua in Italy, described his experiments that led to a new concept of the organization of the human circulatory system.

Prior to Harvey's publication, the established views of blood circulation in humans had been formulated by a Greek named Galen, who had lived 1,400 years earlier and had served as the personal physician for the Roman Emperor Marcus Aurelius. Galen died in 201 A.D., but his views were still accepted in the sixteenth century, and his authority remained unchallenged. A brilliant, although dogmatic, scientist, Galen had made a number of important discoveries. But, he concluded that blood was formed in the liver then flowed to the heart and through the arteries en route to the tissues, where it was entirely absorbed. The

Contrary to the early view that veins and arteries were independent, blood flows through a circuit leading from the heart, into

blood utilized by the tissues was replaced by new blood from the liver. Galen envisioned the veins as a system of vessels independent of the arteries; blood in the veins simply ebbed back and forth within the same vessels, much the way the tide moves in and out along the shore. The fact that this view survived for 1,400 years, is an indication of the stagnation of science throughout the Middle Ages, which were also known as the Dark Ages.

In order to test Galen's hypothesis of blood flow, Harvey attempted to measure just how much blood is actually pumped out of the heart in a given amount of time. He measured the internal volume of the heart of a cadaver and multiplied this figure by the number of times the heart beats per minute. Using this method, Harvey concluded that it took the heart nearly 30 minutes (which is actually a considerable overestimate) to pump out an amount of blood equivalent to the body's entire blood supply. It was inconceivable that the tissues could actually absorb this amount of blood so rapidly, or that the liver could resupply the blood so quickly. Harvey proposed that, rather than being absorbed by the tissues, blood circulated through the body along some type of "circular" pathway. According to his hypothesis, blood left the heart through the arteries, passed into the various tissues, and then returned to the heart through the veins.

To support his hypothesis, Harvey demonstrated that blood could flow in only one direction through a given vessel. In one simple, but convincing, demonstration, Harvey pressed his finger against one of the major veins of the forearm then moved his depressed finger along the vein toward the individual's hand (away from the heart), pushing the blood out of the vein. If blood flowed in both directions, as was the prevailing view of the time, then the vein should rapidly refill with blood. If the vein carried blood in only one direction—back to the heart—the vein should remain empty. The results were clear: The vein remained empty of

blood until Harvey removed his finger. Galen's 1,400-year-old hypothesis of blood flow had been disproved with this simple, but elegant, demonstration.

One of Harvey's great frustrations was his inability to demonstrate just *how* blood flowed from the arteries into the veins. Harvey hypothesized that the tissues contain tiny vessels that complete the circuit from the arteries to the veins, but he had no way of demonstrating the existence of such vessels. The vessels linking the arterial and venous circulation were finally discovered in 1661, 4 years after Harvey's death, by the Italian anatomist Marcello Malpighi, whom we met in Chapter 19 in reference to his experiments on circulation and transport in plants. Using a newly developed instrument, the microscope, Malpighi prepared a thin piece of tissue from the lungs of a frog. He allowed the tissue to hang in the air to dry and then observed the blood vessels with a microscope; the red color of the blood vessels contrasted strikingly against the light background. Malpighi saw the larger arterial vessels branching into smaller vessels and finally giving rise to short, minuscule vessels that merged at the other end into the venous circulation. In this image, Malpighi had discovered the link between the arteries and veins; he named these tiny linking vessels *capillaries,* the Latin word for "hairlike." Malpighi confirmed his observations by examining living tissues, such as the wall of the urinary bladder. He traced the blood as it flowed through the arteries and into capillaries; the blood never spilled out into the spaces of the tissue. Malpighi had demonstrated that blood flowed in a unidirectional, continuous, and uninterrupted cycle.

arteries (red), through microscopic capillaries, into veins (blue), and back to the heart.

*J*ust as the towns and cities of a country are connected by roads and rails, the various parts of your body are connected by an extensive system of "living tubes," or vessels, that provide a continuous route for the movement of blood. Like a road or rail system carrying trucks or trains, the blood picks up and delivers materials as it courses through the body: Nutrients are picked up as the blood passes along the wall of the small intestine and are removed by the liver and, ultimately, delivered to all the tissues of the body; oxygen is picked up as the blood passes through the lungs and is removed by the body's cells; hormones are picked up as the blood passes through the various endocrine glands and are carried to their respective target organs; waste products are swept up as the blood passes through the tissues and are removed by the kidneys, lungs, and liver for disposal.

Blood vessels, the blood that flows through them, and the heart whose contractions propel the blood, make up an animal's **circulatory system.** The circulatory system is more than just a conduit for the movement of materials from one organ to another; it is the means by which complex, multicellular animals maintain homeostasis. If ions or other solutes become too concentrated in the body, for example, the blood carries them to the kid-

ney for elimination. If the tissues become too acidic, the blood provides the buffering agents that lower the hydrogen ion concentration. If the tissues become depleted of oxygen or steeped in carbon dioxide, the blood carries the message to the nervous system to stimulate deeper or more rapid breathing. In birds and mammals, the blood also plays a crucial role in maintaining a constant body temperature; the blood carries heat either to the body surface, where it can be dissipated into the environment, or deep into the body, where it can be conserved. The bloodstream also plays a vital role in an animal's defense against disease-causing organisms (Chapter 30).

The bulk of a complex, multicellular animal is made up of fluid, which is distributed among three different fluid compartments: intracellular, interstitial, and vascular. The **intracellular compartment** is the space present within all of the animal's cells—the fluid of the cytoplasm and nuclei. The **interstitial compartment** is the space between and surrounding all of the cells: the extracellular fluid. The **vascular compartment** is the space present within the vessels and chambers of the circulatory system: the fluid of the bloodstream. The spatial relationship among the three compartments is depicted in Figure 28-1. Because of its position, the interstitial fluid acts as a "middleman" between the other two compartments. Substances such as oxygen, glucose, and hormones are able to reach the cells by diffusing from the blood through the interstitial fluid. Conversely, substances such as carbon dioxide and nitrogenous waste products are able to move in the opposite direction: from the cells, through the interstitial fluid, to the bloodstream. These exchanges are essential for maintaining the homeostatic condition of the organism.

▼ ▼ ▼

THE HUMAN CIRCULATORY SYSTEM: FORM AND FUNCTION

The human circulatory system, or **cardiovascular system** (*cardio* = heart, *vascular* = vessels), consists of blood vessels, the heart, and blood.

BLOOD VESSELS

The human cardiovascular system contains tens of thousands of kilometers of tubing, which assures that every cell of the body is within diffusion distance of a capillary. Blood vessels are divided into five basic types: arteries, arterioles, capillaries, venules, and veins (Figure 28-2), differing in form and function.

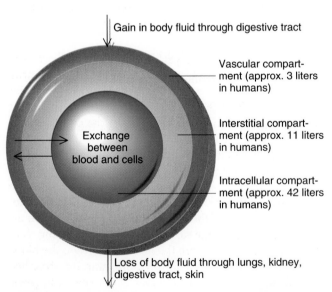

Gain in body fluid through digestive tract

Vascular compartment (approx. 3 liters in humans)

Interstitial compartment (approx. 11 liters in humans)

Exchange between blood and cells

Intracellular compartment (approx. 42 liters in humans)

Loss of body fluid through lungs, kidney, digestive tract, skin

FIGURE 28-1

The spatial interrelationship of the intracellular, interstitial, and vascular fluid compartments. All substances exchanged between the bloodstream and the body's cells must pass through the interstitial fluid that surrounds the cells.

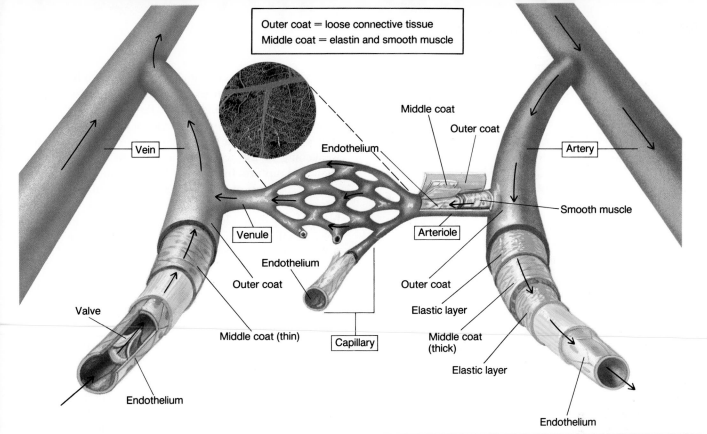

Outer coat = loose connective tissue
Middle coat = elastin and smooth muscle

Vein

Endothelium

Middle coat

Outer coat

Artery

Venule

Smooth muscle

Endothelium

Arteriole

Outer coat

Valve

Outer coat

Elastic layer

Middle coat (thin)

Capillary

Middle coat (thick)

Elastic layer

Endothelium

Endothelium

FIGURE 28-2

The form and function of the body's blood vessels are closely related. Each of the five categories of vessel has a distinct internal anatomy. The inner lining of all the vessels consists of a layer of flattened, "interlocking" endothelial cells which make up the *endothelium*. The walls of the capillaries—the thinnest of the vessels—consist only of a single, thin endothelial cell layer. The inset shows the delicate latticework of a capillary bed. (Note: this drawing is not to scale.)

Arteries: Delivering Blood Rapidly Throughout the Body

Blood is pumped out of the heart into **arteries**—large vessels that function as conduits, rapidly carrying blood to all parts of the body. Arteries typically are large in diameter and have complex walls that contain concentric rings of elastic fibers. When blood is pumped out of the heart, the walls of the arteries are pushed outward, increasing their fluid capacity. As the walls stretch, the rings of elastic fibers respond much as a rubber band does; that is, they recoil and exert pressure against the blood in the arteries. This is what we measure as **blood pressure.** To visualize this effect, imagine filling a rubber balloon with water. If you squeeze one end of the balloon with your hand, you are exerting pressure against the water. Since water can't be compressed into a smaller volume, the fluid pushes outward against the wall of the balloon, forcing it to elongate. This is the same principle underlying blood flow: The heart contracts, forcing blood into the arteries; the stretched arteries push against the blood, forcing it through the smaller vessels farther along the route. As the fluid is pushed out of the arteries, the diameters of the arteries decrease, and the arterial pressure drops, until the next contraction abruptly returns the pressure to its maximum value.

Blood Pressure When you have your blood pressure taken, an inflatable cuff is strapped around your upper arm, providing a measure of the fluid pressure in the major arteries in that limb (Figure 28-3). Blood-pressure readings are expressed as two numbers, one over the other, such as 120/80 (the normal values for a young adult). The first number is the **systolic pressure,** as measured in millimeters of mercury. This is the highest pressure attained in the arteries as blood is propelled out of the heart. The pressure drops rapidly as blood is pushed out of the arteries and the diameter of the arteries decreases. Since the arteries don't have time to return to their resting (unstretched) diameter before the next contraction, the blood in the arteries remains under pressure from the stretched arterial walls. The second number is the **diastolic pressure,** or the lowest pressure in the arteries of the arm recorded just prior to the next heart contraction. If the observed systolic and diastolic values are higher than about 140/90, the person is said to have high blood pressure.

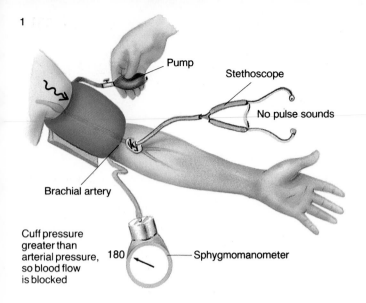

1

Pump

Stethoscope

No pulse sounds

Brachial artery

Cuff pressure greater than arterial pressure, so blood flow is blocked

180

Sphygmomanometer

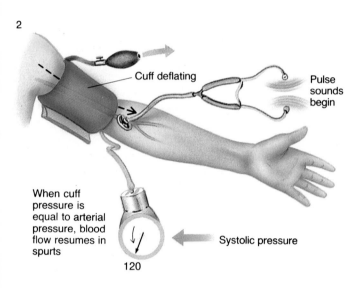

2

Cuff deflating

Pulse sounds begin

When cuff pressure is equal to arterial pressure, blood flow resumes in spurts

120

Systolic pressure

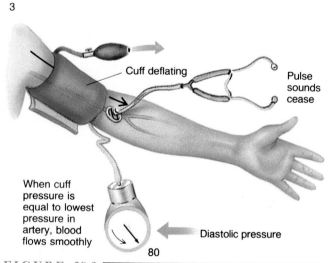

3

Cuff deflating

Pulse sounds cease

When cuff pressure is equal to lowest pressure in artery, blood flows smoothly

80

Diastolic pressure

FIGURE 28-3

Determining blood pressure using a *sphygmomanometer* and a stethoscope. The normal blood pressure in the brachial artery, the major artery of the arm, is 120/80.

While the causes of high blood pressure, or **hypertension,** are still poorly understood (see The Human Perspective: The Causes of High Blood Pressure), the effects of hypertension can become very evident if the condition continues untreated. The excessive pressure leads directly to a weakening of the walls of the arteries, increasing their chance of rupture. Increased arterial pressure also accelerates the buildup of fatty plaques on the walls of arteries (*atherosclerosis*) and a greatly increased likelihood that blood clots will develop in the vessels. Together, these deleterious effects on the body's arteries promote the rupture or blockage of cerebral vessels, causing a stroke and possible subsequent brain damage. Other risks include the rupture of vessels in the kidney, which can cause kidney failure, and blockage of the coronary arteries, which can cause a heart attack.

Arterioles: Regulating Blood Flow to the Tissues

The major arteries branch into smaller and smaller arteries, eventually giving rise to **arterioles:** small vessels whose walls lack elastic fibers but contain a preponderance of smooth muscle cells. The amount of blood that flows into a particular tissue depends largely on the diameter of the local arterioles. Arteriolar diameter is determined by the state of contraction of the smooth muscle cells in the walls of these vessels, which is regulated by the sympathetic nervous system (page 490). If an organ needs more oxygen, as does the heart of a person who is engaged in strenuous exercise, sympathetic stimulation to the muscle cells of the arterioles in the organ *decreases*. As a result, the muscle cells become more relaxed, and the arterioles increase in diameter—a response known as **vasodilation.** Conversely, those organs of the body that operate at a low activity level at a particular time, such as the digestive tract during a time of stress or exercise, receive less blood due to a temporary narrowing of the arterioles. The decrease in arteriole diameter is caused by an *increase* in sympathetic stimulation, causing the arteriole's muscle cells to contract—a response known as **vasoconstriction.**

Capillaries: Exchange between Tissue and Bloodstream

From the arterioles, blood passes through the **capillaries,** the smallest, shortest, and most porous channels of the vascular network. The lumen of these vessels is just large enough for red blood cells to move along in single file (Figure 28-4). The capillaries are the sites of exchange between the cells and the bloodstream. Your body contains about 40,000 kilometers of capillaries—enough to circumscribe the earth at its equator—creating an enormous surface area for the exchange of materials between the blood and the interstitial fluid of the tissues. The movement of water, nutrients, waste products, and dissolved gases in and out of the capillaries is illustrated in Figure 28-5. Red blood cells and most proteins are too large to penetrate the pores

◁ THE HUMAN PERSPECTIVE ▷
The Causes of High Blood Pressure

Approximately 20 percent of all Americans suffer from high blood pressure (hypertension) which, if untreated, can lead to kidney failure, heart attack, or stroke. Even though hypertension is one of the most thoroughly researched medical conditions, the underlying causes are still uncertain. We do know that high blood pressure usually arises as a result of a decreased diameter of arteries and arterioles. Most cardiologists believe that excessive vessel constriction results from a variety of factors that affect people differently, depending primarily on a person's genetic predisposition. Some of these are physiological risk factors discussed below; others are behavioral risk factors associated with stress, obesity, smoking, and consumption of dietary fats and alcohol.

In 1934, Harry Greenblatt of Case Western Reserve University published one of the most important experiments in modern medicine. Having constricted the major *(renal)* artery leading into one of the kidneys of a dog, Greenblatt found that the animal developed a markedly higher blood pressure. This experiment, among others, led to the discovery that the kidneys produced an enzyme, called *renin,* which is secreted into the blood, where it leads to the formation of another protein, called *angiotensin II.* Angiotensin II is a vasoconstrictor—a substance that acts on the smooth muscles of arterioles, causing them to contract, decreasing their diameter, and increasing blood pressure. It was soon shown that injection of renin (or angiotensin II) into a test animal or a human causes temporary hypertension. These studies led to the widespread belief that elevated levels of renin and angiotensin II were a major cause of human hypertension. But humans suffering from chronic hypertension typically lack elevated levels of these blood proteins. The question of the role of these proteins remains unanswered. Some vascular researchers argue that elevated renin levels cause hypertension *early in life* and that even though the renin levels decrease over time, the hypertension remains. Others believe that the renin-angiotensin system is not an important cause of hypertension.

Another factor implicated as a cause of hypertension is salt, specifically sodium chloride (NaCl). The argument is based primarily on studies of sodium intake in different human populations. It appears that people from nonindustrialized societies that have very little sodium in their diet are virtually free of hypertension. When members of these groups change their diet to include sodium, however, as occurred among native Kenyans who joined the Kenyan army and began eating food high in sodium, hypertension appears within the population. It is difficult to extrapolate these findings to industrialized societies, however. Why do some people who ingest large quantities of salt have normal, or even low, blood pressure, while others with similar diets suffer from hypertension? The answer may lie in our genes. It appears that some members of the population are less able to tolerate high-sodium-containing diets than are others. Moreover, once a person has developed hypertension, decreasing the salt intake appears to lower blood pressure only modestly. The question that remains is whether a person who is genetically prone to hypertension could avoid developing the condition if his or her salt intake is limited at an early age.

The latest research on the cause of hypertension has focused on a small protein called *endothelin* which is secreted by the cells that form the inner lining of blood vessels. Endothelin is the most potent vasoconstrictor discovered to date; it is ten times more potent than is angiotensin II, the runner-up. Studies suggest that endothelin functions as a local hormone, causing contraction of the same vessel that is responsible for secreting the substance. Even though endothelin has not yet been shown to be a factor in human hypertension, a number of pharmaceutical companies are racing to develop products that interfere with its action, hoping these drugs will prove to be effective antihypertensive agents.

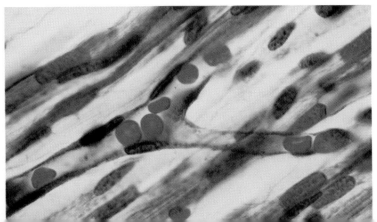

FIGURE 28-4
Red blood cells moving single file through a capillary. The walls of these microscopic vessels typically consist of a single layer of flattened cells, whose edges fit together in a manner resembling the interlocking pieces of a jigsaw puzzle (see Figure 28-2). Between the cells are openings, or pores, which, in most tissues, are large enough to allow the passage of small solutes but small enough to block the outward movement of proteins and blood cells. The thinness of the capillary walls is made evident by the fact that we can see the red blood cells so clearly.

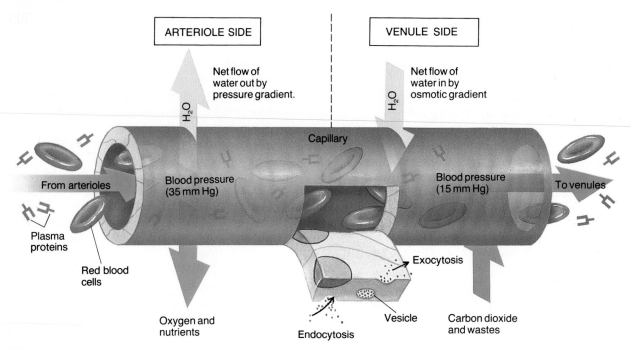

ARTERIOLE SIDE

VENULE SIDE

Net flow of
water out by
pressure gradient.

Net flow of
water in by
osmotic gradient

H_2O

H_2O

Capillary

From arterioles

Blood pressure
(35 mm Hg)

Blood pressure
(15 mm Hg)

To venules

Plasma
proteins

Red blood
cells

Exocytosis

Vesicle

Oxygen and
nutrients

Endocytosis

Carbon dioxide
and wastes

FIGURE 28-5

Action at the capillaries. Some substances, including oxygen, carbon dioxide, and wastes, move between the interstitial fluid and the bloodstream by simple diffusion (red arrows) down their respective concentration gradients. (Water movement is indicated by the yellow arrows.) Water is forced out of the beginning portion of the capillaries in response to the steep pressure gradient between the blood, at about 90 millimeters, and the surrounding fluid, which has virtually no pressure. As blood moves through the capillary, its pressure drops rapidly, forcing less and less water out of the capillary. In contrast, as the blood moves through the capillary, it gains in osmotic pressure due to the loss of water. As a result, water is drawn back into the terminal portion of the capillary by osmosis that results from the greater solute concentration within the blood compared to the surrounding fluid. In addition, some substances move into and out of the cells of the capillary wall by endocytosis and exocytosis.

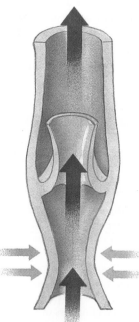

FIGURE 28-6

No turning back. Blood flow is maintained through the veins by one-way valves that open when pressure forces blood toward the heart (indicated by the red arrows) and close shut when the flow is reversed.

Blood
settles

(Closed
valve
prevents
backflow)

When squeezed
by skeletal
muscles

When skeletal
muscles relax

between the cells that form the walls of the capillaries and thus remain in the blood vessel.

The movement of fluid back and forth across the porous walls of the capillaries is a key determinant of the balance of fluid between the interstitial and vascular compartments. Under some conditions, this balance can become greatly disturbed. For example, the rapid swelling that follows a physical injury results from an excess movement of fluid out of the damaged capillaries into the interstitial fluid, creating a condition known as *edema*.

Venules and Veins: The Return Trip to the Heart

Blood from capillaries collects in larger vessels, called **venules,** which empty their contents into large, thin-walled **veins;** from there, the contents return to the heart. Blood pressure in veins is very low, having been dissipated by travel through the narrow capillary passageways. Since most human veins direct blood flow upward to the heart, against the force of gravity, it may seem that blood would simply collect in these large, low-pressure vessels and go nowhere, but this is not the case. Unlike arteries, veins have virtually no elastic fibers or smooth muscle. Instead, they are thin walled and distensible, allowing some pooling of blood during periods of inactivity. However, skeletal muscle activity,

such as that which occurs during walking, combined with the pressure that is exerted during breathing, squeezes veins and force their contents to travel toward the heart. This direction of blood flow is maintained by one-way flaps, or *valves* (Figure 28-6), which project from the walls of these vessels. Without the valves, blood would simply squirt back and forth in the veins as they are squeezed by skeletal muscles. In humans, the veins in the legs receive the greatest weight of blood. This weight can sometimes be so great that the walls of the veins become overdistended and the valves fail, causing the veins to *varicose* (enlarge), which accounts for the characteristic thick, blue appearance of varicose veins.

THE HUMAN HEART: A MARATHON PERFORMER

The major organ of the circulatory system—the **heart**—is a fist-sized, muscular pump that delivers an astonishing performance. The average person's heart works continually for more than 70 years. During this time, the human heart will beat 2.5 billion times and pump approximately 200 million liters (more than 50 million gallons) of blood. Every minute, the heart recirculates the entire volume of blood in the body—approximately 5 liters.

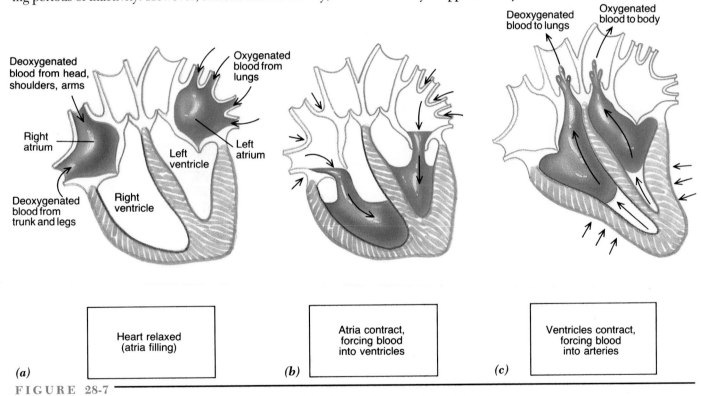

FIGURE 28-7

The movement of blood through the human heart. *(a)* When the heart is relaxed, blood flows from the veins into the two thin-walled atria. Deoxygenated blood from the body flows into the right atrium, and oxygenated blood from the lungs flows into the left atrium. *(b)* In the next stage of the cycle, the atria contract, forcing blood into the relaxed, thicker-walled ventricles. *(c)* In the last part of the cycle, the ventricles contract, forcing blood into the arteries. Oxygenated blood is forced out of the left ventricle into arteries that lead to the tissues of the body, while deoxygenated blood is forced out of the right ventricle into arteries that lead to the lungs. The average duration for the entire cycle in a person at rest is 0.83 seconds (72 beats per minute).

The human heart is a two sided, four-chambered model. Each side has a thin-walled **atrium** and a larger, thicker-walled **ventricle** (Figure 28-7). The left and right atria (plural of atrium) function as receiving stations for the blood that flows into the heart from the veins (Figure 28-7a). Blood from the left atrium flows into the left ventricle, while blood from the right atrium flows into the right ventricle (Figure 28-7b). Contraction of the walls of the ventricles forces the blood out of the heart and into the major arteries (Figure 28-7c).

Pumps and Vascular Circuits

The left and right sides of the heart are separated from each other by partitions that effectively divide the organ into two separate pumps, each of which powers blood through a different vascular *circuit* (Figure 28-8). In the **pulmonary circuit,** deoxygenated blood that has returned from its travels through the body's tissues is pumped from the right ventricle into the pulmonary arteries and on to the lungs, where it is oxygenated. Oxygenated blood returns to the left atrium of the heart via the pulmonary veins. (The

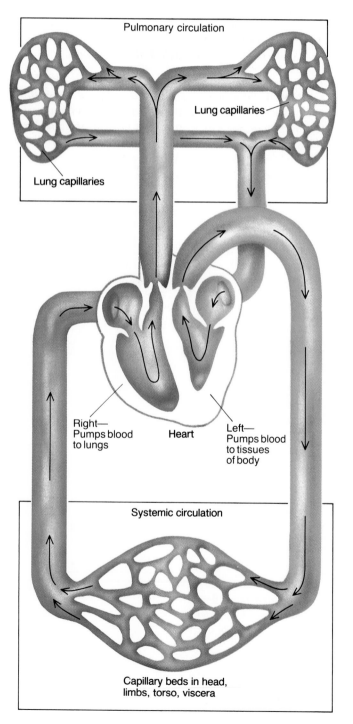

Pulmonary circulation

Lung capillaries

Lung capillaries

Right—
Pumps blood
to lungs

Heart

Left—
Pumps blood
to tissues
of body

Systemic circulation

Capillary beds in head,
limbs, torso, viscera

FIGURE 28-8

Our dual circulatory system. Deoxygenated blood is depicted in blue; oxygenated blood in red. In this schematic diagram, the right side of the heart drives the pulmonary circulation, pumping deoxygenated blood into the capillary beds of the lungs, where the blood picks up oxygen and returns to the left side of the heart. The oxygenated blood is then propelled from the left side of the heart through the systemic circulation. Oxygen is given up in the capillary beds of the tissues, and deoxygenated blood returns to the right side of the heart.

designation of a vessel as an artery or a vein is based on its position relative to the heart, not whether or not the blood is oxygenated or deoxygenated. Arteries carry blood from the heart; veins return it to the heart.) Blood returning to the left atrium from the lungs passes into the left ventricle, whose powerful contraction sends it out into the **systemic circuit,** where it nourishes the tissues of the body. In the systemic circuit (Figure 28-9), blood is initially pumped into

the **aorta,** the largest artery of the body (approximately 2.5 centimeters, or 1 inch, in diameter), which feeds into many of the major arterial thoroughfares of the body.

Blood does not nourish the heart as it travels through its chambers; rather, large **coronary arteries** branch from the aorta immediately after the aorta emerges from the heart, providing cardiac muscle priority access to oxygen-rich blood. Oxygen is required to maintain the oxidation of glu-

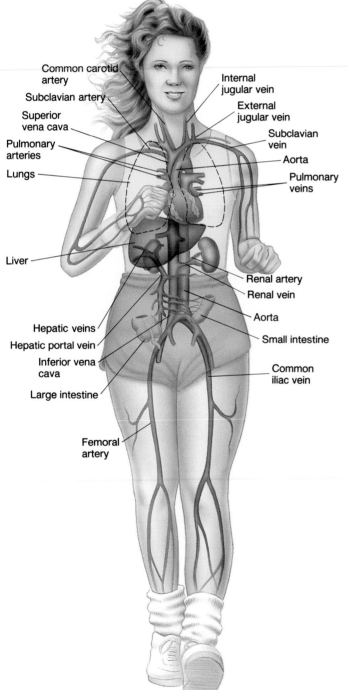

Common carotid artery
Subclavian artery
Superior vena cava
Pulmonary arteries
Lungs
Liver
Hepatic veins
Hepatic portal vein
Inferior vena cava
Large intestine
Femoral artery

Internal jugular vein
External jugular vein
Subclavian vein
Aorta
Pulmonary veins
Renal artery
Renal vein
Aorta
Small intestine
Common iliac vein

FIGURE 28-9

The major arteries and veins in the systemic circulation. As the *aorta* emerges from the heart, it travels in a headward (*anterior*) direction, then makes a sharp "U-turn," known as the *systemic* (or *aortic*) *arch,* and proceeds in a footward (*posterior*) direction through the thorax and into the abdominal cavity. As the aorta extends from the heart, it branches into a number of major arteries that provide blood to the body's organs. Blood from the body's tissues collects in a number of veins that feed into two large trunks, the *inferior* and *superior vena cavae,* which direct the flow of blood back to the heart.

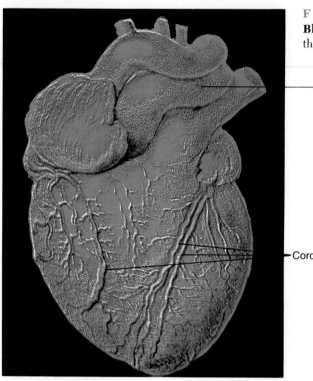

FIGURE 28-10
Blood flow to the heart. Coronary arteries are the first to emerge from the aorta, giving the heart muscle first claim to freshly oxygenated blood.

Aorta

Coronary arteries

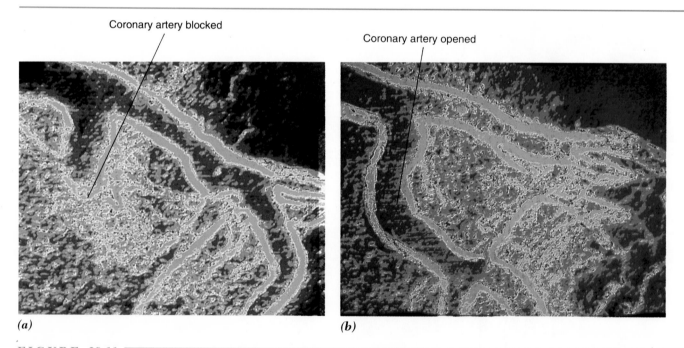

Coronary artery blocked

Coronary artery opened

(a)

(b)

FIGURE 28-11
Coronary artery disease. *(a)* An *angiogram* showing the inability of injected dye to flow normally through an obstruction in a coronary artery. *(b)* Angiogram showing the same coronary artery depicted in part *a* after being opened by angioplasty. *(c)* Instrument utilized in balloon angioplasty. After the tube is threaded into the blocked coronary artery, the tube is inflated near its tip, pushing the walls of the artery outward. *(d)* Diagrammatic results of coronary bypass surgery, whereby a portion of a leg vein has been used to shunt blood directly from the aorta to a coronary artery past the point of an obstruction.

cose, which provides the chemical energy that fuels the heart's contractions. The large coronary arteries are evident in the computer-enhanced view of the heart depicted in Figure 28-10. Obstructions in the coronary arteries can deprive the heart of the oxygen needed to keep heart tissue alive. Such obstructions are the primary cause of heart attacks. If a substantial portion of the heart is deprived of oxygen for a long enough period of time, the heart muscle ceases to function, and the person will die.

A person suffering from insufficient blood flow through the coronary arteries often receives warning signs of the condition, such as an inability to exert themselves or chest pains. The degree to which these arteries are "clogged" is revealed by *angiography,* a procedure in which a dye is injected into the bloodstream and its progress through the heart is monitored radiographically (Figure 28-11*a*). If one or more of the coronary arteries is shown to be largely obstructed, the patient is usually treated either by balloon angioplasty or coronary bypass surgery (Figure 28-11*b*). In *balloon angioplasty,* a tube is threaded into the blocked coronary artery. The tip of the tube is then inflated like a balloon (Figure 28-11*c*), pushing the walls of the artery outward. In *coronary bypass surgery,* a segment of

vein taken from the patient's leg is inserted between the aorta and the coronary artery at a point beyond the site of occlusion (Figure 28-11*d*). In this way, blood is able to flow from the aorta into the heart muscle, bypassing the blocked arterial vessel. Several blocked coronary arteries are usually bypassed in this way during a single operation.

The Heartbeat

The one-way movement of blood through the chambers of the heart and into the appropriate arteries is ensured by the opening and closing of two pairs of strategically located **heart valves** (Figure 28–12). The openings between the atria and the ventricles are guarded by a pair of *atrioventricular (AV) valves,* each of which consists of a flap of tissue that can open in only one direction. When the atria contract, the pressure of the atrial blood rises, causing the fluid to push against the valves. The valves are forced open, allowing blood to flow into the ventricles. When the atrial contractions cease, the pressure of the blood in the ventricles becomes greater than that in the atria. This pressure differential forces the atrioventricular valves to close, preventing the flow of blood back into the atria. The direction of flow out of the ventricles and into the major arteries is main-

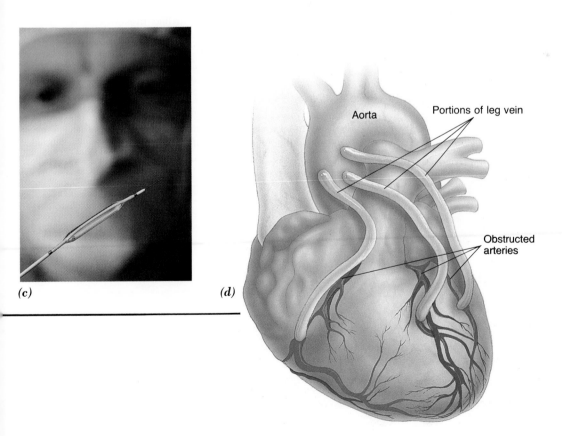

(c)   *(d)*

Aorta

Portions of leg vein

Obstructed arteries

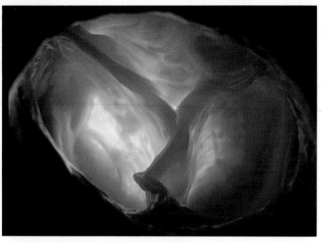

FIGURE 28-12
One of the heart valves that maintains the flow of blood
through the chambers of the heart and into the major arteries.

tained by another pair of valves, the *semilunar valves,* which
are located on each side of the heart.

If you listen to the heartbeat through a stethoscope,
you will hear a "lub-dub" sound that repeats itself with each
beat of the heart. The lower-pitched "lub" is produced by the
closure of the AV valves at the onset of ventricular
contraction. The higher-pitched "dub" is produced by the
closure of the semilunar valves at the onset of ventricular
relaxation. A number of conditions can cause a disruption in

the smooth flow of blood through the heart, creating turbu-
lence, which is heard as a *murmur.* A heart murmur may
result if the opening through a valve is narrower than normal or if the valve is damaged, thereby allowing the backflow of blood. Heart murmurs may or may not be a reflec-
tion of a serious problem. Most serious heart-valve defects
are either congenital or are the result of damage from rheumatic fever. If necessary, the valves can be replaced surgically with artificial ones.

Excitation and Contraction of the Heart

◑ Embedded in the wall of the right atrium is a small
piece of specialized cardiac muscle tissue, called the **sinoatrial (SA) node** (Figure 28-13), whose function is to excite
contraction. If a healthy human heart is isolated, as is done
during a heart transplant, the isolated heart can beat rhythmically on its own, without any outside stimulation. Each
beat of an isolated heart is initiated by a spontaneous electrical discharge that originates every 0.6 seconds or so in the
SA node. Because of its role in determining the rate the
heart beats, the SA node is called the heart's *pacemaker.* A
number of conditions can cause the heart to develop an
abnormal *(arrhythmic)* heartbeat. If the condition is life-
threatening, a surgeon will implant an electronic pacemaker in the chest. Artificial pacemakers deliver an electric
shock at intervals that approximate the intrinsic cardiac
rhythm.

Once an action potential is generated in the SA node, a
wave of electrical activity spreads across the walls of the
atria, passing from one cardiac muscle cell to the next

FIGURE 28-13
Synchronizing the beat. Impulses generated in the SA node
(the pacemaker) sweep through the muscle cells of the two atria,
causing them to contract in unison. When this impulse reaches the
AV node, it launches a wave of electrical activity that is channeled
through the ventricular walls, causing the two ventricles to contract
just as the atria relax.

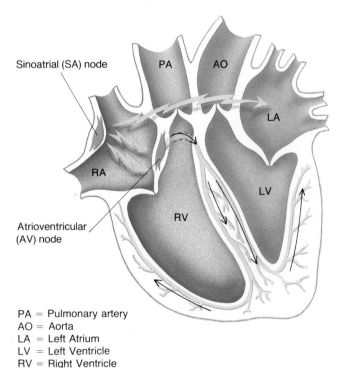

Sinoatrial (SA) node

PA
AO
LA
RA
LV
RV

Atrioventricular
(AV) node

PA = Pulmonary artery
AO = Aorta
LA = Left Atrium
LV = Left Ventricle
RV = Right Ventricle
RA = Right Atrium

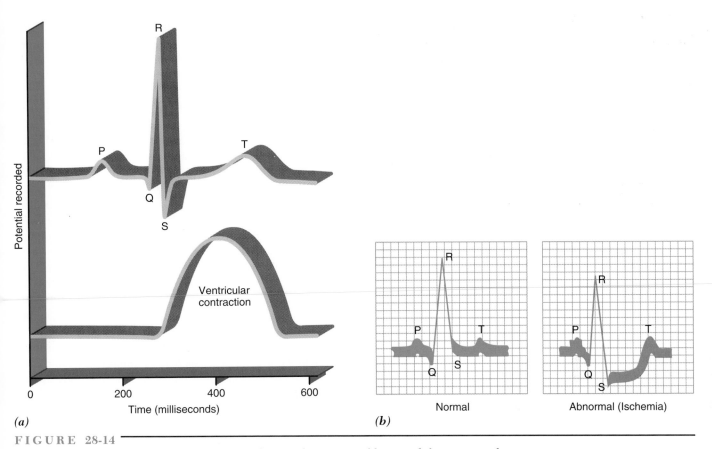

FIGURE 28-14

Diagnosing heart function. *(a)* An electrocardiogram from a normal heart and the corresponding period of ventricular contraction. Three distinct peaks are seen in the electrocardiogram: a P-wave, which results from the flow of current that sweeps across the atria at the start of the heart's contraction; a QRS complex, which results from the flow of current across the ventricles; and a T-wave, which results from the current associated with the return of the ventricles to the resting condition. *(b)* Abnormal cardiac cycles are often revealed by deviations from the normal tracing. Such deviations often show up during a stress EKG that is taken while the person is running on a treadmill. Ischemia results when the heart is not receiving sufficient oxygen due to blockage of the coronary arteries.

through communicating junctions (page 110) that link neighboring cells. The transmission of impulses through such junctions is extremely rapid, causing the entire atrial wall to contract in a synchronous, coordinated manner. As the wave of electrical excitation reaches the boundary of the atria, it cannot pass into the wall of the ventricles because the atria and ventricles are separated by a layer of connective tissue. Only one point of electrical connection exists between the two major parts of the heart—the **AV node** (Figure 28-13). The conduction of the electrical stimulus through the AV node is delayed by about 0.1 second, which allows time for the blood to flow from the atria to the ventricles on each side of the heart. Once the stimulus reaches the AV node, it spreads out over the ventricles, causing the ventricles to contract, forcing blood into the major arteries.

The electrical activity that occurs at different stages of a heartbeat is sufficiently intense to be detected by electrodes placed on the skin of the chest. The recordings of this activity—an **electrocardiogram (EKG)**—constitutes one of the most important diagnostic tools for the analysis of heart disease (Figure 28-14).

Regulating the Heartbeat

The rate at which the heart's pacemaker becomes excited is regulated by a number of factors, primarily by signals that arrive from the autonomic nervous system (Chapter 23). In general, impulses that reach the SA node over *parasympathetic nerves* decrease the rate of heart contraction, while impulses that arrive over *sympathetic nerves* increase it. Thus, at any given moment, the heart rate is determined by the balance between these antagonistic influences. The centers that control the cardiovascular system are housed in the hypothalamus and the medulla of the brainstem (page 486). If we become active, angry, or frightened, this information is passed down the brainstem to the medulla, increasing the sympathetic output to the SA node and decreasing the parasympathetic output, thereby increasing the heart rate.

COMPOSITION OF THE BLOOD

Human blood is a complex tissue (Table 28-1) that is composed of blood cells and cell-like components suspended in a clear, straw-colored liquid called **plasma.**

Plasma: The Soluble Phase of the Blood

The average person contains about 4.7 liters (10 pints) of blood, about 55 percent of which is plasma. Most (about nine-tenths) of plasma is water, the blood's solvent. The rest is composed of various dissolved substances, predominantly three plasma proteins:

- *Albumin,* which helps maintain osmotic conditions that favor recovery of water that has been forced out of capillaries and into the surrounding tissues (see Figure 28-5).

- *Globulins,* such as the *gamma-globulins* (including antibodies that contribute to immunologic defenses) and other globulins that assist in the transport of lipids and fat-soluble vitamins.

- *Fibrinogen,* which provides the protein network necessary for blood clotting.

Salts, vitamins, hormones, dissolved gases, sugars, and other nutrients make up just over 1 percent of plasma's volume.

Red Blood Cells

Erythrocytes, which are often called red blood cells due to their high concentration of red-colored hemoglobin molecules, transport about 99 percent of the oxygen carried by the blood. Oxygenated blood takes on a bright red color, whereas blood that is depleted of oxygen takes on a darker, purplish color that appears blue through blood-vessel walls. Erythrocytes are flattened, disk-shaped cells with a central depression that gives them the appearance of a doughnut with a depression instead of a hole (Figure 28-15). Not only are red blood cells the most abundant cells in the body (there are approximately 5 billion in each milliliter of blood), they are also the simplest. Human erythrocytes lack a nucleus, ribosomes, and mitochondria and constitute little more than nonreproducing sacks of oxygen-binding hemoglobin.

The human body produces red cells at the astonishing rate of 2.5 million every second, or about half a ton during an average person's lifetime. This massive construction project occurs in the red bone marrow, where undifferentiated precursors, called **stem cells,** proliferate and differentiate into erythrocytes. After circulating for about 4 months, aging erythrocytes are engulfed by scavenger cells of the liver and spleen.

TABLE 28-1
COMPONENTS OF BLOOD

Component	Percent	Function
Plasma	55	Suspends blood cells so they flow. Contains substances that stabilize pH and osmotic pressure, promote clotting, and resist foreign invasion. Transports nutrients, wastes, gases, and other substances.
White blood cells	<0.1	Allow phagocytosis of foreign cells and debris. Act as mediators of immune response.
Platelets	<0.01	Seal leaks in blood vessels.
Red blood cells	45	Transport oxygen and carbon dioxide.

Blood settles into three distinct layers when treated with substances that prevent clotting.

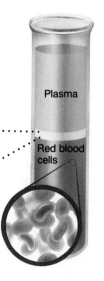

Plasma

Red blood cells

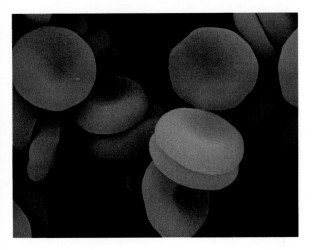

FIGURE 28-15
Red blood cells: biconcave sacks of hemoglobin. The flattened shape of erythrocytes keeps each of the 270 million hemoglobin molecules close to the cell's surface for rapid gas exchange.

The number of erythrocytes produced by the body fluctuates according to availability of oxygen. For example, a person becomes adjusted (*acclimated*) to higher elevations (Figure 28–16), where the air has a lower concentration of oxygen and other gas molecules, by producing more red blood cells, thereby compensating for the reduced amount of oxygen in each breath. The trigger for this response is *erythropoietin,* a hormone secreted by the kidney which initiates the transformation of stem cells in the bone marrow into erythrocytes. Low oxygen concentration stimulates erythropoietin formation, speeding up erythrocyte production until the oxygen supply in the tissues is restored to normal.

Lately, erythropoietin has been at the center of a major controversy in athletics. Now that it is available through recombinant DNA technology, some athletes are injecting themselves with erythropoietin to increase the number of erythrocytes in their blood, thereby supplying more oxygen to their tissues, which they believe improves their performance. Not only is this practice illegal, it may also be dangerous since the effects of excess levels of this highly active protein remain unknown.

White Blood Cells

If you examine a stained blood smear (Figure 28-17*a*), you will notice a small number of white blood cells, or **leukocytes** (*leuko* = white, *cyte* = cell), scattered among the erythrocytes. Leukocytes defend us against foreign microorganisms that could otherwise invade, overwhelm, and destroy our bodies. These blood cells also function as sanitary engineers, cleaning up dead cells and tissue debris that would otherwise accumulate to obstructive levels.

Five classes of leukocytes are clearly recognized: neutrophils, eosinophils, basophils, monocytes, and lymphocytes (Figure 28-17*b*). The first three types—*neutrophils, eosinophils,* and *basophils*—are characterized by their cytoplasmic granules. These cells protect the body by engulf-

FIGURE 28-16
At higher altitudes, an ice climber must become acclimated to the lower concentration of oxygen in the air.

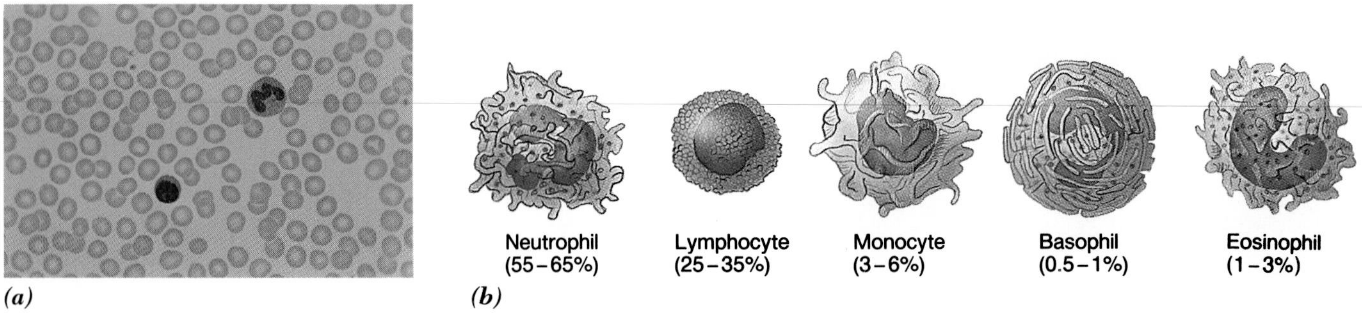

FIGURE 28-17

Human leukocytes and their relative abundance in the blood. *(a)* A blood smear showing two scattered leukocytes. *(b)* The five types of leukocytes. The red dots in the neutrophil, basophil, and eosinophil are granules that can be distinguished by different types of stains.

ing foreign intruders by phagocytosis (page 144). Basophils also function by releasing powerful chemicals that trigger an inflammatory response at the site of the wound or infection. These three white blood cells are distinguished by the types of stains that are used to make their granules visible. In contrast, *monocytes* are nongranular leukocytes that are attracted to sites of inflammation, where they too act as phagocytes, engulfing debris. Monocytes also give rise to phagocytic *macrophages*—huge cells that wander through the body's tissues, ingesting foreign agents. Macrophages also clean up "battlefields" that are littered with the cellular debris of combat between other phagocytes and invading microorganisms. Phagocytosis is not the only mechanism of protection afforded by leukocytes. The *lymphocytes,* a group of nonphagocytic white cells, are the "masterminds" of the immune system, the subject of Chapter 30.

The number and type of leukocytes in the blood can provide an indication of a person's health, which is why your physician may take a "blood cell count" when you are ill. Most infections stimulate the body to release into the bloodstream large numbers of protective leukocytes that are normally held in reserve, causing the "white cell number" to rise. Some disease agents affect only certain types of leukocytes. For example, the viral disease mononucleosis is characterized by an elevated number of monocytes, while infection with the hookworm parasite results in an elevated number of eosinophils. In contrast, certain viruses, including the virus responsible for AIDS, cause infections that *deplete* certain leukocytes to abnormally low levels (Chapter 30).

Factors Necessary for Blood Clotting

A hole in an injured vessel can lead to fatal blood loss if the leak is not quickly patched. In other words, a person could bleed to death. The blood is equipped with its own lines of defense that minimize blood loss and maintain ho-

meostasis. Immediately following injury to a vessel, circulating **platelets**—spiny fragments that are released from special blood cells—become trapped at the injury site. As platelets rapidly accumulate, they plug the leak, providing the first step in damage control. The aggregation of platelets triggers two additional responses at the site of injury: (1) the constriction of the damaged vessel by muscular contraction, and (2) the formation of a clot by coagulation, both of which further seal the vessel.

The resources for clotting circulate in the blood in an inactive form. **Fibrinogen** is a rod-shaped plasma protein which, when converted to the insoluble protein **fibrin,** generates a tangled net of fibers that binds the wound and stops blood loss until new cells replace the damaged vessel (Figure 28-18). The conversion of fibrinogen to fibrin involves the removal of several segments of the fibrinogen molecule, a reaction catalyzed by a proteolytic enzyme, called **thrombin.** Thrombin is also derived from an inactive precursor *(prothrombin)* by a reaction catalyzed by yet another proteolytic enzyme, **Factor X.** As a result of this "reaction cascade," a single molecule of Factor X can stimulate the rapid formation of a large number of fibrin molecules.

In some persons, the cascade fails to complete the formation of a clot. The most common clotting disorder is *hemophilia.* Most hemophiliacs have a defective gene that codes for an inactive version of **Factor VIII,** a protein that normally activates Factor X. Episodes of bleeding in hemophiliacs can now be treated with synthetic preparations of Factor VIII, which are produced by recombinant DNA technology. Such preparations eliminate the need for transfusions of blood products, which may be contaminated with viruses that cause AIDS or hepatitis.

While blood clots are one of the body's primary defense mechanisms, they are also responsible for the majority of deaths in industrialized countries. Most heart attacks are

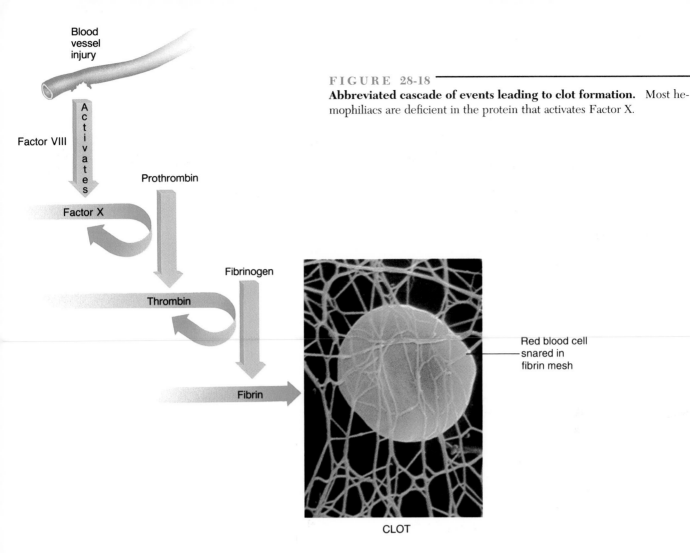

Blood vessel injury

Factor VIII

Activates

Factor X

Prothrombin

Thrombin

Fibrinogen

Fibrin

Red blood cell snared in fibrin mesh

CLOT

FIGURE 28-18

Abbreviated cascade of events leading to clot formation. Most hemophiliacs are deficient in the protein that activates Factor X.

due to the formation of a blood clot *(thrombus)* within a coronary artery at a site where the vessel has already been occluded by the buildup of plaque. While the plaque reduces the flow of blood to the heart, the blood clot totally obstructs the vessel, causing the rapid death of heart tissue which accompanies a heart attack. For this reason, the preferred treatment for heart-attack victims is the injection of a massive dose of a clot-dissolving substance, such as *tissue plasminogen activator (TPA).* It is important that the "clot buster" be administered very soon after the onset of the attack, otherwise damage to the heart tissue due to oxygen deprivation becomes irreversible.

Maintaining human life requires the continual circulation of blood through a vast network of living vessels of varying structure and dimensions. Circulating blood carries substances and cells from one part of the body to another, maintaining the stability of the internal environment and facilitating the exchange of respiratory gases, wastes, and nutrients between the external environment and the living tissues. (See CTQ #1.)

EVOLUTION OF CIRCULATORY SYSTEMS

Before complex animals could become independent of the sea in which they evolved, they had to develop an internal fluid that mimicked the properties of sea water. In other words, they needed a "portable ocean." They also needed a system that circulated this fluid throughout the body, bathing all the cells in a life-sustaining solution, even while the animal lives on dry land. The evolution of this circulatory system did not occur in a single step.

Simpler, multicellular animals, including sponges, jellyfish, and flatworms, attain considerable size without a true circulatory system. This is possible because of the way these animals are constructed; that is, because of their *body plan.* The body plan of these simpler animals places virtually all cells close to the source of oxygen and nutrients. This is particularly well illustrated by the flatworm, whose flattened body shape allows oxygen to diffuse directly from the environment to all of the body's cells and allows waste products to diffuse in the opposite direction (Figure 28-

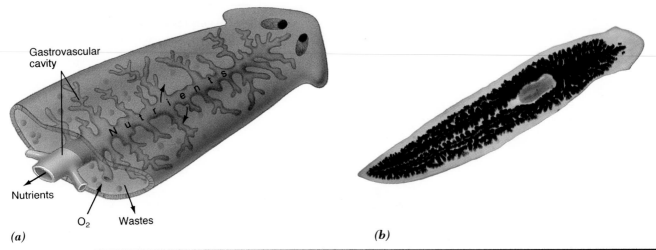

Nutrients

Gastrovascular cavity

Nutrients

O₂ **Wastes**

(a)

(b)

FIGURE 28-19

Flatworm form and function. *(a)* A flatworm's body is highly flattened, which brings every cell close enough to the outside medium to receive oxygen and expel wastes by diffusion. *(b)* The digestive chamber of a flatworm branches to form blind (dead-ended) passageways throughout the body. Nutrients are able to diffuse from the digestive channels directly to the body's cells. Because of its dual role as both digestive and circulatory system, the spacious, internal compartment of the flatworm is called a *gastrovascular cavity.*

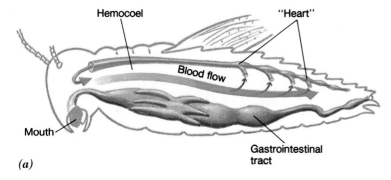

Hemocoel **"Heart"**

Blood flow

Mouth

Gastrointestinal tract

(a)

FIGURE 28-20

Circulatory strategies. *(a)* The open circulatory system of a grasshopper. Blood is pumped out of blood vessels into spaces (which form the *hemocoel*) that bathe the body's tissues. *(b)* The closed circulatory system of an earthworm; blood remains in vessels.

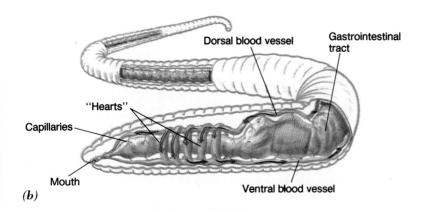

Dorsal blood vessel **Gastrointestinal tract**

"Hearts"

Capillaries

Mouth **Ventral blood vessel**

(b)

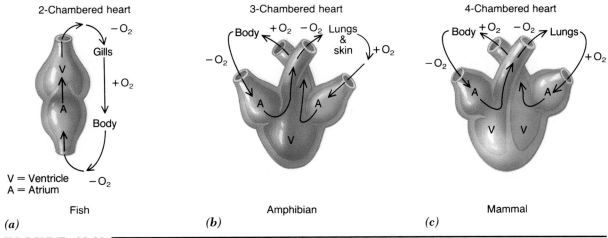

2-Chambered heart

Gills

$-O_2$

$+O_2$

Body

$-O_2$

V = Ventricle
A = Atrium

Fish

(a)

3-Chambered heart

Body $+O_2$ $-O_2$ Lungs & skin

$-O_2$ $+O_2$

A A

V

Amphibian

(b)

4-Chambered heart

Body $+O_2$ $-O_2$ Lungs

$-O_2$ $+O_2$

A A

V V

Mammal

(c)

FIGURE 28-21

Evolution of the vertebrate heart. *(a)* Fishes have a two-chambered heart, consisting of a single atrium and ventricle, which pumps blood through a single vascular circuit. The blood flows from the heart to the gills, where it is oxygenated, and then on to the body, before returning to the heart. At first glance, this single, direct circuit in the fish seems to make more sense than does the more complex, double circuit in mammals, but it has a serious disadvantage: Blood that has passed through the capillary network of the gills will have lost nearly all of the pressure it had when it was pumped from the heart. Blood pressure is an important factor in determining flow rates, fluid exchange with the tissue, kidney function, and so forth, so that the drop in pressure that occurs in the gills seriously reduces the efficiency of the circulatory system. *(b)* Amphibians have a three-chambered heart—two atria and a single ventricle—and are the first vertebrates to have evolved separate pulmonary and systemic circuits. Oxygenated blood from the lungs enters the left atrium, and deoxygenated blood from the body enters the right atrium. Despite the presence of only a single ventricle, physiological studies suggest that the two streams do not become intermixed in the ventricular chamber and that the oxygenated blood from the lungs flows largely to the body, while the deoxygenated blood from the body flows largely to the lungs and skin, a major site of gas exchange in frogs. *(c)* The four-chambered heart of a mammal.

19*a*). Similarly, nutrients can diffuse to the body's cells directly from the highly branched digestive cavity (Figure 20–19*b*).

The evolution of large, more complex animals occurred together with the evolution of a transport system that could deliver oxygen and nutrients to all cells of the body and carry away waste products. Over the course of evolution, two basic types of circulatory systems have appeared: open and closed systems. Some animals, including arthropods (such as insects and spiders) and most molluscs (including snails and clams), have an **open circulatory system** (Figure 28-20*a*), whereby the blood is pumped through vessels that empty into a large, open space (*hemocoel*), in which most of the body's organs are immersed. The cells receive nutrients directly from the fluid in which they are bathed.

In contrast, the human circulatory system is a **closed circulatory system.** In a closed system, blood surges throughout the body in a *continuous* network of closed vessels. Many simpler animals, including the earthworm (Figure 28-20*b*), also have closed systems. In a closed sys-

tem, respiratory gases, nutrients, and waste products are exchanged across the porous walls of the finest vessels in the network—the capillaries.

Vertebrate circulatory systems provide an example of how an entire organ system can undergo gradual evolutionary change. The human heart, with its two separate pumping organs and two distinct circuits, evolved gradually, and some of the presumed intermediate stages have been preserved in the systems of other living vertebrates. The evolution of the vertebrate heart is described in Figure 28-21.

Diffusion is adequate for the movement of substances from one place to another as long as the distances traversed are small—less than a few millimeters. The evolution of larger organisms was accompanied by the gradual development of transport systems that could carry substances back and forth between the environment and the animal's deepest internal tissues. (See CTQ #2.)

THE LYMPHATIC SYSTEM: SUPPLEMENTING THE FUNCTIONS OF THE CIRCULATORY SYSTEM

In addition to the cardiovascular system, humans are equipped with a secondary network of fluid-carrying vessels and associated organs that make up the **lymphatic system** (Figure 28-22). Lymphatic vessels are a series of one-way channels that originate in the tissues as a bed of *lymphatic capillaries.* Lymphatic capillaries absorb excess interstitial fluid that fails to reenter the capillaries. The smaller lymphatic vessels fuse to form larger lymphatic vessels, which ultimately drain into the large veins of the neck. Fats that are absorbed from the intestine following digestion of a fat-containing meal are also collected and delivered to the bloodstream by lymphatic vessels (review Figure 27-6).

Before reentering the bloodstream, lymphatic fluid, or *lymph,* passes through a series of **lymph nodes** (Figure 28-22)—structures that help "purify" the fluid by removing foreign substances and any microorganisms that may be present. During infection, lymph nodes become engorged with white blood cells that are recruited to fight the microbial aggressor. "Swollen glands" (which is actually a misnomer since lymph nodes are not glandular) are therefore common signs of infection. The **spleen** is another lymphoid organ, which filters blood as well as lymph and serves as a reservoir of lymphocytes. Other lymphoid organs include the *tonsils,* which lie on either side of the throat, and the *thymus gland,* which is located at the base of the neck.

The lymphatic system is of great clinical importance because the lymphatic vessels are the common path taken by cancer cells that have been released from a tumor. These malignant cells travel to distant points in the body, where they become lodged and proliferate, forming secondary tumors. This is the reason that lymph nodes that lie in the vicinity of a malignant tumor are usually removed during surgery, along with the tumor itself. Examination of these lymph nodes provides a measure of whether or not the cancer has spread.

The vessels of the lymphatic network provide a specialized pathway for the recovery of excess tissue fluid and the movement of fatty food molecules and the body's protective white blood cells. (See CTQ #3.)

EXCRETION AND OSMOREGULATION: REMOVING WASTES AND MAINTAINING THE COMPOSITION OF THE BODY'S FLUIDS

One day you might eat a large bag of salty potato chips, while the next day you may choose relatively salt-free foods. Even though your diet may change drastically from day to

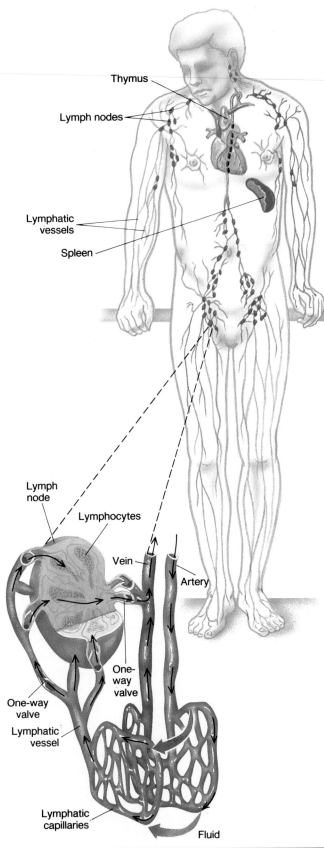

FIGURE 28-22
The lymphatic system. Lymphatic vessels capture and return excess fluid that is leaked from the capillaries of the bloodstream. Lymphoid organs include the lymph nodes, thymus, spleen, and tonsils.

day, the composition of your body fluids remains relatively constant. This, once again, is the essence of homeostasis. Maintaining the proper composition of the body's internal fluids requires two related activities: osmoregulation and excretion.

1. **Osmoregulation** is the maintenance of the body's normal salt and water balance. If you were not able to rid yourself of the excess salt you ingested after eating that bag of potato chips, the concentration of sodium, potassium, and chloride ions could increase to dangerous levels. Elevated potassium concentrations in the blood, for example, can lead to a fatal disruption of the rhythmic beating of the heart. In fact, numerous cases of physician and nurse "mercy killings" in hospitals have been attributed to administration of potassium chloride. Similarly, when you drink more fluid than your body needs to maintain its proper water content, excess water must be eliminated.

2. **Excretion** is the discharge of the body's metabolic waste products. The wastes that tend to accumulate at highest concentration and pose the greatest threat to the delicate physical and chemical balance required for life are products of the metabolic breakdown of nitrogen-containing compounds, notably proteins and nucleic acids. The nitrogen is released during metabolism as highly poisonous ammonia (NH_3). In humans, ammonia is quickly converted by the liver to **urea,** a relatively nontoxic nitrogenous molecule that is tolerated in the tissue fluids until it is eliminated by the kidney.

As we discuss later, many animals convert ammonia to *uric acid,* a nontoxic product that is excreted as an almost dry paste.

Humans have a single system that carries out both osmoregulation and excretion; this is not the case in all animals.

THE HUMAN EXCRETORY SYSTEM: FORM AND FUNCTION

The human excretory system is illustrated in Figure 28-23. The paired **kidneys** are biological cleaning stations that remove nitrogenous wastes as well as excess salts and water from the blood, forming **urine,** a solution destined for discharge from the body. From the kidneys, urine is moved by peristalsis down a pair of muscular tubes, the **ureters,** and into a holding tank, the **urinary bladder.** Discharge of urine from the bladder through the **urethra** is under control of neural commands.

The Structure of the Kidney

The functional unit of the kidney is the renal tubule, or **nephron** (Figure 28-24). Each nephron produces a small volume of fluid, or urine; together, the urine produced by the million or so nephrons of the kidneys makes up the fluid that you void several times a day. The nephron wall consists of a single layer of epithelial cells surrounding a central, hollow lumen. Closely entwined around each nephron is a system of blood vessels, which allows for the exchange of

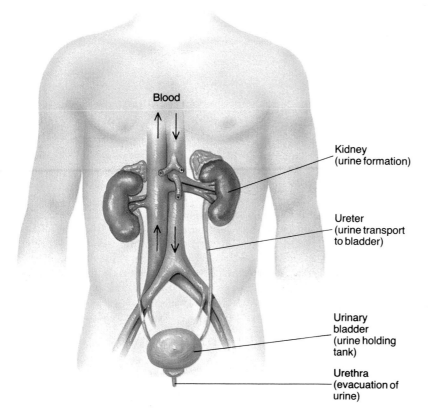

Blood

Kidney
(urine formation)

Ureter
(urine transport
to bladder)

Urinary
bladder
(urine holding
tank)

Urethra
(evacuation of
urine)

FIGURE 28-23
The human excretory system. The kidneys filter the blood, removing waste products and excess salts, forming urine. Ureters provide passageways to the urinary bladder, where urine is held until its release from the body through the urethra.

materials back and forth between the urinary fluid in the tubule and the bloodstream (Figure 28-24).

Even though the weight of both kidneys is only about 1 percent of the body's total weight, these organs receive approximately 25 percent of the body's total blood flow. Blood enters the kidney through the renal artery, which branches to form smaller vessels, each of which leads into a bundle of capillaries, called a **glomerulus.** Each glomeru-

lus is embedded in a blind, cup-shaped end of a nephron, called **Bowman's capsule** (Figure 28-24). From Bowman's capsule, the nephron begins a winding path, first as the **proximal convoluted tubule,** then into a long, U-shaped portion, called the **loop of Henle,** then into the **distal convoluted tubule,** and finally into a **collecting duct,** which drains the nephron's contents into the ureter. Notice in Figure 28-24 that part of each nephron (Bow-

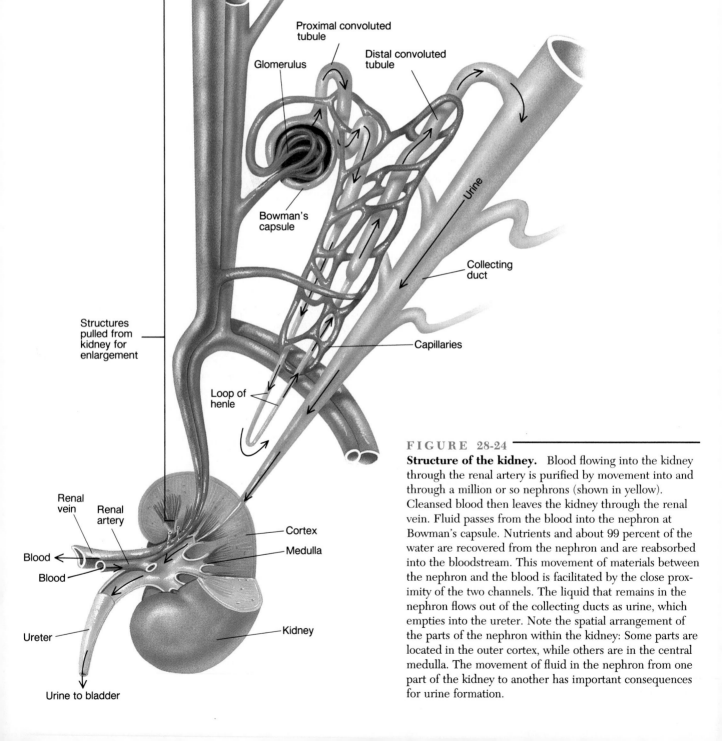

FIGURE 28-24

Structure of the kidney. Blood flowing into the kidney through the renal artery is purified by movement into and through a million or so nephrons (shown in yellow). Cleansed blood then leaves the kidney through the renal vein. Fluid passes from the blood into the nephron at Bowman's capsule. Nutrients and about 99 percent of the water are recovered from the nephron and are reabsorbed into the bloodstream. This movement of materials between the nephron and the blood is facilitated by the close proximity of the two channels. The liquid that remains in the nephron flows out of the collecting ducts as urine, which empties into the ureter. Note the spatial arrangement of the parts of the nephron within the kidney: Some parts are located in the outer cortex, while others are in the central medulla. The movement of fluid in the nephron from one part of the kidney to another has important consequences for urine formation.

man's capsule, and the proximal and distal convoluted tubules) is located in the outer *cortex* of the kidney, while the remainder (the loop of Henle and collecting duct) projects down into the inner *medulla* of the kidney. As you will see shortly, this structural feature of each nephron holds the key to the formation of urine.

Three Processes During Urine Formation

Urine formation occurs as the result of three distinct processes: glomerular filtration, tubular reabsorption, and tubular secretion.

Glomerular Filtration   The capillaries of a glomerulus are particularly porous, and the blood in these vessels is present at unusually high pressure. During **glomerular filtration,** the pressure of the blood in the glomerulus provides the force that pushes fluid out of the capillaries (Figure 28-25, step 1), across the epithelial wall of Bowman's capsule, and into the lumen of the nephron. The first insight into glomerular filtration came in the early 1920s, when the physiologist Alfred Richards of the University of Pennsylvania developed a technique of withdrawing fluid from various portions of a single nephron of a frog kidney, using a

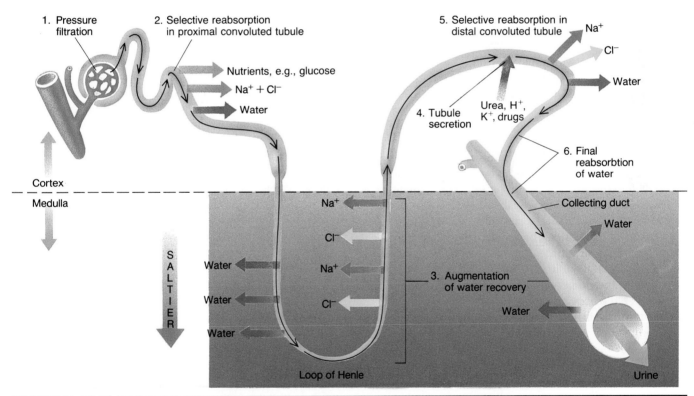

FIGURE 28-25

Urine formation in the nephron.   (1) The fluid portion of the blood is forced into the proximal end of the nephron through Bowman's capsule by pressure filtration. (2) As the fluid passes along the proximal convoluted tubule, nutrients, salts, and water are recovered by reabsorption. The movement of water out of the proximal convoluted tubule by osmosis follows the active transport of salt. (3) The recovery of additional water is augmented by the loop of Henle, whose active transport of sodium establishes a steep osmotic gradient in the extracellular fluid of the medulla of the kidney (indicated by the increased shading). (4) As the fluid moves through the distal convoluted tubule, tubular secretion transfers hydrogen and potassium ions, along with waste products from the blood, into the nephron. (5) Additional salt is reabsorbed from the distal convoluted tubule, along with water, which flows outward by osmosis. (6) The fluid then enters the collecting duct, where it passes through the osmotic gradient established by the loop of Henle. As the fluid in the collecting duct moves into a saltier environment, more and more water is drawn out of the nephron, where it diffuses back into the bloodstream. The fluid that remains in the collecting ducts after it has passed through the medulla passes into the ureters as urine.

very fine pipette. More recently, the technique has been adapted to mammalian kidneys and has provided the best information we have about the processes occurring within the nephron.

Richards determined that the **glomerular filtrate**— the fluid that passes from the glomerulus into the lumen of the nephron—is similar to the blood of the glomerular capillaries in some respects and different in others. Unlike the blood, the glomerular filtrate lacks blood cells and proteins but otherwise has a similar concentration of solutes. These findings suggested that the passage of materials from blood to nephron is nonselective and determined solely by the size of the pores in the walls of the glomerular capillaries and Bowman's capsule. Proteins and cells are simply too large to pass through these pores, but virtually all other blood constituents, including valuable nutrients, water, and wastes, are forced out of the glomerular capillaries and enter the nephron.

The delicate structure of the glomerulus and the nephron is very sensitive to blood pressure. Too low a pressure decreases glomerular filtration, while too high a pressure can produce extensive, microscopic damage. This explains why a person with prolonged high blood pressure often experiences kidney failure. It is also one of the obvious reasons why hypertension should not go untreated.

Tubular Reabsorption The kidney is a remarkable purification plant. Approximately 900 liters of blood passes through the kidneys each day. Of this huge quantity, 20 percent (180 liters) is actually forced out of the million or so glomeruli into the adjacent nephrons. Of course, 180 liters of glomerular filtrate is considerably more fluid than you excrete as urine each day. In fact, the average output of urine is 1 to 2 liters per day, or approximately 1 percent of the daily glomerular filtrate. The remaining 99 percent is *reabsorbed* as the fluid flows down the nephrons. Most of the solutes of the glomerular filtrate are also reabsorbed; if they weren't, the body's store of essential substances would soon be discarded in urine. For example, the loss of glucose, which occurs in untreated diabetes (page 518), deprives a person of his or her readily available fuel supplies. The result is similar to having a hole in the bottom of your car's gasoline tank.

Glomerular filtration is a nonselective process, whereas **tubular reabsorption** is highly selective. The kidneys' "strategy" is to push everything into the nephron at its proximal end and then selectively remove those substances, such as salts, sugars, and water, that the body can't afford to lose, leaving behind waste products to be discharged in the urine. Let's look more closely at how reabsorption can be so selective.

As the water-rich, nutrient-laden glomerular filtrate moves along the proximal convoluted tubule, active transport systems in the membranes of the epithelial cells export glucose, salts, and other valuable nutrients out of the nephron (Figure 28-25, step 2). These substances move through the interstitial fluid and back into the bloodstream. As the solutes move into the interstitial fluid, they create an osmotic gradient that draws water out of the proximal convoluted tubule. In other words, the cells of the wall of the nephron expend energy to recover salts and nutrients, and water follows along passively by osmosis. Approximately 80 percent of the water in the glomerular filtrate is reabsorbed from the proximal portion of the nephron.

The efficiency of water recovery from the glomerular filtrate is boosted by the loop of Henle, whose role it is to establish a salt gradient within the kidney, which is needed to reclaim water from the distal portion of the nephron. To do this, ionized salt (Na^+ and Cl^-) is actively pumped from the ascending side of the loop (Figure 28-25, step 3), making the surrounding fluid very salty. The highest salt concentration develops in the inner medulla of the kidney, while the lowest salt concentration develops in the outer cortex. Establishing a steep salt gradient is an energy-expensive process, requiring ATP hydrolysis. This energy is expended so that water can be reabsorbed from the fluid of the collecting duct at a later stage (Figure 28-25, step 6) as it passes out of the kidney, toward the ureter.

Tubular Secretion The composition of urine is modified in the distal portion of the nephron by **tubular secretion** —the process whereby substances are transported from the blood into the fluid of the distal convoluted tubule (Figure 28-25, step 4). A variety of substances enter the nephron by secretion, including waste molecules, such as urea and urobilin (the yellow pigment derived from hemoglobin breakdown which gives urine its yellow color), certain ions (K^+ and H^+), and a number of medications, such as penicillin and phenobarbitol, and other drugs, such as the active substances found in marijuana and cocaine. Tubular secretion provides one of the final regulatory steps in maintaining homeostasis of the tissue fluids. For example, the kidney regulates blood pH by secreting more hydrogen ions when the blood becomes slightly acidic; this process is reversed when blood becomes too alkaline. The hydrogen ion concentration of the urine can vary more than 1,000-fold, depending on conditions.

Regulation of Kidney Function

Like most physiological activities, urine formation is regulated by both neural and endocrine mechanisms. Two hormones—*aldosterone* and *antidiuretic hormone (ADH)* —that act on distal parts of the nephron are particularly important. If you were to ingest a particularly salty meal, the increased level of sodium in your blood would trigger a decrease in the secretion of aldosterone by your adrenal cortex (page 528), leading to a decreased reabsorption of sodium by the distal convoluted tubule (Figure 28-25, step 5) and increased excretion of salt in the urine. Conversely, a drop in the blood's sodium concentration stimulates an increased secretion of aldosterone, leading to increased salt retention. A person suffering from untreated diseases of the

◁ B I O L I N E ▷
The Artificial Kidney

Kidney failure leads to many deleterious effects, including the retention of waste products, an abnormal concentration of salts and water, an altered pH, and the general deterioration of homeostasis. Until a few decades ago, kidney failure meant certain death. With the development of the *artificial kidney,* however, a person without kidney function can live a long life. The treatment is not without possible side effects, however, including fatigue and impotence (in men). Ideally, the use of the artificial kidney is a stopgap measure that keeps a person alive until he or she can receive a healthy kidney through transplantation.

The modern artificial kidney was developed largely by the efforts of Willem Kolff, a Dutch physician who emigrated to the United States after World War II and eventually settled at the University of Utah. During the use of an artificial kidney, blood is pumped out of an artery and into a long stretch of cellophane tubing. This tubing is immersed in a chamber filled with a salt solution of similar composition

to normal blood plasma. The permeability properties of cellophane tubing, which is the same material that is used as sausage casing, are similar to that of the capillary wall; that is, small-molecular-weight solutes can pass through the pores, but proteins and blood cells are retained within the tubing.

The "purification" of the blood is accomplished by *dialysis,* a process that requires several hours and must be repeated two to three times a week. During dialysis, diffusible molecules pass back and forth between the tubing and the surrounding fluid. The net movements of a particular type of molecule depend on the relative concentration of the substance on the two sides of the cellophane membrane. For example, urea is present in the blood but is absent in the fluid of the dialysis chamber. Consequently, when the blood flows through the dialysis tubing, urea will diffuse out of the blood and into the chamber, cleansing the blood of this waste product. In contrast, the blood is likely to contain a reduced concentration of bicarbonate ions

because these basic ions will have picked up much of the excess acid produced by the body since the last dialysis treatment. During dialysis, bicarbonate ions, which are present in the chamber's fluid, will enter the blood as it flows through the cellophane tubes, restoring the blood's buffering capacity. Blood is kept from clotting during dialysis by treatment with the anticoagulant heparin.

It is interesting to note that both the dialysis machine and the kidney bring about a similar end result but achieve it by very different means. The dialysis machine depends strictly on the nonspecific, passive diffusion of dissolved molecules, whereas the natural kidney utilizes highly specific transport processes and regulated changes in membrane permeability. That is why the natural kidney performs a much better job in a much more compact "apparatus" without having to carry around a large chamber of fluid that has to be changed on a daily basis.

---

adrenal cortex may produce little or no aldosterone and will excrete large quantities of both salt and water.

The volume of the blood is determined by its water content, which is regulated at the collecting ducts (Figure 28-25, step 6). Blood volume decreases when too little water is reabsorbed from the nephrons back into the bloodstream; this causes the blood to become too concentrated. Receptors in the hypothalamus detect the increase in the blood's osmotic strength and direct the posterior pituitary to release ADH, as we described in Chapter 25. ADH increases the permeability of the collecting ducts to water, allowing more water to be reabsorbed from the urine; this dilutes the blood while concentrating the urine. Alcohol interferes with the secretion of ADH, which is why beer induces more trips to the restroom than does drinking a comparable volume of water. In addition to directing the

secretion of ADH, the hypothalamus alerts the cerebral cortex when the water content of blood is low. What you perceive consciously as thirst is your hypothalamus instructing you to consume more water.

EVOLUTION AND ADAPTATION OF OSMOREGULATORY AND EXCRETORY STRUCTURES

Some animals, including sponges, sea anemones, and sea stars, lack specialized organs for osmoregulation and excretion. It is presumed that the ionic regulation and excretion of wastes in these animals are handled by the cell membranes that line the body surface (the *integument*). Most animals—whether earthworms, crabs, or vertebrates—

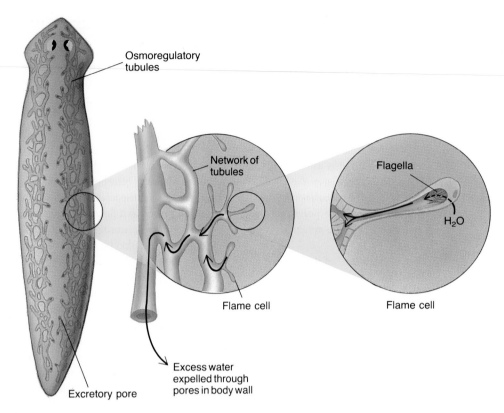

Osmoregulatory
tubules

Network of
tubules

Flagella

H₂O

Flame cell

Flame cell

Excretory pore

Excess water
expelled through
pores in body wall

FIGURE 28-26

The osmoregulatory system of a flatworm. A network of osmoregulatory tubules run the length of the body on each side of the animal. Emerging from the tubules along their length are short canals that end blindly in a *flame cell*. These cells are named for the flickering motion exhibited by a tuft of flagella that projects from the cell into the lumen of the canal. The movement of the flagella is thought to suck fluid across the wall of the flame cell and into the lumen of the tubule. The contents of the fluid are altered by reabsorption and secretion, and the final product flows out of excretory pores located along the body.

possess some type of tubular organ whose mechanism of operation is similar to that found in humans. For example, flatworms contain an interconnected series of tubules that maintain the body's salt and water balance (Figure 28-26). As is the case in many animals, excretion of nitrogenous wastes in flatworms is handled by a separate mechanism.

▐▶ Osmoregulatory and excretory mechanisms are correlated with the type of habitat in which an animal lives. Invertebrates that live in marine habitats, such as a sea anemone or an octopus (Figure 28-27a), have very salty body fluids that are approximately equal in osmotic concentration to that of the surrounding sea. As a result, these animals neither lose nor gain water by osmosis. Therefore, they are not in need of an elaborate osmoregulatory system.

In contrast, marine vertebrates are generally much less salty than is their environment and thus tend to lose water by osmosis. Most marine fishes (Figure 28-27b) regain the

water they lose through osmosis by drinking the sea water they live in and excreting concentrated salt solutions from their gills. These animals produce virtually no urine, and many have very rudimentary kidneys. Marine sea birds and sea turtles (Figure 28-27e,f) must also obtain their water from the sea; these animals drink sea water and excrete concentrated salt solutions from salt glands located in their heads.

Animals that live in freshwater face just the opposite problem: Their environment is very high in water concentration and low in available salts. Consequently, freshwater animals tend to *gain* water by osmosis—water that must be expelled back into the environment. Freshwater fish and frogs (Figure 28-27c,d) have well-developed kidneys and produce a very dilute urine. The most serious problem faced by these animals is the loss of valuable salts, which are inevitably washed away in the large volume of urine. Freshwater animals possess highly effective active-trans-

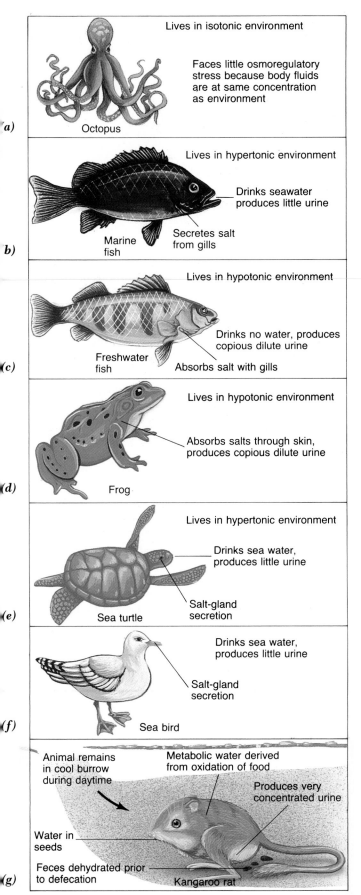

FIGURE 28-27

A diversity of osmoregulatory challenges. After: ANIMAL PHYSIOLOGY by R. Eckert. Copyright © 1988 by W. H. Freeman and Co. Reprinted by permission.

port mechanisms that pump salts back into their bodies, despite the low salt concentration of their environment: Freshwater fish have salt-absorbing gills, while frogs have salt-absorbing skin.

The most serious osmoregulatory problems are faced by terrestrial animals, particularly by those that live in dry, desert habitats. These animals possess physiological and behavioral adaptations that keep water loss to an absolute minimum. One of the best-studied water savers is the desert kangaroo rat (Figure 28-27g). This animal can live its entire life without ever drinking a drop of water: All the water the kangaroo rat needs is either present in the animal's food or formed as a byproduct of metabolic reactions (so-called *metabolic water*). Water conservation is accomplished by the animal's ability to produce an extremely concentrated urine (several times saltier than that which can be excreted by humans) and by its staying out of the desert heat. During the day, these animals remain in their relatively cool burrows; they emerge at night to carry out their feeding and social activities.

Nitrogenous Waste Products: Adaptations to the Environment

Aquatic and terrestrial animals also differ in the type of nitrogenous waste products they excrete. In a metabolic sense, the least expensive nitrogenous waste product is ammonia, which is formed directly when amino ($-NH_2$) groups are removed from amino acids. But ammonia is highly toxic and cannot be allowed to accumulate in body fluids. Aquatic animals are able to excrete ammonia since they can either allow it simply to diffuse into the environment through their body surface or to void it in a dilute urine. In contrast, terrestrial animals must conserve water and excrete urine with concentrated solutes. Ammonia is too toxic for this type of excretory practice. Instead, terrestrial animals convert the ammonia (at the expense of some ATP) to relatively nontoxic products that can be excreted in a concentrated state, such as urea (excreted by adult frogs and most mammals) or uric acid (the white, pasty material found in bird and reptile excrement).

Excretory/osmoregulatory systems maintain the constancy of the internal fluids by ridding the body of nitrogenous waste products, excess water and salts, and other undesirable substances that are produced by the animal or enter from the environment. The formation of an excretory fluid typically occurs within a tubule whose walls are specialized for the reabsorption of substances to be retained and the secretion of substances to be eliminated. (See CTQ #4.)

THERMOREGULATION: MAINTAINING A CONSTANT BODY TEMPERATURE

Humans normally maintain a constant body temperature of 37°C (98.6°F), regardless of whether they are walking on a desert path under a blazing sun or swimming in an ice-cold mountain stream. The homeostatic process by which mammals (and birds) maintain such a high body temperature is called **thermoregulation.** Thermoregulation requires that an animal follow one simple physiological rule: The amount of heat gained by the body must equal the amount of heat lost from the body.

GAINING HEAT TO KEEP THE BODY WARM

The heat present within an animal's body can derive from one of two sources: its own metabolism or the external environment. Animals that use metabolic heat to raise their body temperature are called **endotherms.** In contrast, animals whose body temperature derives from heat absorbed from the rays of the sun, the ground in their burrows, or the water in which they live, are called **ectotherms.** Birds and mammals are the only animals that are classified as true endotherms; all others are ectotherms. The reason for the rarity of endothermy is that it is very expensive metabolically. When you consider the dollar cost of maintaining the temperature of a house at about 70°F in the winter, you can appreciate the metabolic cost for an animal that lives outside during the same winter conditions and maintains its temperature at 98° to 100°F. This is the primary reason that birds and mammals (other than humans) are covered with feathers or fur; these materials provide a layer of insulation that retains body heat in the face of cold environmental temperatures.

Humans are not very well insulated. A slight drop in body temperature is immediately recognized by receptors in the hypothalamus of the brain which send out a message via sympathetic nerves to constrict the arterioles leading to the skin. This constriction causes the skin to become pale and cold, a response that conserves heat by reducing the flow of blood to the body's surface. If our body temperature continues to drop despite such vasoconstriction, more drastic measures are called for, and the brain sends out a signal to the body's muscles to contract involuntarily. This response, called **shivering,** represents the body's attempt to increase the output of metabolic heat in order to raise the body's temperature.

LOSING HEAT TO KEEP THE BODY COOL

Endotherms must also possess mechanisms to rid themselves of body heat as the temperature of the external environment rises or when the individual's level of physical activity increases. The first step in dispersing body heat is dilation of blood vessels in the outer portions of the body, which allows more blood to reach the body surface, where its heat radiates to the environment. If the body temperature continues to rise despite this conservative measure, additional heat must be lost by either sweating or panting. *Sweating*—the loss of body heat as water evaporates from the surface of the skin—is the primary mechanism of cooling in a variety of mammals, including antelopes, camels, and humans. Humans placed under conditions of extreme heat stress can produce approximately 4 liters of sweat in a single hour. It is only after the body becomes severely dehydrated that the flow of sweat decreases and the body temperature rises. Panting—which is the loss of body heat as water evaporates from the moist surfaces of the respiratory tract—is the primary mechanism of cooling in birds, dogs, and bears.

An animal's enzymes function optimally at a particular temperature, which varies from species to species. Endotherms tend to maintain their body at this optimal temperature by balancing the amount of heat gained—largely as a product of metabolism—with the amount of heat lost—largely by shunting warm blood to the body surface or by evaporative cooling. (See CTQ #5.)

REEXAMINING THE THEMES

Relationship between Form and Function

The microscopic structure of a blood vessel is closely correlated with its function. The larger arteries are wide in diameter, promoting rapid flow, and have concentric rings of elastic tissue, which allow the wall to maintain pressure on the blood when the heart is not contracting. Arterioles are endowed with a layer of smooth muscle cells, whose state of contraction determines which tissues will receive the greatest supply of blood. The capillaries are devoid of both elastic and muscle tissue. The extreme thinness and porosity of the capillary walls promote the exchange of substances between the blood and tissues. The veins have large diameters and thin, distensible walls, whereby large volumes of blood can return to the heart with very little external pressure, assisted by the action of one-way valves.

Biological Order, Regulation, and Homeostasis

The circulatory and excretory systems are both primary agents in maintaining homeostasis. The bloodstream serves as the conduit for carrying excess materials and body heat to sites specializing in their disposal. For example, excess heat is carried by the blood to the body surface; carbon dioxide is carried to the lungs; and nitrogenous wastes and excess salts are carried to the kidneys. The concentrations of various substances in the blood are maintained at appropriate levels by the processes of selective reabsorption and secretion that take place in the kidney. Kidneys balance blood pH, maintain the volume and osmotic strength of the blood, and eliminate nitrogenous wastes. Maintaining a constant body temperature, which keeps all of the body's enzyme systems working at maximum efficiency, is another key aspect of homeostasis.

Acquiring and Using Energy

Energy is required for virtually all of the activities described in this chapter. For example, pumping blood through the body requires large amounts of chemical energy. If supplies of chemical energy to the heart are reduced, as occurs when a person's coronary arteries become occluded, the heart cannot increase its pumping activities to meet the body's increased demands for oxygen, curtailing the person's ability to engage in strenuous activities. The reabsorptive and secretory activities of the kidney, which are required to produce urine, also depend on energy, which is used to drive the various ionic pumps located in the plasma membranes of the cells that line the nephrons. Thermoregulation can be the most energy-demanding activity of all. Up to 90 percent of a bird's or mammal's chemical energy may be expended in generating the heat needed to maintain an elevated body temperature.

Evolution and Adaptation

Organs that carry out osmoregulation are highly adapted to an animal's environment and may perform quite differently among closely related animals. For example, most marine fishes lose water to their environment, but they have no difficulty obtaining salts. These animals typically drink the surrounding sea water and excrete the excess salts through their gills; they have poorly developed kidneys. In contrast, freshwater fish tend to gain water and live in an environment where salts are scarce. These animals have well-developed kidneys and excrete a copious, dilute urine. They obtain the salts they need by active transport from their environment.

SYNOPSIS

Circulatory systems deliver a well-oxygenated nutrient solution to the body's tissues and carry away the wastes. The system also contributes to homeostasis by stabilizing pH and osmotic balance, delivering hormones, protecting against foreign intruders, and (in birds and mammals) resisting changes in body temperature.

The anatomy of blood vessels is well suited to their function. The design of the large, elastic arteries allows them to snap back after each surge from the heart, squeez-

ing blood onward. To meet the needs of the tissues they service, muscular arterioles either increase or decrease their diameter. Capillaries are thin and porous and serve as sites of exchange with the surrounding interstitial fluid. Venules collect the blood leaving the capillaries and route it to veins. One-way valves assist the blood's return to the heart. Excess fluid returns to the bloodstream via the lymphatic system. Lymph is filtered through lymphatic organs, notably the spleen and lymph nodes.

The mammalian heart is equipped with one-way valves to keep blood flowing in a single direction. Acting as a double pump, the right side of the heart pumps blood to the lungs for oxygenation; the left side powers blood through the general (systemic) circulation. Although the heart will contract rhythmically without an external stimulus, neural impulses regulate the rate of contraction by stimulating the sinoatrial node, which initiates contraction of the atria and relays the stimulus to the rest of the heart.

Nearly half the volume of the blood consists of cells and cell fragments (platelets). Erythrocytes carry oxygen; granular leukocytes and monocytes phagocytize foreign intruders and tissue debris; platelets aid in clotting; and lymphocytes launch the immune response. The fluid phase (plasma) contains a number of proteins that are activated in a specific sequence, leading to the formation of a clot that consists of a tangled net of insoluble fibers of the protein fibrin.

Excretion and osmoregulation. The mammalian kidney has two roles: It maintains the body's salt and water balance, and it rids the blood of waste products generated by cellular metabolism. Wastes, excess solutes, and water are voided in the urine—the fluid produced in the nephrons of the kidney as a result of three processes: glomerular filtration, selective reabsorption, and secretion. Blood (minus its cells and proteins) is filtered out of the capillaries of the glomerulus into the proximal end of the nephron. From there, the fluid flows down the length of the tubule, while salts, sugars, and other nutrients are transported out of the nephron back into the bloodstream. Nitrogenous wastes, hydrogen ions, and excess potassium ions are secreted in the opposite direction. Water flows out of the collecting duct by osmosis in response to an osmotic gradient that is established in the interstitial fluid of the kidney by the salt-pumping cells of the loops of Henle. The hormones aldosterone and ADH act on the distal portion of the nephrons to regulate the amount of salt and water voided in the urine.

Birds and mammals maintain a constant, elevated body temperature by balancing the amount of heat gained and heat lost. Heat is generated primarily as a result of metabolism and is lost at the body surface. If the body temperature starts to drop, heat is retained by vasoconstriction of peripheral arterioles and by shivering. Heat is lost by vasodilation of peripheral arterioles and by evaporative cooling.

Key Terms

circulatory system (p. 588)
intracellular compartment (p. 588)
interstitial compartment (p. 588)
vascular compartment (p. 588)
cardiovascular system (p. 588)
artery (p. 589)
blood pressure (p. 589)
systolic pressure (p. 589)
diastolic pressure (p. 589)
hypertension (p. 590)
arteriole (p. 590)
vasodilation (p. 590)
vasoconstriction (p. 590)
capillary (p. 590)
venule (p. 593)
vein (p. 593)
heart (p. 593)
atrium (p. 594)
ventricle (p. 594)
pulmonary circuit (p. 594)
systemic circuit (p. 594)
aorta (p. 595)

coronary artery (p. 595)
heart valve (p. 597)
sinoatrial (SA) node (p. 598)
atrioventricular (AV) node (p. 599)
electrocardiogram (EKG) (p. 599)
plasma (p. 600)
erythrocyte (p. 600)
stem cell (p. 600)
leukocyte (p. 601)
platelet (p. 602)
fibrinogen (p. 602)
fibrin (p. 602)
thrombin (p. 602)
Factor X (p. 602)
Factor VIII (p. 602)
open circulatory system (p. 605)
closed circulatory system (p. 605)
lymphatic system (p. 606)
lymph node (p. 606)
spleen (p. 606)
osmoregulation (p. 607)
excretion (p. 607)

kidney (p. 607)
urine (p. 607)
ureter (p. 607)
urinary bladder (p. 607)
urethra (p. 607)
nephron (p. 607)
glomerulus (p. 608)
Bowman's capsule (p. 608)
proximal convoluted tubule (p. 608)
loop of Henle (p. 608)
distal convoluted tubule (p. 608)
collecting duct (p. 608)
glomerular filtration (p. 609)
glomerular filtrate (p. 610)
tubular reabsorption (p. 610)
tubular secretion (p. 610)
thermoregulation (p. 614)
endotherm (p. 614)
ectotherm (p. 614)
shivering (p. 614)

Review Questions

1. Name and discuss five essential functions of the circulatory system.

2. Name three locations where one-way valves are essential to fluid flow.

3. Explain why a two-chambered heart, such as that found in fishes, would be inadequate for humans and other mammals.

4. Discuss how, in closed circulatory systems, tissues get what they need from the blood without ever coming in direct contact with it. Include capillary anatomy and function in your answer.

5. Describe the various ways your kidney helps maintain homeostasis.

6. Draw and label a typical nephron, and relate each part to the three urine-forming processes.

7. Metabolism generates heat. How does blood flow prevent heat from building to lethal temperatures? What happens to heat after it is carried away by the blood?

Critical Thinking Questions

1. Suppose Galen had been right and blood really flowed back and forth in the veins. How would this have affected Harvey's experiment in which he used his finger to stop blood flow? How would the absence of valves in the veins of the arm have affected Harvey's experiment?

2. Redraw Figure 28–2 to represent circulation in an organism such as (a) the jellyfish, *without* a circulatory system, and (b) the crayfish, with an *open circulatory system.* Which body plan would you expect to find in an active, fast-moving animal: no circulatory system, an open system, or a closed system? Why?

3. The circulatory system has vessels that carry blood to the body tissues as well as from the tissues back to the heart. The lymph system, however, has only vessels leading back towards the heart. How can this be? Where does the lymph come from? Why must there be a means for emptying the lymph fluid into the blood system? Without this, what would happen?

4. Excretion in humans begins with nonselective filtration in the glomeruli. If you were designing an excretory system, would you make this part of the system selective or nonselective? What are the advantages of each approach? Do you think that the existing system is the most efficient design? Why, or why not?

5. Until recently, dinosaurs were considered to be ectotherms, like their relatives, the modern-day reptiles. But examination of the structure of fossil dinosaur bones reveals that they have characteristics usually associated with endothermy. What advantages would endothermy have given these reptiles? What disadvantages?

6. After a lengthy visit to a city at a high altitude we adjust to the lowered oxygen content of the air. Why is this capability referred to as *acclimation* rather than *adaptation?*

Additional Reading

Andrade, J. P., ed. 1986. *Artificial organs.* New York: VCH Publishers. (Advanced)

Golde, D. W., and J. C. Gasson. 1988. Hormones that stimulate the growth of blood cells. *Sci. Amer.* July:62–70. (Intermediate)

Golde, D. W. 1991. The stem cell. *Sci. Amer.* Dec:36–43. (Intermediate)

Grantham, J. J. 1992. Polycystic kidney disease I. Etiology and pathology. *Hospital Practice* March:51–59. (Advanced)

Lehrer, S. 1979. *Explorers of the body.* New York: Doubleday. (Introductory)

Robinson, T. F., S. M. Factor, and E. H. Sonneblick. 1986. The heart as a suction pump. *Sci. Amer.* June:84–91. (Intermediate)

Schmidt-Nielsen, K. 1964. *Desert animals: Physiological problems of heat and water.* New York: Oxford University Press. (Advanced)

Valtin, H. 1983. *Renal function: Mechanisms preserving fluid and solute balance in health, 2d ed.,* Boston: Little, Brown. (Advanced)

Weinberger, M. H. 1992. Hypertension in the elderly. *Hospital Practice* May:103–120. (Advanced)

Weisse, A. B. 1992. The alchemy of Willem Kolff, the first successful artificial kidney, and the artificial heart. *Hospital Practice* Feb:108–128. (Introductory)

Gas Exchange:
The Respiratory System

STEPS
TO
DISCOVERY
Physiological Adaptations in Diving Mammals

THE HUMAN PERSPECTIVE

Dying for a Cigarette

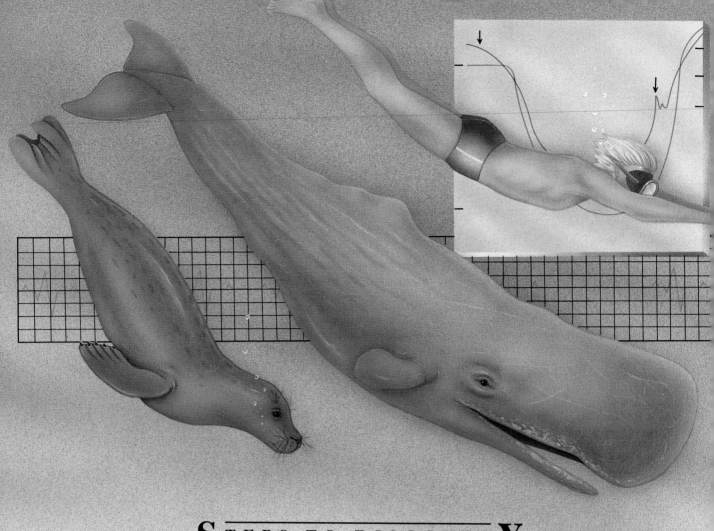

Physiological Adaptations in Diving Mammals

*I*n the South Pacific Islands, native pearl divers plunge into the ocean and search for oysters on the sea floor, often staying beneath the water for several minutes without the help of breathing devices. If you have ever experienced sensations of panic after being underwater for only a minute or so, you can appreciate the magnitude of this feat. But even the most accomplished pearl diver can scarcely approach the talents of a number of other mammalian divers. Consider the Weddell seal of the Antarctic, for example. This animal can remain below water for 70 minutes, searching for food or escaping predators. A sperm whale can dive to depths of 1,500 meters (nearly 1 mile) and remain submerged for as long as 2 hours.

How do these air-breathing mammals accomplish these feats? Perhaps these animals have huge lungs that fill with vast amounts of oxygen before diving. But that would be like our trying to swim to the bottom of a pool holding an inflated air mattress. In fact, just the reverse is true: Weddell seals and sperm whales have modest-sized lungs, and they *exhale* before a dive, thereby reducing their buoyancy. Exhaling also rids the animal of nitrogen-containing air, which can form nitrogen gas bubbles as the animal resurfaces at the end of the dive. Such bubbles can block the flow of blood through vessels, producing a potentially fatal condition known as the *bends*.

Some of the first insights into the physiological adapta-

Once submerged, scientists observed that whales, seals, and humans exhibit a diving reflex. This reflex is an adaptive mechanism

tions of diving mammals emerged in the 1930s from the studies of Per Scholander of the University of Oslo in Norway. Scholander and his colleagues brought seals into the laboratory, where the animals could be connected to instruments that monitored various physiological responses. The scientists had the seals simulate a dive by training the animals to hold their heads under water in a shallow pan. While this simulation is quite different from a true deep-sea dive, the seal's body had to adjust to a lack of fresh oxygen, just as it would have had to do in the sea environment. Scholander soon discovered that this exercise was accompanied by major physiological changes that permitted the diving seal to remain active in the absence of a fresh supply of oxygen.

Almost as soon as the seal's nose was submerged in the water, the rate of the animal's heartbeat fell dramatically, often to as low as one-tenth its normal rate. Because the reaction occurred so quickly, Scholander concluded that it was a reflex response triggered by submersion itself rather than by any metabolic changes that occurred as a result of a lack of oxygen. This cardiac response turned out to be only the initial phase of the so-called *diving reflex*. When blood flow was monitored, the investigators found that only the seal's brain and heart received their normal supply of blood —the same two organs that are most sensitive to oxygen deprivation in humans. The other parts of the seal's body —even the muscles that are required for swimming—were almost totally cut off from fresh oxygen-carrying blood.

How can muscles continue to work if they stop receiving oxygen? Recall from Chapter 9 that skeletal muscle tissue can continue to produce small amounts of ATP in the absence of oxygen as long as the pyruvic acid that is produced by glycolysis is converted to lactic acid. This is one of the paths of fermentation (page 178). Scholander found that the muscles of the submerged seal accumulated large quantities of lactic acid, a clear indication of energy production by fermentation. Normally, when a mammal exerts itself, as when you run long distances, the lactic acid appears in the bloodstream. In this case, however, because the blood flow to the seal's muscles was essentially shut off, the lactic acid remained in the muscle tissue. Once the seal resumed breathing air, blood flow to the muscles returned,

and lactic acid flooded into the general circulation and was oxidized.

After studying the diving reflex in seals, Scholander wondered whether humans might have a similar reflex response. To help answer this question, he recruited the native pearl divers of Australia and found that they showed the same physiological adaptations as did the seals, but to a lesser extent. Within about 20 seconds into their dive, the pearl divers' heart rate dropped, even though the divers were engaged in strenuous activity. As with the seals, lactic acid remained in the divers' muscles until the recovery period, at which time the lactic acid concentration in the blood rose dramatically. These results suggested that the distribution of blood in humans, as in seals, is altered following submersion.

In the 1960s and 1970s, Claude Lenfant of the University of Washington discovered that the blood of Weddell seals had a much greater capacity for oxygen transport and storage than did that of other mammals. In fact, 60 percent of the blood of diving seals is composed of oxygen-carrying red blood cells, as compared to 40 percent in the blood of humans, providing the seals with yet another physiological adaptation for diving.

Advances in electronic technology in recent years has allowed the study of diving mammals to move out of the laboratory and into the field. By attaching miniature computerized devices to seals living in their natural environment, the animals' physiological changes during diving can be monitored. The development of this type of instrumentation was pioneered by Roger Hill of the Massachusetts General Hospital. Although some of the earlier laboratory findings have needed revision, the basic physiological profile obtained from seals who held their noses in a pan of water also applies to those swimming dozens of meters below the water's surface.

that enables the animal to remain alive underwater for extended periods of time.

You can survive for weeks without food and for days without water but without oxygen you will last only a few minutes. Virtually all organisms depend on oxygen to keep the metabolic "fires" burning. Oxygen unleashes the chemical energy that is stored in food molecules, generating more than 90 percent of the ATP that drives energy-consuming activities. In Chapter 9, we referred to this process as *aerobic respiration.*

In this chapter, we will deal with two other aspects of animal respiration: the uptake of oxygen from the environment by a specialized set of organs that make up the **respiratory system,** and the transport of oxygen to the individual cells of the body. We also saw in Chapter 9 how the oxidation of biochemicals generates a waste product, carbon dioxide. The elimination of carbon dioxide from the body accompanies the acquisition of oxygen; therefore, the respiratory system is often called the **gas exchange system.**

▼ ▼ ▼

PROPERTIES OF GAS EXCHANGE SURFACES

The exchange of respiratory gases—oxygen and carbon dioxide—occurs across a **respiratory surface**—a portion of the body surface that has a specialized structure suited for this particular physiological activity. Despite the fact that respiratory organs are found in all shapes and sizes, they all have respiratory surfaces with similar properties:

1. The respiratory surface must remain moist to avoid suffocation, since oxygen must dissolve in fluid before it can move across a cellular membrane. The moisture you see condensing as you exhale on a cold day is water that is lost from the surfaces that line your lungs (Figure 29-1).

2. The epithelial cells that make up the respiratory surface must be extremely thin and permeable; otherwise, oxygen could not cross the barrier from the environment into the body, and carbon dioxide would be unable to move in the opposite direction. The cells that constitute a respiratory surface are usually so flat that they are barely recognizable as cells under the microscope (Figure 29-2).

3. Getting oxygen across the respiratory surface is the first step in respiration, but some mechanism must be present for getting the absorbed oxygen to all the remote cells of the body where it is needed. If an animal is very small or very thin, such as a flatworm, the ab-

sorbed oxygen can simply diffuse through the body to all of the cells. In most larger animals, however, such as birds and mammals, gas exchange surfaces operate in conjunction with the circulatory system. The respiratory tissues of these animals abound with tiny blood vessels (capillaries) that receive oxygen as it diffuses across the respiratory surface from the environment (Figure 29-2). Oxygen is then transported from the site of acquisition, via the circulatory system, to the remainder of the body.

4. All movements of dissolved gases across biological membranes occur by simple diffusion; that is, there are no known transport mechanisms to aid in the passage. Consequently, the rate of oxygen uptake depends on the surface area across which the gas can diffuse. In general, the greater the surface, the more rapid the uptake.

SURFACE AREA AND RESPIRATION

Many animals, including jellyfish, earthworms, and frogs, respire *cutaneously;* that is, across virtually their entire body surface. Since the respiratory surface must be moist and thin, it follows that animals that respire cutaneously tend to be highly vulnerable to environmental conditions. This is evidenced by the familiar sight of a dehydrated earthworm that managed to make it half way across a sidewalk before succumbing to the loss of water across its thin, permeable outer surface. Because of the vulnerability associated with **cutaneous respiration,** evolution has favored animals

FIGURE 29-1

Breathing and water loss. Normally, you are not aware of the water that you lose with every breath, but the loss becomes evident when the exhaled water vapor condenses in cold air.

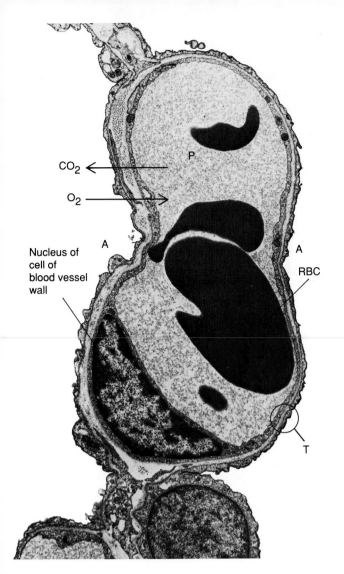

CO₂

O₂

P

A

A

RBC

Nucleus of
cell of
blood vessel
wall

T

FIGURE 29-2

The microanatomy of gas exchange. Electron micrograph of a section through a portion of a human lung, showing the respiratory surface on both sides of a blood capillary. Note the extreme thinness of the barrier (T) between the air space in the lung (A, for alveolus) and the plasma of the blood (P). This barrier consists of a thin, flattened layer of epithelial cells of the lung and an equally thin, flattened layer of cells that line the capillary. Combined, these two cellular layers are approximately 0.2 micrometers thick. In comparison, the red blood cells (RBC) shown in the photograph are about 7 micrometers in diameter. The net movement of oxygen and carbon dioxide is indicated by the arrows.

whose delicate respiratory surfaces are restricted to particular sites on or within the body. This arrangement allows the rest of the body's surface to remain impermeable, as exemplified by the skin that covers your body or the exoskeleton of an insect or spider.

Since the amount of oxygen an animal can absorb from the environment depends on the area of body surface available for gas exchange, restricting the respiratory surface to a small part of the body creates a potential problem. The evolutionary solution has been an increase in the surface area available for absorption without a concurrent increase in the space required to house the enlarged surface. Your lungs, for example, fit conveniently within your chest cavity, yet they have a respiratory surface of 60 to 70 square meters, an expanse equivalent to that of a badminton court. We will see how this is accomplished shortly.

In order to absorb oxygen, an animal must expose a part of its delicate, living surface to the external environment. To facilitate gas exchange, respiratory surfaces are thin, moist, and permeable, and often have large surface areas (See CTQ # 2.)

THE HUMAN RESPIRATORY SYSTEM: FORM AND FUNCTION

The human respiratory system, which is typical of mammals, is depicted in Figure 29-3. The system consists of the following components:

1. a branched passageway, or **respiratory tract,** through which air is conducted, and
2. a pair of **lungs** through which oxygen enters the body and is absorbed into the bloodstream. Even though your lungs reside deep within your chest, like your digestive tract, they are actually a part of your body surface; that is, a surface that is exposed to the external environment.

THE PATH TO THE LUNGS

The entire passageway to the lungs is lined by a mucus-secreting epithelium. The blanket of mucus secreted by these cells keeps the surfaces of the airways moist so that even the

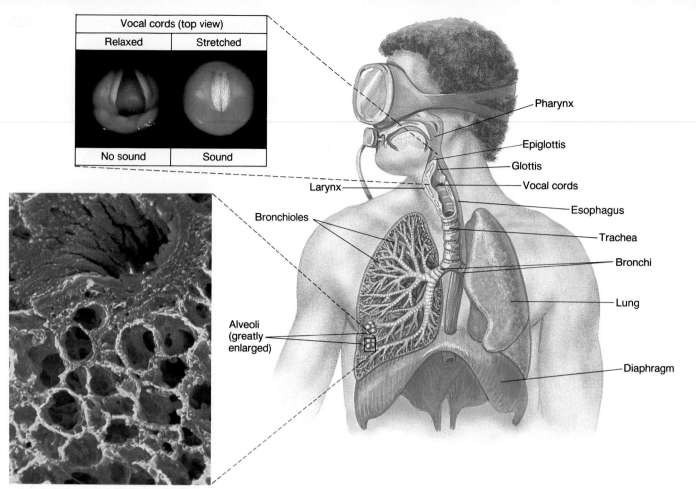

Pharynx

Epiglottis

Glottis

Larynx

Vocal cords

Esophagus

Bronchioles

Trachea

Bronchi

Lung

Alveoli
(greatly
enlarged)

Diaphragm

FIGURE 29-3

The human respiratory system.   The human respiratory system consists of a system of airways that carry air into a pair of lungs. The airways are lined by a ciliated, mucus-secreting epithelium that moistens the air and traps microscopic debris. The top inset shows the vocal cords, which are located in the larynx. Sounds are produced when exhaled air passes through the vocal cords as they are stretched by muscles. The bottom inset shows a cluster of alveoli that fill the interior of the human lung with air pockets. Each lung is a sac that is filled with smaller sacs, thereby increasing the surface area by a factor of hundreds, with no increase in the space required to house it.

driest air is humidified by the time it reaches the gas exchange surface in the lungs. The sticky mucous layer also traps microbes and other dangerous airborne particles in the upper regions of the respiratory tract before they have a chance to enter the lungs and cause serious injury or infection, such as pneumonia. The mucus with its trapped particles is moved by hairlike cilia away from the lungs at the rate of about 2.5 centimeters per minute. Trapped microbes are moved into the throat, where they are swallowed and then killed by stomach acid.

Through the Nasal Cavity and the Airways

Air enters the respiratory system through either the mouth or *nostrils*—the openings into the **nasal cavity.** Air entering through the nostrils passes through a forest of nasal hairs, which traps inhaled particles, preventing them from entering the lungs. Air is warmed and moistened as it passes

through the nasal cavity to the **pharynx** (throat), a corridor that is shared by both the respiratory tract and the digestive tract. Food and liquids are routed to the esophagus and are kept out of the airways by a combination of anatomy and reflexes (Figure 29-3). The passage to the lungs carries the inhaled air through a small opening, called the **glottis,** and into the **larynx,** a short passageway that leads into the lower portion of the respiratory tract. During swallowing, the entire larynx moves up, forcing a flap of tissue, the **epiglottis,** to seal off the entrance to the lower airways. This reaction prevents food and liquids from mistakenly entering the airways on its route to the esophagus and stomach (page 569).

The human larynx is more than just a passageway for air. The sides of the larynx contain a pair of muscular folds —the **vocal cords**—from which the human voice emanates. The vocal cords are operated by the passage of air through the larynx. When we inhale, air silently rushes through the opening between the relaxed vocal cords (top inset, Figure 29-3). As air escapes from the lungs when we exhale, the stretched vocal cords vibrate, creating the sound of the voice. The loudness of the voice is determined primarily by the force of the exhaled air, while the pitch (highness or lowness of the note) is determined by the level of tension on the vocal cords, which is regulated by the contraction of muscles in the larynx. The sound is amplified by the hollow chamber that is formed by the laryngeal cartilage (the "Adam's apple" or "voicebox"). The larynx is also one of the targets of male sex hormones, which is why men have larger larynxes than do women, as evidenced by their deeper voices and more prominent Adam's apple.

Air passes through the larynx and into the tubular **trachea** (or windpipe), which descends into the chest. The trachea divides into a series of smaller and smaller tubes, or *bronchi*, which extend into the various regions of each lung. Both the trachea and the bronchi contain **C**-shaped bands of cartilage that provide structural reinforcement for these passageways, preventing the tubes from collapsing and impairing the flow of air. The bronchi branch further to form a series of even smaller tubules, the **bronchioles.** Bronchioles lack cartilage, but they contain rings of smooth muscle fibers, whose state of contraction regulates the flow of air into the lungs. During an asthma attack or severe allergic reaction, the bronchioles can become clogged with mucus and may constrict due to prolonged muscular contraction.

Into the Lungs

As you have seen by now, the ultimate destination of inhaled air is the lungs. The lungs contain millions of microscopic pouches, called **alveoli** (Figure 29-3, bottom inset), which resemble the air pockets in a kitchen sponge. Due to their structure, alveoli are ideally suited for gas exchange. Each alveolus is a hollow, thin-walled sac that is richly surrounded by a network of capillaries (Figure 29-4). Alveoli are the sites of gas exchange in the lung. Gases readily pass through the thin walls of both the alveolus and the capillaries (see Figure 29-2) so that blood moving through the lung tissue can quickly pick up a fresh supply of oxygen

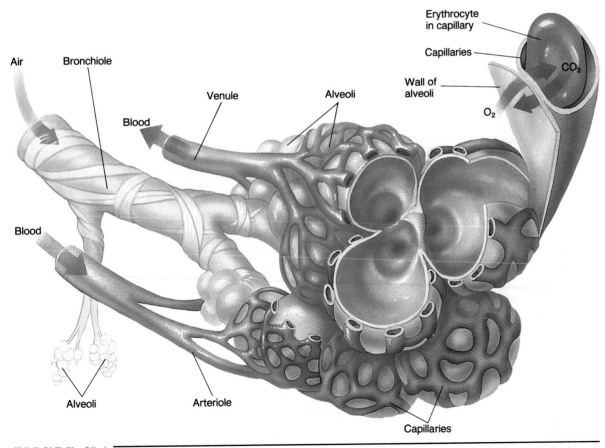

FIGURE 29-4

Alveoli, the lung's trade centers. Each alveolus is a thin-walled, bubblelike chamber (the size of a pinpoint) that is surrounded by capillaries. Oxygen and carbon dioxide gases quickly move in the directions indicated by the arrows. An electron micrograph of the alveolar–capillary interface is shown in Figure 29-2.

and unload its cargo of carbon dioxide. Freshly oxygenated blood (depicted in red in Figure 29-4) travels from the alveolar capillaries into larger vessels, until it enters the pulmonary vein, which carries the blood to the heart; from there, the oxygenated blood is pumped to the remainder of the body. The stale air in the lungs, which has been depleted of a portion of its oxygen, is forced out of the alveoli and back into the bronchioles, where it is expelled from the airways during exhalation.

If your lungs were mere hollow bags, the gas exchange surface would be less than half a square meter (about 5 square feet)—too small a surface to absorb enough oxygen to keep you alive under even the most restful situations. Instead, the spongy interior of your lungs houses more than 300 million alveoli, providing 60 to 70 square meters of an oxygen-collecting, carbon-dioxide-discharging surface. Because of their structure, the lungs are a very delicate tissue that can easily be damaged by inhaled pollutants (see The Human Perspective: Dying for a Cigarette?)

BREATHING: EXPANDING THE CHEST CAVITY

The lungs occupy a space—the **thoracic cavity**—situated within the chest. Cradled in the rib cage, the lungs are sealed in a waterproof, airtight, double-membraned sac, called the **pleura.** Occasionally the pleura becomes inflamed, a condition called *pleurisy,* causing fluid accumulation and painful breathing. A sheet of muscle tissue, the **diaphragm** (Figure 29-5), separates the thoracic cavity from the abdominal cavity. This enclosed configuration not only protects the lungs but also provides a mechanism for breathing.

In general, gases move down a pressure gradient, from a region of high pressure to one of low pressure. This simple principle explains the direction taken by winds, as well as the movement of air into and out of the lungs. The operation of the lungs occurs in a manner analogous to that of a bellows (Figure 29-5). While the pressure of air in the

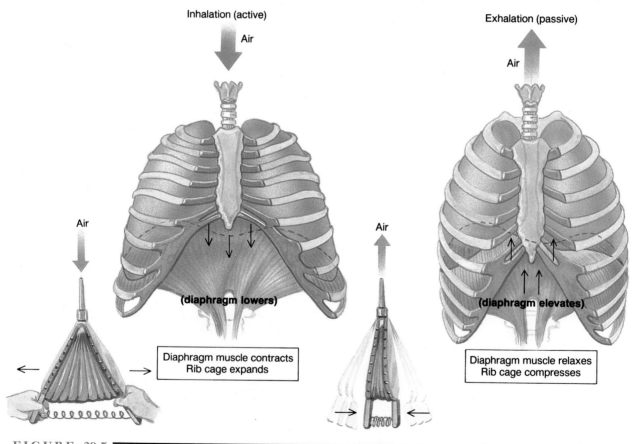

FIGURE 29-5

Operation of the lungs. Contracting the diaphragm muscle (and intercostal muscles) expands the volume of the thoracic cavity, sucking air into the lungs. The inhaled air is expelled when these muscles relax. The operation of the lungs can be compared to that of a bellows, whereby air is forced in and out of the bellow's chamber.

atmosphere remains constant, the pressure within the lungs (or bellows) changes according to the lungs' (or bellow's) volume. When we inhale, the chest cavity expands in volume; the pressure inside each lung drops below that of the atmosphere; and air rushes into the lungs' chambers. The increase in lung volume during inhalation is not a result of the direct expansion of the lungs; rather, it occurs indirectly following the contraction of the muscular diaphragm and the *external intercostal muscles,* which lie between the ribs. The walls of the thoracic cavity contain elastic fibers that stretch during chest expansion. When the diaphragm and intercostal muscles relax, the lungs spring back to their smaller volume, causing the pressure inside each lung to rise above that of the atmosphere, forcing the stale air out of the lungs.

In humans, air is sucked down a pressure gradient through a series of increasingly narrow passageways and into the lungs—organs that contain millions of microscopic, thin-walled chambers, where respiratory gases can be exchanged with the bloodstream. The passageways to the lungs act to warm, moisten, and cleanse the inhaled air of debris. (See CTQ # 3.)

THE EXCHANGE OF RESPIRATORY GASES AND THE ROLE OF HEMOGLOBIN

Two types of gas exchanges are constantly occurring in your body: one between the alveoli and the capillaries in the lung, and the other between the capillaries and the various tissues of the body. These events are summarized in Figure 29-6. The principle underlying both types of exchange is a similar one: the passive diffusion of a dissolved respiratory gas from a region of higher concentration, across a permeable surface, to a region of lower concentration; that is, down a concentration gradient.

GAS EXCHANGE IN THE LUNGS: UPTAKE OF OXYGEN INTO THE BLOOD

Inhaled air is rich in oxygen (about 21 percent) and poor in carbon dioxide (about 0.4 percent). Blood returning from the tissues to the lungs carries the opposite complement of gases: a high concentration of carbon dioxide and a depleted oxygen cargo. As a result of these differences in concentration, oxygen in the lungs diffuses across the thin, cellular walls of both the alveolus and the capillary and into the bloodstream, while carbon dioxide diffuses in the opposite direction (Figure 29-6). This exchange is completed in about a quarter of a second, about the time it takes blood to flow through the site of exchange in a capillary.

If oxygen simply dissolved in the fluid of blood, the blood's oxygen-carrying capacity would be severely limited because oxygen is not very soluble in water. The capacity of the bloodstream to transport oxygen is greatly increased by the presence of the reddish, iron-containing protein **hemoglobin.** Each hemoglobin molecule contains four polypeptide subunits and binds four oxygen molecules, forming *oxyhemoglobin.* Hemoglobin allows human blood to carry 70 times more oxygen than it could otherwise. Put another way, the average person extracts 4,500 grams (about 10 pounds) of oxygen from 50,000 liters of air inhaled every day. Without hemoglobin, this number would plummet to about 60 grams of oxygen, only slightly more than we could obtain by breathing water, with equally fatal results.

The importance of hemoglobin in human blood is readily demonstrated by the fatal effects of carbon monoxide (CO), a gas that is present in high concentration in car exhaust fumes. Carbon monoxide molecules bind to the same sites on the hemoglobin molecule as do oxygen molecules, but they do so 200 times more tenaciously. As a result, the hemoglobin in the blood of a person exposed to high levels of carbon monoxide loses its ability to bind oxygen; if the person is not removed from the toxic environment, he or she will die as the result of oxygen deprivation.

While high concentrations of hemoglobin are needed to carry a large supply of oxygen, the protein cannot simply be dissolved in the blood plasma since it would thicken the fluid, impairing its flow through the vessels. This problem has been solved by encapsulating hemoglobin in **erythrocytes** (red blood cells). Each milliliter of your blood contains about 5 billion erythrocytes; your entire body contains a total of about 25 trillion.

GAS EXCHANGE IN THE TISSUES: RELEASE OF OXYGEN FROM THE BLOOD

Blood leaves the lungs carrying high concentrations of oxygen and low levels of carbon dioxide. When this oxygenated blood reaches a metabolically active tissue, it finds itself in an environment that is low in oxygen and high in carbon dioxide. As a result, oxygen diffuses into the tissues as carbon dioxide enters the blood; both gases move passively from areas of higher concentration to areas of lower concentration (Figure 29-6). The first oxygen molecules to move out of the blood are those that are simply dissolved in the plasma. As these dissolved oxygen molecules move out of the capillary, the concentration of oxygen dissolved in the plasma decreases, promoting the dissociation of oxygen molecules from their binding sites on hemoglobin molecules in the red blood cells. The oxygen molecules released from hemoglobin move out of the erythrocytes and dissolve in the plasma. From there, the dissolved oxygen molecules move out of the capillaries, promoting the release of additional oxygen molecules from the hemoglobin, and so forth. This is the process by which hemoglobin unloads its oxygen cargo in the tissues and readies itself to pick up more oxygen on its upcoming trip through the lungs.

◐ The exchange of gases in the tissues is self-regulating. Those tissues that are metabolizing more actively utilize

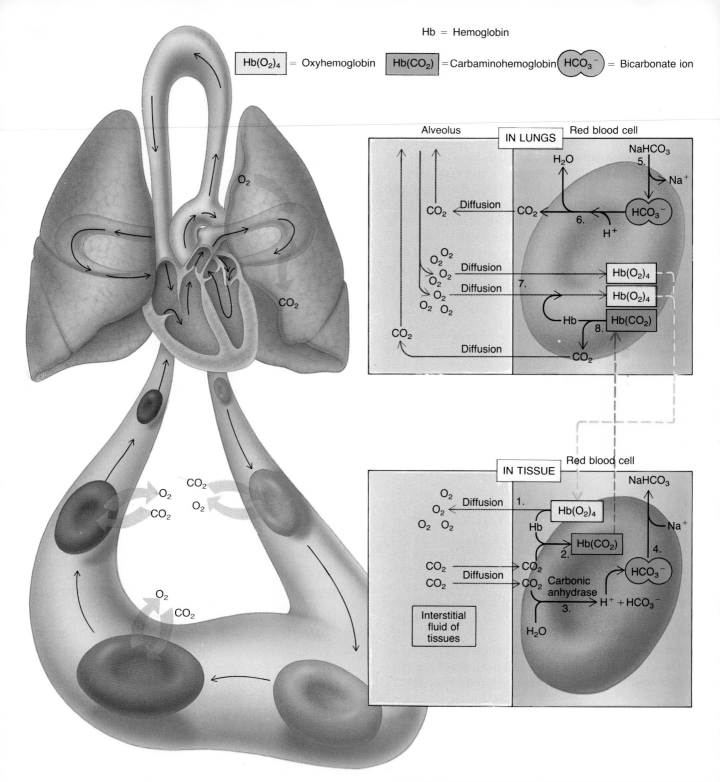

FIGURE 29-6

Transport and exchange of gases. In the tissues: (1) O_2 molecules are released from hemoglobin and diffuse into the cytoplasm of the red blood cell and then into the tissues. (2) As CO_2 diffuses from the tissues into red blood cells, some of it complexes with hemoglobin. (3) Most of the CO_2, however, reacts with water to form bicarbonate ions (HCO_3^-). The protons released by this reaction bind to hemoglobin, which promotes the release of additional oxygen. (4) Some bicarbonate diffuses from the cell and provides the blood with buffering capacity. **In the lungs:** (5) Bicarbonate ions diffuse into the red blood cell and (6) are converted back to CO_2, which diffuses into the alveolus. (7) As O_2 diffuses into the red blood cells, it complexes with hemoglobin, so a steep O_2 concentration gradient is maintained. (8) CO_2 molecules that were bound to hemoglobin molecules are released, freeing the hemoglobin molecules to pick up additional O_2.

more oxygen, thereby producing lower oxygen concentrations. The lower the oxygen concentration in the tissue, the steeper the gradient between tissue and blood, which favors the release of more oxygen from the bloodstream. According to this principle, those tissues most "in need" of oxygen receive the most oxygen from the passing blood.

We noted on page 62 that the blood contained buffers —particularly bicarbonate ions—that kept the blood from becoming too acidic or too alkaline. The bicarbonate ions in the blood are formed as the result of the following reaction between carbon dioxide and water that occurs when carbon dioxide is taken into the blood from the tissues:

$$CO_2 + H_2O \xrightarrow{\text{carbonic anhydrase}} \underset{\substack{\text{carbonic} \\ \text{acid}}}{H_2CO_3} \longrightarrow \underset{\text{proton}}{H^+} + \underset{\substack{\text{bicarbonate} \\ \text{ion}}}{HCO_3^-}$$

This first reaction is catalyzed by the enzyme *carbonic anhydrase*, which is present in red blood cells. Carbonic acid, the product of the reaction, dissociates into two ions, a hydrogen ion (H^+) and a bicarbonate ion (HCO_3^-). Hydrogen ions lower the pH, which affects the shape of hemoglobin molecules, causing them to lose their grip on oxygen. This phenomenon is known as the *Bohr effect*. Those tissues that are metabolizing most actively not only have a lower oxygen concentration but also a higher carbon dioxide concentration, which leads to a lower pH. Since hemoglobin releases more oxygen at a lower pH, metabolically active tissues have a "built-in" mechanism for extracting additional oxygen from the passing bloodstream.

As the blood passes through the tissues, it gives up its oxygen and takes up carbon dioxide (Figure 29-6). Most of the carbon dioxide simply dissolves in the blood, forming bicarbonate, but a portion of the carbon dioxide molecules becomes bound to hemoglobin for the trip back to the lungs. When the blood reaches the lungs, the process of gas exchange is reversed from that which occurs in the tissues. The more oxygen that is removed during the previous passage of the blood through the body, the greater the number of hemoglobin molecules that will be lacking their full complement of oxygen molecules, and the more oxygen that will be picked up from the alveoli. The lungs have a remarkable capability for gas absorption. Even if you are running at top speed, and your blood is virtually depleted of its oxygen content, the blood will be fully resupplied with oxygen during its short, rapid passage through the lungs.

Oxygen and carbon dioxide move into and out of the bloodstream by simple diffusion in response to differences in their respective concentrations. These concentration gradients arise as the result of the consumption of oxygen and the formation of carbon dioxide within the tissues. The uptake of oxygen into the blood as it courses through the lungs is greatly facilitated by the presence of the oxygen-binding protein hemoglobin. (See CTQ # 4.)

REGULATING THE RATE AND DEPTH OF BREATHING TO MEET THE BODY'S NEEDS

During the course of a typical day, the respiratory system must make rapid and dramatic changes in its level of activity. One minute, you might be resting quietly on a park bench, and the next minute you might be running after a bus. This shift from a resting to a running state results in a great increase in the utilization of oxygen. How can the blood increase its rate of oxygen delivery twentyfold or more in less than a minute?

↻ Three major mechanisms are available to supply the body with an increased supply of oxygen, each of which becomes quite evident during strenuous exercise (Figure 29-7): You breathe more rapidly; you breathe more deeply; and the rate of your heartbeat increases. All of these changes are controlled by regulatory centers located in the brain.

Although you can hold your breath for a short time, it is impossible to stop breathing voluntarily to the point of severe oxygen deprivation. Eventually, involuntary regulatory

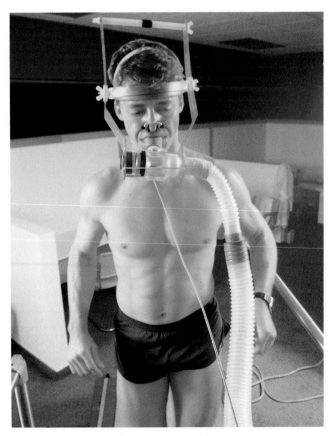

FIGURE 29-7 ——————————————

Measuring the rate of oxygen consumption of a person running on a treadmill.

◁ THE HUMAN PERSPECTIVE ▷
Dying for a Cigarette?

On the average, smoking cigarettes will cut approximately 6 to 8 years off your life, more than 5 minutes for every cigarette smoked! Cigarette smoking is the greatest cause of preventable death in the United States. According to a 1991 report by the Centers for Disease Control (CDC), nearly 450,000 Americans die each year from smoking-related causes. Smoking accounts for 87 percent of all lung-cancer deaths, and smokers are more susceptible to cancer of the esophagus, larynx, mouth, pancreas, and bladder than are nonsmokers. The increased incidence of lung cancer deaths among smokers compared to nonsmokers is shown in Figure 1a, and the benefits attained by quitting is shown in Figure 1b. The effects of smoking on lung tissue is shown in Figure 2. Atherosclerosis, heart disease, and peptic ulcers also strike smokers with greater frequency than they do nonsmokers. For example, long-term smokers are 3.5 times more likely to suffer from severe arterial disease than are nonsmokers. Emphysema (a condition caused by the destruction of lung tissue, producing severe difficulty in breathing) and bronchitis (an inflammation of the airways) are 20 times more prevalent among smokers.

Smokers also endanger other people. Each year, smokers are responsible for the deaths of thousands of "innocent bystanders," nonsmokers who share the same air with smokers. The risks of passive (involuntary) smoking are well known; second-hand smoke can make you seriously ill (Figure 3). Children of smokers have double the frequency of respiratory infections as do children who are not exposed to tobacco smoke in the home. Being married to a smoker is especially hazardous; 20 percent of lung-cancer deaths among nonsmokers are attributable to inhaling other

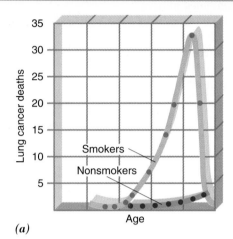

(a)

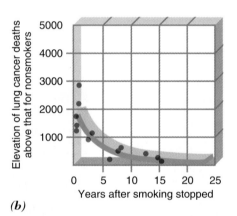

(b)

FIGURE 1

people's tobacco smoke. Another "innocent bystander" is a fetus developing in the uterus of a woman who smokes. Smoking increases the incidence of miscarriage and stillbirth and decreases the birthweight of the infant. Once born, these babies suffer twice as many respiratory infections as do babies of nonsmoking mothers.

Why is smoking so bad for your health? The smoke emitted from a burning cigarette contains more than 2,000 identifiable substances, many of which are either irritants or carcinogens. These compounds include carbon monoxide, sulfur dioxide, formaldehyde, nitrosamines, toluene, ammonia, and radioactive isotopes. Autopsies of respiratory tissues from smokers (and from nonsmokers who have lived for long periods with smokers) show widespread cellular changes, including the presence of precancerous cells (cells that may become malignant, given time) and a marked reduction in the number of cilia that play a vital role in the removal of bacteria and debris from the airways.

Of all the compounds found in tobacco (including smokeless varieties), the most important is nicotine, not because it is carcinogenic, but because it is so addictive. Nicotine is addictive because it acts like a neurotransmitter by binding to certain acetylcholine receptors (page 477), stimulating postsynaptic neurons. The physiological effects of this stimulation include the release of epinephrine, an increase in blood sugar, an elevated heart rate, and the constriction of blood vessels, causing elevated blood pressure. A smoker's nervous system becomes "accustomed" to the presence of nicotine and decreases the output of the natural neurotransmitter. As a result, when a person tries to stop smoking, the sudden absence of nicotine, together with the decreased level of the natural transmitter, decreases stimulation of postsynaptic neurons, which creates a craving for a cigarette—a "nicotine fit." Ex-smokers may be so conditioned to the act of smoking that the craving for cigarettes can continue long after the physiological addiction disappears.

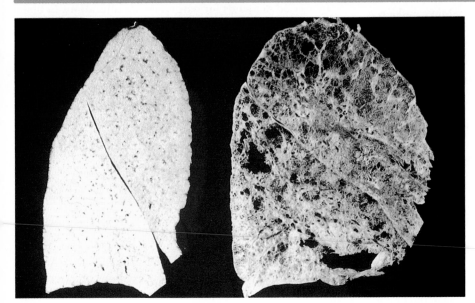

NON SMOKER   SMOKER

FIGURE 2

FIGURE 3

mechanisms restart "automatic" breathing. These same regulatory mechanisms keep you breathing while you sleep or when your attention is on other matters. They also cause you to breathe more often and more deeply as your need for oxygen delivery increases. The breathing control center, or **respiratory center,** is located in the medulla, a portion of the brainstem that regulates automatic activities (page 490). But just how does the respiratory center know how often you need to take a breath and how deep that breath should be?

The respiratory center of the brain receives information from two sources: *peripheral chemoreceptors* located in the walls of the aorta and the carotid arteries of the neck, and *central chemoreceptors* located in the brain itself. Let us consider the role of the more important central chemoreceptors. Recall that an increase in the level of carbon dioxide in the blood leads to a rise in the concentration of hydrogen ions. In other words, the increased carbon dioxide that is generated during more vigorous activity leads to an increased acidity of the blood. Chemoreceptors in the brain are stimulated by even slight increases in acidity, activating the respiratory center of the medulla. This activation, in turn, sends signals to the diaphragm and the intercostal muscles, causing you to breathe more rapidly and deeply. The effect of elevated carbon dioxide levels can be demonstrated in the following ways.

1. If a person is allowed to breathe air that contains a constant level of oxygen but an increasing content of carbon dioxide, the breathing rate will increase markedly, even though the person is not engaged in any physical activity. Since the level of oxygen is held constant, it must be the rising carbon dioxide level that is elevating the respiratory rate.

2. If a person is told to breathe very rapidly and deeply for a period of time and then to breathe in a normal relaxed manner, the breathing rate drops dramatically below that of the normal resting state. This phenomenon is explained on the basis of blood carbon dioxide levels. The period of rapid breathing (known as *hyperventilation*) is not accompanied by increased metabolic activity; thus, there is an increased loss of carbon dioxide from the lungs without a corresponding increase in carbon dioxide production by the tissues. Consequently, blood carbon dioxide concentrations drop. The brain senses the lowered carbon dioxide level—as a drop in hydrogen ion concentration—and interprets the message that further inhalation is not necessary. Underwater divers sometimes use this physiological response to hold their breaths for longer periods of time than would normally be possible. This is actually a very dangerous practice since hyperventilation has no effect on blood oxygen levels. Even though the individual may not feel the need to breathe, the blood oxygen content can become drastically depleted. As a result, the individual may lose consciousness before realizing that he or she is running out of oxygen.

Increased demands for oxygen create alterations in blood chemistry that trigger an increased uptake of oxygen from the environment. As oxygen-saturated blood circulates through the tissues, its delivery is self-regulating; the greater the tissue's need, the more oxygen that is released to the cells of that tissue. (See CTQ # 5.)

ADAPTATIONS FOR EXTRACTING OXYGEN FROM WATER VERSUS AIR

The successful respiratory strategy is one that is adapted to the medium in which the animal lives. Animals that absorb oxygen from water, for example, utilize a different type of respiratory system than do animals that extract oxygen from air.

EXTRACTING OXYGEN FROM WATER

The amount of oxygen dissolved in water is at best 21 times less than that which is dissolved in air. As a result, aquatic animals that acquire their oxygen from the surrounding water have less oxygen available than do their air-breathing counterparts. In addition, water is much heavier than air; thus, water requires considerably more energy in order to be moved over the respiratory surface. We use very little energy to move air in and out of our lungs, whereas a fish uses much of the energy obtained from its food just to pump water through its body to acquire oxygen as needed.

Most animals that extract oxygen from water possess **gills**—outgrowths of the body surface which project into the aqueous environment and are rich with blood vessels. These outgrowths would be ill-adapted in a terrestrial animal since the delicate projections would rapidly dehydrate in air. This is why a fish quickly dies of asphyxiation when it is removed from water even though it is surrounded by a higher concentration of oxygen than that of its normal habitat.

The complexity of an animal's gills correlates with the animal's oxygen requirements. The gills of small or slow-moving aquatic animals, such as sea slugs, sea stars, and aquatic salamanders, are typically simple projections of the body surface (Figure 29-8). Animals with greater oxygen requirements have more complex gills that are capable of harvesting additional oxygen. Such gills are usually covered by a protective flap, such as that found in lobsters or fishes. The form of the complex gills of larger, active animals are well-suited for their function. These gills branch into smaller outgrowths that amplify the surface area for gas exchange without increasing the space required to house them. Gill complexity has reached its current pinnacle in fishes (Figure 29-9a). Each gill consists of fingerlike projections, called **gill filaments,** from which rows of thin, flattened **lamellae** project into the flowing stream of water. The lamellae are the sites of gas exchange and have a rich supply of capillaries into which oxygen molecules diffuse.

Physiologists studying fish gills noted that the flow of blood within the lamellae and the flow of water across the outer surface of the lamellae (over the gills) occurred in opposite directions (Figure 29-9a,b). The movement of the blood and water in opposite directions is imperative to the success of highly active "water breathers" because it maximizes the amount of oxygen that is extracted from water. This principle is called **countercurrent flow,** providing another example of the relationship between form and function. Because the two fluids flow in opposite directions, there will always be a higher concentration of oxygen in the water than in the blood (Figure 29-9b, upper part). This relationship favors the continual diffusion of oxygen from the surrounding water into the bloodstream. Over 80 percent of the oxygen passing through the gills is extracted by this type of respiratory mechanism, much more than that which could be attained if the two currents were flowing in the same direction (Figure 29-9b, lower part).

Not all aquatic animals have gills for extracting oxygen from the water. A variety of aquatic animals, such as porpoises, whales, sea snakes, and some water spiders, have evolved from air-breathing terrestrial ancestors. These animals still rely on air for their oxygen supply, even if they have to bring an external supply of oxygen underwater with them (Figure 29-10).

EXTRACTING OXYGEN FROM AIR

Terrestrial animals live in an environment that is rich in oxygen but low in water content; thus, oxygen is readily available but the loss of some of its precious body water is inevitable. Therefore, the respiratory surfaces of most terrestrial animals—whether a snail, spider, or human— occur as inpocketings of the body surface which are tucked away within the moist, internal environment, thereby minimizing water loss. Two very different types of respiratory systems have evolved in air-breathing animals. Insects, and some of their arthropod relatives, possess *tracheae,* while most other air breathers, including snails, reptiles, birds, and mammals, have *lungs.*

Respiration by Tracheae

If you look at the surface of an insect's abdomen under a microscope, you will find paired rows of openings, called *spiracles* (Figure 29-11a). Each spiracle leads into a tube, which, in turn, branches into a network of finer and finer tubes. These tubes, called **tracheae,** carry air to the most remote recesses of the animal's body (Figure 29-11b). Ultimately, the finer tracheal tubes give rise to microscopic, dead-end, fluid-filled **tracheoles,** which terminate either very close to or actually within the body's cells. Oxygen travels down this complex network of air tubes directly to the cells, without any help from the animal's circulatory system.

While a tracheal respiratory system is very efficient for a small insect, it becomes very limiting in larger species. Even though many insects actively suck fresh air into their larger tracheal tubes (which explains why a grasshopper can sometimes be seen rhythmically contracting its abdomen), the system largely depends on the simple diffusion of air through a network of narrow tubing. This is analogous to a crowd of people trying to breathe at the dead end of a long mine shaft; the available oxygen may be rapidly depleted. The lack of participation of the insect's circulatory system in respiration is one of the key factors that has limited the size

Gills

FIGURE 29-8
Simple gills. Extensions of the external surface of this nudibranch (sea slug) increases the animal's oxygen-absorbing efficiency.

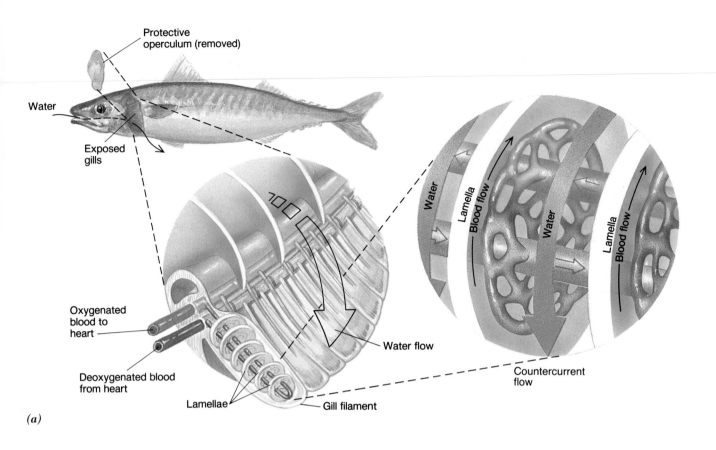

Protective operculum (removed)

Water

Exposed gills

Oxygenated blood to heart

Deoxygenated blood from heart

Lamellae

Gill filament

Water

Lamella

Blood flow

Water

Lamella

Blood flow

Water flow

Countercurrent flow

(a)

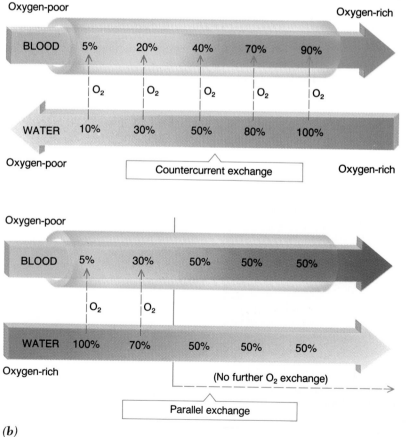

Oxygen-poor

Oxygen-rich

BLOOD   5%   20%   40%   70%   90%

O_2 O_2 O_2 O_2 O_2

WATER   10%   30%   50%   80%   100%

Oxygen-poor

Oxygen-rich

Countercurrent exchange

Oxygen-poor

BLOOD   5%   30%   50%   50%   50%

O_2 O_2

WATER   100%   70%   50%   50%   50%

Oxygen-rich

(No further O_2 exchange)

Parallel exchange

(b)

FIGURE 29-9

Complex gills. Fish have complex gills that have an enormous surface area compacted into a small space *(a)*. In this diagram, blood flowing through each gill filament changes from blue to red as it acquires oxygen from the passing water. The process is enhanced by a countercurrent exchange system, whereby water is forced past the outer surface of the lamellae in a direction counter to that of blood flow within the lamellae. As the blood acquires oxygen, it moves toward water that is even richer in the gas so that at every point along the exchange surface, the oxygen level in the blood is lower than is that of the surrounding water (upper diagram of *b*). This relationship favors the continuing diffusion of oxygen along the entire lamella. If the flow were in the same direction (lower diagram of *b*), absorption would occur only at one end, and much less oxygen could be obtained.

FIGURE 29-10
Carrying more than dinner. This diving spider carries its own oxygen supply in a glistening bubble of air around its abdomen. The spider is hauling its catch home to the larger, submerged air bubble in which it dines, sleeps, and mates.

of these animals, a condition that is undoubtedly fortunate for the rest of the animal life on this planet.

Respiration by Lungs

As we discussed earlier, the lungs are sac-like invaginations of the body surface that contain an extremely thin respiratory surface and a rich supply of microscopically thin blood capillaries. In less active animals, such as snails, frogs, and snakes, the lungs tend to be relatively simple sacs with little internal surface area. In contrast, the more complex lungs of mammals contain millions of microscopic alveoli, which provide an enormous surface area packed into a relatively small space (see Figure 29-3).

Compared to water, air is dry, light-weight, and rich in oxygen. These environmental differences have selected for different types of respiratory organs in air-breathers versus water-breathers. (See CTQ # 6.)

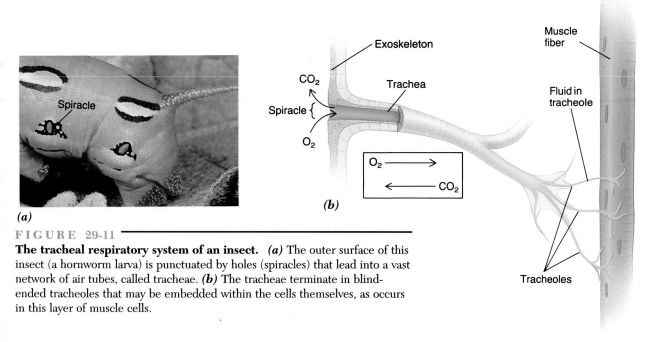

(b)

(a)

FIGURE 29-11
The tracheal respiratory system of an insect. *(a)* The outer surface of this insect (a hornworm larva) is punctuated by holes (spiracles) that lead into a vast network of air tubes, called tracheae. *(b)* The tracheae terminate in blind-ended tracheoles that may be embedded within the cells themselves, as occurs in this layer of muscle cells.

EVOLUTION OF THE VERTEBRATE RESPIRATORY SYSTEM

◼▶ Vertebrates have two very different types of respiratory organs: gills and lungs. Even though they serve the same basic function—gas exchange—these two respiratory organs have different structures and have arisen along two separate evolutionary paths.

The ancestors of fishes were thought to have resembled the tiny, modern-day lancelet (pictured in Figure 29-12). The mouth of a lancelet opens into a pharynx that is punctuated with mucus-covered, microscopic slits, causing it to resemble a woven basket. The animal is a filter feeder; it pumps water into its mouth and through the pharyngeal "basket," while suspended food particles become trapped in the mucus lining the slits. While primarily a feeding organ, this type of pharyngeal apparatus is ideally suited for carrying out gas exchange. A constant stream of oxygen-rich water flows through the pharyngeal basket, which contains a large surface area across which gas exchange may occur. While the early fishes may have been filter feeders like their ancestors, most became adapted to feeding on larger food items. The pharyngeal apparatus lost its food-capturing function and evolved into a system of gills that delivered oxygen to the bloodstream.

The evolutionary shift of the pharynx from a feeding to a respiratory organ can be appreciated by following the development of one of the most primitive vertebrates, the lamprey eel (Chapter 39). The larval lamprey lives on the bottom of streams and feeds on particles strained from the water by its pharyngeal basket. Following metamorphosis, the adult lamprey becomes a predator, but its pharyngeal apparatus is retained as its respiratory organ.

The evolutionary origin of vertebrate lungs is less certain. There are no fossil remains of these soft, spongy

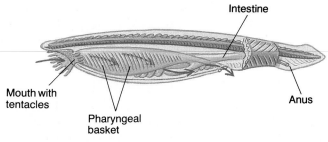

Mouth with tentacles

Pharyngeal basket

Intestine

Anus

FIGURE 29-12

The pharyngeal feeding apparatus of a lancelet. The pharyngeal basket of these small, aquatic invertebrates is thought to resemble that of the ancestors of fishes. The basket is perforated with openings that strain food from the water that flows through the animal's body. While the pharyngeal basket of the lancelet functions primarily as a feeding organ, a similar organ in an ancient ancestor may have given rise to vertebrate gills.

organs, and little has been learned by studying modern-day representatives. Regardless of how they evolved, lungs were probably present in some of the earliest fish. These primitive vertebrates lived in freshwater ponds, and their lungs are thought to have helped them survive periods of drought when the ponds became shallow and unsuitable for aquatic respiration. These air-breathing fishes include the evolutionary predecessors of all of the land vertebrates.

Both the gills and lungs of modern vertebrates have ancient evolutionary roots. Gills are thought to have evolved from feeding baskets present in filter-feeding prevertebrates, while lungs probably evolved in ancient fishes that became dependent on air as their aquatic environments either stagnated or evaporated. (see CTQ # 7.)

REEXAMINING THE THEMES

Relationship between Form and Function

The form of the human respiratory system is closely correlated with its function. This is evident by the hairs in the nasal cavity, which filter the air; the flaplike epiglottis, which seals the airways from food and liquid; the mucus-secreting cells, which line the respiratory tract and keep the air moist; the cartilage bands in the trachea, which prevent the tube from collapsing as we breathe; and the lungs, which consist of millions of thin-walled pouches, whose combined surface area is equivalent to the size of a badminton court yet is packed into a chest cavity smaller than a badminton racket.

Biological Order, Regulation, and Homeostasis

The supply of oxygen to the body's tissues is maintained by a number of homeostatic mechanisms, some of which are self-regulating and occur automatically, without the intervention of the nervous system. The amount of oxygen delivered to a given tissue is determined largely by the oxygen needs of that tissue. More active tissues utilize more oxygen, producing a steeper oxygen gradient between themselves and the blood, causing the release of more oxygen from the blood. In addition, more active tissues produce more carbon dioxide, which increases the acidity of the blood as it passes through the tissues, causing a change in the shape of the hemoglobin molecules and an additional release of oxygen into these areas. The amount of oxygen that is taken into the lungs is regulated by neural mechanisms. As the level of carbon dioxide in the blood rises during periods of physical activity, chemoreceptors in the brain are stimulated, and the respiratory center in the medulla is activated, leading to an increased rate and depth of breathing.

Acquiring and Using Energy

Respiratory systems have evolved as the means by which animals obtain oxygen. Oxygen serves only one major function in animals: It is the terminal acceptor of the electron transport system of the mitochondria (the system by which over 90 percent of the ATP used by most animals is generated). Without oxygen, animals cannot sustain their energy-requiring reactions. Thus, even though oxygen provides no energy itself, the gas is necessary for most animals to obtain energy from food by oxidation.

Unity within Diversity

Respiratory organs come in many different shapes and sizes, ranging from the outer skin of a frog, to the gills of a fish, to the lungs in your chest. Despite this diversity, all of these organs possess a respiratory surface that is extremely thin, highly permeable, and continually moist. In addition, all respiratory surfaces are underlain by a network of thin-walled capillaries and are characterized by a large surface area. Air breathers, from snails to whales, package their respiratory surface within an inpocketing of their body surface, whereas, water-breathing animals, such as starfish and codfish, utilize delicate filaments that project directly into their aquatic environment.

Evolution and Adaptation

The gills of vertebrates evolved from food-trapping pouches, providing an excellent example of how the function of a structure can change over the course of evolution. Since the currents of water that supply food also carried oxygen into the body, this type of feeding strategy set the stage for the evolutionary conversion of a food-collecting organ into oxygen-collecting gills.

SYNOPSIS

The need for gas exchange. All animals must obtain oxygen for aerobic respiration and must dispose of carbon dioxide waste. Most complex animals have impermeable outer surfaces with specialized gas exchange structures (gills in water-breathing animals, tracheae in most terrestrial arthropods, and lungs in air-breathing vertebrates). These respiratory structures contain respiratory surfaces that are thin, permeable, and moist and are characterized by an extensive surface area.

The human respiratory system consists of a branched passageway that leads into a pair of lungs. The lungs consist of millions of microscopic, thin-walled pouches (alveoli) that are underlain by a network of capillaries. The alveoli provide an extensive surface across which gas exchange can occur. Air enters the lungs when the thoracic cavity is expanded by contraction of the diaphragm and intercostal muscles. Expansion of the chest creates a pressure gradient that sucks air into the lungs. When the muscles relax, the walls of the thoracic cavity recoil, forcing stale air out of the lungs.

Gas exchange in the lungs and tissues is driven by the diffusion of oxygen and carbon dioxide down their respective concentration gradients.   Blood entering the lungs from the tissues is relatively high in carbon dioxide and low in oxygen. In the lungs, oxygen diffuses from the alveoli into the blood where over 95 percent of the oxygen molecules bind to hemoglobin in the erythrocytes. The blood becomes completely saturated with oxygen, even when we are undergoing strenuous exercise and the blood is moving rapidly through the lung capillaries. When oxygenated blood is pumped through metabolically active tissues, oxygen molecules diffuse out of the plasma, promoting the release of oxygen molecules bound to hemoglobin. The process of gas exchange in the tissues is self-regulating: The more active the tissue, the lower its oxygen concentration and the steeper the oxygen gradient between the blood and that tissue. This relationship favors the release of additional oxygen. More active tissues also contain higher carbon dioxide levels, producing an increased acidity that alters the shape of the hemoglobin molecules, causing the release of additional oxygen molecules.

The rate and depth of our breathing is regulated by our need for oxygen.   Breathing is controlled by the respiratory center located in the medulla of the brainstem. The most important factor involved in regulating breathing is the acidity of the blood, which is proportional to the carbon dioxide concentration. As carbon dioxide levels rise, the pH drops, and impulses are sent to the diaphragm and intercostal muscles, increasing the rate and depth of expansion of the thoracic cavity.

Water breathers and air breathers have different types of respiratory organs.   Water breathers typically possess gills, which consist of delicate, fingerlike projections that extend into the surrounding water. Air breathers typically possess either lungs—inpocketings of the external surface into the body where the respiratory surface can be kept moist—or tracheae—blind passageways through which air diffuses into the tissues. Tracheae are found in arthropods and carry air to the tissues without the intervention of the circulatory system.

Key Terms

respiratory system (p. 622)
gas exchange system (p. 622)
respiratory surface (p. 622)
cutaneous respiration (p. 622)
respiratory tract (p. 623)
lungs (p. 623)
nasal cavity (p. 624)
pharynx (p. 624)
glottis (p. 624)

larynx (p. 624)
epiglottis (p. 624)
vocal cords (p. 624)
trachea (p. 625)
bronchiole (p. 625)
alveoli (p. 625)
thoracic cavity (p. 626)
pleura (p. 626)
diaphragm (p. 626)

hemoglobin (p. 627)
erythrocyte (p. 627)
respiratory center (p. 632)
gill (p. 632)
gill filament (p. 633)
lamellae (p. 633)
countercurrent flow (p. 633)
tracheae (p. 633)
tracheole (p. 633)

Review Questions

1. Distinguish among aerobic respiration, gas exchange in the lungs and tissues, and breathing.

2. Compare the respiratory strategies of the following: a frog, a snail, a cockroach, a fish, a seal, and a human.

3. Describe the functional interconnection between the human circulatory system and the respiratory system.

4. Describe two mechanisms that help match the amount of oxygen released with a tissue's need for oxygen.

5. Distinguish between the larynx and the trachea; oxygen present in the plasma and in erythrocytes; the pH of blood with a high versus a low carbon dioxide concentration; the mechanism of inhalation and exhalation; the composition of air being inhaled versus air being exhaled; alveoli and gill lamellae; hyperventilation and normal breathing.

Critical Thinking Questions

1. Why do you suppose the diving reflex is more adaptive if it occurs as a response to submersion rather than as a response to metabolic changes resulting from oxygen deprivation?

2. The graph below shows the relationship between the surface area of the alveoli of the lungs of various animals and their rate of oxygen consumption. Explain what the graph shows about this relationship. What characteristics of respiratory surfaces are reflected in this relationship?

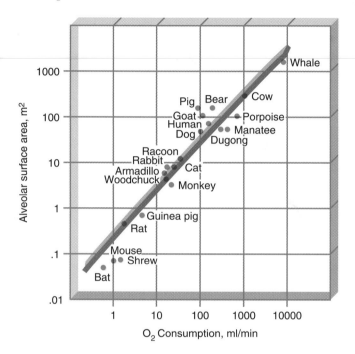

3. Trace the path that a molecule of oxygen would take from the outside atmosphere to a muscle cell in your leg, naming all of the structures through which it would pass.

4. Certain fish living in Antarctic waters have blood that is completely lacking hemoglobin, yet these animals are as active as are related fish that live in warmer waters and have hemoglobin in their blood. Can you suggest any reason that these Antarctic fish can survive without hemoglobin?

5. Suppose you were the subject of a study in which you were exposed to a constant level of carbon dioxide but an increasing level of oxygen. What effect do you think this would have on your rate and depth of breathing? After you have thought about this for a minute, consider the fact that your blood becomes fully saturated with oxygen even when you are exerting yourself and when blood is moving as fast as possible through your lung capillaries.

6. Why is the one-way flow of oxygen-bearing water through the body of a fish better suited for a fish than is the type of movement that occurs in humans, where the oxygen-containing medium is drawn in and out through the same opening?

7. A closed circulatory system would seem to be more efficient than an open circulatory system and, therefore, better adapted to an active way of life. What adaptations enable insects to maintain their high level activity in spite of possessing an open circulatory system?

Additional Readings

American Cancer Society. 1980. *The dangers of smoking; Benefits of quitting.* (Introductory)

Comroe, J. H. 1966. The lung. *Sci. Amer.* Feb:56-71. (Intermediate)

Douville, J. A. 1990. *Active and passive smoking hazards in the workplace.* New York: Van Nostrand Reinhold. (Intermediate)

Powers, S. K., and E. T. Howley. 1990. *Exercise physiology.* Dubuque, Iowa: W. C. Brown. (Intermediate-Advanced)

Sataloff, R.T. 1992. The human voice. *Sci. Amer.* Dec:108–115. (Intermediate)

Scholander, P. F. 1957. The wonderful net. *Sci. Amer.* April:96-110. (Intermediate)

West, J. 1989. *Respiratory physiology: The essentials,* 4th ed. Baltimore: Williams & Wilkins. (Intermediate)

Zapol, W. M. 1987. Diving adaptations of the Weddell seal. *Sci. Amer.* June:100-107. (Intermediate)

Internal Defense:
The Immune System

STEPS
TO
DISCOVERY
On the Trail of a Killer: Tracking the AIDS Virus

NONSPECIFIC MECHANISMS: A FIRST LINE OF DEFENSE

Cellular Nonspecific Defenses

Molecular Nonspecific Defenses

THE IMMUNE SYSTEM: MEDIATOR OF SPECIFIC MECHANISMS OF DEFENSE

The Nature of the Immune System

ANTIBODY MOLECULES: STRUCTURE AND FORMATION

Antibody Specificity: A Unique Combination of Polypeptides

Antibody Formation: Discovery of a Unique Genetic Mechanism

IMMUNIZATION

EVOLUTION OF THE IMMUNE SYSTEM

BIOLINE
Treatment of Cancer with Immunotherapy

THE HUMAN PERSPECTIVE
Disorders of the Human Immune System

S TEPS TO DISCOVER Y

On the Trail of a Killer: Tracking the AIDS Virus

*I*n Fall 1980, a resident at the UCLA Medical Center was visited by a 31-year-old male who was suffering from a persistent fever, weight loss, swollen lymph nodes, and a severe yeast infection in his mouth and throat. This latter condition, known as *candidiasis,* is sometimes observed in patients undergoing chemotherapy during cancer treatment or in babies who are born with an immune system deficiency.

The patient came to the attention of Michael Gottlieb, an immunologist who had recently arrived at UCLA. Gottlieb obtained a sample of the patient's lung fluid and blood. The lung sample revealed the presence of a protozoan,

Pneumocystis carinii, that was known to cause a rare type of pneumonia (called PCP), usually in people suffering from lymphoid cancers. In other words, this patient was suffering from two rare types of infections simultaneously. In a healthy person, pathogenic organisms, such as these yeast and protozoa, are readily attacked and eliminated by certain white blood cells, called T cells. This particular patient lacked an entire class of T cells, called helper T cells, leaving Gottlieb and his colleagues dumbfounded.

In February 1981, Gottlieb was confronted by another patient suffering the same combination of infections and depleted content of helper T cells. Like the previous pa-

AIDS can affect anyone. The disease is caused by a virus that infects T-cells, and reproduces by budding from the infected

tient, this young man was also homosexual. By April, four similar cases had come to Gottlieb's attention, all of which involved male homosexuals. Gottlieb wrote a brief report that was submitted to the weekly newsletter of the U.S. Centers for Disease Control (CDC). This was the first publication concerning the disease that would soon be given the name Acquired Immune Deficiency Syndrome, or AIDS.

One of the physicians who received a copy of the newsletter was Willy Rozenbaum of the Claude Bernard Hospital in Paris. Three years earlier, Rozenbaum had treated a man suffering from a combination of rare diseases, including PCP. The man had recently spent several years in Africa. The next year, two African women were referred to Rozenbaum, both suffering from PCP. By the time Gottlieb's paper was published, all three of Rozenbaum's patients had died from their infections. Rozenbaum had little doubt that the disease plaguing homosexual men in Los Angeles (and simultaneously in New York and San Francisco) was the same malady that had killed his patients from Africa. Furthermore, the fact that the disease was appearing in very different populations on two sides of the world made it very unlikely that the disease was caused by some kind of toxic substance in the environment. Rather, Rozenbaum concluded that the affliction was the result of an infectious agent. Several years had passed between the visit of the African victims and the first patients to exhibit the disease in the United States, suggesting that the disease had been in the human population for several years and, furthermore, that the disease had a long *incubation period:* the time between the infection and the appearance of symptoms. The implications of this finding were frightening: The disease could take hold within a population long before any evidence of its presence would be detected.

Since the African cases of the disease predated those in the United States, it seemed likely that the disease had originated in Africa and had been transported to the United States by an infected male homosexual, who unwittingly passed it into the homosexual community. Soon, the disease began to appear in individuals who had received blood transfusions or blood products, most notably hemophiliacs, and intravenous-drug users who shared hypodermic needles.

As more cases were reported, Don Francis, a scientist at the CDC, began to consider the type of agent responsible for the new plague. Francis had received his doctorate working on a feline leukemia that was caused by a virus that ultimately destroyed the animal's immune system. Like this new human disease, feline leukemia resulted in the animal's contracting a host of unusual parasitic infections and was characterized by a long incubation period. It had been determined that feline leukemia was due to a *retrovirus*—a type of virus whose genetic material is RNA, which is transcribed into a molecule of DNA that is inserted into the host cell's chromosomal DNA. Just 1 year earlier, in 1980, a team led by Robert Gallo of the National Institutes of Health had shown that a retrovirus could cause a human disease—a rare type of leukemia that affected a person's T cells. Taken together, these strands of evidence pointed to a retrovirus as the causative agent of this new disease.

In Spring 1982, Gallo's lab began culturing lymphocytes from patients with AIDS. Meanwhile, another expert on retroviruses, Luc Montagnier, of the Pasteur Institute in Paris, was conducting similar research, searching for a retrovirus in infected lymphocytes. In 1983, both the U.S. and French teams reported the isolation of the retrovirus responsible for AIDS, setting the stage for one of the major scientific controversies of the decade: Which lab should be credited with the discovery of the AIDS virus, later called Human Immunodeficiency Virus (HIV)? For a number of reasons, including the fact that Gallo's virus appears to have been isolated from cells donated by the French scientists, Luc Montagnier is credited with the discovery of HIV.

cell surface.

As you read this sentence, you are being attacked. You are, in fact, fighting for your life. The surface of your body is populated by billions of microorganisms: bacteria and other agents that would invade and use your tissues as their next meal if allowed uncontested entry into your body. With every breath you take, you inhale more than a million microorganisms that are suspended in air, many of which would cause fatal infections if you didn't have an arsenal of weapons to protect your respiratory tract against their invasion. Some microorganisms have already entered your bloodstream, perhaps through cuts in your skin so small you were unaware of them. In your internal tissues and fluids, a microorganism has found a bonanza of nutrients and hospitable environmental conditions. The onslaught never stops.

Repelling this invasion is just as important in maintaining the stability of your internal environment as is the removal of waste products or the gain or loss of heat. The system responsible for this aspect of homeostasis is the **immune system;** it is capable of recognizing breaches in security and effectively eliminating invading microorganisms before they can destroy the biological order required for life to continue. The homeostatic activities that protect humans and other vertebrates from foreign agents are divided into two broad categories: nonspecific mechanisms and specific mechanisms. These two types of defenses are distinguished by the types of cells involved and by whether or not the specific identity of the target is recognized before it is attacked.

▼ ▼ ▼

NONSPECIFIC MECHANISMS: A FIRST LINE OF DEFENSE

The body's most evident protective strategy is to keep viruses, bacteria, and parasites from penetrating the living tissue. As we discussed in Chapter 26, the skin forms a

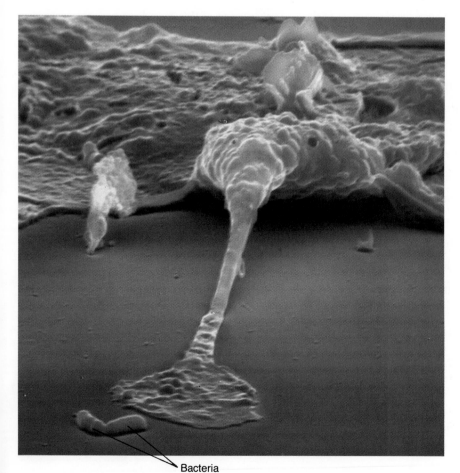

Bacteria

FIGURE 30-1

Death Grip. Among your many weapons against disease are phagocytic cells, which engulf and destroy foreign microbes, such as these doomed bacteria in the clutches of this protective white blood cell.

relatively impregnable outer body layer, as long as it remains undamaged. The mucous membranes that line the respiratory, digestive, and urinary tracts are less formidable barriers than the skin, but they are protected by a sticky layer of mucus, which is continually shed and replenished. The flushing of your urinary tract each time urine is voided is another nonspecific defense.

CELLULAR NONSPECIFIC DEFENSES

If a bacterium or other pathogenic microorganism breaches the body surface and reaches the tissues of the body, it will likely encounter a phagocytic cell that possesses the capacity to ingest and destroy it (see Figure 30-1). These cells are carried to the infected tissues through the blood and lymphatic vessels (Figure 30-2).

Not all potentially dangerous cells arrive from outside the body, however. Cancer cells are spawned within the body and have the same potential for human destruction as do external invading organisms. The transformation of a normal cell into a cancer cell is often accompanied by a change at the cell's surface, which may include the appearance of new or modified membrane proteins. The body contains nonspecific, lymphocyte-like cells, called **natural killer (NK) cells,** which can recognize cells having an altered surface, including cancer cells, and kill them. The precise mechanism by which these cells kill their targets is unclear, but it does not involve phagocytosis.

Inflammation

As important as phagocytic cells are to defending the body, these cells provide little security unless they can be called into the area where protection is needed. The body's overall "strategy" for attracting these and other protective cells to sites of danger, such as an infected wound, is called **inflammation.** Inflammation is initiated when chemicals from cells in the injured tissue are released, causing local blood vessels to dilate, bringing additional blood into the affected region. These chemicals also attract phagocytic leukocytes, such as neutrophils, which escape the vessels by squeezing through the spaces in the vessel wall (Figure 30-2). Protective leukocytes accumulate in the injured or infected region, often forming a yellowish liquid, called *pus*.

The accumulation of fluid and cells in the inflamed area causes painful swelling of the tissues due to stimulation of local neurons. The area grows hot and red from the additional blood. Inflammation creates many of the uncomfortable symptoms we often associate with disease rather than healing. Consequently, inflammation is usually targeted as a problem rather than an ally. Use of drugs to subdue inflammation (such as cortisone) may provide immediate relief from symptoms, but in doing so it suppresses this line of defense.

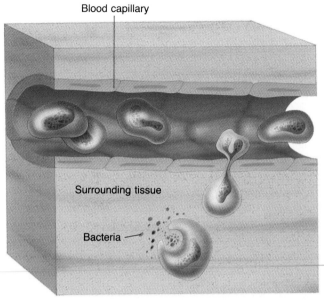

Blood capillary

Surrounding tissue

Bacteria

FIGURE 30-2

Phagocytic leukocytes are able to squeeze through openings between the cells that line the capillaries and enter an inflamed area, where they can carry out phagocytosis.

MOLECULAR NONSPECIFIC DEFENSES

Several *molecular* mechanisms also operate to destroy microbes within the body. Blood contains a group of proteins collectively called **complement.** One of the complement proteins binds to the surface of a bacterium, initiating a series of reactions that poke holes in the plasma membrane of the foreign cell. Fluids and salts can then enter the cell, causing it to burst and die.

The progress of viral infections is blocked by a different type of nonspecific mechanism, spearheaded by the protein *interferon*. Suppose a virus penetrates into the respiratory membranes and infects a cell. Once the virus has entered the cell, the cell responds by secreting interferon into the extracellular fluid. The interferon molecules bind to the surfaces of uninfected, neighboring cells and initiate a series of reactions that block the neighboring cells' ability to manufacture viral proteins. This renders cells incapable of supporting viral reproduction and halts the spread of the infection.

The presence of a foreign cell within the body immediately attracts a nonspecific arsenal of weapons that includes phagocytic cells, blood-borne proteins, and secreted molecules that can attack and destroy the invader. (See CTQ #2.)

THE IMMUNE SYSTEM MEDIATOR OF SPECIFIC MECHANISMS OF DEFENSE

Edward Jenner, an eighteenth-century English physician, was interested in the prevention of smallpox, a deadly and widespread disease characterized by the development of elevated blisters filled with pus. Jenner practiced in the English countryside, where he noticed that the maids who tended the cows were typically spared the ravages of smallpox. Jenner concluded that the milkmaids were somehow "immune" to smallpox because they were infected at an early age with cowpox, a harmless disease they contracted from their cows. Cowpox produces blisters that resemble the pus-filled blisters of smallpox, but the cowpox blisters are localized and disappear, causing nothing more serious than a scar at the site of infection.

In 1796, Jenner performed one of the most famous (and risky) medical experiments of all time. First, he infected an 8-year-old boy with cowpox and gave the boy time to recover. Six weeks later—in an experiment that would be considered unethical today—Jenner intentionally infected the boy with smallpox by injecting pus from a smallpox lesion directly under the boy's skin. The boy remained healthy. Within a few years, thousands of people had become immune to smallpox by intentionally infecting themselves with cowpox. This procedure was termed *vaccination,* after "vacca," the Latin word for cow. Under the leadership of the World Health Organization, vaccination against smallpox has eradicated this once dreaded disease.

You may be wondering how a previous infection with cowpox protects people against a much more serious illness. Infection with cowpox stimulates cells of the immune system to produce **antibodies**—proteins that bind to foreign molecules in a highly specific manner. Any foreign substance that elicits production of an immune response (such as antibody production) is called an **antigen.** Both cowpox and smallpox viruses contain proteins in their outer capsu'‚ (Chapter 36) that act as antigens when they enter the body, evoking antibodies that react with these viral proteins. The proteins present in the cowpox and smallpox viruses have a very similar molecular structure—so similar, in fact, that the antibodies produced by the body's cells against the cowpox virus also react and neutralize the smallpox virus. Once infected with cowpox, these antibodies can be produced rapidly at any time during a person's life, protecting the individual against the subsequent development of smallpox.

THE NATURE OF THE IMMUNE SYSTEM

The immune system works by assembling an arsenal of protective cells and molecules that search out a particular

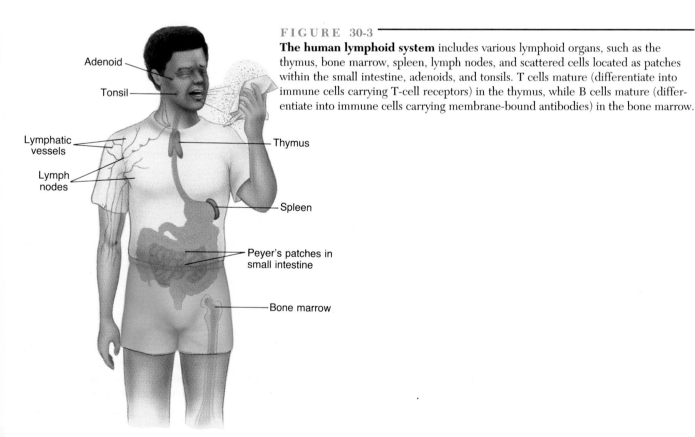

FIGURE 30-3

The human lymphoid system includes various lymphoid organs, such as the thymus, bone marrow, spleen, lymph nodes, and scattered cells located as patches within the small intestine, adenoids, and tonsils. T cells mature (differentiate into immune cells carrying T-cell receptors) in the thymus, while B cells mature (differentiate into immune cells carrying membrane-bound antibodies) in the bone marrow.

Adenoid

Tonsil

Lymphatic vessels

Lymph nodes

Thymus

Spleen

Peyer's patches in small intestine

Bone marrow

antigen and destroy it. The fact that the immune system is not always successful in its endeavors is all too evident by the AIDS virus, which is capable of infecting the very cells that have evolved to destroy such viruses.

The immune system is composed of cells that are scattered throughout the body and particularly concentrated in the lymphoid tissues, which include the thymus, spleen, lymph nodes, bone marrow, and tonsils (Figure 30-3). The most prominent cells of the immune system are lymphocytes, which circulate throughout the blood and lymph. Lymphocytes are aided by *macrophages,* large phagocytic cells that are derived from monocytes (see Figure 28-17*b*). As lymphocytes and macrophages travel throughout the body, they come into contact with foreign materials, which initiates an *immune response.*

The body's immunological arsenal is composed of two major types of lymphocytes: *B lymphocytes* (or simply **B cells**) and *T lymphocytes* (**T cells**). Both types of lymphocytes work in cooperation with each other and with macrophages, but they carry out very different immune responses.

B-Cell Immunity: Protection by Soluble Antibodies

When you come down with a cold or a throat infection, your immune system responds by producing specific antibody molecules that dissolve in the blood, where they circulate throughout the body and bind to the invading pathogenic agents. Once a virus or bacterial cell is covered with these antibody molecules, the pathogen may be inactivated or may become more susceptible to ingestion by a patrolling phagocyte. These soluble, blood-borne antibody molecules are proteins that are secreted by **plasma cells.** Plasma cells are formed from B lymphocytes present within the bone marrow by the process illustrated in Figure 30-4. As we will see below, the B cell recognizes the foreign agent (which acts as an antigen) and responds by transforming into a plasma cell that secretes antibody molecules that can combine with that agent.

Some bacteria (such as those responsible for typhus or meningitis) and all viruses (such as those responsible for colds, measles, or mumps) enter a host cell, reproduce inside the cell, then kill the cell, releasing more bacteria or viruses into the extracellular fluid. It is when these pathogens are *outside* the host cells that they are vulnerable to direct attack by soluble antibodies that are distributed in the body's fluids: tears, milk, nasal and intestinal mucus, interstitial fluid, and blood. In contrast, since soluble antibodies cannot penetrate infected cells, they have little effect on a pathogen while it is inside a host cell. A pathogen growing *inside* the body's cells is eliminated by the other branch of the immune system, T-cell immunity.

T-Cell Immunity: Protection by Intact Cells

While B cells carry out their immune function by secreting antibodies, T cells function by interacting directly with

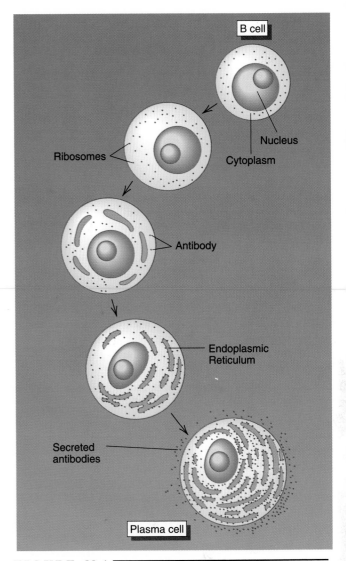

FIGURE 30-4

Differentiation of antibody-secreting plasma cells from B cells. The differentiated plasma cell contains extensive rough ER where antibody proteins are synthesized.

other cells. In other words, T cells function in **cell-mediated immunity.** T cells are able to recognize infected or abnormal target cells by virtue of the antibody-like proteins the T cells carry embedded in their plasma membranes; these membrane proteins are called *T cell receptors.*

There are three functionally distinct subclasses of T cells:

1. **Cytotoxic (killer) T cells** (Figure 30-5) *recognize aged, malignant, or infected cells and release perforins* —proteins that produce holes in the membrane of the target cell, causing its death. By killing infected cells, cytotoxic T cells can eliminate viruses, and intracellular bacteria, yeast, protozoa, and parasites, even after they enter a host cell. Cytotoxic T cells may also play a role in

◁ B I O L I N E ▷
Treatment of Cancer with Immunotherapy

Cancer is currently being treated by three approaches: surgery, radiation, and chemotherapy. Surgery physically removes as much of the malignancy as possible; sometimes all of it. Radiation kills cancer cells in the specific part of the body that contains the malignancy. Chemotherapy poisons cancer cells wherever they might have spread throughout the body. Some forms of cancer, such as childhood leukemias and Hodgkin's disease, have responded well to these treatments, while others, such as lung cancer and pancreatic cancer, have not.

Both radiation and chemotherapy work primarily by killing cells as they divide. Because of the higher rate of division of cancer cells compared to normal cells, cancer cells are more susceptible to the killing action of radiation and chemotherapy. But many normal cells are also killed, and side effects such as nausea and anemia can be debilitating. For several decades, researchers have looked for ways to manipulate the immune system to help in the fight against cancer. Cancer cells often exhibit new proteins at their cell surfaces, which can be recognized by antibodies as foreign antigens. A number of laboratories have been trying to take advantage of the presence of these so-called *tumor-specific antigens* to produce antibodies that can be injected into patients and bring about the death of the tumor cells. While the prospects of antibody therapies have proven disappointing, a new approach has emerged that may prove more promising. This new approach has been pioneered by Stephen Rosenberg of the National Institutes of Health.

Rosenberg's interest in the subject began in 1968 when, as a young surgeon, he removed the gallbladder from a 63-year-old man. It was a routine operation, except for the fact that 12 years earlier this same patient had been diagnosed with ad-

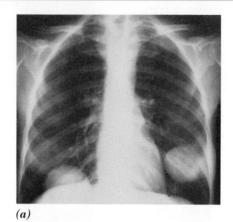

(a)

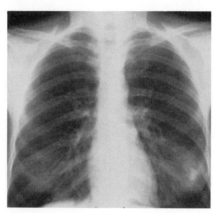

(b)

FIGURE 1 ─────────

One promise of success against cancer utilizes tumor-infiltrating leukocytes (TILs). *(a)* X-ray showing a melanoma that has spread to both lungs. *(b)* Two months after treatment with the patient's own TILs, the tumor masses are greatly reduced in size.

vanced stomach cancer and sent home to die. Within a few months, the cancer had spontaneously disappeared, as had the cancer cells that had spread to other parts of the man's body. Such rare cases of spon-

taneous remission convinced Rosenberg that the immune system has the *potential* to rid the body of its malignant cells.

In the late 1980s, Rosenberg and his colleagues found that human tumors contained cytotoxic T cells that could specifically attack the tumor cells. Under normal circumstances, these T cells are present in too small a quantity and arrive too late to stem the tide of the growing malignancy. Rosenberg devised a plan. What if he were to remove the tumor, isolate the cytotoxic T cells contained in the tumor mass, culture these cells to increase their number, and then reinject them into the same patient? Would these T cells (called **tumor-infiltrating lymphocytes,** or **TILs**) become concentrated in the tumor masses from which they were derived? Would the injected TILs kill the tumor cells? Both of these questions have been answered with an encouraging "yes." When the injected T cells are specially labeled for tracking throughout the body, they are found to accumulate within the tumor. In the first clinical trials on patients with advanced cancer, over half of the treated patients experienced a marked reduction in their tumors (Figure 1); several appeared to have remained in partial or complete remission for years. Moreover, the TILs had no adverse effects on normal tissues, a unique benefit over conventional treatments.

Rosenberg has recently received permission to modify TILs in a way that gives the cells extra killing power. This is done by genetically engineering the cells so that they carry a gene for a protein that is highly toxic to tumor cells. It is hoped that the TILs will be selectively drawn to the tumor and, by delivering the toxic protein, will kill additional tumor cells. If the technique proves effective, within just a few years we may be able to use immunotherapy as a fourth weapon in the war against cancer.

FIGURE 30-5

This large cancer cell is being attacked by numerous, smaller killer T cells. When specific contact is made between the two cells, the T cell releases a protein that kills the target cancer cell by poking holes in its plasma membrane.

inside the sternum. The thymus is quite large in mammals during the period from late fetal life through puberty, after which it gradually shrinks in size and importance. The process by which a T cell matures and produces a particular version of the T-cell receptor occurs within the thymus gland. If the thymus gland fails to develop, T-cell maturation is blocked, and all the various cell-mediated defensive mechanisms fail to appear.

The role of T cells first came to light during studies in the 1960s on mice whose thymus glands had been removed soon after birth. Such mice showed a variety of immunological deficiencies, including the inability to reject tissue and organ grafts from other individuals, which would normally occur. This finding provided one of the first indications that T cells are capable of killing healthy *foreign* cells, which is the basis for the rejection of transplanted human organs, which we will discuss shortly.

More recently, a strain of mice has been developed with a genetic deficiency that leads to the congenital absence of a thymus. These *nude mice* (Figure 30-6), as they are called, are currently playing an important role in cancer research because they accept grafts of malignant human tissue that other mice would reject. Researchers are able to test experimental treatments to stop tumor growth on these mice, which could not be used on human patients. Even more remarkably, when nude mice are given lymphoid tissue that is taken from a human fetus, they actually develop a functioning "human" immune system, complete with both T and B lymphocytes. These "human mice" are currently

destroying cancer cells (see Bioline: Treatment of Cancer with Immunotherapy).

2. **Helper T cells** are regulatory cells, not killers. They regulate immune responses by recognizing and activating other lymphocytes, including both B cells and cytotoxic T cells. Activation is achieved as the helper T cell releases substances that activate the target cell. The best-studied activator substance is *interleukin II*, which is currently being produced by recombinant DNA technology and tested as a treatment for certain types of cancer. Helper T cells are the cells that are hardest hit by AIDS. Ironically, HIV, the virus responsible for AIDS, gains entry to the T cell by binding to the T-cell receptor, the membrane protein normally responsible for detecting the presence of virally infected cells.

3. **Suppressor T cells** also regulate immune responses, but they do so by *inhibiting* the activation of other lymphocytes by a mechanism that remains unknown.

T cells are named for the **thymus gland**—a mass of lymphoid tissue that is situated in the chest cavity, just

FIGURE 30-6

A nude mouse. These homely mice are homozygous for a mutation that leads to the absence of a thymus gland. Since the thymus gland is the site where T cells undergo the process that provides them with a specific T-cell receptor, these animals are lacking immunologically competent T cells; thus, they are unable to reject foreign grafts.

being used as an animal model for AIDS research, since their donated immune systems are susceptible to HIV.

Organ Transplants and Graft Rejection The greatest hurdle that medical research has had to overcome in the transplantation of organs from one person to another is *graft rejection*, which occurs when cells of the transplanted organ are recognized as foreign and are subsequently attacked by the recipient's cytotoxic T cells. Currently, the likelihood of graft rejection is minimized by two factors.

- The first strategy involves matching the donor to the recipient; that is, finding a donor whose cell-surface proteins (called *histocompatibility antigens*) are as closely matched as possible to the recipient. The closer the match, the less likely the rejection. The risk is always present, however, since no two people, other than identical twins, have identical histocompatibility antigens.

- The second strategy requires treating the recipient with drugs (such as the fungal compound cyclosporine) that suppress the person's cell-mediated immunity, thereby reducing the capacity for graft rejection. Even though these drugs are now quite effective, the suppression of the immune system leaves the recipient more vulnerable to infection.

The immune system consists of scattered organs that produce lymphocytes—cells that recognize foreign materials (antigens) and mount a specific response. The immune response may be mediated by soluble antibodies that bind to antigens or by activated T cells that bind to infected cells or to cells with altered surfaces. (See CTQ #3.)

ANTIBODY MOLECULES: STRUCTURE AND FORMATION

B cells are stimulated to produce antibodies by the presence of antigens; that is, by substances that are recognized as foreign. Most antigens are proteins or polysaccharides, such as those present on the surfaces of viruses, bacteria, and parasites. Immune responses are always specific; that is, antigens stimulate production of only those antibodies that can specifically combine with that particular antigen. If you are exposed to the measles virus, for example, the presence of that virus elicits the production of antibodies that will combine only with the measles virus protein (or one with a very similar structure). Therefore, immunity to measles provides no protection against polio, colds, influenza, or any infection other than measles.

You possess the immune capacity to respond to nearly every foreign intruder you will encounter during your lifetime. It is estimated that humans can produce millions of different types of antibodies, which are capable of combin-ing with virtually any molecule of virtually any shape. To appreciate how antibodies possess such specificity and why there are so many different types of antibodies, it is necessary to examine their structure and formation.

ANTIBODY SPECIFICITY: A UNIQUE COMBINATION OF POLYPEPTIDES

Remarkably, all of the millions of different antibodies your body can produce, with their ability to combine with diverse types of antigens, are constructed according to a similar blueprint, using very similar building materials. All antibody molecules (also called *immunoglobulins*) are composed of two basic types of polypeptide chains, called **heavy chains** and **light chains,** based on their relative sizes. The most common class of immunoglobulins, the IgG (immunoglobulin G) class, is composed of two light chains and two heavy chains, linked together by covalent bonds (Figure 30-7). Early studies of a number of different IgG molecules revealed a consistent pattern: Approximately one half of each light chain was *constant* (C) in amino acid sequence among various antibodies, while the other half was *variable* (V) from antibody to antibody. The heavy chains of these antibodies also contained a variable (V) and a constant (C) portion.

Once the basic structure of a number of IgG molecules was compared, the mechanism of antibody function became evident. Each IgG molecule has two antigen-binding sites located at the ends of a Y-shaped molecule. Each antigen-binding site is formed by the association of the variable portion of a light chain with the variable portion of a heavy chain (Figure 30-7). The amino acids that make up the antigen-binding sites give that portion of the protein a three-dimensional structure that is complementary to that of the corresponding antigen (Figure 30-7). Different antibodies contain different light and heavy chains and thus have different shapes and specificities. Since antibody molecules are made up of two different types of polypeptides, an individual is able to produce a tremendous variety of antibodies from a relatively modest number of different heavy and light chains. If, for example, there are 1,000 different light chains and 1,000 different heavy chains, 1 million (1,000 × 1,000) different antibodies—each with a distinct antigen binding site—can be formed.

ANTIBODY FORMATION: DISCOVERY OF A UNIQUE GENETIC MECHANISM

The discovery that antibodies are composed of constant and variable portions raised several basic questions. We can each produce millions of different IgG molecules, all of which have light (or heavy) chains with identical C portions but highly diverse V portions. But just how can different polypeptides have an identical amino acid sequence in one part, and diverse amino acid sequences in another part?

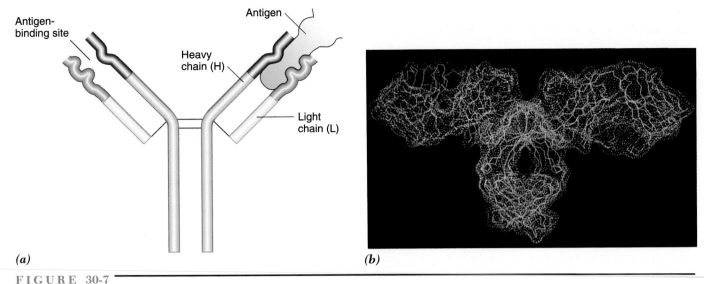

(a)   *(b)*

F I G U R E   30-7

Structure of an IgG immunoglobulin molecule and its interaction with antigen. *(a)* Each IgG antibody molecule consists of two light chains and two heavy chains, each of which has a constant and a variable region. The chains are held together by covalent (—S—S—) bonds. The antigen-combining sites are present at the forked ends of the Y-shaped molecule. Each combining site is made up of the variable (green) portion of both a heavy and a light chain. In this drawing, one site has combined with a complementary-shaped antigen; the other is unbound. *(b)* A computerized, three-dimensional drawing of an IgG molecule.

In 1965, William Dreyer of the California Institute of Technology, and J. C. Bennett of the University of Alabama put forward the "two gene–one polypeptide" hypothesis to account for antibody structure. In essence, Dreyer and Bennett proposed that each antibody chain is coded by two separate genes—a C gene and a V gene—that somehow combine to form one continuous "gene" that codes for a single light or heavy chain. But there was no evidence that such DNA rearrangement was even *possible,* much less that it occurred.

Over a decade later, in 1976, at a Swiss research institute, Susumu Tonegawa provided clear evidence in favor of the DNA rearrangement hypothesis. Using newly developed techniques, Tonegawa and his colleagues measured the distance in the DNA between the nucleotide sequences coding for the C and V portions of a particular antibody chain. They compared this distance in the DNA isolated from two different types of mouse cells, those of embryos and those of antibody-secreting plasma cells. The group found that, while these two DNA segments were widely separated from each other in DNA obtained from embryos, the segments were very close to each other in the DNA obtained from the plasma cells. Tonegawa had shown that parts of the DNA actually rearranged themselves during the formation of antibody-producing cells, earning him a Nobel Prize in 1987.

Later research revealed that the DNA of a particular chromosome contained a single C gene and a large number of different V genes. During the rearrangement process, which occurred while the B cell was maturing in the bone marrow, that single C gene was moved very close to one of the V genes, enabling a single messenger RNA (mRNA) to form from these combined genes (Figure 30-8). The mRNA was then translated into a single polypeptide, containing both a C and a V amino acid sequence. This mechanism of DNA rearrangement is unique to antibody-forming cells and represents an adaptation by which an individual can produce a large number of different antibodies with a minimal amount of genetic information.

The Clonal Selection Theory of Antibody Formation

When you receive a vaccination that contains an inactivated microorganism or you contract a bacterial or viral infection, your body reacts to the presence of the foreign agent by producing a restricted population of antibody molecules, all of which are able to combine with this particular foreign antigen. How does the body know which antibodies to manufacture in response to the presence of a particular antigen?

In 1957, a far-sighted, Nobel Prize-winning proposal was made by an Australian immunologist, Macfarlane Burnet, to answer this question. Burnett's *clonal selection*

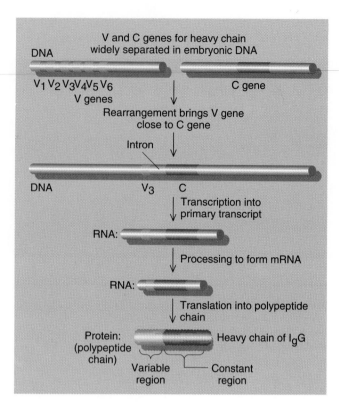

FIGURE 30-8

Antibody genes are formed by DNA rearrangement, which brings the C gene into close proximity with one of the many V genes (in this case, the V_3 DNA segment encoding part of a heavy chain) on the chromosome. The region between the C and V genes after rearrangement represents an intervening sequence (intron). Once DNA rearrangement has occurred, the combined C and V regions are transcribed into one mRNA molecule that codes for a single, combined polypeptide.

theory has gained virtually complete acceptance and remains compatible with all the experimental evidence accumulated since its original proposal. The principles of the clonal selection theory in its present, updated form bring together many different aspects of immune function and are summarized in Figure 30-9. According to this theory, the following events take place during antibody formation.

1. *Each B cell becomes committed to producing one particular antibody.* During embryonic development, B cells are formed from a population of undifferentiated, indistinguishable stem cells. As one of these stem cells becomes a B cell, however, DNA rearrangements occur (see Figure 30-8) which commit the cell to the production of only one particular antibody molecule. Even though a population of B cells remains identical under the microscope, the cells are distinguished by the antibodies they can produce.

2. *B cells become committed to antibody formation in the absence of antigen.* The entire diversity of antibody-producing cells an individual will ever possess is already present within the lymphoid tissues *prior to stimulation by an antigen* (Figure 30-9a) and is independent of the presence of foreign materials. Each B cell "displays" its particular antibody on its surface, with the antigen-reactive portion facing outward; the cell is literally coated with receptors that can fit with one and only one antigen (Figure 30-9a). While most of the cells of the lymphoid tissues will never be called on

to respond during a person's lifetime, the immune system is primed to respond immediately to virtually any type of antigen to which an individual might be exposed.

3. *Antibody production follows selection of B cells by antigen.* Antigens that enter the body trigger the production of complementary antibodies by *selecting* the appropriate antibody-producing cell. When a foreign object, such as a pneumococcus bacterium or the particulate debris from tobacco smoke, enters the body, it is ingested by a phagocytic macrophage. The foreign material is only partially digested, however, generating fragments that make their way to the surface of the macrophage (Figure 30-9b), where they are "presented" to a B cell (Figure 30-9c). This process is called **antigen presentation.** B cells with the appropriate "sample" antibodies at their surface will bind to the antigen on the surface of the antigen-presenting macrophage (Figure 30-10). The binding between the two cells activates the B cell to divide (Figure 30-9d), forming a population, or *clone,* of lymphocytes, all of which are specialized for making the same antibody. Some of these activated cells will then differentiate into plasma cells and begin secreting tens of thousands of antibody molecules per minute.

4. *Immunological memory provides long-term immunity.* Not all of the B lymphocytes that are acti-

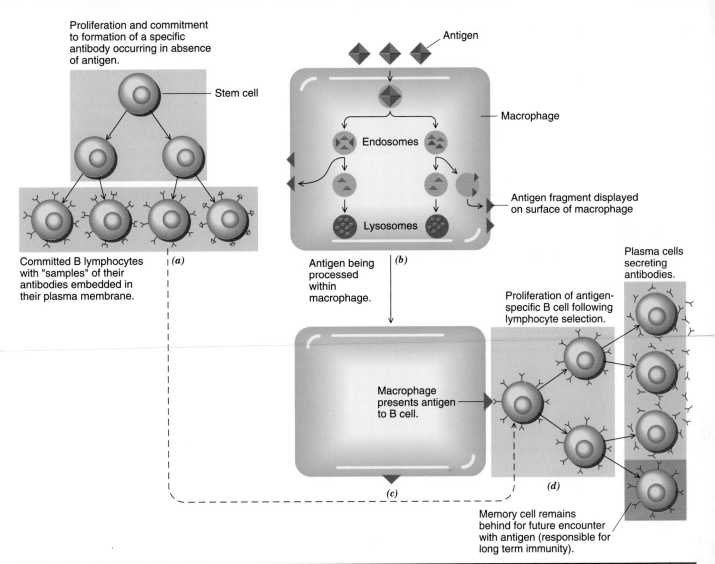

Proliferation and commitment to formation of a specific antibody occurring in absence of antigen.

Stem cell

Antigen

Macrophage

Endosomes

Antigen fragment displayed on surface of macrophage

Lysosomes

Committed B lymphocytes with "samples" of their antibodies embedded in their plasma membrane.

(a)

Antigen being processed within macrophage.

(b)

Plasma cells secreting antibodies.

Proliferation of antigen-specific B cell following lymphocyte selection.

Macrophage presents antigen to B cell.

(c)

(d)

Memory cell remains behind for future encounter with antigen (responsible for long term immunity).

FIGURE 30-9

A modern version of the clonal selection theory. *(a)* Undifferentiated stem cells undergo proliferation, forming a population of B cells that become committed to the formation of a specific antibody. Commitment occurs as a result of DNA rearrangement. Once committed, the B cell carries "sample" antibodies embedded in its plasma membrane. This phase of B-cell development is antigen independent. *(b)* Foreign materials are ingested by wandering macrophages and are packaged into cytoplasmic vesicles (endosomes). Some of the foreign material is completely digested by lysosomal enzymes (the red triangles), while other pieces remain undigested and are moved to the cell surface (the blue triangles), where they can be presented to lymphocytes. *(c)* The activation of a lymphocyte, whose sample antibodies have combining sites that can bind to the processed antigen on the macrophage surface. B cell activation usually requires the participation of "helper" T cells (not shown), which release the soluble factors that stimulate growth and differentiation into antibody producing cells. *(d)* Proliferation of those lymphocytes that have been selected by the antigen form a clone of cells committed to produce antibodies capable of binding to the antigen. Most of these cells go on to differentiate into antibody-secreting plasma cells, but a percentage remain as committed memory cells, which are capable of responding very rapidly should the antigen be reintroduced at a later date.

vated by an antigen differentiate into antibody-secreting plasma cells. Some remain in the lymphoid tissues as **memory cells** (Figure 30-9d), which can respond rapidly at a later date if the antigen reappears in the body. Although plasma cells die off following removal of the antigenic stimulus, memory cells persist, often for years, providing *immunological recall.* When stimulated by the same antigen, some of the memory cells

rapidly proliferate into plasma cells, generating a secondary immune response in a matter of hours rather than the days required for the original response. Thanks to this secondary response, measles, mumps, chickenpox, and many other diseases are suffered only once during a lifetime. The first encounter with the intruder leaves survivors immune to subsequent infection. This type of protection is called **active immunity.**

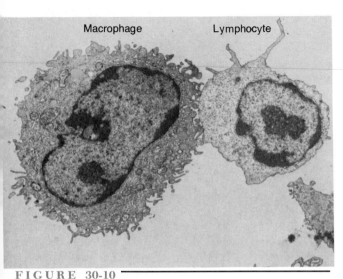

FIGURE 30-10
Interaction between a lymphocyte and macrophage during antigen presentation.

Recognition and Protection of "Self"

Maintaining homeostasis and biological order requires that the immune system be able to distinguish between those substances that have entered the body from the external world and those that "belong" in the body. That is, the system must recognize "self" and refrain from producing antibodies against it, while also recognizing foreign substances ("nonself") and launching an immunological attack against them.

The process by which the immune system becomes *tolerant* of its own tissues is not well understood. During the period of B- or T-cell maturation, if a cell is produced that has antibodies or T-cell receptors capable of reacting with the body's own tissues, that cell is somehow either killed or converted to a nonresponsive state. This process, known as *clonal deletion*, ensures that B cells capable of producing potentially dangerous **autoantibodies**—antibodies against one's own tissues—are rendered incapacitated (see The Human Perspective: Disorders of the Human Immune System). Similarly, formation of cytotoxic T cells capable of interacting with and killing normal, healthy cells is also blocked.

Monoclonal Antibodies: Making Antibodies Available for Use

Antibodies can be very useful proteins. Since they interact so specifically, antibodies can be used to identify or locate particular molecules. For example, the development of prostate cancer in men is associated with the elevation in the blood of a particular protein, called *prostate specific antigen* (PSA), that is shed by prostatic cells. A diagnostic test for prostate cancer measures the level of PSA in the blood by its interaction with a specific antibody. Similar tests are currently being developed for other cancers, including breast cancer.

In earlier decades, antibodies were obtained by injecting a purified substance into an animal, such as a rabbit or goat, and then obtaining samples of the animal's blood, which contained antibodies against the injected, foreign antigen. This approach has severe limitations, however. Today, antibodies are being produced by special cultured cells, called *hybridomas*. Hybridomas are formed by the fusion of two cells: a B cell that is capable of producing a particular antibody, and a malignant cell. Unlike B cells, which have a limited lifespan, hybridomas divide indefinitely in culture, producing an unlimited supply of antibody-producing cells. All of the cells formed from a single fused B cell synthesize and secrete the same antibody molecule into the medium. Different antibodies are obtained by starting with different B cells. Since, in each case, antibody molecules are produced by a clone of identical cells derived from a single fused cell, they are called **monoclonal antibodies.**

The production of monoclonal antibodies has become a billion-dollar biotechnology business, providing antibodies used in a variety of procedures, including cancer screening, pregnancy tests, and basic research. For example, the colorful cells shown in Figure 30-11 were obtained by treating the cells with monoclonal antibodies against specific cytoplasmic proteins. The location of the antibodies—and thus of the cytoplasmic proteins—is made visible by linking the antibody molecules to colored dye molecules.

Antibodies contain sites that are capable of binding molecules that have a complementary shape. The diversity of these binding sites among different antibodies derives from the great variety of polypeptide chains from which antibodies can be constructed. Much of this polypeptide diversity is the result of genetic rearrangements that occur during lymphocyte maturation. Remarkably, an individual possesses the ability to recognize and inactivate millions of different antigens, most of which will never be encountered during his or her lifetime (See CTQ #4.).

IMMUNIZATION

You are probably immune to many diseases from which you have never had to suffer an immunizing infection. Instead, you received a vaccine that "tricked" the immune system into responding and producing immunological memory. Jenner first demonstrated that this was possible when he intentionally stimulated active immunity to smallpox by vaccination with a related cowpox virus. Vaccines have since become one of our most potent weapons against disease. Most vaccines are modified forms of disease-causing agents; these variants have lost the ability to cause the disease but retain some of the same antigens that their danger-

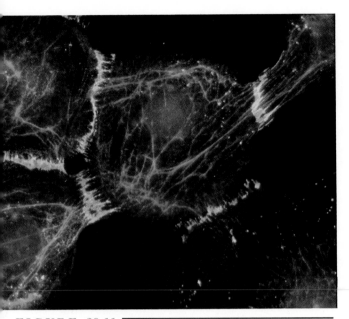

FIGURE 30-11
The use of monoclonal antiodies to locate specific proteins within cells. The antibody molecules are linked to colored dyes to reveal their location.

ous counterparts possess. An immune system exposed to a vaccine, and often several "booster" doses, builds memory cells against a pathogenic agent without the danger of developing the disease. In this fashion, we can now safely promote active immunity against polio, measles, mumps, diphtheria, tetanus, rabies, and many other dangerous diseases.

Sometimes there is no time to wait for an immune system to become sensitized. An attack of diphtheria or tetanus, or the bite of a poisonous snake requires immediate immunity to neutralize the life-threatening antigens. In these cases, antibodies obtained from the blood of an immune individual (or animal) can be injected into the person who needs immediate protection, producing a temporary state of **passive immunity.** For example, the blood of William Haast who, as director of the Miami Serpentarium, had been bitten by a variety of poisonous snakes, has been collected to provide passive immunity to individuals bitten by exotic vipers.

In mammals, passive immunity can be acquired naturally by the transfer of antibodies across the placenta from a mother to her developing fetus. Such passive protection is augmented after birth by the presence of antibodies in breast milk, providing breast-fed babies with more disease resistance—particularly to gastrointestinal infections—than bottle-fed babies. A decline in breast-feeding often results in an increased infant mortality.

Normally, a period of days is required before a person can mount a specific immune response against a toxin or disease-causing microorganism that has entered the body for the first time. A person who has been actively or passively immunized against the foreign intruder can eliminate a foreign agent, before disease develops. (See CTQ #5.)

EVOLUTION OF THE IMMUNE SYSTEM

Dissecting an animal reveals its nerves, vascular channels, digestive tract, and most other systems. It is more difficult to determine whether that same animal has an immune system. The existence of an immune system is determined by the capacity of an animal to respond to foreign materials rather than by the presence of a particular type of organ.
◗ Most invertebrates respond to the injection of foreign material by sending out phagocytic cells in a manner similar to the inflammatory reaction seen in vertebrates. Many invertebrates also appear capable of synthesizing soluble substances that can kill pathogenic organisms, but these substances lack the high degree of specificity that is characteristic of vertebrate antibodies. Most studies indicate that invertebrates lack the ability to reject foreign grafts, suggesting that these animals lack cell-mediated immune systems.

Highly specific, cell-mediated immune mechanisms probably first appeared in vertebrates, as did the ability to produce antibodies. All known species of living vertebrates, from the most primitive, jawless fishes to mammals, have (1) lymphoid tissue that produces lymphocytes, (2) circulating antibodies, and (3) the ability to reject foreign grafts. There is a progression in the complexity of the immune system through the vertebrate classes, however, culminating in the systems found in birds and mammals.

The complexity and effectiveness of immune responses have increased markedly over the course of animal evolution. (See CTQ #6.)

◁ THE HUMAN PERSPECTIVE ▷
Disorders of the Human Immune System

A number of conditions can be traced to a malfunction of the immune system, including allergies, autoimmune diseases, and immunodeficiency disorders.

ALLERGIES

An allergic reaction is triggered when the immune system reacts in an inappropriate manner to a foreign substance, usually one that has been eaten, inhaled, or acquired during injection by an insect sting or a hypodermic syringe. The most common allergic reactions occur in response to exposure to pollen, dust, mold, or certain food products. The allergic individual produces a special class of antibodies (IgE antibodies) against one or more of these *allergens* (Figure 1). The interaction between allergen and antibodies activates various cells to release highly active substances, including prostaglandins (page 532) and *histamine,* which evoke the symptoms of allergy. The

diameter and permeability of blood vessels increase, leading to inflammation; secretion of mucus increases, leading to nasal congestion; and the smooth muscles in the wall of the respiratory airways contract, leading to difficulty in breathing. Allergies are commonly treated with antihistamines, which block the release of histamine. If airways become severely constricted, as can occur during an asthma attack, the quickest relief is obtained by inhaling a mist of epinephrine, which causes a dilation of the airways. Extended periods of asthma may require treatment with anti-inflammatory steroids. People with persistent allergies may be treated by a desensitization program, in which increasing amounts of the allergen are administered by injection over a period of time.

The most severe type of allergic reaction is **anaphylaxis;** it occurs if the substance triggering the reaction is introduced directly into the bloodstream, such as fol-

lowing a bee sting or a shot of penicillin. Under these conditions, allergy-producing substances, such as histamine, are released throughout the body, and the severe reaction, known as *anaphylactic shock,* can be life-threatening.

Not all allergic reactions are triggered by soluble antibodies. The skin rash produced by exposure to poison oak or poison ivy occurs following activation of T cells. In these cases, the allergen is an oil from the plant, which is absorbed into the skin and binds to cells. T cells then release substances that cause inflammation at the site of allergen attachment.

AUTOIMMUNE DISEASES

⟳ Under normal conditions, the immune system does not mount an immune response against its own tissues. Occasionally, however, the discriminatory ability of the system breaks down, and the lymphoid

(a)

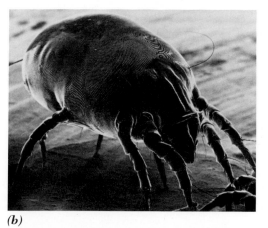

(b)

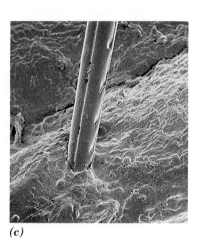

(c)

FIGURE 1

Common allergens. Scanning electron micrographs of selected materials to which many people are allergic: ragweed pollen *(a)* and the house dust mite *(b),* a minute arthropod that lives in dust. *(c)* The sting of a honeybee is, at the worst, painful for most people. People who are allergic to bee venom, however, suffer allergic responses that can be fatal within 10 minutes of the sting.

tissues attack normal components of the body. This phenomenon is responsible for some very serious **autoimmune diseases,** including thyroiditis, rheumatic fever, and systemic lupus erythematosus (often called "lupus"). In all autoimmune diseases, the victim is literally rejecting his or her own tissues.

In the case of thyroiditis, the body produces antibodies that react with tissues of the thyroid gland. The disease probably occurs after some of the protein that is normally stored in the thyroid gland leaks into the bloodstream, stimulating the immune system to produce antibodies against the protein. Rheumatic fever may occur following the body's recovery from infections by the streptococcal bacteria that cause strep throat. These bacteria possess cell-surface molecules that coincidentally are similar in structure to those found on some people's heart valves. In the case of rheumatic fever, antibodies produced against the bacteria *cross react* with the heart tissue after the microbe has been eliminated, causing heart-valve damage. Systemic lupus erythematosus is a debilitating disease in which many different tissues are attacked by antibodies that react with various molecules, including DNA. The reason for the production of these antibodies is unknown. The disease may result in death, usually as a result of kidney damage.

IMMUNODEFICIENCY DISORDERS

Until recently, diseases that left a person with a seriously deficient immune system were very rare and were usually a result of an inherited disorder. AIDS has changed that pattern dramatically, and today immunodeficiency is a growing cause of death. The biology of AIDS is described in Chapter 36. We will restrict the present discussion to the current attempts to halt further spread of the disease and to the treatment of those people already infected.

The offensive against AIDS consists of a three-pronged attack. Education is the first, and currently most effective, weapon. People must be taught how to minimize the risk of contracting AIDS and how to keep from spreading the disease if infected. The virus (Figure 2) is known to be spread by three avenues: sex, blood, and passive transmission from mother to fetus. The spread of AIDS would be greatly reduced if sexually active individuals followed safe sex practices (such as using a condom) and intravenous drug abusers stopped sharing needles.

The second weapon in the fight against AIDS is research to find a treatment that will kill or control the virus with minimum toxicity to the infected person. Presently, AZT (3'-azido-3'-deoxythymidine) is the primary drug that is widely prescribed for use against AIDS, but other drugs have been approved or are under development. AZT inhibits replication of the AIDS virus (HIV) by competing with thymidine (one of the four nucleotide building blocks of DNA) for the active site of the viral DNA-synthesizing enzyme. AZT has prolonged the survival of patients with AIDS and has delayed the onset of the

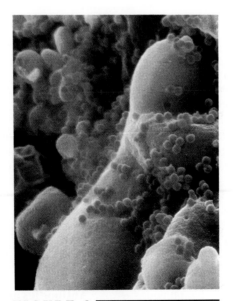

FIGURE 2

Human Immunodeficiency Virus (HIV) particles coating the surface of a white blood cell.

disease in infected people who are asymptomatic. Recent research suggests that treatment of patients with a combination of antiviral drugs may prove more effective than do single drug regimens. Effective drugs have also been developed against most of the secondary infectious diseases that strike AIDS patients, including pneumocystis pneumonia (caused by the protist *Pneumocystis*), toxoplasmosis (caused by the protist *Toxoplasma*), and oral yeast infections (caused by the fungus *Candida*). Controlling these secondary infections allows AIDS patients to live much longer with less suffering. Still, finding a cure for AIDS remains a distant goal, despite the enormous efforts of researchers around the world.

The third approach is the development of a vaccine to immunize people against the virus, creating an AIDS-resistant population. Most vaccines against viral diseases, such as polio, measles, and mumps, consist of attenuated viral particles—viruses that are still active and infectious but are no longer capable of injuring the body—to which the body mounts an immune response. But these viruses do not integrate their DNA into human chromosomes as does HIV. The consensus is that it would be too dangerous to use whole HIV as part of a vaccine, no matter how inactivated or altered it might be, since the virus has the potential to become reactivated as a result of interaction with a person's own DNA. Instead, it is hoped that a vaccine can be produced using parts of proteins from the virus's outer coat, rather than the virus itself. A similar approach has been used to develop a vaccine against hepatitis B, and tests for AIDS vaccines using this approach are ongoing. To date, the results of these tests have not been encouraging since the antibody response has been neither effective nor sustained. Instead, researchers have recently turned to techniques whereby genes that code for proteins of HIV are incorporated into other, safer viruses. It is hoped that infection with a "combination virus" may cause mild infections that stimulate immunity to AIDS.

REEXAMINING THE THEMES

Relationship between Form and Function

The hallmark of immune function is specific molecular interaction. The functioning of both soluble antibodies and T cells—the two weapons of the immune system—depends on their ability to interact specifically with a target. This specificity depends on a complementary shape existing between the target and the combining site of the antibody molecule or the T-cell receptor.

Biological Order, Regulation, and Homeostasis

The maintenance of internal order requires that foreign materials, whether they are disease-bearing pathogens or simple debris, must be eliminated from the body. This is the job of the immune system: to keep the "nonself" out of the "self." This is accomplished by first recognizing the "nonself" antigens and then inactivating or destroying them. Foreign agents can usually be eliminated, even if they are hiding within the body's own cells. The importance of the immune system in maintaining homeostasis is readily revealed by the effects of diseases that impair the system, such as AIDS.

Unity Within Diversity

All T cells are basically alike, as are all B cells and all IgG molecules, yet each group is incredibly diverse. Different B cells contain different membrane-bound antibodies; different T cells contain different T-cell receptors; and different IgG molecules contain different antigen-binding sites. Such diversity allows the body to react to virtually any type of foreign agent and stems from subtle differences in the amino acid sequences of a handful of different categories of polypeptide chains.

Evolution and Adaptation

A sophisticated immune system is a relatively recent evolutionary arrival. While invertebrates possess the ability to resist infectious agents, their defenses lack the high degree of specificity that is found in the human system. Immune systems that are capable of producing specific antibodies and cellular responses are found throughout the vertebrates. Even among vertebrates, however, there is a marked progression in complexity of the immune system, reaching its peak in birds and mammals.

SYNOPSIS

Homeostatic mechanisms that afford protection against invading pathogens are divided into nonspecific and specific defenses. Nonspecific mechanisms include external body surfaces that prevent entry to pathogens; cells that either engulf or poison pathogens that have breached the body's surface; and molecules, such as complement and interferon, that attach to the surface of pathogens or infected cells.

The immune system mediates specific mechanisms of defense, whereby foreign substances are specifically recognized and destroyed. The immune system is composed of cells that are scattered throughout the body and concentrated in lymphoid tissues, such as the lymph nodes and bone marrow. Immune responses may be mediated by either (1) soluble antibody molecules secreted by plasma cells that differentiate from B cells, or (2) cell-mediated immunity mediated by T cells. Antibodies attach to specific targets that are present outside of host cells, leading to their destruction, while T cells interact with other cells. Immune

responses are stimulated by foreign substances called antigens.

Antibodies contain binding sites that combine specifically with the antigen responsible for production of that antibody. Antibodies are constructed of both light and heavy polypeptide chains. Part of the polypeptide is constant from chain to chain, while the remainder is highly variable. Both heavy and light antibody chains are encoded by genes that form as a result of DNA rearrangement. Antigen-binding sites are formed by the association of the variable portions of a heavy and light chain.

Antibodies are formed by clonal selection. Lymphoid tissues contain a population of B cells, each of which is committed to forming a particular antibody. Samples of these antibodies are present in the B-cell plasma membrane. When an antigen binds to a cell's membrane-bound antibody, it stimulates the lymphocyte to proliferate, forming a clone of cells that is capable of producing that antibody. Some of the cells differentiate into plasma cells that secrete the antibody; others become memory cells that can mount a rapid response if the antigen reappears at a later time. This process of antigen-mediated lymphocyte stimulation is mimicked by vaccination. In this case, an antigen, such as an inactivated virus, is injected into the body, where it stimulates the proliferation of lymphocytes that are able to produce antibodies against the injected material.

T cells differentiate in the thymus gland by producing a particular membrane-bound T-cell receptor, which mediates the interaction of the T cell with another cell. Cytotoxic T cells recognize infected or altered host cells and kill them. Helper and suppressor T lymphocytes have a regulatory function; they either specifically activate or inhibit other lymphocytes.

The immune system is able to distinguish self from nonself. The body must be able to suppress the formation of T cells and antibodies capable of interacting with the body's own cells and tissue components. The body becomes tolerant of its own tissues by a process whereby cells capable of reacting with "self" are either suppressed or destroyed. When this process breaks down, the body may produce autoantibodies against itself, causing serious disease.

Key Terms

immune system (p. 644)
natural killer cell (p. 645)
inflammation (p. 645)
complement (p. 645)
antibody (p. 646)
antigen (p. 646)
B cell (p. 647)
T cell (p. 647)

plasma cell (p. 647)
cell-mediated immunity (p. 647)
cytotoxic (killer) T cell (p. 647)
helper T cell (p. 649)
suppressor T cell (p. 649)
tumor-infiltrating lymphocyte (TIL) (p. 648)
thymus gland (p. 649)
heavy chain (p. 650)
light chain (p. 650)

antigen presentation (p. 652)
memory cell (p. 653)
active immunity (p. 653)
autoantibody (p. 654)
monoclonal antibody (p. 654)
anaphylaxis (p. 656)
autoimmune disease (p. 657)
passive immunity (p. 655)

Review Questions

1. Distinguish between B cells, plasma cells, and memory cells; complement and interferon; normal mice and nude mice; C genes and V genes; light chains and heavy chains; active and passive immunity; T-cell receptors and histocompatibility antigens.

2. Describe the elements of the clonal selection theory.

3. Describe the steps that occur between the time a cold virus penetrates the nasal epithelium and the time it is eliminated by the immune system.

4. Why does infection with cowpox protect a person from smallpox?

5. How is it that you have only one C gene per haploid set of chromosomes for a light antibody chain, yet you are able to synthesize hundreds of thousands of different antibody molecules?

6. What is the role of the thymus gland? How does this role change following receipt of an organ by transplantation?

Critical Thinking Questions

1. When AIDS first appeared, some clinicians speculated that the condition was due to the male homosexual practice of inhaling amyl nitrate, a stimulant of the heart. What evidence might you seek either to support or to refute this suggestion about the cause of AIDS?

2. Why are the nonspecific defenses of the body insufficient? How is this characteristic illustrated by disorders that affect the immune system, such as AIDS or thyroiditis?

3. How does the immune system illustrate each of the following characteristics, and how does each contribute to the successful defense of the body against disease: decentralization, division of labor, specificity?

4. One might argue that it is wasteful for B cells to produce antibodies by a process that is independent of the presence of antigens since most of these antibodies will never be needed. Do you agree or disagree? Why?

5. If you were to see your doctor after stepping on a rusty nail, you might receive two very different types of shots, both of which would protect you against tetanus. What types of materials do you suspect might be in these different types of shots, and how would the type of protection they provide differ?

6. Mammals have evolved the most complex and effective immune systems among animals, and yet some disease organisms, such as the HIV that cause AIDS and the protozoa that cause sleeping sickness elude the immune system. HIV do so by a rapid rate of mutation; the sleeping sickness protozoa do so by a reshuffling of genes that code for surface proteins. What do you think are the prospects for humans to evolve mechanisms for fighting these diseases successfully? What do you think are the prospects for humans to develop medical responses to these?

Additional Readings

Boon, T. 1993. Teaching the immune system to fight cancer. *Sci. Amer.* March:82–89. (Intermediate)

Cohen, I. R. 1988. The self, the world and autoimmunity. *Sci. Amer.* April:52–60. (Intermediate)

Grey, H. M., A. Sette, and S. Buus. 1989. How T cells see antigen. *Sci. Amer.* Nov:56–64. (Intermediate)

Hoth, D. F., and M. W. Myers. 1991. Current status of HIV, therapy. *Hospital Practice* Jan:174–197, Feb:105–113. (Advanced)

Koff, W. C. 1991. The prospects for AIDS vaccines. *Hospital Practice* April:99–106. (Advanced)

Mills, J., and H. Masur. 1990. AIDS-related infections. *Sci. Amer.* Aug:50–57. (Intermediate)

Radetsky, P. 1993. Magic missiles. *Discover* March: 42–47. (Intermediate)

Rennie, J. 1990. The body against itself. *Sci. Amer.* Dec:106–115. (Intermediate)

Rosenberg, S. A. 1990. Adoptive immunotherapy for cancer. *Sci. Amer.* May:62–69. (Intermediate)

Rothman, D. J., and H. Edgar. 1991. AIDS, activism, and ethics. *Hospital Practice* July:135–142. (Introductory)

Shilts, R. 1987. *And the band played on.* New York: St. Martin's Press. (Introductory—Excellent prize-winning story of the early history of AIDS)

von Boehmer, and P. Kesielow. 1991. How the immune system learns about self. *Sci. Amer.* Oct:74–81. (Intermediate)

Young, J. D., and Z. A. Cohn. 1988. How killer cells kill. *Sci. Amer.* Jan:38–44. (Intermediate)

CHAPTER
◄ 31 ►

Generating Offspring:
The Reproductive System

STEPS
TO
DISCOVERY
The Basis of Human Sexuality

The Basis of Human Sexuality

*D*uring the eighteenth and nineteenth centuries, physiologists learned a great deal about the workings of the circulatory system, the pathways of the nerves, and the microscopic structure of the kidneys and lungs. Because of the sensitivity of the subject, however, our knowledge of human reproductive physiology lagged woefully behind. One popular hypothesis of the late 1800s even held that a variety of reproductive system dysfunctions were due to disorders of the epithelium lining the nose. One prominent professor at Johns Hopkins University went so far as to compare the anatomy of the nose to that of the penis, concluding that men who masturbated frequently suffered from various nasal diseases, including chronic nosebleeds.

Interest in human sexuality underwent a major revival following the publication of Charles Darwin's second major work, *The Descent of Man, and Selection in Relation to Sex,*

in 1871. In this work, Darwin applied his basic theory of evolution via natural selection to the evolution of humans; in doing so, he created even more furor. Darwin considered the evolution of the human mind as well as the body. For example, he noted that apes and monkeys showed many of the same emotions as humans, including fear, aggression, surprise, sexual desire, and even jealousy.

Darwin became interested in child behavior and closely followed the behavior of his own son during the child's first few years of life. His observations led Darwin to suggest that very young children exhibit signs of sexual behavior. Darwin's writings on this matter had a great influence on the psychiatric world of the late nineteenth century. Darwin called attention to the presence of the instinctive qualities of human behavior, which had been inherited from our less inhibited ancestors. The Darwinian view of

Anonymous questionnaires uncover the social and psychological aspects of human sexuality, while physiological experimentation

human evolution led many psychotherapists to conclude that human behavior was molded largely by two basic, innate drives: self-preservation and sexual gratification.

Just 6 years after the publication of *The Descent of Man,* a paper appeared in a German journal of biology on the structure of the nervous system of the larval stage of a lamprey eel, one of the most primitive of all living vertebrates. The paper was authored by a young medical student at the University of Vienna, who was interested in becoming a neuroanatomist (a biologist who studies the anatomic organization of the nervous system). The author of the paper was Sigmund Freud. Freud received his medical degree a few years later and attempted to secure a position in neuroanatomy at the Physiology Institute. He was discouraged by his mentor Ernst Brucke, who argued that the prospects there were slim and that Freud would be better off pursuing a career in clinical practice. Eventually, Freud opened his own practice in Vienna.

In 1895, Freud and his friend and colleague Josef Breuer published a book entitled *Studies on Hysteria,* in which the two men evaluated hundreds of hours of conversations with patients who were suffering from various forms of mental disorders *(neuroses).* Freud concluded that the basis of these neuroses could be traced to childhood sexual trauma and repressed sexual desires. He recognized a conflict between the human sexual drive, which is inherited from our prehuman ancestors, and the restrictions placed on this drive by modern civilization. Over the following years, Freud became more and more convinced of the pervasive importance of sexuality, particularly infantile sexuality, in shaping our adult personalities.

During the first half of the twentieth century, Freud and other psychotherapists continued to probe the role of sexuality in shaping individual human personalities, but no attempt was made to study sexuality in the human population as a whole. This would soon be changed by Alfred Kinsey of Indiana University, a prominent *taxonomist*—a biologist who studies the anatomic, physiological, and behavioral characteristics that are used to classify a particular group of organisms. During his extensive studies of a group of tiny insects called gall wasps, Kinsey was struck by the genetic variation that existed among the individual insects, noting, for example, that wing length could vary by a factor of over tenfold from one member of a local population to another. As an insect taxonomist, Kinsey was an unlikely student of human sexuality.

In 1944, students at Indiana University petitioned for a biology course on sexuality and marriage. Kinsey was named to head a seven-member faculty to deliver the lectures. At the end of the course, Kinsey passed out a questionnaire asking students for their evaluation of the course as well as personal questions regarding their sexual activities; he planned to discuss the answers to the questions when the course was next offered. From this unlikely beginning, Kinsey realized that by using interviews and questionnaires he could learn about human sexuality, an important subject about which the scientific community was almost totally ignorant. Within a few years, Kinsey had trained a number of interviewers, and the group had collected several thousand sexual case histories from people of all different ages, economic groups, races, religions, and sexual orientations. The data from the sexual case histories were compiled into two large volumes, entitled *Sexual Behavior in the Human Male* and *Sexual Behavior in the Human Female* and were published in 1948 and 1953, respectively.

Kinsey's books provided a wealth of insights into human sexuality. They revealed the striking variation in sexual practices among different individuals; the importance of socioeconomic factors in sexual habits, particularly among males; the differences and similarities in male and female sexuality; the importance of religious convictions in shaping human sexual behavior; the ages at which males and females reach peak sexual interest; and more. Not surprisingly, the books captured the attention of the public as well as the behavioral scientists, bringing both praise and scathing criticism. Although there were shortcomings in Kinsey's method of selecting the individuals to interview, the methods of interviewing, and the handling of data, the Kinsey reports remain among the most cited of all works in the behavioral sciences to this day.

For the past 2 decades, biologists have been investigating the neural and hormonal basis of human sexuality. Particular attention has been focused on the hypothalamus, a region of the brain known to govern sexual responses and to contain distinct anatomic differences between men and women. For example, one cluster of hypothalmic nerve cells, called INAH-3, typically is more than twice as large in men as it is in women. While the reasons for these gender-related differences are not understood, evidence indicates they are caused by the levels of sex hormones (estrogen and testosterone) that circulate in the fetus and newborn. In 1991, Simon LeVay, a neuroanatomist at the Salk Institute, dropped a bombshell when he reported that the INAH-3 cluster in male homosexuals who had died from AIDS tended to be much smaller than that of heterosexual males; in fact, the cluster in homosexuals was about the same size as that in women. Whether or not LeVay's findings are confirmed remains to be seen; regardless, they have created considerable controversy. For example, the findings affirm the conviction of many homosexuals that their sexual orientation is biologically determined and cannot be altered behaviorally. On the other hand, many homosexuals fear that the information could be used as a basis for screening the population in order to assess an individual's sexual orientation.

revealed the importance of the hypothalamus in regulating sexual function.

*O*f all animals, humans are the only species apparently aware of their own mortality. We all know that we will die, but we also know that our lineage can continue by having children. Consider for a moment that the genes we possess have been passed down, generation to generation, from distant ancestors. A living descendent of Andrew Jackson, for example, may have the same genetic information for Jackson's distinctive eyebrows. Some of the information encoded in our genes may be billions of years old, such as that which directs the formation of semipermeable cell membranes or oxygen-dependent respiratory chains. Some of these genes have been preserved—with changes—for millions of generations.

The transmission of genetic information from one generation to the next is a crucial part of **reproduction:** the process by which new offspring are generated. All organisms, from bacteria to redwood trees to humans, have a finite life span. Reproduction is the process by which the species is perpetuated. But reproduction is more than simply a continuation of life; it is also a process of rejuvenation. As each individual grows older, it shows increasing signs of age. Yet the parents' aging is not passed on to the offspring; each new generation begins life with a "fresh start."

▼ ▼ ▼

REPRODUCTION: ASEXUAL VERSUS SEXUAL

Some animals can reproduce by **asexual reproduction;** that is, they produce more of themselves without the participation of a mate, gametes, or fertilization. As we discussed in Chapter 10, since neither meiosis nor fertilization is part of the process, asexual reproduction produces offspring that are genetically identical to the parent. However, most animals produce offspring by **sexual reproduction,** which requires

1. The formation of two different types of haploid gametes—eggs and sperm—by a process that includes meiosis, and

2. the union of a single egg and sperm at fertilization to form a zygote.

You will recall that gametes are formed in reproductive organs called *gonads.* Sperm are produced in the male gonads (testes), whereas eggs are produced in the female gonads (ovaries).

TYPES OF ASEXUAL REPRODUCTION

Asexual reproduction occurs in various ways, depending on the species. During **fission** (Figure 31-1a), an animal splits into two or more parts, each of which becomes a complete individual. Fission is common among sea anemones and various groups of worms. Some animals reproduce asexually by **budding,** whereby offspring develop as an outgrowth of some part of the parent. Budding is common in sponges, hydras, and corals (Figure 31-1b), particularly in those species that form large colonies in which individual members remain connected to one another by a common "pipeline." In such species, each colony arises from a single founder that buds repeatedly, generating new individuals that remain physically joined to one another (as in Figure 1-3a, on page 8).

One type of reproduction is neither strictly asexual or sexual. Although gametes are produced, **parthenogenesis** requires only one parent—a female—who produces eggs that develop into adults without fertilization. Parthenogenesis commonly occurs among some insects, such as the aphids shown in Figure 31-1c, and less commonly in reptiles, amphibians, and fishes.

THE ADVANTAGE OF SEXUAL REPRODUCTION

Why do organisms engage in sexual reproduction? Why don't all organisms simply reproduce asexually, forming an exact, yet younger, version of themselves? After all, asexual reproduction has some obvious adaptive advantages. First, it generates progeny without the greater investment of energy and resources associated with gamete production, mate seeking, and fertilization. Second, asexual reproduction is very efficient; one individual, living by itself, can generate large numbers of offspring very rapidly.

➠ But sexual reproduction is the predominant mode of reproduction among animals. Even among animal species that reproduce asexually, most do not do so exclusively. Rather, individuals in a population typically reproduce asexually for a period of time; then, in response to some environmental trigger, such as food depletion or overcrowding, these same animals switch to a sexual mode of reproduction. Clearly, there must be some selective disadvantage to a total reliance on asexual reproduction. That disadvantage is presumed to be **genetic monotony;** that is, generation after generation, progeny will be genetically the same. In contrast, sexual reproduction combines traits from two genetically distinct parents in a single individual so that each offspring acquires a unique genetic mix. In addition to gene mixing during fertilization, variation is boosted even further by independent assortment and crossing over during meiosis (Chapter 11).

Asexual reproduction is advantageous only in certain situations. If a parent is successful in surviving and reproducing in a particular habitat, identical offspring are also

(a)   *(b)*   *(c)*

FIGURE 31-1

Types of asexual reproduction in animals *(a)* During fission, this sea anemone splits into two individuals. *(b)* In this hydra, a new individual forms as a bud that grows out of the body wall. *(c)* This brood of aphids developed parthenogenetically from unfertilized eggs within the larger female.

likely to succeed. Asexual reproduction is a successful means of reproducing in uniform environments where conditions do not change much from one place to another or from one year to the next. Most of the earth's habitats are not uniform, however. Conditions change over time, as exemplified by the appearance of new diseases, such as AIDS in humans, or the repeated ice ages that have occurred over the past few million years. If an individual member of a species is ill-equipped for a particular change in environmental conditions, then all the identical offspring produced by asexual reproduction will be similarly ill-equipped. The entire species could be wiped out by a single adverse development in the environment. In contrast, sexually reproducing species produce offspring that have different characteristics; consequently, chances are improved that some individuals may be able to survive and reproduce in new habitats, changing environments, or in the face of new diseases. In this way, genetic variation is the foundation of evolutionary change.

STRATEGIES FOR SEXUAL REPRODUCTION

Sexual reproduction requires that eggs and sperm of the same species come together at the same time and place (see Bioline: Sexual Rituals). The simplest strategy, employed by most aquatic animals, is simply to spew sperm and eggs into the surrounding water, which leads to **external fertilization:** fertilization that takes place outside the body (Figure 31-2*a*). With this strategy, the likelihood that a sperm and egg will cross paths is largely a matter of probability; the more gametes produced, the greater the chance of fertilization. Animals that utilize external fertilization typically possess extensive gonads and produce enormous numbers of gametes. A single oyster, for example, releases more than 100 million eggs each season. Only a tiny fraction of these eggs will ever be fertilized, and only a tiny fraction of the resulting zygotes will actually develop into adults. The chance of an egg being fertilized externally is greatly improved when the male discharges his sperm directly onto the eggs, as occurs in many amphibians (Figure 31-2*b*).

External fertilization occurs only in watery environments, where sperm can swim to an egg and fertilize it. Therefore, animals that mate on dry land rely on **internal fertilization** (Figure 31-2*c*), which occurs inside a chamber in the female's body, where a local aquatic environment can be maintained. Internal fertilization has an important selective advantage: It improves the odds that gametes will encounter one another. Animals that utilize internal fertilization typically produce fewer eggs, each of which has a better likelihood of being fertilized.

In many internal fertilizers, sperm is transferred using an intrusive structure called a **penis.** Some internal fertiliz-

(a)

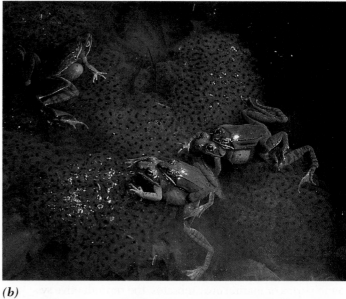

(b)

(c)

FIGURE 31-2

Sexual variety. *(a) External fertilization.* A sponge (lower right) discharging a cloud of gametes into the surrounding sea water. *(b) Coordinated external fertilization.* Among these wood frogs, the female of the species discharges eggs (the black circles surrounded by jelly capsules), which are fertilized by sperm released simultaneously by the male. *(c) Internal fertilization* between male and female damselflies. *(d) Hermaphroditic internal fertilization.* These two entwined slugs exchange sperm that is discharged from the blue penis that extends from each partner's head. This is an example of cross fertilization between individuals that possess both male and female reproductive systems.

(d)

ers lack these copulatory organs. For example, the male octopus produces packages of sperm, which it transfers to the female with one of its eight arms. When inserted into the female, part of the arm breaks off and eventually ruptures, showering the eggs with sperm. Afterward, the male regenerates the tip of the lost appendage.

In many invertebrate species, one individual possesses both male and female reproductive systems; such individuals are called **hermaphrodites** (Figure 31-2*d*). When isolated from other members of their species, some hermaphrodites are capable of reproducing by fertilizing their own eggs. This is particularly true of parasites, such as tapeworms, that may find themselves socially isolated inside a host's digestive tract.

UNITY AND DIVERSITY AMONG REPRODUCTIVE SYSTEMS

⬡ Regardless of the type of animal, all sexual reproductive systems have a clearly defined, universal function: the production of eggs and sperm. Animal gametes may be so similar in appearance that the living eggs or sperm of a jellyfish, snail, sea urchin, or human may be nearly indistinguishable when seen through a microscope.

Despite their similar function, reproductive systems are probably more diverse in structure than is any other group of organs. In some species, such as those of parasitic roundworms or marine invertebrates, the reproductive systems are so complex and voluminous that they fill much of the internal space within the animal (Figure 31-3*a*). For example, the reproductive system of a parasitic flatworm (Figure 31-3*b*) contains a bewildering variety of tubes, chambers, glands, and receptacles, none of which bears any obvious evolutionary relationship to the parts of the reproductive systems of other groups of animals. Rather than attempting to discuss the evolution or diversity of reproductive systems, we will focus on the complex reproductive organs within our own bodies.

Sexual reproduction, which includes meiosis and fertilization, is the predominant mode of reproduction among animals. Although sexual reproduction makes greater demands on animals than does asexual reproduction, it generates genetically diverse offspring, which gives a population a survival advantage. (See CTQ #2.)

HUMAN REPRODUCTIVE SYSTEMS

Both the male and female human reproductive systems consist of a pair of gonads, in which gametes are formed, and a reproductive tract possessing various accessory functions that are discussed later in the chapter. The vastly different structures of the human male and female reproductive systems reflect the difference in their roles during

(a)

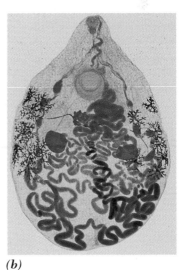

(b)

FIGURE 31-3

Simple animals, complex reproductive systems. *(a)* Huge reproductive organs dominate the anatomy of this jellyfish, testifying to the large number of gametes that must be produced in a species whose eggs are fertilized in the open sea. *(b)* Virtually all of the structures seen in this parasitic flatworm belong to the animal's reproductive system.

◁ B I O L I N E ▷
Sexual Rituals

FIGURE 1

"Timing is the thing; it's true. It was timing that brought me to you."

These words from a popular love song are far more universal than is the "boy-meets-girl" situation the songwriter is describing. Sexual reproduction among virtually all animal species requires some form of timing that increases the likelihood that sperm will meet egg; this is often accomplished by increasing the likelihood that male will meet female. In other words, reproductive timing coordinates the courtship behavior of many animals. The songs of crickets and cicadas, the blinking lights of fireflies, the aromas of prometheus moths, and the nocturnal croaking of frogs are all inherited mechanisms that help mates find and identify one another as members of the same species and opposite genders—in other words, as potential sex partners.

The right smell or the correct approach not only promotes fertilization but can also protect a mate from being mistaken for a potential meal. A male spider, which tends to be smaller than the female of the species, could be consumed by his potential mate before fertilizing her eggs. Not only must the male spider approach the female cautiously, he must also exhibit some behavior that identifies him as a mate, not a meal. Some spiders do so by vibrating the female's web in a rhythm that the female recognizes as an approaching male. Others placate their potentially dangerous mates by providing a "gift," usually a captured meal wrapped in silk (Figure 1). The female spider usually accepts the gift and allows the male to copulate with her. Males occasionally try to trick a female into accepting an empty package. When their deceit is discovered, the "empty-handed" male may become the gift himself.

♻ Younger members of a species, whether a floating sea urchin larva or a newborn colt, are less hardy than are their adult counterparts. For this and other reasons (such as the availability of food), reproduction usually occurs at a particular season of the year, a time that is most favorable for offspring survival. The onset of seasonal reproductive behavior is often coordinated by some specific environmental change that informs the animal's nervous system that the mating season has arrived. For many animals, this environmental factor is *photoperiod,* the length of daylight or darkness. Spring is the most common season for reproduction; the physiological and behavioral changes that lead up to reproduction are often triggered by the increasing length of daylight at this time of year.

In many mammals, the increasing spring daylight stimulates hormonal changes that bring about the onset of *estrus,* or sexual "heat," a period of sexual receptivity. For example, in dogs, as in many other mammals, females in estrus produce a sexual attractant, the odor of which stimulates males to begin courtship activities at a time when eggs are present to be fertilized. The odds of fertilization can be further increased by mechanisms that guarantee that eggs will be available when sperm are deposited. In rabbits, for example, the egg is retained in the ovary until after mating, at which time it is released into the reproductive tract for fertilization.

Reproductive behavior in humans and other primates, such as monkeys and apes, is not limited by such estrus restraints, and sexual arousal follows no seasonal pattern. In fact, the songwriter quoted earlier might find it ironic that the human mating behavior is one of the few behaviors *not* governed by biologically determined timing mechanisms.

reproduction. The male produces huge numbers (trillions during a lifetime) of tiny motile gametes, called **spermatozoa** (singular *spermatozoon*), commonly referred to as sperm, and delivers these cells into the female's reproductive tract. In contrast, the female produces a small number (a few hundred over a lifetime) of relatively large gametes, called **ova** (singular *ovum*), commonly referred to as eggs. The female is responsible for more than just gamete production. She also provides an environment in which the eggs can be fertilized and another environment in which the zygote can develop into a fully formed human infant.

The reproductive activities of both sexes are regulated by a battery of hormones. These hormones are responsible for stimulating the initial development of the reproductive system in the embryo; for causing its maturation during puberty; and for maintaining its day-to-day operation throughout the reproductive years.

THE MALE REPRODUCTIVE SYSTEM

In human males, the paired testes produce sperm throughout adult life at the average rate of 30 million sperm per day. The male reproductive tract is primarily a hollow conduit, equipped with accessory glands, that moves sperm out of the body and into the reproductive tract of the female.

Male External Genitals: From and Function

The male external genitals (Figure 31-4) consist of the penis and a sac, called the **scrotum,** which houses the testes, or testicles. The temperature within the scrotum is several degrees cooler than is the rest of the body; this cooler temperature is necessary for the formation of sperm. If the testes do not descend into the scrotum during fetal development, as occurs on rare occasion, the elevated gonadal temperature results in sterility (the inability to produce offspring) after about 4 years inside the body. Surgically maneuvering the testes from the body cavity into the scrotum prevents such loss of fertility. A passageway, called the **urethra,** runs through the length of the penis, conveying both sperm and urine, although not simultaneously.

In order to deposit sperm into the vagina, the penis must have penetrating capacity, which is accomplished when the organ is engorged with blood. The interior of the penis contains three long cylinders of spongy *erectile tissue.* During periods of sexual excitement, impulses travel along parasympathetic nerves from the brain to the smooth muscle cells that line the arterioles of the penis. Relaxation of the smooth muscles leads to vasodilation of the arterioles, causing blood to engorge the spaces of the erectile tissue, which expands as it fills. As a result, the penis enlarges and becomes rigid enough for vaginal penetration; this is known as an erection.

The failure to develop an erection (known as *impotence*) is a common type of male sexual dysfunction. In younger men, the cause is usually psychological; in older men, it is often a failure of the organ to receive sufficient blood, which may be due to diabetes, atherosclerosis, or as a side effect of medication taken for treatment of high blood pressure. Whether the root cause is psychological or physiological, the condition can, in most cases, be corrected by counseling or medical treatment.

The surface of the penis, especially the tip, or *glans* (Figure 31-4), is richly endowed with sensory receptors that increase tactile sensitivity. In the flaccid (nonerect) penis, the glans is surrounded by the *prepuce,* or "foreskin," which is often removed shortly after birth by a process called **circumcision.** Circumcision is usually performed for religious or hygenic reasons, but it is also used to treat a painful condition called *phimosis*, in which the prepuce cannot retract over the glans as normally occurs during erection.

The Formation of Sperm

A cross section through a testis reveals that the structure is composed almost entirely of tubular elements, called **seminiferous tubules** (Figure 31-5), whose combined length is equivalent to about seven times the length of a football field. Within these tubules, **spermatogenesis,** the formation of the male gametes, takes place. Between the tubules are scattered clusters of *interstitial cells:* the endocrine cells responsible for the production of testosterone, the male sex hormone.

Spermatogenesis takes place in stages, which are revealed in an examination of a cross section of a seminiferous tubule (Figure 31-5). Cells enter spermatogenesis at the outer edge of the tubule and are gradually moved toward its inner lumen as the process continues. The outer edge of the tubule contains a self-perpetuating layer of **spermatogonia,** germ cells that have not yet begun meiosis. Some of these spermatogonia are about ready to enter meiosis and develop into sperm. Others continue to divide by *mitosis* (Figure 31-6), producing more spermatogonia, replacing those that have gone on to form gametes. In the first step of spermatogenesis, a spermatogonium grows in size and enters meiosis, forming a **primary spermatocyte.** Each diploid primary spermatocyte subsequently undergoes the first meiotic division to form two **secondary spermatocytes,** each of which undergoes the second meiotic division to form a total of four haploid **spermatids.** Each spermatid undergoes a dramatic transformation as it is converted into a sperm cell, one of the most specialized cells in the body.

The central lumen of a seminiferous tubule contains germ cells that have almost completed spermatogenesis and resemble, in outward appearance, mature sperm (inset, Figure 31-5). These cells are not yet ready to fertilize an egg, however. Sperm complete their maturation only after they are pushed out of the seminiferous tubule and into the

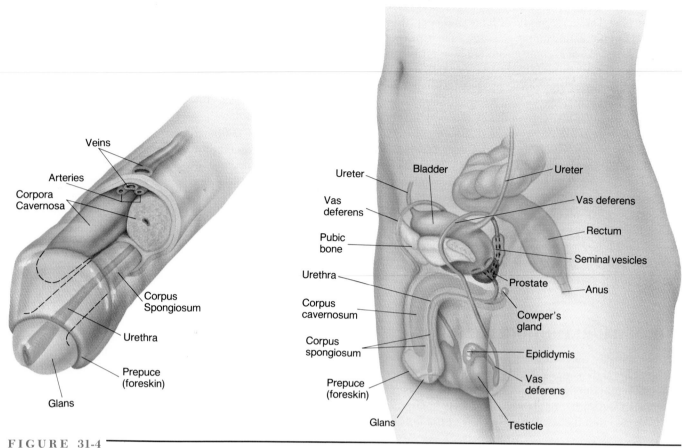

FIGURE 31-4

The male human reproductive system. The purple structures trace the route sperm travel from the testes to the penis. When the arterioles of the penis become dilated, and the erectile tissues (composed of the *corpus spongiosum* and paired *corpora cavernosa*) become engorged with blood, the penis becomes erect.

epididymis. Each **epididymis** (Figure 31-5) is a tubule, approximately 6 meters (20 feet) long; it is coiled into a compact mass attached to each testis. Sperm spend several weeks in the epididymis and then move into the **vas deferens,** a tubule that transports the male gametes to the urethra.

Form and Function of Sperm

A sperm is a compact, streamlined cell (Figure 31-5), whose function it is to move up the female reproductive tract and fuse with an egg. The entire structure of a sperm can be correlated with this function. As it differentiates, the sperm loses nearly all of its cytoplasmic and nuclear fluid, which accounts for more than 90 percent of the volume of the spermatid. The head of a sperm consists of two parts: the nucleus and the acrosome. The nucleus of a sperm is the ultimate in chromosome compactness; the chromosomal material is condensed to a virtual crystalline state. The

acrosome, which forms a cap over the nucleus, contains digestive enzymes that are released as the sperm "digests" its way through the protective layers that surround an egg. The middle portion of the sperm contains tightly packed mitochondria that, by virtue of the ATP they produce, will power the sperm's movements (see Figure 5-14*d*). The tail of a sperm contains a flagellum that whips against the surrounding fluid, driving the sperm toward and into the egg. Sperm are launched like self-propelled torpedoes with a limited range. If they do not reach the egg within the allotted time (24 to 48 hours in humans), the sperm exhaust their fuel supply and die.

The Role of the Male Accessory Glands

Before their release from the penis, sperm are mixed with secretions from several glands (see Figure 31-4) to form **semen,** the sperm-containing fluid that is expelled from the body as the result of strong muscle contractions that occur

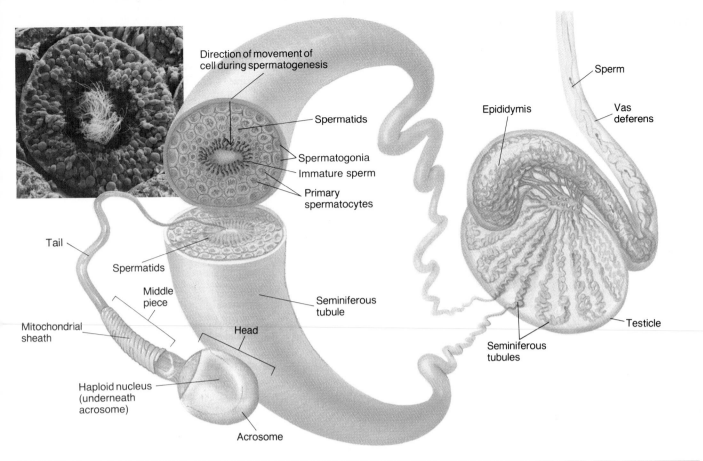

FIGURE 31-5

Spermatogenesis. The cross section through a seminiferous tubule shows cells in different stages of spermatogenesis. The outer layer contains spermatogonia that have not yet begun spermatogenesis. The next layer contains primary spermatocytes that have entered meiosis. Further inward are the secondary spermatocytes and the spermatids—the products of meiosis—and within the lumen are nearly differentiated sperm. The structure of a single sperm is also shown, illustrating several major features such as the head, the mitochondria-rich middle piece, and the tail. The inset shows a scanning electron micrograph of a cross section through a seminiferous tubule; tails of the sperm are clearly seen extending into the tubule's lumen.

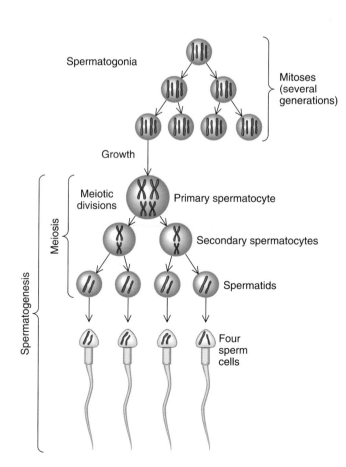

FIGURE 31-6

The stages of spermatogenesis.

during **ejaculation.** Sperm are only a small percentage of the semen's final volume; 5 billion sperm—a quantity large enough to repopulate the world with humans—would fit into a space the size of an aspirin tablet.

Together, the **prostate gland** and the paired **seminal vesicles** and **Cowper's glands** produce most of the ejaculatory fluid. These secretions are rich in fructose, a sugar that serves as an energy source for the sperm. The fluid also contains prostaglandins (page 532), which stimulate the contraction of the smooth muscles of the female reproductive tract, helping to propel sperm cells toward their encounter with an egg. The complete semen is ejaculated from the penis by strong muscular contractions that occur during **orgasm,** a brief period of peak excitement that constitutes the sexual climax. The ejaculatory fluid nourishes and protects the sperm; it provides a liquid medium needed for sperm motility; and, because of its alkaline pH, it temporarily neutralizes vaginal acidity that might otherwise impair sperm-cell activity.

Hormonal Control of Male Reproductive Function

♻ Proper functioning of the testes is maintained by the presence of two hormones that are secreted by the anterior

pituitary: **follicle-stimulating hormone (FSH)** and **luteinizing hormone (LH).** These hormones were originally named because of their effects on the female reproductive system, as we will discuss later in the chapter. The fact that the identical hormones are present in both genders but elicit different responses in males and females illustrates an important principle of endocrine function: The nature of the *target cell* determines the type of response, while the *hormone* itself acts simply as a stimulus or trigger.

FSH and LH are described as *gonadotropins* because they stimulate the activities of the gonads (Figure 31-7). LH acts primarily on the interstitial cells of the testes, stimulating the production and secretion of testosterone. FSH is required for spermatogenesis. The secretion of both LH and FSH by the pituitary is, in turn, regulated by the level of **gonadotropin-releasing hormone (GnRH)** that is secreted by the hypothalamus.

While the gonadotropins act on the gonads, testosterone, the hormone produced by the gonads, acts on the other tissues associated with male sexuality. Testosterone secretion stimulates the differentiation of the male reproductive tract in the embryo, the descent of the testes into the scrotum, the further development of the reproductive tract and penis during puberty, and the development of male *second-*

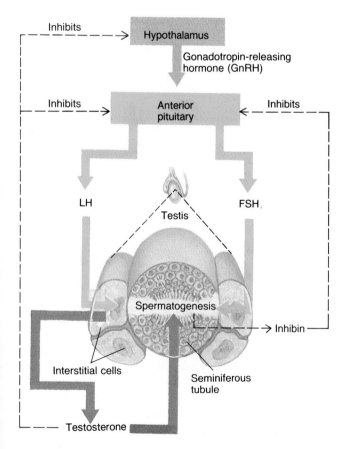

FIGURE 31-7

Production and regulation of male sex hormones. As testosterone concentrations in the bloodstream increase, the release of GnRH by the hypothalamus and LH by the anterior pituitary is inhibited. The reduction in LH concentration, in turn, causes a decrease in testosterone secretion by the interstitial cells of the testes, which, in turn, releases the inhibition on the secretion of GnRH and LH. As a result of this negative-feedback mechanism, a balanced level of all hormones is achieved. FSH production is also believed to be regulated by a negative-feedback mechanism involving a protein hormone called *inhibin,* which is produced by the seminiferous tubules and acts on the anterior pituitary. Both testosterone and FSH are needed for spermatogenesis to occur within the seminiferous tubules.

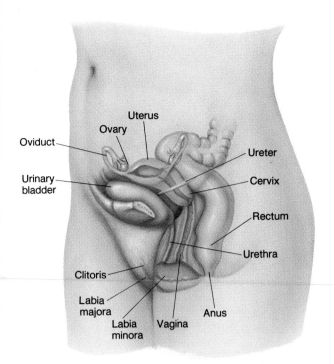

The female human reproductive system. Eggs are formed in the ovary and swept into the oviducts. Meanwhile, sperm enter the body in the vagina and travel through an opening in the cervix up to the oviducts, where fertilization takes place. The resulting embryo passes down the oviduct and into the uterus, where it is implanted into the uterine wall and develops.

ary sex characteristics, including a beard and chest hair, enlargement of the larynx, and increased muscle mass. Testosterone also plays a role in the development and maintenance of the male *libido,* or sexual desire.

THE FEMALE REPRODUCTIVE SYSTEM

A woman's body performs many activities that are essential to reproductive success: Her ovaries are the sites of **oogenesis,** or formation of ova, and her reproductive tract nourishes, houses, and protects the developing fetus. After birth, a woman's body provides breast milk, which nourishes the developing infant. As in the male, each aspect of reproductive activity is under the complex control of hormones.

Female External Genitals and Reproductive Tract: Form and Function

The female's external genitals (Figure 31-8) are collectively known as the **vulva.** The most prominent features of the vulva are an outer and inner pair of lips, the **labia majora** and **labia minora,** respectively. The **clitoris** protrudes from the point where the labia minora merge. Rich in sensory neurons and erectile tissue, the clitoris resembles the penis in its sexual sensitivity and erectile capacity. Unlike the penis, however, the clitoris has no urinary or ejaculatory function; its sole function is to receive sexual stimulation,

which it transmits to the central nervous system. Thus, the clitoris is the only human structure dedicated exclusively to the enhancement of sexual pleasure.

The urethra, which, in the female, is involved solely with the release of urine, opens between the labia minora. Also located between the labia minora is the opening to the **vagina,** an elastic channel that receives the sperm that are ejaculated from the penis and forms the birth canal through which an infant leaves its mother's body during childbirth. The vagina leads to the remainder of the female reproductive tract (Figure 31-8), which consists of the uterus and paired oviducts. The vagina is separated from the uterus by the **cervix,** which contains the opening through which sperm must pass on their way to fertilize an egg. The **uterus** (or *womb*) is a pear-shaped, thick-walled chamber that houses the embryo and fetus during pregnancy. The **oviducts** (or *fallopian tubes*) emerge from the uterus and serve as the sites where sperm and egg become united during fertilization.

The Formation of Eggs

A cross section through an ovary (Figure 31-9) shows an appearance very different from that of a testis. There are no tubules in an ovary. Instead, the **oocytes**—the germ cells that have the potential to form eggs—are housed within spherical compartments, called **follicles,** which are scattered throughout the tissue of the ovary. Each follicle con-

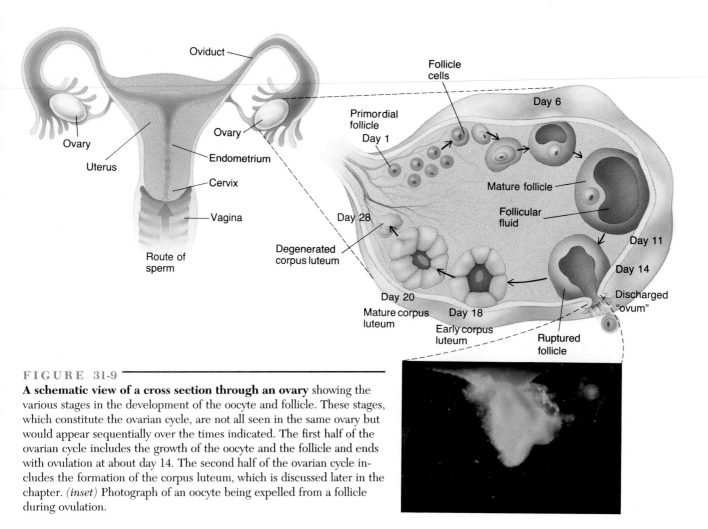

FIGURE 31-9

A schematic view of a cross section through an ovary showing the various stages in the development of the oocyte and follicle. These stages, which constitute the ovarian cycle, are not all seen in the same ovary but would appear sequentially over the times indicated. The first half of the ovarian cycle includes the growth of the oocyte and the follicle and ends with ovulation at about day 14. The second half of the ovarian cycle includes the formation of the corpus luteum, which is discussed later in the chapter. *(inset)* Photograph of an oocyte being expelled from a follicle during ovulation.

tains a single oocyte, surrounded by one or more layers of **follicle cells,** which provide the materials that support the growth and differentiation of the enclosed germ cell. Unlike the male gonad, the ovary of a woman contains no **oogonia,** or premeiotic germ cells. All of the oogonia that are produced during embryonic development have already entered prophase of meiosis I (page 216) by the time of birth. Oocytes will remain suspended in meiotic prophase for years, some for several decades.

The vast majority of the follicles of the adult ovary are small and consist of an undifferentiated oocyte surrounded by a single layer of follicle cells. These *primordial follicles* are storehouses of oocytes that will provide the eggs produced during the reproductive life of the woman. A pair of human ovaries contains approximately 400,000 primordial follicles at the time of birth. Once a female reaches reproductive maturity, a few oocytes will undergo oogenesis during each monthly **ovarian cycle.** During oogenesis, which occurs in the first half of the ovarian cycle, the oocyte increases in size from about 25 to 100 micrometers in diameter. Like the sperm, the architecture of the egg is correlated

with its function. Whereas the sperm is shaped for movement and activation, the egg is packed with nutrients that will be utilized by the embryo during the first days following fertilization. Nutrients take up considerable space and, for the most part, account for the large size of the egg.

The changes in the oocyte that occur during oogenesis are accompanied by a dramatic transformation of the entire follicle. By the time it completes its growth, a *mature follicle* is large enough to be seen as an obvious bulge at the surface of the ovary. Although several follicles typically undergo maturation within each ovarian cycle, one usually outpaces the others. Ultimately, the wall of this follicle suddenly ruptures, and the enclosed oocyte is released from the ovary (inset, Figure 31-9). This is **ovulation.** The ovulated oocyte is swept into the broad, funnel-shaped opening of the oviduct, where fertilization occurs.

Meiotic Divisions of the Female Germ Cell As in the male, the reduction of the chromosome number by meiosis is necessary during the formation of the female gamete. Unlike spermatogenesis, where meiosis occurs before the differentiation of the sperm, meiotic divisions in the female occur after the entire process of growth and differentiation of the ovum has essentially been completed (Figure 31-10).

The first meiotic division is completed in the oocyte just before ovulation. Unlike spermatogenesis, where meiosis I produces two equal-sized cells, the meiotic division in the female produces one large cell and one tiny cell, called a **polar body,** that eventually deteriorates. The second meiotic division begins while the oocyte is in the oviduct, but it does not run to completion. Instead, the meiotic process stops after the chromosomes have lined up for metaphase II. At this stage, the human egg is fertilized. The second meiotic division is completed after fertilization. Meiosis II is also a highly unequal cell division, producing a single, large cell that goes on to develop into a new individual, and another polar body that disintegrates. Thus, unlike meiosis in the male, which produces four equal-sized gametes, meiosis in the female produces only one large egg, which conserves all the cytoplasmic material in one package.

Hormonal Control of Female Reproductive Function

As in the male, the development, maturation, and function of the female reproductive system is under hormonal control. The maturation of the female system during puberty is thought to be initiated within the brain by the production of the hypothalamic hormone GnRH, which stimulates the secretion of LH and FSH by the anterior pituitary. In the female, the gonadotropins LH and FSH act cooperatively on the follicle cells of the ovary, stimulating them to produce the primary female sex hormone, estrogen, which has numerous target tissues. Increased estrogen levels at puberty stimulate maturation of external genitals and the development of secondary sex characteristics, such as the enlargement of breasts and the growth of hair in the armpits and pubic regions. Internally, elevated estrogen levels induce the maturation of the tissues of the uterus, enabling this organ to house and nurture a developing fetus.

Puberty is also the time when the ovary begins to produce mature ova on a cyclical basis, a process that will continue until *menopause,* the cessation of a woman's reproductive cycle. Menopause typically occurs in a woman's late forties; it is probably the result of a drop in GnRH secretion rather than an aging ovary. This is suggested by an experiment carried out by Terry Parkening and co-workers at the University of Texas. If the ovary from an aging mouse is transplanted into the body of a young mouse, the ovary continues to produce viable eggs. This suggests that the brain and pituitary of the young mouse are still able to stimulate oogenesis in an ovary that would ordinarily have stopped functioning if left in the older mouse. In contrast, a "young" ovary transplanted into an older mouse's body stops producing and ovulating oocytes.

The complex relationship among changing levels of hormones, the ovarian cycle, and the menstrual cycle is shown in Figure 31-11. The first phase of the ovarian cycle is characterized by the growth and differentiation of an oocyte. This maturation process is stimulated by both LH and FSH as well as by estrogen, which is produced within the follicle itself. As the growth of one or a few ovarian

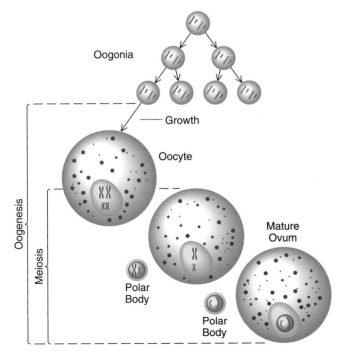

FIGURE 31-10

Gamete formation in the female. Oogonia are premeiotic cells that produce more of their own kind by mitosis. All of these oogonia enter oogenesis during fetal development, generating a large number of primary oocytes. Each month, a few of these primary oocytes (and their surrounding follicles, not shown) undergo growth and differentiation, after which meiosis takes place. Meiosis in the female produces only one viable egg containing the nutrient reserves (yolk) that support early embryonic development. The other products of meiosis are polar bodies that disintegrate. This series of events can be contrasted to that occurring in the male (31-6).

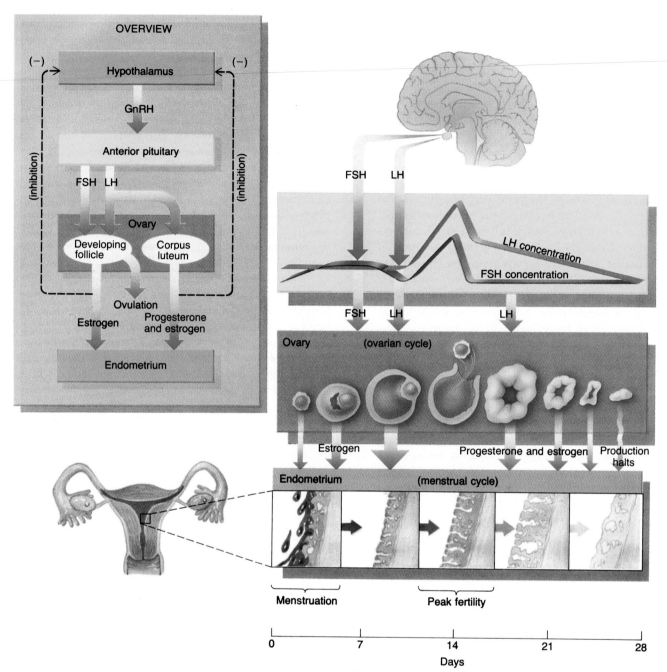

FIGURE 31-11

Synchronizing the ovarian and menstural (uterine) cycles. FSH is released from the anterior pituitary and promotes development of a follicle and the oocyte it contains. The maturing follicle secretes estrogen which stimulates endometrium development in the uterus. Presumably, when estrogen reaches a critical concentration, it stimulates the hypothalamus and anterior pituitary to flood the body with LH. This surge of LH triggers ovulation. LH also transforms the follicle into a corpus luteum. Progesterone (and some estrogen) produced by the corpus luteum maintain the enriched endometrium, so that it is receptive to implantation. Progesterone from the corpus luteum also inhibits GnRH release, so LH secretion stops, and the gonadotropin declines over the next 14 days. In the absence of implantation, the corpus luteum deteriorates (due to declining LH concentrations) until progesterone and estrogen production halts. Without these two hormones, the extra endometrial tissue cannot be maintained and is sloughed as menstrual fluid. Menstruation signals the beginning of a new cycle. With the disappearance of progesterone, inhibition of GnRH is relaxed and FSH is again released and another follicle begins to mature.

follicles nears completion, there is a marked surge in the pituitary's secretion of LH, which triggers ovulation.

The Menstrual (Uterine) Cycle

As the follicles are growing during each ovarian cycle, the uterus undergoes cyclical changes in form that are related to its function as a potential residence for the embryo. These dramatic changes in the uterus constitute the **menstrual cycle,** which takes an average of 28 days to complete (Figure 31-11). The first day of the menstrual cycle (which is defined as the day on which menstruation begins) is characterized by the flow of blood and discarded tissue from the uterus through the vagina. Menstruation takes place when the body "becomes aware" chemically that no fertilization or pregnancy has occurred following the last ovulation. As a result, the interior lining of the uterus, which had been prepared to receive a developing embryo, is broken down and rebuilt. The destruction of the bulk of the uterine wall is initiated by the constriction of arterioles, thereby cutting off the blood supply to the thickened *endometrium,* the inner lining of the uterus. The dead and dying tissue is then expelled from the uterus during menstruation by the contraction of the *myometrium* (uterine muscles).

After the period of menstruation is over, the rebuilding of the uterus begins, preparing the endometrium for the possible reception of a fertilized egg in the upcoming ovarian cycle. The first phase of uterine reconstruction produces the dramatic thickening of the inner glandular endometrium (Figure 31-11). This period is marked by the growth of the endometrial blood vessels and the formation of many new secretory glands. These changes are induced by rising levels of estrogen, produced by the follicle.

Following ovulation, on about day 14 of the menstrual cycle, the ruptured follicle remaining in the ovary is rapidly converted into a yellowish, glandular structure, called the **corpus luteum** (*corpus* = body, *luteum* = yellow). The corpus luteum (see Figure 31-9) secretes estrogen and large quantities of a second ovarian steroid hormone, progesterone. The increasing levels of estrogen and progesterone in the second half of the menstrual cycle act on the uterus to further its growth and development. If the ovulated ovum is fertilized, the resulting embryo will implant itself in the uterus by the eighth day following fertilization. The implanted embryo produces a gonadotropic hormone, called *human chorionic gonadotropin (HCG),* that acts on the ovary to maintain the activity of the corpus luteum. Biochemical tests for detecting HCG in a woman's blood or urine provide a reliable means of determining pregnancy.

If implantation occurs, the sustained corpus luteum continues to produce estrogen and progesterone, which maintain the endometrium where the embryo is developing. Since progesterone also inhibits the release of FSH, follicles cannot mature during pregnancy, making it highly unlikely that ovulation will occur while a woman is pregnant. Some birth control pills prevent ovulation by artificially increasing the concentrations of these hormones.

What happens if the ovulated ovum is not fertilized? Since the egg is not fertilized, there is no embryo available to secrete HCG. In the absence of HCG production, there is a rapid deterioration of the corpus luteum, which quickly loses its ability to produce estrogen and progesterone. When these hormones drop below a critical level, the uterus sloughs its extra endometrial tissue, and menstruation begins.

Women who exercise strenuously, such as marathon runners and gymnasts, often stop having their periods and become temporarily infertile. This response is thought to be due to a shutdown in the secretion of GnRH. Other causes of female infertility are discussed in The Human Perspective: Overcoming Infertility. The Bioethics box entitled "Frozen Embryos and Compulsive Parenthood" discusses the ethical implications of one method of overcoming infertility.

Gamete formation in the male and female are very different processes; each process is suited to the functions of the particular gamete. The role of the ovary is to produce a small number of large cells that contain the nutrients and storage materials needed to carry the fertilized egg through its early stages of development. In contrast, the role of the testis is to produce a large number of small, motile cells, one of which will fuse with the egg and contribute half of the genetic material (See CTQ #3.)

CONTROLLING PREGNANCY: CONTRACEPTIVE METHODS

Throughout history, people have devised many ways to prevent reproduction. *Contraceptive* methods have ranged from inserting elephant and crocodile manure into the vagina in order to block the entrance of sperm into the uterus to "fumigating" the vagina over a charcoal burner after intercourse to kill the sperm. Though primitive, these methods employed the same strategies as do some of the most effective modern contraceptive approaches; that is, physically blocking the uterine entrance or killing the sperm in the vagina.

Modern methods of avoiding pregnancy prevent ovulation, halt spermatogenesis, trap gametes in the oviducts or vas deferens, or interfere with implantation of the fertilized ovum in the uterus. A number of pharmaceutical companies are currently working on contraceptive *vaccines* for both men and women, whereby a person is stimulated to produce antibodies against proteins present on the surfaces of sperm or eggs. Questions have been raised as to the safety of inducing these types of autoimmune responses (page 657) and to the reversibility of the procedure. In the following discussion, *effectiveness* is expressed as the percentage of sexually active women who do not become pregnant in

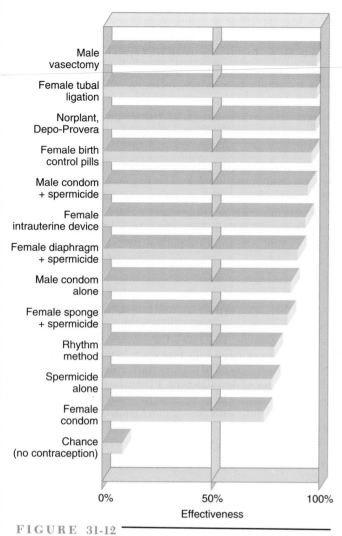

FIGURE 31-12 ——————————
Effectiveness of various forms of contraception. Effectiveness is expressed as the percentage of sexually active women who fail to become pregnant in one year when a particular form of contraception is practiced diligently.

FIGURE 31-13 ——————————
Norplant: an implantable contraceptive. These capsules, which are implanted under the skin, contain a synthetic progesteronelike hormone called levonorgestrel. Six of these capsules provide contraception for about 5 years.

the first year of practicing that form of birth control. In the absence of birth control, "effectiveness" would be approximately 10 percent. The effectiveness of various methods is summarized in Figure 31-12.

CHEMICAL INTERVENTION: ORAL CONTRACEPTIVES, HORMONES, AND SPERMICIDES

Synthetic estrogens and progesterones inhibit the release of FSH and LH which, in turn, prevents ovulation. Taken orally, either alone or in combination, these two hormones constitute *birth control pills*, the most effective (97 percent) *temporary* contraceptive method known today. Birth control pills must be taken daily to be effective. In 1990, the FDA approved the use of a new contraceptive called Nor-

plant, whereby thin, rubber capsules (Figure 31-13) are implanted beneath the skin of a woman's arm. The capsules release a synthetic hormone for a period of 5 years, blocking ovulation and preventing pregnancy. Another new hormonal contraceptive that provides extended protection is Depo-Provera, which consists of hormonal injections every 3 months.

To date, hormone-based contraceptives have been limited to women. Use of hormones to switch off sperm production in men also inhibits testosterone production, causing feminization and a reduction in libido. The World Health Organization has been testing weekly shots of testosterone as a spermatogenesis inhibitor, but excess levels of this hormone may carry unacceptable risks (page 528).

The most controversial chemical contraceptive is a drug called RU486, which is produced in France and interferes with the production of progesterone. Because RU486

◁ THE HUMAN PERSPECTIVE ▷
Overcoming Infertility

Approximately 10 percent of couples who try to have children are unable to do so because of physiological problems. This condition, known as **infertility,** may be due to a number of causes. Infertility in females often results from an inability to ovulate or from a blockage of the reproductive tract, while infertility in males typically results from either the production of abnormal sperm or a low sperm count. On the average, a male's ejaculate contains about 100 million sperm per milliliter of semen. When this number drops below about 20 million sperm per milliliter, fertility is markedly decreased.

Failure of a woman to ovulate is often due to an insufficient surge of LH secretion in the middle of the menstrual cycle. Women with this problem can be administered the hormone prior to the time ovulation would be expected to occur. Since the clinical procedure is not as "fine-tuned" as is the natural surge of LH, several ova are frequently released from enlarged follicles. If more than one of these cells should be fertilized, as is often the case in such situations, multiple births can occur.

When the cause of infertility is a blockage in the female reproductive tract, a couple may try to conceive by *in vitro fertilization,* a more invasive and expensive procedure. In this procedure, oocytes are removed by suction from the enlarged follicles of the ovary prior to ovulation, and the isolated oocytes are fertilized *in vitro* —outside the body in a laboratory dish; babies conceived in this manner are commonly referred to as "test-tube babies." Eggs fertilized *in vitro* are allowed to develop for a few days in culture medium. One of the embryos is then transferred to the woman's uterus, where it has about a 20 percent chance of implanting and developing to term. In recent years, eggs that have been fertilized *in vitro* have been kept alive in a frozen state so that subsequent attempts at implantation can be made without having to operate again to remove additional oocytes.

In a small percentage of cases when the wife is infertile, couples have turned to a *surrogate:* a woman who is willing to be artificially inseminated with the husband's sperm, carry the resulting offspring to term, and give the baby to the couple after birth. Ethical questions concerning the practice of surrogacy were raised in the late 1980s when a surrogate named Mary Beth Whitehead decided she wanted to keep the baby to whom she had given birth, despite having signed a contract in which she agreed to give up the baby in exchange for payment. In 1987, the New Jersey courts awarded custody to the Sterns, the couple who had paid Whitehead for bearing the child; Whitehead was granted surrogate visitation rights. The state court also outlawed the practice of surrogacy, ruling that it amounted to the sale of babies. Since the ruling, a number of other states have passed laws governing surrogacy contracts. Adoption remains the most common choice among couples who are not able to bear children of their own.

renders the uterus incapable of receiving an embryo for implantation, the drug has been used in Europe as a "morning after" pill; that is, a pill that blocks pregnancy (which, by most definitions, begins at implantation) for several days after unprotected sex. RU486 also acts as an "abortion" pill by causing the uterus to reject an implanted embryo. The drug has been prescribed in Europe to terminate pregnancies through about the seventh week. Banned in the United States by the FDA, RU486 became the center of a storm of controversy in 1992 when a woman attempted to bring the drug into the United States aboard a flight from Paris. The incident reopened discussion as to whether or not RU486 should be approved for use in the United States.

Spermicides are sperm-killing chemicals that are inserted into the vagina before intercourse. When used alone, spermicides are not very reliable (79 percent effective) and thus are often used in combination with a diaphragm or condom.

MECHANICAL INTERVENTION

There are various forms of mechanical intervention to prevent pregnancy. The most popular include the IUD, diaphragm, condoms, and the sponge.

Intrauterine Device (IUD)

The *intrauterine device,* or *IUD,* is a plastic or metal device that is inserted by a physician into the uterus, where it prevents implantation of a developing embryo. In spite of its effectiveness (94 percent) and convenience, once inserted, the IUD may cause serious side effects that negate its usefuless for some women. Uterine puncture, a potentially fatal complication, occurs in about 1 in every 1,000 women. The IUD may also aggravate existing low-grade infections of the uterus and lead to a serious and painful condition called *pelvic inflammatory disease.* The most common side effects are painful cramping and irregular

◁ B I O E T H I C S ▷
Frozen Embryos and Compulsive Parenthood
By ARTHUR CAPLAN
Director of the Center for Biomedical Ethics at the University of Minnesota

The Tennessee Supreme Court issued a ruling on June 1, 1992, concerning the fate of seven embryos frozen in a liquid nitrogen tank in a Knoxville, Tennessee, fertility clinic. The ruling has important implications not only for the fate of the more than 22,000 frozen embryos at clinics around the nation but also for the national debate about abortion.

Mary Sue and Junior Lewis Davis were married on April 26, 1980. They wanted children. Mary Sue got pregnant five times, but each time she suffered a tubal pregnancy. She could conceive, but the embryos kept getting stuck in her oviducts instead of implanting in her uterus. Finally, Mary Sue had to have her tubes removed, leaving her able to make eggs but unable to bear a child.

In 1985, the Davises went through six attempts at in vitro fertilization, but they had no luck. They decided after their last attempt, in December 1988, that instead of giving up they would freeze the seven embryos and try again at some future time. Junior Davis filed for divorce in February 1989, however, claiming that he had known his marriage was falling apart for some time, but he had hoped that the birth of a baby might have saved it. The Davises were divorced, but there was one unre-solved issue: who would get custody of the frozen embryos? Mary Sue wanted custody, initially so that she could try to have a baby. Junior Lewis objected. Mary Sue went to court to obtain sole custody of the embryos.

The first court to hear the case awarded Mary Sue custody of the embryos on the grounds that the embryos were "human beings" from the moment of fertilization and ought to have the "opportunity . . . to be brought to term through implantation." Junior Lewis appealed the decision, and a Tennessee appellate court reversed the trial court, assigning joint custody over the embryos on the grounds that Junior had a "constitutionally protected right not to beget a child" without his consent.

Mary Sue decided to appeal this ruling to the Tennessee Supreme Court. She no longer wanted to use the embryos herself, however; she wanted to donate them to another childless couple, rather than have them destroyed. Junior Lewis remained adamantly opposed, preferring that the embryos be destroyed.

In its ruling, the Tennessee Supreme Court scolded fertility clinics for not making the disposition of unused or unwanted embryos part of the consent process. The court rejected the view that legal or moral rights begin at the point of conception. It found no legal basis for assigning personhood to an eight-cell embryo or for giving an embryo the same legal standing as a child. Instead, the court held that there is a fundamental right to privacy, according to the laws of Tennessee and the U.S. Constitution, that forbids compelling Junior Lewis Davis to become a parent against his will. The court concluded that "Mary Sue Davis' interest in donating [the embryos] is not as significant as the interest Junior Davis has in avoiding parenthood" and forbade the implantation of the seven embryos into any woman unless Junior Lewis agrees. In 1993, the U.S. Supreme Court upheld the Tennessee ruling.

The significance of the case goes further than the court ruling. The Tennessee Supreme Court has now dismissed the claim that rights begin at conception. It has also asserted the view that no man should be forced to parent against his will. These two rulings have obvious implications for the future of abortion policy in this country. For example, the U.S. Supreme Court will find it very difficult to ignore the Davis case in deliberating the fate of *Roe* v. *Wade,* the case that legalized abortion.

bleeding, especially immediately after the device is inserted. Although infrequent, these complications created enough concern to cause IUDs to be taken off the market in the United States in 1987. A few types of IUDs are again being made available, notably those that have few negative side effects.

Diaphragm with Spermicide

Legend has it that the famous lover Casanova used half of a hollowed out lemon to cover the cervix of a lover in an attempt to reduce the number of new little Casanovas he sired. The modern *diaphragm,* a thin rubber dome, works in the same way, although with considerably more success (90 percent effectiveness) because it is precisely fitted to the woman's cervix and is used in conjunction with a spermicide. The diaphragm is inserted just before intercourse and must remain in place for 8 hours following intercourse to be fully effective.

Condoms

Condoms are thin sheaths that are worn over the penis and trap sperm, preventing fertilization. Condoms have the additional advantage of preventing sexually transmitted diseases, which is why they are also called "prophylactics," meaning "disease preventing." Condoms have become an important line of defense in the war against AIDS. Used alone, condoms are 88 percent effective as contraceptives; their effectiveness increases to 95 percent when used with a spermicide. Inconvenience and diminished sensations are two drawbacks to condoms.

Recently, a female "condom" has become available, which fits into the vagina, forming a plastic lining that cannot be penetrated by sperm. The sheath is held in place by a ring that remains outside the vagina. The effectiveness of the female condom remains to be determined.

Vaginal Sponge with Spermicide

The spermicide-saturated, cup-shaped *sponge* fits over the cervix prior to intercourse, much like a diaphragm but without the need for fitting by a physician. This advantage may be offset by a serious concern: a slight increase in the incidence of *toxic shock syndrome.* This potentially fatal disease is usually due to excessive growth of a particular strain of toxin-producing bacteria in the vagina. When the sponge is used in conjunction with spermicide, the effectiveness is around 85 percent.

PERMANENT STERILIZATION

In a *vasectomy,* a surgical procedure for men, a portion of each vas deferens is removed, and the cut ends of the tubes are tied. Although spermatogenesis continues, sperm are blocked from reaching the penis, so the man's ejaculate contains no sperm cells, making the procedure 99.85 percent effective. The volume of semen and the sensations of orgasm are unaffected by vasectomy.

In a *tubal ligation,* the woman's oviducts are surgically cut and sealed, preventing an egg from reaching the uterus or from even coming in contact with sperm but allowing ovulation to continue. Tubal ligation is 99.6 percent effective.

NATURAL METHODS

There are two natural methods of birth control: rhythm and coitus interruptus. The *rhythm method* requires abstinence from intercourse during a woman's fertile period, about 12 hours before and 48 hours after ovulation. A major problem with the rhythm method is the difficulty in predicting when ovulation occurs, especially in women with irregular ovarian cycles. (Ovulation occurs 14 days *before* the end of the cycle, not 14 days after the beginning of the cycle, making the prediction difficult.) Even when practiced diligently, failure rates range from 15 to 35 percent; thus, effectiveness is variable, at best. Effectiveness can be improved by keeping a daily record of the woman's body temperature, which rises about half a degree at the onset of her peak fertility days. Another clue is the change in consistency of cervical mucus from thick and sticky to thin and clear just before ovulation.

During *coitus interruptus,* also known as withdrawal, the man removes his penis from the vagina before ejaculation. Because it requires willpower and reduces gratification, failures are very likely.

Prevention of pregnancy can be accomplished by interfering with either the process of sperm–egg union or implantation. There are diverse ways of preventing pregnancy, each with varying degrees of effectiveness and safety. (See CTQ #5.)

SEXUALLY TRANSMITTED DISEASES

No discussion of human sexual reproduction is complete without considering those diseases that are transmitted from one sexual partner to another, a group of disorders called **sexually transmitted diseases (STDs).** These diseases include AIDS, syphilis, gonorrhea, genital herpes and more than a dozen others. To understand STDs, it is necessary to be familiar with the properties of the viruses and bacteria that cause most of these disorders. For this reason, we have deferred the discussion of STDs until Chapter 36, the chapter in which the structure and function of viruses and bacteria are considered (see The Human Perspective, page 794).

REEXAMINING THE THEMES

Relationship between Form and Function

The architecture of a sperm cell is tailored by evolution to accomplish the following tasks: swimming to the egg, penetrating the egg's surface barriers, and delivering a nucleus that contains the paternal genes. These functions are evident in the sperm's streamlined shape, its virtual lack of cytoplasmic and nuclear fluid, its cap of hydrolytic enzymes that are capable of digesting their way through the egg's outer layers, its compact sheath of mitochondria that provide the chemical energy needed to fuel the sperm's journey, its single flagellum that provides the motile force for locomotion, and its condensed nucleus that contains the full haploid number of chromosomes packed into the least possible volume. While the egg lacks such elaborate adaptations, its large size and nutrient content make the egg an ideal package for supporting the needs of embryonic development.

Biological Order, Regulation, and Homeostasis

Virtually every aspect of reproductive activity in mammals—from the development of the system in the embryo, to the maturation of the system at puberty, to the continuing day-to-day functioning of the system in the adult—is under tight regulation by the body's endocrine system. The master controls are located in the hypothalamus. The neuroendocrine cells located in this part of the brain secrete hormones that control the activity of the pituitary gland which, in turn, secretes hormones that control the activities of the gonads. The gonads secrete hormones of their own, particularly steroids, such as estrogen, progesterone, and testosterone. The timely secretion of these steroids determines the ability of the individual to reproduce successfully.

Unity within Diversity

The process of sexual reproduction is basically the same in all animals: Males produce very small, motile sperm, while females produce much larger, immotile eggs. Both types of gametes form in conjunction with meiosis, which reduces their chromosome number in half. These characteristics of sexual reproduction must have appeared at an early stage in the evolution of eukaryotes and have been retained ever since. Despite the similarities in gamete formation and structure, there is a great variety in the structure of reproductive systems and in the strategies used to bring the gametes together. Sex itself provided the genetic variety that allowed this diversity to develop.

Evolution and Adaptation

Sexual reproduction is one of life's most important evolutionary developments. Mixing genes from two parents has generated a vast diversity of organisms, introducing combinations of traits that never would have existed had asexual reproduction remained the only option. Asexual reproduction creates sameness, whereas sexual reproduction generates diversity, creating new types of organisms subject to natural selection; those with favorable adaptations proliferated. Today's rich diversity of life would not have been possible without sexual reproduction.

SYNOPSIS

Both asexual and sexual reproduction have advantages. Asexual reproduction is a biologically economical way to generate identical copies of oneself rapidly. In contrast, sexual reproduction is more costly energetically, but it produces offspring with variable characteristics, which makes it much more likely that a species population will be able to survive changing environmental conditions.

Animals employ several strategies that increase the likelihood of fertilization. Animals that depend on external fertilization release large numbers of sperm and eggs at about the same time, in the same location. External fertilizers use various mechanisms to help coordinate the release of gametes. Internal fertilizers produce fewer eggs, each with a much higher probability of developing to adulthood. Some animals are hermaphrodites; individuals that possess both male and female reproductive systems.

Gametes are produced by gonads. In humans, spermatozoa are generated in the seminiferous tubules of the male's testes, and ova are generated in the follicles of the female's ovaries. Sperm are highly differentiated cells that derive from spermatids that have formed from primary spermatocytes by meiosis. Primary spermatocytes are derived from spermatogonia, whose mitotic divisions provide a continuous source of germ cells throughout life. Sperm produced in the testes move into the adjacent tubules of the epididymis and, ultimately, into the vas deferens, where they are mixed with fluids from the seminal vesicles, Cowper's glands, and prostate gland to form semen, which can be ejaculated during orgasm.

In the female, all germ cells have entered meiosis by the time of birth. Most of these oocytes are located in small, primordial follicles. During each ovarian cycle, a number of follicles enlarge, and their contained oocyte undergoes enlargement and differentiation to form a cell that contains the nutrients needed to support embryonic development. One of these oocytes is released into an oviduct during each cycle. The oocyte then continues to the metaphase of the second meiotic division and awaits fertilization. Meiosis in the female includes highly unequal divisions, producing one cell that retains virtually all of the cytoplas-

mic material and forms the egg, while the others contain little more than nuclei and eventually disintegrate.

Sexual development and gamete formation in both sexes are regulated by hormones. In both males and females, the hypothalamus produces gonadotropin-releasing hormone (GnRH), which controls the production and release of two gonadotropins, LH and FSH, by the anterior pituitary. In the male, FSH promotes spermatogenesis, while LH stimulates the interstitial cells of the testes to produce testosterone, which maintains male reproductive function. In the female, both FSH and LH control the ovarian cycle and stimulate the production of estrogen. These three hormones, in addition to the progesterone that is produced by the corpus luteum (the follicle from which the oocyte is ovulated) control the menstrual (uterine) cycle. During each cycle, the lining of the uterus is thickened and vascularized in anticipation of an implanted embryo. If the embryo implants itself, HCG is produced and the corpus luteum and uterus are maintained. If the ovulated egg is not fertilized, the failure to produce an implanted embryo leads to the deterioration of the corpus luteum, a drop in progesterone and estrogen production, and the breakdown and sloughing of the uterine wall.

Key Terms

reproduction (p. 664)
asexual reproduction (p. 664)
sexual reproduction (p. 664)
fission (p. 664)
budding (p. 664)
parthenogenesis (p. 664)
external fertilization (p. 665)
internal fertilization (p. 665)
penis (p. 665)
hermaphrodite (p. 667)
spermatozoa (p. 669)
ova (p. 669)
scrotum (p. 669)
urethra (p. 669)
circumcision (p. 669)
seminiferous tubule (p. 669)
spermatogenesis (p. 669)
spermatogonia (p. 669)

primary spermatocyte (p. 669)
secondary spermatocyte (p. 669)
spermatid (p. 669)
epididymis (p. 670)
vas deferens (p. 670)
acrosome (p. 670)
semen (p. 670)
ejaculation (p. 672)
prostate gland (p. 672)
seminal vesicle (p. 672)
Cowper's gland (p. 672)
orgasm (p. 672)
follicle-stimulating hormone (FSH) (p. 672)
luteinizing hormone (LH) (p. 672)
gonadotropin-releasing hormone (GnRH) (p. 672)
oogenesis (p. 673)

vulva (p. 673)
labia majora (p. 673)
labia minora (p. 673)
clitoris (p. 673)
vagina (p. 673)
cervix (p. 673)
uterus (p. 673)
oviduct (p. 673)
oocyte (p. 673)
follicle (p. 673)
follicle cell (p. 674)
oogonia (p. 674)
ovarian cycle (p. 674)
ovulation (p. 674)
polar body (p. 675)
menstrual cycle (p. 677)
corpus luteum (p. 677)
infertility (p. 679)

Review Questions

1. Describe three differences in gamete production by men and women.

2. Trace the odyssey of a sperm cell from the spermatogonial stage to fertilization.

3. Explain why LH and FSH have such different effects in human males and females. Describe these effects and how the levels of these gonadotropins are controlled in males and females.

4. Compare and contrast the vas deferens and oviducts; penis and clitoris; testes and ovaries; testosterone and estrogen; parthenogenesis and fission; interstitial cells and follicle cells; ovarian cycle and menstrual cycle; uterus and cervix.

Critical Thinking Questions

1. Speculate on the differences in the use and value of scientific information obtained from the studies of individuals (Freud's method) and the studies of populations (Kinsey's method).

2. Compare the advantages and disadvantages of external fertilization versus internal fertilization and asexual reproduction versus sexual reproduction.

3. What, if any, is the difference in chromosome composition between a spermatogonium and a primary spermatocyte; a spermatid and a sperm; an oocyte in a primordial follicle, an ovulated egg, and an egg awaiting fertilization?

4. Explain why a woman who is taking birth control pills must stop taking them for several days every month to allow menstruation to occur.

5. The United States has one of the highest teenage pregnancy rates in the world. What do you think could be done to reduce the number of unwanted pregnancies among teens? (Obviously, there is no right or wrong answer to this question.)

Additional Readings

Barinaga, M. 1991. Is homosexuality biological? *Science* 253:956–957. (Intermediate)

Bullough, V. L., and B. Bullough. 1990. *Contraception: A guide to birth control methods.* Prometheus Books. Buffalo, NY: (Introductory)

Christenson, C. V. 1971. *Kinsey: A biography.* Indiana University Press. (Introductory)

Daly, M., and M. Wilson. 1983. *Sex, evolution, and behavior.* New York: Wadsworth. (Intermediate)

Duellman, W. E. 1992. Reproductive strategies of frogs. *Sci. Amer.* July:80–87. (Intermediate)

Knobil, E., and J. D. Neill. 1988. *The physiology of sex.* New York: Raven. (Advanced)

Marx, J. L. 1988. Sexual responses are—almost—all in the brain. *Science* 241:903–904. (Intermediate)

Silber, S. J. 1987. *How not to get pregnant.* New York: Scribners. (Introductory)

Symons, D. 1979. *The evolution of human sexuality.* New York: Oxford University Press. (Intermediate)

Sulloway, F. J. 1979. *Freud, biologist of the mind.* New York: Basic Books. (Introductory)

Witters, W., and P. Witters. 1980. *Human sexuality, a biological perspective.* New York: Van Nostrand. (Introductory–Intermediate).

Animal Growth and Development: Acquiring Form and Function

STEPS
TO
DISCOVERY
Genes that Control Development

THE HUMAN PERSPECTIVE

The Dangerous World of a Fetus

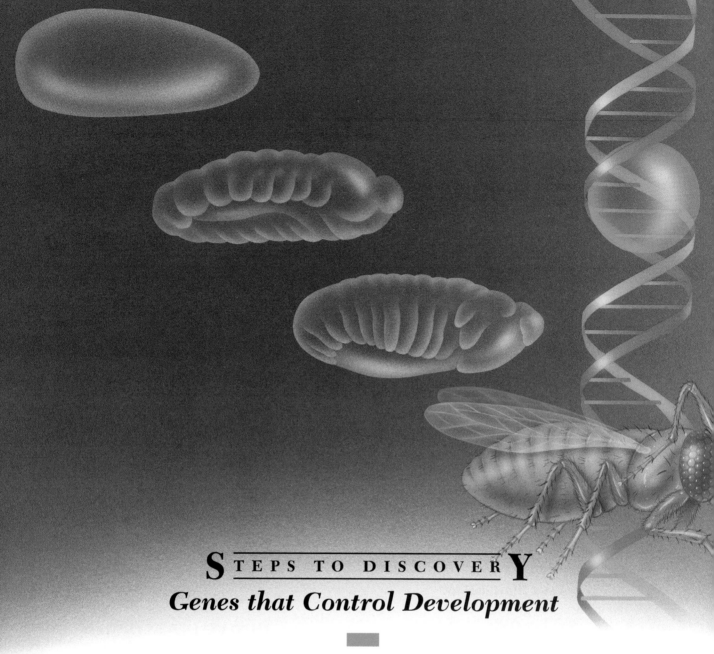

Genes that Control Development

During the 1940s, Edward B. Lewis, a geneticist at the California Institute of Technology, studied a mutant fruit fly with an abnormal body organization. An insect is composed of a head, thorax, and abdomen, each part containing a defined number of segments. The last segment of the thorax of a fruit fly is usually wingless, but Lewis's mutant fly had a second pair of wings on this segment; thus, the mutant was named *bithorax*.

In following years, other mutant fruit flies were isolated that showed even more profound disturbances in body organization. For example, the mutant *antennapedia,* studied

by John Postlethwait and Howard Schneiderman of Case Western Reserve University in the late 1960s, has a pair of legs growing out of its head in the place where antennae are normally found. Genes such as these, that affect the spatial arrangement of the body parts, are called **homeotic genes.** While humans are not known to suffer such drastic homeotic mutations as bithorax and antennapedia, there are many examples of serious developmental malformations that could be the result of mutations in homeotic genes. In addition, human embryos with drastic developmental defects tend to abort spontaneously, so the existence of mu-

Mutations in a crucial gene can "throw a switch" in the stages of development of a fruit fly that causes a pair of legs to develop

tant homeotic alleles might easily go undetected.

Homeotic genes are thought to play a role in the basic process by which each part of an embryo becomes committed to developing along a particular pathway—toward forming a leg rather than an antenna, for example. One way homeotic genes might exert such profound influence on the course of development is by acting as a type of "master" gene. As such, homeotic genes would control the transcriptional activity of other genes, whose products actually form the various tissues. The antennapedia gene, for example, might code for a protein that normally switches on the genes required for antenna formation in the appropriate cells of the developing head. If the antennapedia gene becomes defective, a different cluster of genes may be switched on, and the cells of that part of the body differentiate into a leg instead of an antenna.

In 1983, Walter Gehring, a Swiss biologist, discovered that a number of homeotic genes in the fruit fly contained a common sequence of about 180 nucleotides; this sequence was named the **homeobox.** Once the homeobox DNA from the fruit fly had been isolated, Gehring and others were able to search the DNAs of other organisms to see if they contained similar DNA segments. The homeobox sequence was soon found to exist within the DNA of many different animals, from worms to humans; in fact, it was even present in plants. This finding suggested that similar types of genetic processes take place during the development of very different types of organisms, but it left an important question unanswered: What was the function of the homeobox?

Recall from Chapter 17 that deciphering the amino acid sequence encoded by a gene is a relatively simple matter, once the nucleotide sequence of the DNA has been determined. When the amino acid sequence of the homeobox was deciphered and compared to the sequence of other known proteins, it was found to be very similar to a gene regulatory protein found in yeast that was known to bind to DNA. This correlation suggested that homeotic genes encoded DNA-binding proteins. By binding to a specific portion of a particular chromosome, the product of a homeotic gene could activate or repress transcription of nearby genes, much like a repressor protein in bacterial operons or a steroid receptor protein in eukaryotes (Chapter 15). In this way, homeotic genes might control the course of development. This concept was confirmed in 1988, when Patrick O'Farrell and his colleagues at the University of California demonstrated that products of two of the homeobox-containing genes in the fruit fly actually bound to DNA and altered the rate of transcription of nearby genes.

In the past few years, a number of investigators studying homeotic genes have turned their attention from fruit flies to mice. In doing so, they have invited speculation on the role of these genes in human development. In 1991, for example, Osamu Chisaka and Mario Capecchi of the University of Utah produced transgenic mice (page 319) that carried a genetically engineered version of a homeotic gene in place of the normal gene. The mice that were born with this altered homeotic gene exhibited a variety of severe abnormalities, including deformations of the throat and heart. Interestingly, a similar complex of abnormalities occurs in DiGeorge's syndrome, a rare human disorder that usually causes the death of the affected infant within the first few months of life. It is possible that this human condition is a result of a mutation in a homeotic gene whose expression is required during human development.

where a pair of antennae would normally be located.

CLEAVAGE AND BLASTULATION: DIVIDING A LARGE ZYGOTE INTO SMALLER CELLS

A fertilized egg is an unbalanced cell; it has a huge amount of cytoplasm but only two sets of homologous chromosomes. This situation changes very rapidly during early development, as the egg undergoes a succession of mitotic divisions, called **cleavage.** Cleavage is not a time of growth but a period when the oversized egg is divided into a large number of smaller cells, known as **blastomeres** (Figure 32-4). Cleavage ends with **blastulation,** the formation of a ball of cells, called a **blastula.** In most animals, the blastula contains an internal, fluid-filled chamber, called the **blastocoel,** whose relative size and location depends primarily on the amount of yolk in the egg (Figure 32-5). For example, both sea urchin and mammalian embryos are essen-

tially devoid of yolk; both develop via a blastula stage that has a large, central blastocoel. In contrast, frogs and birds produce larger eggs with considerably more yolk. Consequently, the relative size of the blastocoel in these animals is dramatically reduced.

The Developmental Potential of Cleaving Cells

During cleavage, the fertilized egg divides into many cells, each with a specific developmental "fate." A cell may become part of the liver, the brain, or some other part of the body. One of the first questions developmental biologists asked was: Can the cells of an embryo be experimentally "tricked" into developing into structures other than those they would normally form? In other words, how rigidly is the fate of a cell determined? The answer depends on sev-

First cleavage furrow

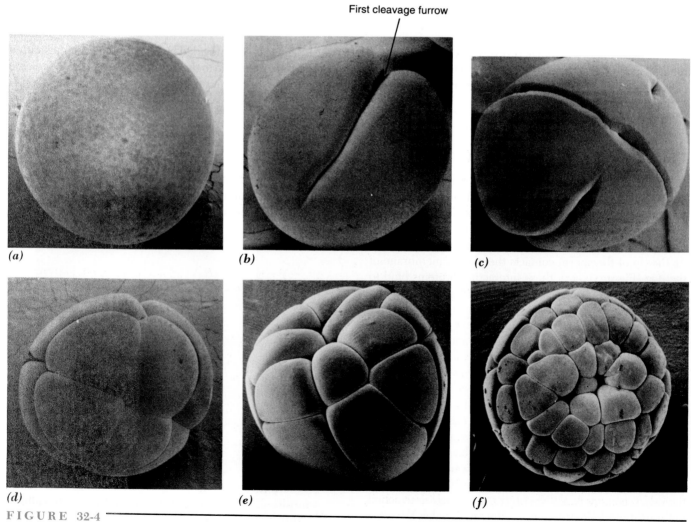

(a) *(b)* *(c)*

(d) *(e)* *(f)*

FIGURE 32-4

Cleavage of a frog egg divides the large, single-celled zygote into a number of smaller cells, called blastomeres. Although the egg contains more cells following cleavage, the total volume of cellular material is the same as is that of the original zygote.

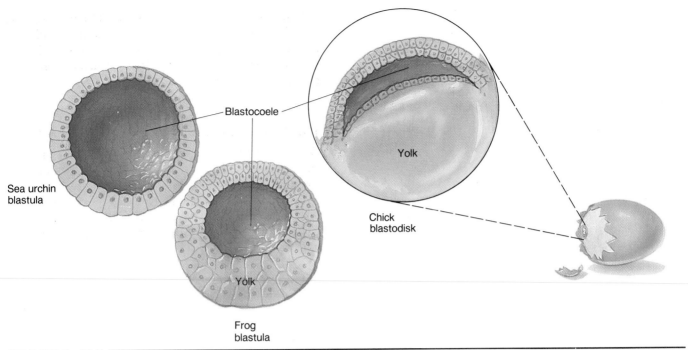

FIGURE 32-5

The blastula stage of a sea urchin, frog, and chick. A sea urchin develops rapidly from a small, relatively yolk-free egg into a feeding larva. The sea urchin's blastula has a large blastocoel, surrounded by a single layer of cells. Frogs and other amphibians have eggs that contain a relatively larger content of yolk, which leaves relatively less room inside the blastocoel. Reptiles and birds—animals whose embryos develop on dry land—must supply their developing offspring with enough food and water to last until they hatch from the shells that enclose them. The enormous yolk and albumen (egg white) provide these resources and occupy most of the egg space. Because of the presence of so much yolk, the growth of early embryonic cells is restricted to a small area on the yolk surface, where an embryonic disk, or *blastodisk*, is formed.

eral factors, particularly the species and the stage of development being studied. We will examine the question at an early stage in the development of a frog.

In most cases, the first cleavage furrow in the frog (see Figure 32-4) divides the egg into two cells that form the left and right halves of the animal. This can be demonstrated by injecting a fluorescent dye into one of the first two blastomeres; only the cells on one half of the body receive the dye. When the first two cells of a frog embryo are separated from each other with a fine needle, each cell is able to develop *independently* into a complete embryo and larva (Figure 32-6, upper). We can conclude that, when separated from its neighbor, each cell somehow responds to the fact that it is now alone and forms a complete individual rather than the half of an individual it would have formed if left attached to the other cell. This is a remarkable observation. How can a part of an embryo have a "sense" of the whole, "knowing" it is now alone and must form the entire organism? How can a cell that is to form a part of an embryo suddenly regulate

its development and form additional parts not in its normal repertoire? These remain among the most basic, unanswered questions in developmental biology.

The Roles of the Genes and the Cytoplasm

Genes and cytoplasmic materials work together to generate the biological complexity that characterizes embryonic development. The importance of both components—genes and cytoplasm—is readily demonstrated. For example, if a fertilized sea urchin, frog, or mouse egg is treated with a drug that inhibits the transcription of genes, the egg will develop relatively normally up to the blastula stage, at which time development stops. This simple experiment illustrates two important aspects of development:

1. *An embryo doesn't need its genes for the very first stages of embryonic development* Whatever new proteins are needed to carry an embryo from the form of a fertilized egg to the blastula stage are synthesized using

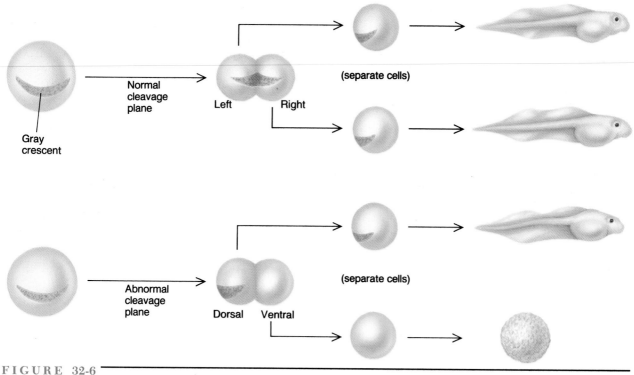

FIGURE 32-6

The developmental fate of isolated frog blastomeres. In the upper panel, the first cleavage plane divides the right and left halves of the egg, as indicated by the bisection of the *gray crescent;* each cell is able to develop independently into an intact tadpole. In the lower panel, the first division is altered so it doesn't bisect the gray crescent; only the dorsal cell containing the gray crescent develops normally.

mRNAs that were stored in the egg at the time it was released from the ovary.

2. *An embryo must use its genes to get past the blastula stage* Before an embryo can begin the next stage of development, its genes must spring into action, producing new mRNAs that direct the sweeping changes that are to occur during later development.

Further experiments provide evidence of the importance of the egg's cytoplasm. Occasionally, the first cleavage plane of an amphibian egg fails to divide an egg into left and right halves. Instead, the plane divides the cell into blastomeres that will give rise to the dorsal and ventral halves of the animal (Figure 32-6, lower). (The dorsal side of the egg will become the dorsal side of the animal—the side containing the animal's backbone. The opposite, or ventral, side will become the belly of the animal.) The division of the egg into dorsal and ventral blastomeres, rather than left and right blastomeres, can be detected by the failure of the cleavage plane to pass through the **gray crescent,** a distinctive, crescent-shaped, cytoplasmic landmark (shown in Figure 32-6) that always forms after fertilization on the *dorsal* surface of a frog egg. One might expect that a dorsal or ventral blastomere would have the same developmental

potential as a right or left blastomere; that is, the potential to form an entire embryo. This is not the case, however. Whereas, the isolated dorsal blastomere will develop into a normal embryo, the isolated ventral blastomere forms little more than a ball of cells. Only the blastomere that contains the gray crescent has all the cytoplasmic components needed for development, even though both cells contain identical genetic information. The important role of the gray crescent in frog development is discussed further below.

GASTRULATION: REORGANIZING THE EMBRYO

Gastrulation is the process whereby the undistinguished blastula is transformed into a more complex stage of development, called a **gastrula.** Gastrulation is characterized by an extensive series of coordinated cellular movements, whereby regions of the blastula are displaced into radically different locations. The dramatic events that take place during gastrulation in a frog (Figure 32-7) provide an overview of the process.

▲ Even though gastrulation takes place in markedly different ways in different types of embryos, in principle, the

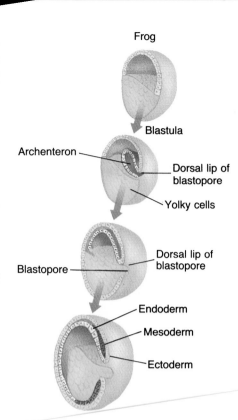

Frog

Blastula

Archenteron

Dorsal lip of blastopore

Yolky cells

Blastopore

Dorsal lip of blastopore

Endoderm

Mesoderm

Ectoderm

FIGURE 32-7

Gastrulation in the frog. The first indication of the onset of gastrulation in the frog is the appearance of a groove on the dorsal side of the embryo, just below the gray crescent. The opening into the interior of the embryo is the *blastopore,* and the fold above the groove is the *dorsal lip* of the blastopore. During gastrulation, cells at the rim of the blastopore migrate into the interior of the embryo and are replaced by new cells that move over the surface toward the blastoporal lip. Once inside the embryo, the cells move deeper into the interior, away from the blastopore, forming interior walls of an increasingly spacious cavity called the *archenteron.* The walls of the archenteron consist of endodermal cells (in purple) that will give rise to the digestive tract. The archenteron remains open to the outside through the blastopore, which corresponds in position to the future anus. Between the ectoderm and the endoderm and above the archenteron, a third group of cells develop into the mesoderm (in green), which will give rise to the skeleton, muscles, and other mesodermal derivatives. Once gastrulation is complete, the entire external layer of the embryo is composed of ectodermal cells. (Note that gastrulation in other animals, such as the sea urchin or chick, occurs by a very different pathway.)

process achieves a similar end result. Regardless of the animal, by the time gastrulation has been completed, the embryo (gastrula) can be divided into an inner, middle, and outer layer, which correspond to the three embryonic *germ layers:* the endoderm, mesoderm, and ectoderm. In vertebrates (such as the frog of Figure 32-7), the inner layer of the gastrula, or **endoderm,** gives rise to the digestive tract

and its derivatives, including the lungs, liver, and pancreas; the outer layer, or **ectoderm,** gives rise to the epidermal layer of the skin and to the entire nervous system; and the middle layer, or **mesoderm,** gives rise to the remaining body components, including the dermal layer of the skin, bones and cartilage, blood vessels, kidneys, gonads, muscles, and the inner linings of the body cavities (Figure 32-8).

FIGURE 32-8

A schematic illustration of the fate of the three germ layers in vertebrates.

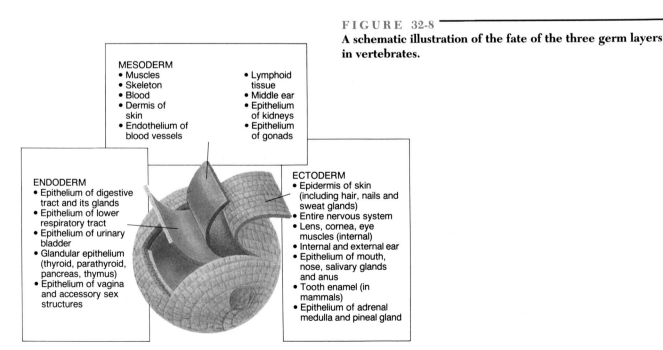

MESODERM
• Muscles
• Skeleton
• Blood
• Dermis of skin
• Endothelium of blood vessels
• Lymphoid tissue
• Middle ear
• Epithelium of kidneys
• Epithelium of gonads

ENDODERM
• Epithelium of digestive tract and its glands
• Epithelium of lower respiratory tract
• Epithelium of urinary bladder
• Glandular epithelium (thyroid, parathyroid, pancreas, thymus)
• Epithelium of vagina and accessory sex structures

ECTODERM
• Epidermis of skin (including hair, nails and sweat glands)
• Entire nervous system
• Lens, cornea, eye muscles (internal)
• Internal and external ear
• Epithelium of mouth, nose, salivary glands and anus
• Tooth enamel (in mammals)
• Epithelium of adrenal medulla and pineal gland

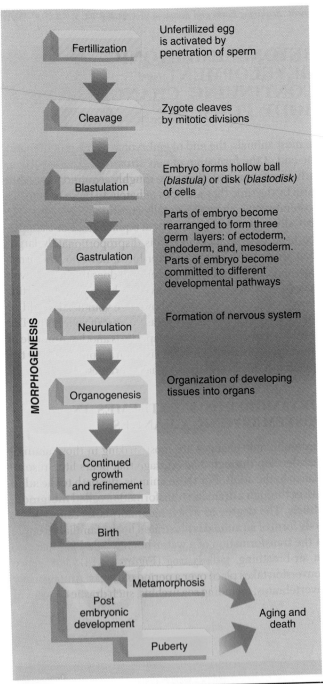

FIGURE 32-15
Summary of the events of growth and development in animals.

form can occur in a matter of minutes. A generalized summary of the stages of animal development is illustrated in Figure 32-15.

Developmental changes do not stop with the formation of an embryo. Animals continue to develop as they achieve a larger form and undergo sexual maturation. Most animals gradually their basic adult body form as the result of morphosis; others do so by a rapid, dramatic metamorphosis. (CTQ #3.)

(a)

HUMAN DEVELOPMENT

The development of a human embryo begins with fertilization; 6 to 8 days later, the embryo "burrows" into the prepared endometrium of the uterus (Figure 32-16), rupturing uterine blood vessels as it penetrates and implanting itself in the lining. By the time the entry site reseals itself, the embryo is surrounded by a pool of maternal blood, which is continually replenished by fresh blood from maternal vessels. Branching projections, called *villi*, sprout from the **chorion,** the embryo's outer membrane covering. These chorionic villi provide a large surface for the exchange of respiratory gases, nutrients, and wastes between the embryo and its mother, until a **placenta,** constructed from the embryo's chorion and the mother's uterine lining, takes over the exchange. Veins and arteries develop in the villi, connecting the embryo's early circulatory system with the blood-filled space, serving as a temporary life support system. These connecting vessels soon become channeled within an **umbilical cord.** One end of the cord is attached to the embryo's belly, the other end to the placenta. In the placenta, the embryo's vascularized chorionic villi are bathed in maternal blood. The two blood supplies are separated by a thin layer of cells, permeable to nutrients, gases, and wastes.

THE STAGES OF HUMAN DEVELOPMENT

During the first 2 months following fertilization, all the major organs appear, many in a functioning capacity. By the end of the eighth week, the embryo has acquired a distinctively human form. Up to this point, the developing offspring is still called an embryo. During the last 7 months in the uterus, however, it is referred to as a **fetus.** In the fetal stage, organ refinement accompanies overall growth, as the fetus develops the ability to survive outside the uterus.

The 266 or so days between conception and birth are traditionally grouped into three stages, each called a *trimester.* The most dramatic changes occur during the first of these periods (Figure 32-17a–d).

The First Trimester

During the first 3 months of development, the embryo is transformed from a single-celled zygote into an ensemble of organ systems that are only slightly less complex than are those found in a full-term baby. Cleavage, blastulation, and implantation take place during the first week or so following fertilization. At the time of implantation, the embryo is called a **blastocyst,** and it contains two different groups of cells (Figure 32-16) with very different fates. Inside the spacious cavity of the blastocyst is a clump of cells called the **inner cell mass,** which will give rise to the tissues of the embryo itself. The outer wall of the blastocyst is called the **trophoblast;** it secretes the enzymes that allow the blasto-

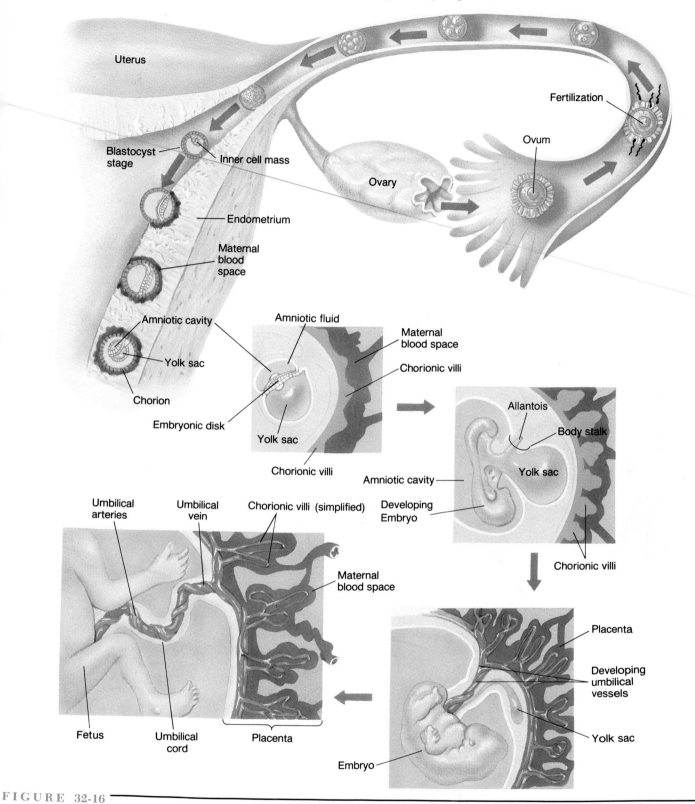

FIGURE 32-16

Implantation and placenta formation.   Following fertilization, cleavage, and blastulation a blasto-cyst is formed that consists of a single outer layer of cells—the trophoblast—and an inner cell mass. Soon after the embryo has implanted in the uterine wall, the inner cell mass gives rise to the embry-onic disk, from which the embryo will develop. The amnion contains the liquid in which the embryo floats as it develops, while the chorion and allantois of the body stalk help form the placenta and um-~~bilical cord~~. Notice that fetal and maternal blood do not mix but are separated by the selectively per-~~meable membranes of~~ the chorion and the capillaries.

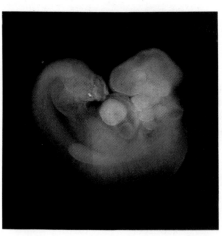

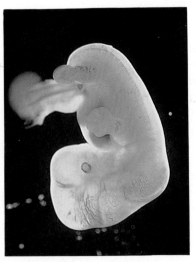

(a) *3 weeks:* Organogenesis begins with neural tube formation, seen here as enlarged crests that delineate the neural groove from which the spinal cord and brain develop.
Length = 0.25 cm (1/10 in).

(b) *4 weeks:* A pumping heart, developing eye, and arm and leg buds are evident. The embryo has a long tail, gill arches, and a relatively enormous head.
Length = 0.7 cm (1/3 in).

(c) *5 weeks:* Internal organ development is well under way. Fingers are faintly suggested. The fetal circulatory system is evident, and the body stalk is now an umbilical cord.
Length = 1.2 cm (1/2 in).

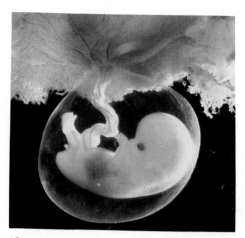

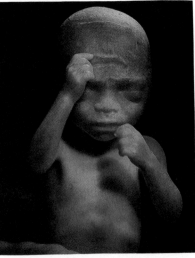

(d) *2.5 months:* The fetus, seen here floating in the amniotic cavity, now has all its major organ systems. The umbilical blood vessels and placenta are well defined, and the tail and gill arches have disappeared.
Length = 3 cm (1-1/2 in).

(e) *5 months:* The fetus looks very much as she will at birth. Bone marrow is assuming more of the blood-producing duties. The mother begins to feel fetal movements, even hiccuping.
Length = 25 cm (10 in).

(f) *5.5 months:* The vernix caseosa is accentuated by a groove the fetus has scraped away with its thumb. Internal organs now occupy their permanent positions (except for the testes in males).
Length = 30 cm (12 in).

FIGURE 32-17

Some stages in human embryonic and fetal development.

cyst to penetrate the uterine wall and then differentiates into the **chorion,** the outermost of four *extraembryonic membranes.* These membranes are termed "extraembryonic" because they do not become part of the embryo and are eventually discarded prior to birth. In addition, the chorion secretes human chorionic gonadotropin (HCG), the hormone that stimulates the corpus luteum to continue to produce progesterone (page 677), which maintains the uterine lining, where the embryo is developing.

■▶ The innermost extraembryonic membrane, the **amnion,** envelops the young embryo and encloses the *amniotic fluid* that suspends and cushions the developing body throughout its duration in the uterus. A third membrane forms the **yolk sac** which, in humans, simply contains fluid (in birds and reptiles, the yolk sac is filled with nutrient-containing yolk). The wall of the human yolk sac is a source of the embryo's blood cells and germ cells, the latter of which will migrate from this extraembryonic tissue into the gonads of the developing embryo. The fourth membrane, the **allantois,** is rich in blood vessels and eventually helps form the vascular connections between mother and fetus. These same four extraembryonic membranes (chorion, amnion, yolk sac, and allantois) are also present in the eggs of birds and reptiles, revealing the evolutionary relationship among the three classes of "higher" vertebrates: reptiles, birds, and mammals.

A few days after implantation, the embryo begins to gastrulate. The cells of the inner cell mass become rearranged to form a double layer of cells, called an *embryonic disk,* which is separated from the overlying chorion by a newly formed space, called the *amniotic cavity* (Figure 32-16). Your entire body, with the exception of the germ cells that migrate in from the yolk sac, is derived from the cells that make up the embryonic disk. By the end of the first month, the embryo, though still less than half a centimeter (3/16 of an inch) long, has begun forming its nervous system, lungs, liver, and several other internal organs. At this stage, the heart is a four-chambered pump; the first signs of eyes and a nose appear; and four buds protrude from the side of the "**C**-shaped" embryo. By the end of the following week, these buds will be vaguely recognizable as developing arms and legs.

During the second month, the liver takes over its temporary job as the main blood-producing organ of the embryo, while an intricate cartilage-containing skeleton begins to change into bone. The brain develops cerebral hemispheres, and spinal nerves grow out from the spinal cord. The face acquires a distinctively human look, with its slate-colored eyes and the beginnings of eyelids. Muscles begin to form and assume their permanent relationships. Fingers and toes become evident. The embryo has become a fetus.

The third month is devoted primarily to the growth and development of existing structures, with the exception of the formation of genitals, which reveals the sex of the fetus. The limbs are clearly recognizable, with well-sculpted fingers and toes, complete with nails. The fetus begins reflex movements that go undetected by the mother.

The first trimester is unquestionably the most dangerous time for the developing embryo. The magnitude of the developmental changes that occur during this time renders the embryo particularly vulnerable to pathogens and chemicals that would be relatively harmless to an older fetus or a newborn human (see The Human Perspective: The Dangerous World of a Fetus).

The Second Trimester

As the 7.5-centimeter fetus enters its fourth month, it develops sucking and swallowing reflexes and soon starts kicking. The mother does not usually feel these movements until the fifth month. The developing skeleton is clearly distinguishable by X-ray examination, and the bone marrow begins blood-cell production. The fetus acquires a downy coat of soft body hair, called *lanugo,* that covers the skin. Lanugo is believed to help the fat droplets that are secreted by fetal sebaceous glands stay on the skin. These droplets form a "greasy" protective coating, the *vernix caseosa,* which conditions the fetal skin during its aquatic tenure in the amniotic fluid. Midway through the pregnancy, two heartbeats become discernible in the mother: her own (about 72 beats per minute) and that of the fetus (up to 150 beats per minute). Convolutions appear in the cerebral cortex of the brain, and sense organs begin supplying the fetus with limited information about its environment.

Although most of the organ systems are at least partially functional by the end of the second trimester, the fetus has little chance of survival if it is born before the seventh month. Even with expert medical care, a 23-week-old fetus has only a 10 percent chance of surviving outside the uterus. If born a month later, its chances of survival jump to 50 percent.

The Third Trimester

During the last 3 months of pregnancy, the fetus increases its body weight by 500 to 600 percent. The brain and peripheral nervous system enlarge and mature at an especially rapid rate during the final trimester. Neurological performance later in life, including intelligence, depends on the availability of protein needed for fetal brain development. Women who suffer protein deficiency during this time tend to have babies that are mentally slower throughout life. Furthermore, this condition cannot be reversed by providing the child with a protein-rich diet. By the end of the third trimester, the fetus is capable of regulating its temperature and can control its own breathing. This latter ability, together with the degree of lung maturation, generally determines a premature infant's chances of survival.

The formerly lean fetus changes in appearance, as fat deposits form under the skin, imparting the rounded, chubby shape characteristic of many babies. In males, the testes descend into the scrotum. Most fetuses change posi-

◁ THE HUMAN PERSPECTIVE ▷
The Dangerous World of a Fetus

The "womb," or uterus, is often used as a metaphor for warmth, protection, and security—a world free of fear and danger. Yet, a fetus may be exposed to a great many threats while residing in its mother's uterus, including nutrient deficiency, infectious microbes, radiation, alcohol, and the chemicals found in drugs, cigarettes, and environmental pollutants. The placenta screens out most harmful substances, but some dangerous agents may pass through the placental barrier and enter the fetal circulation. For example, some radiation passes through *all* living barriers, bombarding the fetus with potential *teratogenic* (embryo-deforming) consequences. Large doses of X-rays during the first trimester of pregnancy may cause mental retardation, skeletal malformations, and reduced head size, and may predispose the unborn child to the development of cancer later in life. When a cell is genetically impaired during the first trimester, all the millions of cells that develop from the affected cell inherit the impairment, amplifying the effect. These cells may fail to differentiate normally and, consequently, produce organs that are malformed, some so severely that the fetus or newborn child has little chance of survival.

The virus that causes *rubella,* or "German measles," illustrates the vulnerability of the embryo to infectious microbes during its first trimester. The disease is so mild that a pregnant woman with rubella may be unaware that she has the disease. The virus can pass through the placenta, however, and infect the emerging embryonic organs, interfering with their normal development. The result can be physical malformations, mental retardation, deafness, or death in about half of the babies exposed during their first 6 weeks of development. Yet, second- and third-trimester fetuses exposed to rubella suffer no impairment. Routine vaccination against rubella has dramatically reduced these tragic occurrences. In recent years, the virus presenting the greatest epidemiologic threat to newborns is HIV, the virus responsible for AIDS. Roughly half of all newborns delivered by infected mothers will be infected with HIV; most of these babies will develop the disease and die within the first few years of life. It is not clear why some babies resist becoming infected in the womb, while others succumb.

Another microbe that can reach the fetus is the streptococcus bacterium, which may infect as many as 12,000 babies in the United States each year. While the bacteria may cause no symptoms in the mother, in the infant, the infection can cause permanent brain damage, cerebral palsy, lung and kidney damage, blindness, and even death. All pregnant women are urged to be tested for Group B streptococcus, the chief cause of preventable newborn infections. Fortunately, most microbes cannot cross the placenta, so the fetus usually remains healthy while the mother is combating a cold, the flu, or most other microbial onslaughts.

Many chemicals to which the mother may be intentionally or accidentally exposed can also compromise fetal development. Some of these chemicals may be found in medications. In the 1950s, thousands of women who took the mild tranquilizer thalidomide during early pregnancy delivered babies whose arms and legs had failed to develop. Although stricter drug regulations resulted from this tragic episode, many pharmaceutical products that are potentially dangerous to fetuses are still available on the market today. These include some vitamins, which can be dangerous when taken in large doses (especially D, K, and C), cortisone, antibiotics, birth control pills, tranquilizers, anticoagulants, and thyroid drugs. Drugs such as LSD, marijuana, cocaine, and alcohol expose the fetus to proven toxins or teratogenic agents. Alcohol consumption by a pregnant woman during a critical period of fetal development can lead to *fetal alcohol syndrome,* a collection of childhood deformities that includes mental retardation, reduced head size, facial irregularities, and learning disabilities later in life. Children born to mothers addicted to crack cocaine also experience severe emotional and learning disabilities.

Tobacco smoke constitutes yet another serious danger to the developing fetus and may contribute to infant mortality. Cigarette smoking during pregnancy tends to impair fetal growth and reduce resistance to such respiratory infections as bronchitis and pneumonia during the first year of life. The carbon monoxide in tobacco smoke competes with oxygen for hemoglobin binding sites, reducing the amount of oxygen available to the fetus. Nicotine constricts blood vessels, further reducing the blood supply. Tobacco smoke may also contain teratogens and may contribute to the development of heart and brain defects, a cleft palate, or sudden infant death syndrome (SIDS).

tion in the uterus, becoming aligned for a head-first delivery through the birth canal.

One system that does not mature by the time of birth is the immune system; human babies are born immunologically deficient. During the final fetal month, however, antibodies from the mother's blood cross the placenta and temporarily fortify the baby's defenses against infectious diseases until its own immune arsenal becomes competent. In breast-fed babies, this immunologic gift is supplemented by the maternal antibodies and immune cells found in breast milk.

BIRTH

After 9 months of development, a human fetus is 6 billion times larger than was the original zygote. Yet, this astonishing size increase probably plays no direct role in inducing **birth,** or *parturition.* In fact, the fetus is believed to have little to do with initiating the uterine contractions of birth. Although the birth-triggering factor(s) remains a mystery, the onset of labor is preceded in most women by a drop in the concentration of progesterone, the hormone that inhibits contractions of the smooth uterine muscles during pregnancy. Once labor begins, a cascade of events promote uterine muscle contractions of increasing strength and frequency. The posterior pituitary releases oxytocin, which stimulates muscle contraction in the uterus. This response then triggers a reflex release of even more oxytocin, establishing a positive feedback loop that intensifies contractions. Prostaglandins (page 532) are also believed to participate in labor; this contention is supported by the ability of aspirin, a prostaglandin antagonist, to delay birth. Both oxytocin and prostaglandins may be administered by injection to induce labor when its delay threatens the health of the mother or fetus.

Birth occurs in three distinct stages. During the first stage, the mucous plug that blocks the cervical canal and prevents microbial invasion of the intrauterine environment is expelled. The amniotic sac ("bag of waters") ruptures during this stage, and its fluid is discharged through the vagina. Contractions during this stage force the cervical canal to dilate until its diameter enlarges to about 10 centimeters (3.9 inches), marking the onset of the second stage of labor: the expulsion of the fetus. Powerful contractions move the fetus out of the uterus and through the vagina, usually head first. Fewer than 4 percent of all deliveries are "breech births," whereby the buttocks or legs come out first. Contractions during the final stage expel the detached placenta, called the *afterbirth,* from the uterus. Additional contractions help stop maternal bleeding by closing vessels that were severed when the placenta detached.

A newborn infant quickly acclimates to its terrestrial environment. With the umbilical cord clamped and severed, the baby rapidly depletes its source of maternal oxygen; still unable to breathe, the baby's respiratory waste (carbon dioxide) accumulates in the blood. The brain's respiratory center responds to this increase by activating the breathing process, and the lungs inflate with air for the first time. But before the lungs can oxygenate the blood, a circulatory modification is required because the unneeded fetal lungs were bypassed by the bloodstream in the uterus. The septum between the right and left atria of the *fetal* heart is perforated by a large hole, called the *foramen ovale,* that allows blood to flow directly from one side of the heart to the other, rather than traveling through the pulmonary circulation. At birth, the oval window is immediately closed by a hinged flap, forcing blood to travel through the lungs to get to the other side of the heart. At the same time, blood vessels to and from the now useless umbilical cord are sealed by muscular contractions, as are vessels that bypassed the fetal lungs and liver.

Humans progress through the same recognizable stages of early development—fertilization, blastulation, gastrulation, and neurulation—as do embryos of other species of vertebrates. As in reptiles, birds, and other mammals, only a small part of a human egg's contents is converted into embryonic tissues; most is used in the formation of extraembryonic membranes which play a key role in providing the human embryo and fetus with oxygen, nutrients, and waste removal. Most human organs are formed within the first 2 months of development; the remaining 7 months are devoted primarily to growth and refinement of structure. (See CTQ #4.)

EMBRYONIC DEVELOPMENT AND EVOLUTION

▐▶ It would be hard to mistake a fish for a bird, or a turtle for a human, yet the embryos of these vertebrates are remarkably similar (Figure 32-18). Embryos tend to change much more slowly over evolution than do the corresponding adults. For this reason, similarities in embryonic development are often used as evidence for evolutionary relatedness. For example, mollusks (such as snails) and flatworms (such as tapeworms) are so different as adults that there is no reason to think the two groups are more closely related to one another than are mollusks and vertebrates. Yet, the pattern of cleavage of certain mollusks and flatworms is so similar that evolutionary schemes typically show mollusks as descendants of flatworm-like ancestors.

The various parts of the embryos depicted in Figure 32-18 appear so similar because they are *homologous* structures; that is, they are derived from the same structure that was present in a common ancestor (Chapter 34). We have seen in this chapter how the resemblance between the four extraembryonic membranes of reptiles, birds, and mam-

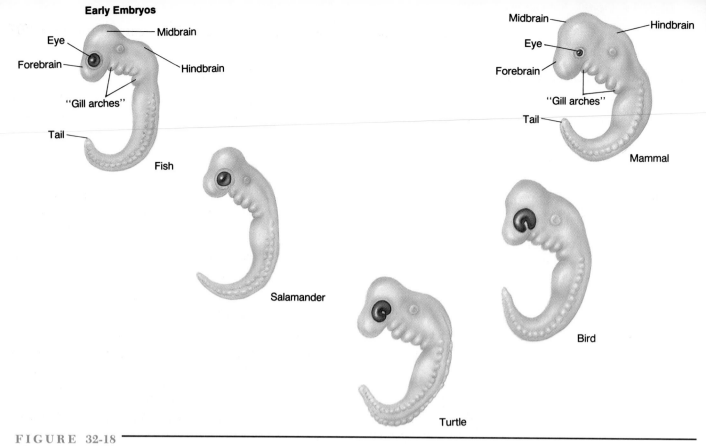

FIGURE 32-18

The striking similarities in the structure of the early embryos of vertebrates reveals their common ancestry.

mals testifies to the common ancestry of all three groups. Even when the original function of a membrane is no longer needed by a particular species, the membrane remains as a vestigial reminder of the organism's evolutionary origins. For example, humans and other mammals have no yolk in their embryo, yet they still develop a yolk sac—a vestigial remnant left over from an early "yolked" ancestor. Similarly, all vertebrate embryos, including those of humans and other air-breathing mammals, develop pharyngeal gill slits (page 636), even though these structures never become functional, even in the embryo. Instead, the gill slits of the human embryo are only transient structures, appearing

rapidly in an early stage and then giving rise to totally different structures, including parts of the jaw, ear, and thyroid gland. The entire jaw apparatus of vertebrates—from fishes to humans—is derived from part of the gills that appear in the embryo.

Embryos of diverse animal species tend to be more similar than are the corresponding adults. As a result, the study of development often provides insights into evolutionary relationships among animals and into the evolutionary pathway by which particular structures arose. (See CTQ #6.)

REEXAMINING THE THEMES

The Relationship between Form and Function

Each of the several hundred cell types found in a human or other mammal has a specialized function supplied by that cell's peculiar structure and metabolic machinery. During embryonic development, each type of cell acquires its specialized morphology. Muscle cells synthe-

size contractile proteins, such as actin and myosin, which become organized into cytoplasmic filaments; salivary gland cells develop an elaborate, interconnected, membranous network that promotes the synthesis and export of secretory proteins; nerve cells develop highly elongated axonal processes, which facilitate cell-to-cell communication; and so forth.

Biological Order, Regulation, and Homeostasis

⟳ Embryonic development is characterized by a striking increase in biological order. A fertilized egg, whether a sea urchin or a human, has one of the least specialized internal structures of any animal cell. Yet, from this relatively simple beginning emerges an animal of striking complexity. Even though a fertilized egg may not show obvious evidence of its awesome potential, the cell has a developmental program that unfolds according to a predetermined plan. This program is encoded within the DNA of the chromosomes and as mRNAs and other cytoplasmic materials. The expression of this program requires such a high degree of regulation that we are now only *beginning* to understand the underlying mechanisms.

Acquiring and Utilizing Energy

☀ The increasing complexity that characterizes embryonic development is driven by the chemical energy that is consumed by the embryo. In some cases, as in birds, the energy is provided in the form of yolk, which is packaged into the egg as it forms in the ovary. In other cases, as in sea urchins, the individual begins life with very little energy reserves and rapidly develops into a free-swimming larva that must obtain food for itself. Mammals produce eggs with very little yolk, but the growth of the embryo and fetus is fueled by the nutrients delivered by the maternal bloodstream.

Unity within Diversity

◬ Although adult animals show great diversity, there is an undeniable similarity in the underlying mechanisms of embryonic development in all species. Virtually all embryos pass through stages of cleavage, blastulation, and gastrulation, even though the embryos themselves may bear little superficial resemblance to one another. The existence of homeotic genes that contain virtually identical homeobox sequences suggests a similarity in the types of genes that control development. Changes in the diverse form of embryos are accomplished by similar morphogenetic mechanisms, including changes in cell shape, cell growth and division, cell death, and cellular migrations.

Evolution and Adaptation

⏩ Since the embryos of animals tend to be much more similar morphologically to one another than are the corresponding adults, the study of embryonic development provides one of the best tools to understanding the evolutionary relatedness among animals. In addition, the pathway by which a particular structure develops often reveals insights into the evolutionary path by which the structure arose. For example, the origin of the vertebrate jaw from part of the pharyngeal gills is hardly evident by examining an adult vertebrate but becomes immediately apparent by observing the formation of jaws in an early vertebrate embryo.

SYNOPSIS

Embryonic development is a programmed course of events that carries an animal from fertilization through cleavage, blastulation, gastrulation, and organ formation. This program is encoded within the fertilized egg in both the genes and the organization of the cytoplasm.

A fertilizing sperm activates an unfertilized egg and donates a set of homologous chromosomes. Unfertilized eggs respond to sperm contact by surface changes that prevent penetration by additional sperm and by the release of calcium, which triggers various responses, including the formation of an extracellular membrane.

Cleavage divides the unbalanced egg into a large number of smaller blastomeres. Cleavage leads to the formation of a blastula, a stage that contains an internal chamber, or blastocoel, whose size and location depend primarily on the amount of yolk in the egg. Cleavage can occur in the absence of transcription, indicating that newly synthesized proteins are formed using stored mRNAs. De-

velopment beyond the blastula requires the activation of embryonic genes. Some blastomeres may be able to form an entire embryo in isolation, while others have a much more restricted potential. In frogs, only those blastomeres possessing a portion of the gray crescent can develop normally.

During gastrulation the various parts of the blastula become rearranged to form an embryo with three defined germ layers. The outer ectoderm gives rise to the skin and nervous system; the inner endoderm to the digestive tract and related organs; and the middle mesoderm to the remainder of the embryo. After gastrulation in vertebrates, the dorsal strip of ectoderm becomes thickened into the neural plate, which rolls into a tube that ultimately gives rise to the entire nervous system. This transformation requires induction from the underlying chordamesoderm, those cells that will form the notochord.

The formation of organs depends on a number of processes. Cells receive chemical and physical signals from their surroundings which inform them of their relative postion within the embryo. The shape of the organ formed by developing cells depends on morphogenetic processes, including changes in the rate of cell division, changes in cell shape, changes in cell adhesion, and programmed cell death. During formation of an organ, the internal architecture of the cells assumes a differentiated state, characteristic of that cell type. Each type of cell transcribes a restricted set of genes, forming a characteristic set of proteins.

Embryonic development provides a window to evolution. Embryos are typically much more similar to one another than are the corresponding adults; thus, embryos are useful in establishing evolutionary relatedness. Following the course of development of a particular structure may reveal information about the evolution of that structure.

Human development occurs in the uterus. The blastocyst implants itself in the endometrium and gastrulates to form an embryonic disk. Of the four extraembryonic membranes, the chorion and allantois contribute to placenta formation; the amnion protects the fetus; and the yolk sac temporarily manufactures blood cells. By the end of the second month, the formation of virtually all embryonic organs has begun. During the remaining 7 months, the fetus refines these structures and grows in size until uterine contractions expel it through the vagina.

Key Terms

homeotic gene (p. 686)
homeobox (p. 687)
embryo (p. 688)
yolk (p. 688)
larva (p. 688)
cleavage (p. 690)
blastomere (p. 690)
blastulation (p. 690)
blastula (p. 690)
blastocoel (p. 690)
gray crescent (p. 692)
gastrulation (p. 692)
gastrula (p. 692)

endoderm (p. 693)
ectoderm (p. 693)
mesoderm (p. 693)
neural plate (p. 694)
neural tube (p. 694)
chordamesoderm (p. 694)
notochord (p. 694)
induction (p. 695)
organogenesis (p. 696)
positional information (p. 696)
morphogenesis (p. 696)
cell differentiation (p. 698)
metamorphosis (p. 699)

chorion (p. 700)
placenta (p. 700)
umbilical cord (p. 700)
fetus (p. 700)
blastocyst (p. 700)
inner cell mass (p. 700)
trophoblast (p. 700)
amnion (p. 703)
yolk sac (p. 703)
allantois (p. 703)
birth (p. 705)

Review Questions

1. Arrange these terms in the order of development and briefly describe the role of each process in embryonic development: neural tube formation, birth, blastulation, fertilization, gastrulation, gametogenesis (meiosis).

2. What is the relationship between the amount of yolk in an egg and the relative size of the blastocoel?

3. Discuss the role of each of the four extraembryonic membranes in human development and indicate their evolutionary significance.

4. Distinguish between inner cell mass and trophoblast; primary and secondary induction; neural plate and neural tube; optic vesicle and optic cup; homeotic and hemoglobin genes.

Critical Thinking Questions

1. Different vertebrae along the backbone can be distinguished by their shape. Occasionally, an infant is born with vertebrae in his or her lower back shaped like those normally found in the neck. In other words, the infant possesses cervical-type vertebrae in the lumbar region of the spine. Do you think this condition could be due to a homeotic mutation? Why or why not?

2. How does embryonic induction help determine the sequence of organogenesis? Why is sequence so important in embryonic development?

3. The graph (right) shows the growth of the body, heart, and brain of a person from birth to age 30. What does this show about the relative growth of the parts of the body? Why is the brain not much larger at age 30 than at age 5, but the heart is?

4. Prepare a timeline, to scale, of the embryonic development of a human. Your timeline should go from 0 to 266 days and should indicate the approximate times of significant events in development. You can indicate these events with words, drawings, or photographs from magazines.

5. Suppose you treated a fertilized snail egg with a drug that is assumed to inhibit transcription and found that the egg developed normally into a snail. What conclusion might you draw? Are there any experiments you might run to be sure of your conclusion? (Consider

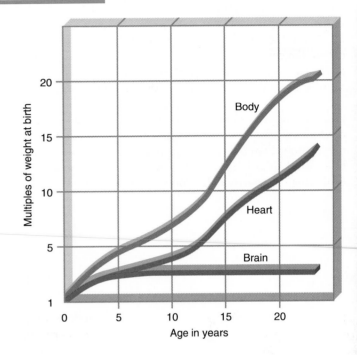

drug permeability and the actual ability of the drug to block transcription in your eggs.)

6. How does the study of comparative embryology contribute to the understanding of evolution? Give examples.

Additional Readings

Beardsley, T. 1991. Smart genes. *Sci. Amer.* Aug:86–95. (Intermediate)

Alberts, B. M. 1989. *Molecular biology of the cell,* 2d. ed. New York: Garland. (Advanced)

Browder, L., C. Erickson, and W. Jeffrey. 1991. *Developmental biology,* 3d ed. Philadelphia: Saunders. (Advanced)

Carlson, B. 1988. *Patten's foundations of embryology,* 5th ed. New York: McGraw-Hill. (Advanced)

Cherfas, J. 1990. Embryology gets down to the molecular level. *Science* 250:33–34. (Advanced)

DeRobertis, E. M., G. Oliver, and C. V. E. Wright. 1990. Homeobox genes and the vertebrate body plan. *Sci. Amer.* July:46–52. (Intermediate)

Gilbert, S. F. 1991. *Developmental biology,* 3d ed. Sunderland, MA: Sinauer. (Advanced)

Hoffman, M. 1990. The embryo takes its vitamins. *Science* 250:372–373. (Intermediate)

Melton, D. A. 1991. Pattern formation during animal development. *Science* 252:234–241. (Advanced)

Steinmetz, G. 1992. Fetal alcohol syndrome. *Nat'l Geog.* Feb:36–39. (Introductory)

Wassarman, P. M. 1988. Fertilization in mammals. *Sci. Amer.* Feb:78–84. (Intermediate)

Evolution

Evolution leads to
characteristics that improve
a species' chances of surviving and
producing offspring. This camouflaged
coralline sculpin blends in with its
environment and with its oversized jaw
can seize unwary victims its
own size, providing the nutrients needed
for growth, development, and
reproduction.

PART
· 6 ·

CHAPTER

33

Mechanisms of Evolution

STEPS TO DISCOVERY
Silent Spring Revisited

THE BASIS OF EVOLUTIONARY CHANGE

The Genetic Composition of Populations

Factors That Cause Gene Frequencies To Change Over Time

SPECIATION: THE ORIGIN OF SPECIES

What Are Species?

Reproductive Isolating Mechanisms

Paths of Speciation

PATTERNS OF EVOLUTION

EXTINCTION: THE LOSS OF SPECIES

THE PACE OF EVOLUTION

BIOLINE

A Gallery of Remarkable Adaptations

STEPS TO DISCOVERY
Silent Spring Revisited

*I*n 1939, Paul Muller, a researcher at a Swiss pharmaceutical company, discovered that the compound dichloro-diphenyl-trichloro-ethane (DDT) was a very effective insecticide. When the United States entered World War II, two insect-borne diseases presented a serious threat to troops: typhus fever, which was common in areas of Europe and was spread by lice; and malaria, which was common in the Pacific and was spread by mosquitos. When an epidemic of typhus threatened to break out in Naples, Italy, in 1943, a powdered preparation of DDT was sprayed under the clothing of over 3 million troops and civilians. For the first time in human history, the spread of this deadly disease

was arrested; DDT was hailed as a "miracle" compound. The insecticide proved equally effective against mosquitoes in the Pacific, where it was sprayed from the air over entire islands.

After the war, DDT became available in large quantities to civilians. In 1957, planes flying over marshy areas of Massachusetts, spraying DDT in order to kill mosquitoes, happened to spray the property of a resident who kept a 2-acre bird sanctuary. In addition to killing the insects in the area, the insecticide killed the birds, leaving the landscape ghostly silent. The resident wrote a letter to a friend, Rachel Carson, a biologist and author of several widely acclaimed

Although spraying DDT killed most mosquitoes, the natural genetic variation in the population allowed a small percentage of

books on the sea and its inhabitants. Carson decided to look into the matter. The deeper she became immersed in the scientific literature on the effects of DDT and related pesticides, the more convinced she became that she had to warn the public of their dangers.

After 4 years of research, Carson wrote *Silent Spring*, a book that documented the devastating effects pesticides were having on the wildlife of the world and the problems they were creating for future insect eradication programs. The book inspired President Kennedy to establish a commission to regulate the use of pesticides. Congress began holding hearings on the subject, and environmental concern groups were established. These events culminated in the establishment of the Environmental Protection Agency in 1970, which banned the use of DDT in 1972.

In addition to changing environmental politics, *Silent Spring* provided documented evidence of the role of natural selection in shaping the characteristics of animal populations. Carson described how insects were changing over time to become resistant to pesticides, particularly DDT. She wrote: "Darwin himself could scarcely have found a better example of the operation of natural selection than is provided by the way the mechanism of [pesticide] resistance operates."

A pesticide is a powerful selective agent. In any given population of insects, some individuals have a combination of genes that makes them less susceptible to harmful chemicals than do other individuals. Those individuals that are susceptible to the pesticide die off, removing the genes that confer susceptibility from the population and leaving resistant individuals to repopulate the species' ranks. Initially, only a small fraction of the insect population had the specific combination of genes that made them resistant to potent pesticides. Consequently, when DDT was first used, most of the insects died off. Since insects can produce tremendous numbers of offspring in very short periods, however, it was only a matter of a few years before highly resistant individuals dominated the population.

Moreover, favorable traits, such as pesticide resistance, need not stop at a population's boundaries. A 1991 study investigating pesticide resistance in mosquitoes of the species *Culex pipiens* found that individual insects from around the world carry precisely the same resistance-promoting genetic alteration. This finding strongly suggests that resistance in different populations is not due to independent mutations within each population; rather, resistance can spread rapidly from population to population when individuals carrying the beneficial genes migrate to new environments.

Species that lack the ability to cope with environmental changes will shrink in number or even become extinct, an event chronicled in Carson's book. Carson noted that, while some insect populations were evolving pesticide resistance, many bird populations were being decimated. Studies showed the birds were eating pesticide-contaminated prey (including insects, earthworms, and fish), and the toxins were building to high concentrations in the birds' fatty tissues. The accumulating DDT prevented many birds from producing healthy offspring: Birds that ingested DDT laid fewer eggs; the eggs had thinner shells and were often broken in the nest; and the chicks that hatched were so loaded with pesticide residues that they often failed to survive. Unlike insects, none of the members of the bird populations possessed genes that conferred resistance to pesticides; therefore, there were no resistant individuals for natural selection to favor. Many people became concerned that some of these bird species, including the American eagle and the peregrine falcon, were being pushed to extinction. Fortunately, since 1970, these bird species have actually grown in number, largely as a result of a ban on the use of DDT.

Concern over the effects of DDT is not limited to birds. Even though DDT has been banned in many countries, the chemical residues of the pesticide can remain stored for decades in the fatty tissues of the human body. A report in 1993 indicated that women with high levels of DDT in their body had four times the risk of contracting breast cancer compared to women with the lowest levels of the pesticide. The earlier use of DDT may be one of the reasons for the puzzling rise in breast cancer rates in the past few decades.

the mosquitoes to survive and repopulate the species.

"A fish out of water" can't survive very long, or so we might think. Yet the mudskipper is a fish that not only survives long treks across mud flats, it even climbs trees (Figure 33-1). In water, the mudskipper's fins and gills work just like those of a typical fish: Its fins propel and steer the mudskipper through the water, and its gills extract dissolved oxygen. But how does the mudskipper remain alive on land with a body and breathing machinery that are so unmistakably adapted for life in the water?

Unlike other fishes, the mudskipper's gills and fins have modifications that enable it to survive and move on land. For its respiration, the mudskipper packs a supply of water into its bulging gill pouches, from which it extracts oxygen; the pouches act like a scuba tank in re-verse. For its motility, the mudskipper's reinforced forefins serve as stubby arms for crawling across the mud or shimmying up tree trunks in search of snails for a meal.

A hobbling mudskipper with water-engorged pouches illustrates how structures originally adapted for one way of life can become refashioned for new functions. The mudskipper's makeshift legs and water bags are modifications of structures that were originally adapted for aquatic life. The fact that the mudskipper possesses these structures is evidence that its ancestors lived strictly underwater and that those ancestors were something other than mudskippers. Every species has a history of ancestors that possessed features and behaviors different from those the existing species possess. To trace the history of change in an organism's ancestors is to follow its course of evolution.

▼ ▼ ▼

FIGURE 33-1

The mudskipper is a fish that can live for hours out of water. Strengthened forefins and modified gill chambers are evidence that the mudskipper evolved from ancestors that were strictly aquatic.

THE BASIS OF EVOLUTIONARY CHANGE

How did there come to be so many kinds of organisms, each with unique characteristics that enable it to survive in one or more of the earth's many diverse habitats? How can we explain the fact that there were organisms living in the past that possessed characteristics very different from those of organisms living today? And why were the earth's first organisms much simpler than are many of today's organisms?

△ **Evolution**—the theory that a population of organisms changes as the generations pass—provides the answers to these questions. Evolution explains how modern animals and plants evolved from more primitive ancestors, which, in turn, evolved from even more primitive types, and so on, back to the first appearance of life, billions of years ago. The fact that all the diverse organisms present on earth today arose from a common ancestor explains why they have the same basic mechanism for the storage and utilization of genetic information, many of the same types of cellular organelles, and similar types of enzymes and metabolic pathways. These shared characteristics were present in the earliest organisms and were retained among their descendants.

Contrary to popular belief, the concept of evolution did not originate with Darwin. In the eighteenth century, a few naturalists had considered the *possibility* of descent with modification, but no one had proposed a convincing explanation as to *how* evolution could occur. Furthermore, strong religious opposition to the concept of evolution contributed to a general lack of interest. Consequently, the theory of evolution was not widely accepted until the middle of the nineteenth century, at which time Charles Darwin suggested a plausible mechanism—natural selection—and collected substantial evidence to support the contention that evolution had indeed occurred (page 21). In the twentieth century, as the principles of genetics became better understood, biologists discovered mechanisms other than natural selection that cause organisms to change from generation to generation. The synthesis of Darwin's original ideas on evolution by natural selection and those of modern genetic theory is referred to as either *neodarwinism* or the *modern synthesis*.

THE GENETIC COMPOSITION OF POPULATIONS

Individual organisms are born, mature, and eventually die. Along the way, an individual may change, but it does not evolve. Rather, it is the **species**—a group of interbreeding organisms—that evolves. The members of a species form groups, or **populations,** that occupy a particular region. Some species consist of just a single population living in one area, such as a small lake or an island. Other species are made up of more than one population, each in a different locality. For evolution to take place, change must occur in the genes that are present in the members of a population. These changes are passed on to the next generation during reproduction and are spread throughout the population by interbreeding. In order to understand this process, biologists have investigated the genetic changes in populations that generate evolutionary change.

You will recall from Chapter 12 that each individual receives one copy of a gene from each of its parents. Recall, also, that genes can occur in different forms, or alleles. Most alleles are either dominant or recessive, and an individual can be either homozygous (two identical alleles) or heterozygous (two different alleles) at any particular gene locus. Not all individuals that make up a population have the same alleles; consequently, there is *genetic variation* in the population. In humans, this variety is reflected by differences in pigmentation, facial characteristics, and blood types among different individuals and ethnic populations (Figure 33-2).

The sum of all the various alleles of all the genes in all of the individuals that make up a population is called the population's **gene pool.** If we could count every allele of every gene in every individual of a population, we could measure the genetic variation in a gene pool. The *relative* occurrence of an allele in the gene pool is expressed as an **allele frequency.**

To illustrate how allele frequencies are determined, we will examine a genetic trait that is prevalent in persons of central African descent: sickle cell anemia. Recall from Chapter 4 that sickle cell anemia is caused by a mutation that results in a substitution of one amino acid for another in one type of polypeptide chain that makes up a hemoglobin molecule (page 81). The mutant allele is denoted as S, while the normal allele is denoted as A. Approximately eight out of every 100 African Americans are heterozygous carriers of sickle cell anemia (genotype SA), and one out of every 500 have sickle cell anemia (genotype SS). Consequently, out of every 500 African Americans, there will be an average of 40 carriers (with a total of 40 copies of the S allele) and one person with the disease (with two copies of the S allele). Thus, out of 500 African Americans, having 1,000 copies of the gene, there will be an average of 42 copies of the sickle cell allele and 958 copies of the normal allele. The frequency for the S allele is $^{42}/_{1,000}$ or 0.042 (4.2 percent). Conversely, the frequency for the normal A allele is 0.958 (95.8 percent). (Calculations of this type can be done more quickly using the equation described in Appendix C.)

△ While differences in allele frequency exist between different human races (as illustrated in Figure 33-2), genetic analysis of a large variety of traits indicates that the overall differences are remarkably small. In the words of Richard Lewontin, "If everyone on earth became extinct except for the Kikuyu of East Africa, about 85 percent of all human variability would still be present in the reconstituted species."

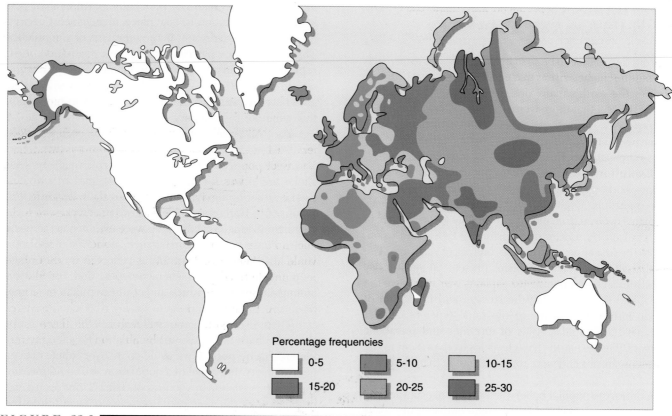

FIGURE 33-2

Frequency of the blood-type allele *I^B* in aboriginal (native) populations of the world. The frequency ranges from a high of about 30 percent (0.3) in Northern India and central Asia to a low of less than 5 percent (0.05) among American Indians and aboriginal Australians. From: *Modern Genetics,* 2nd ed. by: Ayala and Kiger. Copyright © 1984 by The Benjamin/Cummings Publishing Company. Reprinted by permission.

FACTORS THAT CAUSE GENE FREQUENCIES TO CHANGE OVER TIME

▶ Evolution occurs when the composition of the gene pool changes. Therefore, a basic component of the evolutionary process is the change of allele frequencies over time. What causes allele frequencies to change? Sometimes the easiest way to understand a process is first to construct an artificial system or model in which the process does *not* occur. In this case, by uncovering the conditions that are necessary to keep allele frequencies *constant*, we automatically learn what forces will cause them to change.

In 1908, the British mathematician G. H. Hardy and the German biologist W. Weinberg independently discovered that under certain ideal conditions, allele frequencies will remain constant from generation to generation in sexually reproducing populations. Their demonstration is now known as the **Hardy-Weinberg Law** and is discussed in detail in Appendix C. Populations that are not changing — that is, that have the same allele frequencies from one generation to the next — are said to be at *Hardy-Weinberg equilibrium,* or **genetic equilibrium.**

Five "ideal conditions" must exist if a population is to remain at genetic equilibrium:

1. There must be an absence of mutation so that no new alleles appear in the population.

2. Individuals cannot migrate into or out of the population so that no new alleles enter, or existing alleles leave, the population.

3. The population must be very large so that it is not affected by *random* changes in allele frequency.

4. All individuals in the population must have an equal chance of survival; that is, there are no genetic traits that give individuals a survival advantage.

5. Mating must combine genotypes at random; that is, no preference is shown in the selection of a mate.

Based on these five conditions, we can identify those factors that disrupt genetic equilibrium and cause changes in the frequency of alleles in a population's gene pool:

1. *Mutation:* randomly produced inheritable changes in DNA that introduce new alleles (or new genes, as the result of chromosome rearrangements) into a gene pool.

2. *Gene flow:* the addition or removal of alleles when individuals exit or enter a population from another locality.

3. *Genetic drift:* random changes in allele frequency that occur solely by chance.

4. *Natural selection:* increased reproduction of individuals that have phenotypes that make them better suited to survive and reproduce in a particular environment.

5. *Nonrandom mating:* Increased reproduction of individuals that have phenotypes that make them more likely to be selected as mates.

These five forces, alone or in combination, determine the course and rate of evolutionary change. The role of each agent is slightly different but, in general, mutation and gene flow introduce new genetic material into a population, while genetic drift, natural selection, and nonrandom mating determine which alleles will be passed on to the next generation.

Mutation: The Source of New Alleles

A mutation is a random change in the DNA of an organism (page 286). Mutations occur spontaneously in all the cells of the body, but only those that occur in germ cells contribute to evolutionary change because only these cells can become gametes and pass the mutation on to the next generation. Mutations add new alleles to the gene pool, supplying the genetic foundation on which the other evolutionary forces operate.

When we see how quickly some insects have developed resistance to pesticides, such as DDT, it is tempting to propose that the appropriate mutations were stimulated to arise when the insects were *first* exposed to the pesticide, as a direct response to a change in the environment. This is not the case, however; mutations are random and unpredictable.

At any point in time, some mutations are beneficial, some are detrimental, and others are "neutral" and have no apparent effect on the survival or reproductive capacity of an organism. Many harmful mutations are immediately removed from the gene pool because they disrupt the structure and function of a protein whose activity is required for life to continue. Individuals with such lethal mutations typically die during embryonic development. (For example, many human zygotes fail to develop because of lethal mutations.) Other harmful mutations are masked by a dominant allele. For example, each of us is believed to carry an average of 7 to 8 lethal recessive genes. The fact that we are alive testifies to the role of the dominant allele on the homologous chromosome.

Whether a mutation is beneficial, detrimental, or neutral often depends on the environment in which the organism is living at the time. If the environment changes, the effects of the mutation on survival and reproduction can also change. For example, a mutation that causes an enzyme to function optimally at a higher temperature will be beneficial if the environmental temperature rises and will be detrimental if the temperature falls. It is likely that resistance to DDT was originally a neutral, or perhaps mildly beneficial, mutation that spread in low numbers throughout the insect populations through interbreeding. Only when the pesticide was sprayed did the DDT-resistance allele provide a strong survival advantage to the individuals that possessed it. Had no such allele already been present in the population when the individuals were exposed to the pesticide, no insects would have survived, and the population would have been wiped out. In fact, many bird species have been unable to adapt to the presence of the pesticide because the appropriate mutation is not present in their population. As humans continue to modify the earth's environments, it is important to remember that there is no guarantee that organisms will be able to adapt to environmental change.

Gene Flow: Exchanges of Alleles between Populations

It is common for animals or their larvae to migrate over large distances and for the seeds and pollen of plants to be dispersed by the wind or carried by birds to distant locations. Consequently, individuals from one population of a species are moved to another population, creating the opportunity for the transfer of alleles from one population's gene pool to another. The transfer of alleles between populations through interbreeding is called **gene flow.** Immigrants into a population may add new alleles to the population's gene pool, or they may change the frequencies of alleles that are already present. Emigrants out of a population may completely remove alleles, or they may reduce the frequencies of alleles in the remaining pool.

The amount of gene flow between populations varies greatly, depending on a number of factors, including the number of migrating individuals, the ease of movement, the harshness of the environment to be traversed, and the amount of interbreeding that actually takes place when migrants come in contact with a new population.

As described in the chapter opening section, gene flow is one of the factors responsible for the widespread resistance among insects to pesticides. Resistant individuals from one population emigrate into new populations, spreading resistance-conferring alleles into new geographic areas. The importance of gene flow can also be illustrated in humans. The fact that 70 percent of the alleles for cystic fibrosis in the United States can be traced to a single northern European (page 342) reveals how the influx of alleles can affect the genetic composition of a human population.

Genetic Drift: Random Changes in the Gene Pool

Genetic drift is a change in allele frequency that results simply by chance. Chance can affect allele frequency in several ways, but it is especially important during genetic recombination. When gametes are formed by meiosis, the segregation of chromosomes into any particular egg or sperm occurs by chance. When mating takes place, a great many of the gametes are wasted. Only a few happen to combine to form new individuals, representing a random sample of the parents' genes. Genetic drift may be caused by the spread or removal of alleles due to chance segregation into gametes that happen to participate in formation of offspring.

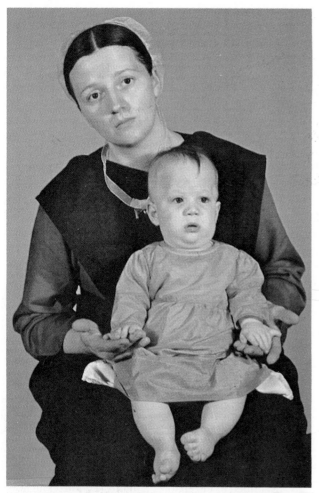

FIGURE 33-3

The founder effect. This Amish woman and her child are descendants of a small group of founding families who immigrated to Pennsylvania in the mid-eighteenth century. As a result of intermarriage between individuals within the community, a recessive allele for Ellis-van Creveld syndrome, which was present in one of the founders, has been able to pair with the same allele, producing homozygotes with the disorder. The child pictured here has the shortened limbs and extra fingers that characterize this syndrome.

Genetic drift occurs in populations of all sizes, but the effects of genetic drift are much more pronounced in small populations; in this case, the genetic composition of a few individuals has a significant impact on the gene pool. In large populations, chance effects tend to be averaged out. The same is true when flipping a coin. It is not unlikely that you will come up with heads or tails 75 percent, or even 100 percent, of the time if you flip a coin only four times, but the chance of this happening if you flip the coin 100 times is very remote. The chance becomes infinitesimally small if you flip the coin 1,000 times.

Even species that normally have large populations may pass through occasional periods when only a small number of individuals survive. During these so-called population **bottlenecks,** allele frequencies can change dramatically due to chance. Climatic changes, disease, predation, and natural catastrophes may reduce the size of a species to a very small number in a small area. During the last Ice Age, for example, the southward movement of glaciers in North America and Europe squeezed many plant and animal species into small areas, reducing population sizes to very low levels.

In a small population with only a few breeding individuals, complete mixing of the gene pool is possible. To take an extreme case, if there are only four individuals in a population, one of whom possesses a unique but selectively neutral trait, such as a dimpled chin, there is a good chance that this trait will be able to spread through the population in just a few generations. It is much less likely that the same trait will spread through a large population of thousands of individuals by chance alone.

An interesting account of genetic drift is provided by a study of the fishes that occupy warm springs in the Death Valley region of California and Nevada. Remarkably, one species, *Cyprinodon diabolis,* is completely confined to a single spring in Nevada, called Devil's Hole. This spring was formed about 12,000 years ago, after the close of the last continental glaciation, when the region was covered by a large lake. Devil's Hole is over 60 meters (200 feet) deep, but the fish are largely confined to a shallow shelf about 20 meters deep, giving these animals the smallest known range (area of occurrence) of any vertebrate species. In fact, it is possible for every fish to be in view at the same time. The number of individuals in the entire species population varies over time, but it is often as low as 50. Because of the small population size, random genetic drift is believed to have been important in the evolution of this species, resulting in a fish that is very different from its relatives.

The Founder Effect When a species expands into another region, a new population may be started by a small number of pioneering individuals. The founders are not likely to possess all the alleles found in the original parental population; even if they do, the proportion of each allele is likely to be different from that of the original population. Since the pioneers represent a small number of individuals, the new population that develops is likely to be strongly affected by genetic drift. This phenomenon is known as the **founder effect.**

Many examples of the founder effect are seen in isolated locations, particularly on oceanic islands of relatively recent geologic origin. When new islands appear, they tend to become colonized by a few members of a species that arrive on the island by chance and are affected by genetic drift. Among animals, a single female arriving on an island carrying fertilized eggs or embryos is all that is required to found a new population. Since many plants can reproduce by either self-fertilization or asexual reproduction, a single seed can colonize a new environment by itself. Consequently, the founder effect has been prominent in plant evolution. As the following example illustrates, the founder effect has also been documented in studies of human populations.

In the 1770s, a small number of Germans of the Amish sect emigrated to the United States and founded a community in Lancaster, Pennsylvania. For over 200 years, this population has remained, for the most part, reproductively isolated, with little intermarrying. One of the members of this founding group apparently carried a recessive allele for a rare form of dwarfism and polydactylism (extra fingers and toes), called the Ellis-van Creveld syndrome (Figure 33-3). A study carried out in the 1960s revealed that of the approximately 8,000 Amish living in the Lancaster area, 43 individuals were homozygous recessive for this allele and exhibited Ellis-van Creveld syndrome, representing more cases of this disorder than in the rest of the world combined! This study provides dramatic evidence of how the founder effect can generate populations whose allelic frequencies may be very different from those of the original population from which the founders arose.

Natural Selection: The Driving Force behind Adaptation

As a young man on the *Beagle,* Charles Darwin became convinced that organisms evolve over time. It wasn't until years later that he conceived of a mechanism that could actually cause that change. This mechanism was **natural selection.**

Ironically, one of the key observations that led Darwin to his conclusions about natural selection came from the practice of *artificial selection,* which has been used for thousands of years by plant and animal breeders to produce strains of crop plants and domestic animals. In this practice, offspring with desirable traits are selected from each generation for breeding purposes, while offspring lacking such traits are prevented from reproducing. The breeder continues to select in a particular direction, generation after generation, until he or she obtains the desired results. This practice often produces varieties of individuals that differ significantly from the original breeding stock. One of the most extreme examples of artificial selection can be seen in dog breeds, all of which are derived from an animal similar to the modern wolf (Figure 33-4*a*). In a few thousand years, artificial selection has produced varieties as different as the two pictured in Figure 33-4*b*.

Artificial selection demonstrated to Darwin and his contemporaries that continued selection was powerful enough to bring about large-scale changes within a species.

(a)

(b)

FIGURE 33-4

Products of artificial selection. *(a)* The wolf, *Canis lupus*, is the wild ancestor of the domestic dog *(b).* After generations of artificially selecting the traits they want to emphasize, breeders have produced dogs as different as the pair depicted here. Despite these differences, all domestic dogs are members of the same species and are capable of interbreeding to produce viable offspring.

one extreme is repeatedly selected, the frequency distribution gradually shifts in the direction of the favored phenotype (Figure 33-7*b*). Directional selection occurs when there is a change in the environment such that the phenotype at one extreme loses its selective advantage, while individuals possessing the phenotype at the other extreme increasingly survive and reproduce. We have already discussed two examples of directional selection in this chapter:

the shift in frequency from the light-colored peppered moth to the dark form during the Industrial Revolution in England, and the increased resistance of mosquitoes to DDT. During human evolution, increased brain size and loss of body hair represent directional changes in phenotype.

In **disruptive selection,** extreme phenotypes become more frequent from generation to generation because indi-

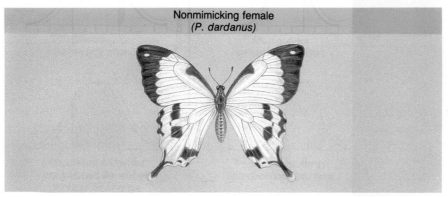

FIGURE 33-9

Altered states. Female African swallowtail butterflies mimic the appearance of local, foul-tasting butterfly species, creating strikingly different female phenotypes, even though they all belong to the same species. These different forms of females provide an example of disruptive selection and polymorphism.

(a)

FIGURE 33-10

Sexual selection. Although they may actually hinder the individual's mobility, brilliant displays of feathers *(a)* or racks of antlers *(b)* are products of sexual selection for characteristics that increase an individual's chances of mating.

(b)

viduals with intermediate phenotypes die or fail to reproduce (Figure 33-7c). Disruptive selection promotes **dimorphism** (two forms of a trait) or even **polymorphism** (several forms of a trait) in a population. This may happen in a diverse or cyclically changing habitat, where different individuals are adapted at different times or in different parts of the environment.

An example of disruptive selection is found in female African swallowtail butterflies (*Papilio dardanus*). Although these butterflies are all members of the same species, the species is widespread, and individuals from one locale are strikingly different in appearance (phenotype) from those of other areas (Figure 33-9). Why would this occur? Some species of butterflies combat predation by concentrating noxious chemicals in their bodies from the plants on which they feed. After one or two nauseating bites, birds learn to recognize these distasteful butterflies and leave them alone. The female African swallowtail butterflies lack these chemicals and would make a tasty meal for a bird, were it not for the fact that they closely resemble (mimic) distasteful species. Distasteful butterflies tend to live in small populations, however. Female African swallowtails will only be protected by mimicking the species of distasteful butterfly that lives in their own small geographic area. Consequently, natural selection has favored the evo-

lution of several distinct color patterns. Intermediate phenotypes between two local groups would not resemble any distasteful butterflies and would be devoured by the birds.

Nonrandom Mating

When individuals choose mates on the basis of their phenotypes, **nonrandom mating** occurs. Nonrandom mating can be caused by a number of factors. It frequently occurs when there is a preference for a particular type of mate or when the population becomes so small that there is no choice except to mate with a close relative.

Sexual Selection Not all characteristics favored by natural selection improve an individual's chances of *survival;* rather, some increase its chance of *reproducing.* The spectacular tail feathers of a peacock and the spreading antlers of a male deer (Figure 33-10) appear as if they could actually impede the animal's pursuit of food and escape from predators. Since these characteristics improve the chances of attracting females and reproducing, however (and in natural selection, passing on your genes is all that matters), they will be strongly selected for. This form of natural selection is called **sexual selection.**

Sexual selection often leads to *sexual dimorphism;* that is, visible differences between the male and female of the

◁ B I O L I N E ▷
A Gallery of Remarkable Adaptations

The most striking result of evolution by natural selection is adaptation, the ways in which organisms seem to fit exactly with the world in which they live. Natural selection has resulted in some truly remarkable adaptations, some of which are morphological, such as sharp teeth and claws, horns, and trichomes. As the following examples illustrate, however, adaptations can also be behavioral or reproductive.

a. *A fishy lure—a morphological adaptation.* The scorpion decoy fish *(Tricundus signifer)* has a dorsal fin

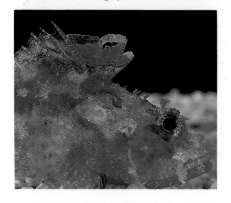

that resembles a smaller fish, complete with its own "dorsal fin," "eye," and "mouth." The scorpion fish swishes its dorsal fin back and forth, luring small, would-be predators. Within a tenth of a second, the hopeful diner quickly becomes the dinner, fatally fooled by a very artful angler.

b. *Disguise—a morphological adaptation.* The tiger swallowtail butterfly

(Papilio glaucus) progresses through a series of larval and pupal stages, each with its own deceptive morphological adaptations. The first larval stage resembles bird droppings. (What predator would eat that?) Three stages later (photo), the green larval caterpillar blends in with the leaves it eats. The caterpillar also has large, false "eyes" that frighten away predators. The pupal stage masquerades as a broken

same species (Figure 33-11). Sexual selection is common among animals because a female's reproductive success is limited by the number of eggs she can produce in her lifetime, and a male's reproductive success is limited by the number of females he can inseminate. Therefore, it is to the female's advantage to choose the most fit male as her mate, and it is to the male's advantage to attract as many females as possible. This leads to natural selection of certain male characteristics, either through male competition with one another or through female choice. On the one hand, males may compete directly by fighting, or they may compete for territory, the possession of which attracts females. Consequently, in these species, males develop characteristics that

enable the animal to fight or intimidate other males, like the antlers of a deer or the huge body size of the male elephant seal. On the other hand, females choose a mate, so natural selection favors those characteristics that females prefer. As a result, characteristics such as the bright-colored plumage of male birds, which is perhaps best exemplified by the gaudy tail feathers of peacocks, become exaggerated.

FIGURE 33-11
Sexual dimorphism in elephant seals and wood ducks. The larger elephant seal and the more brightly colored duck are the males.

twig on a tree trunk, camouflaged from hungry predators.

c. *Playing dead—a behavioral adaptation.* When threatened, the ringed snake *(Natrix natrix)* feigns death by dropping its head, dangling its tongue out of its mouth, and lying completely motionless. These actions help secure the snake's safety because most predators avoid dead organisms.

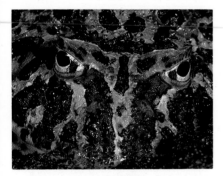

d. *Safe as the ground you walk on—a behavioral adaptation.* The camouflaged horned toad *(Ceratophrys ornata)* buries its body in mud, leaving only its eyes and large jaws protruding. Unwary prey quickly disappear as they move over the concealed head.

e. *Torpedo seeds—a reproductive adaptation.* The seeds of red mangrove trees germinate while they are on the tree. When released, the streamlined radicle slices through the water, planting the seedling upright. Seedlings that don't reach the bottom are able to float for months, until they run aground in shallow water and take root.

f. *Two in one—a reproductive adaptation.* Life as an independent organism is over almost as soon as it begins for the male deep-sea angler fish *(Edriolychunus schmidti).* As an adaption to allow members of the opposite sex to find each other in the blackness of the deep sea, the newly hatched male angler fish permanently attaches itself to the female by sinking its jaws into her body. The female's skin grows over the male's body, and the individuals' circulatory systems connect. The male becomes incorporated into the female body and is reduced to nothing more than a small sperm factory.

FIGURE 33-12

The cheetah population has lost most of its genetic variability due to inbreeding after the population was drastically reduced in size within the past 20,000 years.

Inbreeding   Most harmful alleles originate as rare mutations and are limited to a small percentage of the population. Usually, these mutations exist as recessive alleles. If inbreeding does *not* occur, the chances are slight that two unrelated organisms will have the same harmful recessive alleles. Thus, it is unlikely that many offspring with the homozygous recessive phenotype will be generated, only to die of a genetic disease. If the two organisms are closely related and have received their alleles from a common ancestor, however, the chances of their carrying the same harmful recessive alleles are much greater. If these related organisms mate, chances are even greater that they will produce offspring with the defective phenotype. Obviously, when these offspring fail to reproduce and die, *all* of their genes (beneficial and deleterious) are removed from the gene pool. Therefore, since inbreeding increases the likelihood of death due to genetic defects, the population will lose more and more of its variability. If the environment changes, the population may lack sufficient variability to adapt to the change and may become extinct.

Smaller populations are affected more dramatically by nonrandom mating (and inbreeding, in particular) than are large ones. The effects of inbreeding in a small population are illustrated by the cheetah (Figure 33-12), the world's fastest land animal. At the present time, there are approximately 20,000 cheetahs left in the world, a number that normally would not indicate any danger of extinction. But cheetahs are different. Biologists who have studied these cats have discovered that cheetahs are not a healthy species. Cheetah cubs have a much lower survival rate than do the cubs of other species of large cats. Cheetah cubs are more susceptible to diseases, such as distemper, and the adult males typically produce a small number of sperm, most of which have an abnormal shape.

The reason for the poor health of cheetahs became apparent when a group of biologists led by Stephen O'Brien of the National Institutes of Health analyzed the gene pool of the species to find out why cheetahs were so difficult to breed in captivity. The group began by examining 50 different blood proteins from a variety of individuals, expecting to find a variety of allelic forms of these proteins. To the biologists' surprise, no differences were found in the proteins among the population; all of the cheetahs were homozygous for all of the genes coding for these proteins. Subsequent studies indicated that cheetahs are so genetically similar that they won't even reject tissue grafts from one another, a phenomenon unknown among other mammalian species (page 650).

What has happened to the cheetah? Several million years ago, cheetahs were found in abundant numbers among the African fauna and presumably exhibited a normal variety of alleles within their gene pool. Sometime within the past 20,000 years, conditions arose that apparently decimated the animal's numbers. As the population of cheetahs dwindled to a few individuals, the gene pool was drastically reduced. In other words, the cheetah went through a genetic bottleneck. As the few survivors interbred, the offspring became increasingly homozygous. Furthermore, harmful recessive mutations were paired more frequently by inbreeding, producing homozygous recessive individuals that died, taking with them some of the desirable alleles at other loci, further reducing genetic variability.

Today, the cheetah population probably lacks the variable phenotypes that are needed to ensure the survival of the species. If the cheetah population is to be maintained, new genetic variability will have to come from mutation; the alternative is extinction. The fate of the cheetah illustrates the genetic effects of extreme inbreeding. Virtually all human societies have taboos against incest—sexual relations between parents and their offspring or between brothers and sisters. Approximately half of the states in the United States have laws that prohibit marriages of first cousins. On the average, offspring from marriages between first cousins are about twice as likely to be born with a serious inherited disorder than are offspring from unrelated parents.

➡ The examples we have discussed up to this point—pigmentation in peppered moths, sickle cell anemia in humans, loss of genetic variation in cheetahs, pesticide resistance in mosquitoes—are considered examples of **microevolution** because they result from changes in the allele frequency of a species' gene pool but they have not resulted in the appearance of new species, a phenomenon referred to as **macroevolution.** Microevolution reveals the process of evolutionary change over a short enough period of time so that it can be documented and studied. The occurrence of microevolution allows biologists to study the underlying mechanisms—mutation, gene flow, genetic drift, natural selection, and nonrandom mating—which, given sufficient time, lead to macroevolution, the subject of the remainder of this chapter.

Evolution is the greatest unifying concept in biology. Evolution explains why there are so many different types of organisms on earth; why each species is so well adapted to its particular habitat and lifestyle; and why a species shares many basic features with distant relatives while possessing unique features that distinguish it from all other species. (See CTQ #2.)

SPECIATION: THE ORIGIN OF SPECIES

Biologists have hypothesized that there are more than 5 million species of organisms alive today, even though only about 1.8 million have been described and named so far. Evidence indicates that all of these species, plus all those that lived in the past and have become extinct, descended from a single ancestor that lived approximately 3.5 billion years ago. **Speciation**—the process by which new species are formed—occurs when one population splits into separate populations that diverge genetically from one another to the point where they become separate species. Speciation has produced the millions of species that have inhabited the earth throughout time.

WHAT ARE SPECIES?

Species is a Latin word meaning "kind." Kinds, or species, of organisms were originally identified by their appearances because members of a species typically look alike. In some cases, identifying an individual as a member of a particular species is not so simple, however, because individuals from distinct species may appear very similar. Even among large, familiar vertebrates, such as giraffes, elephants, or camels, there are usually two or more species that closely resemble one another (Figure 33-13a). When we consider the more numerous smaller animals and plants, the problem of identification becomes even more apparent. To distinguish closely related species of insects, for example, a biologist may have to resort to the examination of microscopic bristles.

FIGURE 33-13

Are these animals members of the same or different species? *(a)* These two elephants, one from Africa *(Loxodonta africana)*, the other from India *(Elephas maximus)*, are members of two different species. The Indian elephant (right) has smaller ears and more pronounced "bumps" on its head. *(b)* All of these sea stars are members of the same species, even though they show marked differences in coloration.

(a)   *(b)*

Distinguishing between species of similar morphology may require the analysis of biochemical, ecological, and behavioral traits, as well as those visible to the eye. Some species may be difficult to identify because the members of the species population have different appearances due to a high degree of genetic variation. This is illustrated by the polymorphic African butterflies depicted in Figure 33-9. Similarly, one of the common sea stars of the Pacific coast occurs in a wide variety of colors (Figure 33-13b), yet all of these animals belong to a single species.

Why should two insects that differ slightly in bristle pattern be considered separate species, while two sea stars of totally different color (or two dogs of totally different body shape) are included in the same species? The most important criterion for defining a species is that members are capable of producing other members by mating within the community. One definition of species that incorporated this concept of shared reproduction was given by Ernst Mayr of Harvard University in 1940 and is now known as the **biological species concept.** According to this definition, "Species are groups of actually or potentially interbreeding natural populations which are reproductively isolated from other such groups." By including the phrase "actually or potentially," Mayr acknowledged that although distance, time, or geographic barriers may separate some individuals, the individuals are still members of the same species if they can interbreed once the barrier is removed.

Furthermore, the interbreeding must be natural. Individuals that do not normally breed in the wild are sometimes mated in captivity. For example, zoos sometimes display "tiglons" (offspring from the mating of a tiger and a lion) or similar hybrids. These animals do not occur in nature; thus, they have no real effect on the evolutionary history of groups. While Mayr's definition works well in defining animal species, it does not always hold for plants (see Steps to Discovery, Chapter 38). Among shrubs and trees, in particular, closely related species may interbreed and form fertile hybrids that then give rise to a population of hybrid individuals.

REPRODUCTIVE ISOLATING MECHANISMS

According to the biological species concept, members of one species are *reproductively isolated* from members of all other species. Accordingly, reproductive isolation, which prevents the exchange of genes between populations, is the first step leading to the formation of new species. Once the gene pools are isolated, the separated populations inevitably diverge because of differences in mutation, mating patterns, genetic drift, and natural selection. Over time, the isolated populations amass morphological, physiological, and behavioral differences that prevent them from interbreeding. Consequently, even if the original cause of isolation is removed, the populations remain reproductively isolated; they have become different species.

Barriers that prevent the exchange of alleles between populations (gene flow) are called **isolating mechanisms.** Isolating mechanisms are divided into two categories, depending on whether the isolation prevents a zygote from forming **(prezygotic isolating mechanisms)** or eliminates the success of such crosses as they occur **(postzygotic isolating mechanisms).** Examples of the two categories of isolating mechanisms are presented in Table 33-2.

PATHS OF SPECIATION

The millions of different species that exist today did not emerge by any single sequence of events but have come into existence by a number of different paths of speciation.

Phyletic Speciation

Darwin entitled his great book *On the Origin of Species,* but in it he discussed only how populations could change under the influence of natural selection, as opposed to the other mechanisms described above, such as gene flow and genetic drift. For Darwin, speciation was the simple, gradual accumulation of changes in a lineage through time, until the group was distinct enough to be considered a new species. This process is now called **phyletic speciation.** Although phyletic speciation undoubtedly occurs, speciation is more often the result of one coherent reproductive group splitting into two or more new, discrete species.

Allopatric Speciation

Allopatric speciation (*allo* = other, *patri* = habitat) is believed to be the most common type of speciation. Allopatric speciation typically occurs when a physical barrier, such as a mountain range, a river, or even an oil pipeline, geographically separates a population from its parental population, thereby cutting off gene flow between the two. While isolated, the separated population develops a number of genetic differences, including a reproductive barrier, that distinguish it from the main population. At this point, the two populations can be considered separate species. For example, the dozen or so species of finches discovered by Darwin on the Galapagos Islands are thought to have evolved as the result of their geographic separation from the parental species in Panama and from one another on different islands (see Figure 1-10). Another example of allopatric speciation is provided by several hundred species of fruit flies living on the Hawaiian Islands; all of these fruit flies are believed to have arisen from a single parental species.

Parapatric Speciation

Parapatric speciation (*para* = beside) is thought to occur in populations that lie adjacent to one another. Gene pools diverge because the environment varies sufficiently in the different locales. As a result, different traits are selected in each population. In one study of grasses growing in regions of abandoned mines, for example, investigators found that populations living in areas of toxic mine wastes had devel-

TABLE 33-2
ISOLATING MECHANISMS

1. Prezygotic Isolating Mechanisms

Ecological Isolation Different habitat requirements separate groups, even though the inhabitants may exist in the same general location. Example: Head and body lice are morphologically very similar, yet they live in different "habitats" on a single human body. Head lice live and lay eggs in the hair on the head of a human, whereas body lice live and lay their eggs in clothing. Both suck blood for nutrition.

Geographical Isolation Emerging mountains, islands, rivers, lakes, oceans, moving glaciers, and other geographic barriers keep groups isolated. Example: Different tortoises are found on different Galapagos Islands; surrounding oceans keep tortoise populations isolated.

Seasonal Isolation Differences in breeding seasons prevent gene flow, even when populations are found in the same area. Example: Two populations of bigberry manzanitas grow close together in the mountains of southern California, yet the populations do not interbreed because one completes blooming 2 weeks before the other begins to bloom.

Mechanical Isolation Physical incompatibility of genitalia. Example: Genital structures differ in shape for alpine butterfly species, even though these butterflies look nearly identical in all other ways.

Behavioral Isolation Differences in mating behavior prevent reproduction. Example: Many animals have evolved complicated courtship activities before breeding. Some species of fruit flies (*Drosophila*) are indistinguishable to our eyes, yet they do not mate with each other because of differences in courtship behavior.

Gamete Isolation Sperm and egg are incompatible. Gamete isolation is a common isolating mechanism in many plant and animal species.

2. Postzygotic Isolating Mechanisms

Hybrid Inviability Zygotes or embryos fail to reach reproductive maturity. Example: Hybrid embryos formed between two species of fruit flies fail to develop.

Hybrid Sterility Fertilization is successful between two species, but hybrid progeny are sterile. Example: A mule is a sterile hybrid produced from a mating between a horse and a donkey.

---

oped a tolerance to heavy metals which was not present in populations growing in adjacent, nonpolluted areas. In addition to differences in their sensitivity to metals, the two populations have diverged in time of flowering, resulting in reproductive isolation.

Sympatric Speciation

Sympatric speciation (*sym* = same) occurs in populations where individuals continue to live among one another, even though some type of *biological* difference, such as the time of the year when gonads mature, has divided the members into different reproductive groups. The best-accepted cases of sympatric speciation occur in plants as a result of **polyploidy**—an increase in the number of sets of chromosomes per cell.

The appearance of tetraploid (4N) offspring from diploid (2N) parents is not an uncommon occurrence among certain types of plants. Once formed, the tetraploid cannot interbreed with diploid members of the population because of chromosome incompatibility (page 260). Consequently, the tetraploid plant must either engage in fertilization with another tetraploid individual in the population, or it must produce a population of tetraploid plants on its own (either by self-fertilization or asexual reproduction) which can then interbreed. Regardless of the specific pathway, one interbreeding population is converted into two reproductively isolated populations, setting the stage for speciation. Since many plants are self-fertile and capable of asexual reproduction, polyploidy has been very important in plant evolution and speciation. More than 40 percent of flowering plant species living today are polyploids.

Hybridization

Rapid speciation by **hybridization** occurs when two distinct species come into contact, mate, and produce hybrid offspring that are often reproductively isolated from either parent but not from one another. In just one generation, an entirely new species can be generated by hybridization. At first glance, it may seem unlikely that hybridization could possibly produce viable offspring; in fact, viable offspring are rare in cases of animal hybridization. Plants are different, however, because they are more tolerant of polyploidy than are animals and because a single individual with a unique complement of chromosomes can generate a new population by self-fertilization or asexual reproduction.

Common wheat (*Triticum asestivum*) is believed to have evolved by hybridization. The original stock was probably similar to an ancient crop plant we now call einkorn wheat (*Triticum monococcum*). This wheat appears to have been cross-pollinated by a wild grass (*Aegilops speltoides*) that grows abundantly on the edges of wheat fields in southwestern Asia. Each of these species has seven pairs of chromosomes, but their hybrid offspring have 14 pairs. This hybrid offspring is similar to what we now call emmer wheat (*Triticum durum*). Subsequently, emmer wheat hybridized with goat grass (*Aegilops squarrosa*), which has seven pairs of chromosomes and is found in the mediterranean area. The result is our modern species of wheat, which has 21 pairs of chromosomes.

Most new species are thought to have resulted from the separation of a population into two groups by a physical barrier that prevented interbreeding. The separated populations are influenced differently, causing them to diverge from one another and to become separate species. (See CTQ #4.)

PATTERNS OF EVOLUTION

As new species form and adapt to their environments through natural selection, different patterns of evolution may emerge. The most common pattern, **divergent evolution,** occurs when two or more species evolve from a common ancestor and then become increasingly different over time (Figure 33-14a). Divergent evolution forms the basis for phylogenetic branches, whereby one ancestral species gives rise to two distinct lines (*lineages*) of organisms that continue to diverge. Monkeys and apes, for example, diverged from a common ancestor, as did apes and humans (Chapter 34).

Sometimes, when members of a species move into a new area with many diverse environments, new species form and rapidly diverge, producing a variety of related species that are adapted to different habitats. This rapid divergent evolution is referred to as **adaptive radiation** (Figure 33-14b). One of the most astonishing examples of adaptive radiation occurred on the isolated continent of Australia, where the diversity of habitats and the absence of competitors sparked an adaptive radiation of marsupials (pouched mammals). Beginning with a small, opossum-like marsupial that lived about 100 million years ago, a diverse array of marsupials evolved, ranging from kangaroos that hop across the open Australian prairies to koalas that cling to branches in the forest's trees.

There are a limited number of solutions to any environmental problem. For example, rapid movement through

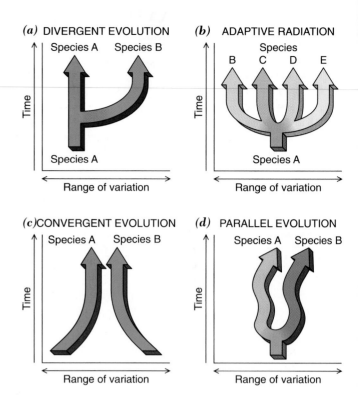

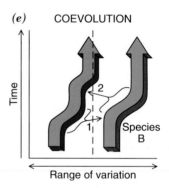

1 Change in species A creates new conditions that cause a change in species B

2 Change in species B creates new conditions that cause a change in species A

FIGURE 33-14

Patterns of evolutionary change. (*a*) Divergent evolution: One species splits into two species. (*b*) Adaptive radiation: One species gives rise to many new species that are adapted to different types of habitats and/or food sources. (*c*) Convergent evolution: Unrelated species evolve similar characteristics as the result of similar selective pressures. (*d*) Parallel evolution: Two related species remain similar over long periods of time. (*e*) Coevolution: Two species evolve in such a way so that changes in one causes reciprocal changes in the other.

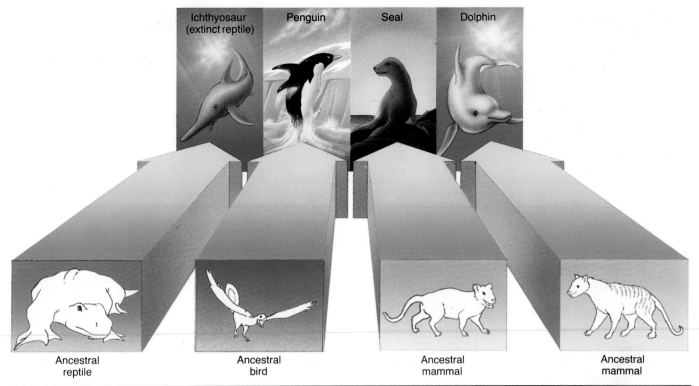

Ichthyosaur (extinct reptile)   Penguin   Seal   Dolphin

Ancestral reptile   Ancestral bird   Ancestral mammal   Ancestral mammal

FIGURE 33-15

Convergent evolution. These marine animals are descended from different ancestors. They have all developed similar streamlined bodies and paddlelike front limbs that adapt them to life in the water.

the water requires a streamlined body shape, while movement through the air requires wings. Therefore, when species with different ancestors colonize similar habitats, they may independently acquire similar adaptations and resemble one another superficially. This phenomenon is called **convergent evolution** (33-14c). For example, each of the four marine animals depicted in Figure 33-15 is descended from a different terrestrial ancestor, but they all share certain common adaptations, such as streamlining and paddlelike forelimbs.

The similarity among many Australian marsupials and placental mammals on other continents is another example of convergent evolution. The Austiralian marsupials evolved independently from their placental counterparts and are only distantly related, yet the two groups include animals with strikingly similar characteristics. For example, there are both marsupial and placental "anteaters," "wolves," and "flying squirrels" (Figure 33-16). Fossils indicate that there was even a marsupial saber-toothed tiger.

Parallel evolution occurs when two species that have descended from the same ancestor remain similar over long periods of time because they independently acquire the same evolutionary adaptations (Figure 33-14d). Parallel evolution occurs when genetically related species adapt to similar environmental changes in similar ways. For example, the ancestral arthropod had a segmented body with a pair of legs on each segment. In all three major arthropod lineages that have descended from this ancestor (the crus-

taceans, insects, and spiders), the number of legs has decreased, and the body segments have become fused, forming larger structures with specialized functions.

Since organisms form part of the natural environment, they can also act as a selective force in the evolution of a species. In nature, species frequently interact so closely that evolutionary changes in one species may cause evolutionary adjustments in others. This evolutionary interaction between organisms is called **coevolution** (Figure 33-14e). Flowering plants and their insect pollinators have coevolved for millions of years, leading to many finely tuned structural and behavioral relationships between flowers and pollinators (Chapter 20). Another example of coevolution is found in predator–prey interactions, where improvements in the hunting ability of a predator favor the survival of a prey with characteristics that increase its ability to escape. Parasites and their hosts also coevolve. Over time, parasites tend to become less destructive of their hosts (a dead host means a dead parasite), and hosts tend to become more resistant to the parasite.

Over long periods of time, several different patterns of evolution may emerge. Most organisms living today are the result of an adaptive radiation whereby an ancestral species gave rise to a number of descendant lines containing organisms that are adapted to different habitats. (See CTQ #5.)

Wolverine

Tasmanian Devil

Southern Flying Squirrel

Sugar Glider

FIGURE 33-16

Living examples of convergent evolution. Australian marsupial mammals and placental mammals on different continents have similar features because they have adapted to similar habitats. Placental mammals are shown on the left, and the Australian marsupial counterpart is on the right.

EXTINCTION: THE LOSS OF SPECIES

Extinction, or the loss of a species, is an important part of evolutionary history. When you consider that approximately 99 percent of the species that have existed since the beginning of life on earth are no longer alive, it becomes apparent that extinction is the ultimate fate of most, if not all, species.

Extinction can occur gradually over a period of tens of thousands of years, or quite suddenly in just one to a few generations. Rapid extinctions are more common among organisms that live in small populations or in geographically restricted areas, such as a single lake or forest. One period of local drought or forest fire can mean the extinction of the entire species. At any point in time, extinction may be limited to just one or a few species, or it may involve the sudden, simultaneous extinction of a multitude of species in a **mass extinction** event (Chapter 35).

Species may become extinct when they lack genetic variability or when they find themselves in the wrong place at the wrong time. In other words, extinction is due to either bad genes or bad luck. In the first case, a species can become extinct when the environment changes and none of the species' members has the genetic makeup that will enable the organism to survive under the new conditions. In the second case, a species may face an unusual catastrophe that essentially eliminates all life in its habitat. Some of the causes that have been proposed for mass extinctions in the past include asteroid impact, volcanic eruptions, drastic changes in sea level, and radical shifts in the earth's climate. Today, organisms on earth are faced with a new cause of mass extinction: The unbridled destruction of natural habitats by humans has increased the extinction rate from a long-term average of about one species each 1,000 years to hundreds, and perhaps thousands, of species in a single year.

Most of the species living today have evolved in recent times. While some of the species that are no longer here were gradually transformed into other species by phyletic speciation, most were unable to adapt to changing conditions and became extinct. (See CTQ #6.)

THE PACE OF EVOLUTION

Darwin viewed evolution by natural selection as a steady, uninterrupted process. He believed that just as natural selection adapted a population to its environment, the process could also turn that population into a new species and eventually found a whole new order, class, or phylum. The discovery of the importance of mutation, gene flow, genetic drift, and nonrandom mating, as well as natural selection, seemed to confirm Darwin's view that most evolution occurs in small, adaptive steps. Under such a model, evolution proceeds by **gradualism** (Figure 33-17a). This view has been criticized by some paleontologists (biologists who study the fossil remains of animals that lived in the past), who contend that fossil evidence does not show a gradual succession of forms. Rather, the analysis of fossils of numerous groups indicates that long periods without significant change (periods of "stasis") are interspersed with short periods of very rapid change.

In 1972, paleontologists Niles Eldredge of the American Museum of Natural History in New York and Stephen Jay Gould of Harvard University proposed a hypothesis called **punctuated equilibrium** (Figure 33-17b) to explain this pattern of evolution. The punctuated equilibrium model includes two separate proposals. The first states that speciation, when it occurs, is a rapid process. We have already seen how allopatric speciation can lead to new species and how small populations are changed more rapidly than are larger ones. When both phenomena occur together (allopatric separation of small populations), spurts of speciation can occur.

The second proposal is that, once formed, species exist for long periods of time without change, unless the environment is altered in some way. This stasis occurs because, even in semistable environments, species often reach a population size large enough for stabilizing selection and gene flow to operate, preventing the species from changing into a new species.

Although gradual evolution and punctuated equilibrium are alternative explanations for the evolution of new species, one does not necessarily exclude the other. The questions now being debated among many biologists are whether one phenomenon occurs more frequently than the other, and which one most likely occurred in the evolution of a particular group.

Evolution does not necessarily progress at a constant pace but may take place in spurts, separated by periods where little change occurs. (See CTQ #7.)

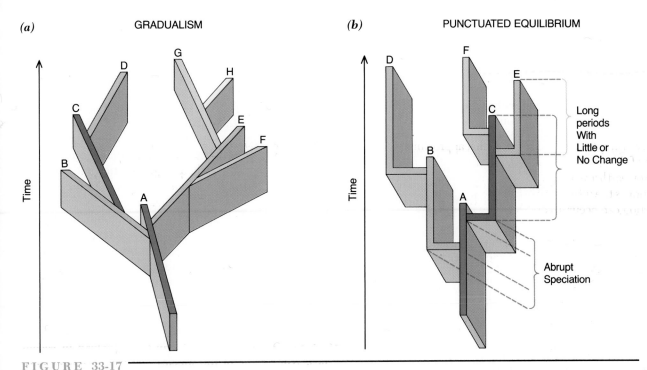

FIGURE 33-17

Gradualism versus punctuated equilibrium. (*a*) During gradualism, species arise through gradual, steady accumulation of changes. (*b*) During punctuated equilibrium, species arise as a result of the rapid accumulation of changes. Once formed, the species remains relatively unchanged for long periods of time.

REEXAMINING THE THEMES

Relationship between Form and Function

Form, function, and natural selection are inseparably linked to one another. Genes code for proteins that determine the form of most structures. Normally, when mutations arise that disrupt the structure of an essential protein, the organism fails to survive and reproduce. As a result, natural selection has eliminated the organism's genes from the species' gene pool. This is not always the case, however, as exemplified by the sickle cell allele (S), in which the alteration of red blood cell structure makes a person carrying one S allele resistant to the ravages of malaria. Consequently, the S allele has been selected for rather than being selected against.

Unity within Diversity

Evolution accounts for both the unity and diversity of life on earth. Genetic changes in populations leads to speciation, which increases the variety of species on earth. Over longer periods of time, evolutionary lineages tend to diverge from one another, leading to greater diversity. At the same time, since all organisms are related by common descent, they share many of the same basic properties of life, including a common system of storage and utilization of genetic information and a common set of metabolic pathways. Unity and diversity are also seen in the characteristics of a single species. Although all members of the species are unified by their capability for interbreeding, the individuals may exhibit a diverse morphology. This is illustrated by the variations in coloration among sea stars and female swallowtail butterflies; the body shape among dogs; and facial appearances, pigmentation, and blood type among humans.

Evolution and Adaptation

Changes in the characteristics of organisms result from changes in the frequency of particular alleles in a population. Changes in allele frequency result from several agents: mutation, genetic drift, gene flow, natural selection, and nonrandom mating. Of these various agents, only natural selection leads to the formation of organisms that are better adapted to their environment and, thus, are more likely to survive to reproductive age.

SYNOPSIS

Evolution is the process whereby species become modified over generations. Evolution occurs when the frequency of alleles in the gene pool of a population changes from one generation to the next. Changes in allele frequency can result from five identifiable factors: mutation (the introduction of new alleles); gene flow (the addition or removal of alleles when individuals move from one population to another); genetic drift (alterations in allele frequency due to chance); natural selection (increased reproduction of individuals with phenotypes that make them better suited to survive and reproduce in a particular environment); and nonrandom mating (increased reproduction of individuals with phenotypes that make them more likely to be selected as mates).

Of the factors listed above, mutation and gene flow introduce new genetic material into a population, while genetic drift, natural selection, and nonrandom mating determine which alleles will be passed on to the next generation. Genetic drift is particularly important in small populations, where chance events can have a major impact on a population's gene pool. Consequently, genetic drift becomes most important when a population shrinks during a bottleneck or when a small group of individuals break away from the main population and colonize a new habitat. Natural selection is the only factor that can cause a species to adapt to its environment. Natural selection is particularly important when environments change, allowing those individuals that possess favorable phenotypes to survive and reproduce, thereby passing their alleles on to the next generation. Natural selection can have a stabilizing, directional, or disruptive effect on the gene pool of a population.

The diversity of life on earth has arisen through repeated speciation events. For speciation to occur, a population must split into two or more separate populations that can no longer interbreed. Reproductive isolation usually occurs as a result of the formation of a geographic barrier. Following reproductive isolation, the separate populations tend to diverge from one another, until they are no longer able to interbreed, and they become different species.

Several identifiable patterns of evolution can be discerned. The most common pattern, divergent evolution, occurs when one species gives rise to two or more species that become increasingly different from one another. When divergent evolution occurs in a new area with diverse environments and an absence of competitors, adaptive radiation may occur, whereby new species form and diverge, producing a variety of related species that are adapted to different habitats. In contrast, when unrelated species colonize similar habitats, they may acquire similar adaptations that cause them to resemble one another. Changes in one species can influence the course of evolution of another species. Extinction is the ultimate fate of most, if not all, species. Extinction occurs when a species lacks the genetic variability needed to adapt to a changing environment ("bad genes") or when a sudden catastrophe occurs that essentially eliminates all life in a particular habitat ("bad luck").

The pace of evolution need not be constant. Evolution within a lineage of organisms may occur gradually in small, adaptive steps, or it may occur in spurts, in which species form and remain unchanged for long periods, followed by a period of rapid change.

Key Terms

evolution (p. 717)
species (p. 717)
population (p. 717)
gene pool (p. 717)
allele frequency (p. 717)
Hardy-Weinberg law (p. 718)
genetic equilibrium (p. 718)
gene flow (p. 719)
genetic drift (p. 720)
bottleneck (p. 720)
founder effect (p. 721)
natural selection (p. 721)
stabilizing selection (p. 724)
directional selection (p.725)

disruptive selection (p. 726)
dimorphism (p. 727)
polymorphism (p. 727)
nonrandom mating (p. 727)
sexual selection (p. 727)
microevolution (p. 731)
macroevolution (p. 731)
speciation (p. 731)
biological species concept (p. 732)
isolating mechanism (p. 732)
prezygotic isolating mechanism (p. 732)
postzygotic isolating mechanism (p. 732)
phyletic speciation (p. 732)
allopatric speciation (p. 732)

parapatric speciation (p. 732)
sympatric speciation (p.733)
hybridization (p. 733)
divergent evolution (p. 734)
adaptive radiation (p. 734)
convergent evolution (p. 735)
parallel evolution (p. 735)
coevolution (p. 735)
extinction (p. 736)
mass extinction (p. 736)
gradualism (p. 737)
punctuated equilibrium (p. 737)

Review Questions

1. Match the term with its definition.
   ____1. allopatric speciation
   ____2. disruptive selection
   ____3. sympatric speciation
   ____4. gene pool
   ____5. gene flow
   ____6. genetic drift
   ____7. extinction
   ____8. convergent evolution

   a. formation of a species by geographic isolation
   b. formation of a species by ecological isolation
   c. result of emigration and immigration
   d. change in gene frequencies due to chance
   e. organisms resemble each other because of similar adaptive pressures, not common ancestry
   f. extreme phenotypes in a species leave more offspring than do average phenotypes
   g. the death of every member of a species
   h. all of the alleles in all of the members of a species.

2. Of all the factors that cause allele frequency to change over time, why is natural selection the only one that leads to increased adaptation to the environment?

3. Why must gene flow stop before speciation can occur?

4. Why do genetic drift and gene flow have a greater impact in changing the gene frequencies of a small population than of a large one?

5. Did speciation occur in the peppered moth populations of England? Under what conditions might speciation occur in the moth?

2. No genetic drift Therefore, it may be changing, but slowly.

3. If a father had a detrimental disease, it would spread like wildfire.

4. Diff. — They are isolated but only 3 reasons why they would Same — Mating produce fertile offspring together.

Critical Thinking Questions

1. Of the five factors that can affect allele frequencies in a population, which could have been important for insects in developing resistance to pesticides? Which could have been unimportant? Why?

2. Populations that are not changing must meet the five conditions identified by Hardy and Weinberg. Consider each of these conditions for the case of sickle cell anemia among African Americans. Do any of the five apply? If so, which one(s)? Based on your analysis, would you predict that change in the frequency of the sickle cell is or is not occurring among African Americans?

3. New reproductive technologies, such as improved artificial insemination, have revolutionized the management of domestic animals. For example, many of the dairy cows in the United States have the same father or grandfather. What is the evolutionary disadvantage of having such a small number of fathers for the population?

4. Two very similar squirrels are found on the north and the south rims of the Grand Canyon. The Kaibab squirrel of the north rim is distinctly darker than the Abert squirrel of the south rim, however. Interbreeding occurs rarely, if ever, in nature, but could occur between members of the two squirrel species, producing fertile offspring. Would you consider these squirrels members of the same or different species? Why, or why not?

5. Match the examples below with the following patterns of evolution: divergent evolution; adaptive evolution; phyletic evolution; convergent evolution; coevolution; parallel evolution. For each one, explain why the example fits the pattern. (1) bears and pandas (2) wolves and foxes (3) the yucca plant and the yucca moth (4) *Homo erectus* and *Homo sapiens* (5) ostrich and emu (6) 16

species of Hawaiian honeycreepers evolved from a common ancestor, each with a different niche.

6. The graph below shows the numbers of extinct species and subspecies of vertebrates from 1760 through 1979. What factors have caused the tremendous increase in extinctions? How do these extinctions differ from the usual extinctions that are a natural part of evolution?

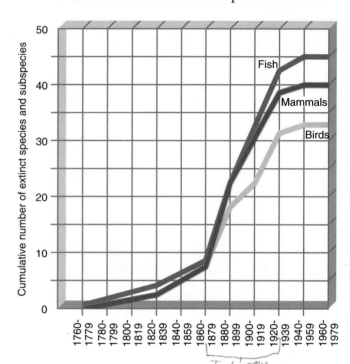

7. How would a scientist who supports gradualism explain gaps in the fossil record? How would a scientist who supports punctuated equilibrium explain the gaps? What sort of evidence would you need to convince you to accept gradualism rather than punctuated equilibrium, or vice versa?

5. 1) convergent evo. 2) parallel evo. 3) coevolution 4) phyletic evo. 5) div. evo evolution 6) adaptive evo

6. Extinction due to humans, industrially, not naturally.

Additional Readings

Avers, C. J. 1989. *Process and pattern in evolution.* New York: Oxford University Press. (Intermediate)

Carson, R. 1962. *Silent spring.* Boston: Houghton Mifflin. (Introductory)

Cook, L. M., G. S. Mani, and M. E. Varley. 1986. Postindustrial melanism in the peppered moth. *Science* 231:611–613. (Advanced)

Dodson, E. O., and P. Dodson. 1985. *Evolution: Process and product,* 3rd ed. Belmont, CA: Wadsworth. (Advanced)

Futuyma, D. J. 1986. *Evolutionary biology,* 2nd ed. Sunderland, MA: Sinauer. (Advanced)

Gould, S. J. 1992. What is a species? *Discover* Dec:40–44. (Intermediate)

Marco, G. J., R. M. Hollingworth, and W. Durham. 1987. *Silent spring revisited.* Washington, D. C.: American Chemical Society. (Introductory)

O'Brien, S. J., D. E. Wildt, and M. Bush. 1986. The cheetah in genetic peril. *Sci. Amer.* 254:84–95. (Intermediate)

Shell, E. R. 1993. Waves of creation. *Discover* May:54–61. (Intermediate)

Wills, C. 1989. *The wisdom of the genes: New pathways in evolution.* New York: Basic Books. (Intermediate)

Evidence for Evolution

STEPS
TO
DISCOVERY

An Early Portrait of the Human Family

STEPS TO DISCOVERY
An Early Portrait of the Human Family

Scientists use the term **hominid** to refer to humans and the various groups of extinct, erect-walking primates that were either our direct ancestors or their relatives. The first fossil remains of a hominid were unearthed in 1856 in caves of the Neander Valley in Germany. After much debate, the remains were dismissed as the bones of a deformed Russian soldier who had died in an earlier war with France.

After the discovery of similar bones in other locations around Europe, it became apparent that the earth had been inhabited at one time by "people" that resembled humans but possessed noticeable differences. They were called Neanderthals, after the site where they were first discovered. Their skulls had a shape different from that of modern humans, with heavy, bony ridges over the eyes, and their bones were much thicker, with indications of larger attached muscles. The Neanderthals were depicted in the

popular press as brutish-looking, grunting, stooped-over cavemen. In reality, if you were to see one of these beings walking down the street in jeans and a T-shirt, you probably wouldn't turn and take notice. Neanderthals lived between 35,000 and 135,000 years ago.

The first evidence of a hominid that would cause you to take notice if you saw one walking down the street was discovered in 1891 by Eugene Dubois, a doctor in the Dutch army stationed in the Dutch East Indies. Soon after arriving on the island of Java, Dubois found the remains of extinct mammals. One day, he found a back (molar) tooth that he thought must have belonged to an ape. A meter away, he discovered a skull that possessed characteristics of both human and ape anatomy. The next year, approximately 15 meters from where he had found the skull, Dubois unearthed a thigh bone (femur) that was very similar to that of

Archaeologists search fossil beds in Africa for remains of early humans illustrated in the time bubble.

a modern human. Most importantly, the shape of the femur indicated that the owner had walked erect. Dubois concluded that he had found the "missing link." He packed the pieces of his "Java Man" into a box and returned triumphantly to Europe.

Most of the scientific world greeted Dubois' claims with skepticism. Sir Arthur Keith, one of the most prominent paleontologists of the time, had a different opinion. After examining the fossils closely, Keith concluded that, even though the size of the Java Man's braincase (the part of the skull that covers the brain) was not much larger than that of an ape, the skull showed definite human features. Keith was so convinced of the similarities that he recommended the Java Man be placed into the same genus as are modern humans. Eventually, Dubois' find became designated *Homo erectus* (see Figure 34-8). However, Dubois never accepted Keith's view that Java Man should be classified as *Homo* (which, in essence, describes it as a human). In response, Dubois buried the bones of his missing link under the floorboards of his dining room, where they remained for the next 30 years.

Over the next 30 to 40 years, a number of other fossils were found that were similar to that of Java Man and were also assigned to the species *H. erectus*. The most important find was Peking Man, discovered in a cave near Peking, China. Like Java Man, Peking Man had a small, apelike braincase; thick, heavy bones; a prominent, bony ridge above the eyes; and a humanlike lower jaw with humanlike teeth. Most importantly, it was demonstrated that Peking Man had walked with an erect posture, used stone tools, and cooked his dinner over a fire. Both Java Man and Peking Man lived about half a million years ago.

Two fossil finds did not fit the profile of *H. erectus*, however. One was a remarkably complete skull that was discovered by an amateur fossil hunter in 1912 near the town of Piltdown in England. The skull of this so-called Piltdown Man had a large braincase (as large as that of a modern human) and an apelike jaw, characteristics in direct contrast to those of Java Man and Peking Man. Piltdown Man presented a serious problem for interpreting the path of human evolution. Some paleontologists dismissed Piltdown Man as an anomaly. Others, including Sir Arthur Keith and the British anthropological establishment, embraced Piltdown Man as an important fossil and suggested that the development of a large brain may have been one of the earliest characteristics to appear along the path of human evolution.

The other perplexing fossil discovery was made in 1924 by Raymond Dart, an Australian on the faculty of a medical school in Johannesburg, South Africa. Dart heard that fossils were being uncovered at a limestone quarry in an area of South Africa called Taung. He asked the owner of the quarry if he might see some of the fossils; two large boxes were shipped to his house. As he was pouring through the contents, Dart spotted a dome-shaped piece of stone. As a neuroanatomist, he immediately recognized the stone as the cast of a brain, complete with indications of convolutions and blood vessels. Sand and lime-containing water had seeped into the skull of an ancient inhabitant of the quarry and hardened, forming a cast of what had once been the creature's brain. Although the brain was the size and form of an ape's, it revealed distinct humanlike characteristics. Dart began searching for the skull that had recently surrounded the cast, believing that it must have been blasted away during the mining operation.

Among the contents of the box, Dart found the remains of the lower jaw and skull, the front of which was covered by an encrusted material, making it impossible to see the face. For the next couple of months, Dart carefully picked away at the crust and slowly revealed an astonishing visage; it was the face of a young "ape," with teeth that showed striking human characteristics (see Figures 34-8 and 34-9). The cranium was slightly larger than that of an ape, and the opening in the skull that allowed entry of the spinal cord was in a position different from that of an ape, suggesting that the individual had walked erect. Based on the other fossils in the box, Dart concluded that the skull was about 1 million years old. He named the creature *Australopithecus africanus* (*Australo* = southern, *pithecus* = ape), but it became known as the Taung Child.

Without delay, Dart wrote up a paper on his skull and sent it to the prestigious British journal *Nature*, in which it was published. Once again the scientific world was very skeptical. Even Keith, who was a friend of Dart's, clung to the notion (based primarily on Piltdown Man) that enlargement of the brain was one of the first steps in hominid evolution and declared that the Taung Child was not a hominid but an extinct ape.

In 1931, Dart traveled to London to attend an anthropological meeting in hopes of convincing his colleagues of his claim. Dart's talk followed a dazzling presentation of the findings that were emerging from China concerning Peking Man. In addition, Dart was a poor speaker, whose evidence was limited to a single skull; he failed to make much of an impact. Discouraged, he went off to dinner with friends while his wife brought the Taung Child back to the hotel. As if the day hadn't gone badly enough, his wife left the infamous skull (wrapped in cloth) in the back seat of the cab, where it traveled around London most of the night. The cab driver finally saw the package and handed it over to the police. Fortunately, Dart was able to recover his package before the police had time to wonder what type of skullduggery they had on their hands.

*T*he "theory of evolution" is no less a fact of life to biologists than the "atomic theory" is to chemists or the "theory of gravitation" is to physicists. For a theory of such importance to have gained such widespread acceptance, it must be backed by a tremendous body of evidence. We will begin by sampling a small portion of this evidence, taken from a wide variety of different fields. The entire matter can be summed up in a single sentence written by the biologist Theodosius Dobzhansky: "Nothing in biology makes sense except in the light of evolution."

▼ ▼ ▼

DETERMINING EVOLUTIONARY RELATIONSHIPS

Given that closely related species share a common ancestor and often resemble one another, it might seem that the best way to uncover evolutionary relationships would be to make comparisons of overall similarity between organisms. In other words, out of a group of species, if two are most similar, can we reasonably hypothesize that they are the closest relatives? Surprisingly, this is not always the case (Figure 34-1). Overall similarity may be misleading because there are actually two reasons why organisms may have similar characteristics, only one of which is due to evolutionary relatedness.

HOMOLOGOUS VERSUS ANALOGOUS FEATURES

Two species that share a similar characteristic they inherited from a common ancestor are said to share a **homologous feature,** or **homology.** The even-toed foot of deer, camels, cattle, pigs, and hippopotamuses, for example, is a homologous feature because all of these animals inherited the characteristic from a common extinct ancestor (Figure 34-2). When *unrelated* species evolve a similar mode of existence, however, their body parts may take on similar functions and end up resembling one another due to convergent evolution (page 735). This type of shared characteristic is called an **analogous feature,** or **homoplasy.** The paddlelike front limbs and streamlined bodies of many aquatic animals (see Figure 33–15) are examples of analogous features.

➭ Homologous similarity is the only evidence that proves that two species are evolutionarily related. But how do biologists tell whether a similarity is homologous or homoplasious? Years of experimentation and observation have resulted in a set of criteria that are used to identify homologies. These criteria include: (1) similar in detail, (2) similar position in relation to neighboring structures or organs, (3) similarity in embryonic development, and (4) agreement with other characters (related animals usually share more than one homology).

These criteria of homology can be illustrated by examining a variety of mammalian forelimbs (Figure 34-3). At first glance, the wing of a bat, the leg of a cat, the flipper of a whale, the arm of a human, and the leg of a horse may not seem very similar, but they are actually homologous. All of these limbs contain the same type of bones (similar in detail); the forelimb always attaches to the shoulder girdle (similar position in relation to neighboring structures); the forelimb develops from the same tissues in each of the embryos (similar embryonic development); and, in addition to the forelimb, all these animals have hair and mammary glands (other shared homologies).

Evolutionary relationships cannot be reconstructed just by grouping species together by their number of shared homologies, however. For example, the hand of the first vertebrates to live on land had five digits (fingers). Many terrestrial vertebrates (such as humans, turtles, lizards, and frogs) also have five digits, which they inherited from this common ancestor. This feature is then homologous in all of these species. In contrast, horses, zebras, and donkeys have only a single digit with a hoof. But, clearly, humans are more

(a)   *(b)*

FIGURE 34-1

Deceptive similarities. These two "palms" could easily be mistaken for closely related plants based on their similar appearances, but the cycad *(a)* is not a palm, or even a flowering plant. It is no more related to the true palm *(b)* than is a pine tree to a rose.

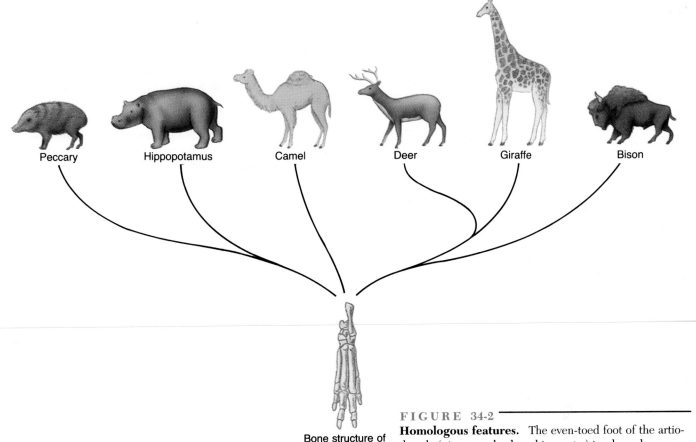

Peccary

Hippopotamus

Camel

Deer

Giraffe

Bison

Bone structure of
extinct ancestor's
foot

FIGURE 34-2

Homologous features. The even-toed foot of the artio-dactyla (pigs, camels, deer, bison, etc.) is a homologous feature because it was inherited from a common ancestor.

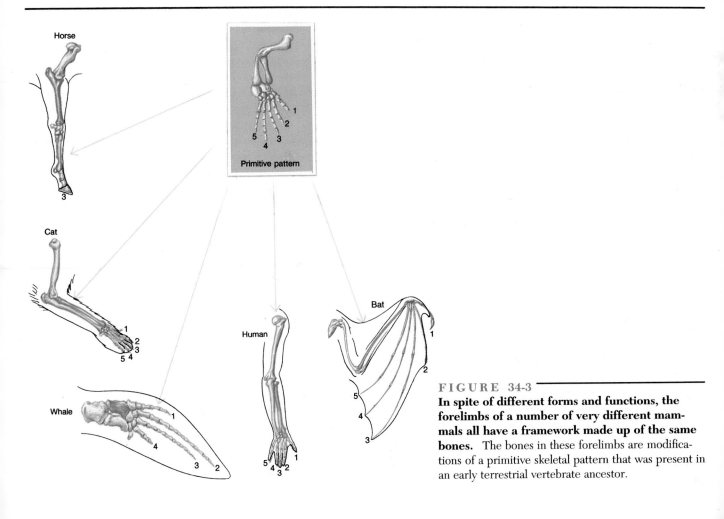

Horse

Cat

Whale

Primitive pattern

Human

Bat

FIGURE 34-3

In spite of different forms and functions, the forelimbs of a number of very different mammals all have a framework made up of the same bones. The bones in these forelimbs are modifications of a primitive skeletal pattern that was present in an early terrestrial vertebrate ancestor.

closely related to horses than they are to lizards! The key point is that the five-digit condition is the *primitive* (original) pattern for the number of digits (shown in Figure 34-2). This primitive feature has been retained in the line of ancestors leading from early amphibians to humans, but it has been modified and reduced to just one digit in the common ancestor of horses, donkeys, and zebras. While the *derived* (modified) trait tells us that horses, zebras, and donkeys share a very recent common ancestor, the primitive form (five digits) tells us only that species are at least distantly related.

Let us now turn to the types of data that biologists use to distinguish between homologies and homoplasies and between primitive homologies, which may be present in rather distantly related species, and shared derived homologies, which indicate closer relationships.

Appearances can be deceiving. Biologists must apply a number of criteria to determine if similarities in structural features are the products of an evolutionary relationship or the consequence of similar selective pressures exerted on unrelated organisms. (See CTQ #2.)

EVIDENCE FOR EVOLUTION

According to the theory of evolution, organisms living today have arisen from earlier types of organisms by a process of genetic change that has occurred over a period of several billion years. The fact that all organisms have arisen from a common ancestor explains why they have the same mechanism for the storage and utilization of genetic information, many of the same types of cellular organelles, and similar types of enzymes and metabolic pathways. At the same time, evolution also explains how a single species can give rise to numerous other species, leading to great biological diversity. Evidence supporting the theory of evolution has accumulated from a variety of different biological disciplines, including comparative anatomy, paleontology, comparative embryology, biochemistry and molecular biology, and biogeography.

FOSSIL RECORDS

A fossil is any trace of life from the past. Many different types of fossils exist, ranging from preserved "footprints" of animals that walked along a trail (see Figure 34-12) to complete remains, such as frozen mammoths or entombed insects (Figure 34-4a), to actual hard parts (teeth and bones), to pieces of petrified trees (Figure 34-4b) and even preserved excrement (*corpolites*). Fossils of body parts may be formed in several ways: Organisms may be buried in sediments, where they harden and mineralize; trapped in tree sap, which hardens into amber; covered in tar or other natural preservatives, such as the liquid found in peat bogs; or frozen in arctic regions or at high altitudes.

The **fossil record** consists of an entire collection of such remains from which paleontologists attempt to reconstruct the biology of the organisms whose remains were left behind. In addition to providing a glimpse of the kinds of organisms that lived in the past, the fossil record provides data evolutionary biologists can use to describe the pathways by which various groups may have evolved. For example, without fossils, we would not realize the close relationship between dinosaurs and birds and the fact that birds are more closely related to reptiles than they are to mammals.

Fossil Dating

For many years, scientists had no way of directly determining the age of a fossil. Instead, scientists would establish the sequence of fossils embedded in the layers of rock from top to bottom and then apply the *law of stratigraphy*. According to this law, in beds of rocks that have not been tilted or folded, the oldest rocks are always on the bottom, and the youngest rocks are always on top. In some sites, rocky strata covering tens of millions of years are present in layers, containing the remains of organisms in chronological order, much like a giant filing cabinet in which the more recent documents are found in drawers situated closer to the top. For example, while remains of ancient fishes are common in rocks 400 million years old, no evidence of reptiles has ever been found in such sediments. Similarly, remnants of reptiles may be unearthed in 285-million-year-old rocks, but we never find evidence there of a bird or mammal. In H. G. Wells's words: "The order of descent is always observed." In the uppermost sediments on earth, biologists may find the remains of animals that have only recently become extinct, including creatures with apelike faces that walked on two feet and crafted primitive stone tools.

While stratigraphy still plays an important role in determining which types of organisms coexisted, the absolute age of fossils can now be determined with a high degree of accuracy using radiodating techniques. As we discussed on page 54, these techniques depend on the existence of naturally occurring radioisotopes that disintegrate into other elements at a predictable rate. Using radioisotopes, scientists have shown that the oldest rocks on earth are over 4 billion years old, much older than Charles Darwin could even have imagined when he first realized that the earth was older than that determined by strict interpretation of the Bible.

Archaeopteryx: An Example of Fossil Evidence

One of the best known fossils was discovered in 1861 in a limestone quarry in Bavaria, Germany. The skeleton of the fossil (Figure 34-5a) suggested that the animal had been a small bipedal dinosaur, but the fine-grained limestone slate also revealed the unmistakable imprint of wings with feathers. Of all the vertebrates, only birds possess feathers; in fact, feathers are a defining characteristic that unites all

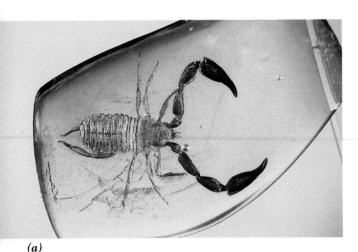

(a)

FIGURE 34-4

Types of fossils. *(a)* This ancient pseudoscorpion (a distant relative of spiders) was trapped in a drop of resin that became transformed into hardened amber. *(b)* A scene from the Petrified Forest of Arizona.

(b)

birds. This animal, which was given the name *Archaeopteryx lithographica* (*archaeo* = old, *pteryx* = wing), was determined to have been a bird that lived 150 million years ago. Yet, unlike all modern birds, *Archaeopteryx* had teeth, a long tail containing over 20 vertebrae, free-floating ribs, and wings containing movable fingers with claws, all characteristics of the small, carnivorous reptiles called *theropods*. Therefore, *Archaeopteryx* provides one of the many pieces of fossil evidence of an evolutionary pathway leading from reptiles to birds.

Although the skeleton of *Archaeopteryx* has a wishbone (two collarbones fused together), which is typical of birds, it lacks the broad, bony breastbone to which the large flight muscles of modern birds are attached. This and other skeletal characteristics reveal a great deal about the lifestyle of this ancient vertebrate. The lack of a breastbone suggests that *Archaeopteryx* was not a strong flyer and may have been primarily a glider. The claws on the toes suggest that the animal could perch on a limb, while the claws on the "fingers" suggest that it may have been able to climb up the trunks of trees (Figure 34-5*b*). The structure of the pelvis and hind legs suggests that *Archaeopteryx* was capable of running over the ground, its long tail acting as a counter-

weight to maintain its balance. The teeth of this pigeon-sized animal would have been suitable for the capture of insects or other small prey.

Cautions Regarding the Fossil Record

The fossil record provides an imperfect view of ancient life because not all organisms and environments are represented equally. For example, fossils of organisms with hard parts, such as shells or skeletons, or the woody parts of plants are more common than are fossils of soft-bodied organisms. Thus, arthropods, which possess hardened exoskeletons, are much more abundant in the fossil record than are jellyfish. Similarly, the likelihood of fossilization is much greater when an organism lives in an environment in which bodies or impressions of bodies can be covered quickly with sediments (such as in shallow seas and river beds) than in other environments.

Furthermore, there is only a remote chance that a fossil will ever be found because only a small proportion of the fossil-bearing rocks are accessible to us. Most fossil-bearing rocks have been eroded away, buried deep beneath the continents or the ocean floor, or broken and destroyed by the earth's movement. Even when a fossil is found, it is

(a)

FIGURE 34-5

Archaeopteryx, a bird that lived 150 million years ago, has many features in common with small, bipedal dinosaurs—features, such as teeth and a long tail, that were eliminated during the evolution of modern birds. ***(a)*** Photograph of the fossil imprint of *Archaeopteryx* in limestone slate. ***(b)*** An artist's rendition of the long-extinct bird as it may have appeared in life.

(b)

A. Snake

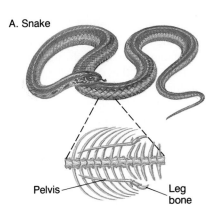

Pelvis — — Leg bone

often only a fragment of a bone or shell, inviting speculation regarding its significance. In general, the further back in time we go, the less complete the fossil record.

Interpreting the fossil record is often difficult, and conclusions are open to speculation. Although some organisms in the fossil record are indeed ancestors of living organisms, an older fossil is not necessarily the ancestor of a younger fossil, and we can't assume an evolutionary lineage or rate of speciation from a fossil sequence. This point is illustrated by the story of human evolution. Most of the fossils that have been found are probably an *offshoot* from the direct line of descent of modern human beings. While there is only one erect-walking, tool-making, word-speaking species of hominid living today, 2 million or 3 million years ago, there may have been quite a number of such species. It is impossible to establish a straight, ladderlike lineage leading from this "bush" of species to a single living representative.

B. Horse

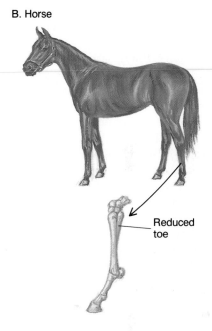

Reduced toe

THE ANATOMY OF LIVING ORGANISMS

▐▶ Comparing the structure of the parts of the bodies of different organisms is probably the most commonly used evidence of evolution. In order to gather comparative evidence for evolution, biologists study external characteristics, examine bones and teeth, dissect organ systems, study sections of tissue under the light microscope, and peer at the finer details of cells and tissues under the electron microscope. Along the way, much is added to our knowledge about the basic biology of different organisms.

Vestigial Structures

The underdeveloped pelvis and leg bones in snakes, the diminished toe bones in horses, and the appendix in humans are all structures that have little or no function in these organisms (Figure 34-6). Each of these **vestigial structures,** as they are called, bears an unmistakable stamp that shows its relationship to a more fully developed, functional structure present in other animals. For example, your appendix is a dwarfed version of the cecum, a part of the digestive tract of many mammals, where food is stored and digested by microorganisms. Similarly, your tailbone is a remnant of the tail present in many of your relatives.

C. Human

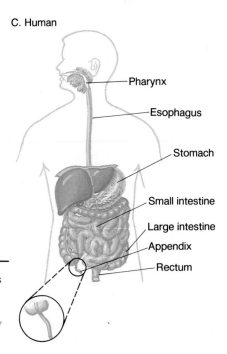

Pharynx

Esophagus

Stomach

Small intestine

Large intestine

Appendix

Rectum

FIGURE 34-6

Vestigial organs: A legacy of ancestors. *(a)* Even though they are legless, snakes retain degenerated leg and pelvic bones inherited from four-legged ancestors. *(b)* Horses still possess degenerated toe bones left over from their three- and four-toed ancestors. *(c)* The human appendix is a degenerated cecum, a chamber used by our vegetarian ancestors for housing cellulose-digesting microorganisms.

Both the plant and animal kingdoms are filled with examples of structures that are in the process of disappearing but still remain as vestiges in living organisms. If species had been created as they are today, there would be no explanation for such structures, but natural selection accounts for the elimination of structures that are no longer needed by an organism exploiting a different way of life.

COMPARATIVE EMBRYOLOGY

Adult fishes, salamanders, turtles, birds, and humans bear virtually no resemblance to one another. Yet these animals are virtually indistinguishable as embryos (see Figure 32-18). Why should animals that have markedly different adult forms and functions develop from such similar embryos? The best scientific explanation is that far back in vertebrate history, fishes, salamanders, turtles, birds, and humans all had a common ancestor, probably some type of primitive fish, that developed from a similar type of embryo. As the various types of vertebrates evolved, they each retained this basic vertebrate embryo as part of their life cycle, even though its parts gave rise to different adult organs.

This example illustrates how developmental evidence can be particularly useful in uncovering evolutionary relationships among diverse adult forms. For example, embryologic similarities between widely separated animal groups (such as snails and flatworms) have been used to reconstruct lines of evolutionary descent within the animal kingdom.

There is another way to illustrate the constraints of embryonic development. Recently, Harvard paleontologist and essayist Stephen J. Gould noted that the question "Why do men have nipples?" heads the list of inquiries from his readers. Gould argues that nipples in men are not adaptive, nor did they evolve from structures that were adaptive in an ancestor. Rather, Gould concludes that nipples in men are the result of a constraint imposed by embryonic development. Male and female mammals of a species pass through identical stages as early embryos; it is not until the secretion of sex hormones that male and female sexual development diverge. Nipples are present in human embryos *prior* to the time of sexual differentiation. Later sexual maturation leads to changes in the female breast and nipple that allow these structures to nourish a newborn infant. In contrast, the male nipple remains as the vestige of an embryonic structure that is simply carried along "for the ride" in the adult. According to Gould, "Male mammals have nipples because females need them . . ."

BIOCHEMISTRY AND MOLECULAR BIOLOGY

Since the characteristics of an organism are determined by its genetic content, changes in organisms over the course of evolution are reflected in changes in the nucleotide sequence of DNA (genes) and the amino acid sequences of proteins (gene products). For the most part, the longer the period since two species have diverged from a common ancestor, the greater the number of substitutions that are found in corresponding genes and proteins between the two species. Common ancestry can now be demonstrated just as forcefully by homologous molecular information as by homologous anatomic structures.

We saw on page 329 how this type of data has been used to determine that humans are more closely related to chimpanzees than are chimpanzees to gorillas. Molecular data of this type allow biologists to determine phylogenetic relationships among organisms based on the degree of nucleotide or amino acid sequence similarity. These diagrams are generally in keeping with conclusions based on anatomic data. As illustrated in the following example, molecular data have also been used to resolve evolutionary controversies.

What Is a Bat? The Use of Molecular Data in Studying Evolution

Biologists recognize two major groups of bats. The microchiroptera include the numerous bats commonly seen swooping up insects at night as well as the more exotic frog-eating bats and vampire bats of Central and South America (Figure 34-7a). The megachiroptera, also called flying foxes because of their foxlike faces, are large fruit-eating bats that live in the tropics (Figure 34-7b). Bats are highly specialized for flight, and many of the features that might reveal their closest relatives have been modified almost beyond recognition by natural selection, resulting in a debate among morphologists and paleontologists over the evolution of bats. One group of scientists determined that the skeletal evidence pointed to a close relationship between microchiroptera and megachiroptera and that both groups were distantly related to the primates (Figure 34-7c). Other scientists claimed that evidence derived from studying the nervous system indicated that the megachiroptera and the primates were closest relatives (Figure 34-7d). There seemed to be no data to solve the debate until late 1991, at which time molecular biologists determined the sequence of the DNA that codes for ribosomal RNA of all three groups. The molecular evidence showed that the microchiroptera and the flying foxes were indeed closest relatives, as reflected in the phylogenetic scheme of Figure 34-7c.

BIOGEOGRAPHY: THE GEOGRAPHICAL DISTRIBUTION OF ORGANISMS

Among the evidence that convinced Darwin of the occurrence of evolution were the observations he made on the Galapagos Islands (Chapter 1). Darwin noted that species present on oceanic islands were not found anywhere else in the world. In fact, many were found only on a particular

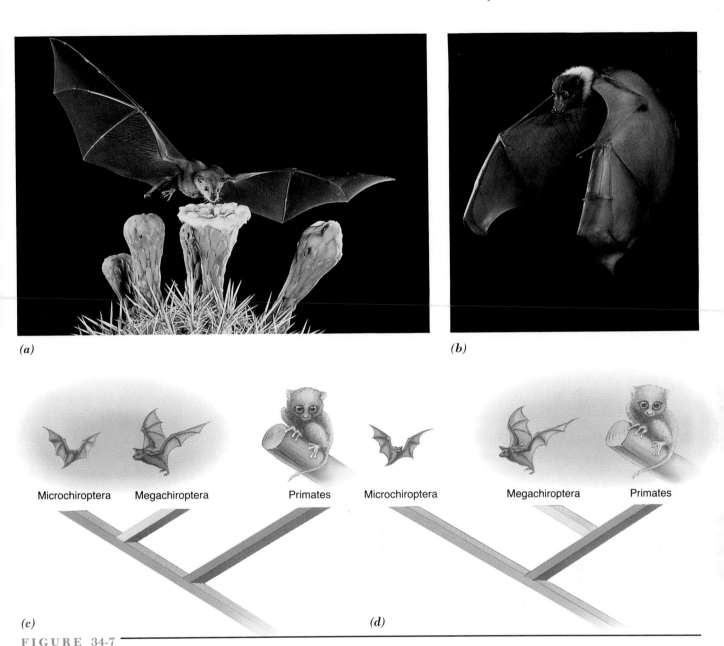

(a)

(b)

Microchiroptera Megachiroptera Primates Microchiroptera Megachiroptera Primates

(c) *(d)*

FIGURE 34-7

Evolution of bats. Evolutionary relationships of bats has been the subject of recent debate. The microchiroptera, such as the long-nosed bat *(a)*, have been thought to be related to the megachiroptera, or flying foxes *(b)*, but flying foxes have also been linked to primates. This conflict was resolved by molecular data that indicates that the megachiroptera are indeed more closely related to the microchiroptera (as depicted in *c*) than they are to the primates *(d)*.

island in the chain and varied from one island to the next. Recall from Chapter 1 that Darwin found different species of finches with distinct anatomic differences living on different islands. Although the various finches possessed different-shaped beaks, which were adapted for obtaining different types of foods (see Figure 1-10), the birds were unmistakably similar in overall anatomy, both to one another and to a species found on the mainland. Darwin concluded that individuals from the mainland species had migrated to the islands, where, given the absence of competition from other birds, they had evolved into a variety of different species adapted to different local conditions and food sources. A common origin also explained why the plants and animals of the Galapagos were generally so similar to those species living on the mainland, even though the two regions had totally different climates and terrain.

These types of biogeographical observations are not restricted to the Galapagos. Plants and animals living in nearby areas typically are similar, regardless of differences in climate and terrain, because they are closely related. In fact, island and mainland species are often placed in the same genus. In contrast, plants and animals living in similar environments on different continents tend to be quite different.

The evidence for evolution comes from such diverse fields as paleontology, embryology, comparative anatomy, biochemistry, molecular biology, and biogeography. This body of evidence is both diverse and overwhelming. While biologists may argue over the mechanisms of evolution, they agree that all life descended with modification from a single common ancestor. (See CTQ #3.)

THE EVIDENCE OF HUMAN EVOLUTION: THE STORY CONTINUES

When we left the discussion of fossil hominids in the chapter-opening vignette, the story had become confused by the presence of conflicting data. On the one hand, we were confronted with *Homo erectus* and the more primitive, less well-accepted, *Australopithecus africanus* (Figure 34-8). These fossils had small, apelike brains and humanlike jaws and teeth. On the other hand, we learned of the Piltdown Man, who had a large, humanlike brain and apelike jaws and teeth.

It was not until the late 1940s and early 1950s that the matter was finally resolved. At that time, a careful analysis was conducted of the jaws of a variety of *Australopithecus* specimens that had been found in South Africa, including Dart's Taung Child (Figure 34-9) and others that had come to light over the intervening years. The analysis was performed under the leadership of Sir Wilfrid Le Gros Clark, who had become the foremost British paleontologist of the time. Le Gros Clark established 11 characteristics that clearly distinguished the teeth of humans from those of modern apes. Examination of the *Australopithecus* fossils indicated that, despite the fact that their braincase was so small and apelike, their teeth were similar to humans in every one of the 11 criteria. There was no longer any doubt that the australopithecines, as they are called, were hominids. In addition, a new radiodating technique revealed that the australopithecines were very old—up to 2 million years old. Even Keith (who was nearly 80 by this time) publicly admitted that he had been wrong and that Dart (who was still actively searching for fossils) had been right.

A second important revelation came from a more careful scrutiny of the Piltdown Man. Radiodating provided

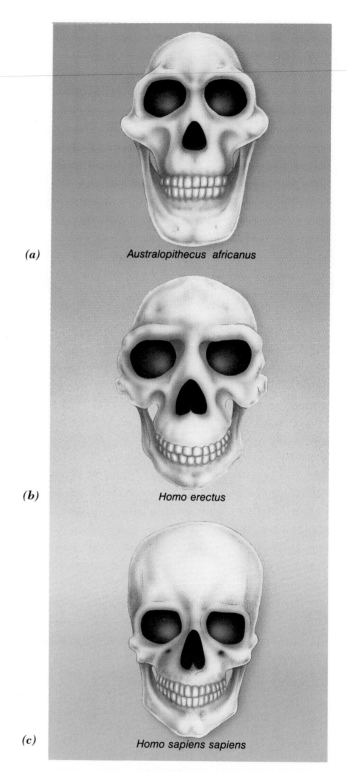

(a) **Australopithecus africanus**

(b) **Homo erectus**

(c) **Homo sapiens sapiens**

FIGURE 34-8

A comparison of the skulls of modern humans and two hominids. *(a)* Skull of the extinct species *Australopithecus africanus* originally represented by the Taung child and later by several other fossils found in South Africa. *(b)* Skull of the extinct species *Homo erectus* that includes Java Man and Peking Man. *(c)* Skull of a modern human, *Homo sapiens sapiens*.

FIGURE 34-9

The skull of the Taung Child in the hands of its discoverer Raymond Dart. Dart died in 1988 at the age of 95.

conclusive evidence that the Piltdown skull was actually of very *recent* vintage—a matter of a few hundred years. In other words, Piltdown Man was a hoax. Someone had taken the skull of a modern human and the lower jaw of a modern orangutan, treated them with chemicals to make them look old, filed down the ape teeth so that they resembled those of a human, broken them into fragments, and buried them alongside one another.

With Piltdown Man out of the way, it was evident that the enlargement of the brain occurred during a late stage of human evolution, not an early stage. In the past 25 years, study of Piltdown Man has shifted from the character of the bones to the perpetrator of the hoax, and fingers have been pointed at some of the leading paleontologists of the time.

HOMO HABILIS AND THE USE OF TOOLS

By the end of the 1950s, the primary scene of hominid excavation had shifted from Europe and Asia to Africa. To understand the more recent evolutionary discoveries, we need to introduce another cast of paleontologists, the most famous of which are Louis and Mary Leakey.

The Leakeys came to East Africa in the 1930s, looking for fossil hominids. They focused their attention on the now-famous Olduvai Gorge, located in Tanzania. The gorge is situated in the Serengeti Plain, on what once was a lake-bed. Over time, sediments were deposited on the bottom of the lake, creating layer upon layer, the deepest sediments

being the oldest. The gorge, which is about 100 meters (330 feet) deep, was created by a river that wound through the area, digging deeper and deeper into the layers of sediments. The sides of the gorge reveal the stratifications, while the bottom corresponds to the bottom of the ancient lake as it existed approximately 2 million years ago. The Leakeys were first drawn to Olduvai by the large numbers of primitive tools that were strewn over the bottom of the gorge. It was the maker of these tools for whom the Leakeys were searching.

According to the Leakeys, the use of tools is just as important (if not more important) than is brain size or tooth structure in describing a fossil hominid as a human (*Homo*), as opposed to some other genus, but not all anthropologists agree. The Leakeys were convinced that the genus *Homo* was older than was generally accepted and that the australopithecines were not our ancestors but our cousins. In other words, they believed that the members of the genus *Homo* went as far back as *Australopithecus* and that the australopithecines were an offshoot that were not on the line leading to modern humans. In fact, the Leakeys argued that members of the two genera lived side by side, a proposal that has since been strengthened by considerable evidence.

During the early 1960s, after 30 years of being scoured, Olduvai Gorge began to reveal bits and pieces of a new hominid. Based on the increased size of the braincase and certain other characteristics, the Leakeys concluded that these hominids were not australopithecines and named them *Homo habilis* ("the handy man"), implying that these hominids were responsible for making the primitive stone tools found at the bottom of the gorge. Their conclusion was supported by radiodating studies that showed that these fossil remains were 1.75 million years old. The age of humans had just been pushed back in time from approximately 500,000 years for *H. erectus* to nearly 2 million years for *H. habilis.*

The argument over whether or not these specimens were actually humans (as opposed to australopithecines) continued for many years. It was finally settled to the satisfaction of most hominid specialists in 1972, with the discovery of a well-preserved skull of the same time period that was unambiguously *Homo*. The discoverer was, appropriately enough, Richard Leakey, the son of Louis and Mary, who had only recently become immersed in the hominid-hunting family passion. A body of evidence now suggests that *Homo habilis* was indeed walking the earth approximately 2 million years ago, probably in the company of several different species of australopithecines. But what type of ancestor had given rise to *Homo habilis*?

THE DISCOVERY OF LUCY

Our best clue as to the nature of that ancestor came in 1974, when Donald Johanson of the Cleveland Museum of Natural History was searching for hominid fossils at a remote site

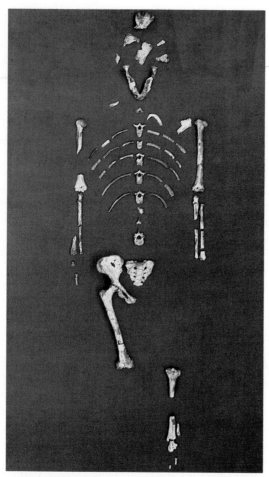

FIGURE 34-10

"**Lucy,**" nearly 40 percent intact, has been assigned to the species *Australopithecus afarensis*. Lucy and other members of "The First Family" may represent a species of hominid that gave rise to the various other species of *Australopithecus* as well as to *Homo*.

in northern Africa, known as Hadar. Like Olduvai, Hadar was an ancient lake bed. Inhabitants of the area had died and become buried by sediment, only to be unearthed at a later time by torrents of water rushing through newly formed gullies. One morning, just as he was planning to return to camp, Johanson noticed a bone projecting out of the ground. On closer inspection, he identified it as the arm bone of a hominid. Nearby, he saw parts of a skull and a thigh bone. In his words, "An unbelievable, impermissible thought flickered through my mind. Suppose all these fitted together? Could they be parts of a single, extremely primitive skeleton?"

That is exactly what they were. Within a few weeks, all of the bones had been recovered. Together they constituted approximately 40 percent of the skeleton of an extremely primitive, small-brained female hominid who had stood only about 3.5 feet tall (Figure 34-10). Most importantly,

the skeleton indicated that the hominid had walked erect, suggesting that bipedal locomotion was one of the first humanlike traits to evolve. Radiodating techniques established the age of the hominid to be 3.5 million years. She was the oldest, most complete, and best-preserved hominid fossil that has yet to be recovered. Johanson named her "Lucy," after the Beatles' song "Lucy in the Sky with Diamonds," which was playing in camp on the night of the discovery. The next year, the remains of 13 additional members of the species were found at a nearby site and have become known as "The First Family."

Johanson puzzled over the species name he should assign to Lucy and The First Family. The hominids were too primitive to be considered "human" and assigned to the genus *Homo*, particularly since there was no evidence that they had used tools. At the same time, Johanson was hesitant to place them into the genus *Australopithecus* since he considered Lucy's species an ancestor of modern humans and the australopithecines were seen by many as an evolutionary offshoot. Johanson could have established a new genus name, but that might have made matters even more controversial and confusing. He finally settled on the name *Australopithecus afarensis* (after the Afar region of Ethiopia, where the species had been discovered) and proposed that it was a common ancestor of the other australopithecines and humans. Others have argued that Lucy's species was not the only one present in east Africa 3.5 million years ago and that some of the fossils from that period are those of the genus *Homo* as well. Johanson's proposal for the evolutionary relationships among the various known species of hominids is illustrated in Figure 34-11.

What about information in the opposite direction? Can Lucy be traced back to an even more primitive ancestor? Since estimates from molecular biology suggest that the common ancestor of humans and chimpanzees lived no longer than 5 million to 7 million years ago, *A. afarensis* would not be very far removed from this common ancestor (Figure 34-11). This contention is supported by the fact that Lucy's arms are particularly long, relative to her legs, suggesting that her species had not yet lost this characteristic of an arboreal (tree-dwelling) ancestor. Somewhere in this period of a few million years, hominids had evolved the anatomic features (changes in the skull, pelvis, and leg bones) that allowed them to walk erect and to use their hands for increasingly complex activities.

These are just some of the highlights of our search for a better understanding of our origins. Only time will tell whether or not the interpretations of the few precious fossils described in these pages will hold up or will be replaced by new proposals that more accurately describe our evolutionary roots. We have just one more "fossil" to describe, not because it tells us what our ancestors looked like but because it provides a mental picture of an event that occurred one rainy day in east Africa nearly 4 million years ago. The fossil in question is a trail of footprints (Figure

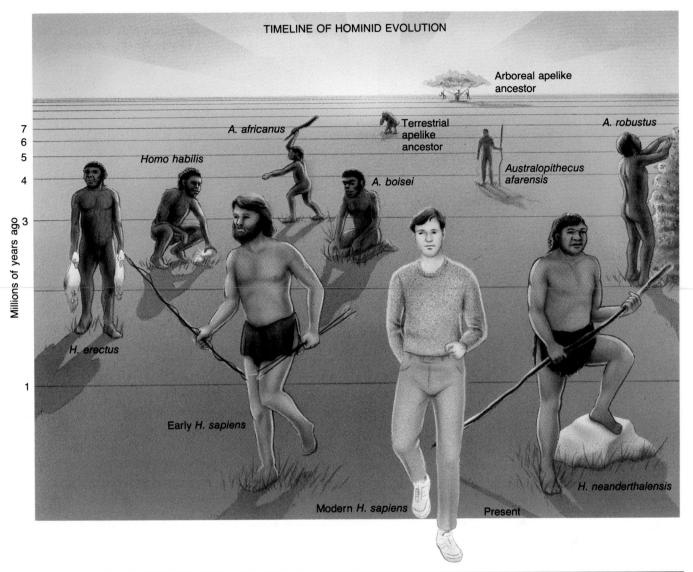

TIMELINE OF HOMINID EVOLUTION

Arboreal apelike ancestor

A. africanus

Homo habilis

Terrestrial apelike ancestor

A. robustus

A. boisei

Australopithecus afarensis

H. erectus

Early *H. sapiens*

Modern *H. sapiens*

Present

H. neanderthalensis

Millions of years ago

7 6 5 4 3 1

F I G U R E 34-11

The human family.   According to this scheme, the earliest hominid was *A. afarensis*, represented by Lucy and the First Family. Lucy walked erect and stood about 3.5 feet tall. She had an apelike skull that housed a brain about 500 cubic centimeters in volume, only slightly larger than that of similar-sized apes. She had exceptionally long arms and showed no evidence of having used tools. This species gave rise to the other australopithecines as well as the genus of humans *(Homo)*. Three species of australopithecines are depicted in the illustration. *A. africanus* had the slightest build and was probably primarily a carnivore. *A. robustus* was the most heavily boned species, with huge (nickel-sized), thickly enameled molars, suggesting that they lived primarily on a diet of coarse vegetation. *A. boisei* was an australopithecine discovered by Mary Leakey, which helped focus attention on Africa as the cradle of human evolution. The first humans, *H. habilis*, appeared about 2 million years ago. *H. habilis*, like *A. afarensis*, was very small in stature and had unusually long arms, reflecting the arboreal habits of their ancestors. Their brain size was about 700 cubic centimeters, and they were able to make simple tools. *H. erectus*, which was very advanced over its predecessors, may have stood as tall as modern humans, had arms of shorter proportion than its ancestors, had a brain capacity of about 850 to 1,100 cubic centimeters, and was able to make sophisticated tools, including hand axes. There is evidence that these hominids hunted in groups and cooked their prey. While originating in Africa more than 1.6 million years ago, *H. erectus* spread over much of Europe and Asia. By about 500,000 years ago, fossils appear that are intermediate between *H. erectus* and *H. sapiens*. By 300,000 years ago, fossils are found which are unmistakably *H. sapiens*. These early *H. sapiens* had larger brains (1,200 to 1,400 cubic centimeters) and used more sophisticated tools. Modern humans *(H. sapiens sapiens)* arose between 100,000 and 200,000 years ago and are thought to have coexisted with another group of *H. sapiens*, the Neanderthals *(H. sapiens neanderthalensis)*, who disappeared about 35,000 years ago.

FIGURE 34-12
Footprints made by a pair of hominids walking together across a plain of wet ash in eastern Africa approximately 4 million years ago.

34-12) that were made by a pair of fully erect hominids walking together through ash that had been spewed from a nearby volcano and then dampened by a light rain. The wet ash quickly hardened like cement and was covered by additional layers of ash, preserving the footprints (and even a few small craters left by the falling raindrops) until their discovery in 1976 by a team led by Mary Leakey. In the words of Tim White, a member of the team: "They are like modern human footprints. There is a well-shaped modern heel with a strong arch and a good ball of the foot in front of it. . . . to all intents and purposes those . . . hominids walked like you and me."

The fossil remains of hominids are fragmentary, leaving questions concerning the path that led to the evolution of modern human beings. At the same time, these remains establish without any doubt the fact that creatures—unlike any living today, with apelike faces and humanlike jaws — roamed the earth over a million years ago, walking erect and using tools. (See CTQ #4.)

REEXAMINING THE THEMES

Relationship between Form and Function

Most fossils are a structural remnant of an organism that lived many years ago. Yet, they tell us more than just how the organism was constructed; they reveal many aspects of the organism's function as well. For example, the discovery of a leg bone or skull tells us whether a primate walked on two legs or four. Similarly, the imprint of *Archaeopteryx's* wings and feathers tells us that this creature was capable of flight, but the lack of a bony breastbone indicates that the flight muscles were poorly developed, suggesting that this primitive bird probably used its wings mostly for gliding. The well-developed claws that were present on the "fingers" as well as the toes suggest that the animal may have been able to climb trees, using its clawed "fingers" to dig into the bark.

Unity within Diversity

The study of evolution encompasses the analysis of many diverse lineages that led to the appearance of many millions of different species. While the organisms generated by evolution are diverse, the underlying mechanisms are similar. Consequently, our understanding of various evolutionary pathways can be obtained by similar approaches, including the study of fossils, comparative anatomy, comparative embryology, molecular biology, and biogeography. Regardless of the particular group whose evolution is being studied, whether humans or gymnosperms, in most cases, each of these approaches can be used to provide relevant data.

Evolution and Adaptation

Evidence for the occurrence of biological evolution is revealed in virtually every area of the biological sciences. We see it when we compare the embryos of animals whose adult forms are highly diverse or when we examine structures that are vestigial and functionless in one organism but well-developed and functional in a related organism. We see it in our own bodies, where virtually every part is homologous to a corresponding part of a chimpanzee or a gorilla. We see it when we compare the types of organisms that live in a particular region of the world to similar organisms that live elsewhere. We see it in the increasing divergence of amino acid and nucleotide sequences among organisms who are thought to be more distantly related. The most direct evidence comes from studying the fossil remains of ancient organisms themselves.

SYNOPSIS

Evolution is the greatest unifying concept in biology. While biologists argue over which mechanisms may have been most important in the evolution of a particular group, there is virtual agreement that all living organisms have arisen by evolution from a common ancestor. The theory of evolution is supported by a mass of evidence gathered from several distinct biological disciplines; no credible scientific evidence has been obtained to suggest that evolution has not occurred.

When using similar characteristics to determine evolutionary relationships, care must be taken to distinguish between homologous features (those that are inherited from a common ancestor) and analogous features (those that result from convergent evolution among unrelated animals that have a similar mode of existence). Homologous and analogous features can be distinguished by applying a number of criteria, including (1) resemblance in detail, (2) similar position in relation to neighboring structures or organs, (3) similarity in embryonic development, and (4) agreement with other characters. Using these criteria, the wings of a bat and the arms of a human are homologous features, while the paddlelike limbs and streamlined bodies of fishes and whales are analogous features.

The evidence for evolution is based on the following: the study of the anatomy of living organisms, comparative embryology, the fossil record, biogeography, and biochemical and molecular data. Comparing the structure of parts of the bodies of different organisms provides evidence of evolutionary relationships. Vestigial structures (ones with little or no apparent function, such as your appendix) are considered to be remnants of structures that had a function in ancestral species that are in the process of evolutionary disappearance. Fossil analysis provides information on the types of organisms that lived in the past and evidence of the pathways by which various groups might have evolved. Comparisons of the embryos of animals can reveal homologies that are not apparent in the adults. Homologies among organisms are also revealed by comparing nucleotide and amino acid sequences of different organisms. For the most part, the longer the amount of time that has passed since two species have diverged from a common ancestor, the greater the number of substitutions found in corresponding genes and proteins between the two species. Geographic information about organisms also helps determine evolutionary relationships. Plants and animals living in nearby areas are more likely to be related than are those living far apart.

Our current view of the evolution of humans from an ancestor common to both apes and humans is based on fragmentary fossil evidence. According to this view, the first known hominid is *Australopithecus afarensis*, represented by Lucy and The First Family, who lived about 3.5 million years ago. These hominids were small in stature, had small brains, and showed no evidence of having used tools, but they walked erect and had jaws with humanlike features. *A. afarensis* may have given rise to a number of other australopithecines as well as to humans (genus *Homo*). The first known humans (*H. habilis*) appeared in Africa about 2 million years ago, and members of the species *H. erectus* appeared about 1.5 million years ago. *H. erectus* survived for over a million years, migrating to diverse regions of the earth. Modern humans (*H. sapiens sapiens*) date back 100,000 to 200,000 years.

Key Terms

hominid (p. 742)
homology (p. 744)
homologous feature (p. 744)

analogous feature (p. 744)
homoplasy (p. 744)

fossil record (p. 746)
vestigial structure (p. 749)

Review Questions

1. Name the major anatomic characteristics that underwent change during the course of human evolution over the past few million years. How are these characteristics illustrated in various fossil hominids?

2. What criteria are used to distinguish homologous and analogous features?

3. Why are vestigial structures evidence for the occurrence of evolution?

4. Compare and contrast Java Man, Peking Man, Taung Child, and Piltdown Man; arboreal and ground-dwelling hominid adaptations; primitive and derived traits; *Archaeopteryx* and fossil reptiles.

5. Why was increased brain size naturally selected during the evolution of modern humans?

Critical Thinking Questions

1. The criteria used to distinguish *Homo* from other genera, such as *Australopithecus*, have never been universally accepted. Is there one (or a few) criterion, that you feel should be the most important determinant of a "human" hominid?

2. Explain how both analogous and homologous structures provide clues to evolution. Why are homologous structures used to infer common ancestry, but analogous structures are not?

3. List one piece of evidence from each of the categories listed below that links humans with apes and/or other primates: comparative anatomy, fossils, vestigial structures, comparative embryology, comparative biochemistry.

4. Which of the four proposed family trees for hominids in the diagram below most closely resembles that described in this chapter? Why are so many different schemes proposed by different scientists working in this field?

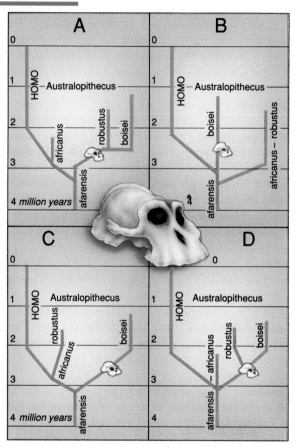

Additional Readings

Avise, J. C. 1989. Nature's family archives. *Nat. Hist.* (March: 24–27). (Intermediate)

Gould, S. J. 1980. *The panda's thumb.* New York: Norton. (Intermediate)

Gould, S. J. 1991. *Bully for brontosaurus.* New York: Norton. (Intermediate)

Lewin, R. 1987. *Bones of contention: Controversies in the search for human origins.* New York: Simon & Schuster. (Introductory)

Johanson, D., and J. Shreeve. 1989. *Lucy's child: The discovery of a human ancestor.* New York: Morrow. (Introductory)

Leakey, R. E. 1983. *One life: An autobiography.* Salem, NH: Salem House. (Introductory)

Shreeve, J. 1993. Human origins. *Discover* Jan:24–28. (Intermediate)

Wilson, A. C. 1985. The molecular basis of evolution. *Sci. Amer.* Oct: 164–173. (Advanced)

The Origin and History of Life

**STEPS
TO
DISCOVERY**
Evolution of the Cell

BIOLINE
The Rise and Fall of the Dinosaurs

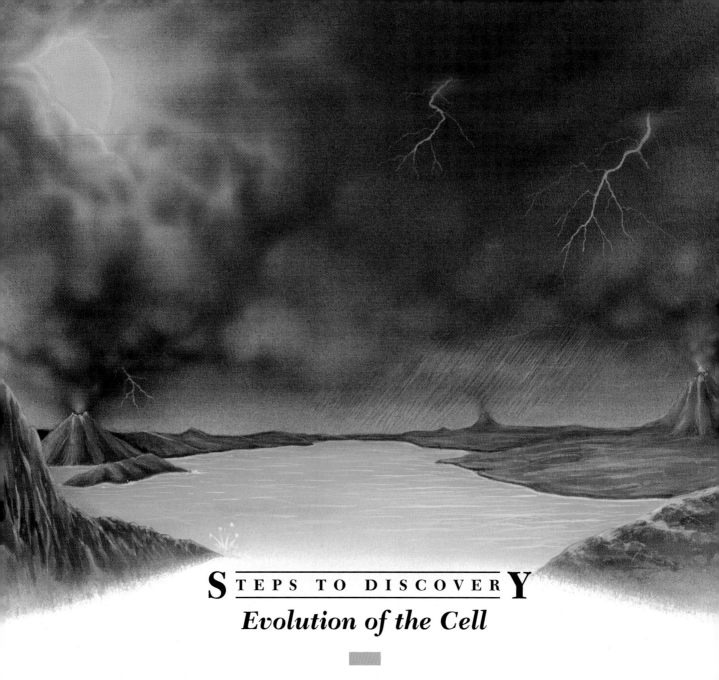

When Louis Pasteur's research finally laid to rest the idea that living organisms could arise from inanimate materials (Chapter 2), he settled one nagging controversy in biology and began a new one. If life could only arise from life, how did living organisms *initially* appear on the planet?

This question was first tackled in the 1920s by the Russian biochemist Aleksandr Oparin, who proposed that life could not have arisen in a single step but only over a long and gradual process of **chemical evolution**—the spontaneous synthesis of increasingly complex organic compounds from simpler organic molecules. In Oparin's view,

the formation of life occurred in several distinct stages.

The first step was the formation of simple molecules, such as ammonia (NH_3), methane (CH_4), hydrogen cyanide (HCN), carbon monoxide (CO), carbon dioxide (CO_2), hydrogen gas (H_2), nitrogen gas (N_2), and water (H_2O) during the formation of the earth's crust and atmosphere. Free molecular oxygen (O_2), which might have destroyed the earth's first delicate life forms (page 174), did not appear until the first photosynthetic organisms had evolved hundreds of millions of years later.

The second step involved the spontaneous interaction

Within the primordial seas, the first life forms appeared over 4 billion years ago by the process of chemical evolution.

of these simple molecules to form more complex organic molecules, such as amino acids, sugars, fatty acids, and nitrogenous bases—the building blocks of the macromolecules that characterize life as we know it today. The energy needed to drive the formation of these organic molecules was derived from various sources, including the sun's radiation, electric discharges in the form of lightning, and heat that emanated from beneath the earth's crust. As these organic compounds accumulated to higher concentrations in the shallow lakes and seas that dotted the primitive earth, they formed an "organic soup," in which additional reactions could take place. Some of the simpler organic molecules polymerized to form macromolecules consisting of chains of subunits, similar in basic structure to proteins and/or nucleic acids. As we discussed in Chapter 15, the first nucleic acids are thought to have been made of RNA rather than DNA.

Molecules eventually arose that could accomplish functions that were necessary for life. In this step, some of these molecules catalyzed certain chemical reactions, forming the basis for a primitive form of metabolism. Others had the capacity for self-replication and somehow formed copies of themselves. Insoluble lipids that were formed in the organic soup were forced together by hydrophobic interactions (page 59), forming lipid complexes, including membranous walls made up of lipid bilayers. Some of these self-assembling membranes formed around clusters of macromolecules, encapsulating them and forming "precells."

In the final step, precells concentrated organic molecules, allowing the molecules to react more frequently. Within the precell, catalysts directed the synthesis of specific organic polymers; the precell conducted metabolism. The eventual evolution of a genetic code enabled the precell to pass on naturally selected codes for metabolism, allowing it to reproduce itself. At this point, the precell possessed three of the basic characteristics of life— metabolism, growth, and reproduction—crossing the line between precell and living cell.

Acceptance of Oparin's theory of chemical evolution received a boost in the early 1950s as a result of a series of experiments by Stanley Miller, a graduate student at the University of Chicago. Miller demonstrated that many of the simpler compounds characteristic of life could be produced in the laboratory under conditions that were thought to have existed soon after the formation of the earth. In a sealed glass vessel, Miller repeatedly jolted a mixture of hydrogen gas, ammonia, methane, and water vapor with electric discharges in order to simulate lightning strikes. Within a matter of days, a number of organic compounds appeared in the reaction vessel, including several amino acids commonly found in proteins. Soon after, Juan Oro of the University of Houston showed that more complex biochemicals, including the nitrogenous bases that form the building blocks of nucleic acids, could be formed under similar conditions.

In 1969, support for the idea that organic compounds could be synthesized under *abiotic* (nonbiological) conditions came from an unlikely source. That September, a series of soniclike booms were heard throughout the town of Murchison, Australia, as a huge meteorite fell from outer space, exploding into pieces. Analysis of the meteorite fragments revealed the presence of a large variety of organic compounds, including amino acids, pyrimidines, and molecules resembling fatty acids. The fact that these compounds could appear under abiotic, extraterrestrial conditions made it even more likely that similar compounds had been able to form on the primitive earth.

*I*t is one of the greatest detective stories of all time. There were no eyewitnesses, and the detectives did not come upon the scene until long after the incident had occurred. Although a host of clues had been left behind, they had all become modified over the years; virtually everything had changed.

What is this baffling case? It is nothing less than the origin and history of life on earth. The detectives in this investigation are scientists from the fields of astronomy, geology, chemistry, physics, and biology. After years of study, the *origin* of life on earth is no less a mystery today than it was decades ago. In contrast, the *history* of life on earth, from the formative beginnings of simple, prokaryotic cells to the present bewildering complexity, is becoming better understood all the time. The story begins nearly 4.6 billion years ago with the formation of the earth.

▼ ▼ ▼

FORMATION OF THE EARTH: THE FIRST STEP

Billions of years ago, our solar system was part of a massive cloud of interstellar gases and cosmic dust (Figure 35-1). As particles of matter were pulled together by gravitational attraction, the vast cloud condensed into a gigantic, spinning disk. Nearly 90 percent of the matter gravitated to the center, causing temperatures to rise high enough to ignite thermonuclear reactions. It was in this scorching center that our sun began to shine.

At the same time the sun was forming, nearly 5 billion years ago, smaller eddies of leftover gases and dust were condensing into the planets of our solar system. Heavier elements, such as iron and nickel, sank inward to form the cores of the planets, while light gases floated near the surface. Most of the lighter gases in the planets nearest the sun (Mercury, Venus, Earth, and Mars) were blasted away when the sun's thermonuclear reactions began, leaving shrunken, dense planets with virtually no gaseous atmospheres.

FIGURE 35-1

Formation of our solar system.

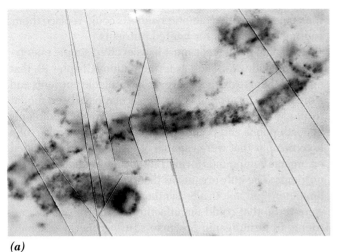

FIGURE 35-2

Imprints of the world's most ancient life. *(a)* A 3.5-billion-year-old cast of a filamentous cyano-bacterium from western Australia. *(b)* These stromatolites, which dot the shore in western Australia, consist of dense masses of prokaryotic cells and mineral deposits. Prokaryotic cells have been found in stromatolites that are 3.5 billion years old.

☀ Immense quantities of thermal energy from gravitational contraction, radioactive decay, meteorite impacts, and solar radiation turned the primitive earth into a red-hot, molten orb. As time passed, collisions with meteorites became less frequent, and the heat from radioactive decay and the earth's contraction lessened. The earth's surface slowly cooled, and a thin crust of crystalline rock formed. Below the crust, the enormous heat of the earth's interior produced massive buildups of hot gases, sparking violent volcanic eruptions that thrust molten rock and gases out through the crust. Repeated eruptions gradually built the earth's rugged land masses and filled the once empty atmosphere with clouds of hot gases and steam, first collecting into small ponds and later forming the earth's oceans, lakes, rivers, and streams. It was in these bodies of water that life emerged.

The size of the earth and its position relative to the sun were ideal for setting the stage for life, as we know it, to appear. Had the earth been much smaller, it would have lacked the mass necessary to generate enough gravitational force to hold onto its atmospheric gases. Had the earth been much closer to the sun, the scorching temperatures on the planet's surface would have prevented the condensation of steam into liquid water. Conversely, had the earth been much farther from the sun, the freezing temperatures would have kept any water in a solid, frozen state.

The first step in the formation of life on earth was the formation of the earth itself, an event that began more than 4.5 billion years ago with the condensation of cosmic matter. (See CTQ #2)

THE ORIGIN OF LIFE AND ITS FORMATIVE STAGES

In the opening pages of this chapter, we saw that the building blocks of macromolecules are readily formed abiotically, both in the laboratory and on extraterrestrial bodies. But no one has demonstrated that these simpler molecules can spontaneously assemble into reaction-catalyzing proteins or information-containing nucleic acids by the process of chemical evolution outlined by Oparin and others. Since it is impossible to recreate the origin of life in the laboratory, many different hypotheses have been offered to explain the course of events that created life. Each of these hypotheses is controversial and has been met with skepticism by some members of the "origin of life" scientific community.[1]

▮▶ The first fossilized evidence of living cells is found in rocks from Australia and South Africa which date back 3.5 billion years. These rocks reveal the presence of prokaryotic cells that are not noticably different in appearance from prokaryotes living today. One such fossil consists of chains of cells (Figure 35-2a) that resemble modern photosynthetic cyanobacteria. Others consist of rocks formed from *stromatolites*—dense masses of bacteria and mineral deposits that grow today in warm, shallow seas (Figure 35-2b). The fact that such "advanced" prokaryotic cells had already appeared 3.5 billion years ago suggests that the process of chemical evolution that led to the first life forms took place relatively rapidly, probably within the earth's first 600 mil-

[1] Those interested in reading about some of these proposals might consult the article entitled "In the beginning," by J. Horgan in the February 1991 issue of *Scientific American*.

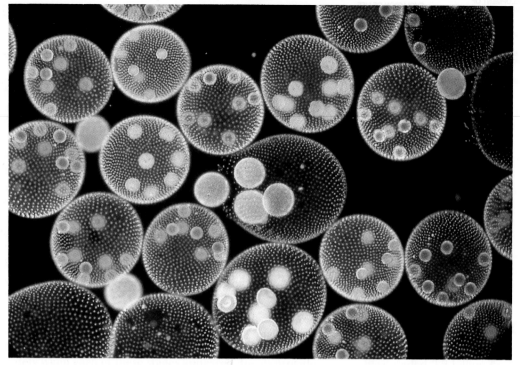

***Volvox*, a colonial protist.** The cells of this colonial protist arise by mitosis from a single cell. Within the colony are cells specialized for locomotion, photoreception, and sexual reproduction. It has been proposed that a colonial organism such as *Volvox* gave rise to all multicellular organisms. The dense green masses inside the large spheres are daughter colonies that will be released into the surrounding environment.

state. The best studied is the colonial protist *Volvox* (Figure 35-4).

▶ By the time the Proterozoic Era ended, members of all the major animal phyla had appeared. We know this to be the case since at the beginning of the next era, fossils representing the various phyla are "suddenly" found in considerable numbers. The lack of Proterozoic fossils probably reflects the fact that the animals living in this era lacked hardened skeletal parts that are usually required for fossil-

ization. A few valuable samples of fossilized Proterozoic life remain, however. The best samples come from a region of sandstone in South Australia, where a group of marine organisms, including jellyfish (cnidarians), segmented worms (annelids), and the soft-bodied ancestors of arthropods became deposited in the fine mud and silt of the Proterozoic ocean bottom about 650 million years ago. Even though the animals lacked skeletons, the imprints of their bodies were preserved in the forming sandstone (Figure 35-5).

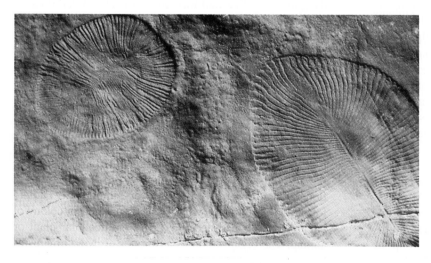

The imprint of a soft-bodied, multicellular animal that lived approximately 650 million years ago. The animal appears to have been segmented but lacked a head or appendages.

THE PALEOZOIC ERA: LIFE DIVERSIFIES

The **Paleozoic Era** lasted approximately 345 million years. Conditions on earth changed dramatically a number of times during this long era, causing episodes of mass extinctions, followed by diversification of organisms via adaptive radiation, whereby a single ancestor gives rise to a diverse array of organisms adapted to different habitats (page 734). Many of these changes were the results of shifts in the positions of the earth's continents. At times, the drifting continents created broad, shallow seas; at other times, land masses became entirely submerged and later rose to form mountain ranges. These changes had enormous impact on the earth's fauna and flora.

Continental Drift

In 1968, an expedition to Antarctica unearthed a collection of fossil reptiles that were virtually identical to fossils found in Africa and India, locales situated thousands of miles apart. This finding raises important questions: How can we explain the presence of the same species of terrestrial ver-tebrates living on several different continents? And how could these "cold-blooded" reptiles have lived in Antarctica, which is so cold that it is home to only a few species of heavily insulated, "warm-blooded" birds and mammals?

In 1912, Alfred Wegener, a German geologist, hypothesized that the continents were not always situated in the same place on the earth's surface as they are today. Wegener observed that the continents of Africa and South America, which are situated on opposite sides of the Atlantic Ocean, have complementary outlines and might be able to fit together spatially. He studied the geologic formations on the opposing coastlines and realized that many of the features that ended abruptly on one continent continued on the opposite continent. Wegener spent the remaining 20 years of his life amassing evidence for his theory that the continents were once situated close together and had then drifted apart, a phenomenon known as **continental drift.**

The occurrence of continental drift is now explained by the theory of *plate tectonics.* According to this theory, the solid crust of the earth, which is estimated to be about 100 kilometers (62 miles) thick, consists of a number of rigid plates (Figure 35-6) that rest on an underlying layer of

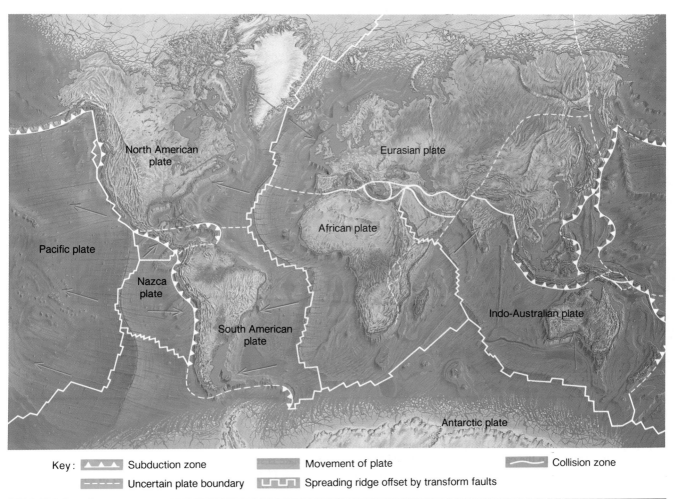

Key: Subduction zone   Movement of plate   Collision zone

Uncertain plate boundary   Spreading ridge offset by transform faults

FIGURE 35-6

The geographic distribution of the major tectonic plates that make up the earth's crust.

◁ B I O L I N E ▷
The Rise and Fall of the Dinosaurs

The dinosaurs, or "terrible lizards," began their adaptive radiation about 225 million years ago, approximately 100 million years after the first reptiles had evolved from amphibians. The earliest species were relatively small carnivores with a bipedal (two-legged) gait that likely carried the animals in swift pursuit of prey. From these modest beginnings, a variety of carnivores of huge dimensions evolved. Perhaps the largest of these was *Tyrannosaurus rex* (Figure 1), which weighed over 6 tons and stood over 4.5 meters (15 feet) tall. The dinosaur's powerful jaws contained large, swordlike teeth. The animal had huge hindlimbs, but its forelimbs were so poorly developed, they were unable to reach its own mouth.

Tyrannosaurus was dwarfed compared to some of the herbivorous dinosaurs, such as *Brachiosaurus*, which weighed 80 tons and reached 15 meters in height, and *Apatosaurus* (formerly *Brontosaurus*), which reached lengths of 25 meters. The herbivorous dinosaurs had massive bodies with disproportionately small heads and brains and very long necks and tails (Figure 2). The animal's front teeth were similar in shape to our own incisors and were used to crop the plants on which the dinosaur browsed. Since they had no rear teeth, herbivorous dinosaurs were probably unable to chew their food and relied instead on some type of grinding mechanism located within their digestive tract (as in modern birds, which evolved from one of the dinosaur lines). Unlike the carnivores, these larger herbivores walked on four legs.

Reptiles dominated the land, air, and water for most of the Mesozoic Era, a reign

exceeding 125 million years. Then, about 65 million years ago, the dinosaurs disappeared from the fossil record, along with a wide variety of other unrelated organisms, including many species of plants, marine invertebrates, and single-celled protists. Many hypotheses have been presented over the decades to explain the extinction of the dinosaurs and other life forms. Taken together, these hypotheses fall into two distinct categories: those that suggest a gradual process of extinction that took place over a period from 2,000 to 3 million years, versus those that propose a catastrophic extinction that might have occurred over a period of months to years.

You might think that the fossil record would enable paleontologists to distinguish easily between a "gradual" versus a "sudden" course of extinction, but this has not been the case. Depending on which organisms have been scrutinized, paleontologists have argued for years over whether or not the loss of certain fossil species is consistent with one or the other hypothesis.

In 1980, a paper was published by Luis Alvarez, a Nobel laureate in physics, and his colleagues. This paper, and the research it stimulated, has gradually shifted opinion over to the side of catastrophic extinction. Alvarez and colleagues were examining the composition of the rocks that

FIGURE 1

A skull of *T. rex* emerging from South Dakota sandstone.

were formed about the time the dinosaurs became extinct, a time that separates the Mesozoic Era from the Cenozoic Era, 65 million years ago. The scientists found a thin layer of rock that contained highly elevated levels of a platinumlike element called *iridium*. Iridium is normally present at low levels in the earth's own crust and at high levels in extraterrestrial bodies, such as asteroids and meteorites. Alvarez interpreted this distinction as evidence that a huge asteroid or meteorite must have struck the earth 65 million years ago, sending a massive cloud of iridium-containing dust into the atmosphere. The dust was dispersed around the world before settling back onto the earth, forming a thin layer that has been preserved in the earth's rocks. In the past few years, corroborating evidence for such an impact has come to light from the discovery of tiny, glasslike particles called *tektites* at the Mesozoic/Cenozoic boundary. Tektites are generated by the tremendous pressure of a meteorite as it hits the ground at speeds of 15 kilometers per second. Current evidence points to the Caribbean as the most likely site of impact.

How can the impact of a meteorite in the Caribbean affect the survival of a dinosaur living thousands of kilometers away on the other side of the earth? Scientists generally agree that such an impact could have radically altered the climate on earth for a period of several years; they disagree on the type of climatic changes that resulted, however. Alvarez initially proposed that the impact would have generated a cloud of dust in the atmosphere, similar to that caused by the eruption of a large volcano. The dust would have blocked the sun's vital rays, greatly lowering the levels of photosynthesis on which virtually all organisms on earth ultimately depend.

Other possible effects of a meteorite have been suggested, including a period of elevated warming due to the greenhouse effect (resulting from increased levels of carbon dioxide in the atmosphere due to global wildfires), the formation of acid rain (due to increased levels of nitrous oxide in the atmosphere), and the release of natural gas (methane) trapped in the earth's crust. We may never know precisely what it was that killed the dinosaurs, but it remains a fascinating topic that captures the imagination of children and geochemists alike.

FIGURE 2

A skeleton of the herbivorous dinosaur, *Diplodocus*. While the accompanying human skeleton is useful in revealing the size of these dinosaurs, closer examination of the limbs, pelvic girdle, and ribcage exposes numerous homologies between the two vertebrates.

One group that was not as severely affected by extinction consisted of small, probably nocturnal animals that had evolved earlier from a group of reptiles known as *therapsids* (Figure 35-11) but had remained in the "shadows" as the reptiles underwent their diversification. These animals were the mammals, and they are thought to have had a number of important biological innovations: They maintained a constant, elevated body temperature that allowed them to remain active at night; they were covered with hair that insulated them from colder temperatures; and they gave birth to their young, rather than hatching them from eggs.

Birds, insects, and mammals survived whatever it was that caused the mass extinction of the ruling reptiles, and flowering plants emerged and thrived. Once the dinosaurs and other large reptiles had become extinct, birds and mammals quickly filled their vacated habitats, giving rise to a number of groups that would dominate animal life during the following era, the Cenozoic.

THE CENOZOIC ERA: THE AGE OF MAMMALS

The present era, the **Cenozoic Era**, began about 65 million years ago and is known as the Age of Mammals (Figure 35-12). Insulating hair and thermoregulatory abilities adapted mammals for the colder climates that became more prevalent during the Cenozoic Era. By the first 10 million to 15 million years of this era, representatives of most of the modern orders of mammals had appeared. During the first half of the Cenozoic Era, the order Primates was represented by the lemurs and tarsiers, small arboreal forms that resembled some of the earliest mammals. Monkeys appear in the fossil record approximately 35 million years ago, and the direct ancestors of the four groups of modern apes (gibbons, orangutans, chimpanzees, and gorillas) appeared about 20 million years ago. The Cenozoic Era also saw the explosive radiation of flowering plants.

The Pleistocene Epoch, which began about 2.5 million years ago, was characterized by intermittent ice ages that caused mass extinctions and migrations, remolding the distribution of life on earth. Once vastly distributed, tropical vegetation became restricted to habitats with mild climates, near the equator. During the last ice age, giant mammoths, ground sloths and numerous other large mammals became extinct (Figure 35-13). Following this last ice age, which ended about 10,000 years ago, semiarid and arid areas developed, and the surviving plants and animals began the most recent period of adaptive radiation.

Now that we have described the major trends of evolution that have occurred on earth over the past several billion years, we are ready to take a closer look at the diverse forms of life that have been generated by evolution. In the following section of the text, we will explore the major characteristics of the organisms that comprise each of life's five kingdoms.

> The history of life on earth has been marked by progression, decimation, and renewal. During the first 3.5 billion years or so, the earth was the site of the evolution of increasingly complex life forms, culminating in multicellular plants and animals. The past 700 million years have seen waves of mass extinctions, which, in some cases, nearly decimated plant and animal life on earth, followed by periods in which new groups rose to prominence, replacing those that either receded in diversity or disappeared altogether. (See CTQ #4.)

FIGURE 35-11

A therapsid reptile of the type that gave rise to mammals. These animals' teeth, jaws, and "upright," four-legged posture made them more like mammals than other reptiles.

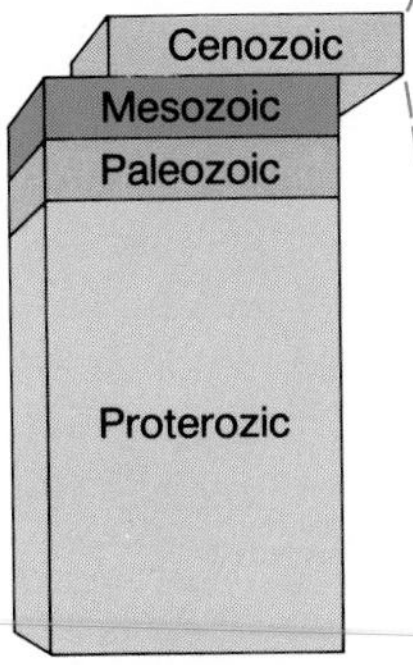

PERIOD	EPOCH	PLANT EVOLUTION	ANIMAL EVOLUTION
Quaternary	Recent	Increase of herbaceous plants	Appearance of *Homo sapiens*
Quaternary	Pleistocene	Repeated glaciation leads to mass extinction	Repeated glaciation leads to mass extinction First *Homo*
Tertiary	Pliocene	Decline of forests, spread of grasslands	Appearance of hominids
Tertiary	Miocene		Appearance of first apes
Tertiary	Oligocene		All modern genera of mammals present In seas, bony fish abound
Tertiary	Eocene		
Tertiary	Paleocene	Explosive radiation of flowering plants	Rise of mammals First placental mammals

2.5

65

Millions of years ago

Cenozoic
Mesozoic
Paleozoic
Proterozic

FIGURE 35-12

The Cenozoic Era is divided into two periods, the tertiary and quaternary, which together are subdivided into seven epochs. Mammals and flowering plants evolved rapidly during the Cenozoic Era. Intermittent ice ages continually remolded the distribution of life and caused the extinction of many organisms. Since mammals became the dominant land animal at this time, the Cenozoic Era is also called the Age of Mammals.

REEXAMINING THE THEMES

Acquiring and Using Energy

☀ During the earliest period of chemical evolution on earth, the energy required for the formation of complex organic chemicals from simpler compounds was derived from a number of sources, including the ultraviolet radiation from the sun, the disintegration of radioactive atoms, the impact of meteorites, molten lava from the earth's core, and electric discharges from lightning. The first cells on earth were presumably heterotrophs. They fueled their activities with the chemical energy present in the compounds that made up the organic soup in which they lived. Ultimately, autotrophs evolved that could capture energy from sunlight and convert it to the chemical energy needed to synthesize their own organic compounds.

Evolution and Adaptation

▕▶ Evolution of life on earth has progressed hand in hand with evolutionary changes in the earth itself. During the

first billion or so years after the earth's formation, the earth's atmosphere was very different from how it is today. It probably contained nitrogen, methane, ammonia, and hydrogen gas but was virtually devoid of molecular oxygen. This atmosphere was conducive to chemical evolution. It allowed for the abiotic formation of organic compounds and eventually led to the formation of life. By about 3 billion years ago, the cyanobacteria began filling the atmosphere with molecular oxygen, promoting the preponderance of aerobic life on earth. The evolution of life has also been markedly affected by geographic and climatic changes. Continents have come together and drifted apart, while global temperatures have varied from that of tropical environments to ice ages. Like the appearance of oxygen, these changes have had dramatic effects on the earth's flora and fauna, resulting in waves of extinction, followed by periods of great biological resurgence, in which newly evolved organisms adapted to the new prevailing conditions.

FIGURE 35-13

A gallery of Ice Age animals. From left to right: Jefferson's mammoth, Conkling's pronghorn antelope, giant heron vulture, and a great short-faced bear.

SYNOPSIS

The earth condensed out of a massive cloud of dust and gases, approximately 4.6 billion years ago. Life subsequently appeared in less than a billion years, the result of spontaneous chemical evolution. The first life forms are presumed to have been chemical heterotrophs that obtained their energy and materials from organic compounds in the surrounding medium. The evolution of prokaryotic autotrophs about 3 billion years ago was essential for the continuation of life on earth. The most advanced autotrophs were the cyanobacteria, which were able to utilize the energy in sunlight to split water, gaining electrons as reducing power and releasing molecular oxygen. Unicellular eukaryotes evolved about 2 billion years ago, paving the way for the evolution of multicellular organisms—presumably from colonial protists—about 700 million years ago.

The geologic time scale divides the earth's lifespan into four great eras, beginning with the Proterozoic Era, during which life evolved. Repeated changes in the shape, size, and location of the earth's continents drastically changed the earth's climate and sea level, affecting the evolution of organisms. By the end of the Proterozoic Era, members of all of the major animal phyla had appeared, even though the fossil record of these groups is very limited.

Vertebrates evolved during the Paleozoic Era, first in the sea and then on land. Arthropods also underwent an adaptive radiation on land, beginning with scorpionlike forms, followed by a variety of wingless and winged insects. Mosses, ferns, and seed plants appeared successively on land. The Paleozoic Era ended with the greatest episode of mass extinction of organisms to have ever taken place, extinguishing the vast majority of marine organisms and many terrestrial forms.

During the Mesozoic Era, reptiles and gymnosperms dominated the land. Adaptive radiation among reptiles produced a diversity of organisms, including the dinosaurs. Flowering plants and the first mammals and birds evolved during this era. The Mesozoic Era ended with another episode of mass extinction, eliminating the dinosaurs and setting the stage for the adaptive radiation of mammals.

The present Cenozoic Era has been marked by the evolution of all of the modern orders of mammals and the explosive radiation of flowering plants. In the past 2.5 million years, a series of intermittent ice ages has caused mass extinctions and reshaped the distribution of plants and animals across the earth.

Key Terms

chemical evolution (p. 760)
Proterozoic Era (p. 764)
cyanobacteria (p. 764)

Paleozoic Era (p. 767)
continental drift (p. 767)

Mesozoic Era (p. 771)
Cenozoic Era (p. 774)

Review Questions

1. Retrace the presumed steps of chemical evolution that led to the appearance of the first living cell.

2. How have the sources of energy important for biological evolution changed from the time the earth formed to the present day?

3. Which of the four major geologic eras saw the appearance of the first living cells? The first eukaryotes? The first protists? The first gymnosperms? The first flowering plants? The first vertebrates? The first reptiles? The first mammals? The first humans?

4. How has the distribution of land masses changed over the past 200 million years?

Critical Thinking Questions

1. What characteristics qualify a structure as a precell? Why are these characteristics so important as steps in the formation of a living cell?

2. Thus far, no signs of life, past or present, have been found on Venus or Mars. Research the environmental conditions on these planets and develop an explanation for why this is so and why the environment on earth is more hospitable to life.

3. According to the second law of thermodynamics, the universe continues to increase toward a state of increasing disorder. How is it that events could take place on earth that led to the evolution of increasingly more complex structures—from simple molecules to organic compounds to simple prokaryotic cells to eukaryotic cells and, finally, to multicellular plants and animals of increasingly complex structures?

4. Develop a graphic illustration *to scale* for the geologic time line, showing the events listed below: origin of earth; origin of first prokaryotic cells; emergence of first autotrophs; rise of the first eukaryotic cells; appearance of the first multicellular organisms; appearance of all the major animal phyla; the dinosaur era; beginning of the age of mammals; appearance of first primates; evolution of Australopithecines; first humans; first appearance of *Homo sapiens*. Indicate the boundaries of the four major geologic eras and the time of the major extinctions.

5. Try to envision the future, 10, 50, and 100 years from now. Where do you think cultural evolution will have led humans? How will humans have affected the evolution *and* extinction of other organisms on earth? What priorities should biologists establish today to circumvent any negative changes in the future?

Additional Readings

Colbert, E. H. 1989. *Digging into the past.* New York: Dembner. (Introductory)

Edey, M. A., and D. C. Johanson. 1989. *Blueprints: Solving the mystery of evolution.* Boston: Little, Brown. (Intermediate)

Hively, W. 1993. Life beyond boiling. *Discover* May:86–91. (Intermediate)

Horgan, J. 1991. In the beginning. *Sci. Amer.* Feb:116–125. (Intermediate)

Knoll, A. H. 1991. End of the Proterozoic Era. *Sci. Amer.* Oct:64–73. (Intermediate)

Paul, G. S. 1989. Giant meteor impacts and great eruptions: Dinosaur killers? *Bioscience.* March:162–172. (Intermediate)

Wilson, E. O. 1992. *The diversity of life.* Cambridge, MA: Harvard University Press. (Intermediate) (If you read only one other book on biology, consider this one.)

Woese, C. R. 1984. *The origin of life.* Carolina Biological Supply Co., Burlington, North Carolina. (Intermediate)

York, D. 1993. The earliest history of the earth. *Sci. Amer.* Jan:90–96. (Intermediate)

The Kingdoms of Life: Diversity and Classification

Two hundred years ago biologists divided all organisms as either plants or animals—the two prominent kingdoms represented in this photo. But closer examination of soil, water, and air revealed organisms that did not fit into either of these categories, necessitating three new kingdoms of life.

PART
· 7 ·

The Monera Kingdom and Viruses

**STEPS
TO
DISCOVERY**

The Burden of Proof: Germs and Infectious Disease

BIOLINE

Living with Bacteria

THE HUMAN PERSPECTIVE

Sexually Transmitted Diseases

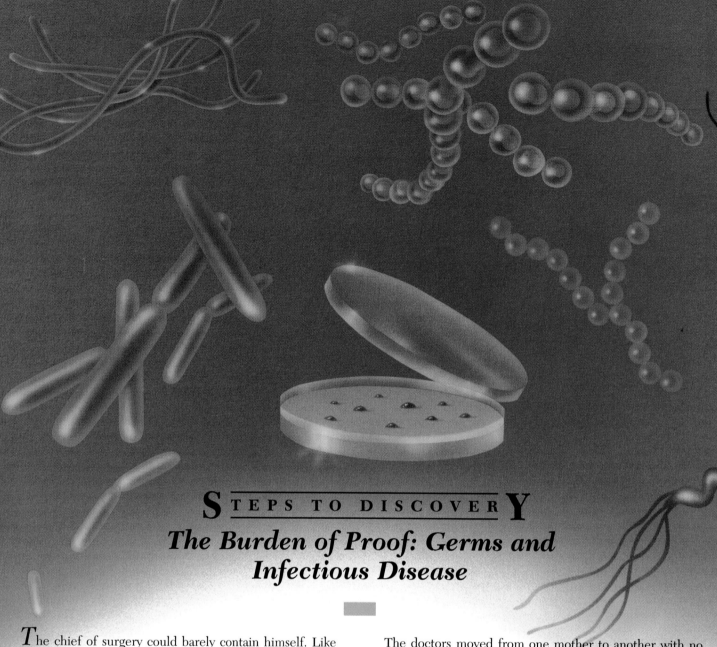

The Burden of Proof: Germs and Infectious Disease

*T*he chief of surgery could barely contain himself. Like others in his position in 1850, he felt helpless, as a tragedy of staggering proportions transpired in his hospital. Nearly half of the women who had their babies delivered by hospital doctors contracted "childbed fever," a massive fatal infection of the uterus and bloodstream caused by bacteria introduced into the birth canal during delivery. The disease's cause was unknown at the time, since no microorganism had ever been proven to cause human disease.

The alarmed chief of surgery, a Hungarian physician named Ignaz Semmelweis, aggressively attacked the situation. He began by observing doctors and patients, trying to identify some factor that might be responsible for the disastrous epidemic in the hospital. He soon found a pattern:

The doctors moved from one mother to another with no intervening sanitary precautions, a common practice since sanitation was not yet recognized as having any practical value. In fact, the doctors often went directly from postmortem dissections to the maternity ward without washing their hands. Semmelweis suspected that the doctors were unwittingly transferring the agent of the fatal disease from one person to another.

Semmelweis conducted a simple but powerful test of his hypothesis. He ordered all physicians in his hospital to wash their hands in a strong chlorine solution before performing each delivery (chlorine kills bacteria). In spite of protests from the insulted doctors, the physicians all complied under the threat of dismissal. The incidence

It was long suspected that microscopic organisms caused certain types of diseases. Only after a cascade of scientific developments,

of childbed fever dropped from 50 percent to 1 percent.

Although Semmelweis published his dramatic results in 1850, the scientific and medical community largely ignored his findings. Evidence that disease could be spread from person to person was hardly unqualified scientific proof that microorganisms were the culprits. A sensitive man, Semmelweis eventually collapsed under the criticism and was committed to a sanitarium, where he died of an infection caused by the same bacteria he tried so hard to fight against in his own hospital.

For 17 additional years, doctors continued their unsanitary "healing" practices. Then, in 1867, an English surgeon, aware of Semmelweis's success, tackled a similar dilemma. Patients in his hospital were also dying of "postsurgical disease." This doctor, Joseph Lister, believed that microorganisms introduced into the surgical wound were responsible. He recommended using an antimicrobial chemical (phenol) to sanitize surgical instruments, surgeons' hands, and the wound itself. He even sprayed a mist of the chemical in the air of the surgical room. The incidence of postsurgical disease soon plummeted to less than 5 percent. Unlike Semmelweis, a persuasive Lister toured the lecture circuit and convinced a reluctant medical community of the value of sanitation. The number of hospital-acquired infections decreased dramatically in all the hospitals that adopted his sanitary precautions. Lister's work provided evidence that microbes were responsible for certain diseases, but a conclusive demonstration that germs actually *cause* disease was still lacking.

Although many scientists and physicians had observed that people suffering from a particular disease always seemed to have microorganisms in their bodies, the mere presence of a microorganism in a person with the disease does not prove that it *causes* the disease. In fact, the microorganism's presence may be the *result* of the disease. The challenge was to establish a cause-and-effect relationship between a specific microorganism and a particular disease. Another 15 years went by without any demonstration that microorganisms cause human disease.

Then, in 1882, a German physician named Robert Koch stepped into the picture. Koch's knowledge of the scientific method ultimately provided incontrovertible proof of what we call today *the germ theory of infectious disease*. Before he could execute his plan, however, Koch had to figure out a way to obtain "pure cultures" of bacteria, cultures that contain only one type of microorganism. Only then could he limit his experimental variables to one microorganism since specimens removed from people contain many different types of organisms. Somehow, Koch had to purify the suspected microorganism so that it alone would be responsible for results when inoculated into experimental animals. The problem was finding a suitable <u>solid</u> medium for growing isolated colonies—groups of cells that are all identical descendants of the one cell deposited on the surface at the spot where the colony would grow. Potato slices and nutrient gelatin proved inadequate because many microbes digested the solid surfaces, turning them to liquid, which allowed cells from different colonies to wash together and different species to mix with each other. Koch needed a solidifying agent that microbes could not digest. The solution was discovered in a household kitchen and delivered to Koch by an alert American woman, Fanny Hesse.

Mrs. Hesse, the wife of one of Koch's associates, suggested using *agar,* a polysaccharide extracted from red seaweed, which she used for hardening jelly. Bacteria could absorb the nutrients added with the agent but could not digest the agar itself, which remained hard. Agar provided Koch with both a solidifying agent and an easy method for obtaining pure cultures of bacteria. Working with anthrax, a rapidly fatal disease transmitted from cattle to humans, Koch used solid media to isolate colonies of the suspected bacterium *(Bacillus anthracis)* from a person suffering from the disease. He then injected a small amount of these bacteria into a susceptible healthy animal free of the anthrax bacteria. The inoculated animal soon developed anthrax, while noninoculated animals did not. From the diseased animal, Koch then reisolated huge numbers of *Bacillus anthracis* in pure culture. Robert Koch had proven the germ theory of infectious disease. In so doing, he provided an approach that is still used today to establish the cause of many infectious diseases.

*T*his chapter is devoted to two groups of agents that many people have branded as pernicious villains. The first group, members of the Monera kingdom, contains *bacteria* and their relatives, organisms that most people automatically associate with human suffering and disease. The other group, the viruses, receives even worse press than do bacteria, especially in the shadow of the virus responsible for today's "modern plague," AIDS. Some of these agents clearly deserve their nasty reputations, but most do not. This is especially true of the Monerans.

Apart from a few disease-causing species, very few species of bacteria are harmful to humans. In fact, some bacteria are essential for the continued survival of life on earth. The following phenomena illustrate just a few of the remarkable abilities and influences of the Monera kingdom's microscopic members.

- A new life form emerges on the young earth, an organism that uses sun energy to run its biological machinery. Its waste product, oxygen gas, permeates the oceans and atmosphere, paving the way for the eventual emergence of plants and animals. A member of the Moneran kingdom had changed the earth more radically than any type of organism before or since.

- One mile beneath the ocean's surface, in total darkness, cracks in the ocean floor spew out superheated sea water at temperatures of 350°C (662°F), which is well above the burning point of paper. This is one of the few spots on earth that would seem too hostile for life to exist. Yet, in the seemingly prohibitive heat and the crushing pressure at that depth, bacteria thrive.

- The Red Sea turns reddish orange, the color for which it was named. This periodic phenomenon accompanies the arrival of nutritional and physical conditions that support the growth of enormous numbers of orange-pigmented monerans in the sea water.

- Deadly diphtheria bacteria fail to gain a fatal foothold in a child's respiratory tract because billions of "friendly" bacteria living in the child's pharynx produce a chemical that inhibits the invader's growth.

- The snow-white polar bears in the San Diego Zoo turned green (Figure 36-1), the color persisting despite vigorous washing. A member of the Monera kingdom was responsible—photosynthetic cyanobacteria that invade the hollow centers of the bear's guard hairs.

FIGURE 36-1

Green polar bears? Normally pure white, these polar bears are green because photosynthetic prokaryotes (cyanobacteria) are living inside the hollow core of the bears' outer hairs, just one of the unusual habitats occupied by members of the Monera kingdom. San Diego zookeepers discovered that bathing the bears in salt water inhibited the cyanobacteria, restoring the bears' snow-white coats.

All of the above phenomena were caused by bacteria and cyanobacteria, the two groups of organisms that make up the Monera kingdom. Although they are the earth's smallest and simplest cells, they are neither the smallest nor the simplest "living" entities. That distinction belongs to the viruses.

Viruses are not usually represented in the five-kingdom classification scheme (page 13) because they are noncellular, and as such, they are considered nonliving by some biologists and are not classified as organisms at all. As we will see later in this chapter, viruses have properties that justify their classification as living entities, although they more appropriately belong in a gray zone, somewhere between the living and the inanimate.

▼ ▼ ▼

KINGDOM MONERA: THE PROKARYOTES

All members of the Monera kingdom are single-celled prokaryotes. They are distinguished from the eukaryotes of the other four kingdoms by some basic characteristics:

- the presence of a **nucleoid** instead of a true nucleus (Chapter 5);
- their unique ribosomes; and
- the absence of membrane-bound organelles (such as mitochondria, chloroplasts, endoplasmic reticulum, and the Golgi apparatus).

Although composed of the same groups of chemicals as are eukaryotes (proteins, nucleic acids, lipids, and carbohydrates), prokaryotes possess several biochemical differences. The most universal distinction is the presence of **peptidoglycan,** a compound found in the prokaryotic cell wall. This polymer extends around the cell, surrounding it with a cross-linked matrix that reinforces the cell wall. Because bacteria have no means of evacuating excess water, the peptidoglycan wall provides enough strength to protect the fragile cell from swelling and bursting due to the osmotic influx of water. Those bacteria surrounded by many layers of peptidoglycan can withstand internal osmotic pressures 25 times that of the normal atmosphere without bursting.

The rigidity and shape of the cell wall determine the shape of the bacterial cell. Three shapes are most prominent in the kingdom: spheres *(cocci)*, rods (bacilli), and spirals (Figure 36-2). Many variations on these three shapes are somewhat common. More recent additions to the catalog of shapes are the square bacteria (genus *Arcula*) that were discovered in a salt pond along the shore of the Red Sea in 1980 and the flower-petal-shaped *Simonsiella*, the cells of which combine to form a cluster resembling a rose.

EVOLUTIONARY TRENDS

Of all the organisms alive today, bacteria most resemble our concept of the earth's original life forms, especially anaerobic bacteria, such as *Clostridium* (Figure 36-3). A comparison of modern bacteria with ancient fossils reveals that bacteria have not changed much in over 3.5 billion years. ⫸ Modern prokaryotes retain a rich variety of the metabolic strategies employed by their primitive ancestors. Today's eukaryotes possess similar metabolic pathways, which are variations on some of the original prokaryotic pathways. This suggests that eukaryotes evolved from one or more prokaryote ancestor. Figure 36-4 depicts this evolutionary process, incorporating the proposition that modern eukaryotic organelles were originally "endosymbiotic

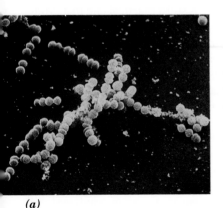
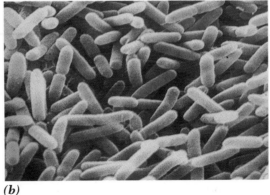
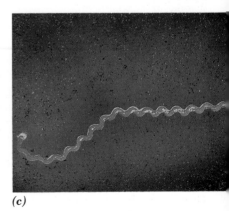

(a) | *(b)* | *(c)*

FIGURE 36-2

Typical bacterial shapes. *(a)* Spheres, *(b)* rods, and *(c)* spirals. Bacterial cells often remain attached to one another after dividing, producing a characteristic arrangement, such as the clusters formed by the *cocci* in part *a*.

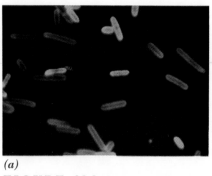
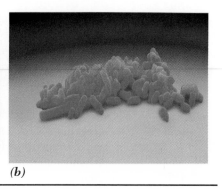

(a)   *(b)*

FIGURE 36-3 ──────────────────────────────

The earth's first life forms may have been very similar to these anaerobic bacteria. *(a)* These species of *Clostridium* (stained here with fluorescent dye) are commonly found in soil and can grow in oxygen-devoid environments, such as injured flesh (causing tetanus and gangrene) and canned foods (causing botulism, a fatal form of food poisoning). *(b)* These *Bacteroides* are among the many bacteria that inhabit your intestine. Gas produced in the large intestine is a byproduct of bacterial metabolism rather than human digestion.

bacteria" that invaded a pre-eukaryotic cell, teaming up to establish a mutually beneficial relationship (see page 111).

In spite of their relative simplicity, prokaryotes are extraordinarily successful organisms. Bacteria populate virtually all the earth's habitats, and, in some especially inhospitable places, such as concentrated sulfuric acid pools, they are the *only* inhabitants. One of the cornerstones of their success as relatively simple organisms is their extremely rapid growth rates, their ability to produce enormous numbers of individuals.

Adaptive Advantage of Rapid Growth Rates

Most bacteria reproduce by *prokaryotic fission,* asexually dividing into two daughter cells of equal size (page 197). These cells, in turn, divide into four cells, which then produce eight cells, and so on, until millions of new cells appear with each doubling of the population. Since bacteria may divide every few minutes, the rate of growth can be astonishingly fast. For example, the common bacterium *Escherichia coli* doubles in number every 20 minutes; in less than 48 hours, a single *E. coli* cell could theoretically produce a population of progeny equal to the volume of the earth. (As we discuss in Chapter 43, the bacterial population's growth is halted long before such an eventuality, usually as the nutrient supply in the local environment becomes exhausted.) Such rapid growth rates often allow bacteria to outcompete competitors that grow more slowly, many of which are simply overgrown and displaced.

Adaptive Advantage of Enormous Numbers

Because of their rapid growth rates, bacteria are often present in very large numbers. The greater the number of organisms, the greater the diversity of each species since the mutations that generate new traits occur as the cells'

DNA is replicated during cell division. Furthermore, many bacteria transfer genetic material to other bacteria (Chapter 14), creating new combinations of traits. This increased diversity improves the likelihood that some members of the species will possess a combination of traits that will enable them to survive a catastrophic change, even one severe enough to kill all the other inhabitants in the area. With a rapid growth rate, the survivors can then quickly repopulate the region with their progeny.

OVERVIEW OF THE KINGDOM

Although fossilized prokaryotes have been found, the fossil record is not sufficiently detailed to allow us to construct a phylogenetic classification scheme without relying on other criteria, such as similarities in DNA sequences and comparisons of ribosomal nucleic acid among living species. Although these techniques continue to modify our understanding of evolutionary relationships among bacteria, our current taxonomic system is still a practical one that groups living bacteria according to properties.

Because of their structural simplicity, bacteria lack the visual detail needed to categorize them solely on the basis of morphology (form), the principal approach used for classifying protists, fungi, plants, and animals. Thus, bacteriologists must supplement morphological observations with information about the organism's biochemistry, metabolic pathways, growth requirements, and genetic composition.

As with other organisms, the starting point in characterizing bacteria is an observation of their appearance on both a microscopic (Figure 36-5) and a macroscopic (Figure 36-6) level. This information is then combined with more subtle criteria, such as those described in Table 36-1. Based on these criteria, taxonomists have described more

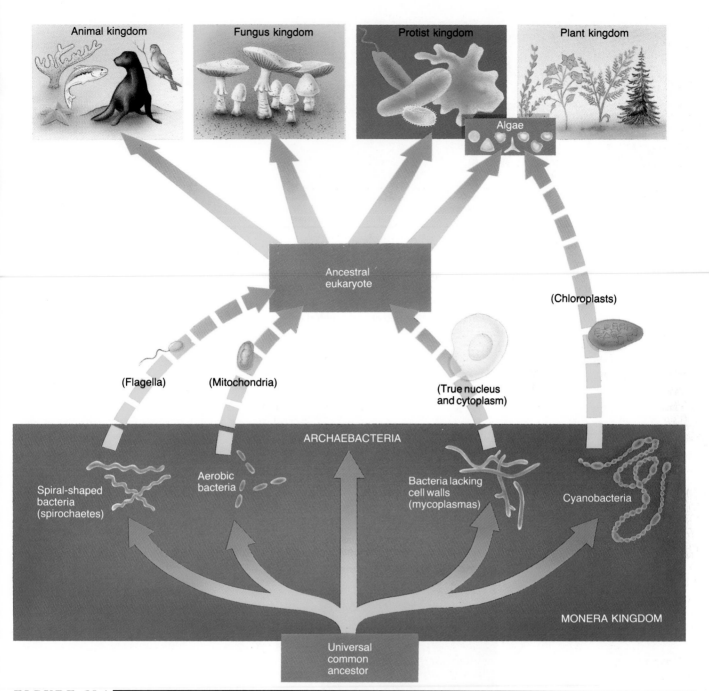

Animal kingdom

Fungus kingdom

Protist kingdom

Plant kingdom

Algae

Ancestral eukaryote

(Chloroplasts)

(Flagella)

(Mitochondria)

(True nucleus and cytoplasm)

ARCHAEBACTERIA

Spiral-shaped bacteria (spirochaetes)

Aerobic bacteria

Bacteria lacking cell walls (mycoplasmas)

Cyanobacteria

MONERA KINGDOM

Universal common ancestor

FIGURE 36-4

One proposal of evolutionary descent of all organisms from early prokaryotic cells. The dotted lines represent the evolution of eukaryotic organelles from bacteria that formed endosymbiotic relationships with an early prokaryote. Archaebacteria (an extremely primitive form of prokaryotic organisms) probably evolved as a separate lineage, rather than descending directly from other prokaryotes.

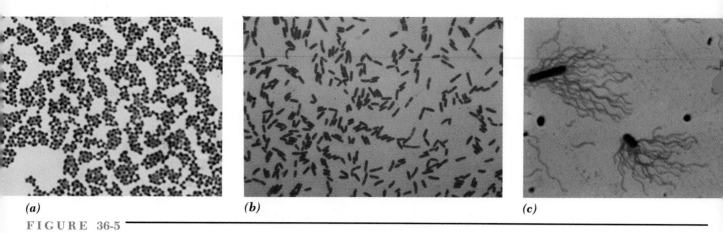

(a) *(b)* *(c)*

FIGURE 36-5

Microscopic views of bacteria. Taxonomists use stains to reveal characteristics that help classify and identify normally colorless bacteria. The most important stain, the gram stain, involves the application of purple dye to the cells, which are washed with alcohol, and a pink dye is applied. The stain identifies the bacterial species as either gram-positive *(a)* in which the cells retain the purple color when washed with alcohol or gram negative *(b)* in which the cells lose the purple color when washed and turn pink. Flagella *(c)* and other cell structures can be detected through the use of special stains that specifically enhance their appearance.

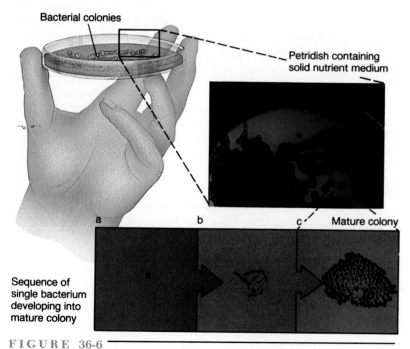

FIGURE 36-6

Colony characteristics are also used by taxonomists to distinguish different species of bacteria from one another. Colonies begin invisibly, as either a single cell or a small group of cells, and then proliferate into billions of bacteria, forming a macroscopic (visible) mass.

TABLE 36-1

CLASSIFYING AND IDENTIFYING BACTERIA

Criteria for Identification and Classification	Examples
Microscopic appearance	
Cell shape and arrangement	Spheres in clusters; rods in chains
Differential staining	Gram-positive or gram-negative
Distinguishing structures	Endospores, flagella, capsule
Macroscopic characteristics	Colony texture, size, shape, and color
Biochemical properties	Unique amino acids in cell wall; Composition of capsule
Physiological activities	Motility; oxygen requirements; carbon and energy requirements; ability to use specific sugars; metabolic byproducts; obligate intracellular growth
Immunologic specificity	Unique antigens of the cell, determined by reacting with antibody preparations of known specificity
Genetic analysis	Extraction of DNA, followed by determination of guanine-cytosine (G-C) content and nucleotide sequence
Ribosomal analysis	Comparison of similarities in one of the RNAs in the small subunit of ribosomes. Closely related organisms have similar ribonucleotide sequences.

than 2,000 species of monerans.[1] Although the Monera kingdom contains the fewest species of any kingdom, the organisms in this kingdom are the most numerous on earth.

The 2,000 members of the Monera kingdom are divided into two fundamental types of cells: (1) **eubacteria** (*eu* = true), the group classically called "true bacteria"; and (2) **archaebacteria** (*archeo* = ancient), a prokaryotic group that resembles eubacteria but differs from the true bacteria in their biochemical constitution. One group of organisms that was incorrectly called "blue-green algae" for years is now recognized as photosynthetic true bacteria. These organisms have been renamed **cyanobacteria** and reclassified as eubacteria.

Eubacteria: True Bacteria

Eubacteria comprise well over 90 percent of the Moneran species, most of which are "typical bacteria," much like the one shown in Figure 36-7. These cells have the prokaryotic anatomy that characterizes the monerans and distinguishes these cells from eukaryotic cells. These features were first observed with the advent of the powerful electron microscope.

Eubacteria are virtually everywhere, from the rod-shaped *E. coli* that inhabits your intestine, to the spherical *Streptococcus* that causes strep throat or scarlet fever, to the corkscrew shaped agent of syphilis, *Treponema pallidum.* These "typical" bacteria vary little in their external form, but they display tremendous diversity in their more subtle features. The scope of these differences is revealed in the following partial list of features.

- *Diversity of metabolic strategies.* Each bacterium's metabolic pathways are suited for the organism's particular habitat. Bacteria show broad variations in their tolerance to environmental stresses, including extremes in osmotic pressure, pH, temperature, water availability, and oxygen deprivation. In terms of oxygen deprivation, for example, aerobic bacteria can thrive only when oxygen is abundant, whereas some other types, *obligate anaerobes,* are killed by oxygen and thrive only in oxygen-devoid environments, such as your intestines or the sediment found at the bottom of a stagnant pond. Still other bacteria are *facultative anaerobes*—organisms that use the more efficient oxygen-dependent pathways in aerobic habitats and then switch to fermentation or anaerobic respiration (Chapter 9) when oxygen is depleted.

- *The ability to conduct photosynthesis.* Photosynthetic bacteria use **bacterial chlorophyll** to convert radiant energy to chemical energy for use as food. Since bacterial chlorophyll absorbs longer wavelengths of light energy than does the chlorophyll of eukaryotic photo-

---

[1] As described in Bergey's *Manual of Determinative Bacteriology* (9th ed.), the authoritative manual on bacterial classification. (William & Wilkins, 1993.)

synthesizers, aquatic bacteria typically inhabit lower depths than do unicellular algae, which grow only near the surface.

- *The formation of endospores.* A few types of eubacteria produce a single spore that enables the species to survive dehydration, exposure to ultraviolet light and caustic chemicals, and extreme heat. This is why boiling medical instruments and foods before canning is an unreliable means of sterilization: Any *endospores* that are present survive boiling and germinate into actively growing bacteria once conditions have returned to normal. Some of these bacteria can cause serious infections or produce dangerous toxins in canned or vacuum-wrapped foods. One such organism, *Clostridium botulinum,* causes botulism, the most dangerous form of food poisoning. Another *Clostridium* species infects wounds and quickly destroys surrounding tissue, causing the notorious disease gas gangrene. Gangrene often necessitates amputation of the affected limb to prevent the bacteria from invading the whole body and killing the individual. Instead of boiling, pressure cookers are now used to superheat steam to 121°C, a temperature high enough to kill the most heat-resistant endospores. Modified pressure cookers, called *autoclaves,* are highly efficient sterilization instruments that are used in hospitals and microbiological laboratories. Endospores are also resistant to the effects of aging, as illustrated by the unearthing of bacterial endospores at an archaeological excavation site; the dormant spores remained alive after 1,300 years of "suspended animation."

- *The formation of a capsule.* Some eubacteria form a capsule around themselves for nutrient storage, protection from drying, or escape from the protective defenses that would eliminate the bacterium from an infected host. Many types of bacteria that cause human pneumonia, for example, can do so only if they are encapsulated. Without capsules, these bacteria are quickly engulfed and destroyed by the body's protective white blood cells.

- *Motility.* Many bacteria are capable of propelling themselves through their environment. Although most bacteria use flagella for motility, some spiral-shaped bacteria spin on their axis and swim by "corkscrewing" through their liquid medium. Other types of cells, called gliding bacteria, move by an unexplained mechanism along the slime tracts they secrete. Even flagellated bacteria are unusual in the way their locomotor appendage works. At the base of the flagellum, just beneath the bacterium's plasma membrane, is a *basal body* that functions as a motor, rotating the flagellum like a helical-shaped propeller that either pushes or pulls the organism (Figure 36-7). This is the only known freely rotating structure in any organism.

- *Pathogenic (disease-causing) bacteria.* About 50 species of bacteria can cause human disease. These bacteria live either in or on host organisms, where they injure or kill the host. Some bacteria, such as the streptococci, infect the throat (producing strep throat). From there, the bacteria can invade the bloodstream, causing scarlet fever, or, in other cases, "blood poisoning." Some bacteria can cross the blood–brain barrier and infect the central nervous system, causing encephalitis or meningitis (inflammation of the brain or its covering). Most of these bacteria live in the tissue fluids between host cells. Some are **obligate intracellular parasites,** organisms that can succeed only if they enter the host cells. One such group, the **rickettsia,** causes several serious diseases, one of which has redirected human history. This disease, epidemic typhus, has incapacitated whole armies. For example, Napoleon lost more soldiers to typhus than to combat-related injuries. Another group of intracellular bacteria, the **chlamydia,** are even smaller than rickettsia and are barely visible using light microscopes. Their small size does not diminish their impact, however; chlamydia cause a number of diseases, such as trachoma, the world's leading cause of human blindness. Chlamydia infections of the genitals also represent one of the most common forms of sexually transmitted diseases in the Western world (see The Human Perspective: Sexually Transmitted Diseases, at the end of this chapter).

In addition to sexual contact, disease-causing bacteria may be transmitted by respiratory droplets, dust, contaminated objects (such as eating utensils), and the bite of an infected animal. One of the newest human diseases is Lyme disease, a tick-borne infection that is caused by the bacterium *Borrelia burgdorferi.* Lyme disease was named for Lyme, Connecticut, where the first case occurred in 1975. It is characterized by extreme exhaustion, joint pain, facial paralysis, and heart ailments that often necessitate installation of a pacemaker. Victims typically experience a large-diameter rash that resembles a bullseye at the site where the organism has entered. Almost all bacterial diseases are treatable with antibiotics.

▐▶ Some of the photosynthetic members of the Moneran kingdom are the *cyanobacteria.* Originally called "blue-green algae," cyanobacteria possess chlorophyll *a* and release oxygen as a waste product of photosynthesis, two properties characteristic of eukaryotic algae. Nonetheless, their lack of chloroplasts (and other organelles) and the presence of peptidoglycan in their cell walls places the cyanobacteria in the Monera kingdom since they are clearly not algae (which belong to the Protist and Plant kingdoms). Instead of chloroplasts, cyanobacteria possess a network of membranes in which the photosynthetic pigments are embedded. These internal membrane systems may have been the evolutionary forerunners of chloroplasts (Figure 36-8).

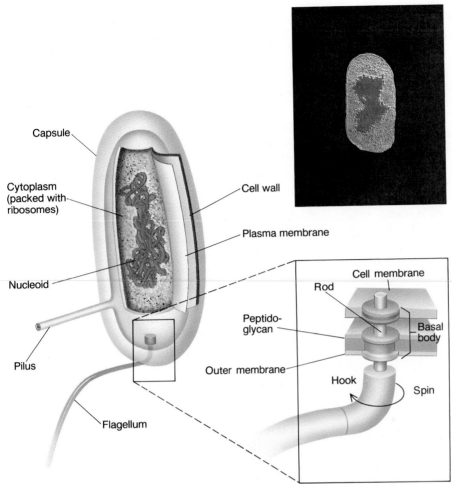

FIGURE 36-7

Anatomy of a typical eubacterium. Compare the simplicity of this prokaryotic cell with the complexity of eukaryotes in Chapter 5. The blow-up details the basal-body connection to a flagellum. The basal body functions like a motor, rotating the shaft of the basal body, which causes the corkscrew shaped flagellum to spin along its long axis.

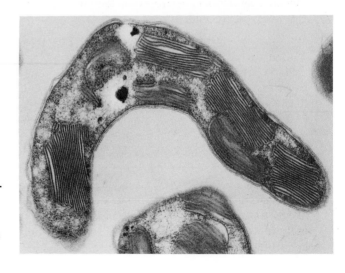

FIGURE 36-8

Forerunner of chloroplasts? The internal membranes of this cyanobacterium contain photosynthetic pigments. This network loosely resembles the membrane stacks (grana) of modern chloroplasts.

◁ THE HUMAN PERSPECTIVE ▷
Sexually Transmitted Diseases

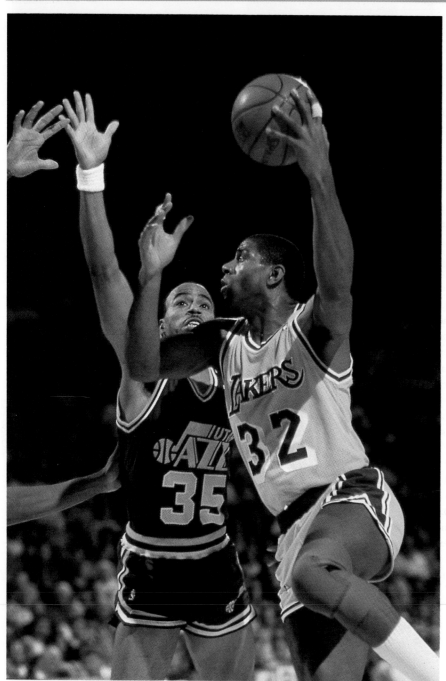

FIGURE 1

Although physical contact between athletes on a basketball court poses no risk of AIDS transmission, Magic Johnson, perhaps the most celebrated person to have tested positive for HIV, permanently retired from his sport in 1992. His retirement was precipitated by his concern that competitors afraid of contracting the virus might not play as aggressively against him. Johnson says he contracted the virus through heterosexual contact with an HIV-infected woman.

As we prepare to enter a new century, we are confronting one of the most frightening diseases to have ever threatened the human population, a new scourge that was unknown just 2 short decades ago. AIDS kills more people than does any other disease transmitted by sexual activity and is caused by one of the most dangerous of all human pathogens. Epidemiologists believe that millions of people in the industrialized world are infected by the human immunodeficiency virus (HIV) and will likely develop AIDS as a result (see Figure 1). Millions more are already infected in Africa, the continent of HIV origin. No cure or effective vaccine is even licensed for experimental testing in humans, much less is on the near horizon for use in fighting this inevitably fatal disease. Yet, as hardy as the virus seems once it has infected a person, it is actually a relatively fragile virus that survives poorly unless it is directly inoculated into a person's tissues through tiny breaks in the skin, such as those that can occur during sexual intercourse.

AIDS is an example of a group of diseases that cause uncontrolled epidemics. These are the **sexually transmitted diseases (STDs)** that can infect a person during sexual contact. Many STDs continue to reach epidemic proportions, in spite of available control measures. For example, even though gonorrhea is curable, it is still ranked as America's number one reportable disease. More cases of gonorrhea are reported to U.S. public health officials than are all other infectious diseases combined.[1] Every 12 seconds, a person contracts this disease, making it possibly even more prevalent than the common cold among sexually active persons with more than one sex partner. Fortunately, many sexually

[1] A reportable (or notifiable) disease is one that must be reported to health authorities anytime a physician diagnoses it. These data are tallied by the Centers for Disease Control (CDC) in Atlanta, Georgia, and are published weekly.

transmitted diseases respond well to anti-microbial drugs and, if treated early enough, can be cured without doing permanent damage to the host. The best strategy for protection, however, is to prevent infection.

Some of the factors that aggravate the continual spread of STDs are listed below:

- Asymptomatic carriers of gonorrhea, syphilis, genital herpes, chlamydia, and human immunodeficiency virus (HIV) serve as hidden reservoirs of infection. These individuals are usually unaware they have the disease until they are informed by a sexual partner they have infected.

- Fear of painful treatment, public disclosure, or social condemnation creates reluctance to seek medical attention and prolongs the amount of time that the pathogen is shed.

- Failure to notify the infected individual's sexual contacts that they have been exposed to a person with STD increases the probability that an unsuspecting individual will transmit the disease to others.

- Failure of the patient's sex partner to seek treatment often leads to reinfection of the cured person by the untreated partner.

- More convenient methods of birth control have decreased the use of condoms, which are the only effective physical barriers against venereal disease (although fear of acquiring AIDS has increased condom use somewhat in the previous decade).

- Sexual activity with multiple partners enhances the probability of exposure to a variety of sexually acquired pathogens.

- No protective immunity follows recovery from the major sexually transmitted diseases. Scientific efforts to develop effective vaccines have proved fruitless thus far.

- The emergence of drug-resistant pathogens represents a potential problem in treating gonorrhea with penicillin. Another disease, recurrent genital herpes, is currently incurable.

Until vaccines against the major STDs are available, the best preventive measures are those that are enacted on an individual basis. These include avoiding sexual contact as long as genital lesions are present, seeking prompt medical attention whenever one suspects possible venereal disease, and, if diagnosed as having an STD, waiting until medical confirmation of cure before resuming sexual activity with another person. All sexual contacts subsequently exposed to an STD should immediately be notified and treated. Even in the absence of symptoms, a sexually active person can determine through medical checkups if he or she is a possible carrier.

GONORRHEA

Although close to 1 million new cases of gonorrhea are reported to the CDC each year, most cases go unreported, and the annual incidence of this disease is estimated to be over 3 million cases. Gonorrhea is caused by the bacterium *Neisseria gonorrhoeae,* which is introduced through sexual contact, usually through the genitals but sometimes through the anus or the oral tract. An infected person may either develop symtoms characteristic of classical gonorrhea or may become a carrier who shows no symptoms. Carrier states are especially common among infected women in the early stages of disease. The large number of gonorrhea carriers in both sexes presents a major obstacle to the control of this venereal disease because carriers often unknowingly transmit gonorrhea to their sex partners.

The bacterium responsible for gonorrhea typically enters through a man's urethra and a woman's vagina. Between 3 and 5 days following exposure, most infected males experience burning during urination and discharge of thick, white pus from the penis (painful urination is *not* a symptom in infected women). If untreated, the pathogen may produce scarring in the vas deferens, which may result in permanent sterility. In women, untreated gonorrhea may cause an obstruction in the fallopian tubes, which may result in infertility. An infected mother can transmit the bacteria to her newborn baby as it travels through the birth canal. These babies most often suffer eye infections or permanent blindness. To prevent infected newborns from developing blinding eye infections, antibiotics are dropped into the eyes of every newborn delivered in U.S. hospitals.

CHLAMYDIA INFECTIONS

Each year, 5 million to 10 million Americans become infected with an obligate intracellular bacterium *Chlamydia trachomatis,* making it the most common sexually transmitted organism in the United States. In symptomatic males, *C. trachomatis* causes burning urination and puslike discharge, similar to the early stages of gonorrhea. Because of its different cause, however, this disease is termed **nongonococcal urethritis (NGU)** to distinguish it from gonorrhea, which responds to different antibiotic treatment. Although not a reportable disease, NGU is the most commonly occurring STD among men in the United States.

Chlamydia can cause all the complications that gonorrhea can: sterility in untreated men and women, and blindness in newborns (gonorrhea and chlamydia infections are believed to be the primary cause of sterility in the United States). Only rapid diagnosis and antibiotic treatment can prevent complications of these diseases. The antibiotic now used in the eyes of newborns to prevent gonorrhea blindness also prevents eye infections caused by chlamydia.

SYPHILIS

After several years of untreated infection, syphilis leaves its victims mentally degenerated, neurologically incompetent, consumed by destructive lesions, and, at least 33 percent of the time, dead. Although these outcomes are far less common today than they were in the past because of accurate diagnosis and effective antibiotic treatment, people still die of syphilis, and babies continue to be born with tragic congenital anomalies as a result of exposure to the syphilis bacterium during gestation in an infected mother.

Syphilis is caused by an especially fragile spirochete bacterium, *Treponema pallidum.* Although between 30,000 and 40,000 cases are reported each year in the United States, many cases go unreported, and the actual number may be as high as 500,000. Today, only a small percentage of these cases progress to the potentially fatal final stage. Most victims (and their partners) are identified and treated while in the early or middle stages. Campaigns to detect the disease in pregnant women help reduce the incidence of congenital syphilis, and antibiotic treatment very early in the pregnancy can also help avert this tragedy. The mandatory blood test people must take before getting a marriage license is solely for the detection of syphilis.

GENITAL HERPES

Genital herpes is characterized by painful blisters on the genitals and surrounding area. It may be transmitted by oral, vaginal, or anal sexual contact. The actual incidence rate of genital herpes is undetermined because physicians are not required to report cases to the CDC. Nonetheless, it is known that the incidence of genital herpes has dramatically escalated since the late 1960s. More than 20 million Americans currently have the disease, and as many as 500,000 join their ranks each year. Genital herpes is caused by *herpes simplex virus* (HSV). The majority of genital infections are caused by the Type 2 genital virus, although at least 20 percent are caused by HSV Type 1 (the virus that usually causes oral fever blisters). It is not uncommon for a person with oral fever blisters to transmit the disease to the genitals of a sex partner (and vice versa). Like oral herpes, genital herpes tends to recur, often following sexual intercourse, menstruation, stress, common cold, or fever. The disease reoccurs when the viruses that have been "hiding" in local nerve fibers (a latent infection) are reactivated.

The lesion that characterizes genital herpes is a fluid-filled vesicle that erupts and becomes an ulcer. These open sores are teeming with virus and are very infectious. They can appear anywhere on the penis in men and on the vagina, vulva, and cervix in women. The anus, perineum, buttocks, and thighs of either sex may also become ulcerated. Although the disease is most communicable while active lesions are present, the virus may also be shed in urine and genital secretions when no lesions are present.

One of the most serious complications of venereal herpes is infection of the newborn by its mother. Such infections are acquired during normal delivery through the birth canal of a woman with open lesions. The consequences of neonatal herpes are much more severe than are those of adult herpes. Infected infants may develop blindness, brain infections, death of much of the skin, or other potentially fatal disorders. Cesarean delivery is necessary to reduce the risk to the newborn whenever the mother has active genital sores.

As with any viral disease, antibiotics do not cure or even shorten the duration of the genital herpes infection. Acyclovir is a chemotherapeutic drug that can prevent the establishment of latency if used early in the course of the primary infection.

AIDS

Perhaps the most frightening modern medical development emerged for the first time in 1979 with the sudden appearance of a new disease, one that eventually kills everyone who contracts it. The disease, called **Acquired Immune Deficiency Syndrome (AIDS),** has become a serious international public health problem. Approximately 10 million people were infected with AIDS as of February 1992. The incidence of AIDS is increasing at an alarming rate. The victims are worldwide and include heterosexuals, homosexuals, males, females, adults, and children. Transmission among adults is primarily by sexual contact; the virus enters open wounds or lesions, usually small breaks in mucous membranes caused by sexual activity. Children usually acquire the disease

---

The more than 200 species of cyanobacteria are diversified in form and physiology. Some grow as single cells, although hundreds of cells may sometimes be held together in a gelatinous matrix. Others form long filaments or clusters of cells that are bound together by cell-to-cell linkages. In some species, an occasional cell is enlarged, forming a **heterocyst** (Figure 36-9). These specialized structures "fix" molecular nitrogen, converting it from its diatomic form (N_2) to a reduced state, such as ammonia (NH_3) or amino groups ($—NH_2$). In this way, cyanobacteria and nitrogen-fixing eubacteria make nitrogen available to other organisms, a process that is critical to the continuation of life on earth. (The nitrogen cycle is discussed in Chapter 41.)

Archaebacteria

Archaebacteria (*archae* = ancient) differ from eubacteria (and from other prokaryotic cells in general) in several ways: They are shaped differently, they possess unusual lipids in their cell membranes, and their cell walls lack peptidoglycan. In fact, archaebacteria resemble eukaryotes in only a few biochemical processes, pigments, and proteins. In addition, some archaebacteria live in areas where conditions are too extreme even for eubacteria. For example, *halophilic* archaebacteria flourish in the Dead Sea's supersaturated salt water. *Thermophilic* archaebacteria thrive in hot springs, in superheated marine waters a mile deep on the ocean floor, and in hot volcanoes. Sulfate-dependent archaebacteria are found in strong sulfuric acid solutions, surviving pHs as low as 1.0, which is acidic enough to dissolve metal.

Other archaebacteria live side by side with eubacteria in habitats devoid of oxygen, such as mammalian intestinal tracts, raw or partially processed sewage, and the smelly ooze found at the bottom of stagnant bodies of water. This last group of archaebacteria, called *methanogens*, often

across the placenta from an infected mother. Other nonsexual routes of transmission include transfusion with contaminated blood or sharing needles with an infected person, an all-too-common practice among drug abusers. There is no evidence that the disease can be transmitted by casual contact. Family members and caretakers of AIDS patients do not become infected.

The disease is caused by the human immunodeficiency virus (HIV). HIV infects cells of the immune system, including macrophages, helper T cells and certain B lymphocytes (page 649). It also can infect glial cells in the brain. Once inside infected cells, the retrovirus RNA is transcribed into a molecule of DNA that is then integrated into the host cell chromosome. After a brief period of replication, which includes a host antibody response, the virus establishes a latent infection. Individuals in this stage of the infection can still transmit the virus. The factors that trigger latency's end are unknown, but some people remain in this latent "healthy" state for as long as 10 years. The resumption of viral replication is accompanied by a decline in the number of helper T cells, which leads to a serious deficiency in the cell-mediated immune system. The infected person becomes vulnerable to cancer or to pathogens that are normally eliminated when the immune system is healthy. When the help T cell population falls to 200 cells per cubic millimeter, the person is diagnosed with the disease AIDS (concentrations of T cells in a healthy person exceed 800 cells per cubic millimeter).

A person with AIDS is especially prone to deelop otherwise rare opportunistic diseases. *Pneumocystis* pneumonia is by far the most common disease, although tuberculosis, candidiasis, toxoplasmosis, and cytomegalovirus (CMV) infections are also prevalent. Many individuals also develop AIDS-related dementia, as the virus replicates in the brain. About one-third of infected males contract a formerly rare cancer called Kaposi's sarcoma, which affects blood vessels in the skin. These opportunistic diseases are severe and are often the first indication that HIV infection has progressed to AIDS. They are also usually the cause of death in persons with AIDS.

HIV infection can usually be diagnosed by an antibody test. There is a period of time, usually 3 months but often more than a year, between infection and the development of an antibody response detectable by current diagnostic assays. More expensive tests can detect viral proteins or integrated DNA copies of the virus genome in individuals who test negative for the antibody. *HIV positive* individuals are those who are infected with the virus and can transmit it to others. They do not have clinically defined AIDS, however, until the helper T cell population falls below 200 cells per cubic millimeter of blood or until they have acquired Kaposi's sarcoma or one of the characteristic opportunistic infections.

Because there is currently no vaccine and no cure, the threat of AIDS can be controlled only by reducing the transmission of the virus. Nonsexual transmission has been effectively slowed in developed countries by screening blood supplies for the virus. It is not clear whether developing countries around the world will be able to afford such vigilance. Reducing intravenous drug use will also protect many individuals from infection. Sexual transmission of the virus can be reduced by the use of latex condoms with a spermicide that contains nonoxynol-9, which kills HIV as well as sperm. Such safer sex practices are not absolute safeguards, however, because HIV may pass through defects in the condoms. The only safe sex is sex with an uninfected partner. Although AIDS is a notifiable disease, results of HIV testing are not reported to public health authorities. A person who learns that he or she is HIV positive, however, should notify all sexual contacts in order to combat the spread of infection.

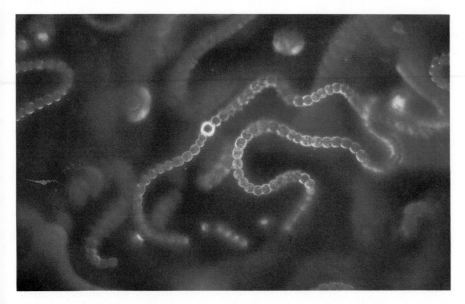

FIGURE 36-9

Swollen heterocysts enable chains of *Anabaena*, a cyanobacterium, to "fix" molecular nitrogen (N_2); that is, to convert it to a biologically usable form.

generate methane, a flammable organic gas known by many common names, including natural gas or swamp gas. Such methane-producing archaebacteria live in the gastric tracts of cattle, which contribute to the greenhouse effect by belching 100 tons of methane into the atmosphere each year. Like carbon dioxide, methane is a "greenhouse gas" that warms the planet by preventing much of the sun's radiant energy from escaping back into space. Excessive global warming may produce such severe climate changes as drought in important agricultural areas and the melting of the polar icecaps, which would raise sea level and flood coastal communities. (see Chapter 41).

▐▶ Because of the unique properties of archaebacteria, some biologists have proposed creating a new kingdom for these organisms. These biologists maintain that a common cellular ancestor (called a urkaryote) gave rise to *three* lines of cellular types: the prokaryotes, the archaebacteria, and the eukaryotes. Evidence for such a proposal comes from the many "unprokaryotic" properties of archaebacteria: Their plasma membranes are unlike any prokaryote; their ribosomal RNA resembles that of eukaryotic cells more than that of prokaryotic cells; and their chromosomal structure contains histones and introns (which eubacteria cells lack). Based on these differences, some biologists believe there should be only three kingdoms rather than the current five. These kingdoms would be the archaebacteria, the prokaryotes, and the eukaryotes. Such a scheme would group humans, plants, fungi, and protists together in the same eukaryote kingdom.

ACTIVITIES OF PROKARYOTES

In addition to fixing nitrogen, monerans perform other activities that have a profound impact on all organisms in the biosphere (see Bioline: Living with Bacteria). Many of these activities are the exclusive domain of prokaryotes, allowing them to occupy habitats that are uninhabitable by eukaryotes. For example, **chemosynthetic bacteria** use carbon dioxide as a carbon source (as do plants, algae, and other photosynthetic autotrophs). Yet, these bacteria do so in areas devoid of light, the usual energy source of autotrophs. As we described in Chapter 8, chemosynthetic bacteria extract their energy from inorganic minerals. The organisms that live near deep, continually dark ocean trenches depend on chemosynthetic bacteria as their ultimate source of food (see "Living on The Fringe," page 167).

◉ The vast majority of bacteria are heterotrophic, consuming organic molecules for energy and carbon. Most of these are *saprophytic decomposers;* they extract their energy and carbon from dead organic matter by disassembling organisms and their wastes, molecule by molecule, soon after they die. In addition, bacteria engage in some very novel biological activities. Some have phosphorescent pigments that glow when energized by respiration (Chapter 9). One unusual bacterium carries a magnetic "compass" that helps it home in on a nutrient-rich, oxygen-free habitat (Figure 36-10). Another bacterium, *Hyphomicrobium vulgare,* initially puzzled scientists because of its ability to grow

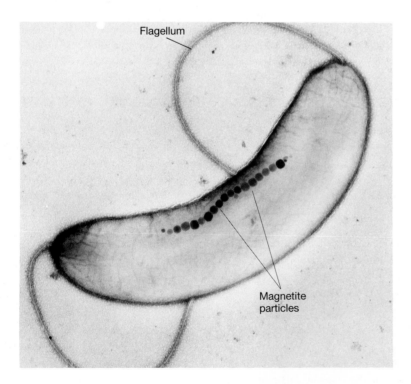

Flagellum

Magnetite
particles

FIGURE 36-10

The "compass bacterium" (*Aquaspirillum magnetotacium*) lives in swampy waters. Particles of magnetite (a component of iron ore) align in response to geomagnetic forces, allowing the bacterium to distinguish up from down. The bacterium rotates like a compass needle, pointing down toward the nutrient-rich sediment at the bottom of a swamp. Homing pigeons and honeybees also possess magnetic crystals, which may help them locate their destinations in a similar manner.

◁ B I O L I N E ▷
Living With Bacteria

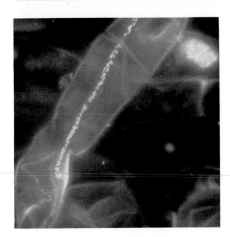

FIGURE 1
An orderly invasion of alfalfa root cells by beneficial nitrogen-fixing bacteria.

No organism can avoid sharing its environment with bacteria. We even share our bodies with them. We are learning how to live with bacteria more successfully today than in the past, however. Compare your awareness of hygiene to that of the people of King Henry VIII's day. At that time, it was customary for a person (including the king himself) to be bathed only three times: once at birth, a second time upon marriage, and the last time after death. The contents of chamber pots (indoor "toilets") were emptied through windows, often onto sidewalks below or into the streets. No one suspected that typhoid fever, cholera, and many other fatal infections were transmitted by bacteria shed in the feces and urine of infected people. Flies carried disease-causing microorganisms from open cesspools to people's lips and food, and runoff from flooded cesspools contaminated drinking water. Food decomposed very rapidly, presenting most people with a difficult dilemma: Eat rotted food or starve. Rats patrolled the streets, harboring such notorious killers as the bubonic plague bacterium, the agent of the "Black Death." The average person's life span was

about half of today's average, a grim testimony to the total lack of microbial awareness of the time.

Bacteria thrive in every habitat that supports life. A single gram of fertile garden soil harbors more than 2 billion bacteria. Each time you inhale, millions of bacteria enter your respiratory tract. In fact, there are ten times more bacteria in your large intestine than there are cells in your body; billions more occupy your skin. But bacteria aren't necessarily harmful. These bacteria constitute your body's *normal flora*. You benefit from their presence every time they successfully outcompete the disease-causing bacteria that could otherwise infect your body.

Many bacteria benefit the entire biosphere:

* Decomposers release organic nutrients from dead organisms and their wastes, recycling the earth's limited nutrients (Chapter 41).

* Nitrogen-fixing bacteria and cyanobacteria provide the biosphere with usable nitrogen. Some of these bacteria live in root nodules of legumes (Figure 1), such as alfalfa and soybeans. The bacteria that reside in the roots free these plants from the need for nitrogen fertilizer. This is why planting a crop of legumes improves the fertility of agricultural fields that have been exhausted by other crops.

* Cattle and sheep are unable to digest cellulose, the chief constituent of their grassy diet. Bacteria (and protozoa) that live in the animals' rumen, a special stomach chamber, digest cellulose for them.

* Bacteria (*Lactobacilli*) are used to make yogurt, some cheeses, vinegar, and several other food products (see Chapter 9).

* Modern waste-water treatment plants employ bacteria to process raw

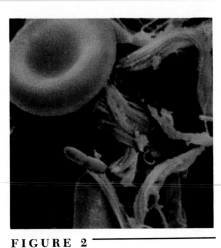

FIGURE 2
Deadly opportunist. The presence of otherwise harmless microbes, such as these bacteria (blue rods) that have colonized a burn wound, poses a constant threat of infection to a person with reduced defenses against disease. With our modern microbial awareness, we can combat such threats with a battery of control measures.

sewage to a safe substance, free of disease-causing microorganisms.

* The newest applications use genetically engineered bacteria to produce antibiotics and other compounds of medical and industrial importance. Chapter 16 discusses many of the modern strategies for employing the microbial "workforce."

In spite of these benefits, a small minority of bacteria cause disease (Figure 2), spoil our food, foul our water, and deteriorate useful products. These bacteria represent a formidable foe, one that requires sustained efforts to minimize its detrimental impact. Considering their benefits and the many critical roles they play in the biosphere, however, bacteria more than compensate for the inconvenience and suffering they may bring. Life on earth could not continue without them.

on the surface of purified water, with no apparent source of nutrients or energy. Scientists eventually discovered the bacterium's source of energy and nutrients: formaldehyde and other small volatile organic molecules that give the air of some laboratories their peculiar smell. These molecules dissolve in water and feed the organism. The bacterium's ability to concentrate such a dilute source of nutrients from air is unmatched by any other heterotroph.

The monerans (bacteria and cyanobacteria) are the smallest and the simplest living cells. Members of the Monera kingdom are found in virtually all the earth's habitats. Their rapid multiplication rates produce enormous numbers of individuals, which contributes to their remarkable adaptability as a group. Generally beneficial, some monerans are essential to the continuation of life on earth. A few types devastate and destroy, causing disease, spoilage, fouled water supplies, and deterioration of useful products. (See CTQ #3.)

VIRUSES

Although the cell is the fundamental unit of life, there are certain "organisms," the **viruses,** that are not made up of cells at all; they have no cytoplasm, no organelles, and no plasma membrane. A fully assembled virus is as inanimate as is a crystal of organic chemicals. Some are nothing more than nucleic acid and protein. Once inside a host cell, however, the virus's biological apparatus "awakens." There, the virus adopts its "living mode," exhibiting two key characteristics of life: reproduction and heredity. How viruses alternate between these two seemingly paradoxical states is a function of their simplicity in structure and life cycle. In general, viruses are characterized by the following properties:

1. *They are obligate intracellular parasites* that require a host cell to perform virtually all biological functions that lead to the production of new viruses.

2. *They are incapable of independent metabolism.* The host cell does virtually everything the virus needs for its successful reproduction.

3. *They are smaller than the tiniest bacteria;* so small, in fact, that they can be seen only with an electron microscope.

4. *They possess only one type of nucleic acid,* either DNA or RNA, but never both. The DNA or RNA of the virus directs the host cell to produce a crop of progeny viruses. The host cell obeys the viral genes as faithfully as it would its own genes.

5. *While in a host cell, viruses undergo an "eclipse phase,"* during which they disintegrate into their molecular constituents. During eclipse, the virus's nucleic acid commandeers the cell's genetic machinery, instructing it to produce new copies of the virus's nucleic acid and the proteins needed for assembling new viruses. Eclipse ends when these components assemble into progeny viruses, sometimes hundreds of them in one cell.

6. *Some viruses can be crystallized and still retain their infectivity.* One such crystal is shown in Figure 36-11, a purified collection of polioviruses. The viruses in the gemlike crystal are still capable of infecting host cells and causing severe life-threatening disease.

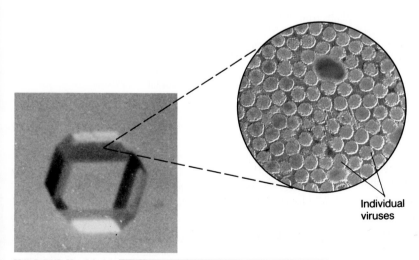

Individual viruses

FIGURE 36-11

"Living gems." This crystal consists of trillions of symmetrically arranged polio viruses. Even when crystallized, the viruses retain their ability to infect cells.

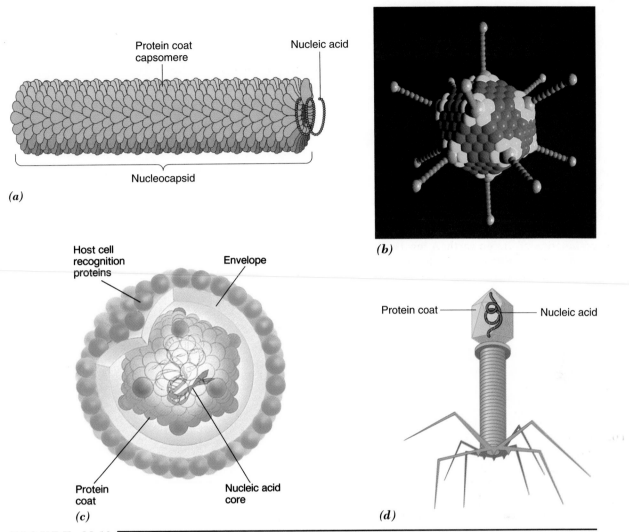

Protein coat
capsomere

Nucleic acid

Nucleocapsid

(a)

(b)

Host cell
recognition
proteins

Envelope

Protein
coat

Nucleic acid
core

(c)

Protein coat

Nucleic acid

(d)

FIGURE 36-12

Four basic forms of viruses. All viruses contain a nucleic acid core housed in a protective protein coat, or capsid. *(a)* Some of these are rod-shaped (proteins attached to a tight spiral of nucleic acid); *(b)* others are shaped like faceted spheres. *(c)* Some are surrounded by an outer membrane that is acquired as the virus escapes the host cell, enclosing itself in a modified portion of the cell's plasma membrane. This envelope is needed by these viruses to infect and enter their host cell. *(d)* Bacteria-infecting viruses, called bacteriophages, are the most complex viruses. Regardless of its structure, a virus must have molecules on its outer surface which specifically fit with and attach to the surface molecules on the particular host cell.

These unusual properties of viruses are a function of their structural simplicity (Figure 36-12). The least complex viruses consist of nothing more than a molecule of nucleic acid (the viral genes) surrounded by a protein coat, called a **capsid.** The capsid protects the genes and allows the virus to attach to and infect its specific host cell. The most complex viruses are those that infect bacteria (called *bacteriophages*). These viruses possess a head (containing the viral nucleic acid) and a tail with fibers that recognize and attach to the cell wall of its particular host cell. The tail then

contracts and forcefully injects a hollow tube into the cell, through which the viral nucleic acid enters the cell, initiating the process of viral infection.

HOW VIRUSES WORK

Regardless of the type of virus, part of the capsid's function is to get the virus into a susceptible host cell, where a single virus can instruct the cell to produce hundreds of "carbon copies" of the parent virus. All it takes for a virus to succeed

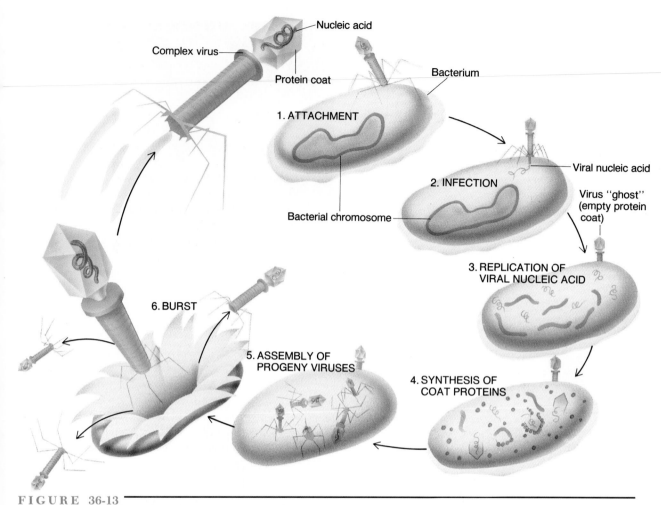

FIGURE 36-13

The life cycle of a typical virus is illustrated here by viral infection of a bacterial cell. Although some variation exists from virus to virus, the basic productive strategy is the same as that shown here. The host cell is not always killed, however. Some infected animal cells "leak" mature viruses through intact plasma membrane. In addition, some viruses insert their nucleic acid into the host cell's chromosome and remain "hidden" for long periods before switching to the productive cycle shown in this figure.

are a set of genes with the information needed for producing more viruses and a way of getting those genes into a host cell. Once the virus abandons its capsid and exists as a naked strand of nucleic acid inside the cell, its eclipse phase has begun. During this "invisible" phase, the virus produces offspring and does its damage to the infected cell (Figure 36-13). Although this figure illustrates bacteriophage infection, it typifies the dynamics of most viral infections, including those that result in human disease.

Once the virus attaches itself to the host cell's surface and injects its nucleic acid (Figure 36-13, parts 1 and 2), the viral genes are transcribed and translated by the host cell's enzymes and ribosomes. The cell's protein synthesis machinery doesn't distinguish between the virus's and the cell's nucleic acid; it simply follows the dictates of the newly acquired genetic instructions (Figure 36-13, part 3). In this way, the virus commandeers the cell's metabolic machinery, redirecting it so that instead of producing new cell material, the cell becomes a virus-producing factory.

Host cell polymerases replicate the viral nucleic acid, sometimes generating several hundred copies of the viral genes (Figure 33-13, part 3). These genes also instruct the cell to produce huge quantities of coat proteins (Figure 33-13, part 4), which automatically aggregate around the new strands of viral nucleic acid, assembling hundreds of progeny viruses (Figure 33-13, part 5). In this case, the cell eventually bursts (Figure 33-13, part 6), liberating viral offspring, each with the capacity to infect another cell.

VIRAL DISEASES

Most human viruses produce infections that rapidly run their course and stimulate permanent immunity in survivors. Immunity results from the production of immune cells and antibodies that specifically recognize and inhibit subsequent infection by the same types of viruses. This explains why chickenpox , measles, mumps, and several other common childhood diseases occur only once in a lifetime; recovery produces long-term immunity. A number of viral diseases pose special medical problems, however. For example, the common cold may be caused by more than 100 distinct strains of *rhinoviruses* (*rhino* = nose). Since immunity is specific, infection with any one strain of rhinovirus fails to induce host immunity to the other strains. Similarly, the *influenza virus* causes recurrent epidemics because of its tendency periodically to alter its surface antigens. This new specificity is no longer recognized and neutralized by the immune defenses of a person previously exposed to earlier strains of the same virus. *Human immunodeficiency virus* (HIV), the virus responsible for AIDS, infects and destroys cells of the immune system itself, thereby rendering the infected individual defenseless against the virus as well as against many other types of potential pathogens.

Some viruses escape elimination by the immune response by establishing **latent infection,** or hidden infection. These viruses remain in the host even after the disease symptoms disappear and are generally undetectable during these latent periods. Evidence suggests that the DNA of these viruses actually inserts itself into the chromosomal DNA of the host cells, where it resides in a dormant state, replicating only with each division of the host cell. When the viral DNA eventually leaves the host chromosome and begins its viral-productive cycle (similar to that depicted in Figure 36-13), symptoms of the disease reappear. The disease may be reactivated periodically by various stimuli. For example, recurrent episodes of Type 1 (oral) herpes or Type 2 (genital) herpes may be triggered by emotional stress, sunburn, menstruation, pregnancy, common colds, or other diseases that cause fever (hence the common terms "fever blisters" and "cold sores"). Between episodes, the host appears "cured," as the virus silently resides within the chromosomes of regional nerve cells, poised for its next attack.

Some RNA viruses insert their nucleic acid into host cells and cause delayed, or latent, infections. These viruses include some cancer viruses (including one that causes a type of leukemia in humans) and HIV, the virus responsible for AIDS. These viruses cannot insert RNA into DNA, however; instead, they make a DNA copy of themselves, which readily integrates into the host's chromosome. These viruses are called **retroviruses** (*retro* = reverse) because they reverse the typical flow of genetic information from DNA to RNA (Chapter 14), using their RNA genomes as a template to transcribe a copy of DNA.

BIOLOGICAL AGENTS EVEN SIMPLER THAN VIRUSES

Two other groups of infectious agents, viroids and prions, are even simpler than viruses in their structure. **Viroids,** are associated with certain diseases of plants. Each viroid consists solely of a small, single-stranded circle of RNA that is unprotected by a protein coat. Viroid-infected plants are stunted in growth and abnormal in development. For example, viroids are the agent of cadang-cadang disease, which infects coconut palms (the disease derives its name from the sound of falling coconuts). A major epidemic of cadang-cadang disease has virtually destroyed the coconut palm population of the Philippines.

A group of slowly progressive neurological diseases are caused by agents called **prions.** Kuru and Creutzfeld-Jakob syndrome, two human prion diseases, are characterized by incubation periods of months to years, followed by fatal attacks on the victim's nervous system. Once symptoms appear, these diseases progress rapidly, and death is inevitable within 1 year after the victim shows signs of illness. Although their infectious natures and small size suggest that prions are similar to conventional viruses, no viruslike particles have been observed in prions, not even by electron microscopy. Furthermore, prions are resistant to physical and chemical treatments that inactivate conventional viruses, and nucleic acid has yet to be detected in preparations of these infectious agents. In fact, prions appear to consist solely of protein.

EVOLUTIONARY SIGNIFICANCE OF VIRUSES

Since viruses lie somewhere between the living and the inanimate, it may be tempting to view them as some kind of "missing link"—the ancient transitional form that evolved from inanimate chemicals into the first cells. Their absolute requirement for a host cell disqualifies viruses as candidates for the earth's initial life forms, however. Instead, most contemporary biologists believe that viruses are "renegade genes," pieces of DNA (or RNA) that were once part of a cell's chromosome (or messenger RNA). These pieces of nucleic acid somehow acquired the information for self-replication using the host cell's machinery and may have been transferred directly from one cell to its neighbors through adjacent membranes. Eventually, the genetic fragments evolved new genes for making their coat proteins, allowing them to escape their host cells and assisting them in infecting new host cells.

■■▶ The evolutionary impact of viruses on other forms of life is significant in determining which organisms will survive. Viruses themselves are selective pressures, much as is a predator. As disease-causing "menaces," viruses eliminate individuals or species that have little resistance to their lethal attacks. The evolutionary significance of viruses may be more profound than is their role as selective pressures, however. Viruses play an important role in gene transfer between bacteria, often carrying advantageous genes that confer traits on recipient cells which allow the cells to survive or better compete in a hostile habitat, such as a human body that is being treated with antibiotics. Viruses may transfer antibiotic-resistant genes to recipient bacteria, which are no longer killed by the drug and continue to proliferate in spite of antimicrobial therapy. Most biologists believe that viruses may have played similar gene-transfer roles in higher organisms, supplementing sex as a means of increasing diversity within a species.

SUCCESSES AND FAILURES IN THE WAR AGAINST VIRUSES

None of the earth's nearly 2 million known species is free of viral infection. Although not always harmful, some of these viruses can cause serious diseases in plants and animals. Viral diseases in humans range in severity from common colds to fatal cases of smallpox, rabies, and AIDS. Even a seemingly innocuous virus that causes diarrhea in children poses a deadly threat in populations that are ill-equipped to provide adequate care. All viral diseases have one thing in common: No antibiotics are effective in hastening recovery.

Cures for viral diseases still elude medical science, mainly because viruses become a functional part of the host cell. As a result, it is difficult to find a chemical that selectively attacks the virus without similarly destroying the host. However, the body's natural immunity eliminates most viral infections if the infected person survives long enough to mobilize defenses and to control the proliferation of the virus.

Despite the difficulty involved in treating viral infections, most of the more serious viruses have been controlled, largely with *vaccines*—inactivated or weakened forms of disease-causing viruses that specifically stimulate and sensitize the immune system (page 655). Vaccination has reduced or eliminated the threat of polio, smallpox, rabies, yellow fever, measles, rubella, and mumps. Some viral diseases still elude attempts to control them, however. The most exasperating of these is the common cold; the most frightening is AIDS.

Viruses have few equals in terms of their parasitic efficiency. Their simple structure provides viruses with everything they need to infect, replicate in, and escape host cells. Viruses depend entirely on the host cell for almost every biological function in their replication cycle; they actually become a new cell component that seizes control of the host cell and directs it to produce progeny virus. (See CTQ #5.)

REEXAMINING THE THEMES

Relationship between Form and Function

Nowhere in nature is the relationship between form and function more elegantly exemplified than in viruses. The extreme simplicity of viral structure reflects the limited number of tasks a virus must accomplish to propagate its kind successfully. Its outer protein coat protects the virus and specifically fits with the surface molecules to which it attaches when infecting a host cell. The nucleic acid structure of a virus is indistinguishable by the host cell's enzymes; consequently, the virus chromosome directs the cell's protein and nucleic acid synthesis machinery to make more viruses. The exquisite structure of bacterial viruses enables the virus to inject its nucleic acid forcefully into a host cell that is protected by a thick cell wall.

Unity within Diversity

The structural differences between prokaryotic and eukaryotic cells are considerable: Prokaryotes lack the mitochondria, endoplasmic reticulum, complex nucleus, and other task-specific organelles that characterize eukary-

otes. Nonetheless, prokaryotic cells carry out the identical biological processes as eukaryotic cells do. These tasks include energy acquisition, genetic control of traits, inheritance of ancestral properties, and cell reproduction. Furthermore, eukaryotes and prokaryotes utilize the same genetic code and many of the same metabolic pathways (e.g., glycolysis, Krebs cycle, respiration, chemiosmosis). These are just a few of the similarities that reveal the unity between such diverse organisms as humans and the bacteria that live in the human colon.

Acquiring and Using Energy

Although most members of the Moneran kingdom acquire energy by heterotrophic or photosynthetic means, some members of this kingdom are capable of chemosynthesis, using chemical energy, yet fixing carbon dioxide, as do autotrophs. In deep oceanic trenches, these chemosynthetic organisms are the primary producers that support a community of organisms. This is the only known ecosystem that does not depend on sunlight and photosynthesis to provide energy to all its organisms.

Evolution and Adaptation

▶ Some living bacteria may be very similar to the original cells from which all life descended. Alternatively, there may have been a primitive cell from which prokaryotes and eukaryotes evolved along separate lines, a cell that also gave rise to archaebacteria. Natural selection favored "team adaptations" that created typical eukaryotic organelles, which originally were independent bacterial cells that later formed mutually dependent intracellular relationships with ancient, pre-eukaryotic cells. Monerans are perhaps the most adaptable organisms, thriving in habitats that are too hostile for any eukaryotic members of the other four kingdoms. Their great metabolic diversity, rapid growth rates, and enormous numbers are responsible for the adaptability of the monerans.

SYNOPSIS

All monerans are unicellular prokaryotes, the simplest living cells. Prokaryotic cells are smaller than are eukaryotic cells and lack a nucleus and other membrane-bound organelles.

Dividing by binary fission, monerans can produce enormous numbers of offspring within a relatively short period. Such rapid growth rates virtually assure the production of genetic variants, greatly enhancing the likelihood that some individuals will survive environmental changes. In addition, rapid growth rates may give bacteria the competitive edge over other organisms in certain habitats.

The Moneran Kingdom contains two fundamental types of prokaryotes. eubacteria (true bacteria) and archaebacteria ("ancient" bacteria). Eubacteria contain all disease-causing bacteria, typical "normal flora" bacteria, and the photosynthetic cyanobacteria. Archaebacteria contain organisms that have adapted to harsher environments, such as extremely hot, salty, or acidic habitats.

Monerans are found in virtually all habitats, including environments that are too severe for the survival of eukaryotes.

Monerans are essential to the continuance of life on earth. As decomposers, bacteria recycle essential nutrients. As nitrogen fixers, they transform molecular nitrogen into a form that can be used by all organisms. As chemosynthetic autotrophs, they form the basis for some marine food chains.

A few bacteria are notorious for their ability to cause infectious disease, spoil food, or decompose useful products. Some bacteria cause severe diseases of people, animals, and plants, extracting an enormous cost in lost revenues and lost lives. Other bacteria spoil food and foul our drinking water. Each year, we spend billions of dollars to protect our water supply and to prevent or retard microbial spoilage and deterioration.

Viruses are noncellular entities composed of DNA or RNA, protein, and sometimes a surrounding envelope. In a host cell, the viral nucleic acid commandeers the cell and directs it to make a new crop of viruses.

Key Terms

nucleoid (p. 787)
peptidoglycan (p. 787)
eubacteria (p. 791)
archaebacteria (p. 791)
bacterial chlorophyll (p. 791)
obligate intracellular parasite (p. 792)
rickettsia (p. 792)

chlamydia (p. 792)
heterocyst (p. 796)
chemosynthetic bacteria (p. 798)
virus (p. 800)
capsid (p. 801)
latent infection (p. 803)
retrovirus (p. 803)

viroid (p. 803)
prion (p. 803)
sexually transmitted disease (STD) (p. 794)
nongonococcal urethritis (p. 795)
Acquired Immune Deficiency Syndrome (AIDS) (p. 796)

Review Questions

1. Describe three features that characterize all the members of the Monera kingdom.

2. The simplicity of prokaryotic cells creates classification and identification difficulties. Name six criteria that are frequently used for classifying and identifying bacteria.

3. Discuss the evolutionary relationship between eubacteria and archaebacteria.

4. Compare the action of eukaryotic and prokaryotic flagella.

5. Other than wiping out infectious disease, what are three consequences of eliminating all bacteria from the biosphere?

6. Why would the antiviral substance pure interferon (Chapter 30) be more effective for treating viral diseases than would antibiotics?

Critical Thinking Questions

1. In 1919, a worldwide epidemic of influenza killed 20 million people in less than a year. In nearly every fatal case, the lungs were infected with a bacterium called *Haemophilus influenza,* which was mistakenly identified as the cause of influenza. Design a procedure that would prove or disprove this bacterium as the cause of influenza.

2. Explain the following statement: Even though an individual bacterium has little ability to adapt to dramatic changes in its environment, bacteria as a group are perhaps the most adaptable organisms on earth.

3. One serious problem in modern medicine is the overuse of antibiotics. Considering the adaptability of bacteria (rapid growth, huge numbers), why would such a practice threaten to decrease the value of antibiotics as treatments of infectious disease.

4. Of the six properties of viruses listed in this chapter, only three are unique to viruses. Identify these three properties, and explain why the other three might be found in a few "organisms" other than viruses.

5. Viruses face competition from the information in the DNA in their host cell's chromosome. Propose a mechanism by which a newly arrived virus particle might shut down or destroy the host chromosome, leaving the viral nucleic acid sole director of that cell.

6. Having read this chapter, do you consider viruses living organisms or inanimate collections of crystals? Justify your answer.

Additional Readings

Dolittle, W. 1987 The evolutionary significance of the archaebacteria. *Ann. N. Y. Acad. of Sci.* 503:72 (Intermediate)

Gallo, R. 1987. The AIDS virus. *Sci. Amer.* Jan.:47–56. (Intermediate)

Kushner, D. 1978. *Microbial life in extreme environments.* New York: Academic Press (Intermediate to Advanced)

McKane, L., and J. Kandel. 1985. *Microbiology, essentials and applications.* New York: McGraw-Hill. (Intermediate)

Simons, K., et al. 1982. How an animal virus gets into and out of its host cell. *Sci. Amer.* Feb.:58–66. (Introductory.)

Shapiro, J. 1988 Bacteria as multicellular organisms. *Sci. Amer.* New York: Freeman. (Intermediate)

Temin, H. 1972. RNA-directed DNA synthesis. *Sci. Amer.* Jan.:24–33. (Intermediate)

CHAPTER
◄ 37 ►

The Protist and Fungus Kingdoms

**STEPS
TO
DISCOVERY**
Creating Order From Chaos

*I*t's an imperfect world, and few people are more aware of this fact than are taxonomists. One of the jobs of a taxonomist is to create meaningful order out of the overwhelming variety of organisms to be studied. It might seem like the easiest job would be defining kingdoms since there are fewer kingdoms than any other category. Indeed, 150 years ago, defining kingdoms was relatively simple: Everything was classified into three categories. Nonliving things were placed in the mineral class, and everything else was assigned to either the Animal or Plant kingdom. But even then, disturbing exceptions intruded on the tidiness of this sys-

tem. For example, mushrooms are plantlike but, like animals, they have no chlorophyll and cannot photosynthesize. Nonetheless, these heterotrophs were forced into the Plant kingdom. Such arbitrary decisions greatly disturbed such taxonomists as Germany's Ernst Haeckel. So many "in-between" organisms had been discovered that Haeckel sought to have the plant–animal dichotomy declared a flop.

Haeckel himself delivered the first crippling blow to the two-kingdom system just over 130 years ago. Growing weary of trying to "squeeze round objects into square holes," he proposed the creation of another "hole," a third

tists and fungi supplied much of the scientific understanding that has expanded our view of the living world from one consisting entire

kingdom called "the Protists." This kingdom contained all those organisms that shared both plant and animal characteristics. Although this three-kingdom system solved the problem of borderline organisms, even Haeckel's scheme would eventually prove inadequate. His Protist kingdom contained organisms made up of two fundamentally different kinds of cells that were too different to belong in the same kingdom.

The problems with Haeckel's system began to surface with the recognition that the nucleus, one of the most prominent features of most cells, was never found in bacterial cells. Camillo Golgi discovered the membrane apparatus that bears his name, another feature found in most types of cells but not in bacteria. Mitochondria, although only faintly visible in plant and animal cells using the light microscopes of the late nineteenth century, were never visible in bacterial cells. Chloroplasts, the most striking microscopic feature of plant cells, were conspicuously absent from photosynthetic bacteria. It became apparent that the relatively simple, phylogenetically primitive bacteria clearly didn't belong in the same kingdom as did the more complex organisms whose cells contained these internal organelles. In 1930, even before the electron microscope detailed the ultrastructural differences between these two types of cells, H. F. Copeland coined the terms *prokaryotic* and *eukaryotic* to distinguish the two. A new kingdom was created to house the more primitive prokaryotic cells.

The new four-kingdom system identified plants, animals, protists,[1] and monerans (the prokaryotes). But one group of organisms still intruded on the tidiness of the system. This group was the fungi, organisms that had cell walls (as do plant cells) but were not photosynthetic. Fungi were different from the other protists, which could all be grouped according to their resemblance to (and probable ancestry of) organisms in the "higher" Plant and Animal kingdoms. In the absence of evidence that fungi were ancestors of either plants or animals, these organisms were assigned their own kingdom in 1969 by Robert Whitaker, who proposed the five-kingdom system that is generally accepted today (page 13). An organism that is neither plant nor animal is assigned to one of three additional kingdoms: the monerans, protists, or fungi.

Even today, new information continues to upset the view that these five kingdoms represent natural lines of division. The discovery of a third fundamental cell type (archaebacteria) has prompted C. R. Woese, G. E. Fox, and others to argue in favor of a three-kingdom system: the Monera kingdom, which would include classical prokaryotic cells; the Archaebacteria, which would occupy their own kingdom; and plants, animals, protists, and fungi, which would be all be lumped together in a single kingdom (called Eukaryota), based on their eukaryotic cell anatomy.

---

[1] Copeland called the Protist kingdom *Protoctista,* a name still preferred by some biologists over *Protists.*

of plants and animals to a world populated by organisms from at least 5 distinct kingdoms.

THE PROTIST KINGDOM

Since Leeuwenhoek first glimpsed microorganisms through his simple microscope more than 300 years ago, biologists have been enamored by the tiny "animalcules" he described. Most of the organisms Leeuwenhoek discovered belong to two modern kingdoms, the protists and the fungi (Figure 37-1). But Leeuwenhoek couldn't even imagine the magnitude of importance these organisms have to humans and to all other organisms in the biosphere. Both kingdoms contain a diverse assortment of eukaryotic organisms, most of which are microscopically small. One kingdom, the protists, presents a special challenge to taxonomists.

▼ ▼ ▼

Many **protists** are "taxonomic misfits" that continue to create problems for those attempting to classify them. This is not to say that protists are unfit. In fact, protists are very well-suited organisms, exceeded only by bacteria in the number of environments to which they have adapted. Nonetheless, as a group, protists are still a hodgepodge of eukaryotic organisms that, rather than sharing broad taxonomic similarities, have been assigned to the Protist kingdom more or less by default. They simply don't fit into any of the other four kingdoms.

Most protists are unicellular, but some are colonial. Most are microscopic, but some are as large as 5 millimeters (about a quarter of an inch). Some are photosynthetic, others are heterotrophic. Some are funguslike, others are plantlike, still others are animal-like, and many combine characteristics of two or more of these. For example, a

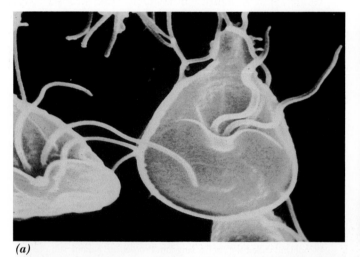
(a)

(b)

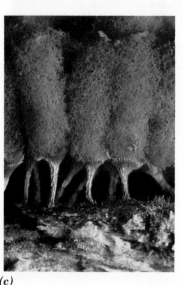
(c)

(d)

FIGURE 37-1

A gallery of protists and fungi: (**a**) Giardia: an animal-like protist. (**b**) Diatoms: plantlike protists. (**c**) Slime mold: a funguslike protist. (**d**) A stinkhorn fungus: a macroscopic member of the kingdom fungi.

FIGURE 37-2

***Euglena*: animal-like and plantlike.** This actively swimming protozoon possess at least one flagellum that imparts motility, and protozoa lack cell walls—both animal cell characteristics. As long as they are confined to darkness, *Euglena* remain in their animal-like state, heterotrophs that absorb dissolved nutrients from their medium. When exposed to light, however, these versatile protists become autotrophs. Their chloroplasts proliferate, and the cell turns into a green photosynthetic, plantlike cell. This taxonomic "fence straddler" is neither plant nor animal; it is a protist.

photosynthetic microorganism that actively swims is neither a plant nor an animal. It is simply a protist (Figure 37-2).

EVOLUTIONARY TRENDS

▮▶ Protists are credited with several evolutionary advances, not the least of which is sexual reproduction (Figure 37-3). For some protists, sex is not a reproductive strategy. These organisms merely conjugate and exchange part of their genetic library with each other. This gene mixing in-

creases genetic diversity of the species, but reproduction continues to be asexual. In contrast, other protists couple sex with reproduction, a process similar to that which occurs in plants and animals. For example, a diploid protist undergoes meiosis, dividing into gametes of opposite mating types, which then fertilize each other. The resulting diploid zygote develops into a progeny individual.

A major evolutionary "breakthrough" of the protists is the development of eukaryotic cellular features, such as a nucleus, mitochondria, chloroplasts, and other double-membraned organelles (Chapter 5). Three varieties of early

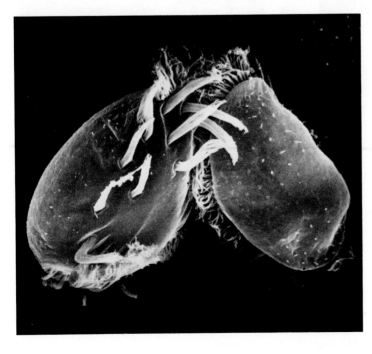

FIGURE 37-3

Sexual interactions are common among some protists. Here, two cells (*Euplotes*) are joined in a preconjugal embrace, holding each other with tufts of fused cilia, called *cirri*. This "recognition behavior" precedes the exchange of genetic material.

eukaryotes evolved and are the ancestors of what we today call the "higher" kingdoms: Fungus, Plant, and Animal (Figure 37-4). Modern protists still retain most of the features of these three ancestral types.

Protists also represent a selective pressure that influences the evolution of other organisms. As either competitors or predators, protists have a significant impact on the species that share their particular microscopic niche. As members of the food chain, protists are essential to many larger organisms that feed on them either directly or indirectly. Protists have also played a major role in human evolution, such as in the development of human sickle cell anemia. People living in areas endemic for malaria, a disease caused by a protist, have an advantage over those who have no sickle cell traits because they are more resistant to the fatal ravages of malaria. Even in malaria-free societies, such as the United States, many citizens have sickle cell anemia, but here the disease confers only disadvantages.

PROTOZOA: PROTISTS THAT RESEMBLE ANIMAL CELLS

Protozoa (*proto* = first, *zoa* = animals) are unicellular protists, most of which lack cell walls, ingest food particles (or absorb organic molecules), move about freely, and produce no multicellular spore-bearing structures. These animal-like traits distinguish protozoa from the other two groups of protists (the algae and slime molds).

Although most protozoa are harmless to humans, like bacteria and viruses, they are often feared because of a few

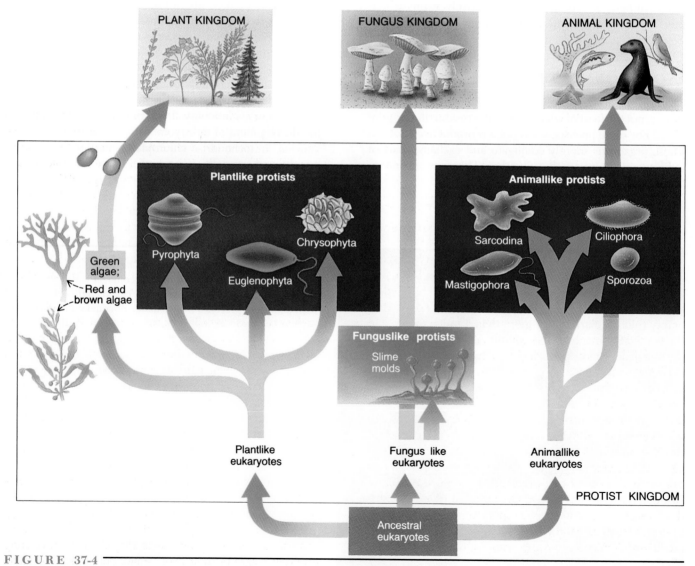

FIGURE 37-4

Evolution of the three major groups of protists. In this proposed lineage, the first eukaryotic cells (as described in Figure 36-4) gave rise to all three major groups of protists.

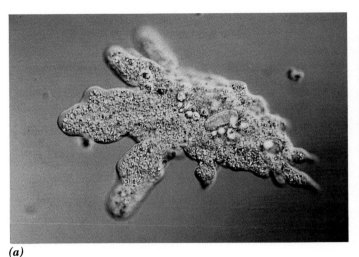

(a)

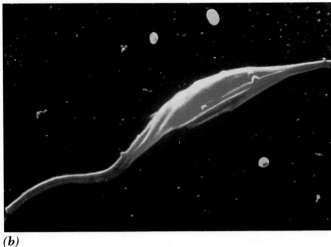

(b)

FIGURE 37-5

The delicate and the deadly. This delicate-looking *Amoeba proteus* (**a**) is a harmless resident of stagnant water, a predator that preys on other microbes that it engulfs by phagocytosis. In contrast, *Trypanosoma gambesia* (**b**) is one of the most deadly protozoa. It is an insect-borne flagellate that causes African sleeping sickness (a fatal blood infection that eventually attacks the human brain). Trypanosomes have an "undulating membrane" attached to their flagellum, forming a broad surface that helps these protozoa swim through thick body fluids.

pathogenic species (Figure 37-5). More than half the people on earth will contract some type of protozoan infection during their lifetimes. Among the most notorious protozoa are four species of *Plasmodium* that cause malaria. These mosquito-borne protozoa kill more people than does any other microorganism. Another protozoon (*Trypanosoma*) causes deadly encephalitis ("sleeping sickness"), an inflammation of the brain that develops after the protozoa enter the body through the bite of an infected insect.

Not all protozoan infections are transmitted by insects. Some spread through food and water that have been contaminated with feces. For example, *Entamoeba histolytica*, a gastrointestinal amoeba, is responsible for amoebic dysentery (bloody diarrhea) in more than a million people in the United States alone. The condition becomes fatal when the amoeba forms cysts in the brain. Another gastrointestinal infection is often contracted by hikers and campers who drink from a "pure" mountain stream. Streams are often contaminated with the *Giardia lamblia* cysts that are shed in the feces of infected muskrats and beavers, neither of which are affected by the organism. (Boiling water before drinking it kills the cysts of this protozoon and protects against the disease.) This organism may cause more infections than does any other protozoon in the United States. Just one infected person can spread the disease through a day-care center, a school, or a hospital, especially if hygienic practices (such as hand washing after diaper changes) are lax. *Giardia's* microscopic appearance is depicted in Figure 37-1.

Recently, another protozoon, *Pneumocystis,* has been thrust into the limelight. Until the early 1980s, this organism and the disease it caused were considered rare. Unfortunately, this is no longer the case. As we discussed in the previous chapter, the disease AIDS cripples the immune system, rendering its victim especially susceptible to fatal pneumonia caused by this protozoon (page 797). *Pneumocystis* infections have been the cause of death in about 63 percent of the people who have died of AIDS.

Yet, protozoa do far more good than harm. They join with other microorganisms to decompose dead organisms and recycle nutrients. Many of these unicellular heterotrophs constitute part of **zooplankton,** an important link in aquatic food chains. Feeding on microscopic algae, zooplankton convert the nutrients of algae (some of which, such as cellulose in cell walls, are nutritionally inaccessible to animals) into protozoan tissue that is digestible by virtually all consumers. Some herbivores, notably cattle, goats, and other ruminant animals, house their own supply of "zooplankton" in their digestive tracts. These protozoa (along with cellulose-digesting bacteria) enable the animal to utilize grass and other high-cellulose food sources that it would otherwise be unable to digest (page 73).

Most of the 40,000 species of protozoa live in water, moist soil, or inside other organisms. Dehydration is therefore not a threat to these protists. Yet, some protozoa actually require temporary exposure to drying conditions or other stressful environments (such as stomach acid) to trigger the events that complete their life cycles. Most of these protozoa are *polymorphic;* that is, they change form during different stages of their lives. When conditions are ideal, these protozoa exist as **trophozoites,** the actively feeding, growing form of the organism. As conditions become drier, hotter, or otherwise less favorable, the microbes transform into protective **cysts:** dehydrated, heavily encased, dormant forms of the organism. Cysts can withstand adverse conditions that would kill the trophozoite of the same organism (Figure 37-6).

Some polymorphic protozoa pass through several hosts during the completion of their life cycle. These protozoa usually change forms as they change hosts, adopting the morphology suited for life inside each animal. The malaria parasite's form in a human, for example, bears no resemblance to its form in a mosquito (Figure 37-7). The existence of many forms of one organism complicates the process of identifying and classifying these species.

Classifying protozoa is traditionally based on the organism's mode of motility (Table 37-1). Protozoa move by *cilia, flagella, or* **pseudopodia** (*pseudo* = false, *pod* =

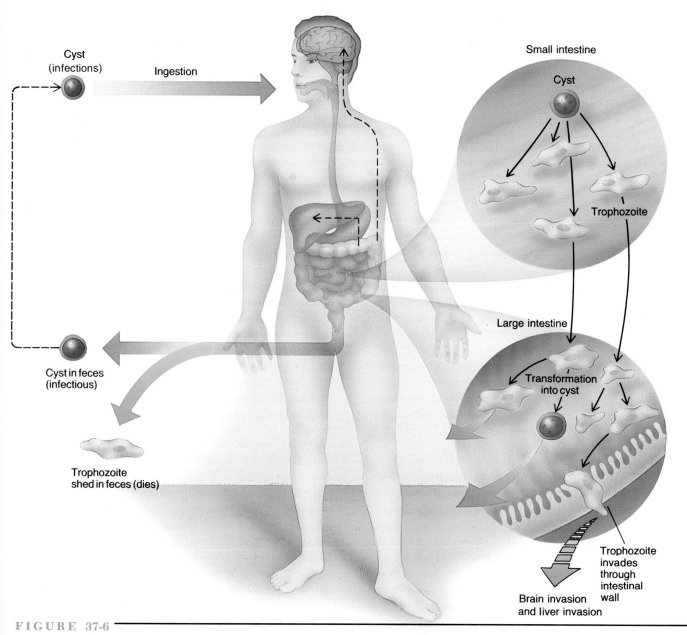

FIGURE 37-6

Example of polymorphism. The life cycle of *Entamoeba histolytica* (the gastrointestinal pathogen that causes amoebic dysentery) shows how the organism exists in two forms: a protective cyst (for surviving outside the host and the enduring trip through the caustic stomach acid), and an actively growing trophozoite (for proliferation in the hospitable intestine). Most trophozoites that remain in the colon are transformed into resistant cysts just before they are discharged in the feces. The cysts live up to 12 days outside the body and can be transferred to a new host by contaminated feces carried on fingers, flies, cockroaches, or in food and water.

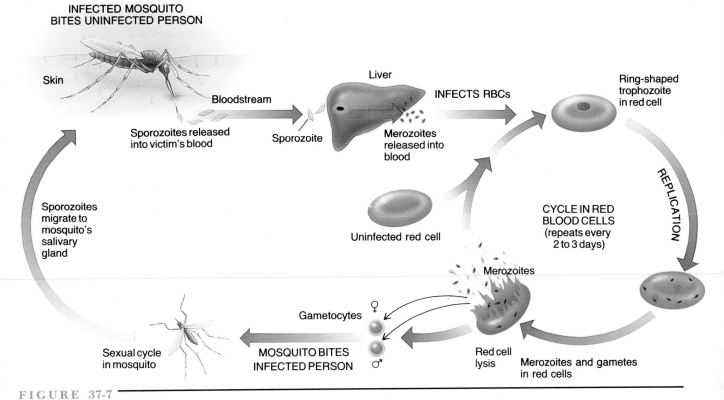

FIGURE 37-7

The world's most deadly microorganism is *Plasmodium*, the agent of malaria. *Plasmodium* kills more people each year than does any other group of microorganisms. During its life cycle, this mosquito-borne parasite alternates in form, depending on its location. Inside red blood cells, the organism proliferates as trophozoites and an intermediate form called merozoites, some of which undergo meiosis and become sex cells (gametocytes). The red blood cells burst, releasing both merozoites (which infect other red blood cells) and gametocytes. When sucked into a female *Anopheles* mosquito who bites a malaria victim, the gametocytes fertilize one another, and the resulting zygotes change into sporozoites, which migrate to the mosquito's salivary glands. The sporozoites are then injected into the mosquito's next victim.

TABLE 37-1

CLASSIFICATION OF PROTOZOA ACCORDING TO MOTILITY

Motility group	*Organelles of Locomotion*	*Characteristics*	*Example*
Mastigophora	Flagella	Heterotrophic or, in some cases, photosynthetic; excess water expelled through contractile vacuole; binary fission along long axis; no sexual reproduction	*Giardia* (agent of intestinal infection)
Ciliata	Cilia	Cilia sweep food particles into mouth; complex cell morphology; many types expel excess water through contractile vacuole; genetic transfer by conjugation; asexual reproduction by fission across long axis	*Paramecium* (predator of smaller organisms)
Sarcodina	Pseudopodia	Amoeba move by extending "fingers" of cytoplasm; feed by phagocytosis; reproduce asexually by binary fission; some species encased in shells of silica or calcium	*Amoeba* (predator of smaller organisms)
Sporozoa	None	Trophozoite stage nonmotile; intracellular parasites; complex life cycles that alternate between sexual and asexual reproductive modes; one species may require two or more different hosts to complete life cycle	*Plasmodium* (agent of malaria)

foot). Pseudopodia are fingerlike extensions of cytoplasm that flow forward from the "body" of an amoeba; the rest of the cell then follows. Even those amoebas that are encased in protective shells (Figure 37-8) can use their pseudopodia by extending them through holes in the encasement.

Pseudopodia, cilia, and flagella often double as a food-gathering instrument. *Paramecium* and some other types of ciliated protozoa use their cilia to create currents that sweep food particles into the cytostome (analogous to a "mouth"). Some flagellated protozoa use their flagella as harpoons to spear prey. The role of pseudopodia in an amoeba's phagocytic engulfment of food particles is described in Chapter 7.

ALGAE: PROTISTS THAT RESEMBLE PLANTS

Taxonomists typically have trouble agreeing on the classification of algae. In fact, they have yet to create a uniformly acceptable definition of them. Some biologists propose doing away with the "algae" label once and for all, insisting that these organisms have too little in common to belong to the same group. For our purposes, **algae** are any unicellular or simple colonial photosynthetic eukaryote. Like other definitions, this one meets with some objections, mainly because it places the multicellular algae in the Plant kingdom.

The problems with classifying algae do not undermine our recognition of their importance, however. Their most important role is as **phytoplankton,** microscopic photosynthesizers that live near the surface of seas and bodies of fresh water. Every breath you take depends on these algae; their photosynthetic activities generate 75 percent of the molecular oxygen available on earth. Virtually every animal in the sea depends either directly or indirectly on phytoplankton for food (Figure 37-9), making algae the most important members of the aquatic food chain. About half the world's organic material is produced by algae.

Perhaps the one characteristic that distinguishes algae from other protists is their photosynthetic capacity, but even some algae lack this trait. These rare nonphotosynthetic algae are heterotrophs. They either feed on particulate organic matter or, in at least one case, parasitize a living host. One parasitic genus, *Prototheca,* infects humans, causing a crippling skin infection of the feet.

Algae fall into one of six divisions, three of which contain unicellular protists and are described below. The other three divisions contain multicellular algae that are considered members of the Plant kingdom by most taxonomists. (These divisions are discussed in the following chapter.) All algae are composed of eukaryotic cells. In fact, there is no such thing as prokaryotic algae. Organisms once called "blue-green algae" are now called cyanobacteria (page 792).

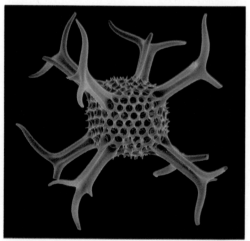

(a)

(b)

FIGURE 37-8

Shelled amoebas. (*a*) The scanning electron microscope captures the beautiful intricacy of some radiolaria. (Complexity doesn't require multicellularity.) Pseudopodia protrude through the holes in the shells of these amoeba to gather food particles. The intricate shells are composed of silica; they are literally glass. (*b*) Foramnifera enclose themselves in "snail-like" shells of calcium carbonate. Geologic upheaval raised sediments of enormous numbers of shells deposited millions of years ago, when these amoebas abounded in oceans. One striking result of this deposition is the White Cliffs of Dover in England, composed entirely of foramnifera shells.

FIGURE 37-9

The world's largest animals dine on microscopic organisms. This whale shark gulps down algae and other microscopic organisms that are filtered from huge mouthfuls of sea water.

Divisions of Algae

Chrysophytes are golden brown and yellow-green algae. They are the major producers of food for all animals living in marine waters. Although yellow-green algae are included in this group, many taxonomists place these algae in their own division, Xanthophyta.

Diatoms are golden-brown algae that are distinguished most dramatically by their intricate silica shells (Figure 37-10). Diatoms secrete their delicate "glass" coats of armor, each of which is perforated with thousands of tiny holes to allow contact between the enclosed photosynthetic cell and its environment. Diatom shells are of great economic significance, ideal for use in polishing, filtering, and insulating materials. Commercially, they are referred to as "diatomaceous earth" and are harvested from enormous geological deposits that resemble fine, white powder. These

FIGURE 37-10

Gallery of diatoms reveals the beauty of the silica shells that are secreted by these golden brown algae. The living cells that occupied them have long since died.

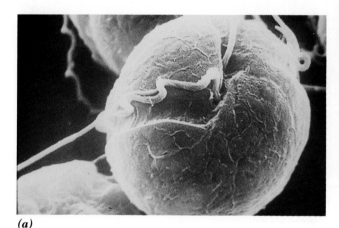

(a)

(b)

FIGURE 37-11

Intriguing but deadly. Some dinoflagellates, such as *Gonyaulax* (*a*), have killed countless fish and hundreds of people. Dinoflagellate blooms tint the ocean red (*b*). Some dinoflagellates are bioluminescent at night, flashing brightly with each breaking wave, inspiring their division name, the "fire algae."

deposits accumulated for millions of years, the diatom shells forming thick sediment layers on the ocean floor. Many of these deposits have been thrust to the surface by geological activity.

The golden color of chrysophytes is provided by accessory pigments that assist chlorophyll in photosynthesis. These pigments are precursors of the fat-soluble vitamins A and D, which are concentrated in the liver oils of fishes that dine on chrysophytes. For years, people have harvested fish liver extract as rich sources of these vitamins. In fact, children used to be given cod liver oil as a vitamin supplement.

Pyrrophyta (fire algae) are **dinoflagellates,** single-celled photosynthesizers that have two flagella. One flagellum moves the alga through its medium, and the other spins the cell on its axis. Dinoflagellates are exceeded only by diatoms in their importance to the marine food chain. Ironically, periodic dinoflagellate "blooms" are often hazardous to many members of the food chain. **Blooms** are massive growths of algae that occur when conditions are optimal for algae proliferation. During the bloom called *red tide*, for example, growth of one reddish brown dinoflagellate species is so extensive that it tints the coastal waters and inland

lakes a distinctive red color (Figure 37-11). Some red tide dinoflagellates produce a powerful nerve toxin that kills any fish that eats them. During red tide, the water's surface may be blanketed with dead fish.

In contrast, the nervous systems of shellfish are not susceptible to the toxin. In fact, to shellfish, red tide is a nutritional gold mine. The toxin becomes concentrated in the tissues of oysters, mussels, and clams. Although the shellfish are unharmed, people who eat them may consume enough toxin to cause paralysis, which typically sets in within an hour of consuming the toxin. Within 8 hours, the victim may die of *paralytic shellfish poisoning*, even if he or she receives medical treatment.

Euglenophyta are a group of unicellular algae that includes *Euglena* and similar protists (called euglenoids) that possess characteristics of both plant cells and animal cells. Exposure to light induces their plantlike property, the production of chlorophyll *a* and *b* (supplemented with carotenoids). The typical euglenoid (Figure 37-2) retains its flagella in its alga form; its movement is directed by its response to light. An "eyespot" (photoreceptor) at the base of the flagellum senses the direction of the light source,

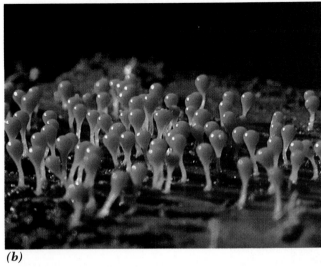

(a)   (b)

FIGURE 37-12

Plasmodial slime molds. (*a*) The protozoalike plasmodium of *Physarium* is an enormous multinu-cleated "cell" that consumes dead organic matter. (*b*) Spore-producing structures of slime molds resemble those of fungi.

enabling the organism to swim toward or away from the light to reach the optimal brightness.

Most euglenoids are freshwater protists, although a few are found in the gastrointestinal tracts of animals. Like some ciliated protozoa, euglenoids eliminate excess intercellular water by collecting it in a contractile vacuole and periodically squeezing out its contents.

SLIME MOLDS: PROTISTS THAT RESEMBLE FUNGI

Although you may never guess it from their name, **slime molds** are distinctly beautiful protists. In their sexually mature stage, these organisms are intricate and attractive, displaying delicate reproductive structures similar to those of molds (a true mold is a fungus, as discussed in the following section).

Slime molds are either *cellular* or *plasmodial*. **Cellular slime molds** exist as microscopic amoebalike cells when food and water are plentiful. These slime molds patrol their habitat for bacteria, spores, and organic debris, which they engulf and digest, rapidly multiplying as a result. When their food is depleted, the "amoebas" release one of the most developmentally important chemicals, *cyclic AMP*. This substance is used by virtually all organisms to direct development or to govern gene expression. Hundreds of amoebas migrate up the cAMP concentration gradient to the point of highest concentration. Here, they aggregate into a single slime-covered "slug" that crawls for a while (helping to disperse the species) and then transforms itself

into a funguslike structure. In this form, the slime mold produces a **fruiting body,** a spore-producing structure that extends upward in an elevated position. When the spores are released, they are carried by the wind or on the surface of animals to new locations. If conditions are suitable, the spores germinate, each forming another amoebalike cell, and the cycle repeats.

Plasmodial (*acellular*) **slime molds** also have a funguslike spore-producing stage in their life cycle. These organisms have no migrating "amoebas," however. Instead, they produce a **plasmodium,** a huge multinucleated "cell" that feeds on dead organic matter. A single plasmodium can grow large enough to cover an entire log (Figure 37-12.) This giant, growing mass then forms fruiting bodies, which release spores that germinate into gametes, provided they can reach a location with suitable conditions. Fertilization between gametes yields a zygote that develops into another multinucleated plasmodium.

The Protist kingdom contains eukaryotic organisms that have few other taxonomic properties in common. Some protists resemble single-celled animals; others resemble single celled plants; and still others pass through a multicellular stage that resembles fungi. All protists are probably very much like the evolutionary forerunners of the "higher" kingdoms they resemble. Protists contribute to the health of the biosphere by decomposing dead organisms and recycling nutrients and by providing critical links in aquatic food chains. (see CTQ #2.)

THE FUNGUS KINGDOM

In the mid-1800s, thousands of farmers in Ireland began to notice a terrifying transformation. Potatoes, their major source of nutrition, were turning black. The Irish helplessly watched their potatoes rot. The following spring, the plants themselves were affected, and the food crop of an entire nation failed. The disaster of the potato blight continued for 14 years, causing the deaths of more than a million Irish people due to starvation. Today, we know the killer to be a member of the Fungus kingdom.

The Irish potato blight is one of several large-scale disasters that have been inflicted on the human population by **fungi.** The vast majority of fungi are beneficial, however, not just to people but to the biosphere as a whole. Here are just a few of their contributions:

- decomposition of dead organisms and recycling of the nutrients;
- soil formation from solid rock;
- formation of *mycorrhizae* ("fungal roots"), filaments that grow in intimate association with plant roots, increasing the plant's absorption of nutrients and water (page 387);
- production of foods, such as cheese, mushrooms, and soy sauce;
- fermentation of sugar to ethyl alcohol, generating beer, wine, and spirits;
- synthesis of valuable chemicals, such as industrial solvents, the chemicals used in the formation of plastics, and compounds used for soap production;
- manufacture of important therapeutic compounds, most notably penicillin and many other antibiotics that continue to save lives that would otherwise be lost to bacterial infections (Figure 37-13).

OVERVIEW OF THE KINGDOM

Because of their cell walls, fungi were once considered members of the Plant kingdom. A lack of photosynthesis and a heterotrophic nutritional strategy disqualifies them as plants, however, and their eukaryotic cellular structure eliminates them as candidates for the Monera kingdom. In 1969, a new and separate kingdom was recognized exclusively for the fungi.

Some members of the Fungus kingdom, notably the **yeast,** are unicellular, forming colonies similar to those of bacteria. **Filamentous fungi** constitute the multicellular members of the kingdom; these organisms are composed mostly of living threads that grow by division of cells at their tips. These filaments, called *hyphae,* are composed of a linear group of cells. The filamentous fungi, most of which

FIGURE 37-13

Destroyer and savior. *Penicillium,* the blue-green mold you find on old oranges, bread, and cheddar cheese, spoils millions of tons of food each year. Yet, the organism also produces the life-saving antibiotic penicillin. In this photograph, the mold's asexual spores are clearly visible at the tips of special "brushlike" spore-forming structures.

are called **molds,** typically grow as fluffy masses. Molds are highly efficient exploiters of available nutrients. During peak growth, a mold colony can grow more than half a mile of new hyphae in a single day.

The members of one group of filamentous fungi are not considered molds, however. Instead, these organisms are called macrofungi, signifying the large size of their fleshy sexual structures. For example, a mushroom is so large and elaborate that it is often mistaken for the whole organism. Yet, most of the fungus grows as an unseen filamentous mass under the ground or in the tissues of a tree, with only its large reproductive structures showing.

⚠ As you can see, fungi range in size from microscopic single-celled yeasts to large, fleshy macrofungi. There are thousands of species in between, including the fungi that collaborate with algae to form associations called **lichens.** Each member of the lichen contributes to the ability of these "composite organisms" to thrive in a wide range of habitats. In dry environments, or those that are poor in organic nutrients, for example, the lichen's photosynthetic alga uses light energy to generate organic nutrients from inorganic compounds, while the fungal filaments gather and conserve what little water is available. Together, the alga–fungus team readily grows in conditions in which neither the fungus nor the alga could survive alone (Figure 37-14). In some harsh environments, lichens support entire food chains, sustaining such large consumers as the reindeer caribou of frozen northern Alaska.

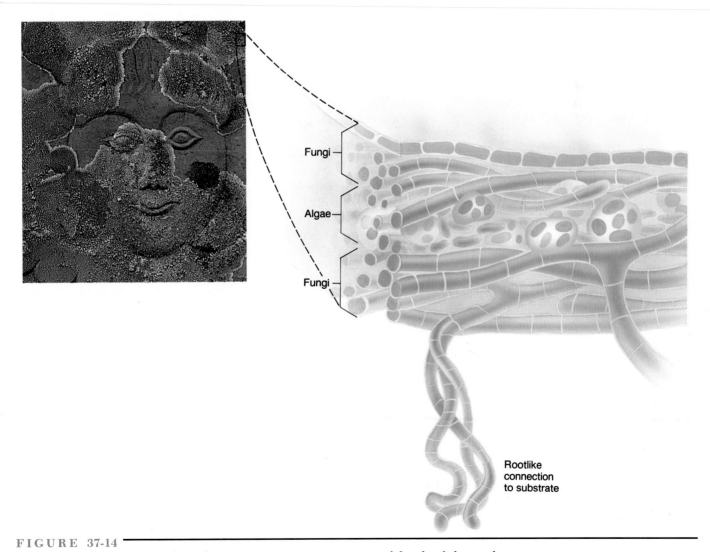

Fungi

Algae

Fungi

Rootlike connection to substrate

FIGURE 37-14

Joining forces to live where few other organisms can. Growing on solid rock, a lichen exploits a niche so inhospitable that competition comes only from other lichens. Acids excreted by the lichen, and penetration of the stone by fungal hyphae, contribute to the eventual breakdown of rock to soil. The biological complex that comprises a lichen is both fungal and algal. The lichen covering and deteriorating this face carved into marble is typical. Its alga is sandwiched inside two outer layers of fungal filaments.

FUNGAL NUTRITION

Fungi are heterotrophs, most of the kingdom's members being **saprobes** that obtain their nutrients by decomposing dead organisms. Yet, a few types of fungi don't wait until an organism is dead before they start consuming it. Most of these parasitic fungi are pathogenic, causing diseases in plants or animals. In fact, fungal diseases constitute some of the most common and persistent human diseases that continue to challenge modern medicine. More than 20 percent of the world's population suffers from fungal diseases, ranging from irritating maladies of the skin and mucous membranes (e.g., athlete's foot, ringworm, and vaginal yeast infections) to life-threatening systemic infections (e.g., histoplasmosis and valley fever). Systemic fungal infections are especially destructive, invading the body from their entry site in the respiratory tract and affecting virtually any organ of the body, including the brain.

Plant pathogens can also have disruptive consequences, both to the plant itself and to organisms that depend on the plant for survival. In addition to the disastrous Irish potato blight, for example, rust and smut fungi continue to reduce human food supplies by attacking agricultural grains. The devastating Dutch elm disease that has downed so many beautiful trees is also caused by a fungus that is introduced into the tree by the bark beetle.

FUNGAL REPRODUCTION

Except for mushrooms and other macrofungi, most members of the Fungus kingdom resort to sexual reproduction infrequently, relying primarily on asexual reproduction to increase their numbers. Yeast, for example, form small "buds" that enlarge and finally break away from the genetically identical parent. Asexual reproduction is more varied among filamentous fungi. Fragments of hyphae may break away from the parent mold and grow mitotically, forming new individuals. Another reproductive strategy is the production of asexual spores, which are then released from sporulating structures and germinate into actively growing hyphae.

When fungi sexually reproduce, the diploid stage is short-lived. The zygote produced by fertilization immediately undergoes meiosis to form haploid spores. All the cells in many types of fungi are therefore haploid, having descended from one haploid spore.

CLASSES OF TERRESTRIAL FUNGI

Fungi have a number of sexual reproductive strategies, most of which are characterized by the formation of a fruiting body that bears the sexual spores. The type of sexual spore produced provides a common criterion for categorizing the fungi. Four of these classes (Table 37-2) are discussed in the following sections. Another group of fungi, commonly called "water molds," have generated controversy over their classification as fungi,[2] and therefore are not included in the table, even though they contain important animal and plant pathogens, including the fungus responsible for the Irish potato blight.

Zygomycetes

The black mold growing on an old loaf of bread typifies the **zygomycetes.** These organisms begin as a white, cottony mass of *nonseptate* hyphae, continuous tubes of cytoplasm with no crosswalls separating nuclei of adjacent cells. A day later, the filaments produce sporulating structures that resemble balloons on the end of sticks, held upright by a

---

[2] Water molds have cell walls atypical of fungi. This has lead some taxonomists to consider them as members of the Protist kingdom.

TABLE 37-2
THE FOUR CLASSES OF TERRESTRIAL FUNGI

Class	Sexual Spores	Asexual Spores	Hyphae	Representative Genera
Zygomycetes	Zygospores	Sporangiospores	Nonseptate[a]	*Rhizopus* (bread mold)
Ascomycetes	Ascospores	Many different types (conidia, arthrospores, etc.)	Septate	*Saccharomyces* (brewer's yeast) *Penicillium* (antibiotic producer), several genera of morels and truffles
Basidiomycetes	Basidiospores	Virtually none	Septate	*Agaricus* (supermarket mushrooms), plus hundreds of other mushroom genera; rusts and smuts (plant pathogens)
Deuteromycetes	None	Conidia; arthrospores, etc.	Septate	*Candida* (causes vaginitis and other human infections); many genera of pathogenic fungi that cause serious diseases of people

[a] Nonseptate hyphae are continuous tubes of cytoplasm, whereas septate hyphae have crosswalls that divide the adjacent cells from one another.

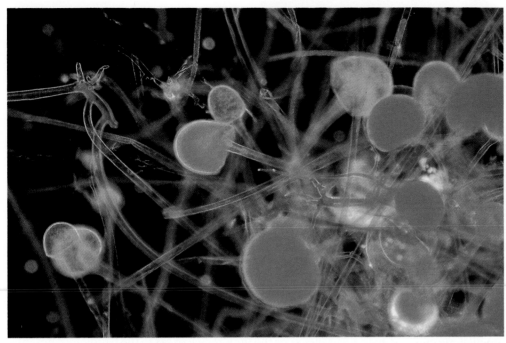

FIGURE 37-15

Packed with sporangiospores, the sporangia of this zygomycete (*Rhizopus*) are ready to burst open and discharge their cargo of reproductive spores. The nonseptate hyphae lack partitions (septa) between adjacent nuclei.

cluster of *rhizoids* that resemble roots. Each "balloon" is really a sac, or *sporangium,* that is filled with black asexual **sporangiospores** (Figure 37-15). It is the production of these spores that darkens the mold as it matures. When sporangia rupture, millions of haploid spores are released, some blowing to new locations by the slightest breeze. Spores that settle in favorable conditions germinate into actively growing hyphae, establishing new mold colonies.

Although zygomycetes produce no sexual fruiting bodies, they do reproduce sexually. When hyphae of opposite mating types grow in contact with each other, fertilization generates a temporary diploid state. The diploid cell forms a dark warty coat around itself, as it develops into a **zygospore.** The zygospore then undergoes meiosis (often after months of dormancy), germinates, and develops into a typical sporangium containing haploid sporangiospores.

Zygomycetes not only spoil food, they can also cause serious, often fatal, human diseases, especially in people with compromised natural defenses. Fortunately, several human adversaries are also on the fungal "hit list," notably destructive insects and parasites. At least 50 species of fungi, most of them zygomycetes, are quite efficient at trapping or snaring these animal pests for dinner. Victims include small worms (nematodes) that are trapped when they crawl through special loop-shaped hyphae of the fungus *Arthrobotrys.* The loop clamps shut when stimulated, snaring the worm, which is then digested by fungal enzymes and absorbed into the hyphae. Houseflies, grasshoppers, and other insects also fall prey to "carnivorous" zygomycetes. Scientists are now evaluating the feasibility of using such predatory fungi as biological insecticides against many types of insect pests, from aphids that destroy millions of dollars in citrus fruit each year to cockroaches that invade the home.

Ascomycetes

Ascomycetes are often referred to as "sac fungi" because they house their sexual **ascospores** in a sac, called an *ascus* (plural = *asci*). In unicellular ascomycetes, such as the common yeast *Saccharomyces cerevisiae* (brewer's and baker's yeast), the cell itself becomes an ascus filled with four ascospores. But most ascomycetes are multicellular and filamentous, some forming very complex fruiting bodies, called ascocarps, that house the spore-filled asci (Figure 37-16). The linear alignment of ascospores in *Neurospora crassa* has allowed geneticists to study genetic crossovers and meiotic distribution of genes. The speed with which the fungus sexually reproduces helped make *Neurospora* a favorite of research geneticists, who sometimes refer to it as "the microbiological fruit fly." Investigators merely have to observe the distribution of spores with particular traits (such as color) in an ascus to obtain results of genetic crosses.

To others, the ascomycetes seem more like assassins than assistants. One species caused the blight that has virtually eliminated chestnut trees from the American land-

FIGURE 37-16
An ascomycete. The inner surface of these cup fungi are lined with thousands of sacs (asci) that contain sexual ascospores that are released when the sacs rupture.

scape. Other ascomycetes have downed many other stately creatures, including millions of elms that succumbed to Dutch elm disease. Humans also fall victim to ascomycetes, which include the agents of such dangerous diseases as histoplasmosis (a systemic infection that initially infects the lungs) and some forms of common "ringworm" infection.

Basidiomycetes

Also known as "club fungi," members of the **basidiomycetes** class are not only the most familiar fungi (Figure 37-17), they are also the most sexually active. The size and complexity of their sexual reproductive structures are paralleled by the importance of sex to the basidiomycetes. For example, mushrooms and puffballs are "fleshy" structures of compactly intertwined hyphae. The hyphae that constitute the partitions ("gills") in the cap of a mushroom form millions of club-shaped sporulating structures called *basidia*. Each basidium bears four **basidiospores,** which are formed by the process shown in Figure 37-18. When mature spores escape the basidiocarp (the fruiting body), they are carried by the wind or animals to new locations, thereby dispersing the species.

FIGURE 37-17
Basidiomycetes: fleshy fungi. (*a*) These leaf fungi are basidiomycetes that decompose dead organic matter. (*b*) Sexual basidiospores are often so abundant that they form a smoky discharge, as from this puffball.

(a)   *(b)*

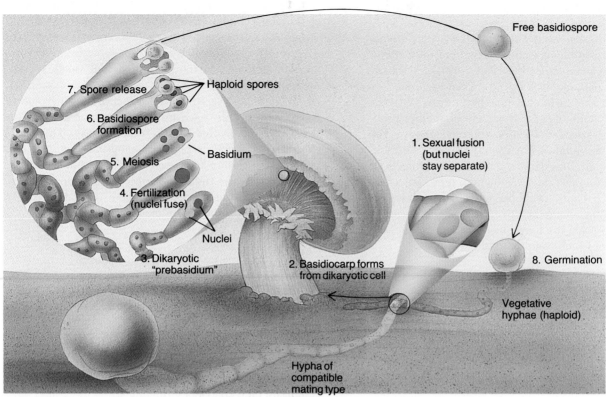

Free basidiospore

7. Spore release
Haploid spores
6. Basidiospore
   formation
5. Meiosis
Basidium
4. Fertilization
   (nuclei fuse)
Nuclei
3. Dikaryotic
   "prebasidium"
2. Basidiocarp forms
   from dikaryotic cell
Hypha of
compatible
mating type

1. Sexual fusion
   (but nuclei
   stay separate)

8. Germination

Vegetative
hyphae (haploid)

FIGURE 37-18

The mushroom's sexual cycle. A mushroom's fruiting bodies (basidiocarps) develop following the fusion of sexually compatible hyphae, creating a *dikaryotic* cell (two nuclei per cell). Proliferation of this cell gives rise to a basidiocarp that is composed of dikaryotic cells. The two nuclei fuse during fertilization in the gills of the basidiocarp, forming a single diploid nucleus. The resulting diploid cell (the basidium) immediately undergoes meiosis, forming four haploid nuclei, each of which ends up in a basidiospore. The cycle begins again when hyphae from a germinated basidiospore come in contact with hyphae from another sexually compatible basidiospore.

The fleshiness of basidiocarps is often accompanied by a delicious taste, but a few are also endowed with deadly toxins for which there are no antidotes. Distinguishing between safe and poisonous varieties sometimes requires microscopic examination. In addition to poisonous mushrooms, detrimental basidiomycetes include the food-robbing plant pathogens wheat rust and smut, as well as tree-killing bracket fungi.

Deuteromycetes

True **deuteromycetes** live their lives without sex. By definition, these are fungi in which sexual reproduction has not (yet) been discovered. Often referred to as "the imperfect fungi," deuteromycetes lack a *perfect stage,* as the sexual cycle is often called. Some species of deuteromycetes probably lack the genetic mechanism needed for sexual reproduction. Others will remain in this class only until scientists discover their elusive sexual stage. Each year, more fungi

are removed from the ranks of the deuteromycetes and assigned to the ascomycetes or basidiomycetes, according to the type of sexual spore produced. The genus *Penicillium,* for example, is traditionally presented as an important example of a deuteromycete. Yet *P. chrysogenum* (the species that produces the antibiotic penicillin) and *P. roqueforti* (the blue mold that imparts the colorful ribbons that flavor Roquefort and blue cheese) were both recently discovered to be ascospore-producing, sexually active fungi.

The most common deuteromycete is *Aspergillus,* perhaps the most ubiquitous organism in the world. It has been found even in the stratosphere, at an altitude of 100 miles. To most people, *Aspergillus* is a pest that spoils food and accelerates the deterioration of useful products. For a few people, it can be deadly. A person with inadequate immunity often cannot combat this organism. Since exposure to the spores of this fungus is virtually unavoidable, people with compromised immune systems often die of overwhelming *Aspergillus* infections. Aspergillus also compounds the misery of hay-fever sufferers, many of whom are allergic to the fungus's spores.

Another troublesome ascomycete is the yeast, *Candida albicans,* which commonly causes vaginal yeast infections, diaper rash, and an oral infection called "thrush." Ordinarily, the yeast is held in check by the normal bacterial flora that resides on the surface of healthy humans. When a person receives antibiotics for treatment of another problem, however, the resident bacteria are killed, allowing the yeast (which is not inhibited by bacterial-specific antibiotics) to grow to enormous numbers, causing local infections in the mouth or vagina. *Candida* infections represent a fatal threat to AIDS patients, alcoholics, leukemia patients, and people undergoing chemotherapy for treatment of cancer. The yeast can invade the lungs, bloodstream, and central nervous system of such compromised individuals.

EVOLUTIONARY TRENDS

Although Figure 37-4 represented fungi as evolving in a single lineage from a eukaryotic ancestor, many biologists believe that the septate and nonseptate fungi actually evolved as separate lines from a pre-eukaryotic cell, some prokaryotic ancestor. If this is true, fungi would have been among the first eukaryotes on earth; indeed, some fossil filaments from precambrian rock formations support this

contention. The similarities that exist between modern ascomycetes and basidiomycetes (such as the formation of dikaryotic cells) suggests that their common ancestry is the most recent of all fungi, the two diverging from the same line that produced the more primitive (but usually septate) deuteromycetes.

▶ Biologists do agree on one aspect of fungal evolution: Fungi were not the ancestors of plants, nor did they descend from plants that lost their chloroplasts. The earlier classification of fungi as plants was a mistake that has left its legacy in our terminology, especially among medically important fungi. The fungi that cause "ringworm" and athlete's foot, for example, are still called "dermatophytes," a word literally meaning "skin plant."

The yeasts and filamentous fungi that compose the Fungus kingdom superficially resemble plants in that they have a cell wall. But fungi are non-photosynthetic; they have chitin instead of cellulose cell walls; and they produce no highly differentiated structures, such as roots or leaves. Most fungi are saprobes, absorbing dead organic matter; therein lies their greatest contribution. Other fungi have killed millions of people by causing disease or by destroying human food resources. (See CTQ #4.)

REEXAMINING THE THEMES

Acquiring and Using Energy

☀ The protists are second only to the monerans in the diversity of energy-acquiring strategies. The Protist kingdom consists of heterotrophic, animal-like predators, heterotrophic decomposers of dead organisms and wastes, and photosynthesizers. The photosynthetic protists are the energy providers in all aquatic habitats.

Unity within Diversity

⬟ Broad differences distinguish members of the protist kingdom from one another. Some are photosynthetic, others are heterotrophic. Some have cell walls, others don't. Some are stationary, others are actively motile. Beneath these differences, however, lie fundamental properties shared by all protists: mitochondria, endoplasmic reticula,

and other organelles that characterize eukaryotic cells. Fungi possess these same organelles, as do members of the other two higher kingdoms: plants and animals. Such unity exists because all eukaryotic organisms retained many of the features they inherited from the ancient ancestor that launched the eukaryotic lineage.

Evolution and Adaptation

▶ 9The similarity of the protists to individual cells of the higher kingdoms suggests how animals, plants, and fungi acquired the adaptations that characterize them. Direct ancestors of each kingdom probably resembled some of the protists alive today. Animals, for example, evolved from an ancient eukaryotic cell that may have been similar to today's flagellated protozoa.

SYNOPSIS

Members of the Protist kingdom. Protists have few obvious properties in common; all are eukaryotic, but some are unicellular, and others are relatively simple multicellular complexes. Protozoa are "animal-like"; algae are "plant-like"; and slime molds are "fungus-like."

Protozoa are protists that resemble animal cells. Some protozoa cause serious human diseases, such as amoebic dysentery, giardiasis, malaria, and encephalitis. Most members of this group contribute immeasurably to the biosphere, however. As zooplankton, protozoa are critical links in food chains, forming a nutritional bridge between the producers (algae) and the animals that cannot digest some of the components of algae. Protozoa also help decompose and recycle nutrients. Many terrestrial livestock animals require cellulose-digesting protozoa in their rumens before they can harvest the energy and nutrients in their grassy diets.

Algae are indispensable to the biosphere. Algae produce most of the world's oxygen and much of the world's food, contributions that greatly outweigh the few detrimental activities associated with algae blooms.

Slime molds are protists that resemble fungi. Cellular slime molds have amoebalike vegetative cells that aggregate and produce a sporulating structure similar to those of fungi. Plasmodial slime molds form a plasmodium, a single giant "cell" with hundreds of nuclei. The plasmodium transforms into spore-filled reproductive fruiting bodies.

The Fungus kingdom consists of yeasts and filamentous fungi. As decomposers, fungi help sustain the earth's many nutrient cycles. In addition to fertilizing soil by decomposing dead organisms, other fungi, those of lichens, actually help make soil out of rock. Lichens are also the fundamental producers in some polar food chains. Fungi are sources of valuable foods, medicines, and other resources for humans. They also cause mild to serious diseases in about a fifth of the world's human population, however, and spoil food. They also cause diseases in crop plants.

Key Terms

protist (p. 810)
protozoa (p. 812)
zooplankton (p. 813)
trophozoite (p. 813)
cyst (p. 814)
pseudopodia (p. 814)
algae (p. 816)
phytoplankton (p. 816)
diatom (p. 817)
dinoflagellate (p. 818)
bloom (p. 818)

red tide (p. 818)
slime mold (p. 819)
cellular slime mold (p. 819)
fruiting body (p. 819)
plasmodial slime mold (p. 819)
plasmodium (p. 819)
yeast (p. 820)
filamentous fungi (p. 820)
hyphae (p. 820)
mold (p. 821)
macrofungi (p. 821)

lichen (p. 821)
saprobe (p. 822)
zygomycete (p. 822)
sporangiospore (p. 823)
zygospore (p. 823)
ascomycete (p. 823)
ascospore (p. 823)
basidiomycete (p. 824)
basidiospore (p. 824)
deuteromycete (p. 825)

Review Questions

1. Explain why both the two-kingdom and three-kingdom systems were taxonomic flops.

2. Determine which of the three protist groups each of these organisms belongs to: *Euglena; Gonyaulax* (the agent of red tide); *Trichomonas vaginalis* (a flagellated heterotroph that causes vaginal infections); and *Trichonympha* (a flagellated, cellulose-digesting heterotroph that lives in the gut of termites).

3. Describe four ways humans benefit from algae or their products.

4. Describe at least one beneficial and one detrimental activity of an organism in each of the following groups: flagellated protozoa; amoeboid protozoa; basidiomycetes; and ascomycetes.

5. How can each of the modes of motility used by protozoa also provide a means of capturing food?

6. Provide four distinctions between true algae and the "blue-green algae" (cyanobacteria) discussed in Chapter 36.

7. Explain the two stages of fertilization that take place in basidiomycetes, accounting for the fact that all the cells in a basidiocarp (e.g., a mushroom) are dikaryotic (containing two nuclei per cell). Describe how the final fertilization stage leads to the production of basidiospores.

8. Discuss the mutual benefits for each organism in the following biological "teams": a "photosynthetic" hydra; mycorrhizae and a pine tree; a lichen; a cow and the cellulose-digesting protozoa in its digestive tract.

Critical Thinking Questions

1. If you were trying to determine in which kingdom to place a newly discovered single-celled organism, what characteristics of the organism would you need to know?

2. Human pollution of coastal sea water and oceans is threatening the growth of algae and other protists that thrive there. Explain how the loss of these microscopic organisms would affect the population of whales; the global economic community; the atmosphere; and the ability of the biosphere to support animal life, including humans.

3. Some biologists classify slime molds as fungi rather than protists. Defend the position of these biologists. Now take the opposite position: What properties warrant placing slime molds in the Protist kingdom?

4. Explorers returning to England from South America fascinated their British sponsors with a novel type of flower they had brought home: the orchid. Unfortunately, none of the orchids would grow in English conservatories. Eventually, it was discovered that if some of the soil in the orchid's original pots were mixed with the European soil, the orchids grew as well as they did in South America. Considering what you've learned about fungal associations with plants, what was lacking in the English soil that was supplied by the soil in which the orchids had already grown?

Additional Readings

Baker, D. 1985. Giardia! *Nat. Wildlife*, August–September. (Introductory)

Botstein, D. and G. Fink. 1988. Yeast: An experimental organism for modern biology. *Science* 240: 1440. (Advanced)

Kessel, R., and C. Shih. 1976. *Scanning electron microscopy in biology: A student's atlas on biological organization.* New York/Berlin: Springer-Verlag. (Introductory–Intermediate)

McKane, L., and J. Kandel. 1985. *Microbiology: essentials and applications.* New York: McGraw-Hill. (Intermediate)

Margulis, L., and K. Schwartz. 1982. *Five kingdoms: An illustrated guide to the phyla.* San Francisco:W. H. Freeman. (Intermediate)

Vidal, G. 1984. The oldest living cells. *Sci. Amer.* 250:48–57. (Intermediate–Advanced)

The Plant Kingdom

STEPS
TO
DISCOVERY
Distinguishing Plant Species: Where Should the Dividing Lines Be Drawn?

MAJOR EVOLUTIONARY TRENDS

Conquering Dry Land

From Prominent Gametophyte to Prominent Sporophyte

OVERVIEW OF THE PLANT KINGDOM

ALGAE: THE FIRST PLANTS

Green Algae: Ancestors of Plants

Brown Algae: Multicellular Giants

Red Algae: Reaching the Greatest Depths

BRYOPHYTES: PLANTS WITHOUT VESSELS

TRACHEOPHYTES: PLANTS WITH FLUID-CONDUCTING VESSELS

Lower (Seedless) Vascular Plants

Seed Plants

BIOLINE

The Exceptions: Plants that Don't Photosynthesize

THE HUMAN PERSPECTIVE

Brown Algae: Natural Underwater Factories

Distinguishing Plant Species:
Where Should the Dividing Lines
Be Drawn?

Ask several biologists "What is a species?" and you are likely to get different answers, depending on the person's specialty. A zoologist distinguishes animal species solely on the basis of whether individuals can successfully interbreed and produce fertile offspring. Unfortunately, this approach, called the *biological species concept,* works poorly for distinguishing plant species since members of some species can hybridize with individuals of other closely related, but morphologically distinct, species. Botanists must rely on different criteria for determining if two plants belong to the

same species. Most botanists use genetically determined morphological traits to define individual plant species. There are two problems with this approach, however. First, are differences in morphology among plants living in different habitats genetically determined or are they induced by differences in environmental conditions? And second, are the genetic differences great enough to justify classifying the two types as separate species?

In the 1920s, Gote Turesson, a Swedish botanist, experimentally attacked the first of these questions by collect-

Plants in the same species often look different when grown in different habitats. Are these distinctions due to genetic changes

ing different individuals of the same species from various habitats in Europe and growing them in uniform conditions in a test garden near Akarp, Sweden. Individuals from environments with more extreme conditions were shorter than were members of the same species from environments with more hospitable conditions. If these differences in height were solely the effect of environmental conditions on gene expression, all plants should be similar when grown in ideal conditions. For example, *Plantago maritima* grows tall and robust in the coastal marshes of Sweden, but it appears dwarfish on the harsh, exposed sea cliffs of the Faeroe Islands.

When grown side by side in Turesson's garden, the two plants did not grow equally tall. They still showed significant height differences, although they were not as extreme as were the differences observed in their native environments. Turesson concluded that these distinctions were due at least in part to *genetic* adaptation to different environments, not just to differences in phenotypic expression of genetically identical plants under different conditions. He coined the term *ecotype* to distinguish genetic varieties within a single species.

Turesson's conclusions were supported by three Stanford scientists, Jens Clausen, David Keck, and William Heisey, who conducted reciprocal transplants of specimens of the perennial herb *Achilla lanulosa* (yarrow). The scientists collected seeds from yarrow specimens growing in dramatically different locations in North America, from the California coasts to the frigid mountain timberlines. They transplanted these seeds in three experimental plots, one representing low coastal habitats (at Stanford), another simulating the moderately high coniferous forest zone (at Mather), and the third simulating the harsh alpine timberline areas (Timberline).

The largest of these plants, an 84-centimeter-tall eco-type from lower altitudes, fared poorly when grown at the Timberline plot. Fewer than half of these plants survived, and those that did grew to an average height of only 15 centimeters (compared to the ecotype native to the timberline habitat, which grew to 24 centimeters). In the Stanford coastal plot, however, the timberline ecotype reached only 21 centimeters, in spite of the more hospitable conditions. This plant actually grew better in its harsh, frigid native alpine area than it did in the "ideal" conditions of coastal California. In fact, the plants from all the different regions sampled by Clausen and his co-workers had adapted to their native habitat. These differences were genetically stable and could not be "undone" by moving plants from hostile habitats to less extreme environments.

These studies clearly illustrated that as a plant species becomes widespread, it develops genetic adaptations to the local conditions. In other words, within a particular species, significant genetic variation exists, some so great that different varieties are no longer capable of interbreeding with other varieties. Although we call these variants *ecotypes*, where do we draw the line and say that the differences are great enough to constitute two separate species? The decision is clearly a judgment call. Had plant taxonomists chosen to draw more lines, the number of plant species would be much greater than the current 400,000, perhaps even approaching that of animals (1.3 million species). In spite of such uncertainties, the system currently used for distinguishing plant species has a clear advantage: It groups together as the same species those plants that are most similar in their evolutionary history.

or to environmental effects? Reciprocal growth experiments proved it to be a combination of both.

A fire rages through a forest. Many animals escape the flames, but the plants cannot flee. Anchored to the ground, they must stand against many life-threatening hazards: violent winds, floods, sleet, herbivores, and fire. Despite what seems to be an enormous handicap, plants not only manage to survive in virtually all habitats, they have become the most prominent form of life on earth.

One of the major reasons for the success of plants is the advantage of photosynthesis. Of the three most diverse kingdoms of organisms, only members of the Plant kingdom are autotrophs. Plants manufacture their own food, through photosynthesis, using sunlight to combine carbon dioxide from the air with water from the soil to form energy-rich sugars (Chapter 9).

With more than 400,000 species, the Plant kingdom is second only to the Animal kingdom in diversity. Plants are structurally simpler than are most animals, but they are more complex than their appearance suggests for many plant adaptations are biochemical. Plants have evolved a battery of chemicals to defend themselves against assaults by animals, fungi, microbes, and even other plants (Figure 38-1). If taxonomists distinguished plant species on the basis of chemical differences in addition to structural differences, the magnitude of diversity in the Plant kingdom would skyrocket.

▼ ▼ ▼

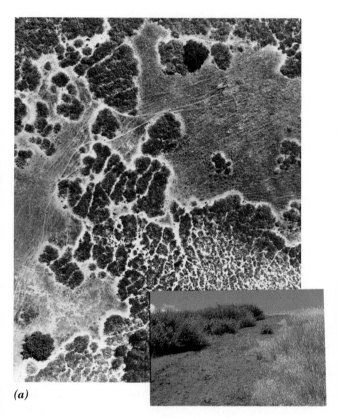

(a)

(b)

FIGURE 38-1

Biochemical warfare. An arsenal of chemicals not only helps protect some plants from foragers, it also enables plants to outcompete rivals for water, space, and sunlight. *(a)* Volatile oils released from the leaves of sage bushes *(Salvia)* help keep the area around the bushes free of competing grasses. These strong-scented oils inhibit seed germination and stunt grass growth as far as 10 meters (30 feet) from the established shrubs. *(b)* The roots of some oaks leak chemicals that inhibit the growth of other oak roots, preventing crowding. The pattern of distribution of oaks in this valley is primarily the result of this form of "crowd control."

MAJOR EVOLUTIONARY TRENDS

Because ancient plants left so few fossils, it is difficult to document the history of plant evolution. Taxonomists and evolutionary botanists must therefore rely on similarities in biochemistry (mainly photosynthetic pigments), cell structure, growth patterns, and gametes to interpret the course of plant evolution. This evidence suggests that plants evolved from a line of filamentous green algae that invaded land over 400 million years ago, during the Silurian period of the Paleozoic Era. Today, only a few plants live in water. Except for the brown, red, and green algae (which some biologists place in the Protist kingdom), most members of the Plant kingdom are terrestrial.

The demands of living on dry land led to a change in the growth pattern of early plants, from single rows of cells that form filaments to sheets of cells that form ground-hugging mats. Lying flat against the soggy ground, water could diffuse into the cells of these early land plants, quickly replacing water used in metabolism or lost to the atmosphere as vapor. Unobstructed light, abundant carbon dioxide, and a virtual lack of competition allowed these early land plants to flourish *as long as they remained close to a continuous supply of water.* Plants were confined to edges of beaches, ponds, or streams until they developed a means of preventing fatal dehydration on dryer land.

CONQUERING DRY LAND

Four types of adaptations enabled plants to make the transition to dry land

1. *Control of water loss.* The above-ground parts of most modern plants are coated with a waxy cuticle that retards evaporation. In addition, a plant's outer surfaces are perforated with stomates (page 368), tiny pores that open when the water supply is adequate, allowing the carbon dioxide needed for photosynthesis to enter. When dehydration threatens the plant, stomates close, reducing water vapor loss.

2. *Vascular tissues.* Plants that lack vascular tissues must grow close to the ground. The evolution of xylem, for transporting water and mineral ions, and phloem, for transporting sugars and other organic molecules, enabled plants to grow taller by providing a means of moving these materials through all parts of the plant (page 370). In addition, thick-walled xylem cells help support large numbers of leaves and heavy stems.

3. *Resistant spores.* Some plants manufacture **spores,** lightweight cells that are specialized for their dispersal and for survival in adverse conditions. With thick walls to impede water loss and a virtual lack of metabolism to consume water, dormant spores can survive for long periods without additional moisture. When water becomes available, the spores regain activity and grow into new plants.

4. *Protective packaging for gametes and embryos.* Gametes and embryos need a defense against dehydration and damage. This protection has been achieved through the evolution of various structures: (1) *Multicellular gametangia* (gamete-producing structures) surround reproductive cells and developing embryos with water-trapping layers of cells. (2) *Pollen grains* encapsulate male gametes in watertight packages that free these plants from the need to use water for transferring sperm to the egg for fertilization. (3) *Seeds* serve as protective, drought-resistant enclosures for plant embryos, enabling the offspring of seed-producing plants to be dispersed to new localities by water, wind, or animals. (4) *Fruits* further clothe the seeds of flowering plants in additional protective layers, enhancing embryo survival and dispersal.

FROM PROMINENT GAMETOPHYTE TO PROMINENT SPOROPHYTE

During sexual reproduction, all members of the Plant kingdom alternate between a diploid, spore-producing generation—the sporophyte—and a haploid, gamete-producing generation—the gametophyte (page 408). Biologists refer to this sequential change from one generation to the next as **alternation of generations** (Figure 38-2).

III▶ A major trend in the evolution of plants has been toward a prolonged sporophyte and a reduced gametophyte generation (Figure 38-3). Complex, advanced plants, such as pines and flowering plants, have a prominent sporophyte generation, whereas more primitive terrestrial plants, like mosses, retain a prominent gametophyte, as did their aquatic ancestors. Diploidy offers an important advantage over the haploid state for complex organisms: Since haploid cells contain only one allele for each gene, the effects of a mutant allele cannot be masked by a dominant allele as it can in a diploid organism. Diploidy not only allows an organism to mask lethal mutations without suffering harmful effects, it also enables an organism to harbor neutral mutations that may later prove to be advantageous. The evolution of the major groups of plants is depicted in Figure 38-4.

Plant evolution spans from the early filamentous green algae, the ancestor of all terrestrial plants, to the modern flowering plants, the most advanced members of the Plant kingdom. No single environmental factor influenced plant evolution more than did the dryness of terrestrial habitats. Evolution eventually produced mechanisms for coping with water loss, for transporting water (and nutrients) throughout plants, and for reproducing in dry terrestrial habitats. Fruits, the most recent evolutionary innovation, improved the dispersal of plants to new locations. The trend from prominent haploid stage to diploidy provided greater protection against the loss of some critical functions due to lethal mutations. (See CTQ # 1.)

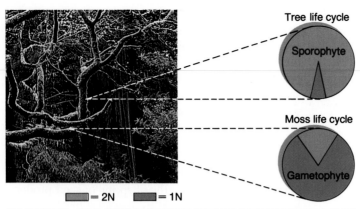

FIGURE 38-2

Two common residents of many forests are mosses and trees.
Both organisms are classified as members of the Plant kingdom, even though the moss is composed entirely of haploid (1N) cells, whereas the tree is made up of diploid (2N) cells. Like all plants, trees and mosses alternate between a diploid sporophyte and a haploid gametophyte generation during their sexual life cycle. Some plants, like trees, have a conspicuous sporophyte, whereas other plants, like this moss, have a conspicuous gametophyte generation.

OVERVIEW OF THE PLANT KINGDOM

All multicellular, tissue-forming eukaryotic photosynthesizers are classified within the Plant kingdom. Those few "plants" that do not photosynthesize are still included in the Plant kingdom because they are direct descendants of photosynthesizing plants (see Bioline: The Exceptions: Plants that Don't Photosynthesize) (page 836). Many botanists also include three divisions of algae (red, brown, and green) in the Plant kingdom, as we have here. Nearly all algae, whether protist or plant, are confined to aquatic environments, as were the ancestors of all plants. Although some of these algae are microscopically small, others are enormous, such as the giant kelp of the brown algae division.

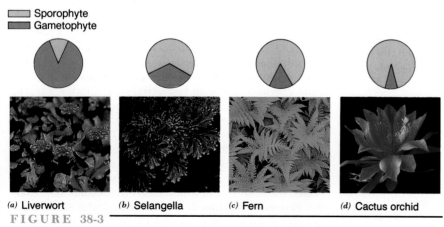

(a) Liverwort   *(b)* Selangella   *(c)* Fern   *(d)* Cactus orchid

FIGURE 38-3

Terrestrial plants: from prominent gametophyte to prominent sporophyte.
Comparing a range of life cycles from primitive to more advanced plants reveals a trend toward a prominent diploid sporophyte.

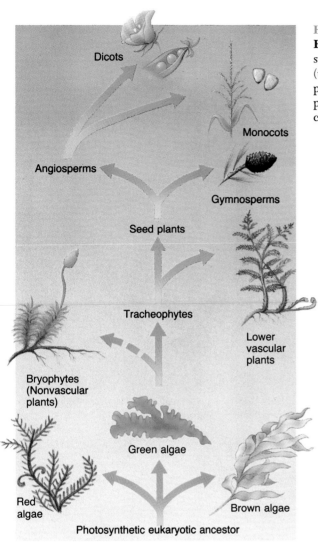

FIGURE 38-4

Evolution of the major groups of plants. From their aquatic photosynthetic ancestors, plants evolved into three divisions of aquatic "plants" (the brown, red, and green algae) and ten divisions of mostly terrestrial plants. The green algae were the ancestors of terrestrial plants. The bryophytes may have evolved directly from green algae or from simple tracheophytes that lost their vascular tissue.

The first plant to colonize the land formed a **thallus,** a flat, ground-hugging plant body that lacks roots, stems, leaves, and vascular tissues. Once plants moved onto the land, the ancestral line branched, producing two major groups of plants, the **bryophytes,** or nonvascular plants, and the **tracheophytes,** or vascular plants (Figure 38-5). Today, there are ten divisions of terrestrial plants, nine of which are vascular plants (see Table 38-1).

ALGAE: THE FIRST PLANTS

Three divisions of algae have been assigned to the Plant kingdom because they contain some organisms that are multicellular or colonial (Figure 38-6). Algae are distinguished from one another by the pigments they use to capture different light rays that penetrate water. These photosynthetic pigments tint the organisms the color that characterizes each group.

GREEN ALGAE: ANCESTORS OF PLANTS

With 7,000 species, green algae (**chlorophyta**) is the largest group of algae. Green algae are the only type found mostly in freshwater habitats, although there are a few marine species. Some green algae are even adapted to terrestrial life and blanket trees, soil, and porous rocks (or bricks). Others live an aquatic life in the snow (Figure 38-7).

▮▮▶ Most biologists believe that the green algae bridge the evolutionary gap between the algae assigned to the Protist kingdom and true plants. Unlike other algae, green algae share the following characteristics with higher plants.

- chlorophyll *a* and *b*,
- carotene accessory pigments,
- the ability to store surplus carbohydrates as starch,
- cell walls of cellulose.

Many green algae are unicellular; others are multicellular or colonial. Colonial forms are intermediate between

(a)

(b)

(c)

(d)

They look like plants, grow like plants, produce flowers, and are multicellular and eukaryotic, but they have no photosynthetic pigments. According to the five-kingdom classification system, these organisms wouldn't fit into any existing kingdom. But these organisms are the inevitable exceptions to any classification system. Although they lack photosynthetic pigments, these organisms are still classified as plants because they evolved from plant ancestors and retain a plantlike anatomy, in spite of their photosynthetic deficiency.

These nonphotosynthetic plants acquire food and nutrients the same way other heterotrophs do: by consuming organic compounds that are produced by other organisms. Nonphotosynthetic plants are either saprobes or parasites (page 972), and some are both. The following is a sampling of some of these atypical members of the Plant kingdom.

Photo (a)   *Corallorhiza trifida* (coralroot) is an orchid that grows in the beech and fir forests of Europe, Siberia, and North America. It is both a parasite and a saprobe, gathering its nutrients either from the roots of other plants or from dead or decaying organic matter.

Photo (b)   *Sarcodes sanguinea* (snow plant) is a striking, red fleshy saprobe that grows in the forests of California and Oregon. The heat of the growing plant melts the snow immediately surrounding it, providing a source of water.

Photo (c)   *Orobanchaceae* is an entire family of plants that parasitize the roots of other plants. *Orobanche,* or broom-rape, is the most widely distributed genus and is found in Europe, Asia, Africa, and North America.

Photo (d)   *Cuscuta* is a parasite that entangles its plant host with its orange, twisting stems. Outgrowths penetrate the host's vascular tissues and tap its fluids. *Cuscuta* grows all over the world and is commonly known as dodder, silkweed plants, boxthorn, or matrimony vine.

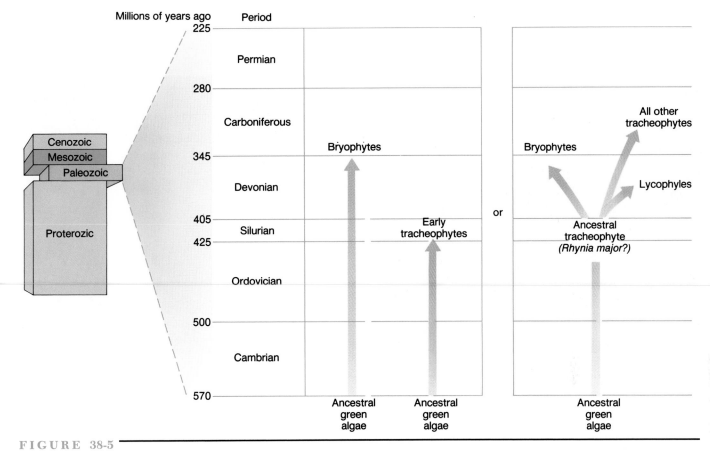

FIGURE 38-5

Plant origins. Two hypotheses have been proposed to explain the origin of plants. Although both hypotheses agree that plants evolved from an ancestral filamentous green alga, one hypothesis contends that the nonvascular plants (bryophytes) and the vascular plants (tracheophytes) arose from different ancestors, whereas the other hypothesis asserts that both bryophytes and tracheophytes arose from the same ancient tracheophyte (possibly *Rhynia major*).

unicellular and multicellular organisms (such as the *Volvox* shown in Figure 38-6a). Individual cells in the colony retain a high degree of functional independence, despite their physical attachments to other (usually identical) cells in the colony. In true multicellular green algae, cell activities are coordinated, and the cells remain tightly bound to adjacent cells, as they are in higher plants. In fact, multicellularity was probably first developed by green algae, the ancestors of higher plants.

One unicellular green alga genus, *Chlorella,* often forms symbiotic associations with animals. For example, green clams provide the algae with protection inside their bodies in exchange for the supplemental food and oxygen the "in-house" algae generate. *Chlorella* is also a favorite of scientists in search of new food and oxygen resources for humans. This alga has even been proposed as a food and oxygen generator for astronauts on long space voyages.

Another well-studied single-celled genus is the flagellated *Chlamydomonas* (a green alga that is not always green, as shown in Figure 38-7). Although it can reproduce asex-ually, *Chlamydomonas* also has different *mating types* (opposite "sexes"). Following union of opposite mating types, the resulting zygospore exists for one cell cycle, during which it divides by meiosis to form the haploid nuclei of new mating types.

Many multicellular green algae undergo a parallel reproductive pattern. Like the mating types of *Chlamydomonas,* gametes of these algae are indistinguishable, regardless of their mating type. Furthermore, the sporophyte and gametophyte stages of some green algae are indistinguishable in form. For example, when the flagellated gametes of *Ulva* (the sea lettuce shown in Figure 38-6b) fuse, the zygote proliferates into a green, leafy seaweed, two cells thick. This diploid sporophyte then forms haploid gametes, each of which develops into a haploid seaweed that looks identical to the diploid "leaf." This gametophyte then produces gametes, which mate with the opposite type. The resulting zygote develops into a new leafy sporophyte.

In other green algae, the female gametes are large and nonmotile, whereas the male gametes are small and motile

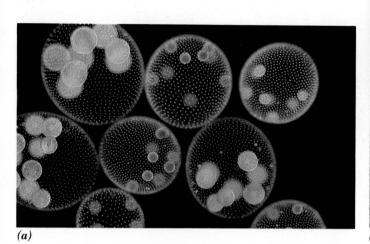

(a)   *(b)*

FIGURE 38-6

Multicellular algae: the simplest members of the Plant kingdom. *(a) Colonial green alga.*
These *Volvox* colonies consist of hundreds of attached cells, each spherical colony functioning as a single organism. Daughter colonies inside the mature "adults" are released when the parental colony disintegrates. *(b) Multicellular green algae* are exemplified by sea lettuce *(Ulva). (c) Brown Algae* are typified by underwater forests of giant kelp *(Macrocystis)* that stay in place by virtue of holdfasts that anchor each organism to the solid ocean floor. *(d) Red algae* live in greater depths than can members of the other divisions.

(analogous to the situation in higher plants and animals). In most of these algae (indeed, in most chlorophytes), the sporophyte is extremely reduced, the more obvious form of the organism being the haploid gametophyte.

BROWN ALGAE: MULTICELLULAR GIANTS

Brown algae (phaeophyta), mostly seaweeds, are distinguished by the presence of an accessory pigment *(fucoxanthin)* that is found in no other group of algae. The brown pigment helps these algae gather blue-green light, the wavelengths that penetrate deep into water. Some brown algae, **phaeophyta,** are giants among algae. Underwater "forests" of giant kelp contain individuals that are long enough to attach firmly to the ocean bottom, with their tops reaching the light-rich ocean surface 300 feet above (Figure 38-6c). Individual plants grow at an extraordinary rate, about 20 centimeters (1 foot) per day. This makes them

particularly productive "crop plants" that can be harvested regularly without depleting their numbers (see The Human Perspective: Natural Underwater Factories).

Brown algae have developed specialized structures and tissues, such as holdfasts, blades, floats, and conducting vessels. *Holdfasts* anchor the alga to rocks on the ocean bottom; leaflike *blades* provide broad surfaces for light absorption, boosting photosynthetic efficiency; hollow, gas-filled *floats* help keep the seaweed upright in the water; and newly synthesized nutrients are shipped from the blades down to the holdfasts and lower parts of the organism through conducting tubes that are reminiscent of the phloem in plants.

The brown algae alternate generations between sporophyte and gametophyte states. The large kelp form is the organism's sporophyte stage, the diploid, spore-producing form. Compared to the sporophyte, the haploid gamete-producing stage is usually very tiny, in both size and duration.

(c)

(d)

TABLE 38-1

CHARACTERISTICS OF DIVISIONS THAT CONTAIN PREDOMINANTLY
TERRESTRIAL PLANTS

Characteristics	Non-vascular	←————————————————Tracheophytes————————————————→								
		←————Lower Vascular Plants————→				←————————Higher Vascular Plants————————→				
	Bryo-phyta (24,000)[a]	*Psilo-phyta* (13)	*Lyco-phyta* (1000)	*Spheno-phyta* (15)	*Ptero-phyta* (12,000)	*Cycado-phyta* (100)	*Ginkgo-phyta* (1)	*Gneto-phyta* (71)	*Conifero-phyta* (700)	*Antho-phyta* (375,000)
Prominent gameto-phyte —										
Prominent sporophyte —										
External water needed for fertilization —										
Dispersal by spores —										
Dispersal by seeds —										
True roots, stems, leaves —										

[a] Number of described species.

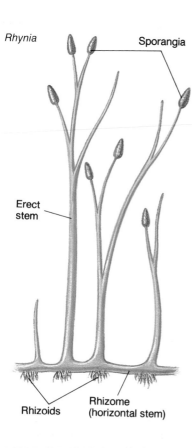

Rhynia

Sporangia

Erect
stem

Rhizoids

Rhizome
(horizontal stem)

FIGURE 38-10

Rhynia major, one of the earliest vascular plants. All rhyniophytes lived between 365 million to 400 million years ago. The sporophyte of *Rhynia major* illustrated here had erect photosynthetic stems that terminated with a sporangium. The plant lacked leaves and roots, attaching itself to the ground by rhizoids.

FIGURE 38-11

A "living fossil." The sporophyte of *Psilotum nudum* (division Psilophyta) closely resembles 400-million-year-old fossils of psilophytes. Round, yellow sporangia are produced on the sides of leafless photosynthetic stems.

heterosporous; they produce two types of spores, a megaspore that grows into a female gametophyte, and a microspore that grows into a male gametophyte. By separating gametangia, heterospory increases the chances of cross fertilization, generating greater genetic variability.

LOWER (SEEDLESS) VASCULAR PLANTS

Seedless vascular plants include four divisions: the Psilophyta, Lycophyta, Sphenophyta, and Pterophyta. Between 300 million and 430 million years ago, members of Psilophyta, Lycophyta, and Sphenophyta were the dominant type of vegetation on earth. Today, only about 1,000 species belong to these three divisions. Even after adding the 12,000 species of ferns (division Pterophyta), the four divisions of lower vascular plants account for less than 3 percent of all known plant species.

Psilophyta: Whisk Ferns and Relatives

Some 300 million years ago, lush green forests were filled with species of whisk ferns, or **psilophytes.** Today, only one family (containing two genera and 13 species) remains of this once diverse division of plants.

Psilotum is the most widespread genus of living psilophytes, extending from tropical and subtropical areas into a few temperate zones. The most common species of *Psilotum,* and the one frequently displayed in introductory biology laboratories, is the naked-looking sporophyte of *Psilotum nudum* (Figure 38-11). *Psilotum nudum* looks "naked" because its stems lack leaves. Instead, the stem bears small, widely spaced scales. Some scales contain three-chambered sporangia that manufacture and disperse spores. The spores grow into tiny, underground gametophytes that possess antheridia and archegonia. The flagellated sperm produced in the antheridia swim through a film of water to fertilize the egg. The resulting zygote grows and develops into another multicellular sporophyte.

Lychophyta: Club Mosses and Relatives

Like the Psilophyta, the division Lycophyta has seen more glorious days. About 350 million years ago, members of the genus *Lepidodendron* grew as tall as 50 meters (165 feet) and dominated the forests of the Carboniferous period (Figure 38-12). Today, all lycopods are small, herbaceous plants.

The lycophyta includes three families, six genera, and about 1,000 species. Most species are tropical, and the majority are *epiphytes*, plants that grow on other plants. The largest genus, *Selaginella,* contains 700 species (Figure 38-3*b*). The most common genus in the United States is Lyco-

FIGURE 38-12

The lush Carboniferous forests were dominated by giant psilophytes, sphenophytes, and lycophytes. *Lepidodendron* was a giant lycopod that produced leaves and sporangia at the ends of its branches. All treelike, seedless vascular plants, including *Lepidodendron,* are now extinct, the victims of changing environmental conditions.

FIGURE 38-13

Lycopodium complanatum sporophytes with strobili. Each strobilus is a tight cluster of spore-producing leaves.

podium, commonly called club moss. These plants are also called ground pines because the sporophytes look like miniature pine trees (Figure 38-13). The spores of lycopods have been used in firecrackers and flash bars (before flashbulbs were developed) because they ignite explosively, emitting light. Ironically, lycopod spores have also been used as baby powder.

Most lycopod sporophytes produce spores in a conelike **strobilus,** a terminal cluster of specialized leaves (called **sporophylls**) that produce sporangia. The majority of lycopods are homosporous. In the few heterosporous lycopods, the megaspore grows into a *megagametophyte* that contains archegonia. Microspores either germinate into a *microgametophyte* that contains antheridia, or they simply release flagellated sperm. (The prefixes "mega" and "micro" refer to the relative sizes of the gametes they produce; a sperm is always smaller than an egg.) Water is required for fertilization in both homosporous and heterosporous lycopods.

Sphenophyta: The Horsetails

Growing alongside the giant lycopods 350 million years ago were woody sphenophytes that grew 15 meters (50 feet) tall. Like the lycopods, all remaining sphenophytes are herbaceous. Today, the Sphenophyta includes 15 species, all of which are classified in one genus, *Equisetum*, the horsetails (Figure 38-14).

The upright sporophytes of horsetails have jointed, hollow stems that arise from a horizontal stem, or rhizome. The vertical stems have distinctive longitudinal grooves and whorls of leaves or branches at the nodes. The leaves of horsetails fuse at the base to form a sheath around each node. Strobili are produced at the ends of stems or from side branches, and homospory is more common in horsetails than is heterospory.

Each *Equisetum* spore is surrounded by coiled *elaters*, elongated flaps that unfurl as they dry. The movement of elaters helps dislodge spores from the sporangium when the air is dry. Dry spores are lighter and remain windborne

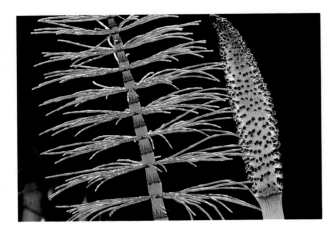

Horsetails. *Equisetum arvensis* has a horizontal rhizome that supports two types of hollow, upright stems. One is topped with a strobilus (on the right in the photo). The other produces orderly whorls of radiating branches (left). Although the stems appear leafless, small leaves on the stem fuse at the base to form sheaths around each node. *Equisetum* is the only genus in the Sphenophyta, a once widely distributed, diverse division.

longer than do wet spores, so dry spores are dispersed over greater distances. The spores remain dormant until moistened, at which time they grow into a tiny, photosynthetic gametophyte about the size of the period at the end of this sentence. It isn't surprising that most gametophytes go completely unnoticed.

Pterophyta: Ferns

Like other seedless vascular plants, ferns are an ancient plant group that arose during the Devonian period between 375 million to 420 million years ago. Two-thirds of the 12,000 ferns species living today are tropical. Because they produce flagellated sperm that must swim to the egg, all ferns are restricted to habitats that are at least occasionally wet.

The life cycle of a typical fern is illustrated in Figure 38-15. The familiar sporophyte of a fern produces *fronds;* leaflike structures that, unlike "true" leaves, have an apical meristem and clusters of sporangia called **sori.** Fern fronds uncoil from tight spirals that arise from rhizomes or upright stems.

Most ferns are homosporous. The spores are released from sporangia that are produced either along the margins or in clusters on the lower surface of the fronds. Clusters of sporangia are covered by an umbrella-shaped **indusium,** whereas the sporangia formed along frond margins are often sheathed by a flap called a **false indusium.** Both types of indusia shrivel and fold back as the air dries, releasing spores during dry periods, when dispersal is favored.

Each fern sporangium is encircled by an **annulus,** a row of specialized cells with thick cell walls on five of its six sides. As water is lost through the thinnest, outward-facing wall, the bonds between the remaining water molecules become strained, and tension builds within each annulus cell. Increasing tension causes the sporangium to bend slowly. When the tension finally exceeds the cohesive force between water molecules, the hydrogen bonds snap, and the thick, elastic cells of the annulus spring back to their original shape, jettisoning the spores into the dry air. (Figure 38-16).

Fern spores grow into a small, heart-shaped gametophyte called a **prothallus.** The photosynthetic, independently growing prothallus lacks vascular tissues. Like many gametophytes of lower vascular plants, the prothallus is reminiscent of primitive ground-hugging plants.

SEED PLANTS

The simple conductive tissues of the seedless vascular plants were more than adequate for life in lowland swamps that were so widespread during the Devonian and Permian periods of the Paleozoic Era. The perpetually saturated soils also provided ample water for sperm and for the plants' ground-hugging, nonvascular gametophytes. Environmental conditions changed dramatically and abruptly at the end of the Permian period, however, shifting the course of plant (and animal) evolution in a new direction.

The new environmental conditions developed when the earth's continents shifted positions (page 767). Swampy lowlands were drained and mountain ranges were thrust high. The land became drier, and deserts formed. The climate began to fluctuate, and yearly seasonal cycles became established on earth. Most lower vascular plants were ill-equipped for these changes and vanished. Among the surviving plants were those that had already acquired a means of surviving less stable, drier climates. These were the earliest seed plants, the primitive **gymnosperms** (*gymno* = naked, *sperm* = seed) that bear naked seeds.

Gymnosperms were able quickly to colonize the new habitats created at the beginning of the Mesozoic Era because they were armed with the following adaptations:

- extensive root systems that reach deeper underground water supplies;

- efficient conducting tissues that deliver water and nutrients to all parts of the plant;

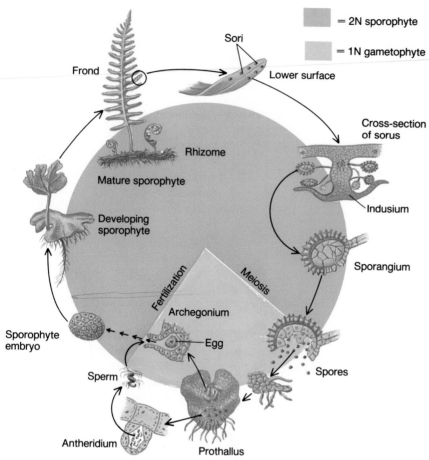

FIGURE 38-15

Fern life cycle. The fern sporophyte and gametophyte live independently, although the young sporophyte extracts food and water during its early development. The heart-shaped fern gametophyte (the prothallus) produces both anteridia and archegonia.

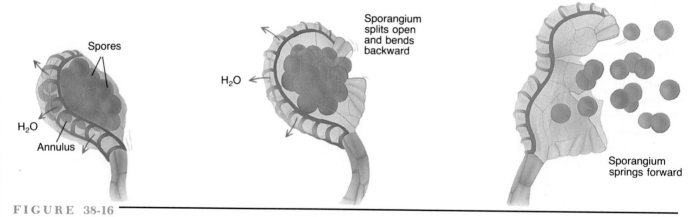

FIGURE 38-16

Engineered for dispersing spores, the sporangium of ferns is encircled by an annulus. As annulus cells lose water, the sporangium is gradually pulled backward. When water bonds snap, the sporangium shoots forward, flinging spores into the air.

- secondary growth that replaces and adds new vascular tissues and increases support (Chapter 20);

- enclosure of gametes in structures that protect the gametes and prevent drying (sperm in pollen grains, and eggs embedded in sporophyte tissues);

- "dry" fertilization (sperm do not require water for dispersal to an egg);

- encasement of the plant embryo in a protective seed that withstands harsh, dry conditions during dispersal and delays germination until conditions become favorable for growth.

Although a great variety of gymnosperms evolved during the Mesozoic Era, some gymnosperms survive today only as relics of the past. These "living fossils" are restricted in their range of distribution because, in most habitats, further environmental changes gave plants with other adaptations the competitive edge. Other gymnosperms still dominate many landscapes, as they did throughout most of the Mesozoic Era.

The Gymnosperms

All modern seed-producing plants are divided into two groups: (1) the gymnosperms (pines, firs, cycads, etc.), or "naked seed" plants, and (2) the angiosperms (*angio* = covered), or flowering plants. The seeds of gymnosperms are referred to as "naked" because they are not surrounded by additional fruit tissues as they are in angiosperms. Instead, the seeds are borne on the upper surface of scales that develop in cones, the gymnosperms' reproductive structures. Cones are tight spirals of sporophylls.

The more than 700 species of living gymnosperms are grouped into four divisions: the Cycadophyta, Ginkgophyta, Gnetophyta, and Coniferophyta.

Cycadophyta **Cycads** were most abundant and widespread during Mesozoic times, the Age of Dinosaurs. The 100 or so species of cycads that still survive today are native only to tropical and subtropical regions, although they are used as ornamental plants in virtually all regions of the world. Some cycads superficially resemble palm trees, but, like all other gymnosperms, cycads produce cones and bear naked seeds, not flowers and fruits as do palms (Figure 38-17).

Cycads are *dioecious;* that is, individual plants produce either male or female reproductive structures but never both. Pollen is produced in male cones and, when released, is dispersed by the wind. When a pollen grain lands on a female cone, it grows a pollen tube that reaches to within a few millimeters of the egg. Flagellated sperm are then released from the pollen tube and swim the remaining distance through a miniature water chamber that is provided by the female gametophyte, a remnant of water fertilization.

Ginkgophyta Like the cycads, a number of **ginkgo** species grew during the Mesozoic Era. Today, only one species of Ginkgophyta remains, *Ginkgo biloba* (the maidenhair tree). *Ginkgo biloba* escaped extinction because it has been cultivated for thousands of years as an ornamental plant in Asian temples and gardens (Figure 38-18). No "wild" or natural populations of ginkgos occur anywhere in the world today; they survive only where tended by humans.

Gnetophyta There are 70 species of **gnetophyta,** grouped into three genera: *Welwitschia, Ephedra,* and *Gnetum. Welwitschia* is the oldest genus; its fossils date back more than 280 million years. The only living species, *Welwitschia mirabilis,* is found in southwestern African

FIGURE 38-17

Mistaken identity. Because of their appearance, cycads are sometimes mistaken for palms. Unlike flowering palms, however, cycads are gymnosperms that produce either male or female cones. Cycads were abundant during the time of the dinosaurs, between 100 million and 200 million years ago.

FIGURE 38-18

Saved from extinction, the maidenhair tree (*Ginkgo biloba*) has been cultivated for thousands of years in gardens, parks, and other tended sanctuaries; there are no natural populations anywhere in the world today. The maidenhair tree is the only living species in the division Ginkgophyta.

FIGURE 38-19

Less than meets the eye. Although *Welwitschia mirabilis* appears to have many leaves, each plant produces only two broad leaves that shred and curl as they grow. *Welwitschia* grows flat against the hot, sandy soil in the Kalahari Desert of Africa. Seeds are produced in cones along leaf edges.

deserts and resembles a pile of tattered leaves growing from a wide, blunt stem (Figure 38-19). Yet, *Welwitschia* produces only two leaves during its 100-year lifetime. Each leaf shreds as it grows, giving the appearance of multiple leaves. *Welwitschia's* stubby, succulent stem and deep tap root store water, helping the plant survive long periods of drought.

All 40 species of *Ephedra* are short shrubs, with jointed stems and scale-like leaves. Native Americans made flour and tea from *Ephedra* plants. Today, drugs for treating asthma, emphysema, and hay fever come from extracts of *E. sinica* and *E. equisetina*.

The third genus, *Gnetum*, includes 30 species of vines, woody shrubs, and trees, all of which are found in Asia, Africa, and Central and South America.

Coniferophyta The most familiar gymnosperms are the **conifers** that dominate many Northern Hemisphere forests. The more than 700 species of conifers are classified into 50 genera, including *Pinus* (pines), *Juniperus* (junipers), *Abies* (firs), and *Picea* (spruces). The largest plants, the giant sequoias *(Sequoiadendron gigantea),* and the tallest plants, the redwoods *(Sequoia sempervirens),* are coni-

fers. The leaves of conifers are either long needles or short scales, both of which are covered with a thick cuticle to retard water vapor loss. Since many conifers retain green leaves throughout the year, they are frequently called evergreens.

Most conifers are *monoecious;* that is, both male and female reproductive structures are produced on the same sporophyte individual. During the pine life cycle (Figure 38-20), microspore mother cells in the male cones divide by meiosis to produce microspores. The microspores then divide mitotically to produce the male gametophyte, a pollen grain that contains sperm. The pollen grains of pines develop balloonlike "wings" that help them remain airborne during dispersal. To improve the chances of fertilization, a single pine tree releases billions of pollen grains, sometimes producing clouds of yellow pollen.

Within the female cones, megaspore mother cells divide meiotically inside megasporangia, producing megaspores. Four megaspores are formed, three of which degenerate. The remaining megaspore divides repeatedly by mitosis to produce a multicellular female gametophyte that contains two archegonia, each with a single egg. The combination of the female gametophyte, megasporangium, and

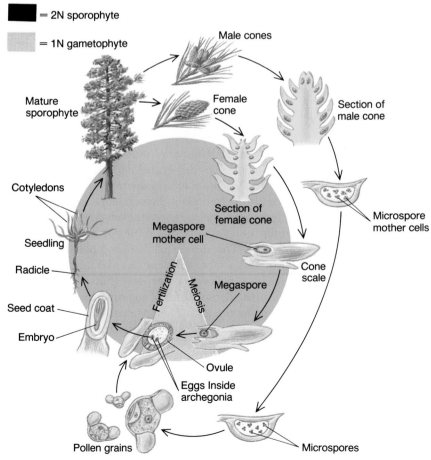

■ = 2N sporophyte

▨ = 1N gametophyte

Male cones

Female cone

Mature sporophyte

Section of male cone

Cotyledons

Seedling

Radicle

Seed coat

Embryo

Megaspore mother cell

Section of female cone

Fertilization

Meiosis

Megaspore

Cone scale

Microspore mother cells

Ovule

Eggs Inside archegonia

Pollen grains

Microspores

FIGURE 38-20
The pine life cycle represents the typical life cycle of the division Coniferophyta.

(a)   *(b)*

FIGURE 38-21

A sampler of success. Flowering plants (division Anthophyta) are now the most diverse and abundant types of plants on earth. The Anthophyta are subdivided into dicots (*a* and *b*) and monocots (*c* and *d*).

surrounding sporophyte tissues forms the pine **ovule,** which functions just as does the ovule of flowers (page 408).

The scales of female cones secrete a sticky substance that captures windborne pollen. Pollen grains then form a pollen tube, which grows through the sporophyte and ovule tissues, delivering a sperm to the egg. Following fertilization, the zygote grows and develops into a pine embryo, which combines with the female gametophyte and the remaining ovule tissues to form the seed, or pine nut. Many seeds are "winged" to aid dispersal by the wind. The entire process normally takes more than 2 years to complete. Fertilization doesn't occur until 13 months after pollination; it takes that long for the pine ovule to develop. Another 12 months or so is then required for the pine embryo and seed to mature.

The Coniferophyta is the most advanced division of gymnosperms. Even with an array of adaptations well suited for life on land, however, conifers are not the most diverse group of plants. The **angiosperms** (flowering plants) far outnumber conifers in number of species and, except in northern and high-altitude habitats, in number of individuals.

The Angiosperms

All plants that produce flowers belong to the division Anthophyta, the only division of angiosperms. Anthophyta is the most recently evolved plant division, appearing about 100 million years ago. The Anthophyta evolved from a now-extinct group of gymnosperms during the middle of the Mesozoic Era. The success of flowering plants was swift. Equipped with a more efficient vascular system and two unique characteristics—flowers and fruits—angiosperms rapidly diversified; thousands of angiosperms appear in the fossil record by the end of the Mesozoic Era. Today, the Anthophyta is the most diverse division of plants; nearly 75 percent of all plant species are flowering plants.

Members of the Anthophyta are called angiosperms (meaning "enclosed seed") because their seeds are surrounded by fruit tissues that form from the mature ovary of the flower (Chapter 21). The more than 350,000 species of angiosperms that have been described to date range in size from tiny, stemless duckweeds to towering eucalyptus trees. Flowering plants grow in virtually every terrestrial and most aquatic habitats.

In general, flowers and fruits are devices for recruiting

(c)

(d)

animals for pollination and seed dispersal. Gymnosperms rely on wind to disperse pollen and seeds indiscriminately. A more efficient means of dispersal is provided by insects and other animals that transfer pollen between individuals of the same species with greater precision. This method not only reduces the need to manufacture large numbers of pollen grains, it also increases the frequency of cross-pollination, which, in turn, increases the genetic variability in a gene pool. Increased variability improves a species' chances of adapting to environmental changes. Seed dispersal by animals also improves the chance that offspring will be deposited in hospitable areas since animals can only live where plants live. In contrast, the wind blows seeds everywhere, to the middle of the ocean or to the top of a glacier, places where no plants can grow.

The division Anthophyta is divided into two classes: (1) the monocotyledons, or monocots, and (2) the dicotyledons, or dicots (page 361). There are more than twice as many dicot as monocot species—250,000 versus 100,000 (Figure 38-21). Dicots include flowering trees and shrubs with wood secondary growth as well as many kinds of herbs.

Most fruits and vegetables that are available in supermarkets are dicots; these include tomatoes, potatoes, various melons, stone fruits (peaches, cherries, apricots), grapes, zucchini, and broccoli. Many important agricultural staples are monocot grasses, however, such as corn, wheat, rye, oats, barley, and rice. Monocots also include orchids, palms, lilies, and ornamental grasses. (The sexual life cycle of flowering plants was discussed fully in Chapter 20.)

The fluid-conducting vessels of tracheophytes allowed them to discard the height restrictions imposed on bryophytes. Lower vascular plants manufacture no seeds, produce only one type of spore, lack true roots, and require standing water for fertilization. Higher vascular plants produce two types of spores that are physically separated from each other in the same plant, improving the chances of cross-fertilization. The latest innovations of plants are found in the flowering plants. These angiosperms house their ovules inside an ovary and produce flowers to promote fertilization and fruits to encourage seed dispersal. (See CTQ # 4.)

REEXAMINING THE THEMES

Relationship between Form and Function

The structure of seeds reflects a reproductive strategy that is tailored for life on dry land. Seeds enclose the plant's embryo in protective epidermal layers that resist drought, encase the embryo in a package readily dispersed to new locals by wind or animals, and provide the germinating sprout with a food supply that enables it to grow large enough to break through the soil and to capture sunlight for photosynthesis.

Acquiring and Using Energy

Plants (along with autotrophic protists and monerans) provide energy and nutrients for all the biosphere's organisms. Most plants capture sunlight that has been filtered only by the atmosphere, but many aquatic plants (mostly algae) must rely on those colors of light that can penetrate their watery habitats. Accessory pigments in red algae capture wavelengths of light that penetrate deepest in water, allowing these plants to occupy deeper regions of the ocean floor.

Unity within Diversity

Plants are second only to animals in the diversity of species found in the kingdom. Their diversity ranges from relatively simple colonies of photosynthetic cells restricted to watery habitats to huge flowering trees with complex tissues and intricate structures for sexual reproduction on dry land. In spite of this diversity, dozens of identical properties are common denominators in all plants: They all use the same photosynthetic pigments (chlorophylls *a* and *b*) they inherited from the photosynthetic eukaryote that was the ancestor of all plants; they all possess green chloroplasts in their photosynthetic cells; and, except for brown and red algae, they all have cell walls made of cellulose, possess the same accessory pigments, and store surplus energy in starch. Mitochondria, endoplasmic reticula, Golgi bodies, and other organelles that characterize eukaryotic cells are also found in all living plant cells.

Evolution and Adaptation

Adaptations to a sedentary life include an arsenal of chemicals that fend off herbivores and competitors. Additional adaptations were required of terrestrial plants, including mechanisms that prevent water loss, seeds that survive fire, and means of transferring gametes and dispersing embryos in nonaquatic habitats. Flowering plants coevolved with animals that serve as pollinators, increasing the likelihood of cross-pollination, promoting diversity, and accelerating the evolution of advantageous traits and new species.

SYNOPSIS

Terrestrial plants evolved from an aquatic filamentous green algae that colonized the land about 400 million years ago. Other algae (brown and red) in the plant kingdom evolved refinements within their divisions but were not ancestors to any other group of plants.

The first land plants grew as ground-hugging mats, remaining close to continuous water sources. The evolution of vascular tissues and adaptations to control water loss enabled early plants to disperse into drier habitats and to throw off their height restrictions. Land plants quickly diversified, becoming the predominant form of terrestrial life. Most modern plants are terrestrial.

The colonization of the land by animals became possible only after plants became established on land, thereby creating the terrestrial habitats and food sources needed by animals.

The roughly 400,000 species of living plants include three divisions of algae, one division of nonvascular bryophytes, and nine divisions of vascular tracheophytes. The latter two groups constitute the 10 divisions of terrestrial plants.

Three divisions of algae constitute the majority of aquatic members of the plant kingdom. Many of the characteristic features of modern plants, such as chlorophyll *a* and *b*, are found in green algae. The first multicellular plants were probably algae, as were the first plants to reside on land.

Bryophytes and lower vascular plants still retain characteristics of semiaquatic ancestors, including the requirement for water as a medium for dispersing sperm to female sex organs that contain an egg.

During sexual reproduction, plants alternate between two multicellular generations: a diploid sporophyte, and a haploid gametophyte. In sporophytes, meiosis produces spores, which grow into sperm and egg-producing gametophytes.

Plants evolved from a prominent gametophyte to a prominent sporophyte, the diploid condition being genetically more advantageous for terrestrial life.

Between 150 million and 350 million years ago, lower (seedless) vascular plants were the predominant plants on earth. Extensive mountain building and shifting continents during the past 150 million years created drier and more variable environmental conditions, causing most seedless vascular plants to become extinct. With their efficient vascular tissues, their protected gametes (eggs in ovules, sperm in pollen grains), and their enclosed embryos, the early seed plants (the gymnosperms) thrived under these new conditions.

The most recently evolved group of plants—the angiosperms—is now the most diverse. All angiosperms produce flowers and develop a fruit from a ripened ovary. Flowers and fruits provide protection for gametes and embryos and enable pollen and seeds to be dispersed economically by insects and other animals.

Key Terms

spore (p. 833)
alternation of generations (p. 833)
thallus (p. 835)
bryophyte (p. 835)
tracheophyte (p. 835)
chlorophyta (p. 835)
phaeophyta (p. 838)
algin (p. 840)
rhodophyta (p. 841)

homosporous (p. 843)
heterosporous (p. 845)
psilophyte (p. 845)
strobilus (p. 846)
sporophyll (p. 846)
sori (p. 847)
indusium (p. 847)
false indusium (p. 847)
annulus (p. 847)

prothallus (p. 847)
gymnosperm (p. 847)
ginkgo (p. 849)
gnetophyta (p. 849)
conifer (p. 851)
ovule (p. 852)
angiosperm (p. 852)

Review Questions

1. Complete the classification scheme for the Plant kingdom using the following terms:

liverworts	bryophytes	gymno-
hornworts	tracheo-	sperms
Sphenophyta	phytes	angiosperms
Cycadophyta	Lycophyta	Psilophyta
higher vascu-	Ginkgophyta	Gnetophyta
lar plants		

2. Explain why there are fewer plant species than animal species.

3. Match the structures with the appropriate plant group.
   ____ a. gymnosperms   1. prothallus
   ____ b. angiosperms   2. strobilus
   ____ c. ferns   3. rhizoids
   ____ d. mosses   4. fruit and flowers
   ____ e. club mosses   5. naked seeds

4. List the divisions of plants that produce the following: pollen, ovules, flowers.

5. How has the need for liquid water to promote fertilization affected the evolution of plants?

Critical Thinking Questions

1. Discuss the role of water in the evolution of modern plants. How are bryophytes analogous to amphibians in the animal kingdom? How are gymnosperms analogous to reptiles?

2. Defend the following statement: "The term *algae* is obsolete and should be discarded. There is simply too much disparity among the organisms that we call algae to lump them into the same group."

3. In some forests of the northern hemisphere, mosses tend to grow in greater profusion on the north-facing surface of the bark. Why do you suppose these mosses do not grow as well on the southernmost faces of trees in these forests?

4. In conifer forests, some gymnosperms rely on periodic episodes of fire to give them a competitive edge over flowering trees. Fire on the forest floor burns the flowering trees, whose lowest branches are readily ignited. The bottom branches of maturing conifers are very high, however, above the level that most fires can reach, and the trunks of these conifers are fairly fire-resistant. Without fire, many conifer forests would soon become forests of flowering trees. What features of these angiosperms would make them so much more successful than these gymnosperms, if not for fire?

5. The perfect plant for a desert environment would possess many of the adaptations discussed in this chapter. List those adaptations plus any additional ones you feel would be beneficial to survival and reproduction in periodically dry, very hot environments. Which adaptations are better suited to very wet habitats, such as a rain forest?

Additional Readings

Banks, H. 1975. Early land plants: proof and conjecture. *Bioscience* 25:730–737. (Intermediate)

Barbour, M., J. Burk, and W. Pitts. 1987. *Terrestrial plant ecology,* 2nd ed. Menlo Park, CA: Benjamin/Cummings. (Advanced)

Briggs, D., and S. Walters. 1984. *Plant variation and evolution,* 2nd ed. Cambridge: Cambridge University Press. (Intermediate)

Chapman, A. 1987. *Functional diversity of plants in the sea and on land.* Boston, MA: Jones and Bartlett. (Advanced)

Gensel, P., and H. Andrews. 1987. The evolution of early land plants. *Amer. Sci.* 75:478. (Intermediate)

Pickersgill, B. 1977. Taxonomy and the origin and evolution of cultivated plants in the New World. *Nature* 268:591–595. (Intermediate)

CHAPTER
◄ **39** ►

The Animal Kingdom

STEPS
TO
DISCOVERY
THE WORLD'S MOST FAMOUS FISH

BIOLINE

Adaptations to Parasitism among Flatworms

*E*ver since the acceptance of Darwin's theory of evolution, biologists have speculated on the origin of four-legged, terrestrial vertebrates, from the earliest amphibians to the more recently evolved reptiles, birds, and mammals. In 1892, Edward Cope of the Academy of Natural Sciences in Philadelphia proposed that the first amphibians evolved during the Devonian period (approximately 370 million years ago) from a particular type of ancient fish called a rhipidistian. Rhipidistians lived at a time when the land was covered by shallow, swampy bodies of water. Warm, swampy waters often become stagnant and oxygen deficient. The rhipidistians were adapted to these conditions; they had lungs that could inhale oxygen from the air under conditions of stress. The rhipidistians also had fins situated at the ends of fleshy lobes that contained an internal bony

skeleton. The front fins of these fish were probably used to support their bodies as they pushed their heads out of the water for a breath of air or dragged their bodies from one shrinking pond to another.

A fish with lungs and bony, lobed fins is *preadapted* to a terrestrial life. In other words, even though these characteristics were adaptations for life in shallow, freshwater ponds, they could also be put to use by animals that were becoming more and more terrestrial. The lungs are preadapted to the respiratory needs of a terrestrial animal, and the bony lobes are preadapted as terrestrial locomotor structures.

The first amphibian known from the fossil record was primarily aquatic and retained numerous fishlike characteristics. (See sketch at end of essay.) This animal had a

---

An air-breathing, lobe-finned fish is believed to be the ancestor of the first four-legged amphibian.

long, fishlike tail, complete with a well-developed tail fin. Its body was partially covered with bony scales typical of fishes living at that time. This animal possessed one obvious feature that is not present in any fish, however: It had four legs and four feet. Cope noticed that the limb bones of this early amphibian were remarkably similar to the bones present within the fleshy lobe of the fins of a rhipidistian. The teeth and skull of these two types of early vertebrates were also similar, including the presence of a peculiar, hingelike joint in their skulls. Cope proposed that rhipidistians were the ancestors of amphibians and, consequently, of all four-legged vertebrates.

The rhipidistians were classified in the subclass Crossopterygii, along with another group of fossil fishes, the coelacanths. These two groups of ancient fishes shared a number of features, including the lobed, bony fins. But while the rhipidistians disappeared hundreds of millions of years ago, their cousins, the coelacanths, straggled through the Mesozoic Era until finally disappearing from the fossil record about 70 million years ago, about the same time that the dinosaurs vanished. It would be nearly 50 years before events occurred that would change this viewpoint.

In December 1938 Marjorie Courtenay-Latimer received a telephone message that a trawler had come into dock with a pile of fish for her to examine. Courtenay-Latimer was the curator of a small museum in East London, South Africa, and had asked the local fisherman for their help in preparing exhibits of the fauna of the area. She took a taxi to the wharf and began rummaging through the fish, most of which were sharks. In the pile was a large, bright blue fish with a symmetric tail and fins situated at the ends of fleshy lobes. Despite her years of experience, Courtenay-Latimer had never seen anything like this creature. She wrapped the unusually smelly fish in a bag and, after much discussion, convinced the taxi driver to allow it in his car. When she returned to the museum, she transported the fish in a hand car to the museum's taxidermist. She then drew a rough sketch of the fish, which she mailed with a note to J. L. B. Smith, a friend and zoologist at a South African university, asking for help in identifying the specimen. Smith received the note 11 days later. He was greatly perplexed by the sketch. With its symmetric tail, thick, armored scales, and limblike fins, the fish looked exactly like a coelacanth. This was impossible, however, since coelacanths had disappeared 70 million years earlier.

It wasn't until the next month that Smith was able to travel to East London to inspect the remains of the fish;

there was no doubt that the animal was indeed a coelacanth. Even the curious intracranial joint characteristic of the skull of rhipidistians and the first amphibians was present in this "living fossil." Smith named the specimen *Latimeria chalumnae,* after its discoverer and the site where it was found. It is considered by many to be the greatest zoological discovery of this century. The finding of a coelacanth was of such extraordinary importance not simply because these fishes were thought to be extinct for so long, but because they were close relatives of the rhipidistians, one of our most important ancestors.

As the years passed, and no other coelacanths were found, Smith distributed thousands of leaflets showing a picture of the fish and offering a large reward for its capture. One day, in 1952, Smith and his wife were talking to the captain of a schooner about the fish; 10 days later, the captain sent the Smiths a cable that he had one on his boat. The fish had been caught by a native fisherman who was attempting to sell it at the local market when it was spotted by a schoolmaster who had seen a copy of the leaflet. In a frantic state, Smith called the prime minister of South Africa, who dispatched a special plane to pick Smith up and fly him to the remote site on a French Comoran Island where the fish was being kept. The subsequent removal of the fish from French territory caused an international uproar. The French government declared the waters off limits to foreign scientists, and for several years French biologists had a monopoly on the study of the next 23 specimens of *Latimeria* to be caught.

In many ways, *Latimeria* is highly specialized for life in the ocean depths and lacks certain features, including lungs, that had adapted ancient coelacanths to shallow ponds. In other ways, *Latimeria* retains many of the aspects of its early ancestors. For example, the structure and movement of the fins is similar to that which would be predicted of a tetrapod (four-legged) ancestor. More recent insights on coelacanths have been made by Hans Fricke of the Max Planck Institute in Germany. From a small submersible, Fricke has filmed coelacanths as they slowly swim and feed in their natural habitat. These studies have also revealed the rarity of the species and have raised some concern that the activities of the scientists studying them might lead to the extinction of these invaluable organisms.

As individuals, animals are greatly outnumbered by plants, bacteria, and even by fungi. Yet, there are more *kinds* of animals than any other type of organism. Nearly 75 percent of all known species (a total of about 1.3 million) are included in the Animal kingdom, and the tally is far from complete. Many biologists believe that millions of animal species remain unnamed and unclassified, most of them insects.

Animals range in size and complexity (Figure 39-1) from the microscopic *Trichoplax adhaerens*, thought to be the most primitive animal yet discovered, to the giant blue whale *(Balaenoptera musculus)* that reaches lengths of nearly 40 meters (130 feet) and weighs more than 160 tons. Between these extremes is an immense assortment of animals that range in body form from worms to giraffes and jellyfish to spiders.

Despite their astounding diversity, members of the Animal kingdom typically share several basic characteristics:

- *Animals are multicellular.* All animals are composed of a large number of cells, some of which are specialized for different functions. Even *T. adhaerens,* the least complex animal known, consists of several different types of cells.

- *Animals are heterotrophic.* Animals are not capable of either photosynthesis or chemosynthesis; thus, they must rely on other organisms to provide raw, organic materials.

- *Animals are motile.* Animals are capable of locomotion at some stage of their life. While many aquatic animals remain in one place as adults (they are said to be *sessile*), such animals invariably develop through a motile larval stage that is capable of swim-

(a)

(b)

FIGURE 39-1

From one extreme to the other. *(a) Trichoplax adhaerens* is the least complex animal known. This organism was originally discovered in 1883 and was classified as the larval stage of an unknown animal. In the early 1960s, the creature was rediscovered as a unique species and is now placed in a separate phylum, Placozoa. *T. adhaerens* consists of several types of cells that form a flattened, amorphous mass. *(b)* A blue whale, the largest animal that has ever lived, is made up of complex tissues, organs, and organ systems. Blue whales are approximately six times larger than the biggest dinosaur. Blue whales can achieve such a massive size because they are supported by the buoyant medium in which they live.

ming away from its parents, thereby dispersing the species population into new environments.

- *Animals engage in sexual reproduction.* With rare exception, animals consist of diploid cells that contain two sets of homologous chromosomes. During meiosis, which occurs in specialized reproductive tissues, haploid gametes are formed. An egg and a sperm (almost always from two different individuals) become united during fertilization to form a zygote that undergoes embryonic development, forming a diploid offspring.

▼ ▼ ▼

EVOLUTIONARY ORIGINS AND BODY PLANS

Like plants, the first animals left no fossils, making it impossible to trace the origin of the Animal kingdom. Biologists believe that animals arose from protozoan ancestors, but it is likely that the transition from a unicellular protozoan to a multicellular animal occurred more than once. For example, sponges contain cells that are virtually identical to those of a peculiar type of flagellated protozoa (a choanoflagellate), leading most zoologists to conclude that sponges arose from these protists. Other animals probably arose from a different ancestor, most likely a flagellated protozoan whose cells lived in close association as a colony (page 766).

THE ELEMENTS OF A BODY PLAN

When comparing various types of motor vehicles, such as motorcycles, sports cars, trucks, vans, motor homes, and so forth, one of the first characteristics you may consider is body plan: the vehicle's basic organization, including the nature and arrangement of its major parts. Animals, too, have *body plans* that describe the layout of their body parts.

Different types of motor vehicles can be distinguished by important structural features, such as the number of wheels or axles, the location and type of transmission or engine, the body size, cargo capacity, and so forth. A motor vehicle's structural features can be explained according to the vehicle's planned use. An animal's structural features can be explained in terms of:

1. its evolutionary ancestry (organisms invariably retain characteristics of their ancestors whose genes they have inherited): and

2. the environment in which it lives (organisms survive because they possess adaptations to their environment; as a result, related animals living in different environments will possess different adaptations).

Three of the most important properties that distinguish the body plans of animals are symmetry, body cavities, and segmentation.

Body Symmetry: The Arrangement of the Body Parts

An object has *symmetry* if it can be bisected into two, mirrored halves. An animal's symmetry often reveals information about its mode of existence. Animals exhibiting **radial symmetry** usually have a cylindrical body composed of similar parts that are arranged in a circular fashion around a single central axis (Figure 39-2a). Radial symmetry is characteristic of groups of animals that are either sessile (sedentary), such as the sea anemone, or slow moving, such as a jellyfish or sea urchin. Radial symmetry is adaptive for these animals since sensory information and food matter usually come toward them from all directions.

More motile animals typically exhibit **bilateral symmetry,** whereby only one plane bisects the animal into mirrored, right and left halves (Figure 39-2b). Bilateral symmetery is found among animals that actively search for food and shelter, such as insects and vertebrates. The evolution of bilateral symmetry was accompanied by the concentration of sense organs and nervous tissue at the leading end of the body to form a "head," an evolutionary phenomenon known as **cephalization** (*cephalic* = head). Cephalization allows a motile animal to gather information about the environment into which it is heading, protecting it from proceeding "blindly" into danger. The various planes that may divide a bilaterally symmetric animal are indicated in Figure 39-3.

Internal Cavities: The Spaces Within

The animal depicted in Figure 39-1a is essentially a flattened mass of cells; it possesses no internal cavities, other than the simple spaces between the cells. Nearly every other type of animal on earth has at least one internal chamber in which food matter is collected. This chamber, or **digestive tract,** may be a blind sac, as in a sea anemone (see Figure 39-8), or a complete internal tube, the *gut*, with openings at both ends, as in an earthworm (see Figure 39-17) or a human.

In addition to the digestive tract, most animals have at least one other major body cavity that develops between the digestive tract and the outer body wall. As we will discuss later in the chapter, this cavity is called by various names (pseudocoelom, coelom, or hemocoel), depending on how it forms in the embryo. Like symmetry, differences in the nature and location of the internal cavities distinguish major groups of animals from one another. In humans, the major body cavity (coelom), is partitioned into several smaller cavities that house the lungs, heart, and digestive organs (Figure 39-4).

(a)   *(b)*

FIGURE 39-2

Symmetry. *(a)* The sea anemone exhibits radial symmetry. Its body can be bisected by a number of planes (dashed lines), as long as the planes pass through a central axis (the round, central dot). The body parts of the sea anemone are arranged in a circle around this central axis. *(b)* A crab exhibits bilateral symmetry. Its body can be bisected by only a single plane, shown by the dashed line.

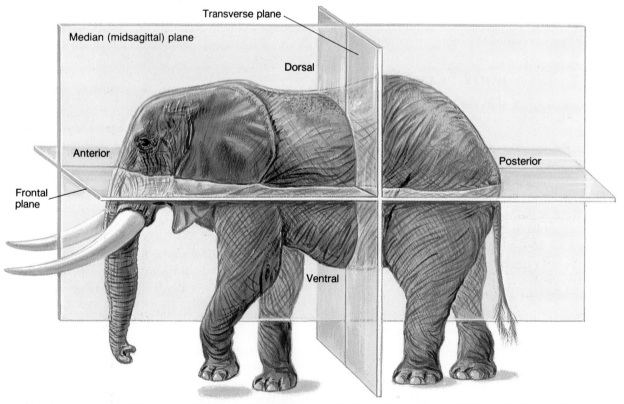

FIGURE 39-3

Sectional planes divide bilaterally symmetric animals into different halves. A *median* (or *midsaggital*) *plane* divides the animal into right and left halves; a *transverse plane* divides it into anterior (front) and posterior (rear) sections; and a *frontal plane* divides it into dorsal and ventral sections.

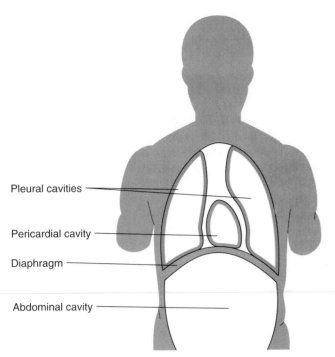

Pleural cavities

Pericardial cavity

Diaphragm

Abdominal cavity

FIGURE 39-4

The coelom is a space that is lined by cells derived from the embryonic mesoderm and situated between the body wall and the digestive tract. In humans, the coelom is partitioned into several distinct cavities: the *abdominal cavity* (which surrounds the visceral organs), the *pleural cavities* (which surround the lungs), and the *pericardial cavity* (which surrounds the heart).

Segmentation: Dividing the Body into Repeating Segments

A quick examination of the internal and external anatomy of an earthworm (see Figure 39-17) reveals that the body of this animal is constructed of repeating subunits, or **segments,** that contain similar organs. The presence or absence of segmentation is one of the most profound characteristics of a body plan.

Three of the largest phyla, the annelids, arthropods, and chordates (which includes vertebrates), consist of segmented animals. As we will discuss later in the chapter, segmentation is thought to have arisen as an adaptation for particular types of locomotion because it allows different parts of the body to engage in different types of activities. Segmentation in the human body is most apparent in your backbone, which consists of a series of bony vertebrae that form from a linear array of segments that develop in the embryo.

DIVIDING ANIMALS INTO PHYLA

🔺 The members of the Animal kingdom are organized by phylum. Even though the members of the same phylum may seem highly diverse—such as a snail and an octopus, or a fish and a human—they are all descended from a common ancestor and are constructed as a variation on a common body plan. Our exploration of the Animal kingdom includes nine of the approximately 35 animal phyla. Their major characteristics are summarized in Table 39-1. We chose these phyla for two reasons: They contain the greatest numbers of species, and they illustrate the principal evolutionary innovations that occurred during animal evolution. An evolutionary tree depicting the relationships among the major phyla is presented in (Figure 39-5).

> **Animals are multicellular, motile, sexually reproducing heterotrophs that are thought to have evolved from protistan ancestors. Differences among animals can be explained by differences in ancestry and in the selective pressures exerted by the environments in which the animals evolved. (See CTQ 1.)**

INVERTEBRATES

For many purposes, particularly college biology courses, animals are divided into two great blocks: **invertebrates,** or animals that lack a backbone, and **vertebrates,** animals that possess a backbone. Considering that there are only about 40,000 vertebrate species and more than a million invertebrate species, this hardly seems like an equitable division, yet it is one that reflects our interests. Humans are vertebrates, as are our pets and domestic animals. All of the 35 or so animal phyla contain invertebrates; only one phylum contains vertebrates. We will begin our discussion with the invertebrates.

TABLE 39-1

SUMMARY OF CHARACTERISTICS FOR ANIMAL PHYLA

Phylum	Cellular Organization	Coelom	Circulatory System	Nervous System	Reproductive System	Distinguishing Characteristics
Porifera (sponges) 10,000 spp.	No tissues	None	None (diffusion)	None	Male, female, or hermaphroditic	Aquatic filter feeders, porocytes, choanocytes
Cnidaria (hydras, corals, jellyfish) 9,000 spp.	Tissues	None	None (diffusion)	Nerve net, sensors over body surface	Male and female medusae, asexual budding	Stinging cnidocytes on tentacles
Platyhelminthes (flatworms) 20,000 spp.	Organs and organ systems	None	None (diffusion)	Ladder-type paired nerve cords, few ganglia	Hermaphroditic	Flatworms with definite head and tail ends
Nematoda (roundworms) 10,000 spp.	Organ systems	Pseudocoelom	None (diffusion)	Simple "brain" dorsal and ventral nerve cords	Male or female, internal fertilization	Tapered body, cuticle covering, tube-within-a-tube digestion
Mollusca (squids, snails, clams) 100,000 spp.	Organ systems	Coelom	Open, 1 heart	Cerebral ganglia, nerve cords	Male, female, or hermaphroditic	Shells in most, free-swimming larvae in some
Annelida (segmented worms) 9,000 spp.	Organ systems	Coelom	Closed, pumping vessels	Simple brain, paired ventral nerve cords	Male or female, external or internal fertilization, or hermaphroditic	Aquatic and terrestrial Repeating segments, but continuous digestive, nervous, and circulatory systems
Arthropoda (spiders, crabs, insects) 1,000,000 spp.	Organ systems	Coelom	Open, 1 heart	Simple brain, ventral nerve cord	Male or female, mostly internal fertilization	Jointed appendages, chitinous exoskeleton, specialized segments
Echinodermata (sea urchins, sea stars, crinoids) 5,300 spp.	Organ systems	Coelom	Open, no heart	Nerve ring, radial nerves to each arm	Male or female, external fertilization	Radially symmetrical adult; bilaterally symmetrical, free-swimming larvae; spines, water vascular system
Chordata (tunicates, lancelets, vertebrates) 44,300 spp.	Organ systems	Coelom	Closed, 1 or no heart	Anterior brain, dorsal hollow nerve cord	Male, female, or hermaphroditic, internal or external fertilization	Notochord, gill slits, tail, dorsal nerve cord

PHYLUM PORIFERA: SPONGES

There are approximately 10,000 species of sponges. Most sponges are marine animals that live in shallow, unpolluted coastal waters, but there are also numerous deep-sea forms and two families that live in freshwater streams and ponds. Adult sponges are sessile and remain attached to rocks, shells, and other submerged surfaces. Sponges are **filter feeders;** they feed on microscopic particles that are swept into their bodies from the external medium.

The simpler sponges have a vase-shaped body (Figure 39-6a,b) containing a central cavity, the **spongocoel,** that opens to the outside through the **osculum,** a large aperture located on the upper surface. A thin **body wall** composed

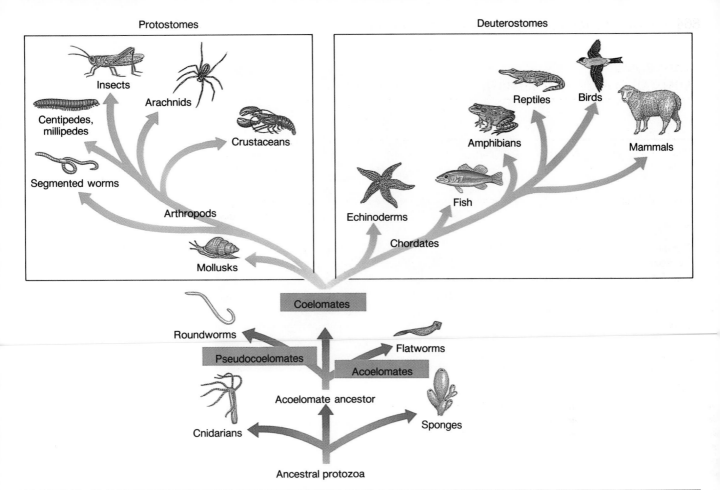

FIGURE 39-5

An overview of the phylogenetic relations among the major groups of animals. The sponges, cnidarians, flatworms, and roundworms are referred to as "lower metazoans," indicating their simpler body structure and the fact these were the first types of animals to evolve. Pseudocoelomates and coelomates have a body cavity that is separate from the digestive tract, while acoelomate animals do not. Two major evolutionary lines emerged from the lower metazoa: the protostomes and the deuterostomes.

of three distinct layers surrounds the spongocoel. The outer layer of the body wall is made up of a sheet of flattened epidermal cells, called *porocytes.* Porocytes are doughnut-shaped cells, each pierced by a microscopic pore. Water enters the sponge through these pores and exits via the single large osculum.

The inner layer of the body wall consists of a loosely organized bed of unique cells, called **choanocytes,** or *collar cells,* each of which has a rounded basal end and an extended collar that surrounds a single undulating flagellum. The combined beating of the flagella of all the choanocytes draws water into the sponge. The fine-mesh collars of the choanocytes then trap the suspended food particles, which are engulfed by phagocytosis and digested in food vacuoles located inside the sponge's cells.

The basal ends of the choanocytes are embedded in a gelatinous middle layer, through which motile cells (*amoebocytes*) wander. The body wall also contains large numbers

of pointed **spicules** that are composed of calcium carbonate (or silica, in some species). The spicules of a sponge function as both a skeleton that supports the animal's soft mass and a protective device that makes the animal less than appetizing to a potential predator.

Most sponges have a more complex internal organization than that depicted in Figure 39-6*b*. The thick body wall of the more complex sponges is riddled with fine channels and hundreds of thousands of microscopic chambers that house the choanocytes (Figure 39-6*c*). This type of internal architecture increases the surface area available for choanocytes and improves the sponge's efficiency in capturing food, exchanging respiratory gases, and eliminating wastes.

Of all the major groups of animals, sponges have the least complex structure: Their bodies often lack symmetry; they lack sensory cells and nerve cells; the layers of the body wall do not constitute bona fide tissues since the component cells do not function in an integrated manner (each cell

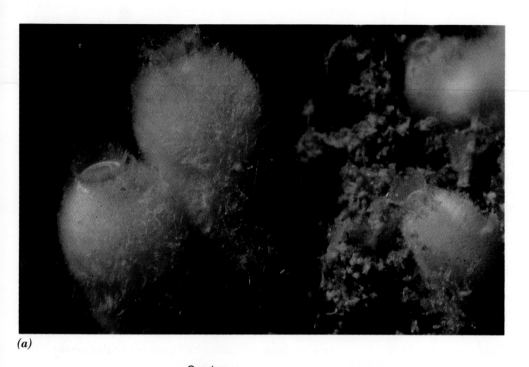

(a)

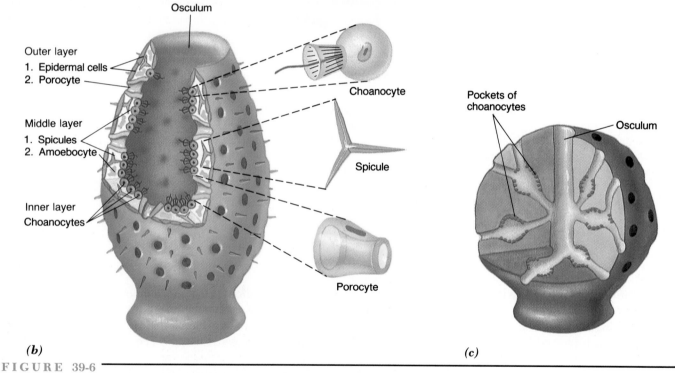

(b)   (c)

FIGURE 39-6

Sponge anatomy. *(a)* One of the simpler, vaselike sponges, *Leucosolenia.* *(b)* Simple sponges, such as *Leucosolenia* have a single, central spongocoel chamber surrounded by a body wall that contains three layers. *(c)* More complicated sponges are composed of many smaller chambers. The presence of these chambers increases the surface area available for choanocytes.

carries out its own activities more or less independently); and there is no mouth—the largest opening into the animal serves as an exit for fluid. In addition, sponges have no digestive tract; the internal chambers are simply part of a pathway that allows the environment to flow through the animal, bringing in food and oxygen and carrying out waste products and gametes.

Most sponges are hermaphrodites; each individual produces both sperm and eggs. Each sponge releases its sperm into the medium, where the sperm cells pass through the pores of another individual, fertilizing its eggs. The fertilized egg develops into a flagellated larva that swims out of the parent through the osculum. Motile larvae are highly adaptive for sessile species since they disperse offspring to more favorable habitats.

Whereas marine sponges usually live in a relatively constant environment, those living in freshwater ponds and streams must be adapted to severe climatic challenges. Under adverse conditions, freshwater sponges produce *gemmules:* clusters of cells surrounded by a hard, protective layer. Gemmules can survive dehydration and very cold temperatures and then regenerate a new sponge when favorable conditions return.

As a group, sponges are diverse and abundant, but they represent an evolutionary dead end; that is, sponges have not given rise to any other form of animal life. Instead, they have remained essentially the same for hundreds of millions of years.

PHYLUM CNIDARIA: HYDRAS, JELLYFISH, SEA ANEMONES, AND CORALS

If you have ever walked along a rocky coastline at low tide, you may have noticed populations of bright, flowerlike sea anemones wedged into the crevices in the rocks. Sea anemones are members of the phylum **Cnidaria,** as are the many types of jellyfish and the microscopic animals that are responsible for building the great coral reefs, such as the Great Barrier Reef off of Australia's eastern coast. The phylum contains about 9,000 species.

Cnidarians are built on a radially symmetric body plan composed of a three-layered body wall surrounding a blind, saclike chamber, called the **gastrovascular cavity.** The simplest cnidarians are the hydras (Figures 39-7a, 39-8a). The outer layer (**epidermis**) and inner layer (**gastoder-**

(a)

(b)

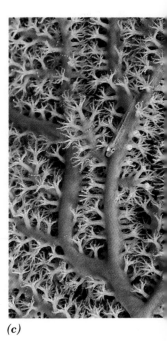
(c)

FIGURE 39-7

Three types of cnidarians, representing the three major classes within the phylum. *(a)* Hydras are sessile, freshwater members of the class Hydrozoa. *(b)* Most jellyfish are members of the class Scyphozoa. The animal moves as a result of pulsating contractions of the muscular bell situated at the top of the animal. The mouth is located at the lower end of the dangling, "frilly" tissue. *(c)* These coral polyps, members of the class Anthozoa, construct an outer skeleton composed of calcium carbonate, into which they can retract when threatened. The mass of a coral reef is made up of successive layers of coral skeletons; the living polyps are restricted to the upper surface.

mis) of the cnidarian are more complex than are those of a sponge; they consist of cells whose activities are coordinated to form a tissue. Thus, cnidarians can be considered as having evolved to the *tissue level of organization,* but they lack true organs, such as a heart or a kidney. Cnidarians do not have such organs because all of their cells are close enough to the outer medium to receive oxygen and dispel waste products by simple diffusion. The middle layer **(mesoglea)** of a cnidarian varies from a thin, noncellular layer in hydras to a thick layer of connective tissue in sea anemones.

All cnidarians are carnivorous. They stun or kill their prey by the use of "projectiles" that are fired from cells, called **cnidocytes,** which are concentrated in the epidermis of the tentacles that surround the cnidarian's mouth. Each cnidocyte (Figure 39-9) contains a capsule *(nematocyst)* that often consists of a coiled thread that is attached to a spiked spear, laden with powerful toxins. When the nematocyst is stimulated, water rushes in, generating hydrostatic pressure capable of ejecting the microscopic spear with a force that can penetrate the body wall of its prey. The threads then ensnare the prey, which is further immobilized by the toxins. Cnidarian toxins are very potent. The Portu-

guese man-of-war produces a venom as potent as that of a cobra; it is capable of raising large welts simply by contact with the skin. The sting of the sea wasp, which lives in the waters of Australia, is powerful enough to kill a human.

Once trapped, the prey is deposited by the tentacles into the gastrovascular cavity. There, the prey is broken down into microscopic particles by the combined action of enzymes that are secreted by the gastrodermis and the muscular activity of the body wall. The appearance of an internal chamber for storing and disassembling food was an important evolutionary innovation that allowed cnidarians to feed on much larger prey than that which could be ingested by animals that relied strictly on phagocytosis, such as sponges.

As a group, cnidarians have two distinct body forms, the polyp and medusa. **Polyps** are cylindrical cnidarians, whose mouth and tentacles are situated at the upward end, as in a hydra, sea anemone, or coral (Figures 39-7*a,c;* 39-8*a*). **Medusas** are umbrella-shaped cnidarians, whose mouth and tentacles are situated on the lower surface, facing downward, as in a jellyfish (Figure 39-8*b*). These two body forms are adapted for different modes of existence. Hydras, sea anemones, and corals occur only in the polyp form, which is

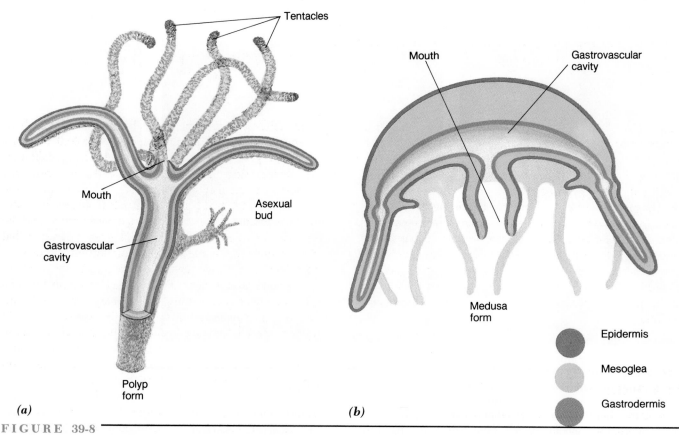

(a)   **(b)**

FIGURE 39-8

Two cnidarian body forms: a polyp *(a)* and a medusa *(b)*.

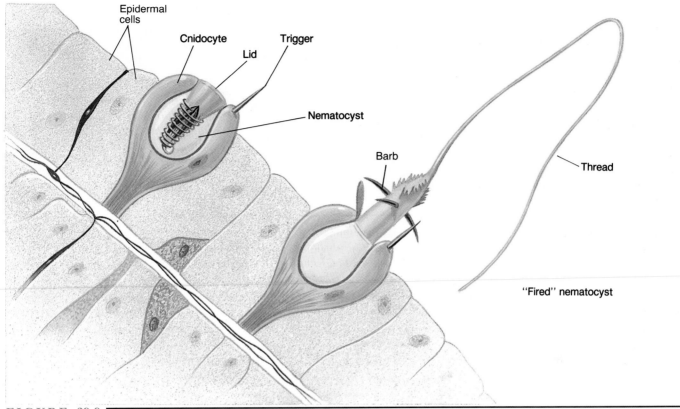

FIGURE 39-9

The tentacles of cnidarians are armed with cnidocytes, each of which contains a nematocyst. When the "trigger" is stimulated, the lid opens and the nematocyst is "fired." Stinging barbs on the nematocyst inject toxins into the prey, immobilizing it. Prey also become entangled in the unraveled threads. Cnidocytes fire only once; they are then absorbed, and new ones are formed.

adapted for a sedentary lifestyle. In contrast, many jellyfish species occur only in the medusa form, which is adapted for a free-swimming lifestyle. Some cnidarian species alternate between the two body types during their lifetimes. In these species, the polyp produces male or female medusae, which then engage in sexual reproduction. Following fertilization, the zygote grows into a flattened, ciliated larva, called a *planula,* that settles and transforms into a new polyp. Many cnidarians live as part of large *colonies* in which individuals are physically attached to one another. In colonial species, including corals (Figure 39-7c), one individual founds the colony and then produces the other members asexually by budding.

PHYLUM PLATYHELMINTHES: FLATWORMS

Among the simplest animals that have a bilaterally symmetric body plan are the flatworms, members of the phylum **Platyhelminthes** (*platy* = flat, *helminthes* = worm). The name of these organisms reflects the flattened body shape (Figure 39-10). The phylum includes about 20,000 species that are divided into three major classes: The *turbellarians* are nonparasitic (*free-living*) worms that live primarily under rocks on the bottom of lakes or ocean beds; the *trematodes,* or flukes, are typically leaf-shaped parasites that live as adults inside the bodies of vertebrates; and the *cestodes,* or tapeworms, are highly elongated parasites that live as adults in the intestines of their vertebrate hosts.

Planaria are common turbellarians that inhabit freshwater ponds and streams (Figure 39-11). Examination of a planarian reveals numerous "evolutionary advances," compared to the sponges and cnidarians discussed above. As in most bilaterally symmetric animals, the planarian's sense organs are concentrated near its anterior end and near a mass of nervous tissue, or brain (Figure 39-11a). The brain and associated nerve cords (which run the length of the animal) are the evolutionary beginnings of a central nervous system. The brain and nerve cords receive information from the sense organs and issue motor commands that activate the muscles responsible for locomotion. The middle layer of

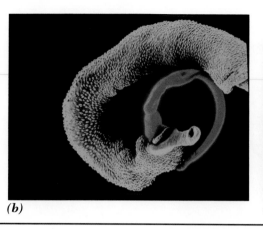
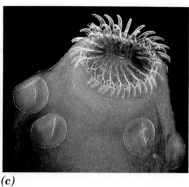

(a) *(b)* *(c)*

FIGURE 39-10

Three types of flatworms, representing the three major classes in the phylum Platyhleminthes, *(a)* A free-living member of the class Turbellaria. Note the flattened body shape. *(b)* A parasitic blood fluke, *Schistosoma.* As adults, the thinner female lives within a groove in the body wall of the stouter male. The male uses its two anterior suckers to hold onto the wall of a human host's blood vessel. *(c)* The anterior end of a tapeworm (class Cestoda) contains a row of hooks and suckers that helps the parasite attach itself to the surface of the host's intestinal wall.

the flatworm is well developed and contains layers of muscle tissue and a variety of differentiated organs that are linked together into organ systems. A planarian, for example, has complex osmoregulatory, reproductive, and digestive systems (Figure 39-11 *b*). Planaria reproduce both sexually and asexually. During asexual reproduction, one individual simply splits into two by the process of fission.

Flatworms are described as **acoelomates,** which means that they lack an internal cavity between their digestive tract and their outer body wall (Figure 39-12). Instead, the spaces between the various organs of the flatworm are filled by a mass of undifferentiated cells. The presence of a bulky middle layer that lacks an internal cavity creates certain problems for an animal. How do the cells of this layer

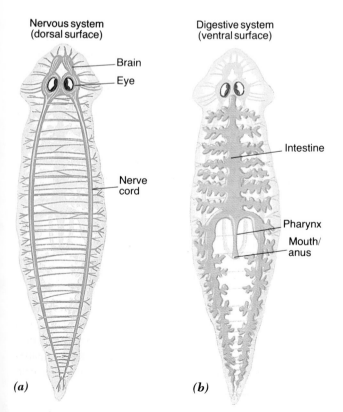

Nervous system
(dorsal surface)

Digestive system
(ventral surface)

Brain
Eye

Nerve
cord

Intestine

Pharynx
Mouth/
anus

(a) *(b)*

FIGURE 39-11

Two systems in a planarian flatworm. *(a)* The ladderlike nervous system has two nerve cords that terminate at a primitive brain. The planarian's eyes respond to light intensity, enabling these free-living flatworms to avoid bright light, which helps them hide from predators. *(b)* The branching digestive system consists of a pharynx that can be everted from the body. At the open end of the pharynx is the mouth that takes in food. The other end of the pharynx leads into the intestine in which food is digested; the nutrients are then distributed via channels of the intestine to the body's cells. Because of the animal's highly branched intestine, cells receive nourishment without the need for a circulatory system.

Acoelomate (Flatworm)

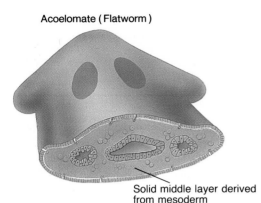

Solid middle layer derived
from mesoderm

FIGURE 39-12
Flatworms are acoelomate animals. As seen in this cross section, flatworms have a "solid" middle layer (derived from the mesodermal cells of the embryo) that lacks a body cavity.

receive nutrients and oxygen? While more complex animals have a circulatory system to meet these metabolic demands, the flatworms manage as a result of their shape, illustrating the relationship between form and function. Because of the animal's flattened shape, all the cells of the body are close enough to the outside environment to allow the exchange of respiratory gases and wastes by direct diffusion through the body wall. The highly branched intestine of a planarian (Figure 39-11) not only acts as a chamber for digestion, it also transports the products of digestion into every nook and cranny of the animal, allowing nutrients simply to diffuse to all of the body's cells.

Trematodes (flukes) and cestodes (tapeworms) are entirely parasitic. Trematodes (Figure 39-10*b*) are not very different in structure from their free-living turbellarian relatives. The life cycle of trematodes involves at least two different host species, one of which is a vertebrate, the other a snail. Among the various flukes that can infect humans, the *schistosomes,* or blood flukes, pose the most serious health hazard, currently infecting as many as 200 million people and causing more fatalities than any other parasites, except for those responsible for malaria.

Humans usually become infected with blood flukes by wading in water that contains schistosome larvae. The larvae penetrate the skin and eventually settle as male and female pairs (Figure 39-10*b*) in the blood vessels of their human hosts. The adult female releases thousands of fertilized eggs that clog the host's blood vessels, interfering with blood flow, destroying tissue, and causing debilitating infections. Ironically, increased modernization in certain areas of Africa has caused the spread of the disease. Prior to the construction of the Aswan Dam in northern Egypt, only about 1 percent of the population was infected by schistosomes. Since the completion of the dam, the irrigation canals and artificial lakes created by the dam have provided a habitat in which schistosome-bearing snails have thrived. In the absence of sewage-treatment facilities, the incidence of infection has risen to nearly 100 percent in some lakeside villages.

Humans may become infected with a tapeworm (Figure 39-10*c*) by eating undercooked fish, pork, or beef that contains a tapeworm larva. Once ingested, the larva develops into an adult within the human intestine. An adult tapeworm may consist of thousands of segments (*proglottids*) and extend several meters in length. Ripe proglottids—those packed with fertilized eggs and developing embryos—typically break off from the posterior end of the worm and pass out of the host with the feces. Some of the specialized characteristics of flukes and tapeworms are discussed in the Bioline: Adaptations to Parasitism Among Flatworms.

PHYLUM NEMATODA: ROUNDWORMS

Members of the phylum **Nematoda,** which includes at least 10,000 species, are the most widespread and abundant animals on earth. We are not aware of the existence of most roundworms because of their microscopic size, but these animals are present in incredible numbers, particularly in the soil. In one count, 90,000 roundworms of several different species were found within a single, rotting apple!

All nematodes have cylindrical, bilaterally symmetrical bodies that are basically constructed as a tube within a tube (Figure 39-13). The inner tube consists of a relatively simple digestive tract, and the outer tube consists of a relatively complex body wall that is covered by a protective, nonliving (noncellular), outer layer, or **cuticle.** Between the tubes is a fluid-filled body cavity that represents an important evolutionary advance over the body plan of flatworms. The nematode body cavity is called a *pseudocoelom* to distinguish it from a *coelom,* which is found in the remaining animals to be discussed in this chapter (compare cross sections in Figures 39-13 and 39-17).

A **coelom** is lined by a epithelial layer that is derived from the embryonic mesoderm (page 693), whereas a **pseudocoelom** ("false coelom") is an unlined cavity that typically develops from the embryonic blastocoel. Regardless of their embryonic origin, the two types of cavities have

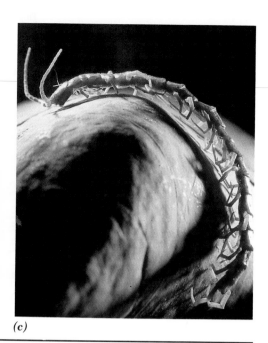

(a)　　　　　　　　　　　　　　*(b)*　　　　　　　　　　　　*(c)*

FIGURE 39-18

The phylum Arthropoda is divided into four subphyla. 1. Trilobita. Although there were more than 10,000 species of marine trilobites during the Paleozoic Era, today they are all extinct. *(a)* A fossilized trilobite. 2. Chelicerata. Chelicerates are composed of segments that are clustered into two major body sections: a cephalothorax (head-thorax), covered with a single exoskeletal plate; and an abdomen. Chelicerates include horseshoe crabs, spiders, scorpions, ticks, and mites. *(b)* A jumping spider. 3. Crustacea. Crustaceans have specialized feeding appendages, called mandibles, and two pairs of antennae (see Figure 39-22). 4. Uniramia. Uniramiates have a pair of mandibles and one pair of antennae. This group includes insects, millipedes, and centipedes. *(c)* A centipede.

receptor cells. The most elaborate of these sensory devices are the compound eyes of crustaceans and insects, which are composed of thousands of photoreceptor elements.

The cuticle of an arthropod is secreted by the underlying epidermis as a single sheet that is molded to the outer contour of the animal; it cannot expand as the animal grows. Being covered by a hardened, nonexpansible, nonliving cuticle is somewhat like living in a suit of armor. As an arthropod grows, it periodically sheds its cuticle and replaces it with a larger one. This process, called **molting,** is under hormonal control. Even though is it made of relatively lightweight materials (consider the weight of an empty crab leg), the weight of an outer body covering greatly increases with increasing body size, which is a major factor that limits the size of arthropods, particularly the terrestrial species. Terrestrial arthropods are also limited in size by the nature of their respiratory system, which requires that air diffuse down long tubules, called **tracheae,** into the depths of the body (page 633).

Arthropods are bilaterally symmetrical animals with segmented bodies and jointed appendages. In many arthropods, adjoining segments are grouped into clusters that serve a common function. In the grasshopper depicted in Figure 39-19*b*, for example, those segments with appendages that are involved in food intake and sensory reception are concentrated in the head, while those with locomotor functions (wings and legs) are concentrated in the middle region, or thorax. The abdominal segments are involved primarily with digestive, excretory, and reproductive functions. In arthropods, the coelom is reduced to small pockets, while the major body cavity (or *hemocoel*) is part of the circulatory system. Blood flows out of the heart into the hemocoel, which directly nourishes most of the body's tissues.

Insects

The largest group of arthropods are the **insects,** which have three thoracic segments, each bearing a pair of legs. Insects can be found at the tops of mountains, in the hottest deserts, the binding of a book, a sack of the driest flour, and nearly every other conceivable terrestrial habitat on earth. The first insects appeared in the Devonian period, approximately 400 million years ago. They are thought to have been

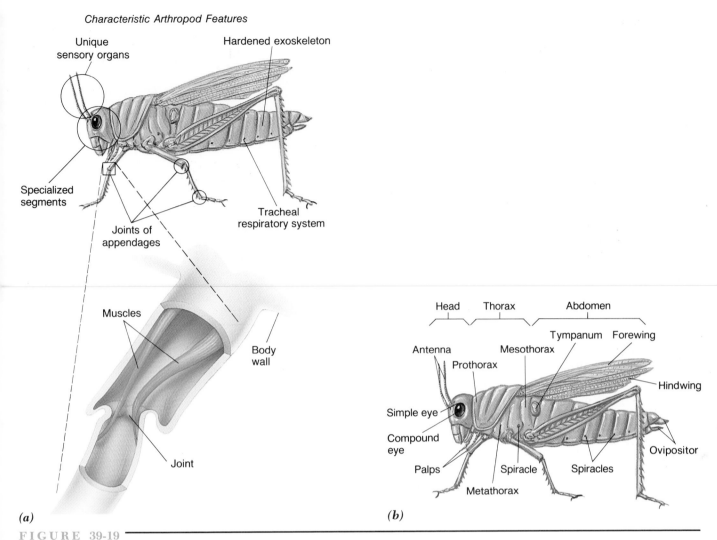

Characteristic Arthropod Features

Unique sensory organs

Hardened exoskeleton

Specialized segments

Joints of appendages

Tracheal respiratory system

Muscles

Body wall

Joint

(a)

Head · Thorax · Abdomen

Antenna

Prothorax

Mesothorax

Tympanum · Forewing

Simple eye

Compound eye

Palps

Metathorax

Spiracle

Spiracles

Hindwing

Ovipositor

(b)

FIGURE 39-19

The arthropod body plan, as illustrated by the grasshopper. *(a)* A number of unique arthropod features can be seen by examining the animal's external anatomy. The inset shows the attachment of bundles of muscle fibers to the inside surface of the exoskeleton on either side of a flexible joint. *(b)* Details of the external anatomy of a grasshopper.

inconspicuous, wingless animals that lived under the leaves, probably resembling modern wingless bristletails (Figure 39-20*a*). These ancestors underwent a remarkable adaptive radiation, spurred by the evolution of flight. The ability to fly is of immeasurable importance; it allows insects to avoid predators, to cross barriers, and to exploit widely scattered resources.

An insect's body is divided into a head, thorax, and abdomen (Figure 39-19*b*). The sensory organs of the head include antennae, which, depending on the species, convey information about touch, specific chemicals, and/or air speed during flight. For example, the antennae of the male silkworm moth are capable of detecting the sex attractant

molecules released by a single female several miles away and guiding the male toward its source. An insect's mouthparts (the appendages around the mouth that are involved in food handling) have shapes and functions that have been modified during evolution so that they are adapted to the mode of feeding of the particular species (Figure 39-21).

This same type of evolutionary modification is seen in the legs and wings of the thoracic segments. For example, the forelegs of a praying mantis (Figure 39-20*b*) grab prey in a lightning-fast attack, while the forelegs of a mole-cricket are used for digging. The hindlegs of both grasshoppers and fleas are modified for jumping. One of the most highly specialized sets of legs is found in honeybees; the legs

(a)

(b)

FIGURE 39-20

This gallery of insects provides a glimpse of the most diverse and perhaps most successful group of animals on earth. Some biologists estimate that there may be as many as 30 million species of insects still to be named and classified. *(a)* A wingless bristletail, one of the most primitive insects. *(b)* A female praying mantis standing atop her egg case. The animal is displaying a threatening posture triggered by intrusion of the camera. *(c)* An ichneumon wasp with its long ovipositor penetrating into a tree where it will deposit its eggs in a woodboring larva.

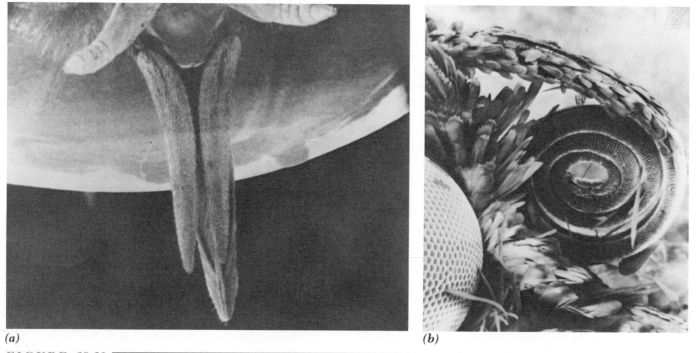

(a)

(b)

FIGURE 39-21

Mouthparts are adapted for specific modes of feeding. The sharp, pointed mouthparts of a horsefly *(a)* are adapted to piercing the skin, while the hollow, tubular mouthparts of a moth *(b)* are adapted to sucking fluid from a flower.

(c)

Ovipositor

of these insects contain structures that collect and store pollen and clean pollen and wax from various parts of the body.

In most insects, both pairs of wings are used in flight, but there are many variations. Some insects are wingless, either because they evolved from primitive wingless insects (as in springtails and bristletails) or because the wings were secondarily lost during evolution (as in lice and fleas). A cockroach's hindwings provide the power for flight, while its forewings are modified as protective covers that fit over the hindwings. In houseflies, the forewings are functional wings, but the hindwings are replaced by *halteres,* which function as balancing organs or stabilizers during flight. The halteres, which resemble a pair of balls mounted at the ends of flexible sticks, swing up and down during flight. If the fly tilts one way or another, the halteres register changes, enabling the fly to readjust its orientation.

The insect's abdomen bears the external genitalia—an *ovipositor* in the female, which lays eggs, or a penis in the male, which transfers sperm during copulation. The shape of the female ovipositor is adapted to the way in which eggs are laid. For example, the very long ovipositor of the ichneumon wasp (Figure 39-20c) projects through the bark of a tree and deposits a single egg into the body of a wood-wasp larva residing within the tree. The location of the larva is determined by sensory receptors that are located in the sheath of the ovipositor. In female bees, the ovipositor has lost its egg-laying function (with the exception of the queen bee, who is the only fertile female) and has become modified into a "stinger."

In some insects, including grasshoppers, the egg develops into a larva that is essentially a small version of the adult. However, most insects progress through a course of development that involves *metamorphosis*—dramatic changes in body form. In these species, the egg develops into a larva that bears no resemblance to the adult. The larva is typically a wormlike form (such as a maggot or a caterpillar) that feeds voraciously and grows rapidly. The larva then transforms into a *pupa,* in which the winged adult form takes shape. Once it emerges from the pupal casing, the adult is specialized for mating, dispersal, and egg laying. In general, each developmental stage has its own habitat, environmental requirements, and adaptations that differ from those of other stages.

Crustaceans, Chelicerates, and Myriapods

If you watch a barnacle in a saltwater aquarium, you will see it repeatedly fan the water with a "net" of hairy-looking appendages (Figure 39-22a). In carrying out this movement, the barnacle is straining its environment for small organisms and other food particles. This is how most crustaceans make a living: as filter feeders. Larger crustaceans, such as crabs and lobsters, are typically bottom-dwelling scavengers or predators. The most numerous crustaceans are copepods that live in the upper layers of lakes and oceans, a habitat referred to as the *plankton* (Figure 39-22b). Copepods and other planktonic crustaceans are important links in aquatic food chains, serving as the primary food source for small fish as well as for giant filter-feeding blue whales and basking sharks. A few crustaceans, such as the familiar "pill pug," which is often mistaken as an insect, have become terrestrial (Figure 39-22c).

The most familiar chelicerates (named after the first pair of appendages, the *chelicerae*) are the spiders (Figure 39-18b), arthropods whose trademark is the use of silk. Silk is produced as a liquid within silk glands located in the abdomen. The liquid silk is converted into a solid thread as the result of a change in the intermolecular arrangement of the silk molecules as they are forced through the fine tubular channels located at the tip of the spider's abdomen. The chelicerae in spiders take the form of a sharp curved fang that is used to pierce the body of its prey and to inject venom. Spiders have no jaws to chew their food, and their mouths do not take in solid material. Instead, spiders are primarily fluid feeders.

In order to fully extract the nutritive contents of their prey, spiders exude digestive enzymes into the prey. The enzymes liquify the tissues, which are then pumped into the digestive tract of the spider with the aid of its sucking stomach. A comparison between the size of an insect when it flies into the web and the size of the corpse that remains attests to the effectiveness of the spider's feeding habits.

Myriapods (*myria* = many; *poda* = feet) are wormlike arthropods with large numbers of legs. Among the group, the centipedes are fast-moving, carnivorous animals with long legs and a long stride (Figure 39-18c). Centipedes feed

(a)

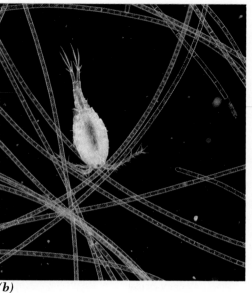
(b)

(c)

Crustaceans. *(a)* A barnacle straining the water for food particles. *(b)* A copepod swimming in the plankton. *(c)* Unlike most crustaceans, this isopod lives in moist terrestrial habitats.

primarily on worms and insects, which they immobilize with venom that is delivered by sharp fangs on one pair of appendages. Millipedes have more numerous shorter legs than do centipedes and are almost exclusively vegetarians. Many of the millipedes are burrowers, whose legs provide the animal with the power to push through the topsoil. Most millipedes protect themselves from predators by secreting noxious substances, including hydrogen cyanide.

PHYLUM ECHINODERMATA: ECHINODERMS

The phylum **Echinodermata** (Figure 39-23) includes about 5,000 species of sea urchins, sand dollars, sea stars, brittle stars, feather stars, and sea cucumbers. Most of the physiological activities of echinoderms are dispersed throughout the organism, rather than being concentrated in particular organs. Echinoderms have no head or brain, and

(a)

(b)

Echinoderms range from sand dollars *(a)* that remove small food particles from the sand to exotic feather stars *(b)* that capture prey with their featherlike arms.

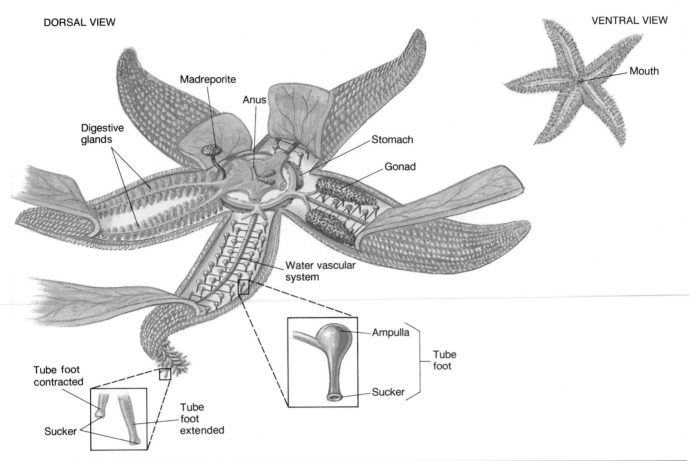

DORSAL VIEW

Madreporite

Anus

Digestive glands

Stomach

Gonad

Water vascular system

VENTRAL VIEW

Mouth

Ampulla

Tube foot

Sucker

Tube foot contracted

Sucker

Tube foot extended

FIGURE 39-24

The anatomy of a sea star, detailing the water vascular, reproductive, and digestive systems. The inset shows a single tube foot. Water enters the water vascular system through the sievelike *madreporite* passes along a network of channels, which connect with the five paired rows of *ampullae.* When muscles in the wall of the ampulla contract, fluid is forced into the hollow, cylindrical tube foot, extending the foot in the direction of movement. When the muscles of the ampulla relax, the foot is withdrawn. When the tip of the foot is pressed against the substrate, it acts like a sucker holding the animal to the surface.

there is no concentration of sensory, respiratory, or excretory structures; all of these functions are carried out by groups of cells that are scattered throughout the animal's body.

Echinoderms possess a number of traits that set them apart from all other invertebrates. Adult echinoderms are basically radially symmetrical but in a different way from the cnidarians, the other major group of radial animals. The radial symmetry of echinoderms is fivefold, or *pentamerous;* that is, parts of the body (such as arms or gonads) tend to be repeated five times around the circle (Figure 39-24). The fossil evidence suggests that echinoderms were originally bilaterally symmetrical and only secondarily evolved a type of radial symmetry adapted to a sessile mode of existence. Unlike their parents, echinoderm larvae are bilaterally symmetrical forms that swim in the bountiful layers of plankton found near the ocean's surface. After a period, they settle to the ocean floor and undergo one of the most

drastic metamorphic transformations found anywhere in the Animal kingdom.

Echinoderms possess a unique endoskeleton that is composed of individual plates of calcium carbonate that lie just beneath the outer epidermis. These plates have jutting spines, inspiring the name echinoderm (*echino* = spiny, *derm* = skin). Another feature unique to echinoderms is their **water vascular system,** which consists of a network of channels and tiny "tube feet" (Figure 39-24) that are used in locomotion, respiration, and sensory perception as well as for clinging to surfaces and attacking prey. The tube feet, which protrude through the skeleton, are present in paired rows and are operated by hydraulic pressure that is generated in the water vascular system. The coordinated extension and withdrawl of thousands of these tiny feet allow the sea urchin or sea star to glide slowly over the substrate or to attach itself very tightly by its suckers to one particular spot.

Sea stars use their tube feet to feed on bivalve mollusks. Anyone who has tried to open a living clam or oyster is familiar with the strength by which the valves are held together. Yet, a small number of sea stars can devastate a bed of oysters. The sea star positions itself over the mollusk, attaches its tube feet to the shell, and pulls on the valves with a pressure of about 20 kilograms. Once the valves come apart a scant 0.1 millimeter, the sea star uses a "trick" unavailable to any other animal: It everts its stomach into the slight crack and secretes a fluid of digestive enzymes onto the soft tissues of the mollusk. The everted stomach then ingests the digested food matter by phagocytosis, after which the stomach is withdrawn back into the sea star's body.

PHYLUM CHORDATA: CHORDATES

Even though vertebrates form a distinct and phylogenetically related group, they are not placed in a phylum of their own. Rather, they are grouped together with a number of invertebrates (called **protochordates**) in the phylum **Chordata**. These remarkable little invertebrates are thought to be similar to the organisms from which we all arose.

Chordates are distinguished from all other animals by a number of characteristics (Figure 39-25), including the following:

1. **Pharyngeal gill slits.** At some stage in their lives, even if only temporarily during embryonic stages, chordates possess perforations in the wall of the **pharynx** (the anterior portion of the digestive tract). In protochordates and fishes, these openings in the embryo give rise to the gills. In terrestrial vertebrates, they close over and disappear during development.

2. **A dorsal hollow tubular nerve cord.** The central nervous system of a chordate develops from a flattened plate of cells located on the dorsal surface of the embryo, which rolls up to form a hollow neural tube (page 694). The wall of the tube develops into the brain and spinal cord, while the central channel becomes filled with fluid.

3. **A notochord.** The notochord is a flexible rod that forms in the embryo just beneath the neural tube. The notochord persists throughout life in many of the protochordates and primitive vertebrates, where it functions as a flexible skeletal rod, stiffening the body during swimming or burrowing. In most vertebrates, the notochord is replaced by a bony vertebral column.

4. **A postanal tail.** The anus is not situated at the posterior tip of chordates as it is in most animals; rather, the anus is followed by a tail. In many vertebrates, the tail disappears during embryonic development.

Protochordates

There are two major groups of protochordates: tunicates and lancelets, both of which provide interesting insights into vertebrate evolution.

An adult **tunicate** (Figure 39-26a,b) consists largely of a sac within a sac. The inner sac, or pharynx, is made up of a network of microscopic gill slits that are supported by a framework of gill bars. The beating of cilia generates a current that moves water through the phyarngeal gill slits. Food particles suspended in the water become trapped in mucus that is secreted in copious amounts by gland cells located in a groove, called the *endostyle*, at one end of the pharynx. The endostyle is thought to be the evolutionary forerunner of the vertebrate thyroid gland. The outer sac, or *tunic*, consists primarily of extracellular material, including cellulose, a polysaccharide that is very rare within the animal kingdom.

�▶ If biologists were able to observe only adult tunicates, it is unlikely that these animals would ever have been classified in the same group as vertebrates. When the development of a tunicate is followed, however, the fertilized egg is seen to transform itself into a larva of the type shown in Figure 39-26c, which possesses all of the basic chordate characteristics. When these "tadpole larvae" settle onto a suitable substrate, they metamorphose into an adult; in the process, they lose their notochord, dorsal nerve cord, and postanal tail. Of the major chordate characteristics, only the pharyngeal gill slits remain in the adult.

Since it is the tunicate larva that resembles vertebrates, rather than the adult, biologists speculate that the *larva* of an ancient tunicatelike organism acquired the ability to produce gametes for sexual reproduction and acted as the ancestor of all other chordate groups. The phenomenon whereby a larval form becomes a sexually mature adult occurs quite often among animals. For example, many salamanders fail to undergo metamorphosis but develop into sexually mature adults in a larval body form.

Lancelets are small, fishlike invertebrates that live in shallow oceans throughout the world. The lancelet's body is flattened and pointed at both ends. These animals are normally found with their posterior end buried in the sediment, and their anterior end, which contains the mouth, protruding into the water. Unlike adult tunicates, adult lancelets (Figure 39-27) show all the major chordate characteristics.

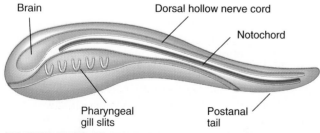

FIGURE 39-25

A generalized chordate showing the four major chordate characteristics: pharyngeal gill slits, a dorsal hollow nerve cord, a notochord, and a postanal tail.

Labels on figure: Brain, Dorsal hollow nerve cord, Notochord, Pharyngeal gill slits, Postanal tail

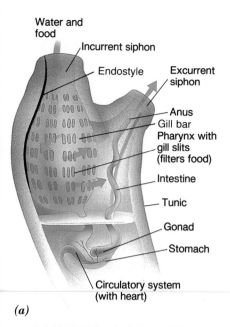

Water and food
Incurrent siphon
Endostyle
Excurrent siphon
Anus
Gill bar
Pharynx with gill slits (filters food)
Intestine
Tunic
Gonad
Stomach
Circulatory system (with heart)

(a)

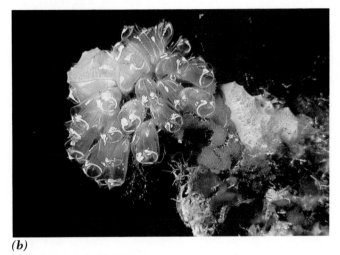

(b)

(c)

FIGURE 39-26

Tunicates. Adult tunicates *(a,b)* are filter feeders that draw sea water in through an incurrent siphon and filter out suspended food particles as the sea water passes through its gill slits. Filtered water, digestive wastes, and gametes exit through an excurrent siphon. The adult has only one chordate characteristic: gill slits. *(c)* The tunicate larva possesses all four of the major chordate characteristics: a dorsal hollow nerve cord, notochord, gill slits, and a postanal tail.

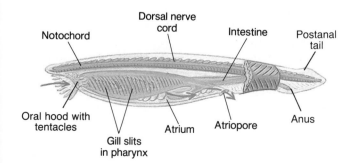

Notochord
Dorsal nerve cord
Intestine
Postanal tail
Oral hood with tentacles
Gill slits in pharynx
Atrium
Atriopore
Anus

FIGURE 39-27

Knifelike adult lancelets wriggle into the seabed, their head protruding for feeding. Sea water enters through the oral hood, passes through the gill slits and into the atrium, exiting via the atriopore. Food particles become trapped in the mucus that lines the pharynx.

Like tunicates, lancelets are filter feeders that pump water through their pharyngeal gill slits, trapping suspended food particles in mucus that is secreted by a glandular endostyle.

The lancelet's notochord is well developed and runs the entire length of the body, providing support for the animal's burrowing activities. On each side of the notochord are blocks of muscle tissue that are separated from one another by thin sheets of connective tissue. This organization of muscle and connective tissue clearly foreshadows the similar arrangement found in fishes. In fact, lancelets are thought to resemble the earliest fishes, which were probably also filter feeders that lived in the bottom sediments of the ocean over half a billion years ago. The division of the lancelet's body musculature (and associated nerve tracts) into blocks is the first evidence of a segmented body plan among chordates. The dorsal nerve cord and postanal tail are also evident in adult lancelets.

Even though the members of a phylum may be extremely diverse, they share certain basic structural features that they all inherited from a common ancestor. Over the course of evolution, the members of each major invertebrate phylum have become increasingly diverse, as they have become adapted to new habitats and to new modes of existence. (See CTQ #2.)

VERTEBRATES

The vertebrates are an immensely successful group of animals, occupying habitats in virtually every terrestrial, marine, and freshwater environment. All vertebrates have a vertebral column—a backbone composed of separate vertebrae that are constructed of bone or cartilage—that surrounds and protects the delicate spinal cord. The vertebral column replaces the embryonic notochord and forms the main axis of the internal skeleton. Anteriorly, the vertebral column connects with the cranium, which encases the brain. Living vertebrates are grouped into seven classes, three of which are composed of fishes.

FISHES

The first traces of vertebrates occur as 550-million-year-old flakes of bone that were found in sediments from an ancient sea that once spread over much of the western United States. The first fossils to reveal the form of early vertebrates date back about 475 million years and consist of small, bottom-dwelling fish called *ostracoderms.* These fish lacked jaws and were covered in heavy, bony plates. Before their demise by the end of the Devonian period, the ostracoderms gave rise to a branch of *jawless* vertebrates that have persisted to the present time.

Living, jawless vertebrates (members of the class Agnatha) include lampreys and hagfish (Figure 39-28), both of which bear a superficial resemblance to eels but lack jaws, fins, and scales. The mouth of a lamprey is surrounded by a round sucker, armed with teeth. The lamprey attaches itself to the outside of another fish, then rasps a hole in the prey's flesh with its tooth-bearing tongue. Along with pollution, lampreys have contributed to the devastation of the fishing industry of the Great Lakes of the United States and Canada, and the eradication of this animal has been one of the primary goals of the industry for several decades.

Hagfish are primarily scavengers that attack dead and dying fish, such as those caught in nets. Hagfish enter the

(a)

(b)

FIGURE 39-28

Living, jawless fishes of the class Agnatha, include *(a)* lampreys that feed on the flesh of other living fishes and *(b)* scavenging hagfish.

mouth or anus of the prey and eat the tissue from the inside, rapidly reducing the animal to a bag of skin and bones. Hagfish are also noted for their prolific mucus production. A single hagfish is able to fill a large bucket with slime in a matter of seconds.

About 415 million years ago, a new type of fish appeared that possessed a revolutionary new adaptation: a pair of jaws. Jaws allowed fishes to become predators, pursuing prey rather than feeding on sediments on the bottom of their aquatic environment. Jaws also provided a means of defense, reducing the need for heavy protective armor and increasing mobility. Jaws evolved from the skeletal bars that lined the anterior gill slits (Figure 39-29).

■■▶ The conversion of gill structures into jaws provides one of the best examples of the change in the function of a structure over the course of evolution; it is clearly revealed by following the embryonic development of a modern fish. The first embryonic gill arch forms in a similar manner to that of the other gill arches, but it then shifts course and develops into the jaws. Along with the evolution of jaws came the appearance of paired **fins** on the side of the body, which must have greatly increased a fish's maneuverability. From these early jawed fishes, two great lines of modern fishes arose: the cartilaginous fishes (class Chondrichthyes), and the bony fishes (class Osteichthyes).

The **cartilaginous fishes** include the sharks, rat fishes, skates, and rays (Figure 39-30). While these fishes arose from ancestral species with heavy bony skeletons, bone formation was lost during evolution, and their skeletons consist entirely of cartilage. The skin contains an enormous number of tiny, sharp *denticles* that become greatly enlarged in the fish's mouth, forming rows of teeth. Most sharks are aggressive predators that possess extremely keen sense organs to help them locate their prey. The whale shark (*Rhincodon typus*) is the largest of all fishes, weighing up to 20 tons and growing to lengths of 18 meters (60 feet). Ironically, this giant shark doesn't consume large prey; a sieve in its pharynx filters minute crustaceans and drifting plankton.

Skates and rays have a flattened shape adapted for living close to the ocean bottom. There, these fish feed primarily on hard-shelled invertebrates, which they crush with their flat, square teeth. Even among skates, the manta ray, the largest of the group, is a filter feeder. Unlike most other fish, the cartilaginous fishes reproduce via internal fertilization, the sperm being delivered into the genital opening of the female by a pair of modified pelvic fins.

The **bony fishes** include species with skeletons that are at least partly composed of bone. The *lobe-finned* crossopterygians described in the chapter opener were primitive, bony fish. While there are only a handful of living species of lobe-finned fishes, another line of bony fish, the *ray-finned teleosts,* has been extremely successful and numbers over 18,000 diverse species today (Figure 39-31).

The structure and function of bony fishes is closely correlated with their aquatic habitat. Since water is much denser than air, it provides an organism with buoyancy

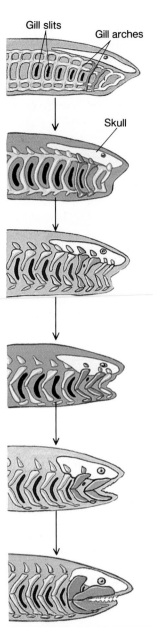

FIGURE 39-29

Jaws evolved from gill arches. This sequence traces the evolutionary changes in the bony skeleton that supports the gills (called gill arches) from an ancient jawless ancestor to a modern shark.

(upward support), eliminating the need for a heavy, supportive skeleton. (This is evident when you try to remove the delicate bones from a trout or salmon steak before eating it.) Although the density of water is high, the density of bone and muscle is even greater; therefore, fish must have some adaptation for keeping them from sinking to the bottom. Most fishes have a gas-filled *swim bladder* that gives the fish an overall density equal to its surrounding, allowing the animal to remain at any desired depth.

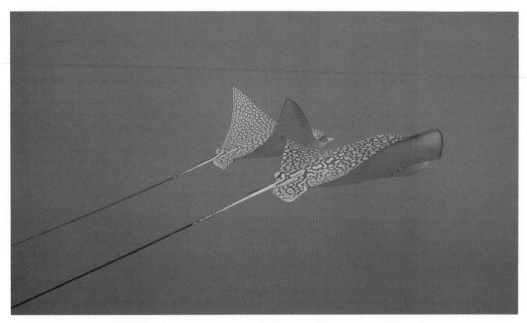

FIGURE 39-30
The class Chondrichthyes includes the manta ray *(photo)*, sharks, and rat fish.

AMPHIBIANS

The class **Amphibia** includes about 3,000 species of frogs, salamanders, and the less familiar wormlike caecilians (Figure 39-32). As we described in the opening section of this chapter, amphibians were the first terrestrial vertebrates, but they never became fully adapted to life in dry terrestrial habitats. Even their descendants, the modern amphibians, remain intermediate in many ways between fully aquatic and terrestrial organisms. For example, unlike reptiles, birds, and mammals, modern amphibians have a moist, thin, permeable skin so they will rapidly lose body water. As a result, adults must remain in damp habitats or stay close to water to prevent dehydration.

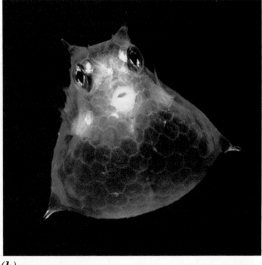

(a)   *(b)*

FIGURE 39-31
Two representatives of the diverse class Osteichthyes (the bony fishes). *(a)* The blenny has eyes on top of its head to scan for danger. *(b)* A juvenile cowfish.

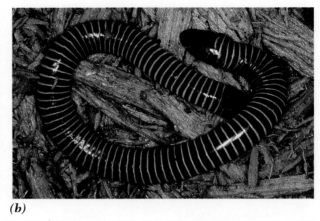

(a)   *(b)*

FIGURE 39-32

Representative amphibians. Amphibians include salamanders and newts, frogs and toads *(a),* and the less familiar caecilians *(b).*

Similarly, amphibian eggs lack a protective impermeable shell and lose water rapidly in dry environments. No matter where they live as adults, most amphibians return to the water to mate and lay their eggs. The eggs develop into free-swimming larvae, such as the familiar frog tadpole, which, in most species, undergoes metamorphosis into a terrestrial adult. It is this dual life—partially aquatic and partially terrestrial—that gave these animals the name "amphibians" (*amphi* = both; *bios* = live). Within the past few years, the numbers of individuals in many amphibian populations have decreased drastically. The reason for the decimation of these species' populations, which is thought to be due to human alteration of the environment, is the subject of current research.

▐▶ Amphibians have played a central role in the history of vertebrates. They were the first vertebrates to leave the water and walk with four legs on the land. Nearly 250 million years ago, a new type of vertebrate appeared that had well-developed lungs and scaly, water-resistant skin. This new animal laid "land eggs" that resisted drying and were equipped with an internal supply of food and water for the developing embryo. Freed of the need to be near water or to return to water for reproduction, these animals were better adapted for life on land than were their amphibian ancestors. These new vertebrates were the reptiles.

REPTILES

The earliest reptiles appeared in the mid-Carboniferous period, approximately 300 million years ago. They remained as small, inconspicuous, lizard-like animals until the Triassic period, at which time they began an adaptive radiation of unprecedented splendor (page 772). Many of them grew to enormous size; if they were alive today, some would be tall enough to peer over the top of a three-story building.

The reptiles of the Mesozoic Era are classified into more than 15 distinct orders. Today, the class **Reptilia** (with approximately 6,000 species) contains only four orders: lizards and snakes, turtles (Figure 39-33*a*), crocodiles and alligators (Figure 39-33*b*), and one lizardlike animal, the tuatara, which is the sole survivor of an ancient order. Of these four groups, the latter three were much more prominent in earlier geologic periods, while the snakes and lizards have increased in recent diversity over the past tens of millions of years.

The most important terrestrial adaptations in reptiles are those that restrict water loss. A reptile's skin is much thicker than that of amphibians, and it is very dry. In addition, the epidermis contains horny *scales,* which protect the skin's surface and help prevent water loss. Reptiles also possess mechanisms that prevent water loss during excretion. For example, reptiles convert nitrogenous wastes to uric acid which, because it is virtually insoluble in water, can be excreted as a nearly dry paste.

Finally, unlike amphibians, reptiles reproduce by internal fertilization, and a reptile's eggs are enclosed in a waterproof outer casing—either a leathery coat or a thin, brittle shell. The embryo that develops within this enclosed environment forms in association with several extraembryonic membranes whose functions we discussed in Chapter 32. One of these membranes, the *amnion,* surrounds a fluid-filled chamber that provides the embryo with a protective, aquatic environment, even though the egg is situated on dry land.

BIRDS

The class **Aves** (Figure 39-34) contains about 9,000 species of birds, ranking this group of vertebrates second only to bony fishes in number of species. The most distinctive characteristic of birds is their possession of feathers, a fea-

(a)

(b)

FIGURE 39-33

Representative reptiles. *(a)* In murky waters, this flattened turtle attracts little attention from an unwary fish, which quickly disappears into the reptile's wide mouth. *(b)* A flick of its head, and this crocodile will soon be dining on a young frog.

ture that is directly involved in two of the most important aspects of avian (bird) biology: endothermy and flight.

Recall from Chapter 28 that endotherms utilize the heat that is generated by their own metabolism to maintain a constant, elevated body temperature. The average body temperature of a bird is 41°C (106°F), several degrees higher than that of mammals. The evolution of feathers may have been a crucial step in the development of endothermy since this feature would have provided insulation, helping to maintain a body temperature higher than that of the environment. Because of their high level of metabolic activity, birds have been able to colonize some of the world's harshest environments, as is exemplified by the emperor penguins that lay their eggs on the bare Antarctic ice in the middle of winter.

Flight gives birds a freedom that is unmatched among other vertebrates. Being airborne, birds have far greater visibility of the terrain and its offerings than do animals that remain on the ground. Flight also allows small, defenseless animals to escape from land-based predators, and it enables birds to take full advantage of distantly separated food sources. When food in one geographic region becomes limited at a particular time of the year, a population of birds is able to migrate to a distant region where conditions are more favorable. For example, the arctic tern *(Sterna paradisaea)* breeds during the summer in the Arctic Circle and spends the winter on the opposite side of the globe, in Antarctica.

The ancestors of birds were probably small, bipedal, insect-eating dinosaurs that walked on their hindlimbs; the forelimbs evolved into wings. This reptilian heritage is evident in many features of modern birds, including their scaly skin, shelled eggs, and *cloaca,* a common chamber that receives discharges from the intestinal, urinary, and reproductive systems.

Flying "machines" must be made to fit strict specifications. This is why most birds have a similar body shape, even though they live in very different habitats and have very different diets. The evolution of flight required sweeping anatomic and physiological changes, including a reduction in the weight of the skeleton, a decrease in the number of bones, and the fusion of many of the bones that remained. For example, the skeleton of a frigate bird weighing 2 kilograms and having a wing span of 2 meters weighs only about 110 grams, or 5 percent of the bird's body weight: less than the weight of its feathers.

The reduction in a bird's skeletal weight is largely a result of the hollow construction of the skeleton (Figure 39-34b). The fusion of individual bones reduces the number of movable joints in the bird's skeleton, making it rigid and better adapted as the internal framework of a flying machine. In order for the flight muscles to deliver the power required to fly, a bird has a very high metabolism; its body contains numerous large air sacs; its circulatory and respiratory systems are extremely efficient; and its digestive system is capable of processing large amounts of food in a very short period of time.

The skull of a bird is also very lightweight (Figure 39-34, inset). It bears a light, horny, toothless *beak* that is composed primarily of the protein keratin. Since the bird's jaws have no teeth to support, the jaw can also be reduced in mass. The beak is a bird's primary "tool." A knowledgeable biologist can look at a bird's beak and tell if it is used for digging, probing, piercing, chiseling, straining, cracking,

(a)

(b)

Interior
of skull

Flight feather

Fingers (fused)

Palm

Thumb

Wrist

Forearm

Vertebrae

Shaft

Barbules

Barb

Upper arm

Shoulder girdle

Ribs

Hip girdle

Sternum with keel
(attachment of
flight muscles)

Upper leg

Lower leg

Heel

Fused ankle

Toes

FIGURE 39-34

Class Aves—birds. *(a)* The numerous hummingbird species in the tropics have brilliant-colored feathers. The diversity of shapes and sizes of their bills enables different species of hummingbirds to suck nectar from various kinds of flowers. *(b)* The anatomy of a bird helps make the animal an efficient flying "machine." The bird's bones are extremely light and are filled with air spaces; its feathers contain lightweight protein and consist of a shaft with projecting barbs and interlocking barbules. The large *flight feathers* play a key role in providing the thrust and lift necessary for flight.

(a)

(b)

FIGURE 39-35

The class Mammalia contains three different groups. *(a)* Monotremes include strange-looking spiny anteaters and this duckbilled platypus, the only mammals that lay eggs. *(b)* Young marsupials are born at an extremely immature stage and complete development in their mother's pouch. A baby koala is born after only 30 days of development. Blind and with no back legs, the underdeveloped koala uses miniature front legs to crawl into its mother's pouch, where it stays for about 6 months. When a young koala becomes too large for the pouch, it clings to its mother's back, returning to the pouch for an occasional drink of milk. *(c)* Most mammals, such as this mountain gorilla, nourish their embryos and fetuses through a placenta, then continue to nourish their newborn with maternal milk from mammary glands.

tearing, pecking, or some other activity. The varying functions of the beak are reflected in the striking differences in its shape among different types of birds.

MAMMALS

The class **Mammalia** includes about 4,500 species, whose body form and lifestyle are highly diversified. Two characteristics distinguish mammals from other vertebrates:

1. Mammalian skin is endowed with hair: thin, elongated filaments made largely of the protein keratin. Most mammals are covered with a relatively dense layer of hair, though it is reduced to patches in humans and is even more sparse in whales, which are hairless except for the sensory whiskers found on their snout. Like feathers, hair traps air, insulating the body surface and reducing the loss of body heat.

2. Mammals nourish their young with milk produced by mammary glands. Milk typically contains large amounts of casein, a phosphate-containing protein, as well as calcium and other salts, lactose, and suspended

globules of fat. The fat composition varies from about 1.5 percent in horses to nearly 40 percent in seals.

▐▶ The ancestors of mammals were the therapsid reptiles, which were abundant in the Permian and early Triassic periods (between 230 million and 270 million years ago). Bona fide mammals, which are distinguished from fossil reptiles by their lower jaw that is composed of a single bone, appeared by the late Triassic period, approximately 210 million years ago. These early mammals remained relatively scarce in number until the end of the Mesozoic Era, at which time the ruling reptiles disappeared from the earth. The earliest mammals are thought to have been small (rat-sized), nocturnal forms that ate insects and depended on a keen sense of smell and an acute sense of hearing. Their need to withstand the cooler night temperatures may have provided strong selective pressure for the evolution of endothermy.

Mammals are classified into three groups (Figure 39-35):

1. **Monotremes.** Unlike other mammals, monotremes lay eggs which, like those of birds, are incubated outside the female's body. Unlike birds, newly hatched

(c)

monotremes are nourished with milk from their mother. Today, monotremes are represented by only the platypus and the spiny anteaters.

2. **Marsupials.** Marsupials give birth to their young at a very early stage of development. The remainder of the developmental process occurs outside the uterus in a special pouch that contains the mammary glands. The most striking feature of the marsupials is the incredible diversity of forms among such a small number of species. Included within the approximately 250 species are carnivores, insectivores, herbivores, and omnivores; terrestrial, arboreal (tree-dwelling), and fossorial (underground) species; and species that run, hop, burrow, swim, climb, and glide. Examples of marsupials include kangaroos, koala bears, and opposums.

3. **Placentals.** Placental mammals nourish developing offspring internally through a placenta, the structure through which nutrients and wastes are exchanged between the mother and fetus. Some types of placental mammals are born at a highly immature stage; they are hairless, blind, deaf, and barely able to move. Mammals of this type, such as many rodents, bears, and foxes, remain in the nest for a considerable period, spending much of their time attached to their mother's nipples. Species whose young are born at immature stages tend to have larger litters with a smaller percentage of survivors. In contrast, other mammals, including horses and sheep, are born at a relatively advanced stage. Most are larger, herbivorous animals that must be able to run with the adults, should a predator appear. For example, a young colt stumbles the first few times it tries to stand, but usually within about an hour's time, it is quite steady on its feet and can run alongside its mother.

Placental mammals are typically divided into about 16 different orders, several of which contain only a single genus. We will confine our discussion to the order in which humans are classified: the order Primates.

Primates

The **primates** are a diverse taxonomic group (Figure 39-36), ranging from primitive insect-eating tree shrews to monkeys, apes, and humans. Primates are one of the oldest mammalian orders, having evolved from small, arboreal ancestors. The primates remained in the trees; this fact, more than any other, seems to have shaped the course of primate evolution. Life in the trees has numerous advantages. Tree-dwelling animals are relatively safe from attack by large, carnivorous animals and are exposed to an abundant supply of insects, leaves, or fruits to eat. But there are also risks to an arboreal life; one serious miscalculation can cause an arboreal animal to fall to the ground and die.

While smaller arboreal animals, such as squirrels, depend on their claws to hold them to the bark as they scamper through the trees, primates have evolved a different approach; they grasp the branches with their hands and feet. The development of a grasping hand, with an opposable thumb that could close to meet the fingertips, has allowed primates to become much larger than other arboreal mammals since they can hold onto the limbs rather than just balance themselves on them. As an added evolutionary "bonus," grasping hands and feet have allowed primates to manipulate objects to a degree unachieved elsewhere among animals.

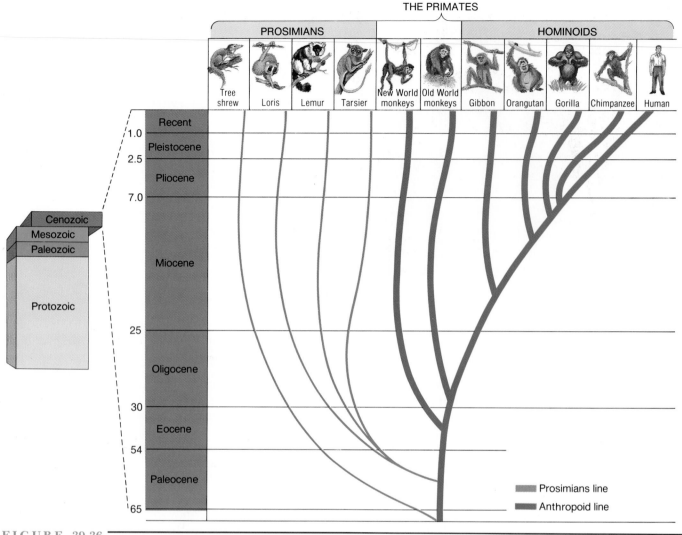

FIGURE 39-36

The primate evolutionary tree. The primates are one of the earliest placental orders still represented by living members. Between 60 million and 70 million years ago, a common ancestral line diverged into two major branches, one leading to the more primitive primates (the *prosimians*), the other leading to the "higher" primates—the *anthropoids,* consisting of monkeys, apes, and humans. The more recently two lines branch apart, the more similar the organisms at the end of each branch. Thus, gorillas and chimpanzees are more closely related to humans than are monkeys.

As they perfected their own unique style of arboreal locomotion, primates became more agile, gaining the ability to hang from branches, to jump from tree to tree, and to perform all the other amazing gymnastic feats performed by monkeys. These feats require a great deal of *hand-eye coordination.* The animal must be able to judge distances with great precision, to learn just how far to jump, and to make instantaneous decisions while in mid-air. Such attributes require a keen sense of vision, a sensitive touch, and a great deal of intelligence and motor control, all characteristics that we recognize in ourselves. If it weren't for the fact that our ancestors lived in trees, it is very unlikely that we would be living in houses today.

The vertebrates are an immensely successful group of animals on the ground, in the air, and under water. The basic vertebrate body plan, which probably first appeared in larval tunicates, has proven highly adaptable as it has been molded by evolution into a highly diverse array of animals. Of all the vertebrates, birds and mammals have evolved the most efficient organ systems, the largest brains and most complex behaviors, the most complex means of locomotion, and the greatest independence from fluctuations in their environment. (See CTQ #3.)

REEXAMINING THE THEMES

The Relationship between Form and Function

☑ With the exception of a few vestigial structures, virtually every part of every animal has a shape that promotes its particular function. We can illustrate this point with sponges, the least complex animals described in this chapter. Sponges possess cells (porocytes) with doughnut-shaped openings that allow microscopic streams of water to enter the animal's body, supplying food and oxygen. Sponges also possess choanocytes, cells that have a single undulating flagellum, whose collective beating generates water currents, and a microscopic collar consisting of a fine mesh that traps suspended particles. Sponges possess delicate, microscopic spicules made of lime or glass, whose pointed shape discourages predators. There are as many examples of the relationship between form and function as there are animals, organs, cell types, and biomolecules.

Acquiring and Using Energy

☀ All animals are heterotrophs, depending on other organisms to provide them with chemical energy in the form of previously-synthesized organic molecules. The energy derived from these food materials is used to fuel the diverse activities described in this chapter.

Unity within Diversity

△ Animals constitute the most diverse kingdom on earth, with over 1.3 million recorded species and millions more that are probably undescribed. Biologists have classified animal life into about 35 different phyla; the members of each phyla have a distinct body plan by which they are all united. Within the major phyla, the ancestral body plan has been modified over evolution, generating animals with strikingly different forms and functions. Nowhere is this more evident than among the chordates. Even though all chordates share a number of basic features that define the phylum, this basic body plan has given rise to animals as diverse as tunicates and birds, fishes, and mammals.

Evolution and Adaptation

▮▶ The characteristics of an organism are shaped by two factors: the genes (and corresponding traits) it inherits from its ancestors, and the environment that selects for animals with adaptations suited to current conditions. For example, both a snail and an octopus have a radula because their ancestors had a radula. The structures of the radula of a snail and that of an octopus are quite different, however. The former may be adapted to scraping algae off rocks, while the latter may be adapted to tearing the flesh from live prey. Similarly, all arthropods have jointed appendages because they evolved from a common ancestor with jointed appendages. The structure and function of the appendages of various arthropods are highly diverse, however. Even among insect forelimbs, we have described appendages adapted for capturing prey, digging, jumping, and collecting pollen.

SYNOPSIS

Members of the Animal kingdom are multicellular, heterotrophic, sexually reproductive, and capable of locomotion at some stage of their life. Animals are thought to have evolved from protists, probably from more than one different protistan group.

An animal's body plan can be explained on the basis of its evolutionary ancestry and the environment in which it lives. Three of the most important characteristics that distinguish body plans are symmetry, body cavities, and the presence or absence of segmentation. Symmetry concerns the arrangement of the body parts relative to an axis or plane through the animal; sessile animals tend to exhibit radial symmetry, while motile animals tend to exhibit bilateral symmetry. All animals but sponges have a digestive cavity. In addition, most animals have a secondary cavity (a pseudocoelom, coelom, or hemocoel) between its digestive tract and body wall. The bodies of annelids, arthropods, and chordates develop from repeating blocks of tissues; they are segmented animals.

Sponges are aquatic animals that circulate water through their bodies, trapping and phagocytizing microscopic food particles. Water enters the body through myriad microscopic pores; food matter is trapped by choanocytes that line the water channels; and the nutrient- and oxygen-depleted water exits through one or several large openings.

Cnidarians include hydras, jellyfish, sea anemones, and coral. Cnidarians are aquatic, radially symmetrical animals that are composed of a three-layered body wall surrounding a blind gastrovascular chamber. There are two basic cnidarian body forms: polyps, which tend to be sessile, and medusae, which tend to be motile. Cnidarians feed on other living animals that are stunned by the nematocysts fired from cells located on the tentacles surrounding the cnidarian's mouth.

Flatworms include planaria, parasitic flukes, and parasitic tapeworms. Flatworms are bilaterally symmetrical animals that exhibit the beginnings of cephalization; they have a head with a concentration of nervous tissues and sense organs. Unlike cnidaria, flatworms have well-developed organ systems. Their flattened shape allows them to nourish and oxygenate their cells without the need for a circulatory system.

Roundworms include the most abundant microscopic organisms of the soil as well as some of the worst human parasites. Roundworms are bilaterally symmetrical, cylindrical animals, whose body contains a fluid-filled pseudocoelom that acts as a hydrostatic skeleton and is covered by a nonliving cuticle.

Mollusks include snails, slugs, oysters, clams, squids, and octopuses. Mollusks are bilaterally symmetrical, nonsegmented animals whose bodies can be divided roughly into a head-foot, a dorsal visceral mass, and a mantle. The ribbon-shaped, tooth-bearing radula is a unique molluskan structure, as is the shell, which contains protein and calcium carbonate and is lined by a layer of "mother of pearl."

Annelids include earthworms, bristleworms (polychaetes), and leeches. Annelids are bilaterally symmetrical, segmented animals with a well-developed head and organ systems. Their segmented body provides flexibility during swimming or burrowing, while their internal, fluid-filled coelom acts as a hydrostatic skeleton. Earthworms have setae to anchor them during burrowing, while polychaetes have paired appendages that function as swimming paddles and sites of gas exchange.

Arthropods include insects, crustaceans, spiders, ticks, and mites. Arthropods constitute the most diverse phylum. They are bilaterally symmetrical, segmented animals that are covered by a complex, jointed cuticle (exoskeleton), and jointed appendages. The exoskeleton provides support, protection, locomotion, food handling, gas exchange, gamete transfer, and sensory reception. It is shed periodically and replaced during periods of growth. The coelom is reduced in size, and the hemocoel (fluid-filled spaces of the circulatory system) comprises the major body cavity.

Echinoderms include sea urchins, sand dollars, and sea stars. Echinoderms are radially symmetrical animals that have a unique water vascular system complete with tiny tube feet used for attachment and locomotion, and an endoskeleton that consists of calcareous plates beneath the epidermis.

Chordates include a small number of invertebrate protochordates and the vertebrates (fishes, amphibians, reptiles, birds, and mammals). Chordates are bilaterally symmetrical animals that are characterized by pharyngeal gill slits, a dorsal hollow tubular nerve cord, a notochord, and a postanal tail.

Key Terms

radial symmetry (p. 861)
bilateral symmetry (p. 861)
cephalization (p. 861)
digestive tract (p. 861)
segment (p. 863)
invertebrate (p. 863)
vertebrate (p. 863)
filter feeder (p. 864)
spongocoel (p. 864)
osculum (p. 864)
body wall (p. 864)
choanocyte (p. 865)
spicule (p. 865)
gastrovascular cavity (p. 867)
epidermis (p. 867)

gastrodermis (p. 867)
mesoglea (p. 868)
cnidocyte (p. 868)
polyp (p. 868)
medusa (p. 868)
acoelomate (p. 870)
coelom (p. 871)
pseudocoelom (p. 871)
protostome (p. 873)
deuterostome (p. 873)
mantle (p. 874)
mantle cavity (p. 874)
radula (p. 874)
nacre (p. 874)
bivalve (p. 874)

parapodia (p. 876)
setae (p. 876)
jointed appendage (p. 876)
exoskeleton (p. 877)
molting (p. 878)
tracheae (p. 878)
water vascular system (p. 883)
protochordate (p. 884)
pharyngeal gill slit (p. 884)
pharynx (p. 884)
dorsal hollow tubular nerve cord (p. 884)
notochord (p. 884)
postanal tail (p. 884)
fin (p. 887)

Review Questions

1. Match the following structures with the appropriate phylum:

   ____notochord   a. Cnidaria
   ____cnidocytes   b. Chordata
   ____radula   c. Porifera
   ____tube feet   d. Mollusca
   ____choanocytes   e. Annelida
   ____setae   f. Arthropoda
   ____halteres   g. Echinodermata

2. What characteristics distinguish mammals from other chordates? Birds from other chordates? Amphibians from reptiles?

3. What characteristics are shared by the arthropods and vertebrates, which enable them to be fully terrestrial?

4. Compare the adaptiveness of radial versus bilateral symmetry.

5. Would you expect sessile animals to have motile larva? Why or why not?

6. Which animals lack tissues? Which lack organs? Which lack a complete digestive system with a mouth and an anus?

7. Which animal phylum probably has the greatest number of individuals? The greatest number of species? The largest-sized members?

Critical Thinking Questions

1. Divide a sheet of paper into two columns, one labeled "Reflect Ancestry," the other "Reflect Environment." Fill in each column with as many characteristics of *Latimeria* as you can. Compare your answers with those of a classmate. If there are any differences, discuss these until you have two lists upon which you both agree.

2. Select one of the phyla listed below and describe the diversity found among its major groups. List the basic structural features held in common by the major groups within the phylum that are evidence of their common ancestry. Explain how the differences among them are adaptations to new habitats and lifestyles. If you prefer, you may present your answers through pictures and/or diagrams:

   Cnidaria, Mollusca, Arthropoda, Echinodermata, Chordata.

3. Most reptiles, birds, and mammals are adapted to life out of the water, as are the flowering plants. Which adaptations to living out of water are similar in these animals and plants? Which are different?

4. Why do you suppose the largest fishes (and the largest mammals) are filter feeders? Why do you suppose the largest arthopods (a species of crab) are aquatic? Why do you suppose the largest sponges contain large numbers of internal chambers rather than simple channels?

5. Which would you expect to be more likely to have a dormant stage in their life cycle—freshwater or marine animals? Why?

Additional Readings

Barnes, R. D. 1987. *Invertebrate zoology*, 5th ed. Philadelphia: Saunders. (Advanced)

Cherfas, J. 1991. New hope for vaccine against schistosomiasis. *Science* 251:630–631. (Advanced)

Hildebrand, M. 1988. *Analysis of vertebrate structure*, 3d ed. New York: Wiley. (Advanced)

Lutz, P. E. 1986. *Invertebrate zoology*. Reading, MA: Addison-Wesley. (Intermediate–Advanced)

Noble, E. R., et al. 1989. *Parasitology*, 6th ed. Baltimore: Lea & Febiger. (Intermediate)

Pearse, V., et al. 1987. *Living invertebrates*. Pacific Grove, CA: Boxwood Press. (Intermediate)

Pough, F. H., et al. 1989. *Vertebrate life*. New York: Macmillan. (Intermediate)

Smith, J. L. B. 1956. *Old fourlegs: The story of the coelacanth*. London: Longman. (Introductory)

Thomson, K. S. 1991. *Living fossil: The story of the coelacanth*. New York: Norton. (Intermediate)

Young, J. Z. 1981. *Life of vertebrates*, 3d ed. New York: Oxford University Press. (Intermediate–Advanced)

Ecology and Animal Behavior

The two major arenas of life are the earth's lands and waters. Lichens slowly transform solid rock into soil that supports a multitude of organisms, including the trees from which these vibrant leaves have fallen. The water teams with microorganisms that break down the leaves. By decaying dead organisms, even simple microorganisms form a critical link in the recycling of chemical nutrients in all ecosystems.

PART
· 8 ·

CHAPTER
◄ **40** ►

The Biosphere

STEPS
TO
DISCOVERY
The Antarctic Ozone Hole

BOUNDARIES OF THE BIOSPHERE

ECOLOGY DEFINED

Ecology and Evolution

THE EARTH'S CLIMATES

THE ARENAS OF LIFE

Aquatic Ecosystems

Biomes: Patterns of Life on Land

THE HUMAN PERSPECTIVE

Acid Rain and Acid Snow: Global Consequences of Industrial Pollution

BIOETHICS

Values in Ecology and Environmental Science: Neutrality or Advocacy?

1981

1984

1985

S TEPS TO DISCOVER Y
The Antarctic Ozone Hole

Virtually all the earth's organisms owe their lives to the filtering effects of the ozone layer 20 to 50 kilometers (12 to 30 miles) above. About 99 percent of the sun's lethal ultraviolet rays are absorbed by this invisible layer; as a result, organisms are spared overexposure to these killer rays that destroy many biological molecules, including DNA. Recent studies have revealed a very alarming trend, however: The ozone layer is rapidly being destroyed. As this protective layer becomes depleted, we are bound to see more cases of skin cancer, cataracts, and immune deficiencies, as well as reduced crop yields and other serious consequences.

The prospect that the ozone layer may be at risk was forwarded in 1974 by F. Sherwood Rowland and Mario Molina, two atmospheric chemists at the University of Cali-

fornia, Irvine. Rowland and Molina created laboratory conditions resembling those found in the mid- to outer stratosphere, the region of the earth's atmosphere where the protective ozone layer is located. The chemists examined the effects of a group of manmade compounds called chorofluorocarbons (CFCs) on ozone molecules (O_3). They discovered that CFCs destroyed ozone molecules with alarming efficiency. CFCs had been widely used as propellents in aerosol products, such as deodorants and air fresheners, as refrigerants, and to inflate the bubbles in styrofoam.

When products containing CFCs are used, CFCs are released into the air as a gas. In their research, Rowland and Molina discovered that when CFC molecules were irradiated with ultraviolet light equivalent to that found in sun-

Photographs taken from space by weather satellites revealed the earth's protective ozone layer was deteriorating

light, the CFC molecule broke apart, releasing a free chlorine atom (Cl). The free chlorine reacted with an ozone molecule (O_3), splitting it into $ClO + O_2$. The ClO molecule then combined with free oxygen (O) to form Cl and O_2, thereby rereleasing the chlorine atom. The free chlorine then split yet another ozone molecule, and so on. Based on their observations, Rowland and Molina projected that a single chlorine atom could destroy between 10,000 and 100,000 ozone molecules in the stratosphere before being washed out of the atmosphere by rain. At that rate, Rowland and Molina hypothesized that CFCs could eventually damage the protective ozone layer. In fact, they projected that CFCs could destroy between 20 and 30 percent of the ozone layer, threatening all life on earth.

Many scientists were skeptical of Rowland and Molina's projections, most believing that a decline of only 2 to 4 percent would be more likely, sometime in the next century. In addition, although Rowland and Molina had clearly demonstrated the ozone-destroying capabilities of CFCs under laboratory conditions, they had not examined whether CFCs actually reached the stratosphere, where they would pose a threat to the ozone layer.

In the early 1980s, satellite studies were initiated to assess stratospheric ozone and CFC levels. These studies verified that CFCs indeed reached the stratosphere. Atmospheric balloons, each carrying flasks that sampled the gases in the air of the stratosphere, were also released. When recovered, the flasks contained both CFCs and chlorine gas, supporting Rowland's and Molina's hypothesis. These samples yielded some very disturbing data: The CFC concentration in the stratosphere had doubled in only 10 years. This led EPA (Environmental Protection Agency) scientists to project a 60 percent decline in ozone levels by 2050 if CFC use and production continued to grow at the current annual rate of 4.5 percent.

Once studies established that significant levels of CFCs were reaching the stratosphere, scientists began to measure changes in the thickness of the ozone layer. The British Antarctic Survey collected monthly samples of ozone in the stratosphere. They found that each spring, ozone concentrations fell sharply and that ozone depletion was growing worse each year. The scientists were also able to correlate rapid ozone depletion to periods of increased usage of CFCs.

In 1987, an international panel of more than 100 scientists was convened to study all of the data and make a judgment about the threat to the ozone layer. In their report, the scientists confirmed that ozone had declined by 1.7 percent to 3.0 percent over the Northern Hemisphere since 1969, and from 5 percent to 10 percent over Antarctica during the same period. In September 1987, the United Nations sponsored negotiations to reduce CFC production worldwide; 24 nations signed what became known as the Montreal Protocol, agreeing to cut CFC production in half by 1999. Since CFC-containing products continue to be used and since CFCs persist in the atmosphere for long periods, however, CFCs are expected to remain in the stratosphere for decades to come.

Satellite photographs clearly depict the trend of ozone depletion. In the late 1980s, ozone had become depleted by as much as 60 percent at certain altitudes over Antarctica. By 1990, ozone levels had dropped by as much as 95 percent. In 1992, the World Meteorological Organization reported that there were regions over Antarctica where no ozone could be detected at all. The report summarized the fact that the ozone hole over Antarctica had enlarged to a record size of over 9 million square miles, about three times the size of the continental United States. This was about 25 percent larger than reported in previous years.

These findings are very important because they reveal that the rate of ozone depletion is even more rapid than originally forecasted by scientists, including Rowland and Molina. In an October 1992 interview, Rowland was asked for this response to these new findings. He said: "What we are looking at is ozone depletion caused by CFCs that were released back in 1987 and 1988. The expectation is that it will probably continue to get worse in the stratosphere for another decade or so." Ultimately, life hangs in the balance.

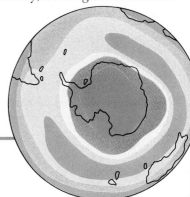

at an alarming rate.

BOUNDARIES OF THE BIOSPHERE

*T*he Apollo astronauts watched the earth recede in the distance as they sped toward the moon (Figure 40-1). The artificial life-support systems that sustained the astronauts' lives while on board the spacecraft temporarily replaced their natural life-support system, the earth.

Without artificial support systems, life as we know it could not exist beyond the earth's atmosphere or deep beneath its solid surface. All life is restricted to a relatively narrow zone of air, water, and land, called the **biosphere,** the thin envelope in which all living organisms are found (see Chapter 1). The biosphere is only 22.4 kilometers (14 miles) thick, from the upper limits of life in the atmosphere to the depths of the dark ocean trenches. In relation to the size of the earth, the biosphere is only about as thick as the skin on an apple. All life as we know it exists within this thin layer that envelops the earth.

▼ ▼ ▼

The biosphere includes portions of the earth's hydrosphere (waters), lithosphere (crust), and atmosphere (gases). Of the three spheres, the **hydrosphere** of oceans, seas, lakes, ponds, rivers, and streams harbors the greatest quantity of life. Most aquatic organisms are found in shallow waters along the shorelines, where sunlight penetrates, providing the energy for photosynthesis. Similarly, most land organisms live near the illuminated surfaces of the **lithosphere,** the rocky crust that forms the earth's rigid plates and terrestrial habitats. In the gaseous **atmosphere,** organisms are found living at altitudes below 7 kilometers (4.3 miles).

There are a few exceptions to these general boundaries. For example, pollen, dormant bacteria, and fungal spores have been found in the atmosphere at altitudes above 62 kilometers (100 miles). In addition, gravity pulls dead organisms and their wastes downward in the hydrosphere, providing food for life in the darkened depths of the earth's waters. Thus, the biosphere is not uniform in width; rather, its thickness varies from one location to another.

Life as we know it is supported within a relatively thin envelope of the earth's air, water, and soil. This life-sustaining envelope is called the biosphere. (See CTQ #2.)

FIGURE 40-1

The earth from space. Portions of Africa, Madagascar, and Antarctica can be seen beneath the swirling cloud layer. Seemingly calm at this distance, the earth teems with millions of species of organisms.

ECOLOGY DEFINED

Ecology is the study of the biosphere and its components. Although the term is a familiar one, the word itself is somewhat new. It was coined a little more than 125 years ago by the German zoologist Ernst Haeckel to refer to the total relations between an animal and its organic and inorganic environment. Haeckel derived the term "oecology" from the Greek, *oikos,* meaning home, and *logos,* meaning the study of. At a national conference held in 1893, American scientists adopted a simpler spelling by dropping the letter "o."

Since the late 1800s, more restrictive definitions have been proposed. For example, in his 1927 book on *Animal Ecology,* Charles Elton defines ecology as "scientific natural history." Nearly 4 decades later, in 1963, in his book *Ecology,* Eugene Odum defined ecology as "the study of the structure and function of nature," a reminder that the unifying theme of form and function applies to ecology as well as to all other biological topics. Many biologists simply define ecology as the "study of ecosystems" because an ecosystem includes both living organisms and the inanimate physical environment.

These three definitions are all somewhat vague. A more exact definition is offered by Charles J. Krebs in his 1985 text, *Ecology.* Krebs wrote, "Ecology is the scientific study of the interactions that determine the distribution and abundance of organisms." Based on Krebs's definition, ecology studies

- where organisms are found,
- how many organisms occur there, and
- why organisms occur where they do.

The distribution and abundance of organisms are affected by the ways in which organisms interact with one another in the biotic community (e.g., competition) and with the surrounding physical abiotic environment (e.g., availability of light or nutrients). These are the topics we will explore in the final five chapters of this text.

ECOLOGY AND EVOLUTION

To study ecology is to study evolution because ecology and evolution are interlinked. Evolution takes place in the ecological arena. Organisms occur where they do because, through evolution, they have acquired adaptations that enable them to survive and reproduce in particular habitats. These adaptations are the result of natural selection, the process that drives evolution. Recall that natural selection is the result of a number of factors, including limited resources, competition for space or mates, adverse changes in the physical environment, the introduction of new organisms into an ecosystem, and so on. These "natural selection factors" are also ecological factors. In other words, natural selection is ecology in action.

No organism can ever be separated from its environment. As a result, organisms are continually affecting and, in turn, are continually being affected by their environment. This constant ecological interplay is the foundation for evolution. (See CTQ #3.)

THE EARTH'S CLIMATES

Climate—the prevailing weather in an area—is the chief physical factor in determining where organisms are distributed within the biosphere. Not only does climate differ greatly over the surface of the earth today, but global climates have changed dramatically over the almost 4 billion years since life originated on earth. These climatic changes have resulted in the formation of a great diversity of biomes as well as the evolution of millions of kinds of organisms with adaptations suited for these unique environments.

The earth's climates are shaped by major circulation patterns that develop in the earth's atmosphere and oceans. Air and ocean currents result from three primary factors: (1) the fact that different parts of the earth receive different amounts of incoming solar radiation; (2) the daily rotation of the earth and its annual orbit around the sun; and (3) the distribution and elevation of the earth's land masses.

Because the earth is round, different parts of the earth receive different amounts of solar radiation (Figure 40-2a). These differences in incoming radiation heat the earth unevenly, producing warm tropics near the equator, where sunlight hits the earth more directly, and progressively cooler regions in higher latitudes, as the intensity of sunlight is reduced. The north and south poles are the two coldest regions in the biosphere because they receive the least amount of sunlight, about five times less than the amount that reaches the equator. This variation in incoming sunlight not only helps produce different climates and biomes at different latitudes, it also affects the activity of organisms that are found there. This explains why tropical areas have much higher productivity than do biomes found in cooler, higher latitudes.

In addition to its round shape, the earth's annual orbit around the sun and its constant 23.5 degree tilt cause annual changes in incoming solar radiation to those parts of the earth farther away from the equator (Figure 40-2b). These annual differences trigger a progression of seasons away from the equator. During the summer months, for example, the Northern Hemisphere of the earth is tipped toward the sun (and therefore receives greater amounts of radiation), while during winter, it is tipped away from the sun. The differences in incoming radiation help explain why summer months are warmer and why organisms are more active during the summer than during the winter months. Because of the earth's constant tilt, the seasons are reversed in the Southern Hemisphere.

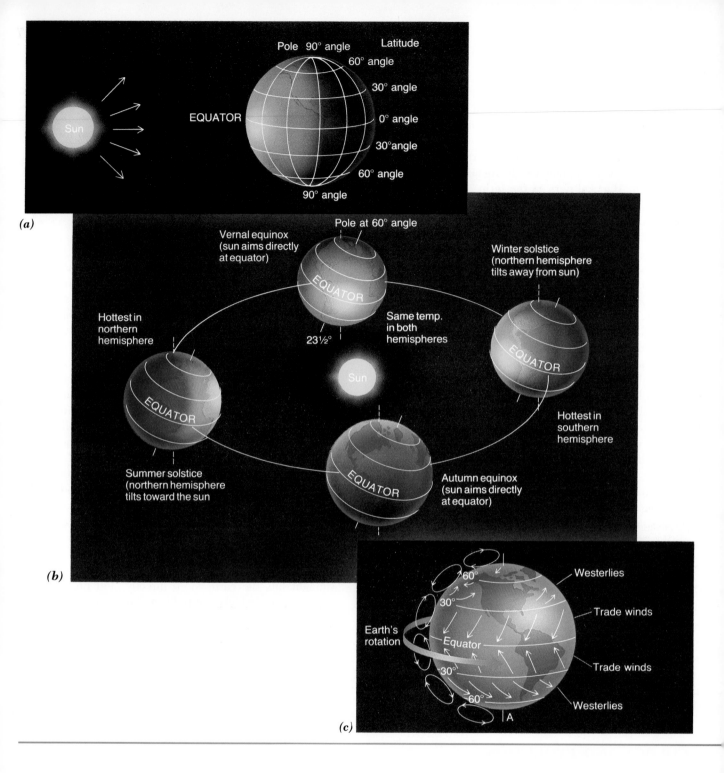

(a)

(b)

(c)

The variation in the amount of sunlight that strikes different parts of the earth at different times of the day and year heats the earth's air and oceans unevenly. Warm air near the equator rises and flows toward the poles, where cooler air sinks. Because of the earth's rotation, however, the moving air mass breaks into six circulating coils—three in the Northern Hemisphere, and three in the Southern Hemisphere (Figure 40-2c).

Intense sunlight at the equator causes the air to rise. As air rises, it cools, releasing its moisture and producing abundant rainfall. The cool poleward-moving air masses sink and become reheated at about 30° north and south latitude, creating zones of decreased rainfall. This is where the earth's largest desert regions are found. As the air in the coils moves over the surface of the rotating earth, prevailing winds are formed. *Trade winds* forms as air moves back toward the equator, and the *westerlies* form as air moves northward from 30° toward 60° latitude. At 60° north and south latitude, the air rises and cools, and, as at the equator, it releases its moisture as rainfall. This region of cool tem-

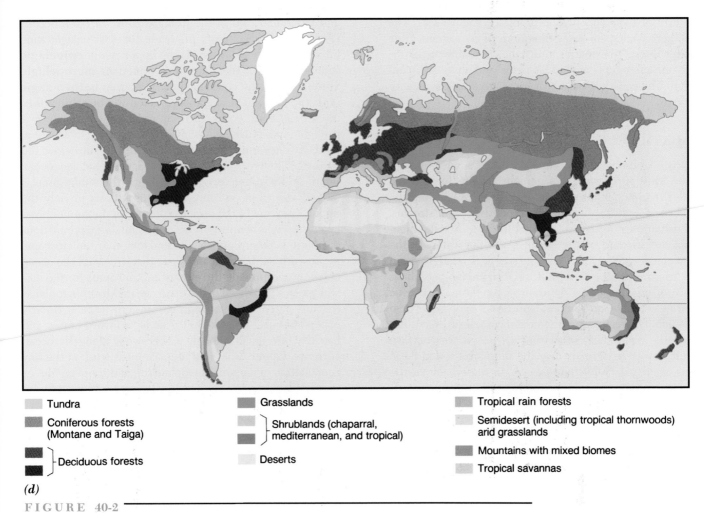

Tundra

Coniferous forests
(Montane and Taiga)

⎫
⎬ Deciduous forests
⎭

Grasslands

⎫
⎬ Shrublands (chaparral,
⎭ mediterranean, and tropical)

Deserts

Tropical rain forests

Semidesert (including tropical thornwoods)
arid grasslands

Mountains with mixed biomes

Tropical savannas

(d)

FIGURE 40-2

Climate and the earth's biomes. *(a)* The round shape of the earth results in differences in the amount of incoming sunlight in different regions. *(b)* The earth's yearly orbit around the sun and its constant tilt result in differences in the amount of incoming sunlight at different times of the year, generating the earth's seasons. *(c)* Six coils of air circulation create the earth's prevailing winds, as well as zones of high and low rainfall. *(d)* Climate is the principal factor in determining where plants grow, which, in turn, is the principal factor in determining where animals live. Not surprisingly, the locations of the earth's biomes closely follow it's 11 main climate types.

peratures and relatively high rainfall produces expansive temperate forests of North and South America, Europe, and Asia.

Prevailing winds blowing over the ocean surface create currents in the sea's upper layers. Continents deflect these water movements, creating slow, circular ocean currents. Circulation mixes the oceans' waters, bringing warmer waters that originated in tropical areas to higher latitudes, and the cooler waters from higher latitudes to tropical areas. These exchanges redistribute heat around the earth, con-

tributing to the formation of the earth's climates (Figure 40-2*d*).

The earth's many climates are the result of the combined effects of air and water currents; the rotation, tilt, and orbit of the earth; and the distribution and elevation of the continents. Climate is the principal factor that affects where organisms are found within the biosphere. (See CTQ #4).

ARENAS OF LIFE

In Chapter 1, we mentioned that the biosphere is the highest level of biological organization (see Figure 1-4). The biosphere consists of many thousands of *ecosystems*, from conifer forests to estuaries to marshes to alpine meadows, each one a functional unit of biological and physical organization. The earth's major ecosystems are either aquatic or terrestrial.

AQUATIC ECOSYSTEMS

Nearly 75 percent of the earth's surface is covered with water. Most of this water (71 percent) is salty, forming the earth's marine habitats of seas and oceans. Other aquatic habitats include freshwater lakes, ponds, rivers, streams, and estuaries of mixed fresh and salt water. Aquatic organisms are affected by the chemistry of the water. In the open oceans, salinity (salt concentration) is relatively even and, as a result, does not limit the distribution and abundance of organisms that naturally occur there. In marine habitats, sunlight and mineral nutrients vary more than does salinity, creating environmental complexity that influences the community of marine organisms. In contrast, salinity levels vary greatly near the shorelines and in those areas where rivers flow into oceans. In these habitats, the salt concentration of the environment has the greatest effect on the organisms that live there.

Marine Habitats

Marine ecologists disagree as to whether distinct ecosystems form in open oceans or whether the ocean is simply one giant ecosystem that is constantly homogenized by fluctuating water currents. The former point of view is supported by studies that clearly reveal that there are specific locations in the oceans where nutrients circulate upward, producing a localized community that has more kinds of organisms, in greater numbers, than are found in adjacent areas. Each of these localized areas displays a unique biological order and regulation. Because of their unique physical and biological composition, some ecologists contend that these areas are distinct ocean ecosystems.

Other studies have established that distinct communities and features also exist in other ocean regions. In a paper written in 1988, however, Richard Barber argues that since these boundaries may easily disappear, perhaps marine communities should not be considered separate ecosystems. For instance, Barber points out that in 1983, at the peak of the El Niño, large-scale changes in temperature and ocean circulation obliterated boundaries. Since an El Niño recurs every 5 years or so, Barber suggests that the concept of ecosystems with well-defined, constant boundaries is more applicable to land than to the oceans. Of course, when the effects of an El Niño subside, marine ecosystem boundaries reappear.

Open Oceans Marine ecosystems are found throughout the earth's oceans, in shallow waters around continents, islands and reefs; in intertidal zones (the area between high and low tides); and in the **pelagic zone,** or open oceans (Figure 40-3). Biologists subdivide the vast pelagic zone into three, vertical layers: (1) the upper, sunlit **epipelagic** *(photic)* **zone;** (2) the dimly lit, intermediate **mesopelagic zone;** and (3) the continually dark, bottom **bathypelagic** *(aphotic)* **zone.** The sea floor is called the **benthic zone.**

In the open ocean two important physical factors—sunlight and nutrients—vary from location to location. Sunlight does not usually penetrate the ocean below 100 meters (320 feet). Considering that the average ocean depth is 3.9 kilometers (12,500 feet), the sunlit zone is indeed very thin, permitting photosynthesis in only the upper 2 percent of the ocean's volume. In the sunlit epipelagic zone, where nutrients are abundant, large populations of *phytoplankton* flourish. Phytoplankton are microscopic photosynthetic bacteria and algae that drift with the ocean currents and provide the energy and nutrients for the animal species that dwell in the open ocean. Most phytoplankton are eaten by *zooplankton,* tiny crustaceans (mostly copepods and shrimplike krill), larvae of invertebrates, and fish that are small enough to be swept along by ocean currents. Larger fishes and other animals feed on the tiny zooplankton or on both phytoplankton and zooplankton (or simply, plankton). An adult blue whale, for example, guzzles an average of 3 tons of plankton in a single day.

Since the dim midwaters of the mesopelagic zone do not receive enough light to power photosynthesis, phytoplankton do not reside there. Inhabitants of this zone must therefore make daily migrations either up to the epipelagic zone or down to the bathypelagic zone to feed. Large fishes, whales, and squid are the principal animal predators of the mesopelagic zone. They eat smaller fishes that feed on the wastes or carcasses of organisms from the sunlit zone.

The benthic zone contains no energy-capturing plants or bacteria, except for a few unusual deep-sea communities of organisms that inhabit the warm waters surrounding fissures and cracks in the ocean floor (see Bioline: Living on the Fringe of the Biosphere, page 167). This pitch-black zone is populated primarily by heterotrophic bacteria and scavengers that feed on a constant rain of organic debris, wastes, and corpses that settle to the bottom as well as predators that eat the scavengers and one another. These bottom dwellers include sponges, sea anemones, sea cucumbers, worms, sea stars, and crustaceans, as well as a collection of odd-looking fish, some with dangling lanterns that light up to attract a meal or a potential mate (Figure 40-4).

Coastal Waters The greatest concentration of marine life inhabits the shallow **coastal waters,** or **neritic** ("near shore") **zone,** along the edges of continents and reefs (Figure 40-5). In addition to abundant light, coastal waters are

◁ B I O E T H I C S ▷

Values in Ecology and Environmental Science: Neutrality or Advocacy?

By ANN S. CAUSEY
Prescott College

Ecology, along with all other branches of science, is usually characterized as a neutral endeavor. Ecologists gather and interpret data and then attempt to describe and explain relationships between organisms and their environments. In recent years, however, some have sought to challenge this picture of ecology as a descriptive, neutral science. Instead, ecology is portrayed as a subversive activity that seeks to impose a conservative (anti-growth) social doctrine on an unsuspecting public. These critics contend that ecology is not merely *descriptive*, but *prescriptive* as well. Ecologists, they say, are not merely gathering facts but are drawing value judgements from their interpretations of those facts and prescribing courses of action that often obstruct technological and economic progress.

Which portrayal is accurate? What is the proper role, if any, of values in ecology? And, how does environmental science enter into this dispute? To understand the source of this controversy, we need to take a closer look at science and values.

The strict separation of facts from values is a key component of the success of science in objectively analyzing reality. The objectivity of science is not a function of the objectivity of scientists however; it is a function of the rules of the game, of the scientific method itself. Granted, the findings of ecology may be particularly relevant to judgements of value made in many arenas of society. However, these findings are obtained by a method designed to separate objective analysis of nature from subjective value judgements; they are, in principle, value-neutral.

Ecologists, then, *in their science*, do not properly become advocates; that is, ecology is not necessarily a thinly disguised form of environmental advocacy, or *environmentalism*, as some have charged. Nevertheless, certain scientific findings, if

highly corroborated, may indeed call for advocacy on particular issues. Thus began a movement, born in the 1960s and maturing through the past 2 decades, to apply ecological principles and theories not only to wild plants and animals in natural environments but to humans, as well.

This application of ecology to human society has become known as *environmental science*. Environmental scientists seek to understand the interactions of humans with other species and with the nonliving environment. Thus, these scientists must integrate knowledge from the natural sciences (ecology, physics, chemistry, biology, geology, etc.) with that from technology and the social sciences (demography, economics, politics, etc.). This interdisciplinary field seeks not only to understand how the various levels of life operate and interact but also to identify, integrate, and apply principles of sustainability to human society.

The blending of ecology and environmentalism to create environmental science has resulted in the charge that science is now value-laden. This characterization may be incorrect. While environmental science rests on ecology as its main scientific foundation, the values reflected by environmental science do not come to it by way of ecology. In other words, ecology can only tell us how nature works, not what humans should to. It may also help us accurately predict the consequences of certain human activities, though it cannot tell us whether or not we should curtail those activities since such decisions always involve weighing competing values and interests. For instance, we know from ecological studies that no population can reproduce uncontrollably. Ultimately, natural resistances, such as disease and starvation, bring a rapidly growing population back into balance with available resources. Environmental scientists assure us that this

mechanism will ultimately check the growth of human populations. No science or scientists can determine whether we should now take steps to control human population growth, however, or whether we should simply wait and let nature take its course. That is a value judgement—an ethical decision.

Environmental scientists use information from the natural and social sciences to tell us how we can best minimize our impact on the natural systems that support humans and other species and how we can develop ecologically and economically sustainable societies. They work under the legitimate *assumption* that such goals are desirable. Their work may even make these goals more appealing and widespread, graphically showing us the undesirable consequences of current unsustainable practices such as deforestation or continued production of ozone-destroying chemicals. Nevertheless, science cannot tell us what objectives to strive for or what our goals should be. These are decisions that reflect larger societal values and standards, standards that are only partly shaped by our scientific understanding.

Do you consider ecology and environmental science advocacy? Some say no. They see ecology as appropriately responsive to environmental crises too urgent to ignore. Modern ecologists and environmental scientists can make a great contribution toward solving many environmental problems by helping us base our policies and judgements on reality, rather than on wish, conjecture, or the political agenda of any particular interest group. According to this view, values do not threaten the objectivity of scientists or the validity of their work. Instead, they help guide modern science in the service of human needs and wants, an increasingly important goal in these times of environmental ignorance and deterioration.

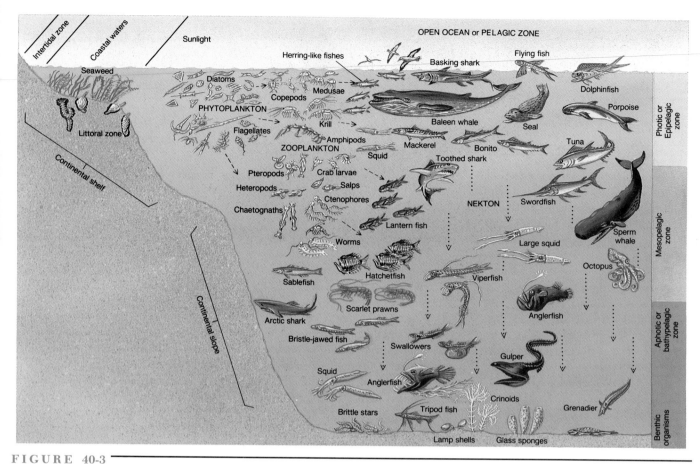

FIGURE 40-3

Marine habitats and representative marine organisms. (Sizes of organisms are not drawn to same scale.)

generally rich in nutrients that are available as a continuous drain from the surrounding land. Waves, winds, and tides constantly stir coastal waters, distributing the nutrients. Bathed in light and nutrients, photosynthetic organisms grow at fantastic rates and in great profusion, providing ample food and habitats for a multitude of fishes, arthropods, mollusks, worms, and mammals.

The richest coastal waters occur in regions of *upwelling,* where nutrient-laden water from below circulates to the surface. Upwellings form along the coasts of Peru, Portugal, Africa, and California and create conditions where light and nutrients are abundant. Under these conditions, it is not surprising that regions of upwellings form the earth's most fertile fishing waters.

Coastal waters present some unique problems for organisms, however. Rough, surging waters can shred or bash organisms against rocks. Many coastal animals have adaptations (both physical and behavioral) that help protect them from the surging waters. For instance, some animals remain in burrows or possess tough, protective shells. Since surging

waters can quickly wash organisms onto the shore or pull them out to sea, a number of coastal organisms have adaptations for clinging to stationary objects. For example, algae use holdfasts to cling to rocks; mussels anchor themselves with powerful cords; and abalone hold tight with their large, muscular foot. Remaining attached to a solid object in churning waters is clearly advantageous, but it also creates problems. How do these sedentary organisms forage for food, find mates, or disperse offspring so that they can colonate new, suitable habitats?

A researcher named I. E. Effort studied sand crabs (*Emerita analoga*) along the Pacific Coast of North America and found, not surprisingly, that the main ocean currents greatly affected the crab's dispersal and, ultimately, its geographic distribution. Sedentary adult sand crabs release tiny larvae for dispersal. The larvae drift along with the plankton from current to current. When the California Current turns sharply westward near Baja, California, all larvae drift out to open sea and perish. As a result, the California Current determines the southern limit of the distribution of this

FIGURE 40-4
Generating its own light, the suspended lantern on this anglerfish attracts this larval fangtooth in the black of the ocean's aphotic zone. Backward-pointing teeth prevent even large fishes from escaping this predator's grasp.

sand crab. Likewise, the northern limit of distribution is affected by a north-flowing current that carries larvae from southern Oregon to Alaska. All larvae produced in Oregon can drift only further north to the Gulf of Alaska, where they all perish in the frigid water. Consequently, Oregon populations are colonized only by larvae from southern populations.

Shallow waters are also found along coral reefs. In these shallow, warm, tropical waters, exceptionally diverse communities flourish. For example, F. H. Talbot and his

FIGURE 40-5
The ocean's bounty. Abundant light and the constant circulation of nutrients make coastal waters the ocean's richest habitats.

colleagues identified nearly 800 species of fish around one small island on the southern edge of the Great Barrier Reef in Australia. At the northern edge of the Great Barrier Reef, over 1,500 species of fish were recorded. Depending on location, coral reefs are classified as one of three types: (1) *fringe reefs*, which extend out from the shores of islands and continents; (2) *barrier reefs*, which are separated from shores by channels or lagoons; or (3) *atolls*, islands of coral with a shallow lagoon in the center (Figure 40-6).

Coral reefs develop only in tropical areas (between 30° north and 30° south latitudes), where water temperatures never fall below 16°C (60°F). Persistent cold currents prevent coral reefs from developing along the western edges of continents.

The Intertidal Zone   Unlike in coastal waters, where organisms are continually submerged in shallow waters, the organisms that inhabit **intertidal zones** must be able to survive in both water and air, as the flow of tides rhythmically submerge and expose their habitats. As its name suggests, the intertidal zone lies between high and low tides at the interface between the ocean and the land. Recurring tides and the geology of the shoreline create a variety of intertidal habitats: rocky shores that often contain *tide pools* (isolated pools that are left behind in rock depressions when the tide recedes) and *mud flats* or *sandy beaches* that form along nonrocky shorelines (Figure 40-7).

The pattern by which organisms inhabit the intertidal zone is a striking example of the dynamic interplay between physical factors (temperature and dehydration) and interactions between organisms (competition and predation). This interplay often produces four distinct strata of organisms in the intertidal zone (Figure 40-8). The uppermost stratum is the **splash zone,** where organisms receive only sprays of water at high tides. Below the splash zone is the **high intertidal zone,** which is followed by the **middle intertidal zone** and the **low intertidal zone.** Organisms in the high intertidal zone must be able to withstand exposure to air for longer periods than do those that inhabit the middle or low intertidal zones.

Estuaries

Estuaries form where rivers and streams empty into oceans, mixing fresh water with salt water (Figure 40-9). An organism's location ion an estuary depends on its ability to tolerate different concentrations of salt. Salinity changes,

(a)

(b)

(c)

FIGURE 40-6

Coral reefs. *(a)* A barrier reef surrounding the island of Taiatea in French Polynesia. *(b)* A fringe reef around Society Island, Tahiti. *(c)* A circular atoll: a ring of coral built around an island that later submerges, creating a central lagoon.

FIGURE 40-7

Part-time oceans: intertidal habitats. *(a) Tide pools* are reservoirs of seawater trapped in depressions during low tide. The pools are filled with flowerlike sea anemones, sea stars, crabs, small fishes, and a variety of seaweeds and other algae. *(b)* A *mud flat* off Point Reyes, California. *(c)* A *sandy beach* below Na Pali sea cliffs, Polihale State Park, Kauai, Hawaii. *(d) Rocky shores.* Large numbers of animals live in the moist tangle of brown seaweeds that drape the shore's rocks at Elkhead Cove in California.

however, as tides bring in new influxes of salty sea water or as storms increase the flow of fresh water. At any one spot in an estuary, salinity may change within minutes, from very low concentrations (equal to that of fresh water) to very high concentrations (the salinity of sea water), creating an ever-changing environment for the organisms that live there.

Despite such fluctuations in salinity, estuaries are tremendously fertile habitats for organisms; estuaries are continually enriched with nutrients from rivers and from the debris that is washed in by tides. Once nutrients enter the estuary, tidal action and the slow mixture of water prevent their escape. Plankton flourishes in the estuary's warm,

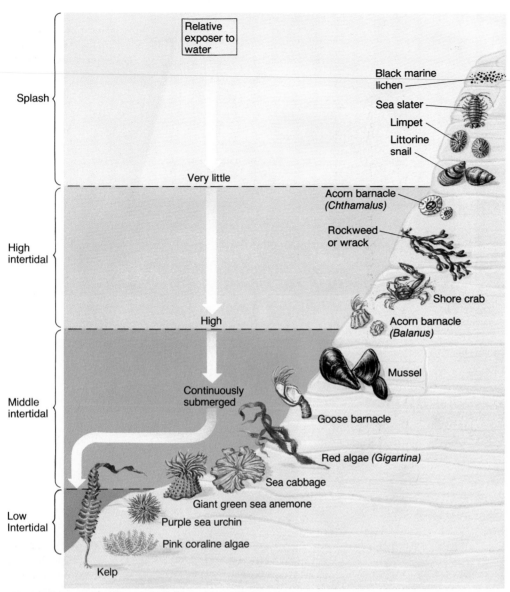

FIGURE 40-8

Intertidal zones. Organisms that live in the *splash zone* must survive with only periodic sprays of moisture, whereas organisms are submerged in water about 10 percent of the time in the *high intertidal zone*, 50 percent of the time in the *middle intertidal zone*, and 90 percent of the time in the *low intertidal zone*.

murky waters. Phytoplankton and rooted plants along the estuary's edges provide food for crustaceans, fishes, shellfishes, and for the young of many open ocean animals that use rich estuaries for spawning.

Freshwater Habitats

Oceans contain 97.2 percent of the earth's total water. The remaining 2.8 percent is fresh water: 2 percent is permanently frozen in ice caps and glaciers; 0.6 percent is groundwater; 0.017 percent is concentrated in lakes and rivers; and 0.001 percent is in the atmosphere as ice and water vapor.

Since the quantity of water on earth is so enormous, even a small percentage like 0.017 percent produces more than 52,000 cubic miles of freshwater habitats on earth. The principal freshwater habitats are flowing rivers and streams and standing lakes and ponds.

Rivers and Streams Size determines whether fresh water flowing over the land forms a **stream** or a **river;** smaller streams converge into larger rivers. The constant flow of water erodes the land, sculpting dramatic landscapes and creating unique habitats that support particular groups of organisms.

FIGURE 40-9

Estuaries are shallow inlets, where freshwater mixes with salty seawater. Nearly one-half of the ocean's living matter are found in such nutrient-rich estuaries. This estuary is on Fraser Island in Australia.

The velocity of a current affects not only erosion but also the deposition of sediments and the supply of oxygen, carbon dioxide, and nutrients, factors that determine which organisms inhabit a stream or river and where organisms occur. Furthermore, a current can vary even in one spot. Friction along the edges and bottom slows water velocity, allowing algae to cling to rocky surfaces, plants to take root, and animals to reside safely without being swept away. Calmer water is also found behind rocks, pebbles, and mounds, creating stable habitats for smaller organisms.

In a river's open waters, the never-ending tug of the current presents unique challenges for animals. Some fishes use their fins to force themselves down to the less turbulent bottom waters. Conversely, powerful trout and salmon can swim fast enough to oppose the current or even to swim upstream.

Lakes and Ponds **Lakes** and **ponds** are standing bodies of water that form in depressions of the earth's crust. As in the case of streams and rivers, the difference between a lake and a pond depends on size: Lakes are larger and usually deeper than ponds.

A lake typically has three zones: a shallow **littoral zone** along the water's edge, a **limnetic zone** that encompasses the lake's open, lighted waters, and a dark **profundal zone** at depths below which light is unable to penetrate. Rooted plants and floating algae are the characteristic photosynthesizers of the littoral zone. In the limnetic zone, phytoplankton and photosynthetic plants provide food for zooplankton, fishes, and other animals. And, as in the deep, dark waters of oceans, heterotrophic predators and scavengers inhabit the lake's profundal zone. Since oxygen often becomes scarce near the bottom of a lake, anaerobic bacteria and other species that can survive anaerobically are also abundant in the profundal zone. For example, some midge larvae that live in lakes have specialized adaptations for withstanding long periods of anaerobic conditions. This is not true for *all* midge larvae, however; those that inhabit streams die quickly without oxygen. Thus, stream and lake midges have evolved along two adaptive lines, and neither species is capable of invading the other's habitat.

In many lakes, bursts of biological activity occur in the spring and autumn, when the lake's waters naturally circulate (Figure 40-10), brining a supply of rich nutrients to the surface. In the spring, the sun warms the upper waters, and the winds stir the lake, driving heat and oxygen down into the deep waters and circulating nutrients up to the surface. This *spring overturn* stimulates a sudden increase in bio-

Spring and autumn overturns

Summer thermal strata

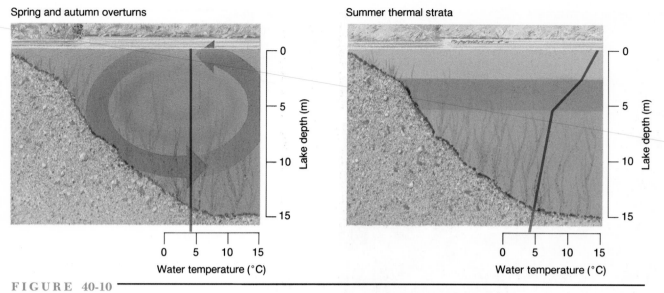

FIGURE 40-10

Seasonal changes in a temperate lake. Water temperature is nearly uniform during spring and autumn overturns. In the summer, warmer surface waters resist mixing, forming distinct temperature zones.

logical activity. As spring passes into summer, heated surface waters become more buoyant and resist mixing. Water circulation stops altogether, producing three temperature layers: a layer of warm surface water; a relatively thin, middle layer called a *thermocline,* in which water temperature declines rapidly; and a bottom layer of cold, relatively still water. Without mixing, oxygen is quickly depleted on the lake's bottom, and nutrients begin to accumulate, slowing down biological activity in the lake throughout the summer.

In the autumn, the lake's thermal layers are disrupted, as heat is dissipated from surface waters to the cold air. As the colder surface waters begin to sink and the autumn winds increase, circulation in the lake is restored, producing a more uniform range of temperature, nutrients, and oxygen. Rising warmer water from the bottom brings nutrients upward and warms the surface waters, producing another burst of biological activity during this *autumn overturn.*

Unlike rivers and streams, lakes and ponds may eventually fill in with sediment and organic matter and become dry land (Figure 40-11). This process may take months, or it may take thousands of years, depending on the size of the lake or pond, the amount of biological activity, and the rate of filling and draining. Thus, biologists speak of a lake's "life span" or a lake's "natural aging process."

A recently formed young lake is termed *oligotrophic* (little nourished) because it contains relatively few nutrients. Oligotrophic lakes support very little life; as a result, they are usually crystal clear. Middle-aged lakes are *mesotrophic* (moderately nourished). Unlike oligotrophic lakes, mesotrophic lakes support large populations of organisms.

Finally, the nutrient-rich waters of old, *eutrophic* (fully nourished) lakes are rich in nutrients and support the largest populations and the greatest diversity of species. As the organisms of eutrophic lakes continually add increasing supplies of organic matter to the water, the filling rate is accelerated. Outside sources of nutrients, such as human sewage or runoff from nutrient-soaked agricultural lands, also promote lake filling.

BIOMES: PATTERNS OF LIFE ON LAND

Large terrestrial (land) ecosystems are called **biomes.** Plants from the bulk of the living mass in a biome; as a result, each biome is characterized by the predominant type of plant that grows there. Dense, tall trees form forest biomes; short, woody plants form shrublands; and grasses and herbs form grasslands. Since the prevailing climate (especially temperature and moisture) is the primary factor in determining the types of plants that grow in an area, the earth's terrestrial biomes tend to follow global climate patterns (Figure 40-2). Climate also changes with elevation, producing successive layers of biomes on a single mountainside (Figure 40-12). Climate affects terrestrial biomes in another, unexpected way: Prevailing climate patterns are spreading pollutants from human activities over very wide distances, causing disruptions in biomes several hundred and thousands of miles away (see The Human Perspective: Acid Rain and Acid Snow: Global Consequences of Industrial Pollution).

(a)

(b)

(c)

FIGURE 40-11

A lake's life cycle. *(a)* Newly formed *oligotrophic* lakes are clear. Oligotrophic lakes eventually begin to fill with sediment, providing nutrients for greater numbers of organisms. *(b)* As sediment and life increases, oligotrophic lakes become *mesotrophic* lakes, *(c)* which then become *eutrophic* lakes.

Studies reveal a pattern in species diversity in terrestrial biomes. The largest number of species are supported in tropical habitats; progressively fewer species are found in temperate and polar areas. For example, in a Malaysian rain forest, biologists counted 227 species of trees in an area of 5 acres (2 hectares), whereas a deciduous forest in Michigan contained only 10 to 15 tree species in an area of the same size. In tropical Mexico, 293 species of snakes were found, compared to 126 in the United States, and only 22 in Canada. Such "global gradients" are found for all organisms, from plants to invertebrates to mammals. In one study, over 150 species of mammals were recorded in Central America, compared to only 15 species in northern Canada.

Forests

Over 30 percent of the earth's land surfaces are covered by forests, or dense patches of trees. Biologists recognize three main forest biomes: (1) lush **tropical rain forests** that grow in a broad belt around the equator; (2) **deciduous forests,** whose trees drop their leaves during unfavorable

seasons; and (3) **coniferous forests** that are dominated by evergreen conifers.

Tropical Rain Forests Warm temperature, abundant rainfall (over 250 millimeters, or 100 inches, a year), and roughly constant daylength throughout the year produce rich, tropical rain forests in Central and South America, Africa, India, Asia, and Australia (Figure 40-2) . Over half of the earth's forests are tropical rain forests. They contain more species of plants and animals than do all other biomes combined. One hectare (10,000 square meters, or 2.5 acres) of tropical rain forest may contain more than 100 species of trees, 300 species of orchids, and thousands of species of animals, most of them insects.

Trees towering over 50 meters (160 feet) form the upper story (*overstory*) of the tropical rain forest (Figure 40-13). Below these giants is a layer of tall trees (the *understory*) so densely packed that very little sunlight penetrates through their canopy. Since much of the light is blocked, plant growth is greatly suppressed below these two layers.

◁ THE HUMAN PERSPECTIVE ▷
Acid Rain and Acid Snow: Global Consequences of Industrial Pollution

Two alarming trends have been recently documented in the biomes of North America and Europe. The first trend is the change in the color of several lakes, from murky green to crystal clear. Sounds good, right? Not really, for a green lake is a biologically active lake, teaming with microscopic algae, which are eaten by small aquatic animals, which, in turn, are eaten by fish. A clear lake is biologically sterile, devoid of aquatic life. How widespread is lake sterility? In eastern Canada, nearly 100 lakes have become sterile, as have more than 1,000 lakes in the northeastern United States and approximately 20,000 lakes in Sweden.

The second trend is the premature death of an excessive number of trees, especially those found on high slopes that face prevailing winds. Huge patches of dead trees, totaling more than 17 million acres (7 million hectares) of trees in North America and Europe, look as if they've been burned, but there have been no fires. More than 50 percent of the forests in Germany alone are affected in this way, impacting more than 1.2 million acres (500,000 hectares).

Just how are dead trees and dead lakes related? The destruction of both trees and lakes is caused by acid deposition from **acid rain** or runoff from **acid snow.** Acid deposition not only destroys forests and

kills lakes, it also damages crops, alters soil fertility, and erodes statues and buildings.

The chemicals that create acid rain and snow (sulfur oxides and nitrogen oxides) come primarily from human activities. Although sulfur oxides are released during volcanic eruptions, forest fires, and from bacterial decay, quantities of sulfur oxides from human activities far exceed those that come from natural sources. Nearly 70 percent of sulfur oxides comes from electrical generating plants, most of which burn coal. Most nitrogen oxides come from motor vehicles and industries, including electrical generation. When sulfur and nitrogen oxides mix with the water in the air, they form acids:

$$SO_2 + H_2O \rightarrow H_2SO_4 \text{ (sulfuric acid)}$$
$$NO_2 + H_2O \rightarrow HNO_3^- \text{ (nitric acid)}$$

Acid rain or acid snow has a pH below 5.7, the pH of unpolluted rain. Over the past 25 years, rains in the northeastern United States maintained an average pH of 4.0. The lowest recorded pH for rainfall was 2.0, reported in Wheeling, West Virginia. The rain in Wheeling was more acidic than lemon juice!

Acids that create acid rain and snow remain airborne for up to 5 days, during which time they can travel over great distances. For example, the acid rainfall that killed many of the lakes in Sweden was

caused by pollutants that were released in England. The acid rain that is damaging trees and lakes in the Adirondack Mountains of New York originated in the upper Mississippi and Ohio River Valleys. Since these acids circulate in large air masses, acid deposition is widespread, spreading from Japan to Alaska, from New Jersey to Canada, to name just a few places.

The rate of destruction caused by acid rain and snow is increasing. A 1988 survey of U.S. lakes lists 1,700 lakes as having high acidity. Another 14,000 lakes were identified as becoming acidified. Scientists estimate that by the turn of the century, over half of the 48,000 lakes in Quebec, Canada, will have been destroyed.

In 1979, the U.S. Congress passed the "Acid Precipitation Act" to *identify* sources of acid deposition, but so far very little has been done to curtail the release of these pollutants. Congress is considering a plan to cut sulfur oxide emissions by nearly 50 percent, and nitrogen oxides by 10 percent by the year 2000. Most other industrialized nations have already taken steps to curb acid deposition; the United States remains of one of a few that has not. Such steps might include: (1) installing scrubbers on power plant smoke stacks; (2) using coal that is low in sulfur; (3) using coal that has been pretreated to remove sulfur; or (4) reducing auto and truck use.

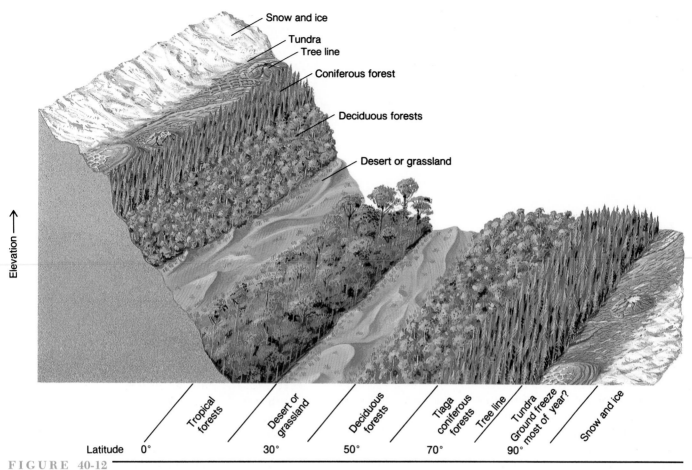

FIGURE 40-12

Elevation and latitude affect the distribution of biomes. Terrestrial biomes change according to elevation, as well as with distance from the equator.

Where a glimmer of sunlight does penetrate, ferns, shrubs, and mosses crowd the forest floor.

🗘 Of all the biomes, the rate of decomposition is fastest in the tropical rain forest. A dead animal or fallen tree can swiftly be cleared from the forest floor by hordes of fungi and bacteria that promptly carry out decomposition. Despite such rapid decomposition, virtually no nutrients accumulate in the soil; the nutrients are either absorbed immediately by plants or washed away by steady rains. Surprisingly, the earth's lushest forests have poor soils, a condition that has contributed to the rapid destruction of the earth's tropical rain forests by humans (see Chapter 21, The Human Perspective, page 428.).

Deciduous Forests In those areas of the earth that have distinct seasons, biological activity follows a seasonal pattern (Figure 40-14). Unfavorable growing conditions during one or more seasons during the year produce deciduous forests with trees, such as maple, beech, hickory, and oak, that produce leaves during warm, wet periods and lose their leaves at the onset of the dry season (in tropical deciduous forests) or the cold season (in temperate deciduous forests).

The trees found in deciduous forests are less dense than are those found in tropical rain forests, enabling sunlight to reach the forest floor. In the presence of adequate light, many layers of plants develop beneath the top layer of deciduous trees. (In both aquatic and terrestrial biomes, vertical layering of plants is associated with a decrease in light.) Plants growing below the upper layer receive varying amounts of sunlight throughout the year, as overstory deciduous trees gradually produce leaf buds, bear fully expanded leaves, and then drop their leaves at the beginning of the unfavorable season. The growth and reproductive cycles of understory plants coincide with brief periods of maximum sunlight and favorable conditions.

In one study, E. Lucy Braun of the University of Cincinnati sampled the deciduous forests of North America and found that tree diversity (the number of different kinds of trees) decreases moving north to colder regions or further west to drier regions. Within a localized area, however,

FIGURE 40-13

A picture of unrivaled diversity. Life in the tropical rain forest is more varied and more abundant than is that found in any other biome on earth. Immense, broad-leafed evergreen trees drip moisture onto masses of clinging epiphytes and enormously long vines, some over 200 meters long. These giant trees have widespreading buttresses that keep them upright, even in shallow tropical soil.

tree diversity was found to be primarily affected by soil moisture and soil calcium levels; drier sites had fewer kinds of trees, as did areas with less calcium.

Coniferous Forests Some of the most extensive forests in the world are populated by evergreen conifers (cone-bearing trees), such as pines, firs, and spruces (Figure 40-15). Belts of coniferous forests girdle the huge continental land masses in the Northern Hemisphere (Figure 40-2) and blanket higher elevations of mountain ranges in North, South, and Central America, and Europe.

The transition from deciduous forests to coniferous forests is a result of colder winters. During the spring and summer, melting snow fills the lakes, forming watery bogs and marshes within many coniferous forests. This is why the coniferous forests of northern latitudes are called **taiga**, which is Russian for "swamp forest." The growing season is relatively short; overall biological activity is restricted to a period of only 3 to 4 months. As a result, decomposition by fungi and bacteria is limited to these relatively brief warm periods. Consequently, the forest floor accumulates a thick layer of needles, producing acidic and relatively infertile soils. The mean temperature is below 0°C (32°F) for more than 6 months of the year.

Coniferous forests typically have two stories of plants: a dense overstory of trees and an understory of shrubs, ferns, and mosses. Unlike the tropical rain forests, coniferous forests are usually not made up of a mixture of many tree species but of expanses of many individuals of a few species.

Tundra

The timberline is a zone in which trees thin and eventually disappear; it marks the boundary between the coniferous forest and the tundra biome (Figures 40-2 and 40-12). In Russian, **tundra** means a marshy, unforested area, describing the frigid, treeless landscape. There are two types of

(a)   *(b)*

FIGURE 40-14

The yearly rhythm of a deciduous forest.   Deciduous trees in temperate areas produce leaves in the spring, when temperatures are warm. During the summer, the overstory trees have fully expanded leaves that shade the plants below. *(a)* A shortened photoperiod and the cool temperatures of autumn trigger the breakdown of chlorophyll, revealing pigments that turn leaves brilliant shades of orange, red, and purple. *(b)* Protected terminal buds enable dormant deciduous trees to withstand the bite of freezing winds and snows during the winter.

FIGURE 40-15

A coniferous forest in the icy grip of winter.   Many animals hibernate or migrate to less severe habitats. Only a few, like this moose, remain active during harsh winter months, forced to travel over large areas in search of scarce food.

FIGURE 40-16

Life is fleeting in the tundra. During the brief summer growing period, ground-hugging tundra plants erupt with new growth, produce flowers, and set seeds, before severe weather returns. In the arctic tundra shown here, permafrost prevents drainage of water from melting snow, producing multitudes of shallow ponds.

tundra: *alpine tundra*, which are found at high elevations of mountain ranges; and *arctic tundra*, which are found to the north at high latitudes in Alaska, Canada, and in northern Europe and Asia (Figure 40-16). Both types look similar, featuring low-growing plants (often only 10 centimeters, or about 4 inches tall) that often belong to the same species. Tundra plants include mosses, lichens, perennial forbs, grasses, sedges, and dwarf shrubs.

◗ The climate that produces a tundra is brutal. The growing season lasts only 50 to 90 days per year, with an average temperature in the warmest month of less than 10°C (50°F). Thick snow covers the ground, and icy winds blow often during winter. During the brief summer period, melting snows create frigid marshes and ponds. In the arctic tundra, only the top 0.5 meters (1.5 feet) thaw, leaving permanently frozen soil, or *permafrost*, that halts root

growth, restricts drainage, and impairs decomposition. Under such harsh conditions, it is not surprising that relatively few plants and animals have evolved adaptations to help them survive the rigorous tundra climate. In fact, there are only about 600 plant species in the entire arctic tundra region of North America, which covers thousands of square miles. This is a smaller number than you would find in a single square mile of tropical rain forest.

Grasslands

Not too long ago, over 30 percent of the earth's lands were covered by **grasslands** of densely packed grasses and herbaceous plants (Figure 40-2 and 40-17). In North America, grasslands were once more widespread than any other biome. But the combination of rich soil and favorable growing (and living) conditions made grassland habitats prime

FIGURE 40-17
Grassland in New South Wales, Australia.

targets for agriculture, livestock grazing, and urbanization. Today, most of the earth's grasslands have been cleared for farming and human development or have been damaged as a result of overgrazing.

Grasslands naturally develop in regions with cold winters, hot summers, and seasonal rainfall (more annual rainfall than in the desert, but not enough to support a forest) and are found worldwide, in places like South Africa, Australia, South America, and the former Soviet Union. Natural fires periodically clear grasslands, opening up space and releasing nutrients for new growth. As in other terrestrial biomes, precipitation patterns greatly affect the nature of the grassland. For example, as precipitation decreases from east to west, *tallgrass prairies*, with plants reaching 2 meters (6.5 feet) in height, gradually give way to *mixed grass prairies*, with grasses growing no more than 1 meter (3.28 feet) tall, which, in turn, give way to *shortgrass prairies*, with bunch grasses less than 0.5 meter (1 foot) in height.

☀ In 1979, a team of researchers from Colorado State University, J. A. Scott, N. R. French, and J. W. Leetham, analyzed herbivory (animals eating plants) in the three types of prairies found in the western United States. The researchers discovered that only a small fraction (less than 15 percent) of the plants were consumed by animals in shortgrass prairies. Herbivore consumption was higher in the tallgrass prairie (about 40 percent) and even higher in mixed grass prairies (50 percent). Since they measured herbivory both above and below ground, the researchers uncovered something very significant and distinctive about the grassland biome: Between 80 and 90 percent of the herbivory occurred underground, mainly by nematodes (page 873.). These results helped confirm the hypothesis that overall growth of grassland plants is primarily limited by root consumption, followed by soil water and competition among the plants for nutrients and light.

Savannas

A combination of grassland and scattered or clumped trees forms a **savanna** biome. *Tropical savannas* are found in South America, Africa, Southeast Asia, and Australia and cover nearly 8 percent of the earth's land (Figure 40-18). In many temperate areas, pockets of savannas are found sandwiched between grasslands and forests.

🐌 Like pure grasslands, savannas are characterized by seasonal rainfall, punctuated with a dry season. During the dry season, the above-ground stems of the grasses and herbs die, providing fuel for fast-moving surface fires. The grasses and herbs recover quickly from fires by resprouting from underground roots and stems. If they are killed, the plants are replaced by fast-growing seedlings. Since ground fires

FIGURE 40-18

The African savanna has flat-topped acacias and dry grassland. Some tropical savannas in central Africa are studded with palm trees.

move rapidly through the savanna, the trees are not usually damaged.

The tropical savannas of Africa support large populations of herbivores, including wildebeest, gazelles, impalas, zebras, and giraffes. Like many grasslands, tropical savannas are now being used for grazing. Overgrazing reduces grass cover and allows trees to invade the area, reducing the number of animals that can inhabit overgrazed savannas.

Shrublands

Woody shrubs predominate in **shrublands.** In regions of the earth with a Mediterranean-type climate (hot, dry summers and cool, wet winters), shrubs grow very close together. The shrubs typically have small, leathery leaves with few stomates and thick cuticles to retard water vapor loss (Chapter 18). Remarkably, similar shrublands develop in all areas that have warm, dry summers and cool, wet winters, from areas around the Mediterranean Sea, to coastal mountains in California and Chile, to the tip of South Africa, to southwestern Australia. In California, this type of shrubland is called a **chaparral** (Figure 40-19).

▐▐▶ Organisms that live in Mediterranean-type shrublands must survive many stressful periods. For example, over the long, hot summers, when water is needed most by organisms, there is little or no rainfall. Since the supply and demand for water are directly out of sync, most plant and animal activity is restricted to the spring, when temperatures are warm and the soil is still moist from winter rains.

Fires are common in Mediterranean-type shrublands. The accumulation of dry, woody stems and highly flammable litter greatly increases the chances of fire during the hot summer. Most plants have evolved adaptations that allow them to cope with recurring fires in these environments, however. In the chaparral of California, for example, a burned area often recovers within just a few years. Even 1 month after an intense fire, many chaparral plants resprout from protected, underground stems, or their seeds germinate, stimulated by the fire itself.

In tropical regions that have a short wet season, another type of shrubland develops: the **tropical thornwood.** Most thornwood plants lose their small leaves during the dry season, reducing transpiration and exposing sharp thorns that discourage even the hungriest browser. One type of thornwood, the *Acacia* plant, has coevolved with certain ants, forming a close partnership that helps both species survive in these habitats. The plant provides food and shelter for the ants, and the ants patrol the ground and stems of the plant, aggressively warding off hungry herbivores and removing flammable debris that may accumulate under the plant (Figure 40-20).

FIGURE 40-19

A chaparral is a shrubland that forms in those regions of California that have a Mediterranean-type of climate. In addition to being tolerant to drought, most chaparral plants are adapted to fire and are able to regrow rapidly after a fire. Layers of charcoal testify that fire is a natural component of the chaparral, and of forests, grasslands, savannas, and other shrublands.

FIGURE 40-20

Mutual aid. Swollen thorns of this acacia tree provide a home and brooding place for ants, while nectar and nutrient-rich plant structures produced at the tips of leaves provide the ants with a balanced diet. Ants protect the acacia by attacking, stinging, or biting herbivores (or scientists). Ants also remove flammable debris from around the base of the tree, forming a natural fire break. Any plant seedling that grows within this "bare zone" is quickly clipped and discarded, protecting the acacia from competing plants, including its own offspring.

Deserts

Intense solar radiation, lashing winds, and little moisture (annual rainfall of less than 25 centimeters or 10 inches) create some of the harshest living conditions in the biosphere. These conditions characterize the **deserts,** which cover about 30 percent of the earth's land and occur mainly near 30° north and south latitude, where global air currents create belts of descending dry air. Some deserts are produced in the rainshadows of high mountain ranges. A **rainshadow** is a reduction in rainfall on the leeward slopes of mountains, slopes that face away from incoming storms. The skies over the deserts are generally cloudless. The sun quickly heats the desert by day, producing the highest air temperatures in the biosphere, the record being 57.8°C (138°F) in Death Valley, California. High daytime temperatures and persistent winds accelerate water evaporation and transpiration of water vapor from plants. High evapo-transpiration and low annual rainfall characterize all deserts, producing sparse perennial vegetation of widely spaced shrubs (Figure 40-21a). Despite such harsh living conditions, only portions of the Sahara Desert in Africa and the Gobi Desert of Asia are totally devoid of organisms, an affirmation of the tenacity of life and the power of evolution to produce a great diversity of plants and animals with adaptations that enable them to survive the desert's severe climate.

Plants have evolved many adaptations for surviving the rigors of the desert (see Bioline: Survival in the Desert, Chapter 19). For example, following brief spring and summer rains, the desert floor often becomes carpeted with masses of small, colorful annual plants (Figure 40-21b). These annuals germinate, grow, flower, and release seeds, all within the brief period when water is available and temperatures are warm. By remaining dormant as seeds the remainder of the year, annuals avoid the most severe stresses of the desert.

The evolution of succulent tissues with cells devoted to storing water allows cacti and other succulents to escape dehydration by accumulating water in specialized roots, stems, or leaves. During dry periods, succulents frugally tap their internal water reserves. Other adaptations of desert plants include long tap roots that siphon deep groundwater supplies. Other plants simply endure extreme heat and dryness as a result of adaptations that enable them to survive various degrees of dehydration.

Many desert animals rely on learned and instinctive behavior to avoid the desert heat and dryness. Some, such as kangaroo rats and ground squirrels, remain in cool, humid underground burrows during the day and search for food at night or in the early morning or late afternoon, when temperatures are lower. If these animals venture out during the day, their underground burrows act as heat sinks, quickly

(a)

(b)

FIGURE 40-21

The desert. *(a)* During dry periods, a few hardy perennial bushes, cacti, and Joshua trees are scattered over the scorching desert floor. Many animals remain underground, in moist, cool burrows. *(b)* Following a spring rain, the desert floor blooms with colorful annual growth.

removing the heat that they acquired while scurrying across the desert in search of food. Many birds avoid stressful desert conditions by simply flying to less hostile areas, while other animals *aestivate;* that is, they sleep through the driest part of the year. With watertight skins, desert snakes and lizards are ale to conserve water even during the heat of the day. The desert toad *(Bufo punctatus)* uses a survival strategy similar to that employed by succulent plants. It stores water in its urinary bladder, carrying its own version of an internal well.

▥▶ Larger animals, like humans, can survive extreme desert heat by sweating. Body heat causes the water in sweat to evaporate, cooling the sweating person. On a very hot day, a human sweats about 1 liter of water per hour, helping to maintain body temperature within a tolerable range. In contrast, camels do not sweat, and, contrary to popular opinion, they do not store water in their humps. With a body size almost five times that of a human, camels absorb, and therefore must dissipate, more heat than a human. How do

they do it, if they neither sweat nor store water? In 1964, a well-respected animal physiologist, K. Schmidt-Nielsen, discovered the answer: First, camels are able to tolerate a wide variation in body temperature, from 35°C (97°F) at night to 40°C (104°F) during the day. Excess heat from the day is stored in the body and then dissipated at night. Second, the camel's dense fur reduces evaporation and cuts down the flow of heat from the hot environment to the skin. Schmidt-Nielsen proved this second point by shearing all the fur off a camel; the animal's evaporation of water increased 50 percent.

There is greater diversity of plants and animals (numbers of species) in land habitats than in aquatic ones, even though water is necessary for all life and is clearly more available in aquatic habitats. Greater numbers of organisms live in aquatic habitats than in terrestrial ones, however. (See CTQ #5.)

REEXAMINING THE THEMES

Relationship between Form and Function

Plants establish the basic structure and dynamics of terrestrial ecosystems or biomes. The growth forms of plants (height, leaf size and shape, and so on) create the vertical and horizontal structure in which all other organisms live. In addition, the rhythm of plant activity affects the functions of all organisms, even other plants. Plant structure and function also modifies the nonliving environment, affecting such factors as light penetration, nutrient cycling, water availability and retention, temperature, and humidity.

Biological Order, Regulation, and Homeostasis

Ecosystems and biomes are levels of biological order that comprise a community of organisms and their physical environment. Both components are permanently interlinked; any change in one will have an affect on the other. A reduction in available water, for instance, will reduce the number of individuals in the community, bringing the size

of the community into balance with water availability. Similarly, changes in a community of organisms can alter the physical environment. For example, habitat destruction by humans in tropical rain forest ecosystems is changing the world weather patterns. Weather changes, in turn, will affect the communities of many ecosystems.

Evolution and Adaptation

▥▶ Natural selection is ecology in action. Not all offspring are equally equipped for acquiring energy and nutrients from their environment, for finding a mate and reproducing, or for surviving under the unique set of environmental conditions present in an ecosystem or biome. Those individuals that are better fit to the existing ecological situations tend to produce more offspring than do those that are less fit. In this way, the more adaptive characteristics of the more fit individuals are passed on to the next generation, increasing the frequency of genes that code for more favorable traits. Natural selection results in adaptation and, under appropriate conditions, in the formation of new species.

SYNOPSIS

The biosphere is composed of the earth's lands, waters, and air, which support all life as we know it. Although organisms are scarce in extreme environments, the biosphere contains all of the resources and conditions necessary for life. Since resources for life are limited, recycling of matter within the biosphere is necessary for the continuation of life throughout time.

The earth's oceans support the greatest quantity of life. Most aquatic life is concentrated in shallow coastal waters along the fringes of land and coral reefs.

Biomes cover vast expanses of the earth's land and are characterized by particular plants. Since prevailing climate is the primary factor that controls plant distributions, terrestrial biomes follow global climate patterns and elevational gradients.

Forests cover more than 30 percent of the earth's land. The major forest biomes include tropical rain forests, deciduous forests, and coniferous forests. Tropical rain forests contain more species of organisms than do all other biomes combined. These rich forests are rapidly being destroyed by humans.

Originally, grasslands covered more than 30 percent of the earth's land. Today, most grasslands are used for agriculture, grazing, and other types of development.

Nearly 8 percent of the earth's lands are covered by savannas of grass and scattered trees. Overgrazing is changing the savannas, reducing their ability to support animal life.

Dense shrublands develop in land areas that have a relatively long dry season. Because fires are frequent in shrublands, many organisms have evolved adaptations for quick recovery from fire.

Nearly 30 percent of the land is covered with hot, dry deserts. Harsh living conditions in the desert have produced many anatomic, physiological, and behavioral adaptations in desert organisms.

Key Terms

biosphere (p. 904)
hydrosphere (p. 904)
lithosphere (p. 904)
atmosphere (p. 904)
ecology (p. 905)
climate (p. 905)
pelagic zone (p. 908)
epipelagic zone (p. 908)
mesopelagic zone (p. 908)
bathypelagic zone (p. 908)
benthic zone (p. 908)
coastal waters (p. 908)
neritic zone (p. 908)
intertidal zone (p. 912)

splash zone (p. 912)
high intertidal zone (p. 912)
middle intertidal zone (p. 912)
low intertidal zone (p. 912)
estuary (p. 912)
stream (p. 914)
river (p. 914)
lake (p. 915)
pond (p. 915)
littoral zone (p. 915)
limnetic zone (p. 915)
profundal zone (p. 915)
biome (p. 916)
acid rain (p. 918)

acid snow (p. 918)
tropical rain forest (p. 917)
deciduous forest (p. 917)
coniferous forest (p. 917)
taiga (p. 920)
tundra (p. 920)
grassland (p. 922)
savanna (p. 923)
shrubland (p. 924)
chaparral (p. 924)
tropical thornwood (p. 924)
desert (p. 926)
rainshadow (p. 926)

Review Questions

1. Complete the chart on aquatic ecosystems by placing the following terms in the appropriate space

pelagic zone	littoral	limnetic
standing	freshwater	aphotic
mesopelagic	ponds	profundal
moving	streams	epipelagic

2. Rank the earth's terrestrial biomes, starting with that which covers the greatest land surface and ending with that which covers the smallest area. Why do some biomes cover more land than others?

3. Rank the earth's terrestrial biomes, starting with that which has the greatest number of species (the highest diversity) and ending with that which has the smallest number of species.

4. Explain why tall trees do not grow in the following biomes:

   a. desert   b. tundra   c. grassland

5. Explain how lakes have alternating periods of low biological activity and high biological activity.

6. List the strata of organisms found in the intertidal zone. What kinds of adaptations are necessary to help organisms withstand increasing lengths of exposure to dry air?

7. List the two most important environmental factors that determine the distribution and abundance of organisms in the ocean.

8. In what ways are the following similar?
   a. lakes and rivers.
   b. profundal zone and desert.
   c. forest understory and bathypelagic zone.
   d. coastal waters and littoral zone.

9. Discuss the changes that would occur in the biosphere and the effect on biodiversity (the numbers of kinds of organisms) of the total destruction of all tropical rain forests. Do you think total destruction is likely?

10. For each of the following, list the major adaptations needed for survival. Are there distinct differences for plants and animals? Make a list of different plant and animal adaptations for each habitat and briefly explain why such differences exist.
    a. prolonged hot and dry season.
    b. surging waters against a rocky coastline.
    c. calm, warm fresh water with abundant nutrients and many species of organisms.
    d. dramatic fluctuations in water salinity.

Critical Thinking Questions

1. Scientists are warning that without the filtering of ultraviolet sunlight by the ozone layer, incidents of skin cancer and crop damage will increase. Explain how increased amounts of UV can increase skin cancer and crop damage. If by the year 2050, the ozone layer is reduced by another 60 percent, as some scientists predict, what other problems may develop?

2. Life cannot be naturally supported outside the biosphere, yet we send astronauts into space. What fundamental requirements must be met for establishing a space station that will maintain life indefinitely? Design a space station capable of supporting life, including humans. Include a list of all types of organisms that would populate the station.

3. Prepare a chart, similar to the one below, showing the

Organism	How It Affects Its Environment	How It Is Affected By Its Environment
earthworm		
moss growing on a rock		
maple tree		
elephant		
reef building coral		
bacterium of decay		

   ways in which each organism affects its environment and how it is affected by the environment.

4. The earth's diverse climates create the myriad habitats that support a tremendous variety of organisms. In what ways would the diversity of climate (and therefore the diversity of life) change if conditions were different? For each of the following conditions describe the impact on climate change and then the impact on life:
   a. the earth does not rotate on its axis once every 24 hours.
   b. the earth is not tilted 23.5°.
   c. all the earth's land mass is contained in a single, large continent. (Is the location of this land mass significant? If so, be sure to state its location and effects.)

5. What environmental and evolutionary factors account for the fact that species diversity is greater in terrestrial biomes, yet the abundance of organisms is greater in aquatic habitats?

6. For each of the human activities listed, describe its effects on the biosphere: deforestation, automobile travel, agricultural use of fertilizers, use of pesticides, heavy industry, destruction of wetlands, building large cities.

Additional Readings

Brewer, R. 1988. *The science of ecology*. New York: Saunders College Publishing. (Intermediate)

Carson, R. 1962. *Silent spring*. Boston: Houghton Mifflin. (Introductory)

Cloud, P. 1983. The biosphere. *Sci. Amer.* SEPTEMBER:176–189. (Introductory)

Disilvestro, R. 1989. *The endangered kingdom: The struggle to save America's wildlife*. New York: Wiley Science Editions. (Introductory)

Newell, N. 1972. The evolution of reefs. *Sci. Amer.* JUNE:54–65. (Introductory)

Schmidt-Nielsen, K. 1964. Desert animals: Physiological problems of heat and water. Oxford, England: Oxford Univ. Press. (Introductory)

Sutton, A., and M. Sutton, 1966. *The life of the desert*. New York: McGraw-Hill. (Introductory)

Sutton, A., and M. Sutton. 1979. *Wildlife of the forests*. New York: Harry N. Abrams. (Introductory)

Ecosystems and Communities

STEPS
TO
DISCOVERY
The Nature of Communities

BIOLINE

Reverberations Felt Throughout an Ecosystem

THE HUMAN PERSPECTIVE

The Greenhouse Effect: Global Warming

STEPS TO DISCOVERY
The Nature of Communities

*C*onflicting views may arise even among brilliant scientists in the same field. Each unresolved conflict heralds a research opportunity for an alert and eager student. When Robert H. Whittaker began graduate school in 1946 at the University of Illinois at Urbana to study insect ecology, he soon found himself embroiled in a 30-year-old conflict among plant ecologists regarding the nature of communities of organisms.

Before Whittaker's time, many scientists had already observed that groups of plants and animals tended to form repeatable, discrete communities of organisms. Indeed, distinct communities have been described at least as far back as 300 B.C. by Theophrastus, the Greek philosopher

who studied with Aristotle and succeeded him as head of the Peripatetic school. Theophrastus' two books on botany, *History of Plants* and *Etiology of Plants*, remained the definitive works on the subject until the Middle Ages. In these books, Theophrastus noted that particular plants tended to group together, forming repeating patterns in similar environments. Between 1800 and the early 1900s, a great deal of research and philosophical discussion centered on the causes of such plant communities.

Carl L. Willdenow, one of the first botanists to study the distributions of plants, noted that similar climates tended to produce similar plant communities, even in regions that are widely separated. Willdenow's findings in-

Hundreds of random samples (small circles) of insects and plants in the Great Smoky Mountains showed that both plants and

trigued Friedrich H. A. von Humboldt, a wealthy Prussian student who was studying botany under Willdenow (along with mathematics and chemistry) at the University of Gottingen in Germany. Upon graduating, von Humboldt traveled to South America, where he collected over 60,000 plant specimens and documented the relationship between climate and plant communities. Von Humboldt eventually coined the term **association** to describe a community that is consistent in the composition of plant species and general appearance, and whose distribution is correlated with a specific climate and other physical factors. Von Humboldt was one of the first scientists to recognize that communities formed as a result of many causes, including elevation, latitude, temperature, rainfall, soils, and so on.

Not everyone shared Von Humbolt's view, however. By the turn of this century, botanists had still not agreed on the principal cause of plant communities. For example, two eminent botanists, Frederick E. Clements of the University of Nebraska and Carnegie Institution of Washington, D.C., and Henry A. Gleason of the New York Botanical Gardens, formulated opposing views on the nature of plant communities. Clements considered a plant community a "super organism"; that is, a well-defined association of plant species that are always found together, much as the cells, tissues, and organs of an organism are always found together. According to Clements, the groups of species in a community all have identical distribution limits, so they recur together in distinct communities. In contrast, Gleason argued that each plant species was distributed individualistically; that is, instead of all species in a community having identical distribution limits, the limits of each species were determined by the species' own genetics that determine its physical and physiological characteristics and tolerances. As a result, each species' distribution changed gradually, making it difficult to divide vegetation into discrete associations or communities. Although Clements proposed his view in 1916, and Gleason proposed his view in 1926, this conflict remained unresolved until the mid-1940s. At that time, Robert Whittaker entered graduate school.

Whittaker referred to the conflict between Clements and Gleason as "exciting confusion" for a new graduate student. Whittaker decided to test these conflicting views on insects. Whittaker sampled the distribution of insects at different elevations in the Great Smoky Mountains National Park in Tennessee and North Carolina. The hypothesis Whittaker formulated was in line with Clements' view. That is, Whittaker proposed that he would find distinct groups of insect species along the continuous elevational gradient, each group more or less separated by abrupt transitions. He discovered that insect populations changed *irregularly* with elevation, however, disproving his hypothesis.

Whittaker decided to retest the same hypothesis, this time on plant species in the same forest. He took 300 *random* samples of plants over the mountain range to see whether natural units of plant species emerged. Again, he failed to find definite groups of plant species. Whittaker finally rejected his hypothesis and reanalyzed his results to see whether Gleason's individualistic view applied to the vegetation in the Great Smoky Mountains. When he plotted plant distributions on a chart with elevation and moisture gradients as axes (an analysis technique he invented), Whittaker verified that plants distributed themselves individualistically and that the vegetation found did not form a mosaic of discrete associations. In his words, Whittaker found "a subtly wrought tapestry of differently distributed species populations variously combining to form the mantle of plant communities covering the mountains."

Whittaker's results resolved the conflict between Clements and Gleason: Clements' group view was rejected, and Gleason's individualistic view was scientifically supported. As often occurs in science, at the same time that Whittaker was completing his studies, other investigators in the United States and in the former Soviet Union had also discovered the validity of the individualistic view of species distributions.

insects distribute themselves independently. Data from different samples are shown by histograms.

*O*rganisms are continually affected by other organisms as well as by the surrounding physical environment.

- A long drought destroys vast stretches of grasses and trees in the African savanna, causing thousands of animals to starve and leaving others frail and malnourished.

- The cold runoff from exceptionally heavy winter snows keeps streams cool throughout most of the summer, slowing hatching larval development of blackflies, mayflies, and caddisflies, which, in turn, reduce the number and size of some fishes that normally eat these larvae.

- Acrid Los Angeles smog blows into mountain forests, where it reduces photosynthesis in pines by as much as 80 percent. Unable to manufacture enough resin (sap) to protect themselves against burrowing insects, thousands of pines are dying from bark beetle infestations in the forests.

These examples illustrate a fundamental ecological principle. Because of this constant interplay, the organisms and the physical environment in a particular area function as an organized unit known as an **ecosystem.** This level of biological organization includes both the community of living organisms and the physical environment.

Biologists have discovered a great deal about many of the levels of biological organization, particularly the cell, tissue, organ, organ system, and organism levels, but relatively little is known about the levels of organization studied in ecology (populations, communities, and ecosystems). This is not surprising because, by definition, ecological levels of organization include all former levels, making them the most complex of all.

In addition to enormous complexity, the timing of ecological events is often difficult to fathom and measure, ranging from a few nanoseconds (a predator capturing its prey) to millions or billions of years (gradual climatic change). Add to these considerations the fact that it is virtually impossible to study whole populations, communities, or ecosystems under controlled conditions, and it is easy to see why ecological studies often require very long periods of time and the collaboration of a number of scientists from many fields.

▼ ▼ ▼

FROM BIOMES TO MICROCOSMS

The variety of ecosystems in the biosphere is enormous. Each expansive biome contains countless ecosystems (Figure 41-1). The boundaries that separate ecosystems are not always sharp delineations, however, since all ecosystems are linked with other ecosystems, to some degree. For example, consider a lake and a cave, two ecosystems that at first seem clearly independent. However, the rate at which the lake fills with sediment may affect the number of beetles that live in a cave, even if the lake and the cave are separated by several miles.

☀ This is possible because bats link the two "separate" ecosystems as these animals forage at night for insects around the lake, where insects are plentiful, and return to the cave during the day to rest. As the lake fills with sediment and organic material, more plants grow along the margins of the lake, providing more food for greater numbers of plant-eating insects, the dietary mainstay for the bats. When well-fed bats return to the cave, they defecate large amounts of feces onto the cave floor. The bat feces, in turn, nourish the growth of mold, a staple of the cave beetles' diet. Thus, the amount of sediment in the lake eventually affects the number of beetles in the cave: More lake sediment means more plants; more plants mean more insects; more insects mean more bat feces; more bat feces means more mold; and more mold means more beetles. Furthermore, since beetles, crickets, flies, springtails, moths, spiders, centipedes—indeed, nearly every organism that makes up the entire cave community—depend on the supply of bat droppings, the lake ecosystem affects many aspects of the cave ecosystem, and the cave affects the lake, as bats regulate the number of insects (Figure 41-2).

Furthermore, a lake may not be linked only to cave ecosystems but to other ecosystems as well. For example, runoff from a nearby forest ecosystem provides nutrients and debris to the lake, while the river ecosystem that drains the lake removes sediments and nutrients. In fact, many biologists view the entire biosphere as one giant global "ecosystem" of tremendous complexity and order.

The earth's vast biomes are subdivided into ecosystems, unique communities of organisms that interact with one another and with the surrounding environment. Although a separate unit of organization, each ecosystem is linked to and affected by other ecosystems. (See CTQ #2.)

THE STRUCTURE OF ECOSYSTEMS

☀ Each ecosystem is a functional unit in which energy and nutrients flow between the physical, nonliving abiotic

FIGURE 41-1

Ecosystems within ecosystems. *(a)* A South American tropical rain forest contains many ecosystems, each with a unique community of organisms and a unique set of physical factors. *(b)* The ecosystem supports colonies of leaf cutter ants that harvest leaves to cultivate a fungus garden ecosystem for food. *(c)* Perched high on a tree branch, the top of a bromeliad collects water, forming a tiny pond ecosystem, the home of this red-eyed tree frog.

FIGURE 41-2

Bats interconnect a cave ecosystem with a lake ecosystem. The more insects found around a lake, the more food available to cave-dwelling bats such as these Mexican freetail bats. When bats return to a cave to rest, they defecate, forming the substrate for the growth of mold. Beetles and a number of other cave-dwelling organisms in turn consume the mold.

◁ B I O L I N E ▷
Reverberations Felt Throughout an Ecosystem

Some great lessons in ecology are stumbled upon accidentally. For example, in the 1950s, the raising of ducks on Long Island, New York, produced some unexpected and economically devastating consequences. No one foresaw that allowing duck farms near the Great South Bay of Long Island would totally destroy the lucrative oyster industry in the bay, but that is just what happened. Before the ducks arrived, the famous blue-point oysters had thrived in the bay, eating a normal mixture of diatoms, unicellular green algae, and dinoflagellates. Shortly after the duck farms were established along the tributaries to the bay, oysters and some other shellfish were found starving to death. Eventually, all the oysters disappeared.

How could such a disaster have happened? The answer lay in the large duck farms that produced enormous quantities of duck droppings that were washed into the tributaries and eventually into the bay. As a consequence of the bay's slow circulation rate, the duck droppings quickly changed the nutrient balance in the bay, drastically altering the prevailing mixture of algae. One species of green algae that was ordinarily present in very small numbers prospered under these new nutrient-rich conditions, while the populations of other phytoplankton species plummeted. Unable to digest the green algae that proliferated, the oysters quickly starved to death. Since the disaster, several attempts have been made to reestablish a normal nutrient balance, in the hopes of reintroducing oysters to the Great South Bay, but all have failed.

The ecological story of "the ducks and the oysters" illustrates how a change in the biotic environment (increasing the number of ducks) can alter the abiotic environment (the levels of inorganic nutrients in the bay), which, in turn, changes the biotic environment (the proportion of algae species), causing another change in the biotic environment (the death of oysters), which produces yet another change in the biotic environment (the loss of food for humans and the bay's other consumers of oysters). The story also illustrates how one seemingly minor alteration can send reverberations throughout an ecosystem, usually with unexpected and destructive results.

---

environment, and a community of living organisms that make up the biotic environment. The abiotic and biotic environments continually affect each other, producing interdependent connections. Even a slight modification in one factor can disrupt an entire ecosystem (see Bioline: Reverberations throughout an Ecosystem).

It is often easier to envision how the abiotic environment changes the biotic community than vice versa. For instance, a long period of freezing temperatures or a devastating fire may kill many organisms outright. Extremely strong winds can uproot plants and blow flying insects and birds far sway from their natural habitat. The relationship between the abiotic and the biotic environment is two-way, however; organisms change the abiotic environment as well. For instance, recall from Chapter 35 that the earth's original atmosphere was devoid of oxygen. Ancient organisms released oxygen during photosynthesis, gradually adding this important gas to the atmosphere. In other words, organisms dramatically altered the abiotic environment, building up the oxygen of the earth's atmosphere to its current level. The biotic environment affects the abiotic environment in various other ways as well: Plants contribute to the formation and fertility of soils; coral reefs change the

flow and temperature of oceans; dense forests modify humidity, temperature, and the amount of light and rain that reaches the forest floor; and today, human activities produce pollutants that are changing the earth's climate.

COMPONENTS OF THE BIOTIC AND ABIOTIC ENVIRONMENTS

Biologists divide ecosystems into five subcomponents, two of which comprise the abiotic environment, and three that comprise the biotic environment (Figure 41-3). The two abiotic subcomponents are:

1. *abiotic resources:* the energy and inorganic substances (nitrogen, carbon dioxide, water, phosphorus, potassium, and so on) needed by organisms for the construction of organic compounds; and

2. *abiotic conditions:* the substrate and/or medium (air, water, and soil) in which organisms live, and the surrounding conditions, such as temperature and water currents.

ABIOTIC
ENVIRONMENT

BIOTIC
ENVIRONMENT

Solar energy

Primary producers

Air

Substrate
or medium

Water

Soil

Consumers

Inorganic nutrients

Decomposers

FIGURE 41-3

The basic components of an ecosystem are the abiotic environment and the biotic community. The abiotic environment includes energy, inorganic substances (essential elements), and the substrate and/or medium in which the community lives. Within the biotic community, primary producers convert energy and inorganic nutrients into organic food molecules, which, in turn, are passed on to consumers and decomposers. Respiration, excretion, and decomposition return essential elements to the abiotic environment. Each year, the earth's primary producers convert an amount of energy equivalent to the output of about 2 billion nuclear power plants, enough to power all life in the biosphere.

The three subcomponents of the biotic environment are:

1. **primary producers:** autotrophs (algae, bacteria, and plants) that use sunlight or chemical energy to manufacture food from inorganic substances;

2. **consumers:** heterotrophs that feed on other organisms or organic wastes; and

3. **decomposers** and **detritovores:** heterotrophs that get their nutrition by breaking down the organic compounds found in waste organic matter and dead organisms (**detritus**). Decomposers are primarily microscopic bacteria and fungi, whereas detritovores are typically larger animals, such as some worms, nematodes, insects, lobsters, shrimp, and birds that feed on detritus.

TOLERANCE RANGE

For each abiotic resource or condition, an organism is able to survive and reproduce only within a certain maximum and minimum limit. For example, the lethal temperature limits for many land animals are a minimum of 0°C (32°F) and a maximum of about 42°C (107°F), whereas the lethal temperature range is often smaller for aquatic animals. This range is known as an organism's **tolerance range** (Figure 41-4). In addition to identifying the tolerance range for a single organism, tolerance ranges can be determined for an entire species by determining the tolerance ranges of all members of the species and then setting the tolerance range limits at the maximum and minimum levels found. The **Theory of Tolerance,** as this concept came to be called, was proposed in 1913 by Victor Shelford, an animal ecologist at the University of Chicago. Since the time the theory was proposed, studies have revealed that tolerance ranges

- differ for each abiotic factor (a marine organism may have a broad range of tolerance for temperature and a narrow range for salinity);

- have an optimum at which conditions are best for the organism (Figure 41-4);

- have some factors that may affect the tolerance range of other factors, especially when conditions approach the maximum or minimum limits for one factor (grasses grown in nitrogen-deficient soils need more water than do those grown in nitrogen-rich soils);

- change as an organism passes through different phases of its life cycle (eggs, gametes, embryos, and young organisms usually have narrower ranges than do adults); and

- may differ slightly from population to population within a species (populations with different genetically fixed tolerance ranges are called **ecotypes** of a species).

LIMITING FACTORS

When the limits of tolerance for one or a few factors are approached, those factors usually take on greater importance than do all the others in determining where or how well an organism will survive. These more critical factors become **limiting factors** because they alone impose restraints on the distribution, health, or activities of the organism. Since each species has a multitude of tolerance ranges (one for each abiotic factor), the ability to identify a few key limiting factors helps ecologists understand otherwise extremely complex ecosystems.

This concept of limiting factors was first proposed in 1840 by Justus von Liebig of the University of Heidelberg and has become known as Liebig's **Law of the Minimum** (Figure 41-5). An agriculturist and physiologist, Liebig discovered that the yield of a crop was restricted by the soil nutrient most limited in amount. His Law of the Minimum states that the growth and/or distribution of a species is dependent on the one environmental factor that is available in the shortest supply.

Examples of Liebig's law are common: Low amounts of phosphate in a lake drastically reduce algae growth, which, in turn, limits the growth of all consumers; the amount of

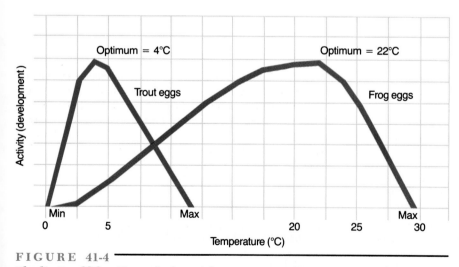

FIGURE 41-4

The limits of life. For each physical factor, organisms have minimum and maximum tolerance limits. Within most tolerance ranges, such as the temperature ranges for the development of trout and frogs, lies an optimum, at which growth is greatest.

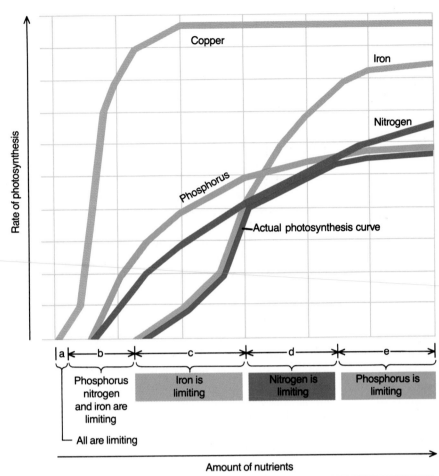

FIGURE 41-5

Nutrient availability and photosynthesis. These graphs illustrate the Law of the Minimum. The rate of photosynthesis in phytoplankton (black line) is limited by the nutrient in the shortest supply. In areas *(a)* and *(b)* on the graph, one or more essential nutrients are lacking, so there is no photosynthesis. Low supplies of iron limit photosynthesis in area *(c)*; of nitrogen in area *(d)*; and of phosphorus in area *(e)*.

zinc in soil is usually so scarce that it alone limits the yield of many agricultural crops; the limited rainfall in deserts severely restricts plant growth; low light intensity on the floor of a tropical rain forest limits photosynthesis in understory plants.

Exceptions to the Law of the Minimum occur when one factor changes the tolerance for another. For example, changes in the oxygen content of water can change the temperature tolerance of American lobsters. When the oxygen content of water is low, lobsters tolerate temperatures only as high as 29°C (84°F), whereas when the oxygen content is high, the temperature tolerance rises to 32°C (89°F). Another exception to the Law of the Minimum is found when one factor substitutes for another, such as when mollusks use strontium to build their shells when calcium is limiting.

Organisms and the nonliving physical environment are inseparably linked; each continually affects the other. Organisms in an ecosystem are primary producers, consumers, or decomposers. Organisms tolerate a certain range of conditions for each abiotic factor; one or a few key limiting factors often determine its growth and distribution. (See CTQ #3).

ECOLOGICAL NICHES AND GUILDS

The organisms that comprise a community possess structural and behavioral adaptations that enable them to grow, reproduce, and survive in a particular ecosystem. Each organism in an ecosystem occupies a specific **habitat,** the physical location in which the organism lives and reproduces (the bottom of a lake, under a rock, inside another organism, and so on). In addition to needing space, organisms also require energy and nutrients. The processes by which organisms acquire these resources partly define their "role(s)" in an ecosystem. Having identified the habitats and roles of organisms that make up ecosystem and investigated what happens when habitats or roles overlap, ecologists have proposed two fundamental concepts to help evaluate and describe the total life history pattern of individuals that occupy the same ecosystem.

NICHES

Together, an organism's habitat, role, requirements for environmental resources, and tolerance ranges for each abiotic factor comprise its **ecological niche.** Since the ecological niche includes all aspects of an organism's existence—its residence, activities, requirements, and effects—it too is the outcome of evolution through natural selection. The full scope of adaptations an organism acquires through natural selection establishes the range and boundaries of an organism's—or a species'—ecological niche.

If only two or three components are considered, it is possible to plot an ecological niche on a graph, each axis representing a different factor (Figure 41-6). It is impossible to draw a graph that includes *all* the components of the ecological niche (all tolerance ranges, the total range of habitats, and all functional roles), however, because such a representation would circumscribe a multidimensional

area, or what ecologists refer to as a **hypervolume.** The hypervolume represents the *potential* niche, or the **fundamental niche.** Most organisms never realize their fundamental niche because interactions with other organisms (such as competition for limited resources or mates) often reduce the range of available habitats or possible functional roles. (We will discuss the range of organism interactions in the next chapter.) The remaining portion of the hypervolume, in which an organism actually exists and functions, is its **realized niche.** In other words, the fundamental niche represents all *possible* ranges under which an organism or species can exist, while the realized niche is the range of conditions in which the organism *actually* lives and reproduces.

Ecologists speak of **niche breadth** as a measure of the relative dimensions of an organism's ecological niche (Figure 41-7). Some species have very broad niches. Hawks, for example, have a relatively broad niche because they visit many habitats over large areas and have evolved adaptations that enable them to feed on several kinds of organisms. Conversely, the cotton boll weevil has a narrow niche because it lives, feeds, and reproduces on cotton plants alone. Species with narrow niches possess adaptations for very specific habitat and environmental requirements. For example, the cotton boll weevil has a long, tubular mouth specifically adapted for feeding on a cotton plant; female boll weevils have adaptations for laying eggs in cotton buds (bolls); and boll weevil larvae can only survive by eating the internal tissues of flower buds and developing cotton bolls.

Although they may not always be apparent, practical benefits sometimes arise from studying an organism's niche breadth. For example, consider the constant battle between farmers and weeds. All farmers want to stop weeds from

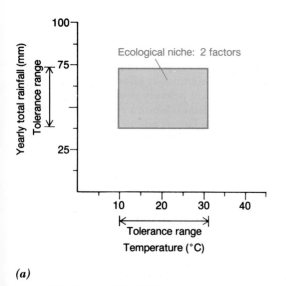

(a)

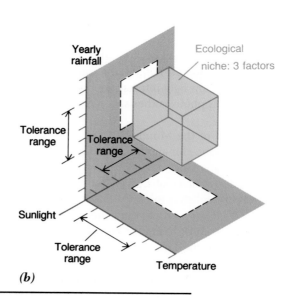

(b)

FIGURE 41-6

An ecological niche based on the tolerance range of two *(a)* and three *(b)* factors.

(a)   *(b)*

FIGURE 41-7

Opposite ends of the spectrum. *(a)* The boll weevil's specific adaptations for living and reproducing only on cotton plants results in a narrow niche breadth, whereas *(b)* adaptations that enable a
hawk to harvest a variety of animals for food from a number of habitats result in a broad niche breadth.

invading and overgrowing their agricultural fields, since
weeds contaminate harvests and cut down yields by competing with the crops for space and limited nutrients. One
way to stop weed invasion is to apply a chemical weed killer
that destroys only the weeds, not the crops. Unfortunately,
such "selective chemical killers" are not available for every
kind of weed. Alternatively, weeds could be stopped by
using a **biological control,** whereby another organism destroys the weeds. Of course, it is important that the organism kill only the weed and not the crop plants; in other
words, the weed-controlling organism must have a narrow
niche.

   Many examples of biological weed control exist. One in
particular involves the unlikely pairing of the klamath weed
(*Hypericum perforatum*) and a leaf-feeding beetle *Chryso-
lina quadrigemina*). After studying the "natural history" of
C. quadrigemina, scientists discovered that the feeding
habits of the beetle were very specific. These beetles fed
only on plants that had leaves of the same surface texture
and margin as that of the klamath weed. Fortunately, the
beetle was active during the same period that the klamath
week germinates and enters its period of rapid growth. The
beetle's life history pattern was ideal for controlling the
klamath weed. Unfortunately, the two organisms did not
naturally occur together; they didn't even exist on the same
continent. Once the leaf-feeding beetle was introduced into
the United States, however, it became a very effective biological control agent against the klamath weed. This example reaffirms the importance of "basic research" in science.
If separate scientists had not been studying the "natural
history" of both of these organisms, and if they had not

shared this information with each other, effective control of
the klamath weed would still be a fantasy.

   ⚠ **Niche overlap** occurs when organisms share the same
habitat, have the same functional roles, or have identical
environment requirements in some other way. Niche overlap leads to competition for the same needed resource; the
greater the overlap, the more intense the competition. For
example, consider two germinating acorns of different species of oak growing side by side in an oak woodland. Both
young plants share the same habitat, require similar
amounts of space, light, water, and nutrients, and play essentially the same roles in the ecosystem as do mature oak
trees. Because of such close similarities in niche requirements (high niche overlap), these plants will likely compete
for needed resources.

   When two species have *identical* niches, competition
can become so intense that both species are unable to coexist in the same ecosystem; one species eventually excludes
the other. This phenomenon is called the competitive exclusion principle and is discussed in more detail in Chapter
42.

GUILDS

Although species with identical niches cannot coexist in the
same ecosystem for long periods, species with *similar*
niches often can. In a forest, for example, all insect-eating
bird species have similar habitats, roles, and nutrient requirements, yet they are able to coexist. Such groups of
species with similar ecological niches form a **guild.** Not
surprisingly, members of a guild interact frequently, often
with intense competition when resources become scarce.

⚠ Species that have similar ecological requirements, yet live in different ecosystems, are called **ecological equivalents**. (The niches of ecological equivalents do not overlap because the species do not share the same habitat.) Ecological equivalents usually look very similar, the product of convergent evolution (page 735). For example, placental mammals in North America and their marsupial counterparts in Australia are ecological equivalents (see Figure 33-16). Although these animals have completely different ancestry and are separated by thousands of miles of ocean, North American mammals and Australian marsupials have evolved strikingly similar adaptations in response to very similar ecosystem characteristics: Remember, evolution takes place on the ecological stage. These animals may be ecologically equivalent, but they are not the same species. With distinctively different ancestry, ecological equivalents contribute to the great diversity of life on earth.

Each organism lives and reproduces in a specific habitat, requires specific environmental resources, has fixed ecological tolerances, and plays a specific role in an ecosystem. Together, these factors define an organism's ecological niche. (See CTQ #4.)

ENERGY FLOW THROUGH ECOSYSTEMS

☀ All organisms must secure a supply of energy and nutrients from their environment in order to remain alive and reproduce. Biologists categorize organisms according to their energy-acquiring strategy: Primary producers harvest energy through photosynthesis or chemosynthesis, forming a direct link between the abiotic and biotic environments, whereas consumers and decomposers obtain energy and nutrients from other organisms.

TROPHIC LEVELS

Tracking the transfer of food (energy and nutrients) among the organisms in an ecosystem is relatively easy if you follow a single feeding path. You simply count how many leaves a caterpillar eats, how much phytoplankton goes into a copepod, or how many prairie dogs a hawk consumes. Energy and nutrients are transferred step by step between organisms. For example, in a streamside community, a hawk eats a snake that ate a frog that ate a moth that sipped nectar from the flower of a periwinkle plant that converted the sun's energy into chemical energy during photosynthesis.

This linear feeding pathway has the same organization as do those of all ecosystems: It begins with a primary producer (the periwinkles in the streamside community), which provides food for a **primary consumer** (the moth), which is eaten by a **secondary consumer** (the frog), which is eaten by a **tertiary consumer** (the snake), and so on, until the final consumer dies and is disassembled by decomposers. Each step along a feeding pathway is referred to as a **trophic level** (*trophic* = feeding).

☀ The sequence of trophic levels maps out the course of energy and nutrient (food) flow between functional groups of organisms (primary producers, primary consumers, secondary consumers, and so on). As Figure 41–8 illustrates, trophic levels are numbered consecutively to indicate the order of energy flow. Trophic level 1 is always populated by primary producers, and Trophic level 2 is always populated by primary consumers. Levels of consumers beyond the primary consumer are then numbered sequentially. The final carnivore, called the **ultimate**, or **top carnivore**, and those organisms that escape being eaten (such as humans), eventually die and are consumed by decomposers.

FOOD CHAINS, MULTICHANNEL FOOD CHAINS, AND FOOD WEBS

The transfer of food from organism to organism forms a **food chain** (Figure 41-8). Like a trophic level diagram, a food chain is a flowchart that follows the course of energy and nutrients through an ecosystem. Unlike trophic level diagrams, food chains name the organisms (rather than the group) that occupy each link. The example we gave earlier of periwinkle plants, moths, frogs, snakes, and hawks is a food chain because the organism that occupies each step is identified.

☀ Most ecosystems contain many food chains. When more than one food chain originates from the same primary producer, a **multichannel food chain** is formed. In Figure 41-9, for example, different parts of a single manzanita shrub provide energy and nutrients for at least five independent food chains.

Often, linkages form between food chains, creating networks of connections. In the streamside ecosystem, for example, frogs may sometimes eat flies instead of moths, and snakes may eat small rodents instead of frogs (the flies and rodents originating from other food chains). When all interconnections between food chains are mapped out for an ecosystem, they form a **food web** (Figure 41-10). A food web illustrates all possible transfers of energy and nutrients among the organisms in an ecosystem, whereas a food chain traces only one pathway in the food web.

Food webs may intersect within ecosystems. For example, in a grassland, energy and nutrients may flow between a *grazing food web* and a *detritus food web*. In the grazing food web, living tissues of photosynthetic grasses are consumed by a variety of herbivores, which may then be consumed by a variety of carnivores. But not all plants, or all herbivores and carnivores are eaten, so when these organisms die, their bodies enter the detritus web. Even before they die, animals excrete organic wastes that also enter the

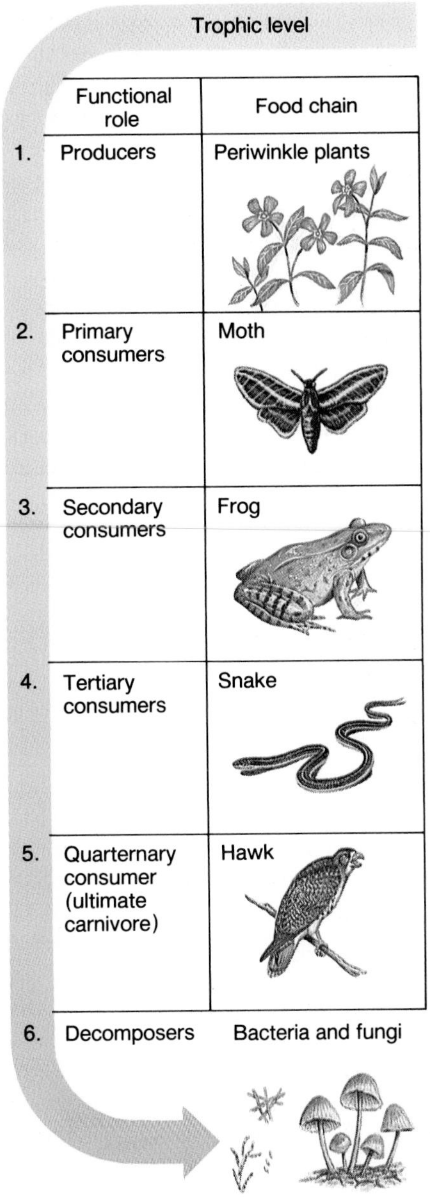

Trophic level

	Functional role	Food chain
1.	Producers	Periwinkle plants
2.	Primary consumers	Moth
3.	Secondary consumers	Frog
4.	Tertiary consumers	Snake
5.	Quarternary consumer (ultimate carnivore)	Hawk
6.	Decomposers	Bacteria and fungi

FIGURE 41-8

A feeding path in a streamside ecosystem. Energy and nutrients flow through the biotic community, as food passes from trophic level to trophic level. Trophic level numbers indicate the order of flow.

detritus web, where earthworms, insects, millipeds, fungi, and bacteria eventually break down organic matter to simple inorganic molecules.

ECOLOGICAL PYRAMIDS

Feeding relationships are graphically represented by plotting the energy content, number of organisms, or **biomass** —the total weight of organic material—at each trophic level. Such graphs are called **ecological pyramids** because of their triangular shape. Each trophic level of a food chain forms a tier on the pyramid; that is, each successive trophic level is stacked on top of the level that represents its food source.

Pyramid of Energy

Energy always flow one way through a food chain; it never recycles. All the energy that enters a food chain is eventually dissipated as unusable heat energy. Since energy does not recycle, there must be an extraterrestrial resource of energy to refuel life continually. For more than 99 percent of the earth's ecosystems, this extraterrestrial energy source is radiant energy from the sun.

A **pyramid of energy** illustrates the rate at which the energy in food moves through each trophic level of an ecosystem (measured in kilocalories of energy transferred per square meter of area per year—$kcal/m^2/year$). A typical pyramid of energy for a food chain in a grassland ecosystem during the spring (excluding decomposers) looks like this:

FIGURE 41-9

Multichannel food chains. This manzanita (*Arctostaphylos pringlei* var. *drupacea*) the primary producer for many food chains, including the five drawn here.

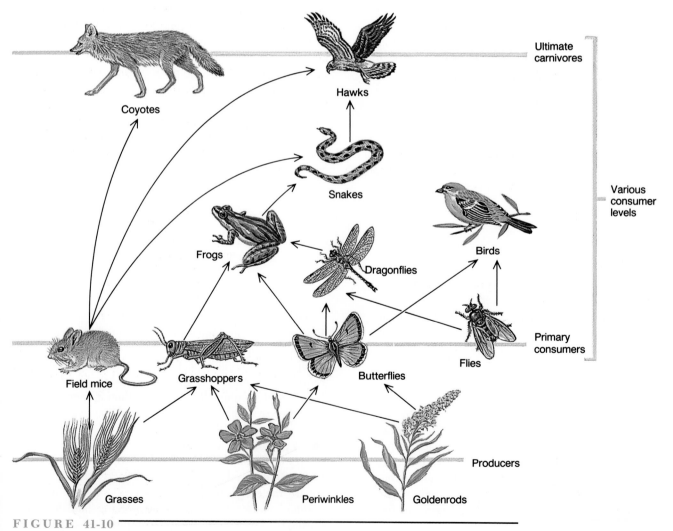

FIGURE 41-10

A simplified food web. A complete food web charts all possible feeding transfers within an ecosystem.

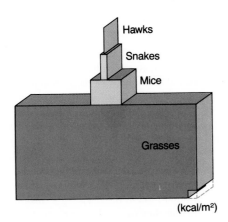

(kcal/m²)

As you can see, there is a dramatic reduction in usable energy as food is transferred from one trophic level to the next. On average, only about 10 percent of the energy in any trophic level is converted into biomass in the next level. The least efficient transfer is between trophic levels 1 and 2; as much as 99 percent of the energy available in the producers is not transferred to the primary consumer level.

What happens to this huge amount of lost energy? Some energy is lost as heat during chemical conversions (the second law of thermodynamics); some energy is used by the organisms in each trophic level for their own metabolism and biological processes; not all food in one trophic level is eaten by organisms in the next trophic level; not all of the food that is eaten is usable (some energy is excreted or defecated as waste), and some energy powers nonbiological phenomena, such as fire. Since such a small fraction of energy gets passed on to successive trophic levels, the number of links in a food chain is limited and rarely exceeds more than four or five. Supporting a food chain with more than five trophic levels would require an enormous energy base at trophic level 1, a situation that does not occur very frequently.

Pyramid of Numbers

The rate of energy flow through a trophic level determines how many individuals can be supported in an ecosystem as well as the overall biomass of the organisms. Because energy decreases with each successive trophic level, the number of organisms and their biomass also decreases. The following **pyramid of numbers** illustrates this principle for a grassland ecosystem:

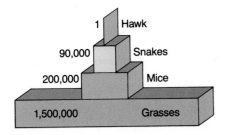

In grasslands, all primary producers are relatively small, compared to primary consumers, so it takes a large number of plants to provide enough food for big consumers, such as wildebeests or bison. In ecosystems where *larger* producers support *smaller* consumers, the pyramid of numbers becomes inverted. In a forest ecosystem, for example, a few large trees provide enough food to support many insects and insect-eating birds. The pyramid of numbers for such a food chain look like this:

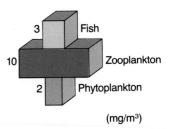

Pyramid of Biomass

A **pyramid of biomass** represents the total dry weight (in grams of dry weight per square meter of area—gm/m²) of the organisms in each trophic level at a particular time. Although most pyramids of biomass are "upright," they may become inverted, depending on when the samples are taken. For example, in the open oceans, where producers are microscopic phytoplankton, and consumers range all the way up to massive blue whales, the biomass of consumers may temporarily exceed that of the primary producers if data are taken when the number of phytoplankton is low. During such sampling periods, the pyramid of biomass could look like this:

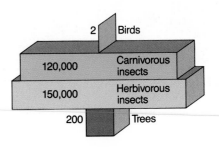

(mg/m³)

If samples are taken during the spring, however, when phytoplankton populations are immensely large, or if multiple generations of phytoplankton are included, the pyramid of biomass assumes the upright pyramid shape:

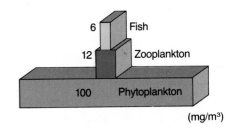

(mg/m³)

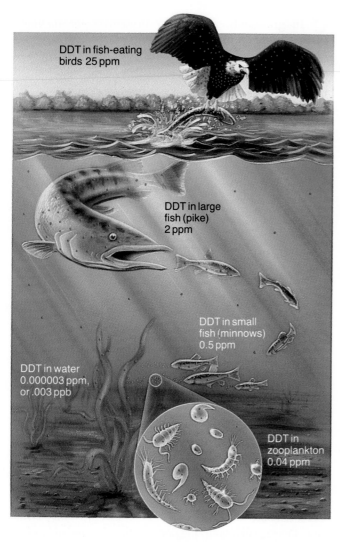

DDT in fish-eating birds 25 ppm

DDT in large fish (pike) 2 ppm

DDT in small fish (minnows) 0.5 ppm

DDT in water 0.000003 ppm, or .003 ppb

DDT in zooplankton 0.04 ppm

FIGURE 41-11

DDT, bioconcentration, and biological magnification. DDT was so effective in killing pests in agricultural fields, it became widely used. Unfortunately, runoff from these fields contaminated streams, rivers, and oceans with DDT. The algae growing in these aquatic ecosystems absorbed the DDT. As the algae were eaten by small aquatic animals, the DDT was passed on. As these animals were eaten by progressively larger fish, the DDT continued to pass along the entire food chain. DDT accumulates at each feeding level because organisms cannot metabolize DDT (such *bioconcentration* is represented by the red dots). When birds eat larger fish, the high doses of DDT interfere with calcium deposition in the birds' eggs, making the egg shells so thin and weak that they can no longer protect developing birds. Because of its harmful effects on wildlife, DDT was banned in the United States in 1968. The chemical is still widely used in other nations, however, and continues to contaminate much of the earth's waters.

Although it is possible for pyramids of biomass and numbers to be inverted, an energy pyramid can never be inverted; there must always be more energy at lower levels to maintain life at higher trophic levels.

Bioconcentration and Biological Magnification

Ecologists often use ecological pyramids to illustrate how toxic chemicals in the environment can gradually accumulate in the bodies of the organisms in an ecosystem, a phenomenon known as **bioconcentration.** Bioconcentration of a toxic chemical may build to a high enough level to kill the organism. Bioconcentration may also lead to **biological magnification,** the buildup of chemicals in the organisms that form a food chain. Biological magnification exposes organisms toward the end of a food chain (at higher trophic levels) to potentially dangerous levels of chemicals.

⚠ The use of DDT as a pesticide provides a clear example of both bioconcentration and biological magnification (Figure 41-11). DDT is fat-soluble, so it accumulates in the body fat of organisms (bioconcentration). Runoff from the use of DDT on agricultural lands mixes the DDT with local water supplies. Small, single-celled animals, like zooplank-

ton, bioconcentrate the DDT in their fat-storage regions. The zooplankton are then eaten by small fish, which are eaten by larger fish, which are eventually eaten by fish-eating birds, each of which also accumulates DDT. The combination of bioconcentration and biological magnification results in DDT concentrations that are several million times greater in the fish-eating birds than in the water. Such magnification has restricted the reproduction of many birds, including peregrine falcons, ospreys, brown pelicans, American eagles, and the California condor, by reducing calcium deposition in their eggshells. Without a hard, protective eggshell, the birds' eggs break easily, and few embryos survive.

The energy for all life in an ecosystem originates from the abiotic environment. The primary producers introduce energy from the sun as well as nutrients from the abiotic environment into the biotic environment. To understand how ecosystems function, ecologists often map and measure the flow of energy and nutrients between organisms, generating food chains, food webs, and ecological pyramids. (See CTQ #5.)

BIOGEOCHEMICAL CYCLES: RECYCLING NUTRIENTS IN ECOSYSTEMS

All organisms are composed of chemical elements (Chapter 4). Of the more than 90 naturally occurring elements found in the biosphere, only about 30 are used by organisms. Some elements, such as carbon, hydrogen, oxygen, phosphorus, sulfur, and nitrogen, are needed in large supplies, whereas others, including sodium, manganese, iron, zinc, copper, and boron, are required in small, or even minute, amounts.

Unlike radiant energy, which showers on the biosphere daily, there is no outside source to supply the elements essential for life. Since there is a finite amount of each, essential elements must be recycled (reused) over and over again for life to continue. Organisms today are using the same atoms that were present in the primitive earth. Some of the atoms that comprise your body may have been part of an ancient bacterium, a long-extinct tree fern, or a dinosaur.

During recycling, elements pass back and forth between the biotic and abiotic environments, forming **biogeochemical cycles.** Primary producers typically introduce elements into the biotic environment by incorporating them into organic compounds, and consumers and decomposers release the elements back into the abiotic environment by breaking down complex organic molecules into simple inorganic forms. Without such biological order and regulation, nutrient recycling could not occur. The rate of nutrient recycling depends primarily on where the element is found in the abiotic environment. For example, elements that cycle through the atmosphere or hydrosphere, forming *gaseous nutrient cycles*, recycle much faster than do those of *sedimentary nutrient cycles*, which cycle through the earth's soil and rocks.

GASEOUS NUTRIENT CYCLES

Gaseous nutrient cycles are those in which the element occurs as a gas at some phase in its cycle, and a large proportion of the element resides in the earth's atmosphere. Although several elements have gaseous cycles, we will discuss only the water (hydrologic), carbon, and nitrogen cycles here because of their central role in living organisms.

The Hydrologic Cycle

Nutrient cycles may involve more than one element. For example, in the **hydrologic cycle,** both hydrogen and oxygen cycle together in the form of water molecules, which often change from one phase to another (liquid water to water vapor, or vice versa) as they move between the earth's oceans, land, organisms, and atmosphere (Figure 41-12).

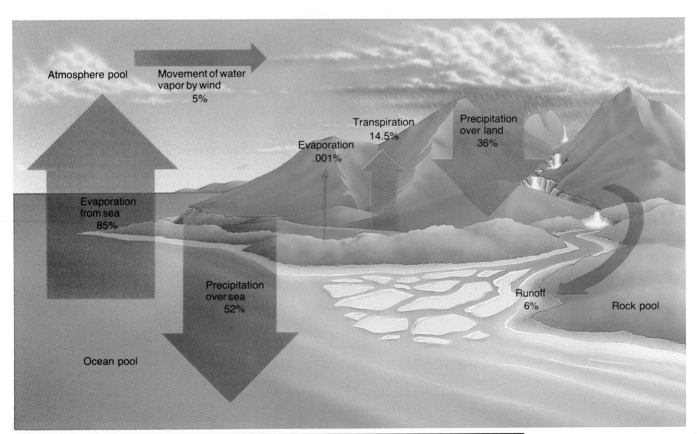

FIGURE 41-12
The hydrologic (water) cycle.

The largest reservoir of water, about 97 percent of the earth's available water, is in the oceans; only about 0.001 percent is present as water vapor in the atmosphere. Yet, despite this huge difference, the greatest quantities of water are exchanged between the oceans and the atmosphere via evaporation and precipitation. Over 80 percent of the liquid water that is evaporated during the hydrologic cycle comes from the oceans and moves into the atmosphere. About 75 percent of water in the atmosphere falls back into the oceans as precipitation, or rain. The remainder either stays in the atmosphere in the form of clouds, ice crystals, and water vapor, or falls back to earth as rain over the land.

Although some of the rain that falls on the land is intercepted by vegetation generally rain reaches ground level and either percolates into the soil or runs off the soil surface into streams and rivers, where it may eventually flow back to the ocean. Some water is pulled downward into the ground by gravity, contributing to ground water supplies, most of which eventually flows back to the ocean. Of course, some of the water in the ground is absorbed by plants. For many plants, more than 90 percent of the water taken in by roots is released back into the atmosphere as water vapor through transpiration. The remainder hydrates plant cells and tissues or is used in biochemical reactions, particularly photosynthesis. In comparison with other biogeochemical cycles, organisms seem to play a relatively minor role in the hydrologic cycle.

The Carbon Cycle

Large amounts of carbon are continually exchanged between the atmosphere and the community of organisms. As a result, the recycling rate of carbon is especially rapid (Figure 41-13). In the **carbon cycle,** primary producers constantly extract carbon dioxide from the atmosphere and use the carbon to form the chemical backbone for building organic molecules. Organic molecules are then disassembled by all organisms during cellular respiration, releasing carbon as carbon dioxide.

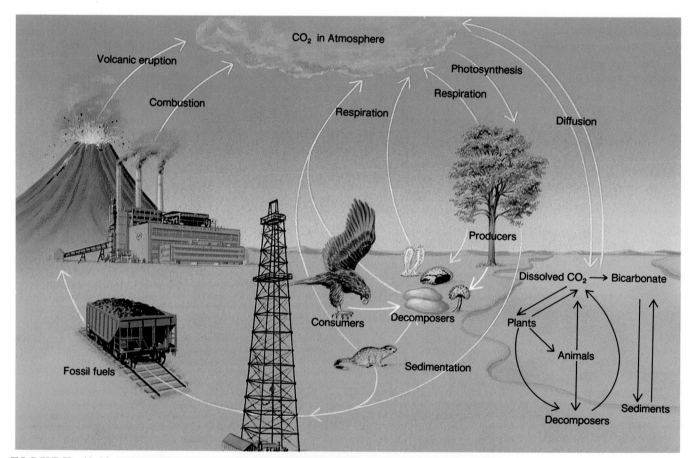

FIGURE 41-13

The carbon cycle. Atmospheric and dissolved carbon dioxide are used by primary producers to make energy-rich organic compounds during photosynthesis. When producers are eaten, carbon-containing organic compounds are passed to consumers and, eventually, to the decomposers. All organisms (producers, consumers, and decomposers) release carbon as carbon dioxide during respiration, most of which is returned to the atmosphere. Organisms that are not decomposed may eventually form fossil fuels. Combustion of fossil fuels also releases carbon as carbon dioxide back into the atmosphere.

On land, producers extract carbon dioxide gas directly from the atmosphere. In water, gaseous carbon dioxide must dissolve before it can be incorporated into organic compounds by aquatic autotrophs. Not all of the carbon dioxide dissolved in water is used in photosynthesis. Resulting carbon-containing bicarbonates and carbonates, compounds with low solubility, eventually settle to the bottom of oceans, streams, and lakes. Additional carbon sediments form as the skeletons and shells of organisms accumulate. The settling of carbonate, shells, and skeletons can tie up carbon for long periods of time in sediment, limestone, and several other forms of rock.

Even on land, dead organisms and organic matter may not be decayed by decomposition, a process that would ordinarily release carbon dioxide back into the atmosphere. Nondecayed organic material may build up deposits that

turn into fossil fuels, producing a carbon reservoir within the earth. Most of the world's fossil fuels were formed during the Carboniferous period, between 285 million and 375 million years ago, when shallow seas repeatedly covered vast forests, preventing decomposition. As we burn fossil fuels, we return this carbon to the atmosphere (see The Human Perspective: The Greenhouse Effect: Global Warming).

The Nitrogen Cycle

Although the earth's atmosphere is 79 percent nitrogen gas (N_2), only a few microorganisms are able to tap this huge reservoir, initiating a series of conversions and transfers that produces the **nitrogen cycle** (Figure 41-14). These "nitrogen-fixing" microorganisms include bacteria and cyanobacteria.

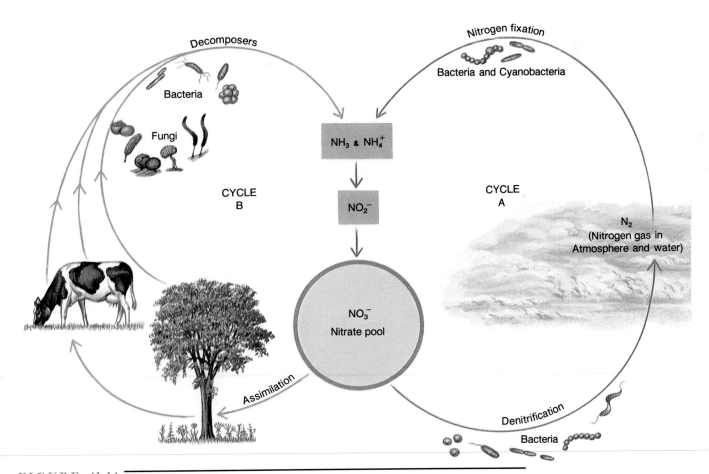

FIGURE 41-14

The nitrogen cycle. A pool of nitrates joins together two cycles that exchange nitrogen between the biotic community and the abiotic environment. In cycle A, nitrogen gas from the atmosphere is converted to ammonia or organic compounds by bacteria during *nitrogen fixation*. The ammonia is converted to nitrites and then nitrates by nitrifying bacteria. The nitrates are either converted by other bacteria into nitrogen gas during *denitrification* (completing Cycle A), or they are absorbed by primary producers and the nitrogen incorporated into nitrogen-containing organic compounds (beginning Cycle B). The ammonia that is excreted by consumers or released during decay is converted into nitrites and then into nitrates, returning nitrogen to the nitrate pool (completing Cycle B).

The **greenhouse effect** is a term that has been coined to describe the rise in global temperature that is triggered by increased amounts of gaseous pollutants, mainly carbon dioxide, that trap heat within the atmosphere. Gaseous pollutants absorb heat-generating infrared radiation from the earth, preventing it from escaping into space at night. The greenhouse effect in the atmosphere is similar to what happens when a car is left parked in the sun with its windows closed. The car's windows are like the gases in the atmosphere: They allow the sun's radiant energy to enter the car, heating the interior but preventing the inside heat from escaping; the car gets hotter and hotter.

As levels of carbon dioxide and pollutants increase in the atmosphere, they trap more and more heat, causing global temperatures to rise gradually, a phenomenon referred to as "global warming." By analyzing air bubbles that formed in the ice of Antarctica and Greenland over the past 1,000 years, scientists have documented an increase in atmospheric carbon dioxide of over 25 percent; 85 percent of this increase

occurred between 1870 and 1989, mostly from the burning of fossil fuels. At the current rate, the global carbon dioxide level is expected to double by 2050, raising the average global temperature between 2°C and 5°C (3.5°F and 9°F). The major consequences of global warming include the following:

- The sea level will rise during the next century. Using computer models, scientists predict that rising sea levels may cover the homes of more than 20 million Americans who live on the East Coast and ruin rice production in Asia. Most of the rice grows in low-lying regions that would be flooded with salt water. Increases in sea temperature will also begin melting the polar ice caps, contributing to rising sea levels. Satellite photographs show that the polar ice caps have shrunk by 6 percent over the past 15 years.

- Reduced rainfall and rising temperatures will amplify the world's hunger crisis. Most of the world's food supply is currently produced in the band of

agricultural land found in North America, Europe, and the Soviet Union. Climatic changes in these regions will cause crop production to fall, shifting farming northward to mountainous areas that are more difficult to farm.

- Many marine organisms breed in coastal estuaries, marshes, and swamps. As seas rise, these vital areas will be flooded, and many plant and animal species will be lost, as will much of our seafood supplies.

Since carbon dioxide has the biggest impact on global warming, reducing activities that produce carbon dioxide will help curb the problem. Much of the carbon dioxide that is released into the atmosphere originates from burning fossil fuels to produce electricity and deliberate burning of huge expanses of tropical rain forests. Thus, reducing energy consumption and saving tropical rain forests (see the Human Perspective, Chapter 21) would make a significant difference.

Nitrogen fixers convert atmospheric nitrogen gas to ammonia (NH_3) in a process called **nitrogen fixation.** Once nitrogen is converted, other groups of soil bacteria, collectively known as the *nitrifying bacteria,* convert ammonia into nitrites (NO_2) and then nitrates (NO_3). Plants absorb nitrates and incorporate the inorganic nitrogen into organic molecules: nucleotides and amino acids, the fundamental building blocks of DNA, RNA, and proteins. When the producers are eaten or die, nitrogen is passed on to the consumers and decomposers in a food chain. The decomposers then convert the nitrogen-containing organic molecules into inorganic ammonia, which is also released directly from the consumers (in urea) as a means of

eliminating the excess nitrogen that would otherwise accumulate and poison the organism.

Once again, ammonia is converted into nitrites and then into nitrates by nitrifying bacteria, making nitrogen available for absorption by plants. Some nitrogen is converted to nitrogen gas during **denitrification** by other bacteria, called **denitrifying bacteria;** nitrogen fixation is therefore needed to renew usable nitrogen resources. Some plants form nodules on their roots, housing nitrogen-reducing bacteria, thereby receiving a direct source of nitrates (refer to Figure 20-1*b*).

↻ The nitrogen cycle illustrates the critical role microorganisms play in the biosphere. Without bacteria, there

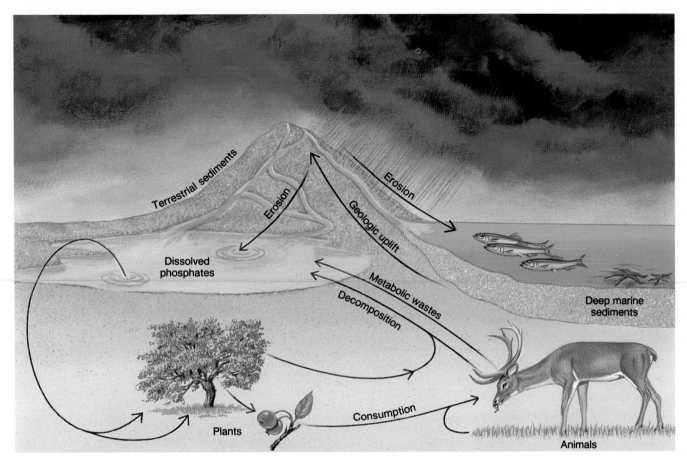

FIGURE 41-15

The phosphorus cycle. Phosphorus has a *sedimentary nutrient cycle* because most of the element is found in sedimentary rocks. Geologic uplift raises phosphate sediment, and erosion caused by waves and rain dissolves the phosphorus. Some dissolved phosphates are absorbed by primary producers, incorporated into organic compounds, and then passed to the other organisms in a food chain. However, the majority of dissolved phosphates runs off and accumulates as precipitated solids at the bottom of streams, lakes, and oceans, where it becomes part of new sediments. Decomposition and wastes from some animals, such as birds, also add to the dissolved phosphate pool.

wouldn't be sufficient nitrogen recycled to support the diversity of life on earth. Without bacteria-dependent nitrogen fixation, there wouldn't be any source of usable nitrogen for producers or the consumers they support. In other words, without bacteria, life on earth would cease.

THE PHOSPHORUS CYCLE: A SEDIMENTARY NUTRIENT CYCLE

Some elements never, or only rarely, exist as a gas. Such elements accumulate in the soil or rocks and, as a result, have a sedimentary nutrient cycle. Elements with sedimentary cycles include calcium, iron, magnesium, sodium, and phosphorus. We use the phosphorus cycle (Figure 41-15) to

illustrate the general features of a sedimentary nutrient cycle.

All organisms require large amounts of phosphorus to construct ATP, DNA, RNA, and cellular membranes. Most phosphorus is contained in rock deposits. Erosion and runoff from rain dissolve the phosphorus in rocks and form phosphates (PO_4^{-2}). Plants and other primary producers absorb phosphates and use the phosphorus to build organic molecules. When these primary producers are eaten, phosphorus is passed to the primary consumer and then to the other organisms in a food chain. Decomposers break down phosphorus-containing compounds and release it back into the environment as phosphates, which are either reabsorbed by plants or leached out of the soil, where they accumulate in sediments. Since phosphorus is easily leached from soil, it is one of the least available essential elements in the biosphere. The large quantities of phosphorus needed by organisms often makes this element a major limiting factor in many ecosystems.

IMPORTANT LESSONS FROM BIOGEOCHEMICAL CYCLES

All nutrient cycles share certain characteristics:

- The abiotic environment provides the principal reservoir of elements in air, water, or earth.

- The introduction of elements into the biotic environment almost always requires primary producers (mainly plants and algae).

- Microorganisms play a crucial role in nutrient recycling.

♻ Nutrient cycles illustrate a very important and fundamental ecological principle: organisms depend on other organisms. Even technologically advanced human societies ultimately depend on plants and simple microorganisms to supply their essential nutrients. Without them, nutrients could not recycle, and all life on earth would quickly grind to a halt. This is one of the reasons why many biologists are so concerned about how human activities have accelerated the extinction rate of many organisms (page 997). Because of their crucial role in nutrient recycling, we must pay just as much attention to the extinction of plants and microorganisms as we do to large, familiar animals.

Unlike energy, nutrients recycle through ecosystems over and over again, enabling life to persist on earth over immense periods of time. Nutrients are exchanged back and forth between the abiotic and biotic environments through the activities of primary producers and decomposers. (See CTQ #6.)

SUCCESSION: ECOSYSTEM CHANGE AND STABILITY

Ecosystems are constantly changing as energy and elements flow from the abiotic environment to the biotic community and as organisms interact within the biotic community. Ecosystems also change as seasonal shifts generate fluctuations in both the abiotic and biotic environments. Changes triggered by regular, seasonal fluctuations make an ecosystem dynamic, but they do not cause *permanent* changes in the composition and organization of organisms in the biotic community. However, permanent changes in the biotic community do occur in newly formed habitats and in areas disturbed by fire, floods, hurricanes, drought, or the activities of humans. Such large-scale changes trigger the process of **succession,** a progression of distinct communities that eventually leads to a community that remains stable and perpetuates itself over time.

THE CLIMAX COMMUNITY

Communities that remain more or less the same over long periods of time, such as an area of mature forest or grassland, are called **climax communities.** The populations of organisms that make up a climax community are in equilibrium with their abiotic environment. Thus, the kinds of organisms and their abundance remain relatively constant over long periods.

♻ Climax communities tend to contain many species in a highly organized trophic structure. Generally, large amounts of organic compounds are manufactured by producers of climax communities, but consumers and decomposers use nearly all of the excess, so the total biomass does not increase. In addition, the majority of plant species that make up the climax community are long-lived, and a comparatively small portion of their energy is used for reproduction; most is diverted for growth.

Severe or long-term changes in either the abiotic or biotic environments can permanently change the organization and composition of the biotic community, however. For example, cycles of declining temperatures during the Pleistocene Epoch produced a series of ice ages that destroyed and permanently altered most communities in the biosphere. Another large-scale factor that often causes permanent changes in a community is fire. Fires not only kill many organisms outright, they also modify conditions so much that different organisms predominate in the affected area.

When a community has been permanently changed, or when an entirely new habitat is formed (such as following a volcanic eruption or when a glacier retreats), a variety of species invade the area, forming a *pioneer community.* As pioneer species take hold, they modify the environment by changing the soil, the temperature of the ground, the amount of light that penetrates to ground level, and many other environmental characteristics. Eventually, the pioneer community changes conditions so much that new species invade the community. The new group gradually displaces the pioneer community and forms its own community. The process continues—one community replacing another—until a stable, climax community develops. This orderly, directional sequence of communities that leads to a climax community is succession. The entire series of successional communities, from pioneer to climax, forms a successional *sere.* Each community in a sere is called a *stage.*

Ecologists recognize two types of succession: 1. **Primary succession** occurs in areas where no community existed before (new volcanic islands, deltas, dunes, bare rocks, or lakes); and 2. **secondary succession** occurs in disturbed habitats where some soil, and perhaps some organisms, still remain after the disturbance. Fires, floods, drought, and many human practices (such as clearing forests for agriculture and construction projects) would prompt secondary succession. Secondary succession also occurs on abandoned farmlands, in overgrazed areas, and in forests cleared for lumber.

PRIMARY SUCCESSION

New habitats do not remain bare for very long. Even a dry rock is soon colonized by lichens, a "compound organism" composed of a fungus and resident algae (Figure 41-16). The hyphae of the fungus are able to grow into even the tiniest rock fissures, prying the rock open. At the same time, hyphae secrete chemicals that help erode the rock. The combination of intrusive growth and chemical erosion, together with abrasion from the wind and water and repeated heating and cooling, gradually crumbles the rock into small fragments, forming sand. When the lichens die, they mix with the sand, initiating the process of soil formation. As sand and organic matter accumulate, moss spores and grass seeds are eventually able to germinate and grow. These organisms continue the process of disassembling and chemically dissolving the rock. When they die, their remains add even more organic matter to the developing soil. Eventually, the soil is rich enough to support the growth of large plants that outcompete the lichens, and the pioneer lichen community becomes replaced by a new community of mosses, grasses, ferns, and other plants. After hundreds of years, a forest climax community may grow in an area that was once nothing but bare rock.

The transformation from bare rock to forest is only one example of primary succession. Primary succession also takes place in lakes (see Figure 40-11), on hardened lava flows, and in any other newly exposed substrate. Primary succession in the sand dunes along the shores of Lake

FIGURE 41-16

Primary succession on a bare rock begins with lichens. Eventually, the lichens crumble the rock, initiating soil formation. The process continues over many years, until what was once bare rock may eventually support a desert, woodland, or even a forest, depending on the climate.

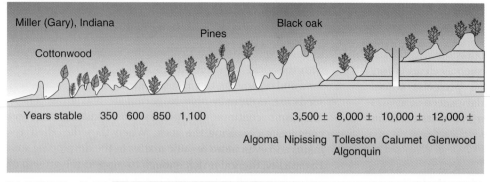

FIGURE 41-17
Primary succession on dunes along Lake Michigan. This diagram illustrates a profile across the sand dunes of Lake Michigan. The level of the lake is gradually falling, uncovering new sand substrate, on which primary succession occurs. Since regions along the shore have emerged most recently, they demonstrate the earliest stages of dune succession. Farther away from the shoreline, successively older stages of dune succession occur. These older dune systems originated along earlier and higher beaches.

(a)

(b)

(c)

(d)

FIGURE 41-18
The eruption of Mount St. Helens in Washington on May 18, 1980, initiated both primary and secondary succession. Near the center of eruption, species of bacteria quickly populated warm, standing pools that formed from rains and melting snows, beginning primary succession in each pool. Away from the eruption center, the force of the eruption severed trees at their base, seared off branches, and blew over burnt trees like toothpicks. Soil remained in both these outer areas, where secondary succession is now occurring. This series of photographs were taken from the same spot *(a)* 3 months, *(b)* 2 years, *(c)* 4 years, and *(d)* 9 years after the May, 1980 eruption.

Michigan has been investigated extensively. Since the retreat of the glaciers that formed the Great Lakes area during the Pleistocene Epoch, the level of Lake Michigan has gradually been falling, creating a nearly ideal situation for studying sand-dune succession. The initial substrate is dune sand (nearest the lake shore), but areas progressively further from the shore are successively older, each the product of an increasing length of time it took for primary succession to occur (Figure 41–17). Bare sand is colonized by grasses, particularly marram grass *(Ammonphila breviligulata)*, which spreads quickly asexually by rhizomes, stabilizing the dune surface. After 6 years or so, marram grass dies out and other grasses become predominant. Cottonwoods *(Populus deltoides)* are the first trees to appear. After 50 to 100 years, pines invade the dunes but are then replaced by black oaks after about 100 to 150 years. The oldest dunes, dated at 12,000 years, still had black oak associations, which were most likely the climax community.

SECONDARY SUCCESSION

Primary succession generally continues over hundreds or thousands of years, but secondary succession is often completed in less time, for two reasons. First, soil is already present, so the process that takes the longest is already completed. Second, in the presence of fully developed soil, plants from the surrounding communities, along with any surviving plants or seeds, quickly colonize the area. As a result, the pioneer community may include individuals of species that are part of the climax community, giving these species an early foothold in the disturbed area.

Of course, rates of secondary succession vary widely, depending on the degree of disturbance, the climate, and the kind of climax community. For example, a cleared area in a tropical rain forest may recover within 10 years, a burned chaparral in 20 years, and a grassland in 40 years; an abandoned farm may revert back to a deciduous forest in 150 years; and a disturbed area in the desert may take several thousands of years to recover, if it recovers at all.

The first eruption of Mount St. Helens in 1980 initiated both primary and secondary succession (Figure 41-18). Near the center of the eruption, primary succession began almost immediately as bacteria populated warm pools. Farther away from the eruption center, secondary succession is currently taking place in areas where tree trunks were severed at their base by the blast and where trees remain standing but were burned during the eruption. The rate of secondary succession is different in each zone because of the varying degree of disturbance.

Although ecosystems bustle with activity as organisms continually interact with each other and with the physical environment, only large-scale changes in the biotic or abiotic environments trigger permanent change in the biotic community. Such change produces a succession of replacement communities that eventually leads to a community that remains stable over long periods. (See CTQ #7.)

REEXAMINING THE THEMES

Biological Order, Regulation, and Homeostasis

Nutrient cycles take place within ecosystems. These critical biogeochemical cycles require order and regulation for the transfer of chemical elements between the abiotic and biotic environments and within the biotic community. Generally, a number of organisms are critical to recycling elements, including some of the simplest microorganisms. Without the participation of essential organisms, particularly primary producers and microorganisms, chemical elements will not recycle. Without recycling of finite amounts of elements, life would eventually grind to a halt.

Acquiring and Using Energy

Unlike chemical nutrients, energy does not recycle through ecosystems. Primary producers, primarily photosynthesizing plants, capture and convert radiant energy from the sun into chemical energy. Only primary producers are capable of acquiring energy from the abiotic environment. All other organisms acquire their energy from the primary producers or from some other organism, forming food chains or food webs if all possible transfers within an ecosystem are considered. In a food chain, organisms incorporate and pass on only a small percentage of the energy they acquire. If the energy transfers between organisms are presented graphically, they form a pyramid shape that illustrates the accentuated loss of energy at each transfer.

Unity within Diversity

Despite the tremendous diversity of organisms found on earth, all are dependent on the physical environment to supply the resources they require for survival and reproduction. All organisms need a habitat in which to live; all organisms have tolerance ranges for each environmental factor; and the distribution and survival of an organism is often determined by a single limiting factor. Despite the tremendous diversity of organisms on earth, however, all organisms are similar in that they depend on one another to secure the resources they need for survival and reproduction.

Evolution and Adaptation

▮▶ An organism's ecological niche is the result of natural selection. Within a specific ecosystem, those traits that en-able an organism to acquire more nutrients, to outrun a predator, to find a mate or any other feature that increases survivorship and reproduction are naturally selected. In this way, natural selection leads to adaptations for a particular habitat, with defined role(s) and tolerance ranges for each abiotic factor in a particular ecosystem. For example, organisms with narrow niches have very specific adaptations for specific habitats and roles, whereas the adaptations of organisms with broad niches enable these organisms to secure resources in a variety of ways and, generally, over a wide range of habitats.

SYNOPSIS

Ecosystems are dynamic, self-sustaining units. They are composed of a community of organisms and the surrounding physical environment. Ecosystems are con-nected to other ecosystems to varying degrees.

Organisms (or species) have a tolerance range for each physical factor. When the maximum or minimum tolerance is approached or exceeded for any given factor, that factor limits the distribution, health, or activities of the organism. Each organism has a suite of adaptations that defines its ecological niche—the combination of an orga-nism's habitat, functional role(s), and total environmental requirements and tolerances.

Energy flows through an ecosystem. Trophic levels, food chains, and food webs track the flow of energy and nutrients between the members of the biotic community.

Essential nutrients cycle between the abiotic and biotic environment, forming biogeochemical cycles. Producers and decomposers are required for exchanging nutrients between the biotic and abiotic environments.

Ecosystems inevitably change. The changes that occur in an ecosystem are triggered by permanent changes that occur in the abiotic or biotic environments.

Key Terms

Review Questions

1. Match the example with the term:

Term	Example
____1. guild	A. fungi and bacteria
____2. primary producers	B. fruit-eating birds and
____3. primary consumers	grass-eating antelopes
____4. decomposers	C. algae and plants
	D. two insect-eating bird species

2. List examples, other than those given in the text, of how the biotic community causes changes in the abiotic environment.

3. Give an example of each of the following:
   a. realized niche
   b. Law of the Minimum
   c. secondary succession
   d. food web
   e. ecological equivalents

4. List as many components as you can think of for the fundamental niche and realized niche of a whale (a relatively broad niche) and a tapeworm (a narrow niche). Now try it for humans.

5. Using your experience with house and garden plants, list some effects of limiting factors that you have observed for yourself.

6. Describe some of the ways humans have triggered secondary succession in ecosystems near where you live. Will the original ecosystems eventually return, or will new ecosystems develop?

7. Refer to the nitrogen cycle on page 949. Describe at least three separate pathways that would enable nitrogen to be recycled through the biotic community.

8. Compare the carbon cycle (a gaseous nutrient cycle) and the phosphorus cycle (a sedimentary nutrient cycle). In what ways are they similar? How do these similarities affect the relative rates of recycling for each?

9. Describe how food (energy and nutrients) is transferred from the abiotic environment to the biotic community of an ecosystem. What eventually happens to the energy? What eventually happens to all of the nutrients?

10. Relative to what you have learned in this chapter, why are biologists so concerned about the increasingly rapid extinction of life on earth?

Critical Thinking Questions

1. Had Robert Whittaker decided to test whether groups of organisms have identical distribution ranges in the Great Smoky Mountains by analyzing bird species instead of insects or plants, would you expect the outcome to be the same or different? What about mammals? Would the lifestyle of the organism being investigated (such as whether the organism is stationary like a plant, has a small range like a rodent, or a wide range like a hawk) have any impact on the results? Explain.

2. Linkage between two ecosystems, even if far apart, is illustrated by the forests of the tropics and North America, which provide winter and summer homes, respectively, for many of our most common songbirds. Read "Why American Songbirds Are Vanishing" (*Scientific American*, May 1992) and list all of the factors that are affecting the songbirds. Are there other ways in which these two ecosystems are linked? (HINT: Think globally!)

3. If organisms are so dependent on their physical environment, how do you explain the following? (1) Many plants and animals thrive when introduced into new areas; for example, many European wild flowers thrive as weeds in North America. (2) When environments undergo change, some plants and animals survive, while others are wiped out; for example, removing maple and beech trees from eastern forests promotes the growth of birch and aspen.

4. Like all biomes, a desert encompasses several ecosystems. One type of desert ecosystem is a "wash." Although dry most of the year or often over several years, a desert wash forms as water from rain is channeled, producing a "river," sometimes only the size of only a small trickle, and other times the size of a large flood. Describe as many components of the ecological niches of five organisms (two plants, two animals, and one protist or fungus) that you would expect to find in a desert wash. Are there similarities in niche breadth between the different types of organisms? In which ways do niches overlap between the plants? Between the animals? Does niche breadth and overlap change

over time as water alternates between abundance and scarcity?

5. All ecosystems depend on a flow of energy through the living system and cycling of material elements. Prepare a diagram showing these characteristics of a generalized ecosystem and the role of producers, consumers, and decomposers in the system.

6. Explain why nutrients must be recycled in an ecosystem. What are the natural recyclers? How are human activities affecting natural cycles? Give two specific examples.

7. The following graph presents data taken over 50 years in a small but unique community of organisms that is completely surrounded by a dense forest. The graph illustrates the change in the number of new species and the change in the average height of primary producers. From these data alone, could you determine whether this unique community was stable or undergoing succession? What are the limitations of these data? If you were conducting this research, what other factors would you investigate? How would you survey these factors, and how would you analyze the data you collect to discern whether changes were cyclic or permanent?

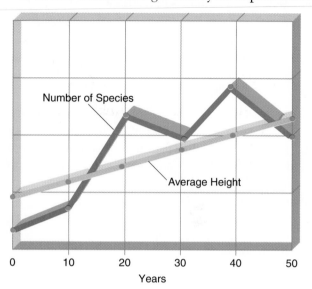

Additional Readings

Brewer, R. 1979. *Principles of ecology.* Philadelphia: Saunders. (Intermediate)

Flannagan, D., and F. Bello (Editors). 1970. *The biosphere.* A Scientific American Book, Scientific American, Inc. San Francisco: W. H. Freeman. (Introductory)

Lewin, R. 1986. In ecology, change brings stability. *Science* 34:1071–1073. (Intermediate)

Odum, E. P. 1983. *Basic ecology.* Philadelphia: Saunders College Publishing. (Intermediate)

Thoreau, H. D. 1937. *Walden.* New York: The Modern Library. (Classic, Introductory)

Community Ecology: Interactions between Organisms

STEPS
TO
DISCOVERY
Species Coexistence: The Unpeaceable Kingdom

SYMBIOSIS

COMPETITION: INTERACTIONS THAT HARM BOTH ORGANISMS

Exploitative and Interference Competition

Competitive Exclusion: Winner Takes All

Resource Partitioning and Character Displacement

INTERACTIONS THAT HARM ONE ORGANISM AND BENEFIT THE OTHER

Predation and Herbivory

Parasitism

Allelopathy

COMMENSALISM: INTERACTIONS THAT BENEFIT ONE ORGANISM AND HAVE NO AFFECT ON THE OTHER

INTERACTIONS THAT BENEFIT BOTH ORGANISMS

Protocooperation

Mutualism

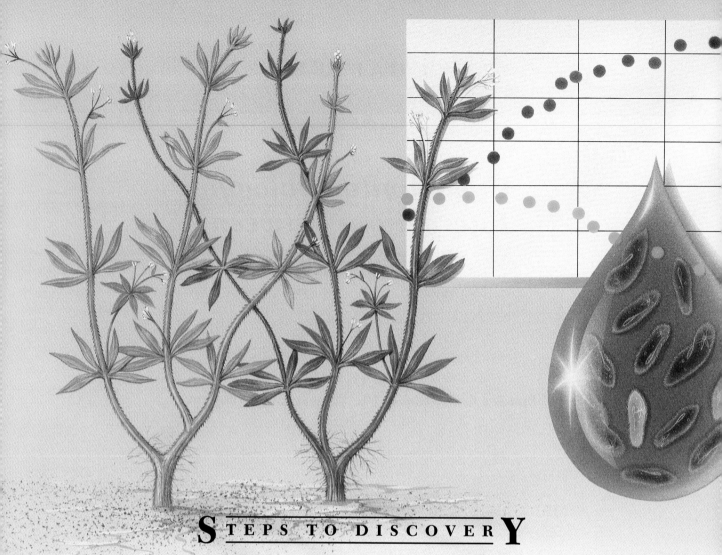

Species Coexistence: The Unpeaceable Kingdom

*C*harles Darwin was an inexhaustible thinker and worker. Not only did he explore the process of evolution, he pursued research in several other areas of biology, including orchid pollination, selective breeding, animal taxonomy, plant movements, and competition. Darwin observed that when two individuals lived in the same area and required the same limited resources—space, food, or whatever—those individuals would compete with one another for that resource. He also noted:

As species of the same genus have usually, though by no means invariably, some similarity in habits and constitution, and always in structure, the struggle will generally be more severe between species of the same genus, when they come into

competition with one another, than between species of distinct genera.

In other words, the more closely related the competing individuals, the more similar their needs and the more intense the competition for limited resources.

Since Darwin's time, many scientists have studied competition. These studies eventually led to the formulation of a mathematical principle regarding competition. At the center of this discovery was G. Gause, a Russian microbiologist at the University of Moscow. As in many scientific investigations, Gause acquired insight not only from his own research efforts but from the research of other biologists as well as two mathematicians and a physicist, L. Boltzmann. In 1905, Boltzmann wrote:

As the graphs illustrate, competition between similar species having identical requirements, whether between bedstraw plants

. . . plants spread under the rays of the sun the immense surface of their leaves, and cause the solar energy before reaching the temperature level of the earth to make syntheses of which as yet we have no idea in our laboratories. The products of this chemical kitchen are the object of the struggle in the animal world.

Although Boltzman refers only to animal competition in this statement, plants also enter into competition with one another, often for the "rays of the sun" to which they expose the "immense surface of their leaves." Indeed, Gause later acknowledged his debt to botanists (plant ecologists, in particular) because their work provided much of the insight and data he needed to formulate his principle. Gause wrote:

Botanists have already recognized the necessity of having recourse to experiment in the investigation of competition phenomena.

One of the first plant ecologists to investigate plant competition was Sir Arthur G. Tansley, who founded the British Ecological Society. In his presidential address to the First Annual General Meeting of the Society, held in 1914, Tansley emphasized the importance of competition among plants in community dynamics and urged fellow ecologists to initiate research on plant competition. Three years later, Tansley reported the results of his own studies on the competition between two species of bedstraw plants, *Galium saxatile* and *G. sylvestre*. Each species of bedstraw is more abundant in different soil types: *G. saxatile* grows best in silica-rich soils, while *G. sylvestre* thrives in limestone soils.

Tansley grew both plants in various soils, including silica-rich soil and limestone, and then monitored germination, seedling survival, and competition. He found that *G. sylvestre* had a higher germination rate, grew more vigorously, and outcompeted *G. saxatile* in lime-rich soils. Eventually, all the *G. saxatile* seedlings in the lime-rich soils died. In contrast, in lime-poor soils, although *G. sylvestre* had a greater germination rate, *G. saxatile* eventually outcompeted *G. sylvestre*. Tansley was one of the first to show that some plants were better able than others to compete in certain soils, suggesting that competition in natural communities influences the distribution and abundance of plants (organisms) in an ecosystem.

From Tansley's work, and that of other plant ecologists, Gause concluded that light, nutrients, water, and pollina-tors were common limiting resources for which plants compete. Sources of animal competition include water, food, mates, nesting sites, wintering sites, and sites that are safe from predators.

During the 1920s, researchers began formulating mathematical models to account for what happens when two species live together and require the same limited resource or when one species preys on or parasitizes another. One model, the *Lotka-Volterra equation,* was derived independently by Alfred J. Lotka at Johns Hopkins University in 1925 and by V. Volterra in Italy in 1926. This model described competition between organisms for food or space by comparing changes in population growth (increases or decreases in numbers of individuals) as competing species affect each other. According to the Lotka-Volterra equation, one possible outcome of competition is for one competitor to displace the other completely, causing the weaker species to become extinct.

To test whether this "winner-takes-all" outcome really occurs in nature, Gause initiated a number of studies in 1932 to test competition between microorganisms, first between competing species of yeast and then between competing species of protozoa. In Gause's best-known experiment, he monitored two species of *Paramecium* (*P. caudatum* and *P. aurelia*). Each species was first grown in a separate culture and then in a mixed culture, where the species competed for a limited food supply. When grown separately, the number of individuals of both paramecia increased rapidly and then leveled off and remained constant. When cultured together, however, competition for limited food supplies resulted in the elimination of *P. caudatum,* which were outcompeted by the more rapidly reproducing "winner," *P. aurelia.*

Gause concluded from this and from other similar experiments that "the process of competition under our conditions has always resulted in one species being entirely displaced by another." Gause's experiments supported the Lotka-Volterra equation and the "winner-take-all" outcome of competition and eventually became known as *Gause's Principle of Competitive Exclusion.*

or paramecia (in droplet), results in one species completely outcompeting the other.

Although seemingly calm, ecosystems actually bustle with activity. During warmer months in a forest, for example, the soil teems with bacteria, fungi, nematodes, springtails, amoebas, mites, slugs, worms, beetles, spiders, and scores of other organisms that churn the ground as they move about, grow, and reproduce. As they erupt through the soil surface, delicate plant seedlings absorb the nutrients recycled by microorganisms and fungi. These seedlings eventually grow into herbs, shrubs, and trees that create habitats and manufacture food for countless herbivores, which are, in turn, devoured by an assortment of carnivores.

Although the participants vary from one ecosystem to the next, all ecosystems are similar to the forest described above in that the organisms that live together often interact with one another. Some of these interactions benefit one or both participants; some have neutral consequences; and some harm either or both participants. The general categories of interactions and the eventual outcome for the participants are previewed in Table 42-1.

▼ ▼ ▼

SYMBIOSIS

Some organisms interact because they physically live together or because they live in very close association with one another. A close, long-term relationship between two individuals of different species is called **symbiosis,** which literally means "to live together" (Figure 42-1). Symbiotic interactions include forms of parasitism, commensalism, protocooperation, and mutualism (Table 42–1).

Many biologists now believe that symbiosis played a critical role in the early stages of the evolution of eukaryotic cells. According to the *endosymbiont theory,* the organelles of eukaryotic cells—mitochondria, chloroplasts, and flagella—are descended from once free-living prokaryotes that developed symbiotic relationships with primitive eukaryotic cells (see Chapters 5 and 35). A great deal of evidence exists to support this theory.

There are many modern examples of symbiotic relationships, involving very different types of organisms, such as between fungi and plants (mycorrhizae, page 386), plants and bacteria (root nodules, page 389), fungi and algae (lichens, page 821), sea anemones and fish, and jellyfish and algae. One interesting example involves an oyster and a crab. The larvae of the crab enters the mantle cavity of the oyster. The crab then grows and resides inside the oysters for its entire life. Since it remains sheltered all its life by

(a)

(b)

FIGURE 42-1

Symbiosis: living together. ***(a)*** *Aphids and ants.* Some ants live with groups of aphids. The aphids feed on the sugary juices of plants, often taking in more than their bodies can hold. The excess juice passes through the aphid's digestive system and out its anus, forming a honeydew drop that is lapped up by the ants. The ants protect the aphids by aggressively keeping predators away (such as ladybird beetles and syrphid fly larvae). ***(b)*** *Anemones and algae.* This anemone is green because algae live in its body. The algae conduct photosynthesis, producing food and oxygen for the anemone. In turn, the algae receive carbon dioxide and a safe habitat inside the animal's tissues.

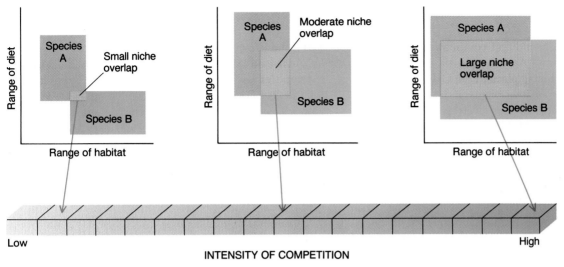

FIGURE 42-2

Niche overlap and competition. In each of the three graphs, habitat and diet requirements are plotted for two species. When niches overlap, species compete for limited resources. The greater the similarity in the requirements of the species, the greater the niche overlap, and the more fierce the competition.

the shell of the oyster, the crab benefits, while the oyster apparently remains unaffected by the crab's presence.

Organisms rely on other organisms. Because of these reliances, some organisms pair up to form a symbiotic relationship, whereby one or both of the organisms provides the needs of the other. (See CTQ #2.)

COMPETITION: INTERACTIONS THAT HARM BOTH ORGANISMS

Recall from Chapter 41 that species can coexist in a community as long as they have slightly different ecological niches, even though their niches may overlap. When a shared resource is abundant, such as oxygen in the air of terrestrial habitats, or water in aquatic habitats, there is more than enough for all. Generally, however, most resources are limited so organisms with overlapping niches enter into **competition.** Competition always harms both participants because each competitor reduces the other's supply of a needed resource. The more similar the requirements of the organisms, the greater their niches overlap; the greater the niches overlap, the more intense the competition (Figure 42-2). Furthermore, since members of the same species require many of the same resources, **intraspecific competition** that occurs between members of one species is often more intense than is **interspecific competition** that occurs between members of different species.

▐▶ Organisms are powerful agents of natural selection. In fact, they are just as powerful as are environmental factors, if not more so. Competition between organisms is also a powerful natural selection force. As we will see, competition can lead to the extinction of one competitor, to the exclusion of one competitor from an ecosystem, or to rapid evolutionary changes in the characteristics of the competitors.

EXPLOITATIVE AND INTERFERENCE COMPETITION

Organisms compete either *directly* or *indirectly* for a limited resource. Indirect competition occurs when competi-

TABLE 42-1

INTERACTIONS BETWEEN ORGANISMS IN A COMMUNITY

Kind of Interaction	Organism 1	Organism 2
Competition	Harmed	Harmed
Predation (including herbivory)	Benefited	Harmed
Parasitism[a]	Benefited	Harmed
Allelopathy	Benefited	Harmed
Commensalism[a]	Benefited	Unaffected
Protocooperation	Benefited	Benefited
Mutualism[a]	Benefited	Benefited

[a] May include symbiotic interactions.

tors have equal access to a limited resource but one species manages to get *more* of the resource, reducing the competitor's supplies. This form of indirect competition is called **exploitative competition.** An example of exploitative competition is currently taking place in the California deserts between deep-rooted native plants and newly introduced tamarisk trees. Tamarisk trees were brought to the California deserts from the Middle East to act as windbreaks along freeways and railroad tracks. The rapidly growing and reproducing tamarisk trees are better able to tap groundwater supplies, interfering with water availability to the deserts' native trees, such as mesquite and desert willows, reducing their populations.

In contrast to exploitative competition, **interference competition** is direct: One species directly interferes with the ability of a competing species to gain access to a resource. Interference competition is exemplified by aggressive behavior in animals, as when hyenas drive away vultures from the remains of a zebra, or by **territoriality,** as when male bighorn sheep establish and defend an area against other males of their species (see Chapter 44 on Animal Behavior).

COMPETITIVE EXCLUSION: "WINNER-TAKES-ALL"

Although species with small niche differences are often able to live in the same community, those with *identical* ecological niches cannot do so, even if they share only one scarce resource and many abundant resources. Competition becomes so intense in this case that one species eventually eliminates the other from the community, either by taking over its habitat and displacing the species from the community or by causing the species' extinction. The "winner" species is successful because it possesses some characteristics (adaptations) that give it a slight advantage over its competitor. This advantage enables members of the winner species to capture a greater share of resources, which, in turn, increases the survival and reproduction of the individuals. Consequently, greater numbers of offspring with better-suited traits gradually displace members of the less efficient species, illustrating natural selection in action. As we mentioned in the introduction to this chapter, this "winner-takes-all" outcome is referred to as Gause's principle of **competitive exclusion.**

One example of competitive exclusion involves the day lily (*Hemerocallis*), a popular cultivated plant. Day lilies often escape gardens and begin multiplying along roadsides and in surrounding communities by forming thick clumps of shoots and roots. Once the day lily becomes established outside a garden, few native plants can compete against it; eventually, they become displaced from their original habitat.

The day lily is an example of a plant that becomes established in a surrounding community that does not naturally include day lilies; it is not an example of competitive exclusion between members of the same community. In natural communities, competitive exclusion is not always apparent because there are so many variables to monitor. For instance, six species of leafhoppers (*Erythoneura*) are able to live on a single sycamore tree, feeding side by side on the same leaves. Not only are the habitats and food source of the insects the same, but the species' life cycles are virtually identical as well. In fact, researchers could not find any niche differences among the six species, a seemingly perfect setup for competitive exclusion. Yet, investigations found no evidence whatsoever that these species harm one another, much less that they exhibit competition that results in exclusion. Perhaps competitive exclusion is avoided in this case because shared resources are abundant.

Competitive exclusion is also often not recorded in ecosystems where environmental conditions change frequently. Under continually changing conditions, there simply is not enough time for one species to displace another during the short periods when resources become limited. This occurs in ocean upwellings and in temperate lakes during seasonal overturns because changes develop too quickly for one species of phytoplankton to grow enough to exclude another, despite intense competition between the species for limited nutrients. Similarly, since steady changes occur during primary and secondary succession (page 953), competitive exclusion is not apparent in communities undergoing succession.

A third explanation for why the process of competitive exclusion may go unnoticed in natural communities is the length of time it takes for one species to exclude another. Researchers are often unable to observe a community continuously so they may miss the process of exclusion entirely. For example, goats were introduced on the island of Abingdon in the Galapagos Islands in 1957. The goats browsed on the same low-growing plants as did the island's native tortoises as well as on the leaves found on higher stems and branches. In the absence of predators, the goats reproduced rapidly and consumed all the low-growing food that could be reached by the tortoises. By the time a research team revisited the island in 1962, all of the tortoises were gone. Competitive exclusion had caused the extinction of the Abingdon tortoise over a 5-year period, but the researchers had missed it.

RESOURCE PARTITIONING AND CHARACTER DISPLACEMENT

Competitive exclusion is not the only outcome of competition. Sometimes, a shared resource becomes partitioned in a way that allows competitors to use different portions of the same resource. For example, five species of North American warblers feed in slightly different zones on the same spruce tree, enabling these very similar birds to coexist with minimum competition (Figure 42-3). In addition to such spatial partitioning, a shared resource may be exploited at different times, producing temporal partitioning. An example of temporal partitioning occurs in a grassland ecosystem, where a species of buttercup (*Ranunculus*) grows only

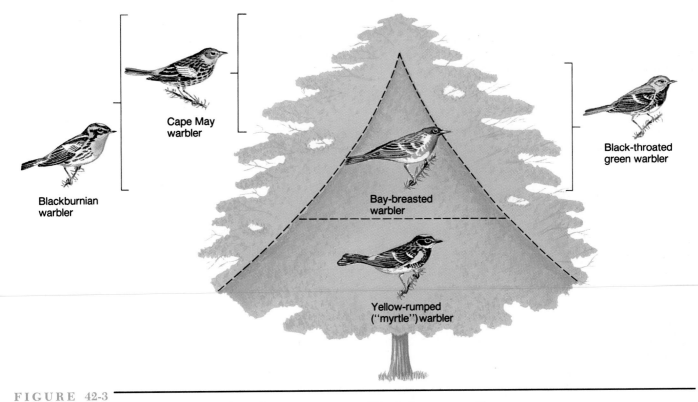

FIGURE 42-3

Resource partitioning: dividing the ecological pie. Five species of North American warblers feed in the same spruce tree, but each feeds in a slightly different zone. Foraging in different areas of a common resource at the same time is one form of resource partitioning. Resource partitioning reduces competition, enabling species with similar ecological niches to coexist in a community.

in early spring, before competing perennial grasses begin to grow. Dividing a resource in time or space is known as **resource partitioning.**

▮▶ Another alternative outcome to competitive exclusion is **character displacement,** whereby intense competition dramatically affects evolution, leading to changes in one or more characteristics of a species. Imagine two bird species that both harvest the same kind and size of fruit. If, as a result of natural selection, one bird species had evolved a different size bill than the other—a bill that can harvest larger fruits, for example—competition between the species would decrease.

The concept of character displacement was formulated in 1956 by William L. Brown and Edward O. Wilson, both of Harvard University, mainly from the observations of the feeding habits of two insect-eating bird species known as nuthatches. The ranges of these two species of nuthatches overlap only in some places. In these overlapping regions, however, the size and coloration of the bills of the two species are very different. In contrast, in regions where the birds' ranges do not overlap, the size and coloration of the bills of both species in adjacent areas are identical. Scientists postulated that, in overlapping regions, interspecific

competition resulted in a selective pressure that resulted in the evolution of different bills. Since there was no such competition in nonoverlapping regions, there was no selective pressure for different bills to evolve.

Competition between organisms is always harmful to both of the individuals involved because each receives less than what it could in the absence of competition. Competition can result in the exclusion of one species from a community or in the species' extinction, or it can lead to behavioral and evolutionary changes that reduce harmful interactions. (See CTQ #3.)

INTERACTIONS THAT HARM ONE ORGANISM AND BENEFIT THE OTHER

All organisms need a source of energy and nutrients to survive and reproduce. When one organism supplies a resource to the other, or is *itself* the resource, the two organisms must interact. Natural selection has produced a bat-

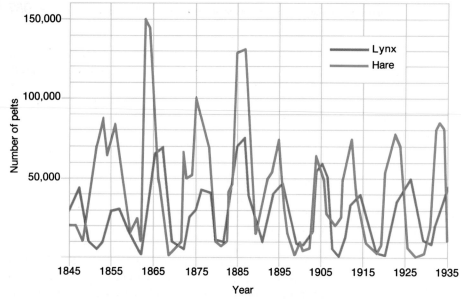

The ups and downs in numbers of predators and prey. This graph plots the number of lynx and snowshoe hare furs sold by trappers to the Hudson's Bay Company in Canada between 1845 and 1930. As you can see, increases and decreases in the number of prey (the hare) triggered increases and decreases in the number of predators (the lynxes), and vice versa. Other factors may also have affected the lynx and hare populations. Overcrowding and fluctuations in plant growth may have altered the availability of food for the hares, while outbreaks of disease and climate changes could have affected the size of the lynx population.

tery of adaptations that help organisms secure needed resources from others and that help organisms defend themselves from becoming a resource.

PREDATION AND HERBIVORY

During **predation,** one organism (the **predator**) acquires its needed resources by eating another organism (the **prey**). If the prey is a primary producer, the interaction is called **herbivory;** plant-eating animals are called **herbivores.** Organisms that eat other animals for energy and nutrients are **carnivores,** and those, like humans, that eat a mixed diet of plants and animals are **omnivores.**

Predator and Prey Dynamics

Although some predators limit their diets to one type of prey, most rely on more than one species for nourishment. The choice often depends on the abundance and accessibility of prey. During the summer, for example, a red fox mainly eats meadow mice. As the availability of meadow mice dwindles in the cooler seasons, however, the fox shifts to eating the more abundant white-footed mice.

As the availability of prey increases in an area, so does the number of predators; more prey feed more predators. More predators consume greater numbers of prey, however, reducing the availability of prey. In turn, the number of predators drops. This reciprocal interaction generates recurring cycles of increases and decreases in predator and

TABLE 42-2
HOW SOME ORGANISMS AVOID BECOMING PREY

Escape Adaptations	Effect
Camouflage	
Cryptic coloration	Hides from predator
Disruptive coloration	Distorts shape and confuses predator
Individual responses	
Startle behavior	Confuses predator
Playing dead	Confuses predator
Shedding body parts	Escapes capture
Outdistancing predators	Escapes capture
Group responses	Warn, protect, and confuse
Defense Adaptations	
Physical defenses	
Armor	Deters an attack
Aposematic	Advertises noxious trait
Mimicry	
Müllerian mimicry	Noxious species avoided
Batesian mimicry	Harmless or palatable species avoided
Chemical Defenses	
Poisons	Kills predators
Hormones	Disrupts predator development
Allelochemicals	Repels predator

prey abundance, resulting in a balance whereby the number of prey remains mostly in balance with the number of predators.

An example of a predator–prey relationship is illustrated in Figure 42-4, where the number of lynxes (the predator) and the number of snowshoe hares (the prey) are plotted for a 90-year period. As you can see, when the number of hares increased, the number of lynxes also increased. As the number of lynxes grew, the lynxes ate more hares, lowering the hare population. A drop in available prey caused a drop in the number of predators, and so on, over the 90-year period. As in other predator–prey relationships, the number of predators (lynxes) lags slightly behind that of the number of prey.

Predator and Prey Adaptations and Defenses

Coevolution between predators and prey over long periods of time has produced an array of remarkable and effective adaptations; some improve the skills of predators in capturing prey, while others improve the prey's chances of escaping predators. Adaptations that aid prey survival can be grouped into two general categories: (1) those that help prey escape being eaten, and (2) those that help a prey defend itself against predator attacks (Table 42-2).

Camouflage Some prey go unnoticed by predators because they blend in with their surroundings or because they appear inanimate (like a dried twig) or inedible (like bird droppings). Such adaptations are called **camouflage** (Figure 42-5) because the color, shape, and behavior of an organism make it difficult to detect, even when in plain sight. Camouflage is not reserved exclusively for prey, however; predators also use camouflage to help conceal themselves while waiting to ambush prey.

(a)   *(b)*

(c)   *(d)*

FIGURE 42-5

Camouflage: nature's masqueraders. *(a)* The "vine" "growing" on this branch is really a grass green whip snake (*Dryophis*). *(b)* This woodcock (directly in the center of the photograph, facing the left) blends in with its surroundings so effectively that is usually escapes detection by predators. *(c)* The Malaysian horned frog has adaptations that help the animal blend in with dead leaves lying on the forest floor. Such adaptations include shading (which conceals the frog's eyes) and curly horns that resemble drying leaf tips. These characteristics help make the horned frog virtually invisible to its prey. *(d)* Transparent wings help the Costa Rican clearing butterfly *Ithomia* virtually disappear.

(a)   *(b)*

FIGURE 42-6

Keeping a "low profile." The fur of the long-tailed weasel (*Mustela frenata*) changes from white in winter *(a)* to brown in summer *(b)*. Both cryptic colors help the weasel integrate into its surroundings and escape detection by predators.

The camouflage of some organisms helps these individuals resemble their background. This type of adaptation is called **cryptic coloration** because the camouflaged organism is hidden from view. The long-tailed weasel exemplifies cryptic coloration. This animal changes color to match the seasonal changes in its surroundings (Figure 42-6). During winter, the weasel's pure white coat helps the animal blend in with the snow (and perhaps conserve heat), whereas during summer, a brown coat helps the weasel blend with the forest floor. The plumage color of several species of grouse and hares also changes seasonally. Laboratory experiments with willow grouse (*Lagopus lagopus*), for example, reveal that the length of daylight (photoperiod) triggers hormonal changes that coordinate color change.

Disruptive coloration disguises the *shape* of an organism, as in the coloration of the moth shown in Figure 42-7. The color pattern breaks up the outline of the moth when the individual is resting on a dark tree trunk, concealing its shape. Disruptive coloration sometimes means camouflaging vital parts, frequently the organism's eyes, since many predators use eyes as an attack target. With the organism's real eyes camouflaged, the predator's attention is often diverted away from a vital part of the prey (the head). If attacked, an individual that has false eyes in a less vital area of the body (such as a wing) usually escapes with only minor damage (Figure 42-8*a*). Furthermore, false eyes may not only divert a predator's attack; it may also threaten or startle a predator by making the prey appear larger than it really is (Figure 42-8*b*).

Cryptic and disruptive color adaptations are not used solely by animals. Cryptic coloration in some plants helps them resemble less palatable plants, and cryptic and disruptive coloration helps camouflage some plants from herbivores (Figure 42-9).

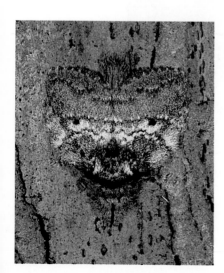

FIGURE 42-7

Disruptive coloration disguises the shape of this moth as it rests on a tree trunk, creating an image that goes unrecognized by its sharp-eyed predators, the birds.

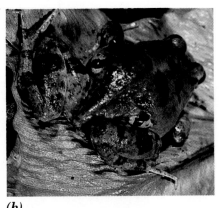

(a)   *(b)*

FIGURE 42-8

An eyefull. *(a)* In addition to false eyes, the Malaysian back-to-front butterfly *(Zeltus amasa)* deflects predator attacks with false legs and fake antennae on its hind wingtips. An attack on this rear end does not damage vital organs, so the butterfly can dart to safety, minus only a wing fragment. *(b)* When confronted with a predator, this South American frog bends over and puffs up its body, revealing large, false eyes. The startled predator is usually frightened away by the intimidating display. If the predator proceeds with the attack, the frog releases an unpleasant secretion from glands located near each false eye.

In addition to coloration, the shape and behavior of an organism also contribute to a successful disguise. The crab spider, African thorn spider, and potoo bird in Figure 42-10 all remain motionless, reducing the chance that they will be detected.

Individual Responses When confronted with a predator, some prey rely on sudden escape responses. Generally, the predator is momentarily stopped by the unexpected response, especially if the escape response seems dangerous. This moment's hesitation may give the prey a chance to escape. Examples of such "last-ditch" responses include the following.

- An owl fluffs its feathers and spreads out its wings, a last-minute bluff that usually startles an attacking hawk.

- A mosquito fish frantically splashes on the surface of a pond when approached by a voracious pickerel (a small fish), making it difficult for the pickerel to launch a pinpoint attack.

- A tiny bombardier beetle sprays hot chemical irritants at a rodent, thwarting the attack.

- An opossum "rolls over and plays dead" when confronted by a coyote; the discouraged predator often searches elsewhere for a fresher meal.

FIGURE 42-9

Appearing more like stones than plants, pebble plants *(Dinteranthus)* usually go unnoticed by passing herbivores.

(a)

(b)

(c)

FIGURE 42-10

Nature's impostors. *(a)* The Borneo crab spider resembles bird droppings, a disguise that lures butterflies and other insect prey that eat genuine droppings. At the same time, birds, the predators of the spider, stay clear of what appears to be their own wastes. *(b)* The "twin thorns" on this acacia branch are really parts of an African thorn spider, resting motionless during the day to avoid predators. At night, the spider spins a web to catch insect prey. *(c)* Unwary animals fall easy prey to this dead tree trunk (really a motionless potoo bird), with its eyes half open and neck outstretched. The adult potoo is not the only one with a perfect disguise; its single-spotted egg also blends with the broken tree stump, camouflaging it from predators.

A few organisms may escape a predator's clutches by releasing the seized part of their body. For example, a lizard quickly detaches its tail, which continues to move for several minutes, keeping the predator occupied while the lizard scurries to safety to begin regenerating a new tail.

Finally, some animals escape becoming prey simply by outrunning their predator. A healthy antelope or impala, for instance, can usually outdistance a lion. Like many predators, lions generally capture the young or the weak. By removing the young and weak from the breeding population, predators act as a powerful natural selection agent for the prey population.

Group Responses Schools, packs, colonies, and herds typically defend themselves more effectively than can a single organism. For example, the first smelt fish to notice an approaching predator releases chemicals into the water, which immediately send the school of smelt fleeing in various directions. The confused predator does not know which way to turn. In grazing herds, stronger individuals generally surround the younger and weaker, protecting them from an advancing predator. There is indeed safety in numbers; a predator is less able to pick out a single target among a swarming group.

Physical Defenses Organisms have evolved an arsenal of anatomic features to help protect themselves against direct attacks. Many have protective shells (mollusks and turtles),

barbed quills (porcupines), needlelike spines (sea urchins), and piercing thorns, spines, and stinging hairs (plants) that can discourage even the hungriest predator. In fact, your skin is a protective armor against the daily invasions of millions of microbes. Imagine how effective the 9-inch-thick hide and blubber of a whale is as a barrier; whale blubber sometimes even prevents the penetration of a high-velocity harpoon.

Many foul-tasting, poisonous, stinging, smelly, biting, or in other ways obnoxious animals ironically have striking colors, or bold stripes and spots. This type of defense, called **aposematic coloring,** or warning coloration, is the opposite of camouflage; it makes an organism stand out from its surroundings. (*Aposematic* refers to anything that serves to warn off potential attackers.) The distinctive aposematic stripes of a skunk, for example, advertise to potential predators that this animal can yield an obnoxiously smelly counterattack. It usually takes only one encounter for a potential predator to avoid any future entanglements with a skunk. In plants, the red and black fruits of poisonous nightshades are examples of warning coloration.

A bad taste is often of little help for the individuals being attacked because at least part of the animal must be eaten before the predator notices its foul taste. The species as a whole profits, however, because individual predators learn to recognize the characteristics of a vile-tasting species and to avoid the distasteful individuals, sparing other members of the species. In evolutionary terms, the advantage is clear: The species survives.

FIGURE 42-11
Equally distasteful, the *Acrea* butterfly *(left)* and the African monarch *(right)* are Müllerian mimics. These butterflies are not even closely related, yet they resemble each other almost exactly in color, pattern, behavior, and flavor.

Color is not the only aposematic defense. Recently, researchers learned that some foul-tasting moth species make a clicking sound when they are being pursued by bats. Bats associate the clicking sound with the bad-tasting moths and learn to pursue quiet, tasty species. Some tasty moths avoid becoming prey for the bats by making the same clicking sound as do the foul-tasting moths. The clicking of tasty moths is an example of aposematic sound defense, as well as of **mimicry,** whereby one species resembles another in color, shape, behavior, or in this instance, sound, as a mechanism for defense or disguise.

Mimicry If similar-appearing species are equally obnoxious, the resemblance is called **Müllerian mimicry.** In the tropics, for example, many species of beetles have bright orange wingcases with bold, black tips. When attacked, the beetles release drops of their own foul-tasting blood; just a taste deters a predator, sparing the beetle. Consequently, predators learn to stay away from other similarly colored beetle species. Examples of Müllerian mimicry abound in many tropical butterfly species, such as between monarch and *Acraea* butterflies (Figure 42-11).

In some cases of mimicry, such as the moths mentioned in the previous discussion, a harmless or palatable species gets a "free ride" by resembling a vile-tasting or stinging species. When a good-tasting or harmless species (the *mimic*) resembles a species with unpleasant, predator-deterring traits (the *model*), the similarly is called **Batesian mimicry** (Figure 42-12).

FIGURE 42-12
The deadly moray eel (left) and the harmless plesiops fish (right) are an example of Batesian mimicry. When pursued by a predator, the plesiops fish swims head first into a rock crevice. The shape, color, pattern, and false eye of its exposed tail strongly resemble the head of the dangerous Spotted Moray eel, frightening predators away.

Chemical Defenses Some plants and animals release **allelochemicals**—chemicals that deter, kill, or in some other way discourage predators. Such defenses must be swift and effective; otherwise, the predator may inflict fatal damage before the allelochemical has a chance to take effect. To accomplish this defense, many animals rely on noxious propellants or painful, sometimes deadly, bites and stings to blunt a predator's attack.

Some tropical toads and frogs secrete extremely poisonous chemicals. South American tribesmen simply need to touch the tip of their arrows to the skin of poisonous toads to produce a lethal missile that can kill an animal (including a human) within minutes. Other swift-killing poisons are manufactured by the Japanese puffer fish, the Asian goby fish, and the American newt.

Poisonous animals frequently exhibit aposematic coloration, which serves as a blatant signal of the consequences of an attack to experienced predators. For example, the conspicuous stripes on a poisonous monarch butterfly larvae make it easy for predators to recognize this species (Figure 42-13). Interestingly, although both the larva and the adult monarch butterfly are poisonous, neither form manufactures the toxic chemicals itself. Instead, larvae are poisonous only because they eat milkweed plants that synthesize the toxic chemicals. Adult monarchs are poisonous only because the toxic milkweed chemicals are passed on from the larvae to the adults during metamorphosis. Some nudibranchs also derive their defenses secondhand. Ironically, these nudibranchs arm themselves with their prey's stinging cnidocytes (see Figure 39-9) and use them to defend themselves against their own predators.

PARASITISM

Parasitism is another type of interaction that benefits one organism and harms the other. A **parasite** secures its nourishment by living on or inside another organism, called the **host.** Although parasitism is sometimes considered a form of predation, the victim of parasitism usually survives the interaction, whereas the victim of predation is almost always killed. The *larvae* of some insect parasites are lethal, however. Such larvae are called **parasitoids.** For example, after a female tarantula wasp captures and paralyzes a tarantula with her sting, she then lays eggs in the spider's flesh. When the eggs hatch, the larvae gorge themselves on fresh tarantula tissues, killing the helpless spider, who is literally eaten alive.

Most parasites are **host specific;** that is, their anatomy, morphology, metabolism, and life history are adapted specifically to those of their host. For example, the human tapeworm lacks eyes, a digestive tract, and muscular systems. The combination of adaptations it evolved are suited for living inside human intestines, however. They include

- an outer cover that protects the tapeworm from powerful digestive enzymes yet allows the absorption of nutrients;

- a long, flat shape that creates a maximum absorptive surface area yet prevents obstruction of the host's intestine;

- hooks on its "head," which anchor it to the host's intestinal lining;

- a reproductive system with both male *and* female parts, allowing for self-fertilization. Self-fertilization is an important reproductive strategy in a location where contact with another tapeworm is highly unlikely. (Internal parasites are often little more than reproduction "machines," producing millions of offspring, increasing the chances of infecting a new host.)

Some internal parasites require more than one host to complete their reproductive cycle. For example, the fox tapeworm requires not only a fox, as its name indicates, but also a rabbit. Inside the intestines of a fox, the tapeworm produces hundreds of eggs that are released into the environment in the fox's feces. When a rabbit eats a plant that is contaminated with fox feces, the tapeworm eggs enter the rabbit's digestive system. Once inside the rabbit, the eggs

FIGURE 42-13

Secondhand poison obtained from eating milkweed plants makes the monarch butterfly and its larva toxic to predators. The bold stripes on the larva (below), and the distinctive color and pattern of the adult, broadcast danger to predators. After just a few tastes and subsequent episodes of vomiting, predators quickly learn to avoid these bold patterns.

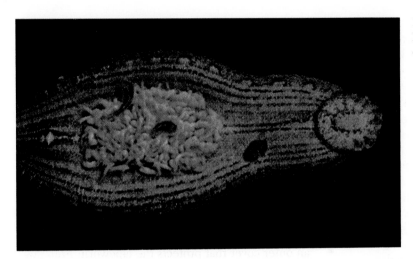

FIGURE 42-14
Even leeches have leeches, illustrating that virtually all organisms—parasites included—have parasites.

hatch, and the larvae bore their way out of the intestine and into the rabbit's muscles. The larvae then form cysts, the resting stage of a parasite. When an infected rabbit is eaten by a fox, the cysts become activated and develop into young tapeworms that attach themselves to the intestines of the new fox host. Within a short time, the fox tapeworm is producing hundreds of eggs a day, and the life cycle begins again.

⬟ Apparently, no organism escapes being parasitized. Even parasites can have parasites (Figure 42-14). Animals are hosts to a huge battery of parasites, including viruses, bacteria, fungi, protozoa, and other animals (flatworms, flukes, tapeworms, nematodes, mites, fleas, and lice, for example). The range of plant parasites is equally broad and includes viruses, bacteria, fungi, nematodes, and other plants, such as mistletoe and dodder.

Most organisms are hosts to a number of different kinds of parasites at the same time. A single bird may have 20 different parasites (Figure 42-15), of which there may be hundreds of individuals. For example, researchers counted more than 1,000 feather lice in the plumage of a single curlew.

Although most parasites use their host only as a source of nutrients, some parasites also use their host as a haven for protection from predators. For example, the pearl fish (*Carapus*) develops and lives in a safe, but very unusual place: the anus of a sea cucumber (*Actinopygia*). As a sea cucumber draws in water through its anus for gas exchange,

Fungus	Tapeworm
Amoeba	Tongue worm
Flagellate	Leech
Plasmodium	Bug
Spirochete	Flea
Trypanosome	Feather louse
Encapsulated tongue worm	Fly larve
Fluke	Louse fly
Roundworm	Mite
Spiny-headed worm	Tick

FIGURE 42-15
The invasion begins immediately after hatching. As many as 20 kinds of parasites can infest a single bird.

a newly hatched pearl fish swims in (Figure 42-16). The pearl fish remains inside the sea cucumber, feeding on the host's tissues and taking periodic excursions outside its host to supplement its diet and to reproduce. It returns to the sea cucumber for protection from predators.

Some parasites exploit the behavior of their host, an interaction called **social parasitism.** Examples of social parasitism are provided by European cuckoos, American cowbirds, and African honey guides. After a host bird builds a nest and lays eggs, the parasite bird destroys one of the host's eggs and replaces it with her own. The egg is usually so similar in size and coloration that the host bird fails to recognize it as an alien egg and incubates the parasite's egg as its own. After hatching, the parasitic baby bird instinctively shoves all solid objects out of the nest, including the host's babies and any unhatched eggs. After clearing the nest of its rivals, the parasite snatches up all of the food brought to the nest by its duped "foster parents."

ALLELOPATHY

Some organisms wage chemical warfare on other members of the community. **Allelopathy** is a type of interaction whereby one organism releases allelochemicals that harm another organism. Although some of the animal chemical defenses we described earlier may also be examples of allelopathy, we will confine this discussion to the harmful chemical defenses of plants.

Some plants manufacture allelochemicals that kill herbivores or competing plants (Figure 42-17). For instance, the chemicals released by some chaparral plants accumulate in the soil beneath the plants, blocking the germination and growth of other plants and reducing competition for scarce water and nutrients.

Sometimes, allelochemicals percolate deep into the soil and may reach high enough concentrations to kill the very plant that produced them. This phenomenon is called *autotoxicity*. Although killing oneself goes against a basic "goal" of life — survival — some biologists argue that autotoxicity has adaptive value for the species as a whole. Since it takes many years for allelochemicals to accumulate to toxic levels, only older plants with low reproductive ability die from autotoxicity, reopening space for new, reproductively vigorous individuals.

▮▶ Some plants defend themselves against herbivores by fatally poisoning them. Members of the crucifer family (cabbages, broccoli, brussels sprouts, mustards, radishes, and so on) produce mustard oils, chemicals that are lethal to many herbivores, and disease-causing fungi and bacteria. In response, some herbivores have evolved *counteradaptations* that detoxify poisonous allelochemicals. For example, cabbage white butterfly caterpillars (*Pieris brassicae*) have been successfully reared on cultures that contain more than ten times the concentration of mustard oils found in cabbage plants. Apparently, what began as a means of protection against herbivores has backfired; cabbage white but-

FIGURE 42-16

One very unusual habitat for a fish is the anus of a sea cucumber. A young pearl fish has poor eyesight and is barely able to swim, making this animal quite vulnerable to predators. Soon after hatching, the pearl fish locates a sea cucumber. When the sea cucumber opens its anus to draw in water, the parasitic pearl fish enters this unusual, but effective, shelter.

FIGURE 42-17

Plant versus plant. Coastal sages (*Salvia leucophylla*) emit oils that prevent the germination and growth of grasses and other herbs, helping to create bare zones that encircle each plant. The bare areas are effective in keeping away other plants that would compete with the sages for scarce water supplies in the southern California chaparral.

terflies now use the scent of mustard oil to locate plants on which they lay more eggs.

During the 1960s, investigators accidentally discovered that plants produce allelochemicals that disrupt the normal growth and development of insect herbivores. While visiting the United States on a sabbatical leave, Dr. Karel Slama, a Czechoslovakian researcher, attempted to continue his investigations on the development of *Pyrrochoris apterus*, an insect he had been studying for a number of years. After having reared thousands of bugs in his native research laboratory, Slama was unable to raise reproductive adults in the United States, even though he was using the exact procedures he had always followed in Czechoslovakia.

After painstakingly reviewing every step, Dr. Slama discovered a single variable that was different. In Czechoslovakia, he had always reared the bug on filter paper, whereas in the United States he was using common laboratory paper towels. When he extracted and analyzed the chemicals found in U.S. laboratory paper towels, Slama discovered a compound that was virtually identical to a critical hormone that triggers metamorphosis in insects. He tested the effect of the chemical by rearing insects on Czechoslovakian filter paper that had been treated with the chemical and compared the results to the development of insects reared on untreated filter paper. Slama's results were conclusive: The plant chemical disrupted the devel-

opment of the insect, preventing the development of adult insects with reproductive organs.

Since Slama's experiments, other researchers have discovered similar allelochemicals in some ferns, conifers, and flowering plants. Once again, these chemicals were virtually identical to those insect hormones that coordinate development during metamorphosis. As larvae consume these plants, the allelochemicals cause premature metamorphosis or produce sterile adults. Either way, herbivore reproduction is disrupted, illustrating a very effective plant adaptation for protection against increasing numbers of herbivores. Some of these hormone-mimicking allelochemicals are being considered for use as natural pesticides because these chemicals would cause considerably less environmental damage than do synthetic insecticides.

Organisms secure needed energy and nutrients from the abiotic environment or from other members of the biotic community. Most species on earth (over 70 percent) secure food by consuming part or all of another organism, leading to a diversity of interactions that benefit one species and harm or kill the other. Through natural selection, organisms have evolved a variety of adaptations that help them secure the food they need. Organisms have also evolved adaptations that help them avoid being eaten themselves. (See CTQ #4.)

COMMENSALISM: INTERACTIONS THAT BENEFIT ONE ORGANISM AND HAVE NO AFFECT ON THE OTHER

The benefits of **commensalism** are one sided: Only one of the participants (the commensal) profits, while the other is virtually unaffected. Nature exhibits many examples of commensalism. For instance, remoras are fish that attach themselves by suckers to the undersides of sharks and gather food scraps as the sharks feed. Remoras benefit from this interaction, but their presence apparently has little or no impact on the shark.

☀ Some commensals simply live in a habitat that is created by another organism. The burrows of large "innkeeper" sea worms, for instance, house an array of "guests" that use the burrow for shelter but do not hinder or benefit the innkeeper worm in any way (Figure 42-18). Epiphytes are commensal plants that grow on the branches of taller plants. Being higher up in the forest canopy, the epiphyte captures more light than it could if it occupied a position lower in the canopy. Barnacles encrusted on a humpback whale are also commensal who gain a habitat as well as a means of transportation to new sources of food.

Organisms inhabit a tremendous variety of habitats, including other organisms. As long as there is no disadvantage for either party, natural selection does not select against such associations. (See CTQ #5.)

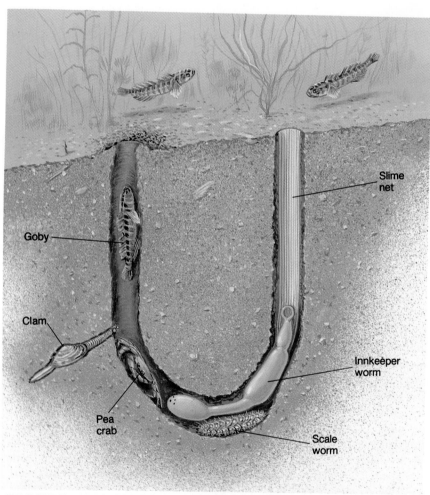

FIGURE 42-18

Commensalism: one benefits, the other is unaffected. The innkeeper worm bores a tunnel in the mud of shallow coastal waters. The worm then spins a slime net that traps minute organisms as the worm pumps water into the tunnel and through the net. When the net is full, the innkeeper gulps down the whole thing—net, trapped food, and all. But the innkeeper worm is not the only occupant of its tunnel. The goby uses the burrow for protection, while the pea crab, clam, and scale worm feast on the innkeeper's leftovers. Although the guests benefit from the association, the innkeeper worm apparently neither profits nor suffers from their presence.

INTERACTIONS THAT BENEFIT BOTH ORGANISMS

Throughout this text, we have seen how natural selection favors characteristics that improve an organism's survival and reproductive success. Interactions between organisms in an ecosystem which provide benefits to both participants are favored by natural selection because the positive interactions contribute to each organism's survival or reproduction. There are many examples of beneficial interactions, some of which are compulsory to both participants, and others that are optional.

PROTOCOOPERATION

Protocooperation interactions benefit both participants but they are noncompulsory. For example, protocooperation between a fungus and algae forms a lichen (page 821). The fungus uses some of the food produced by the algae, while the algae gain a habitat as well as some of the water and minerals absorbed by the fungus. Both organisms benefit from this form of symbiosis, yet they could live successfully on their own as well.

Another example of protocooperation is the relationship between an oxtail bird and a rhinoceros (Figure 42-19). The bird perches on the back of the rhinoceros and removes

FIGURE 42-19

Protocooperation between oxtail birds and rhinos. Although neither animal requires the other for its survival, both benefit from their interaction. The bird removes bloodsucking ticks from the rhino, while the rhino supplies the bird with an abundant food supply and warmth.

pests (bloodsucking ticks and flies). The oxtail benefits by receiving food, warmth, and protection from predators; the rhino benefits by receiving protection against parasites. The sharp-eyed oxtail also alerts the dim-visioned rhino of approaching intruders. Again, the relationship is facultative because both animals are capable of surviving on their own.

MUTUALISM

Mutualism is another form of interaction in which both participants benefit. Unlike protocooperation, however, the mutualistic interaction is essential to the survival or reproduction of both participants. Many mutualistic interactions are symbiotic, involving a close association between the participants. The pollination of some flowers by specific insects, birds, or bats (Chapter 21), and the interaction between ants and the *Acacia* plant found in the tropics (Chapter 40) are examples of mutualism that have been discussed earlier in this book.

⬤ Through coevolution, the adaptations of many mutualistic partners have become functionally interlocked. The partnership between many species of termites and their intestinal protozoa, for instance, goes beyond the termite's simply providing food and housing for the protozoa, and the protozoa's digesting the cellulose in wood for the termite (Figure 42-20). Coevolution has led to synchronized life cycles between these organisms. In fact, the synchrony is so precise that the internal protozoa are transmitted from one developmental stage of the termite to the next during molting. The same hormones that trigger the termite to molt also trigger the protozoa to encyst. When the termite reingests its gut lining after molting, it "reinfects" itself with its mutually beneficial partner.

Interactions between organisms in an ecosystem which enhance both participant's survival and reproduction are strongly favored by natural selection. (See CTQ #6.)

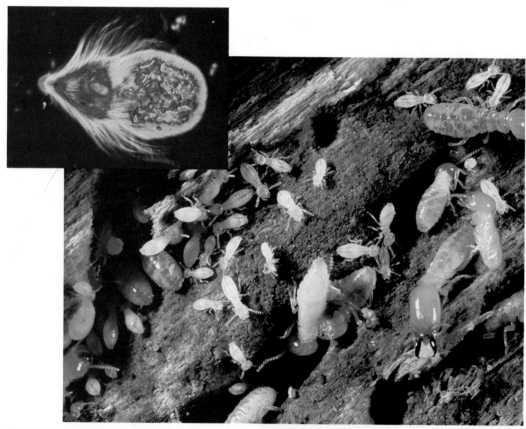

FIGURE 42-20

Mutualism: obligate partnerships with mutual benefits. Without internal protozoa (inset), termites would starve to death because they are unable to digest the wood they consume. Linked through coevolution, protozoa inhabit the gut of termites and obtain a habitat and food supply, while the termites receive a supply of usable nutrients from the digestion of wood by the protozoa.

Relationship between Form and Function

⚡ Organisms must have adaptations that enable them to survive and reproduce, even if the organisms live in very unusual habitats. Internal parasites have evolved a number of adaptations that enable them to live inside another organism. The hooks on a tapeworm's "head" anchors the parasite in an animal's intestine, preventing it from being flushed out in the current of passing food. Since the probability of meeting a mature individual of the opposite sex in the intestine of another animal is so remote, tapeworms are hermaphrodites, possessing both male and female reproductive structures, enabling them to reproduce alone.

Acquiring and Using Energy

☀ Character displacement helps reduce competition for food. For example, competition is reduced if one type of bill is naturally selected in one bird species, while another type is naturally selected in another competing species. Repeated selection of different bills in competing bird species eventually reduces competition because each bird ends up with a bill that harvests food energy in different ways so the species no longer rely on the identical food resource. Resource partitioning also reduces competition, as species share the same food resource.

Unity within Diversity

🔺 Unrelated species sometimes evolve similar adaptations. Müllerian mimicry enables unrelated yet equally repugnant species to benefit from their similar appearance by deterring potential predators. Batesian mimicry allows a palatable species to benefit by resembling an unrelated, yet obnoxious species. All forms of mimicry are examples of similarity in structure and function among diverse species.

Evolution and Adaptation

➠ Camouflage, escape responses, chemical and physical defenses against predators, allelopathy, and mimicry are all examples of adaptations that help organisms survive. Each adaptation evolved as a result of repeated selection of individuals with traits that increase survivability. For example, cryptic coloration allows individuals with coloration and patterns that harmonize best with the background to escape hungry predators more easily than can individuals that stand out. These more cryptically colored individuals will survive and produce more offspring than will less cryptically colored individuals, passing on the adaptive traits. Behaviors can also be adaptive. For example, individual and group escape behaviors are adaptations that result from natural selection.

SYNOPSIS

The organisms that make up the biotic community interact with one another in a variety of ways:

- Competition between organisms harms both participants.
- Predators gain energy and nutrients by consuming prey.
- Parasites live in or on a host organism, damaging or killing the host in the process.
- Commensalism interactions benefit one organism but do not harm the other.
- Both organisms benefit from protocooperation, yet each is able to survive independently.
- Mutualism benefits both interacting organisms, but neither can survive without the other.

When the niches of two species overlap, members of both species compete for the limited resources they require. Intense competition may lead to the sharing of different parts of a resource or of the entire resource at different times or to evolutionary changes in characteristics that reduce competition.

When the ecological niches of two species in a community are identical, competition between the two species results in one rival eliminating or excluding the other from the community. Species can coexist in the same community when they have slightly different ecological niches.

Evolution has resulted in a number of physical and behavioral adaptations that enhance organisms' predatory skills or help organisms escape predators. These adaptations include camouflage, which helps organisms blend in with their surroundings, conceal their shape, or protect vital parts; individual or group behaviors that confuse or distract attackers; anatomic features, such as shells, spines, or armor, that discourage attackers; a foul taste; and chemicals that kill or discourage predators.

Key Terms

symbiosis (p. 962)
competition (p. 963)
intraspecific competition (p. 963)
interspecific competition (p. 963)
exploitative competition (p. 964)
interference competition (p. 964)
territoriality (p. 964)
competitive exclusion (p. 964)
resource partitioning (p. 965)
character displacement (p. 965)
predation (p. 966)
predator (p. 966)

prey (p. 966)
herbivory (p. 966)
herbivore (p. 966)
carnivore (p. 966)
omnivore (p. 966)
camouflage (p. 967)
cryptic coloration (p. 968)
disruptive coloration (p. 968)
aposematic coloring (p. 970)
mimicry (p. 971)
Müllerian mimicry (p. 971)
Batesian mimicry (p. 971)

allelochemical (p. 972)
parasitism (p. 972)
parasite (p. 972)
host (p. 972)
parasitoid (p. 972)
social parasitism (p. 974)
allelopathy (p. 974)
commensalism (p. 976)
protocooperation (p. 977)
mutualism (p. 978)

Review Questions

1. Consider two seedlings of different plant species growing right next to each other in a community. Both grow at about the same rate and develop roots to the same depth. List the resources for which the seedlings will compete as they grow. What is the probable outcome of this situation? What will happen if one plant suddenly outgrows the other?

2. Is there greater opportunity for resource partitioning in a tropical rain forest, a deciduous forest, or a desert? Why? How does each of these terrestrial biomes compare to potential resource partitioning in the pelagic zone of oceans?

3. Match the example with the correct term.

   _____ one bee species chases away another species from flowers.
   _____ similar monkey species with different-size teeth
   _____ similar species of whales visit a feeding bay in different seasons
   _____ a plant releases chemicals that stop other plants from growing
   _____ contrasting colors distort the shape of a fish as it swims through a reef
   _____ a harmless fly looks like a stinging wasp

   a. Müllerian mimicry
   b. Batesian mimicry
   c. disruptive coloration
   d. allelopathy
   e. exploitative competition
   f. interference competition
   g. character displacement
   h. resource partitioning

4. Under what conditions would competitive exclusion not take place in an ecosystem?

5. List some of the reasons why competitive exclusion is rarely observed in natural ecosystems.

6. Use examples to distinguish between cryptic coloration and aposematic coloring. How do these adaptations help prey escape predators?

7. Monarch butterflies and their larvae do not manufacture toxic chemicals, yet both are poisonous to birds. How is this possible? If the poison kills the birds, why wouldn't it kill the larvae or butterfly as well?

8. With the exception of parasitoids, the vast majority of parasites do not kill their host. What advantage is there to killing a host? Must there be an advantage, from a natural selection/evolutionary point of view, in order for there to be any host-killing parasitoids at all?

9. Are herbivores really predators, and are the plants they eat really their prey? If so, why do you think ecologists make this distinction? If not, list the reasons why they should be considered separate.

10. For each of the following pairs of terms, state how they are similar and how they are different.
    a. predation and allelopathy
    b. mutualism and protocooperation
    c. Müllerian and Batesian mimicry

Critical Thinking Questions

1. Each of the six scientists mentioned in the Steps to Discovery (Darwin, Lotka, Boltzman, Volterra, Gause, and Tansley) was an expert in a particular, yet different, scientific discipline (save two). Specifically, what insight was gained from each scientist (discipline) that eventually led to the formulation of the Principle of Competitive Exclusion? Why did this phenomenon become known as a "principle" rather than a "theory" or "hypothesis?"

2. Many symbioses are very specific and permanent. As we discussed in Chapter 20, the pollination of the Spanish dagger *(Yucca whipplei)* by only female pronuba moths is an example of such a relationship. Neither the plant nor the moth can reproduce without the other. From an evolutionary point of view, there are advantages *and* disadvantages to such compulsory and exclusive interactions. List and explain as many advantages and disadvantages as you can. Since there are disadvantages, why would narrow and binding relationships be favored by natural selection at all?

3. In what sense is competition, which is always harmful to the organisms involved, good for the species? How does this concept connect ecology with evolution?

4. More than 70 percent of species obtain their energy and nutrients by consuming all or part of another organism, while only about 30 percent of all species on earth harvest energy and nutrients from the physical environment. These proportions are not always the same for all ecosystems, however. In fact, in some ecosystems (or biomes) the percentages may even be reversed. In which ecosystems would you expect the percentages to be the same, and in which would you expect the percentages to be reversed? Are there any ecosystems in which the biotic community is completely one or the other? With energy being so abundant in the abiotic environment of most ecosystems, explain why 70 percent of species consume other organisms for energy.

5. A biologist who has heard the phrase "nature abhors a vacuum" on many occasions wants to test whether this idea is true or not. As stated, is this a testable hypothesis? If so, design an experiment or series of experiments to test the hypothesis. If not, how could the phrase be reworded so that it could be tested? Design an experiment to test your new hypothesis.

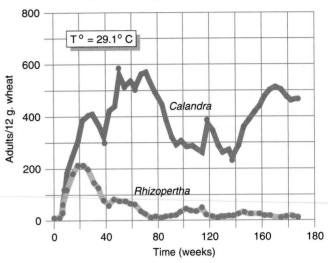

Graph A

6. Study the following graphs on competition between two grain beetles living in wheat at 29.1°C (graph A) and at 32.3°C (graph B). Is the principle of competitive exclusion supported by these data? Altering only one factor (temperature) changed the outcome of competition. Can you offer an explanation for this change?

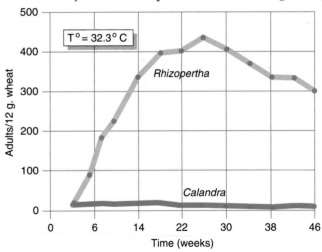

Graph B

Additional Readings

Boltzmann, L. 1905. Populare Schriften. Leipzig. (Advanced)

Barbour, M., J. Burk, and W. Pitts. 1987. *Terrestrial plant ecology.* Menlo Park, CA: Benjamin/Cummings. (Intermediate)

Brewer, R. 1979. *Principles of ecology.* Philadelphia: Saunders. (Intermediate)

Darwin, Charles, 1859. *On the origin of species.* London: John Murphy, p. 76. (Intermediate)

Fogden, M., and P. Fogden. 1974. *Animals and their colors.* New York: Crown Publishers. (Introductory)

Gause, G. F. 1934. *The struggle for existence.* Baltimore: Williams & Wilkins. (Intermediate)

Owen, D. 1980. *Survival in the wild. Camouflage and mimicry.* Chicago: University of Chicago Press. (Introductory)

Putman, R., and S. Wratten. 1984. *Principles of ecology.* Berkeley: University of California Press. (Intermediate)

Tanner, O. 1978. *Animal defenses, wild, wild world of animals.* A Time-Life Television Book. Time-Life Films. (Introductory)

Population Ecology

STEPS
TO
DISCOVERY
Threatening a Giant

Driving across Arizona, you will notice dense forests of giant saguaro cactuses *(carnegiea gigantea)* that extend for miles in all directions. It seems inconceivable that scientists argue that saguaro populations are declining and that this majestic giant may someday disappear from earth. But after conducting a study for the federal government in 1910, Forrest Shreve, a research associate with the Carnegie Institute in Washington, D.C., made just such a conclusion. Based on previous studies that measured growth rates, Shreve estimated the ages of saguaros from their height. He then calculated the age structure of a stand of 240 saguaros on Tumanmoc Hill in central Arizona and noted that the majority of individuals (64 percent) were over 60 years of age. In fact, Shreve found no saguaros younger than 15 years of age and no young seedlings anywhere in the popu-

lation. Shreve concluded that the saguaro was not reproducing enough offspring to replace the older, dying individuals.

Shreve's conclusion verified what other Southwest desert scientists had also observed in other saguaro populations: For some reason, saguaros were dying out. In an effort to understand why this could be happening, and in order to provide a safe haven for saguaros while researchers determined the causes, The Saguaro National Monument was established in 1933 near Tucson, Arizona. Research within the monument also documented population decline. Since the 1930s, a great deal of research has focused on analyzing every aspect of the saguaro's life cycle and population dynamics in an effort to pinpoint the reason(s) why saguaros were failing to reproduce in adequate numbers.

The primary factors that have led to the decreased reproduction of the Saguaro cactus include ants harvesting seeds, rodents

The saguaro, aptly named the giant cactus, is a massive columnar cactus that weighs several tons and grows in the Southwest deserts in Arizona and Sonora, Mexico. An individual saguaro may live for 175 years or more and grow more than 15 meters (45 feet) tall. Although a saguaro may not bloom until it is at least 30 to 50 years old, its reproductive lifetime still stretches well over 100 years, which helps explain why saguaros have such an enormous reproductive potential. Each year, a single saguaro produces an average of 200 fruits, containing a total of 400,000 seeds. Over its reproductive lifetime, an individual saguaro produces some 50 million seeds.

It takes only *one* seed per saguaro to grow, become established, and reach reproductive maturity in order to maintain stable saguaro populations. Despite prodigious seed production, however, this is not happening in many populations. The problem is clearly not with flowering, pollination, or fruit and seed development since each individual generally produces several million seeds in its lifetime, so the explanation for the saguaro's reproductive failure must lie with seed survival, germination, and/or seedling and adult survival.

In 1969, Warren Steenbergh and Charles Lowe of the Saguaro National Monument and the University of Arizona, Tucson, documented the fact that saguaro seeds disappear at a high rate once they fall to the ground. During a 5-week period, mammals, birds, and insects (particularly harvester ants) consumed nearly all of the seeds produced in their study site. Only 4 in 1,000 seeds survived to germination, significantly reducing the saguaro's reproductive potential. To make matters even worse, saguaro seeds do not survive from one year to the next so reproduction is always limited to the current year's seed supply.

Of the seeds that did germinate, all but a few died within the first year. To study the causes of seedling death, Raymond Turner of the U.S. Geological Survey, Stanley Alcorn of the U.S. Department of Agriculture, and George Olin of the National Park Service transplanted 1,600 young saguaro seedlings in the Saguaro National Monument in 1957. The researchers enclosed some of the young saguaros with cages to protect the seedlings from grazing ground squirrels, rodents, and rabbits; other saguaros were left uncaged. All the uncaged plants were killed within just 1 year by grazing; only 1.9 percent of the caged seedlings remained alive after 10 years. The uncaged seedlings were being eaten by rodents, who eat the saguaros for their water content, a very limited resource in the desert.

Turner, Alcorn, and Olin also found that drought caused high losses of saguaro seedlings during the first few years of life. Small saguaros are usually found in the shade of desert trees or shrubs, places where water loss is reduced. Small saguaros have a small water-storage capacity and become dehydrated easily in direct sun. In another study, Turner, Alcorn, and Olin protected seedlings from rodents and studied the effects of shading on these protected seedlings. All 1,200 unshaded seedlings died within 1 year, while 35 percent of 1,200 shaded seedlings survived. The researchers concluded that the survival of young saguaro seedlings is closely tied to that of other perennials that provide the seedlings with shade. This relationship led to speculation that the rapidly growing cattle industry may also be contributing to saguaro population declines since trampling by cattle reduces tree and shrub cover. In fact, a 1965 study led by James Hastings of the Institute of Atmospheric Physics at the University of Arizona and Raymond Turner reported a general deterioration of woody perennial survival in some parts of the Southwest, assuming a combination of changing climate and cattle grazing. Young saguaro seedlings are also crushed outright by grazing cattle.

In 1976, Steenbergh and Lowe identified another important factor associated with saguaro seedling survival: freezing weather. Young saguaros freeze at − 3°C (26°F) to − 12°C (10°F) or when exposed to more than 19 hours of freezing temperatures. This means that during certain years, all saguaro seedlings may be killed by freezing. Steenbergh and Lowe demonstrated that this is precisely what had occurred during a severe January freeze in 1971 and likely occurred in 1894 and 1913. Such vulnerability to freezing temperatures decreases as saguaros grow older.

As you can see, many saguaro populations may be declining primarily because virtually all the seeds produced in a year are quickly eaten by mammals, birds, and ants. If weather conditions are just right, only a few of the remaining seeds germinate and grow, provided that they are in the shade of a tree or shrub and are not eaten by rodents and rabbits or trampled by cattle. Periodic freezing can also kill saguaro seedlings.

Like all organisms, saguaro reproduction and its resulting population dynamics are affected by a number of environmental factors. For the saguaro, some factors, such as shade, low temperature, and camouflage, are more critical than are others. No matter what they are, critical factors affect how quickly or slowly a population grows.

consuming seedlings, cattle trampling small individuals, and decreased protective shade.

*W*hen we consider . . . how soon some species of trees would equal in mass the earth itself, if all their seeds became full-grown trees, how soon some fishes would fill the ocean if all their ova became full-grown fishes, we are tempted to say that every organism, whether animal or vegetable, is contending for the possession of the planet. Nature opposes to this many obstacles, as climate, myriads of brute and also human foes, and of competitors . . . Each suggests an immense and wonderful greediness and tenacity of life. . . .

Henry David Thoreau, journal entry, March 22, 1861.

Elephants are among the slowest reproducers on earth. Over its lifetime, a female elephant can give birth to a maximum of only six babies. Even so, the number of possible descendants from just one pair of mating elephants could total 5 billion (5×10^9) after just 1,000 years. After 100,000 years, the number of potential descendants from one mating pair would theoretically pack the visible universe with elephants.

If such outlandish growth is possible for a slow reproducer like the elephant, imagine what could happen with organisms that have faster reproduction rates, such as house flies. In less than *1 year*, the number of possible descendants from a single pair of house flies would exceed 5.5 trillion (5.5×10^{12})! Consider the magnitude of this number this way: 1 trillion seconds amounts to about 31,700 years. Humans were in the Stone Age only 1 trillion seconds ago.

Clearly, animals have a tremendous capacity to reproduce, and the reproductive potential of plants is often even greater. Yet, the world is not tightly packed with elephants, flies, or any other kind of organism. Disease, parasitism, predation, and limited food and space curb the potential number of individuals, often leading to a balance between the number of individuals living in an area and the availability of resources to support them.

Predator and prey relationships illustrate such a balance. The number of prey in an area is controlled by both the availability of food and the number of predators: More predators eat more prey. In turn, the number of prey determines how many predators can survive: More prey means more predators; fewer prey means fewer predators. As a result, a dynamic balance is often produced between the number of predators and the number of prey in an area (see Figure 42-4).

▼ ▼ ▼

(a)

FIGURE 43-1

Distribution patterns. *(a) Clumped:* A grove of clumped palms. A school of fish or a herd of elephants are other examples of clumped distributions. *(b) Uniform:* Oaks secrete chemicals that prevent growth of nearby oaks, creating more-or-less equal distances between trees. Similarly, when animals defend their territories, the individuals remain separated, producing a uniform distribution. *(c) Random:* Joshua trees may have random distributions in some locations. Random distributions result when the location of one individual has no affect on another individual of the same species.

POPULATION STRUCTURE

Most communities contain many **populations,** each of which consists of the individuals of the same species that live in the same area at the same time. For example, a mountain forest not only contains a population of yellow pine trees but also populations of sugar pine trees, white fir trees, brown bears, Anna's hummingbirds, and more.

To understand the structure and dynamics of each of the populations that make up a community, ecologists examine three fundamental properties of populations:

- *population density* (the size of the population, expressed as the number of individuals in a given area at a particular time);
- *distribution* of individuals throughout the habitat; and
- *growth rate* (increases or decreases in population density per unit of time).

(b)

(c)

POPULATION DENSITY

Population density equals the number of individuals of a species that live in a particular area at the same time. The population density of people in Manhattan is 100,000 per square mile; that of sugar maple trees in Michigan is 300 per hectare (741 acres); and that of dinoflagellates in a red tide is 8 million per liter of ocean water.

Population density can be determined simply by counting every individual in an area. However, ecologists often estimate population density by counting the number of individuals in small, representative areas and then extrapolate that figure to the total area being studied. This technique is called *sampling*. To estimate the number of creosote bushes in California's Mojave Desert, for example, the number of individuals in ten randomly placed 100-square-meter plots were counted and averaged, revealing 30 bushes per sample. Projecting to an acre, the population density is estimated to be about 1,200 bushes per acre.

PATTERNS OF DISTRIBUTION

Although population density reveals the number of individuals in an area, it provides no information about how the individuals are arranged in space. The distribution of individuals is typically categorized into one of three patterns: clumped, uniform, or random (Figure 43-1).

Interactions among individuals often determine how the individuals are distributed within ecosystems. **Clumped patterns** and **uniform patterns** are nonrandom distributions that result when members of a population have some effect on one another (which is almost always the case) or when environmental conditions favor growth in suitable patches. Uniform spacing results when members of a species repel one another, such as when the roots of some plants release chemicals that inhibit the growth of other members of its species, or when animals establish and defend territories. Both chemical inhibition and territoriality maintain an even, maximum distance between members of a species, creating a more or less uniform pattern of individuals.

The most common distribution pattern is clumped, whereby individuals aggregate into groups, forming groves, schools, flocks, herds, and so on. Some animals have a clumped distribution because they cooperate in societies or gain protection from predators by remaining in herds (page 1018). At least three factors can contribute to clumping in plants: (1) favorable conditions for germination and survival occur only in suitable patches; (2) plant seeds are dispersed in groups; and (3) asexual reproduction from runners (stolons), bulbs, branches, or rhizomes concentrates offspring near the parent plant.

Random distribution is the least common pattern in natural populations. For individuals to be randomly distributed, two requirements must be met: (1) the presence of one individual can in no way affect the location of another; and (2) environmental conditions must be more or less the same throughout an area. Both of these prerequisites are very rare.

Not all distribution patterns remain permanent over time. In some animals, for instance, seasonal changes trigger migrations, causing cyclic changes in distribution. Falling temperatures at night and strong winds at the end of summer initiate migratory behavior in several alpine tundra animals (marmots, mountain goats, mountain sheep, pikas, and rosy finches), causing entire populations to move to warmer regions. At the end of winter, when conditions become less severe, the tundra animals migrate back to their previous homes (see Animal Behavior: Migration, Chapter 44).

The distinctiveness of an ecosystem depends on the physical features of the environment and on the nature of the biotic community, including the number of individuals in each population, their distribution, and the rate of growth of all the populations. (See CTQ #2.)

FACTORS AFFECTING POPULATION GROWTH

Like ecosystems, populations inevitably change. Four events trigger increases or decreases in the density of a group of organisms:

1. **Natality** *increases* density, as new individuals are born into a population.
2. **Immigration** *increases* density, as new individuals permanently move into the area.
3. **Mortality** *decreases* density, as individuals in the population die.
4. **Emigration** *decreases* density, as individuals permanently move out of the area.

⟳ If the combination of natality and immigration exceeds that of mortality and emigration, the population grows and density increases. Conversely, when the combination of mortality and emigration exceeds that of natality and immigration, population density decreases. The growth rate of a population equals the rate at which the population size changes. **Zero population growth** occurs when the combined additions and losses to a population are equal; when this happens, population density remains the same.

AGE AND SEX RATIOS

The age and sex of individuals in a population affect population growth. The age of the members of a population is a predictor of both natality and mortality. For example, red

alder trees live to be about 100 years old. If the majority of red alders are older than 95 years, the mortality rate will likely be high and the population of trees will likely decline over the next 5 years. In addition, since each individual reproduces only during part of its lifetime, the ages of the members of a population can also be used to predict natality. In general, the saguaro reproduces between the ages of 40 and 150 years; female humans reproduce between ages 15 and 44.

Population biologists plot the number of individuals of a certain age and sex to determine the **age–sex structure** for a population, which helps them predict future population changes. When a large proportion of individuals in a population are at reproductive age (or younger), the age–sex structure tends to be shaped like a pyramid:

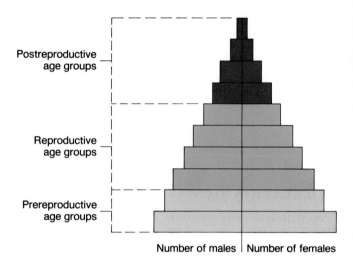

A broad-based population like this one will increase in size; in general, the broader the base, the more rapidly the population will increase. In contrast, a population with an inverted pyramid will decline because most individuals are past reproductive age.

A pyramid for a population with zero population growth would look like this:

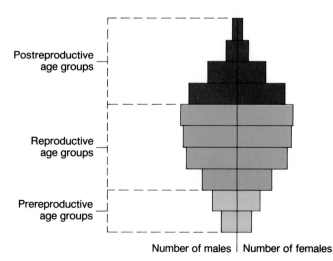

In this stable population, the number of prereproductive individuals balances the number of older individuals, so natality equals mortality. In this stable population, there is either no migration at all, or emigration and immigration are equal.

MORTALITY AND SURVIVORSHIP CURVES

To forecast population changes accurately, ecologists must consider the individual's life expectancy as well as the age–sex structure of the population. When life expectancy is plotted on a graph, a **survivorship curve** is produced. Ecologists identify three general types of survivorship curves: Type I, Type II, and Type III (Figure 43-2).

⚠ In a *Type I curve,* mortality remains low for much of the organisms' lifetimes and then increases sharply as individu-als reach old age. Humans and other animals that provide long-term care for their young often exhibit a Type I survivorship curve. A *Type II curve* is a straight, diagonal line, indicating that the chances of survival remain about the same throughout an individual's lifetime. Many birds and small aquatic animals exhibit a Type II curve. A *Type III curve* is exactly the opposite of a Type I curve: Most individuals die when they are very young, and only a few adults survive to old age. Species that produce enormous numbers of offspring and provide no parental care, such as insects, frogs, and many plants, exhibit a Type III curve.

BIOTIC POTENTIAL

As we mentioned in the beginning of this chapter, all species have the capacity to produce tremendously large numbers of descendants eventually, as long as there are no

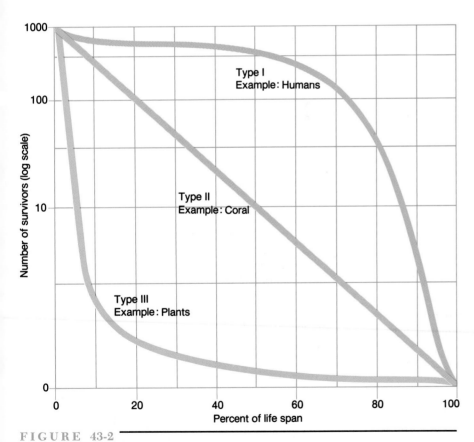

FIGURE 43-2

Three types of survivorship curves are produced by plotting the number of survivors (on a log scale) against the percent of the life span of a species. In the Type I curve, the mortality rate is low in the first years of life and then becomes higher at old age. Humans exhibit a Type I curve. In the Type II curve, the chances of surviving or dying are virtually the same throughout an organism's entire life. In a Type III curve, nearly all of the young die quickly, but the mortality rate is quite low for the few survivors, until they reach old age. Oysters, some insects, and weedy plants exhibit a Type III curve.

FIGURE 43-3
Although it may have seemed like a great idea at the time, importing plants into new habitats without natural population controls has produced some unexpected overpopulation disasters. For example, water hyacinth plants, with their orchid-like flowers and attractive stems and leaves, were brought to the United States from Venezuela to be displayed at the 1884 New Orleans Cotton Exposition. Many of the exposition visitors were given clippings of the charming plant to place in ponds and streams near their homes. With no competitors, few predators, and plenty of available space and nutrients, the growth of the water hyacinth spread rapidly. As this photo dramatically shows, today, water hyacinths are choking streams, rivers, irrigation systems, and hydroelectric installations.

restrictions to curb population growth. This innate capacity to increase in number under ideal conditions is called the **biotic potential** of a population.

▐▶ When conditions are optimal and no limitations exist, the population grows at its **intrinsic rate of increase,** or r_o (r = rate, o = optimum), the maximum increase in numbers of individuals per unit of time under optimal growth conditions. Organisms find themselves in such conditions only on rare occasion, however, such as when organisms colonize a new favorable habitat, when environmental conditions suddenly change for the better, or when organisms are introduced into new habitats in which there are no natural competitors or predators (Figure 43-3).

EXPONENTIAL GROWTH AND THE J-SHAPED GROWTH CURVE

When populations grow at their intrinsic rate of increase, the number of individuals increases exponentially. That is, the number increases by a fixed proportion, such as when a population *doubles* in size with each new generation (2, 4, 8, 16, 32, 64, 128, and so on). Such **exponential growth** can be demonstrated by placing a single *E. coli* bacterium into a nutrient culture (Figure 43-4). The bacterium divides into two cells after 20 minutes; the two cells divide into four cells in another 20 minutes; and the four cells divide into eight cells 20 minutes later. The population continues to double every 20 minutes. After 5 hours, 32,768 bacteria have been produced from the original bacterium. After 7 hours, there are more than 2 million bacteria. And after 36 hours, there would be enough bacteria to blanket the entire earth's surface with 28.8 centimeters (1 foot) of bacteria.

The growth of this bacterial culture illustrates how the number of individuals in a population affects the number of offspring produced; that is, the more reproducing individuals there are in a population, the greater the number of progeny. For example, during a 20-minute period, a population of *E. coli* increases to two when the population contains only the original bacterium, but during the same 20-minute period, the population can jump to 2 million

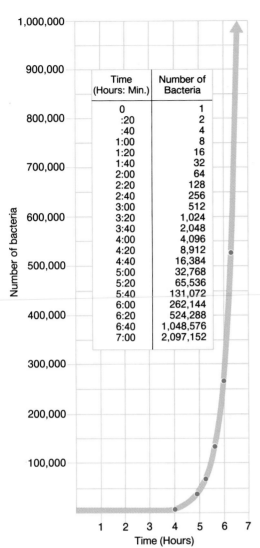

Time (Hours: Min.)	Number of Bacteria
0	1
:20	2
:40	4
1:00	8
1:20	16
1:40	32
2:00	64
2:20	128
2:40	256
3:00	512
3:20	1,024
3:40	2,048
4:00	4,096
4:20	8,912
4:40	16,384
5:00	32,768
5:20	65,536
5:40	131,072
6:00	262,144
6:20	524,288
6:40	1,048,576
7:00	2,097,152

FIGURE 43-4

Exponential growth of an *E. coli* bacterial culture produces a J-shaped growth curve. After rounding the "bend" of the J, the curve gets steeper and steeper, until some limitation, such as depletion of food or the buildup of contaminating waste products, curbs further growth.

when the population starts with 1 million bacteria. Expressed mathematically, the rate of exponential growth ($\Delta N/T$, where Δ = change, N = number of individuals, and T = time) equals the intrinsic rate of increase (r_o) multiplied by the number of individuals in the population (N):

$$\Delta N/T = r_o N$$

The pattern of exponential growth is always the same: The size of a population increases gradually at first and then grows larger and larger in progressively shorter periods of time, as more and more individuals are added to the population. Plotting exponential growth on a graph produces a curve that resembles the letter J; not surprisingly, it is called a **J-shaped curve**. As Figure 43-4 illustrates, once a population "rounds the bend" of a J-shaped curve, the number of individuals added to a population begins to skyrocket.

ENVIRONMENTAL RESISTANCE AND CARRYING CAPACITY

No natural ecosystem can support continuous exponential growth for any species; that is, no ecosystem has unlimited resources, and environmental conditions never remain constantly favorable for limitless growth. Eventually, some environmental limitation imposes a restriction on continued population growth. The factor(s) that eventually limit the size of a population create an **environmental resistance** to population growth. Environmental resistances include competition, predation, hostile weather, limited food or water supplies, restricted space, depleted soil nutrients, and the buildup of toxic byproducts from the organisms themselves.

❖ The combined limitations imposed by the environment establish a ceiling for the number of individuals that can be

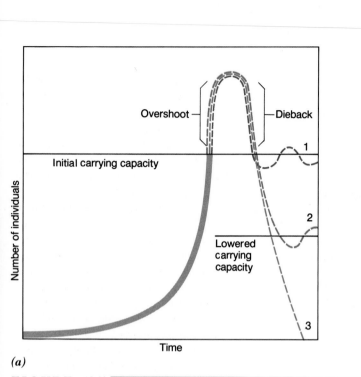

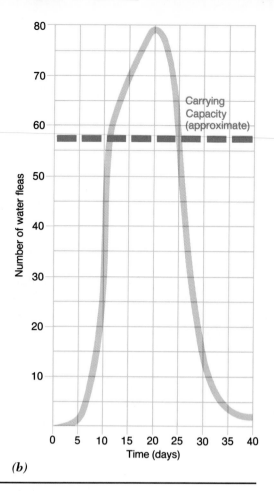

(a)   *(b)*

FIGURE 43-5

The risks of overpopulation. *(a)* A population undergoing exponential growth may overshoot the carrying capacity of the environment. When this happens, the population experiences a dieback to one of three levels: (1) to the original carrying capacity (the least common outcome), (2) to a population density at a lower carrying capacity, or (3) to extinction or near extinction levels. *(b)* A natural population of water fleas *(Daphnia)* illustrates the third fate: extinction.

supported in an area. The size of a population that can be supported indefinitely is called the **carrying capacity,** or **K**, of the environment. Populations undergoing exponential growth sometimes overshoot their carrying capacity before environmental resistances are able to curb their growth. The greater the reproductive momentum, the more likely it is that the population will exceed the environment's carrying capacity for some period of time.

Laboratory experiments and field observations identify three possible fates for populations that overshoot the carrying capacity of their environment (Figure 43-5). All the repercussions begin with a *dieback,* the death of a portion of the population. The least dramatic of the three fates is also the least common: The population simply dies back to the level of the original carrying capacity (curve 1 on the graph). In most cases, however, the excess population damages the environment in some way, which in turn, *reduces* the carrying capacity. In other words, the organisms themselves impact the carrying capacity. For example, an excessively large population of caterpillars could consume all of the leaves on a tree, weakening or possibly killing the plant.

When the carrying capacity is lowered, the population plunges. If the damage caused by overpopulation is not too severe, the population eventually comes into balance with a lower carrying capacity (curve 2). When damage is extensive, however, or when a vital resource is drastically depleted (as in the case of the caterpillars, killing the plant they live on), the population crashes to a very low level or suffers the third fate: It disappears altogether (curve 3).

Occasionally, a population exceeds the carrying capacity because a limiting factor does not take effect until a threshold level is reached. The population continues to grow exponentially until it reaches the threshold level, at which point the factor causes a sharp decrease in population size. For example, the gradual buildup of toxic waste products may have no effect on a population until these chemicals reach a critical level, at which point large numbers of individuals will die. Some biologists warn that this might be the case for the human population, as our air, soil, and water become more and more polluted (see The Human Perspective: Impacts of Poisoned Air, Land, and Water).

◁ THE HUMAN PERSPECTIVE ▷
Impacts of Poisoned Air, Land, and Water

On December 3, 1984, in Bhopal, India, a huge cloud of methyl isocyanate gas leaked out from a storage tank at the Union Carbide chemical plant, killing 3,000 people. It is estimated that an additional 2,000 people will die from side effects by 1995, and 17,000 of the 200,000 people injured from the gas leak have been permanently disabled as a result of lung ailments.

This event dramatically underscores the hazards of toxic substances that surround us in modern society. We are being exposed to more and more toxic substances in the air, in our water, and on the land—chemicals that adversely affect living organisms. In the United States alone, 60,000 chemicals are added to our food or are used to make cosmetics or to combat pests. Hundreds of these chemicals are known to be hazardous. Each year, 700 to 1,000 new chemicals enter the marketplace; fewer than 10 percent are tested to assess their health effects. Over 170 million metric tons (378 billion pounds) of potentially hazardous chemicals are manufactured each year in the United States alone, exposing people to hazardous chemicals in their homes, at schools, at work, and even while playing outdoors.

Toxic chemicals can affect virtually every cell in an organism and can cause cancer, mutations, birth defects, or reproductive impairment. These chemicals can affect cells in several ways: (1) by disturbing enzyme activities that regulate critical chemical reactions (mercury and arsenic inactivate enzymes); (2) by binding directly to cells or to essential molecules in the cell (carbon monoxide binds with hemoglobin in the blood, preventing it from carrying oxygen to cells); or (3) by releasing chemicals that have an adverse effect (in addition to its immediate destructive effects, carbon tetrachloride triggers nerve cells to release large amounts of epinephrine, which is believed to cause long-term liver damage).

Some chemicals gradually accumulate in the bodies of organisms, eventually building to toxic levels. The buildup of chemicals in the organisms in a food chain exposes organisms toward the top of a food chain to potentially dangerous levels of chemicals. The use of DDT as a pesticide is a good example of this phenomenon. Through bioconcentration and biological magnification, DDT concentrations are several million times greater in the fish-eating birds than they are in the water (page 946 and Figure 41-11). Large doses of DDT severely reduced the amount of calcium deposited in eggshells, causing bird eggs to break easily.

Like air pollutants, water pollutants can cause a physical or chemical change that adversely affects life. In the United States, the water in 40 states is already hazardously polluted, and more than half of Poland's water supply is so polluted that it cannot be used even by industry. Water pollutants include poisonous *toxic chemicals,* such as mercury, nitrates, and chlorine; microscopic *pathogens* that cause disease; various *physical agents,* such as soil sediments; and *excess nutrients and organic matter,* such as the remains of plants and animals, feces, debris from food-processing plants, and runoff from feedlots, sewage treatment plants, and fertilized agricultural land.

Pollutants are either dumped into or seep into all the earth's water sources, including surface waters (lakes, ponds, streams, and rivers), ground-water aquifers (which, in the United States, supply more than one-quarter of the annual water demands), and oceans, especially biologically rich coastlines, coastal wetlands (bays, swamps, marshes, and lagoons), and estuaries. Not only are coastal zones rich with myriad organisms, they are also the most vulnerable of the ocean's regions to numerous sources of pollution, including wastes from sewage plants and factories, sediment from erosion, and oil spills. Combine these sources with the fact that many cities may draw huge quantities of fresh water from streams during droughts, and you can see how water flow into important regions has diminished, increasing pollutant concentrations. As a result, coastal zones are being destroyed by pollution, water loss, sedimentation, dredging, and filling, at alarming rates. In the United States alone, more than 40 percent of estuaries have been destroyed, despite the implementation of state and federal laws to protect them.

LOGISTIC GROWTH AND THE SIGMOID GROWTH CURVE

An alternative to unbridled exponential growth occurs when environmental resistance increases as a population approaches the carrying capacity of the environment, slowing growth. When this happens, a **sigmoid growth curve** is produced (Figure 43-6). The "lazy S"-shaped sigmoid curve begins the same as does the J-shaped curve; that is, numbers of individuals increase slowly at first, then, as the reproductive base builds, the population goes into a period of exponential growth.

The rate of growth gradually slows down as the population approaches the carrying capacity, forming the top of the S-shaped curve. Population density eventually fluctuates around the carrying capacity: When the population

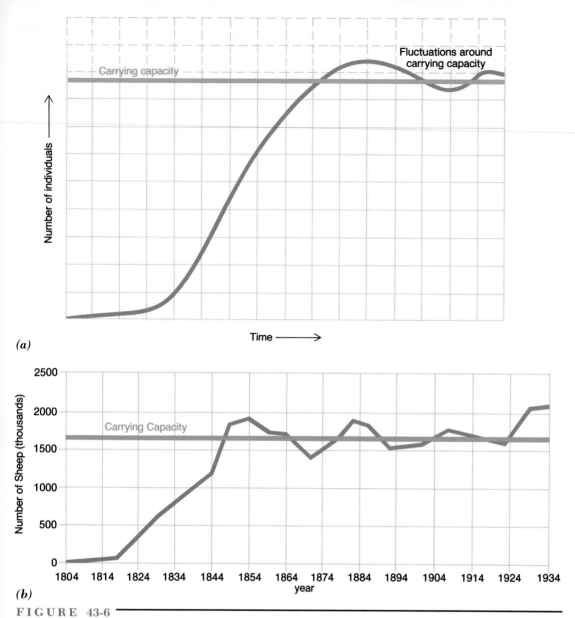

(a)

(b)

FIGURE 43-6

The sigmoid growth curve. *(a)* In a new habitat, a population increases slowly at first and then undergoes exponential growth, until environmental resistances begin to reduce growth rates and stabilize population density around the carrying capacity. This common logistic growth pattern produces a sigmoid growth curve. *(b)* Sheep were introduced on Tasmania in the early 1800s. The population growth pattern produced a sigmoid growth curve, with the population oscillating between 1.5 million and 2.0 million individuals.

slightly exceeds the carrying capacity, environmental resistance intensifies and causes a slight dieback; when the population falls below the carrying capacity, environmental resistance relaxes, and the population gradually increases. In this way, the size of a population remains fairly stable, and the growth curve flattens out at the carrying capacity level. Populations that produce a sigmoid growth curve exhibit **logistic growth.**

Oscillations are not always small, however. In 1957, for example, A. J. Nicholson raised laboratory colonies of Australian sheep-blowflies *(Lucilia cuprina)* by introducing a

small number of blowflies into a cage. All the conditions were kept constant, including food and water supplies. Blowfly populations grew rapidly because of their high biotic potential. Nicholson documented that blowfly populations underwent "violent oscillations," changes that could not have been caused by environmental fluctuations.

Nicholson suggested that since sheep-blowflies "scramble" for their food, overcrowding triggered excessive mortality; as a result, density dropped far below that which would be governed by normal environmental factors, such as seasonal changes in weather. As the population recovered in size, the blowflies' extremely high reproductive potential caused the population to overshoot the carrying capacity once again, triggering overcrowding and excessive mortality again.

THE "r AND K" CONTINUUM

Just as natural selection favors physical traits that improve survival (such as camouflage), traits that increase reproduction are also naturally selected. Over the course of evolution, species have evolved a range of reproductive strategies that offer advantages in different types of habitats. For example, species with the ability to reproduce large numbers of offspring will dominate a disturbed or uninhabited area before slower reproducing species can do so. In stable climax communities, however, where competition for resources is intense, species that provide long-term care to a few offspring (koalas, for example) often have a greater chance of success than do species that produce many offspring and provide no care (such as turtles).

▸ Species that produce many offspring at one time are said to be **r-selected species,** r referring to a high intrinsic rate of increase (r_m; r = rate of increase; m = maximum).

In other words, r-selected species have adaptations that maximize r_m. In contrast, species that produce one or a few well-cared-for individuals at a time are said to be **K-selected species,** K referring to those strategies that are more favorable for populations near the carrying capacity of the environment. These opposite strategies are the extremes in a continuum of reproductive strategies found in organisms. The main components of reproductive strategies and their relation to r-selected and K-selected species are presented in Table 43-1.

Differences between r- and K-selection strategies become very important when we consider the fate of rare, threatened, or endangered species, the most probable candidates for extinction. A rare species is one with only a few individuals remaining. Although a rare species is in no particular danger of extinction, because of its small numbers a rare species could be quickly wiped out. Endangered and threatened species are near extinction as a direct result of human activities. The passenger pigeon is an example of how quickly a species can become extinct when the combination of reproductive strategy and human activities works against the survival of a species. It took less than 100 years for the passenger pigeon to become extinct, even though many millions of pigeons existed in the early 1800s. Since

TABLE 43-1

RANGES OF REPRODUCTIVE STRATEGIES[a]

	r-Selected							K-Selected	
Number of offspring	many								few
Number of times an individual reproduces	once								many times
Size of young	small								large
Rate of development	fast								slow
Parental care	minimal								intensive
Life span	short								long
Survivorship curve	Type III				Type II				Type I
Energy to reproduction	high								low
Energy to increasing body size	low								high
Examples									
Animals	oysters		insects		birds		elephants		humans
Plants	weeds		saguaro		oaks		pears		mangroves

[a] r-selected and K-selected strategies are at the extremes of each continuum.

the passenger pigeon was *K*-selected—females laid only one egg each year—the production of offspring could not keep up with the enormous number of pigeons that were killed by sportsmen. Thus, the species died out.

With an innate ability to reproduce quickly, *r*-selected species may not become extinct as quickly as *K*-selected species do. However, no amount of reproduction can save a species if its environment is destroyed or severely contaminated (see Bioline: Accelerating Species Extinction).

The size of a population and the rate at which the population size increases or decreases depend on a balance between the pace at which new individuals are added to and the pace at which individuals are removed from a population. (See CTQ #3.)

FACTORS CONTROLLING POPULATION GROWTH

A change in the physical environment can have an impact on population growth, leading to a change in the size of a population. For example, global warming and its influence on rainfall patterns is expected to trigger reductions of redwood tree populations along the Pacific coast of North America. In addition to environmental impacts, all interactions within an ecosystem—between organisms and between organisms and their physical environment—can have an impact on population growth. The impact of some population-controlling factors increases or decreases in intensity as the size of the population changes, while the intensity of other population controls remains the same, regardless of the population size.

DENSITY-DEPENDENT FACTORS

As the size of a population increases, the increasing density of the number of individuals within the population sometimes limits the population growth rate. For example, when population density is low, young locusts develop normal-length wings. When locust density is high, however, hormonal changes trigger the development of longer wings in offspring. Long wings increase emigration, which, in turn, reduces the population density in that area. Such factors that are influenced by the number of individuals in the population and ultimately affect population density are called **density-dependent factors.**

There is a direct relationship between the intensity of density-dependent factors and population size: As population density rises, the intensity of density-dependent regulatory mechanisms increases, dampening population growth by reducing natality or by boosting mortality or emigration (as in locust populations). Conversely, when

population density falls, density-dependent mechanisms decrease in intensity, allowing population growth to accelerate. Density-dependent mechanisms explain why there are small fluctuations around the carrying capacity of the environment during logistic growth.

In addition to locust wing length, examples of density-dependent factors include the following:

- *Disease:* As density increases, the number of contacts between members of the population multiplies, intensifying the likelihood of the spread of disease-causing pathogens or parasites.
- *Competition:* Since members of the same species have very similar needs, intraspecific competition intensifies as density increases.
- *Predation:* The number of predators increases as the size of a prey population increases.
- *Stress:* Crowing increases stress in animal populations, which, in turn, increases aggression, infertility, and other growth-limiting factors.

Laboratory studies on overcrowding in rats show that crowding increases aggressive behavior, delays sexual maturation, reduces sperm production in male rats, and causes irregular menstrual cycles in females. Overcrowding also reduces sexual contacts between males and females, and increases homosexual contact between males. Such density-dependent factors quickly curb population growth by reducing natality. Whether such dramatic effects occur in populations outside the laboratory or in other species is not yet known. However, many demographers (scientists who study human population dynamics) believe that increased crime, drug abuse, and suicide may be partly the result of overcrowding, a situation that worsens as the world's human population swells by more than 250,000 people each day (an increase of 3 persons per second).

DENSITY-INDEPENDENT FACTORS

Not all regulatory factors are affected by the density of a population. Factors that are not influenced by population size are called **density-independent factors.** A killer earthquake, such as the earthquake that jolted the city of Erzincan, Turkey, on March 20, 1992, measuring 6.2 and killing over 500 people in less than 1 minute, is an example of a density-independent factor: The population density neither caused the earthquake nor affected the magnitude of the quake or the percentage of individuals killed. Many catastrophic events, including fires, floods, hurricanes, tornadoes, volcanic eruptions, and avalanches, are density-independent factors.

Density-dependent and density-independent factors often combine to regulate population size. For example, the number of aphids feeding on a sycamore leaf is affected not only by competition (a density-dependent factor that determines how many aphids may feed on the same leaf vein) but also by wind velocity (a density-independent factor; as

One alarming consequence of human activity is the diminished diversity of life on earth. One of every five species that was thriving when you were born is now extinct, the direct result of human impact on the environment. The biosphere is an extraordinarily complex structure, much more complicated than the space shuttle. Yet, in both the biosphere and space shuttle, the failure of vital parts will likely lead to the failure of the whole; all occupants will perish. Furthermore, the vital parts may be as small as the rivets that hold the craft together.

The species that comprise the living component of the biosphere can be considered the "rivets" that hold together spaceship Earth. Loss of these species is tantamount to "popping the rivets." If enough rivets are removed, the entire structure will inevitably fall apart. Many biologists predict that we are rapidly approaching the point of popping some critical rivets. If certain key species disappear, the ripple effect could spread to most of the planet's life forms. For example, extinction of the microscopic phytoplankton that supports virtually all life in the oceans and inland waterways would lead to starvation of every fish and aquatic mammal, animals that are an important source of food for humans. The extinction of phytoplankton would also reduce the amount of breathable oxygen in the atmosphere to less than half its current content, which could lead to the deaths of all air-breathing animals.

What would you do if you were an astronaut about to fly on the space shuttle and you noticed a person popping out rivets with a crowbar? You ask him why he is doing such a crazy thing, and he responds that rivets bring a good price and it's a way to make a bigger profit. "But the ship will crash if you remove its rivets," you protest. The profiteer responds, "I have removed 200 rivets on past flights and the shuttle hasn't crashed. What are you so worried about?" Any rational astronaut would refuse to be a passenger on such a compromised space shuttle.

Unfortunately, we see rivets being removed from spaceship Earth in unprecedented numbers, but we don't have the option to take another flight or to decline the trip altogether. All we can do is try to stop the practice, by slowing down species extinction and by giving the biosphere the time and opportunity it needs to repair the damage. Yet, a few countries continue to pop out the rivets in the face of danger, as was evident in 1992, at The Earth Summit held in Rio de Janeiro, Brazil. At the Summit, some countries refused to sign the Biodiversity Treaty, which would help protect the vast variety of life that exists on earth; that is, the earth's **biodiversity.** These countries claimed that scientists don't know for sure what the consequences of reduced biodiversity will be, so why should they endure economic hardship based on unproven predictions? Such arguments often delay scientific investigations into the possible ramifications of continued reductions in biodiversity. These arguments also ignore clear warnings from respected ecologists who continue to remind us of the popping rivets analogy, even though we don't know exactly how many more loosened rivets it will take before serious damage is done to the earth's ability to support life.

---

winds increase, sycamore leaves brush up against one another, scraping off the aphids).

The rate of population growth and the resulting population size is governed by limitations of the physical environment as well as by factors within the biotic community. The intensity of some of these factors changes with the size of the population, while the intensity of others do not. (See CTQ #4.)

HUMAN POPULATION GROWTH

In 1987, the world human population reached *5.0 billion people.* By the end of 1992, there were more than 5.4 billion people, and the number continues to climb quickly. More than 380,000 people are born each day (over four babies every second), and nearly 130,000 people die each day. This means that the world's human population is growing by some 250,000 people every single day. At this rate, nearly 100 million people are added to the human population every year. What will the future be like for those babies born into an exploding human population?

The answer has a great deal to do with geography. If these children are born in the United States in 1993, many will begin kindergarten in 1998, enter college around 2011, and be eligible to retire in the year 2058. By the time their children enter college, the world's human population may have swelled to more than 8 billion. By the time they retire, at the current growth rate, the world's human population will have climbed to 17 billion people, more than three times the number of people living on earth today, unless human endeavors or density-dependent controls can change the human population growth rate.

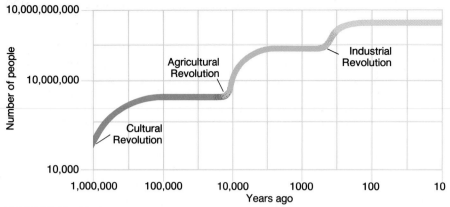

FIGURE 43-7

Surges of human population growth become clear when the size of the human population is plotted against time on a log–log scale (In log–log graphs, both population size and time are plotted on log scales, concentrating huge numbers of individuals and enormous time spans on a single sheet of graph paper.) Bursts of exponential growth occurred on three occasions, corresponding to the cultural, agricultural, and industrial revolutions.

HISTORICAL OVERVIEW

Modern humans appeared on earth less than 1 million years ago. The size of early human populations remained more or less in balance with the available food supply, increasing slowly as our early ancestors spread out and discovered new lands.

When the size of the human population is plotted against time on a log scale (Figure 43-7), three surges of population growth become evident. Each surge is the result of a major technological invention that improved food supply and/or human health.

▐▶ The use of tools and the movement from nomadic tribes to a stationary society were responsible for the first population surge, about 600,000 years ago. The development of agriculture (cultivating food in a village-farming society) produced the second population surge, about 10,000 years ago. The most recent growth spurt occurred about 200 years ago, as humans entered the Industrial Revolution. At the same time, humans continued to make advances in agriculture, medicine, and hygiene, decreasing the death rate and increasing the average life span. These advances lessened some of the former controls on human population growth, mainly food shortages and disease.

The overall growth curve for the world's human population is clearly a J-shaped curve (Figure 43-8). It took hundreds of thousands of years for the world human population to reach 1 billion people. Once a reproductive base of 1 billion people was established in 1800, however, it took progressively less time to add another 1 billion people to the human population: 130 years to reach 2 billion people; 30 years to reach 3 billion; 15 years to reach 4 billion; and finally only 13 years to reach 5 billion people in 1987.

GROWTH RATES AND DOUBLING TIMES

The current human birth rate is 27.7 babies per 1,000 people per year. The death rate is 9.5 people per 1,000 per year,

making the current human population growth rate equal to 18.2 humans per 1,000 people per year:

$$\underset{\text{(Birth rate)}}{27.7} - \underset{\text{(Death rate)}}{9.5} = \underset{\text{(Growth rate)}}{18.2}$$

Population increases are often expressed as the number of people added to the population per 100 individuals, giving the **percent annual increase** in population. The average annual increase for all nations is currently 1.8 percent (see (Table 43-2). This means that it will take less than 56 years for the world's human population to double. That is, if you add 1.8 people to a population of 100 each year, it will take 55.6 years for the population to reach 200 (55.6 × 1.8 = 100). As we discussed earlier, however, population size affects the rate of increase; thus, a growth rate of 1.8 actually has a doubling time of only 39 years instead of 55.6 years because of the large reproductive base. Doubling times for the world human population are given in Table 43-3.

TABLE 43-2

PERCENT GROWTH RATE FOR VARIOUS COUNTRIES

Country	Percent Annual Increase
Developing countries	
India	2.3
Brazil	2.9
Uganda	3.3
Mexico	3.5
Kenya	4.1
Developed countries	
United Kingdom	0.0
United States	0.6
Japan	1.1

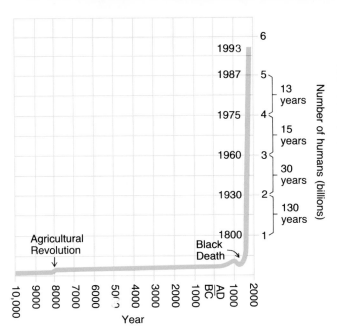

FIGURE 43-8

The human population growth curve. The world's human population grew slowly for 2 million years until the 1800s, at which time population growth began to soar. By the year 2000, there will be an estimated 7 billion to 14 billion people on earth, if our population continues to grow at its present rate.

CURRENT AGE–SEX STRUCTURE

There is a big difference between the age-sex structure of developing nations (many countries in Asia, Africa, and South America) and that of developed countries (Figure 43-9). The populations of developing nations are growing rapidly because large numbers of individuals in these populations are moving into their reproductive years. Since developing countries are heavily populated, they have a large effect on the world's human population growth. As a result, the world's human population will continue to increase at a rapid rate.

FERTILITY RATES AND POPULATION GROWTH

Although age–sex structure diagrams help predict population growth, they do not take into consideration another factor that affects human population growth: **fertility rates.** A fertility rate is the average number of children born to each woman between 15 and 44 years of age (the reproductive years). The average fertility rate for all nations in the world is now slightly below 2.1 births per woman. In developed countries, however, fertility rates average 1.9 births, compared to 4.5 births in developing nations.

Given the current mortality rates, fertility rates of between 2.1 and 2.5 are required to maintain zero population growth. With an average fertility rate of 1.9 in developed countries, populations in these countries will decline. In contrast, with an average fertility rate of 4.5, populations in developing countries will rapidly increase. As a result, the world's human population will continue to increase by greater and greater numbers each year. Some scientists believe that the world's human population will begin leveling off in 30 to 40 years, to between 8 billion and 14 billion people, while others believe that the human population will not reach these levels because we are very close to, or have already exceeded, the earth's carrying capacity for humans.

THE EARTH'S CARRYING CAPACITY AND FUTURE POPULATION TRENDS

Density-dependent controls are already beginning to curb human population growth, as the every-burgeoning human population is reducing the ability of the earth to support life (Figure 43-10). Lowering the earth's carrying capacity affects not only millions of other organisms but ultimately humans as well. We have learned now important photosynthetic plants are to life on earth; they supply chemical energy and nutrients to virtually all heterotrophs, including humans. With only 15 staple plant species standing between humankind and starvation, it is imperative that we minimize the damage to our environment. If just one of these staple crops were to become extinct, millions, or possibly billions, of people might perish.

TABLE 43-3

HUMAN POPULATION SIZE ESTIMATES, DOUBLING TIMES, AND PERCENT GROWTH RATES

Date	Estimated Human Population (bil.)	Doubling Time (yrs.)	Percent Annual Increase
Pre 8000 B.C.			0.0007
800 B.C.	0.005	1500 million	0.0015
1650 A.D.	0.500	200	0.1
1850 B.C.	1.000	80	0.8
1930 A.D.	2.000	45	1.9
1975 A.D.	4.000	35	2.0
1987 A.D.	5.000	39	1.8

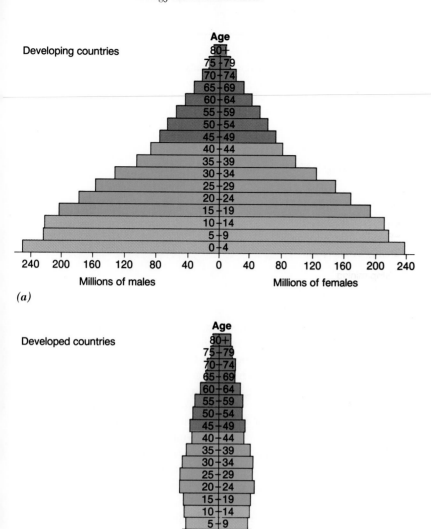

(a)

(b)

FIGURE 43-9

Age–sex structure diagrams for **(a)** developing and **(b)** developed countries differ radically. With a tremendous reproductive base, developing countries will continue to grow rapidly, while developed countries will decline in population. Since 75 percent of humans live in developing countries, the world's human population is expected to continue its perilous increase.

Even simple, microscopic organisms, like bacteria, fungi, and nematodes, are important to human life (and life in general) for they contribute to the recycling of nutrients. Many organisms provide medical and agricultural remedies. The extinction of species not only reduces ecological diversity but also cuts into the genetic bank, a bank of genes we have tapped a number of times to solve medical, industrial, and agricultural problems.

Between 10 million and 20 million people—mostly children—die each year from starvation or malnutrition-related disease. Even in affluent countries, overpopulation has accelerated the rate of environmental deterioration, lowering the quality of life and most likely reducing the environment's carrying capacity. Is it possible that the world's human population is already in overshoot? If so, what lies ahead? A dieback to the original carrying capacity,

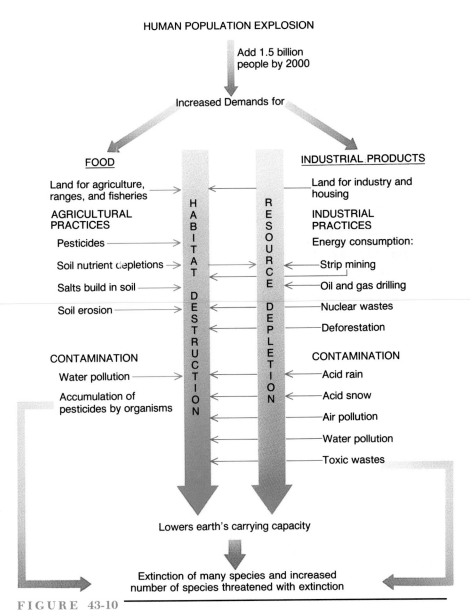

FIGURE 43-10

Impacts of the human population explosion. Increased numbers of people mean increased demands for food and products. These demands deplete resources and destroy habitats, which, in turn, lower the earth's carrying capacity and increase the number of species threatened with extinction.

a dieback to a lower carrying capacity, or a dieback to extinction?

Many biologists believe that we can still avoid these undesirable consequences through efforts to curb population growth and protect the environment; these are the challenges facing us today. Biotechnological advances for increasing food production, methods for disposing of wastes that otherwise contaminate the biosphere, and other advances that would help control growth rates may someday help us avert the disasters of overpopulation.

The world's human population is currently growing exponentially. As with all species, a population ceiling for humans will ultimately be reached. Humans are unique in their ability to analyze and anticipate future trends as well as to modify their environment to their advantage. All of these skills will be necessary in order to solve the problem of the human population explosion. (See CTQ #5.)

REEXAMINING THE THEMES

Biological Order, Regulation, and Homeostasis

↻ The populations that make up an ecosystem interact with one another in many ways. Prey populations provide energy and nutrients for predator populations, just as plant populations provide energy and nutrients for herbivores. A change in the size of one of these interacting populations eventually changes the size of the other. Interacting populations also affect each other's distribution pattern in an ecosystem. For example, when an animal population disperses seeds in groups, the result is a clumped distribution for the plant. When plants produce chemicals that inhibit the growth of other plants, or when animals establish territories, uniform distribution patterns are formed.

Unity within Diversity

⬣ All organisms, from the slowest reproducer to the fastest, have the potential eventually to produce huge numbers of descendants. This capacity for reproduction enables species to colonize new habitats, to recover from disasters that kill a large proportion of the population, and to provide the variability upon which natural selection operates.

Evolution and Adaptation

▶ Reproduction is a basic characteristic of life. Organisms have evolved a range of reproductive strategies, from *r*-selected species that (1) grow fast, (2) reach reproductive maturity early, and (3) invest the greatest amount of energy and nutrients they acquire into a single, large reproductive event to *K*-selected species that (1) grow slowly, (2) delay reproduction, (3) invest the largest proportion of energy and nutrients into growth, increasing competitive abilities, and (4) have small, multiple reproductive events. Between these extremes lies a full range of species with intermediate reproductive strategies, each an adaptation that increases survival in particular ecosystems.

SYNOPSIS

Communities are made up of populations. Each population contains all of the individuals of the same species that occupy an area at the same time.

Every population has the potential to reproduce large numbers of offspring. A population growing at its maximum rate increases in size exponentially and produces a J-shaped growth curve. Eventually, some factor or combination of factors (limited food or the buildup of toxic wastes, for instance) halts continued exponential growth, causing a population dieback, sometimes to extinction.

For most populations, growth starts off slowly, increases rapidly, and then slows down again and levels off. The population levels off at the carrying capacity of the environment, where growth remains in dynamic balance with available nutrients and appropriate conditions.

Populations grow when natality and immigration exceed mortality and emigration. In populations where there is no immigration or emigration, the rate of growth equals the difference between the birth rate and the death rate.

Organisms have evolved a broad range of reproductive strategies. Some species produce enormous numbers of offspring at one time, while others reproduce just one or a few offspring, which they provide with extended care. Each strategy has its advantages in certain environments.

The world's human population exceeded 5.5 million people in 1993. The human population is continuing to grow exponentially, mainly as a result of growth in developing countries.

Key Terms

population (p. 986)
population density (p. 987)

clumped pattern (p. 987)
uniform pattern (p. 987)

random distribution (p. 987)
natality (p. 988)

Review Questions

1. What is the relationship between interspecific competition and density-dependent factors that limit population growth?

2. Match the process (numbered column) with the resulting outcome (lettered column).

   ____1. logistic growth   a. J-shaped growth curve
   ____2. intrinsic rate of   b. sigmoid growth curve
         increase   c. a population dieback
   ____3. exponential growth   d. increased density
   ____4. immigration

3. Check those conditions that would cause a population to *decrease* in density.

   ____1. natality that greatly exceeds mortality, immigration, and emigration
   ____2. pyramid-shaped age–sex structure
   ____3. a Type III survivorship curve, high emigration, and high mortality
   ____4. a population that has greatly exceeded the carrying capacity of the environment

4. In the following habitats, which reproductive strategy —K-selected or r-selected—would be of greater advantage? Give a brief explanation for your answer.

Environment	*r or K* Strategy	Explanation
a. cool lava		
b. recent burn area		
c. climax rainforest		
d. abandoned farm		
e. desert		
f. tundra		
g. streamside		
h. recent flood plain		

5. List two density-dependent and two density-independent factors that would affect human population growth. For example, a collision between two planes as a result of crowded airways would be a density-dependent factor, whereas a plane crash into a mountain as a result of adverse weather would be a density-independent factor.

6. What is the relationship between age–sex ratios and survivorship curves to biotic potential?

7. Describe the reproductive characteristics of an organism that is precisely in the middle of the r and K continuum. In what types of ecosystems would such an organism be at a clear advantage over organisms at either extreme of the continuum?

8. Carrying capacity of the environment, biotic potential, and environmental resistance all affect the rate of population growth. Consider two distinctly different types of organisms in two distinctly different habitats: elephants in an African savanna, and phytoplankton in the open ocean. Describe the similarities and differences between population growth factors for these two organisms. Be sure to consider all aspects.

9. List the conditions that determine the carrying capacity of a prey population. If the prey population exceeded its carrying capacity, would the predator population also exceed its carrying capacity? Explain your answer.

10. Name four factors or events that have blocked or postponed traditional limits on human population growth. Given the current rate of growth, name two ways humans can continue to block traditional limits.

Critical Thinking Questions

1. As in all plants, the critical phases in the life cycle of the saguaro cactus include seed germination, seedling growth and development, growth to reproductive maturity, flowering, pollination, fertilization, fruit and seed development, and seed dispersal. Which critical phases were tested by the research presented in the chapter opening vignette and which phases were not? Of those phases that were tested, are there any other elements that should be looked at to be sure that nothing was missed? Of those phases that were not tested,

1004 • PART 8 / Ecology and Animal Behavior

choose one that you believe may explain why saguaros are not reproducing adequate numbers of offspring to maintain stable populations, and devise an experiment to test whether you are correct.

2. Each ecosystem supports many populations of species, each growing at different rates. Since all populations share the same abiotic environment, explain how the growth rates of each population can be different. Could any populations have the same growth rate? How?

3. The growth rates of populations can be calculated from birth rates and death rates, assuming there is no emigration or immigration. By convention, birth and death rates are given as the number per 1,000 people, whereas growth rate is given as a percent (i.e., per 100). Thus, the calculation for growth rate is

growth rate (%) = birth rate − death rate/10.

You can estimate the time in years that it will take a population to double by dividing 70 by the annual growth rate.

Calculate the growth rate and doubling time of the countries in the table below:

Country	Birth Rate	Death Rate	Growth Rate	Doubling Time
United States	17	9		
India	31	10		
China	21	7		
Somalia	49	19		
Poland	15	10		
France	14	9		

4. Wildebeests and birds migrate from one location to another in search of food and other resources. Barnacles and mussels are sessile and cannot travel to locate richer sources of needed materials. Yet, the populations of both migratory and sessile species are subject to limitations to growth. List some of the differences in the ways limiting factors would affect the size of migratory populations, compared to sessile ones. Are there common features to the limiting factors in each category that may lead you to draw a general conclusion about how environmental resistances differ between migratory and sessile organisms?

5. Describe your opinion concerning the future trends in human population growth. Listen to the news and read a major newspaper over the next week and document which stories support your opinion and which stories do not. Based on these stories, should your projection of future population trends be modified? In what way? Choose one of the following statements that you feel best supports your revised opinion, and devise an experiment for testing whether the statement is correct or not:

a. Humans are fundamentally different from all other organisms (behaviorally, intellectually, and physiologically) and are not subject to the same population controls as are other species.

b. Humans are governed by the same population controls as are all other species. Although humans cannot change the types of natural population controls, their intellect and ingenuity empower them to modify levels of natural controls.

Additional Readings

Ayensu, E., V. Heywood, G. Lucas, and R. Defilipps. 1984. *Our green and living world. The wisdom to save it.* Cambridge, London: Cambridge University Press. (Introductory)

Council on Environmental Quality, 1981. *Global future: Time to act. Report to the President on Global Resources, Environment and Population.* Washington, D.C.: U.S. Government Printing Office. (Introductory)

Ehrlich, P., and A. Ehrlich 1972. *Population, resources, environment. Issues in human ecology.* San Francisco: W.H. Freeman. (Introductory)

Ehrlich, P., and A. Ehrlich. 1981. *Extinction.* New York: Random House. (Intermediate)

Population Reference Bureau, Inc. 1972. *The world population dilemma.* Washington, D.C.: Columbia Books. (Introductory)

World Resources. 1990–91. *A report by the World Resource Institute, in collaboration with The United Nations Environment and Development Programs.* New York: Oxford University Press. (Intermediate)

CHAPTER
◄ 44 ►

Animal Behavior

STEPS
TO
DISCOVERY
Mechanisms and Functions of Territorial Behavior

MECHANISMS OF BEHAVIOR

Primarily Innate Behavior

Genes and Behavior

LEARNING

Habituation

Classical Conditioning

Operant Conditioning

Insight Learning

Social Learning

Play

Learning as an Adaptation

DEVELOPMENT OF BEHAVIOR

EVOLUTION AND FUNCTION OF BEHAVIOR

Optimality and Territoriality

SOCIAL BEHAVIOR

Costs and Benefits of Group Living

Animal Communication

ALTRUISM

Alarm Calls

Helping

Cooperation in Mate Acquisition

Food Sharing

Eusociality

BIOLINE
Animal Cognition

Mechanisms and Functions of Territorial Behavior

Susan Reichert has explored both the mechanisms and function of territorial behavior of the funnel-web building spider *Agelenopsis aperta*. These spiders compete for web sites and defend an area around the web as a territory. The territory must be large enough to provide sufficient food for survival and reproduction. This species occupies a wide variety of habitats, ranging from northern Wyoming to southern Mexico. Spiders living in the relatively lush vegetation along the rivers and lakes in Arizona (riparian populations) allow neighbors to build webs closer to their own than do spiders in desert grassland populations of New Mexico. Furthermore, the intensity of territorial disputes between desert grassland spiders is greater than is that between spiders that live near water. Threat displays of grassland spiders are more likely to escalate into battles, and the fighting more often results in physical injury.

A series of interesting experiments has demonstrated the genetic basis of territorial behavior in these spiders. Spiders were collected from a desert grassland environment in New Mexico and from a riparian environment in Arizona. Pure-bred lines were established by allowing individuals from a particular habitat to mate only with one another. After the spiderlings emerged from the eggs, each was raised separately on a mixed diet of all they could eat. When they were mature, sixteen females from each population line were placed into experimental enclosures where they could build webs. Just as in field populations, the average distance between laboratory raised females from riparian populations was less than that between females from desert grassland populations. Thus, territory size is an inherited characteristic and is not determined by recent feeding history or learned from previous territorial disputes.

With limited resources, competition is greater among funnel spiders living in the desert as opposed to funnel spiders living in

Working with John Maynard Smith, Reichert continued to explore the genetic mechanisms underlying territoriality and aggression in these spiders. Pure-bred lines of grassland spiders were mated with pure-bred lines of riparian spiders. Smith and Reichart found that fights between the hybrid offspring were even more likely to end in injury or death than were those between pure grassland spiders. Why should hybrid spiders be more combative than either of their parents? The simplest explanation for these results is that the behavior of a spider is determined by two conflicting tendencies: "aggression" and "fear." Each tendency is controlled by a gene or, more likely, a gene complex. The allele(s) for high aggression (A) is dominant to that for low aggression (a), and the allele(s) for low fear (B) is dominant to that for high fear (b). Researchers proposed that aggression and fear are low in riparian spiders and high in grassland spiders. Thus, grassland spiders would be homozygous for high aggression (AA) and for high fear (bb). In contrast, riparian spiders would be homozygous for low aggression (aa) and for low fear (BB). As a result, the hybrid offspring would have high aggression and low fear, a situation that would be expected to lead to costly fights. In additional crosses in which the hybrid offspring were mated with one another to create an F_2 generation, and in which they were mated with spiders thought to be homozygous recessive for these traits (backcrosses), the results were consistent with this model. The results of these later crosses also indicated that aggression was inherited on sex chromosomes and that fear was inherited on autosomal chromosomes.

Reichert also examined the degree of genetic diversity between the two populations of spiders. To do this, she used electrophoresis, a technique that reveals the number of alleles that exist for a given gene. Electrophoresis also can be used to estimate the degree of genetic variability among individuals in a population (page 328).

Those genetic differences and the differences in territorial behavior presumably arose as natural selection favored traits in the populations that suited the local environments. In relatively lush riparian areas found along rivers and lakes, for instance, prey are more abundant and there are more suitable web sites than in the desert grassland habitat. As a result, spiders living along a river are able to capture adequate prey in a smaller area. In contrast, the desert grassland environment is severe. Prey are scarce, and the scorching sun makes it difficult for spiders to forage during much of the day. Thus, larger territories are suited to stringent environmental conditions. Evidence suggests that territory size is genetically set. One reason for the intensity of territorial disputes among grassland spiders may be that the scarcity of web sites increases the value of a web. So, differences in genetic diversity between species of spiders is increased as a response to its environment.

Territoriality in the funnel-web spider seems to be an adaptation to the existing ecological conditions. However, there are no data to prove that owners of large territories in desert grassland populations leave more offspring than do individuals with small territories. In another locality, a recent lava bed in central New Mexico, there are data to support the hypothesis that territory quality may influence an animal's reproductive success. Here, spiders with quality web sites have thirteen times the reproductive potential of their neighbors in poorer quality areas.

rich riparian habitats.

"Why is that animal doing that?" This is the fundamental question of ethology, the study of animal behavior. This seemingly simple question has been interpreted in several ways. For example, the Dutch biologist Niko Tinbergen, a corecipient of the Nobel Prize in medicine and psychology in 1973, identified four related questions: (1) What are the mechanisms that cause the behavior? (2) How does it develop? (3) What is its survival value? (4) How did it evolve? Tinbergen believed that the biological study of behavior should "give equal attention to each and to their integration."

To better appreciate the types of questions we may ask about animal behavior, consider those that may be raised regarding a massive herd of caribou in their migratory march across the frozen tundra (Fig. 44-1). (1) How do the caribou "know" when it is time to migrate? How do they find their way along a predictable migratory path? Such questions focus on the mechanisms underlying a behavior. (2) Do those making this journey for the first time learn the route from experienced travelers, or do they inherit a directional tendency from their parents?

Questions such as these concern development. (3) Why do they migrate? How do the advantages they gain outweigh the risks and demands of such a journey? These are questions regarding the survival value, or adaptiveness, of migration. (4) Finally, how did caribou migration begin? This question centers on the evolution of the behavior.

▼ ▼ ▼

MECHANISMS OF BEHAVIOR

When we observe an animal in nature, we generally find that its behavior is adaptive; that is, the behavior enhances the animal's chances of surviving and reproducing. Adaptation includes not only traits with known genetic causes but also the inherited potential for learning and even the learned behaviors themselves. The relative importance of genes and experience may vary tremendously, but neither is ever equal to zero. Genes generally code for a range of potential phenotypes. Sometimes genes specify a precise behavior, leaving little room for modification by learning. Behaviors that are precisely specified by genes are often those that must be expressed in nearly perfect form, even on the very first trial. For example, if an animal fails to respond

FIGURE 44-1

Migrating caribou. Many questions may come to mind when watching the behavior of animals. Some deal with the details of the behavior, others concern its advantages and evolution.

appropriately the first time it encounters a predator, it may not get a second chance to refine its escape response. Genes also play an important role in determining the actions of animals that have little opportunity to learn. For example, fruit fly parents are generally not present when their offspring emerge from the pupal cases. Nonetheless, a male fruit fly who has been isolated from the larval stage until adulthood still exhibits specific courtship behavior. In other cases, the behavioral blueprint is more general so that the behavior is almost entirely shaped by experience. A predator that fails to capture food on the first attempt may be hungry, but it learns from its experience, increasing its chances of success the next time.

PRIMARILY INNATE BEHAVIOR

Innate behaviors are those that are under fairly precise genetic control. Innate behaviors are often species-specific and highly stereotyped.

The Fixed Action Pattern

Among the primarily innate behaviors are **fixed action patterns (FAPs).** These are motor responses that are triggered by some environmental stimulus. Once started, FAPs continue to completion without the help of external stimuli. For example, a brooding female greylag goose will retrieve an egg that has rolled just outside her nest by reaching beyond it with her bill and rolling it toward her with the underside of the bill. Once the rolling behavior has begun, if the egg is experimentally removed, the goose will continue the retrieval response until the now imaginary egg is safely returned to the nest. The egg retrieval response of the female greylag goose illustrates other characteristics that are generally true of most FAPs. An FAP is performed by all appropriate members of a species. Furthermore, in the case of the greylag goose, each time an egg is retrieved, the sequence of actions is virtually identical, modified very little by experience. As evidence, an FAP will be exhibited even in inappropriate circumstances. For example, a brooding female will retrieve a beer bottle or any small object outside the nest, as if it were her egg.

Stimuli and Triggers

A fixed action pattern is produced in response to something in the environment. Ethologists called such a stimulus a **sign stimulus.** If the sign stimulus is given by a member of the same species, it is termed a **releaser.** Releasers are important in communication among animals.

Sign stimuli may be only a small part of any environmental situation. For example, a male European robin will attack another male robin that enters its territory. Experiments have shown, however, that a tuft of red feathers is attacked as vigorously as is an intruding male. Of course, in the world of male robins, red feathers usually appear on the breast of a competitor.

Any of the traits possessed by an animal or an object may serve as a sign stimulus. It may have a certain color, a special shape, or a particular pattern of movement, or it may produce a sound or have an odor. How do we know which character serves as the sign stimulus?

One way ethologists can identify the sign stimuli from the barrage of information reaching an animal is with models in which only one trait is presented at a time. The model is presented to an individual in the appropriate physiological state to see whether it will respond as it would to the normal stimulus. For example, a male stickleback in reproductive condition defends his territory from any intruding males. By constructing dummies of sticklebacks of varying degrees of likeness to the real male and painting them red, pale silver, or green. Niko Tinbergen and his co-workers demonstrated that in male sticklebacks, a red tint on the undersurface of the trespasser releases an aggressive territorial response. A very realistic replica lacking the red color was not attacked, but a model barely resembling a fish, on which the underside was painted red, provoked an assault.

Chain of Reactions

So far, we have considered only relatively simple behaviors; more complex behaviors can also be built from sequences of FAPs. The final product is an intricate pattern called a *chain of reactions,* whereby each component FAP brings the animal into the situation that triggers the next FAP.

An early analysis of a chain of reactions was conducted on the courtship ritual of the three-spined stickleback. This sequence of behaviors culminates in synchronized gamete release, an event of obvious adaptive value in an aquatic environment. Each female behavior is triggered by the preceding male behavior which, in turn, was triggered by the preceding feminine behavior (Fig. 44-2).

FIGURE 44-2
The courtship ritual of sticklebacks is built from a sequence of fixed action patterns.

Sometimes a male stickleback attacks a female entering his territory. If the female displays the appropriate "head-up" posture, in which she hangs in the water, exposing her egg-swollen abdomen, the male will begin his courtship with a zig-zag dance. He repeatedly alternates a quick movement toward her with a sideways turn away. This dance releases approach behavior of the female. Her movement induces the male to turn and swim rapidly toward the nest, an action that entices the female to follow. Once at the nest, the male stickleback lies on his side and makes a series of rapid thrusts with his snout into the entrance to the nest, while raising his dorsal spines toward his mate. This action is the releaser for the female to enter the nest. The presence of the female in the nest, in turn, is the releaser for the male to begin to prod the base of the female's tail with his snout, causing the female to release her eggs. The female then swims out of the nest, making room for the male to enter and fertilize the eggs. We can see that this complex sequence is largely a chain of FAPs, each triggered by its own sign stimulus or releaser.

GENES AND BEHAVIOR

We have seen that some behaviors have a strong genetic basis. Genes do, in fact, influence all behavior to some extent. We might wonder then what it means to say that a behavior has a genetic basis. It simply means that an animal with a certain gene will be able to perform the behavior. If the animal lacks that gene, it may be unable to perform the behavior, or the behavior may be expressed in a different manner.

As you already know, genes direct the synthesis of proteins, proteins that may affect some of the connections within the nervous system, or may act as regulators, such as enzymes that have a regulatory function. In a few cases, we know the link between the protein product of a specific gene and a behavior. For example, the gene behind egg-laying behavior of the sea hare *Aplysia* codes for a long chain of amino acids that is later cleaved into many proteins, three of which are known to be important in egg-laying behavior (Fig. 44-3). One of the three proteins is egg-laying

FIGURE 44-3

The sea hare Aplysia lays eggs in a stereotyped sequence of actions that is controlled by a single gene that codes for a long chain of amino acids. This chain is cleaved into three shorter proteins, three of which are important in orchestrating the behavior.

hormone (ELH), which increases the firing rate of the abdominal ganglion, increases heart and respiration rates, and stimulates the ducts of the reproductive system to contract and expel the egg string. The two other proteins function as neurotransmitters that increase or decrease the activity of neurons involved in egg-laying behavior.

Many animals perform complex behaviors that are either entirely or largely directed by the animal's genes. Such innate behaviors are particularly adaptive in species whose offspring have little or no contact with their parents. Innate behavior often reveals itself by its rigid, stereotypic pattern and by the fact that it can be evoked by highly specific environmental stimuli, such as a red feather or a specific aroma. (See CTQ #1.)

LEARNING

Learning is a process in which the animal benefits from experience so that its behavior is better suited to environmental conditions. As we look at learning in various situations, it appears that the process can occur in several fundamentally distinct ways. Although not everyone agrees on how it should be done, it is often useful to group types of learning into categories, including habituation, classical conditioning, operant conditioning, insight learning, social learning, and play.

HABITUATION

In **habituation,** the simplest form of learning, the animal learns *not* to show a characteristic response to a particular stimulus because the stimulus was shown to be unimportant during repeated encounters.

For an example that illustrates the essential characteristics of habituation we turn to the clamworm, a marine polychaete that lives in underwater burrows which it constructs out of mud. A clamworm partially emerges from the burrow while feeding. However, certain sudden stimuli, such as a shadow that could herald the approach of a predator, cause the clamworm to withdraw quickly for protection. If the stimulus is repeated and there are no adverse consequences, the withdrawal response gradually wanes. If the stimulus is changed, however, say from a shadow to a touch, the clamworm will again respond. Thus, the loss of responsiveness to the shadow is not due to muscular fatigue (page 483).

Habituation is beneficial in that it eliminates responses to frequently occurring stimuli that have no bearing on the animal's welfare, without diminishing reactions to significant stimuli. Obviously, it is important for the clamworm to withdraw to the safety of its burrow when a shadow belongs to an approaching predator. If the shadow is encountered often without a predator's attack, however, it is probably caused by something harmless, perhaps a patch of algae repeatedly blocking the sun. In this case, responding to the shadow each time it appears would waste energy and leave the worm little time for other essential activities, such as feeding or reproducing. Thus, habituation is one of the mechanisms that focuses attention and energy on the important aspects of the environment.

CLASSICAL CONDITIONING

In **classical conditioning,** an animal learns a new association between a stimulus and a response. Because the new stimulus repeatedly occurs before the usual one, it gradually begins to serve as a signal that the usual stimulus will occur. Eventually the new stimulus alone is sufficient to cause the response.

The most familiar example of classical conditioning is that of Pavlov's dogs, who learned to associate the sound of a bell with the presence of food (Fig. 44-4). During training, Pavlov rang a bell immediately before feeding a hungry dog. When the dog saw the food, it began to salivate. The procedure was then repeated many times. Eventually, the bell became a signal that food would be delivered soon, and the dog began to salivate at the mere sound of the bell. This response—salivating at the sound of the bell—is an example of a **conditioned reflex.**

In more general terms, an animal has a particular inborn response to a certain stimulus. This stimulus (e.g., food) is called the **unconditioned stimulus (US)** because the animal did not have to learn the response to it. A second

F I G U R E 44-4
Ivan Pavlov, a Russian physiologist, demonstrated classical conditioning in the salivary reflex of a dog. The dog is prepared so that the saliva it produces can be collected and measured. In this experiment, food serves as the unconditioned stimulus since it triggers the desired response (salivation) without prior training. A tone is sounded immediately before food is offered to the dog. Because of the repeated pairing of the tone and food, the dog eventually begins to salivate in response to the tone alone.

FIGURE 44-5

A male blue gourami that is classically conditioned to cues signaling the approach of a rival has a competitive edge in territorial disputes.

stimulus—one that does not initially elicit the response—is repeatedly presented immediately before the US. After several pairings, the second stimulus is able to elicit the response. The new stimulus (e.g., a bell) is now called the **conditioned stimulus (CS)** since the animal's response has become conditional upon its presentation.

▐▶ Karen Hollis has experimentally tested the hypothesis that the adaptive function of classical conditioning is to prepare animals for important events. Her studies have centered on territorial and reproductive behaviors in blue gouramis (Fig. 44-5). Successful territorial defense is important because female blue gouramis rarely mate with a male without a territory. A conditioned response to signals indicating the approach of a rival gourami might prepare a male for battle and give him a competitive edge. In nature, as the rival approached, he would inadvertently send visual, chemical, or mechanical signals that territorial invasion was imminent. In the laboratory, male blue gouramis that are classically conditioned to a signal (a brief light) that predicts an encounter with a rival are more successful in aggressive contests. It may be that conditioned males are the winners because the light (CS) increased the level of androgens, male sex hormones known to heighten aggressiveness in many species. A more aggressive male has a better chance of winning the battle and defending his territory, thereby increasing his chances of mating.

OPERANT CONDITIONING

When a behavior has favorable consequences, the probability that the act will be repeated is increased. This relationship may result because the animal learns to perform the behavior in order to be rewarded. This type of learning has been named **operant conditioning** to emphasize the fact that the animal operates on the environment to produce consequences. The behavior must be spontaneously emitted, not elicited by a stimulus, as it is in classical condition-

ing, and the favorable result, or reinforcement, must follow the behavior closely. The timing of events is critical. In a sense, a cause-and-effect relationship develops between the performance of the act and the delivery of the reinforcer.

B. F. Skinner devised an apparatus used to study operant conditioning in the laboratory. Typically, a hungry animal is placed into a "Skinner box" and must learn to manipulate a mechanism that yields food. For example, a hungry rat placed in a Skinner box will move about randomly, investigating each nook and cranny. Eventually, the rat will put its weight on a lever provided in the box. When the lever is pressed, a bit of food drops into a tray. The rat will usually press the lever again within a few minutes. In other words, the rat first presses the lever as a random act; then, when the action is rewarded, the probability of its being repeated increases.

INSIGHT LEARNING

Insight learning is a sudden solution to a problem without obvious trial-and-error procedures. For example, captive chimpanzees have been known to stack boxes in order to climb up and reach a banana hanging from the ceiling of their cage. One interpretation of the chimps' problem-solving abilities was that they saw new relationships among events—relationships that were not specifically learned in the past—and that they were able to consider the problem as a whole, not just a stimulus–response association between certain elements of the problem. It has been suggested, for instance, that the chimp forms a mental representation of the problem and then mentally applies trial-and-error patterns to the problem.

Other researchers explain sudden problem solving as the result of associations among previously learned components. It has been argued, for instance, that chimps that moved boxes and then climbed on them to reach a banana had previously acquired two separate behaviors: moving boxes toward targets, and climbing on an object to reach another object.

SOCIAL LEARNING

Some organisms are able to learn from others. The possibility for such **social learning** is much greater in social species since they spend more time close to others.

Although each member of a population may have the capacity to learn appropriate responses for himself or herself, it is often more efficient, and perhaps less dangerous, to learn about the world from others. Individuals of some species may learn to avoid dangerous situations by watching their fellow members. For example, rhesus monkeys can learn to fear and avoid snakes by watching other monkeys show fear of snakes. In other species, interaction with adults is *critical* to learning appropriate behaviors by the young. For example, a juvenile wren must perfect its crude rendition of the song used for territorial defense by countersinging with a neighboring adult.

When a tradition spreads through a population, it is not always because animals have learned the trick simply by seeing it done. For instance, a tradition of washing sweet potatoes in the sea was begun by a young Japanese snow monkey, and it spread rapidly to other members of the troop (Fig. 44-6). Although the habit clearly spread throughout the population, we don't know whether the snow monkeys learned to wash food by imitating others or whether they were trained to do so by the caretaker. Because food washing amuses tourists, the caretakers may have given more sweet potatoes to those members of the troop that were known to wash them. The habit may have

spread because the monkeys near those who washed their sweet potatoes (who, by the way, were likely to be relatives) were also close to the caretaker and the source of reinforcement.

Some traditions are due to social learning but not because the individual observed another performing the activity and then imitated it. For example, rats can learn what to eat, not by watching others, but by smelling the breath of others. In one experiment, a "demonstrator" rat ate food flavored with cocoa or cinnamon. The "demonstrator" was then anesthetized and placed 2 inches away from the wire cage of an awake "observer" rat. Although the "demonstrator" slept through the demonstration, the observer later showed a preference for the food the demonstrator had eaten.

PLAY

Although it is easy to spot "play," it is a difficult term to define. One reason that play eludes a simple definition is that there is no specific behavior pattern or series of activities that exclusively characterize play. Play borrows pieces of other behavior patterns, usually incomplete sequences and often in an exaggerated form.

What function does frolicking serve? One hypothesis is that it is physical training for strength, endurance, and muscular coordination. Indeed, it is thought that the sensory and motor stimulation of play causes the formation of a network of synapses in the cerebellum, a part of the brain responsible for sensory-motor coordination. Another hypothesis maintains that play allows individuals to practice social skills, such as grooming and sexual behavior, that are important in establishing and maintaining social bonds. Hatchling sea turtles, for example, take turns vibrating a foot in front of another hatchling's face, a gesture that will later be part of a male's courtship display. Finally, play may be a mechanism for learning specific skills or improving overall perceptual abilities.

Filial Imprinting in Birds

Young chicks, ducklings, and goslings generally follow their mother wherever she goes. How does such following behavior develop? Konrad Lorenz, an Austrian biologist and corecipient of the 1973 Nobel prize, was the first to systematically study this behavior, working with newly hatched goslings. In one experiment, Lorenz divided a clutch of eggs laid by a greylag goose into two groups. One group was hatched by its mother; as expected, these goslings trailed behind her. The second group was hatched in an incubator. The first moving object these goslings encountered was Lorenz, and the goslings responded to him as they normally would to their mother. Lorenz marked the goslings so that he could determine in which group they belonged and placed them all under a box. When the box was lifted, the goslings streamed toward their respective "parents," nor-

FIGURE 44-6 ——————

Snow monkeys washing food. The tradition of washing sweet potatoes was begun by a young Japanese snow monkey and spread rapidly throughout the troop.

mally reared goslings toward their mother, and incubator-reared youngsters toward Lorenz. The goslings had developed a preference for characteristics associated with their "mother" and expressed this preference through their following behavior. The attachment was unfailing, and from that point on Lorenz had goslings following in his footsteps.

Today, the process by which young birds develop a preference for following their mother is called **filial imprinting.** Imprinting is distinguished from other types of learning by several characteristics: (1) It is relatively quick; (2) it occurs only during a limited time (called the **critical period**); and (3) it occurs without any obvious reward. Presumably, the biological function of filial imprinting is to allow young birds to recognize close relatives and thereby distinguish their parents from other adults that might attack them.

Early experience also has important consequences for development of mate preferences in birds. In many species, experience with parents and siblings early in life influences the sexual preferences that form in adulthood. The learning process in this case is called **sexual imprinting.** Sexual imprinting is typically exhibited in the preferences of sexually mature birds for individuals of the opposite sex. One dramatic demonstration of the importance of early experience to subsequent mate preference came from cross-fostering experiments with finches. Eggs of zebra finches were placed in clutches belonging to Bengalese finches. The Bengalese foster parents raised the entire brood until the young were old enough to feed themselves. From then on, young zebra finch males were reared in isolation until they were sexually mature. When later given a choice between a zebra finch female and a Bengalese finch female, zebra finch males courted Bengalese females almost exclusively.

LEARNING AS AN ADAPTATION

▮▶ Many scientists now accept that evolution shapes the learning "styles" of different species to suit their ecological demands. That is, individuals may learn certain things more easily because, in the natural ecological setting, those that do learn have a better chance of surviving and leaving offspring than those that do not.

Species-specific differences in spatial learning and memory among Clark's nutcrackers, pinyon jays, and scrub jays illustrate how ecology and evolution may influence a species' learning skills. These birds are among those that store seeds and recover them later, when food is more difficult to find. To ensure that their seeds won't be stolen by other animals, the birds hide them in small holes they dig in the ground and then cover over. The seeds are cached in this way in the autumn and are used, as needed, throughout the winter and spring. This means that the birds must be able to return to cache sites months after the seeds are buried. Recovery of the seeds is quite an impressive feat for all three species but particularly for Clark's nutcrackers,

which may have as many as 9,000 caches, covering many square kilometers of ground.

If natural selection shapes the learning ability of species, we might predict that those species that depend more heavily on cached food for survival would be better at recovering caches than would other species that are less reliant on cached food. Clark's nutcracker, pinyon jays, and scrub jays are related species (members of the same family) that differ in their dependence on cached food. The nutcrackers live at high altitudes, where they have little else to eat during the winter and spring but their stored seeds. Their winter diet consists almost entirely of cached pine seeds. Nutcrackers prepare for winter by storing as many as 33,000 seeds. Pinyon jays live at slightly lower altitudes, where food is a little easier to find during the winter. Nonetheless, 70 to 90 percent of the pinyon jays' winter diet consists of some of the 20,000 seeds they cached in preparation for winter. In contrast, winter stresses are not as severe for a scrub jay. Scrub jays are smaller than the other two species, so they require less energy to maintain themselves. Furthermore, scrub jays live at much lower altitudes, where food is somewhat easier to find in the winter. Thus, scrub jays store only about 6,000 seeds a year, and these account for less than 60 percent of the winter diet.

A study comparing cache recovery by Clark's nutcrackers, pinyon jays, and scrub jays found species differences that are correlated with their relative dependence on stored seeds. In these experiments, the birds were permitted to store seeds in sand-filled holes in the floor of an indoor aviary. Each bird's ability to remember where it hid its seeds was tested 1 week later. Pinyon jays and nutcrackers, the species that depend most heavily on finding their stored seeds to survive the winter, remembered where they had hidden their seeds more accurately than did the scrub jays.

Unlike innate behaviors, learned behaviors are the result of prior experience. Learned behaviors range from simple reflex responses that change over time to highly complex behaviors, such as those exhibited by chimpanzees who stack boxes to reach bananas. The types of behavior learned by a species is suited to the ecological demands it faces. For example, those birds that depend on stored seeds to survive the winter are better able to remember seed-hiding places than are birds that live in areas where food remains available. (See CTQ #2.)

DEVELOPMENT OF BEHAVIOR

During the development of a behavior, there is an intimate relationship between the behavior's genetic component and its learned component. Although genes and experience work together throughout the organism's development to

FIGURE 44-7 ───────

Song development in this species depends on both genes and experience. A young male must learn to sing its song by hearing the song of adult males. However, it inherits a "template" of its own species' song.

produce most behaviors, as the organism changes over time, the nature of the interaction will vary. For example, as a result of its genetic makeup, an animal may be more or less responsive to certain environmental stimuli at different times of its life. Conversely, the presence or absence of certain stimuli at critical times during development may alter the expression of the genes.

The interaction of genes and learning is seen in the development of the song of a male white crowned sparrow (Fig. 44-7). Young males must learn to sing by hearing adult males sing. However, if a young male white-crowned sparrow is isolated from other members of its species and is allowed to hear recordings of bird songs, including one of its own species, it will learn to sing correctly. How does the sparrow know which song is correct? It must have inherited a template, or image, of its species' song. The template allows the male to recognize his own song, but the experience of hearing the song is also needed before the male can sing properly.

Most complex behaviors cannot be divided neatly into either the innate or learned categories but, rather, are influenced by both genetics and prior experience. (See CTQ #3.)

EVOLUTION AND FUNCTION OF BEHAVIOR

▐▶ Ecological conditions will determine which traits are favored during evolution. Therefore, as a result of natural selection, we would expect traits of individuals to become better fitted to the environment from generation to generation. The traits that allow individuals to survive and reproduce better than their competitors can are called adaptations. When we ask questions about the adaptiveness of behavior, we are asking about its survival value. The aim in answering such a question is to understand why those animals that behave in a certain way survive and reproduce better than do those that behave in some other way.

Studies of parental behavior among species of gulls provide an example of how behavior may be shaped by natural selection to suit the environmental conditions. Kittiwakes, which have low predation rates, leave eggshell pieces in the nest, while ground nesting gulls, which have high predation rates, generally remove broken eggshells. Niko Tinbergen and his colleagues experimentally tested the hypothesis that eggshell removal reduced predation on chicks. It was noted that the eggs, chicks, and nest were camouflaged and might be difficult for a predator to spot. However, the bright white inner surface of a piece of eggshell might catch a predator's eye and reveal the nest site. The experimenters painted some blackheaded gull eggs white to test the idea that white eggs might be more vulnerable to predators than would the naturally camouflaged eggs. The difference in predation rates supported the hypothesis that the white inner surface of egg shell pieces might endanger nearby eggs or chicks. Tinbergen then painted hen's eggs to resemble those of a gull and placed white pieces of shell at various distances from some of the nests. The broken egg shell bits did attract predators. Furthermore, the risk of predation decreased with increasing distance of eggshell pieces from the nest. Thus, fastidious parents leave more offspring to perpetuate their genes. Eggshell removal is clearly adaptive.

The timing of eggshell removal is also adaptive. Whereas oystercatchers remove eggshell pieces almost immediately after a chick hatches, a blackheaded gull stays near its chick an hour or two after hatching before removing the eggshell pieces. The difference in the timing of eggshell removal between these species is related to their nesting habits. Oystercatchers generally nest alone, so neighbors do not pose a threat. In contrast, blackheaded gulls live in colonies, and their chicks are commonly eaten by neighboring gulls. Furthermore, a newly hatched chick is wet and easier to swallow than is one that has dried and become fluffy. Thus, although removing the shells reduces predation by other species of birds, delaying removal until the chicks are dry decreases the likelihood of the chick's being cannibalized by neighboring gulls while its parents are away from the nest.

By focusing on the survival value of adaptation, we have considered only the benefits gained through certain behaviors. But most actions also have costs. *Optimality theory* views natural selection as "weighing" the costs and benefits of each available alternative. First, natural selection translates all costs and benefits into common units: fitness. Then it chooses the behavioral alternative that maximizes the difference between costs and benefits. The choice that maximizes fitness is the alternative that would continue into the next generation.

OPTIMALITY AND TERRITORIALITY

A **territory** is an area that is defended against intruders, generally in the protection of some resource. Optimality theory predicts that an individual should be territorial if the benefits from enhanced access to the resource are greater than is the cost of defending the resource. The benefits of territoriality include increased mate attraction, decreased predation, protection of young and/or mates, reduced transmission of disease, and a guaranteed food source. The costs of defending the territory include energy expenditure, risk of injury, and increased visibility to predators.

Optimality theory predicts that territoriality will evolve only when the benefits exceed the costs. On the one hand, if

Marine iguanas on the Galapagos Islands. Optimality theory predicts that territoriality will evolve only when its benefits exceed its costs. Female iguanas defend nest sites on Hood Island, where suitable sites are rare, but not on other Galapagos Islands, where nest sites are abundant.

the resource is scarce, an individual may not gain enough to pay the defense bill; it may be economically wiser to look for greener pastures. Accordingly, the golden-winged sunbird will abandon a territory when it no longer contains enough food to meet the energy costs of daily activities as well as defense. On the other hand, if there is more than enough of the resource to go around, it would be energetically wasteful and economically unsound to defend it. Water striders are among the species that will cease to defend territories if supplied with abundant food. Likewise, we find that female marine iguanas don't bother defending territories with nest sites on most of the Galapagos Islands (Fig. 44-8); they defend territories only on Hood Island, the only Galapagos island where nest sites are in short supply.

We have been discussing optimality as if an individual's reproductive success depends solely on its own behavior. There are situations, however, in which the individual is not the sole master of its fate; instead, the reproductive success gained by behaving a certain way depends on what the other members of the population are doing. Such an optimal course of action is called an **evolutionary stable strategy (ESS).** By definition, an ESS is a strategy that cannot be bettered and, therefore, cannot be replaced by any other strategy when most of the members of the population have adopted it. An ESS cannot be bettered because when it is adopted by most of the members of the population, it results in maximum reproductive success for the individuals employing it. As a result, an ESS is both unbeatable and uncheatable.

ESS theory has improved our understanding of the logic of animal combat, a situation in which the best strategy depends on what competitors are doing. Consider, for instance, aggressive games in which there are only two strategies that might be employed. The "hawk" strategy is to fight to win. A hawk will continue to fight either until it is seriously injured or its opponent retreats. The "dove" strategy is to display but never engage in a serious battle. If a dove is attacked, it retreats before it can be injured seriously. Many factors will determine a strategy's success. As the cost of fighting increases, the dove strategy should be favored because the potential cost of injury is greater in species that possess weapons, such as horns or sharp teeth. Therefore, the hawk–dove model predicts that contests between members of well-armed species should rarely escalate from display to battle (Fig. 44-9). Conversely, the hawk strategy should be more common when the risk of injury is low. Accordingly, animals without weapons, such as toads, fight fiercely. In this case, although some individuals are injured, the risk of injury is much less than it would be if the individuals had the prowess of lions.

It seems reasonable to assume that the intensity of fighting should increase with the value of the resource. Offspring—a direct measure of fitness—are obviously of great value. It is not surprising, then, that a mother will fight fiercely to defend her young. Similarly, male elephant seals, for instance, fight brutal and often bloody battles for the right to mate. Battles are titanic because a male's entire reproductive success is at stake. All matings are performed by a few dominant males who defend harems of females. The duels between males are so strenuous that a male can usually be harem master for only a year or two before he dies.

Behaviors are shaped by natural selection to suit the environmental conditions in which the species lives. Such behaviors are adaptive because they increase the chances of survival and reproduction of its members. Even though a particular behavior may have costs, such as an increased risk of predation or the expenditure of energy, the benefits from that behavior are presumed to exceed the costs; otherwise, the behavior would not have evolved. (See CTQ #4.)

FIGURE 44-9

Optimality theory predicts that fights are not likely to escalate when the cost of injury is great. These males could kill one another, but they generally do not bite.

SOCIAL BEHAVIOR

Not all animal species live in groups. Tigers, for example, may live and hunt alone in large territories. Furthermore, not all groups are social (as anyone who has experienced a rush hour traffic jam will testify). Sometimes animals that live alone come together only to share a vital resource, such as a water hole in the desert. But some groups are social. We may wonder, then, what makes the difference?

COSTS AND BENEFITS OF GROUP LIVING

☀ Soon after hatching, a *Holocnemus pluchei* spiderling is faced with a momentous "decision"—build a web of its own or join large individuals of its kind on an existing web. The small spiderlings are generally outcompeted for food by larger individuals whose web they share. Indeed, they catch fewer prey items than do solitary spiderlings. It is energetically expensive to build a web, however, and spiderlings add little or no silk to an existing web. The spiderlings must weigh the value of food lost against that of silk saved.

Living in groups, then, has both costs and benefits. Among the costs may be increased competition for mates, nest sites, or food. Other costs include increased exposure to parasites or disease, increased conspicuousness to predators or prey, and an increased risk of wasted energy in raising offspring that are not one's own.

Group living also has many benefits, particularly those relating to eating and not being eaten. Some predators, for example, engage in cooperative hunting. As a result, they have a higher capture success rate and average energy intake per individual than do solitary hunters. For example, groups of predatory jackfish may work together in catching small prey. Some members of the group form a circle around swarms of prey while other fish make strikes into the midst of the swarm.

In addition, groups are usually more successful than are individuals at detecting, confusing, and repelling predators. With all the members of the group on alert, a predator is likely to be spotted. Many species call to warn others in the group of a predator's approach. Vervet monkeys have a particularly sophisticated warning system. Different calls warn of the approach of an eagle, a leopard, or a poisonous snake. This system makes sense because the best manner of escape would differ depending on whether the predator were striking from the air or land.

During an attack, an individual within a group has a smaller chance of becoming the next meal. In addition, since predators find it easier to attack a single individual, they often swoop at a group, causing the individuals to scatter. Then one member may be singled out. It isn't surprising then that many prey species, such as flocks of starlings or schools of fish, counter this "divide and conquer"

predator strategy by forming even tighter clusters when a predator is detected.

Many prey species will engage in mobbing behavior, whereby adults harry a predator that is frequently far superior to any individuals of the group. For example, small birds will frequently mob a predator, such as an owl. Baboons and chimpanzees use a similar strategy against leopards. Screaming, the baboons or chimpanzees charge and retreat; they may even throw sticks at the leopard. Finally, living in groups may aid in defending young, food, or space against other members of one's species.

An increase in the density of reproducing animals may also facilitate reproduction by providing mutual stimulation. The sights, sounds, and smells of other courting individuals appears to enhance and synchronize breeding. It is thought that this may have been a factor in the evolution of group courtship displays. Such group "advertising" is observed in frogs and insects. Although it is relatively rare in higher vertebrates, the *lek* displays of some bird species, such as the sage grouse, may be thought of in this way as well. In a lek display, as many as 50 or 60 male sage grouse may congregate at dawn. As they congregate for a communal display, the males may stimulate the production of reproductive hormones in one another as well as in the females about to choose a mate from among the madly displaying cocks.

ANIMAL COMMUNICATION

Communication is essential to social life. It underlies cooperative as well as aggressive behaviors.

Reasons For Communication

There are numerous functions for communication. Species recognition is important to ensure that reproductive efforts are not wasted on members of the wrong species and that aggression is directed toward those individuals who are competing for the same resources.

Sexual reproduction is often dependent on communication. First a mate must be attracted and then courted. During courtship, the male often advertises himself, and the female assesses his qualities. Aggression is reduced, and the behavior and physiology of the mates is coordinated. Following mating, the offspring may communicate its desire to be fed, or the parent may have to indicate its willingness to feed the young.

Communication is also important in aggressive interactions. Most animal disputes are settled without fighting; rather, the animals threaten one another with stereotyped displays. Threat displays may allow competitors to assess one another's abilities. In this way, the weaker individual may accept its loss and leave without risking injury. The display is often tied to a physical characteristic, such as size or strength, that cannot be faked. Thus, rivals are kept honest. Male red deer stags, for example, challenge each

FIGURE 44-10
Male red deer roaring. Male red deer challenge one another in a roaring contest. Because roaring is strenuous, it provides an honest cue for assessing a rival's fighting ability.

other in a vocal duel in which each male takes a turn roaring at the other (Fig. 44-10). As the pace of the bellowing increases, one contestant usually gives up. Because roaring is strenuous, it is a reliable indicator of the rival's fighting ability.

Other signals allow animals to keep in touch with one another. Sometimes, it is important to remain in contact so that future association is possible. At other times, signals bring animals together immediately. *Recruitment* occurs when individuals are brought together to perform a specific duty, as occurs in social insects. For example, fire ants leave an odor trial that will guide recruits to a food source.

Honeybees have a remarkable system of recruitment, which was discovered by the Austrian biologist Karl von Frisch, the third corecipient of the 1973 Nobel prize. When a scout bee finds food, she recruits nestmates to help in the harvest by dancing on the vertical surface of the comb. When the food is close to the hive, the bee performs a *round dance,* in which she walks in circles, first to the right and then to the left (Fig.44-11a). By feeling the dancer with their antennae, other bees follow the dance. The dancer informs recruits that food is within a certain distance of the hive. The *waggle dance* is performed when food is farther from the hive (Fig. 44-11b). A path similar to a figure eight

is traced on the comb. During the straight part of her run, the dancer vigorously shakes her abdomen. The speed of the dance, the number of waggles on the straightaway, and the duration of the bee's buzzing sound correlate well with the distance to the food source. The orientation of the straight part of the run indicates the direction of the food source. The angle a forager leaving the hive must assume with the sun as it flies to the food source is indicated by the angle of the waggle run relative to gravity. If the scout bee dances straight up on the comb, recruits should fly directly toward the sun. A dance oriented straight downward indicates a food source directly away from the sun. To indicate other directions, the dancer changes the angle of the waggle run relative to gravity so that it matches the angle between the sun and the food source.

Types of Communication Signals

Any sensory modality may be used for communication. The particular sensory modality that has been naturally selected during evolution to be used to communicate in a given species will have been influenced by the nature of the message as well as by the ecological situation of the organism. What factors may influence the nature of the communication? One factor is how well the animal sending the

(a)

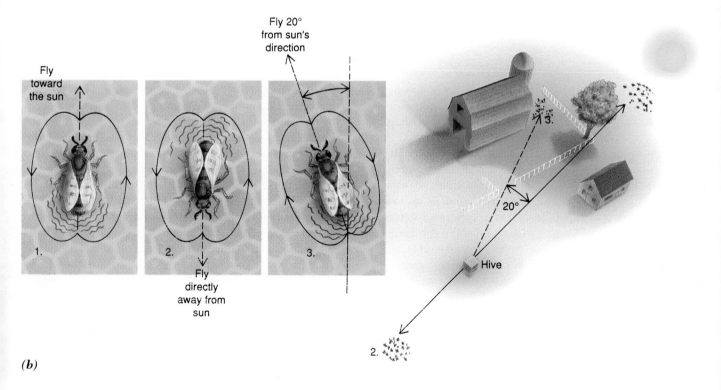

(b)

◀ F I G U R E 44-11 ─────────────

Round dance of honeybee scout. *(a)* After finding food close to the hive, a scout returns and does a round dance on the vertical surface of the comb. The dance consists of circling alternately to the left and right. The dance informs recruits that food can be found within a certain distance of the hive. *(b)* Waggle dance of honeybee is performed when a scout finds food at some distance from the hive. The dancer traces the pattern of a figure eight and waggles her abdomen during the central straight part of the dance. Aspects of the dance correlate with the distance and direction to the food source. The dancer indicates direction by the orientation of her waggle run relative to gravity. The direction in which recruits should fly is equal to the angle of the waggle run relative to vertical. When the food is in the direction of the sun, the dancer waggles straight up. When the recruits should fly directly way from the sun, the waggle run is oriented straight down. A food source located 20 degrees to the right of the sun would be indicated by a dance oriented so that the waggle run were 20 degrees to the right of vertical.

message can be localized. Visual signals that may be enriched with color and brightness are easiest to localize. Therefore, visual displays are frequently used by animals that are active during the day for short-range communication in open environments. Sound can also be localized and is generally employed by animals that are active at night or that live in dense vegetation, where vision is limited. As an example, consider the melodies of insect and amphibian calls that are so obvious on summer nights in the country.

Another important consideration is the distance over which the signal is effective. Pheromones—chemical signals secreted by animals which influence the behavior of other members of the same species—may be the best signal when long distance communication is desired. Perhaps the best known pheromones are sex attractants, such as those produced by moths. For instance, the gypsy moth sex attractant is carried by the wind and may attract males more than a mile away from the female. Sound is not usually as good as are chemicals for long-distance communication, particularly when the sound must be transmitted through the air. Sound travels much farther in water, however. The songs of whales can be heard hundreds of miles away, for example. Light is rapidly attenuated in water, so if a signal must be sent over a distance, visual displays are generally not used by aquatic animals. Likewise, in any environment where long-distance vision is limited, such as in a dense forest, auditory signals are often employed.

A third important factor is the duration of the signal. Chemical signals are generally among the most durable, which is why chemicals are often used to mark territories. Not all chemicals are long-lived, however; some, like the alarm signals of some ants, fade within 30 seconds.

Bodily Contact

Social bonds may be cemented by physical contact. Some species of animals reduce tensions by touching. Greeting ceremonies are common in social animals and often involve touching and sometimes even embracing. It is not uncommon to observe even nonhuman primates sitting with their arms around one another. Some animals, such as black-tailed prairie dogs, may kiss to establish community membership. Sea lions and chimps may kiss one another as a greeting. The members of a wolf pack may surround the dominant male and lick his face and poke his mouth with their muscles. This ceremony occurs at those times when it is useful to reinforce social ties, such as when the wolves awake in the morning, when they have been separated for a period of time, and when they are ready to go out on a hunt (Fig. 44-12).

F I G U R E 44-12 ─────────────

The wolf greeting ceremony illustrates the importance of bodily contact in cementing social bonds.

FIGURE 44-13
Grooming in primates. Although grooming originally functioned only for skin care, its social functions have now become more important. Primates spend a major portion of their day grooming; it helps form and maintain social bonds.

Social grooming (Fig. 44-13) is a form of social behavior that is found in a variety of animals, but it is especially prominent among higher primates. One effect of such behavior is to rid the animal being groomed of parasites, hardened skin secretions, and debris on the skin. Equally important, grooming allows animals to cement social bonds.

Living in a group may provide significant benefits for an individual over a solitary lifestyle. Group living provides safety in numbers, available mates, and an increased opportunity to obtain food. Social life depends on communication between its members, allowing essential activities to be coordinated. Different types of signals—visual, auditory, tactile, or chemical—may be suited for communicating different types of information in different environments. (See CTQ #5.)

ALTRUISM

Altruism is the performance of a service that benefits a *conspecific* (another member of the species) at some cost to the *altruist* (the one who does the deed). Strictly speaking, the benefits and costs of altruism are measured in units of fitness (the reproductive success of a gene, organism, or behavior). Since changes in fitness are nearly impossible to ascertain, however, the gains and losses are usually arbitrarily defined by researchers as certain goods or services that seem to influence the participants' chances of survival.

On the surface, the existence of altruism seems to contradict evolutionary theory. Evolution involves a change in the frequency of certain alleles in the gene pool of a population. If aiding a conspecific costs the altruist, then the altruist should be less successful in leaving offspring that bear copies of its alleles than are the recipients of its services. As a result, the alleles for altruism would be expected to decrease in the population.

The hypotheses for the evolution of altruism can be arbitrarily classified into several overlapping classes:

1. *Individual selection.* The general thrust of these hypotheses is that, when the interaction is examined closely enough, the altruist will be found to be gaining, rather than losing, as a result of its actions. The benefit may not be immediate; sometimes the gain is in the individual's *future* reproductive potential.

2. *Kin selection.* One of the most cogent explanations for altruism is kin selection. Kin selection is based on the theoretical work of the British geneticist W. D. Hamilton. Central to kin selection is the idea of **inclusive fitness** which considers all the adult offspring of an individual as well as those of its relatives that are alive because of the actions of that individual. If family members are assisted in a way that increases their reproductive success, the alleles that the altruist has in common with them are also duplicated, just as they would be if the altruist reproduced personally. In other words, since common descent makes it likely that a certain percentage of the alleles of family members are identical, assisting kin is another way to perpetuate one's own alleles. Since an individual shares more alleles with certain relatives by common descent, the possibility of genetic gain increases with the closeness of the relationship. For example, aiding a cousin, who shares an average of only one-eighth of the same alleles, is less productive than is assisting a brother or sister, who is likely to share half of its alleles with the altruist. Thus, the fitness gained through family members must be devalued in proportion to their genetic distance (diminished relatedness).

How can relatives be identified? There are several possibilities. One way might be to use location as a cue: The individuals who share one's home are likely to be kin. Alternatively, individuals might be identified as kin because they are recognized from prior social contact or as a result of their association with a known relative. Another possible way to recognize a relative is by certain traits that characterize family members. In other words, an image of a family member may be developed that is matched or compared to the appearance of a stranger. Finally, recognition may be genetically programmed. Perhaps there are alleles that, in addition to labeling relatives with a noticable characteristic, may cause the altruist to assist others who bear the label.

3. *Reciprocal Altruism.* Altruism may evolve—despite the initial cost to the altruist—if the service is repaid with interest by other members of the population. In other words, altruism will be favored if the final gain to the altruist exceeds the initial cost. In order for reciprocal altruism to work, however, individuals who fail to make restitution must be discriminated against. Because of this latter requirement, certain conditions make reciprocal altruism more likely in particular species: First, there should be a good chance that an opportunity for future repayment will arise; second, the individuals must be able to recognize one another.

There is no one explanation for the evolution of altruism that applies to every example. Different life histories and ecological conditions may alter the relative importance of a particular evolutionary mechanism. Thus, similar behaviors may evolve by different mechanisms in different species. Furthermore, as we will see, the mechanisms suggested are not mutually exclusive and may be working simultaneously. Let's look at several examples of altruism.

ALARM CALLS

Belding's ground squirrels are often victims of aerial predators, such as hawks, or terrestrial predators, such as coyotes, long-tailed weasels, badgers, and pine martens. The alarm calls warning of these classes of predators are different, and the selective forces behind the evolution of these two types of alarm calls appear to be different. Whereas the alarm calls warning of aerial predators appear to promote *self-*preservation, those warning of terrestrial predators do not. When a hawk is spotted overhead or when an alarm whistle is heard, near pandemonium breaks out in a Belding's ground squirrel colony. Following the first warning, other squirrels whistle a similar alarm, and all scurry to shelter. As a result, a hawk is rarely successful in attracting a group of Belding's ground squirrels. In those cases where the hawk is successful, the victim is most likely to be a noncaller. Thus, it seems that the alarm whistles given at the sight of a predatory bird directly benefit the caller by increasing its chances of escaping predation. This behavior would appear to be an example of individual selection.

In contrast, kin selection seems to be behind the evolution of ground squirrel alarm trills, which are issued in response to terrestrial predators. In this case, the caller is truly assuming risk; significantly more callers than non-callers are attacked. However, those saved by the warning are likely to be the caller's relatives. Because daughters tend to settle and breed near their birthplace, the females within any small area are usually genetically related to one another. In contrast, the sons set off independently before the first winter hibernation. When a terrestrial predator appears, females are more likely to sound an alarm than are males. This is consistent with kinship theory since females are more likely to have nearby relatives who would benefit from the warning. Reproductive females are even more likely to call than are nonreproductive females. Furthermore, reproductive females with living relatives call more frequently than do reproductive females with no living family!

HELPING

Another form of altruism is helping. A helper is an individual who assists in the rearing of offspring that are not its own, usually by providing food or by protecting the young. In most species, helpers are offspring who are helping their parents raise their siblings. Thus, kin selection seems to be a reasonable explanation for these occurrences of helping. The helpers are not always relatives, however. In these cases, helping may be a means of maximizing individual fitness in the future. Helping commonly accompanies ecological conditions that make reproduction difficult or costly. Under such conditions, helping may be a means of obtaining permission to remain in a high-quality territory, of maintaining group or territory cohesiveness, of earning the future assistance of those helped, of obtaining a mate, or of protecting young from predators.

COOPERATION IN MATE ACQUISITION

Another apparently altruistic behavior is helping another individual acquire a mate. Both individual selection and kin selection seem to have played a role in the evolution of this behavior among lions. The males of a pride are usually related. The group of males, called a coalition, generally consists of brothers, half-brothers, and cousins who left their natal pride as a group. These lions challenge the males of other prides (Fig. 44-14). In such contests, the larger coalition usually wins, and the reward is a harem of lionesses. Females of a harem often come into reproductive condition simultaneously. During the 2 to 3 days of estrus (page 668), any of the males in the coalition may be the first to mate with a female and thereby gain fitness directly. When another male takes over, it is likely to be a relative. At that point, the first male may still gain fitness indirectly.

◁ B I O L I N E ▷
Animal Cognition

Many people have wondered what it is like to be an animal. Do nonhuman animals have thoughts or subjective feelings? Such musings have led some investigators to consider whether or not nonhuman animals are cognitive, conscious, aware beings.

Donald Griffin has suggested that tapping animal communication lines is a way that we might find out whether animals have conscious thoughts or feelings. After all, the only way we know about the thoughts or feelings of other people is when they *tell* us what they are thinking or feeling, through either verbal or nonverbal communication. So, if nonhuman animals have thoughts and feelings, they probably communicate these to others via their communication signals as well. If we could learn to speak their language, we could "eavesdrop" and thereby glimpse into the animal mind.

Most people agree that one sign of cognition is the ability to form mental representations of objects or events that are out of sight. We might ask, then, whether animal signals are symbolic; that is, whether they refer to things that are not present. Certain apes can learn a language that uses symbols. Kanzi, the pygmy chim-

panzee, for instance, can communicate by using a computer keyboard that has over 250 symbols, called lexigrams. In addition, Alex, an African gray parrot, is able to request more than 80 different items vocally, even if they are out of sight. In addition, Alex can quantify and categorize those objects. He has shown an understanding of the concepts of color, shape, and same versus different with both familiar and novel objects. But these animals have been taught to use language.

Are the communication signals that animals use in nature symbolic? The apparent simplicity of this question is deceptive because observations are often open to alternative interpretations. For example, the waggle dance of the honeybee that was described earlier is symbolic in that it contains information regarding the distance and direction to a distant food source. However, most scientists agree that this behavior is not evidence of thought since the dances are genetically preprogrammed; bees can perform and understand dances without previous experience.

Another sign of cognition might be whether or not animals adjust signals according to conditions at the time. One way we might see such an adjustment in signal-

ing is if an individual determined whether or not to signal on the basis of the composition of its audience. In other words, if an individual sees a predator, does it always sound an alarm, or does this action depend on the individual's present company? The company does seem to affect the likelihood of alarm calling among domestic chickens. A cock is more likely to call in the presence of an unaltered companion. This has been interpreted by some as an indication that the cock chooses whether or not to call. The choice of calling or withholding the call would be taken as evidence of cognition.

Learning studies may also shed some light on the issue of animal cognition. Some scientists believe that insight learning shows that the animal is *thinking,* an animal that thinks about objects or events can be said to experience a simple level of consciousness. An animal that thinks must also form mental representations of objects or events. Therefore, insight has been used as evidence of animal awareness or cognition. But not everyone agrees that animals, or even *some* animals, might be aware. Some might be willing to accept the idea of awareness in a chimp but not in a pigeon that shows similar behavior.

Roughly half of all male coalitions contain at least one unrelated male. Unrelated males may be accepted because the larger the coalition, the greater a male's reproductive success. Larger coalitions have a better chance of ousting the current coalition in a pride, of maintaining control of that pride, and perhaps even of gaining residence in a succession of prides. A solitary male has little chance of reproducing and, therefore, much to gain by joining another coalition. A small coalition may also gain by accepting an unrelated male because the extra member may help it take over prides.

FOOD SHARING

☀ Vampire bats share food with needy familiar roostmates even if they are not related. This generosity may mean the difference between life and death for the recipient. If a vampire bat fails to find food on two successive nights, it will starve to death, unless a bat that has successfully fed regurgitates part of its blood meal. The hungry bat begs for food first by grooming, which involves licking the roostmate under the wings, and then by licking the donor's lips. A receptive donor will then regurgitate blood. The regurgitated food must be plentiful enough to sustain the bat until the next night, when it may find its own meal. Although the benefit to the recipient is great, the cost to the donor is small. Since a bat's body weight decays exponentially following a meal, the recipient may gain 12 hours of life and, therefore, another chance to find food. But the donor loses less than 12 hours of time until starvation and usually has about 36 hours—another 2 nights of hunting—before it would starve.

Generally, only individuals who have had a prior association share food. In one experiment, a group of bats was formed from two natural clusters in different areas and maintained in the laboratory. Aside from a grandmother and granddaughter, all the bats were unrelated. The bats were fed nightly from plastic measuring bottles in order to determine the amount of blood consumed by each bat. Each night, one bat was chosen at random, removed from the cage, and deprived of food. When it was reunited with its cagemates the following morning, the hungry bat would beg for food. In almost every instance, blood was shared by a bat that came from the starving bat's population in nature. Furthermore, there seemed to be pairs of unrelated bats that regurgitated almost exclusively to one another, suggesting a system of reciprocal exchange.

EUSOCIALITY

Eusocial species are those that have sterile workers, engage in cooperative care of the young, and have an overlap of generations so that the colony labor is a family affair. The eusocial insects, (e.g., ants, bees, and wasps) behave altruistically in several ways: Food is shared; those colony members specialized for defense often die performing their duty; and some members of the colony are sterile but care for the young of the colony's royalty.

Eusociality is also found in African naked mole-rats, burrowing rodents that look like rats but act like moles. Naked mole-rat societies are similar to honeybee societies in several ways. First, in both societies, breeding is restricted to a single female, the queen. Second, colonies contain overlapping generations of offspring. Third, there is differentiation of labor among individuals within the colony. During the first 12 or so days after a honeybee emerges from her pupal case, she specializes almost entirely on cleaning nest cells. For the next week, she performs a variety of tasks associated with brood and queen care, nest maintenance, and food storage. When she is about 3 weeks old, she begins to gather pollen and nectar. The duties assumed by the nonbreeding members of mole-rat colonies seem to depend on both their size and age. The duties of the smaller members generally include gathering food and transporting nest material. As they grow, they begin to clear the elaborate tunnel system of obstruction and debris. Larger members dig tunnels and defend the colony.

Both honeybees and naked mole-rats may have been predisposed to a eusocial lifestyle by a close genetic relationship among colony members along with ecological factors that maximize the benefits of group life or the costs of dispersal. The close genetic relationship among honeybees results from their system of sex determination, called **haplodiploidy,** in which fertilized eggs develop into females

F I G U R E   44-14

Male lions from different prides. Coalitions of male lions, most of them relatives, fight other coalitions for control of a harem.

and nonfertilized eggs develop into males. Haplodiploidy results in a closer relationship between sister workers and their siblings than between females and their own offspring. The female workers are likely to share 75 percent of their alleles with the reproductively capable siblings they helped to raise, but they share only 50 percent of their alleles with the offspring they produced. As a result, a female worker makes greater gains in inclusive fitness by raising siblings than she would by producing offspring.

There is also an unusually high degree of genetic relatedness among the members of any single mole-rat colony. This similarity is thought to be a consequence of the extreme inbreeding within a colony. Naked mole-rats live for about 15 years and are prolific breeders. Generally, only two or three males mate with the queen mole-rat, who may give birth to a litter consisting of up to 12 young every 70 to 80 days. Furthermore, there is little mixing of genes between colonies. In fact, members of different colonies are quite aggressive toward one another; intruders may even be killed.

The high degree of genetic relatedness cannot, by itself, explain the evolution of eusociality in mole-rats. Another predisposing factor may be the mole-rats subterranean lifestyle. Within their underground tunnels, naked mole-rats are fairly safe from predators. Furthermore, the dry regions they inhabit have many plants with subterranean roots, tubers, and bulbs. Since naked mole-rats feed primarily on these parts of the plants, they need not leave the safety of the tunnel system to forage. The tunnel system can also be expanded easily as the colony grows. Finally, dispersal is risky because the tubers and bulbs the mole-rats eat are distributed unevenly throughout the habitat. The natal colony may have access to a patch of food, but a group that sets off on its own may have to burrow extensively before encountering another rich area. Burrowing uses quite a bit of energy so members of a small group might die of starvation before locating a new food resource.

The discovery that, in certain species, individuals will exhibit behaviors that help other members of the species, even if that behavior increases the risk to themselves, at first seemed to contradict accepted evolutionary theory. Several explanations for the evolution of altruistic behavior were formulated and are now generally accepted. Altruistic behavior reaches its greatest expression among eusocial species, such as bees, wasps, and naked mole-rats, where many or all of the members of a colony are closely related. (See CTQ #6.)

REEXAMINING THE THEMES

Relationship between Form and Function

The form of an object—the shape of an egg or the color of a part of the body—may play a key role in triggering a particular innate behavior. For example, a tuft of red feathers introduced into the territory of a male red robin provides a sufficient sign stimulus to evoke a "full-blown" attack by the territory's occupant. Similarly, the sight of a female stickleback in a "head-up" posture with an exposed, egg-swollen abdomen, or a facsimile thereof, provides the sign stimulus for triggering the courtship dance of a male member of the species. Similarly, the form of a communication signal—whether visual, vocal, chemical, or tactile—can be correlated with the function of the signal and the ecology of the organism.

Acquiring and Using Energy

Many diverse behaviors exhibited by animals can be explained on the basis of conservation of energy. For example, a clamworm that stops withdrawing into its burrow with every passing shadow saves energy that would otherwise be wasted. Similarly, a male bird or mammal that defends a territory against outside invaders is able to utilize the resources available on that territory to feed itself and any mate that he might attract. Some types of social behavior may be explained, at least in part, in terms of energy conservation. A small spiderling, for example, saves energy if it shares a web with other individuals of its species rather than bearing the considerable expense involved in constructing a web of its own.

Evolution and Adaptation

As in the case of anatomic traits or physiologic activities, specific behaviors evolve because they are adaptive; that is, they increase the likelihood of an individual's surviving and reproducing. The adaptive behavior of simpler behaviors, such as habituation, which prevents an animal from repeatedly continuing an inappropriate physiologic response, or filial imprinting, which increases the likelihood that a newly hatched bird will remain in the company of a parent, is usually evident. While the adaptive quality of altruistic behaviors is less evident, these behaviors are also thought to provide selective advantage, if not always for the individual that displays the behavior, then for close relatives that share the same genes.

SYNOPSIS

Tinbergen's four questions about animal behavior ask about the mechanisms that underlay the behavior, its development, its survival value and its evolution.

Behavior is adaptive, whether it is primarily innate or learned. Innate behaviors are largely controlled by genes and are generally stereotyped species-specific actions. An FAP is an innate behavior that, once triggered, will continue to completion without further stimulation. The stimulus for an FAP, called a sign stimulus or a releaser, is usually only a small part of the total environmental situation. One FAP may bring an animal into a stimulus situation that triggers another FAP. The sequence of FAPs that results is called a chain of reactions.

Learning is a process through which an animal benefits from its experience. Habituation is a simple form of learning, whereby the animal learns not to show a characteristic response to a stimulus because the stimulus was shown to be unimportant during repeated encounters. Thus, habituation is a mechanism that focuses attention on important aspects of the environment. In classical conditioning, the animal learns a new association between a stimulus and response. Classical conditioning begins with an unlearned association between an unconditioned stimulus (US) and a response. A new stimulus repeatedly occurs immediately before the US. Eventually, the new stimulus alone is sufficient to cause the response. The new stimulus is then called a conditioned stimulus (CS), and the response is called a conditioned reflex. In operant conditioning, the probability of an animal repeating an action increases because the action met with favorable consequences. Insight learning is the sudden solution to a problem without obvious trial-and-error procedures. Social learning occurs when an animal learns from others. A proposed function of play is to serve as a mechanism for learning specific skills or improving overall perceptual abilities. Imprinting is a form of learning that occurs quickly during a restricted interval of time, called a critical period, and without any obvious reinforcement.

Many scientists accept that the inherited potential for learning is an adaptation. Studies comparing the spatial memory of three species of seed-caching birds support this hypothesis. Pinyon jays and Clark's nutcrackers rely more heavily on cached seeds to survive the winter than do scrub jays; the former species also remember where they hide their seeds more accurately than do scrub jays.

Genes and experience interact during development to produce each behavior. Song development in male white-crowned sparrows provides a good example of this interaction. Young male white-crowned sparrows must learn the correct song by hearing adult males of its species sing. The young male inherits a template, or image, of its species' song that allows him to recognize the correct song, but the experience of hearing the song is needed for its development.

Comparisons of parental behavior among certain gull species have supported the hypothesis that behavior is adaptive. The white inner surface of eggshell pieces may catch a predator's attention. Species, such as the black-headed gull, that experience high predation rates remove eggshell pieces from the nest after the chicks hatch, but those that experience low predation rates, such as kittiwakes, leave eggshell pieces in the nest. Furthermore, oystercatchers, which generally nest alone, remove the eggshell pieces immediately. However, blackheaded gulls, which nest in colonies, and may cannibalize the chicks of absent neighbors, delay removing the eggshell pieces until the chicks have dried and are more difficult to swallow.

Optimality theory views natural selection as "weighing" the costs and benefits of each available behavioral alternative. For example, optimality theory predicts that an animal would defend a territory when the benefits from enhanced access to the resource within the territory are greater than the costs of defending the territory.

When an individual's reproductive success depends on what others in the population are doing, the optimal course of action is called an evolutionarily stable strategy (ESS). An ESS is a course of action that cannot be bettered and, therefore, cannot be replaced by any other strategy when most of the members of a population have adopted it.

Communication is essential to social life. The functions of communication include: species recognition, mate attraction and courtship, aggressive interactions, and recruitment. Some displays allow individuals to assess the qualities of a potential mate or competitor. Any sensory channel can be used for communication. The evolutionary choice of a sensory channel for a particular communication signal will be influenced by both the nature of the message and the ecological situation of the organism.

Altruism is the performance of a service that benefits a conspecific at some cost to the altruist. Hypotheses for the evolution of altruism can be grouped into several overlapping classes: (1) individual selection, in which there is a net gain directly to the altruist; (2) kin selection, in which the altruist gains indirectly through increased reproductive success of the altruist's relatives; and (3) reciprocal altruism, in which the altruist is later repaid for its service by other members of the population. These hypothetical mechanisms for the evolution of altruism may work simultaneously, and none applies to every example of altruism. There are several actions that are often considered to be examples of altruism. Many species alert their companions to danger. Another form of altruism is helping, whereby an individual assists in rearing offspring that are not its own. In some species, some individuals may assist others in acquiring a mate. Eusociality is the highest form of altruism. Eusocial species are those that have sterile workers, engage in cooperative care of the young, and have an overlap of generations. The close genetic relationship among eusocial insects is thought be a result of their form of sex determinism, called haplodiploidy, in which fertilized eggs develop into females and unfertilized eggs develop into males. As a result of haplodiploidy, sister workers are more closely related to their siblings than they would be to their own offspring.

Key Terms

fixed action patterns (FAPs) (p. 1009)
sign stimulus (p. 1009)
releaser (p. 1009)
habituation (p. 1011)
classical conditioning (p. 1011)
conditioned reflex (p. 1011)
unconditioned stimulus (p. 1011)

conditioned stimulus (p. 1012)
operant conditioning (p. 1012)
insight learning (p. 1012)
social learning (p. 1013)
filial imprinting (p. 1014)
critical period (p. 1014)
sexual imprinting (p.1014)

territory (p. 1016)
evolutionarily stable strategy (ESS) (p. 1017)
inclusive fitness (p. 1022)
eusocial species (p. 1025)
haplodiploidy (p.1025)

Review Questions

1. Define and give an example of each of the following: fixed action pattern, releaser, and chain of reactions. Explain how the three are related.

2. What is habituation? Explain how habituation may be adaptive.

3. What is the difference between an unconditioned stimulus (US) and a conditioned stimulus (CS)? Describe a situation in which a conditioned reflex might be beneficial to an animal.

4. What characteristics of imprinting are often used to distinguish it from other types of learning?

5. Explain how genes and experience interact during the development of song in male white-crowned sparrows.

6. What are some costs and benefits of defending a territory? What conditions would favor territorial defense? What conditions would make territorial defense economically unsound?

7. Define (1) evolutionarily stable strategy (ESS); (2) "hawk" strategy; and (3) "dove" strategy. What factors might favor the "dove" strategy? What factors might favor the "hawk" strategy?

8. List three factors that will influence the evolutionary choice of the nature of a communication signal.

9. Explain the three hypotheses for the evolution of altruism. Explain why each of the following may be considered examples of altruism: alarm calling by ground squirrels; helping; male lions cooperating in acquiring a mate; food sharing among vampire bats; and eusociality.

10. Explain how the close genetic relationship among honeybees within a colony develops. How is this different from the way that a close genetic relationship among naked mole rats within a colony develops?

Critical Thinking Questions

1. Herring gull chicks peck at the adult's beak, which is yellow with a red spot near the tip, until the adult regurgitates food into the chick's beak. Describe how you would determine which characteristic(s) of an adult herring gull's head serve(s) to release pecking behavior in the chicks. Would you expect this behavior to be innate or learned?

2. In *The Life and Times of Archie and Mehitabel* by Don Marquis, Archie the cockroach says, "as a representative of the insect world I have often wondered on what man bases his claims to superiority. Everything he knows he has had to learn whereas insects are born knowing everything we need to know." On the basis of what you have learned in this chapter, write a reaction to this statement.

3. Chimpanzees in the wild can be observed stripping leaves from a stem and poking the stem into termite or ant hills. They then withdraw the stem and lick off any insects clinging to it. Describe how you would determine whether this behavior is genetically determined or learned.

4. Discuss the probable survival value of each of the following behaviors: (a) Butterfly courtship involves a series of signals in a particular order between male and female. Failure to produce the right signal at the right time interrupts the courtship. (b) Toads exhibit a striking and swallowing reflex to any elongated shape moving lengthwise. (c) Male bower birds, which lack colorful feathers, decorate nests with brightly colored objects. (d) Tawny owls, which are long-lived, territorial

predators, can lay up to four eggs a year. Typically, however, not all pairs in an area breed every year, and some that do breed fail to incubate the eggs.

5. Ostriches practice cooperative breeding; that is, most nests contain eggs from more than one female. During the breeding season, males scrape out many depressions in the ground, only some of which will be used as nests. The eggs—the largest of any bird (weighing about 1,600 grams)—are laid every 2 days for about 24 days. The female who lays the first egg in a nest becomes the one to incubate all the subsequent eggs, including those laid by other females. The incubating female contributes 8 to 16 eggs, while other females contribute 3 to 20 eggs to a nest. Outsiders tend to lay eggs in a nest on the "off" days of the incubating female. Nocturnal carnivores, such as jackals and hyenas, destroy 40 percent of the nests, while daytime predators, such as vultures, destroy about 10 percent. Discuss the reproductive advantages and disadvantages of this pattern of breeding and nesting behavior.

6. Incubating eggs laid by other females appear to be a case of altruism. However, scientists have found that the degree of relatedness between a chick and the incubating female among ostriches is very low. Furthermore, females are able to recognize their own eggs, and those that are pushed out of the nest are more likely to belong to a female other than the one incubating the nest. How does this evidence argue against altruism as an explanation of communal nesting in ostriches?

Additional Readings

Caro, T. M. 1988. "Adaptive significance of play. Are we getting closer?" *Trends Ecol. and Evol.* 3:50–54.

Carter, C. S. and Getz, L. L. 1993. Monogamy and the prairie vole. *Sci. Amer.* June:100–106. (Intermediate)

Goodenough, J. E. 1984. "Animal Communication." *Carolina Biological Readers.* Carolina Biological Supply Co. Burlington, N.C.

Goodenough, J. E., B. McGuire, and R. A. Wallace. 1992. *Perspectives on Animal Behavior.* John Wiley & Sons, NY.

Hailman, J. P. 1969. "How an instinct is learned." *Sci. Am.* Dec:98–106.

Heinsohn, R. G., A. Cockburn, and R. A. Mulder. 1990. "Avian cooperative breeding: Old hypotheses and new directions." *Trends Ecol. and Evol.* 5:403–407.

Hoage, R. J., and L. Goldman. 1986. *Animal Intelligence. Insights into the Animal Mind.* Smithsonian Institution Press. Washington, D.C.

Honeycutt, R. L. 1992. "Naked mole rats." *Am. Sci.* 80:43–53.

Linden, E. 1992. "A curious kinship: Apes and humans." *National Geographic* 181(3):2–45.

Linden, E. 1993. Can animals think? *Time* March 22:55–61. (Introductory)

Ostfeld, R. S. 1990. "The ecology of territoriality in small mammals." *Trends Ecol. and Evol.* 5:411–415.

Reichert, S. E. 1986. "Spider fights as a test of evolutionary game theory." *Am. Sci.* 74:604–610.

Scheller, R. H., and R. Axel. 1984. "How genes control an innate behavior." *Sci. Am.* Mar:54–62.

Schneider, D. 1974. "The sex-attractant receptor of moths." *Sci. Am.* Jul:28–35.

The Marvels of Animal Behavior. National Geographic Press, Washington, D.C.

Tinbergen, N. 1952. "The curious behavior of the stickleback." *Sci. Am.* Dec:22–26.

Tinbergen, N. 1960. *The Herring Gull's World.* Doubleday, Garden City, NY.

Wilkinson, G. S. 1990. "Food sharing in vampire bats." *Sci. Am.* Feb:76–82.

Wilson, E. O. 1963. "Pheromones." *Sci. Am.* May: 100–114.

Wilson, E.O. 1972. "Animal Communication." *Sci. Am.* Sept:52–60.

Winston, M. L. and K. N. Slessnor. 1992. "The essence of royalty: Honey bee queen pheromone." *Am. Sci.* 80:374–385.

A P P E N D I X

◄ **A** ►

Metric and Temperature Conversion Charts

Metric Unit (symbol)		Metric to English	English to Metric
Length			
kilometer (km)	$= 1{,}000\ (10^3)$ meters	1 km = 0.62 mile	1 mile = 1.609 km
meter (m)	$= 100$ centimeters	1 m = 1.09 yards	1 yard = 0.914 m
		= 3.28 feet	1 foot = 0.305 m
centimeter (cm)	$= 0.01\ (10^{-2})$ meter	1 cm = 0.394 inch	1 inch = 2.54 cm
millimeter (mm)	$= 0.001\ (10^{-3})$ meter	1 mm = 0.039 inch	1 inch = 25.4 mm
micrometer (μm)	$= 0.000001\ (10^{-6})$ meter		
nanometer (nm)	$= 0.000000001\ (10^{-9})$ meter		
angstrom (Å)	$= 0.0000000001\ (10^{-10})$ meter		
Area			
square kilometer (km²)	$= 100$ hectares	1 km² = 0.386 square mile	1 square mile = 2.590 km²
hectare (ha)	$= 10{,}000$ square meters	1 ha = 2.471 acres	1 acre = 0.405 ha
square meter (m²)	$= 10{,}000$ square centimeters	1 m² = 1.196 square yards	1 square yard = 0.836 m²
		= 10.764 square feet	1 square foot = 0.093 m²
square centimeter (cm²)	$= 100$ square millimeters	1 cm² = 0.155 square inch	1 square inch = 6.452 cm²
Mass			
metric ton (t)	$= 1{,}000$ kilograms	1 t = 1.103 tons	1 ton = 0.907 t
	$= 1{,}000{,}000$ grams		
kilogram (kg)	$= 1{,}000$ grams	1 kg = 2.205 pounds	1 pound = 0.454 kg
gram (g)	$= 1{,}000$ milligrams	1 g = 0.035 ounce	1 ounce = 28.35 g
milligram (mg)	$= 0.001$ gram		
microgram (μg)	$= 0.000001$ gram		
Volume Solids			
1 cubic meter (m³)	$= 1{,}000{,}000$ cubic centimeters	1 m³ = 1.308 cubic yards	1 cubic yard = 0.765 m³
		= 35.315 cubic feet	1 cubic foot = 0.028 m³
1 cubic centimeter (cm³)	$= 1{,}000$ cubic millimeters	1 cm³ = 0.061 cubic inch	1 cubic inch = 16.387 cm³
Volume Liquids			
kiloliter (kl)	$= 1{,}000$ liters	1 kl = 264.17 gallons	
liter (l)	$= 1{,}000$ milliliters	1 l = 1.06 quarts	1 gal = 3.785 l
			1 qt = 0.94 l
			1 pt = 0.47 l
milliliter (ml)	$= 0.001$ liter	1 ml = 0.034 fluid ounce	1 fluid ounce = 29.57 ml
microliter (μl)	$= 0.000001$ liter		

TEMPERATURE

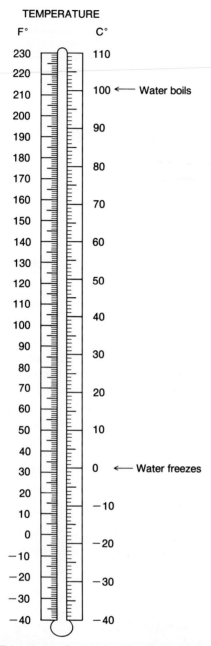

Fahrenheit to Centigrade: $°C = \frac{5}{9}(°F - 32)$
Centigrade to Fahrenheit: $°F = \frac{9}{5}(°C + 32)$

A P P E N D I X
◄ B ►

Microscopes: Exploring the Details of Life

Microscopes are the instruments that have allowed biologists to visualize objects that are vastly smaller than anything visible with the naked eye. There are broadly two types of specimens viewed in a microscope: whole mounts which consist of an intact subject, such as a hair, a living cell, or even a DNA molecule, and thin sections of a specimen, such as a cell or piece of tissue.

THE LIGHT MICROSCOPE

A light microscope consists of a series of glass lenses that bend (refract) the light coming from an illuminated specimen so as to form a visual image of the specimen that is larger than the specimen itself (a). The specimen is often stained with a colored dye to increase its visibility. A special phase contrast light microscope is best suited for observing unstained, living cells because it converts differences in the density of cell organelles, which are normally invisible to the eye, into differences in light intensity which can be seen.

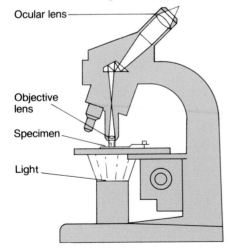

- Ocular lens
- Objective lens
- Specimen
- Light

(a)

All light microscopes have limited *resolving power*—the ability to distinguish two very close objects as being separate from each other. The resolving power of the light microscope is about 0.2 μm (about 1,000 times that of the naked eye), a property determined by the wave length of visible light. Consequently, objects closer to each other than 0.2 μm, which includes many of the smaller cell organ-

elles, will be seen as a single, blurred object through a light microscope.

THE TRANSMISSION ELECTRON MICROSCOPE

Appreciation of the wondrous complexity of cellular organization awaited the development of the transmission electron microscope (or TEM), which can deliver resolving powers 1000 times greater than the light microscope. Suddenly, biologists could see strange new structures, whose function was totally unknown—a breakthrough that has kept cell biologists busy for the past 50 years. The TEM (b) works by shooting a beam of electrons through very thinly sliced specimens that have been stained with heavy metals,

(b)

such as uranium, capable of deflecting electrons in the beam. The electrons that pass through the specimen undeflected are focused by powerful electromagnets (the lenses of a TEM) onto either a phosphorescent screen or high-contrast photographic film. The resolution of the TEM is so great—sufficient to allow us to see individual DNA molecules—because the wavelength of an electron beam is so small (about 0.0005 μm).

THE SCANNING ELECTRON MICROSCOPE

Specimens examined in the scanning electron microscope (SEM) are whole mounts whose surfaces have been coated with a thin layer of heavy metals. In the SEM, a fine beam of electrons scans back and forth across the specimen and the image is formed from electrons bouncing off the hills and valleys of its surface. The SEM produces a three-dimensional image of the surface of the specimen—which can

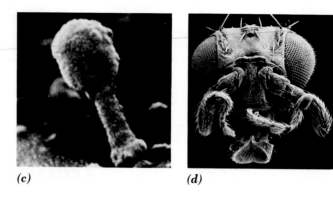

(c)   *(d)*

range in size from a virus to an insect head (c,d)—with remarkable depth and clarity. The SEM produces black and white images; the colors seen in many of the micrographs in the text have been added to enhance their visual quality. Note that the insect head (d) is that of an antennapedia mutant as described on p 687.

APPENDIX
◄ **C** ►

The Hardy-Weinberg Principle

If the allele for brown hair is dominant over that for blond hair, and curly hair is dominant over straight hair, then why don't all people by now have brown, curly hair? The **Hardy-Weinberg Principle** (developed independently by English mathematician G. H. Hardy and German physician W. Weinberg) demonstrates that the frequency of alleles remains the same from generation to generation unless influenced by outside factors. The outside factors that would cause allele frequencies to change are mutation, immigration and emigration (movement of individuals into and out of a breeding population, respectively), natural selection of particular traits, and breeding between members of a small population. In other words, unless one or more of these forces influence hair color and hair curl, the relative number of people with brown and curly hair will not increase over those with blond and straight hair.

To illustrate the Hardy-Weinberg Principle, consider a single gene locus with two alleles, A and a, in a breeding population. (If you wish, consider A to be the allele for brown hair and a to be the allele for blond hair.) Because there are only two alleles for the gene, the sum of the frequencies of A and a will equal 1.0. (By convention, allele frequencies are given in decimals instead of percentages.) Translating this into mathematical terms, if

p = the frequency of allele A, and
q = the frequency of allele a,

then $p + q = 1$.

If A represented 80 percent of the alleles in the breeding population ($p = 0.8$), then according to this formula the frequency of a must be 0.2 ($p + q = 0.8 + 0.2 = 1.0$).

After determining the allele frequency in a starting population, the predicted frequencies of alleles and genotypes in the next generation can be calculated. Setting up a Punnett square with starting allele frequencies of $p = 0.8$ and $q = 0.2$:

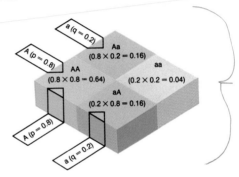

The chances of each offspring receiving any combination of the two alleles is the product of the probability of receiving one of the two alleles alone. In this example, the chances of an offspring receiving two A alleles is $p \times p = p^2$, or $0.8 \times 0.8 = 0.64$. A frequency of 0.64 means that 64 percent of the next generation will be homozygous dominant (AA). The chances of an offspring receiving two a alleles is $q^2 = 0.2 \times 0.2 = 0.04$, meaning 4 percent of the next generation is predicted to be aa. The predicted frequency of heterozygotes (Aa or aA) is 0.32 or $2pq$, the sum of the probability of an individual being $Aa(p \times q = 0.8 \times 0.2 = 0.16)$ plus the probability of an individual being $aA(q \times p = 0.2 \times 0.8 = 0.16)$. Just as all of the allele frequencies for a particular gene must add up to 1, so must all of the possible genotypes for a particular gene locus add up to 1. Thus, the Hardy-Weinberg Principle is

$$p^2 + 2pq + q^2 = 1$$
$$(0.64 + 0.32 + 0.04 = 1)$$

So after one generation, the frequency of possible genotypes is

$$AA = p^2 = 0.64$$
$$Aa = 2pq = 0.32$$
$$aa = q^2 = 0.04$$

Now let's determine the actual allele frequencies for A and a in the new generation. (Remember the original allele frequencies were 0.8 for allele A and 0.2 for allele a. If the Hardy-Weinberg Principle is right, there will be no change in the frequency of either allele.) To do this we sum the frequencies for each genotype containing the allele. Since heterozygotes carry both alleles, the genotype frequency must be divided in half to determine the frequency of each allele. (In our example, heterozygote Aa has a frequency of 0.32, 0.16 for allele A, plus 0.16 for allele a.) Summarizing then:

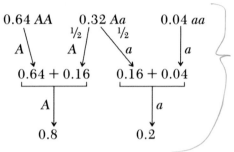

As predicted by the Hardy-Weinberg Principle, the frequency of allele A remained 0.8 and the frequency of allele a remained 0.2 in the new generation. Future generations can be calculated in exactly the same way, over and over again. As long as there are no mutations, no gene flow between populations, completely random mating, no natural selection, and no genetic drift, there will be no change in allele frequency, and therefore no evolution.

Population geneticists use the Hardy-Weinberg Principle to calculate a starting point allele frequency, a reference that can be compared to frequencies measured at some future time. The amount of deviation between observed allele frequencies and those predicted by the Hardy-Weinberg Principle indicates the degree of evolutionary change. Thus, this principle enables population geneticists to measure the rate of evolutionary change and identify the forces that cause changes in allele frequency.

A P P E N D I X
◄ D ►

Careers in Biology

Although many of you are enrolled in biology as a requirement for another major, some of you will become interested enough to investigate the career opportunities in life sciences. This interest in biology can grow into a satisfying livelihood. Here are some facts to consider:

- Biology is a field that offers a very wide range of possible science careers

- Biology offers high job security since many aspects of it deal with the most vital human needs: health and food

- Each year in the United States, nearly 40,000 people obtain bachelor's degrees in biology. But the number of newly created and vacated positions for biologists is increasing at a rate that exceeds the number of new graduates. Many of these jobs will be in the newer areas of biotechnology and bioservices.

Biologists not only enjoy job satisfaction, their work often changes the future for the better. Careers in medical biology help combat diseases and promote health. Biologists have been instrumental in preserving the earth's life-supporting capacity. Biotechnologists are engineering organisms that promise dramatic breakthroughs in medicine, food production, pest management, and environmental protection. Even the economic vitality of modern society will be increasingly linked to biology.

Biology also combines well with other fields of expertise. There is an increasing demand for people with backgrounds or majors in biology complexed with such areas as business, art, law, or engineering. Such a distinct blend of expertise gives a person a special advantage.

The average starting salary for all biologists with a Bachelor's degree is $22,000. A recent survey of California State University graduates in biology revealed that most were earning salaries between $20,000 and $50,000. But as important as salary is, most biologists stress job satisfaction, job security, work with sophisticated tools and scientific equipment, travel opportunities (either to the field or to scientific conferences), and opportunities to be creative in their job as the reasons they are happy in their career.

Here is a list of just a few of the careers for people with degrees in biology. For more resources, such as lists of current openings, career guides, and job banks, write to Biology Career Information, John Wiley and Sons, 605 Third Avenue, New York, NY 10158.

A SAMPLER OF JOBS THAT GRADUATES HAVE SECURED IN THE FIELD OF BIOLOGY°

Agricultural Biologist	Bioanalytical Chemist	Brain Function	Environmental Center
Agricultural Economist	Biochemical/Endocrine	Researcher	Director
Agricultural Extension	Toxicologist	Cancer Biologist	Environmental Engineer
Officer	Biochemical Engineer	Cardiovascular Biologist	Environmental Geographer
Agronomist	Pharmacology Distributor	Cardiovascular/Computer	Environmental Law Specialist
Amino-acid Analyst	Pharmacology Technician	Specialist	Farmer
Analytical Biochemist	Biochemist	Chemical Ecologist	Fetal Physiologist
Anatomist	Biogeochemist	Chromatographer	Flavorist
Animal Behavior	Biogeographer	Clinical Pharmacologist	Food Processing Technologist
Specialist	Biological Engineer	Coagulation Biochemist	Food Production Manager
Anticancer Drug Research	Biologist	Cognitive Neuroscientist	Food Quality Control
Technician	Biomedical	Computer Scientist	Inspector
Antiviral Therapist	Communication Biologist	Dental Assistant	Flower Grower
Arid Soils Technician	Biometerologist	Ecological Biochemist	Forest Ecologist
Audio-neurobiologist	Biophysicist	Electrophysiology/	Forest Economist
Author, Magazines & Books	Biotechnologist	Cardiovascular Technician	Forest Engineer
Behavioral Biologist	Blood Analyst	Energy Regulation Officer	Forest Geneticist
Bioanalyst	Botanist	Environmental Biochemist	Forest Manager

Forest Pathologist
Forest Plantation Manager
Forest Products Technologist
Forest Protection Expert
Forest Soils Analyst
Forester
Forestry Information Specialist
Freeze-Dry Engineer
Fresh Water Biologist
Grant Proposal Writer
Health Administrator
Health Inspector
Health Scientist
Hospital Administrator
Hydrologist
Illustrator
Immunochemist
Immunodiagnostic
   Assay Developer
Inflammation Technologist
Landscape Architect
Landscape Designer
Legislative Aid
Lepidopterist
Liaison Scientist,
   Library of Medicine
   Computer Biologist
Life Science Computer
   Technologist
Lipid Biochemist
Livestock Inspector
Lumber Inspector

Medical Assistant
Medical Imaging Technician
Medical Officer
Medical Products Developer
Medical Writer
Microbial Physiologist
Microbiologist
Mine Reclamation Scientist
Molecular Endocrinologist
Molecular Neurobiologist
Molecular Parasitologist
Molecular Toxicologist
Molecular Virologist
Morphologist
Natural Products Chemist
Natural Resources Manager
Nature Writer
Nematode Control Biologist
Nematode Specialist
Nematologist
Neuroanatomist
Neurobiologist
Neurophysiologist
Neuroscientist
Nucleic Acids Chemist
Nursing Aid
Nutritionist
Occupational Health Officer
Ornamental Horticulturist
Paleontologist
Paper Chemist
Parasitologist

Pathologist
Peptide Biochemist
Pharmaceutical Writer
Pharmaceutical Sales
Pharmacologist
Physiologist
Planning Consultant
Plant Pathologist
Plant Physiologist
Production Agronomist
Protein Biochemist
Protein Structure & Design
   Technician
Purification Biochemist
Quantitative Geneticist
Radiation Biologist
Radiological Scientist
Regional Planner
Regulatory Biologist
Renal Physiologist
Renal Toxicologist
Reproductive Toxicologist
Research and Development
   Director
Research Technician
Research Liaison Scientist
Research Products Designer
Research Proposal Writer
Safety Assessment Sanitarian
Scientific Illustrator
Scientific Photographer
Scientific Reference Librarian

Scientific Writer
Soil Microbiologist
Space Station Life Support
   Technician
Spectroscopist
Sports Product Designer
Steroid Health Assessor
Taxonomic Biologist
Teacher
Technical Analyst
Technical Science Project
   Writer
Textbook Editor
Theoretical Ecologist
Timber Harvester
Toxicologist
Toxic Waste Treatment
   Specialist
Urban Planner
Water Chemist
Water Resources Biologist
Wood Chemist
Wood Fuel Technician
Zoning and Planning
   Manager
Zoologist
Zoo Animal Breeder
Zoo Animal Behaviorist
Zoo Designer
Zoo Inspector

°Results of one survey of California State University graduates. Some careers may require advanced degrees

Glossary

◄ **A** ►

Abiotic Environment Components of eco-systems that include all nonliving factors. (41)

Abscisic Acid (ABA) A plant hormone that inhibits growth and causes stomata to close. ABA may not be commonly involved in leaf drop. (21)

Abscission Separation of leaves, fruit, and flowers from the stem. (21)

Acclimation A physiological adjustment to environmental stress. (28)

Acetyl CoA Acetyl coenzyme A. A complex formed when acetic acid binds to a carrier coenzyme forming a bridge between the end products of glycolysis and the Krebs cycle in respiration. (9)

Acetylcholine Neurotransmitter released by motor neurons at neuromuscular junctions and by some interneurons. (23)

Acid Rain Occurring in polluted air, rain that has a lower pH than rain from areas with unpolluted air. (40)

Acids Substances that release hydrogen ions (H^+) when dissolved in water. (3)

Acid Snow Occurring in polluted air, snow that has a lower pH than snow from areas with unpolluted air. (40)

Acoelomates Animals that lack a body cavity between the digestive cavity and body wall. (39)

Acquired Immune Deficiency Syndrome (AIDS) Disease caused by infection with HIV (Human Immunodeficiency Virus) that destroys the body's ability to mount an immune response due to destruction of its helper T cells. (30, 36)

Actin A contractile protein that makes up the major component of the thin filaments of a muscle cell and the microfilaments of nonmuscle cells. (26)

Action Potential A sudden, dramatic reversal of the voltage (potential difference) across the plasma membrane of a nerve or muscle cell due to the opening of the sodium channels. The basis of a nerve impulse. (23)

Activation Energy Energy required to initiate chemical reaction. (6)

Active Site Region on an enzyme that binds its specific substrates, making them more reactive. (6)

Active Transport Movement of substances into or out of cells against a concentration gradient, i.e., from a region of lower concentration to a region of higher concentration. The process requires an expenditure of energy by the cell. (7)

Adaptation A hereditary trait that improves an organism's chances of survival and/or reproduction. (33)

Adaptive Radiation The divergence of many species from a single ancestral line. (33)

Adenosine Triphosphate (ATP) The molecule present in all living organisms that provides energy for cellular reactions in its phosphate bonds. ATP is the universal energy currency of cells. (6)

Adenylate Cyclase An enzyme activated by hormones that converts ATP to cyclic AMP, a molecule that activates resting enzymes. (25)

Adrenal Cortex Outer layer of the adrenal glands. It secretes steroid hormones in response to ACTH. (25)

Adrenal Medulla An endocrine gland that controls metabolism, cardiovascular function, and stress responses. (25)

Adrenocorticotropic Hormone (ACTH) An anterior pituitary hormone that stimulates the cortex of the adrenal glands to secrete cortisol and other steroid hormones. (25)

Adventitious Root System Secondary roots that develop from stem or leaf tissues. (18)

Aerobe An organism that requires oxygen to release energy from food molecules. (9)

Aerobic Respiration Pathway by which glucose is completely oxidized to CO_2 and H_2O, requiring oxygen and an electron transport system. (9)

Afferent (Sensory) Neurons Neurons that conduct impulses from the sense organs to the central nervous system. (23)

Age-Sex Structure The number of individuals of a certain age and sex within a population. (43)

Aggregate Fruits Fruits that develop from many pistils in a single flower. (20)

AIDS See Acquired Immune Deficiency Syndrome.

Albinism A genetic condition characterized by an absence of epidermal pigmentation that can result from a deficiency of any of a variety of enzymes involved in pigment formation. (12)

Alcoholic Fermentation The process in which electrons removed during glycolysis are transferred from NADH to form alcohol as an end product. Used by yeast during the commercial process of ethyl alcohol production. (9)

Aldosterone A hormone secreted by the adrenal cortex that stimulates reabsorption of sodium from the distal tubules and collecting ducts of the kidneys. (23)

Algae Any unicellular or simple colonial photosynthetic eukaryote. (37,38)

Algin A substance produced by brown algae harvested for human application because of its ability to regulate texture and consistency of products. Found in ice cream, cosmetics, marshmellows, paints, and dozens of other products. (38)

Allantois Extraembryonic membrane that serves as a repository for nitrogenous wastes. In placental mammals, it helps form the vascular connections between mother and fetus. (32)

Allele Alternative form of a gene at a particular site, or locus, on the chromosome. (12)

Allele Frequency The relative occurrence of a certain allele in individuals of a population. (33)

Allelochemicals Chemicals released by some plants and animals that deter or kill a predator or competitor. (42)

Allelopathy A type of interaction in which one organism releases allelochemicals that harm another organism. (42)

Allergy An inappropriate response by the immune system to a harmless foreign substance leading to symptoms such as itchy eyes, runny nose, and congested airways. If the reaction occurs throughout the body (anaphylaxis) it can be life threatening. (30)

Allopatric Speciation Formation of new species when gene flow between parts of a population is stopped by geographic isolation. (33)

Alpha Helix Portion of a polypeptide chain organized into a defined spiral conformation. (4)

Alternation of Generations Sequential change during the life cycle of a plant in which a haploid (1N) multicellular stage (gametophyte) alternates with a diploid (2N) multicellular stage (sporophyte). (38)

Alternative Processing When a primary RNA transcript can be processed to form more than one mRNA depending on conditions. (15)

Altruism The performance of a behavior that benefits another member of the species at some cost to the one who does the deed. (44)

Alveolus A tiny pouch in the lung where gas is exchanged between the blood and the air; the functional unit of the lung where CO_2 and O_2 are exchanged. (29)

Alzheimer's Disease A degenerative disease of the human brain, particularly affecting acetylcholine-releasing neurons and the hippocampus, characterized by the presence of tangled fibrils within the cytoplasm of neurons and amyloid plaques outside the cells. (23)

Amino Acids Molecules containing an amino group ($-NH_2$) and a carboxyl group ($-COOH$) attached to a central carbon atom. Amino acids are the subunits from which proteins are constructed. (4)

Amniocentesis A procedure for obtaining fetal cells by withdrawing a sample of the fluid that surrounds a developing fetus (amniotic fluid) using a hypodermic needle and syringe. (17)

Amnion Extraembryonic membrane that envelops the young embryo and encloses the amniotic fluid that suspends and cushions it. (32)

Amoeba A protozoan that employs pseudopods for motility. (37)

Amphibia A vertebrate class grouped into three orders: Caudata (tailed amphibians); Anura (tail-less amphibians); Apoda (rare worm-like, burrowing amphibians). (39)

Anabolic Steroids Steroid hormones, such as testosterone, which promote biosynthesis (anabolism), especially protein synthesis. (25)

Anabolism Biosynthesis of complex molecules from simpler compounds. Anabolic pathways are endergonic, i.e., require energy. (6)

Anaerobe Organism that does not require oxygen to release energy from food molecules. (9)

Anaerobic Respiration Pathway by which glucose is completely oxidized, using an electron transport system but requiring a terminal electron acceptor other than oxygen. (Compare with fermentation.) (9)

Analogous Structures (Homoplasies) Structures that perform a similar function, such as wings in birds and insects, but did not originate from the same structure in a common ancestor. (33)

Anaphase Stage of mitosis when the kinetochores split and the sister chromatids (now termed chromosomes) move to opposite poles of the spindle. (10)

Anatomy Study of the structural characteristics of an organism. (18)

Angiosperm (Anthophyta) Any plant having its seeds surrounded by fruit tissue formed from the mature ovary of the flowers. (38)

Animal A mobile, heterotrophic, multicellular organism, classified in the Animal kingdom. (39)

Anion A negatively charged ion. (3)

Annelida The phylum which contains segmented worms (earthworms, leeches, and bristleworms). (39)

Annuals Plants that live for one year or less. (18)

Annulus A row of specialized cells encircling each sporangium on the fern frond; facilitates rupture of the sporangium and dispersal of spors. (38)

Antagonistic Muscles Pairs of muscles whose contraction bring about opposite actions as illustrated by the biceps and triceps, which bends or straightens the arm at the elbow, respectively. (26)

Antenna Pigments Components of photosystems that gather light energy of different wavelengths and then channel the absorbed energy to a reaction center. (8)

Anterior In anatomy, at or near the front of an animal; the opposite of posterior. (39)

Anterior Pituitary A true endocrine gland manufacturing and releasing six hormones

when stimulated by releasing factors from the hypothalamus. (25)

Anther The swollen end of the stamen (male reproductive organ) of a flowering plant. Pollen grains are produced inside the anther lobes in pollen sacs. (20)

Antibiotic A substance produced by a fungus or bacterium that is capable of preventing the growth of bacteria. (2)

Antibodies Proteins produced by plasma cells. They react specifically with the antigen that stimulated their formation. (30)

Anticodon Triplet of nucleotides in tRNA that recognizes and base pairs with a particular codon in mRNA. (14)

Antidiuretic Hormone (ADH) One of the two hormones released by the posterior pituitary. ADH increases water reabsorption in the kidney, which then produces a more concentrated urine. (25)

Antigen Specific foreign agent that triggers an immune response. (30)

Aorta Largest blood vessel in the body through which blood leaves the heart and enters the systemic circulation. (28)

Apical Dominance The growth pattern in plants in which axillary bud growth is inhibited by the hormone auxin, present in high concentrations in terminal buds. (21)

Apical Meristems Centers of growth located at the tips of shoots, axillary buds, and roots. Their cells divide by mitosis to produce new cells for primary growth in plants. (18)

Aposematic Coloring Warning coloration which makes an organism stand out from its surroundings. (42)

Appendicular Skeleton The bones of the appendages and of the pectoral and pelvic girdles. (26)

Aquatic Living in water. (40)

Archaebacteria Members of the kingdom Monera that differ from typical bacteria in the structure of their membrane lipids, their cell walls, and some characteristics that resemble those of eukaryotes. Their lack of a true nucleus, however, accounts for their assignment to the Moneran kingdom. (36)

Archenteron In gastrulation, the hollow core of the gastrula that becomes an animal's digestive tract. (32)

Arteries Large, thick-walled vessels that carry blood away from the heart. (28)

Arterioles The smallest arteries, which carry blood toward capillary beds. (28)

Arthropoda The most diverse phylum on earth, so called from the presence of jointed limbs. Includes insects, crabs, spiders, centipedes. (39)

Ascospores Sexual fungal spore borne in a sac. Produced by the sac fungi, Ascomycota. (37)

Asexual Reproduction Reproduction without the union of male and female gametes. (31)

Association In ecological communities, a major organization characterized by uniformity and two or more dominant species. (41)

Asymmetric Referring to a body form that cannot be divided to produce mirror images. (39)

Atherosclerosis Condition in which the inner walls of arteries contain a buildup of cholesterol-containing plaque that tends to occlude the channel and act as a site for the formation of a blood clot (thrombus). (7)

Atmosphere The layer of air surrounding the Earth. (40)

Atom The fundamental unit of matter that can enter into chemical reactions; the smallest unit of matter that possesses the qualities of an element. (3)

Atomic Mass Combined number of protons and neutrons in the nucleus of an atom. (3)

Atomic Number The number of protons in the nucleus of an atom. (3)

ATP (see **Adenosine Triphosphate**)

ATPase An enzyme that catalyzes a reaction in which ATP is hydrolyzed. These enzymes are typically involved in reactions where energy stored in ATP is used to drive an energy-requiring reaction, such as active transport or muscle contractility. (7, 26)

ATP Synthase A large protein complex present in the plasma membrane of bacteria, the inner membrane of mitochondria, and the thylakoid membrane of chloroplasts. This complex consists of a baseplate in the membrane, a channel across the membrane through which protons can pass, and a spherical head (F_1 particle) which contains the site where ATP is synthesized from ADP and P_i (8, 9)

Atrioventricular (AV) Node A neurological center of the heart, located at the top of the ventricles. (28)

Atrium A contracting chamber of the heart which forces blood into the ventricle. There are two atria in the hearts of all vertebrates, except fish which have one atrium. (28)

Atrophy The shrinkage in size of structure, such as a bone or muscle, usually as a result of disuse. (26)

Autoantibodies Antibodies produced against the body's own tissue. (30)

Autoimmune Disease Damage to a body tissue due to an attack by autoantibodies. Examples include thyroiditis, multiple sclerosis, and rheumatic fever. (30)

Autonomic Nervous System The nerves that control the involuntary activities of the internal organs. It is composed of the parasympathetic system, which functions during normal activity, and the sympathetic system, which operates in times of emergency or prolonged exertion. (23)

Autosome Any chromosome that is not a sex chromosome. (13)

Autotrophs Organisms that satisfy their own nutritional needs by building organic molecules photosynthetically or chemosynthetically from inorganic substances. (8)

Auxins Plant growth hormones that promote cell elongation by softening cell walls. (21)

Axial Skeleton The bones aligned along the long axis of the body, including the skull, vertebral column, and ribcage. (26)

Axillary Bud A bud that is directly above each leaf on the stem. It can develop into a new stem or a flower. (18)

Axon The long, sometimes branched extension of a neuron which conducts impulses from the cell body to the synaptic knobs. (23)

◄ **B** ►

Bacteriophage A virus attacking specific bacteria that multiplies in the bacterial host cell and usually destroys the bacterium as it reproduces. (36)

Balanced Polymorphism The maintenance of two or more alleles for a single trait at fairly high frequencies. (33)

Bark Common term for the periderm. A collective term for all plant tissues outside the secondary xylem. (18)

Base Substance that removes hydrogen ions (H^+) from solutions. (3)

Basidiospores Sexual spores produced by basidiomycete fungi. Often found by the millions on gills in mushrooms. (37)

Basophil A phagocytic leukocyte which also releases substances, such as histamine, that trigger an inflammatory response. (28)

Batesian Mimicry The resemblance of a good-tasting or harmless species to a species with unpleasant traits. (42)

Bathypelagic Zone The ocean zone beneath the mesopelagic zone, characterized by no light; inhabited by heterotrophic bacteria and benthic scavengers. (40)

B Cell A lymphocyte that becomes a plasma cell and produces antibodies when stimulated by an antigen. (30)

Benthic Zone The deepest ocean zone; the ocean floor, inhabitated by bottom dwelling organisms. (40)

Bicarbonate Ion HCO_3^-. (3, 29)

Biennials Plants that live for two years. (18)

Bilateral Symmetry The quality possessed by organisms whose body can be divided into mirror images by only one median plane. (39)

Bile Salts Detergentlike molecules produced by the liver and stored by the gallbladder that function in lipid emulsification in the small intestine. (27)

Binomial A term meaning "two names" or "two words". Applied to the system of nomenclature for categorizing living things with a genus and species name that is unique for each type of organism. (1)

Biochemicals Organic molecules produced by living cells. (4)

Bioconcentration The ability of an organism to accumulate substances within its' body or specific cells. (41)

Biodiversity Biological diversity of species, including species diversity, genetic diversity, and ecological diversity. (43)

Biogeochemical Cycles The exchanging of chemical elements between organisms and the abiotic environment. (41)

Biological Control Pest control through the use of naturally occurring organisms such as predators, parasites, bacteria, and viruses. (41)

Biological Magnification An increase in concentration of slowly degradable chemicals in organisms at successively higher trophic levels; for example, DDT or PCB's. (41)

Bioluminescence The capability of certain organisms to utilize chemical energy to produce light in a reaction catalyzed by the enzyme luciferase. (9)

Biomass The weight of organic material present in an ecosystem at any one time. (41)

Biome Broad geographic region with a characteristic array of organisms. (40)

Biosphere Zone of the earth's soil, water, and air in which living organisms are found. (40)

Biosynthesis Construction of molecular components in the growing cell and the replacement of these compounds as they deteriorate. (6)

Biotechnology A new field of genetic engineering; more generally, any practical application of biological knowledge. (16)

Biotic Environment Living components of the environment. (40)

Biotic Potential The innate capacity of a population to increase tremendously in size were it not for curbs on growth; maximum population growth rate. (43)

Blade Large, flattened area of a leaf; effective in collecting sunlight for photosynthesis. (18)

Blastocoel The hollow fluid-filled space in a blastula. (32)

Blastocyst Early stage of a mammalian embryo, consisting of a mass of cells enclosed in a hollow ball of cells called the trophoblast. (32)

Blastodisk In bird and reptile development, the stage equivalent to a blastula. Because of the large amount of yolk, cleavage produces two flattened layers of cells with a blastocoel between them. (32)

Blastomeres The cells produced during embryonic cleavage. (32)

Blastopore The opening of the archenteron that is the embryonic predecessor of the anus in vertebrates and some other animals. (32)

Blastula An early developmental stage in many animals. It is a ball of cells that encloses a cavity, the blastocoel. (32)

Blood A type of connective tissue consisting of red blood cells, white blood cells, platelets, and plasma. (28)

Blood Pressure Positive pressure within the cardiovascular system that propels blood through the vessels. (28)

Blooms are massive growths of algae that occur when conditions are optimal for algae proliferation. (37)

Body Plan The general layout of a plant's or animal's major body parts. (39)

Bohr Effect Increased release of O_2 from hemoglobin molecules at lower pH. (29)

Bone A tissue composed of collagen fibers, calcium, and phosphate that serves as a means of support, a reserve of calcium and phosphate, and an attachment site for muscles. (26)

Botany Branch of biology that studies the life cycles, structure, growth, and classification of plants. (18)

Bottleneck A situation in which the size of a species' population drops to a very small number of individuals, which has a major impact on the likelihood of the population recovering its earlier genetic diversity. As occurred in the cheetah population. (33)

Bowman's Capsule A double-layered container that is an invagination of the proximal end of the renal tubule that collects molecules and wastes from the blood. (28)

Brain Mass of nerve tissue composing the main part of the central nervous system. (23)

Brainstem The central core of the brain, which coordinates the automatic, involuntary body processes. (23)

Bronchi The two divisions of the trachea through which air enters each of the two lungs. (29)

Bronchioles The smallest tubules of the respiratory tract that lead into the alveoli of the lungs where gas exchange occurs. (29)

Bryophyta Division of non-vascular terrestrial plants that include liverworts, mosses, and hornworts. (38)

Budding Asexual process by which offspring develop as an outgrowth of a parent. (39)

Buffers Chemicals that couple with free hydrogen and hydroxide ions thereby resisting changes in pH. (3)

Bundle Sheath Parenchyma cells that surround a leaf vein which regulate the uptake and release of materials between the vascular tissue and the mesophyll cells. (18)

◄ **C** ►

C₃ Synthesis The most common pathway for fixing CO_2 in the synthesis reactions of photosynthesis. It is so named because the first detectable organic molecule into which CO_2 is incorporated is a 3-carbon molecule, phosphoglycerate (PGA). (8)

C_4 Synthesis   Pathway for fixing CO_2 during the light-independent reactions of photosynthesis. It is so named because the first detectable organic molecule into which CO_2 is incorporated is a 4-carbon molecule. (8)

Calcitonin   A thyroid hormone which regulates blood calcium levels by inhibiting its release from bone. (25)

Calorie   Energy (heat) necessary to elevate the temperature of one gram of water by one degree Centigrade (1° C). (6)

Calvin Cycle   The cyclical pathway in which CO_2 is incorporated into carbohydrate. See C_3 synthesis. (8)

Calyx   The outermost whorl of a flower, formed by the sepals. (20)

CAM   Crassulacean acid metabolism. A variation of the photosynthetic reactions in plants, biochemically identical to C_4 synthesis except that all reactions occur in the same cell and are separated by time. Because CAM plants open their stomates at night, they have a competitive advantage in hot, dry climates. (8)

Cambium   A ring or cluster of meristematic cells that increase the width of stems and roots when they divide to produce secondary tissues. (18)

Camouflage   Adaptations of color, shape and behavior that make an organism more difficult to detect. (42)

Cancer   A disease resulting from uncontrolled cell divisions. (10,13)

Capillaries   The tiniest blood vessels consisting of a single layer of flattened cells. (28)

Capillary Action   Tendency of water to be pulled into a small-diameter tube. (3)

Carbohydrates   A group of compounds that includes simple sugars and all larger molecules constructed of sugar subunits, e.g. polysaccharides. (4)

Carbon Cycle   The cycling of carbon in different chemical forms, from the environment to organisms and back to the environment. (41)

Carbon Dioxide Fixation   In photosynthesis, the combination of CO_2 with carbon-accepting molecules to form organic compounds. (8)

Carcinogen   A cancer-causing agent. (13)

Cardiac Muscle   One of the three types of muscle tissue; it forms the muscle of the heart. (26)

Cardiovascular System   The organ system consisting of the heart and the vessels through which blood flows. (28)

Carnivore   An animal that feeds exclusively on other animals. (42)

Carotenoid   A red, yellow, or orange plant pigment that absorbs light in 400-500 nm wavelengths. (8)

Carpels   Central whorl of a flower containing the female reproductive organs. Each separate carpel, or each unit of fused carpels, is called a pistil. (20)

Carrier Proteins   Proteins within the plasma membrane that bind specific substances and facilitate their movement across the membrane. (7)

Carrying Capacity   The size of a population that can be supported indefinitely in a given environment. (43)

Cartilage   A firm but flexible connective tissue. In the human, most cartilage originally present in the embryo is transformed into bones. (26)

Casparian Strip   The band of waxy suberin that surrounds each endodermal cell of a plant's root tissue. (18)

Catabolism   Metabolic pathways that degrade complex compounds into simpler molecules, usually with the release of the chemical energy that held the atoms of the larger molecule together. (6)

Catalyst   A chemical substance that accelerates a reaction or causes a reaction to occur but remains unchanged by the reaction. Enzymes are biological catalysts. (6)

Cation   A positively charged ion. (3)

Cecum   A closed-ended sac extending from the intestine in grazing animals lacking a rumen (e.g., horses) that enables them to digest cellulose. (27)

Cell   The basic structural unit of all organisms. (5)

Cell Body   Region of a neuron that contains most of the cytoplasm, the nucleus, and other organelles. It relays impulses from the dendrites to the axon. (23)

Cell Cycle   Complete sequence of stages from one cell division to the next. The stages are denoted G_1, S, G_2, and M phase. (10)

Cell Differentiation   The process by which the internal contents of a cell become assembled into a structure that allows the cell to carry out a specific set of activities, such as secretion of enzymes or contraction. (32)

Cell Division   The process by which one cell divides into two. (10)

Cell Fusion   Technique whereby cells are caused to fuse with one another producing a large cell with a common cytoplasm and plasma membrane. (5, 10)

Cell Plate   In plants, the cell wall material deposited midway between the daughter cells during cytokinesis. Plate material is deposited by small Golgi vesicles. (5, 10)

Cell Sap   Solution that fills a plant vacuole. In addition to water, it may contain pigments, salts, and even toxic chemicals. (5)

Cell Theory   The fundamental theory of biology that states: 1) all organisms are composed of one or more cells, 2) the cell is the basic organizational unit of life, 3) all cells arise from pre-existing cells. (5)

Cellular Respiration   (See **Aerobic respiration**)

Cellulose   The structural polysaccharide comprising the bulk of the plant cell wall. It is the most abundant polysaccharide in nature. (4, 5)

Cell Wall   Rigid outer-casing of cells in plants and other organisms which gives support, slows dehydration, and prevents a cell from bursting when internal pressure builds due to an influx of water. (5)

Central Nervous System   In vertebrates, the brain and spinal cord. (23)

Centriole   A pinwheel-shaped structure at each pole of a dividing animal cell. (10)

Centromere   Indented region of a mitotic chromosome containing the kinetochore. (10)

Cephalization   The clustering of neural tissues and sense organs at the anterior (leading) end of the animal. (39)

Cerebellum   A bulbous portion of the vertebrate brain involved in motor coordination. Its prominence varies greatly among different vertebrates . (23)

Cerebral Cortex   The outer, highly convoluted layer of the cerebrum. In the human, this is the center of higher brain functions, such as speech and reasoning. (23)

Cerebrospinal Fluid   Fluid present within the ventricles of the brain, central canal of the spinal cord, and which surrounds and cushions the central nervous system. (23)

Cerebrum   The most dominant part of the human forebrain, composed of two cerebral hemispheres, generally associated with higher brain functions. (23)

Cervix   The lower tip of the uterus. (31)

Chapparal   A type of shrubland in California, characterized by drought-tolerant and fire-adapted plants. (40)

Character Displacement   Divergence of a physical trait in closely related species in response to competition. (42)

Chemical Bonds   Linkage between atoms as a result of electrons being shared or donated. (3)

Chemical Evolution   Spontaneous synthesis of increasingly complex organic compounds from simpler molecules. (35)

Chemical Reaction   Interaction between chemical reactants. (6)

Chemiosmosis   The process by which a pH gradient drives the formation of ATP. (8, 9)

Chemoreceptors   Sensory receptors that respond to the presence of specific chemicals. (24)

Chemosynthesis   An energy conversion process in which inorganic substances (H, N, Fe, or S) provide energized electrons and hydrogen for carbohydrate formation (9, 36)

Chiasmata   Cross-shaped regions within a tetrad, occurring at points of crossing over or genetic exchange. (11)

Chitin Structural polysaccharide that forms the hard, strong external skeleton of many arthropods and the cell walls of fungi. (4)

Chlamydia Obligate intracellular parasitic bacteria that lack a functional ATP-generating system. (36)

Chlorophyll Pigments Major light-absorbing pigments of photosynthesis. (8)

Chlorophyta Green algae, the largest group of algae; members of this group were very likely the ancestors of the modern plant kingdom. (38)

Chloroplasts An organelle containing chlorophyll found in plant cells in which photosynthesis occurs. (5, 8)

Cholecystokinin (CCK) Hormone secreted by endocrine cells in the wall of the small intestine that stimulates the release of digestive products by the pancreas. (27)

Chondrocytes Living cartilage cells embedded within the protein-polysaccharide matrix they manufacture. (26)

Chordamesoderm In vertebrates, the block of mesoderm that underlies the dorsal ectoderm of the gastrula, induces the formation of the nervous system, and gives rise to the notochord. (32)

Chordate A member of the phylum Chordata possessing a skeletal rod of tissue called a notochord, a dorsal hollow nerve cord, gill slits, and a post-anal tail at some stage of its development. (39)

Chorion The outermost of the four extraembryonic membranes. In placental mammals, it forms the embryonic portion of the placenta. (32)

Chorionic Villus Sampling (CVS) A procedure for obtaining fetal cells by removing a small sample of tissue from the developing placenta of a pregnant woman. (17)

Chromatid Each of the two identical subunits of a replicated chromosome. (10)

Chromatin DNA-protein fibers which, during prophase, condense to form the visible chromosomes. (5, 10)

Chromatography A technique for separating different molecules on the basis of their solubility in a particular solvent. The mixture of substances is spotted on a piece of paper or other material, one end of which is then placed in the solvent. As the solvent moves up the paper by capillary action, each substance in the mixture is carried a particular distance depending on its solubility in the moving solvent. (8)

Chromosomes Dark-staining structures in which the organism's genetic material (DNA) is organized. Each species has a characteristic number of chromosomes. (5, 10)

Chromosome Aberrations Alteration in the structure of a chromosome from the normal state. Includes chromosome deletions, duplications, inversions, and translocations. (13)

Chromosome Puff A site on an insect polytene chromosome where the DNA has unraveled and is being transcribed. (15)

Cilia Short, hairlike structures projecting from the surfaces of some cells. They beat in coordinated ways, are usually found in large numbers, and are densely packed. (5)

Ciliated Mucosa Layer of ciliated epithelial cells lining the respiratory tract. The beating of cilia propels an associated mucous layer and trapped foreign particles. (29)

Circadian Rhythm Behavioral patterns that cycle during approximately 24 hour intervals.

Circulatory System The system that circulates internal fluids throughout an organism to deliver oxygen and nutrients to cells and to remove metabolic wastes. (28)

Class (Taxonomic) A level of the taxonomic hierarchy that groups together members of related orders. (1)

Classical Conditioning A form of learning in which an animal develops a response to a new stimulus by repeatedly associating the new stimulus with a stimulus that normally elicits the response. (44)

Cleavage Successive mitotic divisions in the early embryo. There is no cell growth between divisions. (32)

Cleavage Furow Constriction around the middle of a dividing cell caused by constriction of microfilaments. (10)

Climate The general pattern of average weather conditions over a long period of time in a specific region, including precipitation, temperature, solar radiation, and humidity. (40)

Climax Final or stable community of successional stages, that is more or less in equilibrium with existing environmental conditions for a long period of time. (41)

Climax Community Community that remains essentially the same over long periods of time; final stage of ecological succession. (41)

Clitoris A protrusion at the point where the labia minora merge; rich in sensory neurons and erectile tissue. (31)

Clonal Selection Mechanism The mechanism by which the body can synthesize antibodies specific for the foreign substance (antigen) that stimulated their production. (30)

Clones Offspring identical to the parent, produced by asexual processes. (15)

Closed Circulatory System Circulatory system in which blood travels throughout the body in a continuous network of closed tubes. (Compare with open circulatory system). (28)

Clumped Pattern Distribution of individuals of a population into groups, such as flocks or herds. (43)

Cnidaria A phylum that consists of radial symmetrical animals that have two cell layers. There are three classes: 1) Hydrozoa (hydra), 2) Scyphozoa (jellyfish), 3) Anthozoa (sea anemones, corals). Most are marine forms that live in warm, shallow water. (39)

Cnidocytes Specialized stinging cells found in the members of the phylum Cnidaria. (39)

Coastal Waters Relatively warm, nutrient-rich shallow water extending from the high-tide mark on land to the sloping continental shelf. The greatest concentration of marine life are found in coastal waters. (40)

Coated Pits Indentations at the surfaces of cells that contain a layer of bristly protein (called clathrin) on the inner surface of the plasma membrane. Coated pits are sites where cell receptors become clustered. (7)

Cochlea Organ within the inner ear of mammals involved in sound reception. (24)

Codominance The simultaneous expression of both alleles at a genetic locus in a heterozygous individual. (12)

Codon Linear array of three nucleotides in mRNA. Each triplet specifies a particular amino acid during the process of translation. (14)

Coelomates Animals in which the body cavity is completely lined by mesodermally-derived tissues. (39)

Coenzyme An organic cofactor, typically a vitamin or a substance derived from a vitamin. (6)

Coevolution Evolutionary changes that result from reciprocal interactions between two species, e.g., flowering plants and their insect pollinators. (33)

Cofactor A non-protein component that is linked covalently or noncovalently to an enzyme and is required by the enzyme to catalyze the reaction. Cofactors may be organic molecules (coenzymes) or metals. (6)

Cohesion The tendency of different parts of a substance to hold together because of forces acting between its molecules. (3)

Coitus Sexual union in mammals. (31)

Coleoptile Sheath surrounding the tip of the monocot seedling, protecting the young stem and leaves as they emerge from the soil. (21)

Collagen The most abundant protein in the human body. It is present primarily in the extracellular space of connective tissues such as bone, cartilage, and tendons. (26)

Collenchyma Living plant cells with irregularly thickened primary cell walls. A supportive cell type often found inside the epidermis of stems with primary growth. Angular, lacunar and laminar are different types of collenchyma cells. (18)

Commensalism A form of symbiosis in which one organism benefits from the union while the other member neither gains nor loses. (42)

Community The populations of all species living in a given area. (41)

Compact Bone The solid, hard outer regions of a bone surrounding the honey-combed mass of spongy bone. (26)

Companion Cell Specialized parenchyma cell associated with a sieve-tube member in phloem. (18)

Competition Interaction among organisms that require the same resource. It is of two types: 1) intraspecific (between members of the same species); 2) interspecific (between members of different species). (42)

Competitive Exclusion Principle (Gause's Principle) Competition in which a winner species captures a greater share of resources, increasing its survival and reproductive capacity. The other species is gradually displaced. (42)

Competitive Inhibition Prevention of normal binding of a substrate to its enzyme by the presence of an inhibitory compound that competes with the substrate for the active site on the enzyme. (6)

Complement Blood proteins with which some antibodies combine following attachment to antigen (the surface of microorganisms). The bound complement punches the tiny holes in the plasma membrane of the foreign cell, causing it to burst. (28)

Complementarity The relationship between the two strands of a DNA molecule determined by the base pairing of nucleotides on the two strands of the helix. A nucleotide with guanine on one strand always pairs with a nucleotide having cytosine on the other strand; similarly with adenine and thymine. (14)

Complete Digestive Systems Systems that have a digestive tract with openings at both ends—a mouth for entry and an anus for exit. (27)

Complete Flower A flower containing all four whorls of modified leaves—sepals, petals, stamen, and carpels. (20)

Compound Chemical substances composed of atoms of more than one element. (3)

Compound Leaf A leaf that is divided into leaflets, with two or more leaflets attached to the petiole. (18)

Concentration Gradient Regions in a system of differing concentration representing potential energy, such as exist in a cell and its environment, that cause molecules to move from areas of higher concentration to lower concentration. (7)

Conditioned Reflex A reflex ("automatic") response to a stimulus that would not normally have elicited the response. Conditioned reflexes develop by repeated association of a new stimulus with an old stimulus that normally elicits the response. (44)

Conformation The three-dimensional shape of a molecule as determined by the spatial arrangement of its atoms. (4)

Conformational Change Change in molecular shape (as occurs, for example, in an en-zyme as it catalyzes a reaction, or a myosin molecule during contraction). (6)

Conjugation A method of reproduction in single-celled organisms in which two cells link and exchange nuclear material. (11)

Connective Tissues Tissues that protect, support, and hold together the internal organs and other structures of animals. Includes bone, cartilage, tendons, and other tissues, all of which have large amounts of extracellular material. (22)

Consumers Heterotrophs in a biotic environment that feed on other organisms or organic waste. (41)

Continental Drift The continuous shifting of the earth's land masses explained by the theory of plate tectonics. (35)

Continuous Variation An inheritance pattern in which there is graded change between the two extremes in a phenotype (compare with discontinuous variation). (12)

Contraception The prevention of pregnancy. (31)

Contractile Proteins Actin and myosin, the protein filaments that comprise the bulk of the muscle mass. During contraction of skeletal muscle, these filaments form a temporary association and slide past each other, generating the contractile force. (26)

Control (Experimental) A duplicate of the experiment identical in every way except for the one variable being tested. Use of a control is necessary to demonstrate cause and effect. (2)

Convergent Evolution The evolution of similar structures in distantly related organisms in response to similar environments. (33)

Cork Cambium In stems and roots of perennials, a secondary meristem that produces the outer protective layer of the bark. (18)

Coronary Arteries Large arteries that branch immediately from the aorta, providing oxygen-rich blood to the cardiac muscle. (28)

Corpus Callosum A thick cable composed of hundreds of millions of neurons that connect the right and left cerebral hemispheres of the mammalian brain. (23)

Corpus Luteum In the mammalian ovary, the structure that develops from the follicle after release of the egg. It secretes hormones that prepare the uterine endometrium to receive the developing embryo. (31)

Cortex In the stem or root of plants, the region between the epidermis and the vascular tissues. Composed of ground tissue. In animals, the outermost portion of some organs. (18)

Cotyledon The seed leaf of a dicot embryo containing stored nutrients required for the germinated seed to grow and develop, or a food digesting seed leaf in a monocot embryo. (20)

Countercurrent Flow Mechanism for increasing the exchange of substances or heat from one stream of fluid to another by having the two fluids flow in opposite directions. (29)

Covalent Bonds Linkage between two atoms which share the same electrons in their outermost shells. (3)

Cranial Nerves Paired nerves which emerge from the central stalk of the vertebrate brain and innervate the body. Humans have 12 pairs of cranial nerves. (23)

Cranium The bony casing which surrounds and protects the vertebrate brain. (23)

Cristae The convolutions of the inner membrane of the mitochondrion. Embedded within them are the components of the electron transport system and proton channels for chemiosmosis. (9)

Crossing Over During synapsis, the process by which homologues exchange segments with each other. (11)

Cryptic Coloration A form of camouflage wherein an organism's color or patterning helps it resemble its background. (42)

Cutaneous Respiration The uptake of oxygen across virtually the entire outer body surface. (29)

Cuticle 1) Waxy layer covering the outer cell walls of plant epidermal cells. It retards water vapor loss and helps prevent dehydration. (18) 2) Outer protective, nonliving covering of some animals, such as the exoskeleton of anthropods. (26, 39)

Cyanobacteria A type of prokaryote capable of photosynthesis using water as a source of electrons. Cyanobacteria were responsible for initially creating an O_2-containing atmosphere on earth. (35, 36)

Cyclic AMP (Cyclic adenosine monophosphate) A ring-shaped molecular version of an ATP minus two phosphates. A regulatory molecule formed by the enzyme adenylate cyclase which converts ATP to cAMP. A second messenger. (25)

Cyclic Pathways Metabolic pathways in which the intermediates of the reaction are regenerated while assisting the conversion of the substrate to product. (9)

Cyclic Photophosphorylation A pathway that produces ATP, but not NADPH, in the light reactions of photosynthesis. Energized electrons are shuttled from a reaction center, along a molecular pathway, back to the original reaction center, generating ATP en route. (8)

Cysts Protective, dormant structure formed by some protozoa. (37)

Cytochrome Oxidase A complex of proteins that serves as the final electron carrier in the mitochondrial electron transport system, transferring its electrons to O_2 to form water. (9)

Cytokinesis Final event in eukaryotic cell division in which the cell's cytoplasm and the new nuclei are partitioned into separate daughter cells. (10)

Cytokinins Growth-producing plant hormones which stimulate rapid cell division. (21)

Cytoplasm General term that includes all parts of the cell, except the plasma membrane and the nucleus. (5)

Cytoskeleton Interconnecting network of microfilaments, microtubules, and intermediate filaments that serves as a cell scaffold and provides the machinery for intracellular movements and cell motility. (5)

Cytotoxic (Killer) T Cells A class of T cells capable of recognizing and destroying foreign or infected cells. (30)

 ◀ **D** ▶

Day Neutral Plants Plants that flower at any time of the year, independent of the relative lengths of daylight and darkness. (21)

Deciduous Trees or shrubs that shed their leaves in a particular season, usually autumn, before entering a period of dormancy. (40)

Deciduous Forest Forests characterized by trees that drop their leaves during unfavorable conditions, and leaf out during warm, wet seasons. Less dense than tropical rain forests. (40)

Decomposers (Saprophytes) Organisms that obtain nutrients by breaking down organic compounds in wastes and dead organisms. Includes fungi, bacteria, and some insects. (41)

Deletion Loss of a portion of a chromosome, following breakage of DNA. (13)

Denaturation Change in the normal folding of a protein as a result of heat, acidity, or alkalinity. Such changes result in a loss of enzyme functioning. (4)

Dendrites Cytoplasmic extensions of the cell body of a neuron. They carry impulses from the area of stimulation to the cell body. (23)

Denitrification The conversion by denitrifying bacteria of nitrites and nitrates into nitrogen gas. (41)

Denitrifying Bacteria Bacteria which take soil nitrogen, usable to plants, and convert it to unusable nitrogen gas. (41)

Density-Dependent Factors Factors that control population growth which are influenced by population size. (43)

Density-Independent Factors Factors that control population growth which are not affected by population size. (43)

Deoxyribonucleic Acid (DNA) Double-stranded polynucleotide comprised of deoxyribose (a sugar), phosphate, and four bases (adenine, guanine, cytosine, and thymine). Encoded in the sequence of nucleotides are the instructions for making proteins. DNA is the genetic material in all organisms except certain viruses. (14)

Depolarization A decrease in the potential difference (voltage) across the plasma membrane of a cell typically due to an increase in the movement of sodium ions into the cell. Acts to excite a target cell. (23)

Dermal Bone Bones of vertebrates that form within the dermal layer of the skin, such as the scales of fishes and certain bones of the skull. (26)

Dermal Tissue System In plants, the epidermis in primary growth, or the periderm in secondary growth. (18)

Dermis In animals, layer of cells below the epidermis in which connective tissue predominates. Embedded within it are vessels, various glands, smooth muscle, nerves, and follicles. (26)

Desert Biome characterized by intense solar radiation, very little rainfall, and high winds. (40)

Detrivore Organism that feeds on detritus, dead organisms or their parts, and living organisms' waste. (41)

Deuterostome One path of development exhibited by coelomate animals (e.g., echinoderms and chordates). (39)

Diabetes Mellitus A disease caused by a deficiency of insulin or its receptor, preventing glucose from being absorbed by the cells. (25)

Diaphragm A sheet of muscle that separates the thoracic cavity from the abdominal wall. (29)

Diastolic Pressure The second number of a blood pressure reading; the lowest pressure in the arteries just prior to the next heart contraction. (28)

Diatoms are golden-brown algae that are distinguished most dramatically by their intricate silica shells. (37)

Dicotyledonae (Dicots) One of the two classes of flowering plants, characterized by having seeds with two cotyledons, flower parts in 4s or 5s, net-veined leaves, one main root, and vascular bundles in a circular array within the stem. (Compare with Monocotylenodonae). (18)

Diffusion Tendency of molecules to move from a region of higher concentration to a region of lower concentration, until they are uniformly dispersed. (7)

Digestion The process by which food particles are disassembled into molecules small enough to be absorbed into the organism's cells and tissues. (27)

Digestive System System of specialized organs that ingests food, converts nutrients to a form that can be distributed throughout the animal's body, and eliminates undigested residues. (27)

Digestive Tract Tubelike channel through which food matter passes from its point of ingestion at the mouth to the elimination of indigestible residues from the anus. (27)

Dihybrid Cross A mating between two individuals that differ in two genetically-determined traits. (12)

Dimorphism Presence of two forms of a trait within a population, resulting from diversifying selection. (33)

Dinoflagellates Single-celled photosynthesizers that have two flagella. They are members of the pyrophyta, phosphorescent algae that sometimes cause red tide, often synthesizing a neurotoxin that accumulates in plankton eaters, causing paralytic shellfish poisoning in people who eat the shellfish. (37)

Dioecious Plants that produce either male or female reproductive structures but never both. (38)

Diploid Having two sets of homologous chromosomes. Often written 2N. (10, 13)

Directional Selection The steady shift of phenotypes toward one extreme. (33)

Discontinuous Variation An inheritance pattern in which the phenomenon of all possible phenotypes fall into distinct categories. (Compare with continuous variation). (12)

Displays The signals that form the language by which animals communicate. These signals are species specific and stereotyped and may be visual, auditory, chemical, or tactile. (44)

Disruptive Coloration Coloration that disguises the shape of an organism by breaking up its outline. (42)

Disruptive Selection The steady shift toward more than one extreme phenotype due to the elimination of intermediate phenotypes as has occurred among African swallowtail butterflies whose members resemble more than one species of distasteful butterfly. (33)

Divergent Evolution The emergence of new species as branches from a single ancestral lineage. (33)

Diversifying Selection The increasing frequency of extreme phenotypes because individuals with average phenotypes die off. (33)

Diving Reflex Physiological response that alters the flow of blood in the body of diving mammals that allows the animal to maintain high levels of activity without having to breathe. (29)

Division (or Phylum) A level of the taxonomic hierarchy that groups together members or related classes. (1)

DNA (see **Deoxyribonucleic Acid**)

DNA Cloning The amplification of a particular DNA by use of a growing population of bacteria. The DNA is initially taken up by a bacterial cell—usually as a plasmid—and then replicated along with the bacteria's own DNA. (16)

DNA Fingerprint The pattern of DNA fragments produced after treating a sample of DNA with a particular restriction enzyme and separating the fragments by gel electrophoresis. Since different members of a population have DNA with a different nucleotide sequence, the pattern of DNA fragments produced by this method can be used to identify a particular individual. (16)

DNA Ligase The enzyme that covalently joins DNA fragments into a continuous DNA strand. The enzyme is used in a cell during replication to seal newly-synthesized fragments and by biotechnologists to form recombinant DNA molecules from separate fragments. (14, 16)

DNA Polymerase Enzyme responsible for replication of DNA. It assembles free nucleotides, aligning them with the complementary ones in the unpaired region of a single strand of DNA template. (14)

Dominant The form of an allele that masks the presence of other alleles for the same trait. (12)

Dormancy A resting period, such as seed dormancy in plants or hibernation in animals, in which organisms maintain reduced metabolic rates. (21)

Dorsal In anatomy, the back of an animal. (39)

Double Blind Test A clinical trial of a drug in which neither the human subjects or the researchers know who is receiving the drug or placebo. (2)

Down Syndrome Genetic disorder in humans characterized by distinct facial appearance and mental retardation, resulting from an extra copy of chromosome number 21 (trisomy 21) in each cell. (11, 17)

Duodenum First part of the human small intestine in which most digestion of food occurs. (27)

Duplication The repetition of a segment of a chromosome. (13)

◀ **E** ▶

Ecdysis Molting process by which an arthropod periodically discards its exoskeleton and replaces it with a larger version. The process is controlled by the hormone ecydysone. (39)

Ecdysone An insect steroid hormone that triggers molting and metamorphosis. (15)

Echinodermata A phylum composed of animals having an internal skeleton made of many small calcium carbonate plates which have jutting spines. Includes sea stars, sea urchins, etc. (39)

Echolocation The use of reflected sound waves to help guide an animal through its environment and/or locate objects. (24)

Ecological Equivalent Organisms that occupy similar ecological niches in different regions or ecosystems of the world. (41)

Ecological Niche The habitat, functional role(s), requirements for environmental resources and tolerance ranges for each abiotic condition in relation to an organism. (41)

Ecological Pyramid Illustration showing the energy content, numbers of organisms, or biomass at each trophic level. (41)

Ecology The branch of biology that studies interactions among organisms as well as the interactions of organisms and their physical environment. (40)

Ecosystem Unit comprised of organisms interacting among themselves and with their physical environment. (41)

Ecotypes Populations of a single species with different, genetically fixed tolerance ranges. (41)

Ectoderm In animals, the outer germ cell layer of the gastrula. It gives rise to the nervous system and integument. (32)

Ectotherms Animals that lack an internal mechanism for regulating body temperature. "Cold-blooded" animals. (28)

Edema Swelling of a tissue as the result of an accumulation of fluid that has moved out of the blood vessels. (28)

Effectors Muscle fibers and glands that are activated by neural stimulation. (23)

Efferent (Motor) Nerves The nerves that carry messages from the central nervous system to the effectors, the muscles, and glands. They are divided into two systems: somatic and autonomic. (23)

Egg Female gamete, also called an ovum. A fertilized egg is the product of the union of female and male gametes (egg and sperm cells). (32)

Electrocardiogram (EKG) Recording of the electrical activity of the heart, which is used to diagnose various types of heart problems. (28)

Electron Acceptor Substances that are capable of accepting electrons transferred from an electron donor. For example, molecular oxygen (O_2) is the terminal electron acceptor during respiration. Electron acceptors also receive electrons from chlorophyll during photosynthesis. Electron acceptors may act as part of an electron transport system by transferring the electrons they receive to another substance. (8, 9)

Electron Carrier Substances (such as NAD^+ and FAD) that transport electrons from one step of a metabolic pathway to the next or from metabolic reactions to biosynthetic reactions. (8, 9)

Electrons Negatively charged particles that orbit the atomic nucleus. (3)

Electron Transport System Highly organized assembly of cytochromes and other proteins which transfer electrons. During transport, which occurs within the inner membranes of mitochondria and chloroplasts, the energy extracted from the electrons is used to make ATP. (8, 9)

Electrophoresis A technique for separating different molecules on the basis of their size and/or electric charge. There are various ways the technique is used. In gel electrophoresis, proteins or DNA fragments are driven through a porous gel by their charge, but become separated according to size; the larger the molecule, the slower it can work its way through the pores in the gel, and the less distance it travels along the gel. (16)

Element Substance composed of only one type of atom. (3)

Embryo An organism in the early stages of development, beginning with the first division of the zygote. (32)

Embryo Sac The fully developed female gametophyte within the ovule of the flower. (20)

Emigration Individuals permanently leaving an area or population. (43)

Endergonic Reactions Chemical reactions that require energy input from another source in order to occur. (6)

Endocrine Glands Ductless glands, which secrete hormones directly into surrounding tissue fluids and blood vessels for distribution to the rest of the body by the circulatory system. (25)

Endocytosis A type of active transport that imports particles or small cells into a cell. There are two types of endocytic processes: phagocytosis, where large particles are ingested by the cell, and pinocytosis, where small droplets are taken in. (7)

Endoderm In animals, the inner germ cell layer of the gastrula. It gives rise to the digestive tract and associated organs and to the lungs. (32)

Endodermis The innermost cylindrical layer of cortex surrounding the vascular tissues of the root. The closely pressed cells of the endodermis have a waxy band, forming a waterproof layer, the Casparian strip. (18)

Endogenous Plant responses that are controlled internally, such as biological clocks controlling flower opening. (21)

Endometrium The inner epithelial layer of the uterus that changes markedly with the uterine (menstrual) cycle in preparation for implantation of an embryo. (31)

Endoplasmic Reticulum (ER) An elaborate system of folded, stacked and tubular membranes contained in the cytoplasm of eukaryotic cells. (5)

Endorphins (Endogenous Morphinelike Substances) A class of peptides released from nerve cells of the limbic system of the brain that can block perceptions of pain and produce a feeling of euphoria. (23)

Endoskeleton The internal support structure found in all vertebrates and a few invertebrates (sponges and sea stars). (26)

Endosperm Nutritive tissue in plant embryos and seeds. (20)

Endosperm Mother Cell A binucleate cell in the embryo sac of the female gametophyte, occurring in the ovule of the ovary in angiosperms. Each nucleus is haploid; after fertilization, nutritive endosperm develops. (20)

Endosymbiosis Theory A theory to explain the development of complex eukaryotic cells by proposing that some organelles once were free-living prokaryotic cells that then moved into another larger such cell, forming a beneficial union with it. (5)

Endotherms Animals that utilize metabolically produced heat to maintain a constant, elevated body temperature. "Warm-blooded" animals. (28)

End Product The last product in a metabolic pathway. Typically a substance, such as an amino acid or a nucleotide, that will be used as a monomer in the formation of macromolecules. (6)

Energy The ability to do work. (6)

Entropy Energy that is not available for doing work; measure of disorganization or randomness. (6)

Environmental Resistance The factors that eventually limit the size of a population. (43)

Enzyme Biological catalyst; a protein molecule that accelerates the rate of a chemical reaction. (6)

Eosiniphil A type of phagocytic white blood cell. (28)

Epicotyl The portion of the embryo of a dicot plant above the cotyledons. The epicotyl gives rise to the shoot. (20)

Epidermis In vertebrates, the outer layer of the skin, containing superficial layers of dead cells produced by the underlying living epithelial cells. In plants, the outer layer of cells covering leaves, primary stem, and primary root. (26, 18)

Epididymis Mass of convoluted tubules attached to each testis in mammals. After leaving the testis, sperm enter the tubules where they finish maturing and acquire motility. (31)

Epiglottis A flap of tissue that covers the glottis during swallowing to prevent food and liquids from entering the lower respiratory tract. (29)

Epinephrine (Adrenalin) Substance that serves both as an excitatory neurotransmitter released by certain neurons of the CNS and as a hormone released by the adrenal medulla that increases the body's ability to combat a stressful situation. (25)

Epipelagic Zone The lighted upper ocean zone, where photosynthesis occurs; large populations of phytoplankton occur in this zone. (40)

Epiphyseal Plates The action centers for ossification (bone formation). (26)

Epistasis A type of gene interaction in which a particular gene blocks the expression of another gene at another locus. (12)

Epithelial Tissue Continuous sheets of tightly packed cells that cover the body and line its tracts and chambers. Epithelium is a fundamental tissue type in animals. (22)

Erythrocytes Red blood cells. (28)

Erythropoietin A hormone secreted by the kidney which stimulates the formation of erythrocytes by the bone marrow. (28)

Essential Amino Acids Eight amino acids that must be acquired from dietary protein. If even one is missing from the human diet, the synthesis of proteins is prevented. (27)

Essential Fatty Acids Linolenic and linoleic acids, which are required for phospholipid construction and must be acquired from a dietary source. (27)

Essential Nutrients The 16 minerals essential for plant growth, divided into two groups: macronutrients, which are required in large quantities, and micronutrients, which are needed in small amounts. (19)

Estrogen A female sex hormone secreted by the ovaries when stimulated by pituitary gonadotrophins. (31)

Estuaries Areas found where rivers and streams empty into oceans, mixing fresh water with salt water. (40)

Ethology The study of animal behavior. (44)

Ethylene Gas A plant hormone that stimulates fruit ripening. (21)

Etiolation The condition of rapid shoot elongation, small underdeveloped leaves, bent shoot-hook, and lack of chlorophyll, all due to lack of light. (21)

Eubacteria Typical procaryotic bacteria with peptidoglycan in their cell walls. The majority of monerans are eubacteria. (36)

Eukaryotic Referring to organisms whose cellular anatomy includes a true nucleus with a nuclear envelope, as well as other membrane-bound organelles. (5)

Eusocial Species Social species that have sterile workers, cooperative care of the young, and an overlap of generations so that the colony labor is a family affair. (44)

Eutrophication The natural aging process of lakes and ponds, whereby they become marshes and, eventually, terrestrial environments.

Evolution A process whereby the characteristics of a species change over time, eventually leading to the formation of new species that go about life in new ways. (33)

Evolutionarily Stable Strategy (ESS) A behavioral strategy or course of action that depends on what other members of the population are doing. By definition, an ESS cannot be replaced by any other strategy when most of the members of the population have adopted it. (44)

Excitatory Neurons Neurons that stimulate their target cells into activity. (23)

Excretion Removal of metabolic wastes from an organism. (28)

Excretory System The organ system that eliminates metabolic wastes from the body. (28)

Exergonic Reactions Chemical reactions that occur spontaneously with the release of energy. (6)

Exocrine Glands Glands which secrete their products through ducts directly to their sites of action, e.g., tear glands. (26)

Exocytosis A form of active transport used by cells to move molecules, particles, or other

cells contained in vesicles across the plasma membrane to the cell's environment. (5)

Exogenous Plant responses that are controlled externally, or by environmental conditions. (21)

Exons Structural gene segments that are transcribed and whose genetic information is subsequently translated into protein. (15)

Exoskeletons Hard external coverings found in some animals (e.g., lobsters, insects) for protection, support, or both. Such organisms grow by the process of molting. (26)

Exploitative Competition A competition in which one species manages to get more of a resource, thereby reducing supplies for a competitor. (42)

Exponential Growth An increase by a fixed percentage in a given time period; such as population growth per year. (43)

Extensor Muscle A muscle which, when contracted, causes a part of the body to straighten at a joint. (26)

External Fertilization Fertilization of an egg outside the body of the female parent. (31)

Extinction The loss of a species. (33)

Extracellular Digestion Digestion occurring outside the cell; occurs in bacteria, fungi, and multicellular animals. (27)

Extracellular Matrix Layer of extracellular material residing just outside a cell. (5)

◀ **F** ▶

F_1 First filial generation. The first generation of offspring in a genetic cross. (12)

F_2 Second filial generation. The offspring of an F_1 cross. (12)

Facilitated Diffusion The transport of molecules into cells with the aid of "carrier" proteins embedded in the plasma membrane. This carrier-assisted transport does not require the expenditure of energy by the cell. (7)

FAD Flavin adenine dinucleotide. A coenzyme that functions as an electron carrier in metabolic reactions. When it is reduced to $FADH_2$, this molecule becomes a cellular energy source. (9)

Family A level of the taxonomic hierarchy that groups together members of related genera. (1)

Fast-Twitch Fibers Skeletal muscle fibers that depend on anaerobic metabolism to produce ATP rapidly, but only for short periods of time before the onset of fatigue. Fast-twitch fibers generate greater forces for shorter periods than slow-twitch fibers. (9)

Fat A triglyceride consisting of three fatty acids joined to a glycerol. (4)

Fatty Acid A long unbranched hydrocarbon chain with a carboxyl group at one end. Fatty acids lacking a double bond are said to be saturated. (4)

Fauna The animals in a particular region.

Feedback Inhibition (Negative Feedback) A mechanism for regulating enzyme activity by temporarily inactivating a key enzyme in a biosynthetic pathway when the concentration of the end product is elevated. (6)

Fermentation The direct donation of the electrons of NADH to an organic compound without their passing through an electron transport system. (9)

Fertility Rate In humans, the average number of children born to each woman between 15 and 44 years of age. (43)

Fertilization The process in which two haploid nuclei fuse to form a zygote. (32)

Fetus The term used for the human embryo during the last seven months in the uterus. During the fetal stage, organ refinement accompanies overall growth. (32)

Fibrinogen A rod-shaped plasma protein that, converted to fibrin, generates a tangled net of fibers that binds a wound and stops blood loss until new cells replace the damaged tissue. (28)

Fibroblasts Cells found in connective tissues that secrete the extracellular materials of the connective tissue matrix. These cells are easily isolated from connective tissues and are widely used in cell culture. (22)

Fibrous Root System Many approximately equal-sized roots; monocots are characterized by a fibrous root system. Also called diffuse root system. (18)

Filament The stalk of a stamen of angiosperms, with the anther at its tip. Also, the threadlike chain of cells in some algae and fungi. (20)

Filamentous Fungus Multicellular members of the fungus kingdom comprised mostly of living threads (hyphae) that grow by division of cells at their tips (see molds). (37)

Filter Feeders Aquatic animals that feed by straining small food particles from the surrounding water. (27, 39)

Fitness The relative degree to which an individual in a population is likely to survive to reproductive age and to reproduce. (33)

Fixed Action Patterns Motor responses that may be triggered by some environmental stimulus, but once started can continue to completion without external stimuli. (44)

Flagella Cellular extensions that are longer than cilia but fewer in number. Their undulations propel cells like sperm and many protozoans, through their aqueous environment. (5)

Flexor Muscle A muscle which, when contracted, causes a part of the body to bend at a joint. (26)

Flora The plants in a particular region. (21)

Florigen Proposed A chemical hormone that is produced in the leaves and stimulates flowering. (21)

Fluid Mosaic Model The model proposes that the phospholipid bilayer has a viscosity similar to that of light household oil and that globular proteins float like icebergs within this bilayer. The now favored explanation for the architecture of the plasma membrane. (5)

Follicle (Ovarian) A chamber of cells housing the developing oocytes. (31)

Food Chain Transfers of food energy from organism to organism, in a linear fashion. (41)

Food Web The map of all interconnections between food chains for an ecosystem. (41)

Forest Biomes Broad geographic regions, each with characteristic tree vegetation: 1) tropical rain forests (lush forests in a broad band around the equator), 2) deciduous forests (trees and shrubs drop their leaves during unfavorable seasons), 3) coniferous forest (evergreen conifers). (40)

Fossil Record An entire collection of remains from which paleontologists attempt to reconstruct the phylogeny, anatomy, and ecology of the preserved organisms. (34)

Fossils The preserved remains of organisms from a former geologic age. (34)

Fossorial Living underground.

Founder Effect The potentially dramatic difference in allele frequency of a small founding population as compared to the original population. (33)

Founder Population The individuals, usually few, that colonize a new habitat. (33)

Frameshift Mutation The insertion or deletion of nucleotides in a gene that throws off the reading frame. (14)

Free Radical Atom or molecule containing an unpaired electron, which makes it highly reactive. (3)

Freeze-Fracture Technique in which cells are frozen into a block which is then struck with a knife blade that fractures the block in two. Fracture planes tend to expose the center of membranes for EM examination. (5)

Fronds The large leaf-like structures of ferns. Unlike true leaves, fronds have an apical meristem and clusters of sporangia called sori. (38)

Fruit A mature plant ovary (flower) containing seeds with plant embryos. Fruits protect seeds and aid in their dispersal. (20)

Fruiting Body A spore-producing structure that extends upward in an elevated position from the main mass of a mold or slime mold. (37)

FSH Follicle stimulating hormone. A hormone secreted by the anterior pituitary that prepares a female for ovulation by stimulating the primary follicle to ripen or stimulates spermatogenesis in males. (31)

Functional Groups Accessory chemical entities (e.g., —OH, —NH₂, —CH₃), which help determine the identity and chemical properties of a compound. (4)

Fundamental Niche The potential ecological niche of a species, including all factors affecting that species. The fundamental niche is usually never fully utilized. (41)

Fungus Yeast, mold, or large filamentous mass forming macroscopic fruiting bodies, such as mushrooms. All fungi are eukaryotic nonphotosynthetic heterotrophics with cell walls. (37)

◄ **G** ►

G₁ Stage The first of three consecutive stages of interphase. During G₁, cell growth and normal functions occur. The duration of this stage is most variable. (10)

G₂ Stage The final stage of interphase in which the final preparations for mitosis occur. (10)

Gallbladder A small saclike structure that stores bile salts produced by the liver. (27)

Gamete A haploid reproductive cell—either a sperm or an egg. (10)

Gas Exchange Surface Surface through which gases must pass in order to enter or leave the body of an animal. It may be the plasma membrane of a protistan or the complex tissues of the gills or the lungs in multicellular animals. (29)

Gastrovascular Cavity In cnidarians and flatworms, the branched cavity with only one opening. It functions in both digestion and transport of nutrients. (39)

Gastrula The embryonic stage formed by the inward migration of cells in the blastula. (32)

Gastrulation The process by which the blastula is converted into a gastrula having three germ layers (ectoderm, mesoderm, and endoderm). (32)

Gated Ion Channels Most passageways through a plasma membrane that allow ions to pass contain "gates" that can occur in either an open or a closed conformation. (7, 23)

Gel Electrophoresis (See **Electrophoresis**)

Gene Pool All the genes in all the individuals of a population. (33)

Gene Regulatory Proteins Proteins that bind to specific sites in the DNA and control the transcription of nearby genes. (15)

Genes Discrete units of inheritance which determine hereditary traits. (12, 14)

Gene Therapy Treatment of a disease by alteration of the person's genotype, or the genotype of particular affected cells. (17)

Genetic Carrier A heterozygous individual who shows no evidence of a genetic disorder but, because they possess a recessive allele for a disorder, can pass the mutant gene on to their offspring. (17)

Genetic Code The correspondence between the various mRNA triplets (codons, e.g., UGC) and the amino acid that the triplet specifies (e.g., cysteine). The genetic code includes 64 possible three-letter words that constitute the genetic language for protein synthesis. (14)

Genetic Drift Random changes in allele frequency that occur by chance alone. Occurs primarily in small populations. (33)

Genetic Engineering The modification of a cell or organism's genetic composition according to human design. (16)

Genetic Equilibrium A state in which allele frequencies in a population remain constant from generation to generation. (33)

Genetic Mapping Determining the locations of specific genes or genetic markers along particular chromosomes. This is typically accomplished using crossover frequencies; the more often alleles of two genes are separated during crossing over, the greater the distance separating the genes. (13)

Genetic Recombination The reshuffling of genes on a chromosome caused by breakage of DNA and its reunion with the DNA of a homologue. (11)

Genome The information stored in all the DNA of a single set of chromosomes. (17)

Genotype An individual's genetic makeup. (12)

Genus Taxonomic group containing related species. (1)

Geologic Time Scale The division of the earth's 4.5 billion-year history into eras, periods, and epochs based on memorable geologic and biological events. (35)

Germ Cells Cells that are in the process of or have the potential to undergo meiosis and form gametes. (11, 31)

Germination The sprouting of a seed, beginning with the radicle of the embryo breaking through the seed coat. (21)

Germ Layers Collective name for the endoderm, ectoderm, and mesoderm, from which all the structures of the mature animal develop. (32)

Gibberellins More than 50 compounds that promote growth by stimulating both cell elongation and cell division. (21)

Gills Respiratory organs of aquatic animals. (29)

Globin The type of polypeptide chains that make up a hemoglobin molecule.

Glomerular Filtration The process by which fluid is filtered out of the capillaries into the proximal end of the nephron. Proteins and blood cells remain behind in the bloodstream. (28)

Glomerulus A capillary bundle embedded in the double-membraned Bowman's capsule, through which blood for the kidney first passes. (28)

Glottis Opening leading to the larynx and lower respiratory tract. (29)

Glucagon A hormone secreted by the Islets of Langerhans that promotes glycogen breakdown to glucose. (25)

Glucocorticoids Steroid hormones which regulate sugar and protein metabolism. They are secreted by the adrenal cortex. (25)

Glycogen A highly branched polysaccharide consisting of glucose monomers that serves as a storage of chemical energy in animals. (4)

Glycolysis Cleavage, releasing energy, of the six-carbon glucose molecule into two molecules of pyruvic acid, each containing three carbons. (9)

Glycoproteins Proteins with covalently-attached chains of sugars. (5)

Glycosidic Bond The covalent bond between individual molecules in carbohydrates. (4)

Golgi Complex A system of flattened membranous sacs, which package substances for secretion from the cell. (5)

Gonadotropin-Releasing Hormone (GnRH) Hypothalmic hormone that controls the secretion of the gonadotropins FSH and LH. (31)

Gonadotropins Two anterior pituitary hormones which act on the gonads. Both FSH (follicle-stimulating hormone) and LH (luteinizing hormone) promote gamete development and stimulate the gonads to produce sex hormones. (25)

Gonads Gamete-producing structures in animals: ovaries in females, testes in males. (31)

Grasslands Areas of densely packed grasses and herbaceous plants. (40)

Gravitropisms (Geotropisms) Changes in plant growth caused by gravity. Growth away from gravitational force is called negative gravitropism; growth toward it is positive. (21)

Gray Matter Gray-colored neural tissue in the cerebral cortex of the brain and in the butterfly-shaped interior of the spinal cord. Composed of nonmyelinated cell bodies and dendrites of neurons. (23)

Greenhouse Effect The trapping of heat in the Earth's troposphere, caused by increased levels of carbon dioxide near the Earth's surface; the carbon dioxide is believed to act like glass in a greenhouse, allowing light to reach the Earth, but not allowing heat to escape. (41)

Ground Tissue System All plant tissues except those in the dermal and vascular tissues. (18)

Growth An increase in size, resulting from cell division and/or an increase in the volume of individual cells. (10)

Growth Hormone (GH) Hormone produced by the anterior pituitary; stimulates protein synthesis and bone elongation. (25)

Growth Ring In plants with secondary growth, a ring formed by tracheids and/or vessels with small lumens (late wood) during periods of unfavorable conditions; apparent in cross section. (18)

Guard Cells Specialized epidermal plant cells that flank each stomated pore of a leaf. They regulate the rate of gas diffusion and transpiration. (18)

Guild Group of species with similar ecological niches. (41)

Guttation The forcing of water and mineral completely out to the tips of leaves as a result of positive root pressure. (19)

Gymnosperms The earliest seed plants, bearing naked seeds. Includes the pines, hemlocks, and firs. (38)

◀ **H** ▶

Habitat The place or region where an organism lives. (41)

Habituation The phenomenon in which an animal ceases to respond to a repetitive stimulus. (23, 44)

Hair Cells Sensory receptors of the inner ear that respond to sound vibration and bodily movement. (24)

Half-Life The time required for half the mass of a radioactive element to decay into its stable, non-radioactive form. (3)

Haplodiploidy A genetic pattern of sex determination in which fertilized eggs develop into females and non-fertilized eggs develop into males (as occurs among bees and wasps). (44)

Haploid Having one set of chromosomes per cell. Often written as 1N. (10)

Hardy-Weinberg Law The maintenance of constant allele frequencies in a population from one generation to the next when certain conditions are met. These conditions are the absence of mutation and migration, random mating, a large population, and an equal chance of survival for all individuals. (33)

Haversian Canals A system of microscopic canals in compact bone that transport nutrients to and remove wastes from osteocytes. (26)

Heart An organ that pumps blood (or hemolymph in arthropods) through the vessels of the circulatory system. (28)

Helper T Cells A class of T cells that regulate immune responses by recognizing and activating B cells and other T cells. (30)

Hemocoel In arthropods, the unlined spaces into which fluid (hemolymph) flows when it leaves the blood vessels and bathes the internal organs. (28)

Hemoglobin The iron-containing blood protein that temporarily binds O_2 and releases it into the tissues. (4, 29)

Hemophilia A genetic disorder determined by a gene on the X chromosome (an X-linked trait) that results from the failure of the blood to form clots. (13)

Herbaceous Plants having only primary growth and thus composed entirely of primary tissue. (18)

Herbivore An organism, usually an animal, that eats primary producers (plants). (42)

Herbivory The term for the relationship of a secondary consumer, usually an animal, eating primary producers (plants). (42)

Heredity The passage of genetic traits to offspring which consequently are similar or identical to the parent(s). (12)

Hermaphrodites Animals that possess gonads of both the male and the female. (31)

Heterosporous Higher vascular plants producing two types of spores, a megaspore which grows into a female gametophyte and a microspore which grows into a male gametophyte. (38)

Heterozygous A term applied to organisms that possess two different alleles for a trait. Often, one allele (A) is dominant, masking the presence of the other (a), the recessive. (12)

High Intertidal Zone In the intertidal zone, the region from mean high tide to around just below sea level. Organisms are submerged about 10% of the time. (40)

Histones Small basic proteins that are complexed with DNA to form nucleosomes, the basic structural components of the chromatin fiber. (14)

Homeobox That part of the DNA sequence of homeotic genes that is similar (homologous) among diverse animal species. (32)

Homeostasis Maintenance of fairly constant internal conditions (e.g., blood glucose level, pH, body temperature, etc.) (22)

Homeotic Genes Genes whose products act during embryonic development to affect the spatial arrangement of the body parts. (32)

Hominids Humans and the various groups of extinct, erect-walking primates that were either our direct ancestors or their relatives. Includes the various species of *Homo* and *Australopithecus*. (34)

Homo the genus that contains modern and extinct species of humans. (34)

Homologous Structures Anatomical structures that may have different functions but develop from the same embryonic tissues, suggesting a common evolutionary origin. (34)

Homologues Members of a chromosome pair, which have a similar shape and the same sequence of genes along their length. (10)

Homoplasy (see **Analogous Structures**)

Homosporous Plants that manufacture only one type of spore, which develops into a gametophyte containing both male and female reproductive structures. (38)

Homozygous A term applied to an organism that has two identical alles for a particular trait. (12)

Hormones Chemical messengers secreted by ductless glands into the blood that direct tissues to change their activities and correct imbalances in body chemistry. (25)

Host The organism that a parasite lives on and uses for food. (42)

Human Chorionic Gonadotropin (HCG) A hormone that prevents the corpus luteum from degenerating, thereby maintaining an ade-quate level of progesterone during pregnancy. It is produced by cells of the early embryo. (25)

Human Immunodeficiency Virus (HIV) The infectious agent that causes AIDS, a disease in which the immune system is seriously disabled. (30, 36)

Hybrid An individual whose parents possess different genetic traits in a breeding experiment or are members of different species. (12)

Hybridization Occurs when two distinct species mate and produce hybrid offspring. (33)

Hybridoma A cell formed by the fusion of a malignant cell (a myeloma) and an antibody-producing lymphocyte. These cells proliferate indefinitely and produce monoclonal antibodies. (30)

Hydrogen Bonds Relatively weak chemical bonds formed when two molecules share an atom of hydrogen. (3)

Hydrologic Cycle The cycling of water, in various forms, through the environment, from Earth to atmosphere and back to Earth again. (41)

Hydrolysis Splitting of a covalent bond by donating the H^+ or OH^- of a water molecule to the two components. (4)

Hydrophilic Molecules Polar molecules that are attracted to water molecules and readily dissolve in water. (3)

Hydrophobic Interaction When nonpolar molecules are "forced" together in the presence of a polar solvent, such as water. (3)

Hydrophobic Molecules Nonpolar substances, insoluble in water, which form aggregates to minimize exposure to their polar surroundings. (3)

Hydroponics The science of growing plants in liquid nutrient solutions, without a solid medium such as soil. (19)

Hydrosphere That portion of the Earth composed of water. (40)

Hydrostatic Skeletons Body support systems found usually in underwater animals (e.g., marine worms). Body shape is protected against gravity and other physical forces by internal hydrostatic pressure produced by contracting muscles encircling their closed, fluid-filled chambers. (26)

Hydrothermal Vents Fissures in the ocean floor where sea water becomes superheated. Chemosynthetic bacteria that live in these vents serve as the autotrophs that support a diverse community of ocean-dwelling organisms. (8)

Hyperpolarization An increase in the potential difference (voltage) across the plasma membrane of a cell typically due to an increase in the movement of potassium ions out of the cell. Acts to inhibit a target cell. (23)

Hypertension High blood pressure (above about 130/90). (28)

Hypertonic Solutions Solutions with higher solute concentrations than found inside the cell. These cause a cell to lose water and shrink. (7)

Hypervolume In ecology, a multidimensional area which includes all factors in an organism's ecological niche, or its' potential niche. (41)

Hypocotyl Portion of the plant embryo below the cotyledons. The hypocotyl gives rise to the root and, very often, to the lower part of the stem. (20)

Hypothalamus The area of the brain below the thalamus that regulates body temperature, blood pressure, etc. (25)

Hypothesis A tentative explanation for an observation or a phenomenon, phrased so that it can be tested by experimentation. (2)

Hypotonic Solutions Solutions with lower solute concentrations than found inside the cell. These cause a cell to accumulate water and swell. (7)

◀ **I** ▶

Immigration Individuals permanently moving into a new area or population. (43)

Immune System A system in vertebrates for the surveillance and destruction of disease-causing microorganisms and cancer cells. Composed of lymphocytes, particularly B cells and T cells, and triggered by the introduction of antigens into the body which makes the body, upon their destruction, resistant to a recurrence of the same disease. (30)

Immunoglobulins (IGs) Antibody molecules. (30)

Imperfect Flowers Flowers that contain either stamens or carpels, making them male or female flowers, respectively. (20)

Imprinting A type of learning in which an animal develops an association with an object after exposure to the object during a critical period early in its life. (44)

Inbreeding When individuals mate with close relatives, such as brothers and sisters. May occur when population sizes drastically shrink and results in a decrease in genetic diversity. (33)

Incomplete Digestive Tract A digestive tract with only one opening through which food is taken in and residues are expelled. (27)

Incomplete (Partial) Dominance A phenomenon in which heterozygous individuals are phenotypically distinguishable from either homozygous type. (12)

Incomplete Flower Flowers lacking one or more whorls of sepals, petals, stamen, or pistils. (20)

Independent Assortment The shuffling of members of homologous chromosome pairs in meiosis I. As a result, there are new chromosome combinations in the daughter cells, which later produce offspring with random mixtures of traits from both parents. (11, 12)

Indoleatic Acid (IAA) An auxin responsible for many plant growth responses including apical dominance, a growth pattern in which shoot tips prevent axillary buds from sprouting. (21)

Induction The process in which one embryonic tissue induces another tissue to differentiate along a pathway that it would not otherwise have taken. (32) Stimulation of transcription of a gene in an operon. Occurs when the repressor protein is unable to bind to the operator. (15)

Inflammation A body strategy initiated by the release of chemicals following injury or infection which brings additional blood with its protective cells to the injured area. (30)

Inhibitory Neurons Neurons that oppose a response in the target cells. (23)

Inhibitory Neurotransmitters Substances released from inhibitory neurons where they synapse with the target cell. (23)

Innate Behavior Actions that are under fairly precise genetic control, typically species-specific, highly stereotyped, and that occur in a complete form the first time the stimulus is encountered. (44)

Insight Learning The sudden solution to a problem without obvious trial-and-error procedures. (44)

Insulin One of the two hormones secreted by endocrine centers called Islets of Langerhans; promotes glucose absorption, utilization, and storage. Insulin is secreted by them when the concentration of glucose in the blood begins to exceed the normal level. (25)

Integumentary System The body's protective external covering, consisting of skin and subcutaneous tissue. (26)

Integuments Protective covering of the ovule. (20)

Intercellular Junctions Specialized regions of cell-cell contact between animal cells. (5)

Intercostal Muscles Muscles that lie between the ribs in humans whose contraction expands the thoracic cavity during breathing. (29)

Interference Competition One species' direct interference by another species for the same limited resource; such as aggressive animal behavior. (42)

Internal Fertilization Fertilization of an egg within the body of the female. (31)

Interneurons Neurons situated entirely within the central nervous system. (23)

Internodes The portion of a stem between two nodes. (18)

Interphase Usually the longest stage of the cell cycle during which the cell grows, carries out normal metabolic functions, and replicates its DNA in preparation for cell division. (10)

Interstitial Cells Cells in the testes that produce testosterone, the major male sex hormone. (31)

Interstitial Fluid The fluid between and surrounding the cells of an animal; the extracellular fluid. (28)

Intertidal Zone The region of beach exposed to air between low and high tides. (40)

Intracellular Digestion Digestion occurring inside cells within food vacuoles. The mode of

digestion found in protists and some filter-feeding animals (such as sponges and clams). (27)

Intraspecific Competition Individual organisms of one species competing for the same limited resources in the same habitat, or with overlapping niches. (42)

Intrinsic Rate of Increase (r_m) the maximum growth rate of a population under conditions of maximum birth rate and minimum death rate. (43)

Introns Intervening sequences of DNA in the middle of structural genes, separating exons. (15)

Invertebrates Animals that lack a vertebral column, or backbone. (39)

Ion An electrically charged atom created by the gain or loss of electrons. (3)

Ionic Bond The noncovalent linkage formed by the attraction of oppositely charged groups. (3)

Islets of Langerhans Clusters of endocrine cells in the pancreas that produce insulin and glucagon. (25)

Isolating Mechanisms Barriers that prevent gene flow between populations or among segments of a single population. (33)

Isotopes Atoms of the same element having a different number of neutrons in their nucleus. (3)

Isotonic Solutions Solutions in which the solute concentration outside the cell is the same as that inside the cell. (7)

◀ **J** ▶

Joints Structures where two pieces of a skeleton are joined. Joints may be flexible, such as the knee joint of the human leg or the joints between segments of the exoskeleton of the leg of an insect, or inflexible, such as the joints (sutures) between the bones of the skull. (26)

J-Shaped Curve A curve resulting from exponential growth of a population. (43)

◀ **K** ▶

Karyotype A visual display of an individual's chromosomes. (10)

Kidneys Paired excretory organs which, in humans, are fist-sized and attached to the lower spine. In vertebrates, the kidneys remove nitrogenous wastes from the blood and regulate ion and water levels in the body. (28)

Killer T Cells A type of lymphocyte that functions in the destruction of virus-infected cells and cancer cells. (30)

Kinases Enzymes that catalyze reactions in which phosphate groups are transferred from ATP to another molecule. (6)

Kinetic Energy Energy in motion. (6)

Kinetochore Part of a mitotic (or meiotic) chromosome that is situated within the centromere and to which the spindle fibers attach. (10)

Kingdom A level of the taxonomic hierarchy that groups together members of related phyla or divisions. Modern taxonomy divides all organisms into five Kingdoms: Monera, Protista, Fungi, Plantae, and Animalia. (1)

Klinefelter Syndrome A male whose cells have an extra X chromosome (XXY). The syndrome is characterized by underdeveloped male genitalia and feminine secondary sex characteristics. (17)

Krebs Cycle A circular pathway in aerobic respiration that completely oxidizes the two pyruvic acids from glycolysis. (9)

K-Selected Species Species that produce one or a few well-cared for individuals at a time. (43)

◀ **L** ▶

Lacteal Blind lymphatic vessel in the intestinal villi that receives the absorbed products of lipid digestion. (27)

Lactic Acid Fermentation The process in which electrons removed during glycolysis are transferred from NADH to pyruvic acid to form lactic acid. Used by various prokaryotic cells under oxygen-deficient conditions and by muscle cells during strenuous activity. (9)

Lake Large body of standing fresh water, formed in natural depressions in the Earth. Lakes are larger than ponds. (40)

Lamella In bone, concentric cylinders of calcified collagen deposited by the osteocytes. The laminated layers produce a greatly strengthened structure. (26)

Large Intestine Portion of the intestine in which water and salts are reabsorbed. It is so named because of its large diameter. The large instestine, except for the rectum, is called the colon. (27)

Larva A self-feeding, sexually, and developmentally immature form of an animal. (32)

Larynx The short passageway connecting the pharynx with the lower airways. (29)

Latent (hidden) Infection Infection by a microorganism that causes no symptoms but the microbe is well-established in the body. (36)

Lateral Roots Roots that arise from the pericycle of older roots; also called branch roots or secondary roots. (18)

Law of Independent Assortment Alleles on nonhomologous chromosomes segregate independently of one another. (12)

Law of Segregation During gamete formation, pairs of alleles separate so that each sperm or egg cell has only one gene for a trait. (12)

Law of the Minimum The ecological principle that a species' distribution will be limited by whichever abiotic factor is most deficient in the environment. (41)

Laws of Thermodynamics Physical laws that describe the relationship of heat and mechanical energy. The first law states that energy cannot be created or destroyed, but one form can change into another. The second law states that the total energy the universe decreases as energy conversions occur and some energy is lost as heat. (6)

Leak Channels Passageways through a plasma membrane that do not contain gates and, therefore, are always open for the limited diffusion of a specific substance (ion) through the membrane. (7, 23)

Learning A process in which an animal benefits from experience so that its behavior is better suited to environmental conditions. (44)

Lenticels Loosely packed cells in the periderm of the stem that create air channels for transferring CO_2, H_2O, and O_2. (18)

Leukocytes White blood cells. (28)

LH Luteinizing hormone. A hormone secreted by the anterior pituitary that stimulates testosterone production in males and triggers ovulation and the transformation of the follicle into the corpus luteum in females. (31)

Lichen Symbiotic associations between certain fungi and algae. (37)

Life Cycle The sequence of events during the lifetime of an organism from zygote to reproduction. (39)

Ligaments Strong straps of connective tissue that hold together the bones in articulating joints or support an organ in place. (26)

Light-Dependent Reactions First stage of photosynthesis in which light energy is converted to chemical energy in the form of energy-rich ATP and NADPH. (8)

Light-Independent Reactions Second stage of photosynthesis in which the energy stored in ATP and NADPH formed in the light reactions is used to drive the reactions in which carbon dioxide is converted to carbohydrate. (8)

Limb Bud A portion of an embryo that will develop into either a forelimb or hindlimb. (32)

Limbic System A series of an interconnected group of brain structures, including the thalamus and hypothalamus, controlling memory and emotions. (23)

Limiting Factors The critical factors which impose restraints of the distribution, health, or activities of an organism. (41)

Limnetic Zone Open water of lakes, through which sunlight penetrates and photosynthesis occurs. (40)

Linkage The tendency of genes of the same chromosome to stay together rather than to assort independently. (13)

Linkage Groups Groups of genes located on the same chromosome. The genes of each linkage group assort independently of the genes of other linkage groups. In all eukaryotic organisms, the number of linkage groups is equal to the haploid number of chromosomes. (13)

Lipids A diverse group of biomolecules that are insoluble in water. (4)

Lithosphere The solid outer zone of the Earth; composed of the crust and outermost portion of the mantle. (40)

Littoral Zone Shallow, nutrient-rich waters of a lake, where sunlight reaches the bottom; also the lakeshore. Rooted vegetation occurs in this zone. (40)

Locomotion The movement of an organism from one place to another. (26)

Locus The chromosomal location of a gene. (13)

Logistic Growth Population growth producing a sigmoid, or S-shaped, growth curve. (43)

Long-Day Plants Plants that flower when the length of daylight exceeds some critical period. (21)

Longitudinal Fission The division pattern in flagellated protozoans, where division is along the length of the cell.

Loop of Henle An elongated section of the renal tubule that dips down into the kidney's medulla and then ascends back out to the cortex. It separates the proximal and distal convoluted tubules and is responsible for forming the salt gradient on which water reabsorption in the kidney depends. (28)

Low Density Lipoprotein (LDL) Particles that transport cholesterol in the blood. Each particle consists of about 1,500 cholesterol molecules surrounded by a film of phospholipids and protein. LDLs are taken into cells following their binding to cell surface LDL receptors. (7)

Low Intertidal Zone In the intertidal zone, the region which is uncovered by "minus" tides only. Organisms are submerged about 90% of the time. (40)

Lumen A space within an hollow organ or tube. (28)

Luminescence (see **Bioluminescence**)

Lungs The organs of terrestrial animals where gas exchange occurs. (29)

Lymph The colorless fluid in lymphatic vessels. (28)

Lymphatic System Network of fluid-carrying vessels and associated organs that participate in immunity and in the return of tissue fluid to the main circulation. (28)

Lymphocytes A group of non-phagocytic white blood cells which combat microbial invasion, fight cancer, and neutralize toxic chemicals. The two classes of lymphocytes, B cells and T cells, are the heart of the immune system. (28, 30)

Lymphoid Organs Organs associated with production of blood cells and the lymphatic system, including the thymus, spleen, appendix, bone marrow, and lymph nodes. (30)

Lysis (1) To split or dissolve. (2) Cell bursting.

Lysomes A type of storage vesicle produced by the Golgi complex, containing hydrolytic (digestive) enzymes capable of digesting many kinds of macromolecules in the cell. The membrane around them keeps them sequestered. (5)

◄ **M** ►

M Phase That portion of the cell cycle during which mitosis (nuclear division) and cytokinesis (cytoplasmic division) takes place. (10)

Macroevolution Evolutionary changes that lead to the appearance of new species. (33)

Macrofungus Filamentous fungus so named for the large size of its fleshy sexual structures; a mushroom, for example. (37)

Macromolecules Large polymers, such as proteins, nucleic acids, and polysaccharides. (4)

Macronutrients Nutrients required by plants in large amounts: carbon, oxygen, hydrogen, nitrogen, potassium, calcium, phosphorus, magnesium, and sulfur. (19)

Macrophages Phagocytic cells that develop from monocytes and present antigen to lymphocytes. (30)

Macroscopic Referring to biological observations made with the naked eye or a hand lens.

Mammals A class of vertebrates that possesses skin covered with hair and that nourishes their young with milk from mammary glands. (39)

Mammary Glands Glands contained in the breasts of mammalian mothers that produce breast milk. (39)

Marsupials Mammals with a cloaca whose young are born immature and complete their development in an external pouch in the mother's skin. (39)

Mass Extinction The simultaneous extinction of a multitude of species as the result of a drastic change in the environment. (33, 35)

Maternal Chromosomes The set of chromosomes in an individual that were inherited from the mother. (11)

Mechanoreceptors Sensory receptors that respond to mechanical pressure and detect motion, touch, pressure, and sound. (24)

Medulla The center-most portion of some organs. (23)

Medusa The motile, umbrella-shaped body form of some members of the phylum Cnidaria, with mouth and tentacles on the lower, concave service. (Compare with polyp.) (39)

Megaspores Spores that divide by mitosis to produce female gametophytes that produce the egg gamete. (20)

Meiosis The division process that produces cells with one-half the number of chromosomes in each somatic cell. Each resulting daughter cell is haploid (1N) (11)

Meiosis I A process of reductional division in which homologous chromosomes pair and then segregate. Homologues are partitioned into separate daughter cells. (11)

Meiosis II Second meiotic division. A division process resembling mitosis, except that the haploid number of chromosomes is present. After the chromosomes line up at the meta-phase plate, the two sister chromatids separate. (11)

Melanin A brown pigment that gives skin and hair its color (12)

Melanoma A deadly form of skin cancer that develops from pigment cells in the skin and is promoted by exposure to the sun. (14)

Memory Cells Lymphocytes responsible for active immunity. They recall a previous exposure to an antigen and, on subsequent exposure to the same antigen, proliferate rapidly into plasma cells and produce large quantities of antibodies in a short time. This protection typically lasts for many years. (30)

Mendelian Inheritance Transmission of genetic traits in a manner consistent with the principles discovered by Gregor Mendel. Includes traits controlled by simple dominant or recessive alleles; more complex patterns of transmission are referred to as Nonmendelian inheritance. (12)

Meninges The thick connective tissue sheath which surrounds and protects the vertebrate brain and spinal cord. (23)

Menstrual Cycle The repetitive monthly changes in the uterus that prepare the endometrium for receiving and supporting an embryo. (31)

Meristematic Region New cells arise from this undifferentiated plant tissue; found at root or shoot apical meristems, or lateral meristems. (18)

Meristems In plants, clusters of cells that retain their ability to divide, thereby producing new cells. One of the four basic tissues in plants. (18)

Mesoderm In animals, the middle germ cell layer of the gastrula. It gives rise to muscle, bone, connective tissue, gonads, and kidney. (32)

Mesopelagic Zone The dimly lit ocean zone beneath the epipelagic zone; large fishes, whales and squid occupy this zone; no phytoplankton occur in this zone. (40)

Mesophyll Layers of cells in a leaf between the upper and lower epidermis; produced by the ground meristem. (18)

Messenger RNA (mRNA) The RNA that carries genetic information from the DNA in the nucleus to the ribosomes in the cytoplasm, where the sequence of bases in the mRNA is translated into a sequence of amino acids. (14)

Metabolic Intermediates Compounds produced as a substrate are converted to end product in a series of enzymatic reactions. (6)

Metabolic Pathways Set of enzymatic reactions involved in either building or dismantling complex molecules. (6)

Metabolic Rate A measure of the level of activity of an organism usually determined by measuring the amount of oxygen consumed by an individual per gram body weight per hour. (22)

Metabolic Water Water produced as a product of metabolic reactions. (28)

Metabolism The sum of all the chemical reactions in an organism; includes all anabolic and catabolic reactions. (6)

Metamorphosis Transformation from one form into another form during development. (32)

Metaphase The stage of mitosis when the chromosomes line-up along the metaphase plate, a plate that usually lies midway between the spindle poles. (10)

Metaphase Plate Imaginary plane within a dividing cell in which the duplicated chromosomes become aligned during metaphase. (10)

Microbes Microscopic organisms. (36)

Microbiology The branch of biology that studies microorganisms. (36)

Microevolution Changes in allele frequency of a species' gene pool which has not generated new species. Exemplified by changes in the pigmentation of the peppered moth and by the acquisition of pesticide resistance in insects. (33)

Microfibrils Bundles formed from the intertwining of cellulose molecules, i.e., long chains of glucose molecules in the cell walls of plants. (5)

Microfilaments Thin actin-containing protein fibers that are responsible for maintenance of cell shape, muscle contraction and cyclosis. (5)

Micrometer One millionth (1/1,000,000) of a meter.

Micronutrients Nutrients required by plants in small amounts: iron, chlorine, copper, manganese, zinc, molybdenum, and boron. (19)

Micropyle A small opening in the integuments of the ovule through which the pollen tube grows to deliver sperm. (21)

Microspores Spores within anthers of flowers. They divide by mitosis to form pollen grains, the male gametophytes that produce the plant's sperm. (20)

Microtubules Thin, hollow tubes in cells; built from repeating protein units of tubulin. Microtubules are components of cilia, flagella, and the cytoskeleton. (5)

Microvilli The small projections on the cells that comprise each villus of the intestinal wall, further increasing the absorption surface area of the small intestine. (27)

Middle Intertidal Zone In the intertidal zone, the region which is covered and uncovered twice a day, the zero of tide tables. Organisms are submerged about 50% of the time. (40)

Migration Movements of a population into or out of an area. (44)

Mimicry A defense mechanism where one species resembles another in color, shape, behavior, or sound. (42)

Mineralocorticoids Steroid hormones which regulate the level of sodium and potassium in the blood. (25)

Mitochondria Organelles that contain the biochemical machinery for the Krebs cycle and the electron transport system of aerobic respiration. They are composed of two membranes, the inner one forming folds, or cristae. (9)

Mitosis The process of nuclear division producing daughter cells with exactly the same number of chromosomes as in the mother cell. (10)

Mitosis Promoting Factor (MPF) A protein that appears to be a universal trigger of cell division in eukaryotic cells. (10)

Mitotic Chromosomes Chromosomes whose DNA-protein threads have become coiled into microscopically visible chromosomes, each containing duplicated chromatids ready to be separated during mitosis. (10)

Molds Filamentous fungi that exist as colonies of threadlike cells but produce no macroscopic fruiting bodies. (37)

Molecule Chemical substance formed when two or more atoms bond together; the smallest unit of matter that possesses the qualities of a compound. (3)

Mollusca A phylum, second only to Arthropoda in diversity. Composed of three main classes: 1) Gastropoda (spiral-shelled), 2) Bivalvia (hinged shells), 3) Cephalopoda (with tentacles or arms and no, or very reduced shells). (39)

Molting (Ecdysis) Shedding process by which certain arthropods lose their exoskeletons as their bodies grow larger. (39)

Monera The taxonomic kingdom comprised of single-celled prokaryotes such as bacteria, cyanobacteria, and archebacteria. (36)

Monoclonal Antibodies Antibodies produced by a clone of hybridoma cells, all of which descended from one cell. (30)

Monocotyledae (Monocots) One of the two divisions of flowering plants, characterized by seeds with a single cotyledon, flower parts in 3s, parallel veins in leaves, many roots of approximately equal size, scattered vascular bundles in its stem anatomy, pith in its root anatomy, and no secondary growth capacity. (18)

Monocytes A type of leukocyte that gives rise to macrophages. (28)

Monoecious Both male and female reproductive structures are produced on the same sporophyte individual. (20, 38)

Monohybrid Cross A mating between two individuals that differ only in one genetically-determined trait. (12)

Monomers Small molecular subunits which are the building blocks of macromolecules. The macromolecules in living systems are constructed of some 40 different monomers. (4)

Monotremes A group of mammals that lay eggs from which the young are hatched. (39)

Morphogenesis The formation of form and internal architecture within the embryo brought about by such processes as programmed cell death, cell adhesion, and cell movement. (32)

Morphology The branch of biology that studies form and structure of organisms.

Mortality Deathrate in a population or area. (43)

Motile Capable of independent movement.

Motor Neurons Nerve cells which carry outgoing impulses to their effectors, either glands or muscles. (23)

Mucosa The cell layer that lines the digestive tract and secretes a lubricating layer of mucus. (27)

Mullerian Mimicry Resemblance of different species, each of which is equally obnoxious to predators. (42)

Multicellular Consisting of many cells. (35)

Multichannel Food Chain Where the same primary producer supplies the energy for more than one food chain. (41)

Multiple Allele System Three or more possible alleles for a given trait, such as ABO blood groups in humans. (12)

Multiple Fission Division of the cell's nucleus without a corresponding division of cytoplasm.

Multiple Fruits Fruits that develop from pistils of separate flowers. (20)

Muscle Fiber A muiltinucleated skeletal muscle cell that results from the fusion of several pre-muscle cells during embryonic development. (26)

Muscle Tissue Bundles and sheets of contractile cells that shorten when stimulated, providing force for controlled movement. (26)

Mutagens Chemical or physical agents that induce genetic change. (14)

Mutation Random heritable changes in DNA that introduce new alleles into the gene pool. (14)

Mutualism The symbiotic interaction in which both participants benefit. (42)

Mycology The branch of biology that studies fungi. (37)

Mycorrhizae An association between soil fungi and the roots of vascular plants, increasing the plant's ability to extract water and minerals from the soil. (19)

Myelin Sheath In vertebrates, a jacket which covers the axons of high-velocity neurons, thereby increasing the speed of a neurological impulse. (23)

Myofibrils In striated muscle, the banded fibrils that lie parallel to each other, constituting the bulk of the muscle fiber's interior and powering contraction. (26)

Myosin A contractile protein that makes up the major component of the thick filaments of a muscle cell and is also present in nonmuscle cells. (26)

◄ N ►

NADPH Nicotinamide adenine dinucleotide phosphate. NADPH is formed by reduction of $NADP^+$, and serves as a store of electrons for use in metabolism (see Reducing Power). (9)

NAD⁺ Nicotinamide adenine dinucleotide. A coenzyme that functions as an electron carrier in metabolic reactions. When reduced to NADH, the molecule becomes a cellular energy source. (9)

Natality Birthrate in a population or area. (43)

Natural Killer (NK) Cells Nonspecific, lymphocytelike cells which destroy foreign cells and cancer cells. (30)

Natural Selection Differential survival and reproduction of organisms with a resultant increase in the frequency of those best adapted to the environment. (33)

Neanderthals A subspecies of Homo sapiens different from that of modern humans that were characterized by heavy bony skeletons and thick bony ridges over the eyes. They disappeared about 35,000 years ago. (34)

Nectary Secretory gland in flowering plants containing sugary fluid that attracts pollinators as a food source. Usually located at the base of the flower. (20)

Negative Feedback Any regulatory mechanism in which the increased level of a substance inhibits further production of that substance, thereby preventing harmful accumulation. A type of homeostatic mechanism. (22, 25)

Negative Gravitropism In plants, growth against gravitational forces, or shoot growth upward. (21)

Nematocyst Within the stinging cell (cnidocyte) of cnidarians, a capsule that contains a coiled thread which, when triggered, harpoons prey and injects powerful toxins. (39)

Nematoda The widespread and abundant animal phylum containing the roundworms. (39)

Nephridium A tube surrounded by capillaries found in an organism's excretory organs that removes nitrogenous wastes and regulates the water and chemical balance of body fluids. (28)

Nephron The functional unit of the vertebrate kidney, consisting of the glomerulus, Bowman's capsule, proximal and distal convoluted tubules, and loop of Henle. (28)

Nerve Parallel bundles of neurons and their supporting cells. (23)

Nerve Impulse A propagated action potential. (23)

Nervous Tissue Excitable cells that receive stimuli and, in response, transmit an impulse to another part of the animal. (23)

Neural Plate In vertebrates, the flattened plate of dorsal ectoderm of the late gastrula that gives rise to the nervous system. (32)

Neuroglial Cells Those cells of a vertebrate nervous system that are not neurons. Includes a variety of cell types including Schwann cells. (23)

Neuron A nerve cell. (23)

Neurosecretory Cells Nervelike cells that secrete hormones rather than neurotransmitter substances when a nerve impulse reaches the distal end of the cell. In vertebrates, these cells arise from the hypothalamus. (25)

Neurotoxins Substances, such as curare and tetanus toxin, that interfere with the transmission of neural impulses. (23)

Neurotransmitters Chemicals released by neurons into the synaptic cleft, stimulating or inhibiting the post-synaptic target cell. (23)

Neurulation Formation by embryonic induction of the neural tube in a developing vertebrate embryo. (32)

Neutrons Electrically neutral (uncharged) particles contained within the nucleus of the atom. (3)

Neutrophil Phagocytic leukocyte, most numerous in the human body. (28)

Niche An organism's habitat, role, resource requirements, and tolerance ranges for each abiotic condition. (42)

Niche Breadth Relative size and dimension of ecological niches; for example, broad or narrow niches. (41)

Niche Overlap Organisms that have the same habitat, role, environmental requirements, or needs. (41)

Nitrogen Fixation The conversion of atmospheric nitrogen gas N_2 into ammonia (NH_3) by certain bacteria and cyanobacteria. (19)

Nitrogenous Wastes Nitrogen-containing metabolic waste products, such as ammonia or urea, that are produced by the breakdown of proteins and nucleic acids. (28)

Nodes The attachment points of leaves to a stem. (18)

Nodes of Ranvier Uninsulated (nonmyelinated) gaps along the axon of a neuron. (23)

Noncovalent Bonds Linkages between two atoms that depend on an attraction between positive and negative charges between molecules or ions. Includes ionic and hydrogen bonds. (3)

Non-Cyclic Photophosphorylation The pathway in the light reactions of photosynthesis in which electrons pass from water, through two photosystems, and then ultimately to NADP⁺. During the process, both ATP and NADPH are produced. It is so named because the electrons do not return to their reaction center. (8)

Nondisjunction Failure of chromosomes to separate properly at meiosis I or II. The result is that one daughter will receive an extra chromosome and the other gets one less. (11, 13)

Nonpolar Molecules Molecules which have an equal charge distribution throughout their structure and thus lack regions with a localized positive or negative charge. (3)

Notochord A flexible rod that is below the dorsal surface of the chordate embryo, beneath the nerve cord. In most chordates, it is replaced by the vertebral column. (32)

Nuclear Envelope A double membrane pierced by pores that separates the contents of the nucleus from the rest of the eukaryotic cell. (5)

Nucleic Acids DNA and RNA; linear polymers of nucleotides, responsible for the storage and expression of genetic information. (4, 14)

Nucleoid A region in the prokaryotic cell that contains the genetic material (DNA). It is unbounded by a nuclear membrane. (36)

Nucleoplasm The semifluid substance of the nucleus in which the particulate structures are suspended. (5)

Nucleosomes Nuclear protein complex consisting of a length of DNA wrapped around a central cluster of 8 histones. (14)

Nucleotides Monomers of which DNA and RNA are built. Each consists of a 5-carbon sugar, phosphate, and a nitrogenous base. (4)

Nucleous (pl. nucleoli) One or more darker regions of a nucleus where each ribosomal subunit is assembled from RNA and protein. (5)

Nucleus The large membrane-enclosed organelle that contains the DNA of eukaryotic cells. (5)

Nucleus, Atomic The center of an atom containing protons and neutrons. (3)

◀ **O** ▶

Obligate Symbiosis A symbiotic relationship between two organisms that is necessary for the survival or both organisms. (42)

Olfaction The sense of smell. (24)

Oligotrophic Little nourished, as a young lake that has few nutrients and supports little life. (40)

Omnivore An animal that obtains its nutritional needs by consuming plants and other animals. (42)

Oncogene A gene that causes cancer, perhaps activated by mutation or a change in its chromosomal location. (10)

Oocyte A female germ cell during any of the stages of meiosis. (31)

Oogenesis The process of egg production. (31)

Oogonia Female germ cells that have not yet begun meiosis. (31)

Open Circulatory System Circulatory system in which blood travels from vessels to tissue spaces, through which it percolates prior to returning to the main vessel (compare with closed circulatory system). (28)

Operator A regulatory gene in the operon of bacteria. It is the short DNA segment to which the repressor binds, thus preventing RNA polymerase from attaching to the promoter. (15)

Operon A regulatory unit in prokaryotic cells that controls the expression of structural genes. The operon consists of structural genes that produce enzymes for a particular metabolic pathway, a regulator region composed of a promoter and an operator, and R (regulator) gene that produces a repressor. (15)

Order A level of the taxonomic hierarchy that groups together members of related families. (1)

Organ Body part composed of several tissues that performs specialized functions. (22)

Organelle A specialized part of a cell having some particular function. (5)

Organic Compounds Chemical compounds that contain carbon. (4)

Organism A living entity able to maintain its organization, obtain and use energy, reproduce, grow, respond to stimuli, and display homeostatis. (1)

Organogenesis Organ formation in which two or more specialized tissue types develop in a precise temporal and spatial relationship to each other. (32)

Organ System Group of functionally related organs. (22)

Osmoregulation The maintenance of the proper salt and water balance in the body's fluids. (28)

Osmosis The diffusion of water through a differentially permeable membrane into a hypertonic compartment. (7)

Ossification Synthesis of a new bone. (26)

Osteoclast A type of bone cell which breaks down the bone, thereby releasing calcium into the bloodstream for use by the body. Osteoclasts are activated by hormones released by the parathyroid glands. (26)

Osteocytes Living bone cells embedded within the calcified matrix they manufacture. (26)

Osteoporosis A condition present predominantly in postmenopausal women where the bones are weakened due to an increased rate of bone resorption compared to bone formation. (26)

Ovarian Cycle The cycle of egg production within the mammalian ovary. (31)

Ovarian Follicle In a mammalian ovary, a chamber of cells in which the oocyte develops. (31)

Ovary In animals, the egg-producing gonad of the female. In flowering plants, the enlarged base of the pistil, in which seeds develop. (20)

Oviduct (Fallopian Tube) The tube in the female reproductive organ that connects the ovaries and uterus and where fertilization takes place. (31)

Ovulation The release of an egg (ovum) from the ovarian follicle. (31)

Ovule In seed plants, the structure containing the female gametophyte, nucellus, and integuments. After fertilization, the ovule develops into a seed. (20, 38)

Ovum An unfertilized egg cell; a female gamete. (31)

Oxidation The removal of electrons from a compound during a chemical reaction. For a carbon atom, the fewer hydrogens bonded to a carbon, the greater the oxidation state of the atom. (6)

Oxidative Phosphorylation The formation of ATP from ADP and inorganic phosphate that occurs in the electron-transport chain of cellular respiration. (8, 9)

Oxyhemoglobin A complex of oxygen and hemoglobin, formed when blood passes through the lungs and is dissociated in body tissues, where oxygen is released. (29)

Oxytocin A female hormone released by the posterior pituitary which triggers uterine contractions during childbirth and the release of milk during nursing. (25)

◀ **P** ▶

P680 Reaction Center (P = Pigment) Special chlorophyll molecule in Photosystem II that traps the energy absorbed by the other pigment molecules. It absorbs light energy maximally at 680 nm. (8)

Palisade Parenchyma In dicot leaves, densely packed, columnar shaped cells functioning in photosynthesis. Found just beneath the upper epidermis. (18)

Pancreas In vertebrates, a large gland that produces digestive enzymes and hormones. (27)

Parallel Evolution When two species that have descended from the same ancestor independently acquire the same evolutionary adaptations. (33)

Parapatric Speciation The splitting of a population into two species' populations under conditions where the members of each population reside in adjacent areas. (33)

Parasite An organism that lives on or inside another called a host, on which it feeds. (39, 42)

Parasitism A relationship between two organisms where one benefits, and the other is harmed. (42)

Parasitoid Parasitic organisms, such as some insect larvae, which kill their host. (42)

Parasympathetic Nervous System Part of the autonomic nervous system active during relaxed activity. (23)

Parathyroid Glands Four glands attached to the thyroid gland which secrete parathyroid hormone (PTH). When blood calcium levels are low, PTH is secreted, causing calcium to be released from bone. (25)

Parenchyma The most prevalent cell type in herbaceous plants. These thin-walled, polygonal-shaped cells function in photosynthesis and storage. (18)

Parthenogenesis Process by which offspring are produced without egg fertilization. (31)

Passive Immunity Immunity achieved by receiving antibodies from another source, as occurs with a newborn infant during nursing. (30)

Paternal Chromosomes The set of chromosomes in an individual that were inherited from the father. (11)

Pathogen A disease-causing microorganism. (36)

Pectoral Girdle In humans, the two scapulae (shoulder blades) and two clavicles (collarbones) which support and articulate with the bones of the upper arm. (26)

Pedicel A shortened stem carrying a flower. (20)

Pedigree A diagram showing the inheritance of a particular trait among the members of the family. (13)

Pelagic Zone The open oceans, divided into three layers: 1) photo- or epipelagic (sunlit), 2) mesopelagic (dim light), 3) aphotic or bathypelagic (always dark). (40)

Pelvic Girdle The complex of bones that connect a vertebrate's legs with its backbone. (26)

Penis An intrusive structure in the male animal which releases male gametes into the female's sex receptacle. (31)

Peptide Bond The covalent bond between the amino group of one amino acid and the carboxyl group of another. (4)

Peptidoglycan A chemical component of the prokaryotic cell wall. (36)

Percent Annual Increase A measure of population increase; the number of individuals (people) added to the population per 100 individuals. (43)

Perennials Plants that live longer than two years. (18)

Perfect Flower Flowers that contain both stamens and pistils. (20)

Perforation Plate In plants, that portion of the wall of vessel members that is perforated, and contains an area with neither primary nor secondary cell wall; a "hole" in the cell wall. (18)

Pericycle One or more layers of cells found in roots, with phloem or xylem to its' inside, and the endodermis to its' outside. Functions in producing lateral roots and formation of the vascular cambium in roots with secondary growth. (18)

Periderm Secondary tissue that replaces the epidermis of stems and roots. Consists of cork, cork cambium, and an internal layer of parenchyma cells. (18)

Peripheral Nervous System Neurons, excluding those of the brain and spinal cord, that permeate the rest of the body. (23)

Peristalsis Sequential waves of muscle contractions that propel a substance through a tube. (27)

Peritoneum The connective tissue that lines the coelomic cavities. (39)

Permeability The ability to be penetrable, such as a membrane allowing molecules to pass freely across it. (7)

Petal The second whorl of a flower, often brightly colored to attract pollinators; collectively called the corolla. (20)

Petiole The stalk leading to the blade of a leaf. (18)

pH A scale that measures the concentration of hydrogen ions in a solution. The pH scale extends from 0 to 14. Acidic solutions have a pH of less than 7; alkaline solutions have a pH above 7; neutral solutions have a pH equal to 7. (3)

Phagocytosis Engulfing of food particles and foreign cells by amoebae and white blood cells. A type of endocytosis. (5)

Pharyngeal Pouches In the vertebrate embryo, outgrowths from the walls of the pharynx that break through the body surface to form gill slits. (32)

Pharynx The throat; a portion of both the digestive and respiratory system just behind the oral cavity. (29)

Phenotype An individual's observable characteristics that are the expression of its genotype. (12)

Pheromones Chemicals that, when released by an animal, elicit a particular behavior in other animals of the same species. (44)

Phloem The vascular tissue that transports sugars and other organic molecules from sites of photosynthesis and storage to the rest of the plant. (18)

Phloem Loading The transfer of assimilates to phloem conducting cells, from photosynthesizing source cells. (19)

Phloem Unloading The transfer of assimilates to storage (sink) cells, from phloem conducting cells. (19)

Phospholipids Lipids that contain a phosphate and a variable organic group that form polar, hydrophilic regions on an otherwise nonpolar, hydrophobic molecule. They are the major structural components of membranes. (4)

Phosphorylation A chemical reaction in which a phosphate group is added to a molecule or atom. (6)

Photoexcitation Absorption of light energy by pigments, causing their electrons to be raised to a higher energy level. (8)

Photolysis The splitting of water during photosynthesis. The electrons from water pass to Photosystem II, the protons enter the lumen of the thylakoid and contribute to the proton gradient across the thylakoid membrane, and the oxygen is released into the atmosphere. (8)

Photon A particle of light energy. (8)

Photoperiod Specific lengths of day and night which control certain plant growth responses to light, such as flowering or germination. (21)

Photoperiodism Changes in the behavior and physiology of an organism in response to the relative lengths of daylight and darkness, i.e., the photoperiod. (21)

Photoreceptors Sensory receptors that respond to light. (24)

Photorespiration The phenomenon in which oxygen binds to the active site of a CO_2-fixing enzyme, thereby competing with CO_2 fixation, and lowering the rate of photosynthesis. (8)

Photosynthesis The conversion by plants of light energy into chemical energy stored in carbohydrate. (8)

Photosystems Highly organized clusters of photosynthetic pigments and electron/hydrogen carriers embedded in the thylakoid membranes of chloroplasts. There are two photosystems, which together carry out the light reactions of photosynthesis. (8)

Photosystem I Photosystem with a P700 reaction center; participates in cyclic photophosphorylation as well as in noncyclic photophosphorylation. (8)

Photosystem II Photosystem activated by a P680 reaction center; participates only in noncyclic photophosphorylation and is associated with photolysis of water. (8)

Phototropism The growth responses of a plant to light. (21)

Phyletic Evolution The gradual evolution of one species into another. (33)

Phylogeny Evolutionary history of a species. (35)

Phylum The major taxonomic divisions in the Animal kingdom. Members of a phylum share common, basic features. The Animal kingdom is divided into approximately 35 phyla. (39)

Physiology The branch of biology that studies how living things function. (22)

Phytochrome A light-absorbing pigment in plants which controls many plant responses, including photoperiodism. (21)

Phytoplankton Microscopic photosynthesizers that live near the surface of seas and bodies of fresh water. (37)

Pineal Gland An endocrine gland embedded within the brain that secretes the hormone melatonin. Hormone secretion is dependent on levels of environmental light. In amphibians and reptiles, melatonin controls skin coloration. In humans, pineal secretions control sexual maturation and daily rhythms. (25)

Pinocytosis Uptake of small droplets and dissolved solutes by cells. A type of endocytosis. (5)

Pistil The female reproductive part and central portion of a flower, consisting of the ovary, style and stigma. May contain one carpel, or one or more fused carpels. (20)

Pith A plant tissue composed of parenchyma cells, found in the central portion of primary growth stems of dicots, and monocot roots. (18)

Pith Ray Region between vascular bundles in vascular plants. (18)

Pituitary Gland (see **Posterior and Anterior Pituitary**).

Placenta In mammals (exclusive of marsupials and monotremes), the structure through which nutrients and wastes are exchanged between the mother and embryo/fetus. Develops from both embryonic and uterine tissues. (32)

Plant Multicellular, autotrophic organism able to manufacture food through photosynthesis. (38)

Plasma In vertebrates, the liquid portion of the blood, containing water, proteins (including fibrinogen), salts, and nutrients. (28)

Plasma Cells Differentiated antibody-secreting cells derived from B lymphocytes. (30)

Plasma Membrane The selectively permeable, molecular boundary that separates the cytoplasm of a cell from the external environment. (5)

Plasmid A small circle of DNA in bacteria in addition to its own chromosome. (16)

Plasmodesmata Openings between plant cell walls, through which adjacent cells are connected via cytoplasmic threads. (19)

Plasmodium Genus of protozoa that causes malaria. (37)

Plasmodium A huge multinucleated "cell" stage of a plasmodial slime mold that feeds on dead organic matter. (37)

Plasmolysis The shrinking of a plant cell away from its cell wall when the cell is placed in a hypertonic solution. (7)

Platelets Small, cell-like fragments derived from special white blood cells. They function in clotting. (28)

Plate Tectonics The theory that the earth's crust consists of a number of rigid plates that rest on an underlying layer of semimolten rock. The movement of the earth's plates results from the upward movement of molten rock into the solidified crust along ridges within the ocean floor. (35)

Platyhelminthes The phylum containing simple, bilaterally symmetrical animals, the flatworms. (39)

Pleiotropy Where a single mutant gene produces two or more phenotypic effects. (12)

Pleura The double-layered sac which surrounds the lungs of a mammal. (29, 39)

Pneumocytis Pneumonia (PCP) A disease of the respiratory tract caused by a protozoan that strikes persons with immunodeficiency diseases, such as AIDS. (30)

Point Mutations Changes that occur at one point within a gene, often involving one nucleotide in the DNA. (14)

Polar Body A haploid product of meiosis of a female germ cell that has very little cytoplasm and disintegrates without further function. (31)

Polar Molecule A molecule with an unequal charge distribution that creates distinct positive and negative regions or poles. (3)

Pollen The male gametophyte of seed plants, comprised of a generative nucleus and a tube nucleus surrounded by a tough wall. (20)

Pollen Grain The male gametophyte of conifers and angiosperms, containing male gametes. In angiosperms, pollen grains are contained in the pollen sacs of the anther of a flower. (20)

Pollination The transfer of pollen grains from the anther of one flower to the stigma of another. The transfer is mediated by wind, water, insects, and other animals. (20)

Polygenic Inheritance An inheritance pattern in which a phenotype is determined by two or more genes at different loci. In humans, examples include height and pigmentation. (12)

Polymer A macromolecule formed of monomers joined by covalent bonds.. Includes proteins, polysaccharides, and nucleic acids. (4)

Polymerase Chain Reaction (PCR) Technique to amplify a specific DNA molecule using a temperature-sensitive DNA polymerase obtained from a heat-resistant bacterium. Large numbers of copies of the initial DNA molecule can be obtained in a short period of time, even when the starting material is present in vanishingly small amounts, as for example from a blood stain left at the scene of a crime. (16)

Polymorphic Property of some protozoa to produce more than one stage of organism as they complete their life cycles. (37)

Polymorphic Genes Genes for which several different alleles are known, such as those that code for human blood type. (17)

Polyp Stationary body form of some members of the phylum Cnidaria, with mouth and tentacles facing upward. (Compare with medusa.) (39)

Polypeptide An unbranched chain of amino acids covalently linked together and assembled on a ribosome during translation. (4)

Polyploidy An organism or cell containing three or more complete sets of chromosomes. Polyploidy is rare in animals but common in plants. (33)

Polysaccharide A carbohydrate molecule consisting of monosaccharide units. (4)

Polysome A complex of ribosomes found in chains, linked by mRNA. Polysomes contain the ribosomes that are actively assembling proteins. (14)

Polytene Chromosomes Giant banded chromosomes found in certain insects that form by the repeated duplication of DNA. Because of the multiple copies of each gene in a cell, polytene chromosomes can generate large amounts of a gene product in a short time. Transcription occurs at sites of chromosome puffs. (13)

Pond Body of standing fresh water, formed in natural depressions in the Earth. Ponds are smaller than lakes. (40)

Population Individuals of the same species inhabiting the same area. (43)

Population Density The number of individual species living in a given area. (43)

Positive Gravitropism In plants, growth with gravitational forces, or root growth downward. (21)

Posterior Pituitary A gland which manufactures no hormones but receives and later releases hormones produced by the cell bodies of neurons in the hyopthalamus. (25)

Potential Energy Stored energy, such as occurs in chemical bonds. (6)

Preadaptation A characteristic (adaptation) that evolved to meet the needs of an organism in one type of habitat, but fortuitously allows the organism to exploit a new habitat. For example, lobed fins and lungs evolved in ancient fishes to help them live in shallow, stagnant ponds, but also facilitated the evolution of terrestrial amphibians. (33, 39)

Precells Simple forerunners of cells that, presumably, were able to concentrate organic molecules, allowing for more frequent molecular reactions. (35)

Predation Ingestion of prey by a predator for energy and nutrients. (42)

Predator An organism that captures and feeds on another organism (prey). (42)

Pressure Flow In the process of phloem loading and unloading, pressure differences resulting from solute increases in phloem conducting cells and neighboring xylem cells cause the flow of water to phloem. A concentration gradient is created between xylem and phloem cells. (19)

Prey An organism that is captured and eaten by another organism (predator). (42)

Primary Consumer Organism that feeds exclusively on producers (plants). Herbivores are primary consumers. (41)

Primary Follicle In the mammalian ovary, a structure composed of an oocyte and its surrounding layer of follicle cells. (31)

Primary Growth Growth from apical meristems, resulting in an increase in the lengths of shoots and roots in plants. (18)

Primary Immune Response Process of antibody production following the first exposure to an antigen. There is a lag time from exposure until the appearance in the blood of protective levels of antibodies. (30)

Primary Oocyte Female germ cell that is either in the process of or has completed the first meiotic division. In humans, germ cells may remain in this stage in the ovary for decades. (31)

Primary Producers All autotrophs in a biotic environment that use sunlight or chemical energy to manufacture food from inorganic substances. (41)

Primary Sexual Characteristics Gonads, reproductive tracts, and external genitals. (31)

Primary Spermatocyte Male germ cell that is either in the process of or has completed the first meiotic division. (31)

Primary Succession The development of a community in an area previously unoccupied by any community; for example, a "bare" area such as rock, volcanic material, or dunes. (41)

Primary Tissues Tissues produced by primary meristems of a plant, which arise from the shoot and root apical meristems. In general, primary tissues are a result of an increase in plant length. (18)

Primary Transcript An RNA molecule that has been transcribed but not yet subjected to any type of processing. The primary transcript corresponds to the entire stretch of DNA that was transcribed. (15)

Primates Order of mammals that includes humans, apes, monkeys, and lemurs. (39)

Primitive An evolutionary early condition. Primitive features are those that were also present in an early ancestor, such as five digits on the feet of terrestrial vertebrates. (34)

Prions An infectious particle that contains protein but no nucleic acid. It causes slow diseases of animals, including neurological disease of humans. (36)

Processing-Level Control Control of gene expression by regulating the pathway by which a primary RNA transcript is processed into an mRNA. (15)

Products In a chemical reaction, the compounds into which the reactants are transformed. (6)

Profundal Zone Deep, open water of lakes, where it is too dark for photosynthesis to occur. (40)

Progesterone A hormone produced by the corpus luteum within the ovary. It prepares and maintains the uterus for pregnancy, participates in milk production, and prevents the ovary from releasing additional eggs late in the cycle or during pregnancy. (25)

Prokaryotic Referring to single-celled organisms that have no membrane separating the DNA from the cytoplasm and lack membrane-enclosed organelles. Prokaryotes are confined to the kingdom Monera; they are all bacteria. (36)

Prokaryotic Fission The most common type of cell division in bacteria (prokaryotes). Duplicated DNA strands are attached to the plasma membrane and become separated into two cells following membrane growth and cell wall formation. (10, 36)

Prolactin A hormone produced by the anterior pituitary, stimulating milk production by mammary glands. (25)

Promoter A short segment of DNA to which RNA polymerase attaches at the start of transcription. (15)

Prophase Longest phase of mitosis, involving the formation of a spindle, coiling of chromatin fibers into condensed chromosomes, and movement of the chromosomes to the center of the cell. (10)

Prostaglandins Hormones secreted by endocrine cells scattered throughout the body responsible for such diverse functions as contraction of uterine muscles, triggering the inflammatory response, and blood clotting. (25)

Prostate Gland A muscular gland which produces and releases fluids that make up a substantial portion of the semen. (31)

Proteins Long chains of amino acids, linked together by peptide bonds. They are folded into specific shapes essential to their functions. (4)

Prothallus The small, heart-shaped gametophyte of a fern. (38)

Protists A member of the kingdom Protista; simple eukaryotic organisms that share broad taxonomic similarities. (36, 37)

Protocooperation Non-compulsory interactions that benefit two organisms, e.g., lichens. (42)

Proton Gradient A difference in hydrogen ion (proton) concentration on opposite sides of a membrane. Proton gradients are formed during photosynthesis and respiration and serve as a store of energy used to drive ATP formation. (8, 9)

Protons Positively charged particles within the nucleus of an atom. (3)

Protostomes One path of development exhibited by coelomate animals (e.g., mollusks, annelids, and arthropods). (39)

Protozoa Member of protist kingdom that is unicellular and eukaryotic; vary greatly in size, motility, nutrition and life cycle. (37)

Provirus DNA copy of a virus' nucleic acid that becomes integrated into the host cell's chromosome. (36)

Pseudocoelamates Animals in which the body cavity is not lined by cells derived from mesoderm. (39)

Pseudopodia (psuedo = false, pod = foot). Pseudopodia are fingerlike extensions of cytoplasm that flow forward from the "body" of an amoeba; the rest of the cell then follows. (37)

Puberty Development of reproductive capacity, often accompanied by the appearance of secondary sexual characteristics. (31)

Pulmonary Circulation The loop of the circulatory system that channels blood to the lungs for oxygenation. (28)

Punctuated Equilibrium Theory A theory to explain the phenomenon of the relatively sudden appearance of new species, followed by long periods of little or no change. (33)

Punnett Square Method A visual method for predicting the possible genotypes and their expected ratios from a cross. (12)

Pupa In insects, the stage in metamorphosis between the larva and the adult. Within the pupal case, there is dramatic transformation in body form as some larval tissues die and others differentiate into those of the adult. (32)

Purine A nitrogenous base found in DNA and RNA having a double ring structure. Adenine and guanine are purines. (14)

Pyloric Sphincter Muscular valve between the human stomach and small intestine. (27)

Pyrimidine A nitrogenous base found in DNA and RNA having a single ring structure. Cytosine, thymine, and uracil are pyrimidines. (14)

Pyramid of Biomass Diagrammatic representation of the total dry weight of organisms at each trophic level in a food chain or food web. (41)

Pyramid of Energy Diagrammatic representation of the flow of energy through trophic levels in a food chain or food web. (41)

Pyramid of Numbers Similar to a pyramid of energy, but with numbers of producers and consumers given at each trophic level in a food chain or food web. (41)

◄ **Q** ►

Quiescent Center The region in the apical meristem of a root containing relatively inactive cells. (18)

◄ R ►

R-Group   The variable portion of a molecule. (4)

r-Selected Species   Species that possess adaptive strategies to produce numerous offspring at once. (43)

Radial Symmetry   The quality possessed by animals whose bodies can be divided into mirror images by more than one median plane. (39)

Radicle   In the plant embryo, the tip of the hypocotyl that eventually develops into the root system. (20)

Radioactivity   A property of atoms whose nucleus contains an unstable combination of particles. Breakdown of the nucleus causes the emission of particles and a resulting change in structure of the atom. Biologists use this property to track labeled molecules and to determine the age of fossils. (3)

Radiodating   The use of known rates of radioactive decay to date a fossil or other ancient object. (3, 34)

Radioisotope   An isotope of an element that is radioactive. (3)

Radiolarian   A prozoan member of the protistan group Sarcodina that secretes silicon shells through which it captures food.

Rainshadow   The arid, leeward (downwind) side of a mountain range. (40)

Random Distribution   Distribution of individuals of a population in a random manner; environmental conditions must be similar and individuals do not affect each other's location in the population. (43)

Reactants   Molecules or atoms that are changed to products during a chemical reaction. (6)

Reaction   A chemical change in which starting molecules (reactants) are transformed into new molecules (products). (6)

Reaction Center   A special chlorophyll molecule in a photosystem (P_{700} in Photosystem I, P_{680} in Photosystem II). (8)

Realized Niche   Part of the fundamental niche of an organism that is actually utilized. (41)

Receptacle   The base of a flower where the flower parts are attached; usually a widened area of the pedicel. (20)

Receptor-Mediated Endocytosis   The uptake of materials within a cytoplasmic vesicle (endocytosis) following their binding to a cell surface receptor. (7)

Receptor Site   A site on a cell's plasma membrane to which a chemical such as a hormone binds. Each surface site permits the attachment of only one kind of hormone. (5)

Recessive   An allele whose expression is masked by the dominant allele for the same trait. (12)

Recombinant DNA   A DNA molecule that contains DNA sequences derived from different biological sources that have been joined together in the laboratory. (16)

Recombination   The rejoining of DNA pieces with those of a different strand or with the same strand at a point different from where the break occurred. (11, 13)

Red Marrow   The soft tissue in the interior of bones that produces red blood cells. (26)

Red Tide   Growth of one of several species of reddish brown dinoflagellate algae so extensive that it tints the coastal waters and inland lakes a distinctive red color. Often associated with paralytic shellfish poisoning (see dinoflagellates). (37)

Reducing Power   A measure of the cell's ability to transfer electrons to substrates to create molecules of higher energy content. Usually determined by the available store of NADPH, the molecule from which electrons are transferred in anabolic (synthetic) pathways. (6)

Reduction   The addition of electrons to a compound during a chemical reaction. For a carbon atom, the more hydrogens that are bonded to the carbon, the more reduced the atom. (6)

Reduction Division   The first meiotic division during which a cell's chromosome number is reduced in half. (11)

Reflex   An involuntary response to a stimulus. (23)

Reflex Arc   The simplest example of central nervous system control, involving a sensory neuron, motor neuron, and usually an interneuron. (23)

Regeneration   Ability of certain animals to replace injured or lost limbs parts by growth and differentiation of undifferentiated stem cells. (15)

Region of Elongation   In root tips, the region just above the region of cell division, where cells elongate and the root length increases. (18)

Region of Maturation   In root tips, the region above the region of elongation; cells differentiate and root hairs occur in this region. (18)

Regulatory Genes   Genes whose sole function is to control the expression of structural genes. (15)

Releaser   A sign stimulus that is given by an individual to another member of the same species, eliciting a specific innate behavior. (44)

Releasing Factors   Hormones secreted by the tips of hypothalmic neurosecretory cells that stimulate the anterior pituitary to release its hormones. GnRH, for example, stimulates the release of gonadotropins. (25)

Renal   Referring to the kidney. (28)

Replication   Duplication of DNA, usually prior to cell division. (14)

Replication Fork   The site where the two strands of a DNA helix are unwinding during replication. (14)

Repression   Inhibition of transcription of a gene which, in an operon, occurs when repressor protein binds to the operator. (15)

Repressor   Protein encoded by a bacterial regulatory gene that binds to an operator site of an operon and inhibits transcription. (15)

Reproduction   The process by which an organism produces offspring. (31)

Reproductive Isolation   Phenomenon in which members of a single population become split into two populations that no longer interbreed. (33)

Reproductive System   System of specialized organs that are utilized for the production of gametes and, in some cases, the fertilization and/or development of an egg. (31)

Reptiles   Members of class Reptilia, scaly, air-breathing, egg-laying vertebrates such as lizards, snakes, turtles, and crocodiles. (39)

Resolving Power   The ability of an optical instrument (eye, microscopes) to discern whether two very close objects are separate from each other. (APP.)

Resource Partitioning   Temporal or spatial sharing of a resource by different species. (42)

Respiration   Process used by organisms to exchange gases with the environment; the source of oxygen required for metabolism. The process organisms use to oxidize glucose to CO_2 and H_2O using an electron transport system to extract energy from electrons and store it in the high-energy bonds of ATP. (29)

Respiratory System   The specialized set of organs that function in the uptake of oxygen from the environment. (29)

Resting Potential   The electrical potential (voltage) across the plasma membrane of a neuron when the cell is not carrying an impulse. Results from a difference in charge across the membrane. (23)

Restriction Enzyme   A DNA-cutting enzyme found in bacteria. (16)

Restriction Fragment Length Polymorphism (RFLP)   Certain sites in the DNA tend to have a highly variable sequence from one individual to another. Because of these differences, restriction enzymes cut the DNA from different individuals into fragments of different length. Variations in the length of particular fragments (RFLPs) can be used as genetic signposts for the identification of nearby genes of interest. (17)

Restriction Fragments   The DNA fragments generated when purified DNA is treated with a particular restriction enzyme. (16)

Reticular Formation   A series of interconnected sites in the core of the brain (brainstem) that selectively arouse conscious activity. (23)

Retroviruses RNA viruses that reverse the typical flow of genetic information; within the infected cell, the viral DNA serves as a template for synthesis of a DNA copy. Examples include HIV, which causes AIDS, and certain cancer viruses. (36)

Reverse Genetics Determining the amino acid sequence and function of a polypeptide from the nucleotide sequence of the gene that codes for that polypeptide. (17)

Reverse Transcriptase An enzyme present in retroviruses that transcriibes a strand of DNA, using viral RNA as the template. (36)

Rhizoids Slender cells that resemble roots but do not absorb water or minerals. (36)

Rhodophyta Red algae; seaweeds that can absorb deeper penetrating light rays than most aquatic photosynthesizers. (36)

Rhyniophytes Ancient plants having vascular tissue which thrived in marshy areas during the Silurian period.

Ribonucleic Acid (RNA) Single-stranded chain of nucleotides each comprised of ribose (a sugar), phosphate, and one of four bases (adenine, guanine, cytosine, and uracil). The sequence of nucleotides in RNA is dictated by DNA, from which it is transcribed. There are three classes of RNA: mRNA, tRNA, and rRNA, all required for protein synthesis. (4, 14)

Ribosomal RNA (rRNA) RNA molecules that form part of the ribosome. Included among the rRNAs is one that is thought to catalyze peptide bond formation. (14)

Ribosomes Organelles involved in protein synthesis in the cytoplasm of the cell. (14)

Ribozymes RNAs capable of catalyzing a chemical reaction, such as peptide bond formation or RNA cutting and splicing. (15)

Rickettsias A group of obligate intracellular parasites, smaller than the typical prokaryote. They cause serious diseases such as typhus. (36)

River Flowing body of surface fresh water; rivers are formed from the convergence of streams. (40)

RNA Polymerase The enzyme that directs transcription and assembling RNA nucleotides in the growing chain. (14)

RNA Processing The process by which the intervening (noncoding) portions of a primary RNA transcript are removed and the remaining (coding) portions are spliced together to form an mRNA. (15)

Root Cap A protective cellular helmet at the tip of a root that surrounds delicate meristematic cells and shields them from abrasion and directs the growth downward. (18)

Root Hairs Elongated surface cells near the tip of each root for the absorption of water and minerals. (18)

Root Nodules Knobby structures on the roots of certain plants. They house nitrogen-fixing bacteria which supply nitrogen in a form that can be used by the plant. (19)

Root Pressure A positive pressure as a result of continuous water supply to plant roots that assists (along with transpirational pull) the pushing of water and nutrients up through the xylem. (19)

Root System The below-ground portion of a plant, consisting of main roots, lateral roots, root hairs, and associated structures and systems such as root nodules or mycorrhizae. (18)

Rough ER (RER) Endoplasmatic reticulum with many ribosomes attached. As a result, they appear rough in electron micrographs. (5)

Ruminant Grazing mammals that possess an additional stomach chamber called rumen which is heavily fortified with cellulose-digesting microorganisms. (27)

◄ **S** ►

S Phase The second stage of interphase in which the materials needed for cell division are synthesized and an exact copy of cell's DNA is made by DNA replication. (10)

Sac Body The body plan of simple animals, like cnidarians, where there is a single opening leading to and from a digestion chamber.

Saltatory Conduction The "hopping" movement of an impulse along a myelinated neuron from one Node of Ranvier to the next one. (23)

Sap Fluid found in xylem or sieve of phloem. (20)

Saprophyte Organisms, mainly fungi and bacteria, that get their nutrition by breaking down organic wastes and dead organisms, also called decomposers. (42)

Saprobe Organism that obtains its nutrients by decomposing dead organisms. (37)

Sarcolemma The plasma membrane of a muscle fiber. (26)

Sarcomere The contractile unit of a myofibril in skeletal muscle. (26)

Sarcoplasmic Reticulum (SR) In skeletal muscle, modified version of the endoplasmic reticulum that stores calcium ions. (26)

Savanna A grassland biome with alternating dry and rainy seasons. The grasses and scattered trees support large numbers of grazing animals. (40)

Scaling Effect A property that changes disproportionately as the size of organisms increase. (22)

Scanning Electron Microscope (SEM) A microscope which operates by showering electrons back and forth across the surface of a specimen prepared with a thin metal coating. The resultant image shows three-dimensional features of the specimen's surface. (APP.)

Schwann Cells Cells which wrap themselves around the axons of neurons forming an insulating myelin sheath composed of many layers of plasma membrane. (23)

Sclereids Irregularly-shaped sclerenchyma cells, all having thick cell walls; a component of seed coats and nuts. (18)

Sclerenchyma Component of the ground tissue system of plants. They are thick walled cells of various shapes and sizes, providing support or protection. They continue to function after the cell dies. (18)

Sclerenchyma Fibers Non-living elongated plant cells with tapering ends and thick secondary walls. A supportive cell type found in various plant tissues. (18)

Sebaceous Glands Exocrine glands of the skin that produce a mixture of lipids (sebum) that oil the hair and skin. (26)

Secondary Cell Wall An additional cell wall that improves the strength and resiliency of specialized plant cells, particularly those cells found in stems that support leaves, flowers, and fruit. (5)

Secondary Consumer Organism that feeds exclusively on primary consumers; mostly animals, but some plants. (41)

Secondary Growth Growth from cambia in perennials; results in an increase in the diameter of stems and roots. (18)

Secondary Meristems (vascular cambium, cork cambrium) Rings or clusters of meristematic cells that increase the width of stems and roots when the divide. (18)

Secondary Sex Characteristics Those characteristics other than the gonads and reproductive tract that develop in response to sex hormones. For example, breasts and pubic hair in women and a deep voice and pubic hair in men. (31)

Secondary Succession The development of a community in an area previously occupied by a community, but which was disturbed in some manner; for example, fire, development, or clear-cutting forests. (41)

Secondary Tissues Tissues produced to accommodate new cell production in plants with woody growth. Secondary tissues are produced from cambia, which produce vascular and cork tissues, leading to an increase in plant girth. (18)

Second Messenger Many hormones, such as glucagon and thyroid hormone, evoke a response by binding to the outer surface of a target cell and causing the release of another substance (which is the second messenger). The best-studied second messenger is cyclic AMP which is formed by an enzyme on the inner surface of the plasma membrane following the binding of a hormone to the outer surface of the membrane. The cyclic AMP diffuses into the cell and activates a protein kinase. (25)

Secretion The process of exporting materials produced by the cell. (5)

Seed A mature ovule consisting of the embryo, endosperm, and seed coat. (20)

Seed Dormancy Metabolic inactivity of seeds until favorable conditions promote seed germination. (20)

Secretin Hormone secreted by endocrine cells in the wall of the intestine that stimulates the release of digestive products from the pancreas. (25)

Segmentation A condition in which the body is constructed, at least in part, from a series of repeating parts. Segmentation occurs in annelids, arthropods, and vertebrates (as revealed during embryonic development). (39)

Selectively Permeable A term applied to the plasma membrane because membrane proteins control which molecules are transported. Enables a cell to import and accumulate the molecules essential for normal metabolism. (7)

Semen The fluid discharged during a male orgasm. (31)

Semiconsevative Replication The manner in which DNA replicates; half of the original DNA strand is conserved in each new double helix. (14)

Seminal Vesicles The organs which produce most of the ejaculatory fluid. (31)

Seminiferous Tubules Within the testes, highly coiled and compacted tubules, lined with a self-perpetuating layer of spermatogonia, which develop into sperm. (31)

Senescence Aging and eventual death of an organism, organ or tissue. (3, 18)

Sense Strand The one strand of a DNA double helix that contains the information that encodes the amino sequence of a polypeptide. This is the strand that is selectively transcribed by RNA polymerase forming an mRNA that can be properly translated. (14)

Sensory Neurons Neurons which relay impulses to the central nervous system. (23)

Sensory Receptors Structures that detect changes in the external and internal environment and transmit the information to the nervous system. (24)

Sepal The outermost whorl of a flower, enclosing the other flower parts as a flower bud; collectively called the calyx. (20)

Sessile Sedentary, incapable of independent movement. (39)

Sex Chromosomes The one chromosomal pair that is not identical in the karyotypes of males and females of the same animal species. (10, 13)

Sex Hormones Steroid hormones which influence the production of gametes and the development of male or female sex characteristics. (25)

Sexual Dimorphism Differences in the appearance of males and females in the same species. (33)

Sexual Reproduction The process by which haploid gametes are formed and fuse during fertilization to form a zygote. (31)

Sexual Selection The natural selection of adaptations that improve the chances for mating and reproducing. (33)

Shivering Involuntary muscular contraction for generating metabolic heat that raises body temperature. (28)

Shoot In angiosperms, the system consisting of stems, leaves, flowers and fruits. (18)

Shoot System The above-ground portion of an angiosperm plant consisting of stems with nodes, including branches, leaves, flowers and fruits. (18)

Short-Day Plants Plants that flower in late summer or fall when the length of daylight becomes shorter than some critical period. (21)

Shrubland A biome characterized by densely growing woody shrubs in mediterranean type climate; growth is so dense that understory plants are not usually present. (40)

Sickle Cell Anemia A genetic (recessive autosomal) disorder in which the beta globin genes of adult hemoglobin molecules contain an amino acid substitution which alters the ability of hemoglobin to transport oxygen. During times of oxygen stress, the red blood cells of these individuals may become sickle shaped, which interferes with the flow of the cells through small blood vessels. (4, 17)

Sieve Plate Found in phloem tissue in plants, the wall between sieve-tube members, containing perforated areas for passage of materials. (18)

Sieve-Tube Member A living, food-conducting cell found in phloem tissue of plants; associated with a companion cell. (18)

Sigmoid Growth Curve An S-shaped curve illustrating the lag phase, exponential growth, and eventual approach of a population to its carrying capacity. (43)

Sign Stimulus An object or action in the environment that triggers an innate behavior. (44)

Simple Fruits Fruits that develop from the ovary of one pistil. (20)

Simple Leaf A leaf that is undivided; only one blade attached to the petiole. (18)

Sinoatrial (SA) Node A collection of cells that generates an action potential regulating heart beat; the heart's pacemaker. (28)

Skeletal Muscles Separate bundles of parallel, striated muscle fibers anchored to the bone, which they can move in a coordinated fashion. They are under voluntary control. (26)

Skeleton A rigid form of support found in most animals either surrounding the body with a protective encasement or providing a living girder system within the animal. (26)

Skull The bones of the head, including the cranium. (26)

Slow-Twitch Fibers Skeletal muscle fibers that depend on aerobic metabolism for ATP production. These fibers are capable of undergoing contraction for extended periods of time without fatigue, but generate lesser forces than fast-twitch fibers. (9)

Small Intestine Portion of the intestine in which most of the digestion and absorption of nutrients takes place. It is so named because of its narrow diameter. There are three sections: duodenum, jejunum, and ilium. (27)

Smell Sense of the chemical composition of the environment. (24)

Smooth ER (SER) Membranes of the endoplasmic reticulum that have no ribosomes on their surface. SER is generally more tubular than the RER. Often acts to store calcium or synthesize steroids. (5)

Smooth Muscle The muscles of the internal organs (digestive tract, glands, etc.). Composed of spindle-shaped cells that interlace to form sheets of visceral muscle. (26)

Social Behavior Behavior among animals that live in groups composed of individuals that are dependent on one another and with whom they have evolved mechanisms of communication. (44)

Social Learning Learning of a behavior from other members of the species. (44)

Social Parasitism Parasites that use behavioral mechanisms of the host organism to the parasite's advantage, thereby harming the host. (42)

Solute A substance dissolved in a solvent. (3)

Solution The resulting mixture of a solvent and a solute. (3)

Solvent A substance in which another material dissolves by separating into individual molecules or ions. (3)

Somatic Cells Cells that do not have the potential to form reproductive cells (gametes). Includes all cells of the body except germ cells. (11)

Somatic Nervous System The nerves that carry messages to the muscles that move the skeleton either voluntarily or by reflex. (23)

Somatic Sensory Receptors Receptors that respond to chemicals, pressure, and temperature that are present in the skin, muscles, tendons, and joints. Provides a sense of the physiological state of the body. (24)

Somites In the vertebrate embryo, blocks of mesoderm on either side of the notochord that give rise to muscles, bones, and dermis. (32)

Speciation The formation of new species. Occurs when one population splits into separate populations that diverge genetically to the point where they become separate species. (33)

Species Taxonomic subdivisions of a genus. Each species has recognizable features that distinguish it from every other species. Members of one species generally will not interbreed with members of other species. (33)

Specific Epithet In taxonomy, the second term in an organism's scientific name identifying its species within a particular genus. (1)

Spermatid Male germ cell that has completed meiosis but has not yet differentiated into a sperm. (31)

Spermatogenesis The production of sperm. (31)

Spermatogonia Male germ cells that have not yet begun meiosis. (31)

Spermatozoa (Sperm) Male gametes. (31)

Sphinctors Circularly arranged muscles that close off the various tubes in the body.

Spinal Cord A centralized mass of neurons for processing neurological messages and linking the brain with that part of peripheral nervous system not reached by the cranial nerves. (23)

Spinal Nerves Paired nerves which emerge from the spinal cord and innervate the body. Humans have 31 pairs of spinal nerves. (23)

Spindle Apparatus In dividing eukaryotic cells, the complex rigging, made of microtubules, that aligns and separates duplicated chromosomes. (10)

Splash Zone In the intertidal zone, the uppermost region receiving splashes and sprays of water to the mean of high tides. (40)

Spleen One of the organs of the lymphatic system that produces lymphocytes and filters blood; also produces red blood cells in the human fetus. (28)

Splicing The step during RNA processing in which the coding segments of the primary transcript are covalently linked together to form the mRNA. (15)

Spongy Parenchyma In monocot and dicot leaves, loosely arranged cells functioning in photosynthesis. Found above the lower epidermis and beneath the palisade parenchyma in dicots, and between the upper and lower epidermis in monocots. (18)

Spontaneous Generation Disproven concept that living organisms can arise directly from inanimate materials. (2)

Sporangiospores Black, asexual spores of the zygomycete fungi. (37)

Sporangium A hollow structure in which spores are formed. (37)

Spores In plants, haploid cells that develop into the gametophyte generation. In fungi, an asexual or sexual reproductive cell that gives rise to a new mycelium. Spores are often lightweight for their dispersal and adapted for survival in adverse conditions. (37)

Sporophyte The diploid spore producing generation in plants. (38)

Stabilizing Selection Natural selection favoring an intermediate phenotype over the extremes. (33)

Starch Polysaccharides used by plants to store energy. (4)

Stamen The flower's male reproductive organ, consisting of the pollen-producing anther supported by a slender stalk, the filament. (20)

Stem In plants, the organ that supports the leaves, flowers, and fruits. (18)

Stem Cells Cells which are undifferentiated and capable of giving rise to a variety of different types of differentiated cells. For example, hematopoietic stem cells are capable of giving rise to both red and white blood cells. (17)

Steroids Compounds classified as lipids which have the basic four-ringed molecular skeleton as represented by cholesterol. Two examples of steroid hormones are the sex hormones; testosterone in males and estrogen in females. (4, 25)

Stigma The sticky area at the top of each pistil to which pollen adheres. (20)

Stimulus Any change in the internal or external environment to which an organism can respond. (24)

Stomach A muscular sac that is part of the digestive system where food received from the esophagus is stored and mixed, some breakdown of food occurs, and the chemical degradation of nutrients begins. (27)

Stomates (Pl. Stomata) Microscopic pores in the epidermis of the leaves and stems which allow gases to be exchanged between the plant and the external environment. (18)

Stratified Epithelia Multicellular layered epithelium. (22)

Stream Flowing body of surface fresh water; streams merge together into larger streams and rivers. (40)

Stretch Receptors Sensory receptors embedded in muscle tissue enabling muscles to respond reflexively when stretched. (23, 24)

Striated Referring to the striped appearance of skeletal and cardiac muscle fibers. (26)

Strobilus In lycopids, terminal, cone-like clusters of specialized leaves that produce sporangia.

Stroma The fluid interior of chloroplasts. (8)

Stromatolites Rocks formed from masses of dense bacteria and mineral deposits. Some of these rocky masses contain cells that date back over three billion years revealing the nature of early prokaryotic life forms. (35)

Structural Genes DNA segments in bacteria that direct the formation of enzymes or structural proteins. (15)

Style The portion of a pistil which joins the stigma to the ovary. (20)

Substrate-Level Phosphorylation The formation of ATP by direct transfer of a phosphate group from a substrate, such as a sugar phosphate, to ADP. ATP is formed without the involvement of an electron transport system. (9)

Substrates The reactants which bind to enzymes and are subsequently converted to products. (6)

Succession The orderly progression of communities leading to a climax community. It is one of two types: primary, which occurs in areas where no community existed before; and secondary, which occurs in disturbed habitats where some soil and perhaps some organisms remain after the disturbance. (41)

Succulents Plants having fleshy, water-storing stems or leaves. (40)

Suppressor T Cells A class of T cells that regulate immune responses by inhibiting the activation of other lymphocytes. (30)

Surface Area-to-Volume Ratio The ratio of the surface area of an organism to its volume, which determines the rate of exchange of materials between the organism and its environment. (22)

Surface Tension The resistance of a liquid's surface to being disrupted. In aqueous solutions, it is caused by the attraction between water molecules. (3)

Survivorship Curve Graph of life expectancy, plotted as the number of survivors versus age. (43)

Sweat Glands Exocrine glands of the skin that produce a dilute salt solution, whose evaporation cools in the body. (26)

Symbiosis A close, long-term relationship between two individuals of different species. (42)

Symmetry Referring to a body form that can be divided into mirror image halves by at least one plane through its body. (39)

Sympathetic Nervous System Part of the autonomic nervous system that tends to stimulate bodily activities, particularly those involved with coping with stressful situations. (23)

Sympatric Speciation Speciation that occurs in populations with overlapping distributions. It is common in plants when polyploidy arises within a population. (33)

Synapse Juncture of a neuron and its target cell (another neuron, muscle fiber, gland cell). (23)

Synapsis The pairing of homologous chromosomes during prophase of meiosis I. (11)

Synaptic Cleft Small space between the synaptic knobs of a neuron and its target cell. (23)

Synaptic Knobs The swellings that branch from the end of the axon. They deliver the neurological impulse to the target cell. (23)

Synaptonemal Complex Ladderlike structure that holds homologous chromosomes together as a tetrad during crossing over in prophase I of meiosis. (11)

Synovial Cavities Fluid-filled sacs around joints, the function of which is to lubricate and separate articulating bone surfaces. (26)

Systemic Circulation Part of the circulatory system that delivers oxygenated blood to the tissues and routes deoxygenated blood back to the heart. (28)

Systolic Pressure The first number of a blood pressure reading; the highest pressure attained in the arteries as blood is propelled out of the heart. (28)

◄ **T** ►

Taiga A biome found south of tundra biomes; characterized by coniferous forests, abundant precipitation, and soils that thaw only in the summer. (40)

Tap Root System Root system of plants having one main root and many smaller lateral roots. Typical of conifers and dicots. (18)

Taste Sense of the chemical composition of food. (24)

Taxonomy The science of classifying and grouping organisms based on their morphology and evolution. (1)

T Cell Lymphocytes that carry out cell-mediated immunity. They respond to antigen stimulation by becoming helper cells, killer cells, and memory cells. (30)

Telophase The final stage of mitosis which begins when the chromosomes reach their spindle poles and ends when cytokinesis is completed and two daughter cells are produced. (10)

Tendon A dense connective tissue cord that connects a skeletal muscle to a bone. (26)

Teratogenic Embryo deforming. Chemicals such as thalidomide or alcohol are teratogenic because they disturb embryonic development and lead to the formation of an abnormal embryo and fetus. (32)

Terminal Electron Acceptor In aerobic respiration, the molecule of O_2 which removes the electron pair from the final cytochrome of the respiratory chain. (9)

Terrestrial Living on land. (40)

Territory (Territoriality) An area that an animal defends against intruders, generally in the protection of some resource. (42, 44)

Tertiary Consumer Animals that feed on secondary consumers (plant or animal) or animals only. (41)

Test Cross An experimental procedure in which an individual exhibiting a dominant trait is crossed to a homozygous recessive to determine whether the first individual is homozygous or heterozygous. (12)

Testis In animals, the sperm-producing gonad of the male. (23)

Testosterone The male sex hormone secreted by the testes when stimulated by pituitary gonadotropins. (31)

Tetrad A unit of four chromatids formed by a synapsed pair of homologous chromosomes, each of which has two chromatids. (11)

Thallus In liverworts, the flat, ground-hugging plant body that lacks roots, stems, leaves, and vascular tissues. (38)

Theory of Tolerance Distribution, abundance and existence of species in an ecosystem are determined by the species' range of tolerance of chemical and physical factors. (41)

Thermoreceptors Sensory receptors that respond to changes in temperature. (24)

Thermoregulation The process of maintaining a constant internal body temperature in spite of fluctuations in external temperatures. (28)

Thigmotropism Changes in plant growth stimulated by contact with another object, e.g., vines climbing on cement walls. (21)

Thoracic Cavity The anterior portion of the body cavity in which the lungs are suspended. (39)

Thylakoids Flattened membrane sacs within the chloroplast. Embedded in these membranes are the light-capturing pigments and other components that carry out the light-dependent reactions of photosynthesis. (8)

Thymus Endocrine gland in the chest where T cells mature. (30)

Thyroid Gland A butterfly-shaped gland that lies just in front of the human windpipe, producing two metabolism-regulating hormones, thyroxin and triiodothyronine. (25)

Thyroid Hormone A mixture of two iodinated amino acid hormones (thyroxin and triiodothyronine) secreted by the thyroid gland. (25)

Thyroid Stimulating Hormone (TSH) An anterior pituitary hormone which stimulates secretion by the thyroid gland. (25)

Tissue An organized group of cells with a similar structure and a common function. (22)

Tissue System Continuous tissues organized to perform a specific function in plants. The three plant tissue systems are: dermal, vascular, and ground (fundamental). (18)

Tolerance Range The range between the maximum and minimum limits for an environmental factor that is necessary for an organism's survival. (41)

Totipotent The genetic potential for one type of cell from a multicellular organism to give rise to any of the organism's cell types, even to generate a whole new organism. (15)

Trachea The windpipe; a portion of the respiratory tract between the larynx and bronchii. (29)

Tracheal Respiratory System A network of tubes (tracheae) and tubules (tracheoles) that carry air from the outside environment directly to the cells of the body without involving the circulatory system. (29, 39)

Tracheid A type of conducting cell found in xylem functioning when a cell is dead to transport water and dissolved minerals through its hollow interior. (18)

Tracheophytes Vascular plants that contain fluid-conducting vessels. (38)

Transcription The process by which a strand of RNA assembles along one of the DNA strands. (14)

Transcriptional-Level Control Control of gene expression by regulating whether or not a specific gene is transcribed and how often. (15)

Transduction A type of genetic recombination resulting from transfer of genes from one organism to another by a virus.

Transfer RNA (tRNA) A type of RNA that decodes mRNA's codon message and translates it into amino acids. (14)

Transgenic Organism An organism that possesses genes derived from a different species. For example, a sheep that carries a human gene and secretes the human protein in its milk is a transgenic animal. (16)

Translation The cell process that converts a sequence of nucleotides in mRNA into a sequence of amino acids in a polypeptide. (14)

Translational-Level Control Control of gene expression by regulating whether or not a specific mRNA is translated into a polypeptide. (15)

Translocation The joining of segments of two nonhomologous chromosomes (13)

Transmission Electron Microscope (TEM) A microscope that works by shooting electrons through very thinly sliced specimens. The result is an enormously magnified image, two-dimensional, of remarkable detail. (App.)

Transpiration Water vapor loss from plant surfaces. (19)

Transpiration Pull The principle means of water and mineral transport in plants, initiated by transpiration. (19)

Transposition The phenomenon in which certain DNA segments (mobile genetic elements, or jumping genes) tend to move from one part of the genome to another part. (15)

Transverse Fission The division pattern in ciliated protozoans where the plane of division is perpendicular to the cell's length.

Trimester Each of the three stages comprising the 266-day period between conception and birth in humans. (32)

Triploid Having three sets of chromosomes, abbreviated 3N. (11)

Trisomy Three copies of a particular chromosome per cell. (17)

Trophic Level Each step along a feeding pathway. (41)

Trophozoite The actively growing stage of polymorphic protozoa. (37)

Tropical Rain Forest Lush forests that occur near the equator; characterized by high annual rainfall and high average temperature. (40)

Tropical Thornwood A type of shrubland occurring in tropical regions with a short rainy season. Plants lose their small leaves during dry seasons, leaving sharp thorns. (40)

Tropic Hormones Hormones that act on endocrine glands to stimulate the production and release of other hormones. (25)

Tropisms Changes in the direction of plant growth in response to environmental stimuli, e.g., light, gravity, touch. (21)

True-Breeder Organisms that, when bred with themselves, always produce offspring identical to the parent for a given trait. (12)

Tubular Reabsorption The process by which substances are selectively returned from the fluid in the nephron to the bloodstream. (28)

Tubular Secretion The process by which substances are actively and selectively transported from the blood into the fluid of the nephron. (28)

Tumor-Infiltrating Lymphocytes (TILs) Cytotoxic T cells found within a tumor mass that have the capability to specifically destroy the tumor cells. (30)

Tumor-Suppressor Genes Genes whose products act to block the formation of cancers. Cancers form only when both copies of these genes (one on each homologue) are mutated. (13)

Tundra The marshy, unforested biome in the arctic and at high elevations. Frigid temperatures for most of the year prevent the subsoil from thawing, which produces marshes and ponds. Dominant vegetation includes low growing plants, lichens, and mosses. (40)

Turgor Pressure The internal pressure in a plant cell caused by the diffusion of water into the cell. Because of the rigid cell wall, pressure can increase to where it eventually stops the influx of more water. (7)

Turner Syndrome A person whose cells have only one X chromosome and no second sex chromosome (XO). These individuals develop as immature females. (17)

◄ U ►

Ultimate (Top) Consumer The final carnivore trophic level organism, or organisms that escaped predation; these consumers die and are eventually consumed by decomposers. (41)

Ultracentrifuge An instrument capable of spinning tubes at very high speeds, delivering centrifugal forces over 100,000 times the force of gravity. (9)

Unicellular The description of an organism where the cell is the organism. (35)

Uniform Pattern Distribution of individuals of a population in a uniform arrangement, such as individual plants of one species uniformly spaced across a region. (43)

Urethra In mammals, a tube that extends from the urinary bladder to the outside. (28)

Urinary Tract The structures that form and export urine: kidneys, ureters, urinary bladder, and urethra. (28)

Urine The excretory fluid consisting of urea, other nitrogenous substances, and salts dissolved in water. It is formed by the kidneys. (28)

Uterine (Menstrual) Cycle The repetitive monthly changes in the uterus that prepare the endometrium for receiving and sustaining an embryo. (31)

Uterus An organ in the female reproductive system in which an embryo implants and is maintained during development. (31)

◄ V ►

Vaccines Modified forms of disease-causing microbes which cannot cause disease but retain the same antigens of it. They permit the immune system to build memory cells without diseases developing during the primary immune response. (30)

Vacoconstriction Reduction in the diameter of blood vessels, particularly arterioles. (28)

Vacuole A large organelle found in mature plant cells, occupying most of the cell's volume, sometimes more than 90% of it. (5)

Vagina The female mammal's copulatory organ and birth canal. (31)

Variable (Experimental) A factor in an experiment that is subject to change, i.e., can occur in more than one state. (2)

Vascular Bundles Groups of vascular tissues (xylem and phloem) in the shoot of a plant. (19)

Vascular Cambrium In perennials, a secondary meristem that produces new vascular tissues. (18)

Vascular Cylinder Groups of vascular tissues in the central region of the root. (18)

Vascular Plants Plants having a specialized conducting system of vessels and tubes for transporting water, minerals, food, etc., from one region to another. (18)

Vascular Tissue System All the vascular tissues in a plant, including xylem, phloem, and the vascular cambium or procambium. (18)

Vasodilation Increase in the diameter of blood vessels, particularly arterioles. (28)

Veins In plants, vascular bundles in leaves. In animals, blood vessels that return blood to the heart. (28)

Venation The pattern of vein arrangement in leaf blades. (18)

Ventricle Lower chamber of the heart which pumps blood through arteries. There is one ventricle in the heart of lower vertebrates and two ventricles in the four-chambered heart of birds and mammals. (28)

Venules Small veins that collect blood from the capillaries. They empty into larger veins for return to the heart. (28)

Vertebrae The bones that form the backbone. In the human there are 33 bones arranged in a gracefully curved line along the bone, cushioned from one another by disks of cartilage. (26)

Vertebral Column The backbone, which encases and protects the spinal cord. (26)

Vertebrates Animals with a backbone. (39)

Vesicles Small membrane-enclosed sacs which form from the ER and Golgi complex. Some store chemicals in the cells; others move to the surface and fuse with the plasma membrane to secrete their contents to the outside. (5)

Vessel A tube or connecting duct containing or circulating body fluids. (18)

Vessel Member A type of conducting cell in xylem functioning when the cell is dead to transport water and dissolved minerals through its hollow interior; also called a vessel element. (18)

Vestibular Apparatus A portion of the inner ear of vertebrates that gathers information about the position and movement of the head for use in maintaining balance and equilibrium. (24)

Vestigial Structure Remains of ancestral structures or organs which were, at one time, useful. (34)

Villi Finger-like projections of the intestinal wall that increase the absorption surface of the intestine. (27)

Viroids are associated with certain diseases of plants. Each viroid consists solely of a small single-stranded circle of RNA unprotected by a protein coat. (36)

Virus Minute structures composed of only heredity information (DNA or RNA), surrounded by a protein or protein/lipid coat. After infection, the viral nucleic acid subverts the metabolism of the host cell, which then manufactures new virus particles. (36)

Visible Light The portion of the electromagnetic spectrum producing radiation from 380 nm to 750 nm detectable by the human eye.

Vitamins Any of a group of organic compounds essential in small quantities for normal metabolism. (27)

Vocal Cords Muscular folds located in the larynx that are responsible for sound production in mammals. (29)

Vulva The collective name for the external features of the human female's genitals. (31)

◀ W ▶

Water Vascular System A system for locomotion, respiration, etc., unique to echinoderms. (39)

Wavelength The distance separating successive crests of a wave. (8)

Waxes A waterproof material composed of a number of fatty acids linked to a long chain alcohol. (4)

White Matter Regions of the brain and spinal cord containing myelinated axons, which confer the white color. (23)

Wild Type The phenotype of the typical member of a species in the wild. The standard to which mutant phenotypes are compared. (13)

Wilting Drooping of stems or leaves of a plant caused by water loss. (7)

Wood Secondary xylem. (18)

◀ X ▶

X Chromosome The sex chromosome present in two doses in cells of a female, and in one dose in the cells of a male. (13)

X-Linked Traits Traits controlled by genes located on the X chromosome. These traits are much more common in males than females. (13)

Xylem The vascular tissue that transports water and minerals from the roots to the rest of the plant. Composed of tracheids and vessel members. (18)

◀ Y ▶

Y Chromosome The sex chromosome found in the cells of a male. When Y-carrying sperm unite with an egg, all of which carry a single X chromosome, a male is produced. (13)

Y-Linked Inheritance Genes carried only on Y chromosomes. There are relatively few Y-linked traits; maleness being the most important such trait in mammals. (13)

Yeast Unicellular fungus that forms colonies similar to those of bacteria. (37)

Yolk A deposit of lipids, proteins, and other nutrients that nourishes a developing embryo. (32)

Yolk Sac A sac formed by an extraembryonic membrane. In humans, it manufactures blood cells for the early embryo and later helps to form the umbilical cord. (32)

◀ Z ▶

Zero Population Growth In a population, the result when the combined positive growth factors (births and immigration). (43)

Zooplankton Protozoa, small crustaceans and other tiny animals that drift with ocean currents and feed on phytoplankton. (37, 40)

Zygospore The diploid spores of the zygomycete fungi, which include Rhizopus, a common bread mold. After a period of dormancy, the zygospore undergoes meiosis and germinates. (36)

Zygote A fertilized egg. The diploid cell that results from the union of a sperm and egg. (32)

Photo Credits

Part 1 Opener Norbert Wu. **Chapter 1** Fig. 1.1: Jeff Gnass. Fig. 1.2a: Frans Lanting/Minden Pictures, Inc. Fig. 1.2b: David Muench. Fig. 1.3a: Courtesy Dr. Alan Cheetham, National Museum of National History, Smithsonian Institution. Fig. 1.3b: Manfred Kage/Peter Arnold. Fig. 1.3c: Stephen Dalton/NHPA. Fig. 1.3d: Charles Summers, Jr. Fig. 1.3e: Larry West/Photo Researchers. Fig. 1.3f: Bianca Lavies. Fig. 1.3g: Steve Allen/Peter Arnold. Fig. 1.3h: George Grall. Fig. 1.3i: Michio Hoshino/Minden Pictures, Inc. Fig. 1.6a: CNRI/Science Photo Library/Photo Researchers. Fig. 1.6b: M. Abbey/Visuals Unlimited. Fig. 1.6c: Steve Kaufman/Peter Arnold. Fig. 1.6d: Willard Clay. Fig. 1.6e: Jim Bradenburg/Minden Pictures, Inc. Fig. 1.7: (pages 16-17) Charles A. Mauzy; (page 16, top) Anthony Mercieca/Natural Selection; (page 16, center) Wolfgang Bayer/Bruce Coleman, Inc.; (page 16, bottom) Dr. Jeremy Burgess/Science Photo Library/Photo Researchers; (page 17, top) Dr. Eckart Pott/Bruce Coleman, Ltd.; (page 17, bottom) Art Wolfe. Fig. 1..8: Paul Chesley/Photographers Aspen. Fig. 1.9a: Rainbird/Robert Harding Picture Library. Fig. 1.11: Dick Luria/FPG International. **Chapter 2** Fig. 2.2a: Couresy Institut Pasteur. Fig. 2.3a: Topham/The Image Works. Fig. 2.3b: Leonard Lessin/Peter Arnold. Fig. 2.3c: Bettmann Archive. Fig. 2.5a: Laurence Gould/Earth Scenes/Animals Animals. Fig. 2.5b: Ted Horowitz/The Stock Market. **Part 2 Opener:** Nancy Kedersha. **Chapter 3** Fig. 3.1: Franklin Viola. Fig. 3.2: Courtesy Pachyderm Scientific Industries. Bioline: Jacan-Yves Kerban/The Image Bank. Fig. 3.7: Courtesy Stephen Harrison, Harvard Biochemistry Department. Fig. 3.11: Gary Milburn/Tom Stack & Associates. Fig. 3.12: Courtesy R. S. Wilcox, Biology Department, SUNY Binghampton. **Chapter 4** Fig. 4.1: Alastair Black/Tony Stone World Wide. Fig. 4.5a: Don Fawcett/Visuals Unlimited. Fig. 4.5b: Jeremy Burgess/Photo Researchers. Fig. 4.5c: Cabisco/Visuals Unlimited. Fig. 4.6: Tony Stone World Wide. Fig. 4.7: Zig Leszcynski/Animals Animals. Fig. 4.8: Robert & Linda Mitchell. Human Perspective: (a) Photofest (b) Bill Davila/Retna. Fig. 4.12a: Frans Lanting/AllStock, Inc. Fig. 4.12b: Mantis Wildlife Films/Oxford Scientific Films/Animals Animals. Fig. 4.17: Stanley Flegler/Visuals Unlimited. Fig. 4.18a: Courtesy Nelson Max, Lawrence Livermore Laboratory. Fig. 4.18b: Tsuned Hayashida/The Stock Market. **Chapter 5** Fig. 5.1a: The Granger Collection. Fig. 5.1b: Bettmann Archive. Fig. 5.2a: Dr. Jeremey Burgess/Photo Researchers. Fig. 5.2b: CNRI/Photo Researchers. Fig. 5.4: Omikron/Photo Researchers. Fig. 5.6a: Courtesy Richard Chao, California State University at Northridge. Fig. 5.6c: Courtesy Daniel Branton, University of Berkeley. Fig. 5.7: Courtesy G. F. Bohr. Fig. 5.8: Courtesy

Michael Mercer, Zoology Department, Arizona State University. Fig. 5.9: Courtesy D. W. Fawcett, Harvard Medical School. Fig. 5.10a: Courtesy U.S. Department of Agriculture, Fig. 5.11: Courtesy Dr. Birgit H. Satir, Albert Einstein College of Medicine. Fig. 5.13: Courtesy Lennart Nilsson, from *A Child Is Born*. Fig. 5.14a: K. R. Porter/Photo Researchers. Fig. 5.14c: Courtesy Lennart Nilsson, BonnierAlba. Fig. 5.14d: Courtesy Lennart Nilsson, From *A Child Is Born*. Fig. 5.15a: Courtesy J. Elliot Weier. Fig. 5.16a: Courtesy J.V. Small. Fig. 5.17a: Peter Parks/Animals Animals. Fig. 5.17b: Courtesy Dr. Manfred Hauser, RUHR-Universitat Rochum. Fig. 5.18: Courtesy Jean Paul Revel, Division of Biology, California Institute of Technology. Fig. 5.19a: Courtesy E. Vivier, from *Paramecium* by W.J. Wagtendonk, Elsevier North Holland Biomedical Press, 1974. Fig. 5.19b: David Phillips/Photo Researchers. Fig. 5.20: Courtesy C.J. Brokaw and T.F. Simonick, *Journal of Cell Biology*, 75:650 (1977). Reproduced with permission. Fig. 5.21: Courtesy L.G. Tilney and K. Fujiwara. Fig. 5.22a: Courtesy W. Gordon Whaley, University of Texas, Austin. Fig. 5.22c: Courtesy R.D. Preston, University of Leeds, London. **Chapter 6** Fig. 6.1a: Alex Kerstitch. Fig. 6.1b: Peter Parks/Earth Scenes. Fig. 6.2: Marty Stouffer/Animals Animals. Fig. 6.3: Kay Chernush/The Image Bank. Fig. 6.4b: Courtesy Computer Graphics Laboratory, University of California, San Francisco. Fig. 6.8: Courtesy Stan Koszelek, Ph.D., University of California, Riverside. Fig. 6.10: Swarthout/The Stock Market. **Chapter 7** Fig. 7.5: Ed Reschke. Fig. 7.10a: Lennart Nilsson, ©Boehringer Ingelheim, International Gmbh; from *The Incredible Machine*. Human Perspective: (Fig. 1a) Martin Rotker/Phototake; (Fig. 1b) Cabisco/Visuals Unlimited. Fig. 7.11a: Courtesy Dr. Ravi Pathak, Southwestern Medical Center, University of Texas. **Chapter 8** Fig. 8.1a: Carr Clifton. Fig. 8.1b: Shizuo Lijima/Tony Stone World Wide. Fig. 8.5: Joe Englander/Viesti Associates, Inc. Fig. 8.6: Courtesy T. Elliot Weier. Fig. 8.10: Courtesy Lawrence Berkeley Laboratory, University of California. Fig. 8.13: Pete Winkel/Atlanta/Stock South. Bioline: Courtesy Woods Hole Oceanographic Institution. **Chapter 9** Fig. 9.1: Stephen Frink/AllStock, Inc. Bioline: (Fig. 1) Frank Oberle/Bruce Coleman, Inc.; (Fig. 2) Movie Stills Archive. Fig. 9.6: Topham/The Image Works. Fig. 9.9: H. Fernandez-Moran. Human Perspective: (top and bottom insets) Courtesy MacDougal; (top) Jerry Cooke; (bottom) Richard Kane/Sportchrome East/West. Fig. 9.11: Peter Parks/NHPA. Fig. 9.12a: Grafton M. Smith/The Image Bank. **Chapter 10** Fig. 10.1: Sipa Press. Fig. 10.2a: Institut Pasteur/CNRI/Phototake. Fig. 10.3a: Dr. R. Vernon/Phototake.

Fig. 10.3b: CNRI/Science Photo Library/Photo Researchers. Fig. 10.6: Courtesy Dr. Andrew Bajer, University of Oregon, Fig. 10.7: CNRI/Science Photo Library/Photo Researchers. Fig. 10.8: Dr. G. Shatten/Science Photo Library/Photo Researchers. Fig. 10.9: Courtesy Professor R.G. Kessel. Fig. 10.10: David Phillips/Photo Researchers. **Chapter 11** Fig. 11.3: Courtesy Science Software Systems. Fig. 11.5a: Courtesy Dr. A.J. Solari, from *Chromosoma*, vol. 81, p. 330 (1980), Human Perspective: Donna Zweig/Retna. Fig. 11.7: Cabisco/Visuals Unlimited. **Part 3 Opener:** David Scharf. **Chapter 12** Fig. 12.1: Courtesy Dr. Ing Jaroslav Krizenecky. Fig. 12.6a: Doc Pele/Retna. Fig. 12.6b: FPG International. Fig. 12.9: Sydney Freelance/Gamma Liaison. Fig. 12.11 Hans Reinhard/Bruce Coleman, Inc. **Chapter 13** Fig. 13.2a: Robert Noonan. Fig. 13.6: Biological Photo Service. Bioline: Norbert Wu. Fig. 13.9: Historical Pictures Service. Fig. 13.11a: Courtesy M.L. Barr. Fig. 13.11b: Jean Pragen/Tony Stone World Wide. Fig. 13.12: Courtesy Lawrence Livermore National Laboratory. **Chapter 14** Fig. 14.1b: Lee D. Simon/Science Photo Library/Photo Researchers. Fig. 14.4: David Leah/Science Photo Library/Photo Researchers. Fig. 14.5: Dr. Gopal Murti/Science Photo Library/Photo Researchers. Fig. 14.6b: Fawcett/Olins/Photo Researchers. Fig. 14.7: Courtesy U.K. Laemmli. Fig. 14.10a: From M. Schnos and.R.B. Inman, *Journal of Molecular Biology*, 51:61-73 (1970), ©Academic Press. Fig. 14.10b: Courtesy Professor Joel Huberman, Roswell Park Memorial Institute. Human Perspective: (Fig. 2) Courtesy Skin Cancer Foundation; (Fig. 3) Mark Lewis/Gamma Liaison. Fig. 14.17: Courtesy Dr. O.L. Miller, Oak Ridge National Laboratory. **Chapter 15** Fig. 15.1: Courtesy Richard Goss, Brown University. Fig. 15.2a: (top left) Oxford Scientific Films/Animals Animals; (top right) F. Stuart Westmorland/Tom Stack & Associates. Fig. 15.2b: Courtesy Dr. Cecilio Barrera, Department of Biology, New Mexico State University. Fig. 15.3b: Courtesy Michael Pique, Research Institute of Scripps Clinic. Fig. 15.7b: Courtesy Wen Su and Harrison Echols, University of California, Berkeley. Fig. 15.8a: Courtesy Stephen Case, University of Mississippi Medical Center. Fig. 15.9: Roy Morsch/The Stock Market. **Chapter 16** Fig. 16.1a: David M. Dennis/Tom Stack & Associates. Fig. 16.1b: Courtesy Lakshmi Bhatnagor, Ph.D., Michigan Biotechnology Institute. Fig. 16.2: Art Wolfe/All Stock, Inc. Fig. 16.3: Ken Graham. Fig. 16.4a: Courtesy R.L. Brinster, Laboratory for Reproductive Physiology, University of Pennsylvania. Fig. 16.4b: John Marmaras/Woodfin Camp & Associates. Fig. 16.5: Courtesy Robert Hammer, School of Veterinary Medicine, University of

Index